Biology of
Marine Mammals

BIOLOGY OF MARINE MAMMALS

EDITED BY JOHN E. REYNOLDS III
AND SENTIEL A. ROMMEL

SMITHSONIAN INSTITUTION PRESS
Washington and London

Copy editor: Danielle Ponsolle
Production editor: Duke Johns
Designer: Janice Wheeler

Library of Congress Cataloging-in-Publication Data
Biology of marine mammals / edited by
John E. Reynolds III and Sentiel A. Rommel.
p. cm.
Includes bibliographical references (p.) and index.
ISBN 1-58834-250-6 (alk. paper)
1. Marine mammals. I. Reynolds, John Elliott, 1952–
II. Rommel, Sentiel A.
QL713.2.B54 1999
599.5—dc21 98-27808

British Library Cataloguing-in-Publication Data available

Manufactured in the United States of America
06 05 04 03 02 01 00 99 5 4 3 2 1

♾ The paper used in this publication meets the minimum requirements of the American National Standard for Information Sciences—Permanence of Paper for Printed Library Materials ANSI Z39.48-1984.

Contents

Preface

My introduction to marine mammal science and policy occurred in the autumn of 1974 when I took a class taught by Dan Odell at the University of Miami's Rosenstiel School of Marine and Atmospheric Science. At the time little did I appreciate the rare opportunity I had to take one of the few such courses in existence. Even today there are so few formal courses on the subject that a survey carried out by the Society for Marine Mammalogy showed that the majority of marine mammalogists with Ph.D.s have never taken one.

Between 1975 and 1980 I worked as a teaching assistant in Dan's course. Although my primary responsibility was teaching marine mammal anatomy laboratories, I began to consider how I would teach different aspects of marine mammalogy; my musings crystallized in 1980 when I was hired by Eckerd College and was charged with precisely that task.

Development of a coherent course that dealt with both science and policy was hindered in 1980 by lack of a comprehensive textbook. The same holds true as I write this in 1998, despite the existence of more and more books on marine mammals. Some contain excellent species accounts; others are highly technical discussions of particular topics; a few are appropriate for use as texts, except that they deal with only a subset (e.g., cetaceans) of the marine mammals or with only those species in a particular geographic area; many are nontechnical "coffee-table books" with stunning photographs but inadequate text to be useful for students. In short, people who teach, or want to teach, marine mammalogy continue to be frustrated that a comprehensive textbook for upper-level undergraduate and graduate students is still unavailable.

This deficiency befuddles many marine mammalogists. The field itself matured remarkably during and after the 1970s. Marine mammal science historically involved a lot of field observations, some excellent anatomical studies, and some physiological investigations. The first Marine Mammal Conference, held in 1975 in Santa Cruz, California, attracted only some 150 scientists. A decade later, the field had grown sufficiently that the Society for Marine Mammalogy was created, along with its own publication, *Marine Mammal Science.* Topics covered in the journal still address traditional scientific questions (e.g., estimates of abundance, causes of mortality, physiological adaptations, social behavior, functional anatomy, stock discreteness, and effects of human activities), but do so using technology (e.g., telemetry, polymerase chain reaction, electron microscopy, geographic information systems) that allows greater resolution. The 11th Biennial Conference on the Biology of Marine Mammals, held in 1995 in Orlando, Florida, attracted some 1500 participants, an order of magnitude more than the conference two decades earlier.

The growth of scientific research has been paralleled in terms of policy and conservation. In 1972 the Marine Mammal Protection Act was adopted by the U.S. Congress. The Act was noteworthy both for its focus on marine mammals and its approach to management from an ecosystem point of view. This and other legislation (e.g., the Endangered Species Act of 1973) provide a legal framework for conserving and managing marine mammal stocks and habitats. Over the years recovery plans, which consider both science and socioeconomics, have been developed for a number of endangered or threatened species. A 1994 workshop sponsored by the U.S. Marine Mammal Commission articulated a set of guiding principles for the worldwide conservation of marine mammals and other living natural resources.

In addition the lay public grew increasingly interested in marine mammals during the past three decades. Marine mammals have become symbolic—even mystical. Marine zoological parks and aquaria display marine mammals and develop educational and conservation messages for the public; people who might never see an ocean, and yet as voters could impact issues involving marine mammals, can see and learn firsthand about these animals. Television documentaries feed the appetite of a public that wants to know more about marine mammals and how to "save" them. Special interest groups have taken up the cause over the years; for example, such groups contributed to efforts to establish a moratorium on worldwide commercial whaling.

Despite the whirlwind of scientific advances, promulgation of policy, and public interest, there has continued to be a dearth of college-level courses to either groom future marine mammal managers and scientists or to enlighten future professionals in related fields (e.g., marine ecology, fisheries biology, epidemiology) for whom an overview of marine mammalogy would be useful. The few existing courses have been handicapped by the lack of a comprehensive text.

Biology of Marine Mammals provides a straightforward, integrated approach to understanding these animals. Most chapters consider topics one would expect in a biology textbook, including functional morphology, physiology, sensory biology, energetics, reproduction, communication, behavior and ecology. A chapter that is perhaps less typical, but which is very useful for students to understand marine mammal biology, is the final chapter on toxicants and their possible effects on marine mammal health.

Chapters have been written so that they may either stand alone or be used as a part of a coherent whole. Terminology has been standardized wherever possible, and unnecessary redundancy eliminated. However, some repetition of principles and concepts remains for two reasons: (1) not all students respond equally to particular pedagogical approaches; therefore to reach everyone, it is helpful to present key ideas from at least a couple of points of view, and (2) it is simply impossible to pigeonhole topics in an integrated subject such as biology with no overlap.

This is not to suggest that the text is only for students. *Biology of Marine Mammals* should be very useful to professionals other than educators, but the text may trick readers who expect only to learn about marine mammals. These animals (often called "charismatic megafauna") represent a "hook" that captures many peoples' interest. However, if one attempts to view them dispassionately, marine mammals are simply living organisms; they are components of ecosystems; they are resources that people use. In short, despite their amazing attributes, they are fundamentally not unlike many other animals.

I tell friends that marine mammal scientists are the luckiest people in the world because they combine vocation and avocation. I hope that *Biology of Marine Mammals* helps students to better understand what the field really entails so that they can decide whether their avocation should actually become their vocation. In that regard, students should note topics about which few data exist and consider why there are such gaps. Although I hope students never lose their sense of wonder, I do hope that the book helps to demystify marine mammals. Ultimately I hope it provides a useful resource to help create a generation of informed, concerned individuals who will succeed better than my generation in understanding and conserving marine mammals and other living resources.

Biology of Marine Mammals is the product of hard work by a number of people. I am especially grateful to my co-editor, Sentiel (Butch) Rommel. I thank all the authors who contributed their time and expertise. I also thank Peter Cannell, director of Smithsonian Institution Press, and the Press's staff for their unflagging support of this project and remarkable tolerance of many delays. I am very grateful to reviewers of chapters who helped ensure thoroughness, accuracy, and appropriateness for students; reviewers included Greg Bossart, Phil Clapham, Samantha Eide, John Eisenberg, Melissa Etheridge, Frank Fish, Ted Grand, David Grove, John Heyning, Aleta Hohn, Brenda Jensen, Lynn Lefebvre, Lloyd Lowry, Chris Marshall, Jim Mead, Bill McLellan, Paul Nachtigall, Dan Odell, Galen Rathbun, Andy Read, Roger Reep, Dan Rubenstein, Laela Sayigh, Michael Scott, John Stein, Hans Thewissen, Jeanette Thomas, Ted Van Vleet, Mike Walsh, and Graham Worthy. Finally, I appreciate the editorial assistance of Alison Kirk Long, Jan Sechrist, Suzanne Montgomery, and Samantha Eide; for her cheerful daily assistance, Samantha merits particular thanks.

John E. Reynolds III

1

JOHN E. REYNOLDS III, DANIEL K. ODELL, AND SENTIEL A. ROMMEL

Marine Mammals of the World

Although marine mammals of various types are often considered as a group by biologists, managers, and legislators, marine mammals actually represent three different orders of mammals: the Carnivora, Cetacea, and Sirenia. We note here that the major groups of marine mammals arose independently from very different ancestors: the carnivorous marine mammals (polar bears, otters, seals, sea lions, and walruses) are derived from ursid (bear) and mustelid (mink, weasel, etc.) ancestors, the cetaceans (whales, dolphins, and porpoises) evolved from an artiodactyl (even-toed ungulates, such as cows or pigs) stock, and the sirenians (manatees and dugongs) are related to elephants and other subungulates.

With the obvious lack of phylogenetic links, why are marine mammals so unified in many people's minds? To be sure, the various marine mammals do have some common ground. First, they are mammals, and they demonstrate the anatomical and physiological features typically associated with that class of organisms. In addition, they occupy or rely on aquatic, if not strictly marine, habitats, and have evolved similar anatomical features, including large body size, streamlined shape (compared to terrestrial relatives), insulation in the form of blubber and dense fur, and in most cases, a modified appendicular skeleton resulting in reduction in the size of appendages. These and other morphological features are described in detail by Pabst, Rommel, and McLellan, Chapter 2, this volume. Marine mammals also possess some similar physiological adaptations (e.g., for diving, thermoregulation, osmoregulation, communication, and orientation; see Elsner, Chapter 3, this volume) to permit them to exploit the aquatic environment. Such commonalities have led biologists to consider marine mammals as a group.

There are other similarities among marine mammals as well. They are united by habitat requirements, in that all are dependent on an aquatic ecosystem for survival. This dependency makes them visible indicators of habitat degradation; the occurrence of more than 1000 dead striped dolphins (*Stenella coeruleoalba*) in the Mediterranean Sea and adjacent bodies of water in the early 1990s (Marine Mammal Commission 1995a) sent an important warning to people dependent on the marine ecosystem for their livelihood, especially when it was found that the dolphins that died carried significantly higher pesticide loads than "normal" (Aguilar and Borrell 1994). Management of marine mammals (and indeed of all living resources) involves maintenance of the habitat on which a species depends, not simply maintaining the species itself; therefore, managers may tend to consider marine mammals as a group that has some common habitat requirements and perhaps common vulnerabilities.

Another unifying feature is that many species have been

deliberately exploited, some (e.g., Steller's sea cow [*Hydrodamalis gigas*]) to the point of extinction, and others, such as right whales (*Eubalaena* spp.), to the point where populations have not been able to recover, even after decades of legal protection. Other species (e.g., the vaquita [*Phocoena sinus*] and the baiji [*Lipotes vexillifer*]) have been decimated by incidental takes and habitat destruction or modification, and their survival over the next few decades is in jeopardy. In fact, two species (the Caribbean monk seal [*Monachus tropicalis*] and the Steller's sea cow) and stocks of other species have become extinct in the past 250 years, and a number of others are classified as endangered, threatened, or depleted (Table 1-1) (Marine Mammal Commission 1995a). Regardless of the reason behind the endangered or threatened status of certain marine mammals, people in many nations have become concerned about the status of marine mammals in general.

The reasons why many people care about marine mammals are numerous and include aesthetic appreciation of the animals, concern for the health of ecosystems that support both marine mammals and humans, the possibility that our species can benefit from studies of marine mammals (e.g., through biomedical or physiological research), and a feeling

Table 1-1. Marine Mammal Species and Populations Listed as Endangered (E) or Threatened (T) under the U.S. Endangered Species Act, and Depleted (D) under the U.S. Marine Mammal Protection Act, as of 31 December 1994[a]

Species	Common Name	Status	Range
Manatees and Dugongs			
Trichechus manatus	West Indian manatee	E/D	The Atlantic coast and rivers of North, Central, and South America from southeast U.S. to Brazil; Puerto Rico and other Greater Antilles Islands
T. inunguis	Amazonian manatee	E/D	Amazon river basin of South America
T. senegalensis	West African manatee	T/D	West African coast and rivers; Senegal to Angola
Dugong dugon	Dugong	E/D	Northern Indian Ocean from Madagascar to Indonesia; Philippines; Australia; southern China; Palau
Otters			
Lutra felina	Marine otter	E/D	Western South America; Peru to southern Chile
Enhydra lutris nereis	Southern sea otter	T/D	Central California coast
Seals and Sea Lions			
Monachus schauinslandi	Hawaiian monk seal	E/D	Hawaiian Archipelago
M. tropicalis	Caribbean monk seal	E/D	Caribbean Sea and Bahamas (probably extinct)
M. monachus	Mediterranean monk seal	E/D	Mediterranean Sea; Atlantic coast of northwest Africa
Arctocephalus townsendi	Guadalupe fur seal	T/D	Baja California, Mexico, to southern California
Callorhinus ursinus	Northern fur seal	D	North Pacific Rim from California to Japan
Eumetopias jubatus	Steller sea lion	T/D	North Pacific Rim from California to Japan
Phoca hispida saimensis	Saimaa seal	E/D	Lake Saimaa, Finland
Whales, Porpoises, and Dolphins			
Lipotes vexillifer	Baiji	E/D	Changjiang (Yangtze) River, China
Platanista minor	Indus river dolphin	E/D	Indus River and tributaries, Pakistan
Phocoena sinus	Vaquita	E/D	Northern Gulf of California, Mexico
Stenella attenuata	Northeastern offshore spotted dolphin	D	Eastern tropical Pacific Ocean
S. longirostris orientalis	Eastern spinner dolphin	D	Eastern tropical Pacific Ocean
Tursiops truncatus	Bottlenose dolphin	D	Atlantic coastal waters from New York to Florida
Eubalaena glacialis	Northern right whale	E/D	North Atlantic, North Pacific oceans; Bering Sea
E. australis	Southern right whale	E/D	All oceans in the southern hemisphere
Balaena mysticetus	Bowhead whale	E/D	Arctic Ocean and adjacent seas
Megaptera novaeangliae	Humpback whale	E/D	Oceanic, all oceans
Balaenoptera musculus	Blue whale	E/D	Oceanic, all oceans
Eschrichtius robustus	Western North Pacific gray whale	E/D	Okhotsk Sea to South China Sea
Balaenoptera physalus	Fin whale	E/D	Oceanic, all oceans
B. borealis	Sei whale	E/D	Oceanic, all oceans
Physeter macrocephalus	Sperm whale	E/D	Oceanic, all oceans

Data from Marine Mammal Commission 1995a, reprinted with permission.

[a]From Fish and Wildlife Service Regulations at 50 C.F.R. §17.11 and National Marine Fisheries Service Regulations at §216.15.

Terms are defined as follows: endangered = species in danger of extinction throughout all or a significant portion of their range (Clark 1994); threatened = species likely to become endangered in the forseeable future (Clark 1994); depleted = a species or population stock considered to be below the optimum sustainable population, or one listed as endangered or threatened under the Endangered Species Act (Marine Mammal Commission 1995b).

of stewardship that extends to endangered or threatened species, especially if they are mammals like ourselves. Domning (1991) provides a well-articulated discussion of reasons why people do (or should) value manatees and, by extension, all marine mammals. The similar bases for caring and concern have promoted common legislation (e.g., the U.S. Marine Mammal Protection Act of 1972) and other conservation actions. In fact, a number of nonprofit conservation groups, including but not limited to the Center for Marine Conservation, Friends of the Sea Otter, Greenpeace, and the Save the Manatee Club, focus much, if not all, of their efforts on marine mammal conservation.

Of course, one of the overarching reasons why humans care about something involves money. Certainly this is the case for marine mammals. Interestingly, both ends of a financial spectrum exist; that is, some people make money from marine mammals, and other people lose money due to marine mammals. For example, marine zoological parks and aquaria, as well as whale-watching ventures, care about marine mammals, in part, as major attractions for the public; Reeves and Mead (Twiss and Reeves 1999) point out, however, that marine zoological parks such as SeaWorld are not simply focused on maximizing profits as they use income from admissions to support research, conservation, and education programs. Some fishermen, on the other hand, consider marine mammals as effective competitors and destroyers of fishing gear. In some societies, direct harvest of marine mammals provides products for sale and trade. Uses and values of marine mammals are discussed extensively in various chapters of Twiss and Reeves (1999); suffice it to say, when finances are involved, people tend to be very interested!

Clearly, a number of attributes cause marine mammals to be considered as a group by scientists, managers, legislators, and the lay public. However, as indicated at the outset of this chapter, marine mammals actually represent several diverse lineages. This chapter separates the marine mammals into the various phylogenetic groups and provides an overview of diagnostic traits. Although the chapter provides some information on individual species, it is not intended to be a detailed species identification or field guide; there exist several excellent, recent, affordable, and readily available references for this purpose, including Leatherwood et al. (1983), Reidman (1990), Reeves et al. (1992), Geraci and Lounsbury (1993), Jefferson et al. (1993), and Reijnders et al. (1993). In addition, Brownell et al. (1995) provide a brief, nontechnical review of marine mammal biodiversity. For people who want to focus on particular species (as opposed to the comparative approach used in this book), the volumes of the *Handbook of Marine Mammals,* edited by Ridgway and Harrison (1981a,b, 1985, 1989, 1994, 1999), provide thorough coverage, and there exist excellent volumes devoted to particular species (e.g., *The Bottlenose Dolphin* by Leatherwood and Reeves 1990). The purpose of this chapter is to discuss general features of marine mammals at the ordinal and family levels to create a framework and a working vocabulary to support concepts developed in subsequent chapters.

Readers should note that some species are readily recognized as marine mammals by almost all marine biologists of the world. However, it is also important to note that there are people who would argue on scientific, political, religious, or other grounds for more or fewer species than those on the lists we provide. The marine mammals we list in this chapter are essentially the species that the United States recognizes as marine mammals under the Marine Mammal Protection Act of 1972. We do not include, for example, the greater bulldog bat (*Noctilio leporinus*), the fishing bat (*Myotis vivesi*), and the North American river otter (*Lontra canadensis*) that are, in part, dependent on the marine ecosystem (J. G. Mead, pers. comm.). If we did include the bats, we would need to add the mammalian order Chiroptera to the group we call marine mammals, an addition with which many marine mammal scientists would disagree. We would also have to include animals that do not dive and change general statements we make regarding marine mammals (e.g., that they have large body size and are well insulated).

A very different example relates to religious beliefs; marine biologists consider dolphins to be marine mammals and yet Buddhist fishermen in Sri Lanka have, until recently, considered them fish and harvested them for food.

Order Carnivora

Five families of carnivorous mammals have marine representatives. The families Otariidae (sea lions and fur seals), Phocidae (true seals), and Odobenidae (walruses) are collectively grouped (by most current systematists) in the suborder Pinnipedia, a group that, until recently, was classified as a separate order of mammals. There are 34 extant and recently extinct species of pinnipeds (Reijnders et al. 1993). The family Ursidae (bears) is represented by a single marine species, the polar bear, and the family Mustelidae (weasels, otters, minks, etc.) by two marine otter species.

Despite being tied to land or ice to breed, the pinniped species are better adapted to a marine lifestyle than are the other carnivorous marine mammals. Pinnipeds are united by some obvious anatomical and behavioral features such as large body size (more than 5 m in length and 5000 kg in weight in the case of male southern elephant seals [*Mirounga leonina*]), presence of insulation (fur or blubber, or both), modified appendicular skeletons, the distal aspects of which form the framework for the characteristic flippers ("pinniped" means feather- or fin-footed), fusiform (spindle-shaped) bodies, ability to exist and to move around on land or ice, delayed implantation of the blastocyst during fetal development, and enhanced diving ability relative to terrestrial mammals.

For many people, the most familiar pinnipeds are members of the family Otariidae, which are common as trained or display animals at zoological parks and aquaria. Otariids are characterized by their ability to rotate the pelvis and bring the hind flippers under the body to permit walking or running on land; the presence of a small external ear or pinna (which has led people to call them "eared seals"); generally temperate-to-subtropical distribution; dense fur that consists of long, coarse, sparse, guard hairs and thick underfur in which air is trapped to promote insulation; large fore flippers that provide propulsion underwater; and sexual dimorphism and polygynous breeding. Some typical otariid features are illustrated in Figures 1-1 and 1-2. The family Otariidae is subdivided (Table 1-2) into two subfamilies: the Otariinae (sea lions) and the Arctocephalinae (fur seals).

The phocids are considered to be the "true seals." They may easily be recognized by the following traits: lack of external ears (hence the occasional use of the name "earless seals"); inability to rotate the pelvis to position the hind limbs under the body, which leads to relatively poor terrestrial locomotion; use of the pelvic appendage (i.e., pelvic flipper) for underwater propulsion; relatively small pectoral appendages that are held tightly against the body and used primarily for steering while swimming; excellent diving ability in some species; blubber for insulation; and, except in pups of

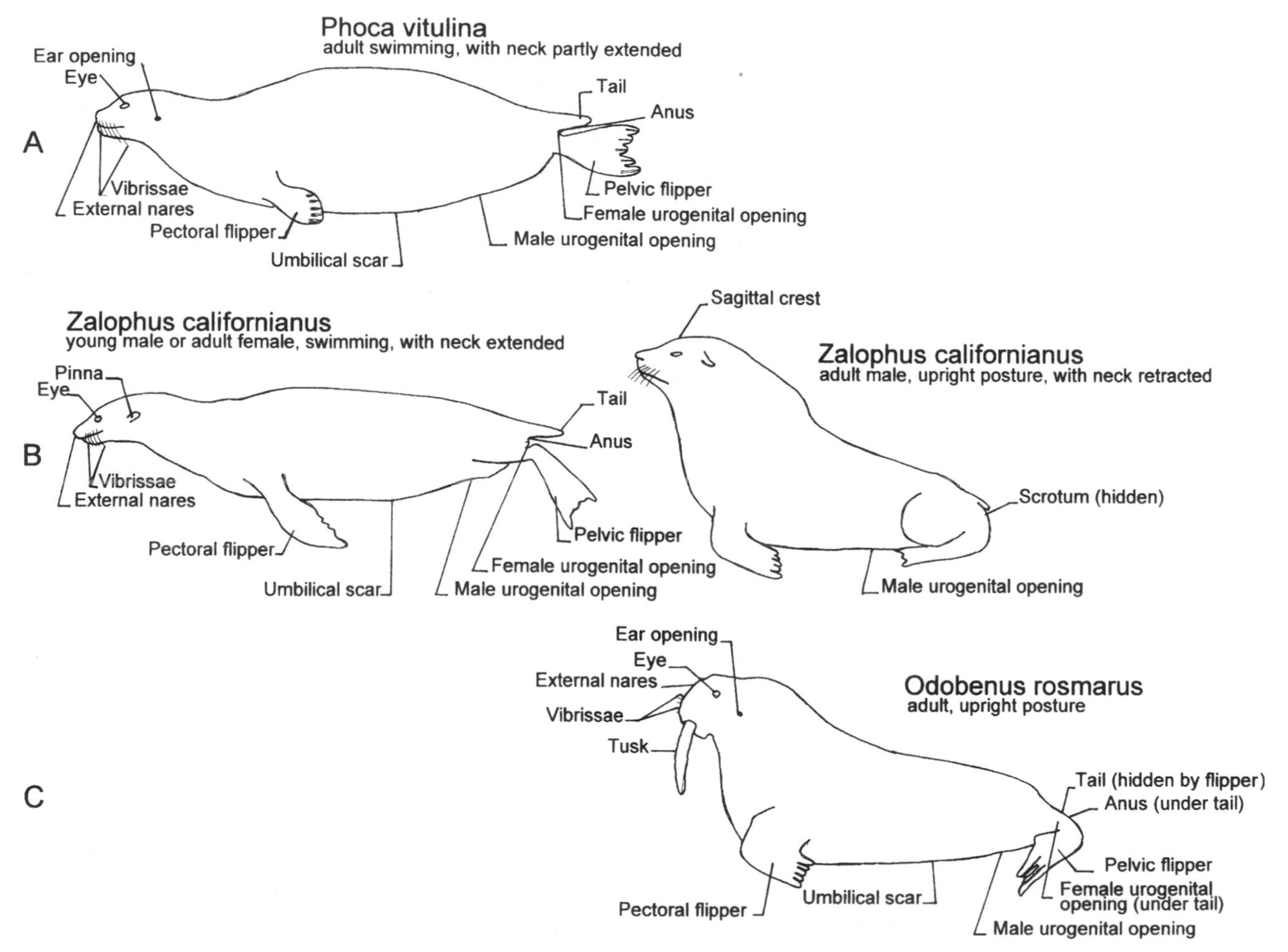

Figure 1-1. Drawings of representative pinnipeds with some diagnostic features labeled. (A) A phocid, or earless seal (some are called hair seals), represented by the harbor seal (*Phoca vitulina*); (B) otariids (fur seals and sea lions) (also called eared seals), represented by the California sea lion (*Zalophus californianus*); (C) the walrus, (*Odobenus rosmarus*). Recall that pinniped means feather- (or fin-) footed. Note that phocid seals change little in appearance when in the water or on land. Phocids commonly tuck their heads back against the thorax, thereby giving the appearance of a short neck, and they locomote in the water by lateral undulation of their pelvic flippers. Otariids and walruses can assume distinctly different postures on land by rotating their pelvises to position their hind flippers under their bodies. Note the enlarged forehead (sagittal crest) of the adult male otariid (sexual dimorphism). Eared seals propel themselves with their pectoral flippers when swimming. (Figure by S. A. Rommel.)

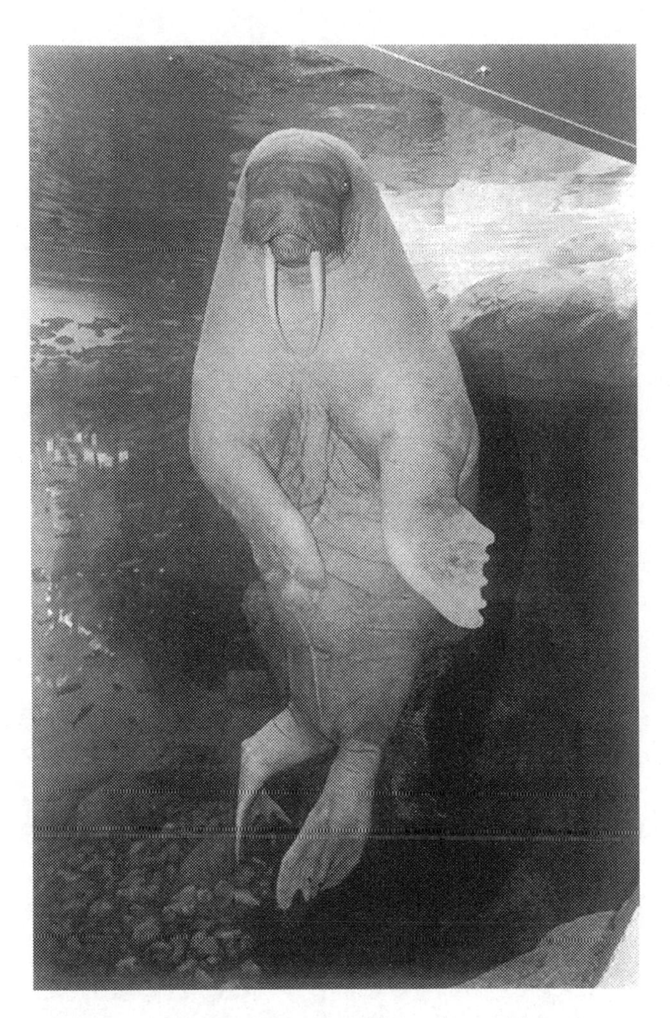

Figure 1-2. The upright posture of the walrus (*Odobenus rosmarus*) in this underwater photograph nicely illustrates the tusks, pectoral, and pelvic flippers, and relatively loose baggy skin. (Photograph courtesy of SeaWorld, Inc.)

some species, thin fur (relative to otariids) that does not trap air and becomes wetted to the skin when the seal is in the water (see Figs. 1-1 and 1-2 for illustration of some typical characteristics). Most phocid species are not sexually dimorphic, although elephant seals are a notable exception to this generality. Phocids, relative to otariids, are more streamlined and better adapted to an aquatic lifestyle. The family is generally divided (Table 1-3) into two subfamilies: the Phocinae (northern phocids) and the Monachinae (southern phocids).

The family Odobenidae (Table 1-4) is represented by one species (*Odobenus rosmarus*), with three subspecies: the Atlantic walrus (*O. r. rosmarus*); the Pacific walrus (*O. r. divergens*); and the Laptev walrus (*O. r. laptevi*) from the Laptev Sea, north of Siberia. Distinctive features of walruses include the presence of large tusks, an ability to rotate the pelvis so that the hind limbs are under the body (although the weight of adult animals cannot be supported to allow them to move on land as easily as otariids), underwater propulsion using either the

Table 1-2. Species and Common Names of Members of the Family Otariidae, the Eared Seals

Species	Subspecies	Common Name(s)
Subfamily Otariinae (sea lions)		
Eumetopias jubatus		Steller sea lion
		Northern sea lion
Neophoca cinerea		Australian sea lion
Otaria byronia		Southern sea lion
Phocarctos hookeri		Hooker's sea lion
		New Zealand sea lion
Zalophus californianus	*californianus*	California sea lion
	japonicus[a]	Japanese sea lion
	wollebaeki	Galapagos sea lion
Subfamily Arctocephalinae (fur seals)		
Arctocephalus australis	*australis*	Falkland fur seal
	gracilis	South American fur seal
A. forsteri		New Zealand fur seal
A. galapagoensis		Galapagos fur seal
A. gazella		Antarctic fur seal
A. philippii		Juan Fernandez fur seal
A. pusillus	*doriferus*	Australian fur seal
	pusillus	South African fur seal
A. townsendi		Guadalupe fur seal
A. tropicalis		Subantarctic fur seal
Callorhinus ursinus		Northern fur seal

Source: Reijnders et al. 1993.

Note that more than one common name is often used and recognized for particular species.

[a]Extinct.

pectoral appendages (similar to otariids) or the pelvic limbs (similar to phocids), absence of an external ear, virtually naked skin (which appears "warty" in adult males) with only a few sparse hairs, and thick blubber and skin (see Figs. 1-1 and 1-2).

The other carnivorous marine mammals resemble terrestrial mammals more than the pinnipeds. The polar bear (*Ursus maritimus*) (Table 1-4) is, in fact, a congener of some terrestrial bears (e.g., brown [*Ursus arctos*] and black bears [*Ursus americanus*]). The polar bear is distinctive due to its generally white pelage, very large size (males reach lengths of about 2.5 m), and a relatively streamlined form (e.g., lacking a shoulder hump, having small ears and head, and possessing a long neck) (see Figs. 1-3 and 1-4), and arctic habitat.

The mustelid representatives (Tables 1-1 and 1-4) are the sea otter (*Enhydra lutris*), found along the Pacific coast of North America and Russia, and the marine otter (*Lutra felina*) from coastal waters of Chile and Peru. There are three subspecies of sea otter recognized (Wilson et al. 1991): *E. l. gracilis* is found in the western north Pacific from the Kamchatka Peninsula through the Kuril Islands; *E. l. lutris* exists in the Commander Islands, the Aleutian Islands, and throughout central and southeastern coastal Alaskan waters; and *E. l. nereis* primarily occupies waters off central Cal-

Table 1-3. Species and Common Names for Members of the Family Phocidae, the Earless or True Seals

Species	Subspecies	Common Name(s)
Subfamily Phocinae (northern phocids)		
Cystophora cristata		Hooded seal
Erignathus barbatus	*barbatus*	Atlantic bearded seal
	nauticus	Pacific bearded seal
Halichoerus grypus		Gray seal
Phoca caspica		Caspian seal
P. fasciata		Ribbon seal
P. groenlandica		Harp seal
P. hispida	*botnica*	Baltic seal
	hispida	Arctic ringed seal
	ladogensis	Ladoga seal
	ochotensis	Okhotsk Sea ringed seal
	saimensis	Saimaa seal
P. largha		Larga seal
P. sibirica		Baikal seal
P. vitulina	*concolor*	Western Atlantic harbor seal
	mellonae	Ungava seal
	richardsi	Eastern Pacific harbor seal
	stejnegeri	Western Pacific harbor seal
	vitulina	Eastern Atlantic harbor seal
Subfamily Monachinae (southern phocids)		
Hydrurga leptonyx		Leopard seal
Leptonychotes weddellii		Weddell seal
Lobodon carcinophagus		Crabeater seal
Mirounga angustirostris		Northern elephant seal
M. leonina		Southern elephant seal
Monachus monachus		Mediterranean monk seal
M. schauinslandi		Hawaiian monk seal
M. tropicalis[a]		Caribbean monk seal
Ommatophoca rossii		Ross seal

Source: Reijnders et al. 1993.

Note that more than one common name may be used and recognized for particular species.

[a]Extinct.

ifornia (*sensu* Jefferson et al. 1993), but note that Reeves et al. (1992) use a somewhat different nomenclature. The sea otter, which is the largest mustelid species (with males measuring almost 1.5 m in length and weighing up to 45 kg), is characterized by extremely dense underfur, with sparse guard hairs, flattened hind feet for propulsion in water, retractile claws on the front feet only, lack of functional anal scent glands, a loose, axillary flap of skin to hold or store food and tools, posterior cheek teeth that lack cutting edges, a small pinna, and a horizontally flattened tail (see Figs. 1-3 and 1-4) (Reeves et al. 1992).

The marine otter is the smallest member of its genus (weighing only about 4.5 kg), which also includes the river otters. In fact, *L. felina* resembles river otters and may be confused with a congener, *L. provocax*, in Chile. Little is known about the biology of marine otters (Jefferson et al. 1993).

Table 1-4. Species and Common Names of Marine Mammals in the Families Odobenidae, Ursidae, and Mustelidae

Species	Subspecies	Common Name
Odobenidae		
Odobenus rosmarus	*divergens*	Pacific walrus
	laptevi	Laptev walrus
	rosmarus	Atlantic walrus
Ursidae		
Ursus maritimus		Polar bear
Mustelidae		
Enhydra lutris		Sea otter
	gracilis	
	lutris	
	nereis	
Lutra felina		Marine otter

Source: Jefferson et al. 1993.

Order Cetacea

With the advent of molecular techniques and the greater availability of specimen materials for biochemical and morphological comparisons, cetacean systematics has become a particularly dynamic field. In fact, although most biologists consider the cetaceans to be a separate and distinct order, Graur and Higgins (1994) used molecular evidence to make an unusual proposal, namely that the cetaceans simply be included as a suborder within the order Artiodactyla (the even-toed ungulates). Exact relationships within the order Cetacea and between cetaceans and other mammals are not entirely clear.

Systematists disagree about the exact number of cetacean species. Part of this uncertainty arises as a result of the availability of new data or application of new techniques for species that have been well known for some time; for example, Heyning and Perrin (1994) recently argued that there are two species of common dolphin (genus *Delphinus*) rather than just one. It is also exciting that part of the uncertainty is attributable to scientists continuing to discover new species, as recently occurred with the pygmy beaked whale (*Mesoplodon peruvianus*) (Reyes et al. 1991). In fact, the field of mammalogy in general is enlivened by the occasional discovery of new terrestrial and aquatic species.

Although we recognize that some controversy exists regarding cetacean systematics, a consensus has been more or less reached (Mead and Brownell 1993, Mead 1995). We list 78 living cetacean species (Tables 1-5 to 1-8) that comprise two suborders, the Mysticeti (baleen whales) and the Odontoceti (toothed whales). Considerable morphological and ecological variability exists among the cetaceans, but all members of the order share certain key characteristics.

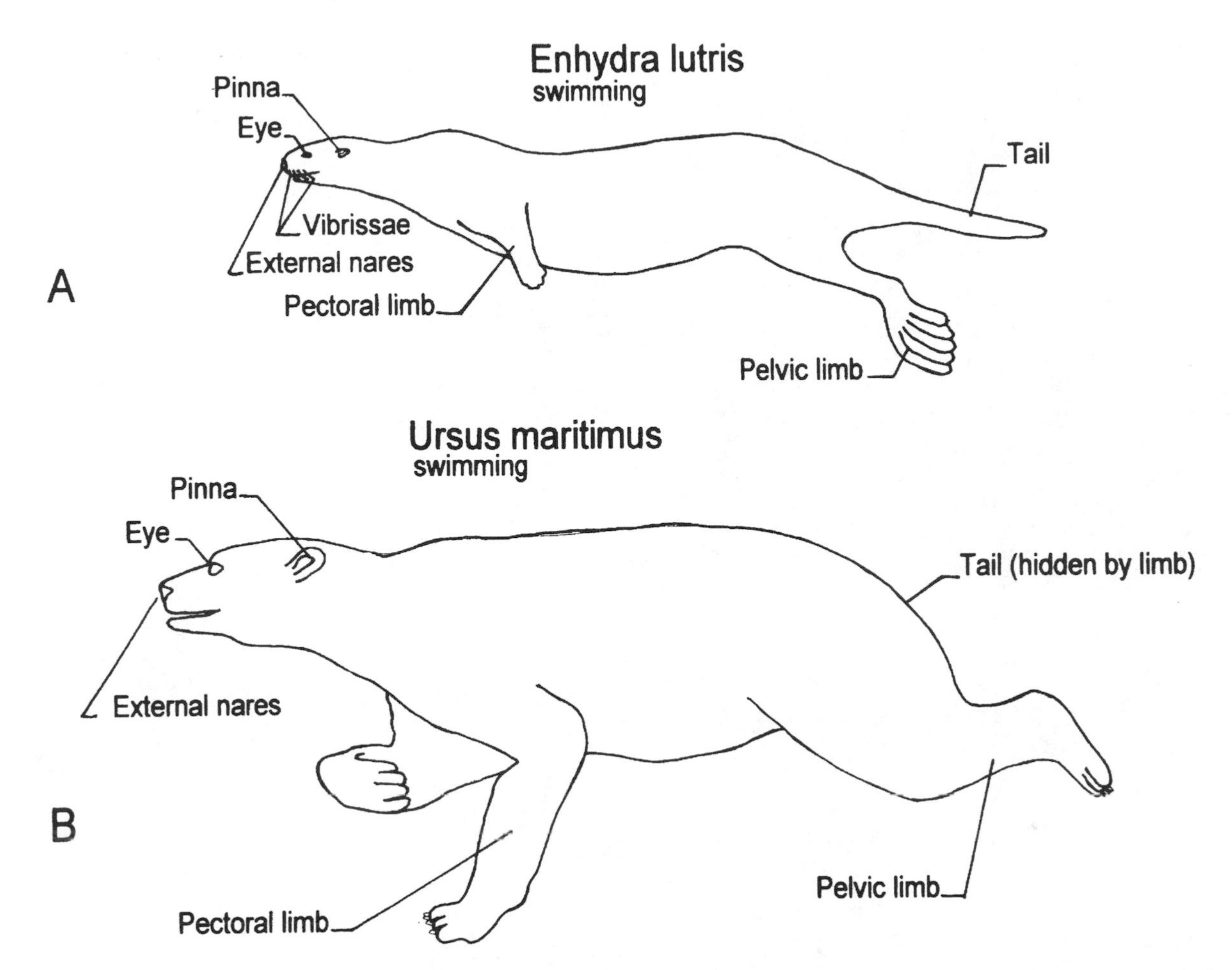

Figure 1-3. Drawings of (A) the sea otter (*Enhydra lutris*) and (B) the polar bear (*Ursus maritimus*), both in swimming postures. Note the small pectoral limbs and large pelvic limbs of the otter. The otter propels itself through the water by lateral undulations of its pelvic limbs. The polar bear typically propels itself through the water with its pectoral limbs, dragging its pelvic limbs passively, occasionally pushing off with the pelvic limbs if near solid objects. The long, loose hair of the polar bear adds to its apparent bulk. Note the absence of prominent vibrissae from the face of the polar bear. (Figure by S. A. Rommel.)

Cetaceans have become modified or extremely derived, relative to the "typical" mammal. This, of course, makes sense; one would expect mammals that are completely aquatic to possess very different adaptations than would exist in terrestrial mammals. Those adaptations range from obvious external morphological features to more subtle physiological attributes (for full discussion of these adaptations, see Pabst, Rommel, and McLellan, Chapter 2; Elsner, Chapter 3; Wartzok and Ketten, Chapter 4; and Costa and Williams, Chapter 5, this volume).

Externally, cetaceans are characterized by features that facilitate survival and free movement in the aquatic medium. All cetaceans are large animals (with adults of different species ranging in size from just under 2 m to more than 30 m in length), an attribute associated, in part, with thermoregulation (for an explanation, see Pabst, Rommel, and McLellan, Chapter 2, this volume). To at least some degree, they possess streamlined, fusiform bodies with minimal protuberances that could cause drag. For example, cetaceans lack external ears, external reproductive organs, pelvic appendages (although vestiges of the pelvic girdle and, rarely, the pelvic limb are present), and generally, hair (although all have some hair during fetal development and some possess a few hairs on the rostrum as adults). The pectoral skeleton is modified from the "typical" mammalian plan to produce paddlelike flippers, and the primary means of propulsion are the powerful tail flukes. A dorsal fin or ridge is present (to facilitate thermoregulation and dynamic stability) in some, but not all, species (see Figs. 1-5 and 1-6 for illustration of some diagnostic features of cetaceans).

Some additional, readily apparent characteristics of cetaceans include an elongated skull with overlapping bones (termed a telescoped skull), positioning of the nares (blowholes) dorsally, and a subdermal layer of blubber that can measure tens of centimeters thick in some species. Less obvious attributes characteristic of the order include po-

A

B

Figure 1-4. Photographs of (A) the sea otter (*Enhydra lutris*) and (B) the polar bear (*Ursus maritimus*). The otter photograph shows the individual in an upright posture; otters commonly float on their backs to groom, prepare and eat food, and play with their young. The polar bear is shown standing on ice in a posture typical of its terrestrial mode. (Polar bear photograph courtesy of SeaWorld, Inc.; sea otter photograph by Richard Bucich, Friends of the Sea Otter.)

rous, oil-filled bones, multiple stomach compartments, and physiological and anatomical adaptations relating to sensory physiology, diving, osmoregulation, and thermoregulation.

Although they share the various characteristics listed above, the mysticetes (Table 1-5) and odontocetes (Tables 1-6 to 1-8) can be readily distinguished by a number of features. The most obvious difference involves their feeding apparatus; mysticetes filter organisms from the water using baleen, which is made of plates of keratin (the substance of which fingernails are made), with a brushlike inner border, suspended from the upper jaw, whereas odontocetes possess teeth in one or both jaws. The size and number of baleen plates and the coarseness of the fringe vary among mysticete species and reflect food habits (see Bowen and Siniff, Chapter 9, this volume). Similarly, the number, shape, and size of odontocete teeth vary, and in females of some species (family Ziphiidae), teeth do not erupt through the gums.

There are other differences between mysticetes and

Table 1-5. Species and Common Names of Members of the Suborder Mysticeti, the Baleen Whales

Species	Common Name
Balaenidae (right whales)	
Balaena mysticetus	Bowhead whale
Eubalaena australis	Southern right whale
E. glacialis	Northern right whale
Neobalaenidae	
Caperea marginata	Pygmy right whale
Balaenopteridae (rorquals)	
Balaenoptera acutorostrata	Minke whale
B. borealis	Sei whale
B. edeni	Bryde's whale
B. musculus	Blue whale
B. physalus	Fin whale
Megaptera novaeangliae	Humpback whale
Eschrichtiidae	
Eschrichtius robustus	Gray whale

Source: Jefferson et al. 1993.

Note that more than one common name may be used and recognized for particular species.

odontocetes. Most odontocetes possess a dorsally asymmetrical skull, whereas mysticetes have a symmetrical one; odontocetes have only a single external blowhole and mysticetes have two. In addition, mysticetes have a nonsymphyseal mandible (it is symphyseal in odontocetes); mysticetes are always extremely large (with the smallest species being the pygmy right whale [*Caperea marginata*], which grows to just over 6 m long), and adult odontocetes vary in size from about 1.7 m long (e.g., the vaquita) to 18 m long (male sperm whales [*Physeter macrocephalus*]); and female mysticetes are larger than males (a trait sometimes referred to as "reverse sexual dimorphism," relating to energetics of reproduction), whereas in odontocetes, males of some social species are larger than females (e.g., in some dolphins, family Delphinidae), members of both sexes are the same size (e.g., in the Kogiidae), or females are larger than males (e.g., in the harbor porpoise, river dolphins, and beaked whales).

As noted above, there is no complete agreement regarding cetacean systematics. Following the identification key and organizational scheme of Jefferson et al. (1993), there are four families of mysticetes (Table 1-5): the Balaenidae (right and bowhead whales), Neobalaenidae (pygmy right whale), Balaenopteridae (rorquals), and Eschrichtiidae (gray whale). These investigators indicate that there are nine families of odontocetes (Tables 1-6 to 1-8): the Physeteridae (sperm whale), Kogiidae (pygmy and dwarf sperm whales), Ziphiidae (beaked and bottlenose whales), Monodontidae (beluga and narwhal), Delphinidae (true dolphins), Phocoenidae

Table 1-6. Species and Common Names of Members of the Families Physteridae, Kogiidae, Ziphiidae, and Monodontidae in the Suborder Odontoceti, the Toothed Whales

Species	Common Name
Physteridae	
Physeter macrocephalus	Sperm whale
Kogiidae	
Kogia breviceps	Pygmy sperm whale
K. simus	Dwarf sperm whale
Ziphiidae	
Berardius arnuxii	Arnoux's beaked whale
B. bairdii	Baird's beaked whale
Hyperoodon ampullatus	Northern bottlenose whale
H. planifrons	Southern bottlenose whale
Mesoplodon bidens	Sowerby's beaked whale
M. bowdoini	Andrews' beaked whale
M. carlhubbsi	Hubbs' beaked whale
M. densirostris	Blainville's beaked whale
M. europaeus	Gervais' beaked whale
M. ginkgodens	Ginkgo-toothed beaked whale
M. grayi	Gray's beaked whale
M. hectori	Hector's beaked whale
M. layardii	Strap-toothed whale
M. mirus	True's beaked whale
M. pacificus	Longman's beaked whale
M. peruvianus	Pygmy beaked whale
M. stejnegeri	Stejneger's beaked whale
Tasmacetus shepherdi	Tasman beaked whale
Ziphius cavirostris	Cuvier's beaked whale
Monodontidae	
Delphinapterus leucas	White whale
	Beluga
Monodon monoceros	Narwhal

Source: Jefferson et al. 1993.

Note that more than one common name may be used and recognized for particular species.

(porpoises), and the Platanistidae, Iniidae, and Pontoporiidae (river dolphins).

The members of the Balaenidae are huge, bulky whales with large heads and distinctly arched mouths containing baleen that can measure more than 5 m long in the bowhead whale (*Balaena mysticetus*); the common name "right whale" reflects the attitude among early whalers that these were the right whales to hunt because they were easily accessible, lived close to shore, dove poorly, swam slowly, produced lots of oil and other products, and floated when dead. Among mysticetes, the Balaenopteridae represent the opposite extreme compared to the Balaenidae; most rorquals are longer, more slender, and faster swimming than are right and bowhead whales. Additional diagnostic features of the rorquals include the presence of numerous ventral pleats (called throat grooves) that may extend from the cranial margin of

Table 1-7. Species and Common Names for Members of the Family Delphinidae in the Suborder Odontoceti, the Toothed Whales

Species	Common Name
Cephalorhynchus commersonii	Commerson's dolphin
C. eutropia	Black dolphin
C. heavisidii	Heaviside's dolphin
C. hectori	Hector's dolphin
Delphinus delphis	Common dolphin[a]
Feresa attenuata	Pygmy killer whale
Globicephala macrorhynchus	Short-finned pilot whale
G. melaena	Long-finned pilot whale
Grampus griseus	Risso's dolphin
Lagenodelphis hosei	Fraser's dolphin
Lagenorhynchus acutus	Atlantic white-sided dolphin
L. albirostris	White-beaked dolphin
L. australis	Peale's dolphin
L. cruciger	Hourglass dolphin
L. obliquidens	Pacific white-sided dolphin
L. obscurus	Dusky dolphin
Lissodelphis borealis	Northern right whale dolphin
L. peronii	Southern right whale dolphin
Orcaella brevirostris	Irrawaddy dolphin
Orcinus orca	Killer whale
Peponocephala electra	Melon-headed whale
Pseudorca crassidens	False killer whale
Sotalia fluviatilis	Tucuxi
Sousa chinensis	Indo-Pacific hump-backed dolphin
S. teuszii	Atlantic hump-backed dolphin
Stenella attenuata	Pantropical spotted dolphin
S. clymene	Clymene dolphin
S. coeruleoalba	Striped dolphin
S. frontalis	Atlantic spotted dolphin
S. longirostris	Spinner dolphin
Steno bredanensis	Rough-toothed dolphin
Tursiops truncatus	Bottlenose dolphin

Source: Jefferson et al. 1993.

Note that more than one common name may be used and recognized for particular species.

[a]Heyning and Perrin (1994) provide evidence that there are two species in the genus Delphinus: D. delphis (short-beaked common dolphin) and D. capensis (long-beaked common dolphin).

the lower jaw to the umbilicus, relatively short baleen, dorsal fins, and a jaw line with virtually no arch.

The Neobalaenidae and Eschrichtiidae are each represented by single species (Table 1-5), whose appearance is intermediate relative to the balaenids and balaenopterids. The pygmy right whale is small for a mysticete, is more streamlined than the balaenid whales, and possesses two throat grooves, a dorsal fin, and a moderately arched jaw. The gray whale is stockier than most other mysticetes (but not the balaenids) and possesses a moderately arched jaw, very coarse baleen, two to five throat grooves, a dorsal hump and several dorsal "knobs," and only four digits in the flipper (Jefferson et al. 1993).

Table 1-8. Species and Common Names of Members of the Families Phocoenidae, Platanistidae, Iniidae, and Pontoporiidae, in the Suborder Odontoceti, the Toothed Whales

Species	Common Name
Phocoenidae (porpoises)	
Phocoena dioptrica	Spectacled porpoise
Neophocaena phocaenoides	Finless porpoise
Phocoena phocoena	Harbor porpoise
P. sinus	Cochito
P. spinipinnis	Burmeister's porpoise
Phocoenoides dalli	Dall's porpoise
Platanistidae[a]	
Platanista gangetica	Ganges Susu
P. minor	Indus Susu
Iniidae[a]	
Inia geoffrensis	Boutu/Boto
Pontoporiidae[a]	
Lipotes vexillifer	Baiji
Pontoporia blainvillei	Franciscana

Source: Jefferson et al. 1993.

Note that more than one common name may be used and recognized for particular species.

[a]Members of these three families are collectively referred to as river dolphins.

The odontocetes are more variable than the mysticetes. The sperm whales, members of the families Physteridae and Kogiidae (Table 1-6), share certain traits, including the presence of a spermaceti organ in the head, extreme skull asymmetry, and small underslung lower jaws; the sperm whale is much larger and has a much larger, squarer head than does *Kogia;* however, the *Kogia* skull is wider than it is long, unusual in mammals. Like the sperm whales, the beaked and bottlenose whales (Table 1-6) are deep divers; on close examination, the latter are easily recognized by the presence of a pair of throat grooves (unusual in odontocetes), a pronounced beak, a small, caudad dorsal fin, and only one or two pairs of teeth that may not penetrate the gums (except in *Tasmacetus shepherdi,* which has 17 to 29 functional teeth in each tooth row of the upper and lower jaws, plus two larger teeth at the tip of the mandible [Leatherwood et al. 1983]).

The members of the Monodontidae (Table 1-6) are intermediate-sized, stocky whales found in arctic and subarctic regions. They have bulbous heads and relatively large unfused cervical vertebrae, which allow more neck flexibility than most other cetaceans. Among the most unusual anatomical features observed among cetaceans is the tusk (which can exceed 3 m in length) of the narwhal (*Monodon monoceros*).

The odontocetes with which most people are familiar include members of the family Delphinidae (Table 1-7). In

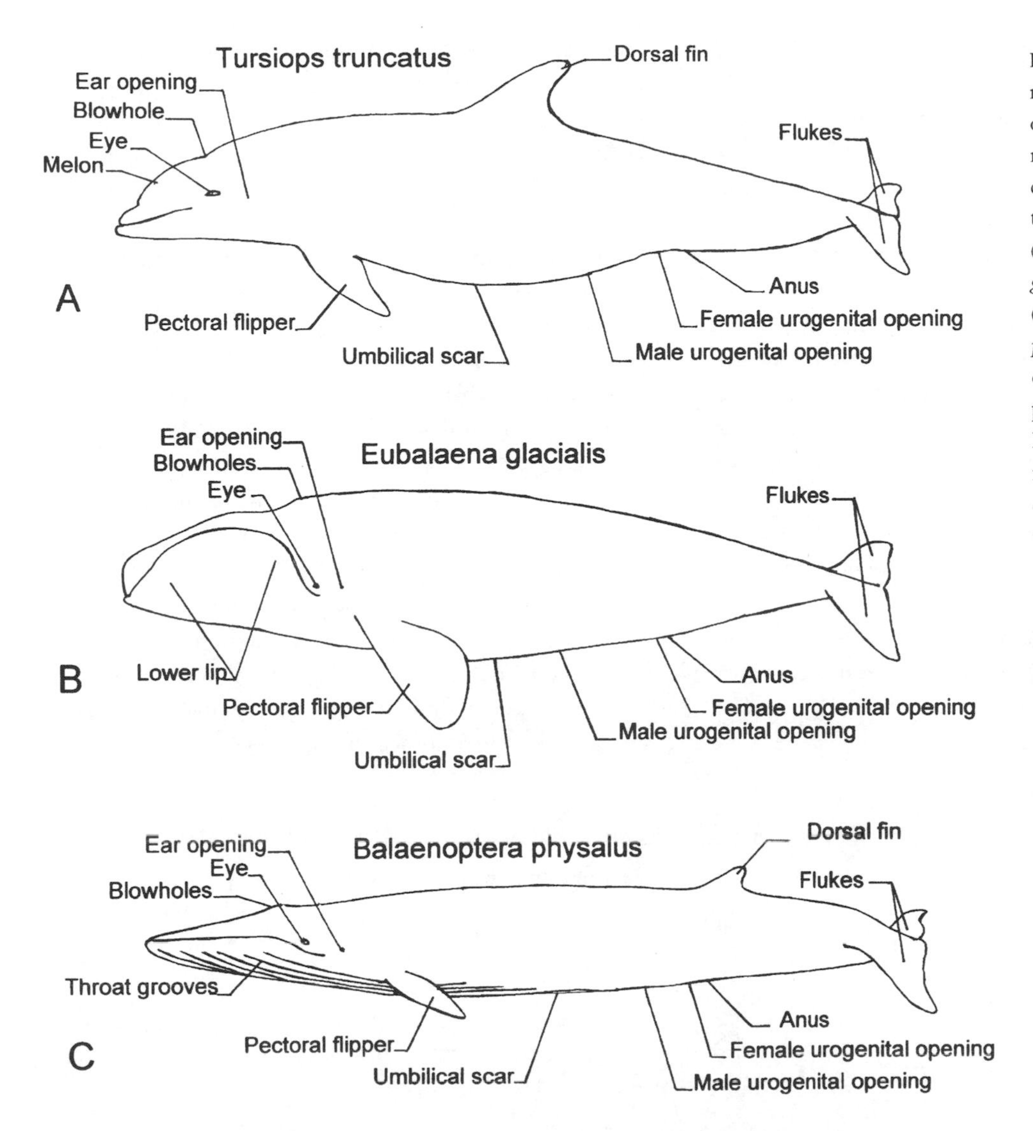

Figure 1-5. Drawings of representative cetaceans. (A) An odontocete or toothed whale represented by the bottlenose dolphin (*Tursiops truncatus*) and two mysticetes or baleen whales (B), the right whale (*Eubalaena glacialis*) and (C) the finback (or finner) whale (*Balaenoptera physalus*). The cetaceans are characterized by the absence of pelvic limbs but are graced with large caudal appendages called flukes. Some cetaceans have dorsal fins, midline fleshy structures that may help stabilize them hydrodynamically. Cetaceans also have a pair of pectoral flippers that help them steer. The rorquals ("gulpers" such as the finner) have throat grooves to allow distention of the ventral throat and thorax to accommodate the large volumes of water they take in when they feed. Compare the finback with the right whale (a "skimmer") that moves its large lower lips to funnel water around its relatively long baleen. The rorquals are more slender and are faster swimmers than the right whales.

Figure 1-6. An example of an odontocete cetacean is the killer whale (*Orcinus orca*). (Photograph courtesy of SeaWorld, Inc.)

cluded in the group are such animals as killer whales (*Orcinus orca*) and bottlenose dolphins (*Tursiops truncatus*). The common name "dolphin" can create confusion because large marine teleosts (bony fish), *Coryphaena* spp., are also called dolphins. Delphinids are medium-sized whales that have conical teeth (see Fig. 1-7), and generally possess a beak and a large falcate dorsal fin located about midway along the back.

Common names, as indicated above, can create confusion; in fact, some delphinids may actually be referred to by laypersons as porpoises (e.g., *T. truncatus* is sometimes referred to as the bottlenose porpoise). The true porpoises, Phocoenidae (Table 1-8), are separate from the delphinids; the former have flattened, spade-shaped teeth (see Fig. 1-7), short, triangular dorsal fins, indistinct, if any, beaks, and more bilaterally symmetrical skulls. Phocoenids are generally small in size.

The various river dolphins (Pontoporiidae, Platanistidae, and Iniidae) (Table 1-8) collectively possess long beaks with numerous teeth; one species, the franciscana (*Pontoporia blainvillei*) is actually coastal in its distribution, rather than confined to rivers. All species are small and have very modified skulls relative to other odontocetes.

Order Sirenia

The members of the order Sirenia (Table 1-9) are totally aquatic, as are the cetaceans, but they inhabit shallow waterways where they feed primarily, if not exclusively, on plants. Because they occupy tropical and subtropical waters, do not dive deeply for food, and apparently lack the sophisticated social behavior characteristic of many cetacean species, the sirenians are considered to show less specialization for an aquatic lifestyle than the whales, dolphins, and porpoises.

That is not to say that sirenians are not extremely derived relative to "typical" mammals. To begin with, their herbivorous diet requires specialized digestive tract anatomy and physiology, more similar to that of an elephant or horse than to that of most other mammals. In addition, sirenians, like other marine mammals, are large (occasionally more than 4 m long in the case of the West Indian manatee, *Trichechus manatus*), fusiform animals that lack external ears and pelvic

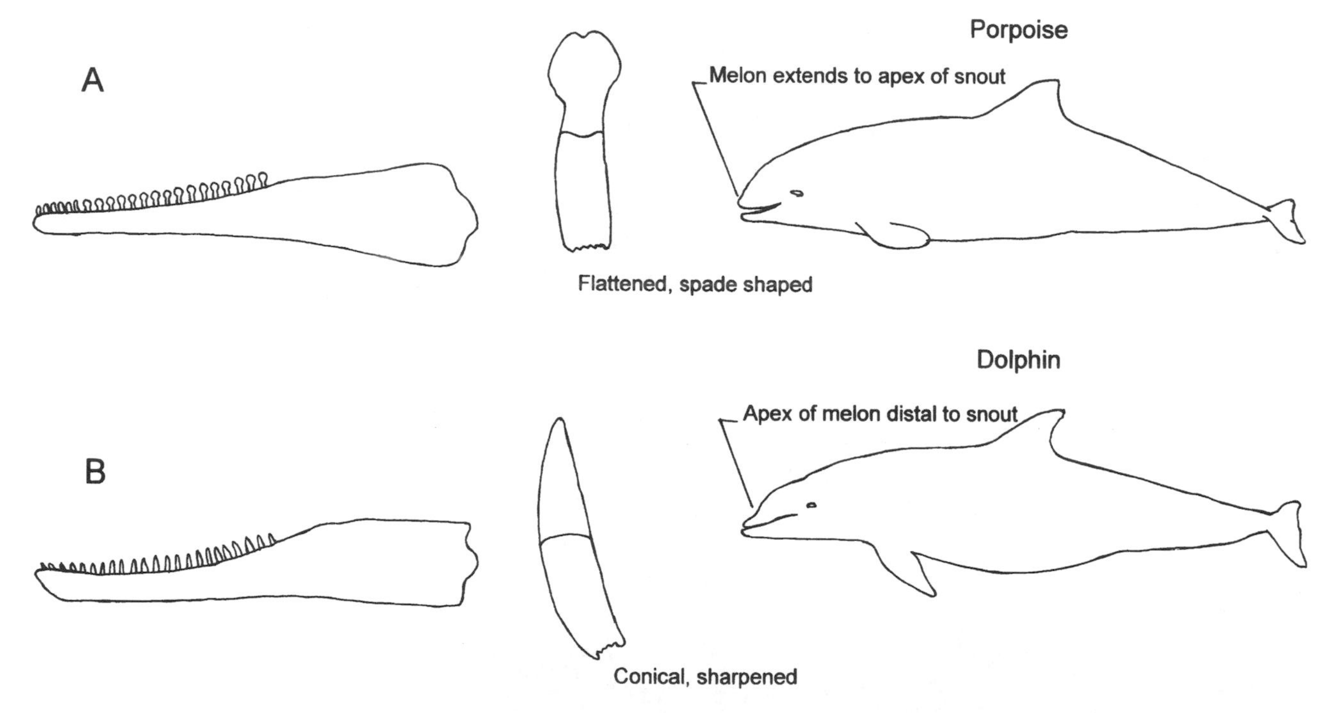

Figure 1-7. Drawings contrasting features of the head and teeth of a representative porpoise and a representative dolphin. One of the most commonly asked questions concerns the difference between dolphins and porpoises. The porpoises have flattened, spade-shaped teeth and the lower, anterior margin of the melon extends all the way to the margin of the upper jaw or beak—there is no "bottle-shaped nose." The dolphins have conical, pointed (when young and unworn) teeth and they have a distinct rostral narrowing of the beak (a bottleneck or bottle-shaped nose) where the apex of the melon does not extend to the tip of the rostrum. It is interesting to note that with old age the teeth of both porpoises and dolphins wear down through abrasion with ingested material and each other. Some individuals have flattened teeth—in fact the name *truncatus* (of the now-named *Tursiops truncatus*) is derived from the truncated (completely flattened) appearance of the teeth in the original specimen (Montague 1821). (Figure by S. A. Rommel.)

Table 1-9. Species and Common Names of Members of the Order Sirenia, the Sea Cows

Species	Subspecies	Common Name
Trichechidae		
Trichechus inunguis		Amazonian manatee
T. manatus		West Indian manatee
	latirostris	Florida manatee
	manatus	Antillean manatee
T. senegalensis		West African manatee
Dugongidae		
Dugong dugon		Dugong
Hydrodamalis gigas[a]		Steller's sea cow

Source: Reynolds and Odell 1991.

Note that more than one common name may be used and recognized for particular species.

[a]Extinct.

limbs, possess paddlelike pectoral appendages and sparse hair, and use a powerful fluke for locomotion (see Fig. 1-8). Sirenian nostrils are located on top or at the tip of the muzzle, but are not as derived as cetacean blowholes. Some other characteristics of sirenians include axillary mammary glands, dense and heavy (i.e., pachyostotic) bones, thick, tough skin, and, like other marine mammals, some morphological and physiological adaptations to facilitate diving, thermoregulation, osmoregulation, and communication in an aquatic medium.

The only extant species in the family Dugongidae is the dugong (*Dugong dugon*), found in tropical and subtropical areas of the Indo-Pacific. The demise of a dugongid, Steller's sea cow, in 1768 is documented by a number of references including Reynolds and Odell (1991). The dugong is easily distinguished from the various species of manatees (family Trichechidae); for example, dugongs possess tusks (lacking in manatees), split dolphinlike flukes (rounded in manatees), and a rostrum that is strongly deflected downward to facilitate bottom feeding. Furthermore, dugongs are much more marine in their distribution than are the various species of manatees.

Conclusion

These, then, are the marine mammals, a group of animals that have responded evolutionarily in similar ways to selective pressures that, to humans, may seem quite foreign. The chapters that follow explore marine mammal evolution and explain those pressures and how marine mammals have become adapted to an aquatic, if not marine, environment. At the end of the text, it may be worthwhile for students to ask the question: altogether, are marine mammals so different biologically from humans and other mammals after all?

Acknowledgments

We thank Dr. James G. Mead and William A. McLellan for their helpful comments on the chapter. We are also grateful to the various individuals and organizations who made their photographs available for our use.

Literature Cited

Aguilar, A., and A. Borrell. 1994. Abnormally high polychlorinated biphenyl levels in striped dolphins (*Stenella coeruleoalba*) affected by the 1990–1992 Mediterranean epizootic. The Science of the Total Environment 154:237–247.

Brownell, R. L., K. Ralls, and W. F. Perrin. 1995. Marine mammal biodiversity. Oceanus 38:30–33.

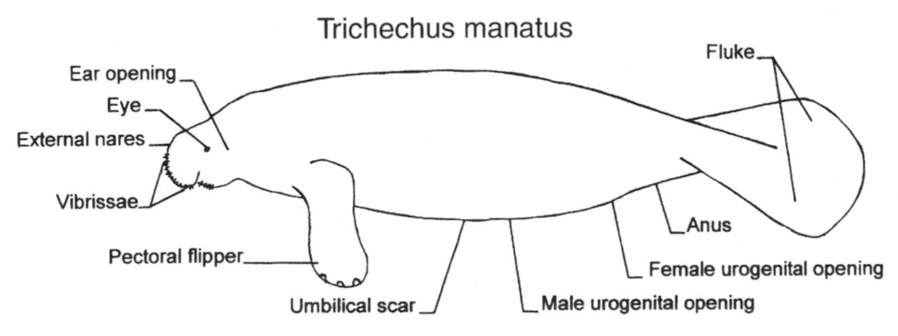

Figure 1-8. Drawing of a manatee (*Trichechus manatus*) showing the lack of hind limbs and a dorsoventrally flattened fluke, not substantially unlike the cetacean flukes. There is no dorsal fin and the pectoral limbs or flippers are much more mobile than those of the cetacea; it is common to see manatees with their flippers folded across their chests or manipulating food into the mouth. Also note the curve of the flippers is more rounded in the front than in the cetacea. The vibrissae are robust but short and the body hairs are fine but sparse to give a nude appearance to the skin of the manatee. The eyes of manatees are small and, unlike other mammals, use a sphincter to close rather than closing with distinct upper and lower eyelids (palpebrae). (Figure by S. A. Rommel.)

Clark, J. A. 1994. The Endangered Species Act: Its history, provisions and effectiveness. Pages 19–43 *in* T. W. Clark, R. P. Reading, and A. L. Clarke (eds.). Endangered Species Recovery. Island Press, Washington, D.C.

Domning, D. P. 1991. Why save the manatee: A scientist's perspective. Pages 167–173 *in* J. E. Reynolds III, and D. K. Odell (eds.). Manatees and Dugongs. Facts on File, Inc., New York, NY.

Geraci, J. R., and V. J. Lounsbury. 1993. Marine Mammals Ashore: A Field Guide for Strandings. Texas A&M Sea Grant Publication TAMU-SG-93-601, Galveston, TX.

Graur, D., and D. G. Higgins. 1994. Molecular evidence for the inclusion of cetaceans within the order Artiodactyla. Molecular Biology and Evolution 11:357–364.

Heyning, J. E., and W. F. Perrin. 1994. Evidence for two species of common dolphins (genus *Delphinus*) from the eastern North Pacific. Contributions in Science, Natural History Museum of Los Angeles County, CA, Number 442:1–35.

Jefferson, T. A., S. Leatherwood, and M. A. Webber 1993. Marine Mammals of the World. FAO Species Identification Guide, Food and Agriculture Organization of the United Nations, Rome.

Leatherwood, S., and R. R. Reeves (eds.). 1990. The Bottlenose Dolphin. Academic Press, San Diego, CA.

Leatherwood, S., R. R. Reeves, and L. Foster. 1983. Sierra Club Handbook of Whales and Dolphins. Sierra Club Books, San Francisco, CA.

Marine Mammal Commission. 1995a. Annual Report to Congress 1994. Marine Mammal Commission, Washington, D.C.

Marine Mammal Commission. 1995b. The Marine Mammal Protection Act of 1972 as Amended. Marine Mammal Commission, Washington, D.C.

Mead, J. G. 1995. Biodiversity in cetaceans. Pages 41–46 *in* R. Arai, M. Kato, and Y. Doi (eds.). Biodiversity and Evolution. The National Science Museum Foundation, Tokyo.

Mead, J. G., and R. L. Brownell Jr. 1993. Order Cetacea. Pages 349–364 *in* D. E. Wilson and D. M. Reeder (eds.). Mammalian Species of the World, 2nd ed. Smithsonian Institution Press, Washington, D.C.

Montague, G. 1821. Description of a species of *Delphinus*, which appears to be new. Memoirs of the Wernerian Natural History Society III:75–82 + Plate III.

Reeves, R. R., B. S. Stewart, and S. Leatherwood. 1992. The Sierra Club Handbook of Seals and Sirenians. Sierra Club Books, San Francisco, CA.

Reidman, M. 1990. The Pinnipeds: Seals, Sea Lions, and Walruses. University of California Press, Berkeley, CA.

Reijnders, P., S. Brasseur, J. van der Toorn, P. van der Wolf, I. Boyd, J. Harwood, D. Lavigne, and L. Lloyd. 1993. Seals, Fur Seals, Sea Lions, and Walrus. Compiled by the IUCN / SSC Seal Specialist Group. International Union for the Conservation of Nature and Natural Resources, Gland, Switzerland.

Reyes, J. C., J. G. Mead, and K. Van Wearebeek. 1991. A new species of beaked whale *Mesoplodon peruvianus* sp. n. (Cetacea: Ziphiidae) from Peru. Marine Mammal Science 7:1–24.

Reynolds, J. E. III, and D. K. Odell. 1991. Manatees and Dugongs. Facts on File, Inc., New York, NY.

Ridgway, S. H., and R. J. Harrison (eds.). 1981a. Handbook of Marine Mammals, Vol. 1: The Walrus, Sea Lions, Fur Seals, and Sea Otter. Academic Press, New York, NY.

Ridgway, S. H., and R. J. Harrison (eds.). 1981b. Handbook of Marine Mammals, Vol. 2: Seals. Academic Press, New York, NY.

Ridgway, S. H., and R. J. Harrison (eds.). 1985. Handbook of Marine Mammals, Vol. 3: The Sirenians and Baleen Whales. Academic Press, New York, NY.

Ridgway, S. H., and R. J. Harrison (eds.). 1989. Handbook of Marine Mammals, Vol. 4: River Dolphins and the Larger Toothed Whales. Academic Press, New York, NY.

Ridgway, S. H., and R. J. Harrison (eds.). 1994. Handbook of Marine Mammals, Vol. 5: The First Book of Dolphins. Academic Press, New York, NY.

Ridgway, S. H., and R. J. Harrison (eds.). 1999. Handbook of Marine Mammals, Vol. 6: The Second Book of Dolphins and of Porpoises. Academic Press, San Diego, CA.

Twiss, J. R., and R. R. Reeves (eds.). 1999. Conservation and Management of Marine Mammals. Smithsonian Institution Press, Washington, D.C.

Wilson, D. E., M. A. Bogan, R. Brownell Jr., A. M. Burdin, and M. K. Maminov. 1991. Geographic variation in sea otters, *Enhydra lutris*. Journal of Mammalogy 72:22–36.

2

D. ANN PABST, SENTIEL A. ROMMEL, AND WILLIAM A. McLELLAN

The Functional Morphology of Marine Mammals

Structure without function is a corpse; function sans structure is a ghost. Vogel and Wainwright 1969

How is the mammalian body adapted to the marine environment? We use this question as our foundation for investigating the diversity of morphological form and function in marine mammals. It is important to remember that marine mammals do not form a monophyletic group. "Marine mammals" is simply a descriptor (see Reynolds, Odell, Rommel, Chapter 1, this volume)—it should be read as "a diverse assemblage of distantly related mammals that spend some or all of their life in the marine environment." It is the diversity of form that is our focus throughout this chapter.

We tend to think about terrestrial mammals moving through the fluid environment of air. In contrast, physical characteristics of the aquatic environment present certain functional challenges to the mammalian body. A few cursory examples illustrate the point. (1) The density of seawater is almost three orders of magnitude greater, and the viscosity approximately 60 times greater, than air at a similar temperature (Vogel 1994). These properties influence drag—the forces that tend to resist movement of a body through a fluid (for review, see Alexander 1983; Vogel 1988, 1994; Fish 1993a). Thus, drag forces are much greater on the mammalian body moving through water than through air. (2) The thermal conductivity coefficient of water (i.e., the ability to transfer heat from a body in physical contact with water) is 25 times that of air at an equivalent temperature (Schmidt-Nielsen 1990). This means that heat can be removed from the mammalian body minimally 25 times faster in water than in air. (3) The aquatic medium severely attenuates light energy at rates much greater than through air (for review, see Au 1993). Thus, a diving marine mammal might literally be in the dark upon reaching depths of as little as a few meters. These three examples suggest that the physical properties of water profoundly influence how marine mammals perceive and respond to their aquatic environment.

The goal of this chapter is to offer an overview of existing morphological "solutions," as we understand them to date, to the functional "problems" posed by a marine existence. Our approach will be comparative—how are cetaceans, sirenians, pinnipeds, polar bears, and sea otters built, and how do their body plans compare to that of a "typical" terrestrial mammal? Because there is no such thing as a typical terrestrial mammal, we have arbitrarily decreed one—the dog—because it is described well anatomically (e.g., Evans and de Lahunta 1980, Evans 1993) and because it is a familiar mammal to most of us. We tour the major organ systems in the body (e.g., integument, musculoskeletal, nervous, respiratory), and determine what functional challenges may be posed to each system by the aquatic environment and what morphological solutions are observed. Because different body systems often function together (e.g., buoyancy control is an integrated function of integument, musculoskeletal, and respiratory systems), we will revisit a number of functions throughout our tour. The reader is encouraged to consult Elsner (Chapter 3, this volume) for another perspective on function. (We have not included the endocrine or lymphatic systems in our survey because few data on the gross morphology of these organs exist for any marine mammal; see a recent review of microscopic structure by Romano et al. 1993.)

A number of different system "designs" (*sensu* Lauder 1982) have evolved in marine mammals. Our goal will be to compare broadly these diverse morphologies within a functional context rather than investigate the detailed anatomy of any one species. In so doing, we put to use techniques that we all use to interpret and interact with our physical world—we are all practicing functional morphologists. As an example, look at Figure 2-1. By simply visually observing the morphologies of these animals you can make functional pre-

Figure 2-1. Representative extant marine mammals. The arrangement of this figure will be the template for many of the figures throughout the chapter. Our "typical" mammal, the dog (*Canis familiaris*), is at the center, with representative species of marine mammals at the periphery. Clockwise from the top are the California sea lion (*Zalophus californianus*), sea otter (*Enhydra lutris*), West Indian manatee (*Trichechus manatus*), right whale (*Eubalaena* sp.), bottlenose dolphin (*Tursiops truncatus*), polar bear (*Ursus maritimus*), and harbor seal (*Phoca vitulina*). We wish to make it clear that the dog should not be considered the "generic" mammal, as it is, in fact, highly specialized. Neither should the dog be considered as a terrestrial "ancestor" to the marine mammals—it is not. The dog is simply a representative mammal familiar to all of us and is well described anatomically (Evans 1993). The figures are intended to illuminate major patterns in systematic morphology displayed in marine mammals. Not all major groups are represented in the figure; for example, walruses, dugongs, porpoises, river dolphins, rorquals, and beaked, gray, sperm, and monodontid whales are absent. We chose the species we depict, in part, because they are better represented in the anatomical literature and are illustrative in a broadly comparative context. (Illustrations by S. A. Rommel adapted from schematics or photographs in Murie 1872, Tarasoff et al. 1972, Nickel et al. 1986, Stirling 1988, Rommel 1990, Reynolds and Odell 1991, Reeves et al. 1992, Evans 1993, and the authors' unpublished data.)

dictions about how each moves through its environment. You can surmise, for example, that the dog and polar bear are ambulatory (although both can swim), whereas cetaceans and sirenians probably would not fare well on land. This is a simplistic example, but it illustrates that we all have a strong intuitive sense of how morphological structure influences function.

Although the goal of this chapter is not to be encyclopedic, we do hope to offer a guide to the voluminous primary literature on marine mammal anatomy. We also suggest three secondary sources from which we have borrowed when preparing this chapter—*Animal Physiology* by K. Schmidt-Nielsen (1983, 1990), *Whales* by E. J. Slijper (1979), and *Seals of the World* by Judith E. King (1983). Schmidt-Nielsen's text is, we believe, the finest synthetic treatise on comparative physiology ever written. Slijper's work, although a bit outdated, is still the most accessible and comprehensive anatomical treatment of cetaceans for the nonspecialist. King's text combines anatomical information with the natural history and evolution of pinnipeds. We would encourage all interested readers to search out these valuable references.

A few notes about our approach. We have attempted to use fundamental physical principles to illustrate the functional "problems" posed by the marine environment. Thus, we have used a number of equations to identify those physical parameters. This approach works especially well with the integumentary and musculoskeletal systems, as they interact directly with the fluid environment. Our approach is less quantitative as we tour other body systems. Throughout the chapter, we also investigate the effect of body size on form and function. Note that although the marine mammals in Figure 2-1 are all illustrated at a similar size, their body masses span three orders of magnitude (Table 2-1).

Table 2-1. Total Mass and Total Lengths for Representative Species

Species	Total mass (kg)	Total length (m)
Dog (Chihuahua "Sticker")	2	0.2
Northern fur seal—female	30	1.4
Franciscana dolphin	32	1.7
Sea otter	45	1.5
Dog (Great Dane "River")	80	1.5
Harbor seal	140	1.9
Amazonian manatee	450	3.0
Bottlenose dolphin	650	4.0
Polar bear	800	2.5
Northern sea lion	1100	3.2
Walrus	1200	3.2
Elephant seal	5000	5.0
Steller's sea cow	10,000	7.0
Sperm whale—male	45,000	18.5
Right whale	90,000	17.7
	increases by four orders of magnitude	increases by two orders of magnitude

Integument

The integument of marine mammals is a multifunctional organ system that sculpts the animal's boundary with its aquatic environment. It forms a protective and dynamically insulative layer, adds buoyancy, and forms propulsive structures, such as the flukes of cetaceans and sirenians.

A view of the typical mammalian integument, a multilayer tissue complex, is schematically represented in Figure 2-2 (for general descriptions, see Sisson and Grossman 1953, Tregear 1966, Freeman and Bracegirdle 1968, Spearman 1973, Schummer et al. 1981). The most superficial layer, the epidermis, is formed by many layers of epithelial cells. The outer cell layers are often keratinized and provide waterproofing to the mammalian skin. Hair, vibrissae, nails, claws, and sebaceous (oil) and sweat glands are all specialized epidermal structures. The underlying dermis is formed by two regions of connective tissue. The papillary dermis projects into the epidermis in fingerlike processes; each papilla carries blood vessels and nerves to a position near the overlying epidermal cells. Oxygen and nutrients diffuse from these vessels into the epidermal cells. The deeper reticular dermis forms a connection with the underlying hypodermis. The hypodermis is a more loosely woven connective tissue layer, often invested with a large number of adipose cells.

Although the skins of all marine mammals conform to this general plan and perform certain common functions, the morphologies of the integument are highly variable across the marine mammals. We focus our comparative investigation on three functions of the integument—thermoregulation, drag reduction, and buoyancy control. What structural features of the generalized mammalian integument are specialized in marine mammals to permit these functional roles?

Thermoregulation

The heat produced by an animal must be transported to the surface before it can be transferred to the environment.
Schmidt-Nielsen 1990

Mammals are homeothermic endotherms, that is, they maintain a high (36 to 38°C) and relatively stable body temperature. (See Schmidt-Nielsen 1990 for a full discussion of body temperature including advantages and disadvantages

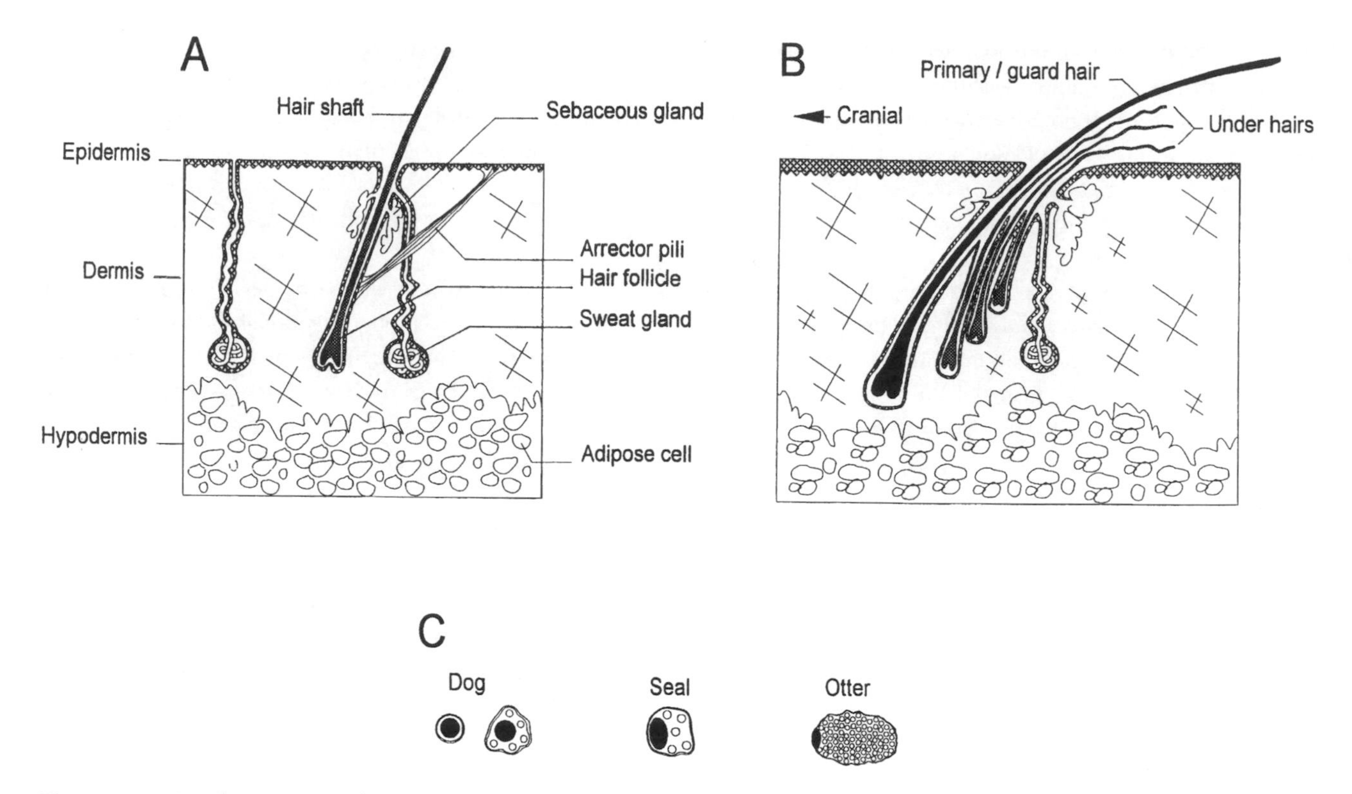

Figure 2-2. (A) Schematic view of typical mammalian integument. The integument is formed by three tissue layers, the epidermis, dermis, and hypodermis. The epidermis, the superficial-most layer, is formed by pigmented, stratified squamous epithelium. The epidermal surface is formed by dead, keratin- (a protective waterproofing protein) filled squamous cells. The underlying dermis is formed by two layers of connective tissue. The papillary dermis projects into the epidermis in fingerlike processes. The deeper reticular dermis forms a connection with the underlying hypodermis. The hypodermis is a more loosely woven connective tissue layer, often invested with a large volume of adipose cells. Specialized features of the integument include the following. Hairs, which are formed by keratinized epidermal cells, grow in canals called follicles. Associated with hair follicles are sebaceous glands that produce a thick oily secretion that lubricates and conditions the hair and skin surface. Sweat glands produce a watery, and sometimes salty, secretion that is used to cool the body surface by evaporation. Arrector pili muscles (smooth muscle) are associated with most hair follicles and act to make the hair stand "on end" and, in addition, may act to squeeze oils from the sebaceous gland. (Illustrations adapted from Schummer et al. 1981.) (B) Schematic representation of hair structure in a variety of marine mammals. The primary or guard hair emerges on the cranial edge of the hair canal. Behind the guard hair emerges multiple softer, smaller diameter, and shorter underhairs. The number of underhairs per hair canal can vary from 0 to 100. Elephant and monk seals, manatees, and cetaceans lack underhairs altogether, whereas fur seals and sea otters have the largest number of underhairs. The large number of underhairs per canal in fur seals is an interesting artifact of their molt (shedding of hair coat). Although the hair canal might contain only 15 follicles, it will often contain 45 to 50 underhairs. The extra underhairs are apparently accumulated over a number of molts (Scheffer and Johnson 1963, Scheffer 1964a,b, 1968). Sea otters can have 100 underhairs per hair canal, but it is not clear from the literature whether sea otters maintain unshed underhairs between molts. (C) Cross-sections taken just above the surface of the epidermis. Notice that the guard hairs are flattened in cross-section in the marine mammals and round in cross-section in the dog.

of homeothermy.) To maintain a constant body temperature, a mammal must produce metabolic heat at the same rate at which heat is removed from its body. Marine mammals live in a fluid environment that conducts heat away from the body 25 times faster than in air. Does this mean that marine mammals must produce metabolic heat at rates 25 times those of their terrestrial counterparts? Although still controversial (see Brodie 1975, Ryg et al. 1993), available physiological data suggest that most marine mammals do not have significantly elevated metabolic rates relative to terrestrial mammals (for review, see Lavigne et al. 1986; Elsner, Chapter 3, this volume). Thus, most marine mammals must cope with the increased thermal conductivity coefficient of water, and therefore, the potential for increased heat losses, by ways other than increased heat production.

Equation 1 defines those variables that influence rate of conductive heat loss, H', measured in watts (W), from a body:

$$H' = (SA)\,C\,(T_b - T_a), \qquad \text{(equation 1)}$$

where SA is surface area, measured in square meters (m^2), across which heat is transferred, *C* is thermal conductance ($Wm^{-2}°C^{-1}$) of a given body, T_b is body temperature and T_a is ambient water temperature (°C) (from Resnick and Halliday 1966, Costa and Kooyman 1982, Schmidt-Nielsen 1990; see Appendix for list of abbreviations). This rather intuitive equation suggests that marine mammals can adjust three variables to decrease their heat flow.

Marine mammals could decrease their body temperatures to that of the surrounding ambient water, thereby decreasing the temperature differential, $T_b - T_a$, and the heat lost to the environment. Of course, this is not an option for a homeothermic mammal living in near-freezing waters, as it must maintain high body temperatures to survive. Marine mammals maintain a high, constant temperature by adjusting surface area, conductance, or both.

A small surface area-to-volume ratio, SA:V, would be advantageous to marine mammals because it would decrease the relative area across which heat is transferred to the aquatic environment. Estes (1989) suggests that marine mammals do have a decreased SA:V relative to terrestrial mammals due to their large size. For example, the sea otter is twicc as large as the largest terrestrial mustelid (Morrison et al. 1974, Estes 1989), and the polar bear is probably the largest of the ursids (Stirling 1988, Reeves et al. 1992). Although there are no closely related terrestrial representatives with which to compare, pinnipeds, cetaceans, and sirenians are all large mammals. The extreme example, the blue whale, is the largest mammal that has ever lived. In addition, when compared to unrelated terrestrial mammals of similar mass, pinnipeds, cetaceans, and sea otters have significantly less surface area (Innes et al. 1990). Thus, decreased SA:V appears to be a morphological "solution" that decreases heat losses from the marine mammal body to the aquatic environment (see also Worthy and Edwards 1990).

Decreasing the thermal conductance of a body in contact with the water will also decrease heat losses. Thermal conductance is the inverse of insulation. Therefore, a good conductive material is a poor insulator. The integument, in contact with the fluid environment, must have the ability to provide insulation sufficient for the maintenance of a relatively constant and elevated body temperature in the highly conductive aquatic environment.

First let us look at terrestrial mammals, which use the epidermal structure, hair (or fur), to insulate their bodies. Fur provides insulation by trapping air (a poor thermal conductor) among individual hairs (e.g., Scholander et al. 1950). The retained air forms an insulative layer in contact with the skin surface. The insulative properties of mammalian fur in air are positively correlated with the hair length (Fig. 2-3), and the density of the hairs (Scholander et al. 1950, Schmidt-Nielsen 1990).

Marine mammals that spend time both on land and in the water do have fur, and for some of those mammals (e.g., sea otters, fur seals, polar bears), it is their primary insulator. But what happens to the insulative properties of fur when it is submerged in water? For terrestrial mammals, wet fur loses most of its insulative properties—perhaps surprisingly, so do the furs of most marine mammals (Fig. 2-3; Scholander et al. 1950, Frisch et al. 1974).

Only the densest fur can maintain a layer of trapped air and thus, remain waterproof and insulative upon submersion (Fig. 2-3). Of all the marine mammals, only the sea otter (the densest fur of any mammal at approximately 130,000 hairs/cm^2) and fur seals (up to 60,000 hairs/cm^2) maintain the skin surface dry and insulated when wetted (Kenyon 1969, Tarasoff 1974, Williams et al. 1992). Why are these furs so dense as compared to other mammals?

Most mammals have a single, relatively coarse, primary hair that emerges from a hair canal (see Fig. 2-2). In polar bears, sea otters, most pinnipeds, and some terrestrial mammals, the canal supports a primary (also called guard) hair and a variable number of smaller, softer underhairs (see Fig. 2-2). In most mammals with multiple underhairs, there are 1 to 10 underhairs per hair canal. In sea otters and fur seals there can be as many as 50 to 100 underhairs per canal. Such dense fur literally makes it impossible for water to reach the skin surface in these animals. Interestingly, although hair density seems critical to the effective insulation properties of sea otter and fur seal, in water, the length of the guard hair is not functionally relevant to thermal insulation (see Fig. 2-2). This is very different from the situation in terres-trial mammals.

The sea otter has an extra feature that helps waterproofing—the fur is richly covered with a hydrophobic lipid, squalene (Williams et al. 1992). The paired sebaceous glands (see Fig. 2-2) associated with each hair bundle in fur seals may also help waterproof the integument (Scheffer 1962). The existence of lipid secretions alone is not enough to waterproof fur. Many phocid seals, with fur densities insufficient to be insulative when wet, also have skins richly supplied with sebaceous glands (Montagna and Harrison 1957, Sokolov 1960).

Marine mammals with less dense fur, such as the walrus (*Odobenus rosmarus*), adult phocid seals, and the bare-skinned sirenians and ce-taceans, obviously do not rely on fur to insulate their bodies from their highly conductive fluid environment. These marine mammals use "blubber," a thickened, adipose-rich hypodermis. Blubber can be characterized as a continuous sheet of adipose tissue, reinforced by a network of collagen and elastic fibers (Parry 1949; Sokolov et al. 1973; Ling 1974; Ackman et al. 1975; Lockyer et al. 1984, 1985). The lipid content can vary between 9% and 82% of the wet weight of blubber, de-

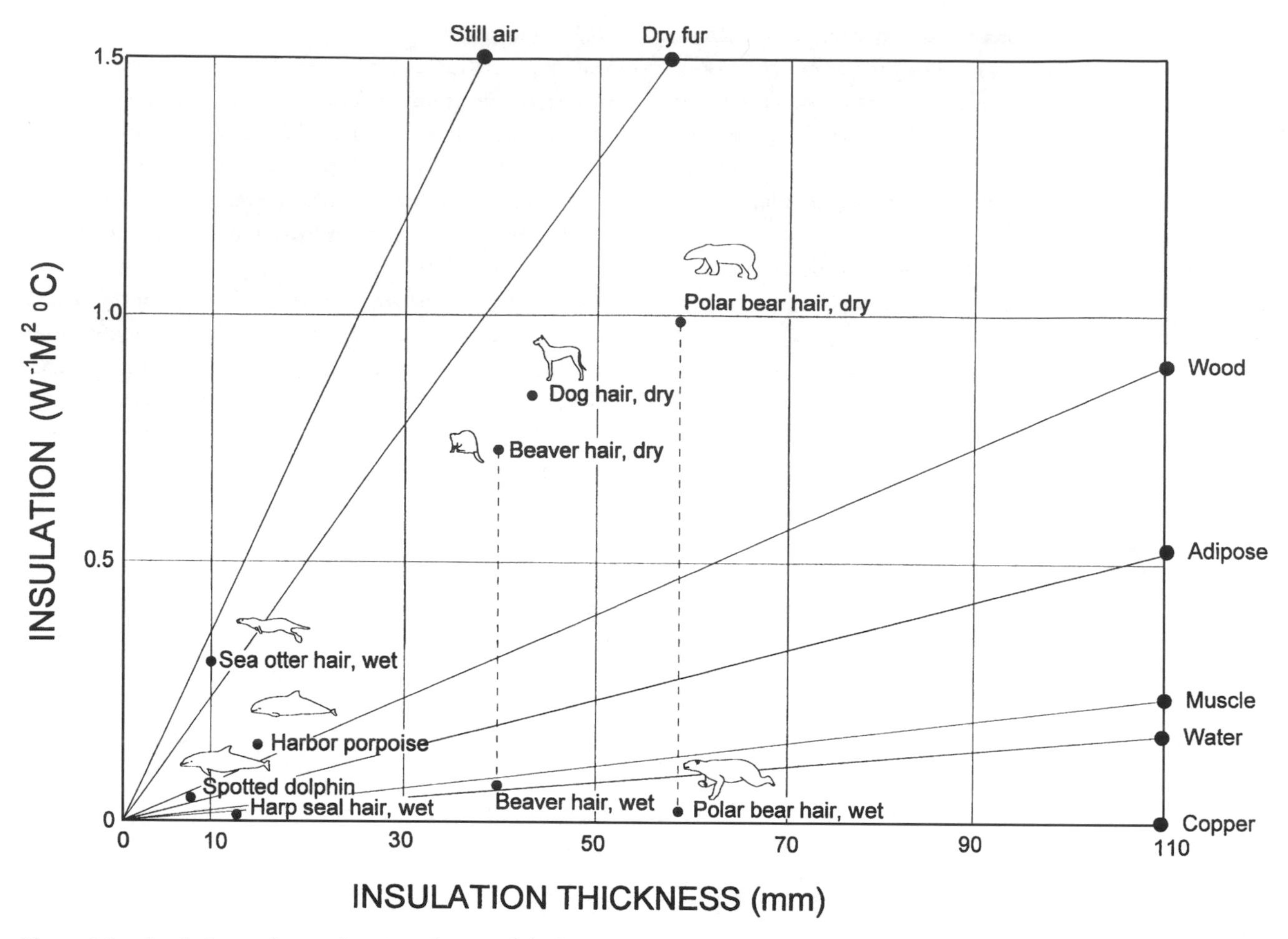

Figure 2-3. Insulative quality as a function of material thickness. Still air is a highly effective insulator. Dry fur, also a high-quality insulator, functions by trapping air between adjacent fibers. Most fur loses insulative value when wet (e.g., polar bear, harp seal, and beaver). In contrast, sea otter hair is a better insulator (even when measured in water) than typical animal fur, and is almost as good as still air. Adipose, an insulative material used by many marine mammals, is far less effective than dry fur. Notice that even wet sea otter fur, at 10 mm in depth, offers insulation equivalent to approximately 70 mm of adipose tissue. The blubber of the harbor porpoise, a small temperate-water cetacean, is a better insulator than that of the Pacific spotted dolphin, a tropical water species. Compare these materials with copper, a very good thermal conductor and thus, a poor insulator. Copper conducts heat at rates 10,000 times higher than that of typical animal fur and about 1,000 times higher than adipose tissue. (Illustration adapted from Scholander et al. 1950 and Schmidt-Nielsen 1990; values for air, typical fur, wood, adipose, muscle, water, and copper, represented by thick straight lines, taken from Schmidt-Nielsen 1990; values for sea otter body conductance in 5°C water from Costa and Kooyman 1982; polar bear, dog, and beaver hair from Scholander et al. 1950; harbor porpoise and Pacific spotted dolphin from Worthy and Edwards 1990; and harp seal from Frisch et al. 1974.)

pending on the species and the sample site (Lockyer et al. 1984, Worthy and Edwards 1990). Blubber can make up a substantial percentage of the total body weight of marine mammals, often in excess of 30% (Bryden 1964, 1969; Kooyman 1973).

Blubber is not nearly as effective an insulating material as dry fur (Fig. 2-3; Scholander et al. 1950); heat is conducted through fat at rates between three to seven times faster than through high-quality, dry fur (Schmidt-Nielsen 1990). Fur, as we have mentioned, loses most of its insulative value when wet. Even the high-density furs of sea otters and fur seals would lose their insulative values if those animals dove to depths that sufficiently compressed the trapped layer of air within the fur. Blubber is the insulating material used by most marine mammals.

The insulative properties of blubber depend on both its thickness and its lipid content (see Parry 1949, Doidge 1990). For example, Worthy and Edwards (1990) demonstrated that blubber of the pan-tropical spotted dolphin, a tropical cetacean, is approximately half the thickness of, and contains less lipid than, blubber of the harbor porpoise, an animal of similar size that inhabits colder temperate waters (Fig. 2-3). The conductance of heat through spotted dolphin blubber is almost four times higher than that of harbor porpoise blubber. Interestingly, harbor porpoise blubber is not homogeneous in its fatty acids constituents, suggesting that its insulative properties also vary regionally (Koopman et al. 1996).

Sea otters and fur seals use fur, and cetaceans, most pinnipeds, and sirenians use blubber as their insulative integumentary structures (Fig. 2-4). Interestingly, insulation of the polar bear is less well understood. Polar bear hair appears to be a poor insulator on land and is not dense enough to maintain a layer of trapped air upon submersion (Scholander et al. 1950). Frisch et al. (1974) determined that polar bear guard hairs were long enough to effectively trap a "dead layer" of water that could act to help insulate their body, much like a wet suit on a human. Unlike guard hairs on most marine mammals, which lie flat against the body (see discussion on drag reduction), polar bear guard hairs stand erect and expose underhairs. The underhairs trap a layer of relatively undisturbed water, which maintains an insulative shell around the bear's body. Although water is a relatively poor insulator (see Fig. 2-3), Frisch et al. (1974) suggested that the fur could provide 30% of the insulation the polar bear required. Some researchers have suggested that polar bears also use their thickened subcutaneous fat deposits as an insulative "blubber" layer (Oritsland 1970). Yet Pond et al. (1992) suggest that there are no morphological or biochemical data to support the hypothesis that polar bear fat has evolved for thermal insulation (instead, they suggest that the fat is primarily used as a metabolic energy store between fasts). The complete story of thermal insulation in polar bears has yet to be determined.

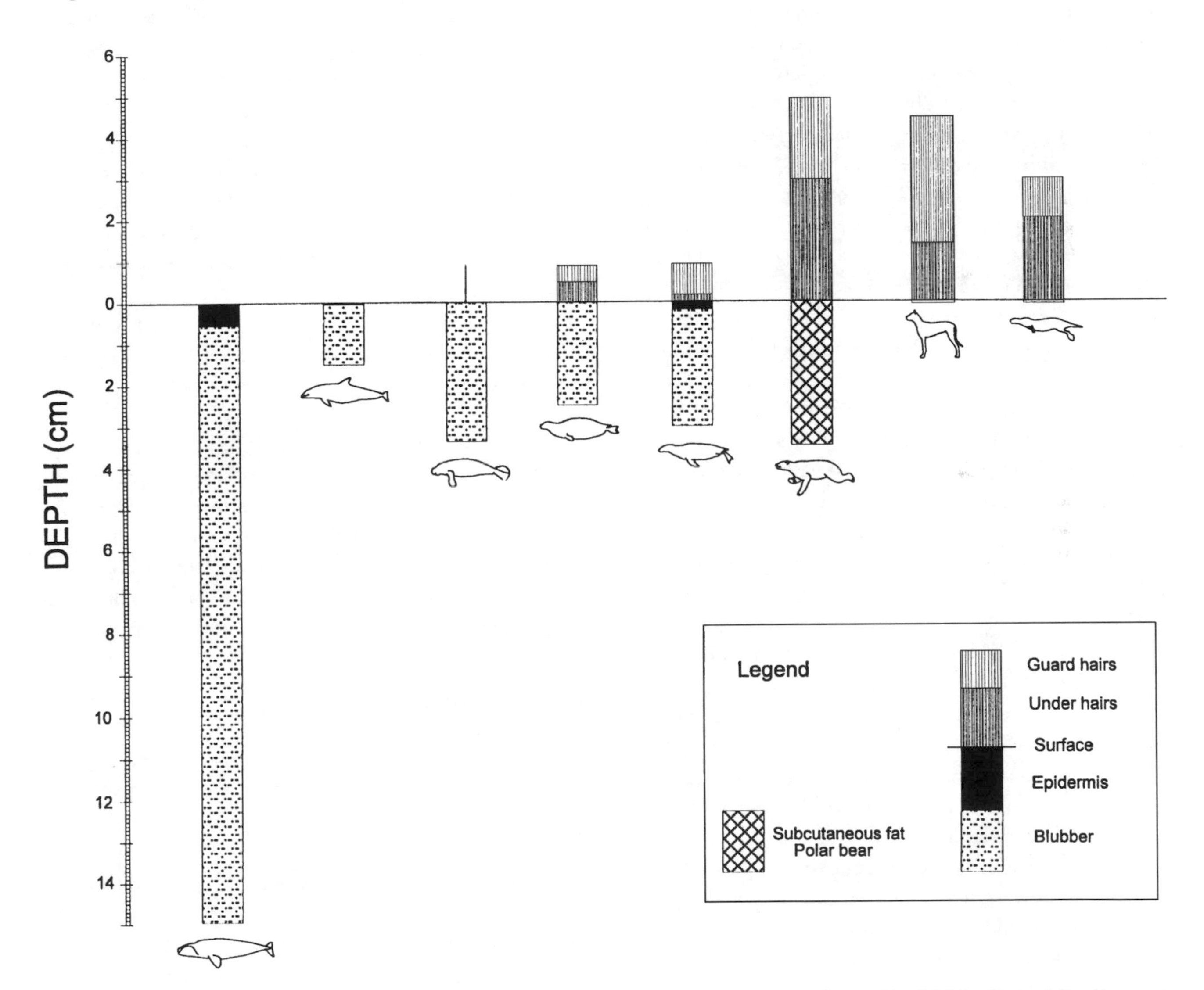

Figure 2-4. Integumentary structures used for insulation in marine mammals. Marine mammals use either blubber (a specialized hypodermal tissue) or fur as their primary insulation. The polar bear may also use subcutaneous fat deposits for insulation although this hypothesis is controversial (Pond et al. 1992). (The insulation values for polar bear hair shown in Figure 2-3 are for hair only; in this study, the subcutaneous fat layers were removed.) Compare thicknesses of each structure used by mammals of diverse sizes. (Thickness measurements from Omura 1958, Sokolov 1960, Oritsland 1970, Bonde et al. 1983, Williams et al. 1992, and authors' unpublished data.)

Thus far our discussion has focused exclusively on the functional requirement of conserving body heat. We have seen that marine mammals use two integumentary "solutions" to deal with potentially large heat losses to their aquatic environment—fur and blubber. But there are conditions under which marine mammals must dump, rather than conserve, body heat. For example, while actively swimming, large whales probably need to dump excess body heat, as do smaller, active marine mammals that live in tropical waters (Brodie 1975, Lavigne et al. 1986, Schmidt-Nielsen 1990, Ryg et al. 1993). Marine mammals that periodically expose themselves to the less conductive medium of air must also be able to dump excess body heat (e.g., Hart and Irving 1959; Oritsland 1969, 1970; Morrison et al. 1974). Perhaps the most extreme example would be that of the Hawaiian monk seal that hauls itself out of the highly conductive aquatic environment and rests for long periods on the hot sands of a tropical beach.

To dump excess body heat, marine mammals must be able to bypass their thermal insulation (Irving and Krog 1955, Scholander and Schevill 1955). For furred mammals, where the insulator is superficial to the heat-conducting surface of the skin, the most effective bypasses available are areas of the body that are naked or thinly covered by hair (Bryden and Molyneux 1978, Schmidt-Nielsen 1990, Bryden 1994). The flippers of pinnipeds and the poorly insulated feet of sea otters act as "thermal windows" to allow excess body heat to be conducted to the environment (Hart and Irving 1959; Tarasoff 1972, 1974; McGinnis 1975; Blix et al. 1983). Heat is transported to the surfaces of these appendages by blood moving through vessels that lie very close to the skin. These appendages generally have large SA:V, thereby increasing conductive heat loss (see equation 1). Thus, heat is transported by the vascular system to the exposed body surfaces where it can be transferred from the body to the aquatic environment.

For mammals that use blubber, there are two mechanisms to bypass their insulative layer. The blubber is richly vascularized (Palmer and Weddell 1964) and carries blood vessels (arterioles) that will eventually reach the papillary dermis, just deep to the epidermal surface of the animal. These arterioles vasoconstrict when the blubber needs to act as an insulative shell, but may vasodilate when the animal needs to dump body heat. Thus, heat is transported by the vascular system through the blubber and to the body surface where it can be conducted away from the body to the aquatic environment (McGinnis et al. 1972, McGinnis 1975, Hampton and Whittow 1976).

Cetaceans also use their uninsulated dorsal fins, flukes, and flippers as thermal windows to dissipate body heat, in much the same way as pinnipeds use their appendages (Scholander and Schevill 1955; McGinnis et al. 1972; Elsner et al. 1974; Hampton and Whittow 1976; Rommel et al. 1992, 1993, 1994; Pabst et al. 1995). In cetaceans the appendages are also invested with large superficial veins that carry warm venous blood to the skin's surface, where the blood can be cooled by exposure to ambient water.

The existence of large surface areas of poorly insulated integument in cetaceans, pinnipeds, and sea otters does raise a question. If these appendages are such good thermal windows, why is body heat not being lost continuously across them? The solution is to have two venous return systems within the appendages—one that lies close to the surface, as described above, and one that forms a countercurrent heat exchanger with the arteries that feed the fins, flippers, and flukes (Scholander and Schevill 1955; Elsner 1969; Irving 1969; Tarasoff and Fisher 1970; Elsner et al. 1974; Tarasoff 1974; Rommel et al. 1992, 1995; and see Fig. 2-5 for explanation of countercurrent heat exchange). In the dorsal fin, flukes, and flippers of cetaceans (Fig. 2-6), the hind limbs of seals, and the feet of sea otters, veins and arteries lie side by side. Cooled venous blood returning from the animal's periphery can be used to selectively precool arterial blood going out to the periphery. Heat is transferred from warm arterial blood to cool venous blood, and little heat is lost to the environment. Although the exact geometry of countercurrent heat exchange systems varies from animal to animal, the physical principle of heat exchange is a shared anatomical "solution" to limit heat loss to the environment (Schmidt-Nielsen 1990).

Marine mammals have highly effective insulation, but they are not simply wrapped like a thermos bottle! The conductance properties of their insulation can be exquisitely adjusted by changes in blood flow patterns through the skin. Body heat may or may not be transported to the skin by the vascular system, depending on environmental conditions and whole body thermoregulation needs of the animal.

Drag Reduction

Optimization of energy by aquatic mammals requires adaptations that reduce drag. Fish 1993b

As stated in the introductory section, drag forces are much higher in the aquatic environment than in air, because of the increased viscosity and density of seawater (Webb 1984, Fish 1993b, Vogel 1994). The rate of energy use, or the power output, P' (W), required of a marine mammal to move through such a medium is proportional to its drag, D (N), multiplied by the speed, u (m/sec), at which it swims:

$$P' \propto Du. \qquad \text{(equation 2)}$$

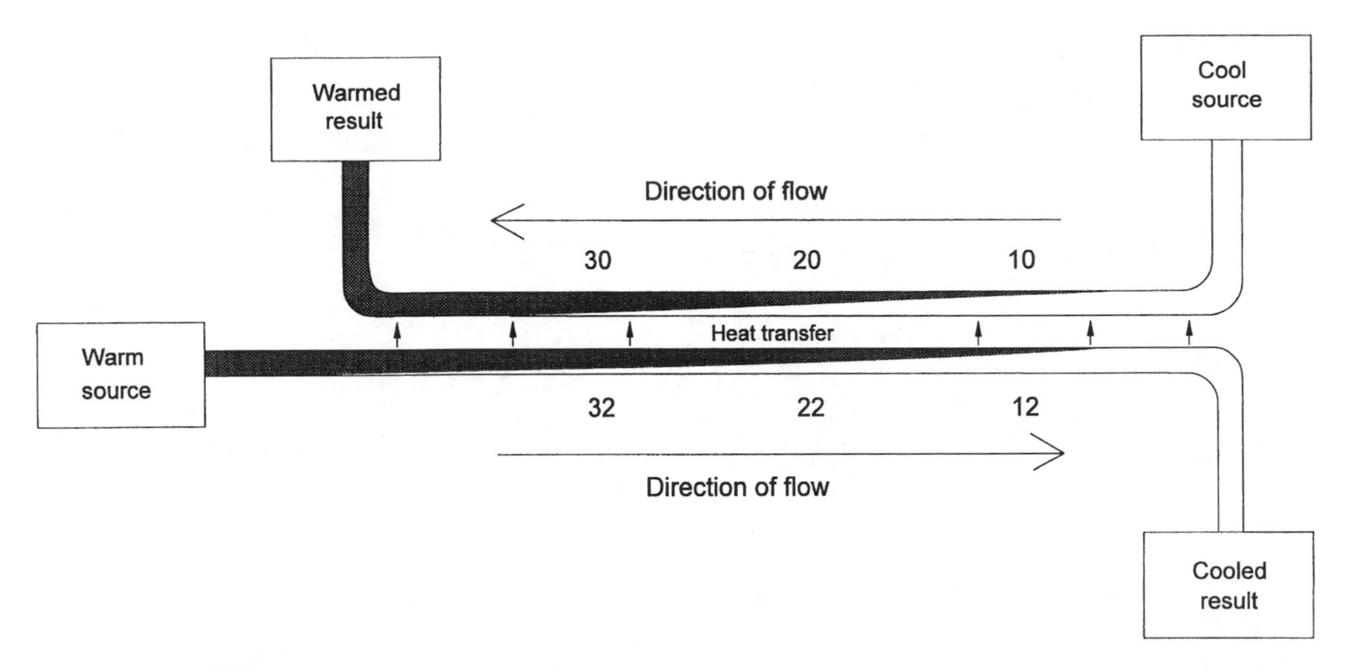

Figure 2-5. Countercurrent heat exchange—a maximally efficient design. How does a countercurrent heat exchanger work? Consider the two pipes in this figure, juxtaposed to each other. The lower pipe is carrying warm fluid that flows in one direction (from left to right in this example). The upper pipe carries cool fluid that is flowing in the opposite direction. Because these pipes are juxtaposed and have opposite (antiparallel) flows, heat is maximally transferred from the fluid in the lower pipe to the fluid in the upper pipe. Almost 100% heat exchange can be achieved with a countercurrent heat exchanger. The result is that by the time the fluid in the lower pipe reaches the point marked "result," it has been significantly cooled. Countercurrent heat exchange is common in animals, and the "pipes" are arteries and veins.

Thus, a marine mammal could reduce its power output (save energy), or alternatively, swim at a higher speed (cover more distance/energy expended), if it could reduce the drag it experiences (Vogel 1994). Although drag forces are easy to understand at an intuitive level (i.e., the forces that resist movement of a body through a fluid), they are quite complex mathematically. We introduce you rather superficially to drag, so that you have enough information to evaluate possible drag reduction adaptations in marine mammals. A more thorough treatment can be found in Vogel (1994), an easy-to-read text on the not-so-intuitive field of fluid dynamics.

We consider two types of drag forces that act on a body, viscous drag (D_v) and pressure drag (D_p). Viscous drag (also called frictional drag or skin friction) results from the viscosity of water (recall, this is approximately 60 times greater in seawater than in air). You can think of viscosity as a measure of the stickiness between thin layers of fluid that slide, or are sheared, past each other (Vogel 1994). Viscous drag results from fluid being sheared across the surface of a body. The fluid in contact with the body surface is not moving relative to that surface, and the fluid close to the body surface is moving more slowly than fluid that is farther away. The thin layer of fluid whose velocity is being affected by the body surface is called the boundary layer (Vogel 1994). The magnitude of the viscous drag is dependent on the amount of body surface that is in contact with—and slowing the flow of—the fluid; as surface area increases, the amount of viscous drag increases. A smooth body surface reduces surface area, and thus, acts to decrease the viscous drag (Vogel 1994).

Pressure (inertial) drag, which is influenced by both the viscosity and density of the fluid, results from the distribution of pressure around a body and its effect on fluid flow. In general, for a body moving through a fluid, pressures are highest at the cranial end and lowest at the widest part. Fluids flow from areas of high to low pressure, therefore the fluid flowing along the front of the body stays close to the surface (Vogel 1988, 1994). As fluid flows past the widest part of the body and toward the caudal end, pressure begins to increase again. Because fluid does not voluntarily flow from areas of low to high pressure, the fluid separates from the surface and heads downstream, leaving a wake behind the body. The larger the wake, the higher the pressure drag. The most effective way to reduce pressure drag is to postpone the separation point until it is further toward the rear of the body and thus, reduce the wake (Vogel 1988, 1994). This is best accomplished by gradually tapering the tail-end of the body, rather than having it terminate abruptly. The resulting fusiform (spindle) shape is referred to as "streamlined." For the flow regimes experienced by marine mammals, pressure drag will be a much larger percentage of the total drag on the body than viscous drag (Vogel 1994).

We can now predict how marine mammals could reduce

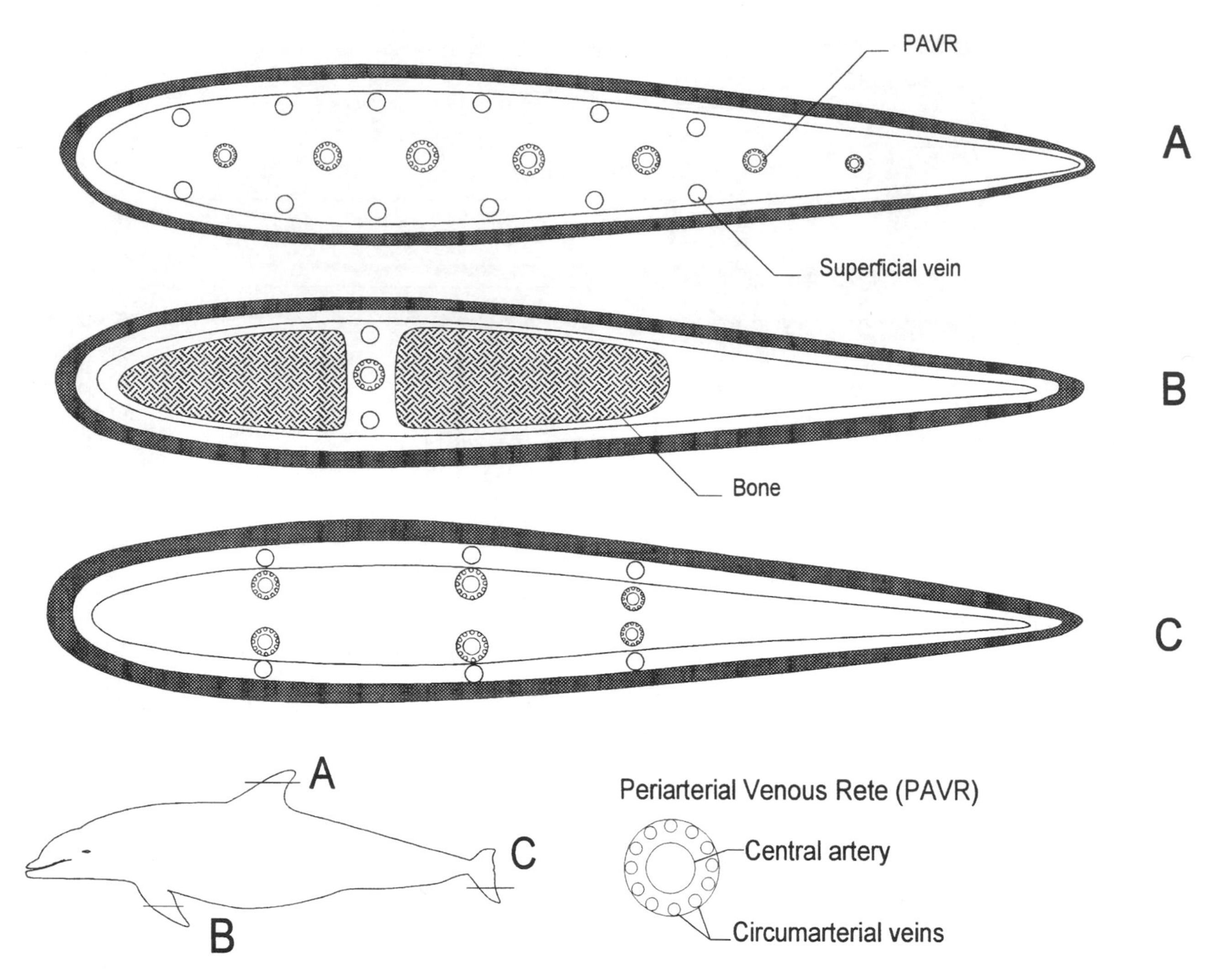

Figure 2-6. Schematic representations of cross-sections through the dorsal fin (A), flipper (B), and flukes (C) of a bottlenose dolphin. Notice that these appendages have two venous systems. The circumarterial venous channels that form a periarterial venous rete (PAVR) lie deep within the appendages and surround central arteries. Blood routed through the circumarterial veins will absorb heat from the relatively warm blood flowing through the deep arteries. The juxtaposed circumarterial veins and arteries form a heat-conserving, countercurrent exchange system. The second system of superficial veins transports warm blood to the surface where body heat would be transferred to the fluid environment (see also Fig. 2-5). Epidermis is hatched; superficial veins are open circles. (Illustrations adapted from Scholander and Schevill 1955, Rommel et al. 1992, and authors' unpublished data.)

the drag they experience while swimming. To reduce viscous drag, they could have low SA:V and smooth surfaces. To reduce pressure drag, they could postpone flow separation around their bodies by streamlining.

We have already seen that marine mammals tend to have a low SA:V. In addition to having a low SA:V, many marine mammals have very smooth surfaces when submerged. In pinnipeds and sea otters the smooth surface is achieved by the morphology and position of the guard hairs. The guard hairs are the leading, or cranial-most, hairs exiting the hair canal (see Fig. 2-2). These hairs, which point caudally and are flattened in cross-section, lie superficial to the underhairs and flat along the body, forming a smooth surface when wetted (Sokolov 1960; Scheffer 1964a,b; Costa and Kooyman 1982; Williams et al. 1992). In contrast, the guard hairs of polar bears are said to stand off the body surface when wetted (Frisch et al. 1974).

Interestingly, pinnipeds and sea otters lack arrector pili muscles (Sokolov 1960, Williams et al. 1992), the small, smooth muscles that are usually associated with each hair follicle (see Fig. 2-2). Each arrector pili muscle acts to raise a hair off the body surface. It has been stated that the absence of arrector pili muscles is an adaptation to maintain the hairs flat along the body while submerged, and thus, to reduce drag (Ling 1974, Fish 1993b). This hypothetical adaptation should be reevaluated as these marine mammals have diffuse smooth muscle bundles running throughout their dermis (Sokolov 1960, Williams et al. 1992), which may function similarly to arrector pili muscles (Williams et al. 1992).

The integument of bare-skinned cetaceans also forms an exceptionally smooth boundary with the fluid environment (for review, see Geraci et al. 1986). The smooth surface may be attributable in part to the very high turnover rate of the outermost layer of epidermal cells (Brown et al. 1983, Hicks et al. 1985). This high turnover rate is not due to increased rates of epidermal cell proliferation, but rather as a result of the relatively large and convoluted surface area of germinal cells that lie on the tall and densely packed dermal papillae of cetaceans. Thus, underlying and replacing a relatively small surface area of superficial skin is a very large surface area of germinal cells (Hicks et al. 1985). In the bottlenose dolphin, the outermost epidermal layer may be replaced every 2 hours (Hicks et al. 1985)! This high turnover rate helps maintain a "smooth, self-cleaning surface" (Hicks et al. 1985) that may reduce opportunities for fouling organisms to settle and attach to the skin surface (Geraci et al. 1986). It is not known yet whether all cetaceans have such high turnover rates of epidermal cells.

Unfortunately, little research has been done on the integument of the sirenians (for review, see Ling 1974). It is interesting to note that the dugong (*Dugong dugon*), which tends to be more marine in its distribution and may be a more accomplished swimmer than the manatee (*Trichechus* spp.), has smoother skin as well (Nishiwaki and Marsh 1985).

Although decreasing SA:V and maintaining smooth surfaces are useful mechanisms for reducing viscous drag, the largest component of drag experienced by marine mammals is pressure drag. Thus, body streamlining will be the single most effective way for marine mammals to decrease total drag (Webb 1984; Fish and Hui 1991; Fish 1993a,b; Vogel 1994). Fineness ratio, a hydrodynamic parameter that reflects streamlined shape, is equal to total body length divided by maximum body diameter (Vogel 1994). An optimally streamlined body has maximum volume and minimum surface area at a fineness ratio of 4.5 (for review, see Fish 1993b). Cetaceans, pinnipeds, sirenians, and the sea otter have fineness ratios between 3.3 and 8, with most species having body shapes incurring drag only 10% above theoretical optimum (data compiled in Williams and Kooyman 1985 and in Fish 1993b). To the best of our knowledge there are no data on the fineness ratio of polar bears, although their external body shape is not obviously streamlined, except, perhaps relative to other ursids.

The integument plays a critical role in defining the streamlined shape of most marine mammals. The blubber of pinnipeds, cetaceans, and sirenians sculpts the external body form and smooths body contours (Fig. 2-7) (for views of cross-sectional shape defined by integument see Slijper 1936, Boice et al. 1964, St. Pierre 1974, Pabst 1990, Domning and de Buffrenil 1991). For example, in the ringed seal (*Phoca hispida*), blubber is distributed throughout most of the body in such a way that the ratio of blubber thickness to body diameter is relatively constant (Ryg et al. 1988). This distribution of blubber, which may be optimal for insulation, is not observed in the caudal portion of the seal, where the blubber is much thicker and the animal appears "overinsulated." The thicker blubber layer in the caudal portion of the animal may play a role in defining a gradually tapering shape, a critical feature of streamlined bodies. Ryg et al. (1988) point out that the distribution of blubber may represent a compromise between the thermoregulatory and streamlining functions of the blubber layer in seals. In cetaceans, blubber also sculpts the caudal peduncle, or tailstock (Fig. 2-7), which is streamlined in the plane of oscillation (e.g., Koopman et al. 1996). This hydrodynamic shape decreases the drag forces on the tailstock as the flukes move through the water (Lighthill 1969, Lindsey 1978, Weihs and Webb 1983, Webb 1984).

The integument not only plays a role in reducing drag by sculpting the body contours of most marine mammals, it also forms the stabilizing dorsal fins of cetaceans, the propulsive flukes of cetaceans and sirenians, and the webbing between the digits of pinnipeds and sea otters. It is a system critical to the effective locomotion of mammals through the aquatic environment.

Of course, the fusiform, streamlined body shapes of many marine mammals are not solely attributable to specializations of the integument. Reduction in the size, change in the shape or absence of limbs, and reduction or absence of ears and external reproductive structures are all morphological characteristics that reduce drag (for review, see Howell 1930, Slijper 1979, Würsig 1989). We will investigate these morphologies further as we tour the musculoskeletal and reproductive systems.

Buoyancy Control

Animals are denser than either fresh or sea water, and therefore tend to sink, unless they have adaptations that give buoyancy. Alexander 1990

Archimedes established that, when a body is submerged, it displaces a volume of water equivalent to its own volume (Schmidt-Nielsen 1990). If the weight of the body is less than

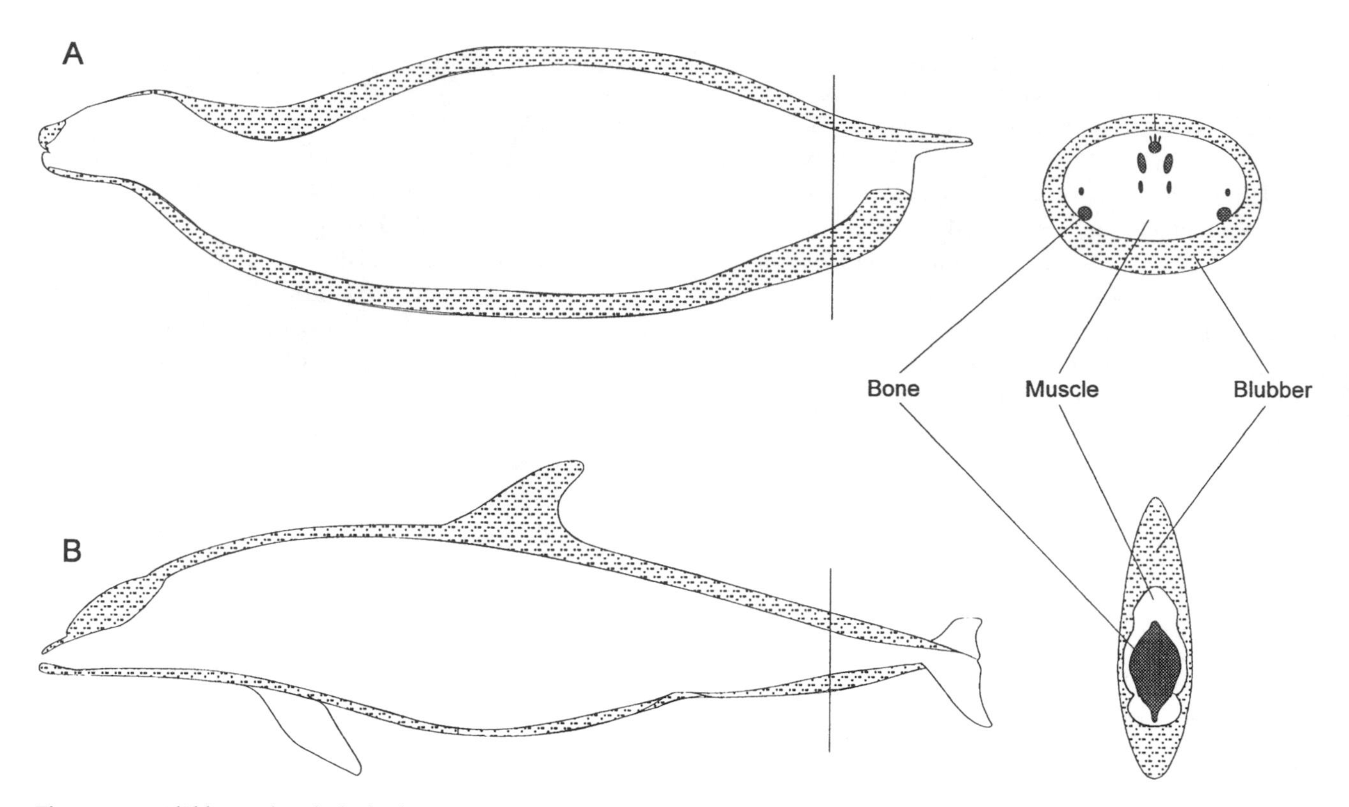

Figure 2-7. Blubber sculpts the hydrodynamically tuned body shape in marine mammals. (A) The caudal portion of the seal body has a relatively thick blubber layer that appears to act to smooth the contours of the body. (B) The dorsal and ventral keels of the caudal peduncle in cetaceans are formed by blubber. The caudal peduncle is streamlined in the plane of oscillation, which decreases drag forces on the body as the flukes move through the water. (Illustrations adapted from Pabst 1990 and authors' unpublished data.)

the weight of the water it displaces, it experiences a net upward force. On the other hand, if the body weighs more than the volume of water it displaces, it will experience a net downward force. This force is called buoyancy, *B*, and it can be expressed as:

$$B = Mg = (\rho_{body} - \rho_{water}) Vg, \qquad \text{(equation 3)}$$

where *M* is the body mass (kg), *g* is the acceleration due to gravity (9.8 $msec^{-2}$), ρ_{body} is the density of the body (kg m^{-3}), ρ_{water} is the density of water (seawater = 1024 kg m^{-3}), and *V* is the volume of the body (for review, see Vogel 1994).

Mammalian bodies are made up of materials that are generally denser than seawater. For example, muscle has a density of 1056 kg m^{-3} (Mendez and Keyes 1960) and most vertebrate bone has a density of about 3200 kg m^{-3} (Schmidt-Nielsen 1990). On the other hand, the air-filled lung is less dense than seawater. Schmidt-Nielsen (1990) suggests two mechanisms available to mammals to reduce their density and thus, avoid sinking. (1) Mammals could reduce the amount of heavy materials in their bodies, especially dense bone (see discussion on buoyancy control in the following section on the musculoskeletal system). (2) Mammals could increase the amount of materials in their bodies that are less dense than water, such as air-filled lungs (see discussion of the respiratory system) and fat (900 to 920 kg m^{-3}) (Schmidt-Nielsen 1990).

How could the morphology of the integument affect buoyancy? The blubber layer, which may be comprised predominantly of fat and may account for a large percentage of the total body weight in many species, may function to reduce the density of some marine mammals. It has been postulated that blubber increases buoyancy (e.g., Slijper 1979, Lockyer et al. 1984, Fish and Stein 1991), although to the best of our knowledge there exists no study that thoroughly substantiates this assertion. The air trapped in the dense, water-repellent fur of sea otters may also aid in maintaining positive buoyancy in these animals (Tarasoff et al. 1972, Kenyon 1981). Fish (1993b) found that the nonwettable fur of the semi-aquatic opossum (*Chironectes minimus*) did add buoyancy, whereas the wettable fur of the terrestrial opossum (*Didelphis virginiana*) did not. Despite the obvious importance of buoyancy regulation in the aquatic environment, there appear to be few systematic studies of the topic in marine mammals.

Musculoskeletal System

The musculoskeletal system acts to define the body shape (Fig. 2-8) and provide support and protection of internal or-

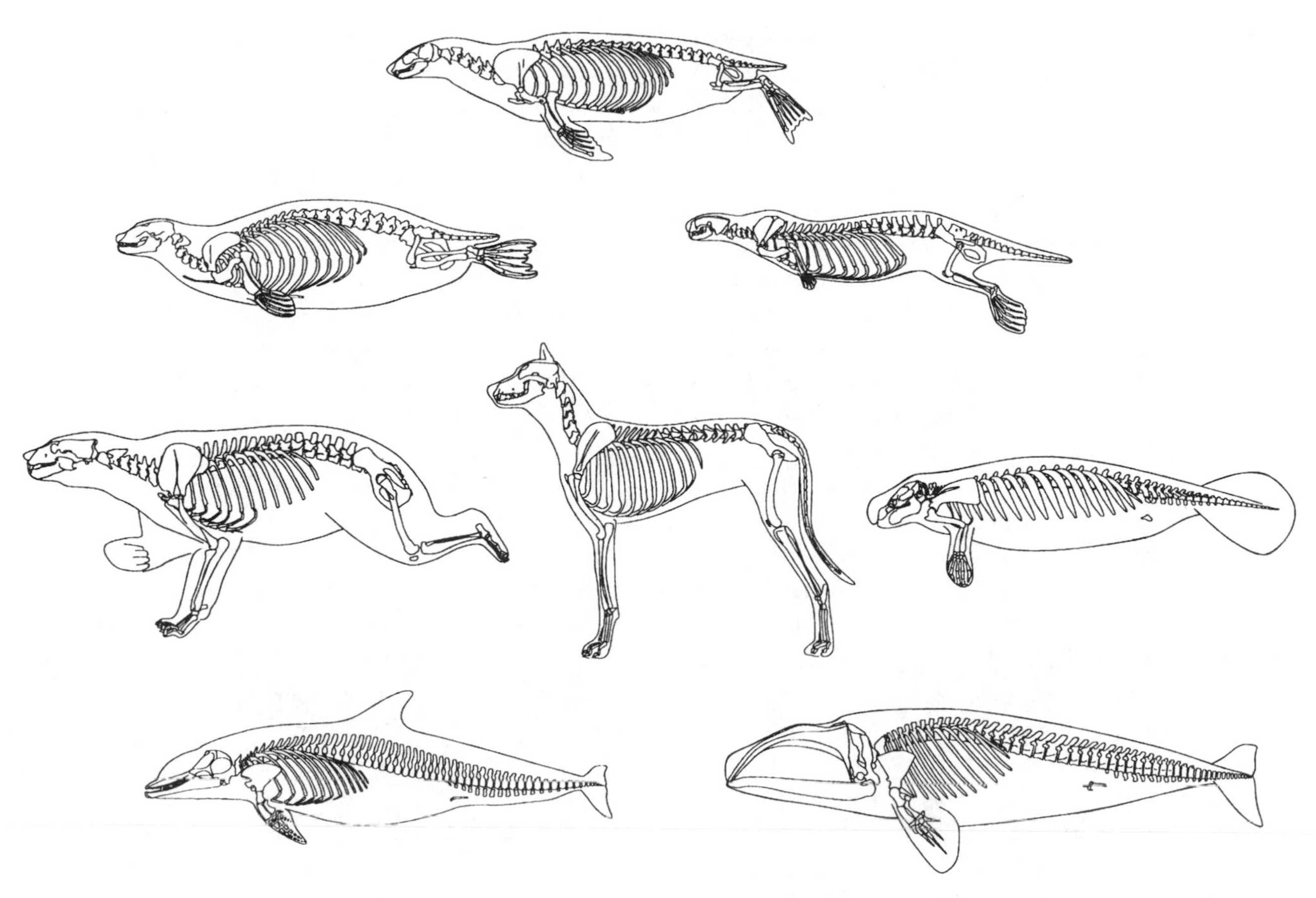

Figure 2-8. Skeletal systems of representative marine mammals. The skeleton helps define the body shape, support and protect internal organs, and transmit forces during locomotion. Note the relative size and shape of the appendicular versus the axial skeleton and how much of the appendicular skeleton lies within the contour of the body wall. (Illustrations adapted from van Beneden and Gervais 1868–1879, Struthers 1881, Flower 1885, True 1904, Howell 1930, Omura et al. 1969, Nishiwaki and Kasuya 1970, Nickel et al. 1986, Rommel 1990, Evans 1993, and authors' unpublished data.)

gans. This system of muscles, bones, tendons, ligaments, and joints also produces and transmits the mechanical forces required for movement through the environment.

In this section we focus most of our attention on the function of aquatic locomotion (i.e., swimming) and compare and contrast the various locomotor designs used by marine mammals. We also briefly touch on two other functions of the musculoskeletal system: (1) providing support for soft tissues of other body systems, specifically the skull's support for the sensory, feeding, and respiratory structures investigated later in the chapter, and (2) controlling buoyancy.

Locomotion

It has taken years for a multitude of highly trained technicians to discover the proper lines for a rigid airplane fuselage and wings. If the latter were propelled by the accompaniment of contortive wrigglings by the hinder end, or convulsive gyrations of the wings by means of a multitude of small engines (i.e., muscle) of unknowable horsepower delivering their power at vague points, one may visualize how little would be known about aerodynamics at the present time. Precisely this situation is encountered by one who would investigate the principles underlying the swimming of mammals.
Howell 1930

Despite its age of over 65 years, Howell's colorful lament is still surprisingly applicable—much less is known about the mechanics of swimming in marine mammals than the mechanics of running in terrestrial mammals. This might seem surprising, especially because we can rather easily define the mechanics of the situation—to swim a mammal must produce propulsive or thrust forces equal to or greater than the drag forces it experiences (Daniel and Webb 1987, Webb 1988, Fish 1993b). A critical function of the musculoskeletal system is to produce those thrust forces.

Recall that the rate of energy expended, or the power output required to swim is a function of the drag force on a body multiplied by its velocity (equation 2). A marine mammal swimming at a constant velocity produces thrust forces, T, that are equal to its drag forces, D. Thus, the power output is equal to

$$P'\eta = Du = Tu, \qquad \text{(equation 4)}$$

where η is the propulsive efficiency. (Notice that this is simply an expansion of equation 2, but we have added an efficiency parameter to the power output side of the equation.)

What do bones and muscles have to do with this equation? The design of the musculoskeletal system profoundly influences the power output the mammal can develop by affecting both thrust and propulsive efficiency. Thrust forces are dependent on muscle morphology and the mechanical design of the skeletal system. The propulsive efficiency of the animal is dependent on the size, shape, position, and behavior of the appendage used to produce thrust (Feldkamp 1987a,b; Webb 1988; Fish 1993a). We describe those aspects of musculoskeletal design that act to increase both the thrust forces generated and the efficiency of propulsion. In so doing, we also use a greatly simplified approach to the hydrodynamics of propulsion. See treatises by Blake (1983), Lighthill (1969), and Webb (1975, 1988) for more thorough discussions of hydrodynamic models of underwater swimming in vertebrates.

Before diving into our investigation of musculoskeletal design, we should offer a brief overview of the diversity of locomotor styles used by marine mammals (Fig. 2-8). As a point of comparison, terrestrial mammals usually use their limbs to swim. Our "typical" mammal swims using the proverbial dog paddle—alternate strokes of the forelimbs (and sometimes hind limbs). Polar bears swim in much the same manner, using only their forelimbs. Their hind limbs trail behind the body, which may help in streamlining the overall body shape (Flyger and Townsend 1968, Fish 1993b). Pinnipeds and sea otters also use their limbs to swim. Unlike most of the other marine mammals, the fully aquatic sirenians and cetaceans swim using their vertebral or axial musculoskeletal systems.

Increasing Thrust Production

An example of a typical mammalian musculoskeletal system is schematically represented in Figure 2-9. Muscles produce contractile force, which is transmitted by tendons to skeletal elements. If the muscular force produced is large enough, the skeletal element changes its position, usually by pivoting at a low friction joint. Thus, the musculoskeletal system is a machine that transmits force from one place in the body to another, and the mechanical design underlying its function is leverage (for review, see Hildebrand 1982). A lever is simply a rigid structure that transmits force by rotating at a pivot point, or fulcrum (Fig. 2-9).

A lever system can be designed to optimize either its force output or the velocity at which it moves (for review, see Kier and Smith 1985). Force output (force-out) is increased by increasing either (1) the force-in, or (2) the mechanical advantage of the lever system, which is defined as the ratio of the length of the lever arm-in to length of lever arm-out (Fig. 2-9).

With this brief overview of musculoskeletal design, we can now make predictions about the morphology of the swimming systems of marine mammals. (1) To produce large thrust forces, the locomotor muscles should have high force-generating potential (i.e., increased force-in). The force that a muscle can generate is equivalent to the number of muscle fibers that are functionally in parallel (e.g., Spector et al. 1980, Sacks and Roy 1982). The functional or physiological cross-sectional area of a muscle is a measure of the force-generating potential, but other measures of muscle morphology are often used to estimate muscle force output. (2) To increase thrust forces, the skeletal elements should also have appropriate proportions to increase the mechanical advantage of the lever system.

Let us first look at the California sea lion, an animal that has been well studied by functional morphologists. Sea lions, like all other otariids, use synchronous movements of their forelimbs to swim, and the primary locomotor muscles are those acting on the brachium (the upper arm) (Howell 1930; English 1976a,b, 1977; Feldkamp 1987a,b; Gordon 1983; Godfrey 1985). English's (1977) thorough and quantitative study describes structural features of the sea lion forelimb that function to increase force output, as compared with terrestrial fissiped carnivores. The muscles acting on the brachium of the sea lion have larger force-generating capacities than do homologous muscles in the terrestrial animals (English 1977). Thus, the force-in of the lever system is enhanced in sea lions. The humerus, the proximal bone of the forelimb, is also greatly shortened relative to the terrestrial fissipeds (Howell 1928, 1930; English 1977). The shortened humerus reduces the length of the lever arm-out, which, in turn, increases the mechanical advantage of the primary locomotor muscles (Fig. 2-9; Howell 1928, English 1977, Gordon 1983). Thus, the appendicular locomotor system of the sea lion is well designed for increased thrust production.

The walrus swims primarily by the alternate action of the hind limbs. Gordon's (1983) study compares the locomotor systems of walrus and sea lion, and illustrates that they share common design features. For example, the cross-sectional areas and thus, force-generating potentials of the hind limb locomotor muscles in the walrus are larger than in the hind limb muscles in the sea lion. In contrast, the forelimb locomotor muscles of sea lions have larger force-generating potentials than the forelimb muscles of walruses. In the walrus,

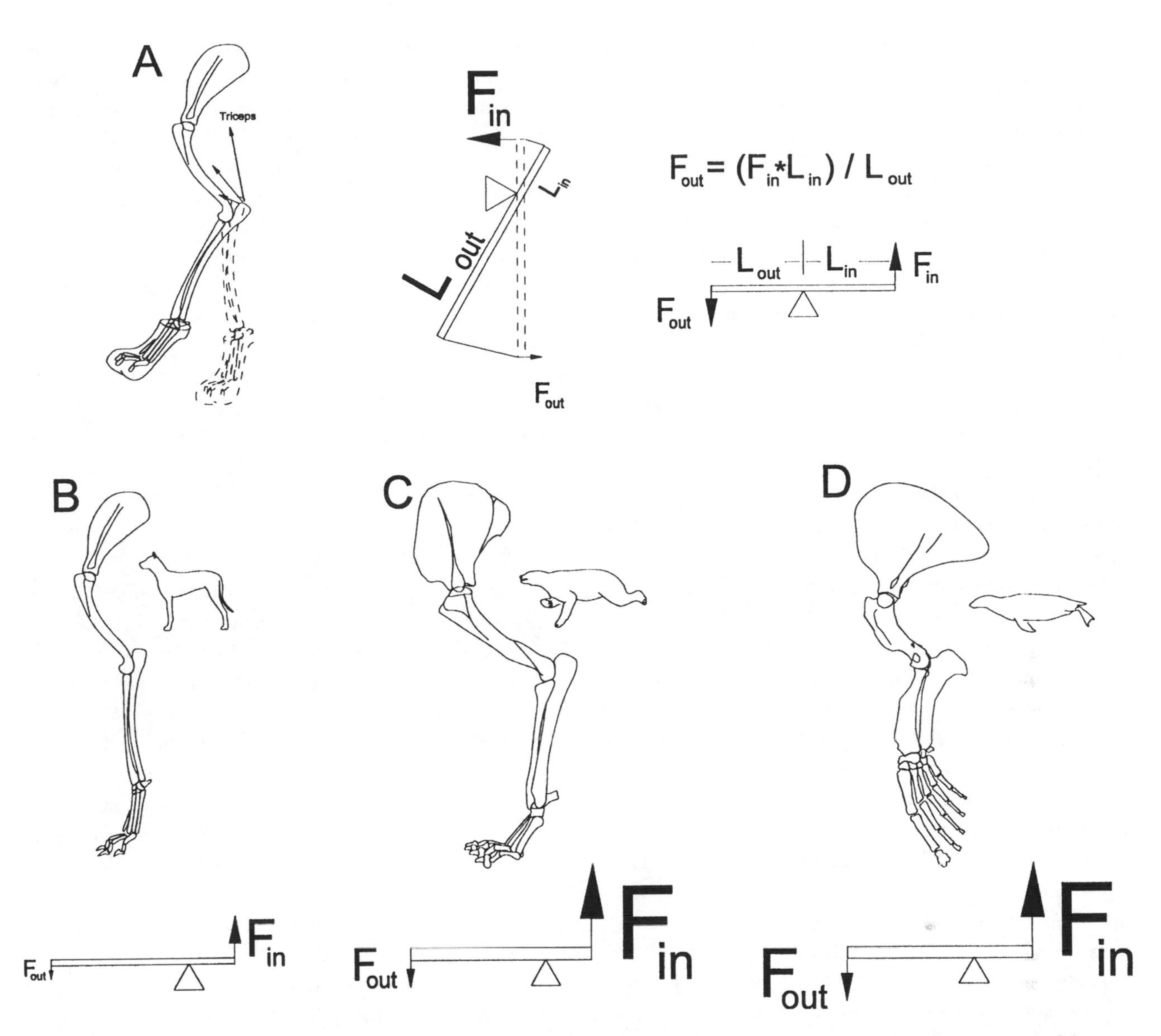

Figure 2-9. The vertebrate musculoskeletal system is a lever system. (A) A lever is a rigid bar that transmits force by rotating at a fulcrum (triangle). A lever system is balanced when force-in (F_{in}) multiplied by length-in (L_{in}) of the lever arm is equal to force-out (F_{out}) times length-out (L_{out}) of the lever arm. A lever system increases F_{out} by increasing F_{in} or increasing the mechanical advantage of the lever (L_{in}/L_{out}). Musculoskeletal design in (B) dog; (C) polar bear; (D) sea lion. The pectoral limb of both polar bears and sea lions produces a large F_{out} relative to the dog. Polar bears increase the F_{in} mainly by increasing the mass of the muscles acting on the brachium. Sea lions increase F_{out} by increasing both F_{in} and the mechanical advantage of the lever (L_{in}/L_{out}).

the femur, the proximal hind limb bone, is also shortened relative to those of terrestrial mammals (based on comparative data presented in Evans 1993).

Unfortunately, there exist no quantitative morphological studies on the appendicular locomotor system for sea otters (descriptive anatomies are found in Taylor 1914 and Howard 1973, 1975). Thus, we cannot comment in as much detail on the design of their musculoskeletal system as we can for sea lions and walruses. Sea otters use their hind limbs in conjunction with dorsoventral undulations of their caudal axial skeleton for high-speed swimming (Kenyon 1969, Tarasoff et al. 1972, Williams 1989). The proximal bones of the sea otter hind limb have been described as "relatively short" (Taylor 1914, Howard 1975) and "heavily muscled" (Howard 1975). Phocid seals also rely on their hind limbs for locomotion, propelling themselves using alternate lateral strokes of their hind limbs (Howell 1928, 1930; Williams and Kooyman 1985; Fish et al. 1988). As in the sea otter, phocid muscles acting on the femur are relatively large, and the femur is also a shortened bone (Taylor 1914; Howell 1928, 1930; Bryden and Felts 1974).

Thus, in mammals that use their limbs to swim, we see two morphological "solutions" to increased thrust production. Proximal locomotor muscles tend to have large cross-sectional areas and thus, large force-in-generating potentials. (This is reflected in the morphology of the proximal limb bones as well; they are large diameter, robust bones that can accommodate the large muscular forces they experience.) Proximal limb bones tend to be shortened, which increases the mechanical advantage of the lever system. The shortened proximal limb bones have an added hydrodynamic benefit. These bones tend to be partially or completely enveloped in the body, which helps reduce drag on the appendage and increase body streamlining (Tarasoff et al. 1972, English 1977, King 1983).

Cetaceans and sirenians use their axial muscles to move their flukes. The action of the axial muscles is to bend a variably flexible beam—the vertebral column. Epaxial muscles bend the vertebral column dorsally in an upstroke; hypaxial muscles, and probably abdominal muscles as well, bend the vertebral column ventrally in a downstroke. In so doing, the muscles generate thrust forces that are delivered to the water by the flukes (for review, see Domning 1977, 1978; Strickler 1980; Pabst 1990).

Axial muscles are long muscle bodies (they can run virtually the entire length of the vertebral column) composed of many shorter muscle fibers. Unlike appendicular muscles that have discrete tendons of insertion, axial muscles have serially arranged tendons, that is, individual tendons insert sequentially on vertebrae along the length of the muscle (e.g., Howell 1927, 1930; Slijper 1936; Parry 1949; Strickler 1980; Pabst 1990, 1993). These anatomical features complicate quantitative, functional morphological studies of axial musculoskeletal systems (but see Arkowitz and Rommel 1985, Bennett et al. 1987, Pabst 1993, Blickhan and Cheng 1994).

The functional cross-sectional areas and thus, the force generated by the axial locomotor muscles of cetaceans, such as short-finned pilot whales (Arkowitz and Rommel 1985) and short-beaked common dolphins (*Delphinus delphis*) and bottlenose dolphins (Pabst 1993), are large. Quantitative data on the force-gen-erating potentials of the axial muscles of sirenians are not available, but images of cross-sections taken through an Amazonian manatee (Domning and de Buffrenil 1991) indicate that these muscles are not as well developed as in the cetaceans. The axial muscles of the manatee do increase in cross-sectional area near the insertion of the flukes (Domning and de Buffrenil 1991), suggesting that there may be relative increases in force-generating potential at the tail in these animals.

The elongated neural spines and transverse processes of cetacean vertebrae also increase the mechanical advantage of the axial muscles relative to terrestrial mammals (compare the height of neural spines in the dog vs. the bottlenose dolphin shown in Fig. 2-8). By inserting far from the point of vertebral column rotation, the lever arm-in is increased, and thus, force output is increased. A novel interaction between the tendons of the epaxial muscles and a subdermal connective tissue sheath that envelops those muscles also increases the work output of the axial musculoskeletal system in cetaceans (Pabst 1993, 1996). Instead of inserting on neural spines, some tendons travel through the subdermal sheath at oblique angles to insert on transverse processes of vertebrae located further caudally. The oblique tendon geometry increases the distance through which these tendons can be pulled by their muscles.

The sirenian axial skeleton does not display elongated neural spines, which would increase the lever arm-in for dorsoventral flexion; the lumbar and cranial-most caudal vertebrae, although, have elongated transverse processes (Domning 1977, 1978).

Increasing Propulsive Efficiency

The design of the musculoskeletal system also influences the efficiency of propulsion by affecting the pattern of movement, and the size and shape of the propulsive structure. For example, the dog paddle, used by most terrestrial mammals, is an adequate, but inefficient swimming mode (Webb 1988). The paddle stroke can be broken into two phases. The power phase produces useful thrust forces, whereas the recovery phase produces no net thrust but rather repositions the limb for the next power phase. Thus, for paddle-based swimming, thrust is generated during about half of the locomotor stroke.

Propulsive efficiency would be increased if the propulsive structures could generate thrust during the entire locomotor stroke. Hydrofoils, oriented appropriately to flow, can produce lift-based thrust throughout the locomotor stroke (Lighthill 1969; Webb 1984; Feldkamp 1987a,b; Fish 1993a; Vogel 1994). Rather than stick a broad, flat "paddle" or "oar" out in the water, marine mammals could use an oscillating "wing" capable of generating lift and thrust continuously to increase propulsive efficiency. A good hydrofoil design is one that has a long span (the length of the wing from base to tip) and short chord (the width of the hydrofoil in the direction of flow) (Webb 1988, Webb and de Buffrenil 1990, Vogel 1994). The ratio of span to chord is called the aspect ratio. A high aspect ratio hydrofoil will produce large lift and thrust forces with low drag, but may compromise maneuverability.

We can now make predictions about the morphology of the musculoskeletal locomotor system that might influence propulsive efficiency. For appendicular locomotors, the distal limb should be shaped like a high aspect ratio hydrofoil. This is just what we see in pinnipeds (Figs. 2-8 and 2-9). In fur seals and sea lions, the bones of the forearm (radius and ulna) are dorsoventrally flattened, and bones of the manus have lengthened to form the winglike flipper used for propulsion (Howell 1928, 1930; English 1976a,b, 1977). Cartilaginous rods at the distal tips of the phalanges increase the effective span of the otariid flipper (Howell 1928, 1930; King 1983). Similarly, the pes of walruses (Gordon 1983, King 1983) and phocid seals (Taylor 1914; Howell 1928, 1930; Tarasoff et al. 1972; King 1983) are expanded, predominantly owing to increases in length of the phalanges and distally placed cartilaginous rods. These appendages form hydrofoils that increase efficiency by creating lift and thrust throughout the locomotor stroke.

Interestingly, the feet of sea otters are also expanded due to lengthening of the phalanges (Taylor 1914, Tarasoff et al. 1972, Howard 1975). The individual foot is not shaped like a hydrofoil, but like a square-shaped oar, which sea otters use for slow paddling (Tarasoff et al. 1972). When swimming at higher velocities otters undulate their hind limbs dorsoventrally and synchronously with the caudal axial skeleton (Kenyon 1969, Tarasoff et al. 1972, Williams 1989). The hind limbs, when positioned together and moved synchronously, form a propulsive surface that approximates a hydrofoil with a lunate caudal border (Lighthill 1969, Tarasoff et al. 1972).

Although accomplished swimmers (Flyger and Townsend 1968), polar bears use a paddle stroke to swim. It is said that the forepaws of polar bears are large and "oarlike" (Stirling 1988, Reeves et al. 1992), although we have been unable to find any anatomical data to suggest that their paws are relatively larger than those of any other bear. In contrast to the pinnipeds and sea otters, the polar bear appendicular locomotor system does not display morphological characteristics suggestive of increased swimming efficiency.

The axially locomoting cetaceans and sirenians use their flukes to produce thrust forces. The lunate flukes of cetaceans are considered highly derived and efficient hydrofoils that decrease the energetic costs of swimming (Webb 1988; Fish and Hui 1991; Williams et al. 1992; Fish 1993a,b). The fluke of the marine dugong is also a lunate-shaped hydrofoil, whereas the fluke of manatees is decidedly less streamlined and oval in shape (Domning 1977, 1978; Nishiwaki and Marsh 1985). Manatees are apparently not sustained fast swimmers, but can produce large accelerations for burst swimming with their high surface area fluke.

Although the locomotor morphologies are diverse in marine mammals, shared patterns in musculoskeletal design can be identified in many species. To increase thrust forces, appendicular locomotors increase their force production by increasing muscle cross-sectional area and the mechanical advantage of their limb skeletal lever system. The cetacean axial locomotor muscles increase thrust forces by increasing their cross-sectional areas and increasing the excursion advantage of those muscles by using novel tendon arrangements. Propulsive efficiency is increased by using lift-based hydrofoils rather than using drag-based paddles.

Skull Morphology

The departure from the generalized type of land mammal skull is most striking in living cetaceans. Kellogg 1928

We introduce you to the diverse skull morphologies displayed in marine mammals for three reasons. First, the skull provides the skeletal support for soft tissues of the respiratory and digestive systems. It forms the bony capsules for the sensory structures, such as eyes, ears, and olfactory equipment, and the protective covering for the brain. Thus, knowing some of the morphology of the skull is of use when we investigate other systems in the body. Second, the skull has undergone massive reorganization in some of the marine mammals. This has profound effects on the relative placement of facial structures (Fig. 2-10). Third, the skull provides a critical set of phylogenetic characters used in the study of the evolution of marine mammals. We hope that this brief overview of skull morphology is of value when delving into the exciting studies of the evolution of this diverse group of mammals.

Marine mammal skulls are formed by no fewer than 20 interlocking and overlapping bones, each with its own complex morphology. The skull can be thought of as two functional units—the face and the vault of the cranium, or braincase. The facial bones form a platform for structures of the special senses of vision, olfaction, and taste, respiratory structures, and feeding structures of the oral cavity. The bones that form the vault of the cranium function to protect the brain and to articulate with the remainder of the axial skeleton (i.e., vertebrae). We have chosen four bony features to illustrate some of the functional diversity within the marine mammals.

First, we focus on the maxilla, a large facial bone that typically bears teeth and forms the floor of the nasal cavity and roof of the oral cavity. (In most species the mandible, or lower jaw, also bears teeth.) The maxillae of seals, sea lions, polar bears, and sea otters are similar in shape and function to that of our "typical" mammal, the dog (Fig. 2-10). Notice that in all these mammals, the maxilla lies predominantly cranial to the orbit (i.e., the region that surrounds the eye). The sirenian maxilla extends far caudal to the orbit (Vrolik 1848–1854; Husar 1977, 1978), perhaps owing to a relatively cranial placement of the orbit.

The cetacean maxilla is extensively remodeled, as compared to other mammals. In general, the cetacean skull is said to have "telescoped" (Miller 1923, Kellogg 1928). Telescop-

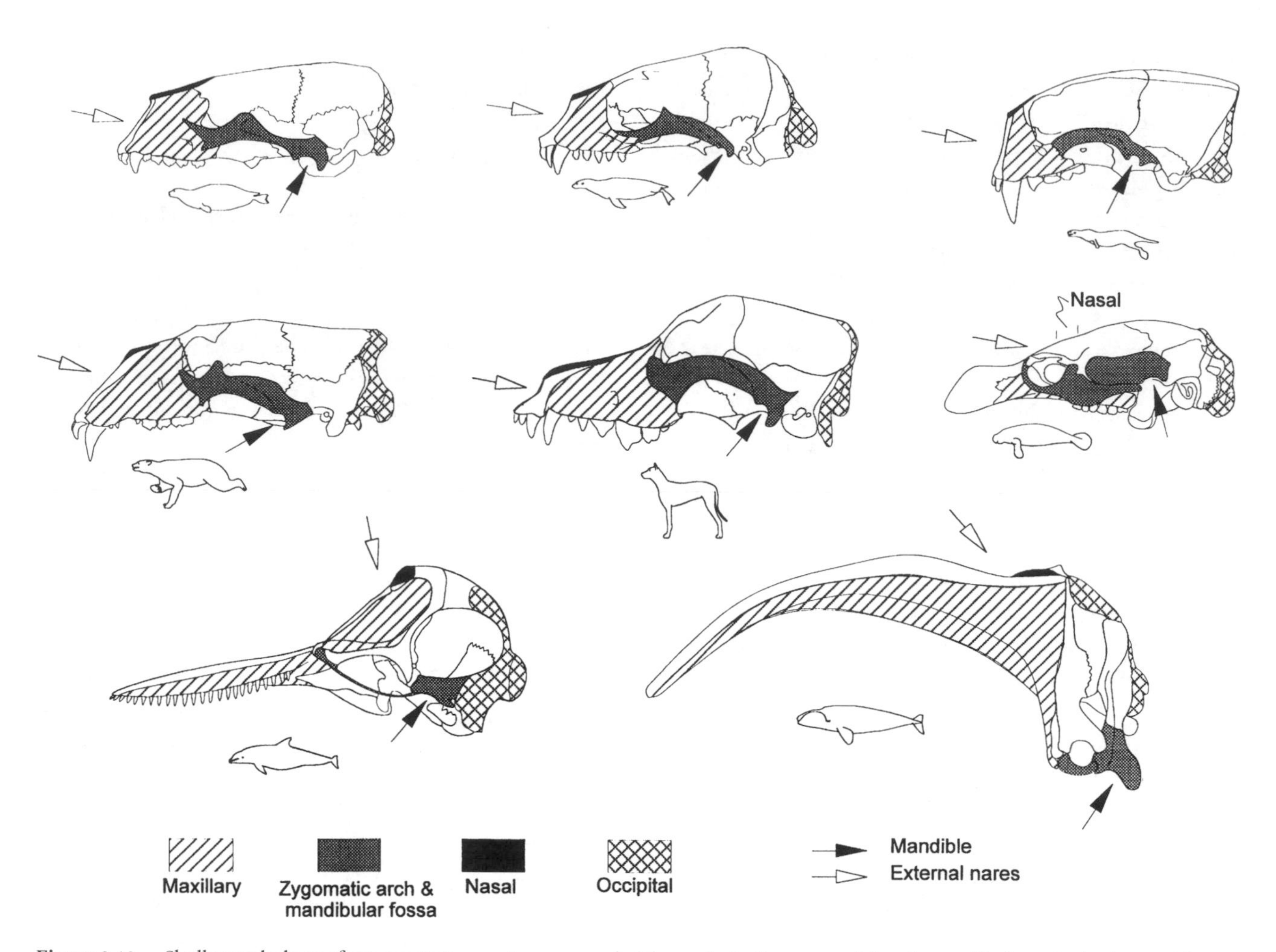

Figure 2-10. Skull morphology of representative marine mammals. The unshaded bones caudal to the maxilla form the braincase. The telescoped skull of the cetacean possesses an elongated maxilla and shortened braincase. The mandible (see Fig. 2-17) forms a lever with its fulcrum at the mandibular fossa of the zygomatic arch. One of the primary masticatory muscles, the masseter, originates on the zygomatic arch. (Illustrations adapted from Lichtenstein 1827-1834; Vrolik 1848-54; van Beneden and Gervais 1868-1879; Flower 1885; Allen 1902; True 1904; Howell 1930; Jacobi 1938; Omura et al. 1969; Nishiwaki and Kasuya 1970; King 1971, 1972; Nickel et al. 1986; Rommel 1990; Evans 1993; and authors' unpublished data.)

ing in both odontocetes and mysticetes is defined as a cranial elongation of the bones that form the rostrum (including the maxilla) and an overlapping of skull bones that tends to produce a shortened braincase. Take a few minutes and look at Figure 2-10. Compare the lengths of the maxillae and positions of the maxillae relative to the frontal bone in the dolphin and the dog. The dog maxilla abuts the frontal bone, whereas in odontocetes, the maxilla slides dorsally over the surface of the frontal. Thus, the maxilla extends both rostral and caudal to the orbit in odontocete cetaceans. Now compare the relative lengths of the braincase (unfilled bones) of the dog versus cetaceans. The cetacean braincase is very foreshortened relative to those of other mammals. Romer (1966) noted that for cetaceans, "osteologically there is no top to the skull; it is all front and back." The skulls of many species of odontocete cetaceans are also strongly bilaterally asymmetric (see Mead 1975, Flower 1885, Heyning 1989, Rommel 1990). This asymmetry cannot be seen in the left lateral view in Figure 2-10, but is illustrated in Figure 2-26 later in this chapter.

In mysticete cetaceans, the maxilla is tremendously elongated. It is also telescoped, but whereas the odontocete maxilla lies over the dorsal surface of the frontal bone, the mysticete maxilla has slid over the ventrally extended frontal bone (Miller 1923, Kellogg 1928). Another unique feature of the mysticete maxilla is that it bears baleen, the keratinous sievelike feeding apparatus of the mysticete cetaceans (see discussion of the digestive system). Mysticetes bear tooth buds early in their ontogeny, but the tooth buds are lost and replaced with baleen as the whales age (Van Valen 1968, Barnes and Mitchell 1978).

A result of skull telescoping in the cetaceans, and the second feature to notice, is that the external nares (or nostrils) no longer lie rostrally, but rather are positioned dorsally and almost centrally in the skull. Notice that in typical mammals, the nasal cavity is a horizontally oriented passage. In cetaceans, the nasal cavity is oriented obliquely vertically and opens dorsally. In typical mammals, the nasal bones are flat, horizontally oriented plates of bone that form the roof of the nasal cavity. In odontocete cetaceans, the nasal bones are small lumps of bone positioned at the caudal edge of the external nares, and do not cover the nasal cavity (see also Figure 2-26 in the section on the nervous system).

In sirenians, the nasal cavity is obliquely horizontal in its orientation and the small nasal bones lie caudal to the nares. The nasal bones are in a similar position in the elephant (Flower 1885).

A third skull feature that is variable across marine mammals is the zygomatic arch, which typically bridges the facial and braincase regions of the skull, forms a part of the ventral margin of the orbit, and is the site of origin for the masseter muscle. The masseter is one of the masticatory (chewing) muscles that closes the jaw. It is extremely robust in herbivores, because it transmits its forces to the chewing and grinding teeth in the caudal oral cavity (Hildebrand 1982). As with the maxillae and nasal bones, the zygomatic processes of seals, sea lions, polar bears, and sea otters are similar in shape to that of the dog. The zygomatic arch of cetaceans is reduced, especially in odontocetes where it exists only as a slender, delicate rod of bone. Sirenians, on the other hand, have a massive, but porous and oil-filled, zygomatic arch that acts to increase the surface area of attachment for the masseter.

The last skull feature we bring to your attention is the occipital bone that articulates with the first cervical vertebra called the atlas. The spinal cord exits the braincase through the foramen magnum of the occipital bone. Notice again that the region of the braincase between the frontal and the occipital bones is greatly reduced in the cetaceans, as compared to other mammals.

Buoyancy Control

Ballast, whether in a submarine or a seacow, can serve three major functions: to decrease buoyancy, to maintain proper trim, and to increase stability, especially in a roll.

Domning and de Buffrenil 1991

In the discussion of the integument system, we argue that there are a number of mechanisms that mammals could use to reduce their density and, thus, avoid sinking. Morphological mechanisms used to decrease density, of course, have the concomitant effect of increasing the energy an animal might require to overcome buoyancy to dive. It has been argued that for a number of marine mammals, increased body density would be advantageous for ease of diving (Fish and Stein 1991). Because bone is one of the densest materials in the mammalian body, increasing the skeletal density has been viewed as an adaptation to increasing whole body density (Wall 1983, Domning and de Buffrenil 1991, Fish and Stein 1991).

Fish and Stein (1991) attempted to determine whether there was a correlation between limb bone densities and degree of aquatic specialization within the mustelids. Although these researchers state that there is an apparent trend toward increased limb bone density in the more aquatic species, the trend is not statistically significant. Their data also did not indicate any consistent pattern in limb bone densities from less to more aquatic forms, as might be expected if bone density changes were adaptations to an aquatic lifestyle. Limb bone density may be correlated to overall body mass, especially in species that spend time on land, as would be expected from scaling models of skeletal support (Alexander 1977, Biewener 1989).

Sirenians, on the other hand, do possess bones of dramatically high density (Domning and de Buffrenil 1991). Domning and de Buffrenil (1991) present interesting anatomical data that suggest that the dense rib bones probably serve as ballast to assist in buoyancy and stability control in the herbivorous, routinely slow-moving sirenians. In contrast, cetaceans and some pinnipeds have reduced bone density (Slijper 1936, Wall 1983, de Buffrenil et al. 1985). Thus, within the marine mammals, there is no simple, consistent pattern of bone density associated with the degree of aquatic specialization. Moreover, very little quantitative data exist on buoyancy regulation for most marine mammal species (Domning and de Buffrenil 1991).

Respiratory System

The respiratory system functions to deliver air to a location in the body where oxygen can diffuse into, and carbon dioxide can diffuse out of, the blood (for review, see Negus 1949, Schmidt-Nielsen 1990). In most mammals the respiratory system is also used to produce sounds or vocalizations.

The "typical" mammalian respiratory system is schematically represented in Figure 2-11. In most mammals there are two entrances to the respiratory system—the nostrils (or external nares) and the mouth. Air travels through these openings and into the pharynx, a short, muscular tube that is partially shared by both respiratory and digestive pathways. From this shared pathway, food preferentially travels through the esophagus (see digestive system) and air through the larynx. The larynx contains a complex set of cartilages and muscles that functions to protect the respiratory system from food and fluids and to house the vocal cords that many mammals use to produce sounds. From the larynx, air

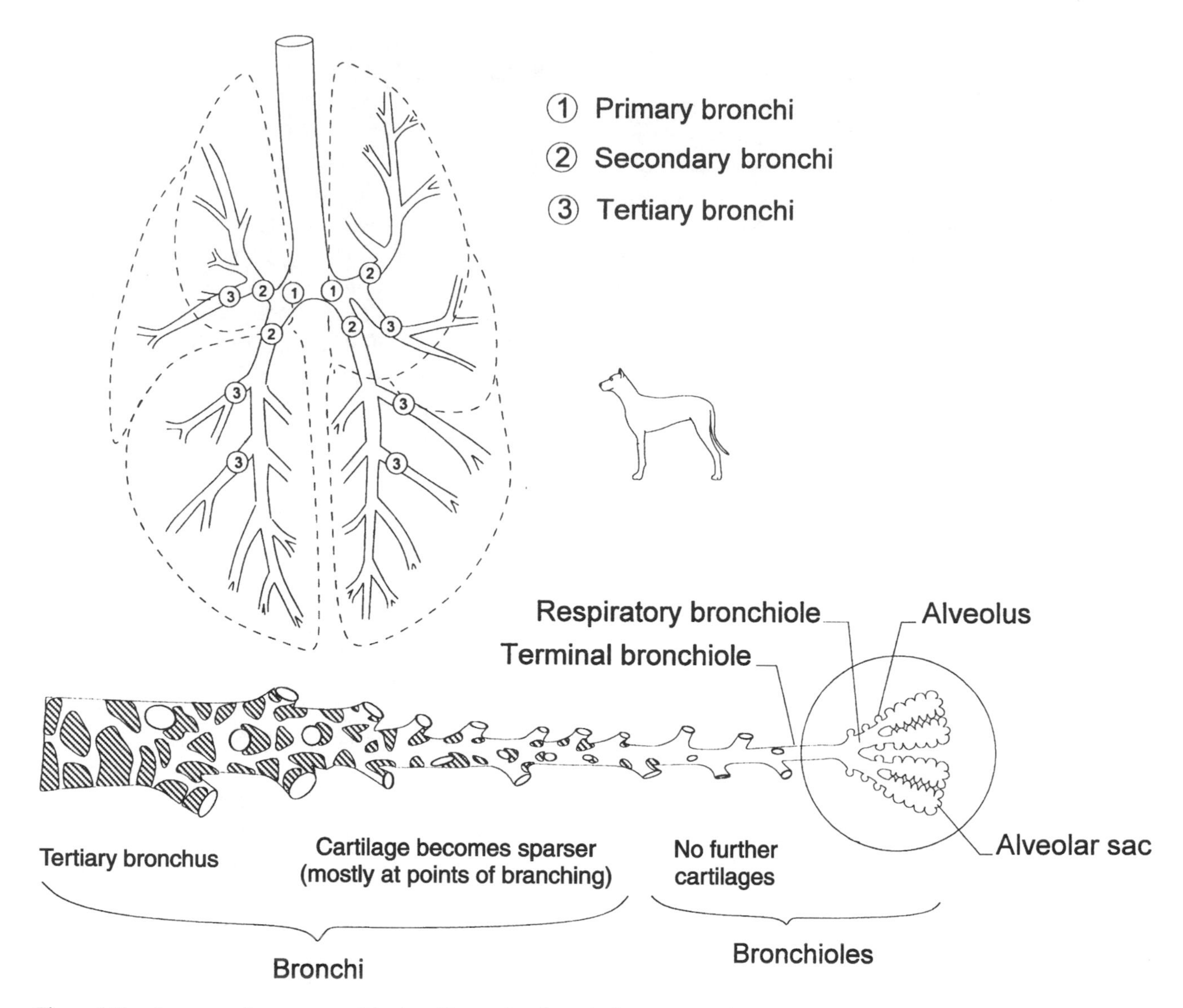

Figure 2-11. Lower respiratory tract of the dog. Air traveling from the larynx enters the cartilage-reinforced trachea. The trachea bifurcates into primary bronchi that enter the lungs. The bronchi continue to branch into the small bronchioles, which lack cartilaginous reinforcement. Bronchioles lead to sites of gas exchange, the respiratory bronchioles and alveoli. (Illustrations adapted from Schummer et al. 1979, Evans 1993.)

enters the trachea, a cartilage-reinforced tube that bifurcates into primary bronchi (Fig. 2-11). The primary bronchi enter the lungs and branch into secondary bronchi that supply individual lobes of each lung. The secondary bronchi continue to branch into cartilage-reinforced conducting tubes of decreasing diameter. The smallest of these conducting tubes are the delicate bronchioles, which lack cartilage reinforcement and are reinforced by a few obliquely oriented muscle fibers (Denison et al. 1971). The bronchioles, in turn, lead to the sites of gas exchange—the thin-walled, membranous respiratory bronchioles, alveolar sacs, and alveoli. The conducting and respiratory tubes from the primary bronchi to the alveoli are collectively named the bronchial tree.

Although the respiratory systems of marine mammals generally conform to this pattern, they also possess morphological specializations apparently adaptive to their aquatic lifestyle. We focus our comparative investigation on the role of the lung in buoyancy control and during diving. We also describe structural features displayed in some marine mammals to produce sounds and to protect the respiratory system from exposure to seawater.

Buoyancy Control

A seal floating upright in the water rides about as high as a man in a life vest. Kooyman 1973

We have seen that blubber and regional variations in bone densities have been implicated as buoyancy regulation mechanisms in some marine mammals. The lung is perhaps the most intuitively obvious and most easily adjustable buoyancy compensator in the mammalian body. You have probably noted how easy it is to adjust your own flotation at the water's surface by changing the volume of air in your lungs. The lung volume of some marine mammals is relatively larger than comparably sized terrestrial mammals (Fig. 2-12; Kooyman 1973). The sea otter, an animal that lacks a buoyant blubber layer, has the largest lung volume-to-body weight ratio of any marine mammal, twice (Fig. 2-12, upper dotted line) that predicted (solid line) for a terrestrial mammal of its body mass (Tarasoff and Kooyman 1973a). Kooyman (1973) points out that this might be particularly important for an animal that feeds while floating, apparently effortlessly, on the surface—especially when one considers that a sea otter uses its abdomen as a resting surface for rocks against which it breaks open abalone shells and urchin tests!

Manatee lungs are unlike those of most other mammals—each is supplied with a single bronchus that runs virtually the entire length of each lung (Murie 1872, Engel 1962). The very elongated and horizontally oriented lungs of sirenians have also been hypothesized to function as buoyancy control structures (Domning and de Buffrenil 1991). Thus, relatively increased lung volumes in some marine mammals (Fig. 2-12), and unique lung morphologies in others, may function to decrease the energetic cost of floating at the surface, relative to terrestrial mammals.

Diving

The environmental condition that has the most profound effect on the respiratory system of marine mammals is the rapid change in hydrostatic pressure which occurs when they dive to depth. Kooyman 1973

Most marine mammals undergo extended periods of apnea, or breath-holding, during dives. During those dives, the lung acts as a site of oxygen storage (for review, see Kooyman 1969, 1973, 1985, 1987; Elsner, Chapter 3, this volume). The atmospheric gases within the lung also pose potential threats to a mammal undergoing prolonged, multiple or deep dives (see Elsner, Chapter 3, this volume). How is the lung constructed to act as an oxygen store while protecting the marine mammal against the damaging effects of some gases at high pressures? We compare lung structure in deep versus shallow divers to investigate this question.

Oxygen Storage

Humans usually breath-hold dive on fully inflated lungs and thus, use the lung as a site of oxygen stores throughout the dive. Humans are neither deep nor prolonged divers. Interestingly, one of the most proficient divers, the Weddell seal, usually exhales before diving (Scholander 1940, Kooyman et al. 1970) and its lung volume per body weight is less than that of comparably sized terrestrial mammals (Kooyman 1973; Fig. 2-12). Apparently, the lung is not the primary site of oxygen storage in the body of a Weddell seal during a dive. Where are other on-board oxygen stores of diving marine mammals?

Oxygen is stored and transported in blood by reversibly binding to hemoglobin in the red blood cells and is also stored in muscle, by reversibly binding to myoglobin found within muscle cells (for review, see Schmidt-Nielsen 1990). Thus, oxygen can be stored in the lung, blood, and muscle tissues (Fig. 2-13). Notice that lung storage is a smaller percentage of total on-board oxygen stores in deep versus shallow diving mammals (Fig. 2-14). Concomitant to decreased oxygen stores in the lung of deep diving mammals are increased oxygen storage capacities of their (1) blood by way of increased blood volume and increased red blood cell concentration, and (2) muscle by way of increased myoglobin stores (e.g., Ridgway and Johnston 1966, Hedrick et al. 1986, Kooyman 1987, Wickham 1989, Thorson and Le Boeuf 1994).

Thus, deep diving mammals have larger on-board oxygen stores in their blood and muscles than in their lungs, relative to both terrestrial and shallow diving mammals. Why might the lung play a less important role in oxygen storage during a deep dive? To investigate that question, we must consider the effects of pressure changes on lung structure and function.

Pressure and the Lung

Mammalian lungs are constructed of numerous gas-filled spaces. The volume of a given mass of gas is inversely related to the pressure it is experiencing, that is, if the pressure exerted on a gas is doubled, its volume will be halved (for review, see Vogel 1988). Water exerts approximately 1 atmosphere of pressure for every 10 m of depth. Therefore, a marine mammal at 10-m depth experiences two times more pressure than it would at the surface, and the air within its lungs occupies one-half its volume at the surface. The lungs occupy just a fraction of their surface volume at the rather heroic depths achieved by many deep diving mammals—600 to 740 m for Weddell seals (Kooyman 1981, Testa 1994), more than 1500 m for elephant seals (DeLong and Stewart 1991), and 2000 m for sperm whales (Heezen 1957, Watkins et al. 1993).

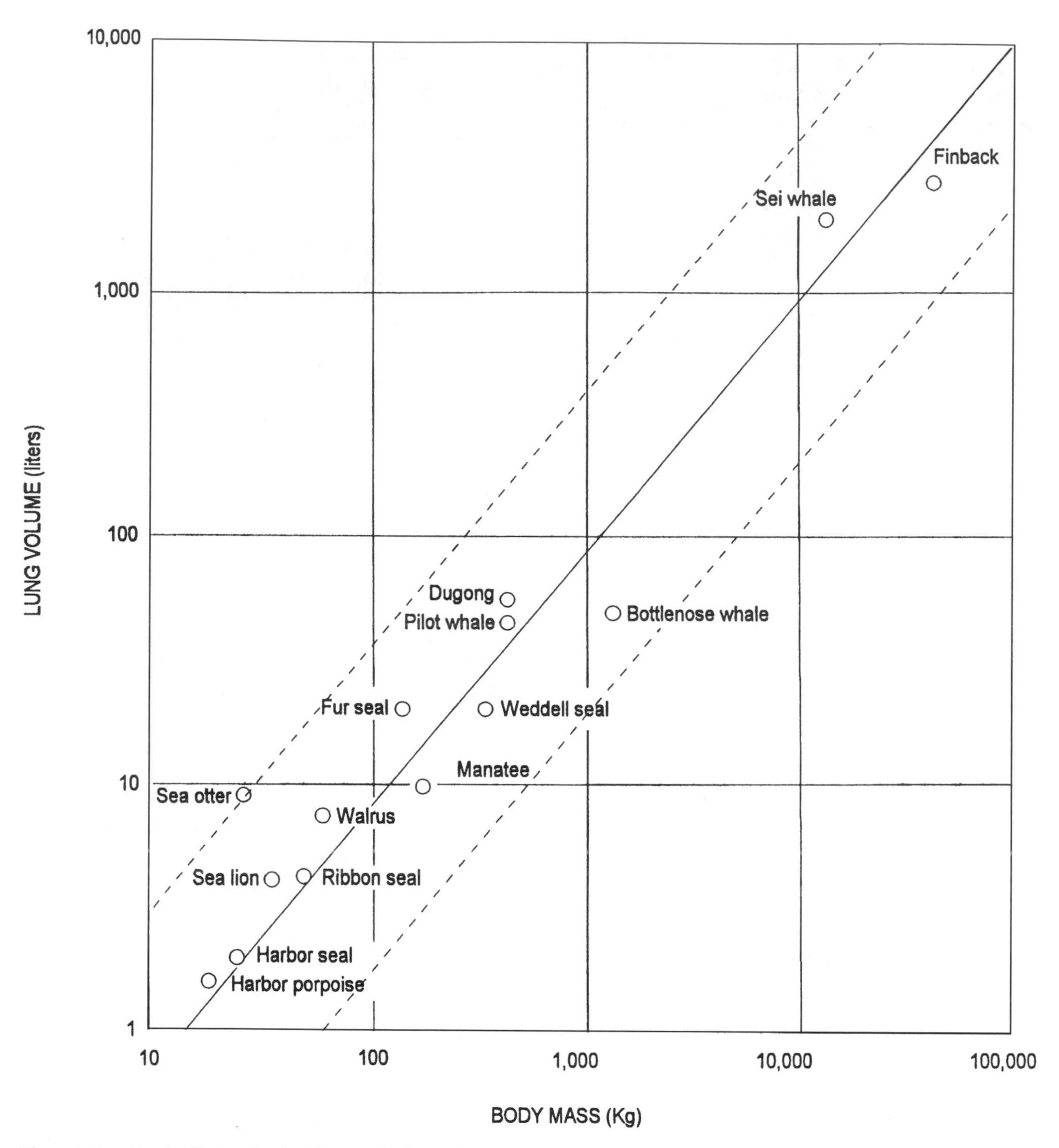

Figure 2-12. Graph of lung volume (in liters) to body mass (in kilograms) for a variety of marine mammals. The solid line is a best-fit regression for a variety of mammals (Tenney and Remmers 1963). The slope of this line is very close to 1, illustrating that most mammals tend to maintain similar relative lung volumes (approximately 5% of their total body volume) no matter their body size (Schmidt-Nielsen 1990). A number of marine mammals fall off the line (data from Kooyman 1973). Most notably, the sea otter, a shallow-diving mammal that spends much of its time floating at the sea surface, has a lung volume twice that predicted for a mammal of its size. The upper dotted line represents twice the values on the solid line; the lower dotted line represents half of the values on the solid line. Contrarily, the lung volumes of the Weddell seal and bottlenose whales, both deep and prolonged divers, are less than would be predicted.

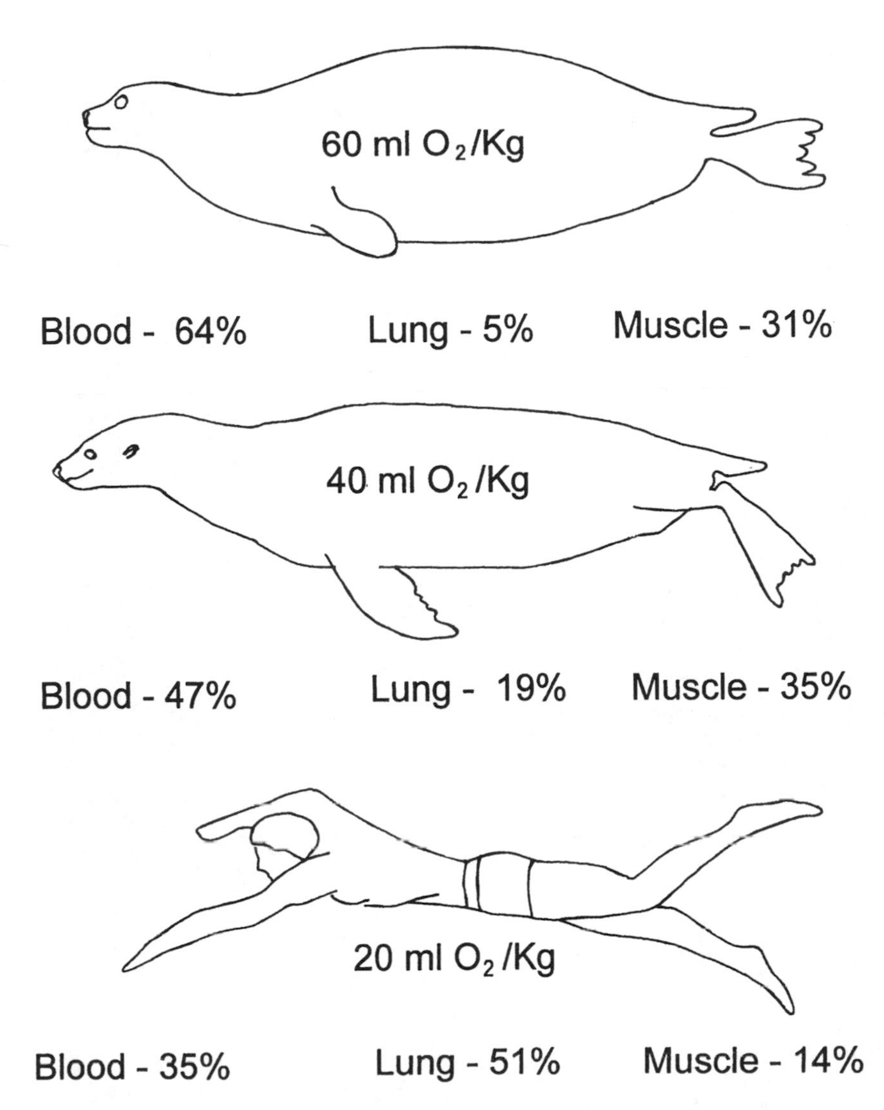

Figure 2-13. Relative oxygen stores in mammals. The deep-diving Weddell seal (top) carries more than 95% of its oxygen reserves in blood and muscle. A fur seal (middle), a shorter duration diver, also carries the majority of its oxygen reserves in blood and muscle, but relies more on the lungs stores than does the seal. The human relies most heavily on the lung as an important oxygen reserve. Note the differences in total oxygen stored per kilogram of body mass between these mammals. (Illustration adapted from Kooyman 1985.)

Data exist to suggest that the entire thoracic cavities of some marine mammals compress during diving. Seals and odontocete cetaceans have flexible rib cages that are easily deformed (Kooyman et al. 1970, King 1983, Rommel 1990). Natural and simulated dive experiments have demonstrated that thoracic cavities of bottlenose dolphins and Weddell and northern elephant seals (*Mirounga angustirostris*) collapse at depths of 10 to 35 m (Ridgway et al. 1969, Kooyman et al. 1970, Ridgway and Howard 1979).

When the lung and thoracic cavity compress during a dive, what happens to the air that is still trapped within the lung? This is not a trivial question because the high partial pressure of nitrogen within that air is potentially dangerous to the mammalian diver. A brief description of Henry's gas law is required to understand what happens to nitrogen under pressure. When gas pressure in contact with a fluid increases, the amount of the gas that will go into solution in the fluid also increases as

$$V_{gas} \propto p_{gas} V_{fluid} \qquad \text{(equation 5)}$$

where V_{gas} is the volume of gas dissolved in the fluid, p_{gas} is the pressure of the gas, and V_{fluid} is the volume of the fluid into which gas is dissolving (Schmidt-Nielsen 1990). Contrarily, decreasing the gas pressure over the solution allows the dissolved gases to come out of solution. A familiar example of Henry's law is provided by a carbonated beverage. Sealed under pressure carbon dioxide remains in solution, but once opened carbon dioxide escapes as effervescent bubbles.

In our diving mammal, increased pressure of air within the lung causes more nitrogen to dissolve into the blood across the alveolar membrane. Mammalian tissues can become saturated with nitrogen at pressures at or above 2 atmospheres; that is, saturation can occur after prolonged deep dives or repeated dives to depths over 10 m. If tissues become saturated, and subsequently the diver surfaces too quickly, the nitrogen can come out of solution and form bubbles in tissues and blood. This condition, known as the bends (also known as caisson disease or aeroembolism), can be extremely painful and even fatal (for review, see Schmidt-Nielsen 1990).

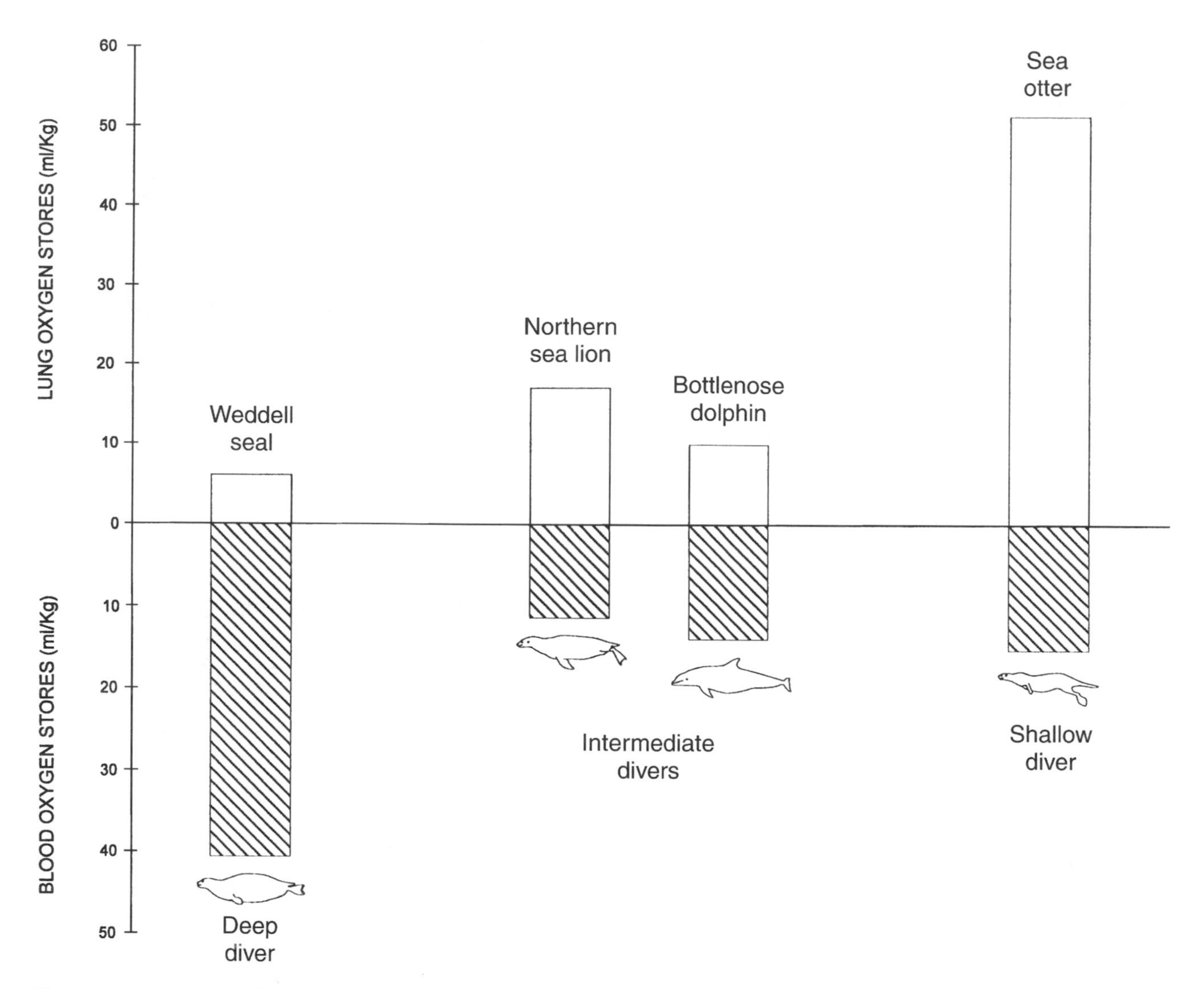

Figure 2-14. Lung versus blood oxygen stores in shallow- and deep-diving mammals. In deeper and longer divers, the lung is a less important oxygen reserve. (Illustration adapted from Kooyman 1973.)

Many marine mammals are deep and continuous divers. How do these diving mammals avoid the bends? (Although marine mammals do not breathe air at pressure while underwater, as do human scuba divers, continuous deep diving free divers are susceptible to the bends [Schmidt-Nielsen 1990].) An obvious solution is simply to avoid exposure of atmospheric gases to blood. One mechanism would be to exhale before diving, as do seals (Scholander 1940, Kooyman et al. 1970). Manatees (Scholander and Irving 1941), sea lions (Dormer et al. 1977), and cetaceans (Ridgway et al. 1969) dive with lungs full of air. Another mechanism to avoid toxic exposure of nitrogen to blood is alveolar collapse. Recall that during a deep dive, the lungs collapse. If the air remaining in the lungs were trapped within the alveoli, nitrogen could easily enter the bloodstream. On the other hand, if the alveoli were collapsed, forcing air into nongas exchanging regions of the respiratory tree, the nitrogen would be mechanically isolated from the blood, and the danger of the bends could be avoided (Scholander 1940). Alveolar collapse occurs in the lungs of diving mammals as a result of unique morphological and mechanical properties of the terminal airways of the bronchial tree.

In the lung of the dog and other terrestrial mammals, the terminal airways leading to the alveoli collapse under compression before the alveoli are empty, thus trapping air at the site of gas exchange (summarized in Denison et al. 1971). The collapse of the airways leading to the alveoli occurs because the bronchioles are thin, delicate tubes (Fig. 2-15). Contrarily, in every diving marine mammal investigated thus far, the terminal airways leading to the alveoli are reinforced with cartilage or thickened muscle (Fig. 2-15; Belanger 1940; Scholander 1940; Engel 1962; Denison et al. 1971; Denison and Kooyman 1973; Kooyman 1973; Tarasoff and Kooyman 1973a,b; Fay 1982; Drabek and Kooyman 1983). Thus, under

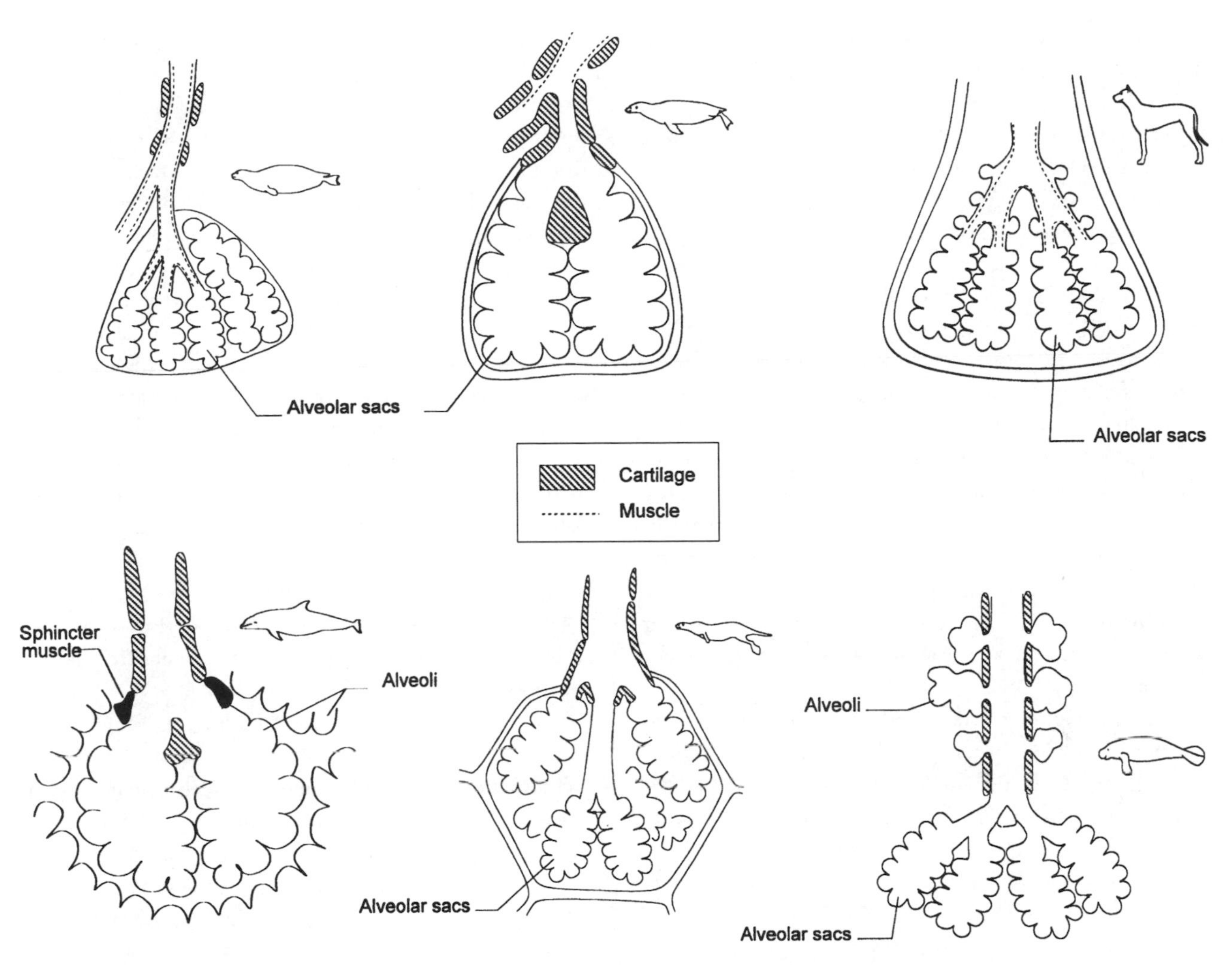

Figure 2-15. Morphology of the terminal airways and alveoli in the lungs of different mammals. The terminal airway of the "typical" mammal is not reinforced with cartilage and can collapse under pressure, trapping air in the alveoli during a dive. The terminal airways of all marine mammals investigated thus far are reinforced with cartilage or thickened muscle (stippled line). When alveoli in the marine mammal lung collapse, air is forced into the reinforced upper airways of the bronchial tree; nitrogen is isolated from the site of gas exchange, and the danger of the bends is avoided. (Illustrations adapted from Belanger 1940, Engel 1962, Denison and Kooyman 1973, Schummer et al. 1979, Drabek and Kooyman 1983, Bergey 1986, Evans 1993.)

compression, the alveoli in the marine mammal lung collapse and air can be forced into the reinforced upper airways of the bronchial tree; nitrogen is isolated from the site of gas exchange and the danger of the bends is avoided. Denison and Kooyman (1973) considered the similar patterns of airway reinforcement displayed in the diverse assemblage of marine mammals a "striking example of convergent evolution."

Although the morphological explanation presented above is accurate, it is also simplified. The terminal airways in all diving marine mammals are reinforced, but there are interesting differences among species (Fig. 2-15). For example, in seals, the penultimate airways are reinforced with cartilage, whereas the terminal airways emptying into the alveoli are reinforced only with muscle. On the other hand, the terminal airways of sea lions and cetaceans are reinforced with cartilage all the way to the entrance to the alveoli. If the muscular reinforcement of the terminal airways is sufficient to allow for alveolar collapse in seals (Kooyman et al. 1970), why do cetaceans and sea lions have "extra" cartilaginous support? Unlike most other mammals, cetaceans and sea lions can maintain high peak expiratory flows at low lung volume (e.g., Kooyman and Sinnett 1979, Kooyman and Cornell 1981). Perhaps cartilage reinforces the terminal airways against collapse during the explosive expiratory phase of respiration in these fast-breathing mammals (Denison and Kooyman 1973, Kooyman 1973).

Another interesting and enigmatic feature of the cetacean bronchial tree is the presence of circular muscular and elastic sphincters at the entrance of the alveoli (e.g., Scholander 1940, Kooyman 1973, Fanning and Harrison 1974,

Drabek and Kooyman 1983). It has been hypothesized that the sphincters function to trap air in the alveoli and to regulate air flow from the alveoli during a dive (for review, see Drabek and Kooyman 1983). Interestingly, if air were trapped within the alveoli during a dive, the body may be exposed to nitrogen under pressure. It is fair to say that this interesting morphological feature, apparently unique to cetaceans, is not well understood.

Although the discussion thus far has centered on reducing the exposure to nitrogen during a dive, some extreme divers may actually use nitrogen to facilitate their diving capabilities. Adult female northern elephant seals make approximately 20-min dives to depths between 500 and 800 m with less than 3.5 min between dives, and repeat this pattern continuously for over 2.5 months while at sea (Le Boeuf et al. 1989)! Le Boeuf et al. (1989) have suggested that elephant seals may use nitrogen as a narcotic during the dive to counteract the effects of high pressure on the nervous system. These extreme divers must have some mechanism for coping with nitrogen saturation, but to date, this is not understood.

This brief overview of diving effects on lung structure and function raises a number of questions. For example, how are the lungs reinflated after extended periods of collapse? Do marine mammals produce more surfactants than terrestrial mammals? What is the morphology of the polar bear respiratory system? Do polar bears share morphological features with other marine mammals? Clearly, there is much left to learn about respiratory structures and functions in diving marine mammals.

Sound Production

The Sirenia, or sea cows, may now be added to the roster of aquatic mammals known to utter underwater sounds, joining the Cetacea and Pinnipedia, albeit with very modest music.
Schevill and Watkins 1965

Most mammals produce sound by vibrating elastic ligaments (vocal cords or folds) in the larynx (Negus 1949, Cartmill et al. 1987). Vibrations are generated by passing air across these ligaments. These sounds can be modulated by actions of the tongue or the teeth, or both. Resonating chambers, such as bony sinuses and pharyngeal air sacs, also influence sound production.

Pinnipeds, sea otters, and polar bears can produce a wide variety of sounds in air, including barking, crying, growling, roaring, and snorting; these sounds are probably produced by the vocal cords (for review, see Kenyon 1981, King 1983). Underwater, some pinnipeds can produce higher frequency sounds including bell-like ringing, "clicks, trains of pulses, whistles, and warbles" (King 1983). These sounds may be produced by structures other than the vocal cords. Air cycled through pharyngeal air pouches found in otariids and male odobenids has been suggested as a source for these underwater sounds (Fay 1960, 1982; Schevill et al. 1966; King 1983). Some sounds may function as echolocation signals (for a review of echolocation, see Au 1993; Wartzok and Ketten, Chapter 4, this volume), which may be particularly useful when these animals are navigating and feeding in deep, poorly lit or turbid water (King 1983). For example, diving Weddell seals produce eerie sounds hypothesized as echolocation signals (Kooyman 1981). Interestingly, phocid seals do not possess extra-pharyngeal structures and it is not yet known how these underwater sounds are produced.

Manatees, once thought to be silent (Murie 1872), produce "squeaky and rather ragged" sounds (Schevill and Watkins 1965). Sirenians apparently possess vocal cords (Murie 1872), and we assume that this is the site of sound production.

Odontocete cetaceans can produce extremely high-frequency sounds (>100,000 Hz) known to function as echolocation signals (for review, see Nachtigall and Moore 1988, Thomas and Kastelein 1990, Au 1993). The larynx was initially implicated as the site of sound production, but experiments on live, phonating dolphins demonstrated that the larynx does not move during high frequency vocalizations (e.g., Norris et al. 1971, Ridgway et al. 1980, Amundin and Anderson 1983, Ridgway and Carder 1988). Rather, structures in the nasal system, including the nasal plug and the elaborate nasal sac system, move when sound is produced, although the exact site of the sound generator is still controversial (Fig. 2-16; Mead 1975, Heyning and Mead 1990, Au 1993). The complex morphology and mechanics of the proposed sound generators lie outside the scope of this review chapter, but the interested reader will find comprehensive reviews in Au (1993), Cranford (1988), Cranford et al. (1996), Heyning (1989), and Mead (1975).

The odontocete nasal apparatus is a highly derived sound-generating system. Equally derived is the path of sound transmitted out of the head: sounds are transmitted to the environment through the fatty melon (Fig. 2-16). Nasal air sacs, the steeply sloping maxillary bones, and the rostrum reflect sounds and help focus the sound beam forward through the melon. Lipids found in various regions of the melon affect the speed of sound transmission (Norris and Harvey 1974). Thus, the melon may function acoustically to couple the sounds generated in the head into the water (Au 1993).

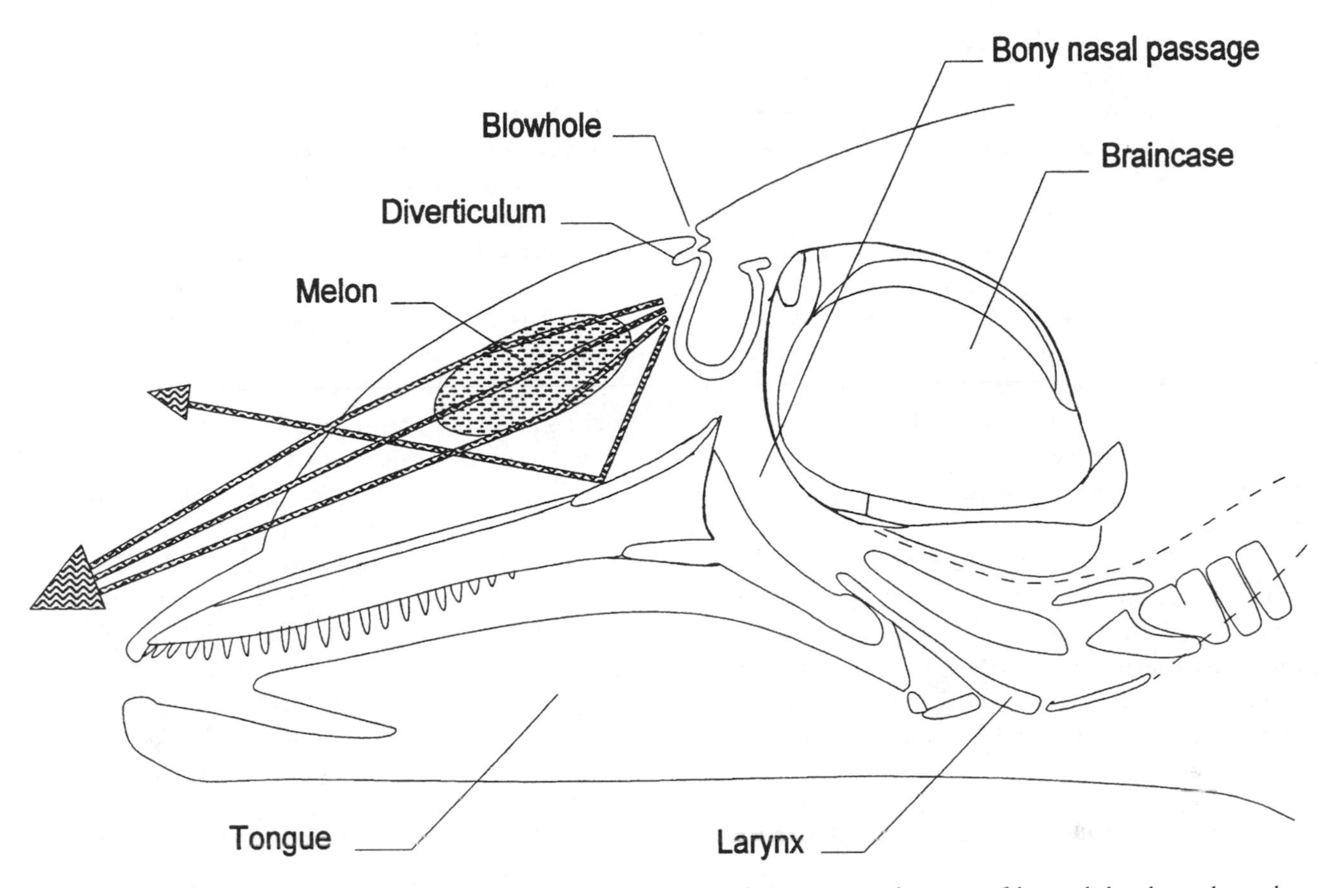

Figure 2-16. Midsagittal view of the head of a bottlenose dolphin. The sound generator is in the region of the nasal plugs located near the diverticulum. Sounds are transmitted and refracted forward through the fatty melon and exit the head. Some sounds are reflected off air-filled nasal sacs, diverticulae, and facial bones. (Illustration adapted from Norris and Harvey 1974, Mead 1975, Au 1993, Cranford et al. 1996.)

Isolating the Respiratory System from the Aquatic Environment

Larynx singularis figurae, aserinum caput refert—the strange larynx resembles the head of a goose.
Bartholinus (1654) commenting on the odontocete larynx (cited and translated in Slijper 1979)

It was once thought that the watery "blow" of large whales was evidence that they inhaled seawater into their lungs to extract oxygen from their fluid environment (Engel 1966). Of course, whales do not inhale water to breathe and must protect their respiratory system from water entering either their blowholes or mouth. In fact, all marine mammals are faced with the same problem. How do deep diving animals keep water from entering their nares at the high pressures they are experiencing? Likewise, how do these animals feed underwater without inspiring seawater?

The nares of marine mammals must close tightly. Mead (1975) points out that pinnipeds have relatively simple nares and suggests that deep diving does not require a specialized nasal apparatus. Perhaps the most specialized feature of pinniped nostrils is that the closed position appears to be their neutral posture (King 1983). For example, northern elephant seals sleep during breath-holding on land with their nostrils closed (Le Beouf et al. 1988, 1989). Delphinid cetaceans possess paired nasal plugs that cover the bony nares. The neutral posture of the nasal plugs is closed. The nasal plugs sit tightly against the bony nares and effectively seal the entrance to the nasal cavity when the nasal plug muscles are relaxed (Lawrence and Schevill 1956, Mead 1975; see detailed anatomy in Cranford et al. 1996). The semicircular nostrils of manatees (Caldwell and Caldwell 1985) and dugongs close with "anteriorly hinged valves" (Nishiwaki and Marsh 1985), although a detailed anatomical description appears to be lacking. Little is known about the morphologies of the nostrils of sea otters and polar bears, although it is likely that they are similar in form to their more terrestrial relatives.

Most marine mammals have laryngeal structures similar to those of terrestrial mammals. In most terrestrial mammals, including the dog, the larynx is located cranially and dorsally in the neck (Negus 1949, Evans 1993). The epiglottis in the breathing position contacts the soft palate and effectively separates the digestive and respiratory passages. This intranarial position of the epiglottis changes during swallowing: the larynx is raised and the epiglottis folds over the entrance to the trachea to protect the respiratory system from food or fluids being swallowed. Thus, terrestrial mammals can swallow fluids that enter their oral cavities without choking.

Pinnipeds possess terrestriallike laryngeal structures, although there are species-specific differences in cartilage form and size (King 1965, 1972, 1983). Seals, for example, possess a large epiglottis, whereas the walrus and sea lion apparently have small epiglottal structures (Negus 1949, Fay 1982). The manatee larynx is also small and appears to lack cartilaginous reinforcement (Murie 1872, Negus 1949). Polar bears have a larynx typical of ursids (Negus 1949), and we know of no description of sea otter laryngeal morphology.

The odontocete cetacean larynx is the most specialized of any marine mammal. The epiglottis and corniculate cartilages form an elongated goosebeaklike structure (Reidenberg and Laitman 1987). The elongated shapes of these cartilages fit rigidly into the elongate nasal passage (Fig. 2-16). The palatopharyngeal sphincter muscle keeps the goosebeak firmly sealed in an intranarial position (Lawrence and Schevill 1965). Wide lateral channels in the esophagus allow food to pass around the larynx during swallowing, and the sphincter muscles ensure that no food or fluids enter the respiratory system (Green et al. 1980). The structure and intranarial position of the larynx "effectively separates the respiratory tract from the digestive tract to a greater extent than that found in any other mammal" (Reidenberg and Laitman 1987). The permanent intranarial position of the larynx also allows odontocetes to simultaneously echolocate and swallow (Reidenberg and Laitman 1987).

Digestive System

The functions of the digestive system can be described in three steps: (1) capture and manipulation of food/prey (feeding), (2) mechanical and chemical digestion, and (3) nutrition (Schmidt-Nielsen 1990). We investigate (1) and (2) and direct the student to Costa and Williams, Chapter 5, this volume, for a review of nutrition and energetic requirements of marine mammals.

In our representative mammalian digestive system, food enters the oral cavity (ingestion), is processed (mastication) and swallowed (deglutition) through the coordinated actions of the lips, muscular tongue, and teeth (Hiiemae and Crompton 1985). The dog, a carnivore, takes relatively large bites of prey (or, more recently, Alpo), chops food with specialized slicing teeth (carnassials), then processes and quickly swallows food in large boluses. Carnivores tend to move their jaws up and down, with little lateral and no rotary movements at the jaw joint (Vaughan 1986). Contrarily, herbivores spend considerable time chewing, grinding, and processing their food to break down the tough cell walls of plants (see Skull Morphology). In both carnivores and herbivores the process of mechanical and chemical digestion (with salivary enzymes) begins in the oral cavity.

Swallowed food passes through the muscular pharynx to the esophagus, then is carried by peristalsis to the stomach, where mechanical and chemical digestion continue. The stomachs of mammals can be simple sacs (e.g., humans, carnivores) or complex, compartmentalized structures (e.g., ruminants) (Hildebrand 1982). The slurry of food and digestive enzymes (now called chyme) passes out of the stomach and enters the small intestine, where bile and digestive enzymes from the liver, gallbladder, and pancreas are added, and where nutrients and water are absorbed. Food then passes into the large intestine where more absorption takes place. Generally, if the diet of mammals is easy to digest, as in carnivores, the intestine length will be relatively short. Contrarily, herbivores, which feed on low-nutrition plants, tend to have very long or voluminous gut tubes (Hildebrand 1982). Because mammals lack enzymes to digest cellulose, commensal bacteria in the gut must help digest plant matter (Vaughan 1986). The undigested material passes through the gut and is eliminated from the body as feces.

Marine mammals exhibit wide variation in their feeding methods and the foodstuffs acquired to meet metabolic needs. This variation ranges from almost total herbivory in sirenians, hunting and capture of individual fish/squid in most small odontocetes and pinnipeds, suction feeding in walruses, physeterids, and beaked whales, to engulfing whole schools of fish or invertebrates in baleen whales. Marine mammals display a wide range of morphologies to acquire and manipulate these varying foodstuffs.

There are two other points to keep in mind regarding feeding. The first is that because mammals must seek their food, they require input from sensory apparati. The second is that how a mammal feeds, especially if it captures live, moving prey, is highly integrated with its mode of locomotion (Hildebrand 1982). Imagine studying feeding in cheetahs without understanding their superb running abilities! The integration of feeding and locomotion is equally important in marine mammals (e.g., Orton and Brodie 1987).

Locating, Capture, and Manipulation of Food

The group of. . . Whalebone Whales embraces all those which are destitute of teeth when adult, and whose palate is lined on each side with rows of horny plates, called whalebone or baleen, which are fringed on their inner edges.
Scammon 1968

The process of finding, localizing, and approaching prey requires specialized sensory systems (see Wartzok and Ketten, Chapter 4, this volume). For example, polar bears possess highly developed nasal turbinates with large surface areas covered with olfactory sensory epithelium (Reeves et al. 1992). These bears can apparently smell seals even through meters of snow (Stirling 1988). Sea otters and pinnipeds have extremely sensitive tactile vibrissae located on the lips and cheek that are used to locate prey (Japha 1912, Yablokov and Klevezal 1962, Ling 1977). Walruses locate benthic invertebrates by probing the ocean sediments with their vibrissae (Fay 1982). Sea cows also possess abundant and sensitive vibrissae that are probably important to integrating feeding behavior (Hartman 1979, Marshall and Reep 1995). Mysticetes possess vibrissae on the dorsal and lateral surfaces of the rostrum (Japha 1912). Yablokov and Klevezal (1962) have suggested that these vibrissae function in determining maximum prey density in the water column and orienting the whale during feeding bouts. In addition, Ogawa and Shida (1950) report sensory tubercles in the lips of mysticetes that help in prey detection. Odontocetes have a highly developed echolocation system (Au 1993) important to prey localization.

Polar bears and sea otters have dentition and jaw structures that are generally similar to those of other carnivores (Fig. 2-17; Kenyon 1981, Nowak 1991). Polar bears use their powerful jaws to capture large prey items, such as cetaceans, seals, and large fish, although they also scavenge from carcasses (for review, see Stirling 1988, Nowak 1991, Reeves et al. 1992). Sea otters feed on benthic invertebrates as well as fish (Kenyon 1969). Otters capture prey items at depth and return to the surface where they use their broad, flat molars to crush the hard shells of their crustacean, mollusk, and urchin prey (Kenyon 1981, Nowak 1991).

Pinniped dentition patterns are similar in general to those of other carnivores, and most species feed on fish or squid (King 1983). There are some interesting species-specific variations in prey and therefore, feeding structures. For example, walrus upper canines develop into large ventrally directed tusks. In male walruses, each tusk can grow to 80 cm in exposed length by 30 years of age (Fay 1982). It had long been believed that the tusks were used to dig, or essentially "root up" mollusks from the benthic sediment (Allen 1880). Fay (1982) concluded that the tusks function as a secondary sexual characteristic in ritualized dominance displays, and that they contribute little to capturing food. This hypothesis is supported by observations of robust and healthy males who had complete loss of tusks (Fay 1982). Nonetheless, walruses display an unusual feeding behavior—they use their muscular tongue to create a vacuum within the oral cavity that is strong enough to remove bivalve mollusks from their shells (Fay 1982, Gordon 1984).

The crabeater seal has been hypothesized to have an equally interesting feeding style. King (1983) states that these seals may feed on krill, sieving these crustaceans from an engulfed "bolus" of water. Crabeaters possess elaborately cusped teeth that interdigitate when the jaws are closed to produce a fine sieve. This dentition may function to strain small schooling krill from the engulfed water, similar to baleen in mysticete whales. It is worth noting that this feeding style has never been observed in the wild (Reeves et al. 1992).

Sirenians use their highly mobile lips and vibrissae to gather aquatic vegetation (Hartman 1979, Nishiwaki and Marsh 1985, Marshall and Reep 1995). Sirenian teeth continuously develop in the caudal maxilla and mandible, and migrate rostrally at a rate of approximately 1 mm per month (Domning and Hayek 1984). This movement allows for the erosion and eventual loss of the rostral-most teeth, which are then replaced by teeth forming caudally. Aspects of this tooth replacement pattern are shared with their relatives, the elephants, and has been hypothesized as an adaptation for continuous grinding of large amounts of abrasive grasses. Domning and Hayek (1984) suggest that the rate of tooth migration in manatees is specific to the mechanical forces experienced by the teeth when grinding tough aquatic grasses.

Odontocete cetaceans generally feed on individual fish and squid. Some large species, such as killer whales and false killer whales, may even feed on other cetaceans (Leatherwood and Reeves 1983). Odontocetes are homodonts, that is, within each species, all teeth have the same morphology (Slijper 1979). The number of teeth is highly variable, ranging from a single tooth per mandible in most beaked whales (Moore 1968) to more than 40 per tooth row in some pelagic delphinid species (e.g., pan-tropical spotted dolphin [Perrin 1975]). Delphinids use their teeth to capture prey items such as fish or squid and to manipulate the prey into a position for swallowing. Many odontocete cetacean species probably use suction to swallow their food (Werth 1992a,b).

Beaked whales (Ziphiidae) are thought of as suction feeders because they lack teeth useful in prey acquisition (J. G. Mead, pers. comm.). The teeth (which generally only erupt

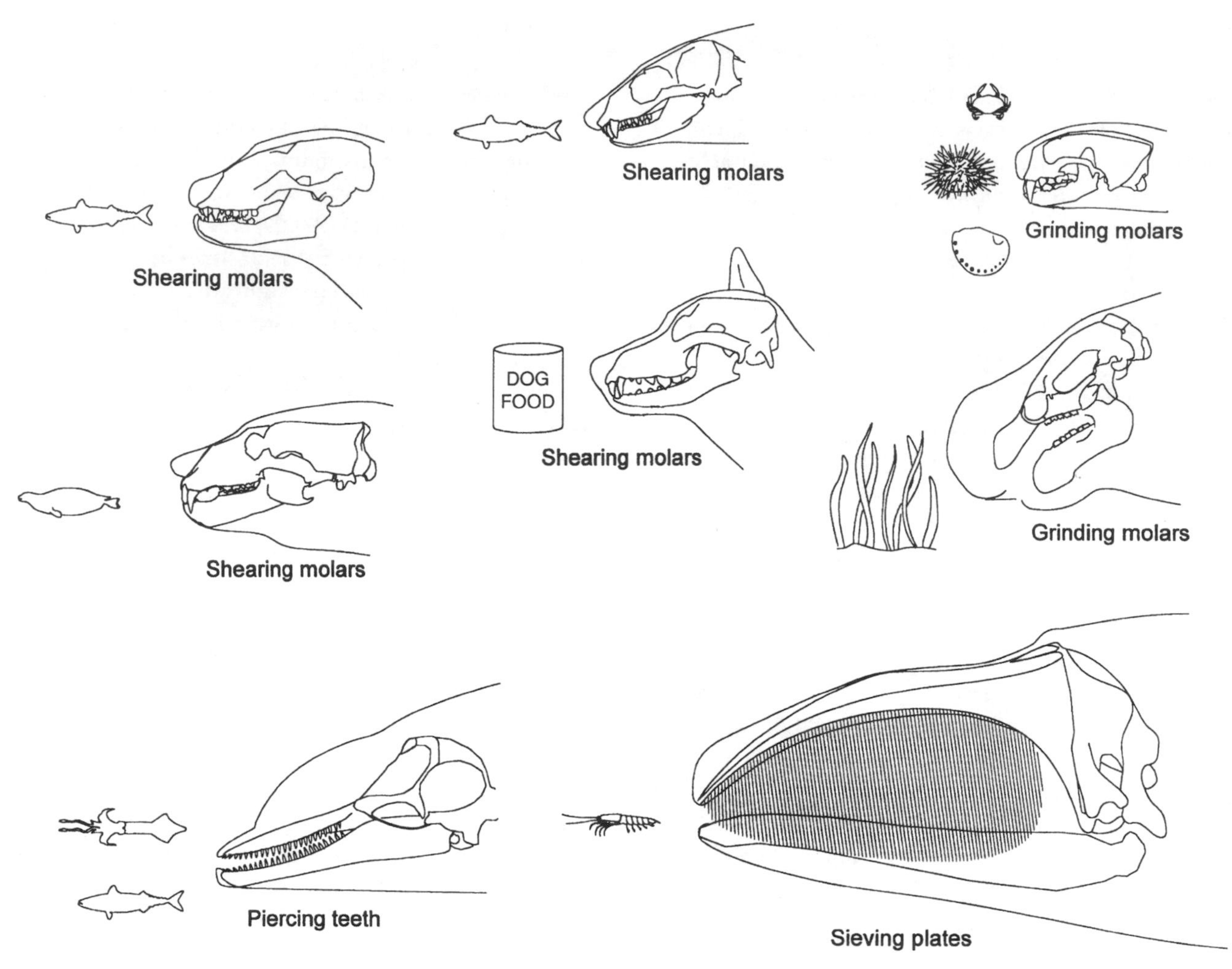

Figure 2-17. Feeding apparatus of representative marine mammals. Polar bears, sea otters, and most pinnipeds have dentition patterns typical of carnivores. Sirenians possess teeth that are replaced throughout most or all of the lifetime of the animal. Odontocete cetaceans are homodonts, with tooth numbers ranging from one (in most beaked whales) to more than 40 (in some pelagic delphinids) per tooth row. Mysticete whales are unique in possessing keratinized plates of baleen that act to filter small fish and invertebrates from engulfed water. (Illustrations adapted from Lichtenstein 1827-1834; Vrolik 1848-1854; van Beneden et Gervais 1868-1879; Flower 1885; Allen 1902; True 1904; Howell 1930; Jacobi 1938; Omura et al. 1969; Nishiwaki and Kasuya 1970; King 1971, 1972; Nickel et al. 1986; Rommel 1990; Evans 1993.)

in males) are used as gouges in aggressive dominance displays (Heyning 1984). Perhaps the most bizarre example of dentition in marine mammals is that of the adult male strapped-toothed whale. Its paired teeth arise from the mandible and cross tightly above the dorsal surface of the rostrum. This configuration apparently keeps the oral cavity from opening more than a few centimeters.

The mysticete whales lack teeth as adults and instead possess baleen—plates of keratinized tissue that hang from their upper jaw (Fig. 2-17; Slijper 1979). The baleen plates grow continuously from the maxilla throughout the life of the individual. Each baleen plate is formed as a sandwich—two outer sheets of keratin surrounding a tubular-keratin "marrow" layer. The action of the tongue passing back and forth across the baleen abrades the adjacent keratin sheets and exposes the tubular marrow layer. These exposed tubules overlap creating a mat that functions as a sieve to trap engulfed prey (Pivorunas 1979).

In bowhead whales, the baleen plates can grow to more than 4 m long (Leatherwood and Reeves 1983) and number up to 300 per side (Lambertsen et al. 1989, Haldiman and Tarpley 1993). In comparison, an adult blue whale may have baleen plates 75 cm long and number up to 380 per side (Tomilin 1967). Right whales have fine fringed baleen plates that function to capture small euphasids, whereas balaenopterid whales have a coarser fringe suitable for capturing fish and larger invertebrates (Pivorunas 1979).

Feeding in balaenopterids is powered by their swimming

(Orton and Brodie 1987) and a feeding bout involves the intake of an immense volume of seawater—in blue whales as much as 60 m³ of water (Pivorunas 1979). The whale begins feeding by lunging (usually on its side) and dropping the mandible to an angle of 90° from the body axis. Water pours into the rapidly expanding oral cavity. The tongue muscles and pleat blubber contain large amounts of elastin fibers (Pivorunas 1977), which can be stretched to four times their resting length (Orton and Brodie 1987). As the oral cavity expands, the soft, flaccid tongue inverts and lines the floor of the oral cavity (Lambertsen 1983). (This function of the tongue is unique to balaenopterids. In most mammals the tongue is a firm, muscular hydrostat that everts from the oral cavity [Kier and Smith 1985].) The whale's mouth is then closed by both the action of the jaw muscles and rotation of the animal. Contraction of muscles and elastic energy stored in the ventral groove blubber help shrink the oral cavity (Orton and Brodie 1987). Seawater is ejected through the baleen plates and food items are filtered in.

In contrast to this "gulping" feeding style, the balaenids are "skimmers," that is, they swim slowly with their mouths open, straining small prey through their baleen filter (for review, see Lowry 1993). The gray whale (*Eschrichtius robustus*) feeds predominantly by engulfing benthic sediments and filtering in associated fauna (for review, see Nerini 1984).

Mechanical and Chemical Digestion

Many biologists, dissecting their first cetacean, are struck by the unexpected resemblance of the stomach to that of ruminants. Gaskin 1978

Food passes from the oral cavity through the muscular pharynx and esophagus to the stomach (Fig. 2-18). Marine mammals either eat large prey items or ingest large volumes of small prey items, and must have highly distensible stomachs to accept the prey. The pinniped stomach is a single, highly distensible chamber (similar to that of other carnivores) and is essentially a dilation of the gut tube (Green 1972, Eastman and Coalson 1974, Olsen et al. 1996). Although the stomach of odontocetes is compartmentalized (discussed later), the forestomach is actually a highly distensible pouch of the esophagus that acts to accept large food items (Slijper 1979). The mysticete whales, although not feeding on large single prey, nonetheless can swallow huge amounts of food at a time. They also possess a nonglandular forestomach that essentially acts as a holding sac for later stomach chambers (e.g., Tarpley et al. 1987, Olsen et al. 1994). We assume that the stomachs of polar bears are similar to those of other carnivores.

The distensible forestomach of cetaceans leads to a glandular main stomach (Fig. 2-18). From there, food passes through connecting channels to the pyloric stomach (Slijper 1979). Mead (1993) points out that this pattern of stomach compartmentalization is found in all cetaceans, save the beaked whales and river dolphins (see also Gaskin 1978, Tarpley et al. 1987, Olsen et al. 1994, Langer 1996). Despite the fact that cetaceans are carnivorous, they retain a complex stomach, superficially more similar to that of herbivorous artiodactyls. Cetaceans also lack a gallbladder.

The single, glandular stomach of sirenians is generally similar in structure to that of other nonruminant herbivores, such as the horse (Reynolds 1980, Caldwell and Caldwell 1985, Reynolds and Rommel 1996). The most unusual feature of the sirenian stomach is an accessory digestive gland, called the cardiac gland, which apparently functions to isolate delicate glandular tissues from abrasive food entering the stomach (Marsh et al. 1977, Reynolds and Rommel 1996).

From the stomach, chyme enters the small intestine and subsequently the large intestine. Pinnipeds have relatively long small intestines and short large intestines (Eastman and Coalson 1974, Helm 1983). The total length of the pinniped intestines is reported to be as long as those of some similarly sized herbivores (Helm 1983). The length of the cetacean intestine is highly variable and some species possess extremely long tracts (for review, see Gaskin 1978). The functional significance of the long intestinal tracts of pinnipeds and cetaceans remains unknown.

Sirenians, like other nonruminating herbivores, have an expanded intestinal tract. The sirenian large intestine, with its rich microbial fauna, may be 20 m in length (Reynolds and Rommel 1996). As in other hindgut fermenters, the large intestine of sirenians is an important site of both cellulolysis (Burn 1986) and nutrient absorption. The sirenian gastrointestinal tract has a number of unique or rare histological features that may accommodate the large volumes of ingested salt water or abrasive plant material (for review, see Reynolds and Rommel 1996).

Cardiovascular System

The mammalian cardiovascular system can be thought of as a plumbing system that uses a pump (heart) to transport a fluid (blood) through a series of pipes (arteries and veins) (Smith and Kampine 1990). The blood is transported to and from exchange sites (capillaries) where dissolved gases, nutrients, waste products, and heat are transferred across the capillary membrane (Vogel 1992).

The four-chambered heart is a double pump with separate pulmonary and systemic circulatory routes. The right half of the heart pumps blood returning from the body to the

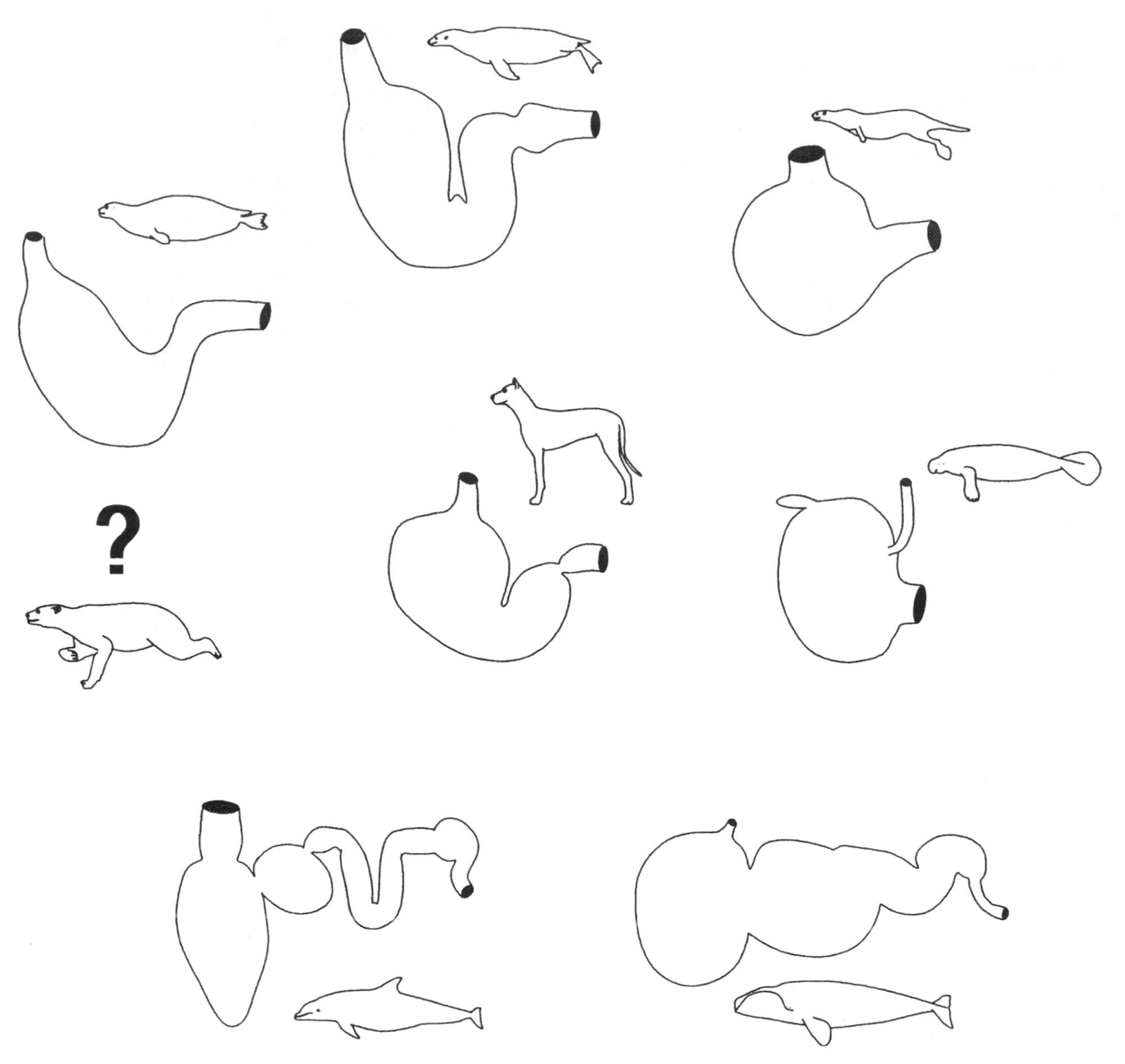

Figure 2-18. Stomach morphology of representative marine mammals. Pinnipeds, sea otters, and polar bears possess stomach morphologies typical of carnivores. The cetacean stomach is compartmentalized, with a highly distensible forestomach. Sirenians possess a single stomach with a specialized cardiac gland. (Illustrations by S. A. Rommel after Harrison et al. 1970, Green 1972, Schummer et al. 1981, Herbert 1987, Tarpley et al. 1987, Reynolds and Rommel 1996.)

lungs; here, carbon dioxide diffuses out and oxygen diffuses into the blood across the alveolar and capillary membranes. The oxygenated blood enters the left half of the heart, where it is pumped, through the aorta, throughout the body.

Because the amount of oxygen that can be stored in any tissue is limited in most terrestrial mammals, the cardiovascular system must provide a constant supply of oxygenated blood. Oxygen-sensitive tissues, such as the heart and brain, can generally only survive brief hypoxic/anoxic periods without experiencing irreversible damage or death (Schmidt-Nielsen 1990, Smith and Kampine 1990). Likewise, some mammalian tissues, such as delicate neural and reproductive tissues, are temperature sensitive. Thus, another important function of the cardiovascular system is to regulate the temperature of body tissues (e.g., Cowles 1958, Taylor and Lyman 1972, Bedford 1977, Baker 1979).

The cardiovascular systems of marine mammals perform the same functions as that of our "typical" mammal. How is oxygen delivered to sensitive neural tissues during extended breath-hold episodes? How do marine mammals regulate the temperature of their reproductive structures? Elsner (Chapter 3, this volume) discusses the complex cardiovascular changes that occur during a dive. We offer some other insights into the functional morphology of the vascular structures specialized for the diving lifestyle. In particular, we investigate (1) the arterial supply to the brain, (2) the role of arteriovenous anastomoses in shunting blood, and (3) thermoregulation of the reproductive structures. Our comparative focus is on the seal and dolphin. Seals possess elaborate venous structures, whereas dolphins display arterial specialization compared to our "typical" mammal.

Arterial Supply to the Brain

Under ordinary circumstances the survival times of the "vital organs" (i.e., the brain and heart) without oxygen are only a few minutes, so even a brief failure of the circulatory or respiratory system may have a serious or fatal outcome.
Smith and Kampine 1990

The "typical" mammalian brain receives its blood supply by way of the internal carotid and vertebral arteries, with some contribution from the external carotid arteries. These vessels enter the cranial cavity and anastomose (form a junction, or join together) at the base of the brain. This circular-shaped anastomosis (sometimes called the circle of Willis) gives rise to all arteries that supply the brain. (In some mammals the carotid vessels form a rete [discussed later] before joining the arterial circle [Baker 1979, Schaller 1992]. Du Boulay and Verity [1973] point out that there exist variations in details with this typical theme, and suggest caution when generalizing arterial patterns to the brain.) Hypothesized functions for the arterial circle include (1) pressure damping so that the delicate neural tissues do not experience potentially damaging pressure pulses (Ask-Upmark 1935), and (2) establishing collateral flow so that damage/ occlusion of one vessel does not necessarily stop blood flow to those areas of the brain supplied by that vessel (Cartmill et al. 1987).

Given that most marine mammals undergo extended breath-hold periods, when blood oxygen stores are not continuously replenished, we might expect some specialized arterial supply to the brain to assure an uninterrupted supply of oxygen. Perhaps surprisingly, those species of seals and sea lions investigated thus far possess a "typical" arterial supply to the brain (du Boulay and Verity 1973, Dormer et al. 1977, King 1977, McFarland et al. 1979). The major arterial path to the brain of these animals is through well-developed internal carotid and vertebral arteries, with some contribution from the external carotid arteries. Although detailed accounts of the arterial supply to the brain of the sea otter are unknown to us, the pattern in the river otter (*Lutra lutra*) is similar to that of a typical mammal (du Boulay and Verity 1973). Bears possess a unique arterial supply to the brain that the polar bear appears to share (du Boulay and Verity 1973). Therefore, pinnipeds, polar bears, and sea otters display "typical" terrestrial arterial blood supplies to the brain.

Unlike these marine mammals, cetaceans possess a highly derived arterial blood supply to the brain. Cetaceans lack vertebral arteries, and their internal carotid arteries are degenerate, often not reaching the cranial cavity (for review, see Walmsley 1938; McFarland et al. 1979; Vogl and Fisher 1981,1982). Thus, the two major paths of arterial blood to the brains of many mammals do not exist in cetaceans. Rather, arterial blood travels to the brain through a series of arterial retia (e.g., Galliano et al. 1966; McFarland et al. 1979; Vogl and Fisher 1981, 1982; Pfeiffer and Kinkead 1990). Retia (plural of rete) are networks of vessels formed when a single vessel splits into multiple smaller channels. Retial vessels can either be organized into multiple, parallel pipes or tortuous, twisting networks (Schmidt-Nielsen 1990).

Arterial blood destined to feed the cetacean brain travels through the thoracic aorta to segmentally arranged intervertebral arteries (Vogl and Fisher 1982). The intervertebral arteries become convoluted, anastomotic bundles that are collectively termed the thoracic rete (e.g., Ommanney 1932a, Barnett et al. 1958; Galliano et al. 1966; McFarland et al. 1979; Vogl and Fisher 1981, 1982). The thoracic rete lies between ribs in the dorsal thoracic cavity. Vessels extending from the thoracic rete enter the neural canal of the vertebral column and form yet another tortuous, anastomotic plexus called the epidural (also named spinal or vertebral) rete (for review, see McFarland et al. 1979). The epidural rete runs the length of the spinal cord and enters the cranial cavity through the foramen magnum. Within the cranial cavity, these arteries form other retia that give rise to arteries that feed the brain.

This serial arrangement of extracranial arterial retia, apparently unique to cetaceans, has been identified as an adaptation to diving, although Vogl and Fisher (1982) suggest that there are "almost as many functions attributed to the cetacean retial system as there are literary references to its anatomy." These researchers tend to support the hypothesis that the thoracic rete acts as a windkessel, that is, a structure designed to dampen pressure fluctuations resulting from the pulsed flows produced by the heart (Vogel 1992). Empirical data support this hypothesis; blood pressures measured at the aorta of anesthetized bottlenose dolphins are higher and more pulsatile in nature than those measured in arteries within the cranium (Nagel et al. 1968). Interestingly, this is the same function ascribed to the arterial circle at the base of the brain of many terrestrial mammals. Such an elaborate structure is also clearly not required for deep diving or extended breath-hold periods, as it is not found in seals or sea lions. Thus, it is fair to say that we do not yet fully understand the functions of the thoracic rete of cetaceans.

Sirenians have not been as extensively investigated as seals and dolphins, and indeed we do not yet completely understand the arterial blood supply to the brain. Manatees do possess internal carotid and vertebral arteries (Murie 1872). Murie (1880) suggests that the manatee circle of Willis "is complete," but he also states that the internal carotid arteries are small; he illustrates a basilar artery, but does not indicate that it is formed by vertebral arteries. Thus, there is equivocal evidence that manatees share the more "typical" arterial supply to the brain.

To complicate matters, sirenians also have an extravagantly elaborate thoracic rete. Unlike the convoluted and anastomosing rete in cetaceans, the sirenian thoracic rete is formed by multiple parallel arranged arteries that tend to fan out like brooms from each intercostal artery (termed vascular bundles by Fawcett 1942). These vascular bundles appear to run peripherally and nourish the thoracic wall, rather than enter the neural canal as they do in cetaceans. Thus, although sirenians do possess a thoracic retial system, there is little evidence that it plays a role in supplying the brain with blood. Our understanding of sirenian arterial structures would greatly benefit from further research.

Marine mammals do not display a consistent pattern of blood supply to the brain, and many possess arterial systems undifferentiated from terrestrial mammals. Thus, perfusion of the brain with oxygenated, arterial blood during a dive does not necessarily require novel arterial *structures*. Rather dynamic changes in *blood flow patterns* through vascular structures occur during diving (for explanation, see Elsner, Chapter 3, this volume). Arterial blood can be shunted from its normal systemic circulatory routes during a dive; perfusion of peripheral tissues is decreased, while blood flow to the brain remains constant (lucidly reviewed in Kooyman 1989). One mechanism to redistribute blood during a dive, elucidated in experiments on harbor seals, is large-scale reduction in the diameter of the aorta and muscular distributing arteries (Bron et al. 1966). Reducing the diameter of the aorta or arteries near the aorta limits the blood supply to peripheral tissues and thus conserves oxygen for vital organs. Another mechanism that can redistribute blood flow and provide a short-circuit to the normal systemic circulation is the arteriovenous anastomosis.

Arteriovenous Anastomoses

In the course of submersion the arteriovenous anastomoses appear to play some important role in blood flow distribution. Kooyman 1989

Arteriovenous anastomoses (AVA) are vascular channels that connect an artery to a vein proximal to the level of the capillary bed (Fig. 2-19). Arterial blood routed through AVA thus bypasses the site of gas and nutrient exchange at the capillaries. The AVA are most commonly observed at the entrance to the microcirculation, that is, the vessels leading directly to and from the capillaries (Smith and Kampine 1990, Vogel 1992). The AVA are also found in more proximal vessels, and can shunt arterial blood into veins higher in the circulatory system/tree (Elsner et al. 1974). The AVA in the flukes of bottlenose dolphins, for example, can be as large as 2 mm in diameter, and are found connecting major arteries and veins (Fig. 2-19; Elsner et al. 1974). These large shunts, also apparently found in the deep-diving Weddell seal, may be important in redistributing blood flow from peripheral tissues to vital organs during a dive (Zapol et al. 1979, Kooyman 1989).

The AVA also perform a critical thermoregulatory function by adjusting blood flow to the surface of the body (for review, see Hales 1985, Grayson 1990). The skins of many mammals are supplied by AVA (Fig. 2-19; Smith and Kampine 1990). Increased blood flow through AVA and capillary beds in the skin increase convective heat loss to the environment (Molyneux and Bryden 1978, Hales 1985, Bryden 1994). The skins of phocid seals are particularly well supplied with AVA, which lie superficial to the insulatory blubber layer and hence are in a position to dump heat to the environment (Molyneux and Bryden 1978, Schmidt-Nielsen 1990, Bryden 1994). The AVA are also found superficial to the blubber layer of bottlenose dolphins (Palmer and Weddell 1964). Unlike seals and dolphins, fur seals have their insulatory fur layer superficial to the skin surface. Interestingly, the skin of their bare flippers is richly invested with AVA (Bryden 1994). Thus, AVA are found in regions of the marine mammal body where they can, if dilated, effectively transfer heat to the environment. Dugongs appear to lack AVA (Bryden, et al. 1978). To the best of our knowledge, AVA have not been described in polar bears or sea otters.

Thermoregulation of the Reproductive System

The body is streamlined and spindle shaped with no sharp protuberances to break the even contour. King 1983

The AVA and countercurrent heat exchangers found in skin, fins, flippers, and flukes of marine mammals are circulatory structures that play a role in whole body thermoregulation (see Figs. 2-6 and 2-19). We now investigate special thermoregulatory requirements of mammalian reproductive tissues.

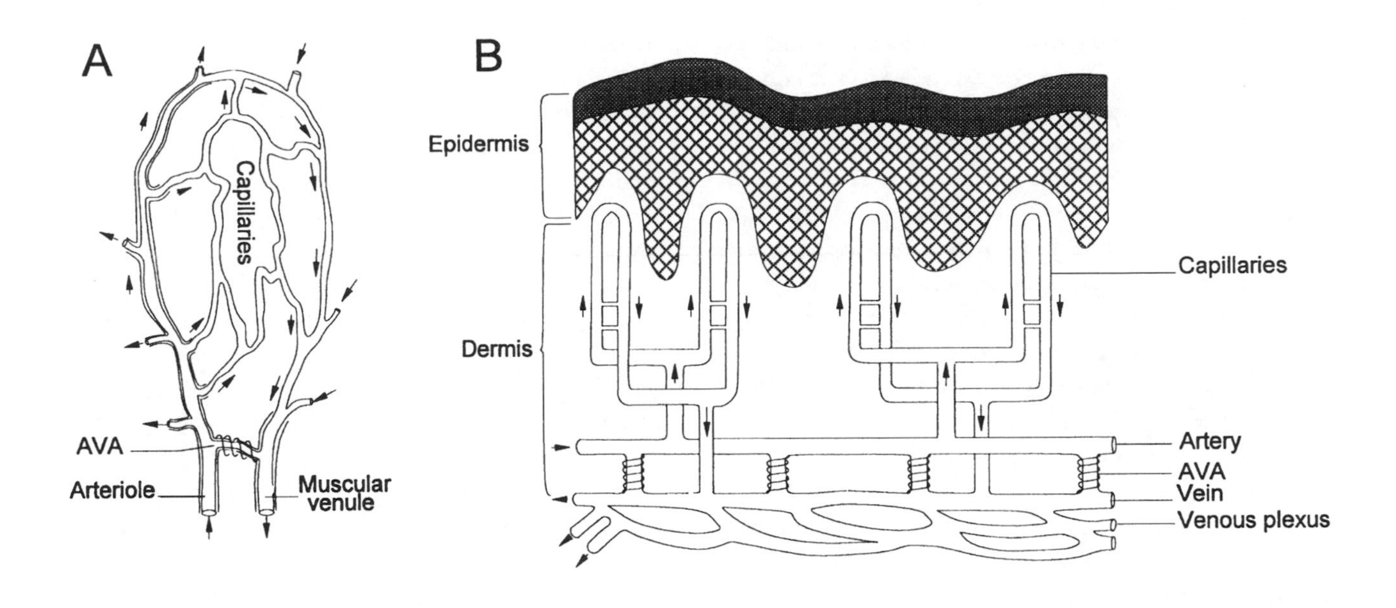

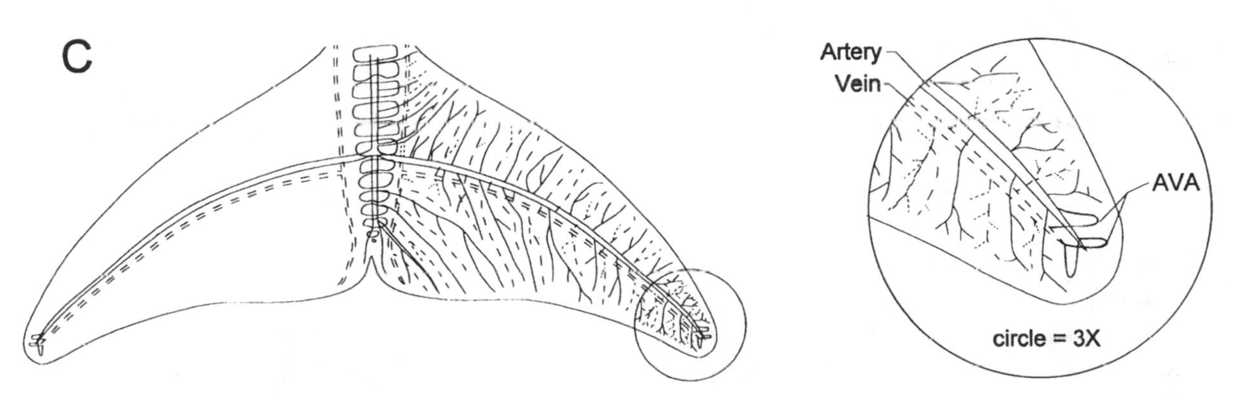

Figure 2-19. Morphologies of various arteriovenous anastomoses (AVA). (A) AVA join arterioles to venules at the entrance to the capillary bed (in this case in the mesentery of stomach). (B) AVA are found in the skin of many mammals and play a role in thermoregulation by adjusting the volume of blood flow to the skin surface. (C) AVA can shunt blood from large arteries to veins higher in the circulatory system, as in the flukes of dolphins. (Illustrations adapted from Elsner et al. 1974, Guyton 1976, Grayson 1990.)

In many mammals, viable sperm production and storage in the epididymis requires temperatures below those found in the body core (Moore 1926; Cowles 1958, 1965; VanDemark and Free 1970; Waites 1970; Bedford 1977; Carrick and Setchell 1977). Physical separation from the body core (e.g., in a scrotum) and vascular countercurrent heat exchangers in the spermatic cord are common mechanisms that maintain below-core temperatures at the mammalian testes and epididymides (Harrison 1948).

Testes of otariids and polar bears lie in scrotal sacs and therefore, are positioned to experience low temperatures (King 1983, Freeman 1990). Sea otter testes are not visible externally (Estes 1980). Walrus and phocid testes lie between the blubber and abdominal musculature, can be invisible externally, and thus, may experience core or above-core body temperatures (Bryden 1967, Fay 1982, King 1983). Yet, testicular temperatures of the elephant seal are 6°C cooler than abdominal or rectal temperatures (Bryden 1967).

Cetacean testes are cryptic (intraabdominal) and are presumably exposed to high core body temperatures (Harrison 1969, Slijper 1979). Indirect measurements of testis temperature in bottlenose dolphins suggest that the testis is also experiencing below-core temperatures, despite its position deep within the abdominal cavity (Rommel et al. 1992, 1994, 1998; Pabst et al. 1995). How can these testes experience below-core body temperatures? Phocid seals and odontocete cetaceans possess specialized vascular structures that regulate the temperature of their testes.

The phocid testis is enveloped in a venous network that lines the inguinal region and pelvic and abdominal cavities (Fig. 2-20). This venous network receives blood from the superficial veins that drain the skin surface of the flipper (Blix et al. 1983, Rommel et al. 1995). Blood returning from the flipper surface is presumably cooled by exposure to ambient water (Galivan and Ronald 1979). The enveloping venous plexus thermally shields the testis from heat generated by adjacent thermogenic muscle.

Odontocete cetaceans possess a countercurrent heat exchanger that cools arterial blood to the testis (Rommel et al. 1992). The countercurrent heat exchanger lies in the caudal abdominal cavity, and is formed by a spermatic arterial plexus juxtaposed to a venous plexus (Fig. 2-20). The blood that supplies the venous plexus returns from superficial veins that drain the dorsal fin and flukes and can be cooled by exposure to ambient water temperatures (Scholander and Schevill 1955). These veins leave the fin and flukes and remain superficial to the body core (i.e., just deep to the blubber layer and superficial to vertebral muscles) until they feed directly into the lateral and caudal borders of the venous plexus. Thus, cooled blood can be introduced directly into the deep caudal abdominal cavity. The juxtaposition of this venous plexus to the spermatic arterial plexus, the vessel dimensions, and the opposite directions of blood flow through the plexuses suggest that dolphins use a countercurrent heat exchanger to regulate the temperature of the blood reaching the testis (Rommel et al. 1992). Noninvasive temperature measurements taken in the region of the countercurrent heat exchanger in bottlenose dolphins support this functional hypothesis (Rommel et al. 1994, 1998; Pabst et al. 1995).

Interestingly, sirenians also possess cryptic testes (Harrison 1969). Manatees have low metabolic rates (Scholander and Irving 1941, Gallivan and Best 1980) and highly variable core body temperatures (for review, see Gallivan et al. 1983) that can be high enough to potentially affect sperm viability detrimentally. The vascular structures associated with the

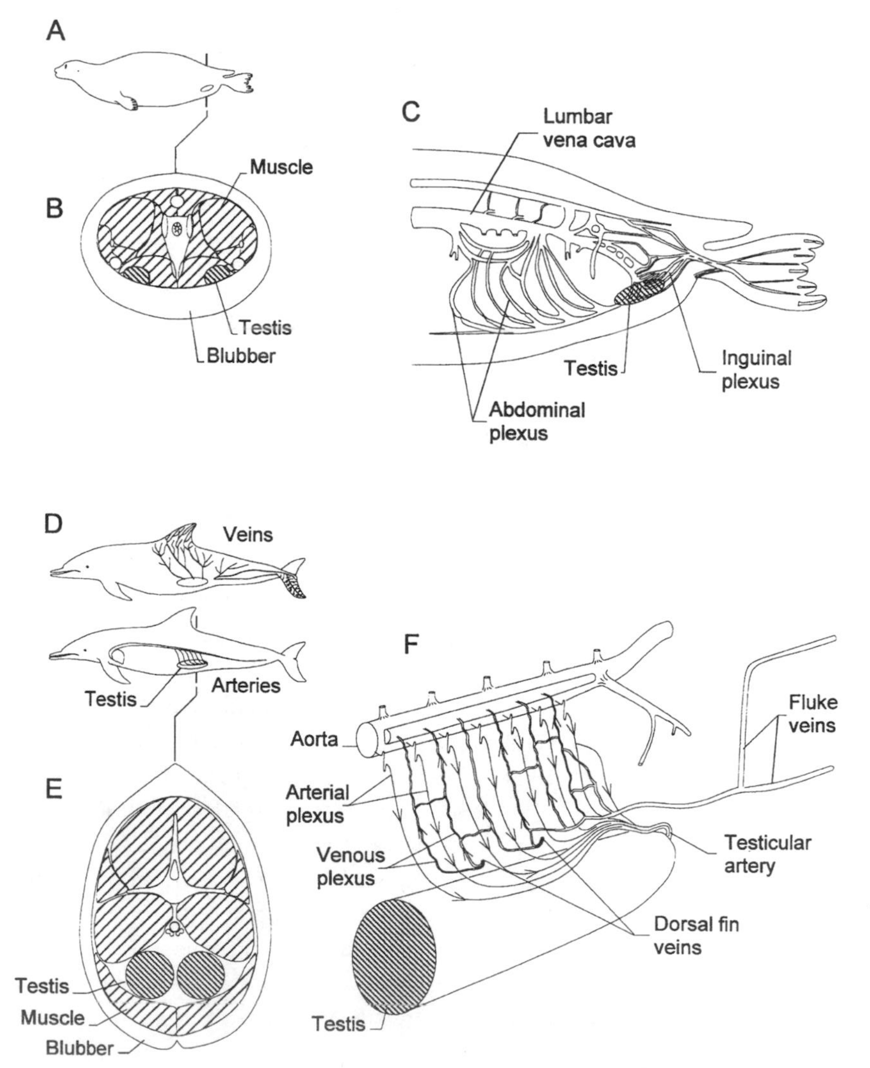

Figure 2-20. The phocid seal testis experiences below core body temperatures by being enveloped in a venous plexus. The plexus receives cooled blood from the superficial veins of the hind flippers. (A) Position of testis in harbor seal. Vertical line indicates position of cross-section seen in (B). (B) The phocid testis lies between the abdominal muscles and blubber layer. (C) Schematic of venous plexus that shields testis. Note extensive anastomoses between venous plexuses in the feet, pelvis, and abdomen of seals. (Illustrations adapted from Rommel et al. 1995 and authors' unpublished data.) Dolphins possess a countercurrent heat exchanger that regulates the temperature of the arterial blood supply to the testis. (D) Superficial veins draining the surfaces of the dorsal fin and flukes supply the venous plexus seen in (F). The arterial blood supply to the testis is through a plexus of small, parallel arteries that arise from the aorta. The vertical line posterior to the dorsal fin indicates position of cross-section seen in (E). (E) The dolphin testis lies within the abdominal cavity surrounded by locomotor muscle. (F) Vascular structures associated with the dolphin testis. The venous blood returning from the dorsal fin and flukes enters a plexus juxtaposed to the arterial plexus. The plexuses form a countercurrent heat exchanger that can cool the arterial blood reaching the testis. (Illustrations adapted from Rommel et al. 1992, 1995.)

manatee testis are poorly understood (Murie 1872). The arterial supply to the dugong testis is through a straight testicular artery and there is no indication of a venous plexus (Marsh et al. 1984a). Thus, it remains a question as to how, or *whether*, sirenians regulate the temperature of their intraabdominal testes.

The developing mammalian fetus is also sensitive to high temperatures. Elevated fetal temperatures can cause detrimental effects including low birth weights, developmental anomalies, and even death (Shelton 1964, Morishima et al. 1975, Bell 1987). The fetus could experience hyperthermic conditions because the reproductive systems of seals and cetaceans is surrounded by thermogenic locomotor muscles and insulating blubber. Similar venous networks in seals (Rommel et al. 1995) and the same countercurrent heat exchange systems in cetaceans described previously (Rommel et al. 1993) have been hypothesized to regulate the temperature of the developing fetus.

Generally speaking, the cardiovascular systems of otariids, and presumably polar bears and sea otters, look very much like their terrestrial relatives. Cetaceans possess specializations of the arterial system (e.g., thoracic retia; see also Shadwick and Gosline 1994, Gosline and Shadwick 1996), whereas seals have elaborate venous structures (e.g., abdominal and pelvic venous plexuses). Sirenians possess bizarre proliferations of both arteries and veins. The function of many of these vascular structures is poorly understood and awaits investigation from future researchers.

Reproductive System

The mammalian reproductive system consists of structures that produce and transport gametes, and provide development and support of offspring (see Table 2-2 for a brief overview of reproductive structures). Although the function of each reproductive structure is shared by all eutherian mammals, there exists great diversity in specific morphologies across the group.

Table 2-2. Reproductive Structures and Functions

Function	Male	Female
Production of gametes	Testis produces sperm	Ovary produces eggs
Transport of gametes from gonads to external environment	Epididymis Ductus deferens Urethra	Uterine tube Uterine horn Cervix Vagina
Site of maintenance of pregnancy		Uterus
Supportive and erectile tissues	Scrotum Penis	Vulva Clitoris

Marine mammals, too, display a tremendous diversity in reproductive morphologies and behaviors (see Boyd, Lockyer, and Marsh, Chapter 6, this volume). Cetaceans, sirenians, and some sea otters give birth to and suckle their young in the aquatic environment, whereas the remaining marine mammals travel to land (or ice) during these times. Are there specializations of the reproductive system of fully aquatic marine mammals? We offer a survey of male and female reproductive structures in marine mammals to investigate this question. We focus on testis size and penis structure in males. In females, we discuss uterus and placenta types, mammary position, and size of offspring at birth.

Male Reproductive Structures

Testes have two major functions—production of sperm and production of steroid hormones. Kenagy and Trombulak 1986

The testis performs similar functions in all mammals, and varies little in its general morphology. Sperm are produced in seminiferous tubules and are stored in a highly convoluted epididymis. The ratio of testis-to-body size does vary across species (Table 2-3 and Fig. 2-21; e.g., Mackintosh and Wheeler 1929, Harrison 1969, Harrison et al. 1972, Kenagy and Trombulak 1986, Read 1990, van Waerebeek and Read 1994). Testis size appears to be related to mating systems; males that compete for females and have high copulatory rates tend to have larger testes (Kenagy and Trombulak 1986).

This pattern is apparent in odontocete species hypothesized to have sperm competition as a mating system (see van Waerebeek and Read 1994; Boyd, Lockyer, and Marsh, Chapter 6, this volume). Odontocete cetaceans possess very high testes-to-body mass ratios, 7 to 25 times higher than would be predicted for "typical" mammals of their size (Kenagy and Trombulak 1986). Dusky dolphins apparently hold the odontocete record for largest ratio of testes-to-body mass at 8.0% (van Waerebeek and Read 1994). (As a point of comparison, male humans invest about 0.08% [i.e., two orders of magnitude lower] of their body weight into gonads [Kenagy and Trombulak 1986].) A similar pattern is seen within the mysticetes. Right whales, which are hypothesized to possess a sperm competition mating system, possess testes more than six times larger than predicted for their body size (Brownell and Ralls 1986). Right whales also have the absolute largest testes—with a combined weight more than 900 kg (Omura 1958, Omura et al. 1969). Comparative data for other marine mammals are less available.

Table 2-3. Brain and Testes Mass as a Percentage of Total Body Mass for Representative Species

Species	Body Mass (kg)	Testes Mass (kg)	%	Brain Mass (kg)	%	References
Eubalaena glacialis	66,778	975.0	1.46	2.75	0.004	Omura et al. 1969
Balaenoptera physalus	50,000	2.8	0.01			Engle 1927
Physeter macrocephalus	37,093			7.8	0.02	Ridgway 1986
Megaptera novaengliae	32,000	1.73	0.01			Engle 1927
Ursus maritimus	365			0.46	0.13	Gittleman 1986
Dugong dugon	ca. 300	0.65	0.2			Marsh et al. 1984a
Tursiops truncatus	ca. 250	0.46	0.2			Mead and Potter 1990
T. truncatus	167			1.6	0.96	Ridgway 1986
Enhydra lutris	28			0.125	0.10	Gittleman 1986
Lagenorhynchus obscurus	114	9.73	8.0			Van Waerbeek and Read 1994
Delphinus delphis	84	1.61	1.9			Harrison, 1969
Phocoena phocoena	58	2.32	4.0			Harrison, 1969
Phoca hispida	49			0.15	0.31	Frost and Lowry 1981, Bryden 1972
Canis lupis	33			0.131	0.39	Gittleman 1986
Callorhinus ursinus	29	0.06	1.1			Fur Seal Investigation 1969

Note: Values for brain mass from Gittleman (1986) were reported in cubic centimeters (cc), which we converted to kg assuming 1 cc = 0.001 kg.

The penis of all marine mammals can be retracted into the body wall (Slijper 1966, 1979; Harrison 1969; King 1983; Marsh et al. 1984a; Pabst et al. 1998). The penis structure of each group tends to be similar to that of their terrestrial relatives. For example, pinnipeds have a vascular penis reinforced with a bone, the os penis or baculum, typical of carnivores (Harrison 1969, King 1983). The vascular penis becomes rigid and increases in length and diameter when blood fills venous sinuses (King 1983). We assume that the penis morphology of the sea otter and polar bear are similar. The cetacean penis is fibroelastic like that of most artiodactyls (Slijper 1966, 1979). The nonerect position of the cetacean penis is curved into a sigmoid flexure within the body wall, as is seen in ruminants (Schummer et al. 1979). Upon erection, the cetacean penis straightens and becomes turgid, but does not dramatically change its length or diameter as does the vascular type. The penis of the sirenians is of the vascular type, is straight and lacks a baculum (Harrison 1969, Marsh et al. 1984a).

Female Reproductive Structures and Calving

Since whales and dolphins cannot rear their young in caves or nests or other sheltered spots, young whales must be able to surface for air, to follow their mothers, and to keep warm . . . the moment they are born. . . . All this implies that, just like a calf or foal, it must have fairly large dimensions at birth.
Slijper 1979

The uterus of pinnipeds, cetaceans, and sirenians is bicornuate, just as in our "typical" mammal (Wislocki 1935; Sinha et al. 1966; Slijper 1966, 1979; King 1983; Marsh et al. 1984b; Pabst et al. 1998). The ovaries are paired, oval organs that lie at the distal portion of the uterine tubes, and are held in place by the broad ligament (e.g., Evans 1993). The placenta type of each group tends to be similar to that of its terrestrial relatives. Thus, cetaceans possess a diffuse type, as do artiodactyls (Wislocki and Enders 1941; Slijper 1966, 1979; Benirschke and Cornell 1987); sirenians possess a zonal type, similar to elephants (Wislocki 1935, Marsh et al. 1984b); and pinnipeds, polar bears, and sea otters possess a zonal type as do carnivores (Turner 1875, Sinha and Mossman 1966, Sinha et al. 1966, Harrison 1969, Ramsey and Stirling 1988).

Most marine mammals have paired mammary glands (although the walrus and polar bear have four glands), usually abdominal in position (Fay 1982). Of the polar bear's four mammaries, the cranial pair lie just behind the axilla (DeMaster and Stirling 1981). In sirenians, the paired mammary glands lie in an axillary position (Hartman 1979, Caldwell and Caldwell 1985, Nishiwaki and Marsh 1985).

Cetaceans and sirenians give birth to their young in the aquatic environment. Cetacean and manatee calves tend to be large (Table 2-4), and must be precocial, at least in their locomotory skills. Pinniped neonates are also large, and their development at birth varies among species. All species spend some amount of time on land before learning to swim and hunt (King 1983). There is evidence that sea otters give birth both on land and at sea (Estes 1980, Kenyon 1981), and Sandegren et al. (1973) point out that these animals are unique in carrying out their maternal activities on the sea surface. The sea otter pup is said to be precocial relative to other otter pups (Estes 1980). Polar bears, like other bears, give birth to altricial young while fasting in winter dens (for

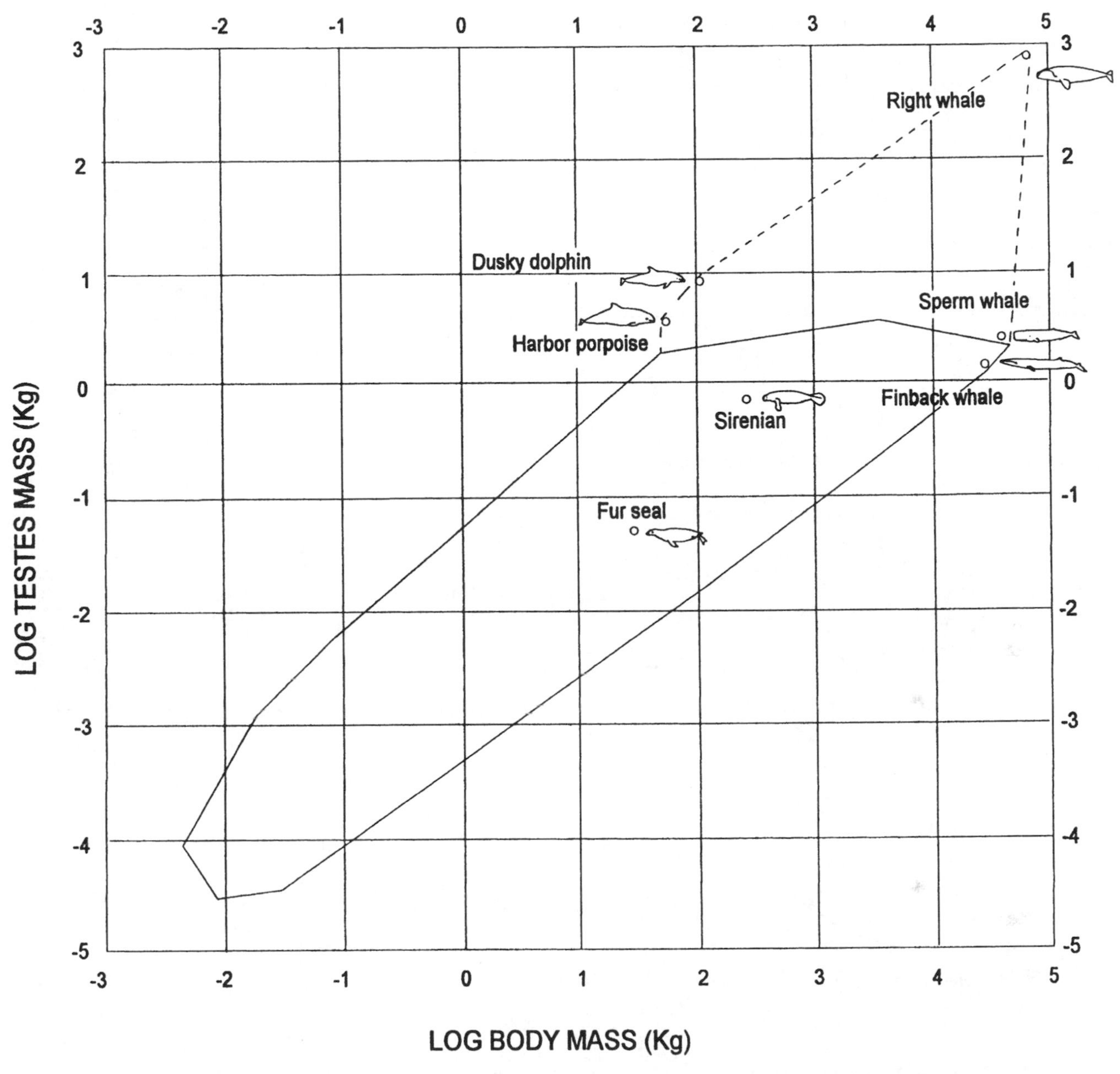

Figure 2-21. Log-log plot of testes mass-to-body mass of a variety of mammals. The ratio of testes-to-body mass is highly variable in mammals as evidenced by the wide-range polygon (solid line for all mammals). Although fur seals and sirenians possess "normal-sized" testes, notice that the values for some odontocete cetaceans are four to six times that predicted for mammals of their body mass and increase the width of the range polygon. The right whale has prodigious testes and extends the range far to the right (dotted line connecting outlying cetaceans). (Data for illustration from Omura 1958, Omura et al. 1969, Kenagy and Trombulak 1986, van Waerebeek and Read 1994.)

review, see Ramsey and Stirling 1988). Marine mammals tend to give birth to one offspring. Twinning can occur in all groups, and triplets have been reported in polar bears (see Hartman 1979; Slijper 1979; King 1983; Ramsey and Stirling 1988; Boyd, Lockyer, and Marsh, Chapter 6, this volume).

Does this survey, albeit superficial, suggest that marine mammals share specialized reproductive morphologies associated with their aquatic lifestyle? We think not. Reproductive structures tend to tell us more about the phylogenetic history of these animals rather than give us insights into common problems associated with reproduction and birthing in the marine environment.

Urinary System

The mammalian urinary system functions to maintain the body's water balance and to excrete metabolic waste products (and soluble foreign substances) in the form of urine

Table 2-4. Adult and Fetus Total Length and Mass for Representative Species

Species	Adult Length (m)	Fetus Length (m)	Adult Mass (kg)	Fetus Mass (kg)	References
Phocoena phocoena	1.55	1.08			Read and Gaskin 1990
Tursiops truncatus	2.40	1.10	190	20	Mead and Potter 1990
Balaenoptera physalus	21.5	6.5			Mackintosh and Wheeler 1929
B. musculus	25	7.5			Mackintosh and Wheeler 1929
Dugong dugon		1.1		15.5	Marsh et al. 1984b
Lagenorhynehus obscurus		0.9		9.6	Van Waerbeek and Read 1994
Phoca vitulina	1.5	0.85	80	11	King 1983
Zalophus californianus	1.8	0.75	110	6	King 1983

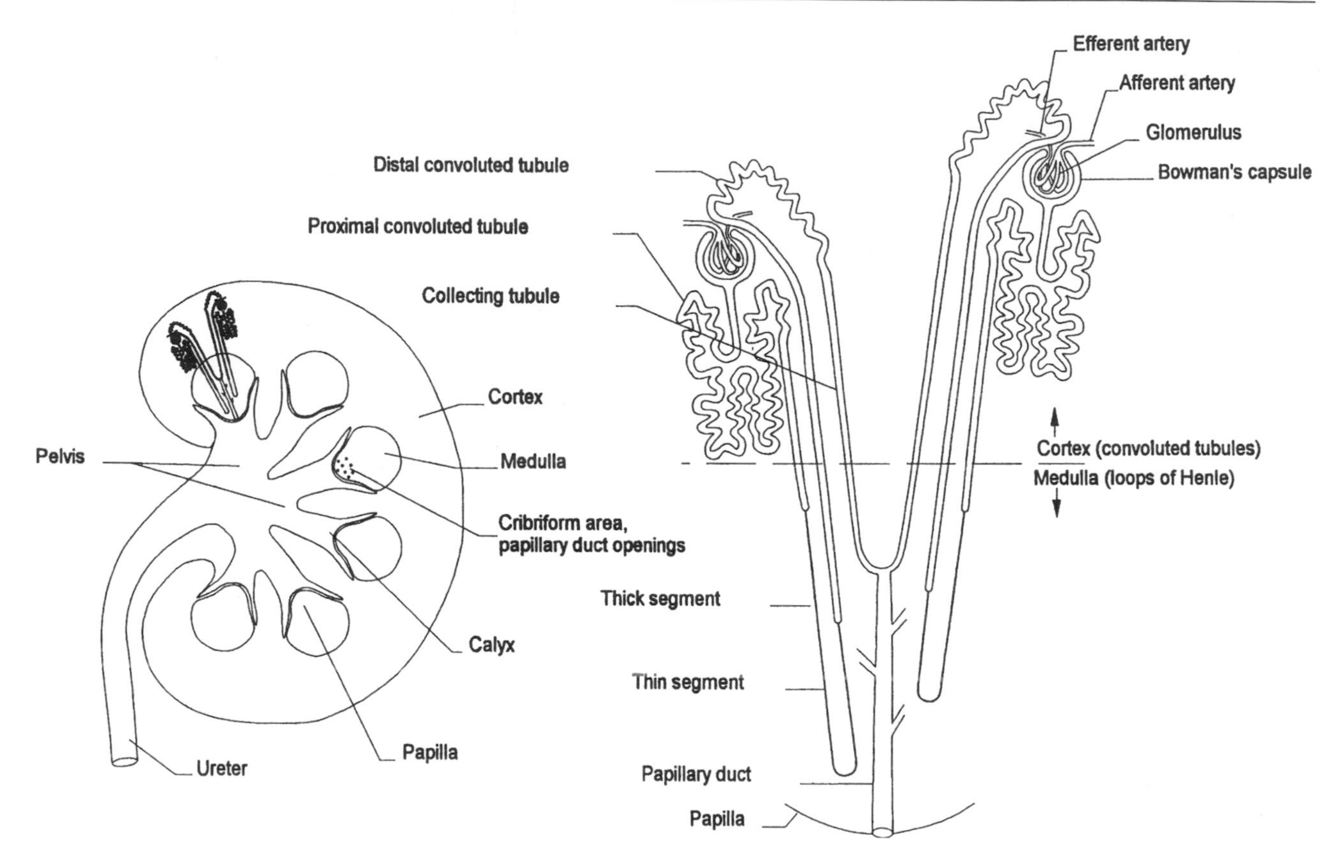

Figure 2-22. Kidney structure of the dog. The kidney is a bean-shaped organ formed by many nephrons. The enlarged view displays two nephrons. A twisted knot of capillaries (glomerulus) is surrounded by the Bowman's capsule (also called a Malpighian body). Water and many soluble, small molecules are filtered out of the blood and into the nephron at the Bowman's capsule. Water, nutrients, and important electrolytes are selectively reabsorbed in the proximal convoluted tubule, the loops of Henle, and the distal convoluted tubule. The urine produced exits the kidney through a ureter. (Illustrations adapted from Freeman and Bracegirdle 1968.)

(Schmidt-Nielsen 1990). Urine is produced as an ultrafiltrate of the blood by the kidney. The kidney of our "typical" mammal is a smooth bean-shaped organ, but mammalian kidneys can be lobulated in their appearance (Schummer et al. 1979).

The functional unit of the kidney is the nephron (Fig. 2-22). Urine is produced in a two-step process. The blood is first filtered across the capillary membrane and the membrane of the Malpighian body (also called the Bowman's capsule) into the nephron. This filtration step filters not only metabolic waste products, but also water, glucose, and important electrolytes out of the blood. Thus, the second step in urine formation is to reabsorb water and nutrients so that they are not lost from the body. Reabsorption occurs along the remainder of the length of the nephron tubule system. This reabsorption step can create a urine more concentrated in some salts and metabolic waste products (predominantly the nitrogenous waste urea) than the blood; urine with a higher osmotic pressure than blood is termed hyperosmotic. From the kidneys, the urine flows through a ureter to the urinary bladder, and exits the body through the urethra.

Notice in Figure 2-22 that the cortex of the kidney contains Bowman's capsule and convoluted tubules, and the loops of Henle are found in the medulla. Mammals with longer loops of Henle, and hence, thicker medullae, tend to be able to create more highly concentrated urine (Schmidt-Nielsen and O'Dell 1961). It has been hypothesized that marine mammals must be able to create a more concentrated urine than terrestrial mammals, as most have no access to freshwater (Schmidt-Nielsen 1990). Is there any morphological evidence to support this hypothesis? We investigate this question by examining kidney shape and relative thickness of the cortex and medulla.

Kidney Shape

The kidney is a composite body. It is made up of a great number of little kidneys bound together by . . . fibrous investments. Ommanney (1932b), describing the kidney of a fin whale

Most marine mammals possess a multiunit kidney—the whole organ is formed by a number of smaller, independently functioning units called reniculi ("little kidneys") (Fig. 2-23; e.g., Ommanney 1932b, Grahame 1959, Van Der Spoel 1963, Cave and Aumonier 1964, Bester 1975, Hedges et al. 1979, Vardy and Bryden 1981). The kidneys of otters, bears, pinnipeds, and cetaceans are "discretely multireniculated,"

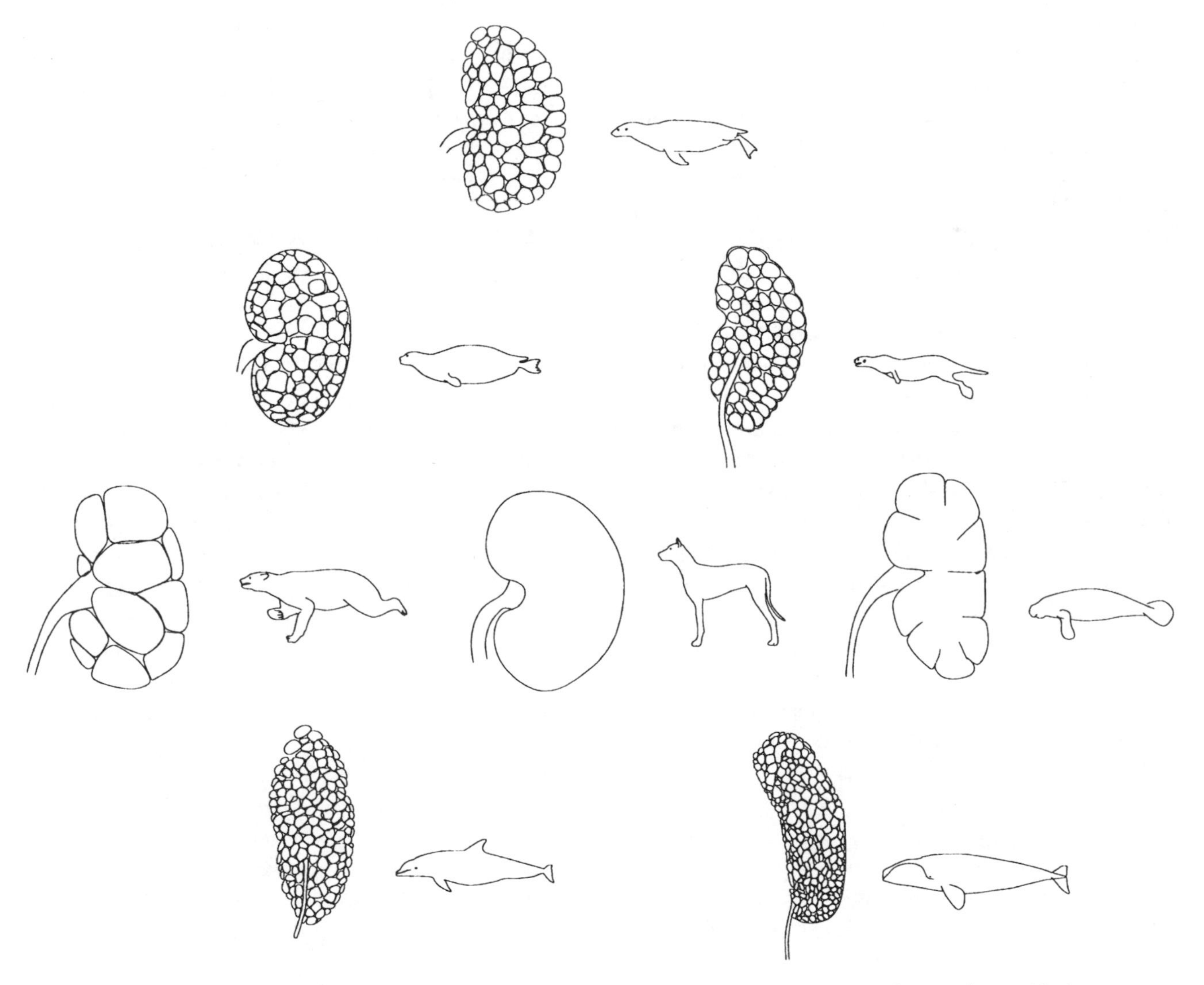

Figure 2-23. Kidney structure of representative marine mammals. The kidneys of most marine mammals are multiunit. The largest number of reniculi are found in the cetacean kidney. (Illustrations after Murie 1874, Harrison and Tomlinson 1956, Slijper 1979, Schummer et al. 1981, Bonde et al. 1985.)

(*sensu* Oliver 1968), that is, each subunit is a fully functional reniculus (Hedges et al. 1979). The number of reniculi varies among species. For example, elephant seal and harbor porpoise kidneys are formed by approximately 300 reniculi (Hedges et al. 1979, King 1983), whereas the fin whale kidney possesses more than 6000 reniculi (Ommanney 1932b).

Manatees possess a "compound multireniculated" kidney (*sensu* Oliver 1968), where each lobe of the kidney is formed by a fusion of a number of reniculi (Schummer et al. 1979, Hill and Reynolds 1989, Maluf 1989). Dugongs are unique within the marine mammals, as they possess an unlobulated kidney (Batrawi 1953, 1957). Steller is apparently the only person to have examined the kidney of the Steller's sea cow (*Hydrodamalis gigas*); he reports that the kidney was multireniculate and similar in form to those of seals and sea otters (reported in Maluf 1989). Interestingly, kidney shape varies more within the sirenians than it does across all other groups of marine mammals.

Does the multiunit kidney of marine mammals influence urine production? The multiunit kidney does possess more surface area than a similarly sized single-lobed kidney (Vardy and Bryden 1981). The relatively increased surface area of this type of kidney may increase both the volume of blood that can be filtered per unit time and the volume of urine that can be produced (Hedges et al. 1979, Costa 1982, King 1983). Similarly, the relatively large size of sea otter and harbor porpoise kidneys may also influence urine production in these species (Hedges et al. 1979, Costa 1982). Yet, large kidney size is not a feature shared by all marine mammals. Phocid seals have kidney weight-to-body weight ratios that are similar to those of terrestrial mammals (Vardy and Bryden 1981). Although the shape and size of multiunit kidneys increase the amount of urine produced, do these features suggest that marine mammals produce a concentrated urine? To date, there is no definitive evidence that the shape of the marine mammal kidney is specifically adapted to osmoregulation in the marine environment. Vardy and Bryden (1981) argue instead that multiunit kidneys may simply be an artifact of large body size. The multiunit shape may increase kidney blood filtration capacity and has been suggested as an adaptation for diving (see Elsner, Chapter 3, this volume).

Kidney Cortex and Medulla

The water requirement of a mammal is made up of that which is used in cooling the body and that required for excretion of metabolic waste. Irving et al. 1935

Recall that the length of the loop of Henle influences urine concentration. Do marine mammals have long loops of Henle? Sirenians and harbor porpoise possess relatively thick medullae, suggesting that they have long loops of Henle and can concentrate urine (Hedges et al. 1979, Hill and Reynolds 1989, Maluf 1989). The extent to which these animals do produce a concentrated urine is unknown. Sea otters can produce a "moderately" concentrated urine (Costa 1982), although we know of no measurement on medullary thickness in these animals. The functional significance of medullary thickness in pinnipeds is controversial and contradictory data regarding their ability to concentrate their urine exist (for review, see Vardy and Bryden 1981). For example, Krogh (1939) suggests that pinnipeds can produce a concentrated urine, whereas later studies offer evidence that at least some pinniped species do not (Irving et al. 1935, Kooyman and Drabek 1968, Vardy and Bryden 1981).

The somewhat confusing data on urine concentration in marine mammals is perhaps understandable. The urinary system functions to regulate precisely the body's water and electrolyte requirements, which is, by its very nature, a highly dynamic process. The variably concentrated urine sampled from mammals in the field probably reflects its extreme dependence on diet, activity, time between feeding or imbibing water, and age of the animals. A comprehensive and comparative study of the functional morphology of marine mammal kidneys would be very beneficial.

Nervous System

The mammalian nervous system functions to detect and respond to changes in both the internal and external environments of the animal. The nervous system is subdivided into the central nervous system (CNS) and the peripheral nervous system (PNS). The CNS is formed by the brain and spinal cord; the PNS includes all other nervous tissues in the body. In the PNS, specialized neurons called receptors transduce mechanical, thermal, chemical, and electromagnetic signals into nervous signals that are transmitted to the CNS through sensory (afferent) nerves. Interneurons (neurons that lie between sensory and motor neurons) in the CNS integrate incoming sensory information and coordinate rapid and appropriate responses by transmitting signals through motor (efferent) nerves to effector organs such as muscles (Evans 1993).

Pairs of spinal nerves, which arise at each intervertebral level, connect the PNS with the spinal cord. Spinal nerves are mixed nerves that carry both sensory and motor neurons. Twelve pairs of cranial nerves connect the PNS with the brain; three are sensory (I, II, and VIII), four are motor (III, IV, VI, and XII), and five are mixed (V, VII, IX, X, XI). A thirteenth cranial nerve was discovered after the standard numbering system was established. This terminal nerve, numbered 0, is closely associated with the olfactory nerve (I) and is sensory in function (Evans 1993).

Marine mammals possess specialized visual, acoustic, and tactile sensory adaptations that are discussed in detail by Wartzok and Ketten in this volume (Chapter 4). We limit our comparative description to brain morphology, particularly relative size, shape, and surface features. We also offer a comparative view of five selected cranial nerves—olfactory (I), optic (II), vestibulocochlear (also called acoustic) (VIII), trigeminal (V), and facial (VII).

Brain Size

Different orders of mammals appear to have characteristically different brain weight–body weight slopes.
Pagel and Harvey 1989

Throughout this chapter we have stressed the relationships between structure and function, but alas, one can glean little information about specifics of brain function by investigating its gross morphology. Nonetheless, humans have long searched for objective measures of function in the gross morphology of the brain. What can the size and the shape of the brain tell us about the animal it controls?

Historically, humans have been interested in "intelligence"—does an animal with a big brain possess greater "intelligence" than one with a smaller brain? For example, is the sperm whale, with absolutely the largest brain of any mammal (7.8 kg; Slijper 1979), really smart? (Note: "Intelligence" is associated with cerebral development; brain size and cerebral size are not equivalent. To read more about sperm whale brain size, see Tyack, Chapter 7, this volume.) This is, of course, a misleading question because we cannot yet accurately measure intelligence in humans (e.g., Hunt 1995), let alone in other animals. As Worthy and Hickie (1986) point out, intelligence "does not occupy space or have physical or chemical correlates," that is, there is little direct relationship between gross brain morphology and intelligence.

There are measurable differences in brain size and shape across different mammalian groups, though, that may offer insights into the mammal's ecology, life history, energetics, and so forth. To elucidate such relationships, one must have a comparable measure of brain size. Researchers use the allometric relationship between brain mass and total body mass (for further information on allometric relationships, see McMahon and Bonner 1983). The approach is theoretically simple—weigh a wide variety of mammal brains and whole bodies and calculate the exponential relationship between the two measured values:

$$M_{BR} = a\,(M)^{b}, \qquad \text{(equation 6)}$$

where M_{BR} is brain mass and M is body mass. A log–log plot of this equation will yield a straight line with slope b and y-intercept a (Fig. 2-24) (see also Appendix C in Schmidt-Nielsen 1990).

Allometric relationships are used to predict the expected brain mass of any mammal based on its total body mass. A value used to compare relative brain size is the encephalization quotient (EQ), calculated as the ratio of brain mass observed-to-brain mass predicted from the allometric equation for mammals as a group. A mammal with an EQ equal to 1 would have exactly the brain mass predicted for its body mass. If your mammal of choice instead had an EQ of 0.5, it would have one-half the brain mass that would be predicted for its body mass. The allometric relationship can be used to identify any systematic deviations from predicted brain mass values within a specific group (i.e., small odontocetes tend to lie above the line [Ridgway and Brownson 1984], sirenians lie below the line [O'Shea and Reep 1990]) (Fig. 2-24). These differences, in turn, may be related to other measurable features of the animal.

Although this approach is theoretically simple, the student delving into relative brain size literature may think otherwise! The primary caution is to recognize, and evaluate accordingly, functional assertions based on absolute versus relative brain size. True, sperm whales have absolutely the largest brain in the mammalian world, but they also possess EQs well below 1 (Fig. 2-24). Another difficulty is that there are at least two significantly different scaling relationships entrenched in the literature. Brain mass scales to $M^{0.67}$ (Jerison 1973, McNab and Eisenberg 1989; represented as the lower dotted line in our Fig. 2-24) or to $M^{0.73}$ to $M^{0.76}$ (Armstrong 1983, Hofman 1983, Pagel and Harvey 1989, Worthy and Hickie 1986; represented as the upper dotted line in our Fig. 2-24). The differences in these scaling relationships result primarily from different samples of mammals used in each study (McNab and Eisenberg 1989). Obviously, the value of EQ that you calculate depends on the scaling relationship you choose. (Use the brain and body masses listed in Table 2-3 to calculate EQs with the two scaling relationships in Figure 2-24. Which marine mammals have the relatively largest and smallest brains? Notice that although the absolute values of EQ depend on the equation used, the pattern of relative brain size does not.)

McNab and Eisenberg (1989) point out another troublesome feature of the relative brain size literature—a similarity in scaling, in and of itself, is not evidence of a functional relationship between two measurable characters. For example, brain mass and basal metabolic rate scale similarly to body mass, and many researchers have asserted that there is a

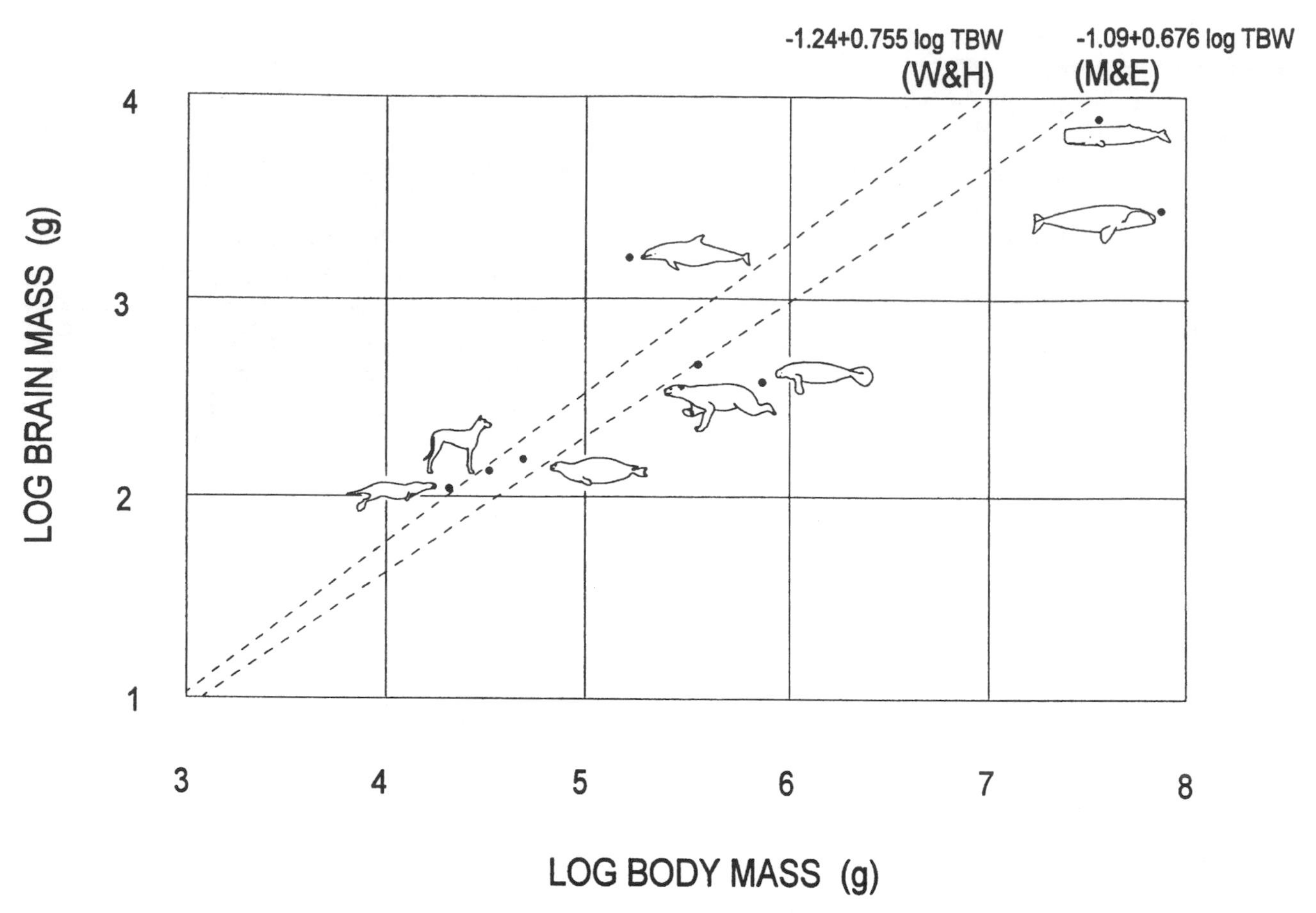

Figure 2-24. Log-log plot of brain mass to body mass of our "typical" mammal and representative marine mammals. The upper (Worthy and Hickie 1986) and lower (McNab and Eisenberg 1989) dotted lines represent two established scaling relationships for mammals. Some small odontocetes have relatively large brains, most pinnipeds have predicted brain sizes, and mysticetes, sperm whales, and sirenians all have relatively small brains. (Data on brain and body masses from Table 2-4. The encephalization quotient for each mammal on the graph can be calculated using data from Table 2-4 and the given logarithmic equations.)

causal relationship between the two (i.e., low metabolic rates somehow constrain mammals to relatively small brains) (e.g., Robin 1973, Armstrong 1983, Hofman 1983). McNab and Eisenberg (1989) analyzed residual variation in brain size in a diverse group of mammals and determined that it is not correlated to relative basal metabolic rate. That is, relatively large brained mammals can have either high or low relative metabolic rates (see also Pagel and Harvey 1990). McNab and Eisenberg (1989) concluded that across a wide diversity of mammals, there is not a detectable functional link between brain mass and metabolic rate.

So what is relative brain size related to? There appears to be no overarching, all encompassing answer to this question. Rather, relative brain size reflects (in a complex manner) a mammal's phylogenetic history, lifestyle, reproductive strategy, and so forth. Thus marine mammals display very different patterns of relative brain size based on their independent histories and diverse aquatic lifestyles. Some small odontocetes have relatively large brains, most pinnipeds have predicted brain sizes, and mysticetes, sperm whales, and sirenians all have relatively small brains (for review, see Worthy and Hickie 1986). How might one explain these differences? Investigators are often left with the less than satisfactory task of creating post-hoc explanations for these patterns, and for every hypothesis, no matter how plausible, there is a mammal that proves the exception.

It has been hypothesized that the relatively large brains of some small odontocete cetaceans may be required to provide sufficient processing for the complex, high frequency sounds used in echolocation (e.g., Worthy and Hickie 1986, Ridgway 1990). Although there exist few quantitative data, some auditory structures are reportedly larger in small odontocetes than in either mysticetes (Breathnach 1960) or humans (for review, see Ridgway 1990). Bottlenose dolphins apparently use a novel area of their dorsal cerebral cortex for auditory processing, but complete mapping of the auditory cortex does not exist for any odontocete cetacean (Ridgway 1990). Thus, to date, there is insufficient evidence to assert that odontocetes use relatively larger areas of cerebral cortex for auditory processing.

It is plausible that the relatively large brain size of small odontocete cetaceans is functionally related to echolocation. There are data to suggest that some auditory structures are large; however, there is no comparative study that substantially supports this assertion. A quantitative survey of specific auditory structures across diverse taxa would test the apparent correlation between relatively large brains and echolocating abilities in odontocetes (for an example of such a study on olfactory capabilities in Carnivora, see Gittleman 1991). Interestingly, microchiropteran bats have relatively low EQs, suggesting that a large brain is not required to echolocate, but Worthy and Hickie (1986) hypothesize that differences in acoustic discrimination capabilities account for this disparity.

Sperm whales and mysticetes are large whales with relatively small brains. Worthy and Hickie (1986) suggest that because these animals invest as much as 25% of their total body length in the head, "they may therefore inflate their body masses without requiring a concomitant increase in neural tissue for their function, resulting in lower EQ's." Although perhaps a plausible hypothesis, the relationship between brain size and head length remains simply a correlation with no evidence of a functional link. These researchers do point out that many other factors (e.g., social structure, feeding styles, vocalizations) may also affect brain size in these large whales.

O'Shea and Reep (1990) provide a systematic investigation of factors that may be functionally related to the unusually small relative brain size of sirenians. Manatees are mosaics, possessing some characteristics often associated with other relatively small brained mammals (e.g., low-quality ubiquitous food and simple foraging strategies), as well as those often associated with relatively large brained mammals (e.g., long gestation period, precocial single offspring, long life span, navigating in a three-dimensional world). Manatees are born with relatively small brains, and their brains grow at very reduced rates, relative to their body mass, after weaning. These two growth traits led O'Shea and Reep (1990) to hypothesize that (1) low maternal metabolic rate constrains the size of the neonate brain (but see Pagel and Harvey 1990, for refutation of this hypothesis in other mammal groups), and (2) selection for large body size at later stages of postnatal development has allowed body and brain growth rates to become uncoupled. O'Shea and Reep's (1990) study is a model for careful use of data sets to reject or support alternative hypotheses of functional relationships between brain size and life history traits.

Brain Shape and Surface Features

A notable gross feature of sirenian brains is the general absence of convolutions on the cortical surface.
Reep and O'Shea 1990

As with relative brain size, marine mammals display great diversity in brain shape and surface folding (Fig. 2-25). Most mammalian brains are longer than they are wide, reflecting the elongated shape of the braincase (see Fig. 2-10). Most mysticete brains are approximately equivalent in length and width, and odontocete brains are actually wider than they are long (Breathnach 1960, Slijper 1979). Cetacean brain shapes mirror the foreshortened braincases of their "telescoped" skulls. The sirenian brain is also relatively short and wide (see also Fig. 2-1 in Reep and O'Shea 1990). King (1983) states that the pinniped brain tends to be more spherical in shape than those of terrestrial carnivores. The brain shape of sea otters is similar to those of other otters (Radinsky 1968), and we presume that polar bears have brain shapes similar to those of their terrestrial relatives.

Although there does not appear to be any functional significance to differences in overall brain shape, folding of the cerebral cortex (convolutedness) has been used to infer brain function. Surface folding increases the area available for cerebral cortex—the more cortex an animal has, the more area available for higher level neuronal processing. Odontocete cetaceans exceed humans, as well as all other animals, in the extent of the convolutedness of their cerebral cortex (Ridgway and Brownson 1984, Ridgway 1990). King (1983) reports that pinniped brains are more convoluted than those of terrestrial carnivores. At the other end of the spectrum are sirenians, which possess exceptionally smooth cerebral surfaces, a condition referred to as lissencephaly (Reep and O'Shea 1990).

The degree of cortical folding is often reported as the gyration index (GI)—the ratio of total perimeter to exposed perimeter of cerebral cortex (see Reep and O'Shea 1990). Sirenian GIs are close to 1, human and ungulate GIs lie between 2 and 3, and odontocete cetacean GIs can exceed 4 (e.g., Ridgway and Brownson 1984, Hofman 1985, Reep and O'Shea 1990). Differences in cortical folding can dramatically influence cerebral surface area. Another way to contrast convolutedness is to compute the ratio of surface area of cerebral cortex (cm^2) to brain mass (grams). Using data from Ridgway and Brownson (1984) and Reep and O'Shea (1990), we calculate that bottlenose dolphins have 2.4 cm^2/g, whereas manatees have 0.4 cm^2/g. That is, per unit weight, there is a sixfold difference in cortical surface area

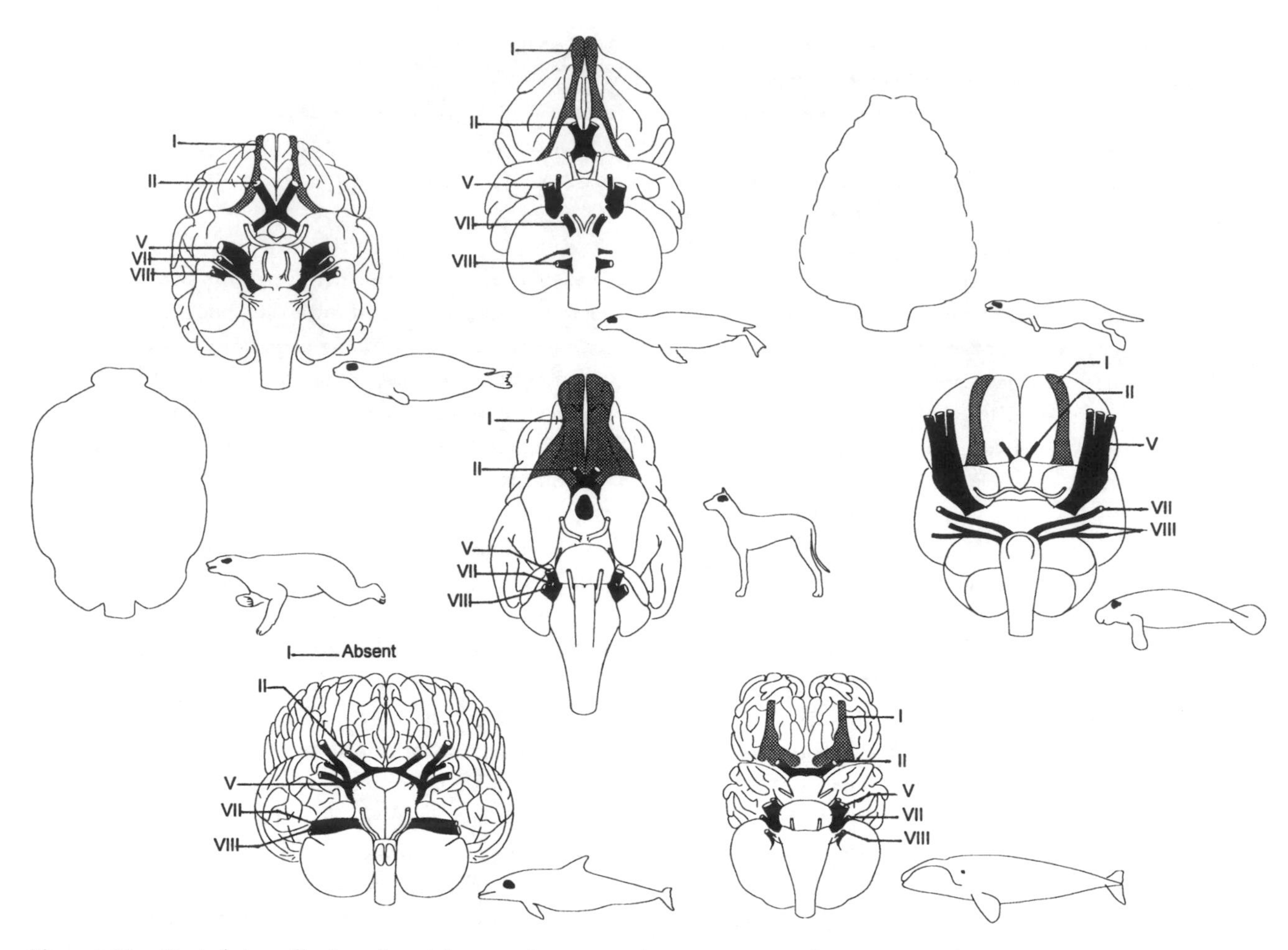

Figure 2-25. Ventral view of brain and cranial nerves of representative marine mammals. Brains are scaled to the same length; lateral views show brain size relative to body size. Sensory nerves I (olfactory), II (optic), and VIII (vestibulocochlear) and mixed nerves V (trigeminal) and VII (facial) are shaded to illustrate relative importance of sensory input (I, II, VIII) and sensory and motor function of the facial region (V, VII). (Illustrations adapted from Gervais 1870; Murie 1874, 1880; Kukenthall 1893; Fish 1896, 1898, 1899; Langworthy 1932; Jelgersma 1934; Gregory 1936; Sisson and Grossman 1953; Breathnach 1960; Pilleri 1964; Radinsky 1968; Anthony 1972; McFarland et al. 1979; Seki 1985; Ridgway 1990; Evans 1993; and the authors' unpublished data.)

between the dolphin and the manatee. Odontocetes possess considerably more (both relatively and absolutely) cerebral cortex than manatees. These numbers allow us to quantify differences between brain structures, but not to explain them. Few comparative data for other marine mammals exist.

Cetaceans and sirenians define "extremes" of mammalian cortex organization (e.g., Glezer et al. 1988, Reep and O'Shea 1990, Welker 1990). These taxa represent outliers from the mainstream of mammalian brain evolution, and as such, offer us insights into the plasticity of mammalian brain organization.

Cranial Nerves

Trigeminal impressions must be considered to be of major importance in the economy of Mysticeti. Breathnach 1960

In Figure 2-25, five of the 12 cranial nerve pairs are compared in representative marine mammals. We have illustrated the olfactory (I), optic (II), and vestibulocochlear (VIII) nerves to offer a connection between the CNS and the special sensory structures discussed in Wartzok and Ketten (Chapter 4, this volume). The trigeminal (V) and facial (VII) nerves are illustrated because they innervate facial structures that have been extensively remodeled in some groups.

Even a cursory glance at Figure 2-25 reveals that cranial nerves vary greatly in relative size within an individual. Odontocete cetaceans have massive vestibulocochlear nerves, yet their olfactory nerves are entirely absent as adults (e.g., Breathnach 1960, Berzin 1972, Slijper 1979, Demski et al. 1990, Ridgway 1990; but for alternative viewpoint, see Behrmann 1989). Likewise, the trigeminal and facial nerves of sirenians are large, but their optic nerves are unremarkable (Murie 1872, 1880). (When viewing Figure 2-25, it is important to note that brains are scaled to the same length. Thus, relative nerve sizes can be compared within an individual, but not among individuals.)

Is it interesting to compare relative nerve size? Yes, because the size of a nerve generally reflects the number of individual neurons it contains or the size of those neurons (e.g., Gao and Zhou 1991). The more neurons there are, the more information can be transmitted along a nerve and the larger the diameter of the axon, the higher the conduction velocity along that nerve (Schmidt-Nielsen 1990).

Gao and Zhou (1991) elegantly demonstrate that the large vestibulocochlear nerve of some odontocete cetaceans is "specialized for high-speed communication with the brain." These investigators record the number and size of neurons (nerve fibers) in the cochlear portion of cranial nerve VIII in three species of odontocetes, and compare these values to those of selected terrestrial mammals and mysticete whales. Odontocetes possess absolutely three to four times more cochlear neurons than do terrestrial mammals. Odontocetes also possess absolutely more vestibulocochlear neurons (combined cochlear and vestibular portions of cranial nerve VIII) than mysticete whales (Jacobs and Jensen 1964). The range, mode, and mean diameter of odontocete cochlear axons is two to five times that found in terrestrial mammals and mysticetes (Gao and Zhou 1991). The maximum axon diameters measured in the cochlear nerve of odontocetes are the largest measured in *any* nerve in *any* vertebrate, implying conduction velocities four to five times faster than in the cochlear nerve of terrestrial mammals. Thus, the large size of cranial nerve VIII in odontocetes reflects morphological specializations for transmitting large quantities of acoustic information at very high speeds.

Few other quantitative studies of cranial nerve structure exist for marine mammals. We can hypothesize the relative "importance" of certain sensory modalities to different marine mammals based on relative sizes of their cranial nerves. For example, compare the olfactory nerve tracts in Figure 2-25. In general, the olfactory nerves and thus, olfactory sensory function, are reduced in most marine mammals. The olfactory nerves are absent in adult odontocetes and are either absent or greatly reduced in adult mysticetes (e.g., Breathnach 1960, Slijper 1979). Interestingly, the terminal nerve, usually associated with the olfactory nerve, is present in odontocetes (Demski et al. 1990, Ridgway 1990). Relative to terrestrial carnivores, the olfactory nerve is apparently reduced in phocid seals, and although less so, is also reduced in otariids (King 1983). The olfactory bulbs (the expanded distal portion of the olfactory nerve tract) of otters are reduced, relative to terrestrial carnivores (for a rigorous quantitative approach, see Gittleman 1991). It is difficult to assess the importance of olfaction in manatees (see Reynolds and Odell 1991), and we know of no information on their olfactory nerve structure. In contrast to other marine mammals, polar bears possess large olfactory nerves, we presume, given their impressive ability to smell seals through meters of snow (Stirling 1988).

Cranial nerve V, the trigeminal nerve, is predominantly sensory to the face, although it also carries motor neurons to the lower jaw (Evans 1993). It is the largest cranial nerve in mysticetes, and Slijper (1979) conjectures that its large size reflects the very large head size of these whales. The large trigeminal nerves of mysticetes may also reflect increased motor innervation required to control their large lower jaws. The trigeminal nerve is the second largest nerve in odontocetes (second only to the vestibulocochlear nerve) and may reflect the high tactile sensitivity of the skin on the head (for review, see Ridgway 1990). The relatively large trigeminal nerves of pinnipeds carry sensory information from their vibrissae, which function as tactile structures and in some species, may respond to sound (e.g., Ling 1977, Renouf 1979, King 1983). The massive trigeminal nerves of manatees probably reflect the highly sensitive vibrissal fields around this animal's mouth (Marshall and Reep 1995). The cerebral cortex of manatees contains unique dense cell aggregates that Reep et al. (1989) hypothesize are related to these tactile hairs. Thus, both specialized morphologies of the PNS and CNS appear to support this sensory function in manatees.

Cranial nerve VII, the facial nerve, is a motor nerve that innervates the facial muscles. The facial nerve of odontocetes is large, apparently reflecting the extensive facial musculature involved with echolocation (for review, see Mead 1975, Ridgway 1990). Comparative studies of this cranial nerve, as well as others, would improve our understanding of sensory and motor functions in marine mammals.

Cranial nerves offer the comparative anatomist one other important investigative tool. Cranial nerves are used to establish the identities of skull bones in diverse taxa because they course through homologous foramina (holes) in homologous bones (e.g., Rommel 1990). For example, a large branch of the trigeminal nerve runs onto the superficial face through holes in the maxilla termed infraorbital foramina (Fig. 2-26). Interestingly, in the dolphin, the maxilla's position dorsal to the orbit (rather than cranial and ventral to the orbit as in the dog) has placed the infraorbital foramena in a dorsal position. Thus, the "telescoping" of the odontocete cetacean skull has rearranged the relative positions of skull bones and cranial nerves, but has not disrupted their developmental and evolutionary relationships.

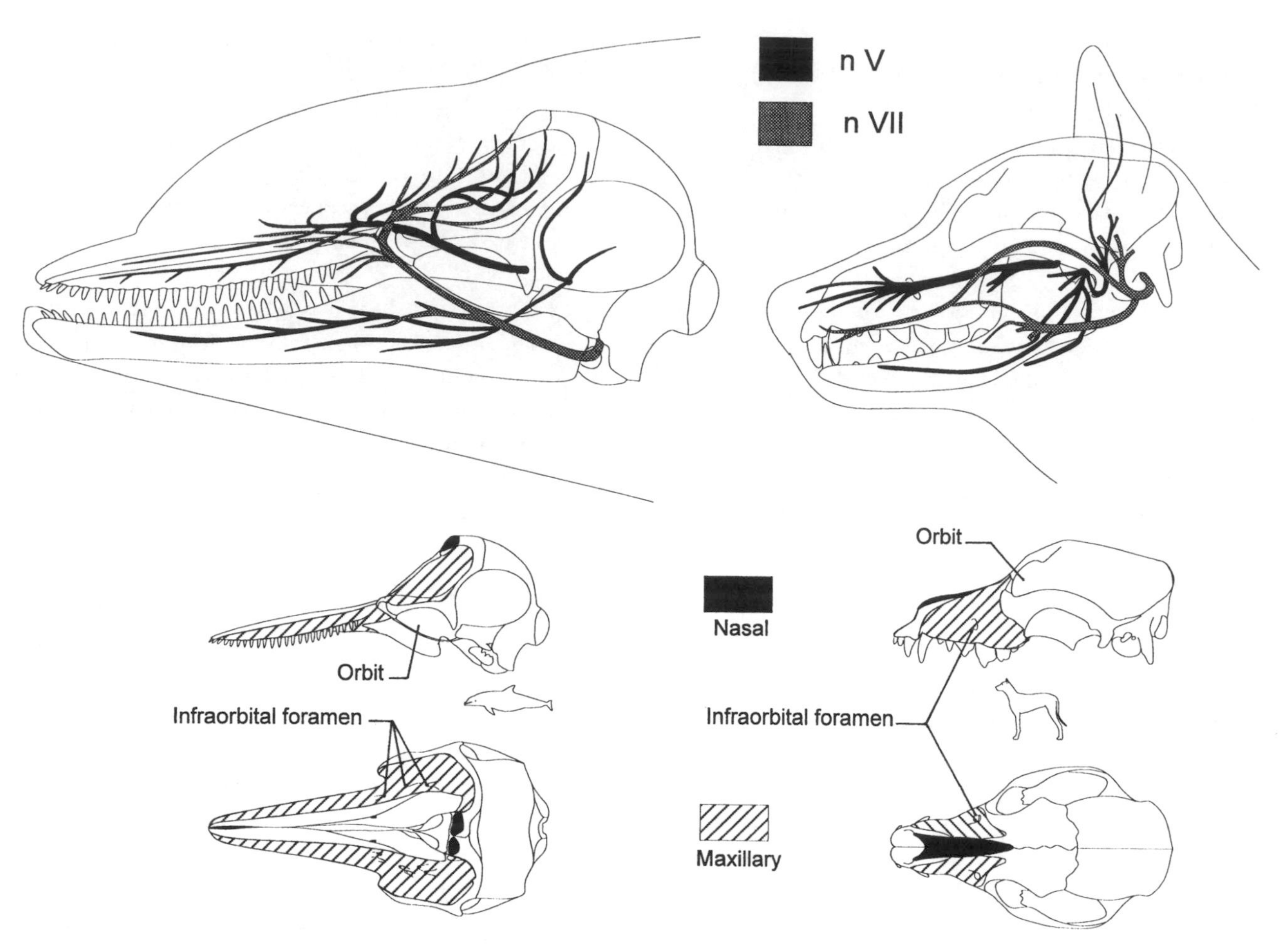

Figure 2-26. Lateral view of major branches of cranial nerves V (trigeminal) and VII (facial) in the bottlenose dolphin and dog. Note that the distribution of the nerves across the face of the dolphin reflects the rearrangement of bony elements in the "telescoped" skull. A major branch of the trigeminal nerve courses through the infraorbital foramen of the maxilla. In the odontocete skull, there are multiple foramina that are dorsal (or superior) to the orbit, rather than ventral to it as in the dog. (Illustrations adapted from Rommel 1990, Evans 1993, and the authors' unpublished data.)

Concluding Remarks

The animals discussed in this chapter define extremes of mammalian function. Some dive to more than 2000 m, some breath-hold for more than 2 hours, some produce and hear sounds at 150,000 Hz, whereas some eat grass underwater. These animals present variations of typical mammalian body forms—with remodeled heads, streamlined limbs, reduced skeletal elements, and wondrous elaborations of their integument. And despite the long and venerable history of marine mammal anatomy, it is abundantly clear that there is much left to be learned. Nearly half of the anatomical citations used in this review are less than 16 years old. The veritable explosion of novel insights into marine mammals has come primarily through the use of new technologies that integrate quantitative morphological data with specific measures of functional performance. The level of sophistication of these technologies, and of our knowledge base, is ever increasing. These are exciting times indeed to be asking questions of form and function in marine mammals.

Acknowledgments

We thank Sharon Gasser at James Madison University for her help with original references, our students and assistants who patiently read and improved earlier versions, James Mead and Charley Potter for their friendship and for getting us all started, and Becky Mead for her unending and gracious support. All figures were created with the FastCAD drawing program. We thank the Office of Naval Research and the National Marine Fisheries Service for their financial support. This is CMSR contribution #220.

List of Abbreviations Used in Text

AVA	arteriovenous anastomoses
B	buoyancy (N)
C	conductance ($Wm^{-2}{}^{o}C^{-1}$)
D	drag (N)
D_p	pressure drag (N)
D_v	viscous drag (N)
g	gravitational acceleration (ms^{-2})
H′	rate of heat lost (W)
n	propulsor efficiency (dimensionless)
L	lift
M	body mass (kg)
M_{BR}	brain mass
PAVR	periarterial venous rete
P′	power output (W)
p	pressure
ρ_{body}	density of body ($kg\,m^{-3}$)
ρ_{water}	density of water ($kg\,m^{-3}$)
SA	surface area (m^2)
T	thrust (N)
T_a	ambient water temperature (°C)
T_b	body temperature (°C)
u	velocity (ms^{-1})
V	volume (m^3)

Literature Cited

Ackman, R. G., J. H. Hingley, C. A. Eaton, V. H. Logan, and P. H. Odense. 1975. Layering and tissue composition in the blubber of the northwest Atlantic sei whale (*Balaenoptera borealis*). Canadian Journal of Zoology 53:1340–1344.

Alexander, R. McN. 1977. Allometry of the limbs of antelopes (Bovidae). Journal of Zoology (London) 183:125–146.

Alexander, R. McN. 1983. Animal Mechanics, 2d ed. Blackwell Scientific Publications, Oxford, London.

Alexander, R. McN. 1990. Size, speed and buoyancy adaptions in aquatic mammals. American Zoologist 30:189–196.

Allen, J. A. 1880. History of North American pinnipeds, A Monograph of the Walruses, Sea-Lions, Sea-Bears and Seals of North America. U.S. Geological and Geographical Survey Terr., Miscellaneous Publications 12.

Allen, J. A. 1902. The hair seals (family Phocidae) of the North Pacific Ocean and Bering Sea. Bulletin of American Museum of Natural History 16:459–499.

Amundin, M., and S. H. Anderson. 1983. Bony nares, air pressure and nasal plug muscle activity during click production in the harbour porpoise, *Phocoena phocoena,* and the bottlenose dolphin, *Tursiops truncatus.* Journal of Experimental Biology. 105:275–282.

Anthony, J. 1972. Le système nerveux des mammifères. Traite de Zoologie, Anatomie, Systematique, Biologie, Tome XVI (Fascicule IV). Masson et Cie, Paris.

Arkowitz, R., and S. A. Rommel. 1985. Force and bending moment of the caudal muscles in the shortfin pilot whale. Marine Mammal Science 1:203–209.

Armstrong, E. 1983. Relative brain size and metabolism in mammals. Science 220:1302–1304.

Ask-Upmark, E. 1935. The carotid sinus and the cerebral circulation. An anatomical, experimental, and clinical investigation, including some observations on the rete mirabile caroticum. Acta Psychiatrica et Neurologica Scandinavia (Suppl.) 6:1–374.

Au, W. W. L. 1993. The Sonar of Dolphins. Springer-Verlag, New York, NY.

Baker, M. A. 1979. A brain-cooling system in mammals. Scientific American 240:130–139.

Barnes, L. G., and E. Mitchell. 1978. Cetacea. Pages 582–601 *in* V. J. Maglio and H. S. B. Cooke (eds.). Evolution of African Mammals. Harvard University Press, Cambridge, MA.

Barnett, C. H., R. J. Harrison, and J. D. W. Tomlinson. 1958. Variations in the venous systems of mammals. Biological Reviews 33:442–487.

Bartholinus, T. 1654. Historiarum anatomicarum rariorum Centuria. I and II (Anatome tursionis) Hafniae *cited in* E. J. Slijper. 1979. Whales. Cornell University Press, Ithaca, NY.

Batrawi, A. 1953. The external features of the dugong kidney. Bulletin of the Zoological Society of Egypt 11:12–13.

Batrawi, A. 1957. The structure of the dugong kidney. Publication of the Marine Biological Station, Al Ghardaqa, Egypt 9:51–68.

Bedford, J. M. 1977. Evolution of the scrotum: The epididymis as the prime mover. Pages 171–182 *in* J. H. Calaby and C. H. Tyndale-Biscoe (eds.). Reproduction and Evolution. Australian Academy of Science, Canberra City.

Behrmann, G. 1989. The olfactory region in the nose of the harbor porpoise, *Phocoena phocoena* (Linne, 1758). Aquatic Mammals 15:130–133.

Belanger, L. F. 1940. A study of the histology portion of the lungs of aquatic mammals. American Journal of Anatomy 67:437–469.

Bell, A. W. 1987. Consequences of severe heat stress for fetal development. Pages 313–333 *in* J. R. S. Hales and D. A. B. Richards (eds.). Heat Stress: Physical Exertion and Environment. Elsevier, Amsterdam.

Benirschke, K., and L. H. Cornell. 1987. The placenta of the killer whale, *Orcinus orca.* Marine Mammal Science 3:82–86.

Bennett, M. R., R. F. Ker, and R. McN. Alexander. 1987. Elastic properties of structures in the tails of cetaceans (*Phocaena* and *Lagenorhynchus*) and their effect on the energy cost of swimming. Journal of Zoology (London) 211:177–192.

Bergey, M. R. 1986. Lung structure and mechanics of the West Indian manatee (*Trichechus manatus*). M.S. thesis. University of Miami, Coral Gables, FL.

Berzin, A. A. 1972. The Sperm Whale (Kashalot). Israel Program for Scientific Translations, Jerusalem.

Bester, M. N. 1975. The functional morphology of the kidney of the Cape fur seal, *Arctocephalus pusillus* (Schreber). Madoqua, Series II 4:69–92.

Biewener, A. A. 1989. Scaling body support in mammals: Limb posture and muscle mechanics. Science 245:45–48.

Blake, R. W. 1983. Fish Locomotion. Cambridge University Press, Cambridge, U.K.

Blickhan, R., and J.-Y. Cheng. 1994. Energy storage by elastic mechanisms in the tail of large swimmers—A re-evaluation. Journal of Theoretical Biology 168:315–321.

Blix, A. S., F. H. Fay, and K. Roland. 1983. On testicular cooling in phocid seals. Polar Research 1:231–233.

Boice, R. C., M. L. Swift, and J. C. Roberts Jr. 1964. Cross-sectional anatomy of the dolphin. Norsk Hvalfangst-Tidende 7:178–193.

Bonde, R. K., T. J. O'Shea, and C. A. Beck. 1983. Manual of Procedures for the Salvage and Necropsy of Carcasses of the West Indian Manatee (*Trichechus manatus*). U.S. Department of the Interior, National Biological Survey. Available from National Technical Information Service, Springfield, VA.

Breathnach, A. S. 1960. The cetacean central nervous system. Biological Reviews 35:187–230.

Brodie, P. F. 1975. Cetacean energetics, an overview of intraspecific size variation. Ecology 56:152–161.

Bron, K. M., H. V. Murdaugh Jr., J. E. Millen, R. Lenthall, P. Raskin, and E. D. Robin. 1966. Arterial constrictor response in a diving mammal. Science 152:540–543.

Brown, W. R., J. R. Geraci, B. D. Hicks, D. J. St. Aubin, and J. P. Schroeder. 1983. Epidermal cell proliferation in the bottlenose dolphin (*Tursiops truncatus*). Canadian Journal of Zoology 61:1587–1590.

Brownell, R. L. Jr., and K. Ralls. 1986. Potential for sperm competition in baleen whales. Reports of the International Whaling Commission (Special Issue 8).

Bryden, M. M. 1964. Insulating capacity of the subcutaneous fat of the southern elephant seal. Nature 203:1299–1300.

Bryden, M. M. 1967. Testicular temperature in the southern elephant seal, *Mirounga leonina* (Linnaeus). Journal of Reproductive Fertility 13:583–584.

Bryden, M. M. 1969. Relative growth of the major body components of the southern elephant seal, *Mirounga leonina* (Linnaeus). Australian Journal of Zoology 17:153–177.

Bryden, M. M. 1972. Growth and development of marine mammals. Pages 1–79 *in* R. J. Harrison (ed.). Functional anatomy of Marine Mammals. Academic Press, London.

Bryden, M. M. 1994. Adaptions in the integument of seals (Abstract). Journal of Morphology 220:330.

Bryden, M. M., and W. J. L. Felts. 1974. Quantitative anatomical observations on the skeletal and muscular systems of four specimens of Antarctic seals. The Anatomical Record 118:589–600.

Bryden, M. M., and G. S. Molyneux. 1978. Arteriovenous anastomoses in the skin of seals. 2. The California sea lion (*Zalophus californianus*) and the northern fur seal (*Callorhinus ursinus*) (Pinnipedia: Otariidae). The Anatomical Record 191:253–260.

Bryden, M. M., H. Marsh, and B. W. MacDonald. 1978. The skin and hair of the dugong, *Dugong dugon*. Journal of Anatomy 126:637–638.

Burn, D. M. 1986. The digestive strategy and efficiency of the West Indian manatee, *Trichechus manatus*. Comparative Biochemistry and Physiology A 85:139–142.

Caldwell, D. K., and M. C. Caldwell. 1985. Manatees: *Trichechus manatus* (Linnaeus, 1758); *Trichechus senegalensis* (Link, 1795); and *Trichechus inunguis* (Natterer, 1883). Pages 33–66 *in* S. H. Ridgway and R. Harrison (eds.). Handbook of Marine Mammals. Volume 3, The Sirenians and Baleen Whales. Academic Press, London.

Carrick, F. N., and B. P. Setchell. 1977. The evolution of the scrotum. Pages 165–170 *in* J. H. Calaby and C. H. Tyndale-Biscoe (eds.). Reproduction and Evolution. Australian Academy of Science, Canberra City.

Cartmill, M., W. L. Hylander, and J. Shafland. 1987. Human Structure. Harvard University Press, Cambridge, MA.

Cave, A. J. E., and F. J. Aumonier. 1964. The reniculus in certain balaenopterids. Journal of the Royal Microscopical Society 83:255–264.

Costa, D. P. 1982. Energy, nitrogen, and electrolyte flux and sea water drinking in the sea otter *Enhydra lutris*. Physiological Zoology 55:35–44.

Costa, D. P., and G. L. Kooyman. 1982. Oxygen consumption, thermoregulation, and the effect of fur oiling and washing on the sea otter, *Enhydra lutris*. Canadian Journal of Zoology 60:2761–2767.

Cowles, R. B. 1958. The evolutionary significance of the scrotum. Evolution XII:417–418.

Cowles, R. B. 1965. Hyperthermia, aspermia, mutation rates and evolution. Quarterly Review of Biology 40:341–367.

Cranford, T. W. 1988. The anatomy of acoustic structures in the spinner dolphin forehead as shown by X-ray computed tomography and computer graphics. Pages 67–77 *in* P. E. Nachtigall and P. W. B. Moore (eds.). Animal Sonar. Plenum Press, New York, NY.

Cranford, T. W., M. Amundin, and K. S. Norris. 1996. Functional morphology and homology in the odontocete nasal complex: Implications for sound generation. Journal of Morphology 228:223–285.

Daniel, T. L., and P. W. Webb. 1987. Physical determinants of locomotion. Pages 343–369 *in* P. De Jours, C. R. Taylor, and E. R. Weibel (eds.). Comparative Physiology: Life in Water and on Land. Liviana Press, New York, NY.

de Buffrenil, V., A. Collet, and M. Pacsal. 1985. Ontogenetic development of skeletal weight in a small delphinid, *Delphinus delphis* (Ceatcea, Odontoceti). Zoomorphology 105:336–344.

DeLong, R. L., and B. S. Stewart. 1991. Diving patterns of northern elephant seal bulls. Marine Mammal Science 7:369–384.

DeMaster, D. P., and I. Stirling. 1981. *Ursus maritimus*. Mammalian Species 145:1–7.

Demski, L. S., S. H. Ridgway, and M. Schwanzel-Fukuda. 1990. The terminal nerve of dolphins: Gross structure, histology, and luteinizing-hormone-releasing hormone immunocytochemistry. Brain Behavior and Evolution 36:249–261.

Denison, D. M., and G. L. Kooyman. 1973. The structure and function of the small airways in pinniped and sea otter lungs. Respiration Physiology 17:1–10.

Denison, D. M., D. A. Warrell, and J. B. West. 1971. Airway structure and alveolar emptying in the lungs of sea lions and dogs. Respiration Physiology 13:253–260.

Doidge, D. W. 1990. Integumentary heat loss and blubber distribution in the beluga, *Delphinapterus leucas*, with comparisons to the narwhal, *Monodon monoceros*. Pages 129–140 *in* T. G. Smith, D. J. St. Aubin, and J. R. Geraci (eds.). Advances in Research on the Beluga, *Delphinapterus leucas*. Canadian Bulletin of Fisheries and Aquatic Sciences 224.

Domning, D. P. 1977. Observations on the myology of *Dugong dugong* (Muller). Smithsonian Contributions to Zoology, Number 226, Washington, D.C.

Domning, D. P. 1978. The myology of the Amazonia manatee, *Trichechus inunguis* (Natterer) (Mammalia: Sirenia). Acta Amazonica 8 (Suppl. 1):1–81.

Domning, D. P., and V. de Buffrenil. 1991. Hydrostasis in the Sirenia: Quantitative data and functional interpretations. Marine Mammal Science 7:331–368.

Domning, D. P., and L.-A. C. Hayek. 1984. Horizontal tooth replacement in the Amazonian manatee (*Trichechus inunguis*). Mammalia 48:105–127.

Dormer, K. J., M. J. Denn, and H. L. Stone. 1977. Cerebral blood flow in the sea lion (*Zalophus californianus*) during voluntary dives. Comparative Biochemistry and Physiology A 58:11–18.

Drabek, C. M., and G. L. Kooyman. 1983. Terminal airway embryology of the delphinid porpoises, *Stenella attenuata* and *S. longirostris*. Journal of Morphology 175:65–72.

du Boulay, G. H., and P. M. Verity. 1973. The Cranial Arteries of Mammals. William Heinemann Medical Books Limited, London.

Eastman, J. T., and R. E. Coalson. 1974. The digestive system of the Weddell seal, *Leptonychotes weddelli*—A review. Pages 253–320 *in* R. J. Harrison (ed.). Functional Anatomy of Marine Mammals, Vol. 2. Academic Press, London.

Elsner, R. 1969. Cardiovascular adjustments to diving. Pages 117–145 *in* H. T. Andersen (ed.). The Biology of Marine Mammals. Academic Press, New York, NY.

Elsner, R., J. Pirie, D. D. Kenney, and S. Schemmer. 1974. Functional circulatory anatomy of cetacean appendages. Pages 143–159 *in* R. J. Harrison (ed.). Functional Anatomy of Marine Mammals, Vol. 2. Academic Press, London.

Engel, E. T. 1927. Notes on the sexual cycle of the Pacific Cetacea of the genera Megaptera and Balaenoptera. Journal of Mammalogy 8:48–54.

Engel, S. 1962. The air-passages of the dugong lung. Acta Anatomica 48:95–107.

Engel, S. 1966. The respiratory tissue of the blue whale and the fin whale. Acta Anatomica 65:381–390.

English, A. W. 1976a. Functional anatomy of the hands of fur seals and sea lions. American Journal of Anatomy 147:1–18.

English, A. W. 1976b. Limb movements and locomotor function in the California sea lion (*Zalophus californinus*). Journal of Zoology (London) 178:341–364.

English, A. W. 1977. Structural correlates of forelimb function in fur seals and seal lions. Journal of Morphology 151:325–352.

Estes, J. A. 1980. *Enhydra lutris*. Mammalian Species 133:1–8.

Estes, J. A. 1989. Adaptions for aquatic living by carnivores. Pages 242–282 *in* J. L. Gittleman (ed.). Carnivore Behavior, Ecology, and Evolution. Comstock Publishing Associates, Ithaca, NY.

Evans, H. E. 1993. Miller's Anatomy of the Dog, 3rd ed. W.B. Saunders, Philadelphia, PA.

Evans, H. E., and A. de Lahunta. 1980. Miller's Guide to the Dissection of the Dog, 2d ed. W.B. Saunders, Philadelphia, PA.

Fanning, J. C., and J. Harrison. 1974. The structure of the trachea and lungs of the South Australian bottle-nosed dolphin. Pages 231–252 *in* R. J. Harrison (ed.). Functional Anatomy of Marine Mammals, Vol. 2. Academic Press, London.

Fawcett, D. W. 1942. A comparative study of blood-vascular bundles in the Florida manatee (*Trichechus latirostris*) and in certain cetaceans and edentates. Journal of Morphology 71:105–134.

Fay, F. H. 1960. Structure and function of the pharyngeal pouches of the walrus (*Odobenus rosmarus* L.). Mammalia 23:361–371.

Fay, F. H. 1982. Ecology and Biology of the Pacific Walrus, *Odobenus rosmarus divergens* Illiger. North American Fauna; no. 74. United States Department of the Interior, Washington, D.C.

Feldkamp, S. D. 1987a. Foreflipper propulsion in the California sea lion, *Zalophus californianus*. Journal of Zoology (London) 212:43–57.

Feldkamp, S. D. 1987b. Swimming in the California sea lion: Morphometrics, drag, and energetics. Journal of Experimental Biology 131:117–135.

Fish, F. E. 1993a. Power output and propulsive efficiency of swimming bottlenose dolphins (*Tursiops truncatus*). Journal of Experimental Biology 185:179–193.

Fish, F. E. 1993b. Influence of hydrodynamic design and propulsive mode on mammalian swimming energetics. Australian Journal of Zoology 42:79–101.

Fish, F. E., and C. A. Hui. 1991. Dolphin swimming—A review. Mammal Review 21:181–195.

Fish, F. E., and B. R. Stein. 1991. Functional correlates of differences in bone density among terrestrial and aquatic genera in the family Mustelidae (Mammalia). Zoomorphology 110:339–345.

Fish, F. E., S. Innes, and K. Ronald. 1988. Kinematics and estimated thrust production of swimming harp and ringed seals. Journal of Experimental Biology 137:157–173.

Fish, P. 1896. A note on the cerebral fissuration of the seal (*Phoca vitulina*). Journal of Comparative Neurology VI:15–19.

Fish, P. 1898. The brain of the fur seal, *Callorhinus ursinus:* With a comparative description of those of *Zalophus californianus, Phoca vitulina, Ursus americanus* and *Monachus tropicalus*. Journal of Comparative Neurology VII:57–91.

Fish, P. 1899. The brain of the fur seal, *Callorhinus ursinus:* With a comparative description of those of *Zalophus californianus, Phoca vitulina, Ursus americanus* and *Monachus tropicalis. in* D. S. Jordan (ed.). The Fur Seals and Fur-Seal Islands of the North Pacific Ocean, Part 3. U.S. Government Printing Office, Washington, D.C.

Flower, W. H. 1885. An introduction to the osteology of the Mammalia, 3rd ed. Macmillan and Co., London (reprinted by A. Asher & Co., Amsterdam, 1966).

Flyger, V., and M. R. Townsend. 1968. The migration of polar bears. Scientific American 218:108–116.

Freeman, S. 1990. The evolution of the scrotum: A new hypothesis. Journal of Theoretical Biology 145:429–445.

Freeman, W. H., and B. Bracegirdle. 1968. An Atlas of Histology, 2d ed. Heinemann Educational Books Ltd., London.

Frisch, J., N. A. Oritsland, and J. Krog. 1974. Insulation of furs in water. Comparative Biochemistry and Physiology A 47:403–410.

Frost, K. J., and L. F. Lowry. 1981. Ringed, Baikal and Caspian seals—*Phoca hispida, Phoca sibirica* and *Phoca caspica*. Pages 29–54 *in* S. H. Ridgway and R. J. Harrison (eds.). Handbook of Marine Mammals. Academic Press, New York, NY.

Galliano, R. E., P. J. Morgane, W. L. McFarland, E. L. Nagel, and R. L. Catherman. 1966. The anatomy of the cervicothoracic arterial system in the bottlenose dolphin (*Tursiops truncatus*) with a surgical approach suitable for guided angiography. The Anatomical Record 155:325–338.

Gallivan, G. J., and R. C. Best. 1980. Metabolism and respiration of the Amazonian manatee (*Trichechus inunguis*). Physiological Zoologist 53:245–253.

Gallivan, G. J., and K. Ronald. 1979. Temperature regulation in freely diving harp seal (*Phoca groenlandica*). Canadian Journal of Zoology 57:2256–2263.

Gallivan, G. J., R. C. Best, and J. W. Kanwisher. 1983. Temperature regulation in the Amazonian manatee *Trichechus inunguis*. Physiological Zoology 56:255–262.

Gao, G., and K. Zhou. 1991. The number of fibers and range of fiber diameters in the cochlear nerve of three odontocete species. Canadian Journal of Zoology 69:2360–2364.

Gaskin, D. E. 1978. Form and function in the digestive tract and associated organs in cetacea, with a consideration of metabolic rates and specific energy budgets. Oceanography and Marine Biology Annual Review 16:313–345.

Geraci, J. R., D. J. St. Aubin, and B. D. Hicks. 1986. The epidermis of odontocetes: A view from within. Pages 4–21 *in* M. M. Bryden and R. Harrison (eds.). Research on Dolphins. Oxford University Press, Oxford.

Gervais, M. P. 1870. Memoire sur les Formes Cerebrales; Propres aux

Carnivores Vivants et Fossiles, Tome sixieme. Nouvelles Archives du Museum D'Histoire Naturelle de Paris, Paris.

Gittleman, J. L. 1986. Carnivore brain size, behavioural ecology, and phylogeny. Journal of Mammalogy 67:23–36.

Gittleman, J. L. 1991. Carnivore olfactory bulb size: Allometry, phylogeny and ecology. Journal of Zoology (London) 225:253–272.

Glezer, I. I., M. S. Jacobs, and P. J. Morgane. 1988. Implications of the "initial brain" concept for brain evolution in Cetacea. Behavioral and Brain Sciences 11:75–116.

Godfrey, S. J. 1985. Additional observations of subaqueous locomotion in the California sea lion (*Zalophus californianus*). Aquatic Mammals 11:53–57.

Gordon, K. R. 1983. Mechanics of the limbs of the walrus (*Odobenus rosmarus*) and the California sea lion (*Zalophus californianus*). Journal of Morphology 175:73–90.

Gordon, K. R. 1984. Models of tongue movement in the walrus (*Odobenus rosmarus*). Journal of Morphology 182:179–196.

Gosline, J. M., and R. E. Shadwick. 1996. The mechanical properties of fin whale arteries are explained by novel connective tissue designs. Journal of Experimental Biology 199:985–997.

Grahame, T. 1959. The arterial circulation of the kidneys of the elephant and grey seals. British Veterinary Journal 115:397–399.

Grayson, J. 1990. Responses of the microcirculation to hot and cold environments. Pages 221–234 *in* E. Schonbaum and P. Lomax (eds.). Thermoregulation: Physiology and Biochemistry. Pergamon Press, New York, NY.

Green, R. F. 1972. Observations on the anatomy of some cetaceans and pinnipeds. Pages 247–297 *in* S. H. Ridgway (ed.). Mammals of the Sea, Biology and Medicine. Charles C. Thomas, Springfield, IL.

Green, R. F., S. H. Ridgway, and W. E. Evans. 1980. Functional and descriptive anatomy of the bottlenosed dolphin nasolaryngeal system with special reference to the musculature associated with sound production. Pages 199–238 *in* R.-G. Busnel and J. F. Fish (eds.). Animal Sonar Systems. Plenum, New York, NY.

Gregory, W. K. 1936. On the phylogenetic relationships of the giant panda (*Ailuropoda*) to other arctoid Carnivora. American Museum Novitates 878:1–29.

Guyton, A. C. 1976. Textbook of Medical Physiology. W.B. Saunders, Philadelphia, PA.

Haldiman, J. T., and R. J. Tarpley. 1993. Anatomy and physiology. Pages 71–156 *in* J. J. Burns, J. J. Montague, and C. J. Cowles (eds.). The Bowhead Whale. Special Publication Number 2. Society for Marine Mammalogy, Lawrence, KS.

Hales, J. R. S. 1985. Skin arteriovenous anastomoses, their control and role in thermoregulation. Pages 433–451 *in* K. Johansen and W. W. Burggren (eds.). Cardiovascular Shunts. Alfred Benzon Symposium 21, Munksgaard, Copenhagen.

Hampton, I. F. G., and G. C. Whittow. 1976. Body temperature and heat exchange in the Hawaiian spinner dolphin, *Stenella longirostris*. Journal of Comparative Biochemistry and Physiology A 55:195–197.

Harrison, R. G. 1948. The comparative anatomy of the blood-supply of the mammalian testis. Proceedings of the Zoological Society of London 119:325–344 + plates I–V.

Harrison, R. J. 1969. Reproduction and reproductive organs. Pages 253–348 *in* H. T. Andersen (ed.). The Biology of Marine Mammals. Academic Press, New York, NY.

Harrison, R. J., and J. D. W. Tomlinson. 1956. Observations on the venous system in certain Pinnipedia and Cetacea. Proceedings of the Zoological Society of London 126:205–233.

Harrison, R. J., F. R. Johnson, and B. A. Young. 1970. The oesophagus and stomach of dolphins (*Tursiops, Delphinus, Stenella*). Journal of Zoology (London) 160:377–390.

Harrison, R. J., R. L. Brownell Jr., and R. C. Boice. 1972. Reproduction and gonadal appearances in some odontocetes. Pages 361–429 *in* R. J. Harrison (ed.). Functional Anatomy of Marine Mammals, Vol. 1. Academic Press, London.

Hart, J. S., and L. Irving. 1959. The energies of harbor seals in air and in water, with special consideration of seasonal changes. Canadian Journal of Zoology 37:447–457.

Hartman, D. S. 1979. Ecology and Behavior of the Manatee (*Trichechus manatus*) in Florida. Special Publication No. 5, American Society of Mammalogists.

Hedges, N. A., D. E. Gaskin, and G. J. D. Smith. 1979. Renicular morphology and renal vascular system of the harbour porpoise *Phocoena phocoena* (L.). Canadian Journal of Zoology 57:868–875.

Hedrick, M. S., D. A. Duffield, and L. H. Cornell. 1986. Blood viscosity and optimal hematocrit in a deep-diving mammal, the northern elephant seal (*Mirounga angustirostris*). Canadian Journal of Zoology 64:2081–2085.

Heezen, B. C. 1957. Whales entangled in deep-sea cables. Deep Sea Research 4:105–115.

Helm, R. C. 1983. Intestinal length of three California pinniped species. Journal of Zoology (London) 199:297–304.

Herbert, D. 1987. The topographical anatomy of the sea otter, *Enhydra lutris*. Master's thesis, The Johns Hopkins University, Baltimore, MD.

Heyning, J. E. 1984. Functional morphology involved in intraspecific fighting of the beaked whale, *Mesoplodon carlhubbsi*. Canadian Journal of Zoology 62:1645–1654.

Heyning, J. E. 1989. Comparative facial anatomy of beaked whales (Ziphiidae) and a systematic revision among the families of extinct odontocetes. Contributions in Science, Natural History Museum of Los Angeles County 405:1–64.

Heyning, J. E., and J. G. Mead. 1990. Evolution of the nasal anatomy of cetaceans. Pages 67–79 *in* J. Thomas and R. Kastelein (eds.). Sensory Abilities of Ceatceans. Plenum, New York, NY.

Hicks, B. D., D. J. St. Aubin, J. R. Geraci, and W. R. Brown. 1985. Epidermal growth in the bottlenose dolphin, *Tursiops truncatus*. Journal of Investigative Dermatology 85:60–63.

Hiiemae, K. M., and A. W. Crompton. 1985. Mastication, food transport, and swallowing. Pages 262–290 *in* M. Hildebrand, D. M. Bramble, K. F. Liem, and D. B. Wake (eds.). Functional Vertebrate Morphology. The Belknap Press of Harvard University Press, Cambridge, MA.

Hildebrand, M. 1982. Analysis of Vertebrate Structure, 2nd ed. John Wiley and Sons, New York, NY.

Hill, D. A., and J. E. Reynolds III. 1989. Gross and microscopic anatomy of the kidney of the west Indian manatee, *Trichechus manatus* (Mammalia: Sirenia). Acta Anatomica 135:53–56.

Hofman, M. A. 1983. Energy metabolism, brain size and longevity in mammals. The Quarterly Review of Biology 58:495–512.

Hofman, M. A. 1985. Size and shape of the cerebral cortex in mammals. 1. The cortical surface. Brain and Behavioral Evolution 27:28–40.

Howard, L. D. 1973. Muscular anatomy of the fore-limb of the sea otter (*Enhydra lutris*). Proceedings of the California Academy of Sciences 39:411–500.

Howard, L. D. 1975. Muscular anatomy of the hind limb of the sea otter (*Enhydra lutris*). Proceedings of the California Academy of Sciences 40:335–416.

Howell, A. B. 1927. Contribution to the anatomy of the Chinese finless porpoise *Neomeris phocaenoides*. Proceedings of the United States National Museum 70 (article 13):1–43 + 1 plate.

Howell, A. B. 1928. Contribution to the comparative anatomy of the eared and earless seals (Genera *Zalophus* and *Phoca*). Proceedings of the United States National Museum 73 (article 15):1–142 + 1 plate.

Howell, A. B. 1930. Aquatic Mammals: Their Adaptions to Life in the Water. Charles C. Thomas, Springfield, IL.

Hunt, E. 1995. The role of intelligence in modern society. American Scientist 83:356–368.

Husar, S. L. 1977. *Trichechus inunguis*. Mammalian Species 72:1–4.

Husar, S. L. 1978. *Dugong dugon*. Mammalian Species 88:1–7.

Innes, S., G. A. J. Worthy, D. M. Lavigne, and K. Ronald. 1990. Surface areas of phocid seals. Canadian Journal of Zoology 68:2531–2538.

Irving, L. 1969. Temperature regulation in marine mammals. Pages 147–174 *in* H. T. Andersen (ed.). The Biology of Marine Mammals. Academic Press, New York, NY.

Irving, L., and J. Krog. 1955. Temperature of skin in the Arctic as a regulator of heat. Journal of Applied Physiology 7:355–364.

Irving, L., K. C. Fisher, and F. C. McIntosh. 1935. The water balance of a marine mammal, the seal. Journal of Cellular and Comparative Physiology 6:387–391.

Jacobi, A. 1938. Der Seeotter. Monographien der Wildersaugetiere VI:1–93.

Jacobs, M. S., and A. V. Jensen. 1964. Gross aspects of the brain and a fiber analysis of cranial nerves in the great whale. Journal of Comparative Neurology 123:55–72.

Japha, A. 1912. The hair of whales. Technical translation Number 1537, National Research Board of Canada, Ottawa (1972). Dier Haare der Waltiere. Zool. Jahrb. Abt. Anat. Ontog. Tiere 32:1–42.

Jelgersma, G. 1934. Das Gehrin der Wassersaugetiere. J. A. Barth, Leipzig.

Jerison, H. J. 1973. Evolution of the Brain and Intelligence. Academic Press, New York, NY.

Kellogg, R. 1928. The history of whales—their adaptation to life in the water. Quarterly Review of Biology 3:29–76.

Kenagy, G. J., and S. C. Trombulak. 1986. Size and function of mammalian testes in relation to body size. Journal of Mammalogy 67:1–22.

Kenyon, K. W. 1969. The Sea Otter in the Eastern North Pacific Ocean. North American Fauna series No. 68. U.S. Government Printing Office, Washington, D.C.

Kenyon, K. W. 1981. Sea otter: *Enhydra lutris* (Linnaeus, 1758). Pages 209–223 *in* S. H. Ridgway and R. J. Harrison (eds.). Handbook of Marine Mammals, Volume 1. Academic Press, New York, NY.

Kier, W. M., and K. K. Smith. 1985. Tongues, tentacles and trunks: The biomechanics of movement in muscular-hydrostats. Zoological Journal of the Linnean Society 83:307–324.

King, J. E. 1965. Swallowing modifications in the Ross seal. Journal of Anatomy (London) 99:206–207.

King, J. E. 1971. The lacrimal bone in the Otariidae. Mammalia 35:465–470.

King, J. E. 1972. On the laryngeal skeletons of the leopard seal, *Hydrurga leptonyx,* and the Ross seal, *Ommatophoca rossi*. Mammalia 36:146–156.

King, J. E. 1977. Comparative anatomy of the major blood vessels of the sea lions *Neophoca* and *Phocarctos;* with comments on the differences between the otariid and phocid vascular systems. Journal of Zoology (London) 181:69–94.

King, J. E. 1983. Seals of the World, 2d ed. Comstock Publishing Associates, Ithaca, NY.

Koopman, H. N., S. J. Iverson, and D. E. Gaskin. 1996. Stratification and age-related differences in blubber fatty acids of the male harbour porpoise (*Phocoena phocoena*). Journal of Comparative Physiology B 165:628–639.

Kooyman, G. L. 1969. The Weddell seal. Scientific American 221:101–106.

Kooyman, G. L. 1973. Respiratory adaptions in marine mammals. American Zoologist 13:457–468.

Kooyman, G. L. 1981. Weddell Seal, Consummate Diver. Cambridge University Press, Cambridge, U.K.

Kooyman, G. L. 1985. Physiology without constraint in diving mammals. Marine Mammal Science 1:166–178.

Kooyman, G. L. 1987. Free diving in vertebrates. Pages 27–37 *in* Dejours (ed.). Comparative Physiology of Environmental Adaptations, Volume 2. 8th ESCP Conference, Strasbourg, Karger, Basel.

Kooyman, G. L. 1989. Diverse Divers: Physiology and Behavior. Springer-Verlag, Berlin.

Kooyman, G. L., and L. H. Cornell. 1981. Flow properties of expiration and inspiration in a trained bottle-nosed porpoise. Physiological Zoology 54:55–61.

Kooyman, G. L., and C. M. Drabek. 1968. Observations on milk, blood, and urine constituants of the Weddell seal. Physiological Zoology 41:187–194.

Kooyman, G. L., and E. E. Sinnett. 1979. Mechanical properties of the harbor porpoise lung, *Phocoena phocoena*. Respiration Physiology 36:287–300.

Kooyman, G. L., D. D. Hammond, and J. P. Schroeder. 1970. Bronchograms and trachograms of seals under pressure. Science 169:82–84.

Krogh, A. 1939. Osmotic regulation in aquatic animals. Cambridge University Press, Cambridge, U.K.

Kukenthall, W. 1893. Comparative anatomical and developmental historical investigations on whales (in German). Gustav Fischer, Jena.

Lambertsen, R. H. 1983. Internal mechanism of rorqual feeding. Journal of Mammalogy 64:76–88.

Lambertsen, R. H., R. J. Hintz, W. C. Lancaster, A. Hirons, K. J. Krediton, and C. Moor. 1989. Characterization of the Functional Morphology of the Mouth of the Bowhead Whale, *Balaena mysticetus,* with Special Emphasis on Feeding and Filtration Mechanisms. Report to North Slope Borough, Contract 87-113.

Langer, P. 1996. Comparative anatomy of the stomach of the Cetacea. Ontogenetic changes involving gastric proportions—mesenteries—arteries. Zeitschrift für Saugetierkunde 61:140–154.

Langworthy, O. R. 1932. A description of the central nervous system of the porpoise (*Tusiops truncatus*). Journal of Comparative Physiology and Psychology 41:111–123.

Lauder, G. V. 1982. Historical biology and the problem of design. Journal of Theoretical Biology 97:57–67.

Lavigne, D. M., S. Innes, G. A. J. Worthy, K. M. Kovacs, O. J. Schmitz, and J. P. Hickie. 1986. Metabolic rates of seals and whales. Canadian Journal of Zoology 64:279–284.

Lawrence, B., and W. E. Schevill. 1956. The functional anatomy of the delphinid nose. Bulletin of the Museum of Comparative Zoology 14:103–151 + 30 figures.

Lawrence, B., and W. E. Schevill. 1965. Gular musculature in delphinids. Bulletin of the Museum of Comparative Zoology 133:5–63.

Leatherwood, S., and R. R. Reeves. 1983. The Sierra Club Handbook of Whales and Dolphins. Sierra Club Books, San Francisco, CA.

Le Boeuf, B. J., D. P. Costa, A. C. Huntley, and S. D. Feldkamp. 1988.

Continuous, deep diving in female northern elephant seals, *Mirounga angustirostris*. Canadian Journal of Zoology 66:446–458.

Le Boeuf, B. J., Y. Naito, T. C. Huntley, and T. Asaga. 1989. Prolonged, continuous, deep diving by northern elephant seals. Canadian Journal of Zoology 67:2514–2519.

Lichtenstein, M. H. K. 1827–1834. Darstellung neuer oder wenig bekannter Saugethiere u. s. w. 2. Abgedruckt in: A. Ermin.

Lighthill, M. J. 1969. Hydromechanics of aquatic animal propulsion. Pages 413–446 *in* W. R. Sears and M. Van Dyke (eds.). Annual Review of Fluid Mechanics, Vol. 1. Annual Reviews Inc., Palo Alto, CA.

Lindsey, C. C. 1978. Form, function, and locomotory habits in fish. Pages 1–100 *in* W. S. Hoar and D. J. Randall (eds.). Fish Physiology, Vol. VII. Academic Press, New York, NY.

Ling, J. K. 1974. The integument of marine mammals. Pages 1–44 *in* R. J. Harrison (ed.). Functional Anatomy of Marine Mammals, Vol. 2. Academic Press, London.

Ling, J. K. 1977. Vibrissae of marine mammals. Pages 387–415 *in* R. J. Harrison (ed.). Functional Anatomy of Marine Mammals, Vol. 3. Academic Press, London.

Lockyer, C. H., L. C. McConnell, and T. D. Waters. 1984. The biochemical composition of fin whale blubber. Canadian Journal of Zoology 62:2553–2562.

Lockyer, C. H., L. C. McConnell, and T. D. Waters. 1985. Body condition in terms of anatomical and biochemical assessment of body fat in north Atlantic fin and sei whales. Canadian Journal of Zoology 63:2328–2338.

Lowry, L. F. 1993. Foods and feeding ecology. Pages 201–234 *in* J. J. Burns, J. J. Montague, and C. J. Cowles (eds.). The Bowhead Whale. Special Publication Number 2. Society for Marine Mammalogy, Lawrence, KS.

Mackintosh, N. A., and J. F. G. Wheeler. 1929. Southern blue and fin whales. Discovery Reports 1:257–540.

Maluf, N. S. R. 1989. Renal anatomy of the manatee, *Trichechus manatus*, Linnaeus. American Journal of Anatomy 184:269–286.

Marsh, H., G. E. Heinsohn, and A. V. Spain. 1977. The stomach and duodenal diverticula of the dugong (*Dugong dugon*). Pages 271–295 *in* R. J. Harrison (ed.). Functional Anatomy of Marine Mammals, Vol. 3. Academic Press, London.

Marsh, H., Heinsohn, G. E., and T. D. Glover. 1984a. Changes in the male reproductive organs of the dugong, *Dugong dugon* (Sirenia: Dugongidae) with age and reproductive activity. Australian Journal of Zoology 32:721–742.

Marsh, H., Heinsohn, G. E., and P. W. Channells. 1984b. Changes in the ovaries and uterus of the Dugong, *Dugong dugon* (Sirenia: Dugongidae), with age and reproductive activity. Australian Journal of Zoology 32:743–766.

Marshall, C. D., and R. L. Reep. 1995. A comparison of sirenian feeding behavior and morphology. American Zoologist 35:59A.

McFarland, W. L., M. S. Jacobs, and P. J. Morgane. 1979. Blood supply to the brain of the dolphin, *Tursiops truncatus*, with comparative observations on special aspects of the cerebrovascular supply of other vertebrates. Neuroscience and Biobehavioral Review 3(Suppl. 1):1–93.

McGinnis, S. M. 1975. Peripheral heat exchange in phocids. *in* K. Ronald and A. W. Mansfield (eds.). Biology of the Seal. International Council for the Exploration of the Sea. Rapports et proces-verbeaux des reunions 169:481–486.

McGinnis, S. M., G. C. Whittow, C. A. Ohata, and H. Huber. 1972. Body heat dissipation and conservation in two species of dolphins. Journal of Comparative Biochemistry and Physiology A 43:417–423.

McMahon, T. A., and J. T. Bonner. 1983. On Size and Life. Scientific American, New York, NY.

McNab, B. K., and J. F. Eisenberg. 1989. Brain size and its relation to the rate of metabolism in mammals. American Naturalist 133:157–167.

Mead, J. G. 1975. Anatomy of the external nasal passages and facial complex in the Delphinidae (Mammalia: Cetacea). Smithsonian Contributions to Zoology 207:1–72.

Mead, J. G. 1993. The systematic importance of stomach anatomy in beaked whales. IBI Reports 4:75–86.

Mead, J. G., and C. W. Potter. 1990. Natural history of bottlenose dolphins along the central Atlantic coast of the United States. Pages 165–195 in S. Leatherwood and R. R. Reeves (eds.). The Bottlenose Dolphin. Academic Press, San Diego, CA.

Mendez, J., and A. Keyes. 1960. Density and composition of mammalian muscle. Metabolism 9:184–188.

Miller, G. S. Jr. 1923. The telescoping of the cetacean skull. Smithsonian Miscellaneous Collection 76:1–71.

Molyneux, G. S., and M. M. Bryden. 1978. Arteriovenous anastomoses in the skin of seals. I. The Weddell seal *Leptonychotes weddelli* and the elephant seal *Mirounga leonia* (Pinnipedia: Phocidea). The Anatomical Record 191:239–252.

Montagna, W., and R. J. Harrison. 1957. Specializations in the skin of the seal (*Phoca vitulina*). American Journal of Anatomy 100:81–113.

Moore, C. R. 1926. The biology of the mammalian testis and scrotum. Quarterly Review of Biology I:4–50.

Moore, J. C. 1968. Relationships among the living genera of beaked whales with classifications, diagnoses and keys. Fieldana: Zoology 53:209–298.

Morishima, H. O., B. Glaser, W. H. Niemann, and L. S. James. 1975. Increased uterine activity and fetal deterioration during maternal hyperthermia. American Journal of Obstetrics and Gynecolology 121:531–538.

Morrison, P., M. Rosenmann, and J. A. Estes. 1974. Metabolism and thermoregulation in the sea otter. Physiological Zoology 47:218–229.

Murie, J. 1872. On the form and structure of the manatee (*Manatus americanus*). Transactions of the Zoological Society of London 8:127–202.

Murie, J. 1874. Researches upon the anatomy of the Pinnipedia. Part 3. Descriptive anatomy of the sealion (*Otaria jubata*). Transactions of the Zoological Society of London 8:501–582.

Murie, J. 1880. Further observations on the manatee. Transactions of the Zoological Society of London 11:19–48.

Nachtigall, P. E., and P. W. B. Moore (eds). 1988. Animal sonar: Processes and performances. Plenum, New York, NY.

Nagel, E. L., P. J. Morgane, W. L. McFarland, and R. E. Galliano. 1968. Rete mirabile of dolphin: Its pressure-damping effect on cerebral circulation. Science 161:898–900.

Negus, V. E. 1949. The Comparative Anatomy and Physiology of the Larynx. W. S. Cowell Ltd., Ipswich, U.K.

Nerini, M. 1984. A review of gray whale feeding ecology. Pages 423–450 *in* M. L. Jones, S. L. Swartz, and S. Leatherwood (eds.). The Gray Whale, *Eschrichtius robustus*. Academic Press, New York, NY.

Nickel, R., A. Schummer, E. Seiferle, J. Frewein, H. Wilkens, and K. -H. Wille. 1986. The locomotor system of the domestic mammals. *in* R. Nickel, A. Schummer, and E. Seiferle (eds.). The Anatomy of the Domestic Animals, Vol. 1. Verlag Paul Parey, Berlin.

Nishiwaki, M., and T. Kasuya. 1970. A Greenland right whale caught at Osaka Bay. Scientific Report of the Whales Research Institute 22:45–62.

Nishiwaki, M., and H. Marsh. 1985. Dugong: *Dugong dugon* (Miller, 1776). Pages 1–31 *in* S. H. Ridgway and R. Harrison (eds.). Handbook of Marine Mammals, Vol. 3: The Sirenians and Baleen Whales. Academic Press, Orlando, FL.

Norris, K. S., and G. W. Harvey. 1974. Sound transmission in the porpoise head. Journal of the Acoustical Society of America 56:659–664.

Norris, K. S., K. J. Dormer, J. Pegg, and G. T. Liese. 1971. The mechanism of sound production and air recycling in porpoises: A preliminary report. Pages 113–129 *in* Proceedings of the Eighth Annual Conference on Biological Sonar and Diving Mammals, Menlo Park, CA.

Northwest and Alaska Fisheries Center. 1969. Fur Seal Investigations. Unpublished Report, 90 p. Northwest and Alaska Fisheries Center, Marine Mammal Division, National Marine Fisheries Service, NOAA, 7600 Sandpoint Way NE, Seattle, WA 98115.

Nowak, R. M. 1991. Walker's Mammals of the World, 5th ed., Vol. II. The Johns Hopkins University Press, Baltimore, MD.

Ogawa, T., and T. Shida. 1950. On the sensory tubercles of lips and of oral cavity in the sei and fin whale. Scientific Reports of the Whales Research Institute 3:1–18.

Oliver, J. 1968. Nephrons and kidneys. A quantitative study of development and evolutionary mammalian renal architectonics. Hoeber Medical Division, Harper & Row, New York, NY.

Olsen, M. A., E. S. Nordoy, A. S. Blix, and S. D. Mathiesen. 1994. Functional anatomy of the gastrointestinal system of Northeastern Atlantic minke whales (*Balaenoptera acutorostrata*). Journal of Zoology (London) 234:55 74.

Olsen, M. A., K.T. Nilssen, and S. D. Mathiesen. 1996. Gross anatomy of the gastrointestinal system of harp seals (*Phoca groenlandica*). Journal of Zoology (London) 238:581–589.

Ommanney, F. D. 1932a. The vascular networks (retia mirablia) of the fin whale (*Balaenoptera physalus*). Discovery Reports 5:327–362.

Ommanney, F.D. 1932b. The urogenital system of the fin whale (*Balaenoptera physalus*) with appendix: The dimensions and growth of the kineys of blue and fin whales. Discovery Reports 5:363–465.

Omura, H. 1958. North Pacific right whale. Scientific Reports of the Whales Research Institute 13:12–52.

Omura, H., S. Ohsumi, T. Nemoto, K. Nasu, and T. Kasuya. 1969. Black right whales in the north Pacific. Scientific Reports of the Whales Research Institute 21:1–78.

Oritsland, N. A. 1969. Deep body temperatures of swimming and walking polar bear cubs. Journal of Mammalogy 50:380–382.

Oritsland, N. A. 1970. Temperature regulation of the polar bear (*Thalarctos maritimus*). Comparative Biochemistry and Physiology 37:225–233.

Orton, L. S., and P. F. Brodie. 1987. Engulfing mechanisms of fin whales. Canadian Journal of Zoology 65:2898–2907.

O'Shea, T. J., and R. L. Reep. 1990. Encephalization quotients and life-history traits in the Sirenia. Journal of Mammalogy 71:534–543.

Pabst, D. A. 1990. Axial muscles and connective tissues of the bottlenose dolphin. Pages 51–67 *in* S. Leatherwood and R. R. Reeves (eds.). The Bottlenose Dolphin. Academic Press, San Diego, CA.

Pabst, D. A. 1993. Intramuscular morphology and tendon geometry of the epaxial swimming muscles of dolphins. Journal of Zoology (London) 230:159–176.

Pabst, D. A. 1996. Morphology of the subdermal connective tissue sheath of dolphins: A new fibre-wound, thin-walled, pressurized cylinder model for swimming vertebrates. Journal of Zoology (London) 238:35–52.

Pabst, D. A., S. A. Rommel, W. A. McLellan, T. M. Williams, and T. K. Rowles. 1995. Thermoregulation of the intra-abdominal testes of the bottlenose dolphin (*Tursiops truncatus*) during exercise. Journal of Experimental Biology 198:221–226.

Pabst, D. A., S. A. Rommel, and W. A. McLellan. 1998. Evolution of thermoregulatory function in cetacean reproductive systems. Pages 379–397 *in* J. G. M. Thewissen (ed.). Emergence of Whales. Plenum, New York, NY.

Pagel, M. D., and P. H. Harvey. 1989. Taxonomic differences in the scaling of brain on body weight among mammals. Science 244:1589–1593.

Pagel, M. D., and P. H. Harvey. 1990. Diversity in the brain sizes of newborn mammals—allometry, energetics, or life history tactics? BioScience 40:116–122.

Palmer, E., and G. Weddell. 1964. The relationship between structure, innervation and function of the skin of the bottle nose dolphin (*Tursiops truncatus*). Proceedings of the Zoological Society of London 143:553–568.

Parry, D. A. 1949. The structure of whale blubber, and a description of its thermal properties. Quarterly Journal of Miscroscopical Science 90:13–25.

Perrin, W. F. 1975. Variation of Spotted and Spinner Porpoise (genus *Stenella*) in the Eastern Tropical Pacific and Hawaii. University of California Press, Berkeley.

Pfeiffer, C. J., and T. P. Kinkead. 1990. Microanatomy of retia mirabilia of bowhead whale foramen magnum and mandibular foramen. Acta Anatomica 139:141–150.

Pilleri, G. 1964. Morphologie des gehirnes des "Southern Right Whale," *Eubalaena australis* Desmoulins 1822 (Cetacea, Mysticeti, Balaenidae). Acta Zoologica XLVI:245–272.

Pivorunas, A. 1977. The fibrocartilage skeleton and related structures in the pouch of balaenopterid whales. Journal of Morphology 151:299–314.

Pivorunas, A. 1979. The feeding mechanisms of baleen whales. American Scientist 67:432–440.

Pond, C. M., C. A. Mattacks, and R. H. Colby. 1992. The anatomy, chemical composition, and metabolism of adipose tissue in wild polar bears (*Ursus maritimus*). Canadian Journal of Zoology 70:326–341.

Radinsky, L. B. 1968. Evolution of somatic specialization in otter brains. Journal of Comparative Neurology 134:495–506.

Ramsey, M. A., and I. Stirling. 1988. Reproductive biology and ecology of female polar bears (*Ursus maritimus*). Journal of Zoology (London) 214:601–634.

Read, A. J. 1990. Reproductive seasonality in harbour porpoises, *Phocoena phocoena,* from the Bay of Fundy. Canadian Journal of Zoology 68:284–288.

Read, A. J., and D. E. Gaskin. 1990. Changes in growth and reproduction of harbour porpoises, *Phocoena phocoena,* from the Bay of Fundy. Canadian Journal of Fisheries and Aquatic Sciences 47:2158–2163.

Reep, R. L., and T. J. O'Shea. 1990. Regional brain morphometry and lissencephaly in the Sirenia. Brain Behavior and Evolution 35:185–194.

Reep, R. L., J. I. Johnson, R. C. Switzer, and W. I. Welker. 1989. Manatee cerebral cortex: Cytoarchitecture of the frontal region in *Trichechus manatus latirostris*. Brain Behavior and Evolution 34:365–386.

Reeves, R. R., B. S. Stewart, and S. Leatherwood. 1992. Seals and Sirenians. Sierra Club Books, San Francisco, CA.

Reidenberg, J. S., and J. T. Laitman. 1987. Postion of the larynx in Odontoceti (toothed whales). The Anatomical Record 218:98–106.

Renouf, D. 1979. Preliminary measurements of the sensitivity of the vibrissae of harbour seals (*Phoca vitulina*) to low frequency vibrations. Journal of Zoology (London) 188:443–450.

Resnick, R., and D. Halliday. 1966. Physics, Part I. John Wiley & Sons, Inc. New York, NY.

Reynolds, J. E. III. 1980. Aspects of the structural and functional anatomy of the gastrointestinal tract of the West Indian manatee, *Trichechus manatus*. Ph.D. dissertation, University of Miami, Coral Gables, FL, 111 pages.

Reynolds, J. E. III, and D. K. Odell. 1991. Manatees and Dugongs. Facts on File, Inc., New York, NY.

Reynolds, J. E., and S. A. Rommel. 1996. Structure and function of the gastrointestinal tract of the Florida manatee, *Trichechus manatus*. The Anatomical Record 245:539–558.

Reynolds, J. E., and S. A. Rommel. 1997. Structure and function of the Florida manatee (*Trichechus manatus latirostris*) diaphragm. Poster presented at the World Marine Mammal Science Conference, Monte Carlo, Monaco, December 1997.

Ridgway, S. H. 1986. Physiological observations on the dolphin brain. Pages 31–59 *in* R. Schusterman, J. Thomas, and F. G. Wood (eds.). Dolphin Cognition and Behaviour: A Comparative Approach. Erlbaum, Hillsdale, NJ.

Ridgway, S. H. 1990. The central nervous system of the bottlenose dolphin. Pages 69–97 *in* S. Leatherwood and R. R. Reeves (eds.). The Bottlenose Dolphin. Academic Press, San Diego, CA.

Ridgway, S. H., and R. H. Brownson. 1984. Relative brain sizes and cortical surface areas in odontocetes. Acta Zoologica Fennica 172:149–152.

Ridgway, S. H., and D. A. Carder. 1988. Nasal pressure and sound production in an echolocating white whale, *Delphinapterus leucas*. Pages 53–60 *in* P. E. Nachtigall and P. W. B. Moore (eds.). Animal Sonar: Processes and Performances. Plenum Press, New York, NY.

Ridgway, S. H., and R. Howard. 1979. Dolphin lung collapse and intramuscular circulation during free diving: Evidence from nitrogen washout. Science 206:1182–1183.

Ridgway, S. H., and D. G. Johnston. 1966. Blood oxygen and ecology of porpoises. Science 151:456–457.

Ridgway, S. H., B. L. Scronce, and J. Kanwisher. 1969. Respiration and deep diving in the bottlenose porpoise. Science 166:1651–1654.

Ridgway, S. H., D. A. Carder, R.F. Green, A. S. Gaunt, S. L. L Gaunt, and W. E. Evans. 1980. Electromyographic and pressure events in the nas / olaryngeal system of dolphins during sound production. Pages 239–249 *in* R.-G. Busnel and J. F. Fish (eds.). Animal Sonar Systems. Plenum, New York, NY.

Robin, E. D. 1973. The evolutionary advantages of being stupid. Perspectives in Biological Medicine 16:369–380.

Romano, T. A, S. Y. Felten, J. A. Olschowka, and D. L. Felten. 1993. A microscopic investigation of the lymphoid organs of the beluga, *Delphinapterus leucas*. Journal of Morphology 215:261–287.

Romer, A. S. 1966. Vertebrate Paleontology, 3d ed. University of Chicago Press, Chicago, IL.

Rommel, S. A. 1990. Oseteology of the bottlenose dolphin. Pages 29–49 *in* S. Leatherwood and R. R. Reeves (eds.). The Bottlenose Dolphin. Academic Press, San Diego, CA.

Rommel, S. A., D. A. Pabst, W. A. McLellan, J. G. Mead, and C. W. Potter. 1992. Anatomical evidence for a countercurrent heat exchanger associated with dolphin testes. The Anatomical Record 232:150–156.

Rommel, S. A., D. A. Pabst, and W. A. McLellan. 1993. Functional morphology of the vascular plexuses associated with the cetacean uterus. The Anatomical Record 237:538–546.

Rommel, S. A., D. A. Pabst, W. A. McLellan, T. M. Williams, and W. A. Friedl. 1994. Temperature regulation of the testes of the bottlenose dolphin (*Tursiops truncatus*): evidence from colonic temperatures. Journal of Comparative Physiology B 164:130–134.

Rommel, S. A., G. A. Early, K. A. Matassa, D. A. Pabst, and W. A. McLellan. 1995. Venous structures associated with thermoregulation of phocid seal reproductive organs. The Anatomical Record 243:390–402.

Rommel, S. A., D. A. Pabst, and W. A. McLellan. 1998. Reproductive thermoregulation in marine mammals. American Scientist 86: 440–448.

Ryg, M., T. G. Smith, and N. A. Oritsland. 1988. Thermal significance of the topographical distribution of blubber in ringed seals (*Phoca hispida*). Canadian Journal of Fisheries and Aquatic Sciences 45:985–992.

Ryg, M., C. Lydersen, L. O. Knotsen, A. Bjorge, T. G. Smith, and N. A. Oritsland. 1993. Scaling of insulation in seals and whales. Journal of Zoology (London) 230:193–206.

Sacks, R. D., and R. R. Roy. 1982. Architecture of the hind limb of cats: Functional significance. Journal of Morphology 173:317–325.

Sandegren, F. E., E. W. Chu, and J. E. Vandevere. 1973. Maternal behavior in the California sea otter. Journal of Mammalogy 54:668–679.

Scammon, C. M. 1968. The marine mammals of the northwestern coast of North America. Dover Publications, New York, NY.

Schaller, O. (ed). 1992. Illustrated Veterinary Anatomical Nomenclature. Ferdinand Enke Verlag, Stuttgart.

Scheffer, V. B. 1962. Pelage and Surface Topography of the Northern Fur Seal. United States Department of the Interior North American Fauna. Number 64.

Scheffer, V. B. 1964a. Hair patterns in seals (Pinnipedia). Journal of Morphology 115:291–304.

Scheffer, V. B. 1964b. Estimating abundance of pelage fibres on fur seal skin. Proceedings of the Zoological Society of London 143:37–41.

Scheffer, V. B. 1968. The Sea Otter in the Eastern Pacific Ocean. United States Department of the Interior North American Fauna. Number 64.

Scheffer, V. B., and A. M. Johnson. 1963. Molt in the Northern Fur Seal. United States Fish and Wildlife Service Special Scientific Report—Fisheries No. 450.

Schevill, W. E., and W. A. Watkins. 1965. Underwater calls of *Trichechus* (Manatee). Nature 205:373–374.

Schevill, W. E., W. A. Watkins, and C. Ray. 1966. Analysis of underwater Odobenus calls with remarks on the development and function of the pharyngeal pouches. Zoologica, Scientific Contributions of the New York Zoological Society 51:103–106.

Schmidt-Nielsen, B., and R. O'Dell. 1961. Structure and concentrating mechanism in the mammalian kidney. American Journal of Phisiology 200:1119–1124.

Schmidt-Nielsen, K. 1983. Animal Physiology: Adaption and Environment, 3d ed. Cambridge University Press, Cambridge, U.K.

Schmidt-Nielsen, K. 1990. Animal Physiology: Adaption and Environment, 4th ed. Cambridge University Press, Cambridge, U.K.

Scholander, P. F. 1940. Experimental investigations on the respiratory function in diving mammals and birds. Hvalradets Skrifter 22:1–131.

Scholander, P. F., and L. Irving. 1941. Experimental investigations on the respiration and diving of the Florida manatee. Journal of Cellular and Comparative Physiology 17:169–191.

Scholander, P. F., and W. E. Schevill. 1955. Counter-current vascular heat exchange in the fins of whales. Journal of Applied Physiology 8:279–282.

Scholander, P. F., V. Walters, R. Hock, and L. Irving. 1950. Body insulation of some arctic and tropical mammals and birds. Biological Bulletin 99:259–269.

Schummer, A., R. Nickel, and W. O. Sack. 1979. The viscera of the domestic mammals. *in* R. Nickel, A. Schummer, and E. Seiferle (eds.). The Anatomy of the Domestic Animals, Vol. 2, 2nd ed. Verlag Paul Parey, Berlin.

Schummer, A., H. Wilkens, B. Vollmerhaus, and K.-H. Habermehl. 1981. The circulatory system, the skin, and the cutaneus organs of the domestic mammals. *in* R. Nickel, A. Schummer, and E. Seiferle (eds.). The Anatomy of the Domestic Animals, Vol. 3. Verlag Paul Parey, Berlin.

Seki, Y. 1985. Anatomical observations on the lower brain stem of the right whale. Scientific Reports of the Whales Research Institute 36:49–87.

Shadwick, R. E., and J. M. Gosline. 1994. Arterial mechanics in the fin whale suggest a unique hemodynamic design. American Journal of Physiology 267:R805–R818.

Shelton, M. 1964. Relation of environmental temperature during gestation on birth weight and mortality in lambs. Journal of Animal Science 23:360–364.

Sinha, A. A., and H. W. Mossman. 1966. Placentation of the sea otter. American Journal of Anatomy 119:521–554.

Sinha, A. A., C. H. Conoway, and K. W. Kenyon. 1966. Reproduction in the female sea otter. Journal of Wildlife Management 30:121–130.

Sisson, S., and J. D. Grossman. 1953. The Anatomy of the Domestic Animals. W. B. Saunders, Philadelphia, PA.

Slijper, E. J. 1936. Die Cetacean: Vergleichend-Anatomisch und Systematisch. Asher and Company, Amsterdam, 1972 reprint.

Slijper, E. J. 1966. Functional morphology of the reproductive system in cetacea. Pages 277–319 *in* K. S. Norris (ed.). Whales, Dolphins, and Porpoises. University of California Press, Berkeley.

Slijper, E. J. 1979. Whales. Cornell University Press, Ithaca, N.Y.

Smith, J. J., and J. P. Kampine. 1990. Circulatory Physiology. Williams and Wilkins, Baltimore, MD.

Sokolov, V. E., M. M. Kalashikova, and V. A. Rodinov. 1973. Micro- and ultrastructure of the skin in the harbor porpoise (*Phocaena phocaena* relicta Abel.) Pages 82–101 *in* K. K. Chapskii and V.E. Sokolov (eds.). Morphology and Ecology of Marine Mammals. Translated by H. Mills. John Wiley and Sons, New York, NY.

Sokolov, W. 1960. The skin structure in pinnipedia of the U.S.S.R. fauna. Journal of Morphology 107:285–296.

Spearman, R. I. C. 1973. The Integument: A Textbook of Skin Biology. Cambridge University Press, Cambridge, U.K.

Spector, S. A., P. F. Gardiner, R. F. Zernicke, R. R. Roy, and V. G. Edgerton. 1980. Muscle architecture and force-velocity characteristics of the cat soleus and medial gastrocnemius: Implications for motor control. Journal of Neurophysiology 44:951–960.

Stirling, I. 1988. Polar Bears. The University of Michigan Press, Ann Arbor, MI.

St. Pierre, H. 1974. The topographical splanchnology and the superficial vascular system of the harp seal *Pagophilus groenlandicus* (Erxleben, 1777). Pages 161–195 *in* R. J. Harrison (ed.). Functional Anatomy of Marine Mammals, Vol. 2. Academic Press, London.

Strickler, T. L. 1980. The axial musculature of *Pontoporia blainvillei,* with comments on the organization of this system and its effect on fluke-stroke dynamics. American Journal of Anatomy 157:49–59.

Struthers, J. 1881. Bones, articulations, and muscles of the rudimentary hind-limb of the Greenland right-whale. Journal of Anatomy and Physiology XV:1–58 + 3 plates.

Tarasoff, F. J. 1972. Comparative aspects of the hind limbs of the river otter, sea otter and seal. Pages 333–359 *in* R. J. Harrison (ed.). Functional Anatomy of Marine Mammals, Vol. 1. Academic Press, London.

Tarasoff, F. J. 1974. Anatomical adaptations in the river otter, sea otter and harp seal with reference to thermal regulation. Pages 111–141 *in* R. J. Harrison (ed.). Functional Anatomy of Marine Mammals, Vol. 2. Academic Press, London.

Tarasoff, F. J. and H. D. Fisher. 1970. Anatomy of the hind flippers of two species of seals with reference to thermoregulation. Canadian Journal of Zoology 48:821–829.

Tarasoff, F. J., and G. L. Kooyman. 1973a. Observations on the anatomy of the respiratory system of the river otter, sea otter, and harp seal. I. The topography, weight, and measurements of the lungs. Canadian Journal of Zoology 51:163–170.

Tarasoff, F. J., and G. L. Kooyman. 1973b. Observations on the anatomy of the respiratory system of the river otter, sea otter, and harp seal. II. The trachea and bronchial tree. Canadian Journal of Zoology 51:171–177.

Tarasoff, F. J., A. Bisaillon, J. Pierard, and A. P. Whitt. 1972. Locomotory patterns and external morphology of the river otter, sea otter, and harp seal (Mammalia). Canadian Journal of Zoology 50:915–929.

Tarpley, R. J., R. F. Sis, T. F. Albert, L. M. Dalton, and J. C. George. 1987. Observations on the anatomy of the stomach and duodenum of the bowhead whale, *Balaena mysticetus*. American Journal of Anatomy 180:295–322.

Taylor, C. R., and C. P. Lyman. 1972. Heat storage in running antelopes: Independence of brain and body temperatures. American Journal of Physiology 222:114–117.

Taylor, W. P. 1914. The problem of aquatic adaption in the carnivora, as illustrated in the osteology and evolution of the sea-otter. University of California Publications, Bulletin of the Department of Geology 7:465–495.

Tenny, S. M., and J. E. Remmers. 1963. Comparative quantitative morphology of the mammalian lung: Diffusing area. Nature 197:54–56.

Testa, J. W. 1994. Over-winter movements and diving behavior of female Weddell seals (*Leptonychotes weddellii*) in the southwestern Ross Sea, Antarctica. Canadian Journal of Zoology 72:1700–1710.

Thomas, J. A., and R. A. Kastelein (eds). 1990. Sensory Abilities of Cetaceans: Laboratory and Field Evidence. Plenum, New York, NY.

Thorson, P. H., and B. J. Le Boeuf. 1994. Developmental aspects of diving in northern elephant seal pups. Pages 271–289 *in* B. J. Le Boeuf and R. M. Laws (eds.). Elephant Seals: Population Ecology, Behavior, and Physiology. University of California Press, Berkley.

Tomilin, A. G. 1967. Mammals of the U.S.S.R. and Adjacent Countries: Cetacea, Volume 4. Israel Program for Scientific Translations, Jerusalem.

Tregear, R. T. 1966. Physical Functions of Skin. Academic Press, London.

True, F. W. 1904. The Whalebone Whale of the Western North Atlantic. Smithsonian Institution, Washington, D.C.

Turner, W. 1875. On the placentation of the seals. Transactions of the Royal Society of Edinburg 27:275–302.

Twiss, J. R., and R. R. Reeves. 1999. Conservation and Biology of Marine Mammals. Smithsonian Institution Press, Washington, D.C.

van Beneden, P. J., and F. L. P. Gervais. 1868–1879. Osteographie des Cetaces vivants et fossiles, comprenant, la description et l'iconographie, du squelette et du systeme dentaire, de ces animaux. Par MM., Atlas, Plates XXIV and XXXV, Librarire de la Societe de Geographie, Paris.

VanDemark, N. L., and M. J. Free. 1970. Temperature regulation and the testis. Pages 233–312 *in* A. D. Johnson, W. R. Gomes, and N. L. VanDemark (eds.). The Testis, Vol. III. Academic Press, New York, NY.

Van Der Spoel, S. 1963. The vascular system in the kidneys of the common porpoise. Bijdr. Dierkd. 33:71–82.

Van Valen, L. 1968. Monophyly or diphyly in the origin of whales. Evolution 22:37–41.

Van Waerbeek, K., and A. J. Read. 1994. Reproduction of dusky dolphins, *Lagenorhynchus obscurus,* from coastal Peru. Journal of Mammalogy 75:1054–1062.

Vardy, P. H., and M. M. Bryden. 1981. The kidney of *Leptonychotes weddelli* (Pinnipedia: Phocidae) with some observations on the kidneys of two other southern phocid seals. Journal of Morphology 167:13–34.

Vaughan, T. A. 1986. Mammalogy, 3rd ed. Saunders College Publishing, Flagstaff, AZ.

Vogel, S. 1988. Life's Devices: The Physical World of Animals and Plants. Princeton University Press, Princeton, NJ.

Vogel, S. 1992. Vital Circuits: On Pumps, Pipes, and the Workings of Circulatory Systems. Oxford University Press, New York, NY.

Vogel, S. 1994. Life in Moving Fluids: The Physical Biology of Flow, 2d ed. Princeton University Press, Princeton, NJ.

Vogel, S., and S. A. Wainwright. 1969. A Functional Bestiary: Laboratory Studies About Living Systems. Addison-Wesley, Reading, MA.

Vogl, A. W., and H. D. Fisher. 1981. The internal carotid artery does not directly supply the brain in the Monodontidae (order Cetacea). Journal of Morphology 170:207–214.

Vogl, A. W., and H. D. Fisher. 1982. Arterial retia related to supply of the central nervous system in two small toothed whales—narwhal (*Monodon monoceros*) and beluga (*Delphinapterus leucas*). Journal of Morphology 171:41–56.

Vrolik, W. T. 1848-1854. Bijdrage tot de natuur- en ontleedkundige kennis van den *Manatus americanus.* Bijdragen tot de Dierkunde 1:53–80.

Waites, G. M. H. 1970. Temperature regulation and the testis. Pages 241–279 *in* A. D. Johnson, W. R. Gomes, and N. L. VanDemark (eds.). The Testis, Vol. I. Academic Press, New York, NY.

Wall, W. P. 1983. The correlation between high limb-bone density and aquatic habits in recent mammals. Journal of Paleontology 57:197–207.

Walmsley, R. 1938. Some observations on the vascular system of a female fetal finback. Contributions to Embryology 164:109–178 + 5 plates and 27 figures.

Watkins, W. A., M. A. Daher, K. M. Fristrup, and T. J. Howald. 1993. Sperm whales tagged with transponders and tracked underwater by sonar. Marine Mammal Science 9:55–67.

Webb, P. W. 1975. Hydrodynamics and energetics of fish propulsion. Bulletin of the Fisheries Research Board of Canada 190:1–159.

Webb, P. W. 1984. Body form, locomotion and foraging in aquatic vertebrates. American Zoologist 24:107–120.

Webb, P. W. 1988. Simple physical principles and vertebrate aquatic locomotion. American Zoologist 28:709–725.

Webb, P. W., and V. de Buffrenil. 1990. Locomotion in the biology of large aquatic vertebrates. Transactions of American Fisheries Society 119:629–641.

Weihs, D., and P. W. Webb. 1983. Optimization of locomotion. Pages 339–371 *in* P. W. Webb and D. Weihs (eds.). Fish Biomechanics. Praeger, New York, NY.

Welker, W. 1990. Why does the cerebral cortex fissure and fold? A review of determinations of gyri and sulci. Pages 3–136 *in* E. G. Jones and A. Peters (eds.). Cerebral Cortex, Vol. 8B. Plenum, New York, NY.

Werth, A. J. 1992a. Suction feeding in odontocetes: Water flow and head shape. American Zoologist 29:92A.

Werth, A. J. 1992b. Anatomy and evolution of odontocete suction feeding. Ph.D. thesis, Harvard University, Boston, MA.

Williams, T. D., D. D. Allen, J. M. Groff, and R. L. Glass. 1992. An analysis of California sea otter (*Enhydra lutris*) pelage and integument. Marine Mammal Science 8:1–18.

Williams, T. M. 1989. Swimming by sea otters: Adaptations for low energetic cost locomotion. Journal of Comparative Physiology A 164:815–824.

Williams, T. M., and G. L. Kooyman. 1985. Swimming performance and hydrodynamic characteristics of harbor seals *Phoca vitulina.* Physiological Zoology 58:576–589.

Williams, T. M., W. A. Friedl, M. L. Fong, R. M. Yamada, P. Sedivy, and J. E. Haun. 1992. Travel at low energetic cost by swimming and wave-riding bottlenose dolphins. Nature 355:821–823.

Wislocki, G. B. 1935. The placentation of the manatee. Memoirs of the Museum of Comparative Zoology 54:158–178 + 7 plates.

Wislocki, G. B., and R. K. Enders. 1941. The placentation of the bottle-nosed porpoise (*Tursiops truncatus*). American Journal of Anatomy 68:97–125.

Worthy, G. A. J., and E. F. Edwards. 1990. Morphometric and biochemical factors affecting heat loss in a small temperate cetacean (*Phocoena phocoena*) and a small tropical cetacean (*Stenella attenuata*). Physiological Zoology 63:432–442.

Worthy, G. A. J., and J. P. Hickie. 1986. Relative brain size in marine mammals. American Naturalist 128:445–459.

Würsig, B. 1989. Cetaceans. Science 244:1550–1557.

Yablokov, A. V., and G. A. Klevezal. 1962. Whiskers of the whales and seals and their distribution, structure and significance. Pages 48–81 *in* S. E. Kleinebergh (ed.). Morphological Characteristics of Aquatic Mammals. Translation Number 1335, Fisheries Research Board of Canada, Ste Anna de Bellevue.

Zapol, W. M., G. C. Liggins, R. C. Schneider, J. Qvist, M. T. Snider, R. K. Creasy, and P. W. Hochachka. 1979. Regional blood flow during simulated diving in the conscious Weddell seal. Journal of Applied Physiology 47:986–973.

3

ROBERT ELSNER

Living in Water

Solutions to Physiological Problems

. . . the greatest part of those things which we do know is the least of the things we know not. Harvey 1628

Existence in an aquatic medium would appear to impose many constraints on mammalian life. The required variations on the mammalian theme, those special features that have made possible a vigorous and productive life in the seas, include adaptations for breath-hold diving, temperature regulation in cold water, water and salt balance, underwater navigation, and high pressure at great depth. The subject matter of this chapter will be concerned with these physiological functions without reference to matters such as foraging and reproduction, which are considered in other chapters. The topics can best be elucidated by comparative examples not exclusively restricted to marine mammals. Morphological aspects of the relevant functions are discussed in Pabst, Rommel, and McLellan, Chapter 2, this volume.

Mammalian physiological adaptations to aquatic life have been subjected to intense study in recent years, and the results of these research efforts have been reviewed in several journal articles and monographs (Andersen 1966, 1969; Ridgway 1972; Galantsev 1977; Kooyman 1981, 1989; Kooyman et al. 1981; Butler and Jones 1982, 1997; Blix and Folkow 1983; Elsner and Gooden 1983; Daly 1984, 1986, 1997; Zapol 1987, 1996; Elsner and Daly 1988; Castellini 1991; Kooyman and Ponganis 1997, 1998). Earlier work and historic reviews are presented in several sources (Irving 1939, 1969; Scholander 1940, 1962, 1963, 1964). The reader is referred to the identified reviews and original publications for more detailed considerations. Optimum appreciation of the topics and issues may require consultation of texts that are directed more deeply into the appropriate subjects of physiology: respiration, circulation, metabolism, and their neural and endocrine regulations.

Diversity in size, range, behavior, life history, and degree of marine adaptation characterize the species to be examined. Aquatic mammals range in size from the freshwater shrew (*Solex palustris*) to the blue whale (*Balaenoptera musculus*). Thus they include the largest mammal ever to have lived and one of the smallest. Maximum diving times vary from less than 1 min (freshwater shrew, Calder 1969) to 2 hr (southern elephant seal, Hindell et al. 1991a,b; sperm whale, Watkins et al. 1985). When southern and northern elephant seals go to sea for extended periods, they dive nearly continuously, day and night. Some of these appear to be foraging dives, others may be sleep episodes (Le Boeuf et al. 1986, 1988, 1989). Weddell seals of Antarctica can dive longer than 1 hr (Kooyman 1966, 1981, 1989) and to depths exceeding 700 m (Testa 1994). Beluga whales have been trained to dive in the open ocean to nearly 650 m (Ridgway et al. 1984). Sperm whales are known to have dived to 1000 to 2000 m (Heezen 1957, Watkins et al 1993), northern elephant seals to more than 1500 m (DeLong and Stewart 1991), and hooded seals dive deeper than 1000 m and remain submerged longer than 52 min (Folkow and Blix 1995).

Marine mammals inhabit all of the world's oceans. Although some populations are native to equatorial regions, and tropical seas are visited seasonally by several species of whales, their greater abundance in the polar regions is clearly evident in the enormous concentrations around Antarctica and in the Arctic Ocean and its peripheral seas. The warm water exceptions include the sirenians (manatee and dugong) of some tropical coasts and estuaries, monk seals, and several species of freshwater dolphins that live in the Ganges, Amazon, Yangtze, and other rivers. Some seals thrive in freshwater lakes, and the largest population of these, some 50,000 Baikal seals, live in Lake Baikal, Siberia.

The physiological adaptations supporting this abundant marine life are the subjects of this chapter. Examples will be drawn from several species among the marine mammals, including Cetacea (whales and dolphins); Sirenia (dugong and manatees); the pinniped families Phocidae, Otariidae, and Odobenidae, including the phocid seals, sea lions and fur seals, and walrus, respectively; and Mustelidae, the sea otter. Our present knowledge of the range and extent of related physiological adaptations has come from comparative studies of other aquatic and nonaquatic mammals, such as polar bear, hippopotamus, muskrat, beaver, pig, dog, and human. Many of the features of marine mammal adaptations are shared by those of aquatic birds.

Breath-Holding and Diving

Marine mammals are the champions at breath-holding (apnea) in the mammalian world, an obvious necessity that supports their underwater excursions. Nevertheless, the inexorable onset of progressive asphyxia begins with the cessation of breathing at the start of submergence, and adaptations to protect the animal against this threat dominate the related physiological events. From the beginning of the dive, the oxygen requirement for the organism is satisfied from storage consisting of oxygen in combinations with hemoglobin circulating in blood and with myoglobin bound in muscle tissue. There follows a steady decline of available oxygen (hypoxia) and a corresponding increase in carbon dioxide content (hypercapnia), a condition defined as *asphyxia* (Dejours 1981), and eventually an accumulation of the products of anaerobic metabolism, lactic acid, and hydrogen ions (acidosis). These combined conditions, the triad of asphyxia, also occur when tissues or organs are deprived of circulating blood, in which instance it is referred to as *ischemia*. It is important to recognize the distinction between asphyxia and hypoxia. The latter represents only one member of the asphyxial triad, and it occurs relatively rarely by itself. The hypoxia of high altitude, for instance, is accompanied by hypocapnia and alkalosis, in contrast to the diving condition.

Organs and tissues exposed to progressive asphyxia or to ischemia continue to function until their immediate sources of oxygen (aerobic metabolism) are depleted and their capacity for anaerobic metabolism is exhausted. Eventually their functional limit is reached, and they weaken with reduced capability for revival. However, if the normal metabolic activity of the cells is reduced during this period of declining energy sources, as by lowered temperature, they can continue to function for a longer period. If blood flow is restored before severe tissue disruption has taken place, recovery without permanent damage is likely. The resistance to asphyxia of different tissues varies widely. The visceral organs of habitual divers are especially tolerant of asphyxia and ischemia (Elsner and Gooden 1983).

Aerobic metabolism sustains the frequently occurring brief dives and the early period of longer dives, and it depends on the quantity and availability of stored oxygen (Scholander et al. 1942a, Kooyman et al. 1980). In adult Weddell seals, relatively short dives of less than about 18 min were found to be aerobic, that is, they were sustained by oxidative metabolism. They represented the vast majority, 97%, of natural dives (Kooyman et al. 1980). Similar limitations of most dives to a brief aerobic phase appear to occur in other marine mammal species, although the metabolic characteristics of their dives are not as well studied as are those of Weddell and Baikal seals (Ponganis et al. 1997).

In longer dives, metabolism switches to an anaerobic energy source when aerobic resources are no longer readily available. The well-adapted marine mammal survives long submergence by using its primary adaptations to extend breath-holding time. One of these adaptations is increased oxygen storage, especially noteworthy in phocid seals (Fig. 3-1). Another entails a preferential distribution of circulating blood during the dive for protection of the organs, brain and heart, which are most intolerant of low oxygen levels. Ultimately, survival depends on the animal's enhanced anaerobic capacity. This general scheme, the classic diving response, is elucidated from the results of laboratory and field experiments and is described in detail in the original publications.

The reduced distribution of cardiac output (the product of ventricular stroke volume and heart rate) during dives is the result of regional vasoconstriction within those organs that tolerate a prolonged lack of oxygen. This selective ischemia results in a lowering of the metabolism in the tissues

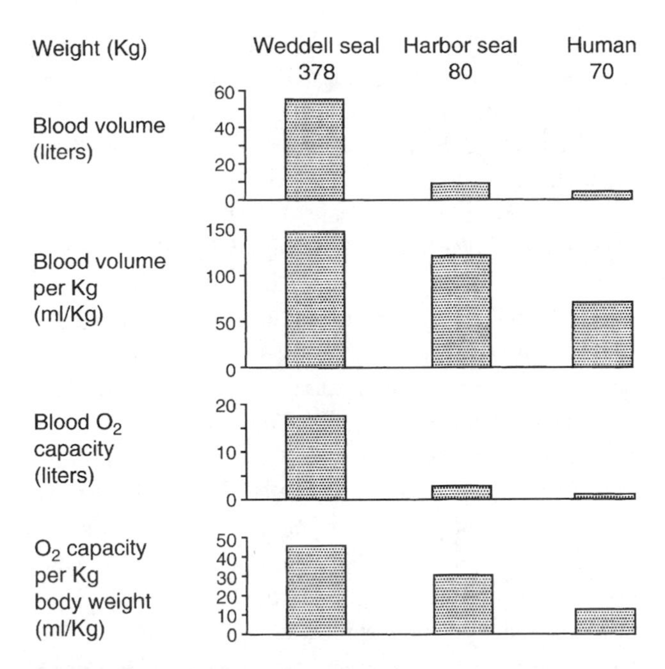

Figure 3-1. Blood oxygen stores of Weddell seal, harbor seal, and human. (Modified from Lenfant et al. 1969b.)

that are deprived of blood and in which the source of oxygen for energy conversion has consequently been temporarily reduced or eliminated. These include the visceral organs, skin, and sometimes muscle. The changes in cardiovascular actions are heralded by a pronounced decline in heart rate (bradycardia). Just as short duration dives are usually aerobic, long dives are supported by eventual recourse to anaerobic metabolism, for which the longest divers among marine mammals are especially well endowed (Scholander 1964, Elsner 1988). The combination of cardiovascular events initiated by submergence and continuing throughout the dive has become known by the term *diving reflex*. This choice of terminology is unfortunate, because we now know that these events are more complex than the single reflex implied. Rather, they are initiated and maintained by a set of reflexes and their interactions acting together to govern the complex array of physiological responses. A more accurate term, coming into general use, is *diving response*.

The reactions described here are not exclusive to aquatic mammals, and they resemble in many respects the responses to asphyxia of many terrestrial birds and mammals (including humans), reptiles, and even some invertebrates (Feinstein et al. 1977, see reviews by Elsner and Gooden 1983, Kooyman 1989). Some of the reactions are unexpected, as recorded in dives performed experimentally on the Australian monotreme, the echidna, *Tachyglossus aculeatus* (Augee et al. 1971), a seemingly unlikely candidate, given its arid habitat and lifestyle. These examples suggest a broader perspective than only that of marine species, a continuum of animals in which the cessation of breathing and its associated asphyxial threat elicits a coordinated response of ancient evolutionary origin. They also suggest that diving responses of marine mammals are specific examples of that more general phenomenon, defense against asphyxia.

Progress in understanding the mechanisms involved in breath-hold diving has been made on two separate research frontiers: experiments performed under controlled laboratory conditions and studies of free-swimming animals in their natural habitats. Questions have arisen concerning the correlation of these two approaches and the relevance of laboratory studies to the results obtained from free-swimming marine mammals.

Studies of animals in situations approaching their natural conditions have opened new vistas. Great strides have been made in the technology of recording and retrieval of physiological and environmental information from marine mammals during natural or near-natural dives. The development and attachment of time–depth recorders, pioneered by Kooyman (1966), has become a highly refined technique and has yielded much important information about the biology and behavior of mammalian divers (see also Ridgway et al. 1984; Gentry and Kooyman 1986; Zapol 1987, 1996; Fedak et al. 1988; Le Boeuf et al. 1989; Guyton et al. 1995; Andrews et al. 1997). Ingenious methods for obtaining depth profile recordings and for the drawing and recovery of blood samples from freely swimming seals (Hill et al. 1987), for determining their swim speeds and body temperatures (Ponganis et al. 1990b, Woakes 1992), for underwater tracking (Wartzok et al. 1992, Watkins and Schevill 1972, Watkins et al. 1993), and for attaching video cameras (Davis et al. 1992) have greatly expanded observational possibilities and have provided new knowledge concerning various events occurring during free dives.

Before these technologies were available, we had only the most superficial understanding of the diving depths and durations of marine mammals, and what we did know suggested more modest capabilities than we now appreciate, as shown by the remarkable feats of the elephant, Weddell, and hooded seals described earlier. These advances in instrumentation have generated new ways of studying diving, but the performance of controlled experiments under these free-swimming conditions is difficult and requires much ingenuity. The comparative results from laboratory and field studies will be described in the following sections and will be discussed where they relate to our knowledge of diving physiology.

Oxygen Availability

The lung capacities of most marine mammals, expressed relative to body weight, are not notably larger than those of land mammals (see Pabst, Rommel, and McLellan, Chapter 2, this volume), and some marine mammals exhale immediately before diving, further decreasing the volume of onboard lung air. Seals begin a dive often with only one-half or less of their total lung capacity (Scholander 1940, Kooyman et al. 1973). An exception to this general rule is the sea otter, *Enhydra lutris,* which is endowed with lungs that are large for a mammal of its size. Lung air provides both oxygen stores and buoyancy in sea otters (Lenfant et al. 1970, Tarasoff and Kooyman 1973). Marine mammals that lack the capability for dives longer than a few minutes, for example, the sea lions and fur seals, often inspire just before diving and depend more on lung capacity for oxygen reserve than do the longer-diving phocid seals (for review, see Snyder 1983).

Major oxygen reserves of hemoglobin and myoglobin are sufficient to provide oxygen storage well above that of their terrestrial counterparts in some species (seals: Scholander 1940; Harrison and Tomlinson 1956; Wasserman and Mackenzie 1957; Bryden and Lim 1969; Lenfant 1969, Lenfant et al. 1969a, 1970; cetaceans: Ridgway and Johnston 1966, Ridgway et al. 1984), but this advantage is by no means universal among marine mammals. Some blood oxygen capacity values are shown in Figure 3-1. In phocid seals the high levels of circulating hemoglobin and their large blood volumes can contribute to a total blood oxygen storage approximately three to five times greater than what might be expected to appear in human blood and other terrestrial species.

Blood volumes vary considerably among marine mammal species, and they correlate roughly with the species' diving capabilities. Among the lower blood volume values are the relatively brief divers: northern fur seal (11% of body weight, Lenfant et al. 1970) and bottlenose dolphin (7.4% of body weight, Ridgway and Johnston 1966). Higher blood volume values exist in the longer divers: southern elephant seal (22% of body weight, Bryden and Lim 1969) and beluga whale (13% of body weight, Ridgway et al. 1984). The comparable human value is about 8% of body weight.

The myoglobin content in muscles of some of these species represents an oxygen storage one order of magnitude greater than the quantity available to terrestrial species (Robinson 1939, Scholander 1940, Blessing and Hartschen-Niemeyer 1969, Lenfant et al. 1970, Castellini and Somero 1981). This amounts to a considerable resource, as skeletal muscle constitutes about 40% to 50% of the fat-free body weight. The bodies of most marine mammals are composed of a high ratio of fat, 20% to 50%, in the form of subcutaneous blubber. These values are higher than in most terrestrial mammals, and they play an important role for insulation, buoyancy, and energy reserve. Despite the increased capacity of the phocid seals to store oxygen, their supplies are not sufficient to provide the energetic basis for the long dives of which they are capable, and it is the cardiovascular accommodations and other adaptations to be described later in this chapter that contribute to the differences in their diving capabilities.

Recent studies of free-diving Weddell seals, in which implanted devices permitted recording of myoglobin oxygen saturation, confirmed the earlier discovery in laboratory dives of harbor seals (Scholander et al. 1942a) that myoglobin functions as an oxygen reserve (Guyton et al. 1995; see Hypoxic Tolerance section). In this free-diving study Weddell seals derived part of their oxygen requirement by myoglobin desaturation, but the maximum desaturations, seen in the earlier laboratory dive study when the harbor seals (*Phoca vitulina*) were pushed to their aerobic limits, did not occur to the same extent in free-swimming Weddell seals. An additional probable role of myoglobin is that of facilitated oxygen diffusion, especially important when cellular oxygen is low (Scholander 1960, Hemmingsen 1965, Wittenberg et al. 1975), as can be expected late in long dives.

As noted earlier, the contributions of oxygen from air contained in the lungs of phocid seals are relatively small, especially as seals often dive on considerably less than their total lung volumes. A further adaptation to long breath-holding dives is the increased capacity for buffering of acid metabolic products of anaerobic metabolism by the hemoglobin of blood and the myoglobin of muscle tissue (Castellini and Somero 1981).

A simple balance sheet of oxygen stores and the seal's estimated metabolic rate has often been used to calculate the available supply for dives. However, given the substantial variations and inaccuracies of the calculations, that procedure can sometimes be misleading. What it does indicate is a confirmation that the available oxygen is insufficient for the duration of very long dives. This is consistent with the importance of the redistributed circulatory perfusion and consequent reduction of metabolism in nonperfused organs, along with enhanced potential for anaerobic metabolism.

Historic Beginnings and Metabolic Reactions

The study of the physiological responses to diving was begun more than a century ago with the experimental studies of the French physiologist Paul Bert (1870). He discovered that when ducks were experimentally submerged, they experienced a dramatic slowing of heart rate. This primary observation initiated a long and still-continuing series of investigations into the diving response signaled by slowing of heart rate in diving birds and mammals. Bert compared the physiological resistance to diving asphyxia in ducks and hens, and he suggested that the superior performance of the ducks was accounted for by their greater blood volumes and related oxygen storage.

A few years later, Richet (1894, 1899) demonstrated that when ducks had been relieved of a portion of their blood volume, making them equivalent in this respect to hens, they still survived immersion longer than might be expected on the basis of differences in relative blood volume alone. He also deduced a decline in metabolic rate during diving on the basis that oxygen stores were insufficient to have lasted throughout the dive. Laurence Irving (1939) was the first to obtain evidence that oxygen was conserved during diving asphyxia by selective circulatory perfusion of the heart and brain, which are essential to survival and require a nearly continuous source of oxygen. The idea is well illustrated by the observation that limbs of mammals can be deprived of blood for long periods, an hour or more, whereas restriction of brain circulation is much more threatening and results in unconsciousness in a few seconds. There followed several years of elaboration of the physiological adaptations of marine mammals by Irving and Scholander, who used the techniques available to them, heart rate and blood pressure recording and indirect blood flow indications, to demonstrate the important mechanisms that have laid the groundwork for studies that continue today.

Much of our present understanding of marine mammal adaptations dates to its origin in laboratory research performed more than a half century ago. The techniques used and the manner of handling the animals require some description. Many of the experimental dives of marine mammals performed by Irving and Scholander in their pioneering studies involved restraint of the animals on a platform that could be immersed in water. These were *forced dives* performed with restrained animals, and the technique has been applied to many subsequent studies. It is important to recognize that the original technique as developed by the early workers involved careful and determined efforts to accustom the animals to the procedure and that considerable effort was made to reduce or eliminate the aspect of fright. The result was that fear and struggling were reduced to a minimum in many of those studies. Nevertheless, the experimental determinations obtained when using this procedure often elicit maximum or near-maximum cardiovascular and other responses.

Early evidence for redistribution of circulation during asphyxia, maintained in brain and reduced in muscle, depended on indirect methodology. Blood flow alterations in several aquatic and nonaquatic mammals were determined by resistance changes in a heated wire flowmeter (Irving 1938). The corresponding reduction in cardiac output was indicated by an abrupt and marked decline in the seal's heart rate. It often decreased drastically to levels of bradycardia amounting to about one-tenth of the predive level. Lactic acid produced by anaerobic metabolism appeared only in low concentrations in the circulating blood of experimentally dived, quiet seals. But in the immediate postdive period the blood levels of lactate increased rapidly. This discovery was interpreted as showing that circulatory perfusion was much reduced during the dives and that the lactic acid accumulated in tissues was flushed out immediately after the dive by the reestablished blood flow (Scholander 1940).

Diving Bradycardia

Diving bradycardia (decreased heart rate) has remained the frequently cited characteristic of the diving response, but not without argument. Evidence for its existence has been surrounded by controversy, largely because of its highly variable nature, and because bradycardia, often profound and long lasting, can be induced by other stimuli such as fright. There is now, after the dust has settled, general agreement regarding its validity and persistence as a component of the diving response. The range of the heart rate response to diving is very great, both among different species and in the same species and individuals in different situations. Scholander (1940) and others have sometimes observed only a steady heart rate or a modest bradycardia during brief free dives. Bradycardia during diving is often less intense during free and trained dives than in forced dives (Elsner 1965, Jones et al. 1973, Kooyman and Campbell 1972, Hill et al. 1987) and it can even be conditioned to occur without diving (Ridgway et al. 1975). Nevertheless, extremely low heart rates, decreasing sometimes to 5% of the predive rate, have been recorded during free dives in several species of phocid seals: harbor seals (Murdaugh et al. 1961a, Jones et al. 1973), free-swimming ringed seals (*Phoca hispida*) (Elsner et al. 1989) (Fig. 3-2), quietly resting, submerged gray seals (*Halichoerus grypus*) (Thompson and Fedak 1993), and northern elephant seals (Andrews et al. 1995). Similar reactions were recorded in an unrestrained, diving hippopotamus (*Hippopotamus amphibius*) (Elsner 1966). Diving bradycardia has been recorded during breath-holding dives in a great variety of species: aquatic and terrestrial mammals, birds, and reptiles.

Modest slowing of heart rate was observed in freely diving Weddell seals (Kooyman and Campbell 1972, Hill et al. 1987), harbor seals, and gray seals (Fedak et al. 1988). These variations of diving bradycardia in the same species and individual animals under different conditions lead one to suspect that they exercise a remarkable voluntary control over the cardiovascular system and that a level is established that is appropriate for the length, effort, and intensity of a particular diving situation. For example, the free-swimming gray seals studied by Fedak et al. (1988) reportedly never showed the

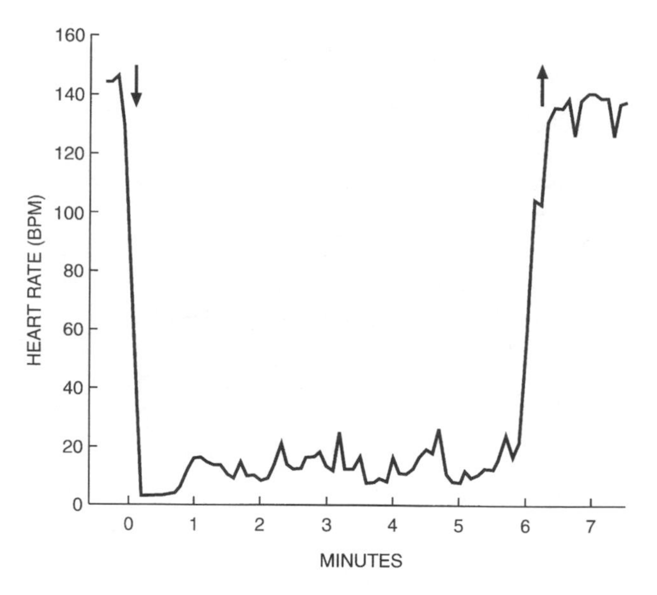

Figure 3-2. Heart rate of a ringed seal during free swimming under ice between breathing holes. Arrows indicate the start and end of the dive. Anticipatory increase in heart rate is evident immediately before surfacing. (Redrawn from Elsner et al. 1989.)

extreme bradycardia seen in forced laboratory dives, yet the same species in quiet resting dives demonstrated heart rates as low as ever determined in that species (Thompson and Fedak 1993). Ringed seals had equally extreme bradycardia whether swimming under ice at a speed of about 1 m/sec or while resting underwater at a breathing hole (Elsner et al. 1989).

Cetaceans also experience bradycardia while diving, but its detection is sometimes more elusive than in seals. Dolphins do not tolerate well conditions of restraint and forcible submersion. They are, however, highly trainable, and this characteristic has been exploited in studies of their diving responses. Bottlenose dolphins were trained to remain touching a target for about 5 min near the bottom of a pool. At the start of the dive the heart rate decreased abruptly from 100 to 12 beats/min, slowly increasing to about 25 beats/min as the trained dive proceeded (Elsner et al. 1966a). In another study a bottlenose dolphin was trained to dive to a target under a boat in the open sea, and by this means an estimate of its aerobic dive duration was obtained. The animal was trained to touch an underwater target for increasingly long periods. After a few minutes it was commanded to exhale underwater into a receptacle for collection of its expired air. The dolphin comfortably tolerated frequently repeated dives up to 2.5 min in duration, but longer dives required a prolonged recovery time, suggesting that it had exceeded its aerobic resource limit (Ridgway et al. 1969).

Recordings demonstrate that the dolphin's heart rate increases each time it surfaces to breathe (Elsner et al. 1966a, Ridgway et al. 1969). This cycling of heart rate is characteristic of sinus arrhythmia, an oscillation of heart rate that accompanies the respiratory excursions in most mammals, showing an increase during inspiration and a decline on exhalation. The excursions are frequently exaggerated in marine mammals. The oscillating heart rate of the dolphins can be interpreted as evidence for a lack of diving bradycardia in this species (Kanwisher and Ridgway 1983). The researchers viewed the lower rate when submerged to be the animal's "normal" heart rate and the increase when taking a breath as tachycardia. The results might just as easily justify the reverse conclusion, as did the earlier recording of persisting low heart rates during prolonged submergence.

Although further experiments on larger cetaceans would be much desired, there are obvious problems with their great size and relative inaccessibility. Simple observations of diving bradycardia have, however, been made on a killer whale (Spencer et al. 1967), a captive pilot whale, and a juvenile gray whale (R. Elsner, unpubl.). Diving bradycardia is sluggish and modest in the Florida manatee (Scholander and Irving 1941) and in the dugong (R. Elsner, unpubl.).

Selective Ischemia

The bradycardia of diving suggests that cardiac output is also reduced, but the extent of the reduction depends in part on what is happening to the heart's stroke volume. Here there are differences in experimental results that are probably accounted for by differing techniques and conditions. Recordings from an implanted ultrasonic flowmeter mounted on the pulmonary artery of a California sea lion trained to immerse its head in water indicated unchanged stroke volume (Elsner et al. 1964). Similar results using other techniques in experimental dives were observed in harbor seals (Murdaugh et al. 1966) and gray seals (Blix et al. 1976). However, two detailed and rigorously controlled studies showed a gradual decline of stroke volume in harbor seals to about one-half of the predive level (Sinnett et al. 1978, Ponganis et al. 1990a). Similarly reduced stroke volume was shown by Blix et al. (1983). It is clear that cardiac output is reduced in dives that are accompanied by intense bradycardia amounting to less than one-half of the nondiving heart rate, because even if stroke volume increased rather than decreased, it would almost certainly be insufficient to compensate for the low heart rate in the determination of cardiac output.

Despite the marked decline in cardiac output in seal dives showing extreme bradycardia, central arterial blood pressure remains relatively constant (Irving et al. 1942, Elsner et al. 1966b); such is the exquisite nature of the baroreceptor reflex mechanism in governing arterial pressure (Fig. 3-3). Consequently, blood pressure is maintained sufficient for

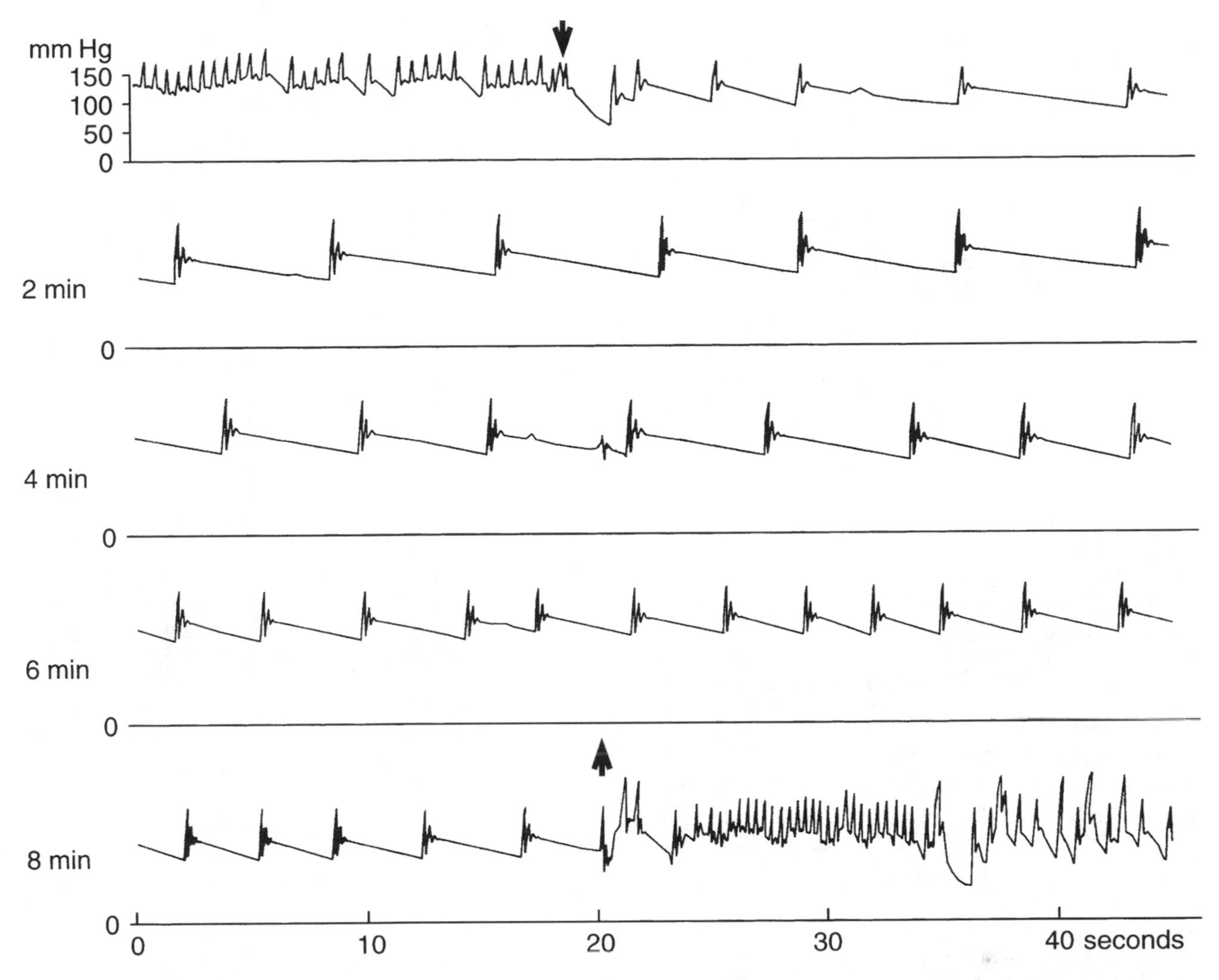

Figure 3-3. Arterial blood pressure of a restraint-dived harbor seal during a dive of 8 min. Sections of the dive record at 0, 2, 4, 6, and 8 min are displayed. Arrows indicate start and end of dive. Note the slow decline of pressure throughout the interbeat (diastolic) period and the maintenance of mean arterial pressure throughout the dive. (Modified from Irving et al. 1942.)

perfusion of the vital organs, brain and heart. The maintenance of arterial pressure throughout the long diastolic interval is accomplished in part by the elastic recoil of the stretched enlargement of the root of the aorta, an aortic bulb roughly equal to the volume of the left ventricle, which is a prominent structure in many marine mammal species (Elsner et al. 1966b, Drabek 1975, Rhode et al. 1986, Shadwick and Gosline 1995). Theoretical analysis indicates that the aortic bulb may also function by its distensibility to decrease left ventricular metabolic cost of pumping against the increased downstream impedance that arises as a consequence of peripheral vasoconstriction (Campbell et al. 1981).

Scholander et al. (1942a) showed that the circulating arterial blood and the muscle microcirculation of harbor seals did not communicate during an experimental dive (Fig. 3-4). This condition was revealed by sampling from the two oxygen storage sites, muscle myoglobin and blood hemoglobin, showing that oxygen bound to myoglobin within the muscle was depleted more rapidly than that in the circulating blood. Because the affinity of myoglobin for oxygen is much greater than that of hemoglobin, circulatory isolation of the muscles was evident (see further discussion under Hypoxic Tolerance). Chronic instrumentation with ultrasonic blood flow transducers (Elsner et al. 1966b) and X-ray angiographic observations (Bron et al. 1966) led to confirmation and elaboration of the predictions derived from the indirect evidence.

The reduction of blood flow varies with the degree to which the diving response is invoked, ischemia being widespread throughout visceral organs, skin, and muscle during its most extreme form. The circulation through the kidneys of harbor and Weddell seal during forced, trained, and free diving is much reduced or ceases entirely (Murdaugh et al. 1961b; Bron et al. 1966; Elsner et al. 1966b, 1970b; Zapol et al. 1979). Cessation of kidney perfusion gives rise to questions

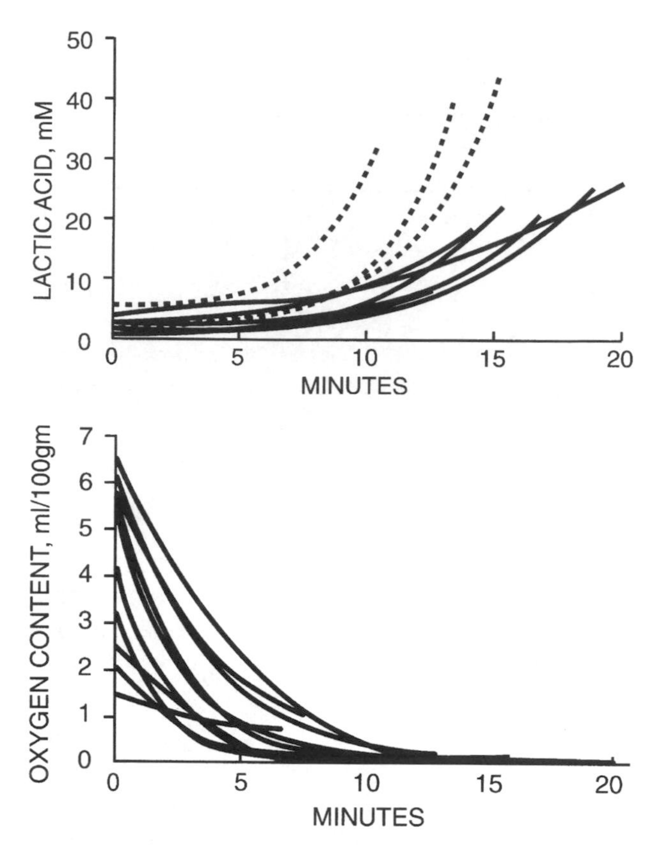

Figure 3-4. Accumulation of muscle lactate (top) and depletion of myoglobin-bound oxygen (bottom) in skeletal muscle of harbor seals during experimental dives. Both are evidence of circulatory isolation of muscle during the dive. The seals were lightly restrained and remained quietly resting, except for those instances identified by dashed lines. (Redrawn from Scholander et al. 1942a.)

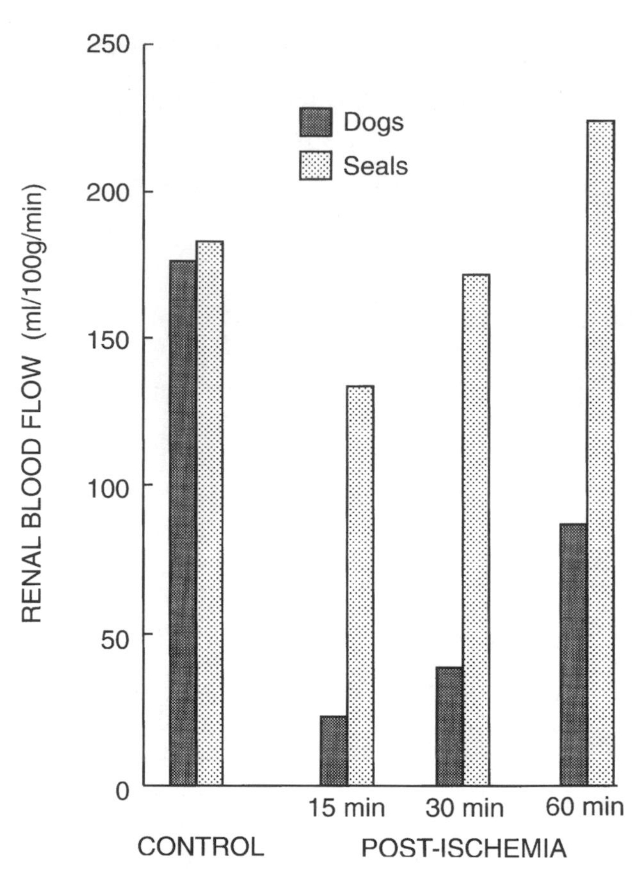

Figure 3-5. Blood flow in isolated harbor seal and dog kidneys before and after 60 min of ischemia. Seal kidneys recovered from the ischemic episode more rapidly and more completely than did dog kidneys. (Redrawn from Halasz et al. 1974.)

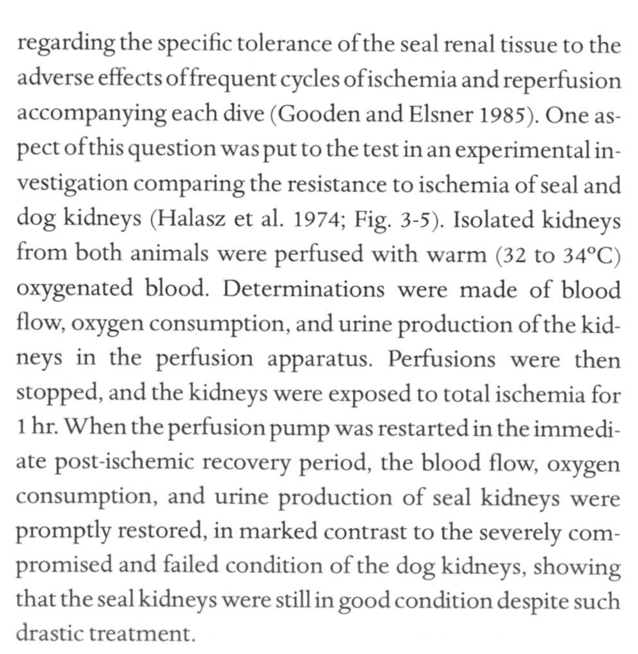

regarding the specific tolerance of the seal renal tissue to the adverse effects of frequent cycles of ischemia and reperfusion accompanying each dive (Gooden and Elsner 1985). One aspect of this question was put to the test in an experimental investigation comparing the resistance to ischemia of seal and dog kidneys (Halasz et al. 1974; Fig. 3-5). Isolated kidneys from both animals were perfused with warm (32 to 34°C) oxygenated blood. Determinations were made of blood flow, oxygen consumption, and urine production of the kidneys in the perfusion apparatus. Perfusions were then stopped, and the kidneys were exposed to total ischemia for 1 hr. When the perfusion pump was restarted in the immediate post-ischemic recovery period, the blood flow, oxygen consumption, and urine production of seal kidneys were promptly restored, in marked contrast to the severely compromised and failed condition of the dog kidneys, showing that the seal kidneys were still in good condition despite such drastic treatment.

Questions have been raised about the possible relevance of the laboratory experiments showing drastic reductions of visceral blood flow in restrained and forcibly dived seals to the situation of natural free swimming. Two studies of Weddell seals shed some light on this problem—and pose new questions. Tests of kidney and liver function were applied to free-swimming seals by administration of inulin into the circulating blood, a procedure used to determine kidney glomerular filtration, and indocyanine green, cleared selectively by the liver. Results suggested that kidney and liver perfusions were maintained during extended bouts of aerobic dives; reduced perfusion occurred in both organs only during long, presumably anaerobic dives (Davis et al. 1983).

In another study of renal and hepatic function obtained by automated blood sampling from Weddell seals during individual dives revealed a much-reduced blood flow in the kidneys and liver during both early and later minutes of what appear to be aerobic and subsequent anaerobic conditions (Guppy et al. 1986). These results suggest contradictory conclusions from those of the earlier study (Davis et al. 1983), but the differences may be obscured by the difficulties of interpretation in these situations in which steady states seldom, if ever, prevail. Furthermore, there is almost certainly a high variability of responses that depend on a multitude of factors relating to the animal's free-diving condition and moment-to-moment changes. Results obtained from blood sampling at a given time during a long dive do not necessarily predict conditions that might occur at the same time in short dives. In addition, seals show that they can "turn on" their maximum diving responses, including widespread vasoconstriction, or invoke more moderate responses, depending on the anticipated duration and intensity of diving activity. On balance, it seems reasonable to conclude that the free-diving seal has the capability of responding physiologically to whatever changing conditions might be, such as pursuit of prey or moments of rest or exploration. What seems clear is that the kidneys and liver show the capacity for intense vasoconstriction and tolerance of its consequences and that the animals sometimes resort to this reaction, even during aerobic dives. The capability for oxygen conservation appears to be available when and if its expression is appropriate for the diving conditions.

The venous system of phocid seals has received attention because of its peculiar structure and large capacity. The inferior vena cava is greatly enlarged and extends into hepatic sinuses of unusual size, and a muscular sphincter innervated by a branch of the phrenic nerve is located where the vena cava passes through the diaphragm (Harrison and Tomlinson 1956, Ronald et al. 1977). The inferior vena cava of northern elephant seals contains roughly one-fifth of the total blood volume, and its capacity can provide considerable oxygen storage (Elsner 1969). Venous blood flowing toward the heart is restricted by the action of the caval sphincter (Elsner et al. 1971b, Hol et al. 1975) and its effect is likely to reduce the volume and pressure of blood returned to the heart, as demonstrated by Sinnett and colleagues (1978) in experimentally dived harbor seals.

Natural Dives Versus Laboratory Studies

The questions that have been raised about the relationship of controlled laboratory dives to what happens in nature have persisted, but a broader comprehension of diving adaptations is emerging. Results from laboratory and field studies are coming increasingly close to resolution of the different approaches involved, and some experimental results once considered to be laboratory artifacts are now understood to be valid phenomena whose clarification contributes to the understanding of diving physiology. Some biologists used to argue that diving bradycardia does not exist, an untenable assertion today. Others have regarded laboratory dives, whether executed with restraint or performed by trained animals, to be aberrations only distantly related to natural diving. Such extreme conclusions appear to be unwarranted in light of what we know about the wide range of natural diving reactions, and they seem unlikely to be useful in the pursuit of solutions to problems in diving physiology that may be best resolved by controlled experimentation.

Precise information about cardiac output cannot be expected by considering heart rate alone. However, some qualitative relationships can be derived. Cardiac stroke volume in mammals rarely varies by more than a factor or two, and therefore, one can be confident that drastic reductions of heart rate, such as occurs in some dives in which it may decline to a small fraction of the nondiving rate, entail similar drastic reductions in cardiac output and severe and widespread vasoconstriction. Without such compensating vasoconstriction, arterial blood pressure, necessary for perfusion of the remaining critical parts of the circulation, including cerebral blood flow, could not be maintained. The close relationship between heart rate and blood flow has been well demonstrated for extreme bradycardia in long, restrained dives (Irving et al. 1941a, Elsner et al. 1966b), less so for moderate bradycardia in shorter trained dives (Elsner et al. 1966b).

The selective restriction of blood flow during the dive is a dynamic and highly variable phenomenon that responds to a variety of competing influences. Some dives are performed with little swimming activity or its absence altogether, as during periods of rest or sleep. In these dives the maximum limitations of blood flow produce generalized ischemia throughout the animal's body except for perfusion of the central nervous system and the requirement to support the reduced mechanical cardiac function of heart muscle. The result, in the latter instance, is a partitioning of circulation to supply fluctuating myocardial blood flow (Elsner et al. 1985). A similar intermittent flow adequate to meet at least minimum needs of skeletal muscle during rest and exercise while diving has been postulated (Gooden and Elsner 1985, Kooyman 1987, Castellini et al. 1992; see section on Hypoxic Tolerance: Vascular Smooth Muscle). Clearly, the cardiovascular response to diving is highly variable—sometimes extreme, with profound bradycardia and intense widespread vasoconstriction, sometimes more modest and showing lesser reactions of the heart and circulatory resistance.

Deprivation of arterial blood flow to selected organs during dives produces a gradual reduction in metabolism of the ischemic tissues, and evidence for that condition has been observed in the decline of deep body temperature that has been noted in diving seals: 2 to 3°C in harbor seals (Fig. 3-6) (Scholander et al. 1942b, Hammel et al. 1977) and Weddell seals (Kooyman et al. 1980, Hill et al. 1987), and 4°C in

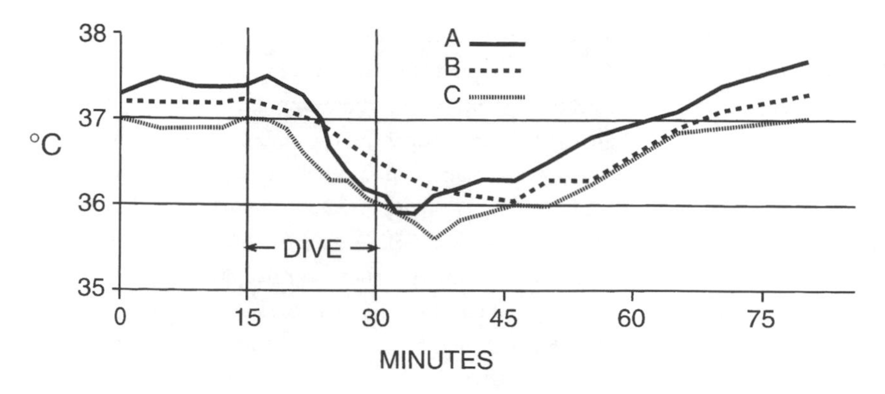

Figure 3-6. Body temperatures of a harbor seal during an experimental dive, evidence of declining metabolism during the dive. A, brain; B, abdominal viscera; C, dorsal musculature. (Redrawn from Scholander et al. 1942b.)

elephant seals (Andrews et al. 1993, 1995). The recent observation of abrupt increases in skin temperatures of northern elephant seals immediately preceding long dives (R. Andrews, pers. comm.) suggests the intriguing possibility that the reduction in metabolism is a *regulated* event, neural in origin, and that it is actuated by the loss of heat from abruptly warmed skin surface, dumping body heat, and resulting in a decline of deep body temperature.

Therefore, there is evidence that cardiovascular events can modulate the level of metabolism in the tissues to which the circulation is varied from full perfusion to ischemia. In the extreme instance, when an organ is made severely ischemic and is totally deprived of blood, obviously its metabolic rate must decline, leading eventually to cellular damage and death. Short of that condition there exists a great range of possibilities for the diving animal, varying from unaltered oxidative metabolism, as may be seen in brief, unstressed dives with little or no slowing of heart rate and uninterrupted circulation, to those longer dives that challenge the animal's homeostasis with drastic reductions in regional circulation. It has also become clear that marine mammal species differ both qualitatively and quantitatively in the timing and expression of their diving responses, depending on many factors. There is no such thing as a "generic" seal or marine mammal, and we are becoming increasingly aware that extrapolations from one species to another can be misleading.

Aerobic Dive Limit

The dependence of harbor seals on oxygen storage was recognized early as defining the limit of aerobic metabolism during experimental dives (Scholander et al. 1942a). The effect, a decline in oxygen resources from myoglobin and hemoglobin, demonstrates the clear priority for oxidative metabolism that precedes an eventual recourse to the more restrictive metabolic condition of anaerobiosis. Lactic acid began to appear in skeletal muscle of quietly resting restraint-dived harbor seals at 10 min. However, this oxygen-supported limit was clearly not uniform, and it is shortened by muscular activity (see Fig. 3-4).

The concept of the aerobic dive limit, ADL in common usage, is consistent with the description of limited aerobic resources shown by Scholander et al. (1942a) in diving harbor seals. It was described and elaborated by Kooyman et al. (1980) in Weddell seals, and among free-swimming species its characteristics have been described in detail in that species and in Baikal seals (Ponganis et al 1997). Caution should be exercised in its extension to other species without knowledge of oxygen stores, metabolic rates, and the onset of anaerobic metabolism and lactate production. The concept has proved to be useful in identifying the demarcation between oxidative and anaerobic metabolic processes in dives. The average length of aerobic dives was about 18 min in the Weddell seals studied by Kooyman et al. (1980) and shorter in younger animals (Kooyman et al. 1983). Variability in the estimation of ADL is indicated by the calculations of 9 to 16 min in adult Weddell seals noted by Castellini et al. (1992). Of course, the ADL endurance will be shortened if the seal exercises vigorously and depletes its available oxygen during the dive. Therefore, the usefulness of the ADL concept is limited to those instances in which the behavior of the animal is restricted to uniform and generally modest activity. Examples of the wide range of ADL estimations from calculated oxygen stores are described in Kooyman (1989). It would seem to be more accurate and useful to apply the concept as a designation of *maximum* aerobic dive limit, ADL_{max}.

As described, most dives by free-swimming Weddell seals are performed within their putative aerobic dive limits (Kooyman 1989). Despite this temporal dominance of oxidative metabolism in most ordinary dives, it is important to remember that the animal's diving capacity is made greater than this limit by its anaerobic and other mechanisms, and that its survival may often depend on its ability to remain submerged for longer than aerobic-limited dives. The investigating biologist may be interested in descriptions of the brief or long diving patterns of a free-swimming marine mammal or, on the other hand, the study of mechanisms by experimental manipulation of the animal's responses determined under controlled laboratory conditions. Sometimes these two approaches to understanding the adaptations of diving animals can be combined to yield new information on physiology and behavior. Both conceptual approaches are required for complete understanding of the diving responses.

Diving Hypometabolism

Whether or not a decline in metabolic rate occurs in marine mammal dives has been the subject of some debate (Elsner and Gooden 1983, Castellini 1985a, Hochachka and Guppy 1987, Kooyman 1989). The numerous estimations of oxygen economy during dives (Richet 1899, Irving 1939, Scholander 1940, Andersen 1966, Elsner and Gooden 1983, Hochachka and Guppy 1987, Kooyman 1989) have all indicated a considerable disparity between oxygen available to the diving mammal from onboard storage and what would be required for sustaining the metabolic rate at anticipated levels predicted from the resting value plus the demands of whatever activity takes place. Clearly, an episodic decline in the level of metabolic rate would allow for longer dives, a temporary *strategic retreat* by the animal into a hypometabolic state that is beneficial in the long run.

Hypometabolism of diving may be detectable only in those instances in which another overriding metabolic activity, for instance, exercise, is reduced or ceases. The phenomenon is much discussed and is the subject of speculation, but hard evidence is difficult to obtain. The original observation of Scholander (1940) that the oxygen consumption after diving, the pay-off of oxygen debt incurred during the dive, was insufficient to account for the anticipated consumption extrapolated from the predive level, and amounted to 50% or less of the predicted value, has withstood the test of time. Little clear evidence, confirming or otherwise, has been produced regarding this question until recently. In a study of trained adult California sea lions, postdive recovery oxygen consumption was 55% to 65% less than could be accounted for by the extrapolated predive value (Hurley 1996).

Other observations, although not direct demonstrations of hypometabolism, lend support to its presence in Weddell and elephant seal dives where the metabolic cost of diving, even including swimming activity, is less than or equal to metabolism of nondiving quiet rest or sleep (Kooyman et al. 1973, Castellini et al. 1992, Andrews et al. 1993). Periods of breath-holding during sleep have been noted in several species: northern elephant seals (Bartholomew 1954; Huntley 1984; Castellini et al. 1986, 1994), harbor seals (S. Ashwell-Erickson and R. Elsner, unpubl.), and Weddell seals (Castellini et al. 1992). These examples, indicative of metabolic conservation, amount to an energetic savings of about 30% in sleeping northern elephant seals (Huntley 1984).

As previously mentioned, body temperatures decline during long dives, and these changes support the concept of diving hypometabolism. The reported declines range from 2°C in harbor seals (Scholander et al. 1942b) to 4°C in elephant seals (Andrews et al. 1993). But the further question remains: Is the reduction in body temperature a result of hypometabolism or is the reduced metabolism a consequence of lowered body temperature? Speculation regarding this question would suggest that the seal might make use of increased skin circulation to accelerate heat loss at the initiation of long dives. Experimental temperature manipulations at brain sites of thermoregulation in the hypothalamus in harbor seals demonstrate the effectiveness of rapidly lowering deep body temperature by increasing skin perfusion (Hammel et al. 1977). Such a maneuver might explain, in part, the remarkable and puzzling ability of elephant seals to make frequent long dives with little recovery times (Le Boeuf et al. 1986, 1988, 1989; Hindell et al. 1991a,b).

Reduced body temperature may, however, be insufficient alone to account for the anticipated overall lowering of metabolic rate during dives. Another suspected mechanism is the effect of progressively increasing tissue acidity as the dive lengthens and which would result in metabolic inhibition (Harken 1976).

Hypometabolism in diving mammals may be thought of as a trade-off in which the advantages of a high and constant body temperature are temporarily abandoned to permit a brief but potentially important extension of diving time. In this example, it resembles other mammalian instances of recourse to hypometabolism for strategic retreats from hostile environments or in fulfilling physiological needs, as in hibernation, estivation, sleep, and certain pathological states.

A form of conscious control of autonomic function, which several observers have remarked, is the increase of heart rate, noticeable immediately before the anticipated termination of dives (Scholander 1940, Murdaugh et al. 1961a, Jones et al. 1973, Casson and Ronald 1975, Fedak 1986, Hill et al. 1987, Elsner et al. 1989, Williams et al. 1991, Wartzok et al. 1992, Thompson and Fedak 1993). Furthermore, some Weddell seal dives begin with heart rates that are apparently linked to the subsequent duration of the dive: the longer the dive, the more intense the initial bradycardia (Kooyman and Campbell 1972, Hill et al. 1987), strongly suggesting that the seal "plans" the extent of cardiovascular preparation that is appropriate for the intended dive duration. Again, as we have seen, there are numerous variations on this general theme. For example, ringed seals free-diving under ice did not use this strategy; their dives, whether brief and quiet or in long excursions, were accompanied by similar lowering of heart rates to about 5% of the predive rates (Elsner et al. 1989; see Fig. 3-2).

Diving and Exercise

An obvious concern in understanding the complexity of free-diving is the question of the combined needs for breath-holding and for the performance of swimming exercise in the same dive. Breath-hold diving precipitates a sequence of reactions designed to provide resistance to progressive asphyxia. These reactions result in an effective conservation of oxygen through reliance on vasoconstriction of as much of the arterial circulation as possible. Exercise, on the contrary, depends on efficient oxygen delivery by vasodilating blood vessels that supply muscle and other active metabolic sites. Thus, requirements for essential underwater activities, such as the pursuit of prey, appear to counter those of diving. Compromises and trade-offs in diving time, exertion, and oxygen economy are the likely explanations (Castellini 1985b, Fedak 1986). The relationships of diving exercise, oxygen consumption, and muscle blood flow necessary for a quantitative description of events are yet to be exposed to scrutiny.

Perhaps one-third to one-half of the skeletal muscle component of the seal's body may be used in locomotion. Studies in terrestrial mammals have shown that muscle blood flow is highly variable, ranging from about 20% of the cardiac output during rest to about 80% of the increased cardiac output during heavy exercise. One might intuitively suspect that a diving mammal would partition metabolic activity to efficiently meet the demands of individual dives, depending on the kind of activity undertaken, and there is observational support for just this kind of grading of dive time and effort (Kooyman 1989, Williams et al. 1993a). The highly variable nature of metabolic demand estimated in natural dives supports this concept (Castellini 1985b; Castellini et al. 1985, 1988, 1992; Fedak et al. 1988; Kooyman 1989).

The range of energetic activity can be expressed as multiples of oxygen consumption extending from quiet rest to maximum exercise (maximum oxygen consumption or aerobic scope; see Costa and Williams, Chapter 5, this volume). The determination of maximum oxygen consumption represents the integration for peak performance of the respiratory and circulatory systems, and it reflects the capacity of the animal for its most intense muscular effort. The determination can be described graphically by construction of a relationship of oxygen consumption as a function of increasing discrete work loads. When the animal's maximum oxygen uptake is reached, an oxygen consumption plateau is reached and further work can only be accomplished anaerobically (Fig. 3-7). Most mammals find it very difficult to increase work loads much beyond the maximum aerobic limit. For many terrestrial mammals, the maximum oxygen consumption reveals an aerobic scope of about 10 times the resting rate, and 20 or more times the resting rate for elite athletes, such as trained dogs, horses, and humans (Seeherman et al. 1981).

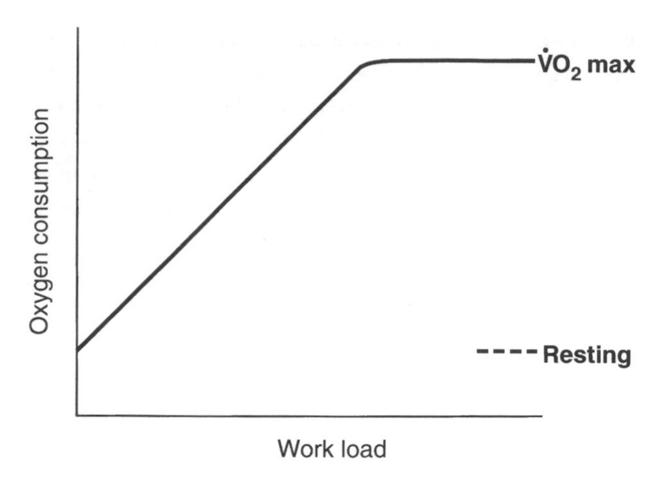

Figure 3-7. Diagrammatic method for determination of the maximum aerobic exercise capacity. The oxygen consumption steadily increases with successively increasing work loads, eventually reaching a level at which further consumption ceases. The subsequent plateau represents anaerobic support (i.e., oxygen debt) of further work load. $\dot{V}O_{2max}$, maximum oxygen consumption.

Maximum oxygen consumption has been determined in three species of marine mammals: harbor seals (Hammel et al. 1977; Ashwell-Erickson and Elsner 1980; Davis et al. 1985; Elsner 1986, 1987; Ponganis et al. 1990b; Williams et al. 1991), California sea lions (Feldkamp 1987), and bottlenose dolphins (Williams et al. 1993b). Two aspects of these records differ significantly from those of terrestrial mammals. First, maximum aerobic scope is relatively low, only four to nine times the resting level, and second, the anaerobic capability for increasing work load beyond the level identified by maximum oxygen consumption is clearly greater in the marine species (Elsner 1987, Williams et al. 1993b). Therefore, the work load that depends on anaerobic reserve could be approximately doubled beyond that defining the aerobic limit, far more than possible for most terrestrial mammals (Seeherman et al. 1981). The exception in anaerobic ability among terrestrial mammals that have been tested is the lion. This suggests that African lions, seals, and dolphins may rely more on anaerobic burst activity than on sustained endurance for intense exercise demands.

The implications for these species of marine mammals, when compared with terrestrial mammals, are twofold: their life-supporting activities probably seldom require them to exercise at a sustained high level, and when they do they can rely in large part on anaerobic metabolism. These limitations indicate a compromise between the contrasting requirements of diving endurance and exercise, and suggest that the energetic costs of mammalian life in the sea could be relatively stable and small (Costa and Williams, Chapter 5, this volume). A highly variable and graded invocation of the diving response, allowing for most dives to be performed aerobically, is consistent with this interpretation.

Evolutionary Considerations, Origin of Aerobic Metabolism

Ready access to oxygen, which we take for granted among terrestrial mammals, is an ever-present problem for diving animals whose breath-holding, underwater excursions sustain their lifestyle. Early forms of life on Earth existed without oxygen. An examination of the ancient conversion from anaerobiosis to aerobic metabolism by the developing primordial oxygen-dependent biota may appear to be remote from matters dealing with present day marine mammals. These early events, however, have left a traceable connection extending throughout the evolutionary history of mammalian physiological adaptations.

Present evidence indicates that the primitive atmosphere of the earth lacked oxygen and that early organisms were obliged to depend on anaerobic energy sources. Atmospheric oxygen appeared about two billion years ago and increased as a by-product of the origin of photosynthetic organisms (Gilbert 1981, 1996). Its appearance brought about subsequent beneficial changes in metabolic efficiency, but it was toxic to existing life of that time. This toxicity is a fundamental by-product of the biological reduction of oxygen leading to the production of toxic intermediates, the oxygen-derived free radicals, and other forms of highly reactive oxygen. These toxic effects had to be controlled for the benefits of oxidative metabolism to be realized. Organisms that depend on oxidation evolved protective mechanisms, including the enzymatic degradation of these potentially damaging oxygen species; hence we find the ubiquitous occurrence in aerobic organisms of the enzymes superoxide dismutase, catalase, and glutathione peroxidase, which catalyze the reduction of highly reactive oxygen species.

The origin of photosynthesis changed the world forever and resulted in the transformation of solar energy into a form that ultimately depended on oxygen for its liberation. This transformation represented an enormous increase in available energy and led eventually to the origin of multicelled and active organisms. But the impetus to the evolutionary advancement of life carried with it the lingering threat of oxygen toxicity. Destructive effects of oxygen are noticeable in mammals during exposure to higher than normal oxygen concentrations and pressures (Clark 1993). These effects can also be produced by conditions in which sensitive tissues are exposed successively to reduced oxygen or cessation of blood flow followed by sudden replenishment of oxygen by reperfusion and the resulting generation of oxygen-free radicals (Gilbert 1963, McCord 1985).

Mammalian life is dependent on oxygen and, accordingly, we may be surprised to be confronted by an *oxygen paradox*, raising by implication the question, "Is oxygen a benefit or a poison?" What seems superficially to be an obscure consideration in the life of mammals will appear again in the form of a problem that must be understood to fully comprehend the extent and the potential of the physiological adaptations to an aquatic environment. When we consider the means by which these species respond to the requirement for long underwater submergence, it turns out that the frequent cyclic episodes of widespread ischemia and reperfusion accompanying most dives would be expected in aterrestrial mammal to be just the appropriate condition for abundant production of highly reactive and potentially damaging forms of oxygen (Weiss 1986). At the cellular level, this form of poisoning is identical to the oxygen toxicity produced by prolonged oxygen breathing in high concentrations or at elevated pressures. The appropriate conditions of ischemia–reperfusion leading to the abundant production of oxygen-free radicals appear to be a necessary accompaniment of the diving habit. The extent and damaging potential of that production and protection against its adverse effects are unknown, but the successful adaptations of diving mammals indicate that they are able to cope with the effects of these products. Whether their protection differs from the naturally occurring defense mechanisms of other animals is unknown (Elsner et al. 1998).

Blood Viscosity

The high hematocrit, that is, the elevated concentration of red blood cells in the circulation of some marine mammals, especially phocid seals, raises the question of a possible increased resistance to blood flow attributable to elevated blood viscosity characteristic of high hematocrits (Wickham et al. 1985, 1989; Hedrick et al. 1986; Hedrick and Duffield 1991). The concern here relates to a potential for impaired circulation caused by the high blood viscosity and the consequent reduced supply to critical tissues in circumstances requiring optimum gas exchange and nutritive perfusion. Blood viscosity is ordinarily directly related to hematocrit value, and the tendency of red blood cells to aggregate accounts for much of the increase in viscosity with increasing hematocrit.

This issue has been examined in recent studies, and a somewhat contradictory picture has emerged. Ringed seal blood shows very little tendency for red cell aggregation, and this results in anomalously lowered viscosity despite high hematocrit (Wickham et al. 1990). Weddell seal blood also shows a high hematocrit, but in contrast to ringed seal blood, it has a very high red cell aggregation and as a consequence, much higher viscosity (Meiselman et al. 1992) (Fig. 3-8). Weddell and elephant seals sequester blood cells, presumably in the spleen and perhaps venous reservoirs, and release them into the general circulation during dives (Kooyman et al. 1980, Qvist et al. 1986, Hurford et al. 1996) and sleep apnea (Castellini et al. 1986, Castellini and Castellini 1993). This storage sequestration occurs in those species in which the oxygen demands of diving and nondiving sleep apnea are supported by intermittent release of blood cells, but is not as prominent in species (e.g., ringed seals) in which viscosity is lowered by reduced cell aggregation (Elsner and Meiselman 1995). This suggestion is supported by indications of differences of spleen size in these species. Postmortem measurements show that the spleen weights, very likely in a fully contracted condition (storing little blood) and expressed as a fraction of body weights, are: Weddell seal, 0.89% (Qvist et al. 1986) and ringed seal, 0.20% (R. Elsner, unpubl.).

It is interesting to note that natural selection has apparently resulted in these two separate mechanisms for dealing with elevated blood viscosity and has endowed these two species at opposite ends of the Earth, Weddell and ringed seals, with differing capabilities for coping with the contradictory requirements for their lifestyles: oxygen conservation for long dives and oxygen delivery for sustained effort.

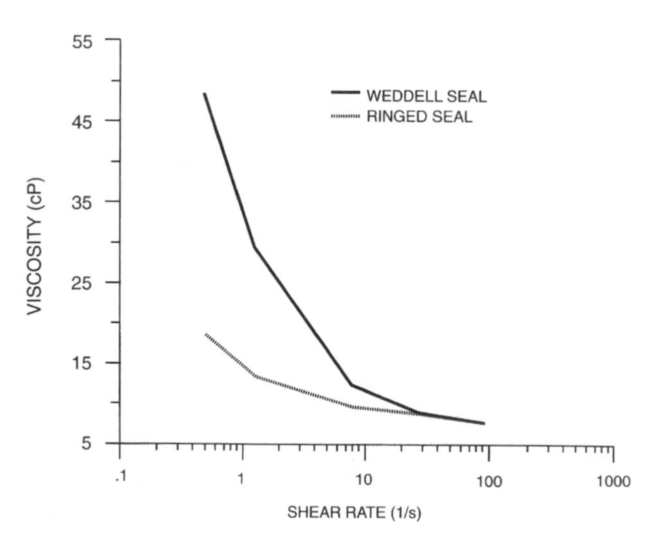

Figure 3-8. Viscosity (in centipoise) versus shear rate (in inverse seconds) of ringed and Weddell seal blood tested at normalized (40%) hematocrit. Human blood viscosity lies roughly midway between these values. Shear rate is an approximate representation of blood flow. (Redrawn from Wickham et al. 1990 and Meiselman et al. 1992.)

Temporary sequestration of red cells in Weddell seals and reduced red cell aggregation in ringed seals achieve the same result—reduced blood viscosity while not sacrificing oxygen storage—evolutionary solutions of a common problem by differing adaptive techniques.

Hypoxic Tolerance

Brain

The mammalian brain functions best when it is well supplied with a near-constant perfusion of oxygen-rich blood, and therefore it is not surprising that a primary adaptation of diving mammals should be the maintenance of cerebral blood flow. Existing evidence shows a relatively well maintained brain perfusion during experimental dives in harbor seals (Kerem and Elsner 1973a) and sea lions (Dormer et al. 1977). In addition, there is evidence to indicate that the brain of certain species of seals is well adapted to periods of very low oxygen that might occur toward the end of long dives. The first indication of this ability was the observation of Scholander (1940) that, at the termination of a long experimental dive, the arterial blood oxygen level of gray seals reached a value of 2 mL/100 mL. (Well-oxygenated blood would have had an oxygen content of 20 or more mL/100 mL of blood.) We can estimate what that value represents in oxygen tension (partial pressure) and, making an assumption about the pH of the blood, we derive a value for oxygen partial pressure of about 10 mm Hg, too low to maintain consciousness in terrestrial mammals. Some years later, determinations of arterial oxygen partial pressure in Weddell seals confirmed that they were able to tolerate the remarkably low value of 8 to 10 mm Hg at the end of long simulated dives (Elsner et al. 1970a). Similar low levels were found in harbor seals (Kerem and Elsner 1973a). Toward the end of a long dive, brain oxygen consumption gradually decreased and was supplanted by anaerobic metabolism as evidenced by lactate production (Kerem and Elsner 1973a) (Fig. 3-9). This represents a capability for anaerobiosis and brain hypoxial tolerance unknown in terrestrial mammals (Kerem and Elsner 1973b; for reviews, see Elsner and Gooden 1983, Lutz and Nilsson 1997). Bottlenose dolphins apparently also tolerate unusual levels of brain hypoxia, as evidenced by collections of expired air containing very low oxygen concentrations after long dives (Ridgway et al. 1969).

Heart

The seal heart provides another example of the ability of some sensitive seal tissues to tolerate low oxygen. Hearts of most terrestrial mammals are dependent almost exclusively on oxidative metabolic resources and when deprived of oxy-

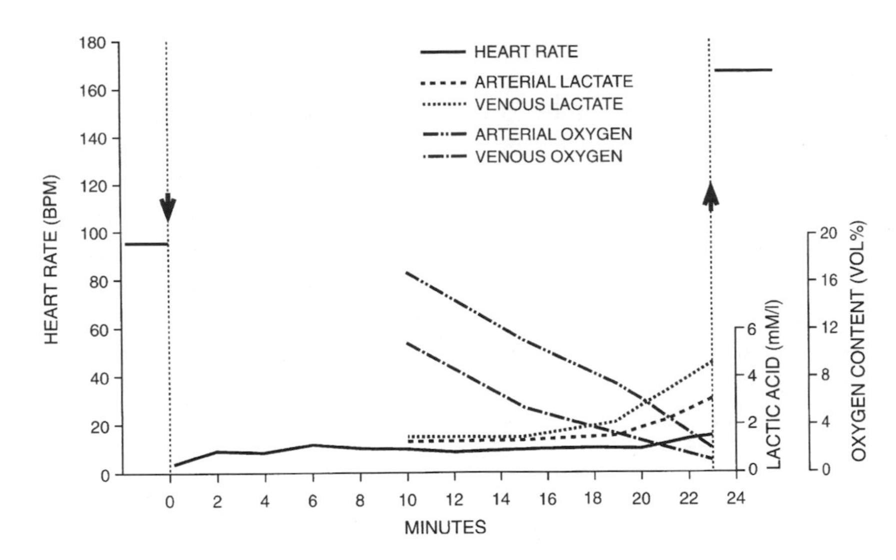

Figure 3-9. Cerebral responses to diving in a harbor seal. Recordings of heart rate, arterial and cerebral–venous oxygen, and lactate contents (blood supplying and draining from the brain). Blood flow to brain remained relatively constant throughout. Lactate release (arterial vs. venous lactic acid) began when oxygen consumption (arterial vs. venous oxygen content) declined. (Modified from Kerem et al. 1973.)

gen their capacity for anaerobiosis is brief, and the contribution from that source is subject to disruption by rapid acidic inactivation of glycolytic enzymes (Rovetto et al. 1975). Several lines of evidence show that certain marine mammals enjoy considerably more cardiac anaerobic capacity and that it can effectively protect the seal heart during long dives (Kjekshus et al. 1982). In a comparison of harbor seals and pigs in which cardiac mechanical function (left ventricular contractile response) was determined during stepwise induction of hypoxia, myocardial oxidative and glycolytic reserves were appreciably higher in seals. The seal hearts tolerated myocardial hypoxia better than did the pig hearts (White et al. 1990). A similar cardiac tolerance of hypoxia exists in the freshwater muskrat, *Ondatra zibethica* (McKean and Landon 1982, McKean 1984).

Laboratory experiments with dogs and pigs have shown that coronary blood flow is regulated to match the metabolic requirements of heart muscle, and myocardial metabolism is closely related to the heart rate. Another determinant of heart work is the downstream or aortic pressure, the ventricular after-load. The myocardium is highly dependent on a constant source of oxygen, and hypoxia imposes a severe stress on the terrestrial mammalian heart (Feigl 1983).

The diving seal's ventricular contractility and metabolic demand decline during experimental dives, and both oxygen consumption and coronary blood flow decrease in approximate proportion to the heart rate. Blood pressure in diving seals is little changed in the dive, and the metabolism of the heart muscle declines along with the diving bradycardia and the accompanying lower work load. Accordingly, the average coronary blood flow of gray and harbor seals is considerably reduced during the dive (Blix et al. 1976, Kjekshus et al. 1982). It is not, however, maintained at a constant lower level, as might be expected. Rather, it fluctuates intermittently, from frequent cessations of flow to high levels approximately equivalent to the nondiving condition (Elsner et al. 1985). The cyclic fluctuations in the seal's coronary perfusion suggest alternate utilizations of both oxidative and glycolytic (i.e., anaerobic) energy sources accomplished by switching between decreased flow in which anaerobic reactions predominate, alternating with conversion to oxidative metabolism accompanying restoration of flow and washout of accumulated end products (Elsner et al. 1992, Elsner and de la Lande 1998).

Net lactate release by the heart occurs as oxygen consumption decreases (Kjekshus et al. 1982). Comparable cardiac lactate production in a human or dog would indicate the presence of a severe cardiac pathological event. The suggestion from this evidence is that the hearts of some marine mammals have advantages over those of terrestrial mammals in being able to continue functioning on anaerobic metabolism supported by abundant myocardial tissue glycogen (Kerem et al. 1973). The universality of the response to diving asphyxia, even in terrestrial mammals, was demonstrated by the observation that dogs trained to immerse their snouts in water for 30 sec showed bradycardia, reduction in coronary perfusion, and reduced myocardial oxygen consumption (Gooden et al. 1974).

Vascular Smooth Muscle

Vasoconstriction is maintained by a persisting contractile state in the vascular smooth muscle of small arteries and arterioles supplying the tissue, a condition normally dependent on some minimal level of oxygen supply. At a value below this critical supply resulting in sufficiently severe local hypoxia, the vascular smooth muscle relaxes, resulting in breakthrough dilation and restoration of blood flow. With resupply of oxygen, the vascular smooth muscle would be reenergized, and vessels would constrict again. The cycle would be repeated providing intermittent perfusion of the organ, as has been demonstrated in seal's cardiac muscle (Elsner et al. 1985) and may well occur in other tissues. Even a trickle of blood from such a mechanism might well be sufficient to maintain the oxidative integrity of the supply blood vessels and perhaps of the tissue itself (Gooden and Elsner 1985, Cherepanova et al. 1993). The recent study of myoglobin–oxygen desaturation during Weddell seal dives (Guyton et al. 1995) lends credence to this concept. Besides its confirmation of oxymyoglobin as a source of oxygen storage, another interesting observation of that study was that there were instances of partial oxygen resaturation, the most likely source of oxygen being occasional brief reperfusions of working skeletal muscle.

Metabolic and Cellular Processes

It has been known for many years that the primary metabolic fuel of the seal, whether diving or not, is fat, as indicated by its respiratory quotient (RQ) of about 0.73 (Scholander 1940, Hart and Irving 1959, Kooyman et al. 1973, Davis 1983). The fish diets of seals are rich in protein and fat, but low in carbohydrates. Study of fuel homeostasis shows that the harbor seal conserves carbohydrate; the fractions of lactate turnover and glucose oxidized were 21% and 3%, respectively, compared with values of 57% to 91% and 41% to 51% in terrestrial mammals. Because most natural dives are of brief duration and therefore are supported by oxidative activity, the normal metabolism of fat was hypothesized to continue while oxygen was available, with a switch to anaerobiosis in longer dives and in tissues rendered ischemic (Davis 1983).

Evidence for unusually high concentrations of glycogen in seal tissues indicates the resources available for anaerobic metabolism (Kerem et al. 1973). The mammalian heart is primarily an aerobic organ, but a considerable capacity for anaerobic metabolism has been demonstrated in the harbor seal heart (Kjekshus et al. 1982). Seal cardiac tissue is rich in myoglobin (Blessing and Hartschen-Niemeyer 1969), and its capacity for acid buffering (Castellini and Somero 1981) probably helps to prevent or delay glycolytic inhibition in the face of the inexorable lowering of pH in long dives. In another study, harbor seal myocardium had fewer mitochondria per gram of tissue and lower respiratory rate than dog heart (Sordahl et al. 1983). This condition suggests lower mitochondrial oxidative activity in seal heart, consistent with unusual dependence on anaerobic metabolism, as in long dives, but it may be more closely related to the lower maximum oxygen transport, as shown in exercising seal and dog studies (Elsner 1986, 1987).

Considering the many indications of an unusual ability to tolerate severe asphyxia in the armament of marine mammals, it would be sensible to look for biochemical manifestations of cellular resistance to the anticipated ill effects of long apneic dives. That search has been undertaken by several investigators, but the results are unclear and unresolved. In their reviews of the subject, Castellini et al. (1981) and Kooyman et al. (1981) argue that clear adaptations of oxidative and glycolytic enzymes have not been demonstrated and that comparisons with similar systems in terrestrial mammals have not shown convincing differences favoring the diving species. There is not universal agreement with that conclusion, and some additional evidence for biochemical adaptations is summarized by Blix (1976), Hochachka and Somero (1984), Hochachka et al. (1988), and Hochachka (1992).

Marine mammals can tolerate anaerobic cellular conditions that result in elevated lactic acid and declining pH that terrestrial mammals would find disruptive (Scholander 1940; Elsner et al. 1970a; Kerem and Elsner 1973a,b). Arterial pH in the nondiving seal runs about the mammalian norm of 7.4, declines during long dives and declines further, sometimes as low as 6.8 (Scholander 1940, Kooyman et al. 1980) in the immediate postdive minutes as lactate is flushed into the general circulation. Inhibition of some enzymes at low pH suggests that the pH optimum of these enzymes may have a greater range in marine mammals to support functioning in acid conditions that would depress their activity in terrestrial species. Indications of tolerance to low pH comes from several sources, including experiments showing that isolated kidney tissues from harbor seals continue to show active transport function despite media pH low enough to inhibit such activity in rat kidney (Koschier et al. 1978, Hong et al. 1982a). However, activities of one glycolytic enzyme, lactate dehydrogenase, from brain, heart, and muscle of rabbits, dogs, and Weddell seals were similar and were little changed over a wide range of low pH (Castellini et al. 1981). This topic requires more study, and it will likely be a fertile field for further investigation.

Pregnant Seals

It might be anticipated that pregnant marine mammals, burdened with a demanding fetus, would be severely challenged by long dives. Weddell seals in advanced pregnancy do not appear to be seriously disadvantaged in this respect, and their free-diving performance did not differ substantially from that of nonpregnant animals (Elsner et al. 1969). Fetal heart rate declined, but more slowly than the maternal rate during both simulated and natural dives (Elsner et al. 1970b, Liggins et al. 1980, Hill et al. 1987). The mechanism governing the fetal heart rate under these conditions remains unknown. Uterine artery blood flow in pregnant Weddell seals (monitored by an implanted flow transducer) was well maintained during simulated dives, contrasting with renal ischemia and indicating a sharing of maternal oxygen reserves with the fetus (Elsner et al. 1970b). A similar finding (by radioactive microsphere distribution) in the same species was demonstrated by Liggins et al. (1980). Interestingly, a protective redistribution of cardiac output, favoring the central nervous system and somewhat resembling that of diving seals, occurs in asphyxia-threatened terrestrial newborn and fetal mammals (Berhrman et al. 1970, Elsner and Gooden 1983), further evidence of the universality and ancient evolutionary origins of asphyxial defenses.

Neural Regulation of Diving Responses

Cessation of breathing, bradycardia, and peripheral vasoconstriction are the triad of fundamental physiological reactions in breath-hold diving. Their initiation in seals is directly related to the onset of breath-holding stimuli, flooding of the face, usually with the lungs in the expired position and thus with relaxation of pulmonary stretch receptors. The lengthening dive results in the progressive decline of blood and cellular oxygen, increasing carbon dioxide and acid metabolic products; and it is these conditions that also influence the regulation, primarily neural, that governs the diving response. These conclusions are based on studies of both marine and terrestrial mammals and birds as well as of humans (for reviews, see Andersen 1966; Angell-James and Daly 1972; Butler and Jones 1982; Blix and Folkow 1983; Elsner and Gooden 1983; Daly 1984, 1986, 1997; Elsner and Daly 1988). Treatment here of this topic will be brief, and the reader is referred to these and other sources for further consideration.

Dogs, Ducks, and Seals

It is clear that the neural and humoral mechanisms operating in the diving response of marine mammals are not exclusive to marine mammals. Rather, they are extensions of regulatory processes common among mammalian species as they respond to the challenges of breath-holding, asphyxia, and diving. They are qualitatively similar but quantitatively elaborated from general vertebrate regulatory reactions. The fundamental neural mechanisms governing mammalian responses to asphyxia were first identified in experimental investigations of terrestrial species (Angell-James and Daly 1969a,b) and were later identified in seals. Although the physiological adaptations to diving are most highly developed in aquatic species, some aspects of the responses and their regulations are detectable in all birds and mammals, including humans, in which they have been sought.

The performances of living systems as they interact with their environments are characterized by orderly and integrated reactions, and diving is no exception to this general rule. We have seen that the acts of breath-holding and diving elicit numerous complex responses, notably in the respiratory and cardiovascular systems. Some of these occur rapidly at the onset of dives and again during restoration of breathing at the termination of the dive. Other reactions take place more slowly, as in the inexorable progressive discomfort leading to the breaking point at which time the animal surfaces and breathing is resumed. Consideration of these contrasting rapid and slow actions tells us something about the nature of the regulatory mechanisms that govern them.

Physiological regulations are of two general types: neural and humoral. Impulse traffic in the nervous system and the circulation of blood-borne chemical–hormonal substances serve the contrasting fast and slow control modes, respectively. These two mechanisms do not operate independently, rather they function by complex interactions whose result is often unpredictable from knowledge of the separate responses alone. Cessation of respiration and onset of bradycardia, for example, begin and end promptly at the beginning and end of the dive, whereas chemical signals, indicating the progressive changes leading to the need to restore breathing, act more gradually. The obvious responses of the diving animal include apnea, bradycardia, and peripheral vasoconstriction. Recognizing that these effects are the result of several neural reflexes and reflex interactions, as well as an integration of neural and humoral communications, we can appreciate the misleading inaccuracy of the term *diving reflex*.

Knowledge derived from study of both terrestrial and aquatic mammals and birds has shown that they are fundamentally similar with regard to the neural controls of respiratory and cardiovascular interactions. Research endeavors have centered on several reflex actions initiated by neural receptors at different locations: (1) the face, nasal mucosa, and pharynx, (2) lungs, (3) peripheral chemoreceptors, and (4) baroreceptors (for reviews, see Andersen 1966; Butler and Jones 1982; Daly 1984, 1986, 1997; Elsner and Gooden 1983). The responses operate through feedback loops; that is, signals generated at the effector end of the reflex arc are transmitted back into the system and indicate the need for increased or decreased action. Negative feedback, for example, provides information that deviation from a target output has occurred and indicates the need for a correction.

Interactions among various reflex circuits vastly increase the complexity and subtlety of the action needed to produce the desired end product. An example of the integrated nature of the response is seen in the exquisite regulation of arterial blood pressure when a seal is experimentally dived (see Fig. 3-3). There is an immediate cessation of breathing and a decrease in heart rate with an associated abrupt decline in cardiac output. The vascular resistance to peripheral blood flow increases to precisely the level required to balance the reduced cardiac output such that the arterial pressure remains nearly constant during the transition from air breathing to diving. The opposite effect, sometimes with overshoot, takes place at the end of the dive when breathing is resumed. The changes at dive onset and termination often occur on a virtual heart beat-to-beat time scale.

We have noted earlier that the diving response operates to protect the animal against the unfavorable effects of asphyxia. The French physiologist Richet initiated the investigation of the neural components of diving in ducks with a simple experiment showing that asphyxia alone was only part of the appropriate stimulus. Water immersion was required to elicit the full and most effective response. He also discovered that the decrease in heart rate was mediated by parasympathetic (vagus nerve) control. Severing the vagus nerve or injecting atropine abolished the bradycardia (Richet 1894, 1899). Many experiments on ducks and seals followed, and the current evidence indicates that the diving response is produced and maintained by the following sequence of events: breath-holding with face immersion and gradual lowering of arterial blood oxygen and increasing carbon dioxide levels (Daly et al. 1979).

The reflexes invoked by diving produce reactions characterized by activation of both the sympathetic and parasympathetic divisions of the autonomic nervous system (Fig. 3-10). Obvious examples are the vigorous peripheral vasoconstriction (sympathetic) and the abrupt bradycardia (parasympathetic), clearly mediated through autonomic activation at the onset of diving. It is likely that catecholamine secretion participates in these reactions, and maintained blood flow in the adrenal gland supports such release (Zapol et al. 1979, Blix et al. 1983). Circulating catecholamine levels increase during voluntary dives in Weddell seals (Hochachka et al. 1995), but their actions may be attenuated in the heart by the corresponding secretions of the parasympathetic-transmitting substance acetylcholine (Vanhoutte and Levy 1980, Cohen et al. 1984).

Face Immersion

Through the experimental work of several investigators we know that the separate effects of breath-holding and face immersion are distinct, that their combination results in an intensified reaction, and that the interactions do not necessarily equal the sum of the two separate components. This principle of unpredictability of the combined contributions from reflex interactions, as considered from the perspective of the separate reflex responses, is a recurring theme in cardiopulmonary neural regulation (Daly 1984, 1986). It has been consistently found that breath-holding with face immersion produces a more profound response than breath-holding alone. The precipitating condition is the combination of apnea accompanied in the dive by the simultaneous stimulation of nerve endings in the region of the nose and mouth (Scholander 1940, Angell-James and Daly 1972, Tanji et al. 1975). Similar effects are produced by stimulating the pharynx. Thus, the primary nerves involved are the trigeminal, glossopharnygeal, and superior laryngeal (for review, see Daly 1984).

In a more general sense than only diving, these reflexes serve the function of airway protection against the intrusion of potentially disturbing substances. It has been known for more than a century that the nasal and pharyngeal mucosa are sensitive to the irritating stimuli of smoke and noxious gases and that such stimuli evoke bradycardia and probably other cardiovascular reactions (Kratschmer 1870, Forster and Nyboer 1955, Angell-James and Daly 1969a, White et al. 1974). Interruption of the trigeminal innervation of diving mammals reduces or abolishes the submersion bradycardia in harbor seals (Dykes 1974) and muskrats (Drummond and Jones 1979). Blockade of the trigeminal nerves also prevents the seal's external nares from opening (R. Elsner, unpubl.), indicating that inspiration requires active muscular effort for dilation of the nasal apertures.

Pulmonary Reflexes

Development of the cardiovascular diving response generally depends on the cessation of lung inflation that occurs along with immersion. Mammals respond to lung inflation with increased heart rate, the cyclic changes referred to as sinus arrhythmia (Hering 1871; Anrep et al. 1936a,b). The lung tissue contains receptors that respond to stretch with tachycardia and peripheral vasodilation. Experimental inflation of the lungs in a dog (Angell-James and Daly 1969a, 1978) or seal (Angell-James et al. 1978, 1981) results in an immediate increase in heart rate. Inflation of the diving harbor seal's lungs with volumes equal to or greater than normal tidal volumes during experimental dives causes prompt increase in heart rate. Deflation reverses the procedure with return to bradycardia (Angell-James et al. 1981). This effect may be involved in the increasing heart rate that sometimes occurs as the seal ascends toward the water surface and its lungs expand with the decrease in external pressure. The neural mediation of pulmonary effects of inflation and deflation, and their influence on cardiovascular responses, is qualitatively similar in terrestrial and marine mammals, but the aquatic species, at least seals, react more vigorously.

Peripheral Chemoreceptors

Another mechanism becomes activated as the dive progress. Progressive asphyxia leads to increasing levels of hypoxia, hypercapnia, and acidosis, and these conditions stimulate the peripheral chemoreceptors. Principal chemoreceptors are the carotid bodies located in the neck at the bifurcation of the internal and external carotid arteries. Chemoreceptor activation would normally stimulate pulmonary ventilation, as it does in the nondiving seal (Daly et al. 1977). That influence, however, is countered by the persisting inhibitory interactions derived from upper airway and facial sensory (trigeminal nerve) input from immersion. The neural inputs from the immersed face are reinforced. This combination results in partial suppression of the need to restore pulmonary ventilation, which would otherwise respond to the induced urge to breathe resulting from hypoxia and hypercapnia (Daly et al. 1977, Elsner et al. 1977a, Daly 1997).

Bradycardia in the experimentally dived seal is reversed by elimination of carotid body activation, but continued carotid body stimulation under diving conditions maintains bradycardia (Daly et al. 1977, Elsner et al. 1977a), reflex peripheral vasoconstriction (M. de B. Daly, R. Elsner, and J. E. Angell-James, unpubl.), and coronary vasoconstriction (Elsner et al. 1992). A similar role of carotid body chemoreceptors in diving ducks was earlier demonstrated by Jones and Purves (1970) by denervation of the carotid bodies and thus eliminating bradycardia.

Arterial Baroreceptors

Little overall change in arterial blood pressure occurs in seals during simulated dives (Irving et al. 1942, Elsner 1969) and on this basis one might suspect that baroreceptor reflexes have little influence on the regulation of diving responses. That view was challenged by Angell-James and Daly (1972) with the suggestion that the relationship between arterial blood pressure and heart rate may be altered during dives such that heart rate at the same reference pressure as in the nondiving condition would be slower. Subsequent experiments with harbor seals demonstrated this effect (Angell-James et al. 1978). Baroreceptor reflexes that would tend to restore heart rate and vascular resistance toward the nondiving condition are readjusted to an increased sensitivity (increased gain) that regulates these outputs at the new level during the dive. Reflex effects are promptly reversed at the termination of the dive (Angell-James et al. 1978).

The regulatory circuitry is summarized in Figure 3-10. This brief summary of the present knowledge regarding neural regulations of the diving phenomenon would be incomplete without mention of the sometimes overriding influence from higher structures in the central nervous system. This is a topic that has received little direct attention because of the obvious difficulties inherent in experimental interventions into the cortical influences operating to alter or control circulatory reflexes. However, there is an abundance of observational information indicating that conditioning and psychological perspective of the animal can drastically influence the reactions of the diving experience. For example, seal heart rates sometimes show anticipatory changes, as when terminating the dive (see Fig. 3-2). In other instances, bradycardia will be apparent in response to an episode of fright or disturbance. A reasonable generalization would suggest that the physiological adaptations to submergence are subject to great variation in timing and intensity, resulting in graded responses depending on the situation and the animal's condition.

Human Divers

We do not ordinarily regard humans as marine mammals, and human adaptations for aquatic life appear trivial and inconsequential when compared with those of truly marine species. Nevertheless, human breath-hold diving has a long and successful history, and physiological adjustments to diving are readily identified. Food gathering from the sea by diving has its origins in prehistory, and the activity persists today in the Ama of Japan and Korea (Hong and Rahn 1967, Hong 1988). The profession is supported by the high market value of harvested invertebrates such as abalone. The modern Ama are usually clothed in wet suits to protect them from excessive cold, and this relatively recent modification permits a full expression of their diving capabilities. The diving pattern of one group of Japanese Ama at Hegura Island in the Sea of Japan reveals a performance of about 100 dives during a 5-hr work day. Dives typically last about 1 min and seldom exceed 90 sec. The emphasis is placed on obtaining maximum useful bottom time while reducing surface recovery time to a minimum. Some divers, known as *funado,* are assisted in descent by weights and during ascent by tethered retrieval from a boat. Their dive durations are similar to those of unassisted *cachido* divers, but dive depths often exceed 20 m, deeper than unassisted dives, which rarely reach 15 m (Mohri et al. 1995).

Whatever overall oxygen conservation might result from human breath-hold dives is almost certain to be modest (Lin and Hong 1996), although some effect has been noted (Elsner et al. 1971a). Clear evidence indicates myocardial oxygen conservation in dogs (Gooden et al. 1974) and humans (Bjertnaes et al. 1984) during experimental dives.

The deepest human breath-hold dive has been recorded

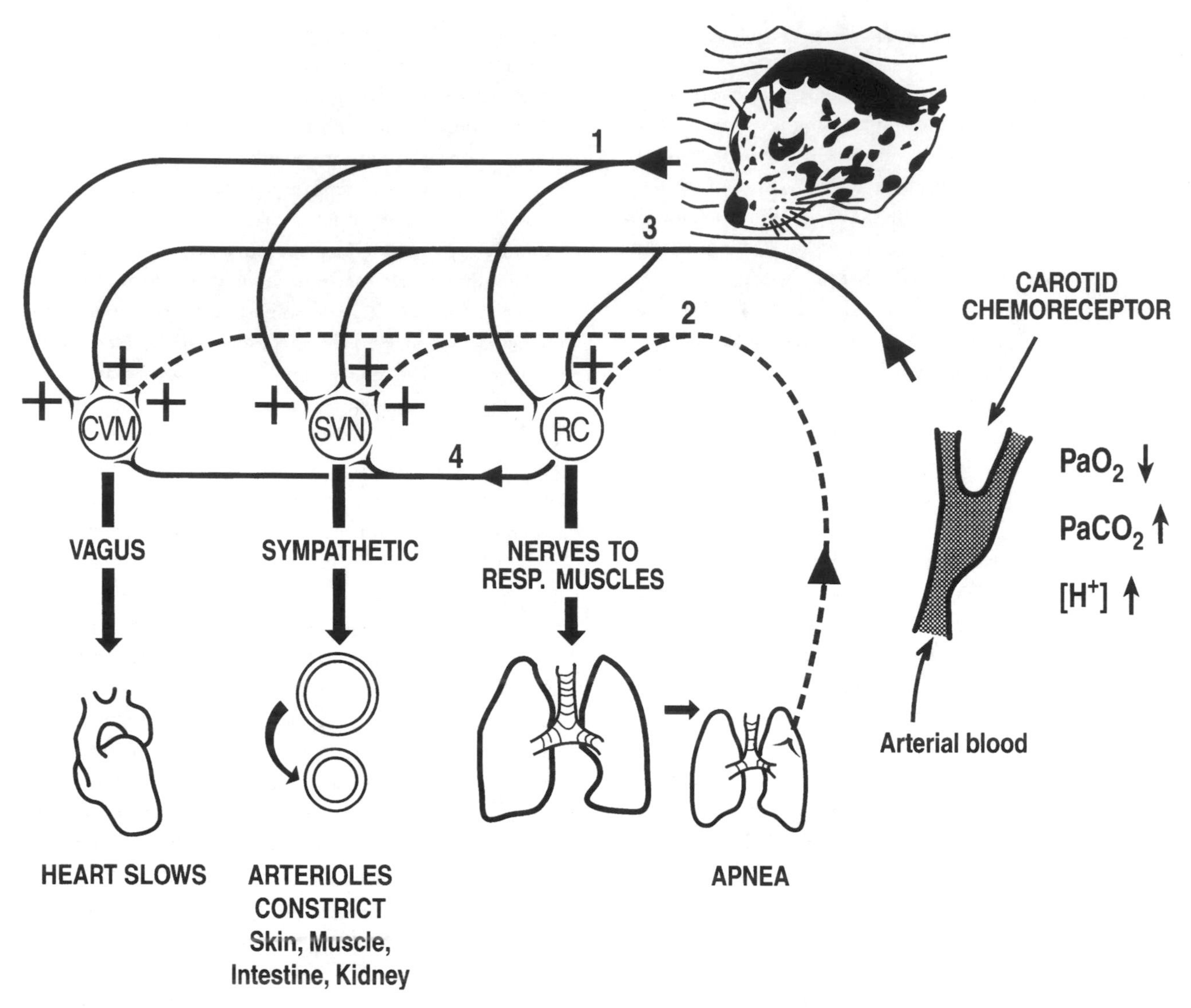

Figure 3-10. Some neural controls operating during experimental dives of harbor seals. The probable sequence is indicated by pathways: (1) stimulation of trigeminal nerve endings of the face by water resulting in apnea with lungs in the expiratory position (relaxing pulmonary stretch receptors), bradycardia, and selective vasoconstriction; (2) reinforcement of bradycardia and vasoconstriction by reduced activity of pulmonary stretch receptors; (3) progressive hypoxia and hypercapnia stimulate carotid body chemoreceptors, also reinforcing bradycardia and vasoconstriction (respiratory drive is suppressed by continuing trigeminal input); (4) cessation of central inspiratory drive. CVM, cardiac vagal motor neurons; $PaCO_2$, arterial carbon dioxide tension; PaO_2, arterial oxygen tension; RC, respiratory "center" neurons; SVN, sympathetic vasomotor neurons. (Redrawn with minor modifications from Daly 1984.)

recently at an astonishing depth of 130 m in a long-standing formal competition (Gessner 1996). The longest apneic dive recorded in the Guinness Book of Records, 5 min 40 sec, has recently been exceeded by a Swedish underwater rugby athlete at just over an equally astonishing 6 min (Schagatay 1996).

Bradycardia and limb vasoconstriction in human breath-hold divers are well documented (for reviews, see Elsner et al. 1966b, Elsner and Gooden 1983, Lin and Hong 1996). During the past 40 years, approximately 1000 human subjects have been studied in various experimental breath-holding dives. Face immersion alone is sufficient stimulus to trigger physiological responses. Nearly all subjects demonstrated some slowing of heart rate, usually of gradual onset. About 2% showed marked bradycardia, even less than 20 beats/min. One healthy subject demonstrated several beat-to-beat intervals equivalent to a remarkably low heart rate of 5 beats/min (Arnold 1985)!

Because asphyxia occurs as a result of or in association with several pathological conditions, there are obvious medical implications. Our appreciation of the mechanisms governing responses in conditions such as neonatal asphyxia, sudden infant death syndrome, and near-drowning and resuscitation has been enhanced by knowledge of diving physiology (Daly et al. 1979, Elsner and Gooden 1983).

Pressure

An inevitable consequence of submergence in water is the increase of pressure with depth. A descent of 10 m in water produces the equivalent of approximately 1 atmosphere of pressure. Thus, the deep-diving Weddell and elephant seals and some species of whales regularly experience 50 to 100 atmospheres of pressure and just as regularly return to 1 atmosphere after each dive. Although marine mammals are not exposed to breathing gases at those great pressures—as is the case with scuba-supported and other engineered human deep dives—they might be exposed repeatedly to some of the cumulative adverse effects of high pressure. A recent review deals with the general subject of gas physiology in deep dives (Lundgren et al. 1996).

Living soft tissue that does not enclose air spaces is composed of substances, in large part water, that are compressible to only a very small extent, and the ambient pressure during the dive is readily distributed throughout the tissues. Only at great pressures when compressibility does become a factor, as in alterations of enzyme kinetics, are they likely to be adversely affected. Some of these effects can occur at pressures experienced by deep-diving marine mammals (Hochachka and Somero 1984), but specific disruptions of cellular functions and adaptive mechanisms are unknown. The air-filled spaces of the animal's body are sites of potential damage due to changes in volume accompanying the frequently repeated cycles of compression and decompression when diving to depths of more than a few atmospheres of pressure. Pressure differences across membranes and other structures make them vulnerable.

Barotrauma

The damage caused by rapid expansion or squeezing of gas spaces that exceeds the structural integrity of tissues is referred to as barotrauma. Boyle's law, relating pressure and volume, expresses these changes, stating that at constant temperature the volume of a gas decreases inversely as the pressure increases. Some of the air-filled spaces of the body, the lungs for instance, are enclosed within tissues of limited distensibility. The human chest wall is more rigid than that of marine mammals, and accordingly the diving human breath-holder cannot exceed a certain depth determined by the flexibility of the chest and pulmonary structures and their resistance to distortion (Agostini 1965, Craig 1968). *Lung squeeze* occurs when these tissues are deformed to the point of distress. Ordinarily this condition limits human breath-hold dives to about 30 m, the depth at which the full lung volume is compressed to its residual volume. Further depth might be accommodated by transfer of blood to the distensible pulmonary veins or by shifting abdominal contents pressing against the diaphragm and thus resulting in displacement of thoracic air (Schaefer et al. 1968). The extraordinary depths reached by the rare human examples reported are likely to have been achieved by some unusual combinations of large lung capacity, small residual volume, and thoracic flexibility.

Deep diving marine mammals have flexible chest walls and other structures that are capable of sufficient collapse to render the lungs virtually airless. Airway structures, including the terminal alveolar ducts leading to the alveoli where gas exchange with the pulmonary capillaries takes place, are reinforced with structural stiffening by cartilage. This morphology permits the alveoli to collapse while the proximal airways are still patent, thus ensuring that pulmonary gas exchange is eliminated as pressure increases and preventing dangerous accumulation of nitrogen in the circulating blood and peripheral tissues (Denison et al. 1971; Denison and Kooyman 1973; Kooyman 1973, 1988). The unusual oblique position of the diaphragm, notable in phocid seals, attached high ventrally and low dorsally, permits its bulging and the displacement of abdominal contents to occupy part of the thorax. Combined with these anatomical features is an unusually small residual lung volume, allowing more complete lung emptying and gas exchange in the respiratory cycle (sea lion: Denison et al. 1971; harbor porpoise: Irving et al. 1941b; pilot whale: Olsen et al. 1969). Resistance to flow in the airways, alveoli to trachea, is reduced (Olsen et al. 1969, Lieth 1976). This adaptation permits rapid exhalation to low lung volumes, convenient for quick restoration of lung gases at the sea surface.

Decompression Sickness ("Bends")

Respiratory gases are absorbed in blood and body fluids according to their partial pressures (Henry's law states that the amount of gas entering solution is proportional to its pressure; for example, doubling the gas pressure doubles the amount of dissolved gas). When these gases are breathed at increased pressure for more than a critical time, the physiologically inert components, primarily nitrogen, are absorbed. Their slow diffusion out of solution during decompression causes bubbles to form in the course of ascent from the dive.

Bends is a disorder resulting from breathing air at elevated pressure, as in scuba diving. The precipitating cause is the diffusion and expansion of inert gas out of critical body regions, nervous tissue and joints, during decompression as the diver rises to the surface from long and deep dives. Increased nitrogen tension in tissues may theoretically gradually reach an increased level during repeated breath-holding dives, depending on depth, frequency, and duration at surface. Such accumulation may expose the diver to decompression sickness by gradual absorption of nitrogen and bubble formation with its complications and potential damage during decompression.

Instances of bends symptoms in human breath-hold divers provide insight into this hazard. The best documented case is that of a Danish submarine medical officer who repeatedly dived to the bottom of an escape training tank (Paulev 1965). He did 60 breath-holding dives to 20 m during 5 hr, a schedule approximating that of some Japanese Ama divers. Shortly after completing this episode, he experienced symptoms of decompression sickness. He was placed in a recompression chamber at 6 atmospheres, whereupon the symptoms promptly disappeared, evidence that he did indeed have incipient decompression sickness. Three similar cases among other divers were also reported, and a description of bendslike symptoms in Tuamoto pearl divers suggests a similar condition (Cross 1965). These examples indicate that the critical conditions are set by the depth, duration, and frequency of the dives, and these are countered by surface intervals sufficient to allow regular reductions in tissue nitrogen or its elimination from tissues.

Decompression sickness may, however, be unlikely in seals and cetaceans because of the earlier described unusual pressure-activated sequential collapse of the airways that removes gas from the alveoli into the larger bronchioles, bronchi, and trachea, regions that lack the dense pulmonary capillary network essential for gas transfer into the circulating blood. Such a possibility was predicted by Scholander in his 1940 monograph. Studies on isolated lungs and on seals in compression chambers have demonstrated the anatomical and physiological characteristics of marine mammal airways that respond in the predicted manner, that is, by alveolar compression at about 3 atmospheres of pressure (Fig. 3-11) that prevents pulmonary gas exchange at greater depth (Kooyman et al. 1972; Kooyman 1973, 1989; Sinnett et al. 1978). Instrumentation of free-swimming Weddell seals confirmed that pulmonary gas exchange was much reduced during deep dives (Falke et al. 1985). The issue still leaves a question of possible additional protective mechanisms operating in marine mammals when their performance is compared with the above-mentioned experience of Paulev (1965) and others.

Evidence for alveolar collapse in a moderately deep diving small cetacean, the bottlenose dolphin, was obtained in a remarkable series of experiments with a highly trained animal. The training consisted in repeated untethered dives to targets and signalling devices in the open ocean at a maximum depth of 300 m. The animal was trained to breath-hold up to 4 min and then to exhale into a collecting container. Analyses of the expired gases suggested that the collapse of alveoli, and hence the prevention of further pulmonary gas exchange, occurred at about 100 m depth. The thoracic region was visibly compressed during descent (Ridgway et al. 1969). Additional experiments involving repetitive diving to 100 m demonstrated that nitrogen tension in the animal's skeletal muscle remained below a level expected to result in bubble formation (Ridgway and Howard 1979). Extension of these experiments on small cetaceans to the great whales would be most interesting, but that enterprise must await development of the techniques required for its study.

Nitrogen Narcosis

A condition of compressed air intoxication in humans with symptoms of neuromuscular incoordination, drowsiness, hallucinations, mental impairment, and alterations of memory is a consequence of dives to about 4 or more atmospheres (Bennett 1993). It is caused by inert gases in addition to nitrogen, and it is understood to be related to the general explanation for the mode of action of anesthetics. Brain synapses are vulnerable to absorption of anesthetic molecules by the lipids of the synaptic membranes, which then expand and disrupt synaptic transmission, resulting in anesthesia. Present theory explains the effects in humans of inert gases under pressure in producing narcosis in a similar manner (Winter and Miller 1985). We have no knowledge of what happens in marine mammals under similar conditions, but it seems unlikely that their deep underwater activities are much disrupted by this problem.

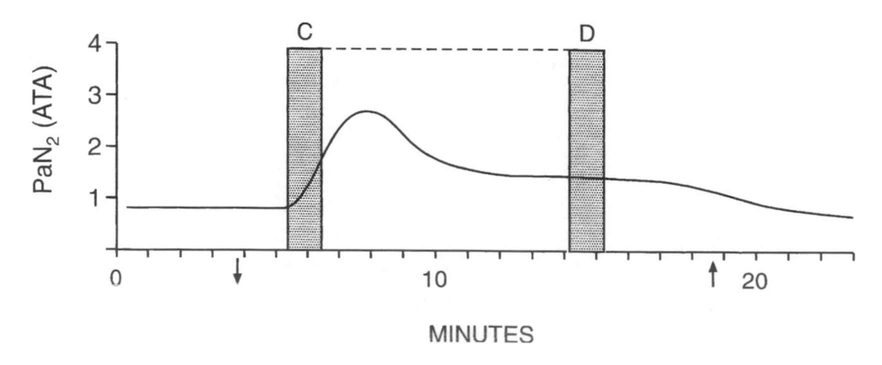

Figure 3-11. Arterial nitrogen tension (PaN_2) during experimental compression (C) and decompression (D) of an elephant seal in a pressure chamber. Blood nitrogen partial pressures remained less than 3 ATA (atmospheres absolute), whereas chamber pressurization varied from 4 to 14.6 atmospheres, indicating lack of lung gas transfer at increasing compressions. (Modified from Kooyman et al. 1972.)

High-Pressure Nervous Syndrome

This condition is experienced by human divers at 15 or more atmospheres of hydrostatic pressure (Bennett and Rostain 1993), a depth equivalent frequently visited by the deeper diving marine mammals. It is not clear whether seals and whales suffer from this condition, although it seems likely that it would be a serious handicap to their normal prey-seeking and underwater navigation. Its symptoms include tremor, dizziness, and nausea. It contrasts with nitrogen narcosis by its onset at greater depths and by its generalized stimulation of the nervous system. Clearly, the diving habits of many marine mammals fall within the depths where serious problems with this pathological effect of exposure to high pressure might occur. We do not know for certain that they are not affected, but as with nitrogen narcosis, it seems unlikely that it can be a major constraint on their diving activities. An understanding of their resistance to these hazards of human diving would be of considerable interest.

Oxygen Toxicity

Prolonged breathing of oxygen in high concentrations results in toxicity noticeable in lung tissue and later in other areas. This toxic effect has fundamentally the same origin as that noted earlier as a consequence of ischemia and reperfusion (Gerschman et al. 1954, Haliwell and Gutteridge 1985). Even in deep dives the alveolar concentration of oxygen in Weddell seal lungs is seldom much elevated (Qvist et al. 1986) as a result of its steady consumption and removal from the lungs. Therefore, this source of oxygen toxicity is an unlikely problem for marine mammals. As noted earlier, however, there is another potential source of toxic oxygen effects: production of oxygen-derived free radicals and other highly reactive forms of oxygen originating from the repeated cycles of ischemia and reperfusion accompanying frequent dives (McCord 1985). How marine mammals might protect themselves against this source of oxygen toxicity is an intriguing question (Elsner et al. 1998).

Hypoxia

Progressive lowering of arterial blood oxygen tension during long dives of seals and dolphins can result in extraordinarily low values at the terminations of long dives, as we have seen. Hypoxia is intensified by a further decline in lung and blood oxygen tension associated with the decreasing ambient pressure experienced during the animal's return to the surface from deeper water. The ensuing cerebral hypoxia rapidly worsens as the diver approaches the surface. It can decrease so much and so rapidly that it can produce fainting in human divers before reaching the surface. This accident is recognized as *shallow-water blackout,* a life-threatening situation and the probable cause of numerous drownings (Craig 1961). The threat seems not to extend to marine mammals, and it has been suggested that their tolerance for extremely low cerebral hypoxia, described earlier, protects them from this frequently encountered situation (Kooyman 1989). One is left with the conclusion that many problems and questions remain unresolved in the matter of marine mammal protection against the adverse effects of their frequent exposures to high pressure in deep dives.

Navigation

Marine mammals engage in prodigious migrations (see Wells, Boness, and Rathbun, Chapter 8, and Bowen and Siniff, Chapter 9, this volume) and the navigational techniques with which these excursions depend remain a mystery. Behavioral skills are certain to be involved, and these will depend on some physiological attributes that will be considered here. Some of the larger baleen whale species travel seasonally between arctic and temperate or tropical regions. Gray whales, for example, migrate annually between the nutrient-rich Bering-Chukchi Seas and the lagoons of Baja California, Mexico, a distance of roughly 8000 km. Their route is well delineated, but we know almost nothing about the mechanisms used by the whales to find their way along this great distance. Other whales, blue, fin, and sei whales, for instance, follow the annual cycle of summer feeding in rich polar seas and winter breeding and giving birth in warmer temperate and tropical regions. Accordingly, the migrations in the northern and southern hemispheres, dictated by the opposite seasons in the north and south, incidentally work to keep the northern and southern populations apart. That is, whales in one hemisphere feed in temperate and polar seas, whereas those in the other hemisphere breed in more tropical waters. Much of this information was obtained in the days of whale hunting, and it was further elaborated by the capture of animals previously marked by Discovery tags (National Institute of Oceanography, U.K.) and others (Dawbin 1966, Mackintosh 1966).

The reasons for the whale migrations are uncertain, but some reasonable speculations have been made. It was pointed out by Kanwisher and Ridgway (1983) that the seasonal massive subcutaneous blubber layer in the great whales is considerably more than would be needed exclusively for an insulation envelope protecting against cold water. They suggest that whale blubber, which can reach 25 cm in thickness in some species, provides sufficient energetic storage to last several months. That capacity of stored fat would likely be enough to supply the whale's need for food during much of the long migratory excursion and its attendant activities, including nurture of the newborn calf. Again, the whaling operations of past years have verified this suggestion by observations of stomach contents, or lack thereof, and blubber thicknesses in whales taken at different locations and seasons. But questions remain regarding what environmental signals they use for guidance and orientation on the long journey, and also what compensatory adjustments there might be for changing buoyancy that would be associated with large changes in body fat content.

Antarctic and northern fur seal populations are larger by several orders of magnitude than their tropical counterparts. The former also engage in long migrations, taking advantage of abundant food in their summer polar sojourn and migrating to temperate waters in winter. The migratory routes and distributions are better known for the northern fur seals, as a result of records from earlier years when they were actively hunted in the open seas (Gentry and Kooyman 1986). Here, too, a great void exists regarding knowledge of their navigational techniques.

Some progress has been made in understanding navigational efforts at local pilotage in the sea ice environment where its performance calls for precision and certainty for finding breathing holes and open water for access to air, and the penalty for failure is swift and fatal. Several seal species inhabit polar oceans dominated by extensive sea ice. The dynamic properties of sea ice formation and breakup are determined by seasonal advance and retreat and by movements of oceanic currents and winds. Leads (cracks in sea ice) open up and later refreeze; ice floes are pressured into ridges and layers of several thicknesses. Winter darkness further complicates this habitat for seals that require ready and assured access to air. Arctic ringed seals and Antarctic Weddell seals are good examples for this consideration, because they prefer continuous shorefast ice for their relatively stable habitats.

Ringed and Weddell seal maximum diving durations are about 20 min and more than 1 hr, respectively. They must depend on their navigation abilities for locating leads and breathing holes. Ringed seals seek ice stable enough for construction of their birth lairs under the snow cover and drifts on the ice (Stirling 1969, Smith and Stirling 1975, Kelly and Quakenbush 1990). Accomplishing this search and navigational effort seems superficially to be an extraordinarily challenging activity. The seals take advantage of weaknesses in the ice for maintaining breathing holes. Holes are created early in the winter season and maintained by continuous attention, in ringed seals by scratching with foreflipper claws, and in Weddell seals by abrasion of the ice with their teeth. Bowhead whales break through thin ice to create breathing holes during under-ice travel (George et al. 1989).

Extensive study of Weddell seal diving characteristics by Kooyman and his colleagues (for reviews, see Kooyman 1981, 1989) has described the species' extraordinary capabilities for under-ice navigation. They can be captured in a seasonal colony, transported to a remote sea ice location far from leads or breathing holes, instrumented with depth recorders and other devices, and released into a hole purposely drilled at that site. They freely dive at the new location, returning regularly for access to air at the newly created breathing hole. They are capable of travelling several kilometers under the ice during their maximum dive limit of more than 1 hr. They are also able to find their way under ice and return unerringly to the breathing hole from which they departed, which is often less than a meter in diameter and located in ice 1 to 3 m thick. Not only that, they must be able to sense the time when their breath-holding capacity—and the half-time of their stored oxygen—will be sufficient for the return journey, altogether a seemingly remarkable feat!

How do they do it? Recent studies designed to determine what sensory modalities might be used have shed some light on the seal techniques. Ringed seals were tested in frozen ponds near Fairbanks, Alaska, and Weddell seals were treated similarly in the fast ice of McMurdo Sound, Antarctica. Breathing holes were drilled at various distances from one another, simulating the natural conditions encountered. Tethered seals swimming freely were able to locate another breathing hole calculated to be close enough for visual identification. Blindfolding of the seals without providing additional cues eliminated their ability to locate another breathing hole, but the blindfolded seals responded to acoustic signals created at holes up to 100 m distant and swam toward that site, locating such holes with a high rate of success (Elsner et al. 1989, Wartzok et al. 1992). The preliminary experiments were followed by a series in which an acoustic tracking procedure was developed to provide a real time display of the seal's location throughout its under-ice traverse, revealing much additional useful information about the navigation process (Wartzok et al. 1992). Both ringed and Weddell seals appeared on occasion to use a ranging procedure for estimating the distance to the acoustic source at the distant hole. That is, at the start of their route they maneuvered some distance at an acute angle to the sound source, as if they were determining the angular separation of the original and new positions, possibly a trigonometric solution to revealing the distance they needed to swim!

In another test the ringed seals willingly explored a route to a breathing hole 100 m distant when the path had been prepared by removal of snowcover, thus admitting light such as might appear through a refrozen lead in sea ice (Wartzok et al. 1992). It is highly likely that such light paths, cracks, and other under ice features provide a chart, even in relative darkness, of routes to be used and identifying locations within the seal's home range. This procedure suggests that the irregular features of the undersurface of the ice make for many identifiable and remembered routes among breathing holes and birth lairs resembling the familiarity that terrestrial mammals enjoy within their home range.

Temperature Regulation and Some Metabolic Implications

Body Temperature and the Environment

Marine mammals conform to the general mammalian endothermic condition, an advantage over ectotherms by having closely regulated chemical and physical reactions relatively undisturbed by alterations in temperature. Rates of biochemical reactions are subject to marked alteration by changing temperature, with a 10°C increase or decrease raising or lowering the rate, respectively, often by a factor of two to three times, the Q_{10} effect (Fig. 3-12).

Many aspects of temperature regulation in marine mammals are poorly understood. As with other kinds of research on these animals, the smaller and more manageable pinnipeds and cetaceans can be studied and experimentally manipulated in the laboratory and in the field. Large whales, for the most part, have been too difficult to study. Much justifiable curiosity has centered on thermal relationships of the great whales, and that knowledge void has been replaced by speculations about their temperature regulation.

Body size has important implications for temperature regulation. Heat is transferred to the environment through the effective body surface, and this consideration has led to definition of a *surface rule* regarding heat flow (Kleiber 1961). The relatively high ratio of surface area to volume of the smaller animal requires more heat production if it is to equal heat loss and thus maintain thermal equilibrium. Although the actual surface area engaged in heat transfer is often difficult to describe, it does not negate the validity of the effective surface-to-volume ratio in its influence on thermal balance.

The topic of temperature regulation in marine mammals has been reviewed in some important earlier literature on the subject, beginning with three classic papers concerned with cold adaptations in arctic animals (Scholander et al. 1950a,b,c). Scholander (1955) also discussed the evolution of homeothermic climatic adaptation in the light of results of these earlier studies. Two reviews by Irving (1969, 1973) dealt specifically with temperature regulation in marine mammals. Blix and Steen (1979) reviewed the special situation of newborn polar homeotherms, including marine mammals. Mathematical models of temperature regulation

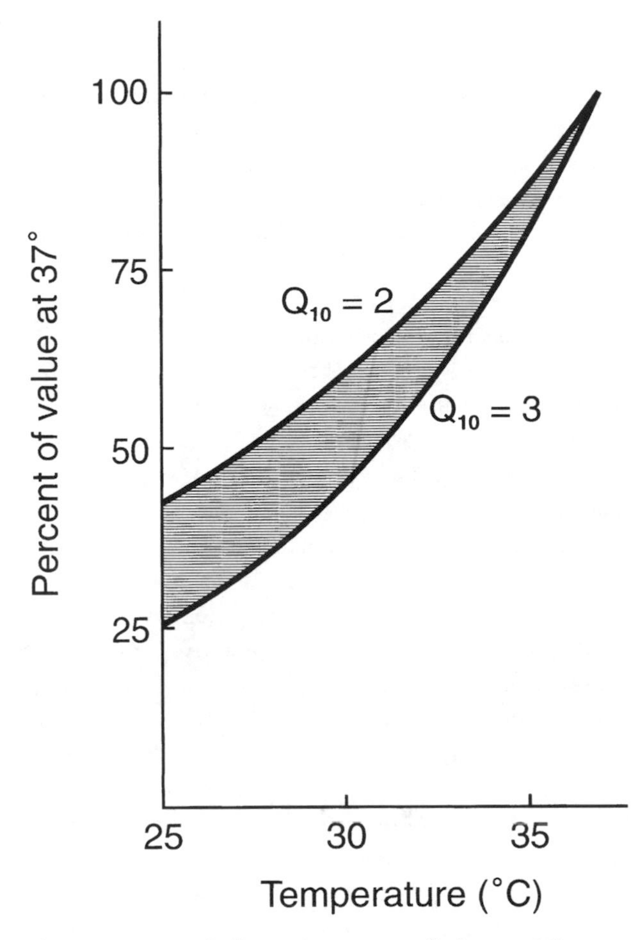

Figure 3-12. Metabolic rate (percentage of value at 37°C) as a function of body temperature showing the effect for two Q_{10} values. Metabolic rate decreases by one-half ($Q_{10} = 2$) to one-third ($Q_{10} = 3$) for a lowering of 10 degrees in temperature. (Redrawn from Willford 1985.)

in marine mammals were considered by Hokkanen (1990). An important review of the more recent literature on thermoregulation in arctic marine mammals appears in Folkow (1992).

Some interesting comparisons can be made with results of thermal studies of water-immersed human subjects (see reviews by Keatinge 1969, Nadel 1984). These include determinations of the importance of subcutaneous fat in reducing heat loss during human immersion in water (Pugh and Edholm 1955, Rennie et al. 1962) and the indication that the layers of muscle underlying the subcutaneous fat can contribute to the insulative envelope during severe cooling (Carlson et al. 1958), a modification that also occurs in marine mammals (see below).

Temperatures in different regions of the marine mammal body vary from the near 37°C value of the body core to a level in the periphery approaching the temperature of the environment. The submerged animal's temperature gradient from core to water is steepest in its insulation, whether fur or subcutaneous blubber. Accordingly, skin temperatures of sparsely furred seals and bare-skinned cetaceans and sirenians approach those of the surrounding water; conversely, the fur-clad sea otter maintains relatively warm skin. When skin temperatures increase, as might occur during overheating when an animal is active, heat is conducted to the adjacent aqueous microclimate and causes it to circulate by convection and to be replaced by cooler water.

Comparing the marine mammal habitats with our own, we can readily suspect that their mechanisms for temperature regulation must differ from ours. We will see, however, that the means by which they keep a relatively constant body temperature are qualitatively like those of terrestrial mammals, including humans. They differ in the extremes of exposure that can be tolerated and in the quantitative expression of their compensatory regulations. An obvious difference is the striking amount of body fat concentrated subcutaneously as insulation, which is a prominent feature of the bare-skinned marine mammals (Bryden 1964). Sea otters depend almost exclusively on insulation provided by air trapped in their dense fur. Fur seals and polar bears also depend partly on this means, as well as subcutaneous blubber, for reducing heat loss (see Pabst, Rommel, McLellan, Chapter 2, Costa and Williams, Chapter 5, this volume).

When an animal in its resting, thermoneutral condition is exposed to gradual cooling, it protects its deep body temperature by reducing heat loss, mainly by autonomic (sympathetic)-mediated peripheral vasoconstriction, thereby increasing internal insulation and widening the thermal gradient from body core to skin and hence to the environment. Although most of this insulative gradient occurs in subcutaneous blubber or in fur, it can sometimes extend deep into muscle tissue. Further cooling eventually requires an increase in metabolic heat production to prevent a decrease in body core temperature. The ambient temperature that produces this increase in metabolic rate is known as the lower critical temperature (Fig. 3-13). The upper critical temperature is less well defined, and has not been clearly established for any marine mammal.

Large populations of marine mammals thrive in the polar regions, and it is obvious that they can be quite comfortable in ice water, an exposure that would be severely cooling and life-threatening for most other mammals. The freezing temperature of seawater varies with the concentration of dissolved salts; the freezing temperature of virtually all of the open seas is –1.8°C. Tropical seas, in which certain marine mammals live or to which they are exposed seasonally, can be as warm as 30°C at the surface, but decrease precipitously to a few degrees above freezing at a depth of a few hundred meters. Thus, most marine mammals encounter cold water throughout their ranges. An exception among marine mammals in their tolerance to cold are the extant sirenians, manatees, and the dugong. They have relatively low metabolic rates (Scholander and Irving 1941, Gallivan and Best 1980, Irvine 1983). Their lower critical temperature in water is about 20° to 23°C, and they prefer warm and shallow water habitats (Gallivan et al. 1983, Reynolds and Odell 1991).

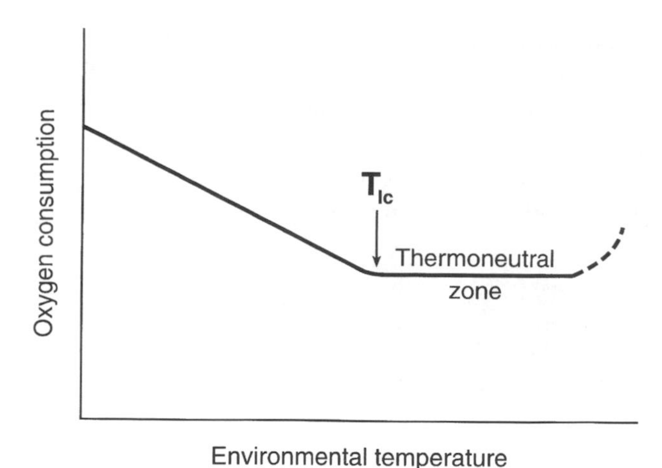

Figure 3-13. Diagrammatic representation of lower critical temperature (T_{lc}) at the environmental temperature value where metabolic rate (oxygen consumption) begins to increase.

Seals experience a wide range of thermoneutrality amounting to about 30°C in air (Irving and Hart 1957). Determinations of the lower critical temperature in adult harbor, gray, and harp seals (*Phoca groenlandica*) in winter showed that they tolerate immersion in ice water without increasing metabolic rate (Gallivan and Ronald 1979, Folkow and Blix 1989). The lower critical temperature of harp seals in still air is about –12°C. (Exposure to lower air temperatures, especially accompanied by wind and overcast skies, might induce the seal to seek the warmth of ice water!) Seasonal variations are, however, evident and appear to be related to the condition of body insulation (Hart and Irving 1959). Insulation is not confined only to blubber in cold-exposed cetaceans, inasmuch as the range of temperatures within thermal gradients from core to skin are still below deep body temperature at the blubber–muscle interface, indicating that insulation includes deeper tissues (Kanwisher and Sundnes 1966, Brodie and Paasche 1985, Vongraven et al. 1990, Folkow and Blix 1992). It should be noted that the conditions for laboratory determinations of critical temperature and other thermal variables cannot be expected to match precisely the complex field environments that include varying conditions affecting heat transfer.

The basis of thermal balance in marine mammals, preventing heat loss in cold water as well as the need to lose heat in response to exercise, can be best understood by an appreciation of the physical laws governing heat transfer. Biological examples do not fit the physical models well, because of the enormous variety of conditions, internal and external, in which living organisms find themselves, and therefore some lack of rigorous application can be anticipated. Nevertheless, an understanding of the physical characteristics of heat transfer clarify the quantitative relationships that determine the thermal state of the animal, and the use of these physical laws is justified, even if the application is simplistic and inexact.

For an endothermic organism to maintain the steady thermal state of a constant deep body temperature, the overall metabolic heat production must equal heat loss. The avenues of heat loss for mammals include conduction, convection, evaporation, and radiation. The relationship is somewhat simplified for animals immersed in water, as radiation becomes vanishingly small, and the conduction–convection avenues can be considered as one, because they are intimately related. For the nonexercising immersed animal, these relationships can be expressed by the heat balance equation:

$$HP = H_c + H_e + H_s,$$

where HP = internal heat production or metabolism, H_c = combined conductive and convective heat loss, H_e = evaporative heat loss (by respiration), and H_s = heat storage (negative for net decrease in body temperature, equal to zero at thermal equilibrium). Radiation, insignificant in water immersion, becomes an important avenue for heat transfer for a seal hauled out on the shore or sea ice, and becomes an added term, H_r (Øritsland 1971, Øritsland and Ronald 1973). Evaporation from the skin surface is then added to the H_e term.

The conductive–convective route of heat transfer is through the skin and therefore its value relates to the surface area of the animal. The thermal gradient of the marine mammal starts with the deep body structures of highest temperature, and from them heat is transferred mainly by the forced convection of blood circulation and by conductance through ischemic tissue ultimately to the skin surface. This circulatory flow is under the control of the sympathetic division of the autonomic nervous system. Blood flow can penetrate the subcutaneous blubber layer, transferring heat directly to the skin surface. Also, the temperature of relatively uninsulated flippers, fins, and flukes responds rapidly for control of heat delivery to the surrounding water. The overall heat transfer from body core to the skin can be expressed by Newton's law of cooling, an empirically derived relationship that holds (approximately) for many biological situations:

$$H = Ak\,(T_c - T_a),$$

where H = heat produced, transferred through the circulation from core to surface, A = surface area, k = heat transfer coefficient, T_c = deep body temperature, and T_a = water temperature. The heat transfer coefficient is determined in part by the conductance (reciprocal of insulation) of the tissues (especially blubber or fur) and blood flow. Relatively small adjustments in peripheral temperatures can maintain deep body thermal equilibrium as heat exchange varies within the thermoneutral range. That is, the body core temperature is protected by insulative cooling of the body shell. Heat production increases when the core heat is encroached on by excessive body cooling below the lower critical temperature. Contrary to one's intuitive perceptions, calculations of thermal balance indicate that large whales may be more likely to have problems with overheating, even in ice water, than with heat conservation, a situation very different from that of the smaller species (Hokkanen 1990). Such information as we have suggests that marine mammals are so well adapted to their aquatic habitats that they remain in a condition almost always above their lower critical temperatures.

Temperature regulation is concerned with the animal's interactions with its thermal environment and the adjustments undertaken to maintain a balance between internal heat production (metabolism) and heat gain or loss. The net result at equilibrium is a steady, or near-steady, body temperature close to the mammalian norm, 37°C. Most marine mammals spend all or the greater part of their lives immersed in water, a medium of high conductivity and with temperatures lower than the body temperatures of the animals; therefore, protection against heat loss is the primary concern. The thermal conductivity of water is about 25 times that of air. Units of heat transfer are expressed in watts, where 1 watt (W) equals 3.6 kJ/hr. These are SI units (International System), which are gradually replacing the older heat transfer designation, kilocalories per hour.

The kilocalorie, the amount of heat required to raise the temperature of 1 kg (very close to 1 liter) of water by 1°C, an expression of specific heat capacity, is still widely used and understood. One watt equals 0.86 kcal/hr. A resting adult human in thermoneutral conditions may have a metabolic heat production of about 60 kcal/hr or 70 W (i.e., about that of a light bulb), depending on factors such as age and body size.

The deep body or core temperatures of some marine mammals are subjected to unusual variations. The effect of diving on deep body temperature is especially noteworthy. The changes that occur during long dives result in a lowering of deep body temperature in seals (see Fig. 3-6) sometimes to as low as 34°C (Scholander 1940; Scholander et al. 1942b;

Hammel et al. 1977; Kooyman et al. 1980; Hill et al. 1987; Andrews et al. 1993, 1995). Such measurements suggest that deep body temperatures of some species of great whales are often a few degrees below those of terrestrial mammals (Morrison 1962, Brodie and Paasche 1985, Vongraven et al. 1990, Folkow and Blix 1992).

From the earlier discussion of diving, we have seen that the deep body temperature of northern elephant seals decreases as much as 4°C during long dives (Andrews et al. 1993, 1995). Skin temperatures increase abruptly just before such dives (R. Andrews, pers. comm.), and the sudden increase in body heat transferred to the surrounding water suggests that the adjustment in diving to a decreased body heat content, resulting in lower body temperature, is in part a neurally regulated phenomenon. As noted earlier, metabolism declines by a factor of two or more for a lowering of temperature by 10°C, the Q_{10} effect (see Fig. 3-12). The combined effects of lowered body temperature and widespread vasoconstriction of internal organs during dives doubtless reduce the seal's oxygen utilization and therefore, extend its aerobic dive duration. Whether or not this extension is sufficient to explain the brevity of surface intervals in northern and southern elephant seal diving bouts is unknown. Nevertheless, it provides at least a partial explanation for the records of long bouts of repeated dives with astonishingly little time for recovery in both of these species (Le Boeuf et al. 1986, 1988, 1989; Hindell et al. 1991a,b). Thermal relationships during dives are intimately related to diving phenomena, and much remains unknown in this important and challenging topic for marine mammal research.

Cetaceans, sirenians, and some pinnipeds are nearly bare-skinned, and they share this distinction with pigs and humans; experiments with pigs have revealed some parallels with their marine mammal counterparts. Swine have thick layers of subcutaneous fat that make for good insulation similar to that of seals (Irving 1956). They tolerate relatively cold skin temperatures as low as 9°C, a skin temperature well below that ordinarily experienced by furred animals and one extremely uncomfortable for humans. Swine also have a lower critical temperature in air of about 0°C, approaching those of seals. (Humans, who originated as tropical mammals, have lower critical temperatures of about 27°C in air, 33°C in water.) It is especially interesting to note that the pigs, when severely cooled, have skin temperatures that regularly oscillate and reveal phasic periods of skin vasodilation (Irving 1956, Irving et al. 1956). These phasic increases in skin temperature resemble the phenomenon of cold vasodilation observed in the appendages of mammals (for review, see Keatinge and Harman 1980), and they suggest a mechanism for intermittent circulation enabling reheating and nutritive perfusion of skin, yet another example of circulatory cyclic behavior. Whether or not this reaction to cold occurs in marine mammals is unknown (Molyneux and Bryden 1978).

Metabolic Heat Production

A possible adaptation of marine mammals to cold water habitats may involve generally increased heat production by elevated metabolic rate. Considerable controversy surrounds this topic (Kasting et al. 1989, Innes and Lavigne 1991, Gallivan 1992). Several earlier determinations indicated that the metabolic rates of marine mammals may be elevated to two or more times the rate of their terrestrial counterparts of similar size (Irving 1969). These early determinations have been challenged by the suggestion that they failed to meet the Kleiber (1961) definitions of a true resting, thermoneutral, and postabsorptive condition in mature animals (Lavigne et al. 1986). The criticism is doubtlessly justified, but it appears that there are exceptions to that general conclusion. Metabolic rate during sleep or basal conditions satisfying the Kleiber criteria has been found to be elevated by a factor of two or less in species such as sea otters (Morrison et al. 1974), mature harbor seals (Ashwell-Erickson and Elsner 1980), and California sea lions (Hurley 1996).

The question of possible elevated metabolic rates in marine mammals cannot yet be regarded as clearly resolved, and other questions remain. The Kleiber specifications do not include such matters as very high body fat content and deviations from normal deep body temperature (Speakman et al. 1993). The vast amount of body fat common to marine mammals, sometimes amounting to 50% of body weight, represents a portion of the total body that has a relatively low circulatory perfusion and probably low metabolic rate and, thus, would dilute the overall value of heat production expressed in terms related to body weight. However, it must be noted that the actual heat production of body fat of any marine mammal has not been determined. It is of interest that fat pigs—whose thermoregulatory similarities to seals have been noted—have anomalously high metabolic rates (Irving et al. 1956).

Determinations of metabolic rate in which the animal is placed in a tank of water and permitted to dive repeatedly during the course of the experiment (seals: Gallivan and Ronald 1979; manatees: Gallivan and Best 1980, Irvine 1983) may be influenced by the decline in metabolic rate that accompanies some dives (discussed earlier). Such measurements would be biased toward lower values if the animals remained submerged for long, quiet dives during metabolic determinations.

The annual pelage molt of seals, which sometimes lasts several weeks, is accompanied by a decline of about 18% in resting metabolic rate in harbor and spotted seals (*Phoca largha*) (Ashwell-Erickson et al. 1986). Metabolic rate can be increased by exercise, shivering, and nonshivering thermogenesis. One source of nonshivering thermogenesis is brown fat, a kind of fat tissue that is much more reactive than ordinary white fat and is capable of sustained high levels of oxygen consumption (see below).

From the foregoing considerations it will be appreciated that the newborns of the smaller marine mammals, because of their size and relatively undeveloped blubber, may be especially vulnerable to the hazards of cold exposure (Davydov and Makarova 1964). Seal pups are born on land or on sea ice, but cetacean and sirenian births take place in water, making for obligatory exposure to the aquatic heat sink from the moment of birth. Many whales migrate to subtropical warm water for parturition. Exceptions are the arctic cetacean species, bowhead, beluga, and narwhal (Doidge 1990).

Weddell seals are born on Antarctic sea ice in the austral spring, often in subfreezing temperatures and wind chill. Metabolism is very high in the first hours after birth, gradually lowering as the fur dries and insulation improves (Elsner et al. 1977b). For the first 5 weeks of life the pups are covered with thick gray fur (lanugo) like the newborn of several other polar seals. It usually dries quickly immediately after birth and provides good insulation, but newborn pup survival is threatened if the drying is prevented by wet weather. Energy transfer from the mother seal to the pup is hastened by the milk's high fat content (42% to 58%; Kooyman and Drabek 1968), and this results in a rapid acquisition of subcutaneous fat by the infant (and also water conservation for the mother). Nevertheless, the pup cannot tolerate immersion in ice water until about 10 days of age, by which time blubber, only 2 mm thick at birth, has increased to about 15 mm (Elsner et al. 1977b). Lanugo is lost about 5 weeks after birth, shortly after weaning.

The smallest newborn marine mammals are especially vulnerable, and they require more protection than is provided by the insulation with which they are born. Ringed seals and Baikal seals are born in subnivean lairs of maternal construction, hollowed out of snow and having access to the water beneath through a hole in the ice floor. Their birth weight is about 5 kg. Dry snow is a good insulator, and the temperature of these lairs is rather steady within a few degrees of freezing, warming a further few degrees when the lair is occupied (Elsner and Pruitt 1959, Blix and Lentfer 1979, B. P. Kelly, pers. comm.) even despite severe outside wind and temperatures to –40°C. Polar bears are born in a similar snow lair on sea ice or in a snow cave on land near the water's edge. The cubs weigh about 500 g and require maternal embrace for the first weeks of life, finally emerging from the den weighing about 10 kg at 3 months of age. By that time they have acquired thick fur insulation, high resting metabolic rate, and a lower critical temperature of –30°C (Blix and Lentfer 1979).

Young harbor seals tolerate cool environmental exposure by virtue of a high resting metabolic rate, body insulation provided by subcutaneous fat, and by the reduction in skin circulation accompanying peripheral insulative cooling (Miller and Irving 1975). Newborn harp and fur seals, and probably other species, can increase heat production by means of thermogenesis originating in brown fat deposits that are strategically situated adjacent to venous circulatory channels, thereby guiding heat from those sources to centrally located organs (Grav and Blix 1976, Blix et al. 1979). The topic has been reviewed by Blix and Steen (1979).

A special circumstance of marine mammal thermal physiology related to body temperature should be pointed out. Deep body temperatures a few degrees lower than those normally recorded in terrestrial mammals can have a considerable influence on lowering the metabolic rate, the Q_{10} effect (see Fig. 3-12). This effect should be considered when determining metabolic rates, and appropriate adjustment should be noted for comparisons with mammals having deep body temperatures at the more normal 37°C. A lower core temperature implies a lesser gradient to the skin surface and thus increased tissue insulation and less heat loss to the environment.

Insulation

The subcutaneous blubber provides an envelope of fixed insulation that is not subject to compression, as is the case with air trapped in fur. The early whalers gave it special attention because of its value when rendered to whale oil. Scammon (1874) measured its thickness in bowhead whales as 25 cm and in gray whales as 18 cm. Melville (1851) devoted a chapter of *Moby Dick* to whale blubber, describing its insulating qualities as a "cozy blanket." The thick marine mammal blubber layer provides, in addition to insulation and buoyancy, an energy resource for long periods of obligatory or facultative fasting. It is, however, a variable fraction of total body mass, changing with season and fasting.

Exceptions to the general rule regarding the thermal protective role of body fat are the sea otters; they depend on thick fur enclosing cells of trapped air to provide its insulation. The air thus trapped is subject to compression by increasing water depth during dives and therefore is insufficient by itself to maintain the necessary insulation to protect deep body temperature in all circumstances. This is the reason why it is necessary for sea otters to consume vast quantities of food, up to one-quarter of their body weight per day, to fuel their high metabolic rate, which is 2.5 times that of similar sized terrestrial mammals (Morrison et al. 1974).

Northern fur seals have modest subcutaneous fat and dense fur that provides sufficient insulation to maintain warm skin temperatures and a steep thermal gradient to the outside of the fur (Irving et al. 1962). At the other end of the energy spectrum are the sirenian species, manatees and dugong. As noted earlier, these tropical-dwelling herbivorous mammals have comparatively low metabolic rates. They depend on a lifestyle limited to shallow and warm water, and they cannot tolerate prolonged exposure to cold (Scholander and Irving 1941).

The migrations of several species of whales, notably gray whales and humpback whales, to warm water regions at the time of parturition raises an interesting question regarding temperature regulation in the newborn. Infant whales often weigh more than 1000 kg at birth, and they are endowed with blubber several centimeters thick. Therefore, they may well be large enough to have a favorable ratio of surface area-to-volume and sufficient blubber to provide thermal protection in colder water than they encounter in their birthplace, avoiding the necessity for the long migrations to the subtropics. Why, then, do the whales migrate to warm water?

A partial answer might relate to the need for warm temperatures at which skin and subdermal tissues of the newborn can grow. Mitotic activity in seal skin occurs optimally at temperatures above 17°C (Feltz and Fay, 1966), and the newborn whale may require warm temperatures for a similar purpose. If so, we are still left with the question of how skin growth in more mature animals immersed in cold polar water takes place at those colder temperatures. We have seen in Chapter 2 a discussion of the remarkable countercurrent structures evident in appendages of marine mammals, which are strikingly developed in cetaceans (Scholander and Schevill 1955, Schmidt-Nielsen 1981). The idea has been reinvented by engineers for heat conservation and depends on heat transfer between the outgoing (arterial) current from the heat source to the incoming (venous) cooler current. The countercurrent structures in the appendages in marine mammals, as well as in other animals and birds, indicates that they function to reduce heat loss from the vulnerable extremities, although we lack adequate quantitative information about the process. We may tend to reach for a conclusion regarding this putative thermal function in whales, despite the theoretical finding that such structures may not be needed for prevention of heat loss in such large animals, even in ice water (Hokkanen 1990). Alternatively, an equally important role could be the occasional perfusion of warm blood to the appendages and skin for the purpose of providing warmth and nutrition for growth. Regrettably, speculation regarding this and other topics of whale physiology must suffice until experimental manipulation is possible.

Speculation, however, is all that we will ever have for consideration of how one member of the sirenians, the extinct Steller's sea cow (*Hydrodamalis gigas*), apparently differed from its extant cousins, manatees and dugongs, in its ability to live comfortably in cold waters of the North Pacific Ocean and the Bering Sea. Its large size (at least 8 m in length and perhaps weighing 10 tons; Reynolds and Odell 1991) may have been enough to provide it with a sufficiently favorable surface-to-volume ratio so it could overcome the general sirenian thermal disadvantage of vulnerability to cold.

Newborn marine mammals undergo phenomenal growth during the first weeks and months of life, and the impacts of the accompanying changes on temperature regulation were discussed earlier in this chapter. Determinations of early growth in a large cetacean were obtained by the study of two gray whale calves in captivity, one of which was followed throughout its first year of life and then released (Wahrenbrock et al. 1974). Growth rate compared well with data collected from wild populations of similar age. The calves weighed probably about 1500 kg at birth, increasing to about 5000 kg at 1 year. The weight gain in the first 8 months was approximately 200 kg per month. A rapid growth phase began at 10 months; on a diet of 550 to 800 kg of squid per day the juvenile whale gained 970 kg per month (1.3 kg/hr!). The increase in length was from 5.5 m at birth to 8.2 m at 1 year, a growth rate of approximately 7 mm per day.

Reducing Respiratory Heat Loss

Respiration is accompanied by warming of the inspired air and a consequent obligatory loss of heat during expiration. The pattern of breathing in marine mammals is adjusted to their dives, but when at the surface, tidal volume and oxygen extraction are relatively high. However, minute volume is lower by 25% to 50% than that of similar-sized terrestrial mammals (Stahl 1967). This lower respiratory volume in relation to body size suggests that marine mammals have lower heat loss by this avenue.

Seals also have another effective mechanism for preventing respiratory heat loss. It involves a systematic cooling of expired air by means of countercurrent heat exchange as it passes over the vast surface area of nasal mucosa. The inspired air cools the mucosa, and subsequent expiration, although warmed in the lungs to deep body temperature, is cooled enough to conserve both heat by conduction and water by condensation (Huntley et al. 1984; Folkow and Blix 1987, 1988). Thus gray seals breathing ambient air at –30°C exhaled that air at 6°C. This amounted to a recovery of 66% of the heat and 80% of the water that would have been lost if the exhaled air were fully saturated and at deep body temperature. The animals were tested in air at environmental temperatures ranging from +20° to –40°C. Folkow and Blix

(1987) calculated that without this mechanism for reducing respiratory heat loss (i.e., if expired air temperature were equal to deep body temperature) a compensatory increase in metabolism of 7% to 15% would have been required for maintenance of overall thermal balance. This mechanism of heat conservation was described originally in desert animals, in some of which (kangaroo rats, for instance) it is also extremely effective in conserving body water (Schmidt-Nielsen 1981, Schmidt-Nielsen et al. 1970), a function important for both desert and marine mammals.

Clearly, behavior occupies a critical part of the thermoregulatory options available to marine mammals. An interesting account reveals some interactions of behavioral and physiological mechanisms used by beached northern elephant seals for temperature regulation by the action of flipping cool, moist sand over their bodies when overheated (White and Odell 1971, see also Wells, Boness, and Rathbun, Chapter 8, this volume).

Water and Salt Balance

Despite the fact that they live in an aquatic medium with a high salt content, marine mammals are able to maintain a body fluid equilibrium of salt concentration not markedly different from that of land mammals. In this respect they resemble some desert mammals that are also limited by reduced water availability (Schmidt-Nielsen 1964). To satisfy their needs for water intake, there are only a few possibilities: extraction from seawater, preformed dietary sources, and metabolic oxidation of food. Therefore, conservation of water becomes an important consideration in their water balance. The mechanisms used by marine mammals for maintaining a well-regulated internal water and salt environment are qualitatively similar to those of land species but with some quantitative differences.

The reader who is unfamiliar with the field of the physiology of body fluids may be confused by the array of terminology that appears to be duplicative or misleading, and therefore a cursory review of the subject's vocabulary might be useful. Salt and water balance depend on osmotic processes, that is, primarily the exchanges of water and electrolytes that take place across semipermeable membranes in the kidney. Terms used to describe solutions are fundamental to the subject. A *molar* concentration of a solute refers to the number of moles per liter of solution. A *molal* concentration of a solute is the number of moles per kilogram of solvent. Because of the unusual complexity of biological fluids, their most reliable description uses the molal nomenclature. Molar terminology is sometimes more convenient, however, because of the simplicity of dealing with the volumes of fluids rather than the weight of the water solvent. Furthermore, the highest specific gravities of plasma and urine seldom exceed 1.04, and therefore a relatively small error is incurred by use of molar designations. In modern physiological literature, however, convention generally prefers the more accurate expression of fluids as molal concentrations.

The terms for describing the salt load of the body fluids are derived from expressions of the concentrations of salts and other substances. The fundamental unit is the *osmole,* defined as the weight of solute in grams dissolved in 1 kg of water providing a solution having an osmotic pressure of 22.4 atmospheres. In biological matters, it is customary to describe solutions in terms of milliosmoles (mOsm). Thus 1000 mOsm equal 1 osmole.

The osmolality of ordinary seawater is approximately 1000 mOsm/kg, and that of terrestrial mammalian plasma is about 290 to 300 mOsm/kg. Plasma of seals tends to be a little higher, about 330 mOsm/kg (Hong et al. 1982b). To maintain osmotic homesostatis in the face of seawater consumption, it is necessary for the kidneys to produce a urine more concentrated than seawater. The maximum urine concentrating ability is approximately 2000 to 2400 mOsm/kg in Weddell seals (Kooyman and Drabek 1968), California sea lions (Pilson 1970), harbor seals (Tarasoff and Toews 1972), and ringed and Baikal seals (Hong et al. 1982b); that is, about seven or eight times the concentration of plasma. In contrast, the maximum concentrating power of the human kidney is only slightly above that of seawater and, with the obligatory loss of water involved in urine formation, it is impossible for humans to maintain salt balance while drinking seawater. This is true without considering the toxic effects of consumption of some salt constituents of the seas, such as magnesium and sulfate. Therefore, some, possibly all, marine mammals have ability for concentrating urine that is somewhat better than the human facility. However, their capability is far exceeded by desert-dwelling rodent species. Kangaroo rats and sand rats can excrete urine with a maximum concentration 14 and 17 times that of their plasma (Schmidt-Nielsen 1964). The champion in this regard appears to be the hopping mouse at 25 times (MacMillen and Lee 1969).

The Kidney

The kidneys are the primary site of conservation of water, electrolytes, and other substances that are essential to life, while also being responsible for excretion of unwanted metabolic products. The kidney cortex contains the glomeruli through which blood plasma is filtered, after which the filtrate is directed in tubules looping through the medulla and finally into collecting ducts and via the kidney pelvis into the ureters. This gross oversimplification of mammalian renal anatomy is modified in cetaceans and pinnipeds by an arrangement of many small individual units, reniculi, contained within the kidney capsule (Vardy and Bryden, 1981). There may be as many as 100 to 300 reniculi in one kidney, and the reasons for such configurations are unknown (see Pabst, Rommel, and McLellan, Chapter 2, this volume).

Approximately one-fifth of mammalian cardiac output flows through the kidneys, one of the highest rates of perfusion of any organ, to provide for high levels of plasma filtration. More than 90% of total renal blood flow perfuses the cortex where glomerular filtration takes place. Only about 0.1% of renal perfusion becomes urine in the typical terrestrial mammal, but whether similar relationships hold for marine mammal kidneys is unknown. Although oxygen extraction by the kidney is low, and in a terrestrial mammal the kidney is only about 0.5% of body weight, its very high blood flow supports high oxygen consumption.

Renal circulation in terrestrial mammals is well regulated to remain steady by a mechanism of autoregulation; that is, a compensatory increase in flow occurs when arterial pressure is lowered, thus maintaining flow over a wide range of blood pressures. The extent to which such a mechanism operates in marine mammal kidneys is not known. As described earlier, renal blood flow in seals can be decreased precipitously to or close to zero flow during dives (Bradley and Bing 1942, Page et al. 1954, Lowrance et al. 1956, Murdaugh et al. 1961b, Bron et al. 1966, Elsner et al. 1966b, Zapol et al. 1979). Renal blood flow and kidney glomerular filtration rate in harbor seals are highly variable, increasing after feeding and decreasing during fasting (Hiatt and Hiatt 1942, Bradley et al. 1954, Page et al. 1954, Schmidt-Nielsen et al. 1959), revealing a mechanism that contributes to water conservation during food deprivation simply by reducing renal activity (Pernia et al. 1989, Adams and Costa 1993). More complete understanding of actual filtration rates and their regulation awaits further research.

Seawater Drinking, Water and Salt Balance

Although some marine mammals appear able to consume seawater without major physiological disturbance, the question of whether they routinely do so as a means of satisfying water balance has been examined in several studies. The food items consumed by marine mammals include fish, invertebrates, and zooplankton for the carnivores and a strictly herbivorous diet for the sirenians. Thus, the possibility of obligatory ingestion of high salt load varies from the high salt content of sea grasses and invertebrates to the relatively low electrolyte intake of an exclusively fish diet. In one study it was found that California sea lions could maintain water balance solely by the food intake of fish, without access to freshwater (Pilson 1970). A similar capability for maintaining water balance without drinking, in this instance dependent on water derived from fat metabolism, has been described in harbor seals (Tarasoff and Toews 1972), fasting northern elephant seal pups (Ortiz et al. 1978), and gray seals (Nordøy et al. 1990).

Depressed metabolism contributes to short-term water conservation by reducing the need for excretion of metabolic end products (Nordøy et al. 1990). Occasional seawater drinking is not unusual for seals and sea lions (Depocas et al. 1971, Gentry 1981), common dolphins (Hui 1981), and sea otters (Costa 1982), but it appears to be rare in northern fur seals (Fadely 1989). Freshwater drinking has been observed in young, postweaned harbor seal pups (Irving et al. 1935), Florida manatees (Reynolds and Odell 1991), and ribbon and spotted seal pups, which also eat snow (R. Elsner, unpubl.).

Administration of water to harbor seals through a stomach tube produces water diuresis and hypotonic urine (Ladd et al. 1951). This effect was inhibited by administration of antidiuretic hormone (ADH), which acts directly on kidney tubules. A source of confusing nomenclature is found in literature designations of the antidiuretic hormone as vasopressin. ADH, or vasopressin, is a peptide composed of nine amino acids having similar structures in most mammals (in pigs leucine is substituted for arginine), hence its occasional designation as arginine vasopressin (AVP) to distinguish it from leucine vasopressin. The commercial product is known as Pitressin.

Water balance is achieved and maintained in terrestrial mammals by interactions originating in the central nervous system and resulting in endocrine activation. For example, drinking a large volume of water (hypo-osmotic condition) operates via neural osmoreceptors and leads to suppression of ADH release from the pituitary gland and consequently a reduction in permeability of the distal tubules and collecting ducts of the kidney, resulting in water diuresis. On the other hand, when blood plasma becomes hyperosmotic as a result of salt intake, more ADH is secreted, and more water is reabsorbed in the kidney. Conservation of essential salts, for example, sodium, is effected by active transport across the membranes of kidney tubules, along with water, which follows by passive osmotic diffusion. Active transport requires metabolic energy, and most of the kidney's large oxygen consumption is devoted to that task. Excess plasma sodium suppresses the secretion of the adrenal hormone aldosterone, which functions to regulate kidney tubular reabsorption; in contrast, sodium deficiency initiates renin secretion by the kidney vasculature, hence by way of angiotensin to adrenal release of aldosterone.

There are contradictory lines of evidence regarding the importance of these mechanisms in marine mammals for achieving water and salt balance. Antidiuretic hormone appears to be involved in the regulation of urine volume in dolphins (Malvin and Rayner 1968, Malvin et al. 1971), but it is not clearly implicated in water conservation of gray seals (Skog and Folkow 1994) or in fasting, postweaned northern elephant seals (Ortiz et al. 1996). However, ADH is likely to be involved in regulation of water balance in Baikal seals (Hong et al. 1982b). The renin–aldosterone system is effective in sodium reabsorbtion in dolphins and sea lions (Malvin et al. 1978) and gray seals and harp seals (Sangalang and Freeman 1976, Engelhardt and Ferguson 1980).

Although some marine mammals can produce urine concentrated enough to indicate their ability to tolerate modest or regular drinking of seawater, the mechanism or mechanisms for accomplishing this task are uncertain. The thickness of the medulla relative to the cortex can contribute to increasing the countercurrent concentrating process (Schmidt-Nielsen and O'Dell 1961), and may be involved here, but definitive experimental or morphological verification of this possibility has been done only in manatee kidneys (Hill and Reynolds 1989, Maluf 1989). Also, the precise role of ADH in the urine-concentrating mechanism in seals has yet to be delineated (Hong et al. 1982b).

The effects of head-out immersion in water produce consistent and sometimes profound responses in humans and raise questions about possible implications for marine mammals. In humans, this condition leading to a blood shift toward the head results from the hydrostatic pressure of the water and by the level of negative pressure breathing, thus facilitating the return of venous blood to the heart and consequently increasing cardiac output. From this sequence of events there follows diuresis and urinary loss of sodium. The responsible relationships remain somewhat obscure, but there appears to be a suppression of ADH and aldosterone activation (for reviews, see Epstein 1978, 1984, 1996; Lin 1984). Similar responses to head-out immersion have been described in experimental animals, and sympathetic innervation of the kidney is involved in these results (Krasney et al. 1984, Miki et al. 1989, Krasney 1996). One can hardly imagine a mammal more accustomed to head-out immersion than the seal, and therefore it is the theoretical model par excellence, yet we know nothing about its responses, if any, to this condition. Possibly its cardiovascular and renal systems are fully adapted in such a way that the hydrostatic effects make little difference. In any event, an examination of the seal's responses to head-out immersion and to gravitational effects of upright posture in air should make good use of this model.

Special Situations

In a comparative study of the renal function of Baikal and ringed seals it was found that the urine-concentrating function of the Baikal seals is similar in magnitude to that of ringed seals. Despite the fact that the Baikal seal has lived in the freshwaters of Lake Baikal, Siberia, for perhaps half a million years (Repenning et al. 1979), its urine concentration after fasting was in excess of 2000 mOsm/kg, comparing well with that of its similarly treated marine cousin, the ringed seal (Hong et al. 1982b).

The sirenians are of special interest in this regard. The Florida manatee occupies a mixed habitat because of its frequent movement between regions of seawater and freshwater, and it is apparently able to consume either freshwater or seawater without osmoregulatory penalty. Its prominent renal medulla suggests that it has the structure that is required for flexibility in salt and water regulation (Hill and Reynolds 1989, Maluf 1989). Studies of both wild and captive members of this species and experimental manipulation by interchanging of animals between seawater and freshwater environments have revealed some of the associated mechanisms. These include involvement of ADH in conserving water in the marine habitat and aldosterone in retaining sodium when exposed to freshwater (Ortiz 1994).

The role of the countercurrent flow of respiratory gases was described earlier in the section on temperature regulation. Water conservation is also achieved by the same mechanism: cooling of expired air and condensation of water contained as saturated vapor in air as it leaves the lungs. This mechanism is effective in the unusually long fast of northern elephant seals by the recovery of about 70% of the water that would otherwise be lost if air were expired at deep body temperature (Huntley et al. 1984). However useful this mechanism, nasal water conservation is neither increased with dehydration nor reduced with water loading in gray seals, which suggests that this function may not be critically responsive as a discrete modifier of water balance but instead may act as a dull instrument in its defense (Skog and Folkow 1994).

Renal Ischemia

As has been mentioned, the seal kidney is remarkably tolerant of repeated cyclic ischemia and reperfusion (Halasz et al. 1974). Harbor seal kidney tissue is resistant to the severe conditions of long dives, and isolated tissues continue certain transport functions even when exposed to hypoxia and low pH. Uptake by seal kidney slices of the organic anion para-aminohippurate (PAH), a test of kidney tubule functional condition (Cross and Taggert 1950), remained effective for 60 min without oxygen, whereas identical treatment of rat kidney slices resulted in marked transport inhibition after 15 min (Koschier et al. 1978). The PAH transport in seal kidney slices was independent of pH over a pH range from 7.5 to 5.0, but was highly dependent on pH and drastically reduced at lowered pH in rat kidney slices (Hong et al. 1982a; Fig. 3-14). A possible explanation may be that anaerobic metabolism supplies the required energy in the seal kidney and that certain glycolytic enzymes that would be sensitive to increasing acidity in the terrestrial mammal are resistant to these effects in the seal. That is, the levels of pH at which the enzymes operate optimally in the seal may extend over a greater range of pH than they do in mammals that lack its tolerance for frequent exposure to asphyxial conditions.

These adaptations for resistance to renal ischemia were reviewed by Hong (1989) in a discussion of cellular and membrane mechanisms. The results indicate that the seal kidney maintained tubular secretory function despite what was for the terrestrial rat an apparently severe hypoxic and acidic insult. Suggested mechanisms include tolerance of the transport system to low oxygen, which presupposes an enhanced anaerobic facility; reduced sensitivity to acidity; maintenance of the transmembrane ionic gradient; and improved efficiency of the membrane sodium–potassium exchange pump (Hong 1989).

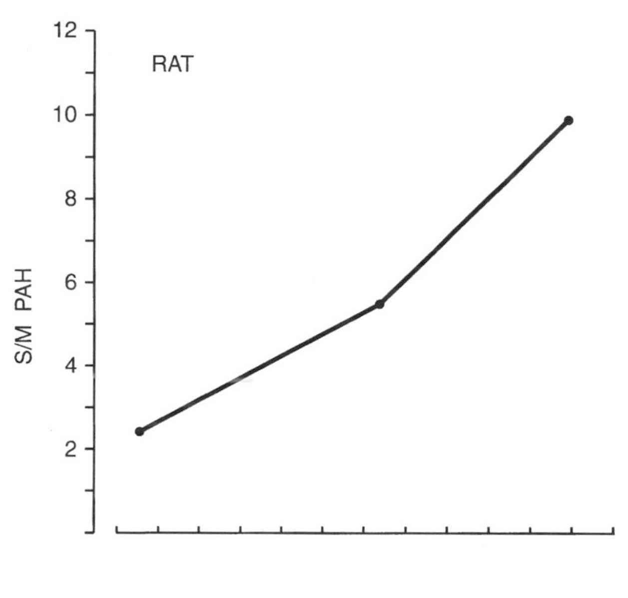

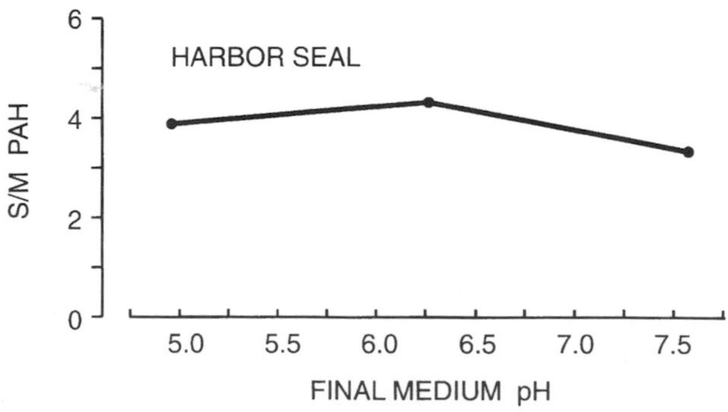

Figure 3-14. Renal transport of para-aminohippurate (PAH) in the presence of acetate in harbor seal and rat kidney slices as affected by pH. Slice-to-medium (S/M) ratios of PAH concentrations show little effect of pH on seal renal transport. (Redrawn from Hong et al. 1982a.)

An additional, and likely important, question deserves consideration. The frequently repeated cycles of ischemia and reperfusion have been cited earlier in this chapter as a well-established source of oxygen-derived free radical generation, and kidney tissue is not immune from this possibility. In terrestrial mammals the degradation of ATP to hypoxanthine in extreme hypoxia leads, with reperfusion, to the production of oxygen-derived free radicals: superoxide radical by way of xanthine oxidase and, in combination with hydrogen peroxide, the destructive hydroxyl radical.

Dogma and New Directions

Our understanding of marine mammal physiological adaptations is derived from field and experimental observations that have resulted from manageable simplifications of the extremely complicated realities of living animals interacting with their environments. The simplifications are necessary for us to be able to characterize the observable events, but we run the risk of assuming that these examples comprise a fixed and unvarying set of physiological responses, dogma that applies to all or most of the situations that the animal encounters in real life. What is required of the observer is the ability to keep one's mind open to the possibility of hidden relationships and subtle clues buried within the simplifications. Alfred North Whitehead (1920) put it well in this excerpt from his collection of essays, *The Concept of Nature:* "The aim of science is to seek the simplest explanations of complex facts. We are apt to fall into the error of thinking that the facts are simple because simplicity is the goal of our quest. The guiding motto in the life of every natural philosopher should be: Seek simplicity and distrust it."

The success of modern marine mammal biology depends on the resources of many disciplines ranging from molecular biology to population studies, using techniques ranging from those of cell chemistry to binoculars and notebook. Just as in other sciences, progress owes much to the piece-by-piece analytical techniques of the reductionist process, the mainstay of the scientific method by which complex problems are broken down and examined as simpler elements. The resulting intellectual power guides the rapid growth of new knowledge, but our ultimate understanding of the whole organism depends on the reassembly of the separate and interacting parts. This essential integrative activity is the province of physiology. Reductive exploitation will continue to reap extraordinary rewards in scientific progress, but our ability to achieve required new understanding of integrated life science is also steadily advancing. We have seen some simple examples in the acknowledgment that the algebraic sum of the parts involved in neural control mechanisms that govern cardiorespiratory and other aspects of the diving response does not equal the whole process. Rather, that complex regulatory process in the intact animal is modulated by influences that are unpredictable from knowledge that is limited to the individual, unassembled parts (Daly 1984).

The message can hardly be better stated than in the recent essay by Stanley Schultz (1996): "The central theme of our message should be that the study of integrated systems cannot be supplanted, or rendered unnecessary or superfluous, by advances at lower levels of organization. . . . The shape of the whole cannot be predicted by knowing only the shapes of the separated parts. Thus, while knowledge of the parts is necessary, it is by no means sufficient. We must also know the rules of association and integration. Deciphering these rules will be the next breathtaking revolution in biology."

Acknowledgments

Colleagues and students have graciously commented on and contributed to the ideas and reviews included in this chapter. I am also grateful to the editors, John E. Reynolds III, and Sentiel A. Rommel, and two anonymous reviewers for useful criticisms and suggestions. Partial support has been provided by the Institute of Marine Science, University of Alaska Fairbanks.

Literature Cited

Adams, S., and D. P. Costa. 1993. Water conservation and protein metabolism in northern elephant seal pups during the postweaning fast. Journal of Comparative Physiology B 163:367–373.

Agostini, E. 1965. Limitations to depths of diving: Mechanics of chest wall. Pages 139–145 *in* H. Rahn and T. Yokoyama (eds.). Physiology of Breath-hold Diving and the Ama of Japan, Publication 1341. National Academy of Sciences, Natural Resources Council, Washington D.C.

Andersen, H. T. 1966. Physiological adaptations in diving vertebrates. Physiological Reviews 46:212–243.

Andersen, H. T. 1969. The Biology of Marine Mammals. Academic Press, New York, NY.

Andrews, R. D., D. R. Jones, J. D. Williams, D. E. Crocker, D. E. Costa, and B. J. Le Boeuf. 1993. Metabolic and Cardiovascular Adjustments to Diving in Northern Elephant Seals *Mirounga angustirostris*. Abstracts, Tenth Biennial Conference, Biology of Marine Mammals, Galveston, TX, 22.

Andrews, R. D., D. R. Jones, J. D. Williams, D. E. Crocker, D. P. Costa, and B. J. Le Boeuf. 1995. Metabolic and Cardiovascular Adjustments to Diving in Northern Elephant Seals (*Mirounga angustirostris*) (abstract). Proceedings of the 4th International Congress, Comparative Physiology and Biochemistry, Birmingham 1995. Physiological Zoology 68:105.

Andrews, R. D., D. R. Jones, J. D. Williams, P. H. Thorson, G. W. Oliver, D. P. Costa, and B. J. Le Boeuf. 1997. Heart rates of northern elephant seals diving at sea and resting on the beach. Journal of Experimental Biology 200:2083–2095.

Angell-James, J. E., and M. de B. Daly. 1969a. Cardiovascular responses in apnoeic asphyxia: Role of arterial chemoreceptors and the modification of their effects by a pulmonary vagal inflation reflex. Journal of Physiology, 201:87–104.

Angell-James, J. E., and M. de B. Daly. 1969b. Nasal reflexes. Proceedings of the Royal Society of Medicine 62:1287–1293.

Angell-James, J. E., and M. de B. Daly. 1972. Some mechanisms involved in the cardiovascular adaptations to diving. Society of Experimental Biology Symposia 26:313–341.

Angell-James, J. E., and M. de B. Daly. 1978. The effects of artificial lung inflation on reflexly induced bradycardia associated with apnoea in the dog. Journal of Physiology 274:349–366.

Angell-James, J. E., M. de B. Daly, and R. Elsner. 1978. Arterial baroreceptor reflexes in the seal and their modification during experimental dives. American Journal of Physiology 234:H730–739.

Angell-James, J. E., R. Elsner, and M. de B. Daly. 1981. Lung inflation: Effects on heart rate, respiration and vagal activity in seals. American Journal of Physiology 240:H190–H198.

Anrep, G. V., W. Pascual, and R. Rössler. 1936a. Respiratory variations of the heart rate. I. The reflex mechanism of the respiratory arrhythmia. Proceedings of the Royal Society, Series B 119:191–217.

Anrep, G. V., W. Pascual, and R. Rössler. 1936b. Respiratory variations of the heart rate. II. The central mechanism of the respiratory arrhythmia and the interactions between the central and the reflex mechanisms. Proceedings of the Royal Society, Series B 119:218–230.

Arnold, R. W. 1985. Extremes in human breath-hold facial immersion bradycardia. Undersea Biomedical Research 12:183–190.

Ashwell-Erickson, S., and R. Elsner. 1980. The energy cost of free existence for Bering Sea harbor and spotted seals. Pages 879–899 *in* D. W. Hood and J. A. Calder (eds.). The Bering Sea Shelf: Oceanography and Resources, Vol. 2. University of Washington Press, Seattle, WA.

Ashwell-Erickson, S., F. Fay, R. Elsner, and D. Wartzok. 1986. Metabolic and hormonal correlates of molting and regeneration of pelage in Alaskan harbor and spotted seals (*Phoca vitulina* and *P. largha*). Canadian Journal of Zoology 64:1086–1094.

Augee, M. L., R. Elsner, B. A. Gooden, and P. R. Wilson. 1971. Respiratory and cardiac responses of a burrowing animal, the echidna. Respiration Physiology 11:327–334.

Bartholomew, G. A. 1954. Body temperature and respiratory and heart rates in the northern elephant seal. Journal of Mammalogy 35:211–218.

Behrman, R. E., M. H. Lees, E. N. Peterson, C. W. de Lannoy, and A. E. Seeds. 1970. Distribution of the circulation in the normal and asphyxiated fetal primate. American Journal of Obstetrics and Gynecology 108:956–969.

Bennett, P. B. 1993. Inert gas narcosis. Pages 170–193 *in* P. B. Bennett and D. H. Elliott (eds.). The Physiology and Medicine of Diving, 4th ed. Williams and Wilkins, Baltimore, MD.

Bennett, P. B., and J. C. Rostain. 1993. The high pressure nervous syndrome. Pages 194–237 *in* P. B. Bennett and D. H. Elliott (eds.). The Physiology and Medicine of Diving, 4th ed. Williams and Wilkins, Baltimore, MD.

Bert, P. 1870. Pages 526–553 *in* Leçons sur la Physiologie Comparée de la Respiration. Baillière, Paris.

Bjertnæs, L., A. Hauge, J. Kjekshus and R. Søyland. 1984. Cardiovascular responses to face immersion and apnea during steady state muscle exercise. Acta Physiologica Scandinavica 120:605–612.

Blessing, M. H., and E. Hartschen-Niemeyer. 1969. Uber den Myoglobingehalt der Herz und Skelettmuskulatur insbesondere mariner Säuger. Zeitschrift für Biologie 116:302–313.

Blix, A. S. 1976. Metabolic consequences of submersion asphyxia in mammals and birds. Biochemical Society Symposia 41:169–178.

Blix, A. S., and B. Folkow. 1983. Cardiovascular responses to diving in mammals and birds. Pages 917–945 *in* J. T. Shepherd and F. M. Abboud (eds.). Handbook of Physiology: Section 2, The Cardiovascular System. American Physiological Society, Washington, D.C.

Blix, A. S., and J. W. Lentfer. 1979. Modes of thermal protection in polar bear cubs—at birth and on emergence from the den. American Journal of Physiology 236:R67–R74.

Blix, A. S., and J. B. Steen. 1979. Temperature regulation in newborn polar homeotherms. Physiological Reviews 59:285–304.

Blix, A. S., J. K. Kjekshus, I. Enge, and A. Bergan. 1976. Myocardial blood flow in the diving seal. Acta Physiologica Scandinavica 96:277–280.

Blix. A. S., L. K. Miller, M. C. Keyes, H. J. Grav, and R. Elsner. 1979. Newborn northern fur seals (*Callorhinus ursinus*)—Do they suffer from cold? American Journal of Physiology 236:R322–R327.

Blix, A. S., R. Elsner, and J. Kjekshus. 1983. Cardiac output and its distribution through capillaries and A-V shunts in diving seals. Acta Physiologica Scandinavica 118:109–116.

Bradley, S. E., and R. J. Bing. 1942. Renal function in the harbor seal *Phoca vitulina* L. during asphyxial ischemia and pyrogenic hyperemia. Journal of Cellular and Comparative Physiology 19:229–237.

Bradley, S. E., G. H. Mudge, and W. D. Blake. 1954. The renal excretion of sodium, potassium, and water by the harbor seal (*Phoca vitulina* L.): Effect of apnea; sodium, potassium and water loading; pitressin; and mercurial diuresis. Journal of Cellular and Comparative Physiology 43:1–22.

Brodie, P., and A. Paasche. 1985. Thermoregulation and energetics of fin and sei whales based on postmortem, stratified temperature measurements. Canadian Journal of Zoology 63:2267–2269.

Bron, K. M., H. V. Murdaugh Jr., J. E. Millen, R. Lenthall, P. Raskin, and E. D. Robin. 1966. Arterial constrictor response in a diving mammal. Science 153:540–543.

Bryden, M. M. 1964. Insulating capacity of the subcutaneous fat of the southern elephant seal. Nature 203:1299–1300.

Bryden, M. M., and G. H. K. Lim. 1969. Blood parameters of the southern elephant seal (*Mirounga leonina*) in relation to diving. Comparative Biochemistry and Physiology 28:139–148.

Butler, P. J., and D. R. Jones. 1982. The comparative physiology of diving in vertebrates. Pages 179–364 *in* O. Lowenstein (ed.). Advances in Comparative Physiology and Biochemistry. Academic Press, New York, NY.

Butler, P. J., and D. R. Jones. 1997. Physiology of diving of birds and mammals. Physiological Reviews 77:837–899.

Calder, W. A. 1969. Temperature relations and underwater endurance of the smallest homeothermic diver, the water shrew. Comparative Biochemistry and Physiology 30:1075–1082.

Campbell, K. B., E. A. Rhode, R. H. Cox, W. C. Hunter, and A. Noordergraaf. 1981. Functional consequences of expanded aortic bulb: A model study. American Journal of Physiology 240:R200–R210.

Carlson, L. D., A. C. L. Hsieh, F. Fullington, and R. Elsner. 1958. Immersion in cold water and body tissue insulation. Journal of Aviation Medicine 29:145–152.

Casson, D. M., and K. Ronald. 1975. The harp seal, *Pagophilus groenlandicus* (Erxleben, 1777). XIV. Cardiac arrhythmias. Comparative Physiology and Biochemistry A 50:307–314.

Castellini, M. A. 1985a. Metabolic depression in tissues and organs of marine mammals during diving: Living longer with less oxygen. Molecular Physiology 8:427–437.

Castellini, M. A. 1985b. Closed systems: Resolving potentially conflicting demands of diving and exercise in marine mammals. Pages 220–226 *in* R. Gilles (ed.). Circulation, Respiration and Metabolism. Springer-Verlag, Berlin.

Castellini, M. A. 1991. The biology of diving: Behavioral, physiological and biochemical limits. Pages 105–132 *in* R. Gilles (ed.). Advances in Comparative and Environmental Physiology. Springer-Verlag, Berlin.

Castellini, J. M., and M. A. Castellini. 1993. Estimation of splenic volume and its relationship to long-duration apnea in seals. Physiological Zoology 66:619–627.

Castellini, M. A. and G. N. Somero. 1981. Buffering capacity of vertebrate muscle: Correlations with potentials for anaerobic function. Journal of Comparative Physiology 143:191–198.

Castellini, M. A., G. N. Somero, and G. L. Kooyman. 1981. Glycolytic enzyme activities in tissues of marine and terrestrial mammals. Physiological Zoology 54:242–252.

Castellini, M. A., B. J. Murphy, M. Fedak, K. Ronald, N. Gofton, and P. W. Hochachka. 1985. Potentially conflicting metabolic demands of diving and exercise in seals. Journal of Applied Physiology 58:392–399.

Castellini, M. A., D. P. Costa, and A. C. Huntley. 1986. Hematocrit variation during sleep apnea in elephant seal pups. American Journal of Physiology 251:R429–R431.

Castellini, M. A., R. W. Davis, and G. L. Kooyman. 1988. Blood chemistry regulation during repetitive diving in Weddell seals. Physiological Zoology 61:379–386.

Castellini, M. A., G. L. Kooyman, and P. J. Ponganis. 1992. Metabolic rates of freely diving Weddell seals: Correlations with oxygen stores, swim velocity and diving duration. Journal of Experimental Biology 165:181–194.

Castellini, M. A., W. K. Milsom, R. J. Berger, D. P. Costa, D. R. Jones, M. Castellini, L. D. Rea, S. Bharma, and M. Harris. 1994. Patterns of respiration and heart rate during wakefulness and sleep in elephant seal pups. American Journal of Physiology 266:R863–R869.

Cherepanova, V., T. Neshumova, and R. Elsner. 1993. Review: Muscle blood flow in diving mammals. Comparative Biochemistry and Physiology A 106:1–6.

Clark, J. M. 1993. Oxygen toxicity. Pages 121–169 *in* P. B. Bennett and D. H. Elliott (eds.). The Physiology and Medicine of Diving, 4th ed. Williams and Wilkins, Baltimore, MD.

Cohen, R. A., J. T. Shepherd, and P. M. Vanhoutte. 1984. Neurogenic cholinergic prejunctional inhibition of sympathetic beta adrenergic relaxation in the canine coronary artery. Journal of Pharmacology and Experimental Therapeutics 229:417–421.

Costa, D. P. 1982. Energy, nitrogen, electrolyte flux and sea water drinking in the sea otter *Enhydra lutris*. Physiological Zoology 55:35–44.

Craig, A. B. 1961. Causes of loss of consciousness during underwater swimming. Journal of Applied Physiology 16:583–586.

Craig, A. B. 1968. Depth limits of breath-hold diving. Respiration Physiology 5:14–22.

Cross, E. R. 1965. Taravana diving syndrome in the Tuamoto diver. Pages 207–219 *in* H. Rahn and T. Yokoyama (eds.). Physiology of Breath-hold Diving and the Ama of Japan. National Academy of Sciences, National Research Council, Washington, D.C.

Cross, R. J., and J. V. Taggert. 1950. Renal tubular transport: Accumulation of p-aminohippurate by rabbit kidney slices. American Journal of Physiology 161:181–190.

Daly, M. de B. 1984. Breath-hold diving: Mechanisms of cardiovascular adjustments in the mammal. Pages 201–245 *in* P. F. Baker (ed.). Recent Advances in Physiology. Churchill Livingstone, London.

Daly, M. de B. 1986. Interactions between respiration and circulation. Pages 529–592 *in* N. S. Cherniack and J. G. Widdicome (eds.). Handbook of Physiology, Section 3, Vol. 2: Control of Breathing, Part 2. American Physiological Society, Bethesda, MD.

Daly, M. de B. 1997. Peripheral Arterial Chemoreceptors and Respiratory–Cardiovascular Integration. Monographs of the Physiological Society, No 46. Clarendon Press, Oxford.

Daly, M. de B., R. Elsner, and J. E. Angell-James. 1977. Cardiorespiratory control by the carotid chemoreceptors during experimental dives in the seal. American Journal Physiology 232:H508–H516.

Daly, M. de B., J. E. Angell-James, and R. Elsner. 1979. Role of carotid body chemoreceptors and their reflex interactions in bradycardia and cardiac arrest. Lancet 1:764–767.

Davis, R. W. 1983. Lactate and glucose metabolism in the resting and diving harbor seal (*Phoca vitulina*). Journal of Comparative Physiology B 153:275–288.

Davis, R. W., M. A. Castellini, G. L. Kooyman, and R. Maue. 1983. Renal glomerular filtration rate and hepatic blood flow during voluntary diving in Weddell seals. American Journal of Physiology 245:R743–R748.

Davis, R. W., T. M. Williams, and G. L. Kooyman. 1985. Swimming metabolism of yearling and adult harbor seals *Phoca vitulina*. Physiological Zoology 58:590–596.

Davis, R. W., D. Wartzok, R. Elsner, and H. Stone. 1992. Attaching a small video camera to a Weddell seal: A new way to observe diving behavior. Pages 631–642 *in* J. A. Thomas, R. A. Kastelein, and A. Y. Supin (eds.). Marine Mammal Sensory Systems. Plenum, New York, NY.

Davydov, A. F., and A. R. Makarova. 1964. Changes in heat regulation and circulation in newborn seals on transition to aquatic form of life. Federation Proceedings 24:T563–T566.

Dawbin, W. H. 1966. The seasonal migratory cycle of humpback whales. Pages 145–170 *in* K. B. Norris (ed.). Whales, Dolphins and Porpoises. University of California Press, Berkeley.

Dejours, P. 1981. Principles of Comparative Respiratory Physiology, 2nd ed. Elsevier, Amsterdam.

DeLong, R. L., and B. S. Stewart. 1991. Diving patterns of northern elephant seal bulls. Marine Mammal Science 7:369–384.

Denison, D. M., and G. L. Kooyman. 1973. The structure and function of the small airways in pinniped and sea otter lungs. Respiration Physiology 17:1–10.

Denison, D. M., D. A. Warrell, and J. B. West. 1971. Airway structure and alveolar emptying in the lungs of sea lions and dogs. Respiration Physiology 13:253–260.

Depocas, F., J. S. Hart, and H. D. Fisher. 1971. Sea water drinking and water flux in starved and in fed harbor seals, *Phoca vitulina*. Canadian Journal of Physiology and Pharmacology 49:53–62.

Doidge, D. W. 1990. Integumentary heat loss and blubber distribution in the beluga, *Delphinapterus leucas,* with comparisons to the narwhal, *Monodon monoceros*. Pages 129–140 *in* T. G. Smith, D. J. St. Aubin, and J. R. Geraci (eds.). Advances in Research on the Beluga Whale, *Delphinapterus leucas*. Canadian Bulletin of Fisheries and Aquatic Sciences 224, Department of Fisheries and Oceans, Ottowa.

Dormer, K. J., M. J. Denn, and H. L. Stone. 1977. Cerebral blood flow in the sea lion (*Zalophus californianus*) during voluntary dives. Comparative Biochemistry and Physiology A 58:11–18.

Drabek, C. M. 1975. Some anatomical aspects of the cardiovascular system of antarctic seals and their possible functional significance in diving. Journal of Morphology 145:85–106.

Drummond, P. C., and D. R. Jones. 1979. The initiation and maintenance of bradycardia in a diving mammal, the muskrat, *Ondatra zibethica*. Journal of Physiology 290:253–271.

Dykes, R. W. 1974. Factors related to the dive reflex in harbor seals: Sensory contributions from the trigeminal region. Canadian Journal of Physiology and Pharmacology 52:259–265.

Elsner, R. 1965. Heart rate response in forced versus trained experimental dives in pinnipeds. Hvalrådets Skrifter, Det Norske Videnskaps-Akadami i Oslo 48:24–29.

Elsner, R. 1966. Diving bradycardia in the unrestrained hippopotamus. Nature 212:408.

Elsner, R. 1969. Cardiovascular adjustments to diving. Pages 117–145 *in* H. T. Andersen (ed.). The Biology of Marine Mammals. Academic Press, New York, NY.

Elsner, R. 1986. Limits to exercise performance: Some ideas from comparative studies. Acta Physiologica Scandinavica (Suppl.) 556:44–51.

Elsner, R. 1987. The contribution of anaerobic metabolism to maximum exercise in seals. Pages 109–114 *in* A. C. Huntley, D. P. Costa, G. A. J. Worthy, and M. A. Castellini (eds.). Marine Mammal Energetics. Society for Marine Mammalogy, Special Publication No. 1, Lawrence, KS.

Elsner, R. 1988. Anaerobic contributions to metabolism in diving seals. Canadian Journal of Zoology 66:142–143.

Elsner, R., and M. de B. Daly. 1988. Coping with asphyxia. News in Physiological Sciences 3:65–69.

Elsner, R., and B. Gooden. 1983. Diving and Asphyxia, A Comparative Study of Animals and Man. Monographs of the Physiological Society No. 40. Cambridge University Press, Cambridge, U.K.

Elsner, R., and I. S. de la Lande. 1998. Heterogeneous cholinergic reactions of ringed seal coronary arteries. Comparative Biochemistry and Physiology A 119:1019–1025.

Elsner, R., and H. J. Meiselman. 1995. Splenic oxygen storage and blood viscosity. Marine Mammal Science 11:93–96.

Elsner, R., and W. O. Pruitt. 1959. Some structural and thermal characteristics of snow shelters. Arctic 12:20–27.

Elsner, R., D. L. Franklin, and R. L. Van Citters. 1964. Cardiac output during diving in an unrestrained sea lion. Nature 202:809–810.

Elsner, R., D. W. Kenney, and K. Burgess. 1966a. Diving bradycardia in the trained dolphin. Nature 212:407–408.

Elsner, R., D. L. Franklin, R. L. Van Citters, and D. W. Kenney. 1966b. Cardiovascular defense against asphyxia. Science 153:941–949.

Elsner, R., G. L. Kooyman, and C. M. Drabek. 1969. Diving duration in pregnant Weddell seals. Pages 477–482 *in* M. Holdgate (ed.). Antarctic Ecology. Academic Press, New York, NY.

Elsner, R., J. T. Shurley, D. D. Hammond, and R. E. Brooks. 1970a. Cerebral resistance to hypoxemia in asphyxiated Weddell seals. Respiration Physiology 9:287–297.

Elsner, R., D. D. Hammond, and H. R. Parker. 1970b. Circulatory responses to asphyxia in pregnant and fetal animals: A comparative study of Weddell seals and sheep. Yale Journal of Biology and Medicine 42:202–217.

Elsner, R., B. A. Gooden, and S. M. Robinson. 1971a. Arterial blood gas changes and the diving response in man. Australian Journal of Experimental Biology and Medical Sciences 49:435–444.

Elsner, R., W. N. Hanafee, and D. D. Hammond. 1971b. Angiography of the inferior vena cava of the harbor seal during simulated diving. American Journal of Physiology 220:1155–1157.

Elsner, R., J. E. Angell-James, and M. de B. Daly. 1977a. Carotid body chemoreceptor reflexes and their interactions in the seal. American Journal of Physiology 232:H517–H525.

Elsner, R., D. D. Hammond, D. M. Denison, and R. Wyburn. 1977b. Temperature regulation in the newborn Weddell seal Leptonychotes weddelli. Pages 531–540 *in* G. A. Llano (ed.). Adaptations Within Antarctic Ecosystems. Proceedings of the 3rd SCAR Symposium on Antarctic Biology. Gulf Publishing, Houston, TX.

Elsner, R., R. W. Millard, J. Kjekshus, F. C. White, A. S. Blix, and S. Kemper. 1985. Coronary circulation and myocardial segment dimensions in diving seals. American Journal of Physiology 249:H1119–H1126.

Elsner, R., D. Wartzok, N. B. Sonafrank, and B. P. Kelly. 1989. Behavioral and physiological reactions of arctic seals during under-ice pilotage. Canadian Journal of Zoology 67:2506–2513.

Elsner, R., M. de B. Daly, A. Maseri, R. W. Millard, and F. C. White. 1992. Coronary circulation in seals. Pages 363–375 *in* S. C. Wood, R. E. Weber, A. Hargens, and R. W. Millard (eds.). Physiological Adaptations in Vertebrates. Marcel Dekker, New York, NY.

Elsner, R., S. Øyasæter, R. Almass, and O. D. Saugstad. 1998. Diving seals, ischemia-reperfusion and oxygen radicals. Comparative Biochemistry and Physiology A 119:975–980.

Engelhardt, F. R., and J. M. Ferguson. 1980. Adaptive hormone changes in harp seals, *Phoca groendanlica,* and gray seals, *Halichoerus grypus,* during the postnatal period. General and Comparative Endocrinology 40:435–445.

Epstein, M. 1978. Renal effects of head-out immersion in man: Implications for an understanding of volume homeostasis. Physiological Reviews 58:529–581.

Epstein, M. 1984. Water immersion and the kidney: Implications for volume regulation. Undersea Biomedical Research 11:113–121.

Epstein, M. 1996. Renal, endocrine, and hemodynamic effects of water immersion in humans. Pages 845–853 *in* M. J. Fregly and C. M. Blatteis (eds.). Handbook of Physiology, Section 4, Vol. 2: Environmental Physiology. Oxford University Press, Oxford.

Fadely, B. S. 1989. Investigations of the water balance and assimilation efficiency of the northern fur seal (*Callorhinus ursinus*). M.S. thesis, University of California, Santa Cruz.

Falke, K. J., R. D. Hill, J. Qvist, R. C. Schneider, M. Guppy, G. C. Liggins, P. W. Hochachka, R. E. Elliott, and W. M. Zapol. 1985. Seal lungs collapse during free diving: Evidence from arterial nitrogen tensions. Science 229:556–558.

Fedak, M. A. 1986. Diving and exercise in seals: A benthic perspective. Pages 11–32 *in* A. Brubakk, J. W. Kanwisher and G. Sundnes (eds.). Diving in Animals and Man, Kongsvold Symposium, 1985. Royal Norwegian Society of Sciences and Letters. Tapir Publishers, Trondheim, Norway.

Fedak, M. A., M. R. Pullen, and J. Kanwisher. 1988. Circulatory responses of seals to periodic breathing: Heart rate and breathing during exercise and diving in the laboratory and open sea. Canadian Journal of Zoology 66:53–60.

Feigl, E. O. 1983. Coronary physiology. Physiological Reviews 63:1–205.

Feinstein, R., H. Pinsker, M. Schmale, and B. A. Gooden. 1977. Bradycardial response in asphyxia exposed to air. Journal of Comparative Physiology 122:311–324.

Feldkamp, S. D. 1987. Swimming in the California sea lion: Morphometrics, drag and energetics. Journal of Experimental Biology 131:117–135.

Feltz, E. T., and F. H. Fay. 1966. Thermal requirements in vitro of epidermal cells from seals. Cryobiology 3:261–264.

Folkow, L. P. 1992. Aspects of thermoregulation in terrestrial and aquatic arctic mammals. Ph.D. dissertation, University of Tromsø, Norway. 150 pp.

Folkow, L. P., and A. S. Blix. 1987. Nasal heat and water exchange in gray seals. American Journal of Physiology 253:R883–R889.

Folkow, L. P., and A. S. Blix. 1988. Anatomical and functional aspects of the nasal mucosal and ophthalmic retia of phocid seals. Journal of Zoology (London) 216:417–436.

Folkow, L. P., and A. S. Blix. 1989. Thermoregulatory control of expired air temperature in diving harp seals. American Journal of Physiology 257:R306–R310.

Folkow, L. P., and A. S. Blix. 1992. Metabolic rates of minke whales (*Balaenoptera acutorostra*) in cold water. Acta Physiologica Scandinavica 146:141–150.

Folkow, L. P., and A. S. Blix. 1995. Distribution and diving behaviour of hooded seals. Pages 193–202 *in* A. S. Blix, L. Walløe, and Ø. Ullyang (eds.). Whales, Seals, Fish and Man. Elsevier, Amsterdam.

Forster, R. P., and J. Nyboer. 1955. Effect of induced apnea on cardiovascular and renal functions in the rabbit. American Journal of Physiology 183:149–154.

Galantsev, V. P. 1977. Evolutionary adaptations of diving mammals (in Russian). Academy of Science USSR, Leningrad.

Gallivan, G. J. 1992. What are the metabolic rates of cetaceans? Physiological Zoology 65:1285–1297.

Gallivan, G. J., and R. C. Best. 1980. Metabolism and respiration of the Amazonian manatee (*Trichechus inunguis*). Physiological Zoology 53:245–253.

Gallivan, G. J., and K. Ronald. 1979. Temperature regulation in freely diving harp seals (*Phoca groenlandica*). Canadian Journal of Zoology 57:2256–2263.

Gallivan, G. J., R. C. Best, and J. W. Kanwisher. 1983. Temperature regulation in the Amazonian manatee (*Trichechus inunguis*). Physiological Zoology 56:255–262.

Gentry, R. L. 1981. Sea water drinking in eared seals. Comparative Biochemistry and Physiology A 68:81–86.

Gentry, R. L., and G. L. Kooyman (eds.). 1986. Fur seals: Maternal strategies on land and at sea. Princeton University Press, Princeton, NJ.

George, J. C., C. Clark, G. M. Carroll, and W. T. Ellison. 1989. Observations on the ice-breaking and ice navigation behavior of migrating bowhead whales (*Balaena mysticetus*) near Point Barrow, Alaska, spring 1985. Arctic 42:24–30.

Gerschman, R., D. L. Gilbert, S. W. Nye, P. Dwyer, and W. O. Fenn. 1954. Oxygen poisoning and x-irradiation: A mechanism in common. Science 119:623–626.

Gessner, B. 1996. Pipin, how deep can you go? Baja Life 5:24–25.

Gilbert, D. L. 1963. The role of pro-oxidants and antioxidants in oxygen toxicity. Radiation Research (Suppl.) 3:44–53.

Gilbert, D. L. 1981. Significance of oxygen on earth. Pages 73–101 *in* D. L. Gilbert (ed.). Oxygen and Living Processes, An Interdisciplinary Approach. Springer-Verlag, New York.

Gilbert, D. L. 1996. Evolutionary aspects of atmospheric oxygen and organisms. Pages 1059–1095 *in* M. J. Fregly and C. M. Blatteis (eds.). Handbook of Physiology, Section 4, Vol. 2: Environmental Physiology. Oxford University Press, Oxford.

Gooden, B. A., and R. Elsner. 1985. What diving animals might tell us about blood flow regulation. Perspectives in Biology and Medicine 28:465–474.

Gooden, B. A., H. L. Stone, and S. Young. 1974. Cardiac responses to snout immersion in trained dogs. Journal of Physiology 242:405–414.

Grav, H. J., and A. S. Blix. 1976. Brown adipose tissue—A factor in the survival of harp seal pups. Canadian Journal of Physiology and Pharmacology 54:409–412.

Guppy, M., R. D. Hill, R. C. Schneider, J. Qvist, G. C. Liggins, W. M. Zapol, and P. W. Hochachka. 1986. Microcomputer-assisted metabolic studies of voluntary diving of Weddell seals. American Journal of Physiology 250:R175–R187.

Guyton, G. P., K. S. Stanek, R. C. Schneider, P. W. Hochachka, W. E. Hurford, D. G. Zapol, G. C. Liggins, and W. M. Zapol. 1995. Myoglobin saturation in free-diving Weddell seals. Journal of Applied Physiology 79:1148–1155.

Halasz, N. A., R. Elsner, R. S. Garvie, and G. T. Grotke. 1974. Renal recovery from ischemia: a comparative study of seal and dog kidneys. American Journal of Physiology 227:1331–1335.

Haliwell, B., and J. M. C. Gutteridge. 1985. Free radicals in biology and medicine. Clarendon Press, Oxford.

Hammel, H. T., R. Elsner, H. C. Heller, J. A. Maggert, and C. R. Bainton. 1977. Thermoregulatory responses to altering hypothalamic temperature in the harbor seal. American Journal of Physiology 232:R18–R26.

Harken, A. H. 1976. Hydrogen ion concentration and oxygen uptake in an isolated canine limb. Journal of Applied Physiology 40:1–5.

Harrison, R. J., and J. D. W. Tomlinson. 1956. Observations on the venous system in certain pinnipedia and cetacea. Proceedings of the Zoological Society (London) 126:205–231.

Hart, J. S., and L. Irving. 1959. The energetics of harbor seals in air and water with special consideration of seasonal changes. Canadian Journal of Zoology 37:447–457.

Harvey, W. 1628. Exercitatio anatomica de motu cordis et sanguinis in animalibus, translated by G. Whitteridge, 1976. Blackwell Scientific Publications, Oxford.

Heezen, B. C. 1957. Whales entangled in deep-sea cables. Deep Sea Research 4:105–115.

Hedrick, M. S., and D. A. Duffield. 1991. Hematological and rheological characteristics of blood in seven marine mammal species: Physiological implications for diving behavior. Journal of Zoology (London) 225:273–283.

Hedrick, M. S., D. A. Duffield, and L. H. Cornell. 1986. Blood viscosity and optimal hematocrit in a deep-diving mammal, the northern elephant seal (*Mirounga angustirostris*). Canadian Journal of Zoology 64:2081–2085.

Hemmingsen, E. A. 1965. Accelerated transfer of oxygen through solutions of heme pigments. Acta Physiologica Scandinavica (Suppl.) 246:1–53.

Hering, E. 1871. Uber den Einfluss der Atmung auf den Kreislauf. Zweite Mittheilung. Ubereine reflectorische Beziehung Zwischen Lunge und Herz. Sitzungs Berichte der Academie der Wissenschaften in Wein 64:333–353.

Hiatt, E. P., and R. B. Hiatt. 1942. The effect of food on the glomerular filtration rate and renal blood flow in the harbor seal *Phoca vitulina* L. Journal of Cellular and Comparative Physiology 19:221–227.

Hill, D. A., and J. E. Reynolds. 1989. Gross and microscopic anatomy of the kidney of the West Indian manatee, *Trichechus manatus* (Mammalia: Sirenia). Acta Anatomica 135:53–56.

Hill, R. D., R. C. Schneider, G. C. Liggins, A. H. Schuette, R. L. Elliott, M. Guppy, P. W. Hochachka, J. Qvist, K. J. Falke, and W. M. Zapol. 1987. Heart rate and body temperature during free diving of Weddell seals. American Journal of Physiology 253:R344–R351.

Hindell, M. A., D. J. Slip, and H. R. Burton. 1991a. The diving behaviour of adult male and female southern elephant seals, *Mirounga leonina*. Australian Journal of Zoology 39:595–619.

Hindell, M. A., D. J. Slip, H. R. Burton, and M. M. Burton. 1991b. Physiological implications of continuous, prolonged, and deep dives of the southern elephant, *Mirounga leonina*. Canadian Journal of Zoology 70:370–379.

Hochachka, P. W. 1992. Metabolic biochemistry and the making of a mesopelagic mammal. Experientia 48:570–575.

Hochachka, P. W., and M. Guppy. 1987. Metabolic arrest and the control of biological time. Harvard University Press, Cambridge, MA.

Hochachka, P. W., and G. N. Somero. 1984. Biochemical Adaptation. Princeton University Press, Princeton, NJ, 537 pp.

Hochachka, P. W., J. M. Castellini, R. D. Hill, R. C. Schneider. J. L. Bengtson, S. E. Hill, G. C. Liggins, and W. M. Zapol. 1988. Protective metabolic mechanisms during liver ischemia: Transferable lessons from long-diving animals. Molecular and Cellular Biochemistry 23:12–20.

Hochachka, P. W., G. C. Liggins, G. P. Guyton, R. C. Schneider, K. S. Stanek, W. E. Hurford, R. K. Creasy, D. G. Zapol, and W. M. Zapol. 1995. Hormonal regulatory adjustments during voluntary diving in Weddell seals. Comparative Biochemistry and Physiology B 112:361–375.

Hokkanen, J. E. I. 1990. Temperature regulation of marine mammals. Journal of Theoretical Biology 145:465–485.

Hol, R., A. S. Blix, and H. O. Myhre. 1975. Selective redistribution of the blood volume in the diving seal (*Pagophilus groenlandicus*). Rapports et Procès-verbaux des Réunions, Conseil International pour L'Exploration de la Mer 169:423–432.

Hong, S. K. 1988. Man as a breath-hold diver. Canadian Journal of Zoology 66:70–74.

Hong, S. K. 1989. Mechanism of tolerance to renal ischemia in harbor seal: Role of membranes. Undersea Biomedical Research 16:381–390.

Hong, S. K., and H. Rahn. 1967. The diving women of Korea and Japan. Scientific American 216:34–43.

Hong, S. K., S. Ashwell-Erickson, P. Gigliotti, and R. Elsner. 1982a. Effects of anoxia and low pH on organic ion transport and electrolyte distribution in the harbor seal, *Phoca vitulina,* kidney slices. Journal of Comparative Physiology 149:19–24.

Hong, S. K., R. Elsner, J. R. Claybaugh, and K. Ronald. 1982b. Renal functions of the Baikal seal, *Pusa sibirica* and ringed seal, *Pusa hispida*. Physiological Zoology 55:289–299.

Hui, C. 1981. Seawater consumption and water flux in the common dolphin *Delphinus delphis*. Physiological Zoology 54:430–440.

Huntley, A. C. 1984. Relationships between metabolism, respiration, heart rate and arousal states in the northern elephant seal. Ph.D. dissertation, University of California, Santa Cruz, 89 pp.

Huntley, A. C., D. P. Costa, and R. D. Rubin. 1984. The contribution of nasal countercurrent heat exchange to water balance in the northern elephant seal, *Mirounga angustirostris*. Journal of Experimental Biology 113:447–454.

Hurford, W. E., P. W. Hochachka, R. C. Schneider, G. P. Guyton, K. E. Stanek, D. G. Zapol, G. C. Liggins, and W. M. Zapol. 1996. Splenic contraction, catecholamine release and blood volume redistribution during diving in the Weddell seal. Journal of Applied Physiology 80:298–306.

Hurley, J. 1996. Metabolic rate and heart rate during trained dives in adult California sea lions. Ph.D. dissertation, University of California, Santa Cruz, 109 pp.

Innes, S., and D. M. Lavigne. 1991. Do cetaceans really have elevated metabolic rates? Physiological Zoology 64:1130–1134.

Irvine, A. B. 1983. Manatee metabolism and its influence on distribution in Florida. Biological Conservation 25:315–334.

Irving, L. 1938. Changes in the blood flow through the brain and muscles during the arrest of breathing. American Journal of Physiology 122:207–214.

Irving, L. 1939. Respiration in diving mammals. Physiological Reviews. 19:112–134.

Irving, L. 1956. Physiological insulation of swine as bare-skinned mammals. Journal of Applied Physiology 9:414–420.

Irving, L. 1969. Temperature regulation in marine mammals. Pages 147–175 *in* H. T. Andersen (ed.). The Biology of Marine Mammals. Academic Press, New York, NY.

Irving, L. 1973. Aquatic mammals. Pages 47–96 *in* G. C. Whittow (ed.). Comparative Physiology of Thermoregulation, Vol. 3. Academic Press, New York, NY.

Irving, L., and J. S. Hart. 1957. The metabolism and insulation of seals as bare-skinned mammals in cold water. Canadian Journal of Zoology 35:497–511.

Irving, L., K. E. Fisher, and F. C. McIntosh. 1935. The water balance of a marine mammal, the seal. Journal of Cellular and Comparative Physiology 6:387–391.

Irving, L., P. F. Scholander, and S. W. Grinnell. 1941a. Significance of the heart rate to the diving ability of seals. Journal of Cellular and Comparative Physiology 18:283–297.

Irving, L., P. F. Scholander, and S. W. Grinnell. 1941b. The respiration of the porpoise, *Tursiops truncatus*. Journal of Cellular and Comparative Physiology 17:145–168.

Irving, L., P. F. Scholander, and S. W. Grinnell. 1942. The regulation of arterial blood pressure in the seal during diving. American Journal of Physiology 135:557–566.

Irving, L., L. J. Peyton, and M. Monson. 1956. Metabolism and insulation of bare-skinned swine in Alaska. Journal of Applied Physiology 9:421–426.

Irving, L., L. J. Peyton, C. H. Bahn, and R. S. Peterson. 1962. Regulation of temperature in fur seals. Physiological Zoology 35:275–284.

Jones D. R., and M. J. Purves. 1970. The carotid body in the duck and the consequences of its denervation upon the cardiac responses to immersion. Journal of Physiology 211:279–294.

Jones, D. R., H. D. Fisher, S. McTaggart, and N. H. West. 1973. Heart rate during breath-holding and diving in the unrestrained harbor seal, *Phoca vitulina richardsi*. Canadian Journal of Zoology 51:671–680.

Kanwisher, J. W., and S. H. Ridgway 1983. The physiological ecology of whales and porpoises. Scientific American 248:111–120.

Kanwisher, J. W., and G. Sundnes. 1966. Thermal regulation in cetaceans. Pages 397–409 *in* K. S. Norris (ed.). Whales, Dolphins and Porpoises. University of California Press, Berkeley.

Kasting, N. W., S. A. L. Adderley, T. Safford, and K. G. Hewlett. 1989. Thermoregulation in beluga (*Delphinapterus leucas*) and killer (*Orcinus orca*) whales. Physiological Zoology 62:687–701.

Keatinge, W. R. 1969. Survival in Cold Water. Blackwell Scientific Publications, Oxford.

Keatinge, W. R., and M C. Harman. 1980. Local Mechanisms Controlling Blood Vessels. Monographs of the Physiological Society, No. 37. Academic Press, London.

Kelly, B. P., and L. T. Quakenbush. 1990. Spatiotemporal use of lairs by ringed seals, *Phoca hispida*. Canadian Journal of Zoology 68:2503–2512.

Kerem, D., and R. Elsner. 1973a. Cerebral tolerance to asphyxial hypoxia in the harbor seal. Respiration Physiology 19:188–200.

Kerem, D., and R. Elsner. 1973b. Cerebral tolerance to asphyxial hypoxia in the dog. American Journal of Physiology 225:593–600.

Kerem, D., D. D. Hammond, and R. Elsner. 1973. Tissue glycogen levels in the Weddell seal: A possible adaptation to asphyxial hypoxia. Comparative Biochemistry and Physiology A 45:731–736.

Kjekshus, J. K., A. S. Blix, R. Elsner, R. Hol, and E. Amundsen. 1982. Myocardial blood flow and metabolism in the diving seal. American Journal of Physiology 242:R79–R104.

Kleiber, M. 1961. The Fire of Life. John Wiley, New York, NY.

Kooyman, G. L. 1966. Maximum diving capacities of the Weddell seal. Science 151:1553–1554.

Kooyman, G. L. 1973. Respiratory adaptations in marine mammals. American Zoologist 13:457–468.

Kooyman, G. L. 1981. Weddell Seal: Consummate Diver. Cambridge University Press, Cambridge, UK.

Kooyman, G. L. 1987. A reappraisal of diving physiology: Seals and penguins. Pages 459–469 *in* P. Dejours, L. Bolis, C. R. Taylor, and E. R. Weibel (eds.). Comparative Physiology: Life in Water and on Land, Vol. 9. Springer-Verlag, Berlin.

Kooyman, G. L. 1988. Pressure and the diver. Canadian Journal of Zoology 66:84–88.

Kooyman, G. L. 1989. Diverse Divers. Springer-Verlag, Berlin.

Kooyman, G. L., and W. B. Campbell. 1972. Heart rates in freely diving Weddell seals *Leptonychotes weddelli*. Comparative Biochemistry and Physiology A 43:31–36.

Kooyman, G. L., and C. M. Drabek. 1968. Observations on milk, blood and urine constituents of the Weddell seal. Physiological Zoology 41:187–194.

Kooyman, G. L., and P. J. Ponganis. 1997. The challenges of diving to depth. American Scientist 85:530–539.

Kooyman, G. L., and P. J. Ponganis. 1998. The physiological basis of diving to depth: Birds and mammals. Annual Reviews of Physiology 60:19–32.

Kooyman, G. L., J. P. Schroeder, D. M. Denison, D. D. Hammond, J. J. Wright, and W. P. Bergman. 1972. Blood nitrogen tensions of seals during simulated deep dives. American Journal of Physiology 223:1016–1020.

Kooyman, G. L., D. H. Kerem, W. B. Campbell, and J. J. Wright. 1973. Pulmonary gas exchange in freely diving Weddell seals (*Leptonychotes weddelli*). Respiration Physiology 17:283–290.

Kooyman, G. L., E. A. Wahrenbrock, M. A. Castellini, R. W. Davis, and E. E. Sinnett. 1980. Aerobic and anaerobic metabolism during voluntary diving in Weddell seals: Evidence of preferred pathways from blood chemistry and behavior. Journal of Comparative Physiology 138:335–346.

Kooyman, G. L., M. A. Castellini, and R. W. Davis. 1981. Physiology of diving in marine mammals. Annual Reviews of Physiology 43:343–356.

Kooyman, G. L., M. A. Castellini, R. W. Davis, and R. A. Maue. 1983. Aerobic dive limits in immature Weddell seals. Journal of Comparative Physiology 151:171–174.

Koschier, F. J., R. Elsner, D. F. Holleman, and S. K. Hong. 1978. Organic anion transport by renal cortical slices of harbor seals (*Phoca vitulina*). Comparative Biochemistry and Physiology 60:289–292.

Krasney, J. A. 1996. Head-out water immersion: Animal studies. Pages 855–888 *in* M. J. Fregly and C. M. Blatteis (eds.). Handbook of Physiology, Section 4, Vol. 2: Environmental Physiology. Oxford University Press, Oxford.

Krasney, J. A., G. Hajduczok, C. Akiba, B. W. McDonald, D. R. Pendergast, and S. K. Hong. 1984. Cardiovascular and renal responses to head-out water immersion in canine model. Undersea Biomedical Research 11:169–183.

Kratschmer, F. 1870. Uber Reflexe von der Nasenchleimhaut auf Athmung und Kreislauf. Sitzungs Berichte der Akademie der Wissenschaften in Wein 62:147–170.

Ladd, M., L. G. Raisz, C. H. Crowder Jr., and L. B. Page. 1951. Filtration rate and water diuresis in the seal, *Phoca vitulina*. Journal of Cellular and Comparative Physiology 38:157–164.

Lavigne, D. M., S. Innes, G. A. J. Worthy, K. M. Kovacs, O. J. Schmitz, and J. P. Hickie. 1986. Metabolic rates of seals and whales. Canadian Journal of Zoology 64:279–284.

Le Boeuf, B. J., D. P. Costa, A. C. Huntley, G. L. Kooyman, and R. W. Davis. 1986. Pattern and depth of dives in northern elephant seals, *Mirounga angustirostris*. Journal of Zoology (London) A 208:1–7.

Le Boeuf, B. J., D. P. Costa, A. C. Huntley, and S. D. Feldkamp. 1988. Continuous, deep diving in female northern elephant seals, *Mirounga angustirostris*. Canadian Journal of Zoology 66:446–458.

Le Boeuf, B. J., Y. Naito, A. C. Huntley, and T. Asaga. 1989. Prolonged, continuous, deep diving by northern elephant seals. Canadian Journal of Zoology 67:2514–2519.

Leith, D. E. 1976. Comparative mammalian respiratory mechanics. Physiologist 19:485–510.

Lenfant, C. 1969. Physiological properties of blood of marine mammals. Pages 95–116 *in* H. T. Andersen (ed.). The Biology of Marine Mammals. Academic Press, New York, NY.

Lenfant, C., R. Elsner, G. L. Kooyman, and C. M. Drabek. 1969a. Respiratory function of the blood of the adult and fetus Weddell seal, *Leptonychotes weddelli*. American Journal of Physiology 216:1595–1597.

Lenfant, C., R. Elsner, G. L. Kooyman, and C. M. Drabek. 1969b. Tolerance to sustained hypoxia in the Weddell seal, *Leptonychotes weddelli*. Pages 471–476 *in* M. W. Holdgate (ed.). Antarctic Ecology. Academic Press, New York, NY.

Lenfant, C., K. Johansen, and J. D. Torrance. 1970. Gas transport and oxygen storage capacity in some pinnipeds and the sea otter. Respiration Physiology 9:277–286.

Liggins, G. C., J. Qvist, P. W. Hochachka, B. J. Murphy, R. K. Creasy, R. C. Schneider, M. T. Snider, and W. M. Zapol. 1980. Fetal cardiovascular and metabolic responses to simulated diving in the Weddell seal. Journal of Applied Physiology 49:424–430.

Lin, Y. C. 1984. Circulatory functions during immersion and breath-hold dives in humans. Undersea Biomedical Research 11:123–138.

Lin, Y. C., and S. K. Hong. 1996. Hyperbaria: Breath-hold diving. Pages 979–998 *in* M. J. Fregly and C. M. Blatteis (eds.). Environmental Physiology, Section 4, Vol. 2: Handbook of Physiology. Oxford University Press, Oxford.

Lowrance, P. B., J. F. Nickel, C. M. Smythe, and S. E. Bradley. 1956. Comparison of the effects of anoxic anoxia and apnea on renal function in the harbor seal *Phoca vitulina* L. Journal of Cellular and Comparative Physiology 48:35–49.

Lundgren, C. E. G., A. Harabin, P. B. Bennett, H. D. Van Liew, and E. D. Thalmann. 1996. Gas physiology in diving. Pages 999–1019 *in* M. J. Fregly and C. M. Blatteis (eds.). Handbook of Physiology, Section 4, Vol. 2: Environmental Physiology. Oxford University Press, Oxford.

Lutz, P. L. and G. E. Nilsson. 1997. The Brain Without Oxygen. Chapman and Hall, Austin, TX, 207 pp.

Mackintosh, N. A. 1966. The distribution of southern blue and fin whales. Pages 125–144 *in* K. S. Norris (ed.). Whales, Dolphins and Porpoises. University of California Press, Berkeley.

MacMillen, R. E., and A. K. Lee. 1969. Water metabolism of Australian hopping mice. Comparative Biochemistry and Physiology 28:493–514.

Maluf, N. S. R. 1989. Renal anatomy of the manatee, *Trichechus manatus* (Linnaeus). American Journal of Anatomy 184:269–286.

Malvin, R. L., and M. Rayner. 1968. Renal function and blood chemistry in cetacea. American Journal of Physiology 214:187–191.

Malvin, R. L., J. P. Bonjour, and S. Ridgway. 1971. Antidiuretic hormone levels in some cetaceans. Proceedings of the Society for Experimental Biology and Medicine 136:1203–1205.

Malvin, R. L., S. Ridgway, and L. Cornell. 1978. Renin and aldosterone levels in dolphins and sea lions. Proceedings of the Society for Experimental Biology and Medicine 157:665–668.

McCord, J. M. 1985. Oxygen-derived free radicals in post-ischemic tissue injury. New England Journal of Medicine 312:159–163.

McKean, T. A. 1984. Response of isolated muskrat and guinea pig hearts to hypoxia. Physiological Zoology 57:557–562.

McKean, T. A., and R. Landon. 1982. Comparison of the response of muskrat, rabbit and guinea pig heart muscles to hypoxia. American Journal of Physiology 243:R245 R250.

Meiselman, H. J., M. A. Castellini, and R. Elsner. 1992. Hemorheological behavior of seal blood. Journal of Clinical Hemorheology 12:657–675.

Melville, H. 1851. Moby Dick or the Whale (reprinted 1930). Random House, New York, NY.

Miki, K., Y. Hayashida, S. Sagawa, and K. Shiraki. 1989. Renal sympathetic nerve activity and natriuresis during water immersion in conscious dogs. American Journal of Physiology 256:R299–R305.

Miller, K., and L. Irving. 1975. Metabolism and temperature regulation in young harbor seals *Phoca vitulina richardsi*. American Journal of Physiology 229:506–511.

Mohri, M., R. Torii, K. Nagaya, K. Shiraki, R. Elsner, H. Takeuchi, Y. S. Park, and S. K. Hong. 1995. Diving patterns of ama divers of Hegura Island, Japan. Undersea and Hyperbaric Medicine 22:137–143.

Molyneux, G. S., and M. M. Bryden. 1978. Arteriovenous anastomoses in the skin of seals. The Weddell seal, *Leptonychotes weddelli*, and the elephant seal, *Mirounga leonina* (Pinnipedia: Phocidae). Anatomical Record 191:239–252.

Morrison, P. 1962. Body temperatures in some Australian mammals. III. Cetacea (*Megaptera*). Biological Bulletin 123:154–169.

Morrison, P., M. Rosenmann, and J. A. Estes. 1974. Metabolism and thermoregulation in the sea otter. Physiological Zoology 47:218–229.

Murdaugh, H. V., J. C. Seabury, and W. L. Mitchell. 1961a. Electrocardiogram of the diving seal. Circulation Research 9:358–361.

Murdaugh, H. V., B. Schmidt-Nielsen, J. W. Wood, and W. L. Mitchell. 1961b. Cessation of renal function during diving in the trained seal (*Phoca vitulina*). Journal of Cellular and Comparative Physiology 58:261–265.

Murdaugh, H. V., E. D. Robin, J. E. Millen, W. F. Drewry, and E. Weiss. 1966. Adaptations to diving in the harbor seal: Cardiac output during diving. American Journal of Physiology 210:176–180.

Nadel, E. R. 1984. Energy exchanges in water. Undersea Biomedical Research 11:149–158.

Nordøy, E. S., O. C. Ingebretsen, and A. S. Blix. 1990. Depressed metabolism and low protein catabolism in fasting grey seal pups. Acta Physiologica Scandinavica 139:361–369.

Olsen, C. R., F. C. Hale, and R. Elsner. 1969. Mechanics of ventilation in the pilot whale. Respiration Physiology 7:137–149.

Øritsland, N. A. 1971. Wavelength-dependent solar heating of harp seals (*Pagophilus groenlandicus*). Comparative Biochemistry and Physiology A 40:359–361.

Øritsland, N. A., and K. Ronald. 1973. Effects of solar radiation and wind-chill on skin temperature of the harp seal. Comparative Biochemistry and Physiology A 44:519–525.

Ortiz, C. L., D. P. Costa, and B. J. Le Boeuf. 1978. Water and energy flux in elephant seal pups fasting under natural conditions. Physiological Zoology 51:166–178.

Ortiz, R. M. 1994. Water flux and osmoregulatory physiology of the West Indian manatee (*Trichechus manatus*). M.S. dissertation, University of California, Santa Cruz.

Ortiz, R. M., S. E. Adams, D. P. Costa, and C. L. Ortiz. 1996. Plasma vasopressin levels and water conservation in fasting, postweaned northern elephant seal pups (*Mirounga angustirostris*). Marine Mammal Science 12:99–106.

Page, L. B., J. C. Scott-Baker, G. A. Zak, E. L. Becker, and C. F. Baxter. 1954. The effect of variation in filtration rate on the urinary concentrating mechanism in the seal, *Phoca vitulina* L. Journal of Cellular and Comparative Physiology 43:257–270.

Paulev, P. 1965. Decompression sickness following repeated breath-hold dives. Journal of Applied Physiology 20:1028–1031.

Pernia, S. D., D. P. Costa, and C. L. Ortiz. 1989. Glomerular filtration rate in weaned elephant seal pups during natural, long term fasts. Canadian Journal of Zoology 67:1752–1756.

Pilson, M. E. Q. 1970. Water balance in California sea lions. Physiological Zoology 43:257–269.

Ponganis, P. J., G. L. Kooyman, M. H. Zornow, M. A. Castellini, and D. A. Croll. 1990a. Cardiac output and stroke volume in swimming seals. Journal of Comparative Physiology B 160:473–482.

Ponganis, P. J., E. P. Ponganis, K. V. Ponganis, G. L. Kooyman, R. L. Gentry, and F. Trillmich. 1990b. Swimming velocities in otariids. Canadian Journal of Zoology 68:2105–2115.

Ponganis, P. J., G. L. Kooyman, E. A. Baranov, P. H. Thorson, and B. S. Stewart. 1997. The aerobic submersion limit of Baikal seals, *Phoca sibirica*. Canadian Journal of Zoology 75:1323–1327.

Pugh, L. G. C., and O. G. Edholm. 1955. The physiology of channel swimmers. Lancet 2:761–768.

Qvist, J., R. D. Hill, R. C. Schneider, K. J. Falke, G. C. Liggins, M. Guppy, R. L. Elliott, P. W. Hochachka, and W. M. Zapol. 1986. Hemoglobin concentrations and blood gas tensions of free-diving Weddell seals. Journal of Applied Physiology 61:1560–1569.

Rennie, D. W., B. G. Covino, B. J. Howell, S. H. Song, B. S. Kang, and S. K. Hong. 1962. Physical insulation of Korean diving women. Journal of Applied Physiology 17:961–966.

Repenning, C. A., C. E. Ray, and D. Grigorescu. 1979. Pinniped biogeography. Pages 357–369 *in* J. Gray and A. J. Boucot (eds.). Historical Biogeography, Plate Tectonics and the Changing Environment. Oregon State University Press, Corvalis.

Reynolds, J. E., and D. K. Odell. 1991. Manatees and Dugongs. Facts on File, Inc., New York, NY.

Rhode, E. A., R. Elsner, T. M. Peterson, K. B. Campbell, and W. Spangler. 1986. Pressure–volume characteristics of aortas of harbor and Weddell seals. American Journal of Physiology 251:R174–R180.

Richet, C. 1894. La résistance des canards à l'asphysie. Journal de Physiologie et de Pathologie Générale 1:244–245.

Richet, C. 1899. De la résistance des canards a l'asphyxie. Journal de Physiologie et de Pathologie Générale 1:641–650.

Ridgway, S. H. 1972. Homeostasis in the aquatic environment. Pages 590–747 *in* S. H. Ridgway (ed.). Mammals of the Sea: Biology and Medicine. Thomas, Springfield, IL.

Ridgway, S. H., and R. Howard. 1979. Dolphin lung collapse and intramuscular circulation during free diving: Evidence from nitrogen washout. Science 206:1182–1183.

Ridgway, S. H., and D. G. Johnston. 1966. Blood oxygen and ecology of porpoises of three genera. Science 151:456–458.

Ridgway, S. H., B. L. Scronce, and J. Kanwisher. 1969. Respiration and deep diving in the bottlenose porpoise. Science 166:1651–1654.

Ridgway, S. H., D. A. Carder, and W. Clark. 1975. Conditioned bradycardia in the sea lion *Zalophus californianus*. Nature 256:37–38.

Ridgway, S. H., C. A. Bowers, D. Miller, M. L. Schultz, C. A. Jacobs, and C. A. Dooley. 1984. Diving and blood oxygen in the white whale. Canadian Journal of Zoology 62:2349–2351.

Robinson, D. 1939. The muscle hemoglobin of seals as an oxygen store in diving. Science 90:276–277.

Ronald, K., R. McCarter, and L. J. Selley. 1977. Venous circulation in the harp seal (*Pagophilus groenlandicus*). Pages 235–270 *in* R. J. Harrison (ed.). Functional Anatomy of Marine Mammals , Vol. 3. Academic Press, London.

Rovetto, M. J., W. F. Lamberton, and J. R. Neely. 1975. Mechanisms of glycolytic inhibition in ischemic rat hearts. Circulation Research 37:742–751.

Sangalang, C. B., and H. C. Freeman. 1976. Steroids in the plasma of the gray seal, *Halichoerus grypus*. General and Comparative Endocrinology 29:419–433.

Scammon, C. M. 1874. The Marine Mammals of the Northwest Coast of North America Together With an Account of the American Whale Fishery. J. H. Carmany, San Francisco, reprinted 1968; Dover, New York, NY, 319 pp.

Schaefer, K. E., R. D. Allison, J. H. Dougherty, C. R. Carey, R. Walker, F. Yost, and D. Parker. 1968. Pulmonary and circulatory adjustments determining the limits of depths in breath-hold diving. Science 162:1020–1023.

Schagatay, E. 1996. The human diving response: Effects of temperature and training. Ph.D. dissertation, University of Lund, Sweden.

Schmidt-Nielsen, B., H. V. Murdaugh, R. O'Dell, and J. Bacsanyi. 1959. Urea excretion and diving in the harbor seal, *Phoca vitulina* L. Journal of Cellular and Comparative Physiology 53:393–411.

Schmidt-Neilsen, B., and R. O'Dell. 1961. Structure and concentrating mechanism in the mammalian kidney. American Journal of Physiology 200:1119–1124.

Schmidt-Nielsen, K. 1964. Desert Animals: Physiological Problems of Heat and Water. Clarendon Press, Oxford, reprinted 1979; Dover, New York, NY.

Schmidt-Nielsen, K. 1981. Countercurrent systems in animals. Scientific American 244:118–128.

Schmidt-Nielsen, K., F. R. Hainsworth, and D. E. Murrish. 1970. Countercurrent heat exchange in the respiratory passages: Effect on water and heat balance. Respiration Physiology 9:263–276.

Scholander, P. F. 1940. Experimental investigations on the respiratory function in diving mammals and birds. Hvalrådets Skrifter, Det Norske Videnskaps-Akademi i Oslo 22:1–131.

Scholander, P. F. 1955. Evolution in climatic adaptation in homeotherms. Evolution 9:15–26.

Scholander, P. F. 1960. Oxygen transport through hemoglobin solutions. Science 131:585–590.

Scholander, P. F. 1962. Physiological adaptations to diving in animals and man. Harvey Lectures 57:93–110.

Scholander, P. F. 1963. The master switch of life. Scientific American 209:92–106.

Scholander, P. F. 1964. Animals in aquatic environments: Diving mammals and birds. Pages 729–739 *in* D. B. Dill, E. F. Adolph, and C. G. Wilber (eds.). Handbook of Physiology, Section 4: Adaptations to the Environment. American Physiological Society, Washington, D.C.

Scholander, P. F., and L. Irving. 1941. Experimental investigations on the respiration and diving of the Florida manatee. Journal of Cellular and Comparative Physiology 17:169–191.

Scholander, P. F., and W. E. Schevill. 1955. Counter-current vascular heat exchange in the fins of whales. Journal of Applied Physiology 8:279–282.

Scholander, P. F., L. Irving, and S. W. Grinnell. 1942a. Aerobic and anaerobic changes in seal muscles during diving. Journal of Biological Chemistry 142:431–440.

Scholander, P. F., L. Irving, and S. W. Grinnell. 1942b. On the temperature and metabolism of the seal during diving. Journal of Cellular and Comparative Physiology 21:53–63.

Scholander, P. F., V. Walters, R. Hock, and L. Irving. 1950a. Body insulation of some arctic and tropical mammals and birds. Biological Bulletin 99:225–236.

Scholander, P. F., R. Hock, V. Walters, F. Johnson, and L. Irving. 1950b. Heat regulation in some arctic and tropical mammals and birds. Biological Bulletin 99:237–258.

Scholander, P. F., R. Hock, V. Walters, and L. Irving. 1950c. Adaptation to cold in arctic and tropical mammals and birds in relation to body temperature, insulation and basal metabolic rate. Biological Bulletin 99:259–271.

Schultz, S. G. 1996. Homeostasis, Humpty Dumpty, and integrative biology. News in Physiological Sciences 11: 238–246.

Seeherman, H. J., C. R. Taylor, G. M. O. Maloiy, and R. B. Armstrong. 1981. Design of the mammalian respiratory system: Measuring maximum aerobic capacity. Respiration Physiology 44:11–24.

Shadwick, R. E., and J. M. Gosline. 1995. Arterial windkessels in marine mammals, Pages 243–252 *in* C. P. Ellington and T. J. Pedley (eds.). Biological Fluid Dynamics, Symposia of the Society for Experimental Biology, Vol. 49. Cambridge University Press, Cambridge, U.K.

Sinnett, E. E., G. L. Kooyman, and E. A. Wahrenbrock. 1978. Pulmonary circulation of the harbor seal. Journal of Applied Physiology 45:718–727.

Skog, E. B., and L. P. Folkow. 1994. Nasal heat and water exchange is not an effector mechanism for water balance regulation in grey seals. Acta Physiologica Scandinavica 151:223–240.

Smith, T. G., and I. Stirling. 1975. The breeding habitat of the ringed seal, *Phoca hispida*. The birth lair and associated structures. Canadian Journal of Zoology 53:1297–1305.

Snyder, G. K. 1983. Respiratory adaptations in diving mammals. Respiration Physiology 54:269–294.

Sordahl, L. A., G. Mueller, and R. Elsner. 1983. Comparative functional properties of mitochondria from seal and dog hearts. Journal of Molecular and Cellular Cardiology 15:1–5.

Speakman, J. R., R. M. McDevitt, and K. R. Cole. 1993. Measurement of basal metabolic rates: Don't lose sight of reality in the quest for comparability. Physiological Zoology 66:1045–1049.

Spencer, M. P., T. A. Gornall, and T. C. Poulter. 1967. Respiratory and cardiac activity of killer whales. Journal of Applied Physiology 22:974–981.

Stahl, W. R. 1967. Scaling of respiratory variables in mammals. Journal of Applied Physiology 22:453–460.

Stirling, I. 1969. Ecology of the Weddell seal in McMurdo Sound, Antarctica. Ecology 50:573–586.

Tanji, D. G., J. Weste, and R. W. Dykes. 1975. Interactions of respiration and the bradycardia of submersion in harbour seals. Canadian Journal of Physiology and Pharmacology 53:555–559.

Tarasoff, F. J., and G. L. Kooyman. 1973. Observations on the anatomy of the respiratory system of the river otter, sea otter and harp seal. I. The topography, weight and measurements of the lungs. Canadian Journal of Zoology 51:163–170.

Tarasoff, F. J., and D. P. Toews. 1972. The osmotic and ionic regulatory capacities of the kidney of the harbor seal, *Phoca vitulina*. Journal of Comparative Physiology 81:121–132.

Testa, J. W. 1994. Over-winter movements and diving behavior of female Weddell seals (*Leptonychotes weddellii*) in the southwestern Ross Sea, Antarctica. Canadian Journal of Zoology 72:1700–1710.

Thompson, D., and M. A. Fedak. 1993. Cardiac responses of grey seals during diving at sea. Journal of Experimental Biology 174:139–164.

Vanhoutte, P. M., and M. N. Levy. 1980. Prejunctional cholinergic modulation of adrenergic neurotransmission in the cardiovascular system. American Journal of Physiology 238:H275–H281.

Vardy, P. H., and M. M. Bryden. 1981. The kidney of *Leptonychotes weddelli* (Pinnipedia: Phocidae) with some observations on the kidneys of two other southern phocid seals. Journal of Morphology 167:13–34.

Vongraven, D., M. Ekker, A. R. Espelien, and F. J. Aarvik. 1990. Postmortem body temperatures in the minke whale *Balaenoptera acutorostrata*. Canadian Journal of Zoology 68:140–143.

Wahrenbrock, E. A., G. F. Maruschak, R. Elsner, and D. W. Kenney. 1974. Respiratory function and metabolism in two baleen whale calves. Marine Fisheries Review 36:3–9.

Wartzok, D., R. Elsner, H. Stone, B. P. Kelly, and R. W. Davis. 1992. Under-ice movements and the sensory basis of hole finding by ringed and Weddell seals. Canadian Journal of Zoology 70:1712–1722.

Wasserman, K., and A. Mackenzie. 1957. Cardiac output in diving seals. Bulletin of the Tulane University Medical Faculty 16:105–110.

Watkins, W. A., and W. E. Schevill. 1972. Sound source location with a three-dimensional hydrophone array. Deep Sea Research 19:691–706.

Watkins, W. A., K. E. Moore, and P. Tyack. 1985. Investigations of sperm whale acoustic behaviors in the southeast Caribbean. Cetology 49:1–15.

Watkins, W. A., M. A. Daher, K. M. Fristrup, and T. J. Howard. 1993. Sperm whales tagged with transponders and tracked underwater by sonar. Marine Mammal Science 9:55–67.

Weiss, S. J. 1986. Oxygen, ischemia and inflammation. Acta Physiologica Scandinavica (Suppl.) 548:9–37.

White, F. C., R. Elsner, D. Willford, E. Hill, and E. Merhoff. 1990. Responses of harbor seal and pig heart to progressive and acute hypoxia. American Journal of Physiology 259:R849–R856.

White, F. N., and D. K. Odell. 1971. Thermoregulatory behavior of the northern elephant seal, *Mirounga angustirostris*. Journal of Mammalogy 52:758–774.

White, S. W., R. J. McRitchie, and D. L. Franklin. 1974. Autonomic cardiovascular effects of nasal inhalation of cigarette smoke in the rabbit. Australian Journal of Biological and Medical Sciences 52:111–126.

Whitehead, A. N. 1920. The Concept of Nature. Cambridge University Press, Cambridge, U.K.

Wickham, L. L., L. H. Cornell, and R. Elsner. 1985. Red cell morphology and blood viscosity: Possible adaptations of phocid seals. Abstracts, 6th Biennial Conference on the Biology of Marine Mammals, Vancouver, B.C.

Wickham, L. L., R. Elsner, F. C. White, and L. H. Cornell. 1989. Blood viscosity in phocid seals: Possible adaptations to diving. Journal of Comparative Physiology 159:153–158.

Wickham, L. L., R. M. Bauersachs, S. Coker, H. J. Meiselman, and R. Elsner. 1990. Red cell aggregation and viscoelasticity of blood from seals, swine and man. Biorheology 27:191–204.

Willford, D. C. 1985. Oxygen transport and utilization during induced hypothermia, Ph.D. dissertation, University of California, San Diego, 174 pp.

Williams, T. M., G. L. Kooyman, and D. A. Croll. 1991. The effect of submergence on heart rate and oxygen consumption of swimming seals and sea lions. Journal of Comparative Physiology B 160:637–644.

Williams, T. M., W. A. Friedl, J. E. Haun, and N. K. Chun. 1993a. Balancing power and speed in bottlenose dolphins. Pages 383–394 *in* I. L. Boyd (ed.). Marine Mammals, Advances in Behavioral and Population Biology, Symposia of the Zoological Society, London, No. 66.

Williams, T. M., W. A. Friedl, and J. E. Haun. 1993b. The physiology of bottlenose dolphins (*Tursiops truncatus*): Heart rate, metabolic rate and plasma lactate concentration during exercise. Journal of Experimental Biology 179:31–46.

Winter, P. M., and J. N. Miller. 1985. Anesthesiology. Scientific American 252:124–131.

Wittenberg, B. A., J. B. Wittenberg, and P. R. B. Caldwell. 1975. Role of myoglobin in the oxygen supply to red skeletal muscle. Journal of Biological Chemistry 250:9038–9043.

Woakes, A. J. 1992. An implantable data logging system for heart rate and body temperature. Pages 120–127 *in* I. G. Priede and S. M. Swift (eds.). Wildlife Telemetry. Ellis Horwood Ltd, London.

Zapol, W. M. 1987. Diving adaptations of the Weddell seal. Scientific American 256:100–107.

Zapol, W. M. 1996. Diving physiology of the Weddell seal. Pages 1049–1058 *in* M. J. Fregly and C. M. Blatteis (eds.). Environmental Physiology, Section 4, Vol. 2: Handbook of Physiology. Oxford University Press, Oxford.

Zapol, W. M., G. C. Liggins, G. C. Schneider, J. Qvist, M. T. Snider, R. K. Creasy, and P. W. Hochachka. 1979. Regional blood flow during simulated diving in the conscious Weddell seal. Journal of Applied Physiology 47:968–973.

4

DOUGLAS WARTZOK AND DARLENE R. KETTEN

Marine Mammal Sensory Systems

Sensory systems evolved to allow animals to receive and process information from their surroundings. To understand how sensory systems operate in any given environment, we must understand how the physical characteristics of that environment affect the available information and its propagation and reception. In a very real sense we need to look at both the medium and the message (McLuhan and Fiore 1967). Signals in the marine environment can be substantially different from those in air, and the oceanic medium itself changes the message in a number of ways.

When their evolutionary paths took them into the oceans, marine mammals had to adapt sensory systems that had evolved in air into ones that were able to detect and process signals in water. The sensory systems of marine mammals are functionally similar to those of terrestrial mammals in that they act as highly selective filters. If every environmental cue available received equal attention, the brain would be barraged by sensory inputs. Instead, sensory organs are filters, selecting and attending to signals that, evolutionarily, proved to be important. Consider how predator and prey are driven to be both similar and different sensorially. Because their activities intersect in place and time, they need to have similar visual sensitivities, but different fields of view. The predator usually has binocular overlap that provides a precise judgment of distance to the prey. The prey may forego binocular vision and accurate visual depth judgments in favor of greater lateral visual fields to detect a predator. Thus, two species may have overlapping sensory ranges, but no two have identical sensory capacities. Consequently, each animal's perceived world is only a subset of the real physical world, that is, it is a species-specific model, constructed from the blocks of data its senses can capture.

In animal behavior, this concept is called the Umwelt (von Uexküll 1934). As a technical term, Umwelt means an animal's perceptually limited construct of the world. In common usage, it simply means the environment. This dual meaning reflects the complex interaction of sensory adaptations and habitat. Senses are tuned to relevant stimuli by evolution but are limited by the physical parameters of the habitat. For example, human sensory systems are geared to diurnal, airborne cues. Humans are highly developed visually, with 38 times more optic nerve fibers than auditory nerve fibers, and a hearing range (20 to 20,000 Hz) that is narrower than that of many other animals. By observing species adapted to different habitats and analyzing their sensory biology, we can learn how they detect and use physical cues that are normally imperceptible to us. If we develop technology that translates those cues into our sensory ranges, we can glimpse at the world as other species perceive it. Marine mammals offer us a very special glimpse. In aquatic environments, our air-adapted senses are out of their element and are effectively detuned. By studying marine mammal sensory systems and abilities, we can understand how land mammal senses were evolutionarily retuned to operate in water. From that knowledge, we gain a valuable window into the oceans, the most extensive and unexplored environment on earth.

In this chapter, we discuss marine mammal audition, vision, chemoreception, tactile sensation, and magnetic detection. We begin with an overview of the basic aspects of sensory receptor systems, and then, for specific sensory systems, examine how water versus air affects the parameters and propagation of related signals and discuss how air-based receptors were adapted to function effectively in an aquatic environment. Different sensory systems and different marine mammal groups (sirenians, cetaceans, pinnipeds, fissipeds, ursids) are discussed in varying detail based on the extent of data available for each.

Generally, the term sensory system refers to the peripheral, as opposed to the brain, or central, components an animal uses to detect and analyze a signal. There are four essential functions for any sensory system: (1) capture an environmental signal, (2) filter it, (3) transduce it to a neural impulse, and (4) send processed information to the central nervous system. Each function may involve more than one form of receptor or peripheral processor. The block diagram in Figure 4-1 compares a generic sensory receptor system with equivalent stages for mammalian eyes and ears. In vision, the first step, signal capture, is accomplished by the refraction of light at the cornea and the pupil's ability to control the light intensity entering the eye. Second, the lens focuses light on the retina while also acting as a first-stage filter, passing only some portions of the full spectrum of light. The tapetum, a reflective layer behind the retina, reflects the photons not captured on the first passage through the retina back through for a second chance at absorption. Third, pigments within the rod and cone receptor cells absorb each wavelength with a different efficiency. Fourth, the rods or cones pass a chemical signal to horizontal or bipolar cells that modify and transmit the signal to amacrine or ganglion cells. Axons of ganglion cells make up the optic nerve, which passes signals to midbrain structures and ultimately to the cortex. In hearing, the first step is the capture of sound by the external ear. The external ear and the ear canal act as first-stage filters, attenuating some sounds based on their direction (pinnal shadowing) and amplifying others according to the resonance characteristics of the outer ear and canal. Second, the middle ear components act as second stage filters. The middle ear bones mechanically transmit vibrations of the eardrum, or tympanic membrane, to the oval window, which is the acoustic entrance to the fluid-filled inner ear. This bony chain acts as a series of levers that provides a nearly 40-dB boost to the incoming signal, which compensates for the loss of acoustic power that would normally occur from a simple transmission of sound in air into fluid. The mass, stiffness, and shape of the middle ear cavity and of the middle ear ossicular chain also influence the efficiency with which different frequencies are transmitted to the inner ear. Third,

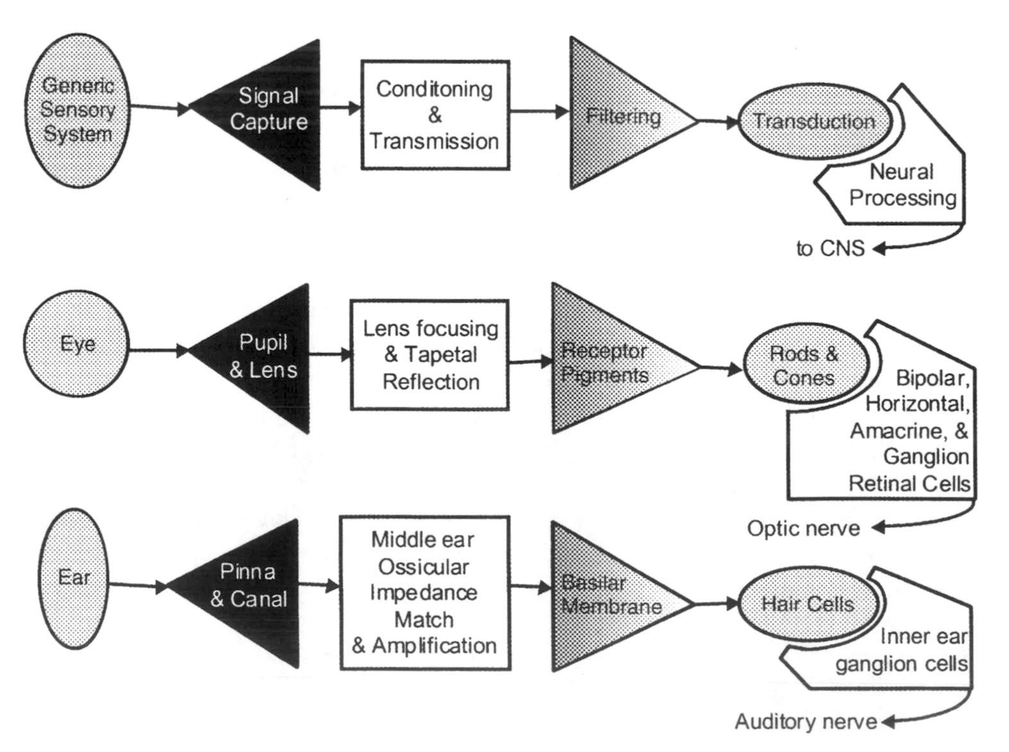

Figure 4-1. A generic sensory system is shown with parallel stages for mammalian visual and auditory systems.

at the level of the inner ear, the basilar membrane acts as a bank of filters that determine the range of frequencies the brain will ultimately process. The detectable sound or "hearing" range is dictated by the stiffness and mass characteristics of this membrane. Fourth, when sensory cells with flexible cilia, the hair cells, are bent through the motion of the basilar membrane, a chemical signal is transmitted via the auditory afferent (inward) fibers to the brainstem. Thus, in both the eye and ear, the signal goes through a minimum of three and as many as five layers of signal processing before it is transformed into a neural impulse.

For both sensory systems, there is extensive central processing as well as efferent (outward) feedback signals from the central nervous system that affect the responses at the receptor. Depending on the stimulus, the behavioral and physiologic state of the animal, and the type of receptor, a stimulus can be perceived but elicit no action, or it can prompt a set of signals to be sent to an effector that modulates the stimulus intensity, as in withdrawal from pain, pupillary contraction in bright light, or rotation of the head or pinna to enhance detection of a particular sound. Now we turn to a more detailed look at individual sensory systems.

Audition

Hearing is simply the detection of sound. "Sound" is the propagation of a mechanical disturbance through a medium. In elastic media, such as air and water, that disturbance takes the form of acoustic waves. The adaptive significance of sound cues is underscored by the ubiquity of hearing. There are lightless habitats on earth with naturally blind animals, but no terrestrial habitat is without sound, and no known vertebrate is naturally profoundly deaf.

Mechanistically, hearing is a relatively simple chain of events: sound energy is converted by biomechanical transducers (middle and inner ear) into electrical signals (neural impulses) that provide a central processor (brain) with acoustic data. Mammalian ears are elegant structures, packing more than 75,000 mechanical and electrochemical components into an average volume of 1 cm^3. Variations in the structure and number of these components account for most of the hearing capacity differences among mammals (for an overview, see Webster et al. 1992).

Normal functional hearing ranges and the sensitivity at each audible frequency (threshold, or minimum intensity required to hear a given frequency) vary widely by species (Fig. 4-2). "Functional" hearing refers to the range of frequencies a species hears without entraining nonacoustic mechanisms. In land mammals, the functional range is generally considered to be those frequencies that can be heard at thresholds below 60 dB SPL. (dB SPL refers to a decibel measure of sound pressure level. The basis for this measure and how it differs in air and water are explained in detail in the next section.) For example, a healthy human ear has a potential maximum frequency range of 0.02 to 20 kHz, but the normal functional hearing range in an adult is closer to 0.04 to 16 kHz.[1] In humans, best sensitivity (lowest thresholds) occurs between 500 Hz and 4 kHz, which is also where most of the acoustic energy in speech occurs (Fig. 4-2; Schuknecht 1993; Yost 1994). To hear frequencies at the extreme ends of any animal's total range generally requires intensities that are uncomfortable, and some frequencies are simply unde-

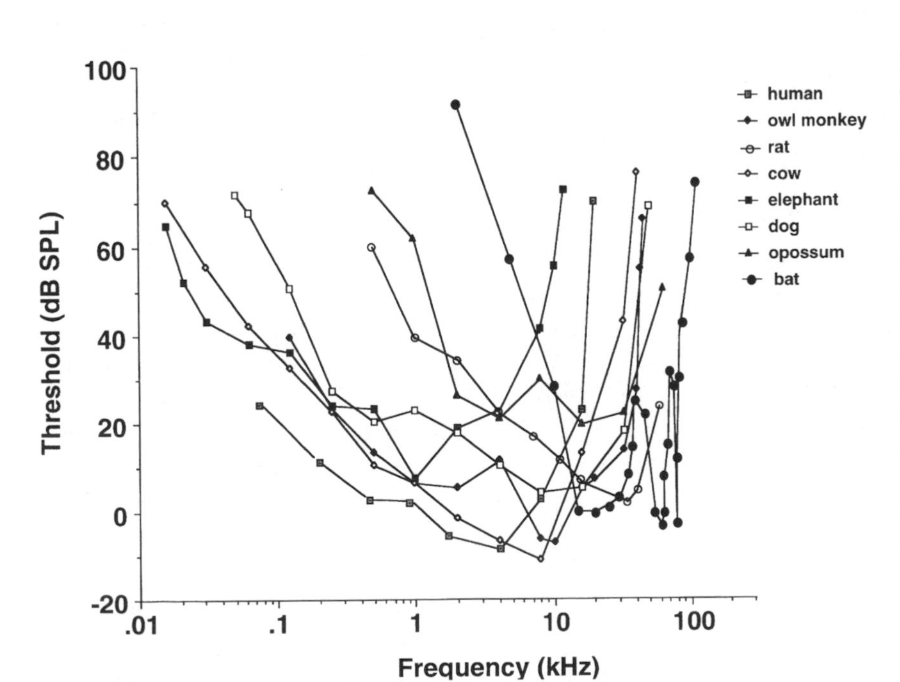

Figure 4-2. Audiograms of representative terrestrial mammals. Note that the ordinate is labeled dB SPL and therefore, thresholds are at or near 0 dB in the regions of best sensitivity for most species (data compiled from Fay 1988). The human curve ends abruptly near 100 Hz at its low-frequency end because most human hearing studies have focused on speech perception and subjects are not tested over the full human hearing range. Compare these audiograms for land mammals in air with the underwater audiograms for cetaceans and pinnipeds in Figure 4-5, taking into consideration the effect that differing reference pressures have on reported threshold values.

tectable because of limitations in the resonance characteristics of the middle and inner ear. Exceptionally loud sounds that are outside the functional range of normal hearing can sometimes be perceived through bone conduction or direct motion of the inner ear, but this is not truly an auditory sensation.

Analyzing how hearing abilities, habitat, and ear anatomy are linked in different species, particularly in animals from diverse habitats, provides insights into how each component in the auditory periphery functions and how different hearing capacities evolved. "Sonic" is an arbitrary term that refers to the maximal human hearing range. Frequencies outside this range are deemed infrasonic (below 20 Hz) or ultrasonic (above 20 kHz). Of course, many animals hear sounds inaudible to humans. Most mammals have some ultrasonic hearing (i.e., can hear well at ultrasonic frequencies) and a few, like the Asian elephant (*Elephas maximus*) hear infrasonic signals (Fig. 4-2).

Hearing ranges are both size and niche related. In general, mammalian ears scale with body size (Manley 1972; Ketten 1984, 1992; West 1985). In land mammals, the highest frequency an animal hears is generally inversely related to body mass; smaller animals typically have good high frequency hearing, whereas larger animals tend to have lower overall ranges (von Békésy 1960, Greenwood 1962, Manley 1972, Ketten 1984). Yet, regardless of size, crepuscular and nocturnal species typically have acute ultrasonic hearing, whereas subterranean species usually have good infrasonic hearing, and, in some cases, can detect seismic vibrations (Sales and Pye 1974, Heffner and Heffner 1980, Payne et al. 1986, Fay 1988).

How well do marine mammals mesh with this general land mammal hearing scheme? Marine mammals evolved from land-dwelling ancestors during the explosive period of mammalian radiation (see Barnes et al. 1985). Today, marine mammals occupy virtually every aquatic niche (freshwater to pelagic, surface to profundal) and have a size range of several magnitudes (e.g., harbor porpoise [*Phocoena phocoena*], 1 m and 55 kg vs. the blue whale [*Balaenoptera musculus*], 40 m and 94,000 kg; Nowak 1991). Water is a relatively dense medium in which light attenuates much faster than sound, therefore marine mammals are, in a sense, de facto crepuscular species, but we also expect to see a wide range of hearing given their diversity of animal size and habitat. Because marine mammals retained the essentials of air-adapted ears, that is, an air-filled middle ear and spiral cochlea, some similarities in hearing mechanism between land and aquatic mammals would not be surprising. In fact, hearing in marine mammals has the same basic size versus auditory structure relationship as in land mammals, but marine mammals have a significantly different auditory *bauplan*, or ear size versus frequency relationship (Solntseva 1971, 1990; Ketten 1984, 1992). Consequently, although some marine mammals, consistent with their size, hear well at low frequencies, the majority, despite their relatively large size, hear best at ultrasonic frequencies because of unique auditory mechanisms that evolved in response to the marine environment.

Land and marine ears have significant structural differences. Because of some of these differences, a common definition of the term ear is somewhat problematic. In this chapter, ear is used in the broadest sense to encompass all structures that function primarily to collect and process sound. As marine mammal ancestors became more aquatic, air-adapted mammalian ears had to be coupled to waterborne sound for hearing to remain functional. Ear evolution took place in tandem with, and in part in response to, body reconfigurations. Just as the physical demands of operating in water exacted a structural price in the locomotory and thermoregulatory systems of marine mammals (see Pabst, Rommel, and McLellan, Chapter 2, this volume), physical differences in underwater sound required auditory system remodeling. In modern marine mammals, the extent of ear modifications parallels the level of aquatic adaptation in each group (Ketten 1984, 1992; Solntseva 1990). The greatest differences from land mammals are found in cetaceans and sirenians. As they evolved into obligate aquatic mammals, unable to move, reproduce, or feed on land, every portion of the head, including the auditory periphery, was modified. Modern cetaceans have the most derived cranial structure of any mammal (Barnes and Mitchell 1978, Barnes et al. 1985). "Telescoping," a term coined by Miller (1923), refers to the evolutionary revamping of the cranial vault as the maxillary bones of the upper jaw were transposed back to the vertex of the skull, overlapping the compressed frontal bones. As the rostrum elongated, the cranial vault foreshortened, and the nares and narial passages were pulled rearward to a dorsal position behind the eyes. Telescoping may have been related primarily to changes that allow respiration with only a small portion of the head exposed, but it also produced a multilayer skull that has a profound effect on how sound enters and leaves the cetacean head. Many land mammal auditory components, like external pinnae and air-filled external canals, were lost or reduced and the middle and inner ears migrated outward. In most odontocetes, the ears have no substantial bony association with the skull. Instead, they are suspended by ligaments in a foam-filled cavity outside the skull. Consequently, they are effectively acoustically isolated from bone conduction, which is important for echolocation. There are also few bony, thin-walled air chambers, which is important for avoiding pressure-related injuries. Specialized fatty tissues (low impedance channels for underwater sound reception) evolved that appear to function in lieu of external air-filled canals.

Mysticete ears are also specialized but they appear to have been shaped more by size adaptations than by special hearing functions. Sirenian ears are not as well understood, but they too appear to have many highly derived adaptations for underwater sound reception. Today, cetacean and sirenian ears are so specialized for waterborne sound perception that they may no longer be able to detect or interpret airborne sound at normal ambient levels. On the other hand, ears of sea otters (*Enhyra lutris*) and some otariids have very few anatomical differences from those of terrestrial mammals, and it is possible these ears represent a kind of amphibious compromise or even that they continue to be primarily air adapted.

That brings us to three major auditory questions: (1) How do marine and terrestrial ears and hearing differ?; (2) How do these differences relate to underwater sound perception?; and (3) How do amphibious species manage hearing in both domains? To address these questions requires collating a wide variety of data. Behavioral and electrophysiological measures are available for some odontocetes and pinnipeds, but there are no published hearing curves for any mysticete, sirenian, or marine fissiped. Anatomical correlates of hearing are fairly well established (Greenwood 1961, 1962, 1990; Manley 1972; for reviews, see Fay 1988, 1992; Echteler et al. 1994). Anatomical data are available on some aspects of the auditory system for approximately one third of all marine mammal species, including nearly half of the larger, non-captive species. Therefore, to give the broadest view of current marine mammal hearing data, both audiometric and anatomical data are discussed. An outline of physical measures of sound in air versus water and of the basic mechanisms of mammalian hearing are given first as background for these discussions.

Sound in Air Versus Water

In analyzing marine mammal hearing, it is important to consider how the physical aspects of sound in air versus water affect acoustic cues. Basic measures of sound are frequency, speed, wavelength, and intensity. Frequency (f), measured in cycles/sec or hertz (Hz), is defined as:

$$f = c/\lambda, \qquad \text{(equation 1)}$$

where c = the speed of sound (m/sec) and λ = the wavelength (m/cycle). The speed of sound is directly related to the density of the medium. Because water is denser than air, sound in water travels faster and with less attenuation than sound in air. Sound speed in moist ambient surface air is approximately 340 m/sec.[2] Sound speed in seawater averages 1530 m/sec, but varies with any factor affecting density. The principal physical factors affecting density in seawater are salinity, temperature, and pressure. For each 1% increase in salinity, speed increases 1.5 m/sec; for each 1°C decrease in temperature, speed decreases 4 m/sec; and for each 100 m depth, speed increases 1.8 m/sec (Ingmanson and Wallace 1973). Because these factors act synergistically, the ocean has a highly variable sound profile that may change both seasonally and regionally (Fig. 4-3). For practical purposes, in-water sound speed is 4.5 times faster than in air and, at every frequency, the wavelength is 4.5 times greater than in air.

How do these physical differences affect hearing? Mammalian ears are primarily sound-intensity detectors. Intensity, like frequency, depends on sound speed and, in turn, on density. Sound intensity (I) is the acoustic power (P) impinging on surface area (a) perpendicular to the direction of sound propagation, or power/unit area ($I = P/a$). In general terms, power is force (F) times velocity ($P = Fv$). Pressure is force/unit area ($p = F/a$). Therefore, intensity can be rewritten as the product of sound pressure (p) and vibration velocity (v):

$$I = P/a = Fv/a = pv \qquad \text{(equation 2)}$$

For a traveling spherical wave, the velocity component becomes particle velocity (u), which in terms of effective sound pressure can be defined as $p/\rho c$ where ρ is the density of the medium. The product ρc is called the charateristic impedance of the medium.

We can then redefine intensity (equation 2) for an instantaneous sound pressure for an outward traveling plane wave in terms of pressure, sound speed, and density:

$$I = pv = p(p/\rho c) = p^2/\rho c \qquad \text{(equation 3)}$$

Recall that for air c = 340 m/sec and for seawater c = 1530 m/sec; for air, ρ = 0.0013 g/cm^3; for seawater, ρ = 1.03 g/cm^3. The following calculations using the intensity–pressure–impedance relation expressed in equation 3 show how the differences in the physical properties of water versus air influence intensity and acoustic pressure values:

$$\begin{aligned} I_{air} &= p^2/(0.0013\ \text{g/cm}^3)(340\ \text{m/sec}) \\ &= p^2/(0.442\ \text{g-m/sec-cm}^3) \\ I_{water} &= p^2/(1.03\ \text{g/cm}^3)(1530\ \text{m/sec}) \\ &= p^2/(1575\ \text{g-m/sec-cm}^3) \end{aligned}$$

To examine the sensory implications of these differences, we will construct a hypothetical mammal, the neffin ("never found in nature"), that hears equally well in water as in air. For this to be true, the neffin, with an intensity-based ear, would require the same acoustic power/unit area in water as in air to have an equal sound percept, or ($I_{air} = I_{water}$):

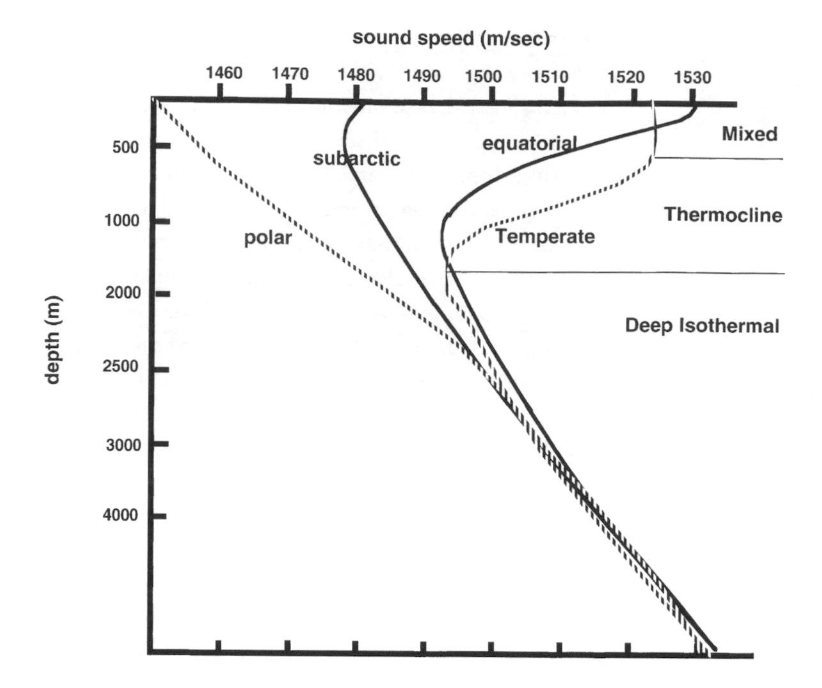

Figure 4-3. Sound speed profiles for temperate, tropical, and polar conditions. In polar waters, there is an inverse trend close to the surface where the water is near freezing; density and therefore, velocity decrease. (Adapted from Ingmanson and Wallace 1973 and Jensen et al. 1994.)

$$I_{air} = p^2_{air}/(0.442 \text{ g-m/sec-cm}^3) = I_{water} = p^2_{water}/(1575 \text{ g-m/sec-cm}^3)$$
$$p^2_{air}(3565.4) = p^2_{water}$$
$$p_{air}(59.7) = p_{water} \quad \text{(equation 4)}$$

This implies that the sound pressure in water must be approximately 60 times that required in air to produce the same intensity and therefore, the same sensation in the neffin ear.

For technological reasons, received intensity, which is measured in watts/m², is difficult to determine. Consequently, to describe hearing thresholds, we capitalize on the fact that intensity is related to the mean square pressure of the sound wave over time (equation 3) and use an indirect measure, effective sound pressure level (*SPL*) (for discussion, see Au 1993). Sound pressure levels are conventionally expressed in decibels (dB), defined as:

$$dB\ SPL = 10 \log (p^2_m/p^2_r) = 20 \log (p_m/p_r), \quad \text{(equation 5)}$$

where p_m is the pressure measured and p_r is an arbitrary reference pressure. Currently, two standardized reference pressures are used. For airborne sound measures, the reference is dB SPL or dB re 20 μPa rms, derived from human hearing.[3] For underwater sound measures, the reference pressure is dB re 1 μPa. Notice that decibels are expressed on a logarithmic scale based on a ratio that depends on reference pressure.

In the earlier hypothetical example, with identical reference pressures, the neffin needed a sound level approximately 35.5 dB greater in water than in air (from equation 4, 10 log 3565.4) to hear equally well in both. However, if conventional references for measuring levels in air versus water are used, the differences in reference pressure must be considered as well. This means that to produce an equivalent sensation in a submerged neffin, the underwater sound pressure level in water would need to be 35.5 dB + 20 (log 20) dB greater than the airborne value. That is, a sound level of 61.5 dB re 1 μPa in water is equivalent to 0 dB re 20 μPa in air. To the neffin, they should sound the same because the intensities are equivalent. Thus, underwater sound intensities with conventional 1 μPa reference pressures must be reduced by 61.5 dB for gross comparisons with in-air sound measures using a 20 μPa reference pressure.

It is important to remember that these equations describe idealized comparisons of air versus waterborne sound. In comparing data from different species, bear in mind that experimental conditions can significantly impact hearing data. Both subtle and gross environmental effects (salinity, temperature, depth, ambient noise, surface reflection, etc.) as well as individual state (motivation, age, pathology) influence results. Comparisons of terrestrial and marine mammal hearing data are particularly difficult because we have no underwater equivalent of anechoic chambers; results are often obtained from few individuals, and test conditions are highly variable.

Basic Hearing Mechanisms

Hearing capacities are the output of the integrated components of the whole ear. All mammalian ears, including those of marine mammals, have three basic divisions: (1) an outer ear, (2) an air-filled middle ear with bony levers and membranes, and (3) a fluid-filled inner ear with mechanical resonators and sensory cells. The outer ear acts as a sound collector. The middle ear transforms acoustic components into mechanical ones detectable by the inner ear. The inner ear acts as a band-pass filter and mechanochemical transducer of sound into neural impulses.

The outer ear is subdivided conventionally into a pinna or ear flap that assists in localization and the ear canal. The size and shape of each component in each species is extraordinarily diverse, which makes any generalized statement about the function of the outer ear debatable. In most mammals, the pinnal flaps are distinct flanges that may be mobile. These flanges act as sound diffractors that aid in localization, primarily by acting as a funnel that selectively admits sounds along the pinnal axis (Heffner and Heffner 1992).

The middle ear is commonly described as an impedance-matching device or transformer that counteracts the approximately 36-dB loss from the impedance differences between air and the fluid-filled inner ear, an auditory remnant of the original vertebrate move from water onto land. This gain is achieved by the mechanical advantage provided by differences in the middle ear membrane areas (large tympanic vs. small oval window) and by the lever effect of the ossicular chain that creates a pressure gain and a reduction in particle velocity at the inner ear.

Improving the efficiency of power transfer to the inner ear may not, however, be the only function for the middle ear. Recent studies on land mammals have led to a competing (but not mutually exclusive) theory called the peripheral filter-isopower function, in which the middle ear has a "tuning" role (for comprehensive discussions, see Zwislocki 1981, Rosowski 1994, Yost 1994). The middle ear varies widely among species in volume, stiffness (K), and mass (M). Each species has a characteristic middle ear resonance based on the combined chain of impedances, which, in turn, depends on the mechanical properties of its middle ear components. For any animal, the sum of impedances is lowest (i.e., middle ear admittance is greatest and energy transmission most efficient), at the middle ear's resonant frequency (f). As expected, this frequency also tends to be at or near the frequency with the lowest threshold (best sensitivity) for that species (Fay 1992).

Stiffness and mass have inverse effects on frequency in a resonant system:

$$f = \left(\frac{1}{2}\pi\right)\sqrt{K/M}\,. \qquad \text{(equation 6)}$$

Put another way, mass-dominated systems have a lower resonant frequency than stiffness-dominated systems. Increasing stiffness in any ear component (membranes, ossicles, cavity) improves the efficiency of transmission of high frequencies. Adding mass to the system (e.g., by increasing cavity volume or increasing ossicular chain mass) favors low frequencies. Consequently, in addition to impedance matching, middle ears may be evolutionarily tuned by different combinations of mass or stiffening agents in each species. Ultrasonic species, like microchiropteran bats and dolphins, have ossicular chains stiffened with bony struts and fused articulations (Reysenbach de Haan 1956, Pye 1972, Sales and Pye 1974, Ketten and Wartzok 1990). Low frequency species, like heteromyid desert rodents, mole rats, elephants, and mysticetes, have large middle ears with flaccid tympanic membranes (Webster 1962; Hinchcliffe and Pye 1969; Webster and Webster 1975; Fleischer 1978; Ketten 1992, 1994).

Inner ears are similarly tuned, in that inner ear stiffness and mass characteristics are major determinants of species-specific hearing ranges. The inner ear consists of the cochlea (primary hearing receptor; Fig. 4-4) and the vestibular system (organs of orientation and balance). Mammalian inner ears are precocial, that is, they are structurally mature and functional at birth and may be active in utero. The cochlea is a fluid-filled spiral containing a primary resonator, the basilar membrane, and an array of neuroreceptors, the organ of Corti (see Fig. 4-7). When the basilar membrane moves, cilia on the hair cells of the organ of Corti are deflected eliciting chemical changes that release neurotransmitters. Afferent fibers of the auditory nerve (cranial nerve VIII) synapsing on the hair cells carry acoustic details to the brain, including frequency, amplitude, and temporal patterning of incoming sounds. Efferent fibers also synapse with the hair cells, but their function is not yet fully understood.

A key component in this system is the basilar membrane. Interspecific differences in hearing ranges are dictated largely by differences in stiffness and mass that are the result of basilar membrane thickness and width variations along the cochlear spiral. Because the cochlea is a spiral with a decreasing radius, the spiral portion with the largest radius (closest to the oval and round windows) is referred to as the base or basal turn; the section with the smallest radius (farthest from the middle ear) is the apex or apical turn. From base to apex, changes in the construction of the basilar membrane in each mammal mechanically tune the ear to a specific set of frequencies. Each membrane region has a particular resonance characteristic and consequently greater deflection than other regions of the membrane for a particular input frequency. For an animal to be sensitive to a sound, its basilar membrane must have resonance capabilities matching that sound at some point along the cochlear spiral.

For any input signal within the hearing range of the animal, the entire basilar membrane responds to some degree. At any one moment, each region of the membrane has a dif-

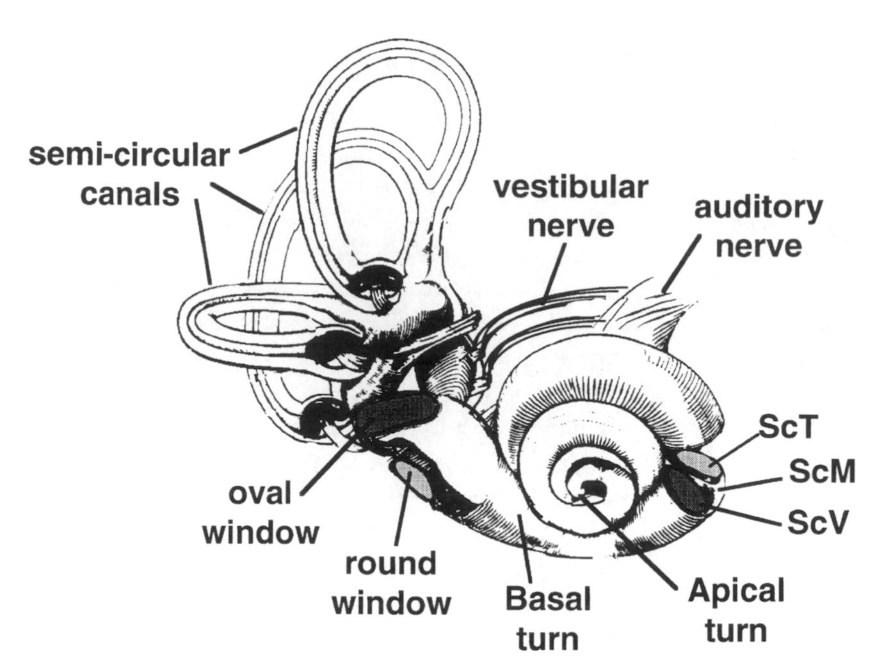

Figure 4-4. The drawing illustrates the fundamental structure of a mammalian inner ear with a 2.5 turn cochlea and 3 semicircular canals. A wedge has been removed from the basal turn to show the three chambers or scalae in the cochlea. (See Fig. 4-7 for additional intracochlear detail.) ScM, scala media; ScT, scala tympani; ScV, scala vestibuli. (Figure was redrawn to scale based on illustrations of human and guinea pig ears in Lewis et al. 1985.)

ferent amount of deflection and a different phase related to the input signal. Over time, changes in amplitude and phase at each point give the impression of a traveling response wave along the cochlea, but because the membrane segments that have resonance characteristics closest to frequencies in the signal have greater displacements than other segments of the membrane, a characteristic profile or envelope develops for the signal.

Basilar membrane dimensions vary inversely, and generally regularly, with cochlear dimensions. The highest frequency each animal hears is encoded at the base of the cochlear spiral (near the oval window), where the membrane is narrow, thick, and stiff. Moving toward the apex of the spiral, as the membrane becomes broader and more pliant, progressively lower frequencies are encoded. Therefore, mammalian basilar membranes are essentially tonotopically arranged resonator arrays, ranging high to low from base to apex, rather like a guitar with densely packed strings graded to cover multiple octaves.[4] The ear, however, is a reverse instrument in that sound energy is the primary input rather than a product, and it is the differential tuning based on the construction of the inner ear membrane "string" array that forms the basis of hearing range differences among species.

Recall from the earlier discussion of animal size in relation to hearing that, in general, small mammals have good high frequency hearing characteristics and large mammals have comparatively low hearing ranges. Early inner ear models were based on the assumption that all mammalian basilar membranes were constructed of similar components that had a constant gradient with length and that length scaled with animal size. On average, smaller animals were assumed to have shorter, narrower, stiffer membranes, whereas larger animals had longer, broader, less stiff membranes (von Békésy 1960; Greenwood 1961, 1990). Given that assumption, frequency distributions in the inner ear of any species could be derived by comparing one parameter, basilar membrane length, with an arbitrary standard, the average human membrane length. For many land mammals, this assumption is correct, but only because length is an indirect correlate of other key features for basilar membrane resonance. For these ears, now termed generalists (Fay 1992, Echteler et al. 1994), basilar membrane thickness and width covary regularly with length; therefore, length can proportionately represent stiffness.

Only recently has it become clear that some species, termed specialists (Echteler et al. 1994), do not have the same thickness–width–length relationship as do generalist land mammals (Manley 1972; Ketten 1984, 1997). Most specialist animals have retuned their inner ears to fit an atypical tuning for their body size by either increasing mass to improve low frequency sensitivity in small ears (as in mole rats) or adding stiffening components to increase resonant frequencies in larger inner ears (as in dolphins) (Hinchcliffe and Pye 1969, Sales and Pye 1974, Webster and Webster 1975, Ketten 1984). The most extreme case of specialization is to be found in some bats, which have relatively constant basilar membrane dimensions for about 30% of the cochlea and thereby devote a disproportionate amount of the membrane to encoding very narrow bands of frequencies related to a component of their echolocation signal (Bruns and Schmieszek 1980, Vater 1988a, Kössl and Vater 1995).

Marine mammal ears fall into both categories and some species have a mix of generalist and specialist traits. Like land mammals, pinnipeds and cetaceans have basilar membranes that scale with animal size. Consequently, because marine mammals are relatively large, most have basilar membranes longer than the human average. If marine mammal ears followed the generalist land mammal pattern, most would have relatively poor ultrasonic hearing. For example, standard land mammal length-derived hearing models (Greenwood 1961, 1990; Fay 1992) predict an upper limit of hearing of approximately 16 kHz for bottlenose dolphins (*Tursiops truncatus*), which actually have a functional high frequency hearing limit of 160 kHz (Au 1993). Before the discovery of dolphin echolocation, it was assumed that these large animals had predominately low functional hearing ranges similar to cows. Hearing is not constrained to low frequencies in marine mammals, because they have radically different inner ear thickness–width gradients than generalist land mammals. In odontocetes, very high ultrasonic hearing is related also to the presence of extensive stiffening additions to the inner ear. These features, discussed in detail later, demonstrate the usefulness of comparative audiometric and anatomical studies for teasing apart sensory mechanisms. In fact, one important outgrowth of marine mammal hearing studies has been the development of multifeature hearing models that are better predictors of hearing characteristics for all mammals than traditional, single-dimension models (Ketten 1994).

Marine Mammal Sound Production

Recordings of naturally produced sounds are available for most marine mammal species (Watkins and Wartzok 1985), and they provide the broadest acoustic framework for hearing comparisons. Sound production data obtained in a wide variety of background noise conditions cannot be used to infer hearing thresholds because it is likely that produced sound levels are elevated over minimum audible levels to override background noise. For example, some recordings of odontocete and mysticete sounds have source levels estimated to be as high as 180 to 230 dB re 1 µPa (Richardson et al. 1991, Au 1993, Würsig and Clark 1993). However, because mammalian vocalizations typically have peak spectra at or near the best frequency for that species, they are generally good indirect indicators of frequencies the animal normally hears well (Sales and Pye 1974, Popper 1980, Watkins and Wartzok 1985, Henson et al. 1990, Ketten and Wartzok 1990, Popov and Supin 1990a). A classic example is the discovery of ultrasonic signal use by dolphins (Kellogg 1959, Norris et al. 1961), which prompted several decades of investigations into echolocation and ultrasonic hearing abilities in marine mammals.

Cetaceans

Cetaceans can be divided into high and low frequency sound producers, which are coincident with the two suborders (Table 4-1). Sound production data for odontocetes are consistent with the audiometric data (i.e., ultrasonic use is common and differences in peak spectra of produced sounds are consistent with best frequency of hearing in species that have been tested) (compare Table 4-1 and Fig. 4-5A). Mysticete sound production data imply they are primarily low frequency animals, and it is likely that many baleen species hear well at infrasonic frequencies.

Odontocetes produce species-stereotypic broadband clicks with peak energy between 10 and 200 kHz, individually variable burst pulse click trains, and constant frequency (CF) or frequency modulated (FM) whistles ranging from 4 to 16 kHz (see Tyack, Chapter 7, this volume). Ultrasonic signals are highly species specific and have been recorded from 21 species, although echolocation (or "biosonar") has been demonstrated in only 11 species of smaller odontocetes (Au 1993). All modern odontocetes are assumed, like bats, to be true echolocators, not simply ultrasonic receptors. That is, they "image" their environment by analyzing echoes from a self-generated ultrasonic signal (Kellogg 1959, Norris et al. 1961, Popper 1980, Wood and Evans 1980, Pilleri 1983, Watkins and Wartzok 1985). Echolocation is a two-way function; to be an effective echolocator, an animal must have a coordinated means of generating a highly directional signal and receiving its echo. For this reason, evidence for high frequency ears alone is not sufficient to determine whether any marine mammal (or fossil species) is an echolocator.

Captive odontocetes routinely vary pulse repetition rate, interpulse interval, intensity, and click spectra, particularly in response to high ambient noise (Schevill 1964, Norris 1969, Au et al. 1974, Popper 1980, Thomas et al. 1988, Moore 1990, Popov and Supin 1990a). Normally, however, each species has a characteristic echolocation frequency spectrum (Schevill 1964, Norris 1969, Popper 1980). Well-documented peak spectra of odontocete sonar signals range from approximately 20 kHz up to 160 kHz with source levels as high as 228 dB, but more commonly in range of 120 to 180 dB (Table 4-1).

There are strong correlations between habitat types, societal differences, and peak spectra (Gaskin 1976, Wood and Evans 1980, Ketten 1984). Considering that frequency and wavelength are inversely related, there is an inverse relationship between frequency and the size of the object or detail that can be detected with echolocation. On the basis of their

Table 4-1. Marine Mammal Sound Production Characteristics

Scientific Name	Common Name	Signal Type	Frequency Range (kHz)	Frequency Near Maximum Energy (kHz)	Source Level (dB re 1 μPa)	References (Partial references only for some species)
Cetacea						
Odontoceti						
Delphinidae						
Cephalorhynchus commersonii	Commerson's dolphin	Pulsed sounds	< 10	0.2–5	—	Watkins and Schevill 1980, Dziedzic and de Buffrenil 1989
		Clicks	—	6	—	Dziedzic and de Buffrenil 1989
		Click	—	116–134	160	Kamminga and Wiersma 1981, Shochi et al. 1982, Evans et al. 1988, Au 1993
C. heavisidii	Heaviside's dolphin	Pulsed sounds	0.8–5[a]	0.8–4.5[a]	—	Watkins et al. 1977
		Click	—	2–5	—	Watkins et al. 1977
C. hectori	Hector's dolphin	Click	—	112–135	150–163	Dawson 1988, Dawson and Thorpe 1990, Au 1993
Delphinus delphis	Common dolphin	Whistles, chirps, Barks	—	0.5–18	—	Caldwell and Caldwell 1968, Moore and Ridgway 1995
		Whistles	4–16	—	—	Busnel and Dziedzic 1966a
		Click	0.2–150	30–60	—	Busnel and Dziedzic 1966a
		Click	—	23–67	—	Dziedzic 1978
Feresa attenuata	Pygmy killer whale	Growls, Blats	—	—	—	Pryor et al. 1965
Globicephala macrorhynchus	Short-finned pilot whale	Whistles	0.5–> 20	2–14	180	Caldwell and Caldwell 1969, Fish and Turl 1976
		Click	—	30–60	180	Evans 1973
G. melaena	Long-finned pilot whale	Whistles	1–8	1.6–6.7[b]	—	Busnel and Dziedzic 1966a
		Clicks	1–18	—	—	Taruski 1979, Steiner 1981
		Click	—	6–11	—	McLeod 1986
Grampus griseus	Risso's dolphin	Whistles	—	3.5–4.5	—	Caldwell et al. 1969
		Rasp/pulse burst	0.1–> 8[c]	2–5	—	Watkins 1967
		Click	—	65	~120	Au 1993
Lagenodelphis hosei	Fraser's dolphin	Whistles	7.6–13.4	—	—	Leatherwood et al. 1993
Lagenorhynchus acutus	Atlantic white-sided dolphin	Whistles	—	6–15[b]	—	Steiner 1981
L. albirostris	White-beaked dolphin	Squeals	—	8–12	—	Watkins and Schevill 1972
L. australis	Peale's dolphin	Pulses (buzz)	0.3–5	0.3	—	Schevill and Watkins 1971
		Clicks	to 12	to 5	Low	Schevill and Watkins 1971
L. obliquidens	Pacific white-sided dolphin	Whistles	2–20	4–12	—	Caldwell and Caldwell 1971
		Click	—	60–80	180	Evans 1973
L. obscurus	Dusky dolphin	Whistles	1.0–27.3	6.4–19.2[b]	—	Wang Ding et al. 1995
Lissodelphis borealis	Northern right whale dolphin	Whistles, tones	1–16	1.8, 3	—	Leatherwood and Walker 1979
Orcinus orca	Killer whale	Whistles	1.5–18	6–12	—	Steiner et al. 1979, Ford and Fisher 1983, Morton et al. 1986
		Click	0.25–0.5	—	—	Schevill and Watkins 1966
		Scream	2	—	—	Schevill and Watkins 1966
		Click	0.1–35	12–25	180	Diercks et al. 1971, Diercks 1972
		Pulsed calls	0.5–25	1–6	160	Schevill and Watkins 1966, Awbrey et al. 1982, Ford and Fisher 1983, Moore et al. 1988
Pseudorca crassidens	False killer whale	Whistles	—	4–9.5	—	Busnel and Dziedzic 1968, Kamminga and van Velden 1987
		Click	—	25–30, 95–130	220–228	Kamminga and van Velden 1987, Thomas and Turl 1990

Table 4-1 continued

Scientific Name	Common Name	Signal Type	Frequency Range (kHz)	Frequency Near Maximum Energy (kHz)	Source Level (dB re 1 μPa)	References (Partial references only for some species)
Sotalia fluviatilis	Tucuxi	Whistles	3.6–23.9	7.1–18.5[b]	—	Wang Ding et al. 1995
		Click	—	80–100	High	Caldwell and Caldwell 1970, Norris et al. 1972, Kamminga et al. 1993
Sousa chinensis	Humpback dolphin	Whistles	1.2–>16	—	—	Schultz and Corkeron 1994
Stenella attenuata	Spotted dolphin	Whistles	3.1–21.4	6.7–17.8[b]	—	Wang Ding et al. 1995
		Whistles	—	—	—	Evans 1967
		Pulse	to 150	—	—	Diercks 1972
S. clymene	Clymene dolphin	Whistles	6.3–19.2	—	—	Mullin et al. 1994
S. coeruleoalba	Spinner dolphin	Whistles	1–22.5	6.8–16.9[b]	109–125	Watkins and Schevill 1974, Steiner 1981, Norris et al. 1994, Wang Ding et al. 1995
		Pulse bursts	Wide band	5–60	108–115	Watkins and Schevill 1974, Norris et al. 1994
		Screams	—	—	—	Norris et al. 1994
S. longirostris	Long-snouted spinner dolphin	Pulse	1–160	5–60	—	Brownlee 1983
		Whistle	1–20	8–12	—	Brownlee 1983
		Click	—	low–65	—	Watkins and Schevill 1974, Norris et al. 1994
		Click	1–160	60	—	Ketten 1984
S. plagiodon	Spotted dolphin	Whistles	5.0–19.8	6.7–17.9[b]	—	Caldwell et al. 1973, Steiner 1981
		Clicks	1–8	—	—	Caldwell and Caldwell 1971b
		Squawks, barks, growls, chirps	0.1–8	—	—	Caldwell et al. 1973
S. styx	Gray's porpoise	Whistles	6–> 24	8–12.5	—	Busnel et al. 1968
Steno bredanensis	Rough-toothed dolphin	Whistles	—	4–7	—	Busnel and Dziedzic 1966b
		Click	—	5–32	—	Norris and Evans 1967
Tursiops truncatus	Bottlenosed dolphin	Whistles	0.8–24	3.5–14.5[b]	125–173	Lilly and Miller 1961, Tyack 1985, Caldwell et al. 1990, Schultz and Corkeron 1994, Wang Ding et al. 1995
		Rasp, grate, mew, bark, yelp	—	—	—	Wood 1953
		Click	0.2–150	30–60	—	Diercks et al. 1971, Evans 1973
		Bark	0.2–16	—	—	Evans and Prescott 1962
		Whistle	4–20	—	—	Caldwell and Caldwell 1967, Evans and Prescott 1962
		Click[d]	—	110–130	218–228	Au et al. 1974, Au 1993
Monodontidae						
Delphinapterus leucas	Beluga	Whistles	0.26–20	2–5.9	—	Schevill and Lawrence 1949, Sjare and Smith 1986a,b
		Pulsed tones	0.4–12	1–8	—	Schevill and Lawrence 1949, Sjare and Smith 1986a,b
		Noisy vocalizations	0.5–16	4.2–8.3	—	Schevill and Lawrence 1949, Sjare and Smith 1986a,b
		Echolocation click	—	40–60, 100–120	206–225	Au et al. 1985, 1987, Au 1993
Monodon monoceros	Narwhal	Pulsed tones	0.5–5	—	—	Ford and Fisher 1978
		Whistles	0.3–18	0.3–10	—	Ford and Fisher 1978
		Click	—	40	218	Møhl et al. 1990
Phocoenidae						
Neophocaena phocaenoides	Finless porpoise	Clicks	1.6–2.2	2	—	Pilleri et al. 1980
		Click	—	128	—	Kamminga et al. 1986, Kamminga 1988

Continued on next page

Table 4-1 continued

Scientific Name	Common Name	Signal Type	Frequency Range (kHz)	Frequency Near Maximum Energy (kHz)	Source Level (dB re 1 μPa)	References (Partial references only for some species)
Phocoenoides dalli	Dall's porpoise	Clicks	0.04–12	—	120–148	Evans 1973, Evans and Awbrey 1984
		Click	—	135–149	165–175	Evans and Awbrey 1984, Hatakeyama and Soeda 1990, Hatakeyma et al. 1994
Phocoena phocoena	Harbor porpoise	Clicks	2	—	100	Busnel and Dziedzic 1966a, Schevill et al. 1969
		Pulse	100–160	110–150	—	Møhl and Anderson 1973
		Click	—	110–150	135–177	Busnel et al. 1965, Møhl and Anderson 1973, Kamminga and Wiersma 1981, Akamatsu et al. 1994
P. sinus	Vaquita	Click	—	128–139	—	Silber 1991
Physeteridae						
Kogia breviceps	Pygmy sperm whale	Clicks	60–200	120	—	Santoro et al. 1989, Caldwell and Caldwell 1987
Physeter catodon	Sperm whale	Clicks	0.1–30	2–4, 10–16	160–180	Backus and Schevill 1966, Levenson 1974, Watkins 1980a,b
		Clicks in coda	16–30	—	—	Watkins 1980a,b
Platanistoidea						
Iniidae						
Inia geoffrensis	Boutu	Squeals	< 1– 12	1–2	—	Caldwell and Caldwell 1970
		Whistle	0.2–5.2	1.8–3.8[b]	—	Wang Ding et al. 1995
		Click	25–200	100	—	Norris et al. 1972
				95–105	—	Kamminga et al. 1989
		Click	—	85–105	—	Diercks et al. 1971, Evans 1973, Kamminga et al. 1993
		Click	20–120	—	156	Xiao and Jing 1989
Platanistidae						
Platanista minor	Indus susu	Clicks	0.8–16	—	Low	Andersen and Pilleri 1970, Pilleri et al. 1971
		Click	—	15–100	—	Herald et al. 1969
Pontoporiidae						
Lipotes vexillifer	Baiji	Whistles	3–18.4	6	156	Jing et al. 1981, Xiao and Jing 1989
Pontoporia blainvillei	Franciscana	Click	0.3–24	—	—	Busnel et al. 1974
Ziphiidae						
Hyperoodon ampullatus	Northern bottle-nose whale	Whistles	3–16	—	—	Winn et al. 1970
		Clicks	0.5–26	—	—	Winn et al. 1970
Hyperoodon spp.	Bottlenose whale	Click	—	8–12	—	Winn et al. 1970
Mesoplodon carlhubbsi	Hubb's beaked whale	Pulses	0.3–80	0.3–2	—	Buerki et al. 1989, Lynn and Reiss 1992
M. densirostris	Blainville's beaked whale	Whistles, chirps	<1–6	—	—	Caldwell and Caldwell 1971a
		Whistles	2.6–10.7	—	—	Buerki et al. 1989, Lynn and Reiss 1992
Mysticeti						
Balaenidae						
Balaena mysticetus	Bowhead whale	Calls	0.100–0.580	0.14–0.16	128–190	Thompson et al. 1979, Ljungblad et al. 1980, Norris and Leatherwood 1981, Würsig and Clark 1993
		Tonal moans	0.025–0.900	0.10–0.40	128–178	Ljungblad et al. 1982, Cummings and Holliday 1987, Clark et al. 1986

Table 4-1 continued

Scientific Name	Common Name	Signal Type	Frequency Range (kHz)	Frequency Near Maximum Energy (kHz)	Source Level (dB re 1 μPa)	References (Partial references only for some species)
Balaena mysticetus	Bowhead whale	Pulsive	0.025–3.500	—	152–185	Clark and Johnson 1984, Würsig et al. 1985, Cummings and Holliday 1987
		Song	0.02–0.50	< 4	158–189	Ljungblad et al. 1982, Cummings and Holliday 1987, Würsig and Clark 1993
Eubalaena australis	Southern right whale	Tonal	0.03–1.25	0.16–0.50	—	Cummings et al. 1972, Clark 1982, 1983
		Pulsive	0.03–2.20	0.05–0.50	172–187	Cummings et al. 1972, Clark 1982, 1983
					181–186	Clark (in Würsig et al. 1982)
E. glacialis	Northern right whale	Call	< 0.400	< 0.200	—	Watkins and Schevill 1972, Clark 1990
		Moans	< 0.400	—	—	Watkins and Schevill 1972, Thompson et al. 1979, Spero 1981
Neobalaenidae						
Caperea marginata	Pygmy right whale	Thumps in pairs	<0.300	0.060–0.135	165–179	Dawbin and Cato 1992
Balaenopteridae						
Balaenoptera acutorostrata	Minke whale	Sweeps, moans	0.06–0.14	—	151–175	Winn and Perkins 1976, Schevill and Watkins 1972
		Down sweeps	0.06–0.13	—	165	Schevill and Watkins 1972
		Moans, grunts	0.06–0.14	0.06–0.14	151–175	Schevill and Watkins 1972, Winn and Perkins 1976
		Ratchet	0.85–6	0.85	—	Winn and Perkins 1976
		Thump trains	0.10–2	0.10–0.20	—	Winn and Perkins 1976
B. borealis	Sei whale	Fm sweeps	1.5–3.5	—	—	Thompson et al. 1979, Knowlton et al. 1991
B. edeni	Bryde's whale	Moans	0.070–0.245	0.124–0.132	152–174	Cummings et al. 1986
		Pulsed moans	0.10–0.93	0.165–0.900	—	Edds et al. 1993
		Discrete pulses	0.70–0.95	0.700–0.900	—	Edds et al. 1993
B. musculus	Blue whale	Moans	0.012–0.400	0.012–0.025	188	Cummings and Thompson 1971, 1994, Edds 1982, Stafford et al. 1994
B. physalus	Fin whale	Moans	0.016–0.750	0.020	160–190	Thompson et al. 1979, Edds 1988
		Pulse	0.040–0.075	—	—	Clark 1990
		Pulse	0.018–0.025	0.020	—	Watkins 1981
		Ragged pulse	< 0.030	—	—	Watkins 1981
		Rumble	—	< 0.030	—	Watkins 1981
		Moans, downsweeps	0.014–0.118	0.020	160–186	Watkins 1981, Watkins et al. 1987, Edds 1988, Cummings and Thompson 1994
		Constant call	0.02–0.04	—	—	Edds 1988
		Moans, tones, upsweeps	0.03–0.75	—	155–165	Watkins 1981, Cummings et al. 1986, Edds 1988
		Rumble	0.01–0.03	—	—	Watkins 1981, Edds 1988
		Whistles[e], chirps[e]	1.5–5	1.5–2.5	—	Thompson et al. 1979
		Clicks[e]	16–28	—	—	Thompson et al. 1979
Megaptera novaeangliae	Humpback whale	Songs	0.03–8	0.1–4	144–186	Thompson et al. 1979,Watkins 1981, Edds 1982,1988, Payne et al. 1983, Silber 1986, Clark 1990
		Social	0.05–10	< 3	—	Thompson et al. 1979
		Song components	0.03–8	0.120–4	144–174	Thompson et al. 1979, Payne and Payne 1985
		Shrieks	—	0.750–1.8	179–181	Thompson et al. 1986
		Horn blasts	—	0.410–0.420	181–185	Thompson et al. 1986
		Moans	0.02–1.8	0.035–0.360	175	Thompson et al. 1986
		Grunts	0.025–1.9	—	190	Thompson et al. 1986

Continued on next page

Table 4-1 continued

Scientific Name	Common Name	Signal Type	Frequency Range (kHz)	Frequency Near Maximum Energy (kHz)	Source Level (dB re 1 μPa)	References (Partial references only for some species)
Megaptera novaeangliae	Humpback whale	Pulse trains	0.025–1.25	0.025–0.080	179–181	Thompson et al. 1986
		Slap	0.03–1.2	—	183–192	Thompson et al. 1986
Eschrichtiidae						
Eschrictius robustus	Gray whale	Call	0.2–2.5	1–1.5	—	Dahlheim and Ljungblad 1990
		Moans	0.02–1.20	0.020–0.200, 0.700–1.2	185	Cummings et al. 1968, Fish et al. 1974, Swartz and Cummings 1978
		Modulated pulse	0.08–1.8	0.225–0.600	—	Dahlheim et al. 1984, Moore and Ljungblad 1984
		FM sweep	0.10–0.35	0.300	—	Dahlheim et al. 1984, Moore and Ljungblad 1984
		Pulses	0.10–2	0.300–0.825	—	Dahlheim et al. 1984, Moore and Ljungblad 1984
		Clicks (calves)	0.10–20	3.4–4	—	Fish et al. 1974, Norris et al. 1977
Fissipedia						
Mustelidae						
Enhydra lutris	Sea otter	Growls[c], whine	3–5	—	—	Kenyon 1981, Richardson et al. 1995
Pinnipedia						
Odobenidae						
Odobenus rosmarus	Walrus	Bell tone	—	0.4–1.2	—	Schevill et al. 1966, Ray and Watkins 1975, Stirling et al. 1983
		Clicks, taps, knocks	0.1–10	< 2	—	Schevill et al. 1966, Ray and Watkins 1975, Stirling et al. 1983
		Rasps	0.2–0.6	0.4–0.6	—	Schevill et al. 1966
		Grunts	≤1	≤L	—	Stirling et al. 1983
Otariidae						
Arctocephalus philippii	Juan Fernandez fur seal	Clicks	0.1–0.2	0.1–0.2	—	Norris and Watkins 1971
Callorhinus ursinus	Northern fur seal	Clicks, bleats	—	—	—	Poulter 1968
Eumetopias jubatus	Northern sea lion	Clicks, growls	—	—	—	Poulter 1968
Zalophus californianus	California sea lion	Barks	<8	<3.5	—	Schusterman et al. 1967
		Whinny	<1–3	—	—	Schusterman et al. 1967
		Clicks	—	0.5–4	—	Schusterman et al. 1967
		Buzzing	< 1–4	< 1	—	Schusterman et al. 1967
Phocidae						
Cystophora cristata	Hooded seal	Grunt	—	0.2–0.4	—	Terhune and Ronald 1973
		Snort	—	0.1–1	—	Terhune and Ronald 1973
		Buzz, click	to 6	1.2	—	Terhune and Ronald 1973
Erignathus barbatus	Bearded seal	Song	0.02–6	1–2	178	Ray et al. 1969, Stirling et al. 1983, Cummings et al. 1983
Halichoerus grypus	Gray seal	Clicks, hiss	0–30, 0–40	—	—	Schevill et al. 1963, Oliver 1978
		6 Calls	0.1–5	0.1–3	—	Asselin et al. 1993
		Knocks	to 16	To 10	—	Asselin et al. 1993
Hydrurga leptonyx	Leopard seal	Pulses, trills	0.1–5.9	—	—	Ray 1970, Stirling and Siniff 1979, Rogers et al. 1995
		Thump, blast	0.04–7	—	—	Rogers et al. 1995
		Ultrasonic	up to 164	50–60	Low	Thomas et al. 1983a

Table 4-1 continued

Scientific Name	Common Name	Signal Type	Frequency Range (kHz)	Frequency Near Maximum Energy (kHz)	Source Level (dB re 1 μPa)	References (Partial references only for some species)
Leptonychotes weddellii	Weddell seal	>34 Calls	0.1–12.8	—	153–193	Thomas and Kuechle 1982, Thomas et al. 1983b, Thomas and Stirling 1983
Lobodon carcinophagus	Crabeater seal	Groan	<0.1–8	0.1–1.5	High	Stirling and Siniff 1979
Ommatophoca rossii	Ross seal	Pulses	0.25–1	—	—	Watkins and Ray 1985
		Siren	4–1–4	—	—	Watkins and Ray 1985
Phoca fasciata	Ribbon seal	Frequency sweeps	0.1–7.1	—	160	Watkins and Ray 1977
P. groenlandica	Harp seal	15 sounds	<0.1–16	0.1–3	130–140	Møhl et al. 1975, Watkins and Schevill 1979, Terhune and Ronald 1986, Terhune 1994
		Clicks	—	30	131–164	Møhl et al. 1975
P. hispida	Ringed seal	Barks, clicks, yelps	0.4–16	<5	95–130	Stirling 1973, Cummings et al. 1984
P. largha	Spotted seal	Social sounds	0.5–3.5	—	—	Beier and Wartzok 1979
P. vitulina	Harbor seal	Clicks	8–150	12–40	—	Schevill et al. 1963, Cummings and Fish 1971, Renouf et al. 1980, Noseworthy et al. 1989
		Roar	0.4–4	0.4–0.8	—	Hanggi and Schusterman 1992, 1994
		Growl, grunt, groan	< 0.1–0.4	< 0.1–0.25	—	Hanggi and Schusterman 1992, 1994
		Creak	0.7–4	0.7–2	—	Hanggi and Schusterman 1992, 1994
Sirenia						
Dugongidae						
Dugong dugon	Dugong	Chirp-squeak[c]	3–8	—	Low	Nair and Lal Mohan 1975
		Sound 1[c]	1–2	—		Marsh et al. 1978
		Chirp[c]	2–4	—	—	Marsh et al. 1978
		All sounds	0.5–18	1–8		Nishiwaki and Marsh 1985, Anderson and Barclay 1995
Trichechidae						
Trichechus inunguis	Amazonian manatee	Squeaks, pulses	6–16	6–16	—	Evans and Herald 1970
T. manatus	West Indian manatee	Squeaks	0.6–16	0.6–5	Low	Schevill and Watkins 1965

Data compiled from Popper 1980, Watkins and Wartzok 1985, Ketten 1992, Au 1993, Richardson *et al.* 1995, Ketten 1997.

[a]Equipment capable of recording to 10 kHz only.

[b]Frequency determined as "mean minimum frequency minus 1 sd . . . to . . . mean maximum frequency plus 1 sd." (*sensu* Richardson et al. 1995).

[c]Recorded in air.

[d]Performance in high background noise (Au 1993)

[e]Few recordings or uncertain verification of sound for species.

ultrasonic signals, odontocetes fall into two acoustic groups: Type I, with peak spectra (frequencies at maximum energy) above 100 kHz, and Type II, with peak spectra below 80 kHz (Ketten 1984, Ketten and Wartzok 1990) (Table 4-1). Type I echolocators are inshore and riverine dolphins that operate in acoustically complex waters. Amazonian boutu (*Inia geoffrensis*) routinely hunt small fish amid the roots and stems choking silted "varzea" lakes created by seasonal flooding. These animals produce signals up to 200 kHz (Norris et al. 1972). Harbor porpoises typically use 110 to 140 kHz signals (Kamminga 1988). Communication signals are rare (or are rarely observed) in most type I species (Watkins and Wartzok 1985); their auditory systems are characterized primarily by ultra-high frequency adaptations consistent with short wavelength signals. Type II species are nearshore and offshore animals that inhabit low object density environments, travel in large pods, and, acoustically, are concerned with both communication with conspecifics and detection of relatively large, distant objects. They may use ultra high frequency signals in high background noise, but typically

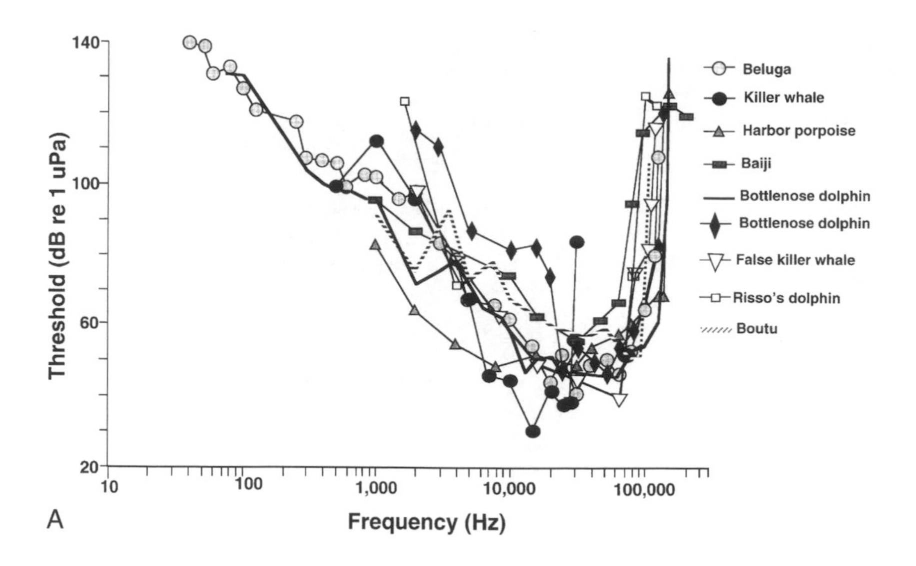

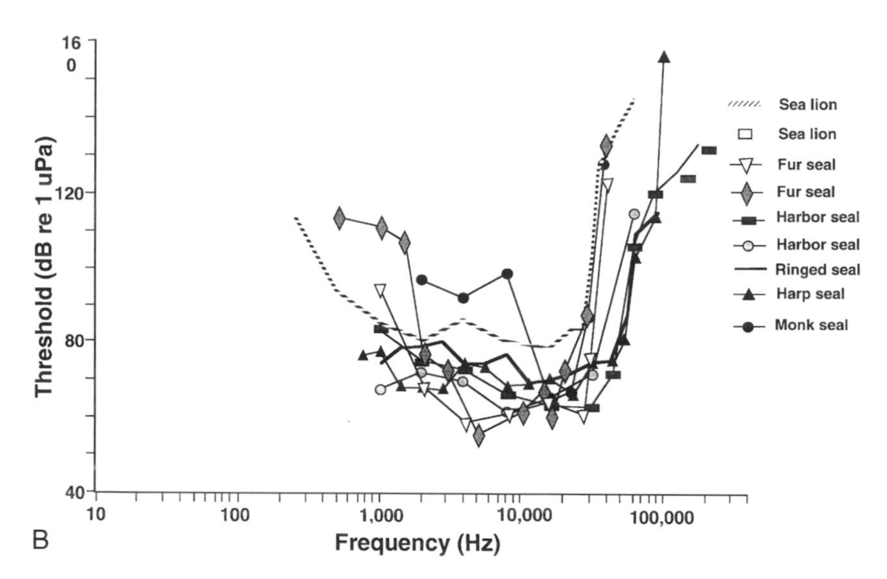

Figure 4-5. Underwater audiograms for (A) odontocetes and (B) pinnipeds. For some species, more than one curve is shown because data reported in different studies were not consistent. Note that for both the bottlenose dolphin and the sea lion, thresholds are distinctly higher for one of the two animals tested. These differences may reflect different test conditions or a hearing deficit in one of the animals. (Data compiled from Popper 1980, Fay 1988, Au 1993, Richardson et al. 1995.)

they use lower ultrasonic frequencies with longer wavelengths that are consistent with detecting larger objects over greater distances and devote more acoustic effort to communication signals than Type I species.

Use of deep ocean stationary arrays has substantially increased our database of mysticete sounds. Recent analyses suggest that mysticetes have multiple, distinct sound production groups, but habitat and functional relationships for the potential groupings are not yet clear (Würsig and Clark 1993; for review, see Edds-Walton 1997). In general, mysticete vocalizations are significantly lower in frequency than those of odontocetes (Table 4-1). Most mysticete signals are characterized as low frequency moans (0.4–40 sec; fundamental frequency well below 200 Hz); simple calls (impulsive, narrow band, peak frequency < 1 kHz); complex calls (broadband pulsatile AM or FM signals); and complex "songs" with seasonal variations in phrasing and spectra (Thompson et al. 1979; Watkins 1981; Edds 1982, 1988; Payne et al. 1983; Watkins and Wartzok 1985; Silber 1986; Clark 1990; Dahlheim and Ljungblad 1990). Infrasonic signals, typically in the 10- to 16-Hz range, are well-documented in at least two species, the blue whale (*Balaenoptera musculus*) (Cummings and Thompson 1971, Edds 1982), and the fin whale (*Balaenoptera physalus*) (Watkins 1981; Edds 1982, 1988; Watkins et al. 1987). Suggestions that these low frequency signals are used for long-distance communication and for topological imaging are intriguing but have not been definitively demonstrated.

Pinnipeds

The majority of pinniped sounds are in the sonic range (20 Hz – 20 kHz), but their signal characteristics are extremely diverse (compare Table 4-1 with Fig. 4-5B). Some species are nearly silent, others have broad ranges and repertoires, and the form and rate of production vary seasonally, by sex, and whether the animal is in water or air (Watkins and Wartzok 1985, Richardson et al. 1995). Calls have been described as grunts, barks, rasps, rattles, growls, creaky doors, and warbles in addition to the more conventional whistles, clicks, and pulses (Beier and Wartzok 1979, Ralls et al. 1985, Watkins and Wartzok 1985, Miller and Job 1992). Although clicks are produced, there is no clear evidence for echolocation in pinnipeds (Renouf et al. 1980, Schusterman 1981, Wartzok et al. 1984).

Phocid calls are commonly between 100 Hz and 15 kHz, with peak spectra less than 5 kHz, but can range as high as 40 kHz. Typical source levels in water are estimated to be near 130 dB re 1 μPa, but levels as high as 193 dB re 1 μPa have been reported (Richardson et al. 1995). Infrasonic to seismic level vibrations are produced by northern elephant seals (*Mirounga angustirostris*) while vocalizing in air (Shipley et al. 1992).

Otariid calls are similarly variable in type, but most are in the 1 to 4 kHz range. The majority of sounds that have been analyzed are associated with social behaviors. Barks in water have slightly higher peak spectra than in air, although both center near 1.5 kHz. In-air harmonics that may be important in communication range up to 6 kHz. Schusterman et al. (1972), in their investigation of female California sea lion (*Zalophus californianus*) signature calls, found important interindividual variations in call structure and showed that the calls have fundamental range characteristics consistent with peak in-air hearing sensitivities.

Odobenid sounds are generally in the low sonic range (fundamentals near 500 Hz; peak < 2 kHz), and are commonly described as bell-like although whistles are also reported (Schevill et al. 1966, Ray and Watkins 1975, Verboom and Kastelein 1995).

Sirenians

Manatee (*Trichechus manatus* and *T. inunguis*) and dugong (*Dugong dugon*) underwater sounds have been described as squeals, whistles, chirps, barks, trills, squeaks, and froglike calls (Sonoda and Takemura 1973, Anderson and Barclay 1995, Richardson et al. 1995) (Table 4-1). West Indian manatee (*Trichechus manatus*) calls typically range from 0.6 to 5 kHz (Schevill and Watkins 1965). Calls of Amazonian manatees (*Trichechus inunguis*), a smaller species than the Florida manatee (a subspecies of *T. manatus*), are slightly higher with peak spectra near 10 kHz, although distress calls have been reported to have harmonics up to 35 kHz (Bullock et al. 1980). Dugong calls range from 0.5 to 18 kHz with peak spectra between 1 and 8 kHz (Nishiwaki and Marsh 1985, Anderson and Barclay 1995).

Fissipeds

Descriptions of otter sounds are similar to those for pinnipeds and for terrestrial carnivores (Table 4-1) (i.e., growls, whines, snarls, and chuckles) (Kenyon 1981). Richardson et al. (1995) state that underwater sound production analyses are not available but that in-air calls are in the 3- to 5-kHz range and are relatively intense.

In Vivo Marine Mammal Hearing Data

As indicated in the introductory sections, hearing capacity is usually expressed as an audiogram, a plot of sensitivity (threshold level in dB SPL in air and dB re 1 μPa in water) versus frequency (Figs. 4-2 and 4-5), which is obtained by behavioral or electrophysiological measures of hearing.

Mammals typically have a U-shaped hearing curve. Sensitivity decreases on either side of a relatively narrow band of frequencies at which hearing is significantly more acute. The rate of decrease in sensitivity is generally steeper above this best frequency or peak sensitivity region than below. Behavioral and neurophysiological hearing curves are generally similar, although behavioral audiograms typically have lower thresholds for peak sensitivities (Dallos et al. 1978). Interindividual and intertrial differences in audiograms may be related to variety of sources, including ear health, anesthesia, masking by other sounds, timing, and anticipation by the subject.

Hearing curves are available for approximately 12 species of marine mammals. All have the same basic U-shaped pattern as land mammal curves (compare Fig. 4-5A,B with Fig. 4-2). As noted earlier, peak sensitivities are generally consistent with the vocalization data in those species for which both data sets are available (compare Table 4-1 with Fig. 4-5A,B). Detailed reviews of data for specific marine mammal groups are available in Bullock and Gurevich (1979), McCormick et al. (1980), Popper (1980), Schusterman (1981), Watkins and Wartzok (1985), Fay (1988), Awbrey (1990), Au (1993), and Richardson et al. (1995). Data discussed here for cetaceans and sirenians are limited to underwater measures. Most pinnipeds are in effect "amphibious" hearers in that they operate and presumably use sound in both air and water; therefore, data are included from both media where available. No published audiometric data are available for mysticetes, marine otters (*Lutra felina* and *Enhydra lutris*), or polar bears (*Ursus maritimus*).

Cetaceans

Electrophysiological and behavioral audiograms are available for seven odontocete species (Au 1993), most of which are Type II delphinids with peak sensitivity in the 40- to 80-kHz range. Data, generally from one individual, are available also for beluga whales (*Delphinapterus leucas*), boutu, and harbor porpoise. There are no published audiograms for adult physeterids or ziphiids, or any mysticete. The available data indicate that odontocetes tend to have at least a 10-octave functional hearing range, compared with 8 to 9 octaves in the majority of mammals. Best sensitivities range from 12 kHz in killer whales (Schevill and Watkins 1966, Hall and Johnson 1971) to more than 100 kHz in boutu and harbor porpoise (Møhl and Andersen 1973, Voronov and Stosman 1977, Supin and Popov 1990).

Until recently, most odontocete audiometric work was directed at understanding echolocation abilities rather than underwater hearing per se. Therefore, much of what is known about odontocete hearing is related to ultrasonic abilities. Acuity measures commonly used in these studies include operational signal strength, angular resolution, and difference limens. The first two are self explanatory. Difference limens (DL) are a measure of frequency discrimination based on the ability to differentiate between two frequencies or whether a single frequency is modulated. Difference limens are usually reported simply in terms of Hz or as relative difference limens (rdl), which are calculated as a percent equal to 100 times the DL in Hz / frequency. Au (1990) found that echolocation performance in bottlenose dolphins was 6 to 8 dB poorer than that expected from an optimal receiver. Target detection thresholds as small as 5 cm at 5 m have been reported, implying an auditory angular resolution ability of 0.5°, although most data suggest 1° to 4° for horizontal and vertical resolutions are more common (Bullock and Gurevich 1979, Popper 1980, Au 1990). Minimal intensity discrimination in bottlenose dolphins (1 dB) is equal to human values; temporal discrimination (≈8% of signal duration) is superior to human abilities. Frequency discrimination in bottlenose dolphins varies from 0.28% to 1.4% rdl for frequencies between 1 and 140 kHz; best values are found between 5 and 60 kHz (Popper 1980). These values are similar to those of microchiropteran bats and superior to the human average (Grinnell 1963, Simmons 1973, Sales and Pye 1974, Long 1980, Pollack 1980, Popper 1980, Watkins and Wartzok 1985). Frequency discrimination and angular resolutions in harbor porpoises (0.1% to 0.2% rdl; 0.5° to 1°) are on average better than those for bottlenose dolphins (Popper 1980).

An important aspect of any sensory system for survival is the ability to detect relevant signals amidst background noise. Critical bands and critical ratios are two measures of the ability to detect signals embedded in noise. In hearing studies, the term masking refers to the phenomenon in which one sound eliminates or degrades the perception of another (for a detailed discussion, see Yost 1994). To measure a critical band, a test signal, the target (usually a pure tone), and a competing signal, the masker, are presented simultaneously. Fletcher (1940) showed that as the bandwidth of the masker narrows, the target suddenly becomes easier to detect. The critical band (CB) is the bandwidth at that point expressed as a percent of the center frequency. If the ear's frequency resolution is relatively poor, there is a broad skirt of frequencies around the target tone that can mask it, and the CB is large. If the ear has relatively good frequency resolution, the CB is relatively narrow. Critical ratios (CR) are a comparison of the signal power required for target detection versus noise power, and are simply calculated as the threshold level of the target in noise (in dB) minus the masker level (dB). Critical bands tend to be a constant function of the CRs throughout an animal's functional hearing range. Consequently, CR measures with white noise, which are easier to obtain than CBs, have been used to calculate masking bandwidths based on the assumption that the noise power integrated over the critical band equals the power of the target at its detection threshold, or,

$$CB\ (\text{in Hz}) = 10^{(CR/10)} \qquad \text{(equation 7)}$$

(Fletcher 1940, Fay 1992). This implies that the target strength is at least equal to that of the noise; however, there are exceptions. Although uncommon, *negative* CRs, meaning the signal is detected at levels below the noise, have been reported for human detection of speech signals[5] and for some bats near their echolocation frequencies (Schuknecht 1993, Kössl and Vater 1995). Critical bands are thought to depend on stiffness variations in the inner ear. In generalist ears, the critical bandwidths are relatively constant at about 0.25 to 0.35 octaves/mm of basilar membrane (Ketten 1984, 1992; West 1985; Allen and Neeley 1992). Although hearing ranges vary widely in terms of frequency, most mammals have a hearing range of 8 to 9 octaves, which is consistent with earlier findings that the number of CBs was approximately equal to basilar membrane length in millimeters (Pickles 1982, Greenwood 1990).

Based on CR and CB data, odontocetes are better than most mammals at detecting signals in noise. Odontocetes have more CBs and the CRs are generally smaller than in humans. Furthermore, odontocete critical bandwidths approach zero and are not a constant factor of the critical ratio at different frequencies. The bottlenose dolphin has 40 CBs, which vary from 10 times the CR at 30 kHz to 8 times the CR at 120 kHz (Johnson 1968, 1971; Moore and Au 1983; Watkins and Wartzok 1985; Thomas et al. 1988, 1990b). Critical ratios for bottlenose dolphins (20 to 40 dB) are, however, generally higher than in other odontocetes measured. The best CRs to date (8 to 40 dB) are for the false killer whale (*Pseudorca crassidens*) (Thomas et al. 1990b), which is also the species that has performed best in echolocation discrimination tasks (Nachtigall et al. 1996).

Sound localization is an important aspect of hearing in which the medium has a profound effect. In land mammals, two cues are important for localizing sound: differences in arrival time (interaural time) and in sound level (interaural intensity). Binaural hearing studies are relatively rare for marine mammals, but the consensus from research on both pinnipeds and odontocetes is that binaural cues are important for underwater localization (Dudok van Heel 1962, Gentry 1967, Renaud and Popper 1975, Moore et al. 1995); however, because of sound speed differences, small or absent pinna, and ear canal adaptations in marine mammals, localization mechanisms may be somewhat different from those of land mammals.

In mammals, the high frequency limit of functional hearing in each species is correlated with its interaural time distance (IATD = the distance sound travels from one ear to the other divided by the speed of sound; Heffner and Masterton 1990). The narrower the head, the smaller the IATD, the higher the frequency an animal must perceive with good sensitivity to detect arrival time through phase differences. For example, consider a pure tone, which has the form of a sine wave, arriving at the head. If the sound is directly in front of the head, the sound will arrive at the same time and with the same phase at each ear. As the animal's head turns away from the source, each ear receives a different phase, given that the inter-ear distance is different from an even multiple of the wavelength of the sound. Therefore, IATD cues involve comparing time of arrival versus phase differences at different frequencies in each ear. Phase cues are useful primarily at frequencies below the functional limit; however, the higher the frequency an animal can hear, the more likely it is to have good sensitivity at the upper end of frequency range for phase cues.

Clearly, IATDs depend on the sound conduction path in the animal and the media through which sound travels. For terrestrial species, the normal sound path is through air, around the head, pinna to pinna. The key entry point for localization cues is the external auditory meatus, and therefore the IATD is the intermeatal (IM) distance measured around the head divided by the speed of sound in air. In aquatic animals, sound can travel in a straight line, by tissue conduction, through the head given that tissue impedances are similar to the impedance of seawater. Experiments with delphinids suggest that intercochlear (IC) or interjaw distances are the most appropriate measure for calculating IATD values in odontocetes (Dudok van Heel 1962, Renaud and Popper 1975, Moore et al. 1995). The IC distances of dolphins are acoustically equivalent to a rat or bat IM distance in air because of the increased speed of sound in water. Supin and Popov (1993) proposed that marine mammals without pinnae were incapable of using IATD cues, given the small interreceptor distances implied by the inner ear as the alternative underwater receptor site. Recently, however, Moore et al. (1995) demonstrated that the bottlenose dolphin has an IATD on the order of 7 μsec, which is better than the average human value (10 μsec) and well below that of most land mammals tested. If IM distances are used for land mammals and otariids in air and IC distances are used for cetaceans and underwater phocid data, marine mammal and land mammal data for IATD versus high frequency limits follow similar trends.

Intensity differences can be detected monaurally or binaurally, but binaural cues are most important for localizing high frequencies. In land mammals, intensity discrimination thresholds (IDT) tend to decrease with increasing sound levels and are generally better in larger animals (Fay 1992, Heffner and Heffner 1992). Humans and macaques commonly detect intensity differences of 0.5 to 2 dB throughout their functional hearing range; gerbils and chinchillas, 2.5 to 8 dB. Behavioral and evoked potential data show intensity differences are detectable by odontocetes at levels equal to those of land mammals and that the detection thresholds, like those of land mammals, decline with increasing sound level. Binaural behavioral studies and evoked potential recordings for bottlenose dolphin indicate an approximate IDT limit of 1 to 2 dB (Bullock et al. 1968, Moore et al. 1995). In harbor porpoise, IDTs range 0.5 to 3 dB (Popov et al. 1986). Thresholds in boutu range from 3 to 5 dB (Supin and Popov 1993), but again, because of small sample size and methodological differences, it is unclear whether these numbers represent true species differences. Fay (1992) points out that the IDT data for land mammals do not fit Weber's Law, which would predict a flat curve for IDT (i.e., intensity discrimination in dB should be nearly constant). Rather, the IDTs decrease with increasing level and increase slightly with frequency.

In the past decade, auditory evoked potential (AEP) or auditory brainstem response (ABR) procedures have been established for odontocetes (Popov and Supin 1990a, Dolphin 1995). These techniques are highly suitable for studies with marine mammals for the same reasons they are widely used for measuring hearing in infants or debilitated humans—namely, they are rapid, minimally invasive, and require no training or active response by the subject. An acoustic stimulus is presented by ear or jaw phones and the evoked neural responses are recorded from surface electrodes or mini-electrodes inserted under the skin. The signals recorded reflect synchronous discharges of large populations of auditory neurons. The ABRs consist of a series of 5 to 7 peaks or waves that occur within the first 10 msec after presentation of click or brief tone burst stimuli. Most mammals have similar ABR patterns, but there are clear species-specific differences in both latencies and amplitudes of each wave (Jewett 1970, Dallos et al. 1978, Achor and Starr 1980, Dolan et al. 1985, Shaw 1990). The delay and pattern of the waves are related to the source of the response. For example, wave I in most mammals is thought to derive from synchronous discharges of the auditory nerve; wave II from the auditory nerve or cochlear nucleus. The ABRs from dolphins show clear species dependence. Typical ABRs from harbor porpoise and bottlenose dolphin have three positive peaks with increasing amplitudes, but those in harbor porpoise have longer latencies (Bullock et al. 1968, Ridgway et al. 1981, Bibikov 1992).

Recent work using continuous amplitude modulated stimuli (AMS) at low frequencies in bottlenose dolphins and false killer whales suggest odontocetes can extract envelope features at higher modulation frequencies than other mammals (Kuwada et al. 1986, Dolphin and Mountain 1992, Dolphin 1995). Supin and Popov (1993) also showed that envelope following responses (EFR) are better measures of low frequency auditory activity than ABR. The anatomical correlates of EFRs have not been identified, but the data suggest auditory central nervous system adaptations in dolphins may include regions specialized for low as well as high frequencies.

Pinnipeds

Pinnipeds are particularly interesting because they are faced with two acoustic environments. Different ways for sensory information to be received and processed are required for equivalent air and water hearing in their amphibious lifestyle. One possibility is that pinnipeds have dual systems, operating independently for aquatic and airborne stimuli. If this is the case, hearing might be expected to be equally acute but possibly have different frequency ranges related to behaviors in each medium (e.g., feeding in water vs. the location of a pup on land). An alternative to the neffinlike dual but equal hearing is that pinnipeds are adapted primarily for one environment and have a "compromised" facility in the other. Renouf (1992) argued that there is an "a priori justification for expecting otariids and phocids" to operate with different sensory emphases, given that phocids are more wholly aquatic. This question cannot be definitively resolved until more pinniped species have been tested. As with cetaceans, present data are limited to a few individuals from mostly smaller species. However, the most recent data suggest there are significant differences among pinnipeds in both their primary frequency adaptations and in their adaptations to air versus water to warrant more widespread species research.

Underwater behavioral audiograms for phocids are somewhat atypical in that the low frequency tail is relatively flat compared to other mammalian hearing curves (compare Figs. 4-2 and 4-5A with Fig. 4-5B; see also Fay 1988 or Yost 1994 for additional comparisons). In the phocids tested (harbor seal [*Phoca vitulina*], harp seal [*Phoca groenlandica*], ringed seal [*Phoca hispida*], and Hawaiian monk seal [*Monachus schauinslandi*]), peak sensitivities ranged between 10 and 30 kHz, with a functional high frequency limit of about 60 kHz, except for the monk seal which had a high frequency limit of 30 kHz (Schusterman 1981, Fay 1988, Thomas et al. 1990a). Low frequency functional limits are not yet well established for phocids, and it is likely that some of the apparent flatness will disappear as more animals are tested below 1 kHz. However, the fact that all phocid plots have remarkably little decrease in overall sensitivity below peak frequency is notable. Currently available data from an on-going study comparing harbor seal and northern elephant seal hearing suggest that the elephant seal has significantly better underwater low frequency hearing thresholds than other pinnipeds tested to date (Kastak and Schusterman 1995, 1996).

In-air audiograms for phocids have more conventional shapes with peak sensitivities at slightly lower frequencies (3 to 10 kHz) (Fay 1988; Kastak and Schusterman 1995, 1996). In-air evoked potential data on these species are consistent with behavioral results (Bullock et al. 1971, Dallos et al. 1978). In-air and underwater audiograms cannot be compared directly; however, when the data are converted to intensity measures, the thresholds for airborne sounds are poorer, on average (Richardson et al. 1995), implying that phocids are primarily adapted for underwater hearing.

Underwater audiograms and aerial audiograms are available for two species of otariids. Underwater hearing curves for California sea lions and northern fur seals (*Callorhinus ursinus*) have standard mammalian shapes. Functional underwater high frequency hearing limits for both species are between 35 and 40 kHz with peak sensitivities from 15 to 30 kHz (Fay 1988, Richardson et al. 1995). As with phocids, otariid peak sensitivities in air are shifted to lower frequencies (< 10 kHz; functional limit near 25 kHz), but there is relatively little difference in the overall in-air versus underwater audiogram shape compared with phocids. The fact that the otariid aerial and underwater audiograms are relatively similar suggests that otariids may have developed parallel, equipotent hearing strategies for air and water or even, in the case of the California sea lion, have "opted" evolutionarily for a slight edge in air.

In frequency discrimination and localization tasks, pinnipeds perform less well than odontocetes. Angular resolution ranges from 1.5° to 9°, with most animals performing in the 4° to 6° range (Møhl 1964, Bullock et al. 1971, Moore and Au 1975). There is wide individual variability and no consistent trend for aerial versus aquatic stimuli. Minimal intensity discrimination (3 dB) by California sea lions is poorer than that of dolphins or humans (Moore and Schusterman 1976); typical frequency discrimination limens for several phocids and the sea lion (1% to 2% rdl) (Møhl 1967; Schusterman and Moore 1978a,b; Schusterman 1981) are similar to some of the bottlenose dolphin data but are on average significantly larger (less sensitive) than those for harbor porpoise.

Critical ratio data are available for only three pinnipeds (Richardson et al. 1995). In the northern fur seal, underwater critical ratios measured over a fairly narrow range (2 to 30 kHz) were on a par with those of most odontocetes at those frequencies (18 to 35 dB). Critical ratios for one harbor seal in air and in water were generally similar but also had anomalously higher values for some data points. Data reported for the ringed seal were consistently 10 dB or more greater than those of the other two species, that is, significantly poorer than those of fur seals, harbor seals, or most odontocetes. Turnbull and Terhune (1993) concluded that equivalent performances in air and water can be explained by having an external reception system (ear canal and middle ear) in which both signal and noise levels produce parallel impedance shifts. However, this implies an identical filter response in air and water, which means either identical processing or parallel but equally efficient paths in the two domains. That is, the ear canal and middle ear transfer functions remain constant regardless of the medium. Given the usual assumptions about the mechanisms underlying CRs, however, the results could also be attributed to a common inner ear response in both media.

Like odontocetes, pinnipeds in water have small acoustic inter-ear distances. It is not known whether they have specialized mechanisms for maintaining the external canal as the sound reception point underwater or if tissue conduction is used. Møhl and Ronald (1975), using cochlear microphonics, determined that in-air reception in the harp seal is through the external canal, but they also found that underwater, the most sensitive region was located below the meatus in a region paralleling the canal. Pinnae allow monaural cues to be used; therefore, eared species may use two different strategies for localizing in air and in water.

Sirenians

Very few audiometric data are available for sirenians, the other obligate aquatic group. Published data for the West Indian manatee consist of one evoked potential study and preliminary reports from on-going work on manatee behavioral audiogram (Patton and Gerstein 1992, Gerstein et al. 1993, Gerstein 1994). Several evoked potential studies of Amazon manatee have been published (Bullock et al. 1980, Klishin et al. 1990, Popov and Supin 1990a), but no behavioral data. No audiometric data are available for dugongs.

Current behavioral data for the West Indian manatee indicate a hearing range of approximately 0.1 to 40 kHz, with best sensitivities near 16 kHz. Functional hearing limits within this range are not yet established. This octave distribution (7 to 8 octaves) is narrower than that of bottlenose dolphins (10.5 octaves: 0.15 to 160 kHz; Au 1993) and phocid seals (8 to 9 octaves: 0.08 to 40 kHz; Kastak and Schusterman 1995, 1996) that have been tested over a wide range of frequencies. Best thresholds for manatees (50 to 55 dB re 1 μPa) are similar to in-water thresholds for several pinnipeds (45 to 55 dB re 1 μPa), but are significantly higher than those for odontocetes tested in similar conditions (30 to 40 dB re 1 μPa). An interesting feature of the manatee audiogram is that it is remarkably flat (i.e., there is less than a 15-dB overall difference in thresholds between 5 and 20 kHz). In terms of level and shape, therefore the West Indian manatee audiogram more closely resembles the "essentially flat" audiograms of phocids noted by Richardson et al. (1995) than it does the sharply tuned curve typical of odontocetes. Bullock et al. (1982), using evoked potential techniques to measure West Indian manatee hearing, found a maximal upper frequency limit (35 kHz), which is similar to the behavioral results but a markedly different peak sensitivity (1.5 kHz). They also reported a sharp decline in response levels above 8 kHz.

Popov and Supin (1990a) found peak responses in evoked potential studies of the Amazon manatee between 5 and 10 kHz with thresholds of 60 to 90 dB re 1 μPa. Klishin et al. (1990) reported best sensitivities to underwater stimuli in the West Indian manatee to be between 7 and 12 kHz, based on auditory brainstem responses from awake animals.

Fissipeds

No conventional audiometric data are available for sea otters. Behavioral measures of hearing in air for two North American river otters (*Lutra canadensis*) (Gunn 1988) indicate a functional hearing range in air of approximately 0.45 to 35 kHz with peak sensitivity at 16 kHz, which is consistent with Spector's (1956) more general description of their hearing.

Noise Trauma

One area that is relatively unexplored audiometrically is marine mammal susceptibility to hearing loss. Particularly when data are obtained from one animal, it is important to consider whether that hearing curve is representative of the normal ear for that species. Age and/or exposure to noise can significantly alter hearing in mammals. Hearing losses are recoverable (TTS, temporary threshold shift) or permanent (PTS) according to the extent of inner ear damage (for reviews, see Lipscomb 1978, Lehnhardt 1986, Richardson et al. 1991). Damage location and severity are correlated with the power spectrum of the signal in relation to the sensitivity of the animal. For narrow-band, high frequency signals, losses typically occur in or near the signal band, but intensity and duration can act synergistically to broaden the loss. Long or repeated exposures to TTS level stimuli without adequate recovery periods can induce permanent threshold shifts. In general, if the duration of intense noise is short and the noise is narrow and not impulsive, hearing is recoverable and the loss is near the signal's peak frequency. If exposure is long, or if the signal is broadband with a sudden onset, some hearing, particularly in the higher frequencies, can be permanently lost.

In humans, PTS results most often from protracted, repeat intense exposures (e.g., occupational auditory hazards from background noise) or sudden onset of intense sounds (e.g., rapid, repeated gun fire). Acoustic trauma induced by sudden onset, loud noise (a "blast" of sound) is not synonymous with blast trauma, nor are noise and blast effects of the same magnitude. Blast injuries generally result from a single exposure to an explosive shock wave that has a compressive phase with a few microseconds initial rise time to a massive pressure increase over ambient followed by a rarefactive wave in which pressure drops well below ambient. Blast damage may be reparable or permanent according to the severity of the single blast exposure. Hearing loss with aging (presbycusis), in contrast, is the accumulation of PTS and TTS insults to the ear. Typically, high frequencies are lost first with the loss gradually spreading to lower frequencies over time.

In experiments, multihour exposures to narrow band noise are used to induce PTS. Most mammals incur losses when the signal is 80 dB over the animal's threshold. Temporary threshold shift has been produced in humans for frequencies between 0.7 and 5.6 kHz (our most sensitive range) from underwater sound sources when received levels were 150 to 180 dB re 1 μPa (Smith and Wojtowicz 1985, Smith et al. 1988). Taking into account differences in measurements of sound pressure in air versus water (equations 4 and 5), these underwater levels are consistent with the 80- to 90-dB exposure levels that induce TTS in humans at similar frequencies in air. Sharp rise-time signals produce broad spectrum PTS at lower intensities than slow onset signals both in air and in water (Lipscomb 1978, Lehnhardt 1986).

Currently, there are insufficient data to determine accurately TTS and PTS exposure guidelines for any marine mammal. Although there is the possibility that dive-related adaptations ameliorate acoustic trauma, recent studies show losses in marine mammals consistent with age-related hearing changes (Ketten et al. 1992, 1995). Significant differences in the hearing thresholds of two California sea lions reported by Kastak and Schusterman (1995) are consistent with age-related hearing differences between the animals. In odontocetes, postmortem examinations of ears from older bottlenose dolphins with known hearing losses found neural degeneration patterns similar to those of older humans, which are consistent with a progressive, profound high frequency loss (Ketten et al. 1995).

It is likely that all marine mammals can be impacted by sound at some combination of frequency and intensity, but those parameters will vary considerably by species. In terms of research, the possibility of a hearing loss from both natural and anthropogenic acoustic impacts should be considered in the interpretation of the data for any animal for which there is little or no history. The possibility of hearing loss over time needs to be tested with long-term studies and should be considered in interpreting data from older animals. Finally, as we expand our use of ocean technologies that have direct and indirect sonic effects, we must be aware of their potential to produce acoustic impacts in ranges important for aquatic species.

Ears

All marine mammals have special adaptations of the external (closure, wall thickening, wax plugs) and middle ear (thickened middle ear mucosa, broad eustachian tubes) consistent with deep, rapid diving and long-term submersion, but they retain an air-filled middle ear and have the same basic inner ear configuration as terrestrial species. Each group has distinct adaptations that correlate with both their hearing capacities and with their relative level of adaptation to water.

Cetaceans

EXTERNAL EAR. Pinnae are absent, although vestigial pinnal rings occur in some individuals. External auditory canals are present in cetaceans, but it is debatable whether they are functional. In odontocetes, the external canal is exceptionally narrow and plugged with cellular debris and dense, waxy cerumen. The canal has no observable attachment to the tympanic membrane or the middle ear. In mysticetes, the canal is narrow along most of its length, but the proximal end flares, cloaking the "glove finger," a complex, thickened membrane capped by a waxy mound in adults (Reysenbach de Haan 1956). Reysenbach de Haan (1956) and Dudok van Heel (1962) were among the first researchers to suggest soft tissue paths as an alternative to conventional external canal sound conduction in odontocetes. Reysenbach de Haan (1956) reasoned that because the transmission characteristics of blubber and seawater are similar, using a canal occluded with multiple substances would be less efficient than conduction through body fat, fluid, or bone. Dudok van Heel (1962) found the minimum audible angle in bottlenose dolphin was more consistent with an interbullar critical interaural distance than with intermeatal distances, and concluded the canal was irrelevant. A passive resonator system involving the teeth of the lower jaw has been suggested for delphinids (Goodson and Klinowska 1990), but this cannot be considered a general explanation because it cannot account for echolocation by relatively toothless species (e.g., the Monodontidae [narwhals and belugas] and Ziphiidae [beaked whales]). Currently, the lower jaw is considered the primary reception path for ultrasonic signals in odontocetes. Norris (1968, 1980) observed that the odontocete lower jaw has two exceptional properties: a fatty core and a thin, ovoid "pan bone" area in the posterior third of the mandible. Norris (1969) speculated this mandibular fat channel acts as a preferential low impedance path to the middle ear and the pan bone as an acoustic window to the middle ear region (Fig. 4-6).

Several forms of data support this hypothesis. The fats in the mandible are wax esters with acoustic impedances close to seawater (Varanasi and Malins 1971). Evoked responses and cochlear potentials in spotted and bottlenose dolphins were significantly greater for sound stimuli above 20 kHz from transducers placed on or near the mandible (Bullock et al. 1968, McCormick et al. 1970). Measurements with implanted hydrophones in severed bottlenose dolphin heads found best transmission characteristics for sources directed into the pan bone (Norris and Harvey 1974). Brill et al. (1988) found that encasing the lower jaw in neoprene significantly impaired performance in echolocation tasks. Some results disagreed, notably those by Popov and Supin (1990b) and Bullock et al. (1968), who found best thresholds for low to sonic frequencies near the external meatus. However, recent computerized tomographic and magnetic resonance imaging of dolphins revealed a second channel of similar fats lateral to the pan bone (Ketten 1994), which may explain the discrepancy in the data as the lateral fatty lobes are near the meatus in delphinids. No discrete soft tissue channels to the ear have as yet been identified in mysticetes.

TYMPANOPERIOTIC COMPLEX. The temporal bones of cetaceans are distinctive in both form and construction. The inner ear is housed in the periotic, which is fused at one or more points to the tympanic, or middle ear bone. This "tympanoperiotic" bullar complex is located outside the skull in a peribullar cavity formed evolutionarily by eliminating the small, pneumatized, mastoid spaces common in terrestrial animals. The periotic and tympanic are bordered laterally and ventrally by the mandible and hyoids. The extracranial position of the tympanic and periotic is important because it increases the acoustic separation of the middle and inner ears, as discussed earlier in the section on localization and interaural distances.

Tympanic and periotic volumes are highly correlated with animal size (Ketten and Wartzok 1990). Mysticete bullae are two to three times larger than those of most odonto-

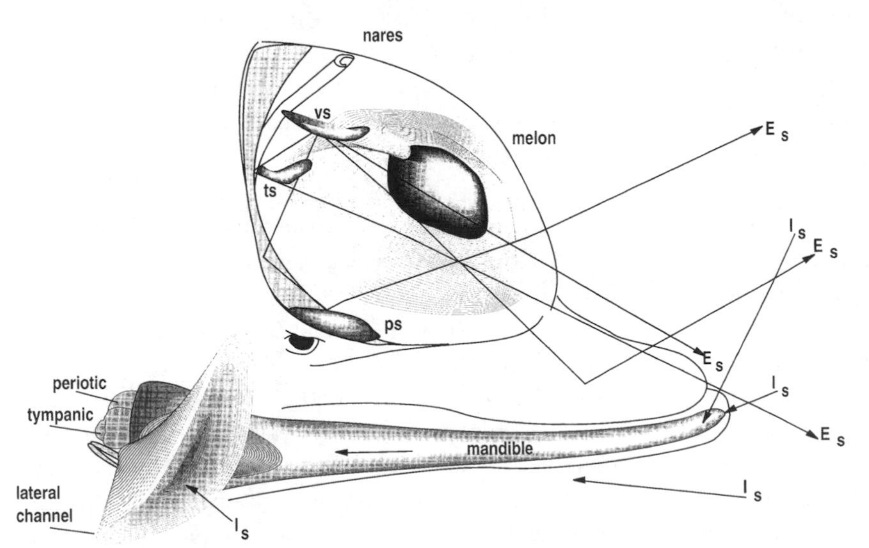

Figure 4-6. Sound paths in the dolphin head. (See also Pabst, Rommel, and McLellan, Chapter 2, this volume.) Outgoing ultrasonic signals (E_s) are generated in the vestibular (vs) and tubular (ts) nasal sac diverticulae and reflected off the cranium and premaxillary sac (ps) (Norris 1980, Au 1993). Incident sounds (I_s) from anterior targets enter the lower jaw where lipids act as preferential low-impedance acoustic conduits to the ear (Varanasi and Malins 1971, Norris 1980). Lateral, trumpetlike fatty lobes (lateral channel) with densities equal to those in the jaw overlie the pan bone and may act as additional input channels for signals from the rear or side of the animal. (Ketten 1994; figure ©Ketten 1992, revised 1997.)

cetes. Mysticete tympanics are nearly twice the volume of the periotics. In odontocetes, periotic and tympanic volumes are nearly equal. The periotic is ovoid, massive, and thick-walled in all cetaceans. The odontocete tympanic is thin-walled and conical, tapering anteriorly. Mysticete tympanics are thick-walled and spherical. This means mysticetes have larger middle ear cavities with relatively low frequency resonance characteristics compared to smaller, higher resonance odontocete middle ear cavities. The fine structure of the tympanoperiotic and the solidity of the tympanoperiotic suture differ among species, but no specific auditory effect has been determined for these differences (Kasuya 1973, Ketten 1984, Ketten and Wartzok 1990).

Odontocete tympanoperiotics are suspended in a spongy mucosa, the peribullar plexus, by five or more sets of ligaments. This mucosal cushion and the lack of bony connections to the skull isolate the ear from bony sound conduction and hold the tympanic loosely in line with the mandibular fatty channels and pan bone. Because peribullar sinuses are most extensive in riverine, ultra-high frequency species, like the boutu, and are poorly developed in pelagic mysticetes, Oelschläger (1986) suggested that the peribullar plexus and expanded sinuses are primarily echolocation-related adaptations, functioning as acoustical isolators for the ear, and were not driven evolutionarily by diving.

In mysticetes, extensive bony flanges wedge the periotic against the skull. The tight coupling of these flanges to the skull suggests both bony and soft tissue sound conduction to the ear occur in baleen whales.

MIDDLE EAR. Ossicles of odontocetes and mysticetes are large and dense, and vary widely in size, stiffness, and shape (Reysenbach de Haan 1956, Belkovich and Solntseva 1970, Solntseva 1971, Fleischer 1978). In odontocetes, a bony ridge, the processus gracilis, fuses the malleus to the wall of the tympanic and the interossicular joints are stiffened with ligaments and a membranous sheath. Mysticete ossicles are equally massive but have none of the high frequency-related specializations of odontocetes. The ossicles are not fused to the bulla and the stapes is fully mobile with a conventional fibrous annular ligament. Furthermore, as noted earlier, the tympanic bone scales with animal size and is on average double the volume of the periotic. Therefore, the mysticete middle ear cavity is substantially larger than that of any odontocete and, in combination with their massive ossicles that are loosely joined, forms a characteristically low frequency ear.

The middle ear cavity in both odontocetes and mysticetes is lined with a thick, vascularized fibrous sheet, the corpus cavernosum. Computerized tomography (CT) and magnetic resonance imaging (MRI) data suggest the intratympanic space is air-filled in vivo (Ketten 1994). If so, a potential acoustic difficulty for a diving mammal is that changing middle ear volumes may alter the resonance characteristics of the middle ear, and, in turn alter hearing sensitivity. Studies are underway with free-swimming beluga whales (S. Ridgway, pers. comm.) to test whether hearing thresholds change with depth. In light of the extensive innervation of the middle ear corpus cavernosum by the trigeminal nerve, one novel task proposed for the trigeminal in cetaceans has been to regulate middle ear volume (Ketten 1992), which could also explain exceptionally large trigeminal fiber numbers in both odontocetes and mysticetes (Jansen and Jansen 1969, Morgane and Jacobs 1972).

There is no clear consensus on how cetacean middle ears function. Both conventional ossicular motion and translational bone conduction have been proposed for cetaceans (McCormick et al. 1970, 1980; Lipatov and Solntseva 1972; Fleischer 1978). On the basis of experiments with anesthetized bottlenose dolphin and a Pacific white-sided dolphin (*Lagenorhynchus obliquidens*) McCormick et al. (1970, 1980) concluded that sound entering from the mandible by bone conduction produces a "relative motion" between the stapes and the cochlear capsule. In their procedure, immobilizing the ossicular chain decreased cochlear potentials, but disrupting the external canal and tympanum had no effect. Fleischer (1978) suggested the procedure introduced an artificial conduction pathway. From anatomical studies, he concluded sound from any path is translated through tympanic vibration to the ossicles, which conventionally pulse the oval window. McCormick's theory assumes fixed or fused tympanoperiotic joints; Fleischer's requires a mobile stapes, distensible round window, and flexible tympanoperiotic symphyses. Both conclusions may have been confounded by experimental constraints: McCormick et al. (1970) had to disrupt the middle ear cavity to expose the ossicles, and Fleischer's data were subject to postmortem and preservation artifacts. In addition, neither theory is completely compatible with the wide structural variability of cetacean middle ears. The question of middle ear mechanisms in cetaceans therefore remains open.

INNER EAR. The cetacean periotic houses the membranous labyrinth of the inner ear, which is further subdivided into auditory and vestibular components.

In all cetaceans, the vestibular system is substantially smaller in volume than the cochlea and may be vestigial (Boenninghaus 1903, Gray 1951, Ketten 1992, Gao and Zhou 1995). Although size is not a criterion for vestibular function, cetaceans are unique in having semicircular canals that are significantly smaller than the cochlear canal (Gray 1951, Jansen and Jansen 1969). Innervation is proportionately re-

Table 4-2. Auditory, Vestibular, and Optic Nerve Distributions

Species	Common Name	Cochlear Type	Membrane Length (mm)	Auditory Ganglion Cells	Density (cells/mm cochlea)	Vestibular Ganglion Cells	Vestibular Auditory Ratio	Optic Nerve Fibers	Optic Auditory Ratio	Optic Vestibular Ratio
Inia geoffrensis	Boutu	I	38.2	104,832	2744			15,500	0.15	
Lipotes vexillifer	Baiji			82,512		3,605	0.04	23,800	0.29	6.60
Neophocaena phocaenoides	Finless porpoise			68,198		3,455	0.05	88,900	1.30	25.73
Sousa chinensis	Humpbacked dolphin			70,226		3,213	0.05	149,800	2.13	46.62
Phocoena phocoena	Harbor porpoise	I	22.5	70,137	3117	3,200		81,700	1.16	25.53
Delphinapterus leucas	Beluga		42	149,386	3557			110,500	0.74	
Delphinus delphis	Common dolphin	II	34.9	84,175	2412	4,091	0.05	165,600	1.97	40.48
Lagenorhynchus obliquidens	White-sided dolphin	II	34.9	70,000	2006			77,500	1.11	
Stenella attenuata	Spotted dolphin	II	36.9	82,506	2236					
Tursiops truncatus	Bottlenosed dolphin	II	38.9	96,716	2486	3,489	0.04	162,700	1.68	46.63
Physeter catodon	Sperm whale		54.3	161,878	2981			172,000	1.06	
Balaenoptera physalus	Fin whale	M	64.7	134,098	2073			252,000	1.88	
Megaptera novaeangliae	Humpback whale	M	58	156,374	2696			347,000	2.22	
Rhinolophus ferrumequinum	Horseshoe bat	T	16.1	15,953	991/1750[a]					
Pteronotus parnellii	Mustached bat	T	14.0	12,800	900/1900[a]					
Cavia porcellus	Guinea pig	T	19.0	24,011	1264	8,231	0.34			
Felis domesticus	Cat	T	28.0	51,755	1848	12,376	0.24	193,000	3.73	15.59
Homo sapiens	Human	T	32.1	30,500	950	15,590	0.51	1,159,000	38.00	74.34

Data compiled from Yamada 1953; Gacek and Rasmussen 1961; Jansen and Jansen 1969; Firbas 1972; Morgane and Jacobs 1972; Bruns and Schmieszek 1980; Dawson 1980; Ketten 1984, 1992; Vater 1988a,b; Nadol 1988; Gao and Zhou 1991, 1992, 1995; Kössl and Vater 1995.

[a]Densities at auditory fovea as described by Bruns and Schmiezek (1980).

duced as well (Table 4-2); on average less than 5% of the cetacean VIIIth nerve is devoted to vestibular fibers (Fig. 4-4), as compared to approximately 40% in other mammals (Ketten 1997). No equivalent reduction of the vestibular system is known in any land mammal. A possible explanation is that fusion of the cervical vertebrae in cetaceans resulted in limited head movements, which resulted in fewer inputs to the vestibular system that led to a reduction of related vestibular receptors. This does not mean that cetaceans do not receive acceleration and gravity cues but rather that the neural "budget" for these cues is less. In land mammals, similar vestibular reductions have been approximated only by experimentation, disease, congenital absence of canals, or, in some extreme cases, through surgery as a cure for vertigo (Graybiel 1964).

All cetacean cochleae have three scalae or chambers like other mammals (Figs. 4-4 and 4-7): scala media (also called the cochlear duct), scala tympani, and scala vestibuli. The scalae are parallel fluid-filled tubes. Scala vestibuli ends at the oval window; scala tympani, at the round window; and scala media, which contains the organ of Corti, is a blind pouch between them. Detailed descriptions of odontocete cochlear ducts are available in Wever et al. (1971a,b,c, 1972), Ketten (1984, 1992, 1997), Ketten and Wartzok (1990), and Solntseva (1971, 1990). We briefly summarize histological observations and discuss in detail only the cochlear features that influence hearing ranges and sensitivity.

Odontocete cochleae differ significantly from other mammalian cochleae by having hypertrophied cochlear duct structures, extremely dense ganglion cell distributions, and unique basilar membrane dimensions. Wever et al. (1971a,b,c, 1972) found all cellular elements of the organ of Corti in *Tursiops* and *Lagenorhynchus* were larger and denser than in other mammals. More recent studies reported hypertrophy of the inner ear in phocoenids and monodontids as well (Ketten 1984, Ketten and Wartzok 1990, Solntseva 1990). Most of the hypercellularity is associated with the support cells of the basilar membrane and with the stria vascu-

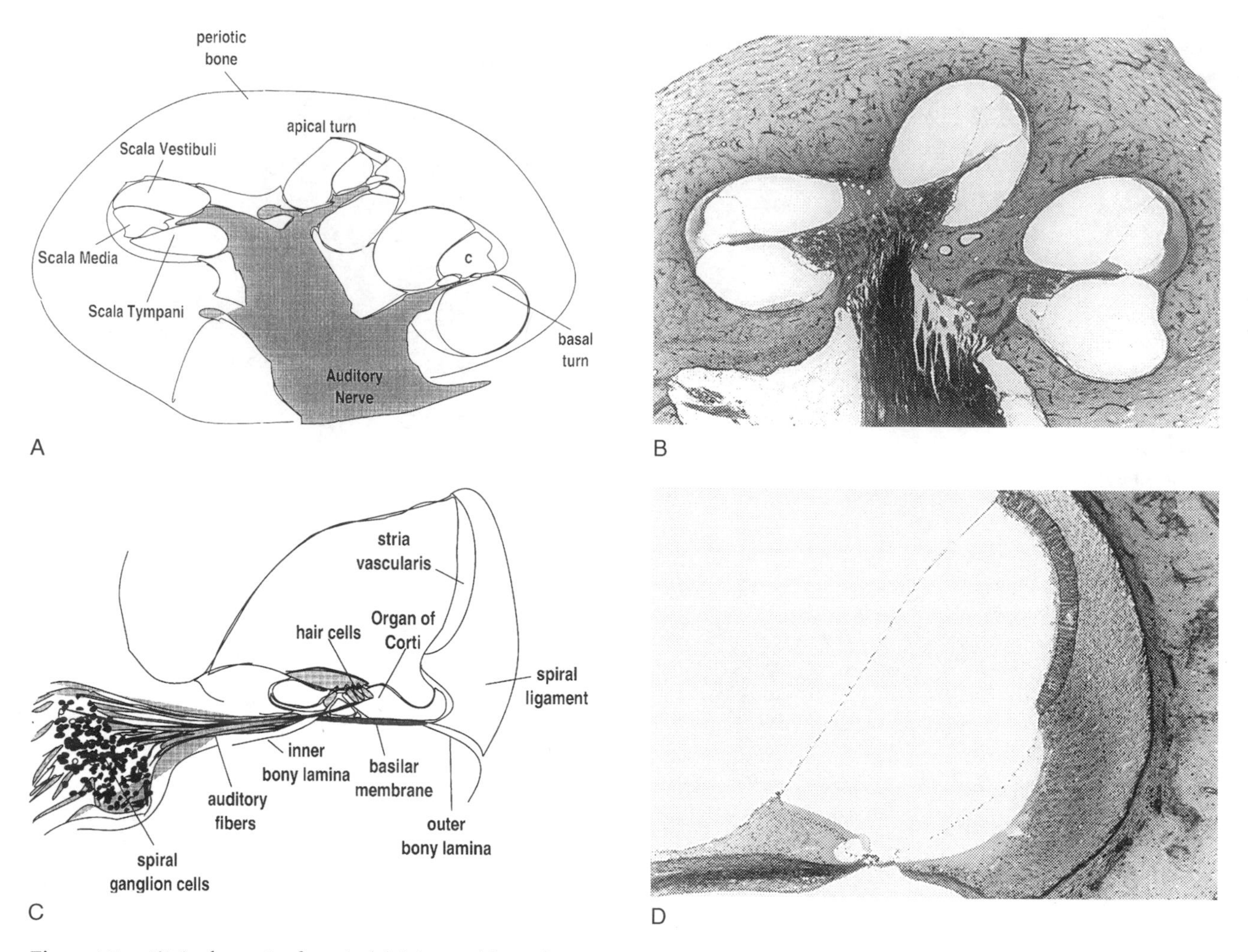

Figure 4-7. (A) A schematic of a typical dolphin cochlea is shown in cross-section with a corresponding photomicrograph (B) from a 25-μm histologic section of a *Phocoena* ear. (C) A schematic of the major features of the cochlear duct in the high-frequency basal region that are equivalent to the magnified view (D) of a 25-μm section of a Atlantic white-sided dolphin inner ear.

laris, which plays a major role in cochlear metabolism. Mysticete ears are less well-endowed cellularly, but this may be a reflection of preservation artifacts that are more common in baleen specimens because of greater difficulties in their collection and generally longer postmortem times before they are preserved.

The fiber and ganglion cell counts for the auditory nerve are exceptional in all cetaceans (Table 4-2). Auditory ganglion cell totals are more than double those of humans in all species, but, more important, the innervation densities (neurons/mm basilar membrane) are two- to threefold greater than in other mammals. Comparisons of the ratios of auditory, vestibular, and optic nerve fibers in cetaceans versus representative land mammals (Table 4-2) underscore the hypertrophy of the cetacean auditory nerve. The vestibular to auditory ratios are approximately 1/10 those of land mammals. Optic to auditory ratios in Type II odontocetes and mysticetes are approximately half those of most land mammals (noting an exception for the exceptionally high human optic value), whereas those of Type I riverine odontocetes are an order of magnitude less.

Auditory ganglion cell densities in Type I odontocetes are particularly notable, averaging more than 3000 cells/mm. Using a mammalian average of 100 inner hair cells/mm (Kiang, pers. comm.) and four rows of outer hair cells/inner hair cell, these data imply a ganglion to hair cell ratio of nearly 6:1 for Type I species. In humans, the ratio is 2.4:1; in cats (*Felis domesticus*), 3:1; and in bats, the average is 4:1 (Firbas 1972, Bruns and Schmieszek 1980, Vater 1988b). Because 90% to 95% of all afferent spiral ganglion cells innervate inner hair cells, the average ganglion cell to inner hair cell ratio is 24:1 for cetaceans, or more than twice the average ratio in bats and three times that of humans. Wever et al. (1971c) speculated that additional innervation is required primarily in the basal region to relay greater detail about ultrasonic signals to the central nervous system in echolocation analyses.

Electrophysiological results are consistent with this speculation. The central nervous system recordings in both porpoises and bats imply increased ganglion cells correspond to multiple response sets that are parallel processed at the central level. Bullock et al. (1968) found three distinct categories of response units in the inferior colliculus of dolphin brains, that is, those that were signal duration specific, those that responded to changes in signal rise time, and those that were specialized to short latencies with no frequency specificity. This division of signal properties among populations of neurons is consistent with, although not identical to, observations in bats of multiple categories of facilitation and analysis neurons (Schnitzler 1983, Suga 1983). The odontocete inner ear neural distribution data imply that equally extensive analyses of signal characteristics are performed by odontocete auditory systems as well. However, although high afferent ratios in odontocetes could be related to the complexity of information extracted from echolocation signals, this theory does not explain similar densities in mysticetes. The similarity of odontocete and mysticete innervations suggests that mysticetes may have equally complex processing but possibly for infra- rather than ultrasonic tasks.

CETACEAN INNER EAR STRUCTURE—HEARING CORRELATES. The cetacean basilar membrane is a highly differentiated structure with substantial variations in length, thickness (T), and width (W) (Table 4-3, Fig. 4-8). Basilar membrane lengths in Cetacea, like those of terrestrial mammals, scale isomorphically with body size. In Cetacea, cochlear length is correlated strongly with animal size ($0.8 < r < 0.95$), but there is no significant correlation for length and frequency (Ketten 1992). Thickness and width, however, are strongly correlated with hearing capacity (Ketten 1984, Ketten and Wartzok 1990). In most odontocetes, basilar membrane width is 30 μm at the base and increases to 300 to 500 μm apically. Basal widths of odontocetes are similar to those of bats and one third that of humans (Firbas 1972, Schuknecht and Gulya 1986). In odontocetes thicknesses

Table 4-3. Cochlear Morphometry in Whales versus Land Mammals

Species	Common Name	Cochlear Type	Turns	Membrane Length (mm)	Outer Lamina (mm)	Base Thickness (μm)	Base Width (μm)	Apex Thickness (μm)	Apex Width (μm)	Basal T:W Ratio	Apical T:W Ratio
Inia geoffrensis	Boutu	I	1.5	38.2	[a]						
Phocoena phocoena	Harbor porpoise	I	1.5	22.5	17.6	25	30	5	290	0.83	0.017
Grampus griseus	Risso's dolphin	II	2.5	41.0	[a]	20	40	5	420	0.50	0.012
Lagenorhynchus albirostris	White-beaked dolphin	II	2.5	34.8	8.5	20	40	5	360	0.50	0.014
Stenella attenuata	Spotted dolphin	II	2.5	36.9	8.4	20	40	5	400	0.50	0.013
Tursiops truncatus	Bottlenosed dolphin	II	2.25	38.9	10.3	25	35	5	380	0.71	0.013
Physeter catodon	Sperm whale		1.75	54.3	[a]						
Balaenoptera acutorostrata	Minke	M	2.25	50.6		—	100	—	1500		
Balaena mysticetus	Bowhead	M	2.25	56.5	< 10[b]	7.5	120	2.5	1670	0.06	0.001
Balaenoptera physalus	Fin whale	M	2.5	64.7		—	100	—	2200		
Eubalaena glacialis	Right whale	M	2.5	54.1	< 8[b]	7	125	2.5	1400	0.06	0.002
Rhinolophus ferrumequinum	Horseshoe bat	Æ	2.25	16.1	[a]	35	80	2	150	0.44	0.013
Pteronotus parnellii	Mustached bat	Æ	2.75	14.0	[a]	22	50	2	110	0.44	0.018
Spalax ehrenbergi	Mole rat	Sb	3.5	13.7		—	120	—	200		
Cavia porcellus	Guinea pig	T	4.25	18.5		7.4	70	2	250	0.11	0.008
Felis domesticus	Cat	T	3	28.0	[a]	12	80	5	370	0.15	0.014
Homo sapiens	Human	T	2.5	33.0		—	120	—	550		

Data compiled from Wever et al. 1971a, b; Firbas 1972; Bruns and Schmieszek 1980; Norris and Leatherwood 1981; Ketten 1984, 1992, 1994; West 1985; Vater 1988a,b; Nadol 1988; Echteler et al. 1994; Kössl and Vater 1995.

[a]Outer osseous lamina length unknown.

[b]Laminar remnant present but not in contact with basilar membrane.

Width (W) = pars arcuata and pectinata; Thickness (T) = pars pectinata maximum.

I = aquatic > 100 kHz; II = aquatic < 90 kHz; M = aquatic < 2 kHz; Æ = æolian > 20 kHz; Sb = subterranean; T = terrestrial.

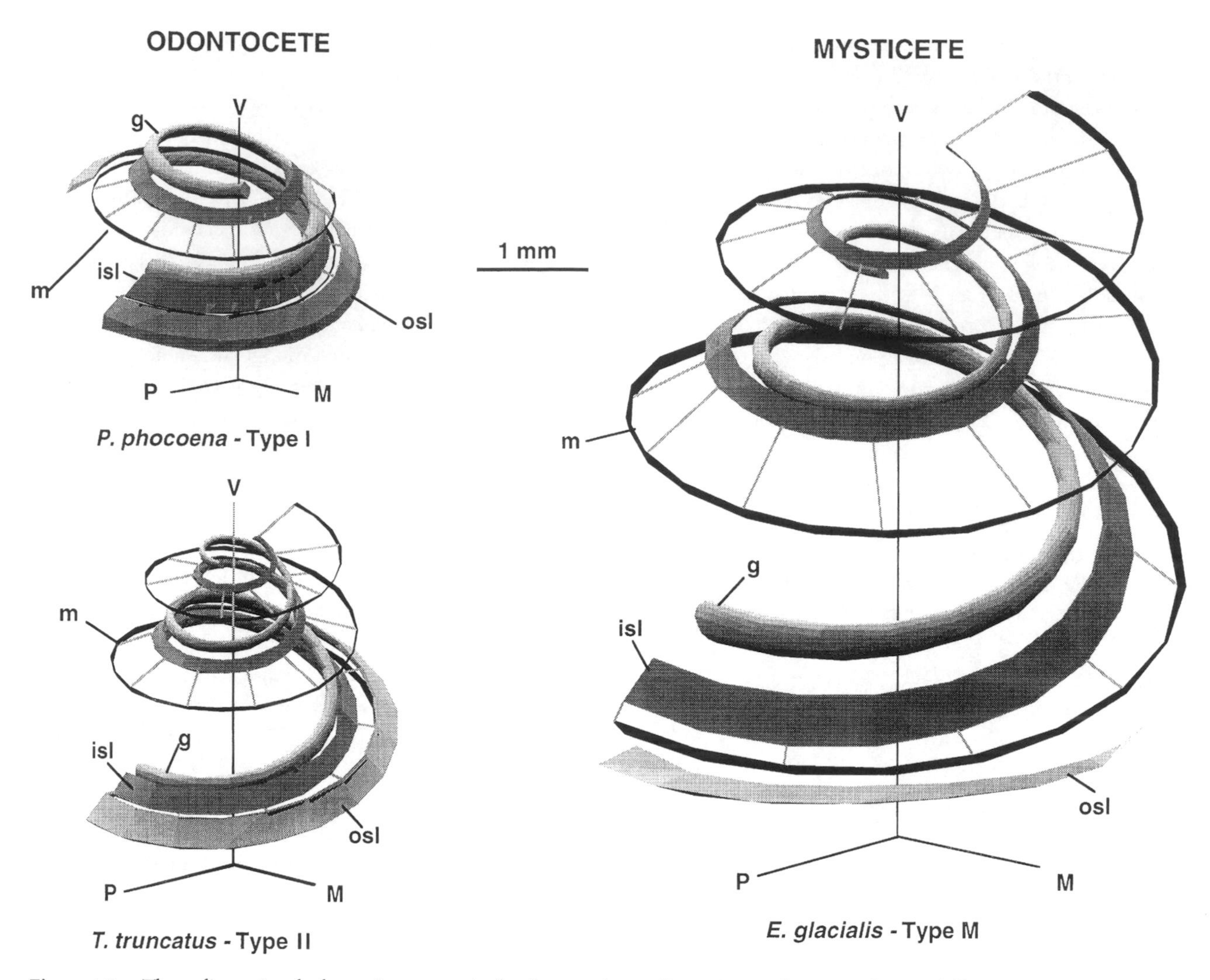

Figure 4-8. Three-dimensional schematics summarize basilar membrane dimensions and support element differences among Type I, Type II odontocete, and Type M mysticete inner ears in cetaceans. The cochlea are inverted from in vivo orientations. The schematics are drawn to the same scale for the three species illustrated (see Table 4-3). (g) spiral ganglion; (isl) inner bony spiral lamina; (m) basilar membrane; (osl) outer bony spiral lamina; (P) posterior; M (medial); (V) ventral.

typically range from 25 μm at the base to 5 μm at the apex (Table 4-3). Therefore, a typical cross-section of an odontocete basilar membrane is square at the base becoming rectangular apically. Mysticete membranes are thin rectangles throughout, varying in thickness between 7 μm at the base to 2 μm at the apex. Width gradients in mysticetes can be as great as in odontocetes with membranes in some species ranging from 100 μm at the base (similar to the base in humans) to 1600 μm at the apex. The apical widths in mysticetes are three times that of human, three to five times those of most odontocetes, and 1.2 times that of elephants, which are known to perceive infrasonics (Payne et al. 1986).

Comparing bat, odontocete, and mysticete basilar membrane thickness to width (T:W) ratios is a good exercise in structure–function relationships. T:W ratios are consistent with the maximal high and low frequencies each species hears and with differences in their peak spectra (Ketten and Wartzok 1990; Ketten 1992, 1997). Echolocators have significantly higher basal ratios than mysticetes, and odontocete ratios are higher than for bats in the basal regions where their ultrasonic echolocation signals are encoded. For example, *Phocoena,* a Type I odontocete, has a basal T:W ratio of 0.9 and a peak frequency of 130 kHz. *Tursiops,* a Type II odontocete, has a T:W ratio of 0.7 and a peak signal of 70 kHz, and *Rhinolophus,* a bat, a 0.3 T:W ratio and a 40 kHz echolocation signal. All three have terminal apical ratios near 0.01. Mysticete T:W ratios range from 0.1 at the base to approximately 0.001 at the apex, that is, the mysticete basal ratios are equivalent to midapical ratios in the three echolocators and decrease steadily to a value one-tenth that of odontocetes at the apex. The exceptionally low apical ratio in mysticeti is consistent with a broad, flaccid membrane that can encode infrasonics.

A striking feature of odontocete basilar membranes is their association with extensive outer bony laminae. In mammals, ossified outer spiral laminae are hallmarks of ultrasonic ears (Yamada 1953, Reysenbach de Haan 1956, Sales and Pye 1974, Ketten 1984). Thick outer bony laminae (Fig. 4-7C) are present throughout the basal turn in all odontocetes, and the proportional extent of outer laminae is functionally correlated with odontocete ultrasonic frequency ranges (Ketten and Wartzok 1990). In the basal, high frequency region of the cochlea, odontocete basilar membranes resemble thick girders, stiffened by attachments at both margins to a rigid bony shelf. In Type I echolocators with peak frequencies above 100 kHz an outer lamina is present for 60% of the cochlear duct (Fig. 4-8). Type II echolocators with lower peak frequencies have a bony anchor for about 30% of the duct. Therefore, the Type I basilar membrane is coupled tightly to a stiff ledge for twice as much of its length as a Type II membrane. If Type I and II membranes have similar thickness:width ratios, a Type I cochlea with longer outer laminae would have greater membrane stiffness and higher resonant frequencies than an equivalent position in a Type II membrane without bony support. Both membrane ratios and the extent or proportion of auxiliary bony membrane support are important mechanistic keys to how odontocetes achieve ultrasonic hearing despite ear size.

Both inner and outer laminae are present in mysticete cochleae but they are morphologically and functionally very different from those of odontocetes. Mysticete outer laminae are narrow spicules located on the outer edge of the spiral ligament. They do not attach to the basilar membrane. The broad, thin mysticete basilar membrane attaches only to a flexible spiral ligament. It is likely that the spikelike outer lamina in mysticetes is a remnant of an ancestral condition rather than a functional acoustic structure and that low basilar membrane ratios and large organ of Corti mass are the principal structural determinants of mysticete hearing ranges. To date, few mysticete species have been analyzed for very low frequency sensitivity, but the inner and middle ear anatomy argues strongly that they are low to infrasonic specialists.

Pinnipeds

EXTERNAL EAR. Pinniped ears are less derived than cetacean ears. The external pinnae are reduced or absent. Ear canal diameter and closure mechanisms vary widely in pinnipeds, and the exact role of the canal in submerged hearing has not clearly been determined. Otariids have essentially terrestrial, broad-bore external canals with moderate to distinctive pinnae. Phocids, particularly northern elephant seals, spend more time in water than otariids and have only a vestigial cartilaginous meatal ring, no pinnae, and narrow ear canals (D. R. Ketten and R. J. Schusterman, unpubl.). Although the phocids have no external pinnae, it is not yet known which species normally have air-filled versus partially to fully blocked external canals. No specialized soft tissue sound paths for underwater hearing have been clearly demonstrated in seals.

An obvious amphibious adaptation in phocid ears is that the external canal is well-developed and has a ring of voluntary muscle that can close the meatus (Møhl 1967, Repenning 1972). It has been suggested that seal middle ears are capable of operating entirely liquid filled (Repenning 1972) and that various soft tissue attachments to the ossicles are related to the operation of a liquid-filled middle ear or for enhancing high frequency sensitivity in water (Ramprashad et al. 1972, Renouf 1992), but neither of these suggestions is consistent with the level of development of the external canal or the size and development of the eustachian tube. Whether the external canal remains patent and air filled, collapses, or becomes flooded during dives remains a heavily debated subject. The ear canal contains a corpus cavernosum (cavernous epithelium) analogous to that in the middle ear, which may close the canal and regulate air pressures during dives (Møhl 1968, Repenning 1972). There are strong theoretical arguments for each position. Flooding the canal would provide a low impedance channel to the tympanic membrane, but then directing sound input to only one window of the cochlea becomes a problem. If the middle ear is fluid filled, the oval and round windows can receive simultaneous stimulation that would interfere with normal basilar membrane response. However, if the canal remains air filled, it poses the problem of an impedance mismatch that could make the canal less efficient for sound conduction to the middle and inner ear than surrounding soft tissues when the animal is submerged. To date, there is no clear evidence for specialized soft tissues, like those found in odontocetes, and no direct measures of the shape of the ear canal when submerged.

The position and attachment to the skull of the tympanic and periotic bones in pinnipeds is not significantly different from that of land mammals. The middle ear space is encased in a tympanic bulla, a bulbous bony chamber with one soft-walled opening, the tympanic membrane. The tympanic bulla is fused to the periotic. Both have partially or fully ossified articulations with the skull. These connections are less rigid than those in some land mammals, but the ears are not as clearly detached (and acoustically isolated) as those of cetaceans.

MIDDLE EAR. Pinniped middle ears have a moderate layer of cavernous tissue, but it is less developed than that of cetaceans (Møhl 1968, Ramprashad et al. 1972, Repenning 1972, Fleischer 1978). Pinniped ossicular chains are diverse. Those in otariids resemble terrestrial carnivores; ossicles of phocids are more massive but with wide species variation in shape (Doran 1879, Fleischer 1978), which suggests a wider range of peak frequencies and more emphasis on lower frequency reception than in otariids. Although some researchers indicate phocids have small eardrums (Repenning 1972), the size is not significantly different from that of equivalent mass land mammals. The oval and round window areas in terrestrial mammals are of approximately the same size. In pinnipeds, the oval window can be one-half to one-third the size of the round window. Eardrum to oval window ratios have been cited frequently as a factor in middle ear gain, but this association is still being debated (Rosowski 1994), and depending on the exact size distributions among these three membranes in each pinniped species, there could be wide differences in middle ear amplification among pinnipeds.

INNER EAR. Relatively few pinniped inner ears have been investigated and published data that are available are largely descriptive (Ramprashad et al. 1972, Solntseva 1990). Most pinnipeds have inner ears that resemble terrestrial high frequency generalists (i.e., multiple turn spirals with partial laminar support). Preliminary data on larger species suggest they may have some low frequency adaptations consistent with their size. There is no indication of extensive adaptation for either high ultrasonic or infrasonic hearing. Pinnipeds have one feature in common with cetaceans—a large cochlear aqueduct. Møhl (1968) suggested that this would facilitate bone conduction, but the mechanism is not clear, nor is it consistent with equally large aqueducts in odontocetes.

Sirenians

Anatomical studies of sirenian ears are largely descriptive and most work has been done only on manatees (Robineau 1969, Fleischer 1978, Ketten et al. 1992). Like Cetacea, they have no pinnae. Also, the tympanoperiotics are constructed of exceptionally dense bone, but like pinnipeds (and unlike odontocetes), sirenian ear complexes are partly fused to the inner wall of the cranium. Neonate ears vary less than 20% in shape and size from adult specimens; consequently, the ear complex is disproportionately large. In young manatees it can constitute 14% of skeletal weight (Domning and de Buffrénil 1991).

EXTERNAL EAR. Exact sound reception paths are not known in sirenians. The unusual anatomy of the zygomatic arch in manatees, combined with its relation to the squamosal and periotic have made the zygomatic a frequent candidate for a sirenian analogue to the odontocete fat channels. The periotic is tied by a syndesmotic (mixed fibrous tissue and bone) joint to the squamosal, which is fused, in turn, to the zygomatic process, a highly convoluted cartilaginous labyrinth filled with lipids. The zygomatic is, in effect, an inflated, oil-filled, bony sponge that has substantial mass but less stiffness than an equivalent process of compact bone (Domning and Hayek 1986, Ketten et al. 1992). In the Amazonian manatee, the best thresholds in evoked potential recordings were obtained from probes overlying this region (Bullock et al. 1980, Klishin et al. 1990), but no clear acoustic function has been demonstrated.

MIDDLE EAR. The middle ear system of sirenians is large and mass dominated but the extreme density of the ossicles adds stiffness (Fleischer 1978, Ketten et al. 1992). The middle ear cavity of manatees, as in other marine mammals, is lined with a thick, vascularized fibrous sheet. The ossicles are loosely joined and the stapes is columnar, a shape that is common in reptiles but rare in mammals and possibly unique to manatees. The manatee tympanic membrane is everted and supported by a distinctive keel on the malleus. Deeply bowed, everted tympanic membranes, epitomized by the fibrous "glove finger" in mysticetes, are common in marine mammals but are relatively rare in nonaquatic species. Like the eardrum of cats, the manatee tympanic membrane has two distinct regions, implying that membrane response patterns are frequency dependent (Pickles 1982). The tympanic-to-oval window ratio is approximately 15:1 in the West Indian manatee, which places it midway between that of humans and elephants (Ketten et al. 1992, Rosowski 1994). Chorda tympani, a branch of the facial nerve (cranial nerve VII) that traverses the middle ear cavity, is relatively large in manatees. It is crosses the middle ear but has no known auditory function. In humans, chorda tympani constitutes approximately 10% of the facial nerve. It conveys taste from the anterior two-thirds of the tongue and carries parasympathetic preganglionic fibers to the salivary glands. In the West Indian manatee, chorda tympani forms 30% of the facial nerve bundle.

INNER EAR. The sirenian inner ear is a mixture of aquatic and land mammal features. Anatomically, *T. manatus* inner ears are relatively unspecialized. The cochlea has none of the obvious features related to ultra- or infrasonic hearing found in cetacean ears. Basilar membrane structure and neural distributions are closer to those of pinnipeds or some land mammals than to those of cetaceans (Ketten et al. 1992). The outer osseous spiral lamina is small or absent, and the basilar membrane has a small base-to-apex gradient. At the thickest basal point, the membrane is approximately 150 μm wide and 7 μm thick; apically it is 600 μm by 5 μm. Therefore, the manatee has a relatively small basilar membrane gradient compared to cetaceans, which is consistent with the audiometric profile and 7-octave hearing range recently reported for the West Indian manatee (Gerstein et al. 1993). Spiral ganglion cell densities (500/mm) are low compared to those of odontocetes, but auditory ganglion cell sizes (20 μm × 10 μm) are larger than those of many land mammals.

The intracranial position of the periotic has important implications for sound localization abilities in manatees. Depending on the sound channels used, the manatee IATD could range from a minimum of 58 μsec (intercochlear distances) to a maximum of 258 μsec (external intermeatal path). If manatees fit the conventional mammalian IATD frequency regression, the potential IATDs imply that the West Indian manatee should have an upper frequency limit of at least 50 kHz, similar to that of a smaller odontocete. To date, there is no indication that any species of manatee has acute ultrasonic hearing (Schevill and Watkins 1965; Bullock et al. 1980, 1982; Klishin et al. 1990; Popov and Supin 1990a). Intensity differences from head shadow could provide some directional cues, but the available anatomical data (Ketten et al. 1992) suggest that manatees may have difficulty using phase cues for sound localization and have poorer localization ability than other marine mammals.

Localization ability is of particular interest in Florida manatees because of the number injured annually from collisions with boats in shallow coastal waters and canals. Since manatee salvage programs began throughout Florida in 1971, more than 2400 dead manatees have been recovered (D. Odell, pers. comm.). Human activities may account, directly or indirectly, for one-third to one-half of all deaths during the past 15 years. Even more important, deaths from collisions doubled in the past decade. In 1991, more than 30% of all deaths were associated with boats (Ketten et al. 1992). Although the sound spectra for most outboard motors are well within what we believe to be the hearing range of Florida manatees, the ability to determine a boat's direction depends on the ability to localize engine noise. Therefore, the collision hazard could be related more to localization abilities than to hearing range or sensitivity per se.

Is the conclusion that sirenians have less localization ability than cetaceans tenable? Did environmental pressures that influenced the evolution of sirenia differ in some important way from those of their terrestrial counterparts and from other marine mammals that are consistent with proposed differences in hearing? The tympanoperiotic complexes of extinct Sirenia are remarkably similar to those of modern West Indian manatee. The structural commonalities between fossil and modern manatee ears imply that few functional changes have occurred in the sirenian auditory periphery since the Eocene. Sirenians first appear in the fossil record in the early Eocene and were already adapted to at least a partially aquatic lifestyle by the early to middle Eocene (Barnes et al. 1985). Their closest affinities are with proboscideans and tethytheres, and, like cetaceans, they arose from unknown ungulate stock. A radiation in the late Eocene led to modern dugongids and trichechids. The West Indian manatee is a docile, slow grazer that lives in fresh or estuarine environments. The Florida manatee averages 400 kg and 3 m, but can exceed 4 m in length (Ridgway 1972, Odell et al. 1981). They cruise at an average speed of 3 km/hr and seldom attain speeds more than 20 km/hr even when alarmed (Reynolds 1981). For most mammals, sounds related to predators, mates, and food sources are important selection pressures. With the exception of Steller's sea cow (*Hydrodamalis gigas*), sirenians have been tropical animals, feeding on marine angiosperms or sea grasses, since the Eocene (Domning 1977, 1981; Eva 1980; Domning et al. 1982). Trichechids therefore developed in a relatively stable environment with few predators, and it is unlikely they use sound to detect food. If acoustic cues carried little or no survival advantage, acoustic developments in ancestral species may have been lost. The limited data on manatees are intriguing because of the diversity of the results. Relatively little is known about masking or localization in any sirenian, and we have no direct data for the dugong. Given their size and salt water habitat, it is likely that dugongs evolved somewhat different sensory abilities from trichechids.

If hearing is not highly evolved in manatees, are there other candidates for a premier sensory system? A highly developed chemosensory apparatus would be useful for a shallow water herbivore, and the dimensions of chorda tympani hint that gustation could be an important sense in sirenians. On the other hand, peripheral targets for chorda tympani in manatees have not been determined, and innervation patterns in marine mammals do not always match those of terrestrial species. Exceptional vibrissal discrimination has been demonstrated for pinnipeds (Dehnhardt 1990, 1994; Kastelein 1991) and suggests similar investigations with sirenians would be worthwhile.

Fissipeds and Ursids

These groups are linked by a lack of data. Remarkably little is known about sea otter hearing even in comparison to the sirenians, and data on polar bear hearing or ear adaptations appear to be nonexistent.

The sea otter has a well-defined external ear flap and a canal that is open at the surface. Kenyon (1981) indicated that the pinnal flange folds downward on dives, which suggests that the canal is at least passively closed during dives, but there are no data on whether specialized valves are associated with the ear canal like those found in pinnipeds. Otter auditory bullae are attached to the skull and resemble those of pinnipeds. CT scans of sea otters (D. R. Ketten, unpubl.)

show that their middle and inner ears are grossly configured like ears of similarly sized terrestrial carnivores, with the same orientation and 2.5 turn distribution. Spector (1956) and Gunn (1988) both indicated an upper frequency limit of 35 kHz for common river otters, which have similar ear anatomy.

It is reasonable to assume that in-air hearing is well-developed in polar bears but there are no published audiograms for air or underwater hearing. Considering their size, polar bears should have moderately good low frequency hearing, but because they are also aquatic, they may have evolved high frequency ear adaptations in parallel with other marine mammals. Anatomically, polar bear ears are interesting because even casual observations show that polar bears have a well-developed, moderately stiff pinnal flange resembling those of other ursids; however, because they dive in cold waters, it is likely polar bears have special mechanisms for preserving the integrity of the ear canal and middle ear. Hearing in this group clearly warrants further research, particularly because they represent an unknown level of auditory adaptation for underwater hearing.

Vision

The characteristics of marine mammal vision can be divided into two major functional categories: visual detection and visual acuity. The first relates to adaptations that maximize the opportunity for marine mammals to detect either predators or prey in an environment that filters and attenuates light very differently than does air. After detection of an object, an animal needs to form a sufficiently clear image of the object to determine what action to take: ignore, flee, capture, etc. Because there are fewer studies of vision than of audition in marine mammals, and because many of the known adaptations are similar, cetaceans and pinnipeds will be considered together in the following discussion. We begin with a consideration of light in water.

Light in the Ocean

As light passes through the water column, it is absorbed, refracted, and scattered. These effects are different depending on the wavelength of the light, the concentration of chlorophyll in the water, and the concentration and type of dissolved organic matter. Based on the differences in transmission of light in different waters, Jerlov (1976) proposed a classification scheme (Fig. 4-9) that shows the wavelength dependence for the transmission of light in different types of oceanic waters. In more coastal waters, light of longer wavelength is transmitted better, whereas in the open ocean, shorter wavelength blue light is transmitted best.

In general, the photon flux density, or downward irradiance, of the light falls off as an exponential function of the depth,

$$I_d = I_s e^{-kd}, \qquad \text{(equation 8)}$$

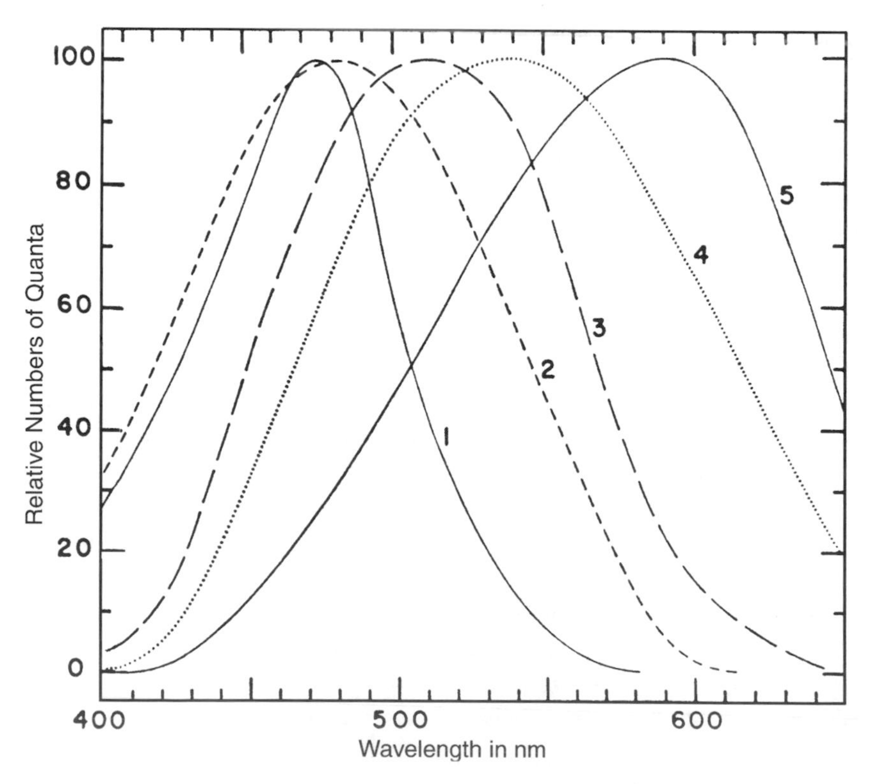

Figure 4-9. Relative spectral distribution of sunlight remaining in different marine waters, after the number of quanta at the respective maxima is reduced to 20% to 25% of the maximum number incident at the surface (wavelength = 600 nm). Curves rescaled to 100 for comparison. Curve 1 represents the spectral distribution of quanta in the clearest oceanic water at 70 m depth; curve 2, average oceanic water at 22 m; curve 3, clearest coastal water at 12 m; curve 4, average coastal water at 5 m; and curve 5, average inshore water at 3 m. (Reprinted from Munz 1965 ©Little Brown and Company.)

where I_d is the downward irradiance at depth d, I_s is the irradiance at the surface, and k is the attenuation coefficient. Light of a wavelength around 475 nm penetrates the best through clear oceanic waters (Fig. 4-9, curve 1). For light of this wavelength in typical oceanic waters (Jerlov type IB, Jerlov 1976), k is approximately 0.033 and the irradiance falls by 90% for approximately every 70 m of depth. In bright daylight, the surface irradiance for photons of 475 nm wavelength is about $1.6 \times 10^{14}/\text{cm}^2/\text{sec}$ or $0.67\ \text{W/m}^2$ (McFarland and Munz 1975).

Visual Detection

With the rapid attenuation of light with depth, how deep can marine mammals use vision to detect predators or prey? Lavigne and Ronald (1972) found that the minimum light level needed for a dark-adapted harp seal to distinguish between light and dark was $1.35 \times 10^{-9}\ \text{W/m}^2$ at light of a wavelength of 475 nm. The bright day surface irradiance decreases to $10^{-9}\ \text{W/m}^2$ at approximately 615 m. Hence, this is the likely depth limit for visual detection in this species. Wartzok (1979) reported a somewhat greater depth value for spotted seals (*Phoca largha*) (670 m). The irradiance of the night sky with a full moon is about six orders of magnitude less than that of bright sunlight. Hence, on a bright moonlight night, irradiance decreases to $10^{-9}\ \text{W/m}^2$ at about 200 m. Thus, the seal that could detect light at 615 m during the day would have to move about 415 m closer to the surface to do equally well on a moonlit night. We do not know the minimum light level actually required to detect prey because the experiments only measured the limits of light detectability. Light levels for prey detection are likely somewhat greater than these minimum levels. Assuming they are 100 times greater, or $10^{-7}\ \text{W/m}^2$, the comparative depths for prey detection under sunlight and moonlight become 476 and 61 m, respectively. The difference in depths remains 415 m because the ratio of intensities in equation 6 at functional depth in sunlight and functional depth in moonlight remains constant.

So far we have been considering only the downwelling light, that is, light that would be sensed by a detector looking straight up. For predator or prey detection, the light coming at other angles is equally important. Figure 4-10 shows the relative intensity of light at other angles. Regardless of water type, the light scattered back toward the surface is only about 1/100 that of the light downwelling. This gives an indication of the significant advantage a predator has when approaching another animal from below. A predator looking up sees the prey as a shadow blocking light that is about two orders of magnitude more intense than the light against which the prey can detect a predator below. Typical countershading of animals minimizes this difference in light intensity. A dark dorsal surface helps the animal reflect no more light toward the surface than upwells naturally from the open ocean at that depth. A pale ventral surface helps camouflage an animal against the brighter light downwelling from above, although the difference in downwelling and upwelling light is so great that an animal which is not bioluminescent cannot match the intensity of the downwelling light and will always appear as a darker object against a lighter background. Many marine predators, such as seals and sea lions (Hobson 1966), typically attack prey from below for this reason.

To maximize the detection of prey at depth, a predator attacking from below should adopt several sensory strategies, including matching receptor pigment sensitivity to downwelling light, increasing density of photoreceptors, and enhancing photon capture by an elaborated tapetum. All of

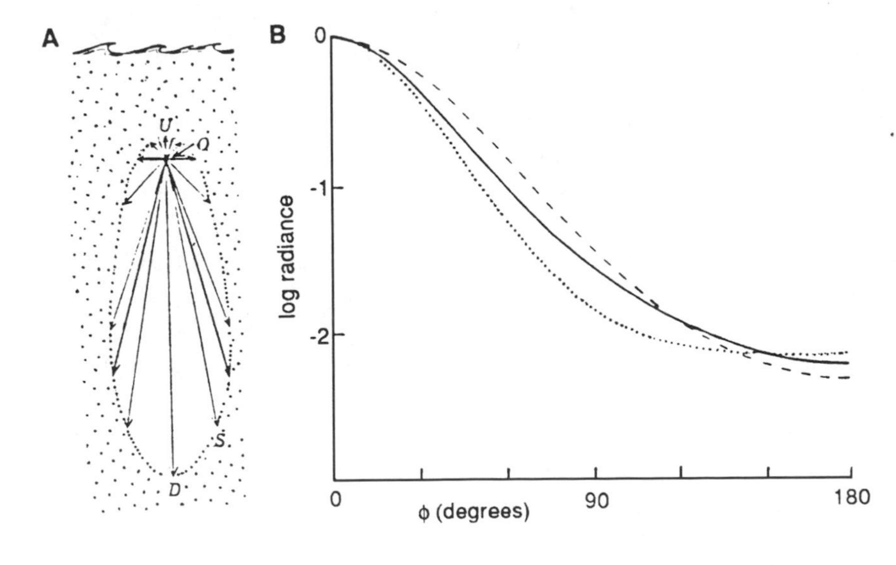

Figure 4-10. (A) Diagram of distribution of radiance found in the sea. For a given point, O, the distribution in three dimensions is given by the surface formed by revolving the dotted line S around the axis UD. The relative radiance in a given direction is the length of the line in that direction joining the point O to a point on the surface S. (B) The log radiance of light is plotted against ϕ, the angle between a given direction and the downward vertical axis. The log radiance for $\phi = 0$ has been made equal to zero. (Solid line) Lake Pend Oreille (after Tyler 1960); (dashed line) from an equation given by Tyler (see Denton et al. 1985); (dotted line) for oceanic Mediterranean waters (after Lundgren 1976, cited by Jerlov 1976). (Reprinted from Denton 1990 ©Cambridge University Press.)

these strategies will be discussed. The photoreceptors should have their greatest visual sensitivity at the same wavelength as the light that is best transmitted through the type of ocean waters in which they typically feed (Munz 1965). For both pinnipeds (Lavigne and Ronald 1975b) and cetaceans (McFarland 1971), pigments in the rod receptors (the retinal receptors specializing in low light level sensitivity) of species that feed in the open ocean have a maximum sensitivity at the blue end of the spectrum, matching the wavelengths of light that best penetrate in the open ocean.

Detection of a visual signal depends on the number of photons captured. Marine mammals maximize their visual sensitivity by having a high density of photoreceptors and a well-developed tapetum. The California sea lion has receptor densities of 200,000 to 260,000 per mm^2 (Landau and Dawson 1970) and the bottlenose dolphin has 400,000 per mm^2 (Dral 1977), compared to 120,000 to 160,000 in humans (Østerberg 1935) and 800,000 in a nocturnal fish, the lota (Walls 1942).

The tapetum, a reflective layer behind the retina, reflects the photons that were not absorbed by the visual pigment as the light initially passed through the visual receptor layer. This reflection gives the visual pigment a second chance to capture a photon and thus increases the probability of detection. Marine mammals have the most developed tapeta of any mammal (Walls 1942, Dawson 1980). The tapeta are thicker, up to 35 layers in the gray seal (*Halichoerus grypus*) (Braekevelt 1986), and back the entire fundus (Nagy and Ronald 1970, Dawson et al. 1972, Braekevelt 1986). Even at the periphery of the tapetum, the average number of cell layers, 15 to 20, is as many as at the center of the tapetum in terrestrial nocturnal predators such as the cat (Pedler 1963). Dawson et al. (1987a) noted that the tapetum of a Risso's dolphin (*Grampus griseus*) (a pelagic species) reflected more blue light than that of a bottlenose dolphin (a coastal species). This difference is consistent with the sensitivity hypothesis for detection, but Dawson et al. (1987a) cautioned against building too much of a case on the basis of one Risso's dolphin specimen. All 11 bottlenose dolphins observed were very similar to each other in spectral light reflected by the tapetum.

In shallower depths, down to approximately 100 m in the open ocean, the downwelling light still has a relatively broad range of wavelengths. In this depth range, a predator can use a second method to detect prey, the contrast method (Lythgoe 1968). For contrast detection, the pigment in the eye needs to have its best sensitivity offset somewhat from the wavelength that best penetrates the ocean. To understand why we need to remember that the wavelength-dependent absorption as light passes through oceanic water will be the same for all light paths. This means that in the first 100 m of the ocean, downwelling light has a range of wavelengths consistent with a 100-m path but an animal looking horizontally would see light that had passed through thousands of meters of ocean, and thus would see light restricted to the narrow band of wavelengths that are best transmitted. A silvery-sided fish that reflected the horizontal light perfectly would be invisible to a predator looking horizontally. However, if the fish also reflected some of the downwelling light, it would be reflecting light of a different spectrum containing longer wavelengths than the horizontal light a predator would see on all sides of the fish. If the predator had a visual pigment with greater sensitivity to longer wavelengths of light, then the difference in the light spectrum reflected from the fish and the general horizontal background light would be detected easily. There are several lines of evidence (discussed below) that show some marine mammals have more than one type of photopigment, and thus can make use of both sensitivity (when attacking prey from below) and contrast (when attacking prey from the side).

The vertebrate photoreceptors that function best at low light levels are the rod photoreceptors. Rod pigments are most accurately measured when they are extracted from a dark-adapted eye preserved immediately after the death of the animal. Such ideal conditions cannot be met for most marine mammals, so the values for extracted pigments are less accurate than for most terrestrial mammals. In addition, many of the measurements are based on only a single specimen. Given these caveats, the current data show a basic correlation between what we know about the diving habits of some cetacean species and the peak sensitivity of the photopigment in their rods. The deep-diving Baird's beaked whale (*Berardius bairdii*) has a pigment maximum at 481 nm, whereas the gray whale (*Eschrichtius robustus*) (a coastal species) has a pigment maximum at 497 nm (McFarland 1971). Recall that shorter wavelengths are transmitted best and reach greater depths. The humpback whale (*Megaptera novaeangliae*), with a typical diving depth intermediate between the beaked whale and the gray whale, has a pigment maximum at 492 nm (Dartnell 1962). The Weddell seal (*Leptonychotes weddelli*) has a pigment maximum at 495 to 496 nm, similar to coastal fishes and terrestrial mammals, but the deep diving southern elephant seal (*Mirounga leonina*) (Slip et al. 1994) has a pigment maximum at 485 to 486 nm (Lythgoe and Dartnall 1970). The elephant seal pigment maximum could be the result of a shift either to the wavelengths best penetrating the depths, or to wavelengths emitted by bioluminescent prey (Young and Mencher 1980). Absorption curves from these whales and seals suggest that there is only one pigment. However, most of these studies are old and did not use sensitive tests presently available. The nocturnal New World owl monkey (*Aotus trivirgatus*) and nocturnal prosimian bush-baby (*Galago garnetti*) used to be considered species with all rod retinas (Walls 1942), but modern monoclonal antibody techniques have identified cones in the retinas of both species (Wikler and Rakic 1990).

A variety of anatomical and behavioral evidence suggests that odontocetes, phocids, otariids, and sirenians have more than one photopigment. Anatomically, both rods and cones have been described in the retinas of bottlenose dolphins (Perez et al. 1972, Dawson and Perez 1973), Dall's porpoise (*Phocoenoides dalli*) (Murayama et al. 1992), harbor porpoise (Kastelein et al. 1990), sperm whales (*Physeter macrocephalus*) (Mann 1946), fin whales (Mann 1946), minke whales (*Balaenoptera acutorostrata*) (Murayama et al. 1992), harbor seals (Jamieson and Fisher 1971), harp seals (Nagy and Ronald 1975), and West Indian manatees (Cohen et al. 1982). Behavioral studies of critical flicker fusion frequency (the fastest flicker rates that can be detected as a flicker rather than as continuous light), indicate that the faster responding cone systems exist in harp seals (Bernholtz and Mathews 1975) and bottlenose dolphins (van Esch and de Wolf 1979). Lavigne and Ronald's (1975a) reanalysis of visual acuity studies in California sea lions conducted by Schusterman and Balliet (1971) and Schusterman (1972) indicate the presence of both rod and cone visual systems. Finally, the demonstration of color discrimination in spotted seals (Wartzok and McCormick 1978) also indicates the presence of more than one visual pigment. The difficulty with which color discrimination was demonstrated in spotted seals (Wartzok and Mc Cormick 1978) and the inability to demonstrate it in bottlenose dolphins (Simons and Huigen 1977, Madsen and Herman 1980) suggest that these marine mammals are using their dual pigment systems primarily for brightness detection and discrimination and did not evolve them mainly for color vision.

Multiple pigment visual systems in marine mammals probably serve three functions: (1) to maximize sensitivity when at depth and approaching prey from below; (2) to maximize contrast when at shallower depths and approaching prey from the side; and (3) to allow vision in up to nine orders of magnitude of changes in light intensity. A rod system sensitive to the lowest levels of light at depth would be bleached out by the much higher intensities at or near the surface. The combination of a rod system for vision at depth and a cone system for vision near the surface, and on land in the case of pinnipeds, allows marine mammals to function visually at almost all light levels and depths. One open question is what adaptations have been made in the rod pigment to allow it to regenerate quickly enough to reach maximum sensitivity as the marine mammal dives to depth. Diving marine mammals often descend at rates of greater than 1 m/sec (Kooyman and Gentry 1986, Martin et al. 1993, Watkins et al. 1993, Asaga et al. 1994), which translates into decreases in light intensity of about one order of magnitude, or one log unit, per min. Initial dark adaptation has only been measured in humans (Hecht and Hsia 1945) where it occurs at a rate of about 0.5 log/min. Clearly the biochemical processes of rod pigment regeneration need to be modified in marine mammals if they are to use the full potential sensitivity of their visual system. Levenson and Schusterman (1998) presented preliminary evidence that northern elephant seals dark adapt completely in 10 minutes. Given their sensitivity range, this suggests that they are adapting at a rate approaching 1 log/min, and thus operating with near maximum sensitivity at most depths.

There is good evidence that freely diving Weddell seals detect novel holes in the ice at ranges where the light from the hole just exceeds their visual sensitivity threshold. A prediction of maximum sensitivity was made from laboratory results with harbor and spotted seals (Wartzok 1979). In spite of the extrapolations from laboratory data to the field situation and from harbor and spotted seals to Weddell seals, the detection distances correspond to those predicted based on the maximum sensitivity at typical states of adaptation, and on the extinction coefficients in the water column (Wartzok et al. 1992).

The fact that the pupils of marine mammals are usually constricted when they are in bright light at the surface is one factor that helps prevent a total bleaching of the rod pigment. The constricted pupil not only helps preserve a partial state of dark adaptation, it is also essential for visual acuity in air.

Visual Acuity

Visual acuity is a measure of how well an animal can resolve features in its visual environment. For example, it can be measured by determining how far apart two points of light must be for the subject to detect that there really are two lights, rather than one; or how broad alternating stripes of black and white must be for the subject to recognize them as stripes, rather than a uniform gray. Because these separations of lights or widths of stripes varies with distance from the subject, visual acuity is usually measured in terms of degrees or minutes of arc. As a point of reference, the diameter of the full moon is about one-half degree, or 30 min of arc. Visual acuity is dependent on the focusing ability of the optics of the eye, the density of receptors in the retina, the connections between the receptors, and the subsequent processing of the signal in the central nervous system.

The extent to which light rays are refracted when they pass through the interface between different media is dependent on the difference in refractive indices of the media and the angle of the light with respect to the interface between the media. If the light ray is perpendicular to the interface, the ray is not refracted. The strength of a homogeneous lens in air, therefore, is dependent on its refractive index and its radius of curvature. A light ray entering a terrestrial eye experiences its greatest difference in refractive index at the air–cornea interface (Fig. 4-11). Depending on the radius of curvature, this can produce a lens with a strength of 25 to 40 diopters. Diopters are a measure of the refractive power of a lens given as the reciprocal of the focal length in meters. In humans, focusing at the air–cornea interface accounts for about two-thirds of the total focusing power of the visual system. The actual lens within the eye provides only about one-third of the focusing power. When terrestrial mammals returned to the sea, they had to compensate for the loss of a major focusing element in their visual systems. The reason they lost the cornea interface lens power is that the refractive index of water is similar to that of the interior of the eye, therefore there is little change in refractive index as light passes from the water into the eye through the thin cornea. To see well underwater, marine mammals have developed much stronger lenses. Refractive measurements of the eyes of several species have shown that they are approximately emmetropic, neither near- nor far-sighted, for underwater vision (bottlenose dolphin: Dawson et al. 1972, Dral 1972; harbor seal: Johnson 1893; harp seal: Piggins 1970; Weddell seal: Wilson 1970).

The lenses in marine mammal eyes are more similar to those in fish eyes than they are to the lenses in terrestrial mammalian eyes. In cetaceans, the lenses are spherical in shape and the ciliary muscles that change the shape of the lens for accommodation, or focusing, appear to be nonexistent (Dral 1972). In pinnipeds, the lenses have a shape intermediate between those of cetaceans and terrestrial mammals, and ciliary muscles are attached to the sclera, an arrangement that has been postulated to allow for a strong pull on the thick lens (Jamieson and Fisher 1974).

The strong lens that compensates for the loss of the air–cornea interface when the animal is underwater leads to severe myopia when the animal is out of the water and regains the air–cornea lens. Exactly how marine mammals overcome myopia in air is not well understood. One way is through extensive pupillary contraction to a slit that can close further to form a pin hole. The slit is oriented vertically in the majority of pinnipeds, and horizontally in cetaceans, and in bottlenose dolphins it takes the shape of a horizontal crescent that closes to two slits at the ends of the crescent (a double slit pupil) (Herman et al. 1975, Dawson et al. 1979). A pin hole camera approaches an infinite depth of field so an eye with a pin hole pupil will be emmetropic and an eye with a slit pupil could provide focused images over a range of distances.

The constricted pupil works well for animals in bright light, but as the light intensity decreases and the pupil opens, visual acuity declines. Schusterman and Balliet (1971) compared visual acuity in California sea lions above and below water as the illumination decreased (Fig. 4-12), and demonstrated that performance degraded much more rapidly in air than underwater. The initial good visual acuity in air is also attributable to an apparently unique feature of the cornea of California sea lions (although no other otariids have been carefully studied). In the center of the cornea there is a 6.6-mm diameter circular flattened area (Dawson et al. 1987b). Because this portion of the cornea has no curvature, it does not act as an additional lens in the air, and the sea lion has equivalent aerial and underwater vision as long as the pupil does not dilate beyond this diameter.

Although the above considerations certainly contribute to functional vision in air, a few problems are not fully ex-

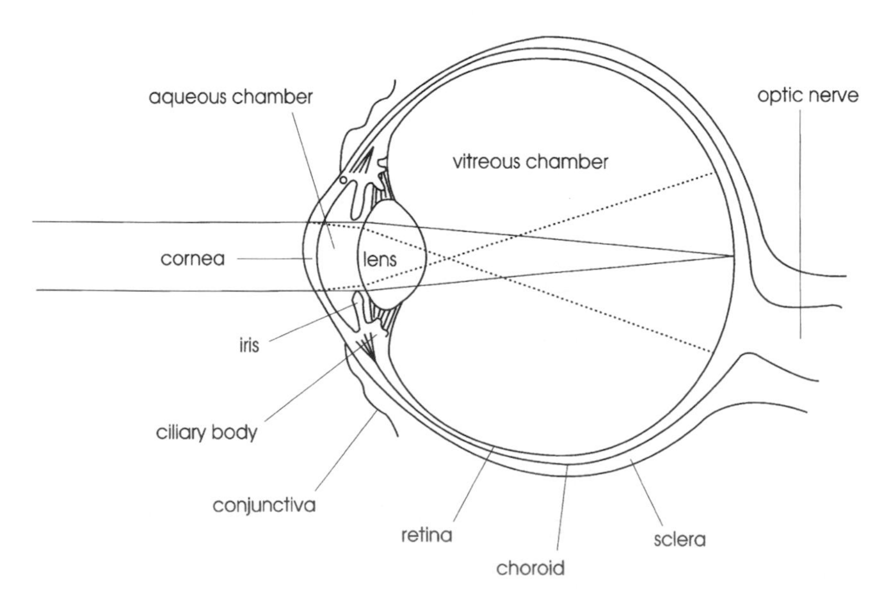

Figure 4-11. Ray drawings of the focusing of a seal's eye in air and underwater. Underwater (solid line rays) there is no refraction at the cornea. The only refraction occurs at the lens. When in air, there is bending of the light at both the cornea and at the lens leading to focusing well in front of the retina (i.e., myopia). The bending at the lens is exaggerated to clearly show the effect.

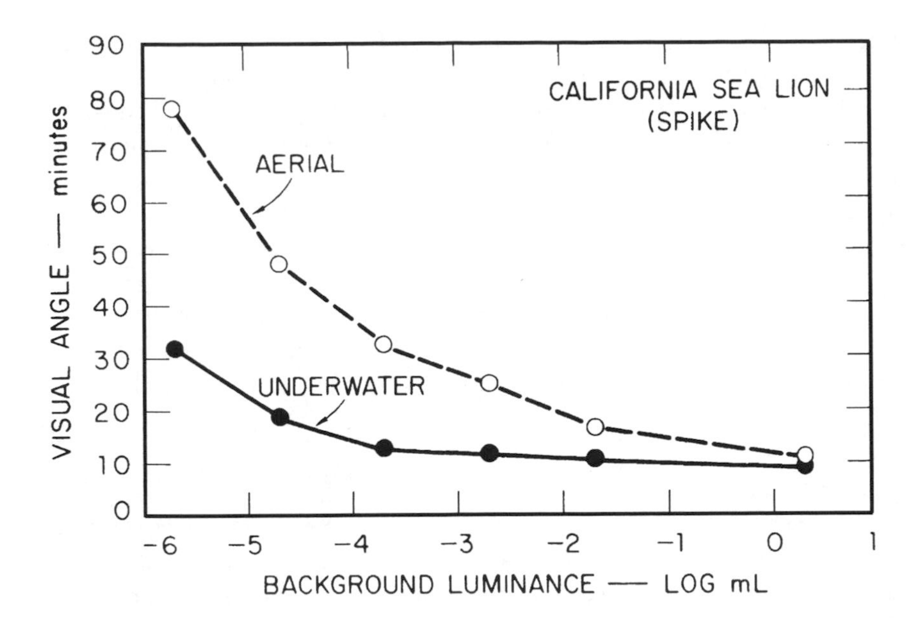

Figure 4-12. Aerial and underwater visual acuity thresholds as a function of luminance. (Reprinted from Schusterman and Balliet 1971, ©New York Academy of Sciences.)

plained. First, Dawson et al. (1979) showed that the time course of pupillary constriction is too slow to allow a bottlenose dolphin to constrict its pupil between the time its head leaves the water on a leap and the time at which it has to make the final trajectory correction before its flukes leave the water. Nevertheless, dolphins can leap accurately to targets placed at variable locations above the surface. Rivamonte (1976) argued that the double slit pupil becomes diffraction limited before the pupil has constricted sufficiently to achieve the acuities reported in air (Herman et al. 1975), and he proposed that the dolphin has a lens that overcorrects for spherical aberration by having the central core of the lens surrounded by a region of lower refractive index. Light passing through the periphery of the lens (i.e., coming from the peripheral field of view), would need the additional strength of the air–cornea interface to focus, and thus would be out of focus underwater, whereas light passing through the lens on axis would be in focus underwater but out of focus in air. There is behavioral evidence that dolphins preferentially use peripheral areas of the retina for forming images in air (Dawson et al. 1972, Pepper and Simmons 1973, Dral 1975).

Two areas of high ganglion cell density are found in the retinas of the following species: bottlenose dolphin (Dral 1975, Mass and Supin 1995); the common dolphin (*Delphinus delphis*) (Dral 1983); the finless porpoise (*Neophocaena phocaenoides*), and the Chinese river dolphin (*Lipotes vexillifer*) (Gao and Zhou 1987); harbor porpoise and gray whale (Mass and Supin 1990, 1997); and Dall's porpoise and minke whales (Murayama et al. 1992). A rostral high-density area could improve vision to the side, whereas a dorsotemporal high-density area could improve frontal vision. However, the peak densities of ganglion cells (500 to 670/mm² in bottlenose dolphins, Dral 1977, Mass and Supin 1995) are well below the peak densities in rabbits (10,000/mm², Rodieck 1973) and cats (4,000/mm², Stone 1965). Mass and Supin (1995) calculate that the cell densities in bottlenose dolphins would limit the retinal resolution to about 9 min of arc underwater and 12 min of arc in air.

Kröger and Kirschfeld (1994) showed that the cornea, rather than being a thin membrane that does not add an optical element to the visual pathway, can be a negative lens for images formed in the periphery of the visual field. Kröger and Kirschfeld (1992, 1994) measured the variable thickness and refractive index of the cornea in harbor porpoises, and argued that the negative corneal lens was required to compensate for a too powerful positive lens in the eye of the porpoise to produce emmetropia for underwater viewing. The same arrangement could help alleviate myopia in air.

Visual acuities of cetaceans and pinnipeds fall in the range of 5 to 9 min of arc both for underwater vision and for in-air vision at high light intensities (Schusterman and Balliet 1970a,b, 1971; White et al. 1971; Herman et al. 1975). These values are similar to those reported for terrestrial hunters such as the cat (Muir and Mitchell 1973).

Sirenians

Sirenians have been reported to have poor vision (Walls 1942), based on early studies of manatees (e.g., Chapman 1875) and dugongs (e.g., Petit and Rochon-Duvigneaud 1929). These studies suggested poor visual acuity both in air and underwater with probably an exclusively rod retina, but no tapetum. On the basis of observations of manatee visual behavior, Hartman (1979) suggested they were not as visually deficient as the older literature indicated. More recent studies (Cohen et al. 1982) indicated that the manatee has both rods and cones in its retina with a central visual area where visual acuity may be enhanced. Although there is little refractive error in the optics of the eye underwater (Piggins et al. 1983), acuity is still considered to be only moderate. The rod-dominated retina, with extensive summing of receptors to ganglion cells, indicates that acuity has been sacrificed for vision at low light levels in murky water, but the lack of a tapetum indicates a less than full adaptation to functioning at these light levels.

Ursids

Polar bears show little adaptation for underwater vision. With the loss of the air–cornea interface, the polar bear loses more than 20 diopters of lens strength underwater (Sivak and Piggins 1975). The peak sensitivity of the dark-adapted eye is at 525 nm, even a bit longer wavelength than typical of terrestrial predators such as the cat. The polar bear has both rods and cones in its retina as indicated by separate scotopic (i.e., dark adapted) and photopic (i.e., light adapted) functions (Ronald and Lee 1981).

Fissipeds

The sea otter has the most spectacular accommodative range of any vertebrate. It is able to see well in air and in water by a unique method of changing the radius of curvature, and hence the strength, of its lens (Murphy et al. 1990). As shown in Figure 4-13, when the sea otter is underwater, the loss of the approximately 60 diopter corneal–air interface lens is compensated for by the action of the ciliary and iris muscles. The ciliary muscle opens a path for fluid to flow from the anterior chamber into the corneoscleral venous plexus. This reduces the pressure in the anterior chamber and causes the lens to be pushed forward by the pressure differential between the posterior and anterior chambers. As the lens is pushed forward, the iris sphincter forces it to assume a new shape with a much smaller radius of curvature, increasing the strength of the lens by 60 diopters. Although the iris does not close to a slit or pin hole, as in dolphins and seals, because the accommodative mechanism is dependent on the iris muscle, there is the same trade off between sensitivity in low light and visual acuity. Although no behavioral studies have been done on sea otter acuity and light levels, Schusterman and Barrett (1973) showed that an Asian "clawless" otter (*Amblonyx cineria cineria*) lost acuity underwater more rapidly than in air when light levels decreased.

Chemoreception

Both air and water are fluids; therefore, the underlying principles of the spread of chemicals through diffusion and convection in laminar and turbulent flow are the same. However, diffusion is approximately 10,000 times slower in water than in air and average water currents are approximately 15 times slower than average air currents (Vogel 1981). Thus, some aspects of chemical communication in water are far less efficient than chemical communication in air.

Smell and taste are closely aligned senses for detection of substances in air and in liquids, respectively. Neither sensory modality has been investigated with much rigor in marine mammals. Olfactory sensation appears to decline in parallel with adaptation to the marine environment. Gustatory sensation is present in dolphins, and presumably in other marine mammals, but there is a paucity of experimental evidence. The difficulty of controlling, measuring, and presenting chemical stimuli, as well as the limited availability of marine mammals for experiments, results in very slow progress in this area. Lowell and Flanigan (1980), Watkins and Wartzok

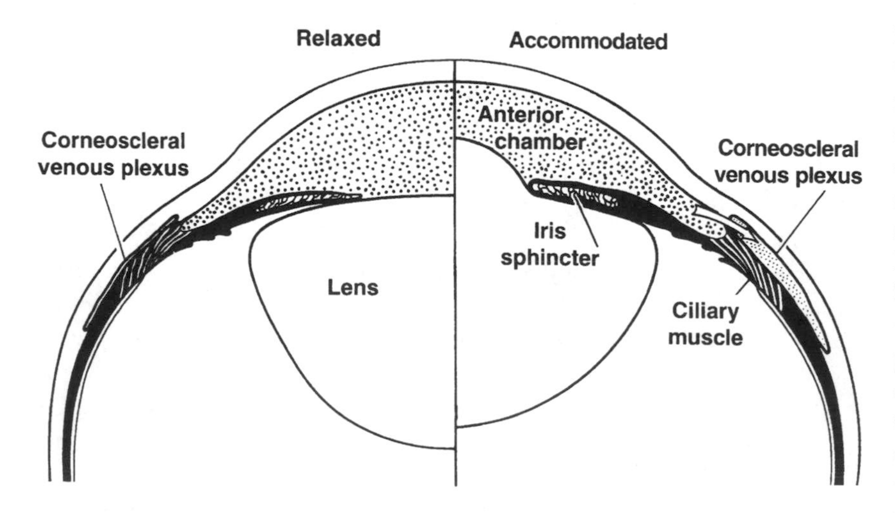

Figure 4-13. Proposed mechanism of visual accommodation for underwater vision in the sea otter. In this model, the ciliary and iridal muscles act to decrease markedly the radius of curvature of the anterior lens surface, thus making a stronger lens to compensate for the loss of the air–corneal interface lens. Contraction of the ciliary muscle will bring about a dilation of the corneoscleral venous plexus, lowering its hydrostatic pressure and causing redistribution of aqueous humor from the anterior chamber into the plexus (arrow). This pressure will bring about an anteriorward movement of the lens which is then deformed by action of the iris sphincter muscle. (Reprinted from Murphy et al. 1990, ©Elsevier Science Inc.)

(1985), and Nachtigall (1986) have reviewed basically the same small set of experimental results.

Olfaction

Anatomically, the olfactory system of mysticetes, odontocetes, and sirenians is less developed than those of terrestrial mammals. Odontocetes lack olfactory bulbs and attendant ganglia or fiber tracts (Breathnach 1960, Pilleri and Gihr 1970, Morgane and Jacobs 1972). Manatees have a very rudimentary olfactory system, and, like cetaceans, lack an important component of the olfactory system, a vomeronasal organ (Mackay-Sim et al. 1985). Pinnipeds have both peripheral (Kuzin and Sobolevsky 1976) and central (Harrison and Kooyman 1968) olfactory structures, although in phocids and walrus they are somewhat reduced compared to those of otariids (Harrison and Kooyman 1968). Both otariids and phocids show behavioral evidence of olfactory sensation. Among many examples that could be cited, males sniff the hindquarters of females to assess the state of estrus in northern fur seals (*Callorhinus ursinus*) (Bartholomew 1953) and South African fur seals (*Arctocephalus pusillus*) (Rand 1955). In southern elephant seals, experienced females begin smelling a newborn pup sooner after birth than do primiparous females (McCann 1982), and learn the identity of their pups sooner.

Gustation

Neuroanatomy (Jacobs et al. 1971) and tongue morphology (Sukhovskaya 1972, Donaldson 1977, Yamasaki et al. 1978) indicate that odontocetes have the ability to taste. Jacobs et al. (1971) suggested that the well-developed olfactory lobes in the rhinencephalon of bottlenose dolphins might be innervated by the trigeminal nerve, in place of the missing olfactory nerve, and thus might be sensitive to substances stimulating receptors on the tongue. Taste buds located in grooves or pits at the root of the tongue have been described in bottlenose dolphins (Sukhovskaya 1972, Donaldson 1977), common dolphins (Sukhovskaya 1972), and the striped dolphin (*Stenella coeruleoalba*) (Yamasaki et al. 1978, Komatsu and Yamasaki 1980). In manatees, taste buds are located in dorsal and posterior lateral swellings on the tongue (Yamasaki et al. 1980). In dugongs, the taste buds are located in pits rather than swellings in the same general dorsal and posterior lateral areas of the tongue (Yamasaki et al. 1980). Although the number of gustatory receptors is greater in sirenians than in cetaceans, they are still poorly endowed compared to herbivorous land mammals.

Kuznetsov and colleagues, using a number of different behavioral and physiological indicators, have reported chemoreception for a number of substances (for review, see Kuznetsov 1990). These studies showed that harbor porpoise, common dolphin, and the Black Sea bottlenose dolphin could detect a variety of chemicals with a well-developed ability to detect bitter chemicals, a less well-developed ability to detect salinity, and an apparent inability to detect sugar. None of the experiments determined detection thresholds by comparing differing response levels with different stimulus concentrations.

Nachtigall and Hall (1984) used a go/no-go operant conditioning technique with a bottlenose dolphin to establish detection thresholds for sour (citric acid), bitter (quinine sulfate), salt, and sweet (sucrose). They showed that the dolphin could detect tastes humans classify as sour and bitter at levels only slightly above the range of human detection thresholds. In contrast to the results of Kuznetsov (1979), the dolphin used by Nachtigall and Hall was able to detect sucrose, although at much higher thresholds than those of humans. Nachtigall and Hall (1984) also showed a sensitivity to salt, but these results were confounded by the animal being held in salt water. In summary, the results show that dolphins with taste receptors located only at the root of the tongue can detect chemicals assigned by humans to all four of the taste groups.

Little work has been done on taste sensation in pinnipeds. Kuznetsov (1982, cited in Friedl et al. 1990) reported that Steller sea lions (*Eumetopias jubatus*) detected sour, bitter, and salty stimuli, but were insensitive to sweet. Friedl et al. (1990) reported on experiments done with a California sea lion using the same techniques used by Nachtigall and Hall (1984). The sea lion was good at detecting citric acid, and discriminated between freshwater and seawater, but had a high threshold for bitter, and appeared not to detect sweet. As Friedl et al. (1990) noted, "much work remains for patient researchers in the area of marine mammal chemoreception."

Tactile Sensation

Cetaceans

Most of the studies of tactile sensation in cetaceans have been done in bottlenose dolphins with the objective to understand how these animals maintain laminar flow over major portions of the body at fast swimming speeds. Gray's paradox (Gray 1936) suggested that dolphin muscles would need to be seven times more powerful than they are if the dolphin experienced the turbulent flow predicted by their body shape and speed of movement through the water. Lang and coworkers (Lang 1966, Lang and Norris 1966, Lang and Pryor

1966) demonstrated that for several odontocetes drag coefficients measured during glides were indicative of turbulent flow over the entire surface of the animal. They suggested that energetically problematic high-speed swimming was limited to short bouts so that reasonable estimates of propulsive power capabilities were not violated. However, drag measured during a glide is not necessarily representative of drag during active swimming (Alexander and Goldspink 1977). If the animals are capable of actively damping turbulence, they could likewise be capable of not damping it or enhancing it based on their intention to move forward or glide to a stop. Support for some form of active damping comes from observations showing that little turbulence is generated over most of the body of a swimming Pacific white-sided dolphin. A number of investigators have speculated, based on the anatomy of the skin and the innervation of the skin, that active damping of incipient turbulence is occurring (Kramer 1965, Surkina 1971, Kayan 1974, Khomenko and Khadzhinskiy 1974, Babenko and Nikishova 1976, Ridgway and Carder 1990).

The first step in determining whether dolphins are capable of active damping is to assess their tactile sensitivity. Kolchin and Bel'kovich (1973) used galvanic skin responses to map areas of greatest sensitivity on the body of common dolphins. They found the area of highest sensitivity within 2.5 cm from the blowhole and in separate 5-cm diameter circles around the eyes. The next most sensitive area was the snout, lower jaw, and melon. The area of the back, both anterior and posterior to the dorsal fin, had a lower sensitivity. Bryden and Molyneux (1986) found large numbers of encapsulated nerve endings in the region of the blowhole in both bottlenose dolphins and false killer whales. They suggested that these mechanoreceptors were involved in detection of the pressure changes associated with the blowhole breaking the water surface. They found more receptors on the anterior lip of the blowhole, the first part of the blowhole that would become exposed when the animal surfaced. These most sensitive areas are not the most critical areas for dampening turbulence.

Using electroencephalograms to map areas of tactile sensitivity on the body of bottlenose dolphins, Ridgway and Carder (1990) expanded on early studies by Lende and Welker (1972). They found highest sensitivity at the angle of the gape, followed by the area around the eyes and melon, followed by the area around the tip of the snout and the blowhole. The least sensitive areas were on the remainder of the dolphin's trunk. Although there are some differences in most sensitive areas, the results of Ridgway and Carder (1990) and Kolchin and Bel'kovich (1973) agree that the head region is the most sensitive. The minimum detectable pressure values of 10 mg/mm^2, obtained by Kolchin and Bel'kovich (1973) using a static, weighted 0.3-mm wire, are similar to those obtained for humans in their most sensitive areas such as the fingertips and lips. Even the 40 mg/mm^2 found on the common dolphin trunk shows a high level of tactile sensitivity over the majority of the body.

Knowing a dolphin is capable of detecting pressure associated with turbulence as it swims is only part of the issue of whether a dolphin can actively damp turbulence. To determine whether there is muscle control sufficient to effect the damping, Haider and Lindsley (1964) and Ridgway and Carder (1990) studied microvibrations in bottlenose dolphins. Haider and Lindsley (1964) showed that these vibrations were at least three times greater in amplitude in

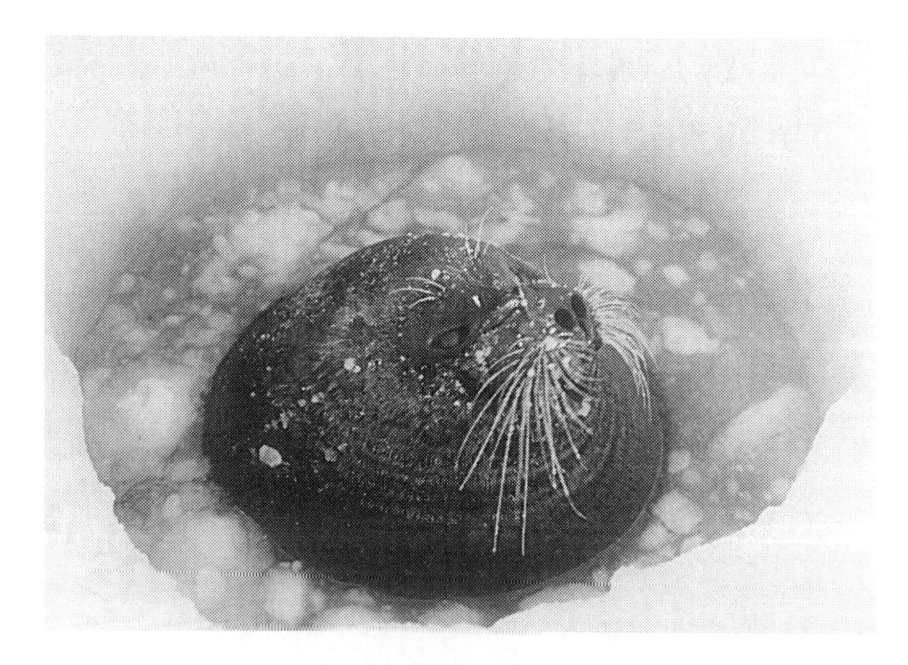

Figure 4-14. A photograph of a ringed seal surfacing through a manmade hole in the ice. The extensively developed vibrissae are obvious.

dolphins than in humans (15 μm, cf. 5 μ maximum) and somewhat faster in frequency (13 Hz, cf. 11 Hz). Ridgway and Carder (1990) showed that the dolphins responded to a vibratory stimulus by entraining their spontaneous microvibrations to the frequency of the stimulus and by increasing the normal amplitude of the microvibrations. This level of control of spontaneous microvibrations suggests a possible mechanism for damping turbulence by flexing the body surface away from regions of higher pressure and toward regions of lower pressure in the water flowing over the dolphin. Currently there is no evidence showing that dolphins actually use this mechanism.

Pinnipeds

The most obvious marine mammal tactile sensory structures are the vibrissae of pinnipeds. Not only are they obvious when viewing the animal (Fig. 4-14), they are also highly innervated peripherally and well-represented cortically with large areas devoted to input from the vibrissae. Anatomical and physiological studies have been more successful in demonstrating their importance than have behavioral studies. Dykes (1975) showed that the vibrissae of harp seals and harbor seals were heavily innervated with slowly and rapidly adapting nerve fibers. Both fiber types had a range of thresholds resulting in a sensory system that could provide detailed information about static displacement and vibratory stimuli. The sensitivity of the vibrissae is enhanced by the rich supply of blood vessels in the dermal sheath (Ling 1966), which could provide adaptive changes in the compliance, or flexibility, and by the histologically prominent circular sinuses around the vibrissae (Stephens et al. 1973, Hyvärinen 1989). In ringed seals, Hyvärinen (1989) found that each vibrissal follicle is innervated by 1000 to 1600 fibers.

Ladygina et al. (1992) mapped the body surface of northern fur seals onto the somatosensory cortex. The fur seal has a duplicate somatosensory projection in the cortex. Although a duplicate projection of the entire body has been observed in some monkeys (Merzenich et al. 1978, Nelson et al. 1980, Sur et al. 1982), it usually only occurs for those parts of the body involved in particularly fine tactile discriminations such as the forepaw of the gray squirrel (*Sciurus carolinensis*) (Sur et al. 1978, Nelson et al. 1979, Krubutzer et al. 1986), the hand of prosimian primates (Carlson and Welt 1980, Sur et al. 1980, Carlson and FitzPatrick 1982, FitzPatrick et al. 1982, Carlson et al. 1986), and the vibrissae of the opossum (*Didelphis virginiana*) (Pubols et al. 1976). Ladygina et al. (1992) calculated magnification factors (the relative distance between two cortical recording places to the distance between the two locations on the body that provided maximum stimulation of the cortical sites) and found that the factor varied from 0.01 to 0.5. On the trunk, the value was 0.01, meaning that a 1-mm shift in recording location in the cortex resulted in a 100-mm shift in the most sensitive body area. For the vibrissae, the factor was as large as 0.5, meaning that a 1-mm shift in recording location in some areas of the cortex corresponded to only a 2-mm shift on the maxillar vibrissal pad. This indicates the extensive cortical space allocated to tactile input from the vibrissae.

Dykes (1975) determined, using electrophysiological recordings from nerves innervating the vibrissae of harbor seals, that the threshold for responses to vibratory stimuli were less than 7 sec of arc. All of the rapidly adapting fibers responded to vibratory stimuli of 256 Hz, but at 1024 Hz only 15% of the fibers followed the vibratory stimulus, and at 1500 Hz, fewer than 1% of the fibers followed the stimulus. In contrast, psychophysical studies by Renouf (1979) and Mills and Renouf (1986) indicated a marked increase in sensitivity to vibratory stimuli as frequencies increased to a peak sensitivity at 1000 Hz, with little decline in sensitivity out to 2500 Hz, the highest frequency tested.

Dykes (1975) postulated that the sensitivity of the slowly adapting nerve fibers to small, but not large, displacements, and the ability of the rapidly adapting nerve fibers to detect very small vibratory stimuli meant that the vibrissae were innervated to provide information about contour and texture. Behavioral results are in line with this prediction.

Kastelein and van Gaalen (1988) showed that a walrus (*Odobenus rosmarus*), fitted with eye cups blocking vision, could use its vibrissae to distinguish between a square and a triangle even when the surface areas of the stimuli were gradually decreased to 0.4 cm^2. Dehnhardt (1990) demonstrated that a California sea lion, deprived of vision by eye cups, could distinguish among five different three-dimensional objects when allowed to touch them with its vibrissae. In a subsequent experiment (Dehnhardt 1994), this same sea lion detected diameter differences as small as 0.33 cm for discs with a mean diameter of 1.12 cm. This threshold led to a Weber's fraction (miniumum detectable diameter difference/mean diameter) of 0.29. The sea lion tactile thresholds followed Weber's law, which states that Weber's fraction should be constant (i.e., ΔD will increase as D increases). Weber's fraction was also 0.29 for discs with a mean size of 8.74 cm. Studies on two harbor seals showed similar values for Weber's fraction at disc sizes of 1.12 cm, but values as low of 0.13 (male) and 0.08 (female) for discs of 5.04 and 8.74 cm (Dehnhardt and Kaminski 1995). Thus, the best tactile size discrimination threshold for the California sea lion was about twice as great as it was for harbor seals. The small number of individuals tested and the individual variability observed preclude definitive species rankings on tactile ability. However, it is interesting to note that the Weber's fractions attained by the harbor seals are similar to those reported for the hands of lower primates such as macaque monkeys (Semmes and Porter 1972, Carlson et al. 1989). Also the Weber fractions for tactile size discrimination by sea lions and harbor seals are similar to those obtained for visual size discriminations by these and other marine mammal species (California sea lion, Schusterman et al. 1965; harbor seal, Feinsein and Rice 1966; spotted seal, Wartzok and Ray 1976; South African fur seal, Busch and Dücker 1987). This similarity between tactile and visual discrimination Weber fractions is an indication of the importance of tactile discrimination for pinnipeds.

All-or-none behavioral experiments, which investigated the transmission of phasic (i.e., not static) information through the vibrissae, showed little difference in the animal's ability to complete the task whether vibrissae were present or not. Renouf (1980) recorded the length of time harbor seals required to capture live fish. Animals with vibrissae clipped off showed no increase in time required in either clear or murky water. Although the vibrissae might not serve a role in fish capture, they could well be important for benthic feeding, particularly at depths where light levels have fallen below threshold. Benthic prey have been observed in fecal samples of harbor seals (Härkónen 1987) and ringed seals (Kelly and Wartzok 1996). Lindt (1956) described Southern sea lions (*Otaria byronia*) swimming near the benthos with their vibrissae erect and touching the sea bottom. Kelly and Wartzok (1996) reported that ringed seals repeatedly dove to the bottom on apparent feeding dives.

As a test of Montagna's (1967) suggestion that one function of the vibrissae might be speed sensing, Renouf and Gaborko (1982) trained a seal to swim through hoops maintaining a constant speed of 6 km/hr. When the animal was able to do this, its vibrissae were cut off. After the removal of the vibrissae, the seal was still able to maintain the same speed. However, this was not a good test of Montagna's hypothesis because the seal was able to see the sequence of hoops even after the vibrissae were removed and thus had other possible speed-sensing cues.

The one all-or-nothing experiment that showed an important role for the vibrissae was a static recognition task. Sonafrank et al. (1983) reported that a blindfolded seal attracted to a hole though the ice using an acoustic cue would ascend directly up through the center of the hole when its vibrissae were unimpeded. However, when the blindfolded seal also had its vibrissae restricted, it still found the hole through the acoustic stimulus just as effectively but bumped into the under side of the ice just lateral to the hole. When finally in the hole, apparently through trial and error, it ascended more slowly to the surface than when the vibrissae were free. A blindfolded seal with free vibrissae but without an acoustic cue would swim directly under an open hole and not detect it. The vibrissae apparently were unable to provide the seal with usable information on the differences in the reflected pressure wave from the ice and the open hole.

Sirenians

The dugong has the most developed sensory hairs of any marine mammal (Kamiya and Yamasaki 1981). These hairs are present over the entire surface of the body, being most dense on the muzzle. A similar pattern is seen in hair distribution in manatees. Marshall and Reep and coworkers (Marshall et al. 1998, Reep et al. 1998) have recently completed studies of the bristles, hair, and bristlelike hair of the facial region of Florida manatees. The bristles appear to be modified vibrissae. They are used in tactile exploration of the environment and appear quite sensitive to touch (Hartman 1979, Marshall et al. 1998). The bristles are different from vibrissae, however, in that they can be actively everted and used in a grasping fashion during feeding and manipulation of objects. The manatee is unique among marine mammals in its ability to use its vibrissae or bristles in a prehensile manner. There appears to be a one-to-one relationship between the number of perioral bristles and the number of neuronal clusters in the region of the cerebral cortex devoted to tactile sensation from the face (Marshall and Reep 1995, Reep et al. 1998).

Magnetic Detection

Water has almost no effect on magnetic flux density. Thus magnetic sensors should be equally effective underwater as in the air. Interest in animal use of the earth's magnetic field was stimulated by Keeton's (1971) finding that the orientation ability of homing pigeons could be disrupted by attaching magnetic coils to the heads of the birds on days when the sun was not visible. A number of studies stimulated by this finding gradually developed methodological and theoretical criteria that should be kept in mind when evaluating the possibility of biomagnetism in marine mammals. Magnetite is the material that has been consistently implicated in magnetic field reception in species easier to work with than marine mammals. Magnetite can be formed through biochemical processes within the body, but unfortunately ferromagnetic materials are also ubiquitous industrial pollutants, and great care must be exercised in determining the origin of any "magnetite" detected in an animal. One way of distinguishing functional magnetite from contamination is through considerations of the size and characteristics of magnetite needed to function as a magnetic detector.

Kirschvink and Walker (1985) argued that single-domain crystals are the most likely form of magnetite to be used in magnetosensation. A second consideration is the chemical composition of the magnetite. Magnetite with few oxides other than iron oxide is more likely derived from biochemical reactions within the animal, whereas magnetite with oxides of rare earth metals, such as titanium and manganese, is more likely of geologic origin (Walker et al. 1985). A third consideration is the consistency of location of magnetite in the same tissues of each specimen from the same species or higher taxa.

The magnetic material found in cetaceans most fully meets the third criterion. Magnetic material was found in the dura matter in the area of the falx cerebri and the tentorium cerebelli (Bauer et al. 1985) in three odontocete species, bottlenose dolphins, Dall's porpoise, Cuvier's beaked whale (*Ziphius cavirostris*), and one mysticete, the humpback whale. The dura is the tough outer membrane covering the brain. The falx cerebri is the region of the dura separating the cerebral hemispheres, and the tentorium cerebelli is the region of the dura separating the cerebrum and the cerebellum. These are the locations where magnetic material has been found in adult cetaceans. Any conclusions regarding the role of this magnetic material in magnetoreception must be tempered by noting that the amount of magnetic material in the dura appears to increase with age (Bauer et al. 1985). The falx cerebri and tentorium cerebelli are the areas of the dura that show increasing ossification with age (Nojima 1988), and magnetite is known to provide structural support by increasing the hardness of chitin in teeth (Kirschvink and Lowenstam 1979). Thus, magnetite in this region may be more related to the ossification of these tissues than to a magnetic sensory system.

The amount of magnetic material isolated from cetaceans to date has not been sufficient to confirm it as single domain magnetite or even to identify it conclusively as magnetite (Bauer et al. 1985). The magnetic material does not, however, appear to be contamination with the probable exception of the first report in *Tursiops* (Zoeger et al. 1981). It is difficult to dissect animals as large as cetaceans using nonmagnetic tools and to carry out the dissection in a clean environment to avoid contamination.

Although the anatomical studies remain equivocal, there is correlational evidence that some cetaceans are using magnetic information to guide their movements. Klinowska (1985, 1986) compared locations in the United Kingdom where dead stranded cetaceans were found and where live strandings took place. She found that all cases of live strandings took place at points where local lows, or valleys, in the magnetic fields intersected the coast or islands. There was no such correlation between strandings of dead cetaceans, which had presumably been washed ashore by currents, and the magnetic field orientation. Because of reversals of the magnetic field over geological time and the spreading of the ocean sea floor, there are alternating bands of magnetic orientation running north to south parallel to the mid-ocean rift where the sea floor spreading originates (Vine 1966). If whales were able to sense these alternations in the magnetic field, they could use these patterns as north–south highways on their annual north–south migrations. An indicator of magnetic field dip would provide information on position along the north–south axis, and if they could keep track of how many reversals they crossed, they could use this information for east–west positioning (Kirschvink and Walker 1985).

Kirschvink and colleagues (Kirschvink et al. 1986, Kirschvink 1990) have extended and confirmed Klinowska's (1985, 1986) United Kingdom study to the United States. The following species were significantly more likely to strand at sites where geomagnetic lows intersected the coastline: long-finned pilot whale (*Globicephala melaena*), short-finned pilot whale (*G. macrorhynchus*), striped dolphin, Atlantic spotted dolphin (*Stenella frontalis*), Atlantic white-sided dolphin (*Lagenorhynchus acutus*), common dolphin, harbor seal, sperm whale, pygmy sperm whale (*Kogia breviceps*), and fin whale.

In contrast to the results of Klinowska and Kirschvink, Brabyn and Frew (1994) found no correlation between magnetic field orientation, minima, or gradients and the locations of live strandings in New Zealand. A possible explanation for the different results is the absence of a consistent orientation of the magnetic contours around New Zealand compared to the sea floor of the North Atlantic off the east coast of the United States and around the United Kingdom.

Attempts to demonstrate sensitivity to magnetic fields in laboratory experiments have been unsuccessful with cetaceans (Bauer et al. 1985). This result is not surprising when compared to studies with more tractable laboratory animals. A key feature of successful experiments is the freedom of the animals to move extensively within the altered magnetic fields. Under these conditions, which are difficult to duplicate for marine mammals, laboratory demonstrations of magnetoreception have been successful in salamanders (Phillips and Adler 1978), honeybees (Walker et al. 1989), and tuna (Walker 1984).

Walker et al. (1992) looked at free-swimming fin whales and found that migrating animals were associated with lows in the geomagnetic gradient or intensity. Because of the great difficulty in obtaining substantial numbers of pure anatomical specimens and in conducting laboratory experiments demonstrating magnetoreception, correlational studies rather than definitive studies will continue to be the norm.

Summary

Marine mammals are acoustically diverse, with wide variations not only in ear anatomy, but also in frequency range and amplitude sensitivity. In general their hearing is as acute as that of land mammals, and they have wider ranges. Although marine mammals exhibit habitat- and size-related hearing trends that parallel those of land mammals in that larger species tend to have lower frequency ranges than smaller species, the majority of species have some ultrasonic capability and there are multiple specialized, auditory adaptations in odontocetes that provide large species exceptional high frequency hearing capabilities. Both mysticetes and odontocetes appear to have soft-tissue channels for sound conduction to the ear. Sirenians may have analogous adaptations. It remains unclear whether pinnipeds use soft-tissue channels in addition to the air-filled external canal for sound reception. Comparisons of the hearing characteristics of otarids and phocids suggest that there are at least two types of pinniped ears, with phocids being better adapted for underwater hearing. Sea otter ears are the most similar to those of land mammals of all marine mammal ears that have been investigated, but they do have some aquatic-related features, and it is not known how well they hear underwater. No data are available on polar bear hearing.

All marine mammals have middle ears that are heavily modified structurally from those in terrestrial mammals in ways that reduce the probability of barotrauma. The end product is an acoustically sensitive ear that is simultaneously adapted to sustain moderately rapid and extreme pressure changes, and which appears capable of accommodating acoustic power relationships several magnitudes greater than in air. It is possible that these special adaptations may coincidentally provide acoustically protective mechanisms that lessen the risk of injury from high intensity noise, but no behavioral or psychometric studies are yet available that directly address this issue.

Visual adaptations parallel the extent to which marine mammals have returned to the marine environment. Although light is extinguished more quickly in water than in air, there is sufficient light that vision is an important sensory modality for marine mammals at almost all depths they inhabit. The lens in the eyes of cetaceans has become more like a fish lens to accommodate for the loss of focusing power of the air–cornea interface. Sirenians, despite of their long history as a fully marine species, have not made as many adaptations to achieve acute underwater vision as have cetaceans. Just as in audition, pinnipeds have compromised full underwater adaptation to allow for functional vision in air as well. One result of these compromises is that although pinnipeds can see well in air in bright light, acuity decreases with declining light levels. The polar bear has a basically terrestrial eye. The sea otter has developed a unique accommodation mechanism to provide good visual acuity in air and underwater. All marine mammals carefully studied have both rod and cone retinas, but few studies have ascertained if they can detect colors. The most likely reasons for the multiple pigment systems is to be able to maximize sensitivity and detection over a wide range of depths and light levels.

Olfaction has declined as adaptation to a marine environment has increased, and is apparently nonexistent in cetaceans, rudimentary in sirenians, and still an important modality, particularly for behavioral interactions, in pinnipeds. The little work that has been done in taste sensation indicates that most tested species can detect the various categories of tastes defined by humans. Gustatory tests are difficult to conduct and quantify for species living in water. Also, it is difficult to define ecologically relevant gustatory discriminations for marine mammals.

Cetaceans have greatest tactile sensitivity near the blowhole where detection of the air–water interface is important. Tactile sensitivity over the trunk of the body appears to be sufficient to provide the sensory input needed to actively damp turbulence, but how the afferent and efferent pathways link and are used to achieve this goal remains an open question. The vibrissae of pinnipeds are extensively developed, richly innervated, and as well-represented cortically as are the hands of some primates. They provide the animal with information about contour and texture, but seem to be less adapted for the transmission of phasic information such as speed sensing.

Magnetic sensation has been investigated primarily in cetaceans. There is evidence that the movements of several cetacean species correlate with geomagnetic patterns. Evidence of a mechanism for detection of geomagnetism is very weak. Magnetite has been identified in the brains of several species, in the same regions of the dura matter, but the increase in concentration with age, which parallels the ossification of these structures, raises questions as to the function of magnetite in these areas. It is likely that most progress on working out the exact mechanisms of magnetic sensation will likely take place in species more tractable than marine mammals.

One irony of sensory system research is that the more tools we invent to explore animals and their senses the greater the hints we receive that our reach is still too short. How extensive is our research arm currently? We know marine mammals use frequencies we cannot hear, but techno logically we can detect and transduce their frequency range into something we can analyze. Tools that help us probe and visualize how marine mammal sounds are produced and processed, like fast biomedical imaging, are helpful but still comparatively limited. All marine mammals carefully studied are sensitive to about the same range of visual wavelengths as humans. Consequently, vision studies can be conducted more straightforwardly on marine mammals because there is no need to transduce the frequencies they see into ones we see. Chemosensation may be reduced in marine mammals, but we should also consider that we may simply not yet have asked the proper questions or developed adequate tools to measure responses in terms we understand. The anatomical sophistication and the extensive cortical space allotted to vibrissal sensory processing implies a more important role than we have yet determined. Our greatest shortcoming is that we cannot yet measure or observe reliably and frequently in the truly relevant environment for marine mammals: at depth in a free-ranging animal. Until we do, we cannot truly understand what is happening in any marine mammal's ear, eye, or brain, and what transpires in the real world of most marine mammals, the open ocean and the deeps, will remain a mystery.

Notes

1. The conventional unit of frequency, the Hertz, abbreviated Hz, is named after a nineteenth century German physicist and is equal to 1 cycle/sec, or 1 cps. kHz is an abbreviation for kiloHertz, or 1000 cps.
2. All physical constants were obtained from the CRC Handbook of Chemistry and Physics, 66th edition (Weast 1985) unless otherwise noted.
3. Like sonic, the airborne sound reference pressure is based on a human metric. The lowest sound level the normal human ear detects at 2 kHz is a diffuse field pressure of 20 μPa, which has an acoustic power density of approximately 1 picowatt/m². Therefore, the common human threshold in air for 2 kHz, which is at or near the most sensitive frequency in a normal human ear, is 20 log (20 μPa / 20 μPa) = 20 log 1 = 0 dB re 20 μPa, or 0 dB SPL. In some older literature, dB SPL is used when decibel values reported were based on ambient pressure as reference, which is generally stated in the text.
4. An octave is a doubling of frequency. For any initial frequency (f), the octave range is 2f, but because the scale is nonlinear, the center frequency is $f*2^{1/2}$. Because octaves are self-referential, two animals may have radically different hearing ranges in terms of frequency, but equal octave spans (e.g., 30–15,000 Hz vs. 100–50,000 Hz; nine octaves in each case).
5. Typical values for human critical ratios (CR) at speech frequencies are 10 to 18 dB.

Literature Cited

Achor, J., and A. Starr. 1980. Auditory brainstem responses in the cat. II. Effects of lesions. Electroencephalography and Clinical Neurophysiology 48:174–190.

Akamatsu, T., Y. Hatakeyama, T. Kojima, and H. Soeda. 1994. Echolocation rates of two harbor porpoises *Phocoena phocoena*. Marine Mammal Science 10:401–411.

Alexander, R. McN., and G. Goldspink. 1977. Mechanics and Energetics of Animal Locomotion. Chapman and Hall, London.

Allen, J. B., and S. T. Neely. 1992. Mircomechanical models of the cochlea. Physics Today 45:40–47.

Anderson, P. K., and R. M. K. Barclay. 1995. Acoustic signals of solitary dugongs: Physical characteristics and behavioral correlates. Journal of Mammalogy 76:1226–1237.

Anderson, S., and G. Pilleri. 1970. Audible sound production in captive *Platanista gangetica*. Investigations on Cetacea II:83–86.

Asaga, T., Y. Naito, B. J. Le Boeuf, and H. Sakurai. 1994. Functional analysis of dive types of female northern elephant seals. Pages 310–327 *in* B. J. Le Boeuf and R. M. Laws (eds.). Elephant Seals: Population Ecology, Behavior, and Physiology. University of California Press, Berkeley.

Asselin, S., M. O. Hammill, and C. Barrette. 1993. Underwater vocalizations of ice breeding grey seals. Canadian Journal of Zoology 71:2211–2219.

Au, W. W. L. 1990. Target detection in noise by echolocating dolphins. Pages 203–216 *in* J. A. Thomas and R. A. Kastelein (eds.). Sensory Abilities of Cetaceans: Laboratory and Field Evidence. Plenum, New York, NY.

Au, W. 1993. The Sonar of Dolphins. Springer-Verlag, New York, NY.

Au, W. W. L., R. W. Floyd, R. H. Penner, and A. E. Murchison. 1974. Measurement of echolocation signals of the Atlantic bottle-nosed dolphin, *Tursiops truncatus* Montagu, in open waters. Journal of the Acoustical Society of America 56:1280–1290.

Au, W. W. L., D. A. Carder, R. H. Penner, and B. Scronce. 1985. Demonstration of adaptation in Beluga whale echolocation signals. Journal of the Acoustical Society of America 772:726–730.

Au, W. W. L., R. H. Penner, and C. W. Turl. 1987. Propagation of beluga echolocation signals. Journal of the Acoustical Society of America 82:807–813.

Awbrey, F. T. 1990. Comparison of hearing abilities with characteristics of echolocation signals. Pages 427–433 *in* J. A. Thomas and R. A. Kastelein (eds.). Sensory Abilities of Cetaceans: Laboratory and Field Evidence. Plenum, New York, NY.

Awbrey, F. T., J. A. Thomas, W. E. Evans, and S. Leatherwood. 1982. Ross Sea killer whale vocalizations: Preliminary description and comparison with those of some Northern Hemisphere killer whales. Report of the International Whaling Commission 32:667–670.

Babenko, V. B., and O. D. Nikishova. 1976. Some hydrodynamic patterns in structure of integument of marine animals. Bionika 10:27–33.

Backus, R. H., and W. E. Schevill. 1966. *Physeter* clicks. Pages 510–527 *in* K. S. Norris (ed.). Whales, Dolphins and Porpoises. University of California Press, Berkeley.

Barnes, L. G., and E. Mitchell. 1978. Cetacea. Pages 582–602 *in* V. J. Maglio and H. B. S. Cooke (eds.). Evolution of African Mammals. Harvard University Press, Cambridge, MA.

Barnes, L. G., D. P. Domning, and C. E. Ray. 1985. Status of studies on fossil marine mammals. Marine Mammal Science 1:15–53.

Bartholomew, G. A. 1953. Behavioral factors affecting social structure in the Alaska fur seal. Transactions of the North American Wildlife Conference 18:481–502.

Bauer, G. B., M. Fuller, A. Perry, J. R. Dunn, and J. Zoeger. 1985. Magnetoreception and biomineralization of magnetite in cetaceans. Pages

489–507 *in* J. L. Kirschvink, D. S. Jones, and B. J. MacFadden (eds.). Magnetite Biomineralization and Magnetoreception in Living Organisms: A New Biomagnetism. Plenum, New York, NY.

Beier, J. C., and D. Wartzok. 1979. Mating behavior of captive spotted seals (*Phoca largha*). Animal Behavior 27:772–781.

Belkovich, V. M., and G. N. Solntseva. 1970. Anatomy and function of the ear in dolphins. U.S. Government Research Development Reports 70:275–282 (read as English summary).

Bernholz, C. D., and M. L. Matthews. 1975. Critical flicker frequency in a harp seal: Evidence for duplex retinal organization. Vision Research 15:733–736.

Bibikov, N. G. 1992. Auditory brainstem responses in the harbor porpoise (*Phocoena phocoena*). Pages 197–211 *in* J. Thomas, R. Kastelein, and A. Y. Supin (eds.). Marine Mammal Sensory Systems. Plenum, New York, NY.

Boenninghaus, G. 1903. Das Ohr des Zahnwales, zyugleich ein Beitrag zur Theorie der Schalleitung. Zoologische Jahrb Åecher (abteilung für anatomie und ontogenie der tiere) 17:189–360 (not read in original).

Brabyn, M., and R. V. C. Frew. 1994. New Zealand herd stranding sites do not relate to geomagnetic topography. Marine Mammal Science 10:195–207.

Braekevelt, C. R. 1986. Fine structure of the tapetum cellulosum of the grey seal (*Halichoerus grypus*). Acta Antarctic 127:81–87.

Breathnach, A. S. 1960. The cetacean central nervous system. Biological Review 35:187–230.

Brill, R. L., M. L. Sevenich, T. J. Sullivan, J. D. Sustman, and R. E. Witt. 1988. Behavioral evidence for hearing through the lower jaw by an echolocating dolphin, *Tursiops truncatus*. Marine Mammal Science 4:223–230.

Brownlee, S. 1983. Correlations between sounds and behavior in the Hawaiian spinner dolphin, *Stenella longirostris*, M.S. thesis, University of California, Santa Cruz.

Bruns, V., and E. T. Schmieszek. 1980. Cochlear innervation in the greater horseshoe bat: Demonstration of an acoustic fovea. Hearing Research 3:27–43.

Bryden, M. M., and G. S. Molyneux. 1986. Ultrastructure of encapsulated mechanoreceptor organs in the region of the nares. Pages 99–107 *in* M. M. Bryden and R. Harrison (eds.). Research on Dolphins. Clarendon Press, Oxford.

Buerki, C. B., T. W. Cranford, K. M. Langan, and K. L. Marten. 1989. Acoustic recordings from two stranded beaked whales in captivity. Page 10 *in* Abstracts of the 8th Biennial Conference on the Biology of Marine Mammals. Pacific Grove, CA, December 1989.

Bullock, T. H., and V. S. Gurevich. 1979. Soviet literature on the nervous system and psychobiology of cetaceans. International Review of Neurobiology 21:47–127.

Bullock, T. H., A. D. Grinnell, E. Ikezono, K. Kameda, Y. Katsuki, M. Nomoto, O. Sato, N. Suga, and K. Yanagisawa. 1968. Electrophysiological studies of central auditory mechanisms in cetaceans. Zeitschrift für Vergleichende Physiologie 59:117–156.

Bullock, T. H., S. Ridgway, and N. Suga. 1971 Acoustically evoked potentials in midbrain auditory structures in sea lions Pinnipedia. Zeitshchrift für Vergleichende Physiologie 74:372–387.

Bullock, T. H., D. P. Domning, and R. C. Best. 1980. Evoked brain potentials demonstrate hearing in a manatee (*Trichechus inunguis*). Journal of Mammalogy 61:130–133.

Bullock, T. H., T. J. O'Shea, and M. C. McClune. 1982. Auditory evoked potentials in the West Indian manatee (Sirenia: *Trichechus manatus*) Journal of Comparative Physiology 148:547–554.

Busch, H., and G. Dücker. 1987. Das visuelle Leistungsvermögen der Seebären (*Arctocephalus pusillus* und *Arctocephalus australis*). Zoologischer Anzeiger 219:197–224.

Busnel, R. -G., and A. Dziedzic. 1966a. Acoustic signals of the pilot whale *Globicephala melaena* and of the porpoises *Delphinus delphis* and *Phocoena phocoena*. Pages 607–646 *in* K. S. Norris (ed.). Whales, Dolphins and Porpoises. University of California Press, Berkeley.

Busnel, R. -G., and A. Dziedzic. 1966b. Caractéristiques physiques de certains signaux acoustiques du Delphinid *Steno bredanensis*. Comptes Rendus Academie du Scientifique (Paris) 262:143–146.

Busnel, R. -G., and A. Dziedzic. 1968. Caractéristiques physiques des signaux acoustiques de *Pseudorcacrassidens* Owen (Cetace Odontocete). Mammalia 32:I–5.

Busnel, R.-G., A. Dziedzic, and S. Anderson. 1965. Rôle de l'impédance d'une cible dans le seuil de sa détection par le système sonar du Marsouin *P. phocaena*. Comptes Rendus des Séances de la Société de Biologie 159:69–74.

Busnel, R. -G., G. Pilleri, and F. C. Fraser. 1968. Notes concernant le dauphin *Stenella styxx* Gray 1846. Mammalia 32:192–203.

Busnel, R. -G., A. Dziedzic, and G. Alcuri. 1974. Etudes préliminaires de signaux acoustiques du marsouin *Pontoporia blainvillei* Gervais et D'Orligny (Cetacea, Plantanistidae). Mammalia 38:449–459.

Caldwell, D. K., and M. C. Caldwell. 1970. Echolocation-type signals by two dolphins, genus *Sotalia*. Quarterly Journal of Florida Academy of Science 33:124–131.

Caldwell, D. K., and M. C. Caldwell. 1971a. Sounds produced by two rare cetaceans stranded in Florida. Cetology 4:1–6.

Caldwell, D. K., and M. C. Caldwell. 1971b. Underwater pulsed sounds produced by captive spotted dolphins, *Stenella plagiodon*. Cetology 1:1–7.

Caldwell, D. K., and M. C. Caldwell. 1987. Underwater echolocaton-type clicks by captive stranded pygmy sperm whales, *Kogia breviceps*. Page 8 *in* Abstracts of the 7th Biennial Conference on the Biology of Marine Mammals, Miami, FL, December 1987.

Caldwell, D. K., M. C. Caldwell, and J. F. Miller. 1969. Three brief narrow-band sound emissions by a captive male Risso's dolphin, *Grampus griseus*. Los Angeles County Museum of Natural History Foundation Technical Report 5. NTIS AD-693157.

Caldwell, M. C., and D. K. Caldwell. 1967. Intraspecific transfer of information via pulsed sound in captive odontocete cetaceans. Pages 879–937 *in* R -G. Busnel (ed.). Animal Sonar Systems: Biology and Bionics II. Laboratoire de Physiologie Acoustique, Jouy-en-Josas, France.

Caldwell, M. C., and D. K. Caldwell. 1968. Vocalization of naive captive dolphins in small groups. Science 159:1121–1123.

Caldwell, M. C., and D. K. Caldwell. 1969. Simultaneous but different narrowband sound emissions by a captive eastern Pacific pilot whale, *Globicephala scammoni*. Mammalia 33:505–508, 2 plates.

Caldwell, M. C., and D. K. Caldwell. 1971. Statistical evidence for individual signature whistles in Pacific whitesided dolphins, *Lagenorhynchus obliquidens*. Cetology 3:1–9.

Caldwell, M. C., D. K. Caldwell, and J. F. Miller. 1973. Statistical evidence for individual signature whistles in the spotted dolphin, *Stenella plagiodon*. Cetology 16:1–21.

Caldwell, M. C., D. K. Caldwell, and P. L. Tyack. 1990. Review of the signature whistle hypothesis for the Atlantic bottlenose dolphin. Pages 199–234 *in* S. Leatherwood and R. R. Reeves (eds.). The Bottlenose Dolphin. Academic Press, San Diego, CA.

Carlson, M., and K. A. FitzPatrick. 1982. Organization of the hand area in the primary somatic sensory cortex (SmI) of the prosimian primates, *Nycticebus coucang*. Journal of Comparative Neurology 204:280–295.

Carlson, M., and C. Welt. 1980. Somatic sensory cortex (SmI) of the prosimian primate *Galago crassicaudatus:* Organization of mechanoreceptive input from the hand in relation to cytoarchitecture. Journal of Comparative Neurology 189:249–271.

Carlson, M., M. F. Huerta, C. G. Cusik, and J. H. Kaas. 1986. Studies on the evolution of multiple somatosensory representations in primates: The organization of anterior parietal cortex in the new world callitrichid, *Sahuinus*. Journal of Comparative Neurology 246:409–426.

Carlson, S., H. Tanila, I. Linnankoski, A. Pertovaara, and A. Kehr. 1989. Comparison of tactile discrimination ability of visually deprived and normal monkeys. Acta Physiologica Scandinavica 135:405–410.

Chapman, H. C. 1875. Observations on the structure of the manatee. Proceedings of the Academy of Natural Sciences, Philadelphia 1872:452–462.

Clark, C. W. 1982. The acoustic repertoire of the southern right whale: A quantitative analysis. Animal Behaviour 30:1060–1071.

Clark, C. W. 1983. Acoustic communication and behavior of the southern right whale (*Eubalaena australis*). Pages 163–198 *in* R. S. Payne (ed.). Behavior and Communication of Whales. AAAS Selected Symposium 76. Westview Press, Boulder, CO.

Clark, C. W. 1990. Acoustic behavior of mysticete whales. Pages 571–584 *in* J. A. Thomas and R. A. Kastelein (eds.). Sensory Abilities of Cetaceans. Laboratory and Field Evidence. Plenum, New York, NY.

Clark, C. W., and J. H. Johnson. 1984. The sounds of the bowhead whale, *Balaena mysticetus,* during the spring migrations of 1979 and 1980. Canadian Journal of Zoology 62:1436–1441.

Clark, C. W., W. T. Ellison, and K. Beeman. 1986. An acoustic study of bowhead whales, *Balaena mysticetus,* off Point Barrow, Alaska during the 1984 spring migration. Report from Marine Acoustics, Clinton, MA, for North Slope Borough Department of Wildlife Management. Barrow, AK.

Cohen, J. L., G. S. Tucker, and D. K. Odell. 1982. The photoreceptors of the West Indian manatee. Journal of Morphology 173:197–202.

Cummings, W. C., and J. F. Fish. 1971. A synopsis of marine animal underwater sounds in eight geographic areas. U.S. Naval Undersea Research & Development Center. NTIS AD-AO68875, 97 pages.

Cummings, W. C., and D. V. Holliday. 1987. Sounds and source levels from bowhead whales off Pt. Barrow, Alaska. Journal of the Acoustical Society of America 82:814–821.

Cummings, W. C., and P. O. Thompson. 1971. Underwater sounds from the blue whale, *Balaenoptera musculus,* Journal of the Acoustical Society of America 50:1193–1198.

Cummings, W. C., and P. O. Thompson. 1994. Characteristics and seasons of blue and finback whale sounds along the U.S. west coast as recorded at SOSUS stations. Journal of the Acoustical Society of America 95(5, Pt. 2):2853.

Cummings, W. C., P. O. Thompson, and R. Cook. 1968. Underwater sounds of migrating gray whales, *Eschrichtius robustus*. Journal of the Acoustical Society of America 44:1278–1281.

Cummings, W. C., J. F. Fish, and P. O. Thompson. 1972. Sound production and other behavior of southern right whales, *Eubalaena australis*. Transactions of the San Diego Society of Natural History 17:1–13.

Cummings, W. C., D. V. Holliday., W. T. Ellison, and B. J. Graham. 1983. Technical feasibility of passive acoustic location of bowhead whales in population studies off Point Barrow, Alaska. T-83-06-002. Report from Tracor Applied Science, San Diego, CA for North Slope Borough, Barrow, AK.

Cummings, W. C., D. V. Holliday, and B. J. Lee. 1984. Potential impacts of man-made noise on ringed seals: Vocalizations and reactions. Outer Continental Shelf Environmental Assessment Program, Final Report Principal Investigator, NOAA, Anchorage, AK 37:95-230. OCS Study MMS 86-0021; NTIS PB87-107546.

Cummings, W. C., P. O. Thompson, and S. J. Ha. 1986. Sounds from Bryde, *Balaenoptera edeni,* and finback, *B. physalus,* whales in the Gulf of California. Fishery Bulletin 84:359–370.

Dahlheim, M. E., and D. K. Ljungblad. 1990. Preliminary hearing study on gray whales *Eschrictius robustus* in the field. Pages 335–346 *in* J. A. Thomas and R. A. Kastelein (eds.). Sensory Abilities of Cetaceans: Laboratory and Field Evidence. Plenum, New York, NY.

Dahlheim, M. E., H. D. Fisher, and J. D. Schempp. 1984. Sound production by the gray whale and ambient noise levels in Laguna San Ignacio, Baja California Sur, Mexico. Pages 511–541 *in* M. L. Jones, S. L. Swartz, and S. Leatherwood (eds.). The gray whale *Eschrictius robustus*. Academic Press, Orlando, FL.

Dallos, P., D. Harris, O. Ozdamar, and A. Ryan. 1978. Behavioral, compound action potential, and single unit thresholds: Relationship in normal and abnormal ears. Journal of the Acoustical Society of America 64:151–157.

Dartnell, H. J. A. 1962. The photobiology of visual processes. Pages 321–533 *in* H. Davson (ed.). The Eye, Vol. 2, Ed. 1. Academic Press, New York, NY.

Dawbin, W. H., and D. H. Cato. 1992. Sounds of a pygmy right whale (*Caperea marginata*). Marine Mammal Science 8:213–219.

Dawson, S. M. 1988. The high frequency sounds of free-ranging Hector's dolphins, *Cephalorhynchus hectori*. Reports of the International Whaling Commission (Special Issue 9):339–344.

Dawson, S. M., and C. W. Thorpe. 1990. A quantitative analysis of the sounds of Hector's dolphin. Ethology 86:131–145.

Dawson, W. W. 1980. The cetacean eye. Pages 54–99 *in* L. M. Herman (ed.). Cetacean Behavior: Mechanisms and Functions. Wiley Interscience, New York, NY.

Dawson, W. W., and J. M. Perez. 1973. Unusual retinal cells in the dolphin eye. Science 181:747–749.

Dawson, W. W., L. A. Birndorf, and J. M. Perez. 1972. Gross anatomy and optics of the dolphin eye (*Tursiops truncatus*). Cetology 10:1–12.

Dawson, W. W., C. K. Adams, M. C. Barris, and C. A. Litzkow. 1979. Static and kinetic properties of the dolphin pupil. American Journal of Physiology 237:R301–R305.

Dawson, W. W., J. P. Schroeder, and J. F. Dawson. 1987a. The ocular fundus of two cetaceans. Marine Mammal Science 3:1–13.

Dawson, W. W., J. P. Schroeder, and S. N. Sharpe. 1987b. Corneal surface properties of two marine mammal species. Marine Mammal Science 3:186–197.

Dehnhardt, G. 1990. Preliminary results from psychophysical studies on the tactile sensitivity in marine mammals. Pages 435–446 *in* J. A. Thomas and R. A. Kastelein (eds.). Sensory Abilities of Cetaceans: Laboratory and Field Evidence. Plenum, New York, NY.

Dehnhardt, G. 1994. Tactile size discrimination by a California sea lion (*Zalophus californianus*) using its mystacial vibrissae. Journal of Comparative Physiology A 175:791–800.

Dehnhardt, G., and A. Kaminski. 1995. Sensitivity of the mystacial vibrissae of harbour seals (*Phoca vitulina*) for size differences of actively touched objects. Journal of Experimental Biology 198:2317–2323.

Denton, E. J. 1990. Light and vision at depths greater than 200 meters. Pages 127–148 *in* P. J. Herring, A. K. Campbell, M. Whitfield, and L. Maddok (eds.). Light and Life in the Sea. Cambridge University Press, Cambridge, U.K.

Diercks, K. J., 1972. Biological sonar systems: A bionics survey. Applied Research Laboratories, ARL-TR-72-34, University of Texas, Austin, TX.

Diercks, K. J., R. T. Trochta, R. L. Greenlaw, and W. E. Evans. 1971. Recording and analysis of dolphin echolocation signals. Journal of the Acoustical Society of America 49:1729–1732.

Dolan, T. G., J. H. Mills, and R. A. Schmidt. 1985. A comparison of brainstem, whole nerve and single-fiber tuning curves in the gerbils: Normative data. Hearing Research 17:259–266.

Dolphin, W. F. 1995. Steady-state auditory-evoked potentials in three cetacean species elicited using amplitude-modulated stimuli. Pages 25–47 *in* R. A. Kastelein, J. A. Thomas, and P. E. Nachtigall (eds.). Sensory systems of aquatic mammals. DeSpil Publishers, Woerden, Netherlands.

Dolphin, W. F., and D. C. Mountain. 1992. The envelope following response: Scalp potential elicited in the Mongolian gerbil using SAM acoustic stimuli. Hearing Research 58:70–78.

Domning, D. P. 1977. An ecological model for late Tertiary sirenian evolution in the North Pacific Ocean. Systematic Zoology 25:352–362.

Domning, D. P. 1981. Sea cows and sea grasses. Paleobiology 7:417–420.

Domning, D. P., and V. de Buffrénil. 1991. Hydrostasis in the Sirenia: Quantitative data and functional interpretations. Marine Mammal Science 7:331–368.

Domning, D. P., and L. C. Hayek. 1986. Interspecific and intraspecific morphological variation in manatees (Sirenia: *Trichechus*). Marine Mammal Science 2:87–144.

Domning, D. P., G. S. Morgan, and C. E. Ray. 1982. North American Eocene sea cows (Mammalia: Sirenia). Smithsonian Contributions to Paleobiology 52:1–69.

Donaldson, B. J. 1977. The tongue of the bottlenose dolphin. Pages 175–198 *in* R. J. Harrison (ed.). Functional Anatomy of Marine Mammals. Academic Press, London.

Doran, A. H. G. 1879. Morphology of the mammalian ossicula auditus. Transactions of the Linnaean Society 1:371–497.

Dral, A. D. G. 1972. Aquatic and aerial vision in the bottle-nosed dolphin. Journal of Sea Research 5:510–513.

Dral, A. D. G. 1975. Some quantitative aspects of the retina of *Tursiops truncatus*. Aquatic Mammals 2:28–31.

Dral, A. D. G. 1977. On the retinal anatomy of cetacea. Pages 86–87 *in* R. J. Harrison (ed.). Functional Anatomy of Marine Mammals, Vol. 3. Academic Press, London.

Dral, A. D. G. 1983. The retinal ganglion cells of *Delphinus delphis* and their distribution. Aquatic Mammals 10:57–68.

Dudok van Heel, W. H. 1962. Sound and cetacea. Netherlands Journal of Sea Research 1:407–507.

Dykes, R. W. 1975. Afferent fibers from mystacial vibrissae of cats and seals. Journal of Neurophysiology 38:650–662.

Dziedzic, A. 1978. Etude experimentale des émissions sonar de certain delphinides et notamment de *D. delphis* et *T. truncatus*. Thèse de Doctorat d'Etat Es-Sciences Appliquées, l'Université de Paris VII.

Dziedzic, A., and V. de Buffrenil. 1989. Acoustic signals of the Commerson's dolphin, *Cephalorhynchus commersonii*, in the Kerguelen Islands. Journal of Mammalogy 70:449–452.

Echteler, S. W., R. R. Fay, and A. N. Popper. 1994. Structure of the mammalian cochlea. Pages 134–171 *in* R. R. Fay and A. N. Popper (eds.). Comparative Hearing: Mammals. Springer-Verlag, New York, NY.

Edds, P. L. 1982. Vocalizations of the blue whale, *Balaenoptera musculus*, in the St Lawrence River. Journal of Mammalogy 63:345–347.

Edds, P. L. 1988. Characteristics of finback *Balaenoptera physalus* vocalizations in the St. Lawrence Estuary. Bioacoustics 1:131–149.

Edds, P. L., D. K. Odell, and B. R. Tershy. 1993. Vocalizations of a captive juvenile and free-ranging adult-calf pairs of Bryde's whales, *Balaenoptera edeni*. Marine Mammal Science 9:269–284.

Edds-Walton, P. L. 1997. Acoustic communication signals of mysticete whales. Bioacoustics 8:47–60.

Eva, A. N. 1980. Pre-Miocene seagrass communities in the Caribbean. Palaeontology 23:231–236.

Evans, W. E. 1967. Vocalizations among marine mammals. Pages 159–186 *in* W. N. Tavolga (ed.). Marine Bio-Acoustics. Pergamon, New York, NY.

Evans, W. E. 1973. Echolocation by marine delphinids and one species of fresh water dolphin. Journal of the Acoustical Society of America 54:191–199.

Evans, W. E., and F. T. Awbrey. 1984. High frequency pulses of Commerson's dolphin and Dall's porpoise. American Zoologist 24:2A.

Evans, W. E., and E. S. Herald. 1970. Underwater calls of a captive Amazon manatee, *Trichechus inuguis*. Journal of Mammology 51:820–823.

Evans, W. E., and J. H. Prescott. 1962. Observations of the sound production capabilities of the bottlenose porpoise: A study of whistles and clicks. Zoologica 47:121–128.

Evans, W. E., F. T. Awbrey, and H. Hackbarth. 1988. High frequency pulses produced by free-ranging Commerson's dolphin *Cephalorhynchus commersonii* compared to those of phocoenids. Reports of the International Whaling Commission (Special Issue 9):173–181.

Fay, R. R. 1988. Hearing Vertebrates: A Psychophysics Databook. Hill-Fay Associates, Winnetka, IL.

Fay, R. R. 1992. Structure and function in sound discrimination among vertebrates. Pages 229–267 *in* D. B. Webster, R. R. Fay, and A. N. Popper (eds.). The Evolutionary Biology of Hearing. Springer-Verlag, New York, NY.

Feinstein, S. H., and C. E. Rice. 1966. Discrimination of area differences by the harbor seal. Psychonomic Science 4:379–380.

Firbas, W. 1972. Über anatomische Anpassungen des Hörorgans an die Aufnahme hîher Frequenzen. Monatszeitschrift Ohrenheilkd Laryngo-Rhinologie 106:105–156.

Fish, J. F., and C. W. Turl. 1976. Acoustic source levels of four species of small whales. NUC TP 547. U.S. Naval Undersea Center, San Diego, CA. NTIS AD-A037620.

Fish, J. F., J. L. Sumich, and G. L. Lingle. 1974. Sounds produced by the gray whale, *Eschrictius robustus*. Marine Fisheries Review 36:38–45.

FitzPatrick, K. A., M. Carlson, and J. Charlton. 1982. Topography, cytoarchitecture, and sulcal patterns in primary somatic sensory cortex (SmI) of the prosimian primate, *Perodicticus potto*. Journal of Comparative Neurology 204:296–310.

Fleischer, G. 1978. Evolutionary principles of the mammalian middle ear. Advanced Anatomy, Embryology and Cell Biology 55:1–70.

Fletcher, H. 1940. Auditory patterns. Reviews of Modern Physics 12:47–65.

Ford, J. K. B., and H. D. Fisher. 1978. Underwater acoustic signals of the narwhal *Monodon monoceros*. Canadian Journal of Zoology 56:552–560.

Ford, J. K. B., and H. D. Fisher. 1983. Group-specific dialects of killer whales *Orcinus orca* in British Columbia. Pages 129–161 *in* R. Payne (ed.). Communication and Behavior of Whales. AAAS Selected Symposium 76. Westview Press, Boulder, CO.

Friedl, W. A., P. E. Nachtigall, P. W. B. Moore, N. K. W. Chun, J. E. Haun, R. W. Hall, and J. L. Richards. 1990. Taste reception in the Pacific bottlenose dolphin (*Tursiops truncatus gilli*) and the California sea lion (*Zalophus californianus*). Pages 447–454 *in* J. A. Thomas and R. A. Kastelein (eds.). Sensory Abilities of Cetaceans: Laboratory and Field Evidence. Plenum, New York, NY.

Gacek, R. R., and G. L. Rasmussen. 1961. Fiber analysis of the statoacoustic nerve of guinea pig, cat, and monkey. Anatomical Record 139:455.

Gao, A., and K. Zhou. 1987. On the retinal ganglion cells of *Neophocaena* and *Lipotes*. Acta Zoologica Sinica 33:316–323.

Gao, G., and K. Zhou. 1991. The number of fibers and range of fiber diameters in the cochlear nerve of three odontocete species. Canadian Journal of Zoology 69:2360–2364.

Gao, G., and K. Zhou. 1992. Fiber analysis of the optic and cochlear nerves of small cetaceans. Pages 39–52 *in* J. A. Thomas, R. A. Kastelein, and A. Supin (eds.). Marine Mammal Sensory Systems. Plenum, New York, NY.

Gao, G., and K. Zhou. 1995. Fiber analysis of the vestibular nerve of small cetaceans. Pages 447–453 *in* R. A. Kastelein, J. A. Thomas, and P. E. Nachtigall (eds.). Sensory Systems of Aquatic Mammals. DeSpil Publishers, Woerden, Netherlands.

Gaskin, D. E. 1976. The evolution, zoogeography, and ecology of Cetacea. Oceanography and Marine Biology: Annual Review 14:247–346.

Gentry, R. L. 1967. Underwater auditory localization in the California sea lion (*Zalophus californianus*). Journal of Auditory Research 7:187–193.

Gerstein, E. R. 1994. Hearing Abilities of the West Indian Manatee, *Trichechus manatus*. Technical Report no. 119, Florida Inland Navigation District.

Gerstein, E. R., L. A. Gerstein, S. E. Forsythe, and J. E. Blue. 1993. Underwater audiogram of a West Indian manatee *Trichechus manatus*. Page 130 *in* Abstract of the 10th Biennial Conference on Marine Mammals, Galveston, TX, November 1993.

Goodson, A. D., and M. Klinowska. 1990. A proposed echolocation receptor for the bottlenose dolphin (*Tursiops truncatus*): Modeling the received directivity from tooth and lower jaw geometry. Pages 255–269 *in* J. A. Thomas and R. A. Kastelein (eds.). Sensory Abilities of Cetaceans: Laboratory and Field Evidence. Plenum, New York, NY.

Gray, J. 1936. Studies in animal locomotion IV. The propulsive powers of the dolphin. Journal of Experimental Biology 13:192–199.

Gray, O. 1951. An introduction to the study of the comparative anatomy of the labyrinth. Journal of Laryngology and Otology 65:681–703.

Graybiel, A. 1964. Vestibular sickness and some of its implications for space flight. Pages 248–270 *in* W. S. Fields and R. R. Alford (eds.). Neurological Aspects of Auditory and Vestibular Disorders. Charles C. Thomas, Springfield, IL.

Greenwood, D. G. 1961. Critical bandwidth and the frequency coordinates of the basilar membrane. Journal of the Acoustical Society of America 33:1344–1356.

Greenwood, D. G. 1962. Approximate calculation of the dimensions of traveling-wave envelopes in four species. Journal of the Acoustical Society of America 34:1364–1384.

Greenwood, D. G. 1990. A cochlear frequency-position function for several species—29 years later. Journal of the Acoustical Society of America 87:2592–2605.

Grinnell, A. D. 1963. The neurophysiology of audition in bats: Intensity and frequency parameters. Journal of Physiology 167:38–66.

Gunn, L. M. 1988. A behavioral audiogram of the north American river otter (*Lutra canadensis*). M.Sc. thesis. San Diego State University, San Diego, CA, 40 pp.

Haider, M., and D. B. Lindsley. 1964. Microvibrations in man and dolphin. Science 146:1181–1183.

Hall, J., and C. S. Johnson. 1971. Auditory thresholds of a killer whale, *Orcinus orca* Linnaeus. Journal of the Acoustical Society of America 51:515–517.

Hanggi, E. B., and R. J. Schusteman. 1992. Underwater acoustic displays by male harbor seals *Phoca vitulina*. Initial results. Pages 449–457 *in* J. A. Thomas, R. A. Kastelein, and A. Y. Supin (eds.). Marine Mammal Sensory Systems. Plenum, New York, NY.

Hanggi, E. B., and R. J. Schusterman. 1994. Underwater acoustic displays and individual variation in male harbor seals, *Phoca vitulina*. Animal Behavior 48:1275–1283.

Härkónen, T. 1987. Seasonal and regional variations in the feeding habits of the harbour seal, *Phoca vitulina*, in the Skagerrak and Kattegat. Journal of Zoology (London) 213:535–543.

Harrison, R. J., and G. L. Kooyman. 1968. General physiology of the pinnipeds. Pages 212–296 *in* R. J. Harrison, R. C. Hubbard, R. S. Peterson, C. E. Rice, and R. J. Schusterman (eds.). The Behavior and Physiology of Pinnipeds. Appleton-Century-Crofts, New York, NY.

Hartman, D. S. 1979. Ecology and behavior of the manatee (*Trichechus manatus*). American Society of Mammalogists, Special Publication No. 5.

Hatakeyama, Y., and H. Soeda. 1990. Studies on echolocation of porpoises taken in salmon gillnet fisheries. Pages 269–281 *in* J. A. Thomas and R. A. Kastelein (eds.). Sensory Abilities of Cetaceans: Laboratory and Field Evidence. Plenum, New York, NY.

Hatakeyama, Y., K. Ishii, T. Akamatsu, H. Soeda, T. Shimamura, and T. Kojima. 1994. A review of studies on attempts to reduce the entanglement of the Dall's porpoise, *Phocoenoides dalli*, in the Japanese salmon gillnet fishery. Report of the International Whaling Commission (Special Issue 15):549–563.

Hecht, S., and Y. Hsia. 1945. Dark adaptation following light adaptation to red and white lights. Journal of the Optical Society of America 35:261–267.

Heffner, R. S., and H. E. Heffner. 1980. Hearing in the elephant (*Elephas maximus*). Science 208:518–520.

Heffner, R. S., and H. E. Heffner. 1992. Evolution of sound localization in mammals. Pages 691–715 *in* D. Webster, R. Fay, and A. Popper (eds.). The Biology of Hearing. Springer-Verlag, New York, NY.

Heffner, R. S., and R. B. Masterton. 1990. Sound localization in mammals: Brainstem mechanisms. Pages 285–314 *in* M. A. Berkley and W. C. Stebbins (eds.). Comparative Perception. John Wiley and Sons, New York, NY.

Henson, O. W. Jr., P. A. Koplas, A. W. Keating, R. F. Huffman, and M. M. Henson. 1990. Cochlear resonance in the mustached bat: Behavioral adaptations. Hearing Research 50:259–274.

Herald, E. S., R. L. Brownell Jr., F. L. Frye, E. J. Morris, W. E. Evans, and A. B. Scott. 1969. Blind river dolphin: First side-swimming cetacean. Science 166:1408–1410.

Herman, L. M., M. F. Peacock, M. P. Yunker, and C. J. Madsen. 1975. Bottlenose dolphin: Double-slit pupil yields equivalent aerial and underwater diurnal acuity. Science 189:650–652.

Hinchcliffe, R., and A. Pye. 1969. Variations in the middle ear of the Mammalia. Journal of Zoology 157:277–288.

Hobson, E. S. 1966. Visual orientation and feeding in seals and sea lions. Nature 210:326–327.

Hyvärinen, H. 1989. Diving in darkness: Whiskers as sense organs of the ringed seal (*Phoca hispida saimensis*). Journal of Zoology (London) 218:663–678.

Ingmanson, D. E., and W. J. Wallace. 1973. Oceanology: An Introduction. Wadsworth Publishing Co., Belmont, CA.

Jacobs, M. S., P. J. Morgane, and W. L. McFarland. 1971. The anatomy of the brain of the bottlenose dolphin (*Tursiops truncatus*). Rhinic lobe (rhinecephalon) I. The paleocortex. Journal of Comparative Neurology 141:205–272.

Jamieson, G. S., and H. D. Fisher. 1971. The retina of the harbor seal, *Phoca vitulina*. Canadian Journal of Zoology 49:19–23.

Jamieson, G. S., and H. D. Fisher. 1972. The pinniped eye: A review. Pages 245–261 *in* R. J. Harrison (ed.). Functional Anatomy of Marine Mammals, Vol. 1. Academic Press, New York, NY.

Jansen, J., and J. K. S. Jansen. 1969. The nervous system of Cetacea. Pages 175–252 *in* H. T. Andersen (ed.). The Biology of Marine Mammals. Academic Press, New York, NY.

Jensen, F. B., W. A. Kuperman, M. B. Porter, and H. Schmidt. 1994. Computational Ocean Acoustics. AIP Press, New York, NY.

Jerlov, N. G. 1976. Marine Optics. Elsevier, Amsterdam.

Jewett, D. L. 1970. Volume conducted potentials in response to auditory stimuli as detected by averaging in the cat. Electroencephalography and Clinical Neurophysiology 28:609–618.

Jing X., Y. Xiao, and R. Jing. 1981. Acoustic signals and acoustic behaviour of the Chinese river dolphin *Lipotes vexillifer*. Scientica Sinica 24:407–415.

Johnson, C. S. 1968. Masked tonal thresholds in the bottlenose porpoise. Journal of the Acoustical Society of America 44:965–967.

Johnson, C. S. 1971. Auditory masking of one pure tone by another in the bottlenose porpoise. Journal of the Acoustical Society of America 49:1317–1318.

Johnson, G. L. 1893. Observations on the refraction and vision of the seal's eye. Proceedings of the Zoological Society of London, No. 48:719–723.

Kamiya, T., and F. Yamasaki. 1981. A morphological note on the sinus hair of the dugong. Pages 193–197 *in* H. Marsh (ed.). The Dugong: Proceedings of the Seminar/Workshop held at James Cook University of North Queensland, May 8–13, 1979. Department of Zoology, James Cook University of North Queensland, Townsville, Queensland, Australia.

Kamminga, C. 1988. Echolocation signal types of odontocetes. Pages 9–22 *in* P. E. Nachtigall and P. W. B. Moore (eds.). Animal Sonar Processes and Performance. Plenum, New York, NY.

Kamminga, C., and J. G. van Velden. 1987. Investigations on cetacean sonar VIII. Sonar signals of *Pseudorca crassidens* in comparison with *Tursiops truncatus*. Aquatic Mammals 13:43–49.

Kamminga, C., and H. Wiersma. 1981. Investigations on cetacean sonar II. Acoustical similarities and differences in odontocete signals. Aquatic Mammals 82:41–62.

Kamminga C., T. Kataoka, and F. J. Engelsma. 1986. Investigations on cetacean sonar VII. Underwater sounds of *Neophocaena phocaenoides* of the Japanese coastal populations. Aquatic Mammals 122:52–60.

Kamminga, C. F., F. J. Engelsma, and R. P. Terry. 1989. Acoustic observations and comparison on wild, captive and open water *Sotalia* and *Inia*. Page 33 *in* Abstracts of the 8th Biennial Conference on the Biology of Marine Mammals, Pacific Grove, CA, December 1989.

Kamminga, C., M. T. van Hove, F. J. Engelsma, and R. P. Terry. 1993. Investigations on cetacean sonar X. A comparative analysis of underwater echolocation clicks of *Inia* spp. and *Sotalia* spp. Aquatic Mammals 19:31–43.

Kastak, D., and R. J. Schusterman. 1995. Aerial and underwater hearing thresholds for 100Hz pure tones in two pinniped species. Pages 71–81 *in* R. A. Kastelein, J. A. Thomas, and P. E. Nachtigall (eds.). Sensory Systems of Aquatic Mammals. DeSpil Publishers, Woerden, Netherlands.

Kastak, D., and R. J. Schusterman. 1996. Temporary threshold shift in a harbor seal (*Phoca vitulina*). Journal of the Acoustical Society of America 100:1905–1908.

Kastelein, R. A. 1991. The relationship between sensory systems and head musculature in the walrus. Page 38 *in* Abstracts of the 9th Biennial Conference on the Biology of Marine Mammals, Chicago, IL, December 1991.

Kastelein, R. A., and M. A. van Gaalen. 1988. The sensitivity of the vibrissae of a Pacific walrus (*Odobenus rosmarus divergens*) Part 1. Aquatic Mammals 14:123–133.

Kastelein, R. A., R. C. V. J. Zweypfenning, and H. Spekreijse. 1990. Anatomical and histological characteristics of the eyes of a month-old and an adult harbor porpoise (*Phocoena*). Pages 463–480 *in* J. A. Thomas and R. A. Kastelein (eds.). Sensory Abilities of Cetaceans: Laboratory and Field Evidence. Plenum, New York, NY.

Kasuya, T. 1973. Systematic consideration of recent toothed whales based on the morphology of tympano-periotic bone. Scientific Reports of the Whales Research Institute (Tokyo) 25:1–103.

Kayan, V. P. 1974. Resistance coefficient of the dolphin. Bionika 8:31–35.

Keeton, W. T. 1971. Magnets interfere with pigeon homing. Proceedings of the National Academy of Sciences 68:102–106.

Kellogg, W. N. 1959. Auditory perception of submerged objects by porpoises. Journal of the Acoustical Society of America 31:1–6.

Kelly, B. P., and D. Wartzok. 1996. Ringed seal diving behavior in the breeding season. Canadian Journal of Zoology 74:1547–1555.

Kenyon, K. W. 1981. Sea otter *Enhydra lutris* (Linnaeus, 1758). Pages 209–223 *in* S. H. Ridgway and R. J. Harrison (eds.). Handbook of Marine Mammals, Volume 1: The Walrus, Sea Lions, Fur Seals and Sea Otter. Academic Press, London.

Ketten, D. R. 1984. Correlations of morphology with frequency for odontocete cochlea: Systematics and topology. Ph.D. dissertation, The Johns Hopkins University, Baltimore, MD, 335 pp.

Ketten, D. R. 1992. The marine mammal ear: Specializations for aquatic audition and echolocation. Pages 717–754 *in* D. Webster, R. Fay, and A. Popper (eds.). The Biology of Hearing. Springer-Verlag, New York, NY.

Ketten, D. R. 1994. Functional analyses of whale ears: Adaptations for underwater hearing. I.E.E.E. Proceedings in Underwater Acoustics 1:264–270.

Ketten, D. R. 1997. Structure and function in whale ears. Bioacoustics 8:103–135.

Ketten, D. R., and D. Wartzok. 1990. Three-dimensional reconstructions of the dolphin cochlea. Pages 81–105 *in* J. A. Thomas and R. A. Kastelein (eds.). Sensory Abilities of Cetaceans: Laboratory and Field Evidence. Plenum, New York, NY.

Ketten, D. R., D. K. Odell, and D. P. Domning. 1992. Structure, function, and adaptation of the manatee ear. Pages 77–95 *in* J. A. Thomas, R. A. Kastelein, and A. Y. Supin (eds.). Marine Mammal Sensory Systems. Plenum, New York, NY.

Ketten, D. R., S. Ridgway, and G. Early. 1995. Apocalyptic hearing:

Aging, injury, disease, and noise in marine mammal ears. Page 61 *in* Abstracts of the 11th Biennial Conference on the Biology of Marine Mammals, Orlando, FL, December 1995.

Khomenko, B. G., and V. G. Khadzhinskiy. 1974. Morphological and functional principles underlying cutaneous reception in dolphins. Bionika 8:106–113.

Kirschvink, J. L. 1990. Geomagnetic sensitivity in cetaceans: An update with live stranding records in the United States. Pages 639–649 *in* J. A. Thomas and R. Kastelein (eds.). Sensory Abilities of Cetaceans: Laboratory and Field Evidence. Plenum, New York, NY.

Kirschvink, J. L., and H. A. Lowenstam. 1979. Mineralization and magnetization of chiton teeth: Paleomagnetic, sedimentologic, and biologic implications of organic magnetite. Earth and Planetary Science Letters 44:193–204.

Kirschvink, J. L., and M. W. Walker. 1985. Particle-size considerations for magnetite-based magnetoreceptors. Pages 243–254 *in* J. L. Kirschvink, D. S. Jones, and B. J. MacFadden (eds.). Magnetite Biomineralization and Magnetoreception in Organisms: A New Biomagnetism. Plenum, New York, NY.

Kirschvink, J. L., A. E. Dizon, and J. A. Westphal. 1986. Evidence from strandings for geomagnetic sensitivity in cetaceans. Journal of Experimental Biology 120:1–24.

Klinowska, M. 1985. Interpretation of the UK cetacean stranding records. Report of the International Whaling Commission 35:459–467.

Klinowska, M. 1986. The cetacean magnetic sense—evidence from strandings. Pages 401–432 *in* M. M. Bryden and R. Harrison (eds.). Research on Dolphins. Clarendon Press, Oxford.

Klishin, V. O., R. P. Diaz, V. V. Popov, and A. Y. Supin. 1990. Some characteristics of hearing of the Brazilian manatee, *Trichechus inunguis*. Aquatic Mammals 16:140–144.

Knowlton, A. R., C. W. Clark, and S. D. Krauss. 1991. Sounds recorded in the presence of sei whales, *Balaenoptera borealis*. Page 40 *in* Abstracts of the 9th Biennial Conference on the Biology of Marine Mammals, Chicago, IL, December 1991.

Kolchin, S. P., and V. M. Bel'kovich. 1973. Tactile sensitivity in *Delphinus delphis*. Zoologichesky Zhurnal 52:620–622.

Komatsu, S., and F. Yamasaki. 1980. Formation of the pits with taste buds at the lingual root in the striped dolphin, *Stenella coeruleoaba*. Journal of Morphology 164:107–119.

Kooyman, G. L., and R. L. Gentry. 1986. Diving behavior of South African fur seals. Pages 142–152 *in* R. L. Gentry and G. L. Kooyman (eds.). Fur Seals: Maternal Strategies on Land and at Sea. Princeton University Press, Princeton, NJ.

Kössl, M., and M. Vater. 1995. Cochlear structure and function in bats. Pages 191–234 *in* R. R. Fay and A. N. Popper (eds.). Hearing by Bats. Springer-Verlag, New York, NY.

Kramer, M. O. 1965. Hydrodynamics of the dolphin. Hydroscience 2:111–130.

Kröger, R. H. H., and K. Kirschfeld. 1992. The cornea as an optical element in the cetacean eye. Pages 97–106 *in* J. A. Thomas, R. A. Kastelein, and A. Y. Supin (eds.). Marine Mammal Sensory Systems. Plenum, New York, NY.

Kröger, R. H. H., and K. Kirschfeld. 1994. Refractive index in the cornea of a harbor porpoise (*Phocoena phocoena*) measured by two-wavelengths laser-interferometry. Aquatic Mammals 20:99–107.

Krubitzer, L. A., M. A. Sesma, and J. H. Kaas. 1986. Microelectrode maps, myeloarchitecture, and cortical connections of three somatopically organized representations of the body surface in the parietal cortex of squirrels. Journal of Comparative Neurology 250:403–430.

Kuwada, S. R., R. Batra, and V. Maher. 1986. Scalp potentials from normal and hearing impaired subjects in response to SAM tones. Hearing Research 21:179–192.

Kuzin, A. Y., and Y. I. Sobolevsky. 1976. Morphological and functional characteristics of the fur seal's respiratory system. Pages 168–170 *in* Proceedings of the 6th All-Union Conference on the Study of Marine Mammals, Kiev October 1–3, 1975. Joint Publication Service, Arlington, VA.

Kuznetsov, V. B. 1979. Chemoreception in dolphins of the Black Sea. Doklady Akadamii Nauk SSSR 249:1498–1500.

Kuznetsov, V. B. 1982. Taste perception of sea lions. Page 191 *in* V. A. Zemskiy (ed.). Izucheniye, okhrana, i ratstional'noye ispol'zovaniye morskikh mlekopitayushchikh. Ministry of Fisheries, USSR, Ichthyology Commission, VNIRO, and the Academy of Sciences, USSR, Astrakhan.

Kuznetsov, V. B. 1990. Chemical sense of dolphins: Quasi-olfaction. Pages 481–503 *in* J. A. Thomas and R. A. Kastelein (eds.). Sensory Abilities of Cetaceans: Laboratory and Field Evidence. Plenum, New York, NY.

Ladygina, T. F., V. V. Popov, and A. Y. Supin. 1992. Micromapping of the fur seal's somatosensory cerebral cortex. Pages 107–117 *in* J. A. Thomas, R. A. Kastelein, and A. Y. Supin (eds.). Marine Mammal Sensory Systems. Plenum, New York, NY.

Landau, D., and W. W. Dawson. 1970. The histology of retinas from the pinnipedia. Vision Research 10:691–702.

Lang, T. G. 1966. Hydrodynamic analysis of cetacean performance. Pages 410–432 *in* K.S. Norris (ed.). Whales, Dolphins and Porpoises. University of California Press, Berkeley.

Lang, T. G., and K. S. Norris. 1966. Swimming speed of a Pacific bottlenose porpoise. Science 151:588–590.

Lang, T. G., and K. Pryor. 1966. Hydrodynamic performance of porpoises (*Stenella attenuate*). Pages 531–533 *in* K. S. Norris (ed.). Whales, Dolphins and Porpoises. University of California Press, Berkeley.

Lavigne, D. M., and K. Ronald. 1972. The harp seal, *Pagophilus groenlandicus* (Erxleben 1777). XXIII. Spectral sensitivity. Canadian Journal of Zoology 50:1197–1206.

Lavigne, D. M., and K. Ronald. 1975a. Evidence of duplicity in the retina of the California sea lion (*Zalophys californianus*). Comparative Biochemistry and Physiology A 50:65–70.

Lavigne, D. M., and K. Ronald. 1975b. Pinniped visual pigments. Comparative Biochemistry and Physiology B 52:325–329.

Leatherwood, S., and W. A. Walker. 1979. The northern right whale dolphin *Lissodelphis borealis* Peale in the eastern North Pacific. Pages 85–141 *in* H. E. Winn and B. L. Olla (eds.). Behavior of Marine Animals, Vol. 3: Cetaceans. Plenum, New York, NY.

Leatherwood, S., T. A. Jefferson, J. C. Norris, W. E. Stevens, L. J. Hansen, and K. D. Mullin. 1993. Occurrence and sounds of Fraser's dolphins *Lagenodelphis hosei* in the Gulf of Mexico. Texas Journal of Science 45:349–354.

Lehnhardt, E. 1986. Clinical Aspects of Inner Ear Deafness. Springer-Verlag, New York, NY.

Lende, R. A., and W. I. Welker. 1972. An unusual sensory area in the cerebral neocortex of the bottlenose dolphin, *Tursiops truncatus*. Brain Research 45:555–560.

Levenson, C. 1974. Source level and bistatic target strength of the sperm whale *Physeter catodon* measured from an oceanographic aircraft. Journal of the Acoustical Society of America 55:1100–1103.

Levenson, D. H., and R. J. Schusterman. 1998. Dark adaptation and visual sensitivity in phocid and otariid pinnipeds. Page 79 *in* Abstracts of the World Marine Mammal Science Conference, Monte Carlo, Monaco, January 1998.

Lewis, E. R., E. L. Leverenz, and W. S. Bialek. 1985. The Vertebrate Inner Ear. CRC Press, Boca Raton, FL.

Lilly, J. C., and A. M. Miller. 1961. Sounds emitted by the bottlenose dolphin. Science 133:1689–1693.

Lindt, C. C. 1956. Underwater behavior of the Southern sea lion *Otaria jubata*. Journal of Mammalogy 37:287–288.

Ling, J. K. 1966. The skin and hair of the southern elephant seal, *Mirounga leonina* (Linn.). Australian Journal of Zoology 14:855–866.

Lipatov, N. V., and G. N. Solntseva. 1972. Some features of the biomechanics of the middle ear of dolphins. Makhachkala 2:137–140 (read as English summary).

Lipscomb, D. M. 1978. Noise and Audiology. University Park Press, Baltimore, MD.

Ljungblad, D. K., S. Leatherwood, and M. Dahlheim. 1980. Sounds recorded in the presence of an adult and calf bowhead whale. Marine Fisheries Review 42:86–87.

Ljungblad, D. K., P. O. Thompson, and S. E. Moore. 1982. Underwater sounds recorded from migrating bowhead whales, *Balaena mysticetus*, in 1979. Journal of the Acoustical Society of America 71:477–482.

Long, G. R. 1980. Some psychophysical measurements of frequency in the greater horseshoe bat. Pages 132–135 *in* G. van den Brink and F. Bilsen (eds.). Psychophysical, Psychological, and Behavioral Studies in Hearing. Delft University Press, Delft.

Lowell, W. R., and W. F. Flanigan Jr. 1980. Marine Mammal Chemoreception. Mammal Review 10:53–59.

Lynn, S. K., and D. L. Reiss. 1992. Pulse sequence and whistle production by two captive beaked whales, *Mesoplodon* species. Marine Mammal Science 8:299–305.

Lythgoe, J. N. 1968. Visual pigments and visual range under water. Vision Research 8:997–1011.

Lythgoe, J. N., and H. J. A. Dartnell. 1970. A "deep sea rhodopsin" in a mammal. Nature 227:955–956.

Mackay-Sim, A., D. Duvall, and B. M. Graves. 1985. The West Indian manatee, *Trichechus manatus*, lacks a vomeronasal organ. Brain, Behavior and Evolution 27:186–194.

Madsen, C. J., and L. M. Herman. 1980. Social and ecological correlates of cetacean vision and visual appearance. Pages 101–147 *in* L. M. Herman (ed.). Cetacean Behavior: Mechanisms and Functions. Wiley Interscience, New York, NY.

Manley, G. A. 1972. A review of some current concepts of the functional evolution of the ear in terrestrial vertebrates. Evolution 26:608–621.

Mann, G. 1946. Ojo y vision de las ballenas. Biologica 4:28–81.

Marsh, H., A. V. Spain, and G. E. Heinsohn. 1978. Physiology of the dugong. Comparative Biochemistry and Physiology A 61:159–168.

Marshall, C. D., and R. L. Reep. 1995. Manatee cerebral cortex: Cytoarchitecture of the caudal region in *Trichechus manatus latirostris*. Brain Behavior and Evolution 45:1–18.

Marshall, C. D., G. D. Huth, D. Halin, and R. L. Reep. 1998. Prehensile use of perioral bristles during feeding and associated behaviors of the Florida manatee (*Trichechus manatus latirostris*). Marine Mammal Science 14:274–289.

Martin, A. R., T. G. Smith, and O. P. Cox. 1993. Behaviour and movements of high arctic belugas. Pages 195–210 *in* I. L. Boyd (ed.). Marine mammals: Advances in behavioural and population biology. Clarendon Press, Oxford.

Mass, A., and A. Supin. 1990. Best vision zones in the retinae of some cetaceans. Pages 505-517 *in* J. A. Thomas and R. A. Kastelein (eds.). Sensory Abilities of Cetaceans: Laboratory and Field Evidence. Plenum, New York, NY.

Mass, A. M., and A. Y. Supin. 1995. Ganglion cell topography of the retina in the bottlenose dolphin, *Tursiops truncatus*. Brain, Behavior and Evolution 45:257–265.

Mass, A. M., and A. Y. Supin. 1997. Ocular anatomy, retinal ganglion cell distribution, and visual resolution in the gray whale, *Eschrichtius gibbosus*. Aquatic Mammals 23:17–28.

McCann, T. S. 1982. Aggressive and maternal activities of female southern elephant seals (*Mirounga leonina*). Animal Behavior 30:268–276.

McCormick, J. G., E. G. Wever, G. Palin, and S. H. Ridgway. 1970. Sound conduction in the dolphin ear. Journal of the Acoustical Society of America 48:1418–1428.

McCormick, J. G., E. G. Wever, S. H. Ridgway, and J. Palin. 1980. Sound reception in the porpoise as it relates to echolocation. Pages 449–467 *in* R. -G. Busnel and J. F. Fish (eds.). Animal Sonar Systems. Plenum, New York, NY.

McFarland, W. N. 1971. Cetacean visual pigments. Vision Research 11:1065–1076.

McFarland, W. N., and F. W. Munz. 1975. The photic environment of clear tropical seas during the day. Vision Research 15:1063–1070.

McLeod, P. J. 1986. Observations during the stranding of one individual from a pod of pilot whales, *Globicephala malaena*, in Newfoundland. Canadian Field-Naturalist 100:137–139.

McLuhan, M., and Q. Fiore. 1967. The Medium Is the Message. Co-ordinated by Jerome Agel. Random House, New York, NY.

Merzenich, M. M., J. H. Kass, M. Sur, and C. -S. Lin. 1978. Double representation of the body surface within cytoarchitectonic areas 3b and 1 in "SI" in the owl monkey (*Aotus trivirgatus*). Journal of Comparative Neurology 181:41–74.

Miller, E. H., and D. A. Job. 1992. Airborne acoustic communication in the Hawaiian monk seal, *Monachus schauinslandi*. Pages 485–531 *in* J. A. Thomas, R. A. Kastelein, and A. Y. Supin (eds.). Marine Mammal Sensory Systems. Plenum, New York, NY.

Miller, G. S. 1923. The telescoping of the cetacean skull. Smithsonian Miscellaneous Collection 76:1–67.

Mills, F., and D. Renouf. 1986. Determination of the vibration sensitivity of harbour seal (*Phoca vitulina*) vibrissae. Journal of Experimental Marine Biology and Ecology 100:3–9.

Møhl, B. 1964. Preliminary studies on hearing in seals. Videnskabelige Meddelelser Fra Dansk Naturhistorisk Forening I Kjobenhaven 127:283–294.

Møhl, B. 1967. Frequency discrimination in the common seal and a discussion of the concept of upper hearing limit. Pages 43–54 *in* V. Albers (ed.). Underwater Acoustics, Vol. II. Plenum, New York, NY.

Møhl, B. 1968. Hearing in seals. Pages 172–195 *in* R. Harrison, R. Hubbard, R. Peterson, C. Rice, and R. Schusterman (eds.). The Behavior and Physiology of Pinnipeds. Appleton-Century, New York, NY.

Møhl, B., and S. Andersen. 1973. Echolocation: High-frequency component in the click of the harbor porpoise (*Phocoena phocoena* L.). Journal of the Acoustical Society of America 57:1368–1372.

Møhl, B., and K. Ronald. 1975. The peripheral auditory system of the harp seal, *Pagophilus groenlandicus* (Erxleben 1777). Rapports et

Procés-Verbaux des Réunions, Conseil Internationale Pour l'Exploration de la Mer 169:516–523.

Møhl, B., J. M. Terhune, and K. Ronald. 1975. Underwater calls of the harp seal, *Pagophilus groenlandicus*. Rapports et Procés-Verbaux des Réunions, Conseil Internationale Pour l'Exploration de la Mer 169:533–543.

Møhl, B., A. Surlykke, and L. A. Miller. 1990. High intensity narwhal clicks. Pages 295–303 *in* J. A. Thomas and R. A Kastelein (eds.). Sensory Abilities of Cetaceans: Laboratory and Field Evidence. Plenum, New York, NY.

Montagna, W. 1967. Comparative anatomy and physiology of the skin. Archives of Dermatology 96:357–363.

Moore, P. W. B. 1990. Investigations on the control of echolocation pulses in the dolphin. Pages 305–317 *in* J. A. Thomas and R. A. Kastelein (eds.). Sensory Abilities of Cetaceans: Laboratory and Field Evidence. Plenum, New York, NY.

Moore, P. W. B., and W. W. L. Au. 1975. Underwater localization of pulsed pure tones by the California sea lion *Zalophus californianus*. Journal of the Acoustical Society of America 58:721–727.

Moore, P. W. B., and W. W. L. Au. 1983. Critical ratio and bandwidth of the Atlantic bottlenose dolphin (*Tursiops truncatus*). Journal of the Acoustical Society of America 74(suppl. 1):s73.

Moore, P. W. B., and R. J. Schusterman. 1976. Discrimination of pure tone intensities by the California sea lion. Journal of the Acoustical Society of America 60:1405–1407.

Moore, P. W. B., D. A. Pawloski, and L. Dankiewicz. 1995. Interaural time and intensity difference thresholds in the bottlenose dolphin (*Tursiops truncatus*). Pages 11–25 *in* R. A. Kastelein, J. A. Thomas, and P. E. Nachtigall (eds.). Sensory Systems of Aquatic Mammals. DeSpil Publishers, Woerden, Netherlands.

Moore, S. E., and D. K. Ljungblad. 1984. Gray whales in the Beaufort, Chukchi, and Bering seas: Distribution and sound production. Pages 543–559 *in* M. L. Jones, S. L. Swartz, and S. Leatherwood (eds.). The Gray Whale *Eschrictius robustus*. Academic Press, Orlando, FL.

Moore, S. E., and S. H. Ridgway. 1995. Whistles produced by common dolphins from the southern California bight. Aquatic Mammals 21:55–63.

Moore, S. E., J. K. Francine, A. E. Bowles, and J. K. B. Ford. 1988. Analysis of calls of killer whales, *Orcinus orca*, from Iceland and Norway. Rit Fiskideilder 11:225–250.

Morgane, P. J., and J. S. Jacobs. 1972. Comparative anatomy of the cetacean nervous system. Pages 117–224 *in* R. J. Harrison (ed.). Functional Anatomy of Marine Mammals, Vol. 1. Academic Press, New York, NY.

Morton, A. B., J. C. Gale, and R. C. Prince. 1986. Sound and behavioral correlations in captive *Orcinus orca*. Pages 303–333 *in* B. C. Kirkevold and J. S. Lockard (eds.). Behavioral Biology of Killer Whales. Alan R. Liss, New York, NY.

Muir, D. W., and D. E. Mitchell. 1973. Visual resolution and experience: Acuity deficits in cats following early selective visual deprivation. Science 180:420–422.

Mullin, K. D., L. V. Higgins, T. A. Jefferson, and L. J. Hansen. 1994. Sightings of the Clymene dolphin *Stenella clymene* in the Gulf of Mexico. Marine Mammal Science 10:464–470.

Munz, F. W. 1965. Adaptation of visual pigments to the photic environment. Pages 27–51 *in* A. V. S. de Reuck and J. Knight (eds.). Ciba Foundation Symposium on Color Vision–Physiology and Exploratory Psychology. Little Brown and Co., Boston, MA.

Murayama, T., Y. Fujise, I. Aoki, and T. Ishii. 1992. Histological characteristics and distribution of ganglion cells in the retinae of the Dall's porpoise and minke whale. Pages 137–145 *in* J. A. Thomas, R. A. Kastelein, and A. Y. Supin (eds.). Marine Mammal Sensory Systems. Plenum, New York, NY.

Murphy, C. J., R. W. Bellhorn, T. Williams, M. S. Burns, F. Schaeffel, and H. C. Howland. 1990. Refractive state, ocular anatomy and accommodative range of the sea otter *Enhydra lutris*. Vision Research 30:23–32.

Nachtigall, P. E. 1986. Vision, audition, and chemoreception in dolphins and other marine mammals. Pages 79–113 *in* R. J. Schusterman, J. A. Thomas, and F. G. Wood (eds.). Dolphin Cognition and Behavior: A Comparative Approach. Lawrence Erlbaum Associates, Hillsdale, NJ.

Nachtigall, P. E., and R. W. Hall. 1984. Taste reception in the bottlenose dolphin. Acta Zoologica Fennica 172:147–148.

Nachtigall, P. E., W. W. L. Au, and J. Pawlowski. 1996. Low-frequency hearing in three species of odontocetes. Journal of the Acoustical Society of America 100:2611.

Nadol, J. B. 1988. Quantification of human spiral ganglion cells by serial section reconstruction and segmental density estimates. American Journal of Otolaryngology 9:47–51.

Nagy, A. R., and K. Ronald. 1970. The harp seal, *Pagophilus groenlandicus* (Erxleben, 1777). VI. Structure of retina. Canadian Journal of Zoology 48:367–370.

Nagy, A. R., and K. Ronald. 1975. A light and electronmicroscopic study of the structure of the retina of the harp seal, *Pagophilus groenlandicus* (Erxleben, 1777). Rapports et Procés-Verbaux des Réunions, Conseil Internationale Pour l'Exploration de la Mer 169:92–96.

Nair, R.V., and R.S. Lal Mohan. 1975. Studies on the vocalisation of the sea cow *Dugong dugong* in captivity. Indian Journal of Fisheries 22:277–278.

Nelson, R. J., M. Sur, and J. H. Kaas. 1979. The organization of the second somatosensory area (SmII) in the grey squirrel. Journal of Comparative Neurology 184:473–490.

Nelson, R. J., M. Sur, D. J. Felleman, and J. H. Kaas. 1980. Representation of the body surface in postcentral parietal cortex of *Macaca fascicularis*. Journal of Comparative Neurology 192:611–644.

Nishiwaki, M., and H. Marsh. 1985. Dugong *Dugong dugon* (Miller, 1776). Pages 1–31 *in* S. H. Ridgway and R. Harrison (eds.). Handbook of Marine Mammals, Vol. 3: The Sirenians and Baleen Whales. Academic Press, London.

Nojima, T. 1988. Developmental pattern of the bony falx and bony tentorium of spotted dolphins (*Stenella attenuata*) and the relationship between degree of development and age. Marine Mammal Science 4:312–322.

Norris, J. C., and S. Leatherwood. 1981. Hearing in the bowhead whale, *Balaena mysticetus*, as estimated by cochlear morphology. Hubbs Sea World Research Institute Technical Report No. 81-132:15.1–15.49.

Norris, K. S. 1968. The evolution of acoustic mechanisms in odontocete cetaceans. Pages 297–324 *in* E. T. Drake (ed.). Evolution and Environment. Yale University Press, New Haven, CT.

Norris, K. S. 1969. The echolocation of marine mammals. Pages 391–423 *in* H. J. Andersen (ed.). The Biology of Marine Mammals. Academic Press, New York, NY.

Norris, K. S. 1980. Peripheral sound processing in odontocetes. Pages 495–509 *in* R. -G. Busnel and J. F. Fish (eds.). Animal Sonar Systems. Plenum, New York, NY.

Norris, K. S., and W. E. Evans. 1967. Directionality of echolocation clicks in the rough-tooth porpoise, *Steno bredanensis* (Lesson). Pages 305–316 *in* W. N. Tavolga (ed.). Marine Bio-Acoustics, Vol. 2. Pergamon Press, Oxford.

Norris, K. S., and G. W. Harvey. 1974. Sound transmission in the porpoise head. Journal of the Acoustical Society of America 56:659–664.

Norris, K. S., and W. A. Watkins. 1971. Underwater sounds of *Arctocephalus philippii*, the Juan Fernandez fur seal. Antarctic Research Series 18:169–171.

Norris, K. S., J. H. Prescott, P. V. Asa-Dorian, and P. Perkins. 1961. An experimental demonstration of echolocation behavior in the porpoise, *Tursiops truncatus*, Montagu. Biological Bulletin 120:163–176.

Norris, K. S., G. W. Harvey, L. A. Burzell, and D. K. Kartha. 1972. Sound production in the freshwater porpoise *Sotalia* cf. *fluviatilis* Gervais and Deville and *Inia geoffrensis* Blainville in the Rio Negro Brazil. Investigations on Cetacea 4:251–262.

Norris, K. S., R. M. Goodman, B. Villa-Ramirez, and L. Hobbs. 1977. Behavior of California gray whale, *Eschrictius robustus*, in southern Baja California, Mexico. Fishery Bulletin 75:159–172.

Norris, K. S., B. Würsig, R. S. Wells, and M. Würsig, with S. M. Brownlee, C. M. Johnson, and J. Solow. 1994. The Hawaiian Spinner Dolphin. University of California Press, Berkeley.

Noseworthy, E., D. Renouf, and W. K. Jacobs. 1989. Acoustic breeding displays of harbour seals. Page 46 *in* Abstracts of the 8th Biennial Conference on the Biology of Marine Mammals, Pacific Grove, CA, December 1989.

Nowak, R. M. 1991. Mammals of the World, Vol. 2, 5th ed. The Johns Hopkins University Press, Baltimore, MD.

Odell, D. K., D. J. Forrester, and E. D. Asper. 1981. A preliminary analysis of organ weights and sexual maturity in the West Indian manatee (*Trichechus manatus*). Pages 52–65 *in* R. L. Brownell and K. Ralls (eds.). The West Indian Manatee in Florida. Proceedings of a Workshop, Orlando, FL.

Oelschläger, H. A. 1986. Comparative morphology and evolution of the otic region in toothed whales Cetacea Mammalia. American Journal of Anatomy 177:353–368.

Oliver, G. W. 1978. Navigation in mazes by a grey seal, *Halichoerus grypus* (Fabricius). Behaviour 67:97–114.

Østerberg, G. 1935. Topography of the layer of rods and cones in the human retina. Acta Ophthalmologica (Suppl.) 6:1–102.

Patton, G. W., and E. Gerstein. 1992. Toward understanding mammalian hearing tractability: Preliminary acoustical perception thresholds in the West Indian manatee, *Trichechus manatus*. Page 783 *in* D. Webster, R. Fay, and A. Popper (eds.). The Biology of Hearing. Springer-Verlag, Berlin.

Payne, K., and R. Payne. 1985. Large scale changes over 19 years in songs of humpback whales in Bermuda. Zeitschrift für Tierpsychologie 68:89–114.

Payne, K. B., P. Tyack, and R. S. Payne. 1983. Progressive changes in the songs of humpback whales (*Megaptera novaeangliae*). Pages 9–57 *in* R. Payne (ed.). Communication and Behavior of Whales. AAAS Selected Symposium 76. Westview Press, Boulder, CO.

Payne, K. B., W. J. Langbauer Jr., and E. M. Thomas. 1986. Infrasonic calls of the Asian elephant (*Elephas maximus*). Behavioral Ecology and Sociobiology 18:297–301.

Pedler, C. 1963. The fine structure of the tapetum cellulosum. Experimental Eye Research 2:189–195.

Pepper, R. L., and J. V. Simmons Jr. 1973. In-air visual acuity of the bottlenose dolphin. Experimental Neurology 41:271–276.

Perez, J. M., W. W. Dawson, and D. Landau. 1972. Retinal anatomy of the bottlenose dolphin (*Tursiops truncatus*). Cetology 11:1–11.

Petit, G., and A. Rochon-Duvigneaud. 1929. L'oeil et la vision de L'*Halicore dugong* Erxl. Bulleting, Société Zoologique de France 54:129–138.

Phillips, J. B., and K. Adler. 1978. Directional and discriminatory responses of salamanders to weak magnetic fields. Pages 325–333 *in* K. Schmidt-Koenig and W. T. Keaton (eds.). Animal Migration, Navigation, Homing. Springer-Verlag, Berlin.

Pickles, J. O. 1982. An introduction to the physiology of hearing. Academic Press, London.

Piggins, D. J. 1970. The refraction of the harp seal, *Pagophilus groenlandicus* (Erxleben 1977). Nature 227:78–79.

Piggins, D. J., W. R. A. Muntz, and R. C. Best. 1983. Physical and morphological aspects of the eye of the manatee *Trichechus inunguis* Natterer 1883: (Sirenia: mammalia). Marine Behaviour and Physiology 9:111–130.

Pilleri, G. 1983. The sonar system of the dolphins. Endeavour (New Series) 7:59–64.

Pilleri, G., and M. Gihr. 1970. The central nervous system of the Mysticete and Odontocete whales. Investigations on Cetacea 2:89–127.

Pilleri, G. C., C. Kraus, and M. Gihr. 1971. Physical analysis of the sounds emitted by *Platanista indi*. Investigations on Cetacea 3:22–30.

Pilleri, G. K., K. Zbinden, and C. Kraus. 1980. Characteristics of the sonar system of cetaceans with pterygoschisis. Directional properties of the sonar clicks of *Neophocaena phocaenoides* and *Phocoena phocoena* (Phocoenidae). Investigations on Cetacea 11:157–188.

Pollack, G. D. 1980. Organizational and encoding features of single neurons in the inferior colliculus of bats. Pages 549–587 *in* R. -G. Busnel and J. F. Fish (eds.). Animal Sonar Systems, Plenum, New York, NY.

Popov, V. V., and A. Y. Supin. 1990a. Electrophysiological studies on hearing in some cetaceans and a manatee. Pages 405–416 *in* J. A. Thomas and R. A. Kastelein (eds.). Sensory Abilities of Cetaceans: Laboratory and Field Evidence. Plenum, New York, NY.

Popov, V. V., and A. Y. Supin. 1990b. Localization of the acoustic window at the dolphin's head. Pages 417–427 *in* J. A. Thomas and R. A. Kastelein (eds.). Sensory Abilities of Cetaceans: Laboratory and Field Evidence. Plenum, New York, NY.

Popov, V. V., T. F. Ladygina, and A. Y. Supin. 1986. Evoked potentials of the auditory cortex of the porpoise *Phocoena phocoena*. Journal of Comparative Physiology 158:705–711.

Popper, A. N. 1980. Sound emission and detection by delphinids. Pages 1–52 *in* L. M. Herman (ed.). Cetacean Behavior: Mechanisms and Functions. John Wiley and Sons, New York, NY.

Poulter, T. C. 1968. Underwater vocalization and behavior of pinnipeds. Pages 69–84 *in* R. J. Harrison, R. C. Hubbard, R. S. Peterson, C. E. Rice, and R. J. Schusterman (eds.). The Behavior and Physiology of Pinnipeds. Appleton-Century-Crofts, New York, NY.

Pryor, T., K. Pryor, and K. S. Norris. 1965. Observations on a pygmy killer whale *Feresa attenuata* Gray from Hawaii. Journal of Mammalogy 46:450–461.

Pubols, B. H., L. M. Pubols, D. J. DePette, and J. C. Sheely. 1976. Opossum somatic sensory cortex: A microelectrode mapping study. Journal of Comparative Neurology 165:229–246.

Pye, A. 1972. Variations in the structure of the ear in different mammalian species. Sound 6:14–18.

Ralls, K. P., P. Fiorelli, and S. Gish. 1985. Vocalizations and vocal mimicry in captive harbor seals, *Phoca vitulina*. Canadian Journal of Zoology 63:1050–1056.

Ramprashad, F., S. Corey, and K. Ronald. 1972. Anatomy of the seal ear *Pagophilus groenlandicus* (Erxleben,1777). Pages 264–306 *in* R. Harrison (ed.). Functional Anatomy of Marine Mammals, Vol. I. Academic Press, London.

Rand, R. W. 1955. Reproduction in the female Cape fur seal, *Arctocephalus pusillus* (Schreber). Proceedings of the Zoological Society of London 124:717–740.

Ray, C. 1970. Population ecology of Antarctic seals. Pages 398–414 *in* M. W. Holdgate (ed.). Antarctic Ecology, Vol. 1. Academic Press, London.

Ray, G. C., and W. A. Watkins. 1975. Social function of underwater sounds in the walrus *Odobenus rosmarus*. Rapports et Procés-Verbaux des Réunions, Conseil Internationale Pour l'Exploration de la Mer 169:524–526.

Ray, G. C., W. A. Watkins, and J. J. Burns. 1969. The underwater song of *Erignathus* (bearded seal), Zoologica 54:79–83, 3 plates, 1 phonograph record.

Reep, R. L., C. D. Marshall, M. L. Stoll, and D. M. Whitaker. 1998. Distribution and innervation of facial bristles and hairs in the Florida manatee (*Trichechus manatus latirostris*). Marine Mammal Science 14:257–273.

Renaud, D. L., and A. N. Popper. 1975. Sound localization by the bottlenose porpoise *Tursiops truncatus*. Journal of Experimental Biology 63:569–585.

Renouf, D. 1979. Preliminary measurements of the sensitivity of the vibrissae of harbour seals (*Phoca vitulina*) to a low frequency vibrations. Journal of Zoology (London) 188:443–450.

Renouf, D. 1980. Fishing in captive harbor seals (*Phoca vitulina concolor*): A possible role for vibrissae. Netherlands Journal of Zoology 30:504–509.

Renouf, D. 1992. Sensory reception and processing in Phocidae and Otariidae. Pages 345–394 *in* D. Renouf (ed.). Behaviour of Pinnipeds. Chapman and Hall, London.

Renouf, D., and L. Gaborko. 1982. Speed sensing in a harbour seal. Journal of the Marine Biological Association of the United Kingdom 62:227–228.

Renouf, D., G. Galway, and L. Gaborko. 1980. Evidence for echolocation in harbor seals. Journal of the Marine Biology Association 60:1039–1042.

Repenning, C. 1972. Underwater hearing in seals. Pages 307–331 *in* R. Harrison (ed.). Functional Anatomy of Marine Mammals, Vol. I. Academic Press, London.

Reynolds, J. E. 1981. Behavior patterns in the West Indian manatee, with emphasis on feeding and diving. Florida Scientist 44:233–242.

Reysenbach de Haan, F. W. 1956. Hearing in whales. Acta Otolaryngologica (Suppl.) 134:1–114.

Richardson, W. J., C. R. Greene Jr., C. I. Malme, and D. H. Thomson. 1991. Effects of noise on marine mammals. USDI / MMA / OCS study 90-0093. LGL Ecological Research Associates, Bryan, TX.

Richardson, W. J., C. R. Greene Jr., C. I. Malme, and D. H. Thomson. 1995. Marine Mammals and Noise. Academic Press, New York, NY.

Ridgway, S. H. 1972. Mammals of the Sea: Biology and Medicine. Charles H. Thomas, Springfield, IL.

Ridgway, S. H., and D. A. Carder. 1990. Tactile sensitivity, somatosensory responses, skin vibrations, and the skin surface ridges of the bottlenose dolphin, *Tursiops truncatus*. Pages 163–179 *in* J. A. Thomas and R. A. Kastelein (eds.). Sensory Abilities of Cetaceans: Laboratory and Field Evidence. Plenum, New York, NY.

Ridgway, S. H., T. H. Bullock, D. A. Carder, R. L. Seeley, D. Woods, and R. Galambos. 1981. Auditory brainstem response in dolphins, Proceedings of the National Academy of Science 78:1943–1947.

Rivamonte, L. A. 1976. Eye model to account for comparable aerial and underwater acuities of the bottlenose dolphin. Netherlands Journal of Sea Research 19:491–498.

Robineau, D. 1969. Morphologie externe du complexe osseux temporal chez les sireniens. Mémoires du Musée Nationale d'Histoire Naturelle, Nouvelle Séries, Série A, Zoologie 60:1–32.

Rodieck, R. W. 1973. The Vertebrate Retina. Freeman, San Francisco, CA.

Rogers, T., D. H. Cato, and M. M. Bryden. 1995. Underwater vocal repertoire of the leopard seal *Hydrurga leptonyx* in Prydz Bay, Antarctica. Pages 223–236 *in* R. A. Kastelein, J. A. Thomas, and P. E. Nachtigall (eds.). Sensory Systems of Aquatic Mammals. DeSpil Publishers, Woerden, Netherlands.

Ronald, K., and J. Lee. 1981. The spectral sensitivity of a polar bear. Comparative Biochemistry and Physiology A 70:595–598.

Rosowski. 1994. Outer and middle ears. Pages 172–247 *in* R. R. Fay and A. N. Popper (eds.). Comparative Hearing: Mammals. Springer-Verlag, New York, NY.

Sales, G., and D. Pye. 1974. Ultrasonic Communication by Animals. John Wiley and Sons, New York, NY.

Santoro, A. K, K. L. Marten, and T. W. Cranford. 1989. Pygmy sperm whale sounds *Kogia breviceps*. Page 59 *in* Abstracts of the 8th Biennial Conference on the Biology of Marine Mammals. Pacific Grove, CA, December 1989.

Schevill, W. E. 1964. Underwater sounds of cetaceans. Pages 307–316 *in* W. N. Tavolga (ed.). Marine Bio-Acoustics. Pergamon Press, New York, NY.

Schevill, W. E., and B. Lawrence. 1949. Underwater listening to the white porpoise *Delphinapterus leucas*. Science 109:143–144.

Schevill, W. E., and W. A. Watkins. 1965. Underwater calls of *Trichechus* (manatee). Nature 205:373–374.

Schevill, W. E., and W. A. Watkins. 1966. Sound structure and directionality in *Orcinus* (killer whale). Zoologica 51:71–76.

Schevill, W. E., and W. A. Watkins. 1971. Pulsed sounds of the porpoise *Lagenorhynchus australis*. Breviora 366:1–10.

Schevill, W. E., and W. A. Watkins. 1972. Intense low frequency sounds from an Antarctic minke whale, *Balaenoptera acutorostrata*. Breviora 388:1–8.

Schevill, W. E., W. A. Watkins, and C. Ray. 1963. Underwater sounds of pinnipeds. Science 141:50–53.

Schevill, W. E., W. A. Watkins, and C. Ray. 1966. Analysis of underwater *Odobenus* calls with remarks on the development and function of the pharyngeal pouches. Zoologica 51:103–106.

Schevill, W. E., W. A. Watkins, and C. Ray. 1969. Click structure in the porpoise, *Phocoena phocoena*. Journal of Mammalogy 50:721–728.

Schnitzler, H. U. 1983. Fluttering target detection in horse-shoe bats. Journal of the Acoustical Society of America 74(suppl. 1):S31–S32.

Schuknecht, H. F. 1993. Pathology of the Ear, 2nd ed. Lea and Febiger, Philadelphia, PA.

Schuknecht, H. F., and A. J. Gulya. 1986. Anatomy of the Temporal Bone With Surgical Implications. Lea and Feibiger, Philadelphia, PA.

Schultz, K. W., and P. J. Corkeron, 1994. Interspecific differences in whistles produced by inshore dolphins in Moreton Bay, Queensland, Australia. Canadian Journal of Zoology 72:1061–1068.

Schusterman, R. J. 1972. Visual acuity in pinnipeds. Pages 469–492 *in* H. E. Winn and B. L. Olla (eds.). Behavior of Marine Animals, Vol. 2: Vertebrates. Plenum, New York, NY.

Schusterman, R. J. 1981. Behavioral capabilities of seals and sea lions: A review of their hearing, visual, learning and diving skills. Psychological Record 31:125–143.

Schusterman, R. J., and R. F. Balliet. 1970a. Conditioned vocalizations as a technique for determining visual acuity thresholds in sea lions. Science 169:498–501.

Schusterman, R. J., and R. F. Balliet. 1970b. Visual acuity of the harbor seal and steller sea lion underwater. Nature 226:563–564.

Schusterman, R. J., and R. F. Balliet. 1971. Aerial and underwater visual acuity in the California sea lion (*Zalophys californianus*) as a function of luminance. Annals of the New York Academy of Sciences 188:37–46.

Schusterman, R. J., and B. Barrett. 1973. Amphibious nature of visual acuity in the Asian "clawless" otter. Nature 244:518–519.

Schusterman, R. J., and P. W. B. Moore. 1978a. Underwater audiogram of the northern fur seal *Callorhinus ursinus*. Journal of the Acoustical Society of America 64: S87.

Schusterman, R. J., and P. W. B. Moore. 1978b. The upper hearing limit of underwater auditory frequency discrimination in the California sea lion. Journal of the Acoustical Society of America 63:1591–1595.

Schusterman, R. J., W. N. Kellog, and C. E. Rice. 1965. Underwater visual discrimination by the California sea lion. Science 147:1594–1596.

Schusterman, R. J., R. Gentry, and J. Schmook. 1967. Underwater sound production by captive California sea lions, *Zalophus californianus*. Zoologica 52:21–24+5 plates.

Schusterman, R. J., R. Balliet, and J. Nixon. 1972. Underwater audiogram of the California sea lion by the conditioned vocalization technique. Journal of Experimental Animal Behaviour 17:339–350.

Semmes, J., and L. Porter. 1972. A comparison of precentral and postcentral cortical lesions on somatosensory discrimination in the monkey. Cortex 10:55–68.

Shaw, N. A. 1990. Central auditory conduction time in the rat. Experimental Brain Research 79:217–220.

Shipley, C., B. S. Stewart, and J. Bass. 1992. Seismic communication in northern elephant seals. Pages 553–562 *in* J. A. Thomas, R. A. Kastelein, and A. Y. Supin (eds.). Marine Mammal Sensory Systems. Plenum, New York, NY.

Shochi, Y., K. Zbinden, C. Kraus, M. Gihr, and G. Pilleri. 1982. Characteristics and directional properties of the sonar signals emitted by the captive Commerson's dolphin, *Cephalorhynchus commersoni* (Gray, 1846). Investigations on Cetacea 13:177–201.

Silber, G. K. 1986. The relationships of social vocalizations to surface behavior and aggression in the Hawaiian humpback whale (*Megaptera novaeangliae*). Canadian Journal of Zoology 64:2075–2080.

Silber, G. K. 1991. Acoustic signals of the vaquita *Phocoena sinus*. Aquatic Mammals 17:130–133.

Simmons, J. A. 1973. The resolution of target range by echolocating bats. Journal of the Acoustical Society of America 54:157–173.

Simons, D., and M. Huigen. 1977. Analysis of an experiment on colour vision in dolphins. Aquatic Mammals 5:27–33.

Sivak, J. G., and D. J. Piggins. 1975. Refractive state of the eye of the polar bear (*Thalarctos maritimus* Phipps). Norwegian Journal of Zoology 23:89–91

Sjare, B. L., and T. G. Smith. 1986a. The vocal repertoire of white whales, *Delphinapterus leucas*, summering in Cunningham Inlet, Northwest Territories. Canadian Journal of Zoology 64:407–415.

Sjare, B. L., and T. G. Smith. 1986b. The relationship between behavioral activity and underwater vocalizations of the white whale, *Delphinapterus leucas*. Canadian Journal of Zoology 64:2824–2831.

Slip, D. J., M. A. Hindell, and H. R. Burton. 1994. Diving behavior of southern elephant seals from Macquarie Island: An overview. Pages 253–270 *in* B. J. Le Boeuf and R. M. Laws (eds.). Elephant Seals: Population Ecology, Behavior, and Physiology. University of California Press, Berkeley.

Smith, P. F., and J. Wojtowicz. 1985. Temporary auditory threshold shifts induced by twenty-five minute continuous exposures to intense tones in water. Naval Medical Research and Development Command. USN, Report 1063:1–13.

Smith, P. F., J. Wojtowicz and S. Carpenter. 1988. Temporary auditory threshold shifts induced by ten-minute exposures to continuous tones in water. Naval Medical Research and Development Command. USN, Report 1122:1–10.

Solntseva, G. N. 1971. Comparative anatomical and histological characteristics of the structure of the external and inner ear of some dolphins. Trudy Atlanticheski Kaliningrad Nauchno-Issledovatel'skogo Instituta Rybnogo Khoziaistva Okeanografii, pages 237–244 (read as English summary).

Solntseva, G. N. 1990. Formation of an adaptive structure of the peripheral part of the auditory analyzer in aquatic, echo-locating mammals during ontogenesis. Pages 363–384 *in* J. A. Thomas and R. A. Kastelein (eds.). Sensory Abilities of Cetaceans: Laboratory and Field Evidence. Plenum, New York, NY.

Sonafrank, N., R. Elsner, and D. Wartzok. 1983. Under-ice navigation by the spotted seal *Phoca largha*. Page 97 *in* Abstracts of Fifth Biennial Conference on the Biology of Marine Mammals, Boston, MA, November 1983.

Sonoda, S., and A. Takemura. 1973. Underwater sounds of the manatees, *Trichechus manatus manatus* and *T inunguis* (Trichechidae). Report of the Institute for Breeding Research, Tokyo University for Agriculture 4:19–24.

Spector, W. S. 1956. Handbook of Biological Data. Saunders, Philadelphia, PA.

Spero, D. 1981. Vocalizations and associated behavior of northern right whales *Eubalaena glacialis*. Page 108 *in* Abstracts of the 4th Biennial Conference on the Biology of Marine Mammals, San Francisco, CA, December 1981.

Stafford, K. M., C. G. Fox, and B. R. Mate. 1994. Acoustic detection and location of blue whales (*Balaenoptera musculus*) from SOSUS data by matched filtering. Journal of the Acoustical Society of America 96(5, Pt. 2):3250–3251.

Steiner, W. W. 1981. Species-specific differences in pure tonal whistle vocalizations of five western North Atlantic dolphin species. Behavioral Ecology and Sociobiology 9:241–246.

Steiner, W. W., J. H. Hain, H. E. Winn, and P. J. Perkins. 1979. Vocalizations and feeding behavior of the killer whale (*Orcinus orca*). Journal of Mammalogy 60:823–827.

Stephens, R. J., I. J. Beebe, and T. C. Poulter. 1973. Innervation of the vibrissae of the California sea lions, *Zalophus californianus*. Anatomical Record 176:421–442.

Stirling, I. 1973. Vocalization in the ringed seal *Phoca hispida*. Journal of the Fisheries Research Board of Canada 30:1592–1594.

Stirling, I., and D. B. Siniff. 1979. Underwater vocalizations of leopard seals *Hydrurga leptonyx* and crabeater seals *Lobodon carcinophagus* near the South Shetland Islands, Antarctica. Canadian Journal of Zoology 57:1244–1248.

Stirling, I., W. Calvert, and H. Cleator. 1983. Underwater vocalizations as a tool for studying the distribution and relative abundance of wintering pinnipeds in the high Arctic. Arctic 36:262–274.

Stone, J. 1965. A quantitative analysis of the distribution of ganglion cells in the cat's retina. Journal of Comparative Neurology 124:337–352.

Suga, N. 1983. Neural representation of bisonar [sic] information in the auditory cortex of the mustached bat. Journal of the Acoustical Society of America 74(suppl. 1):S31.

Sukhovskaya, L. I. 1972. The morphology of the taste organ in dolphins. Investigations on Cetacea 4:201–240.

Supin, A. Y., and V. V. Popov. 1990. Frequency selectivity of the auditory system of the bottlenose dolphin *Tursiops truncatus*. Pages 385–393 *in* J. A. Thomas and R. A. Kastelein (eds.). Sensory Abilities of Cetaceans: Laboratory and Field Evidence. Plenum, New York, NY.

Supin, A. Y., and V. V. Popov. 1993. Direction-dependent spectral sensitivity and interaural spectral difference in a dolphin: Evoked potential study. Journal of the Acoustical Society of America 93:3490–3495.

Sur, M., R. J. Nelson, and J. H. Kaas. 1978. The representation of the body surface in somatosensory area I of the grey squirrel. Journal of Comparative Neurology 179:425–450.

Sur, M., R. J. Nelson, and J. H. Kaas. 1980. Representation of the body surface in somatic koniocortex in the prosimian *Galago*. Journal of Comparative Neurology 189:381–402.

Sur, M., R. J. Nelson, and J. H. Kaas. 1982. Representation of the body surface in cortical area 3b and 1 of squirrel monkeys: Comparison with other primates. Journal of Comparative Neurology 211:177–192.

Surkina, R. M. 1971. Structure and function of the skin muscles of dolphins. Bionika 5:81–87.

Swartz, S. L., and W. C. Cummings. 1978. Gray whales, *Eschrictius robustus,* in Laguna San Ignacio, Baja California, Mexico, MMC-77/04. USMMC Report NTIS PB-276319. San Diego Natural History Museum, Washington, D.C.

Taruski, A.G. 1979. The whistle repertoire of the North Atlantic pilot whale *Globicephala malaena* and its relationship to behavior and environment. Pages 345–368 *in* H. E. Winn and B. L. Olla (eds.). Behavior of Marine Animals, Vol. 3: Cetaceans. Plenum, New York, NY.

Terhune, J. M. 1994. Geographical variation of harp seal underwater vocalizations. Canadian Journal of Zoology 72:892–897.

Terhune, J. M., and K. Ronald. 1973. Some hooded seal *Cystophora cristata* sounds in March. Canadian Journal of Zoology 51:319–321.

Terhune, J. M., and K. Ronald. 1986. Distant and near-range functions of harp seal underwater calls. Canadian Journal of Zoology 64:1065–1070.

Thomas, J. A., and V. B. Kuechle. 1982. Quantitative analysis of Weddell seal *Leptonychotes weddelli* underwater vocalizations at McMurdo Sound, Antarctica. Journal of the Acoustical Society of America 72:1730–1738.

Thomas, J. A., and I. Stirling. 1983. Geographic variation in the underwater vocalizations of Wedell seals *Leptonychotes wedelli* from Palmer Peninsula and McMurdo Sound, Antarctica. Canadian Journal of Zoology 61:2203–2212.

Thomas, J. A., and C. W. Turl. 1990. Echolocation characteristics and range detection threshold of a false killer whale *Pseudorca crassidens*. Pages 321–334 *in* J. A. Thomas and R. A. Kastelein (eds.). Sensory Abilities of Cetaceans: Laboratory and Field Evidence. Plenum, New York, NY.

Thomas, J. A., S. R. Fisher., W. E. Evans, and F. T. Awbrey. 1983a. Ultrasonic vocalizations of leopard seals *Hydrurga leptonyx*. Antarctica Journal U.S. 17:186.

Thomas, J. A., K. C. Zinnel, and L. M. Ferm. 1983b. Analysis of Wedell seal *Leptonychotes wedelli* vocalizations using underwater playbacks. Canadian Journal of Zoology 61:1448–1456.

Thomas, J. A., N. Chun, W. Au., and K. Pugh. 1988. Underwater audiogram of a false killer whale (*Pseudorca crassidens*). Journal of the Acoustical Society of America 84:936–940.

Thomas, J. A., P. W. B. Moore, R. Withrow, and M. Stoermer. 1990a. Underwater audiogram of a Hawaiian monk seal *Monachus schauinslandi*. Journal of the Acoustical Society of America 87:417–420.

Thomas, J. A., J. L. Pawloski and W. W. L. Au. 1990b. Masked hearing abilities in a false killer whale (*Pseudorca crassidens*). Pages 395–404 *in* J. A. Thomas and R. A. Kastelein (eds.). Sensory Abilities of Cetaceans: Laboratory and Field Evidence. Plenum, New York, NY.

Thompson, P. O., W. C. Cummings, and S. J. Ha. 1986. Sounds, source levels, and associated behavior of humpback whales, southeast Alaska. Journal of the Acoustical Society of America 80:735–740.

Thompson, T. J., H. E. Winn, and P. J. Perkins. 1979. Mysticete sounds. Pages 403–431 *in* H. E. Winn and B. L. Olla (eds.). Behavior of Marine Animals, Current Perspectives in Research, Vol. 3: Cetaceans. Plenum, New York, NY.

Turnbull, S. D., and J. M. Terhune. 1993. Repetition enhances hearing detection thresholds in a harbour seal (*Phoca vitulina*). Canadian Journal of Zoology 71:926–932.

Tyack, P. 1985. An optical telemetry device to identify which dolphin produces a sound. Journal of the Acoustical Society of America 7885:1892–1895.

van Esch, A., and J. de Wolf. 1979. Evidence for cone function in the dolphin retina—A preliminary report. Aquatic Mammals 7:35–37.

Varanasi, U., and D. G. Malins. 1971. Unique lipids of the porpoise *Tursiops gilli:* Differences in triacyl glycerols and wax esters of acoustic (mandibular canal and melon) and blubber tissues. Biochemica and Biophysica Acta 231:415.

Vater, M. 1988a. Lightmicroscopic observations on cochlear development in horseshoe bats *Rhinolophus rouxii*. Pages 341–345 *in* P. E. Nachtigall and P. W. B. Moore (eds.). Animal Sonar Processes and Performance. Plenum, New York, NY.

Vater, M. 1988b. Cochlear physiology and anatomy in bats. Pages 225–241 *in* P. E. Nachtigall and P. W. B. Moore (eds.). Animal Sonar Processes and Performance. Plenum, New York, NY.

Verboom, W. C., and R. A. Kastelein. 1995. Rutting whistles of a male Pacific walrus (*Odobenus rosmarus divergens*). Pages 287–299 *in* R. A. Kastelein, J. A. Thomas, and P. E. Nachtigall (eds.). Sensory Systems of Aquatic Mammals. DeSpil Publishers, Woerden, Netherlands.

Vine, F. J. 1966. Spreading of the ocean floor: New evidence. Science 154:1405–1415.

Vogel, S. 1981. Life in Moving Fluids: The Physical Biology of Flow. Willard Grant Press, Boston, MA.

von Békésy, G. 1960. Experiments in Hearing. E. G. Wever (trans.). McGraw-Hill Book Co., New York, NY.

von Uexküll, J. 1934. A stroll through the worlds of animals and men. Pages 5–80 *in* C. Schiller (ed.). A Picture Book of Invisible Worlds, translated *in Instinctive Behavior* (1957). Metheun, London.

Voronov, V. A., and I. M. Stosman. 1977. Frequency-threshold characteristics of subcortical elements of the auditory analyzer of the *Phocoena phocoena* porpoise. Zhurnal Evoliutsionnoi Biokhimii I Fiziologii 6:719 (read as English summary).

Walker, M. M. 1984. Learned magnetic field discrimination in yellow fin tuna, *Thunnus albacares*. Journal of Comparative Physiology A 155:673–679.

Walker, M. M., J. L. Kirschvink, J. Perry, and A. E. Dizon. 1985. Detection, extraction, and characterization of biogenic magnetite. Pages 155–166 *in* J. L. Kirschvink, D. S. Jones, and B. J. MacFadden (eds.). Magnetite Biomineralization and Magnetoreception in Organisms: A New Biomagnetism. Plenum, New York, NY.

Walker, M. M., D. L. Baird, and M. E. Bitterman. 1989. Failure of stationary but not of flying honeybees to respond to magnetic field stimuli. Journal of Comparative Psychology 103:62–69.

Walker, M. M., J. Kirschvink, G. Ahmed, and A. E. Dizon. 1992. Evidence that fin whales respond to the geomagnetic field during migration. Journal of Experimental Biology 171:67–79.

Walls, G. L. 1942 The Vertebrate Eye and its Aadaptive Radiation. Hafner Press, New York, NY.

Wang Ding, B. Würsig, and W. Evans. 1995. Comparisons of whistles among seven odontocete species. Pages 299–325 *in* R. A. Kastelein, J. A. Thomas, and P. E. Nachtigall (eds.). Sensory Systems of Aquatic Mammals. DeSpil Publishers, Woerden, Netherlands.

Wartzok, D. 1979. Phocid spectral sensitivity curves. Page 62 *in* Proceedings of the 3rd Biennial Conference on the Biology of Marine Mammals, Seattle, WA, October 1979.

Wartzok, D., and M. G. McCormick. 1978. Color discrimination by a Bering Sea spotted seal, *Phoca largha*. Vision Research 18:781–785.

Wartzok, D., and G. C. Ray. 1976. A verification of Weber's law for visual discrimination of disc sizes in the Bering Sea spotted seal, *Phoca largha*. Vision Research 16:819–822.

Wartzok, D., R. J. Schusterman, and J. Gailey-Phipps. 1984. Seal echolocation? Nature 308:753.

Wartzok, D., R. Elsner, H. Stone, B. P. Kelly, and R. W. Davis. 1992. Under-ice movements and the sensory basis of hole finding by ringed and Weddell seals. Canadian Joural of Zoology 70:1712–1722.

Watkins, W. A. 1967. The harmonic interval: Fact or artifact in spectral analysis of pulse trains. Pages 15–43 *in* W. N. Tavolga (ed.). Marine Bioacoustics, Vol. 2. Pergamon, Oxford.

Watkins, W. A. 1980a. Acoustics and the behavior of sperm whales. Pages 283–290 *in* R.-G. Busnel and J. F. Fish (eds.). Animals Sonar Systems. Plenum Press, New York, NY.

Watkins, W. A. 1980b. Click sounds from animals at sea. Pages 291–297 *in* R. -G. Busnel and J. F. Fish (eds.). Animal Sonar Systems. Plenum, New York, NY.

Watkins, W. A. 1981. The activities and underwater sounds of fin whales. Scientific Reports of the Whales Research Institute 33:83–117.

Watkins, W. A., and G. C. Ray. 1977. Underwater sounds from ribbon seal, *Phoca (Histriophoca) fasciata*. Fisheries Bulletin 75:450–453.

Watkins, W. A., and G. C. Ray. 1985. In-air and underwater sounds of the Ross seal *Ommatophoca rossi*. Journal of the Acoustical Society of America 77:1598–1600.

Watkins, W. A., and W. E. Schevill. 1972. Sound source location by arrival times on a non-rigid three-dimensional hydrophone array. Deep-Sea Research 19:691–706.

Watkins, W. A., and W. E. Schevill. 1974. Listening to Hawaiian spinner porpoises, *Stenella* cf. *longirostris*, with a three-dimensional hydrophone array. Journal of Mammology 55:319–328.

Watkins, W. A., and W. E. Schevill. 1979. Distinctive characteristics of underwater calls of the harp seal, *Phoca groenlandica*, during the breeding season. Journal of the Acoustical Society of America 66:983–988.

Watkins, W. A., and W. E. Schevill. 1980. Characteristic features of the underwater sounds of *Cephalorhynchus commersonii*. Journal of Mammology 61:738–739.

Watkins, W. A., and D. Wartzok. 1985. Sensory biophysics of marine mammals. Marine Mammal Science 1:219–260.

Watkins, W. A., W. E. Schevill, and P. B. Best. 1977. Underwater sounds of *Cephalorhynchus heavisidii* (Mammalia: Cetacea). Journal of Mammology 58:316–320.

Watkins, W. A., P. Tyack, K. E. Moore, and J. E. Bird. 1987. The 20 Hz signals of finback whales, *Balaenoptera physalus*. Journal of the Acoustical Society of America 82:1901–1912.

Watkins, W. A., M. A. Daher, K. M. Fristrup, T. J. Howald, and G. N. Di Sciara. 1993. Sperm whales tagged with transponders and tracked underwater by sonar. Marine Mammal Science 9:55–67.

Weast, R. C. 1985. CRC Handbook of Chemistry and Physics, 66th ed. CRC Press, Cleveland, OH.

Webster, D. B. 1962. A function of the enlarged middle ear cavities of the Kangaroo rat, *Dipodomys*. American Journal of Anatomy 108:123–148.

Webster, D. B., and M. Webster. 1975. Auditory systems of Heteromyidae: Function, morphology, and evolution of the middle ear. Journal of Morphology 146:343–376.

Webster, D., R. Fay, and A. Popper (eds.). 1992. The Biology of Hearing. Springer-Verlag, New York, NY.

West, C. D. 1985. The relationship of the spiral turns of the cochlea and the length of the basilar membrane to the range of audible frequencies in ground dwelling mammals. Journal of the Acoustical Society of America 77:1091–1101.

Wever, E. G., J. G. McCormick, J. Palin, and S. H. Ridgway. 1971a. The cochlea of the dolphin, *Tursiops truncatus:* General morphology. Proceedings of the National Academy of Sciences 68:2381–2385.

Wever, E. G., J. G. McCormick, J. Palin, and S. H. Ridgway. 1971b. The cochlea of the dolphin, *Tursiops truncatus:* The basilar membrane. Proceedings of the National Academy of Sciences 68:2708–2711.

Wever, E. G., J. G. McCormick, J. Palin, and S. H. Ridgway. 1971c. The cochlea of the dolphin, *Tursiops truncatus:* Hair cells and ganglion cells. Proceedings of the National Academy of Sciences 68:2908–2912.

Wever, E. G., J. G. McCormick, J. Palin, and S. H. Ridgway. 1972. Cochlear structure in the dolphin, *Lagenorhynchus obliquidens*. Proceedings of the National Academy of Sciences 69:657–661.

White, D., N. Cameron, P. Spong, and J. Bradford. 1971. Visual acuity of the killer whale. Experimental Neurology 32:230–236.

Wikler, K. C., and P. Rakic. 1990. Distribution of photoreceptor subtypes in the retina of diurnal and nocturnal primates. Journal of Neuroscience 10:3390–3401.

Wilson, G. 1970. Some comments on the optical system of pinnipedia as a result of observations on the Weddell seal (*Leptonychotes weddelli*). British Antarctic Survey Bulletin 23:57–62.

Winn, H. E., and P. J. Perkins. 1976. Distribution and sounds of the minke whale, with a review of mysticete sounds. Cetology 19:1–12.

Winn, H. E., P. J. Perkins, and L. Winn. 1970. Sounds and behavior of the northern bottle-nosed whale. Pages 53–59 *in* Proceeding of the 7th Annual Conference on Biological Sonar and Diving Mammals. Stanford Research Institute, Menlo Park, CA.

Wood, F. G. Jr. 1953. Underwater sound production and concurrent behavior of captive porpoises, *Tursiops truncatus* and *Stenella plagiodon*. Bulletin of Marine Science of the Gulf and Caribbean 3:120–133.

Wood, F. G., and W. E. Evans. 1980. Adaptiveness and ecology of echolocation in toothed whales. Pages 381–425 *in* R. -G. Busnel and J. F. Fish (eds.). Animal Sonar Systems. Plenum, New York, NY.

Würsig, B., and C. Clark. 1993. Behavior. Pages 157–200 *in* J. Burns, J. Montague, and C. Cowles (eds.). The Bowhead Whale. Society for Marine Mammalogy, Special Publication No. 2. Allen Press, Lawrence, KS.

Würsig, B., C. W. Clark, E. M. Dorsey, M. A. Fraker, and R. S. Payne. 1982. Normal behavior of bowheads. Pages 33–143 *in* W. J. Richardson (ed.). Behavior, Disturbance Responses, and Feeding of Bowhead Whales *Balaena mysticetus* in the Beaufort Sea, 1980-81. USBLM Report NTIS PB86-152170. LGL Environmental Research Associates, Bryan, TX.

Würsig, B., E. M. Dorsey, W. J. Richardson, C. W. Clark, and R. Payne. 1985. Normal behavior of bowheads, 1980–84. Pages 13–88 *in* W. J. Richardson (ed.). Behavioral, Disturbance Responses and Distribution of Bowhead Whales, *Balaena mysticetus*, in the eastern Beaufort Sea, 1980–84. USMMS Report NTIS PB87-124376. LGL Environmental Research Associates, Bryan, TX.

Xiao, Y., and R. Jing. 1989. Underwater acoustic signals of the baiji, *Lipotes vexillifer*. Pages 129–136 *in* W. F. Perrin, R. L. Brownell Jr., K. Zhou, and J. Liu (eds.). Biology and Conservation of the River Dolphins. Occasional Paper. IUCN Species Survival Commission 3. International Union for the Conservation of Nature, Gland, Switzerland.

Yamada, M. 1953. Contribution to the anatomy of the organ of hearing of whales. Scientific Reports of the Whales Research Institute (Tokyo) 8:1–79.

Yamasaki, F., S. Komatsu, and T. Kamiya. 1978. Taste buds in the pits at the posterior dorsum of the tongue of *Stenella coeruleoalba*. Scientific Reports of the Whales Research Institute (Tokyo) 30:285–290.

Yamasaki, F., S. Komatsu, and T. Kamiya. 1980. A comparative morphological study on the tongues of manatee and dugong (Sirenia). Scientific Reports of the Whales Research Institute (Tokyo) 32:127–144.

Yost, W. A. 1994. Fundamentals of Hearing: An Introduction, 3rd ed. Academic Press, New York, NY.

Young, R. E., and F. M. Mencher. 1980. Bioluminescence in mesopelagic squid: Diel color change during counterillumination. Science 208:1286–1288.

Zoeger, J., J. R. Dunn, and M. Fuller. 1981. Magnetic material in the head of the common Pacific dolphin. Science 213:892–894.

Zwislocki, J. 1981. Sound analyses in the ear: A history of discoveries. American Scientist 69:184–192.

5

DANIEL P. COSTA AND TERRIE M. WILLIAMS

Marine Mammal Energetics

Why study energetics? An understanding of energy flow through a single organism or a community made up of multiple organisms provides insight into the relative importance of the many biological and physical processes impinging on an animal. More important, it provides information about the interrelationship between an animal and the environment in which it lives. Energy flow models are analogous to cost–benefit models used in economics. Costs take the form of energy expended to acquire and process prey and to maintain body functions. The energetic benefits are manifest as the food energy used for growth and reproduction.

Because of their high rate of food consumption, marine mammal populations are often major components of energy flow models describing marine communities; a few individuals can have seemingly disparate impacts on their habitat. Examples include sea otter (*Enhydra lutris*) predation on benthic invertebrates (Riedman and Estes 1990), sperm whale (*Physeter macrocephalus*) foraging in the Gulf of Mexico (Davis and Fargion 1996), and the growing conflict between marine mammals and fishermen as they draw from the same fish resource. Models of energy flow enable us to understand the energetic impact of these mammals on marine communities. By measuring the various avenues of energy transfer we can determine how animals organize their daily or seasonal activities and how they prioritize their behaviors.

Patterns of energy acquisition and expenditure also provide insight into the unique strategies developed by marine mammals to survive and reproduce in an energetically challenging environment. To be an effective predator in the aquatic environment, marine mammals have undergone significant modifications of the "typical" mammalian physiology. Extreme examples are the ability of some marine mammals to fast for months at a time, subsisting entirely on energy stored in their extensive blubber. This mammalian group also demonstrates a remarkable ability to conserve thermal energy that might otherwise be lost in maintaining body temperature in the cold polar seas. Furthermore, precise budgeting of energy expenditure is crucial to enabling marine mammals to dive for more than 1 hr to depths in excess of 1500 m.

In this chapter we examine the major pathways of energy flow for marine mammals. The chapter is divided into two sections. In the first section, the basic physiological concepts of energy flow in mammalian systems are presented. We begin with a generalized energy budget, and then describe in detail four major areas of energy expenditure. These areas address the energetic costs associated with (1) maintaining basic biological functions, (2) thermoregulation, (3) activity and work, and (4) growth and reproduction. The ecological implications of energy exchange between the marine mammal and its environment are addressed in the second section. These are discussed in the context of field metabolic rates for free-ranging marine mammals.

Physiology of Mammalian Energetics

Energy Budgets

The basic pattern of energy flow can be described by the balanced growth equation where Ingestion = Egestion + Excretion + Respiration + Somatic growth + Reproductive growth. This relationship applies the first law of thermodynamics, which states that energy and mass are conserved. Consequently, what goes into an animal (ingestion) must come out in the form of growth, waste, or metabolic work. In such models the "currency" for income or expenditure is given in units of total energy (joules), or as a rate of energy production or utilization measured in watts (joules/sec). However, this relationship can be modified and the units adjusted (e.g., 1 calorie = 4.184 joules) accordingly to include all forms of material or energy transfer between an organism and the environment. Thus, we can examine the balance between calories ingested by an animal in different prey items versus the calories of heat generated by the mammal. By adding a rate factor, we can compare metabolic power input in watts to mechanical power output based on drag and swimming speed for an assessment of locomotor efficiency.

The growth equation shows us that marine mammals must achieve a dynamic balance between the costs of existence and their ability to acquire energy. Ultimately, both sides of the equation must balance. Yet, temporal delays in energy acquisition and expenditure are common, particularly among large marine mammals. In the short term an animal may not be able to obtain sufficient food energy when foraging or it may face periodic challenges that require expenditure of more energy than it can obtain. If a marine mammal cannot compensate for decreases in energy acquisition, it must either reduce its overall rate of energy expenditure or use stored energy reserves. Conversely, for growth and reproduction to occur, the animal must obtain considerably more energy than is required for the individual to survive.

How the balance in energy acquisition and expenditure is achieved differs for individual species and environments. For some species (sea otters, sea lions, and fur seals) very high rates of energy expenditure are met by high rates of energy acquisition. These animals may preferentially live in environments near-shore or upwelling regions where food is abundant (Costa 1993a, 1993b). Manatees (*Trichechus* spp.) represent the opposite extreme. These sluggish marine mammals demonstrate comparatively low existence costs and are able to survive on a low-quality diet. They have adapted to a diet of grasses that are in high abundance but of low quality energetically. Fortunately, they live in the climatically benign tropics where existence costs may be kept to a minimum.

These simple relationships suggest that marine mammals living in an energy-rich, benign (thermally neutral) environment have little trouble meeting their energy demands and should have significant amounts of energy available for growth and reproduction. If food is plentiful, then energy can be expended at a rapid rate. In contrast, if food is scarce or difficult to obtain, then the rate of energy expenditure must be reduced to match the lower rate of food energy input. The seasonal migrations of large cetaceans demonstrate this interrelationship between energetic demand, energy availability, and local productivity. Twice a year, gray whales (*Eschrichtius robustus*) migrate more than 4800 km (3000 miles) between the energetically demanding, but highly productive polar environment, to the warm tropics, where productivity is significantly lower. Although existence costs may be elevated in the polar regions, the ability to take advantage of the immense productivity associated with the sea ice during the polar summer appears to more than compensate. When confronted with the high energetic costs of reproduction and winter conditions, the whales opt for the more benign tropics. Prey availability may be low, but so too are the existence costs, especially for a large animal that is able to live for long periods off of stored energy reserves in the blubber.

A representative energy flow diagram for marine mammals is presented in Figure 5-1 and demonstrates the distribution of energy into the various physiological processes of the animal. Ingested energy may be liberated in feces (egestion) and urine (excretion), or used as metabolizable energy (ME). The ME may be stored for somatic and reproductive growth or expended through respiratory processes. Key processes include the energetic costs associated with basal metabolism, digestion (heat increment of feeding), thermoregulation, and activity (locomotion, grooming, feeding, etc.).

Energy Acquisition

Ingested Versus Digested Energy

A large component of the growth equation is related to the processing of the ingested food energy. As this relationship suggests, not all of the ingested material consumed is digestible and energy may be lost as egestion and excretion. The amount of the ingested food energy that can be assimilated across the gut is a function of the composition of the diet. Invertebrate prey (i.e., shrimp, krill, clams, mussels) are

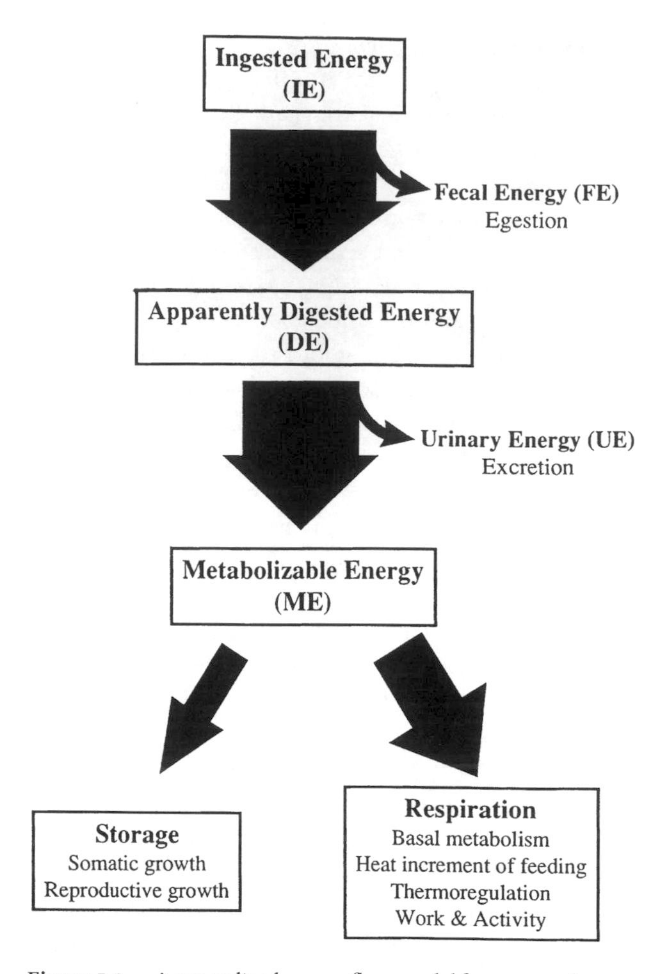

Figure 5-1. A generalized energy flow model for mammals.

important components of the diet of a wide variety of marine mammals from the smallest sea otters to the largest whales. These prey items often contain large amounts of indigestible components such as chitin in their exoskeletons. Foods that are high in fat are the most digestible (Kleiber 1975). Animal material is significantly more digestible than plant material because much of the energy in plants is in the form of cellulose, which vertebrates cannot digest without help from symbiotic microorganisms.

The ability of animals to digest a given prey item is termed the apparent digestibility (AD) and defined by the ratio of energy acquired during digestion and the energy consumed. It is typically presented as the percentage: Apparent digestibility = (Ingested energy − Fecal energy)/Ingested energy. The apparent digestibility of different diets for sea otters (Costa 1982) and pinnipeds has been examined. The ADs for herring (*Clupea harengus*) is 92.6% in gray seals (*Halichoerus grypus*) (Ronald et al. 1984), 93.8% in harp seals (*Phoca groenlandica*) (Keiver et al. 1984), and 97.1% in ringed seals (*Phoca hispida*). This increases slightly to 97.9% for ringed seals fed a diet of capelin (*Mallotus villosus*) (Parsons 1977). For northern fur seals (*Callorhinus ursinus*), AD ranges from 88% for diets consisting of capelin or squid (*Loligo opalescens*), to 90% for pollock (*Theragra chalcogramma*) and 90% to 93% for herring (Miller 1978, Fadely et al. 1990). Harbor seals (*Phoca vitulina*) show an AD of 92.1% and 96.7% for a diet of pollock or herring, respectively (Ashwell-Erickson and Elsner 1981). Polar bears (*Ursus maritimus*) have an AD of 91.6% when they consume ringed seals (Best 1976). Such relatively high AD is to be expected for animals eating prey that is low in fiber or chitin and is easily processed in the gut (Kleiber 1975). Invertebrate prey with a high chitin content, such as shrimp (*Pandalus borealis*), has a lower AD. Values for a shrimp diet average 72.2% for the harp seal (Keiver et al. 1984).

The high fiber and cellulose content of plant material results in a similarly low AD for the grazing manatee (Burn 1986); AD is reportedly 84.6% in the dugong (*Dugong dugon*) feeding on grasses (Murray et al. 1977). Sirenians use hindgut fermentation to fully digest plant material (Murray et al. 1977, Burn 1986, Reynolds and Rommel 1996). Interestingly, the AD of these animals are higher than the 45% to 59% measured for terrestrial mammals, such as the horse (*Equus caballus*), that use hindgut fermentation (Fonnesbeck et al. 1967, Fonnesbeck 1968). The AD of sirenians is similar to that measured for the green sea turtle (Bjorndal 1979). The high AD in sirenians also appears to be related to the low lignin content of aquatic vegetation and to the large body mass of the animal (Burn 1986, Reynolds and Rommel 1996). Large body size provides a comparatively large gastrointestinal tract and a low metabolic rate, which together allow a slow rate of passage of digesta and more efficient breakdown of fibrous plant material.

The ingested food energy (IE) is equivalent to the net chemical energy liberated as heat if food is completely oxidized. Food energy remaining after digestion and elimination of fecal energy (FE) is known as the apparently digested energy (DE) where IE = FE + DE or DE = IE × AD. Fecal energy includes the energy lost through the elimination of nonfood materials (i.e., intestinal secretions, microbes, and cellular debris); it is equivalent to the egestion term of the growth equation. It is important to note that the AD is given as a ratio, whereas the IE, FE, and apparently DE are given in quantitative units such as joules (J).

The chemical energy lost as the end product of metabolic processes is known as urinary energy (UE) and is equivalent to the excretion term in the growth equation. In mammals the most significant component of UE results from the urea formation, the end product of protein catabolism. In general, no UE results from the catabolism of fat or carbohydrate.

Metabolizable energy (ME) is the net energy remaining after fecal and urinary energy loss and represents the energy available to the animal for growth or supporting metabolic processes. In the growth equation ME is described by the energy available for work (locomotion), respiration (thermoregulation, maintenance metabolism, heat increment of feeding), as well as somatic and reproductive growth. Mathematically ME is described by: ME = IE – FE – UE.

The ME has been measured as a percentage of total energy ingested for California sea lions (*Zalophus californianus*) on a diet of herring (88.2%), anchovy (*Engraulis mordax*) (91.6%), mackerel (*Scomber japonicus*) (91.4%), and squid (78.3%) (Costa 1988a). This compares with the ME of 87.1% for harp seals (Keiver et al. 1984), 82.7% for gray seals (Ronald et al. 1984), and 89.8% for ringed seals (Parsons 1977) fed herring. Harbor and spotted seals (*Phoca largha*) fed either pollock or herring diets had a ME of 80.3% (Ashwell-Erickson and Elsner 1981).

Energy Expenditure

The Cost of Maintenance Functions

Maintenance costs are defined as the energy required to maintain constant body mass and composition (homeostasis). After fecal and urinary losses, the remaining energy must be sufficient to meet the energy costs of homeostasis. The cost of homeostasis includes basal metabolism, the heat increment of feeding, heat used for thermoregulation outside the zone of thermoneutrality, and heat liberated as voluntary activity (Fig. 5-1). As mentioned above, these components of energy flow can vary as a function of the proximate composition of the prey consumed and the animal's nutritional history.

BASAL METABOLISM. It has generally been assumed that the basal metabolic rates of aquatic mammals are elevated when compared to terrestrial mammals of similar size. This had been explained as an adaptation for maintaining thermal balance under conditions of high thermal conductivity when in water (Hart and Irving 1959, Kanwisher and Sundnes 1966, Irving 1973, South et al. 1976, Kanwisher and Ridgway 1983). The current view of marine mammal basal metabolic rates is more complex. Lavigne et al. (1986) identified one problem; many past studies assessing metabolic rate in aquatic mammals did not conform to standardized criteria established by Kleiber (1975) for comparing the basal metabolism of animals. These require that the subjects be adults, resting, thermoneutral, and postabsorptive.

Another confounding problem has been the assumption that all marine mammals use identical metabolic responses. Specialization for marine living has occurred in at least five mammalian lineages, the sirenians, cetaceans, pinnipeds, sea otters, and polar bears. On the basis of this diversity, we might expect different metabolic adaptations among the groups (Fig. 5-2). For example, manatees demonstrate basal metabolic rates that are lower than values predicted by the Kleiber relationship based on similarly sized terrestrial mammals (Scholander and Irving 1941, Gallivan and Best 1980, Irvine 1983). The basal metabolism of phocid seals appear to be equivalent to Kleiber predictions (Lavigne et al. 1986).

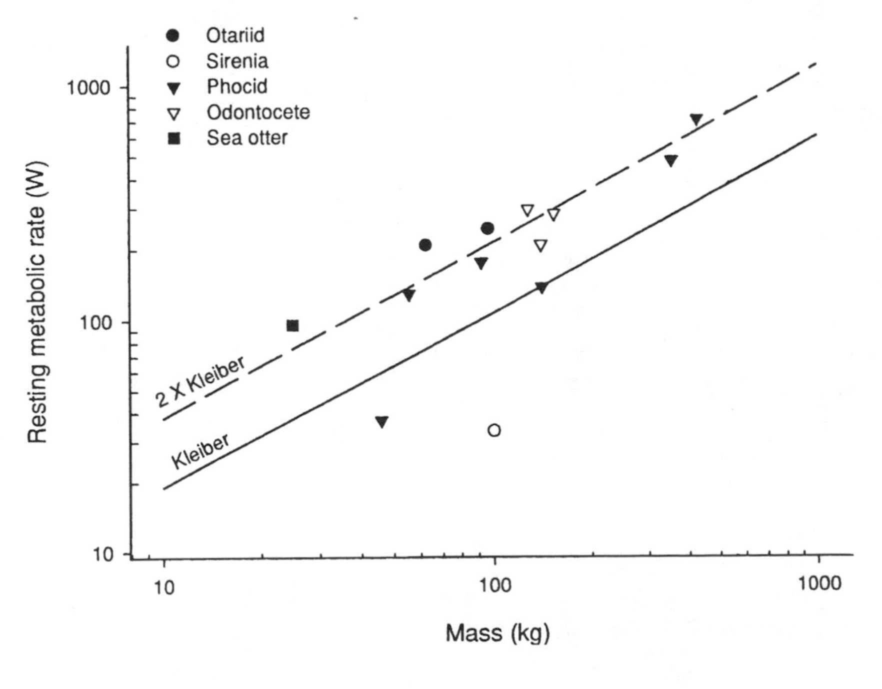

Figure 5-2. Resting metabolic rate (RMR) of marine mammals in relation to body mass. Measurements were made for the animals resting in water. The solid line denotes the predicted metabolic rate from Kleiber (1975); the dashed line represents two times the predicted levels. (The sources of data used in this figure are provided in the text.)

Conversely, data for sea otters, otariids, and odontocetes suggest that the basal metabolic rates of these mammals are greater than predicted for terrestrial mammals (Pierce 1970, Ridgway and Patton 1971, Morrison et al. 1974, Costa and Kooyman 1982, Costa et al. 1989b, Liao 1990, Hurley 1996).

Extreme differences in the body composition of marine mammals introduces another complicating factor when attempting to compare basal metabolic rates. Many species of marine mammal undergo large variations in body fat seasonally or in association with life history patterns (Rice and Wolman 1971, Costa et al. 1986). Some species, such as sirenians and walrus (*Odobenus rosmarus*), have dense bone. Because metabolic rates are conventionally expressed in terms of total body mass, a disproportionate amount of fat or particularly dense bone will lower the apparent metabolic rate. This is attributable to the low metabolic rates of bone and adipose tissue in comparison to lean tissue (Kleiber 1975). Thus, animals of equivalent body mass but different percentages of fat or bone display different metabolic rates. This relationship was confirmed in studies of northern elephant seal (*Mirounga angustirostris*) pups. The fat content of these pups increased from 4% to 48% between birth and weaning. When comparing various sized elephant seal pups, lean mass rather than total body mass was the better predictor of metabolic rate (Rea and Costa 1991).

The difficulty in establishing the basal metabolic rate of marine mammals is attributable in part to the definition of resting. Marine mammals appear at rest while bobbing on the water surface as well as while lying quietly submerged. The metabolic demands of each state may be very different due to physiological changes associated with the diving response (Elsner, Chapter 3, this volume). Recent studies on submerged, resting California sea lions suggest that the metabolic rate of these animals approaches basal values predicted for terrestrial mammals during prolonged periods of apnea (Hurley 1996). In view of this, we may need to redefine the criteria for assessing basal metabolism in marine mammals to conform with their unique lifestyles and physiology.

Costs Associated with the Heat Increment of Feeding

When food is consumed the animal's metabolic rate increases over fasting levels. This increase in metabolism results from the biochemical work associated with the digestion and chemical processing of food. Historically, the process has been referred to as specific dynamic action (SDA), although the recent convention is heat increment of feeding (HIF) (Kleiber 1975). The HIF may be considered the "tax" that is required to process food energy for conversion to ME. The magnitude of the energy allocated to HIF varies between 5% and 17% of the ME and is related to the composition of the diet. In addition, the duration of HIF after a meal will depend on the amount of food consumed and its composition (Table 5-1).

If we continuously monitor the metabolic rate of an animal after a meal, we find that metabolic rate slowly increases, reaches a peak, and then returns to the resting nonfed level. This metabolic bulge is often of sufficient magnitude that it must be taken into account in the analyses of energy metabolism. Expressed in terms of the energy content of the food consumed, HIF is about 6% for carbohydrates, 13% for fats, and 30% for proteins (Bartholomew 1977a).

Under experimental conditions, HIF is usually determined on animals placed in a thermally neutral environment. As a result, the excess heat resulting from HIF is dissipated from the body by thermoregulatory pathways and is considered "wasted" energy (Kleiber 1975). For free-ranging marine mammals or when thermoregulatory demands are high, HIF is an excellent mechanism for defraying high metabolic costs associated with keeping warm. The actively foraging marine mammal must contend with the constant thermal challenge of living in relatively cold water and heating ingested prey items. This can lead to an interesting balancing act between the energetic costs associated with HIF, activity, and thermoregulation. Rather than acting independently, the various components of the energy flow diagram (Fig. 5-1) may act synergistically to reduce the overall cost to the animal. For example, sea otters use heat generated from the digestion of food (HIF) to offset thermoregulatory costs when water temperature is low (Costa and Kooyman 1984).

The Cost of Temperature Regulation

Marine mammals maintain body temperatures within the typical mammalian range of 36° to 40°C. However, this is ac-

Table 5-1. The Heat Increment of Feeding (HIF) in Marine Mammals

	Increase	Duration of	Diet Protein	
	HIF	Above	HIF	Content
Species/Diet	(%)	RMR	(hr)	(%)
Harp seal[a]	17	1.67	7	
Harbor seal[b]	9.0			High
Herring	5.1	1.46	15	Low
Harbor seal[c]	4.7	1.28	10–12	
Elephant seal[d]		1.65	10.4	
Capelin	12.1			15.2
Herring	11.2			17.4
Sea otter[e]		1.54	4.2–5.3	
Clam	13.2			10.5
Squid	10.0			11.1

[a]Gallivan and Ronald 1981; [b]Markussen et al. 1994; [c]Ashwell-Erickson and Elsner 1981; [d]Barbour 1993; [e]Costa and Kooyman 1984.

RMR = resting metabolic rate

complished under conditions of high heat transfer attributable the high thermal conductivity of water in comparison to air. Because heat flow may be 25 times greater in water than in air, marine mammals have evolved a wide variety of morphological and physiological adaptations to control heat loss to the environment. These include (1) an insulating layer comprised of blubber or fur; (2) thermal windows in poorly insulated peripheral sites; (3) complex vascular arrangements, which serve as variable heat exchangers, and (4) in some species, elevated basal metabolic rates (Scholander and Schevill 1955, Hart and Irving 1959, Kanwisher and Sundnes 1966, Irving 1969, Tarasoff and Fisher 1970, Hampton et al. 1971, McGinnis et al. 1972, Irving 1973, Hampton and Whittow 1976, Gallivan and Ronald 1979, Pabst, Rommel, and McLellan, Chapter 2, this volume).

Phocids, otariids, sea otters, and polar bears have the dual problem of thermoregulating in air when hauled out and in water during a wide variety of physiological states from rest to high-speed swimming. Conversely, cetaceans and sirenians are truly aquatic and avoid the conflicting thermoregulatory demands associated with an amphibious lifestyle. For many species of cetacean the primary thermoregulatory problem during rest is heat conservation. Even in waters at tropical temperatures small cetaceans such as Atlantic bottlenose dolphins (*Tursiops truncatus*) and the long-snouted spinner dolphin (*Stenella longirostris*) must guard against excess heat loss during rest (Hampton et al. 1971, McGinnis et al. 1972, Hampton and Whittow 1976). The findings of these investigators lend support to the hypothesis of Parry (1949) who inferred from his studies of the harbor porpoise (*Phocoena phocoena*) that small cetaceans are obliged to remain active, or to elevate their metabolic rate, to maintain thermal balance.

The aquatic environment represents a thermal energetic challenge for marine mammals. Consequently, these animals often depend on the energy produced by activity and feeding as well as marked control over peripheral insulation to maintain thermal balance (Hampton and Whittow 1976). For pinnipeds and cetaceans, a characteristic thermoregulatory feature is the extreme lability of their insulation (Kanwisher and Sundnes 1966, Hampton and Whittow 1976). Dorsal fins, flippers, and flukes act as thermal heat exchangers to afford considerable control in the regulation of body temperature.

THE THERMONEUTRAL ZONE. Surprisingly few studies have examined the thermoneutral zone (TNZ) of marine mammals in water (Fig. 5-3). The thermoneutral zone defines the range of environmental temperatures where metabolic heat production is minimal for the animal (Bartholomew 1977b). Its range is delineated by the lower and upper critical temperatures, T_{lc} and T_{uc}, respectively. Below the T_{lc}, physiological variations in thermal conductance are not sufficient to offset heat loss. Therefore, metabolism increases in accordance with decreases in environmental temperature (Schmidt-Nielsen 1979). Environmental temperatures that exceed the T_{uc} also result in an increase in metabolic rate, this time as a consequence of the additional work necessary to dump excess heat (Bartholomew 1977b).

Studies of the TNZ of marine mammals in water have concentrated on the phocid seals. In general, these pinnipeds demonstrate a broad TNZ when compared to the relatively narrow range of air temperatures representing the TNZ of terrestrial mammals. Irving and Hart (1957) and Gallivan and Ronald (1979) have shown that adult (150 kg) harp seals have a TNZ of at least 28°C. The T_{lc} is below 0°C, whereas the T_{uc} has yet to be defined. In harbor seals, the TNZ is

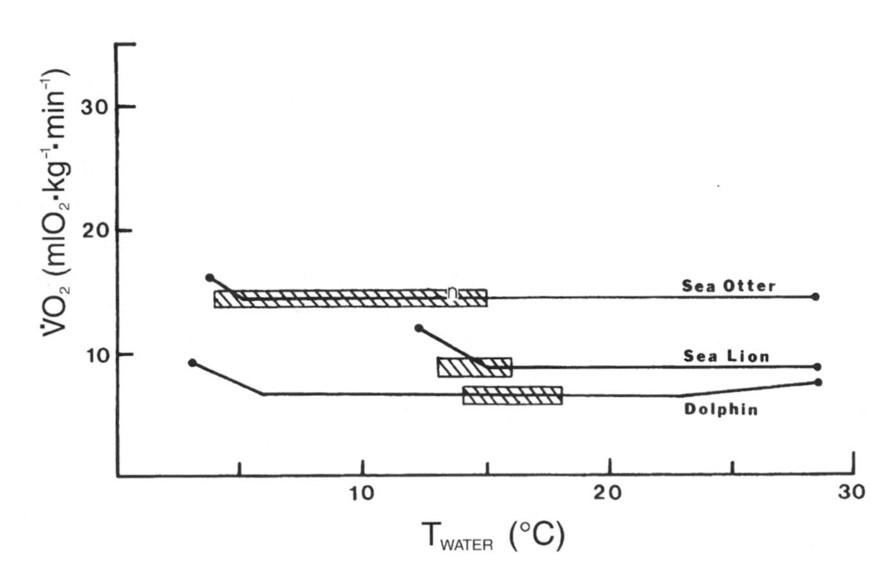

Figure 5-3. Oxygen consumption ($\dot{V}O_2$) in relation to water temperature for three species of marine mammals. Results for sea otters, sea lions, and bottlenose dolphins are compared. The horizontal portion of the line represents the thermoneutral zone (TNZ) for each species. The hatched box denotes the routine water temperature encountered by the animals in the wild. Note that the dolphins were acclimated to conditions in Hawaii, and the sea lions and sea otters to conditions off the California coast. (The sources of data used in this figure are provided in the text.)

dependent on both body size and season. The larger the animal, the lower the surface-to-volume ratio, and hence, smaller the surface area for heat loss. Large size may be associated with changes in the thickness of the blubber layer, which will also contribute to seasonal differences in the TNZ (Irving and Hart 1957, Hart and Irving 1959, Miller and Irving 1975, Miller et al. 1976). Thus, we find that harbor seals show a considerably higher T_{lc} (approximately 20°C) during the summer than when examined during winter (average T_{lc} = 13°C) (Hart and Irving 1959). The simplest explanation for this seasonal change is the 30% increase in insulation during the winter (Irving 1973).

In contrast to phocids, relatively little information exists on the variability of the thermoneutral zone of otariids. Sea lions are temperate water otariids and generally have a thinner blubber layer than arctic phocids of similar size (Bryden and Molyneux 1978). These animals appear to be ill-equipped for dealing with high environmental heat loads, often resorting to behavioral mechanisms for maintaining thermal homeostasis (Whittow et al. 1972). The T_{uc} for sea lions resting in air is between 22°C and 30°C, with smaller animals able to tolerate higher air temperatures than larger ones (Matsuura and Whittow 1973, South et al. 1976, Liao 1990). Once again we find a lower overall insulation and enhanced heat transfer as body size and blubber thickness decrease. With decreases in ambient temperature, sea lions will reduce blood flow to peripheral sites, which leads to an increase in total body insulation (Matsuura and Whittow 1975). This appears to be the primary thermoregulatory mechanism for sea lions until air temperatures decline below 10°C (Matsuura and Whittow 1973, South et al. 1976). Liao (1990) examined the metabolic responses of California sea lions to changes in water temperature (Fig. 5-3). The thermoneutral zone was approximately 20°C for this marine mammal resting in water, although the upper critical temperature was never identified. The T_{lc} of California sea lions was 14.8° to 16.4°C, at least 2°C above the range of water temperatures typically encountered by these animals in the wild (Fig. 5-3). This implies an increased reliance on physiological mechanisms of thermoregulation for these otariids.

Most studies of temperature regulation in small odontocetes have primarily concentrated on resting heat production, body temperature, and heat flow across the body surface (Irving et al. 1941, Kanwisher and Sundnes 1966, Irving 1969, Pierce 1970, Hampton et al. 1971, Ridgway and Patton 1971, Whittow 1987). Observations of elevated heat production and high tissue insulation in captive bottlenose dolphins and spinner dolphins indicate that they must conserve body heat, when at rest even in tropical waters averaging 24°C (Hampton et al. 1971, McGinnis et al. 1972, Hampton and Whittow 1976). Like California sea lions, these animals appear to live close to their critical temperatures and depend on the energy produced from activity and feeding as well as precise control of peripheral insulation to maintain thermal balance (Hampton and Whittow 1976).

Captive bottlenose dolphins acclimated to water temperatures averaging 28°C have a thermoneutral zone of at least 15°C. The upper critical temperature is near 28°C (Fig. 5-3). The T_{lc} depends on the thickness and quality of the blubber layer and the size of the cetacean. Larger animals or individuals with a thicker blubber layer have a lower T_{lc}. Therefore, the T_{lc} was 11°C for a dolphin with an average blubber thickness at more than 20 mm and 15°C for animals with thinner blubber layers (Costa et al. 1989a, Williams et al. 1993). Note that the range for the thermoneutral zone and upper and lower critical temperatures shift for dolphins acclimated to cooler water temperatures, similar to resetting the thermostat of a car for summer and winter conditions (T. M. Williams, pers. obs.).

Similar patterns are observed for wild dolphins. Seasonal changes in the body condition of dolphins in Sarasota Bay, Florida, include considerable variation in blubber thickness and total body fat. These changes undoubtedly reflect the dolphins' response to changes in water temperature. Blubber thickness averaged 18 mm for Sarasota Bay animals in winter acclimated to water temperatures of 16°C (G. A. J. Worthy, D. P. Costa, and R. S. Wells, unpubl.). The similarity in blubber depth for captive Hawaiian dolphins at 28°C and the wild Florida animals at 16°C seems contradictory. However, the quality of insulation as well as its depth need to be considered when assessing the thermal characteristics of the blubber layer (Williams et al. 1992). Worthy and Edwards (1990) found considerable differences in the blubber thickness and blubber fat content of harbor porpoises and spotted dolphins. Harbor porpoises living in cold coastal waters (average water temperature, 12°C) maintained blubber of lower thermal conductivity, $0.10 \pm 0.01\ W \cdot m^{-1} \cdot C^{-1}$. In comparison, the thermal conductivity of blubber from spotted dolphins (*Stenella attenuata*) living in the eastern tropical Pacific Ocean (average surface water temperature, 27°C) was $0.20 \pm 0.02\ W \cdot m^{-1} \cdot C^{-1}$. These variations were attributed to differences in the lipid content of the blubber layer. In addition, blubber thickness varied in harbor porpoises averaging 15 ± 3 mm, whereas spotted dolphin blubber averaged 7.7 ± 1.1 mm (Worthy and Edwards 1990). Consequently, insulating quality and quantity of dolphin blubber can be very different depending on species and routine environmental temperatures.

Unlike those of other marine mammals, the body temperatures of manatees, 35.6° to 36.4°C, are near the lower end of the normal mammalian range. With a T_{lc} of only 20° to 23°C (Gallivan et al. 1983, Irvine 1983), these mammals appear to have a low thermal tolerance to changes in water temperature. This physiological response limits the habitats that manatees can occupy to relatively warm regions. They rely on warm water refuges in the winter and may be susceptible to cold-related mortalities after periods of severe cold weather (Irvine 1983). With the notable exception of Steller's sea cow (*Hydrodamalis gigas*), sirenians as a group tend to have similar physiological limitations and are restricted to tropical habitats (Domning 1978, Marsh et al. 1978).

TEMPERATURE REGULATION DURING ACTIVITY IN WATER. Heat generated as a by-product of activity can alter the thermal energetic demands of the marine mammal. Just as a human running around a track on a cold day feels warmer than a sedentary friend seated in the bleachers, some marine mammals may use exercise-induced thermogenesis to help maintain a stable core body temperature when in cold water. The response depends on the balance between the level of heat production provided by activity and the elevation in convective heat loss associated with movements (Williams 1986).

Several studies on exercising California sea lions have shown conflicting results. Feldkamp (1985) demonstrated comparatively higher metabolic rates over a range of speeds for a juvenile sea lion swimming in water at 18°C than for animals swimming in water at 24°C. The metabolic response of these animals was similar to those observed for humans (*Homo sapiens*) (Nadel et al. 1974) and muskrats (*Ondatra zibethicus*) (Fish 1983) swimming at water temperatures below their TNZ. Conversely, Davis and Williams (1992) found that swimming speeds as low as 1 m/sec generated enough heat to defray the cost of thermoregulation in juvenile sea lions exercising in water temperatures ranging from 5° to 20°C. The conclusions of this study are similar to those presented for sea otters by Costa and Kooyman (1984) in which the level of activity was directly correlated with water temperature. Although these results are inconclusive, they do reveal a potentially important thermoregulatory strategy for marine mammals that warrants further investigation.

The Cost of Work and Activity

Whether a mammal lives on land or ice, or in the water, performing work is energetically expensive. As a result, the cost of locomotion or activities such as grooming, predatory maneuvers, and defending territories can be major components of an animal's total daily energy budget. This is not because these activities necessarily make up a large proportion of time, but because they exact a high metabolic cost. For example, the metabolic rate of an adult sea otter resting in water at 20°C is 12 to 13 mL $O_2 \cdot kg^{-1} \cdot min^{-1}$ (Costa and Kooyman 1982, Williams 1989). Grooming activity including fur pleating and rubbing, and somersaulting raises the otter's metabolic rate to more than 20 mL $O_2 \cdot kg^{-1} \cdot min^{-1}$. Swimming on the water surface at a leisurely 0.8 m/sec increases the metabolic rate to nearly three times resting levels with even higher metabolic rates occurring when the sea otter swims faster (Williams 1989).

The aerobic scope (AS) provides a physiological measure of an animal's upper metabolic and performance limits. Defined by the ratio: Aerobic scope = Maximum oxygen consumption ($\dot{V}O_{2max}$)/Resting oxygen consumption ($\dot{V}O_{2rest}$), the aerobic scope is usually determined by measuring the minimum and maximum rates of energy metabolism under a standard set of conditions (Bartholomew 1977b). For locomotor studies the ratio is often set by the resting metabolic rate and peak metabolic rate during sustained muscular work.

In general, aerobic scope for activity in marine mammals is less than that measured for terrestrial mammals (Fig. 5-4). Running mammals typically show aerobic scopes between 10 and 25. Therefore, the maximum oxygen consumption ($\dot{V}O_{2max}$) during running is 10 to 25 times the level of oxygen consumption for the same animal standing quietly on a treadmill. The better the athlete or training, the greater the aerobic scope (Taylor et al. 1980). Many trained domestic and wild canids and thoroughbred racehorses show aerobic scopes on the high end of this range. In contrast, swimming marine mammals rarely demonstrate aerobic scopes above 6. This includes a variety of pinnipeds exercising in swimming flumes (Davis et al. 1985, Williams et al. 1991), harbor seals and a spotted seal treading water in a tank (Elsner 1987), and bottlenose dolphins swimming against a force platform (Williams et al. 1993). These comparatively low aerobic scopes do not imply that marine mammals are in poorer athletic shape than their terrestrial counterparts. Many factors including the level of maximum oxygen consumption measured during aquatic exercise and the definition of resting metabolic rate for marine mammals may account for the observed differences in AS. The $\dot{V}O_{2max}$ for the bottlenose dolphins exercising against a force platform was 30% to 40% lower than predicted values based on running mammals. Similarly, $\dot{V}O_{2max}$ for the phocid seals treading water was 40% lower than that predicted for a same sized terrestrial mammal. Low $\dot{V}O_{2max}$ levels during aquatic activity are not limited to marine mammal species. Humans achieve a $\dot{V}O_{2max}$ during swimming that is 11% to 19% lower than measured for the same athlete running on a treadmill. The $\dot{V}O_{2max}$ of the American mink (*Mustela vison*), a semiaquatic mammal, is 3% to 14% lower than for the same animals running on a treadmill (Williams 1983). Respiratory constraints and the utilization of smaller skeletal muscle mass during swimming exercise are possible causes for the relatively low $\dot{V}O_{2max}$ of mammals during swimming exercise.

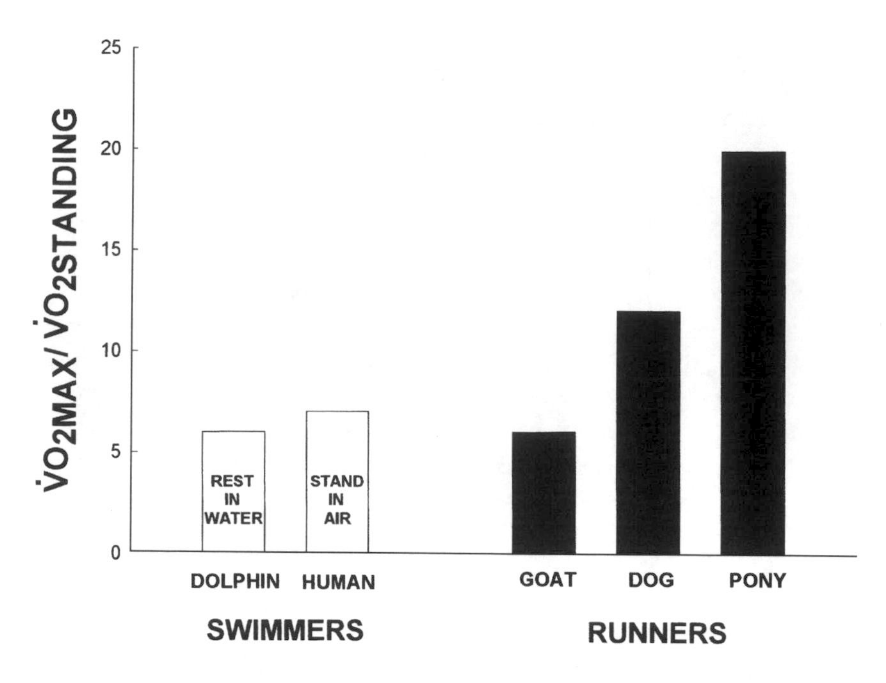

Figure 5-4. The dynamic aerobic scope ($\dot{V}O_{2max}/\dot{V}O_{2std}$) for swimming and running mammals. $\dot{V}O_{2std}$ was measured for dolphins resting in water and for the other species standing in air. Note the similarity in scope for swimming dolphins, swimming humans, and sedentary terrestrial mammals (goats).

Difficulties in defining the standard or resting metabolic rate of marine mammals also lead to complications in calculating aerobic scope in this group of animals. As mentioned in the preceding section, "resting" in marine mammals may range from bobbing on the water surface to lying submerged in apnea and bradycardia. The metabolic rates associated with each of these states may differ two- to threefold resulting in significantly different estimates for aerobic scope. If we use the predicted mammalian standard metabolic rate from Kleiber (1975), the resulting AS for dolphins pushing against a force platform ranges from 7 to 11. These levels are typical of relatively sedentary terrestrial species as described by Taylor et al. (1987), which conflicts with the athletic image of leaping dolphins and breaching whales. The general diving response of marine mammals may also preadapt this group to lower apparent aerobic scopes. In seals adaptations for hypoxia tolerance during submergence, an enhanced capacity to buffer pH shifts associated with lactate accumulation, and high oxygen storage capacities in the skeletal muscles, blood, and lungs may support greater metabolic efforts than indicated by traditional measurements of aerobic scope (Elsner 1986). To cope with a limited aerobic scope, marine mammals may also preferentially rely on anaerobic metabolism during high work loads. In this respect, exercising dolphins and seals appear similar to running lions (Williams et al. 1993). Obviously, further research is needed concerning the appropriate measurement and definition of upper and lower limits of metabolism in marine mammals.

SWIMMING CAPABILITIES. It is important to remember that marine mammals were originally equipped with the same physiological and morphological building blocks for locomotion as terrestrial mammals. The fossil record for cetaceans (Gingerich et al. 1994, Thewissen et al. 1994), pinnipeds (Berta et al. 1989), sea otters (Riedman and Estes 1990), and sirenians (Barnes et al. 1985), demonstrates their terrestrial ancestry. However, marine mammals show a variety of physiological, morphological, and behavioral adaptations that enable them to move quickly and efficiently through water. Consequently, aquatic performance by pinnipeds, cetaceans, and even sea otters far outranks our best Olympic athletes in terms of speed and energetic cost. Most humans rarely swim more than 1.0 m/sec, although elite human athletes attain swim speeds approximately two times faster. During the 1996 Summer Olympics, Amy Van Dyken won the gold medal for the women's 50-m free-style sprint with a swimming speed of 2.0 m/sec. The gold medal winner in this event for men, Aleksandr Popov, swam even faster at 2.3 m/sec. In comparison, the routine cruising speeds of many marine mammals approach the 2.0 m/sec swimming sprint speed of humans (Table 5-2). Northern fur seals, Galapagos sea lions (*Zalophus californianus wollebaeki*), Galapagos fur seals (*Arctocephalus galapagoensis*), and New Zealand sea lions (*Phocarctos hookeri*) demonstrate mean surface swimming velocities ranging from 0.6 to 1.9 m/sec (Ponganis et al. 1990). A similar range is observed for sustained submerged swimming by sea otters (Williams 1989) and northern elephant seals (Le Boeuf et al. 1992). Sustained swimming speeds are somewhat faster for cetaceans. For example, the range of sustained speeds for bottlenose dolphins is 1.4 to 3.1 m/sec depending on the duration of effort, depth, and behavior (Lang and Norris 1966, Würsig and Würsig 1979, Shane 1990, Williams et al. 1993). Free-ranging killer whales (*Orcinus orca*) average 2.4 m/sec (Kriete 1995),

Table 5-2. General Characteristics of Marine Mammal Swimmers

Characteristics	Sea Otter	Otariid	Phocid	Cetacean
Hydrodynamics	Surface	Submerged	Submerged	Submerged/porpoising
Mechanics	Paddle/row, dorsoventral undulate	Pectoral	Lateral thunniform with lunate tail	Dorsoventral thunniform
Stroke frequency[a] (strokes/min)	20–80 (surface) 56 (submerged)	15–50	60–78	60–90
Speed (m/sec)	< 1.4	2.0–6.0	2.0–6.0	2.0–10.0
Energetics				
Measured COT / Predicted COT	12 (surface) 6 (submerged)	4	4	2–3

[a]Values are for adult southern sea otters (Williams 1989), immature sea lions (Feldkamp 1987), harbor seals (Davis et al. 1985), and adult bottlenose dolphins (Williams et al. 1993).

COT = cost of transport.

whereas migrating bowhead whales (*Balaena mysticetus*) average 1.4 m/sec depending on their migratory route (Würsig et al. 1985). Sprinting speeds for both pinnipeds and cetaceans may be three to five times higher than these cruising speeds. Some otariid species reach 4.0 m/sec (Ponganis et al. 1990), whereas trained bottlenose dolphins have been clocked at more than 7.5 m/sec (Lang and Norris 1966), a speed that human swimmers have little hope of ever approaching. Even more remarkable than the high performance capabilities of marine mammals is the comparatively low energetic cost associated with swimming, as addressed in the following sections.

HYDRODYNAMICS. To understand the energetics of swimming in marine mammals, we have to appreciate the underlying hydrodynamics, that is the physics associated with propelling a body through water. A brief description is provided here with more detail found in Pabst et al. (Chapter 2, this volume). Fish (1993) also provides an excellent overview of the effects of body design and propulsive mode on swimming energetics in semiaquatic and marine mammals.

When a body moves through a fluid, a force acts backward on it, resisting its forward motion. Anyone who has waded knee deep in water or mud has experienced the increased effort needed to overcome this force. The resistive force, termed drag, is described by the equation:

$$\text{Drag} = 1/2r\rho V^2 ACd,$$

where ρ = the density of the fluid, V = velocity of the fluid relative to the body, A = body area, and Cd = drag coefficient. For mammalian swimmers, the most critical term in this equation is velocity squared, which causes drag to increase exponentially for every incremental increase in swimming velocity. Thus, the amount of work that the marine mammal must expend to move through the water increases exponentially with its forward velocity. This has a profound effect on both the energetics of swimming and the preferred swimming speeds of marine mammals.

Total body drag of harbor seals, sea otters, and humans has been measured by towing the subjects through water at different velocities. Drag forces for all subjects increased curvilinearly with towing speed (Fig. 5-5). The value of body streamlining in reducing drag is readily apparent if we compare the total body drag of humans to those of marine mammals. Drag on the submerged human form is 3 to 10 times that of the submerged seal (Williams and Kooyman 1985) or sea otter (Williams 1989) measured at comparable speeds. Limb internalization, an overall decrease in surface projections, and the approximation of an elongate, tapered hull contribute to the streamlined character of marine mammals. The result is lower body drag and a decreased energetic demand on the swimmer.

There are several components comprising total body drag of a swimmer. These include viscous drag and pressure drag, which are affected by the swimmer's surface area and body streamlining, respectively (see Pabst, Rommel, and McLellan, Chapter 2, this volume). A third type of drag, wave drag, results from surface waves generated when the swimmer moves on or near the air–water interface. Hertel (1966) measured the changes in drag for a dolphin-shaped model towed on and below the water surface. Near the water surface, drag was four to five times higher than when the model was towed submerged at the identical speed (Fig. 5-6). An in-

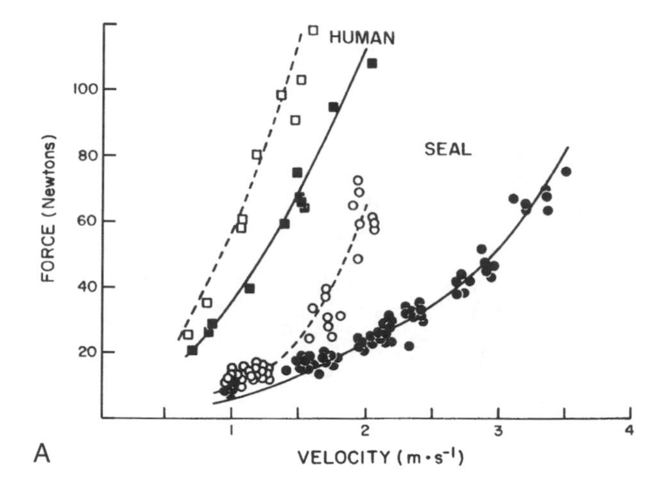

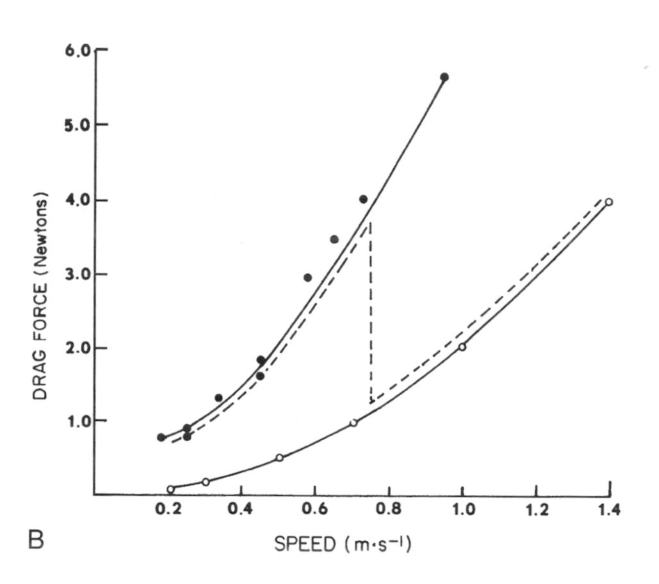

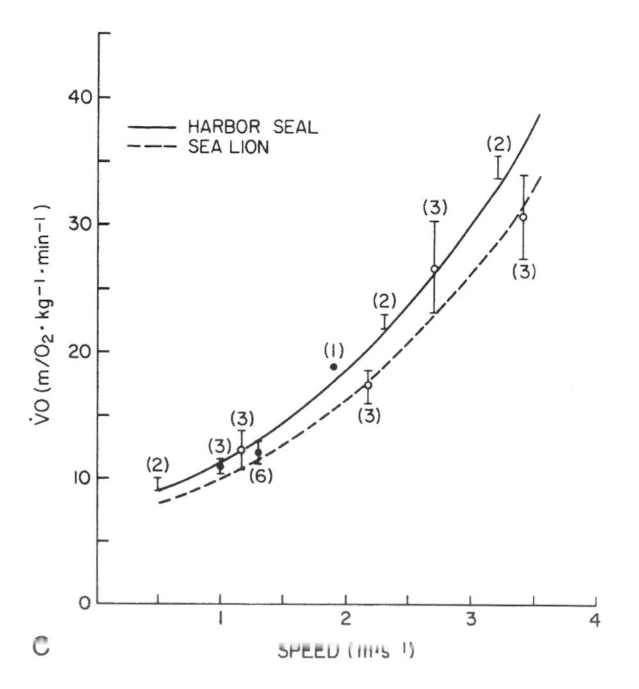

crease in drag was observed until the model was placed at least three body diameters below the water surface. Similar results have been reported for towed seals and sea otters (Williams and Kooyman 1985, Williams 1989); at comparable speeds body drag was consistently higher for the animals towed on the water surface than for the same subject submerged (see Fig. 5-5). Clearly, wave drag plays a dominant role in establishing the total body drag of surface swimmers. In view of this, the practice of removing body hair to achieve a more streamlined body in human swimmers undoubtedly provides more of a psychological or perceptual boost than a hydrodynamic benefit.

Two strategies used by marine mammals to avoid prohibitively high levels of drag are abstinence from high velocity movements when on the water surface and maintenance of a submerged body position (Fig. 5-6). Except for the special case of "porpoising" (see below), aquatic mammals rarely swim at high speed for long periods while near the water surface. Sea otters provide a good example of the interplay between body drag and swimming speed. These mammals routinely use three different modes of swimming depending on their behavior. When eating prey, nursing young, or in the initial stages of escape, the otter swims ventral side up on the water surface using alternate or simultaneous strokes by the hind paws. Slow travel over short distances is conducted ventral side down on the water surface using both hind paws for propulsion. Submerged swimming involves dorsoventral undulation of the caudal half of the body. The particular swimming mode chosen will depend on the desired speed of travel. When swimming slowly the sea otter remains on the water surface, but to travel quickly the otter avoids high drag by submerging (Williams 1989). Analogous to the gait change of terrestrial mammals, the switch from surface to submerged swimming by sea otters occurs at approximately 0.8 m/sec. Likewise, the routine cruising speeds of many pinnipeds and cetaceans appear to

Figure 5-5. Total body drag of harbor seals (submerged = closed circles; surface = open circles) and humans (submerged = closed squares; surface = open squares) (A) and sea otters (submerged = open circles; surface = closed circles) (B) towed at different velocities through seawater. Surface (dashed lines) and submerged (solid lines) body positions are shown for each subject. The dashed line in (B) represents the body drag encountered by a sea otter based on preferred surface and submerged swimming speeds. Graphs are reprinted from Williams and Kooyman (1985) and Williams (1989). (C) The relationship for oxygen consumption in relation to swimming speed for seals and sea lions. Note the similar curvilinear relations for both parameters (drag and oxygen consumption) in relation to speed. (Numbers in parenthese represent number of animals.)

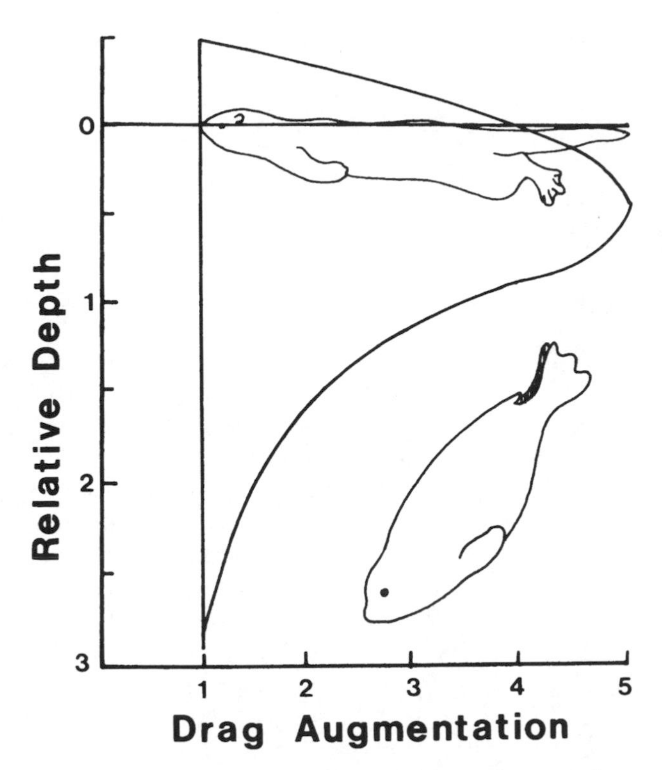

Figure 5-6. The augmentation in body drag in relation to relative depth of the swimmer. The horizontal line can be considered the water surface. Relative depth refers to body diameter of the swimmer. (The sources of data used in this figure are provided in the text.)

be dictated by the relationships between body drag, energetic cost, and velocity.

A submerged swimming position while providing a hydrodynamic benefit to the marine mammal conflicts with the physiological demand to be at the water surface to breathe. A compromise involves the diving response of marine mammals. It is difficult to determine whether the hydrodynamic demands of high-speed swimming or the energetic benefits of being able to forage over a greater range of the water column led to the enhanced breath-holding capabilities of marine mammals. Regardless, marine mammals, by virtue of their adaptations for diving, can swim submerged for prolonged periods and reduce the hydrodynamic and energetic burdens encountered by surface swimmers.

ENERGETIC COST OF SWIMMING. The energetics of swimming in marine mammals reflect the exponential increase in drag that occurs with locomotor speed (see Fig. 5-5). Metabolic rates have been measured for gray seals (*Halichoerus grypus*) (Fedak 1986), harbor seals (Davis et al. 1985), California sea lions (Kruse 1975, Feldkamp 1987), northern fur seals (Feldkamp et al. 1989), and bottlenose dolphins (Williams et al. 1993) swimming over a wide range of speeds. Unlike many terrestrial mammals in which oxygen consumption generally increases linearly with running speed (Taylor et al. 1987), marine mammals often show a curvilinear increase in oxygen consumption with swimming speed. As a result of the nonlinear relationship between energetic effort and speed, the physiological demands associated with swimming at routine speeds may be remarkably low for marine mammals. Bottlenose dolphins provide one example. The metabolic rate, heart rate, respiration rate, and levels of blood lactate of these cetaceans swimming at 2.0 m/sec show only small increases from the values of resting animals (Williams et al. 1993). Compare this with the high respiratory frequency and obvious state of fatigue of the Olympic swimmers after their gold medal sprints at the same swimming speed.

The cost of transport (COT) provides a way to compare the energetic efficiency of locomotion in different animals. The COT is defined as the metabolic cost of moving a unit mass a unit distance (Schmidt-Neilsen 1979), and is calculated by dividing the mass-specific metabolic rate of the animal by its locomotor velocity (VO_2/velocity). The resulting units, mL $O_2 \cdot kg^{-1} \cdot m^{-1}$ can be converted to energy units (joules per kilogram per meter). These are analogous to the gas mileage rating of automobiles. For animal systems we speak of liters per mile or kilometer rather than miles per gallon. The minimum COT indicates the velocity used by the swimmer to travel the greatest distance for the lowest energetic input. The minimum COT occurs in the mid-range of routine swimming speeds and is within the trough of a U-shaped curve that relates COT to swimming speed (Fig. 5-7).

Comparing different types of mammalian swimmers, we find that the minimum COT of aquatic mammals falls into two separate categories, the marine mammals and the semiaquatic mammals (Fig. 5-8). One allometric expression describes total transport costs in relation to body mass for swimming marine mammals ranging in size from 21 to 15,000 kg:

$$COT = 7.94 Mass^{-0.28},$$

where COT is in joules $\cdot kg^{-1} \cdot m^{-1}$ and mass is in kilograms (Williams 1999). This relationship includes otariid and phocid seals, large and small odontocetes, and a mysticete whale. A second allometric equation that includes muskrats, minks, humans, and surface-swimming sea otters describes the COT of semiaquatic mammals. This relationship,

$$COT = 23.9 Mass^{-0.15},$$

where COT is in joules $\cdot kg^{-1} \cdot m^{-1}$ and mass is in kilogram, roughly parallels the regressions for marine mammals and for fish (Williams 1989). However, COT is consistently higher for the surface swimming species. Several factors, in-

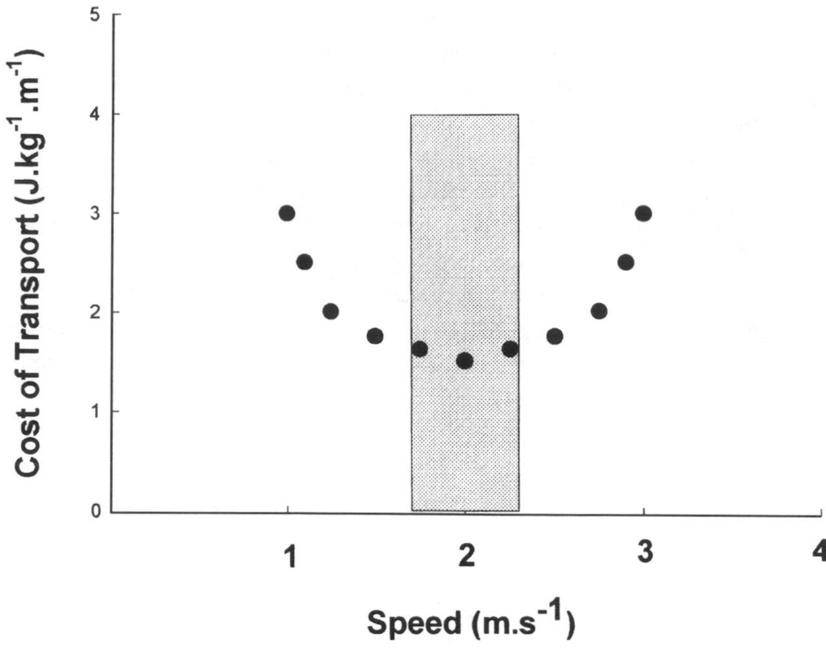

Figure 5-7. The cost of transport (COT) in relation to swimming speed in bottlenose dolphins. The dotted curve was calculated from the values of oxygen consumption determined for animals swimming next to a boat. The shaded area denotes the preferred swimming speeds of wild dolphins in coastal regions. (Reprinted from Williams et al. 1996.)

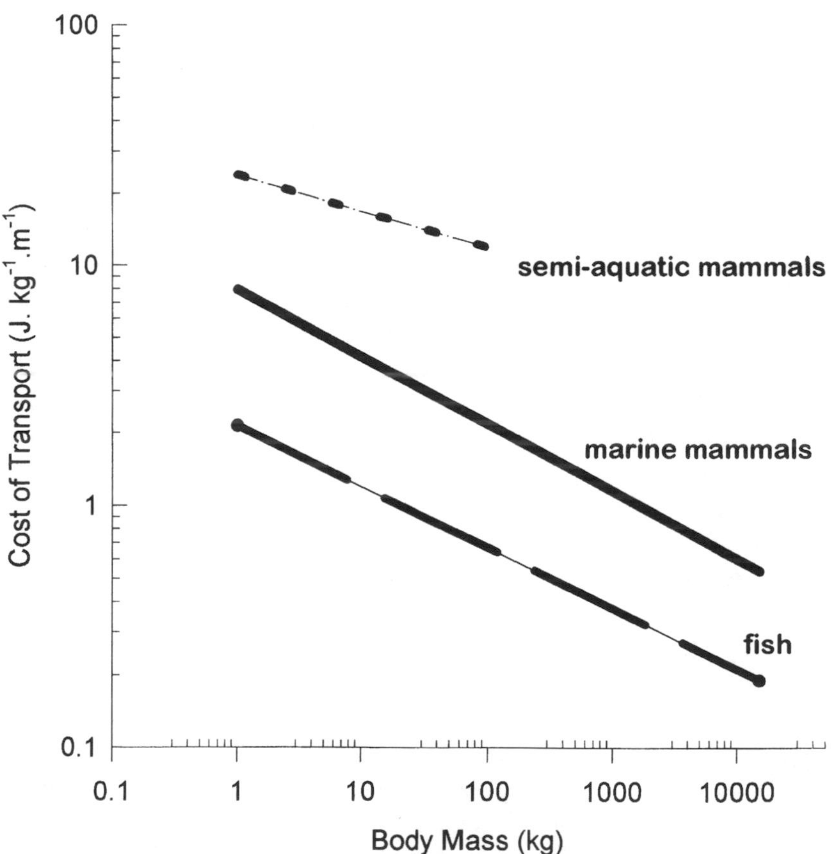

Figure 5-8. Allometric relationships for the minimum COT for swimming vertebrates. The upper dashed line denotes the best fit regression for surface swimmers and includes data for mink, muskrats, and humans. The lower dashed line represents the best fit regression and extrapolation for swimming salmonid fish. The solid line between is the allometric regression for marine mammals ranging in size from a 21-kg juvenile sea lion to a 15,000-kg gray whale.

cluding total hydrodynamic drag, propulsive efficiency, and high maintenance costs, may explain the comparatively high COT of semiaquatic animals. As discussed above, surface-swimming mammals must expend energy to overcome the augmentation in body drag associated with surface wave generation. In addition, semiaquatic mammals typically use paddling modes of swimming rather than the more cost-efficient lift-based propulsion typical of marine mammals (Fish 1993). As a result, the COTs of semiaquatic mammals are 9 to 23 times higher than those predicted for similarly sized fishes, whereas the COTs of marine mammals are only 2 to 5 times higher than such predictions.

Nonetheless, the cost per stroke may not be exceptionally low for marine mammals in comparison to semiaquatic mammals. Figure 5-9 compares the stroking costs for a human performing the front crawl to a dolphin using dorsoventral movements of the flukes to push against a force platform. The remarkable feature of the dolphin's performance is not the animal's metabolic input per se but the amount of work that can be performed for that input. A metabolic input of 60 mL O_2 per stroke allows the human to generate 15 kg of thrust, compared to nearly 100 kg by the dolphin. Differences in efficiency between the different styles of propulsion undoubtedly account in part for different levels of performance (Fish 1993). In addition, the dolphin may be able to take advantage of unique elastic characteristics of the skin and blubber and an arrangement of tendons to generate high levels of thrust for relatively little energetic input (Pabst, Rommel, and McLellan, Chapter 2, this volume).

REDUCING THE ENERGETIC COST OF SWIMMING: MARINE MAMMAL TRICKS. Marine mammals often seem to be in the process of finding the path of least resistance, that is, the most cost-effective way of moving through the marine environment. One relatively simple strategy is to use a submerged mode of swimming. We have already seen that moving under water imparts a hydrodynamic advantage to the swimmer. It also results in a significant physiological benefit.

The ability to swim submerged for sustained periods is an important adaptation for aquatic locomotion in mammals. Marine mammals, by virtue of their physiological responses to diving (i.e., utilization of on-board oxygen stores, bradycardia, and redistribution of blood flow; see Elsner, Chapter 3, this volume) can take advantage of the lower costs associated with submerged swimming. Aquatic species capable of both surface and subsurface modes of swimming (penguins and sea otters) demonstrate the energetic advantage of moving underwater. At 0.8 m/sec the oxygen consumption of sea otters swimming submerged is more than 40% lower than for surface swimming at the same speed (Williams 1989). Little blue penguins (*Eudyptula minor*) show a 40% reduction in COT when changing to a submerged mode of swimming (Baudinette and Gill 1985). In sea otters the COT for surface swimming is more than 12 times the predicted value for a salmonid fish of equal size. This is reduced to six times the predicted fish value when the otter switches to a submerged swimming mode. In view of this, it is little wonder that many marine mammals at sea, whether migrating or moving between prey patches, spend more than 90% of their time below the water surface.

Despite the advantages of remaining submerged, marine mammals must periodically come to the water surface to breathe. Usually, surface intervals are short, comprising less than 1 sec for dolphins and sea lions (Hui 1989). Surface intervals for swimming harbor seals are somewhat longer at 3 to 5 sec depending on swimming speed (Williams et al. 1991). The phocids may take several breaths during this period in contrast to the single breath taken with each surfacing by otariids and cetaceans. The differences in breathing patterns probably reflect morphological differences in lung structure for the different groups of marine mammals. Specifically,

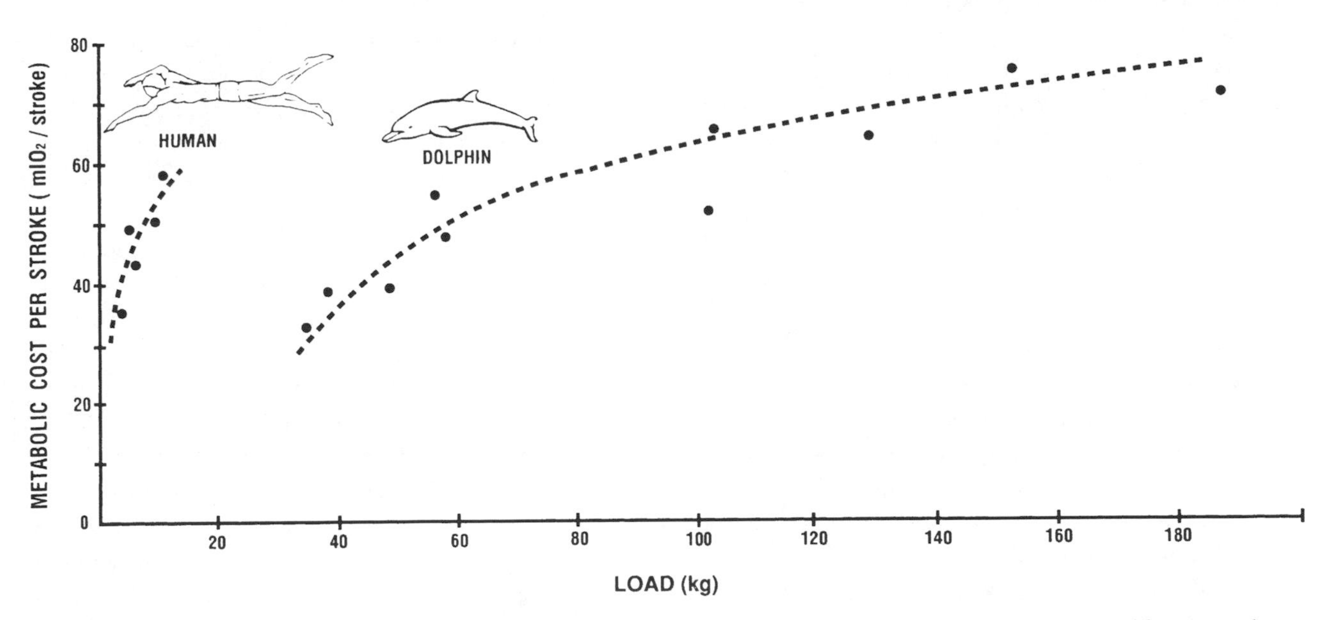

Figure 5-9. The energetic cost per stroke in relation to load for dolphins and humans. Values for dolphins were obtained from animals swimming against a force platform (Williams et al. 1993). Human athletes used swimming strokes to push off submerged force transducers in a pool. (Toussaint 1990.)

cartilaginous reinforcement of the otariid and cetacean lung supports high flow rates of air at low lung volumes during expiration (Drabek and Kooyman 1984). As a result, oxygen loading during the surface interval may be shorter for these animals than for phocid seals and perhaps other marine mammals.

Dolphins also demonstrate a novel solution to the problem of high surface drag; they simply leave the water in a maneuver appropriately termed porpoising. Au and Weihs (1980) and Blake (1983) have constructed theoretical models that predict the cost of swimming on the water surface, swimming submerged, and leaping by dolphins. It is always energetically cheaper to swim submerged at a depth greater than three times body diameter. At low speeds the energetic effort of leaping is greater than that needed to overcome drag encountered on the water surface, and the dolphins should theoretically remain in the water. However, with increasing rates of travel surface wave drag becomes prohibitively expensive. When swimming faster than the calculated transition or crossover speed, the dolphin should take to the air as it surfaces to breathe. To date, these theories have not been tested with actively swimming dolphins. Complicating factors in these tests will be individual and species-specific variations in aptitude and behavioral propensity for leaping.

Wave-riding is another behavioral strategy for reducing the energetic cost of high-speed travel in marine mammals. This behavior is routinely observed from aboard ship when wild dolphins seek the bow and stern wakes of vessels and appear to surf alongside with no apparent tail movements (Scholander 1959). Variations include riding the wake of larger whales, surfing on large wind waves (Woodcock and McBride 1951, Caldwell and Fields 1959), and drifting on currents (Würsig et al. 1985). The energetic benefit of this behavior was investigated for bottlenose dolphins trained by the U.S. Navy to match their speed to that of a moving boat. When the dolphins positioned themselves in the boat's wake their COT at 3.8 m/sec was nearly identical to that of the same animals freely swimming outside of the wake at only 2.1 m/sec. In other words, wave-riding behavior allowed the dolphins to move almost twice as fast for the same energetic cost. In addition, heart rate, respiration rate, and levels of postexercise blood lactate were reduced (Williams et al. 1992). Except for the occasional surfing sea lion or harbor seal, it is unusual to observe species other than small cetaceans take advantage of this energy-saving behavior.

Diving marine mammals may use several other behavioral strategies to reduce activity costs during the descent and ascent portion of their dives. Changes in buoyancy, interrupted modes of swimming, and prolonged periods of gliding during descent affect the energetic cost of a dive (Williams et al. 1996). Because marine mammals must limit their period of submergence to account for on-board oxygen reserves (Kooyman 1981), behavioral adjustments that reduce energetic costs potentially increase the duration of the dive. High-speed ascents and descents, and elevated thermoregulatory costs act to increase the total energetic demand of a dive and significantly reduce the time available for foraging. Conversely, the use of burst-and-glide modes of swimming increases allowable foraging time. The energetic implications of these observations for the actively foraging marine mammal are addressed in detail when we discuss at-sea metabolic rates.

The Cost of Growth and Reproduction

For growth and reproduction to occur an animal must acquire energy and nutrients in excess of that required to support maintenance functions. These additional energy costs vary with the species of marine mammal, sex, and reproductive pattern (for a general overview of all marine mammal reproductive biology, see Boyd, Lockyer, and Marsh, Chapter 6, this volume). An excellent example is the difference in reproductive costs for males and females. In pinnipeds, polar bears, sea otters, and probably mysticetes and sirenians, the cost of reproduction in males is limited to the cost of finding and maintaining access to estrous females. Evolution favors a pattern of energy expenditure that maximizes reproductive success in males. This may be demonstrated by monitoring changes in body mass or body composition of males over the breeding season (Anderson and Fedak 1985, Deutsch et al. 1990, Boyd and Duck 1991, Bartsh et al. 1992, Coltman 1996). Interestingly, the costs associated with reproduction in aquatic and terrestrially breeding males are quite similar when normalized for differences in body mass (Coltman 1996; Fig. 5-10). Larger body size is preferred in terrestrially breeding male pinnipeds as it both confers an advantage in fighting (Haley et al. 1994, Anderson and Fedak 1985). In addition, larger animals can fast longer because they have a lower mass-specific metabolic rate than smaller animals; this allows the male to maintain terrestrial territories longer (Bartholomew 1970). In species that compete for females in the water, we find that the males are comparatively smaller than the species that breed on land. For the aquatic breeders, underwater agility is more important than large size when competing for mates. It is also possible for the male to leave on short foraging trips and then return to the breeding area. Even among these aquatic breeders, large males have an advantage over small males. By reducing the number of foraging trips, large males can establish an advantage over small males in terms of time available for competing for females (Coltman 1996).

The cost of reproduction for females can be broken down into the energy requirements for gestation (the cost of pro-

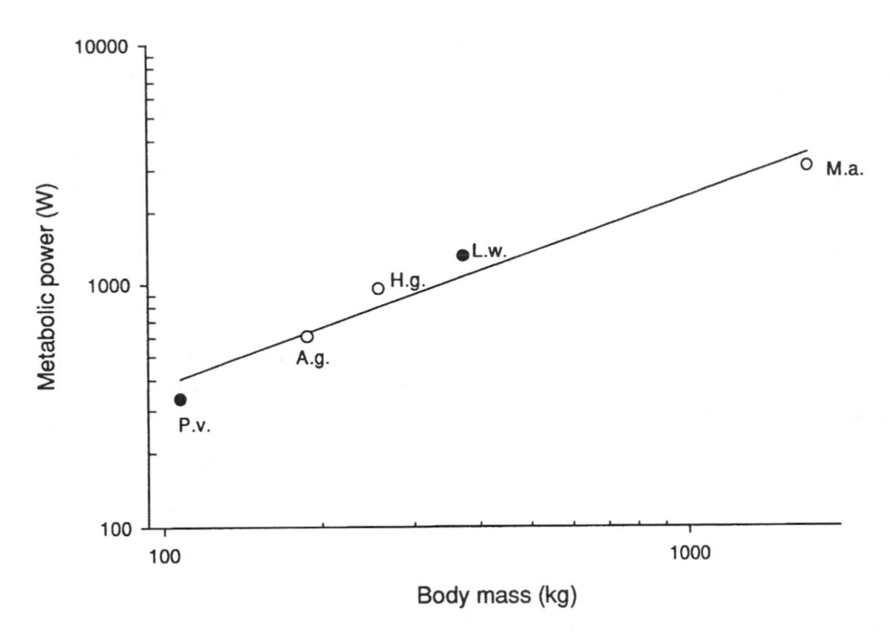

Figure 5-10. Metabolic power of breeding male pinnipeds is shown relative to body mass. Solid circles are for males with aquatic territories and open circles for males with terrestrial territories. Data are for *A.g.*, Antarctic fur seals (Boyd and Duck 1991); *P.v.*, harbor seals (Coltman 1996); *H.g.*, gray seals (Anderson and Fedak 1985); *L.w.*, Weddell seal (Bartsh et al. 1992); and *M.a.*, northern elephant seals (Deutsch et al. 1990). The line represents the least squares linear regression of all data ($R^2 = 0.95$, where $W = 10.45(kg)^{0.78}$).

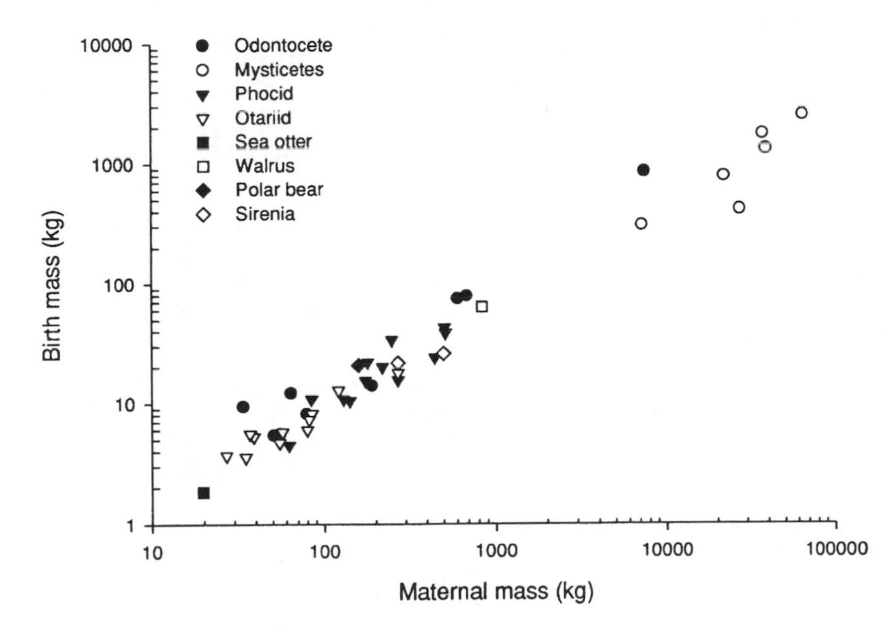

Figure 5-11. Birth mass plotted in relation to maternal mass for marine mammals. Data from Brodie 1971; Rice and Wolman 1971; Lockyer 1977, 1978, 1981a,b, 1984, 1995; Hartman 1979; Best 1982; Fay 1982; Best and da Silva 1984; Christen 1984; Kasuya and Marsh 1984; Perrin and Henderson 1984; Kovacs and Lavigne 1986, 1992b; Ramsay and Stirling 1988; Read 1990; Riedman and Estes 1990; Bowen 1991; Costa 1991a; Kretzman et al. 1991; Derocher et al. 1993; Koski et al. 1993; Martin and Rothery 1993; Perrin and Hohn 1994; Marsh 1995; Marmontel 1995; Rathbun et al. 1995; Reyes and Van Waerebeek 1995.

ducing a fetus) and for lactation (the cost of nursing young until weaning). These costs depend on the rate and intensity of investment by the mother. In marine mammals the pattern of maternal investment is quite variable with profound effects on both the energetics of the individual animal and the species. Maternal investment patterns of marine mammals vary among the taxonomic groups and in many ways dictate life history characteristics.

Although no direct measurements on the cost of gestation are available for marine mammals, investigations of terrestrial eutherians suggests that the cost of producing a fetus is insignificant relative to the costs associated with lactation (Millar 1977). Fetal mass at birth relative to maternal mass can be used as a relative index of the cost of gestation. Among marine mammals, we find that there are no striking differences in birth mass relative to maternal mass (Fig. 5-11). This is not surprising in view of the relatively low energetic cost of producing a fetus. Conversely, lactation strategies vary markedly with maternal body mass among different marine mammal groups (Fig. 5-12). Phocid seals and mysticete whales have significantly shorter lactation durations for their body size than do other marine mammals. The remaining groups are surprisingly similar. A comparison of growth rate as a function of maternal mass indicates that the shorter lactation duration of phocids is compensated for by an increase in growth rate (Fig. 5-13). Otariids, polar bears, sea otters, walruses, and odontocetes have remarkably similar growth rate patterns relative to maternal mass (Boyd, Lockyer, and Marsh, Chapter 6, this volume).

Comparisons of maternal investment patterns must consider differences in the behavior and metabolic rates of the mother and her young. For example, although phocid pups are weaned at an early age, they are not truly nutritionally independent at that time; they rely on maternally derived energy, stored as blubber, for weeks or months after weaning. In contrast, other marine mammals, such as otariids, sea otters, and odontocete cetaceans, wean their offspring much later. As a result, the young are not as reliant on energy reserves stored before weaning. A longer lactation period requires that more energy be supplied to the young in support of its maintenance metabolism. Perhaps for this reason, many marine mammal young begin to feed before weaning with the result that the animal is nutritionally independent when weaning does occur. Obviously, a longer development period requires greater maternal investment.

Longer lactation periods allow the young to acquire a greater proportion of lean tissue growth (Costa 1991a). Growth in phocid young primarily takes the form of adipose tissue with proportionately little growth in lean tissue. Otariids show proportionately greater increases in lean mass during the lactation interval (Bryden 1968; Worthy and Lavigne 1983b; Ortiz et al. 1984; Oftedal et al. 1987; Arnould et al.

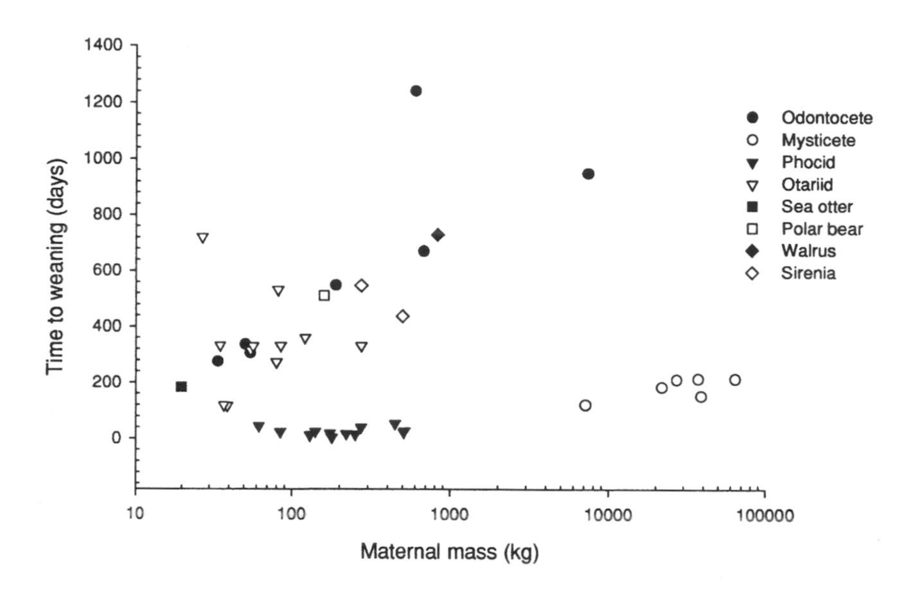

Figure 5-12. Time to weaning plotted as a function of maternal mass for marine mammals. Lactation durations of phocid seals and mysticete whales are shorter than all other marine mammals. Data from Brodie 1971; Rice and Wolman 1971; Lockyer 1977, 1978, 1981a,b, 1984, 1995; Hartman 1979; Best 1982; Fay 1982; Best and da Silva 1984; Kasuya and Marsh 1984; Perrin and Henderson 1984; Kovacs and Lavigne 1986, 1992b; Ramsay and Stirling 1988; Read 1990; Riedman and Estes 1990; Bowen 1991; Costa 1991a; Kretzman et al. 1991; Derochere et al. 1993; Koski et al. 1993; Martin and Rothery 1993; Perrin and Hohn 1994; Marsh 1995; Marmontel 1995; Rathbun et al. 1995; Reyes and Van Waerebeek 1995.

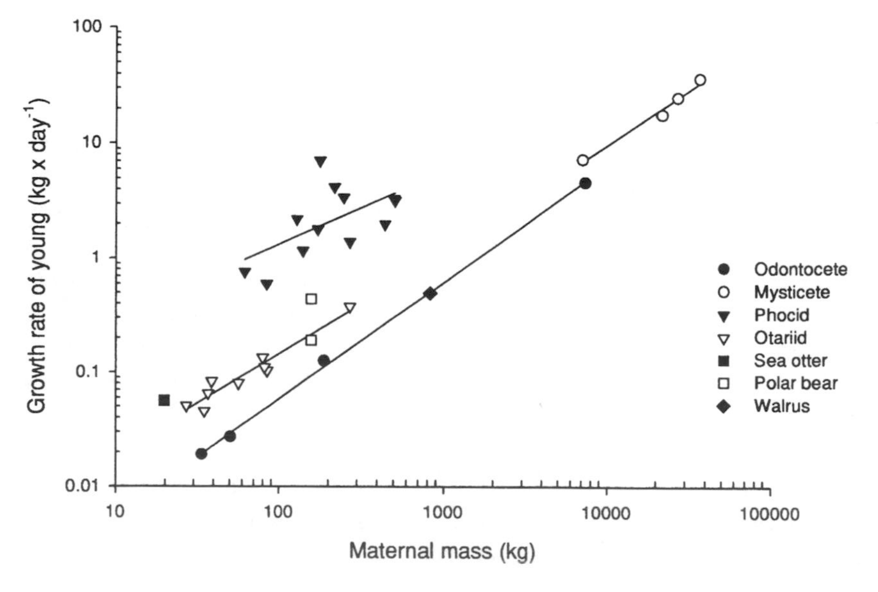

Figure 5-13. Growth rate of suckling marine mammals as a function of maternal mass. Lines represent least squares regressions for each taxonomic group. Data are from Brodie 1971; Rice and Wolman 1971; Lockyer 1977, 1978, 1981a,b, 1984, 1995; Best 1982; Fay 1982; Best and da Silva 1984; Christen 1984; Kasuya and Marsh 1984; Perrin and Henderson 1984; Kovacs and Lavigne 1986, 1992b; Ramsay and Stirling 1988; Read 1990; Riedman and Estes 1990; Bowen 1991; Costa 1991a; Kretzman et al. 1991; Derocher et al. 1993; Koski et al. 1993; Martin and Rothery 1993; Perrin and Hohn 1994; Reyes and Van Waerebeek 1995.

1996a,b). The longer dependency period of otariid pups requires a greater utilization of energy for maintenance functions. This diverts energy from growth and requires that otariid mothers provide more energy for an equivalent relative mass at weaning.

The relationship between lactation duration and weaning mass is well illustrated by a comparison of northern elephant seals and northern fur seals. Northern elephant seal pups are born at 7.5% of maternal mass, are nursed over a 28-day lactation interval, and are weaned at 26% of maternal mass. After weaning, pups remain on the beach fasting for 2.5 months (Reiter et al. 1978) and then go to sea to feed after losing approximately 30% of their mass at weaning (Kretzmann et al. 1993). Pups are not nutritionally independent until they are at least 3.5 months of age (Ortiz et al. 1978) and by this time weigh 18% of maternal mass. Northern fur seal mothers suckle their pups over a 4-month period and wean them at 35% of maternal mass. At or near weaning, northern fur seal pups have begun to feed and may be considered nutritionally independent. Consequently, northern fur seal pups may be proportionately larger at nutritional independence than northern elephant seal pups. In addition to physiological factors, the duration of lactation may be dictated by the habitat of the marine mammals. This is especially apparent for phocids (Kovacs and Lavigne 1986) and is an important factor that permits this group to breed on ice (Stirling 1975, 1983). The shortest lactation interval and fastest growth rates for pups occur in pack ice breeding seals such as hooded (*Cystophora cristata*) and harp seals. The longest lactation intervals are found in species that breed on fast ice such as Weddell (*Leptonychotes weddellii*) and ringed seals. Because fast ice is firmly attached to the shore, it is quite stable. Conversely, pack ice is a very unstable breeding substrate and can disappear at any time. The shortened lactation interval of pack ice seals ensures that the pup is weaned before the break up of the pack (Bowen et al. 1985, Bowen 1991). Island-breeding phocids, including elephant and gray seals, that feed far offshore show an intermediate pattern. These animals may not be able to feed during the lactation period because the food resource is too distant. As a result, there would be a significant advantage in concentrating the investment interval to reduce metabolic overhead (Fedak and Anderson 1982). By shortening the lactation period, a higher proportion of stored maternal resources can go into milk production rather than maternal maintenance metabolism (see below). The comparatively long lactation interval of Weddell seals, ringed seals, and island-breeding harbor seals allows these species to augment their maternal reserves by feeding because prey are nearby (Testa et al. 1989, Boness et al. 1994, Boness and Bowen 1996). With the proximity of food, feeding trips may be shortened during lactation. This results in a concomitant decrease in pressure on the female to shorten the investment interval and reduce metabolic overhead. It is unlikely that such short-duration feeding trips could supply sufficient energy to support the rapid growth rates of phocid pups. Most of the energy and materials supplied to the pup are still derived from maternal body reserves.

BODY SIZE AND MATERNAL RESOURCES: THE ROLE OF MATERNAL OVERHEAD. Fasting during lactation is a unique component of the life history pattern of marine mammals. With the exception of bears, no other mammal is capable of producing milk without feeding. By undertaking this energetic challenge, mysticetes and pinnipeds are able to temporally and spatially separate feeding from breeding. Separation of lactation from feeding allows mysticete whales to feed in the highly productive polar regions of the world's oceans, but retain the thermal advantage of breeding in the calm tropical regions. Migrating to warmer waters for parturition reduces the thermal demands on the newborn calf and provides additional energetic advantages for the mother.

Among pinnipeds, the separation of feeding from lactation allows terrestrial parturition and pup rearing and at-sea foraging by the female. Most phocids store what is required for the entire lactation period, whereas all otariids feed during lactation (Bonner 1984, Kovacs and Lavigne 1986, Oftedal et al. 1987, Costa 1991a, Kovacs and Lavigne 1992a). Phocid mothers remain on or near the rookery continuously from the birth of their pups until they are weaned; milk is produced from body reserves stored before parturition. Although some phocids feed during lactation, most of the maternal investment is derived from body stores. Weaning is abrupt and occurs after a minimum of 4 days of nursing in hooded seals to a maximum of 6 to 7 weeks in Weddell or monk seals (*Monachus* spp.). In contrast, otariid mothers remain with their pups for approximately 1 week after parturition and then periodically go to sea to feed. The pup is suckled intermittently between foraging trips (Bonner 1984).

The ability of a marine mammal female to fast while providing milk to her offspring is related to the size of her energy and nutrients reserves and the rate at which she uses them. When food resources are far from the breeding grounds, as may occur for some phocids and large mysticete whales, the optimal solution is to maximize the amount of energy and nutrients provided to the young and to minimize the amount expended on the mother. Fedak and Anderson (1982) used the term "metabolic overhead" to define the amount of energy a female pinniped expends on herself while ashore suckling her pup (Fig. 5-14). However, this concept can be applied to all marine mammals that lactate while fasting.

One mechanism for reducing metabolic overhead is to attain large body size. This results from the relationship between maintenance metabolism, which scales as $mass^{0.75}$, and fat stores, which scale as $mass^{1.0}$ (Fig. 5-15) (Calder 1984, Millar and Hickling 1990, Lindstedt and Boyce 1985). These relationships demonstrate that as body size increases, energy reserves increase proportionately faster with mass than maintenance metabolism. This implies that larger females have a greater ability to provision their young from stored body reserves than smaller females.

The relationship between metabolic overhead, milk production, and lactation duration has been modeled using data from northern elephant seals (Costa 1993a). The total amount of energy available for maternal investment was assumed to be constant, and metabolic overhead was calculated for each lactation duration. Milk production was calculated as the proportion of maternal resources remaining after the cost of the metabolic overhead was met for a given lactation duration. Not surprisingly, the net amount of energy used for maternal maintenance increases with increasing lactation duration; a concomitant decrease in the energy available for milk production is observed. This model indicates that a shorter lactation period is advantageous because the mother can devote more of her energy stores to milk production than to maternal maintenance.

As stated above, a large female should have a lower metabolic overhead than a small female. If we allow our model to include variations in maternal mass as well as lactation duration, we find that metabolic overhead can be minimized by either increasing body mass or reducing the duration of lactation. The strategy for small phocid seals (i.e., ringed seals, harbor seals) is to maintain short lactation periods or to feed during lactation. Large phocids, such as elephant seals or extremely large mysticete whales, are able to maintain longer lactation intervals as metabolic overhead is minimal because of their large size (Fig. 5-16).

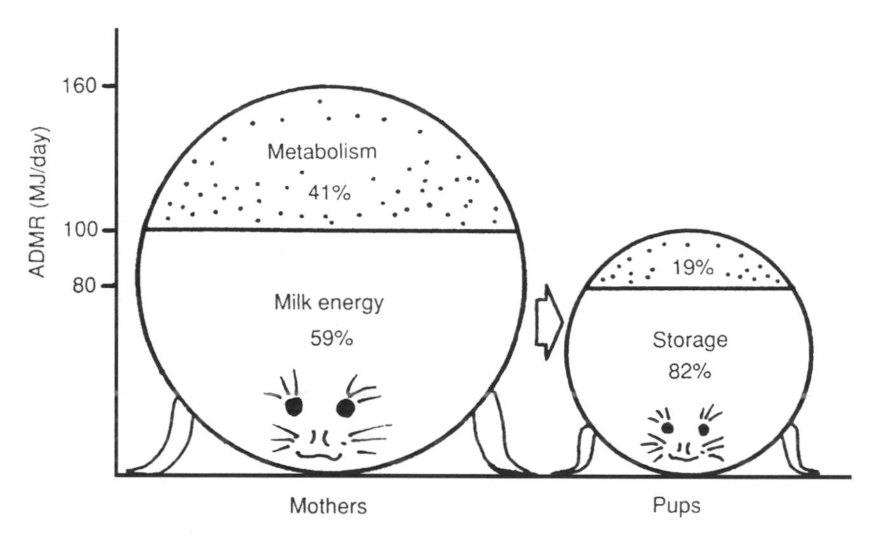

Figure 5-14. Total energy budget of a typical northern elephant seal mother graphically divided between the energy expended on her maintenance metabolism and that contained in the milk fed to the pup. The pup's total energy budget is divided between energy storage and that used for maintenance metabolism. (Figure drawn by A. C. Huntley from data in Costa et al. 1986.)

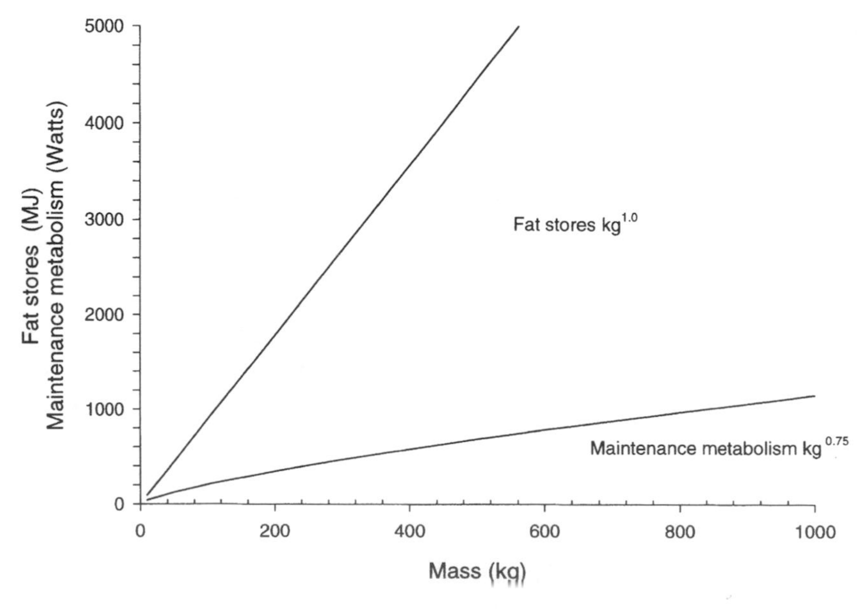

Figure 5-15. The theoretical relationship between fasting metabolic rate and adipose tissue content plotted as a function of maternal body mass for a typical phocid female. Fasting metabolic rate scales with body mass raised to the 0.75 power, whereas fat stores scale linearly with body mass.

Although a comparison of maternal investment strategies for different marine mammal groups would be interesting, there is little information available for species other than pinnipeds. Cetaceans, sirenians, sea otters, and polar bears may spend their entire lives at sea. Therefore, measurements of milk intake, feeding rates during lactation, or other features of reproductive investment are not possible for comparison at this time.

MATERNAL METABOLISM. Variation in maternal metabolism, at least within the otariids, appears to be linked to ambient temperature. Galapagos fur seal females in the warm equatorial environment exhibit fasting metabolic rates on land that are only 1.1 times the predicted basal metabolic rate (Costa and Trillmich 1988). In contrast, northern and Antarctic fur seals (*Arctocephalus gazella*) inhabiting the cold subpolar environment exhibit metabolic rates 3.4 times predicted levels (Costa and Trillmich 1988). Decreases in an animal's onshore metabolism may be achieved by a reduction in activity as observed for phocids and otariids or by periodic breathing as occurs in phocids (Costa et al. 1986, Costa and Trillmich 1988).

The ability to store energy also differs between otariids and phocids. The available data indicate that phocid mothers store significantly more fat than otariids. Values range from 24.5% fat for harbor seals to 47% for harp seals. In comparison, Galapagos, northern, and Antarctic fur seals, and California and Australian sea lions (*Neophoca cinerea*) show 26%, 22%, 19%, 13%, and 8.3% body fat, respectively. With the exception of the Galapagos fur seal, these otariids appear to maintain fasting onshore metabolic rates similar to those of phocids when corrected for body mass (Fig. 5-17) (Costa 1991a, Bowen et al. 1992, Fedak et al. 1996). Overall, the ability of phocid seals to store most if not all of the maternal energy and nutrients before arriving onshore may be attributed to (1) large body size, (2) low metabolic overhead, and (3) greater lipid reserves (Costa 1991a, 1993a).

ENERGY INVESTMENT AND TRIP DURATION. Many phocids fast throughout the lactation interval, whereas otariid females spend 0.5 to 14 days at sea foraging between visits to suckle the pup on the rookery (Gentry and Holt 1986). How do otariid mothers modify the timing and patterning of energy and nutrient investment to accommodate for such different trip durations? If we plot trip duration and milk energy consumed by the pup (normalized for maternal mass$^{0.75}$), we find that otariid mothers making short feeding trips provide their pups with less milk energy than mothers that make long trips (Fig. 5-18). Such a pattern is consistent with the predictions of central place foraging theory (Orians and Pearson 1977), which predicts the optimal behavior of animals foraging at varying distances from a central place, such as a nest or rookery. For example, when foraging a long distance from the rookery, a parent should make few trips of long duration and return with a greater quantity of energy per trip. In contrast, parents feeding close to the

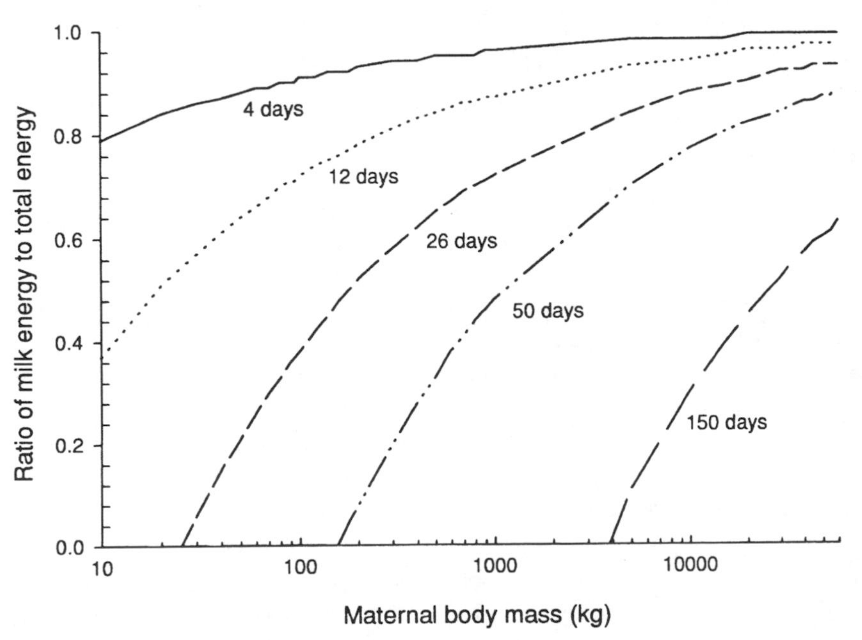

Figure 5-16. The importance of maternal mass and lactation duration on the proportion of total maternal energy available for milk production is detailed in this figure. Lactation durations and maternal mass vary from small phocids (4 days, 100 kg) to mysticete whales (150 days, 60,000 kg). Notice that for small animals it is critical to have a short lactation interval. However, longer lactation intervals are possible with large body size.

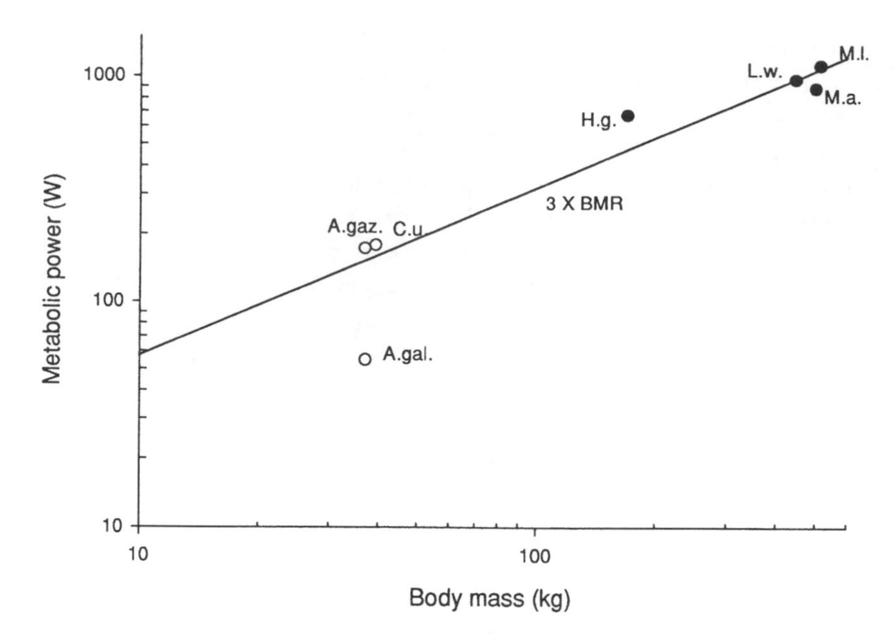

Figure 5-17. Fasting metabolic rate plotted as a function of maternal body mass for four phocids (solid symbols) and three otariids (open symbols). Data were collected using labeled water methodologies (Costa and Gentry 1986, northern fur seals (*C.u.*); Costa and Trillmich 1988, Antarctic fur seals (*A. gaz.*), Galapagos fur seals (*A. gal.*); Costa et al. 1986, northern elephant seals (*M.a.*); Fedak et al. 1996, southern elephant seals (*Mirounga leonina* [*M.l.*]). Measurements derived from mass change data are for gray seals (*H.g.*) Fedak and Anderson 1982 and, Weddell seals (*L.w.*) Tedman and Green 1987. The solid line represents three times the basal metabolic rate for a terrestrial animal of equal size (Kleiber 1975).

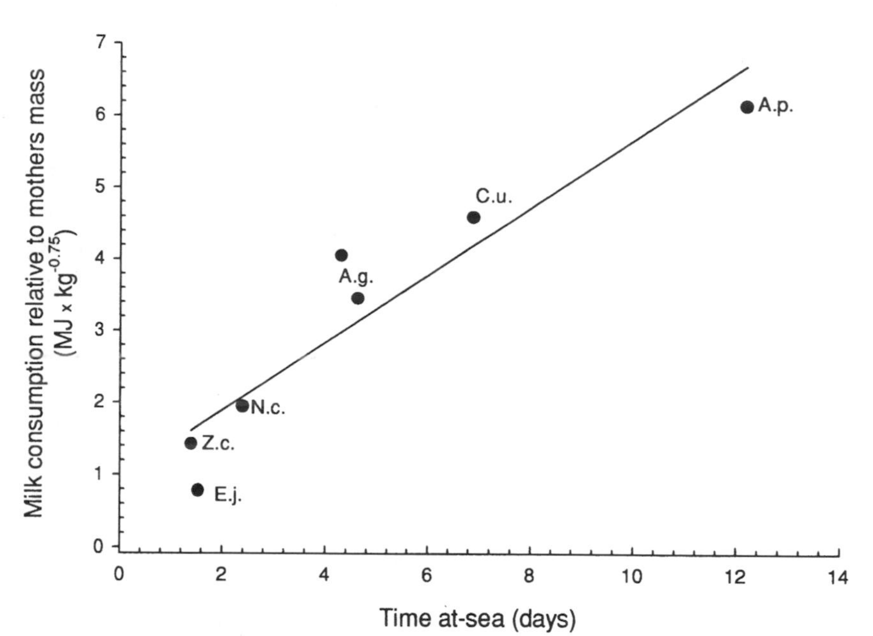

Figure 5-18. Total milk energy consumed by pups per shore visit normalized by dividing by maternal metabolic mass$^{0.75}$ and plotted in relation to the amount of time the mother typically spends at sea feeding before the shore visit. Symbols are for *E.j.*, Steller sea lion (Higgins et al. 1988), *Z.c.*, California sea lion (Oftedal et al. 1987), *A.g.*, Antarctic fur seal (two points, one each from Costa 1991b, Arnould and Boyd 1995a,b), *N.c.*, Australian sea lion (Costa 1991b), *C.u.*, northern fur seal (Costa and Gentry 1986), *A.p.*, Juan Fernández fur seal (Acuña 1995). The line represents the least squares linear regression where milk consumption (MJ × kg$^{-0.75}$) = 8.91(days)$^{0.47}$, $R^2 = 0.883$).

rookery (near-shore) should make many short trips, with a comparatively lower energy return per trip. Otariids, such as the Steller sea lion (*Eumetopias jubatus*), make trips of relatively short duration (approximately 36 hr), feed near shore, and travel short distances to the feeding grounds (Higgins et al. 1988). Northern fur seals feed up to 100 km offshore and make trips of 7 days duration (Loughlin et al. 1987). As predicted by the model, Steller sea lions deliver considerably smaller amounts of milk energy (0.8 MJ × kg$^{-0.75}$) per visit to their pup than northern fur seals (4.6 MJ × kg$^{-0.75}$). A similar pattern is observed for fur seal species. Inshore feeding species (i.e., Galapagos fur seals) forage for less than 24 hr between shore visits to suckle their pups (Trillmich et al. 1986). Offshore feeding species, such as Antarctic, northern, and Juan Fernández fur seals (*Arctocephalus philippii*) may spend anywhere from 4 to 12 days at sea foraging (Doidge et al. 1986, Gentry and Holt 1986, Acuña 1995).

Lactating northern fur seal females consume 80% more food than nonlactating females (Perez and Mooney 1986). Such elevated rates of food intake can only be sustained in the highly productive water characteristic of upwelling environments. Therefore, the absence of a truly tropical otariid may be related to these high reproductive demands and the lower productivity of warmer waters. (The Galapagos fur seal and sea lion are not considered truly tropical species as they exist in a highly productive equatorial upwelling region; Maxwell 1974.)

Optimization of foraging behavior is also observed for phocids. Island-breeding species represent an extreme example of an offshore feeder that uses highly dispersed or distant prey resources and makes as few trips as possible per reproductive event. Elephant seals and gray seals have separated feeding from onshore lactation. The reproductive pattern of these phocids is less constrained by the time it takes to travel and exploit distant prey, which may allow utilization of a more dispersed or patchy food resource. By spreading the acquisition of prey energy required for lactation over many months at sea, northern elephant seal females need to increase their daily food intake by only an estimated 12% to cover the entire cost of milk production and maternal metabolism if we assume that the food energy needed for lactation is consumed over the entire trip to sea (Costa et al. 1986).

In comparison to otariids, phocids may have a reproductive pattern that is better suited for dealing with dispersed or unpredictable prey or prey that is located at great distances from the rookery. The ability of some phocids to forage over long distances is influenced by reducing the importance of feeding during lactation. However, fasting during lactation places a limit on the duration of investment and this limits the total amount of energy that a phocid mother can invest in her pup (Costa 1991a).

What enables otariids and phocids to forage at sea for such variable intervals and yet supply their pups with an appropriate amount of energy? The answer is that the lactation process enables pinnipeds to concentrate the material fed to the young with some independence from prey quality (Pond 1977), or distance or time spent away from the rookery. It is well documented that marine mammals in general, and pinnipeds in particular, produce milk extremely high in lipid (Pilson and Kelly 1962; Bonner 1984; Oftedal et al. 1987, 1996). Although the milk reflects the basic constituents consumed by the mother, she can process, concentrate, or use stored reserves in the production of milk. For example, some species feed on fish, whereas others feed on fish or squid. Yet all of these species provision their offspring with milk of significantly greater energy density than the prey consumed (Fig. 5-19).

Increasing the lipid content of milk to modify the energy density of milk has a disadvantage. The high energy density of pinniped milk is achieved by increasing the lipid content with a reciprocal decrease in water content but no corresponding change in protein content. Therefore, the protein-to-energy ratio of pinniped milk is lowest in the most energy-dense milk. Young may be provided with more than sufficient energy to fuel metabolism but may be limited in their ability to grow due to the reduced protein intake. This is especially important in species that have shortened lactation intervals like hooded seals. In such species, pups receive similar amounts of total energy in smaller quantities of milk. Because the protein content of the milk is independent of lactation duration, these

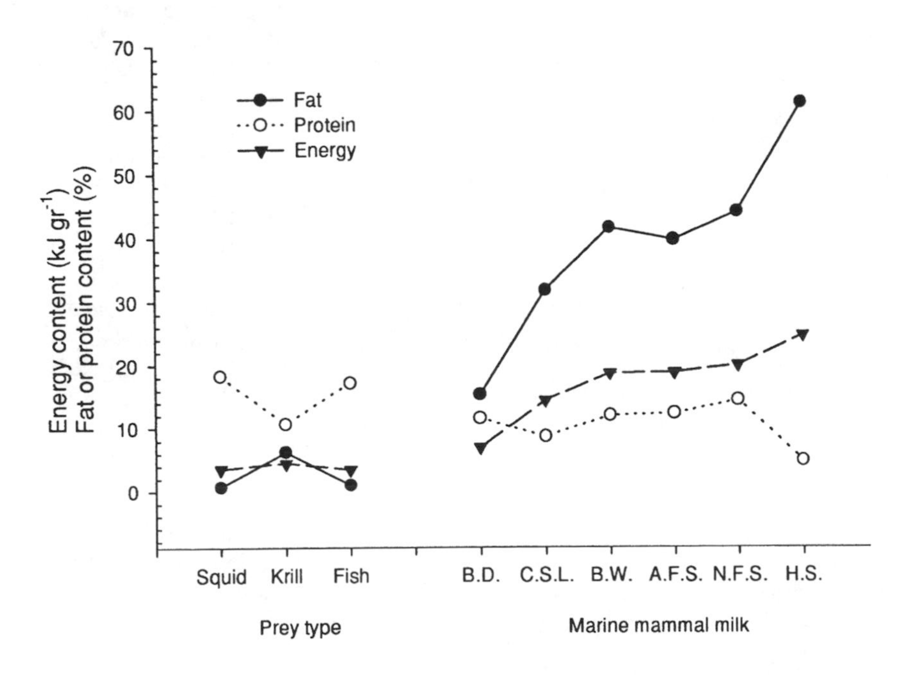

Figure 5-19. The energy density and protein content of squid, notothaenid fish, krill, and milk of bottlenose dolphins (*B.D.*), California sea lions (*C.S.L.*), blue whale (*B.W.*), Antarctic fur seal (*A.F.S.*), northern fur seal (*N.F.S.*), and hooded seal (*H.S.*). Data on squid and fish are from Perez and Mooney (1986); krill from Clarke (1980); California sea lions from Oftedal et al. (1987); Antarctic fur seal, Costa et al. (1985); bottlenose dolphins, D. P. Costa, unpubl.; blue whale, Lockyer (1984); and hoded seals, Oftedal et al. (1988).

pups get less total protein. This constraint can be seen by the fact that most of the postnatal growth of phocid seals is due to the accumulation of adipose tissue stored as blubber with little growth in lean tissue (Bryden 1968, Worthy and Lavigne 1983a, Costa et al. 1986, Oftedal et al. 1987). Thus, northern elephant and harp seal pups are born almost without fat but are composed of approximately 50% lipid at weaning (Worthy and Lavigne 1983a,b; Ortiz et al. 1984).

VARIATION IN MILK COMPOSITION. A consistent adaptation among all marine mammals is that they produce a high-fat and, therefore, energy-rich milk (Fig. 5-20). With a few exceptions, most terrestrial animals produce milk that is fairly low in fat; cows and humans produce milk that contains 4% and 8% milk fat, respectively. The variation in milk fat among marine mammals is correlated with the duration of time that the mother and calf are together. For example, the highest milk fat contents are observed in pinnipeds, particularly phocid seals that spend only a few days to weeks with their pups. A fat, energy-rich milk allows the mother to transfer high levels of energy in a very short period of time. Hooded seals are most impressive with a 4-day lactation interval and a milk fat of 65% lipid (Oftedal et al. 1988). In view of this, it is not surprising that marine mammals with the highest growth rates produce milk with the highest lipid content (Fig. 5-21). However, there is considerable variation in the milk fat content of marine mammals with unexceptional growth rates.

Variations in milk fat content, particularly among otariids, appear to be related to maternal attendance patterns (Costa 1991a). Lipid and, therefore, energy content of the milk of otariids increases with trip duration (Trillmich and Lechner 1986). Correlations between milk fat content and trip duration are complicated by the fact that otariid females with short trip durations are low latitude species and those with long trip durations are high latitude species. High latitude environments are highly seasonal and may force shorter lactation periods than low latitude species (Oftedal et al. 1987).

Ecological Implications of Energy Flow

Field Metabolic Rates

In the preceding sections, we discussed the various components that comprise the energy costs associated with exis-

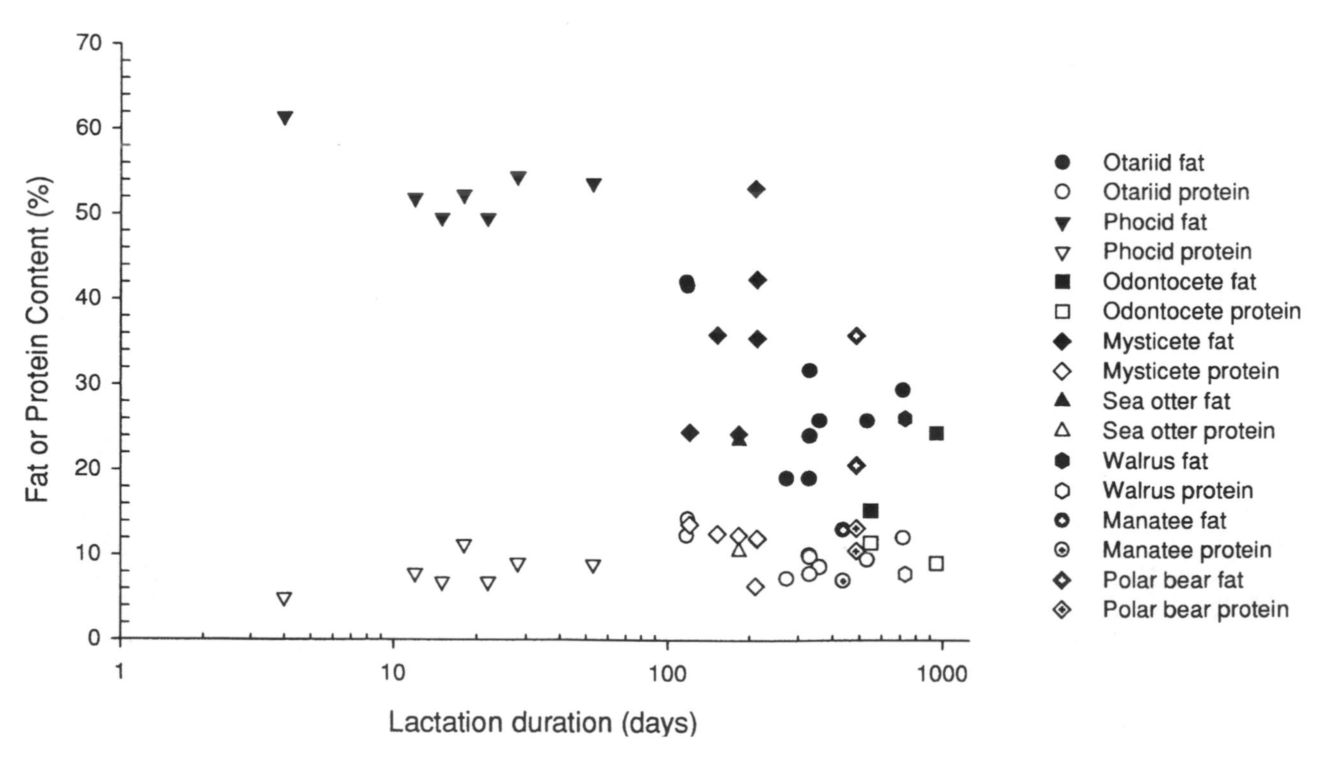

Figure 5-20. Fat (solid symbols) and protein (open symbols) content (%) for marine mammal milk as a function of lactation duration. Data are from Rice and Wolman 1971; Lockyer 1977, 1978, 1981a,b, 1984, 1995; Best 1982; Fay 1982; Best and da Silva 1984; Ramsay and Stirling 1988; Read 1990; Riedman and Estes 1990; Bowen 1991; Costa 1991a; Kretzman et al. 1991; Derocher et al. 1993; Acuña 1995; Marsh 1995; Marmontel 1995; Rathbun et al. 1995.

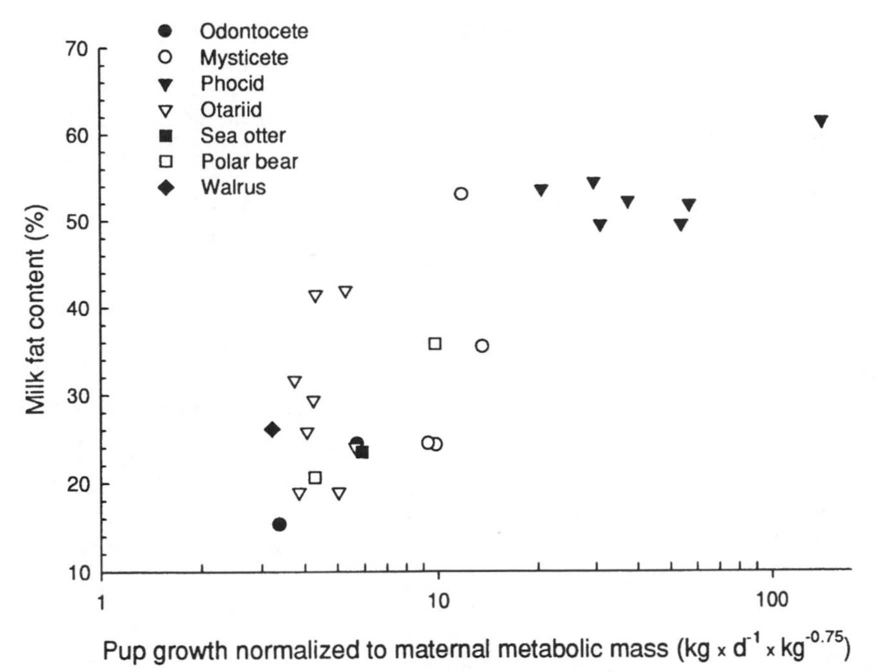

Figure 5-21. Milk fat content of marine mammal milk plotted in relation to the growth rate of the young corrected to maternal metabolic mass ($kg^{0.75}$). Data are from Rice and Wolman 1971; Lockyer 1977, 1978, 1981a,b, 1984, 1995; Best 1982; Fay 1982; Best and da Silva 1984; Ramsay and Stirling 1988; Read 1990; Riedman and Estes 1990; Bowen 1991; Costa 1991a; Kretzman et al. 1991; Derocher et al. 1993; Acuña 1995.

tence for a marine mammal. Although these individual components provide an understanding of energetic adaptations, it is the sum of all of these activities and processes that determine the overall energetic cost of existence, commonly referred to as field metabolic rate. In the following sections, we examine the overall energy budget of marine mammals and how it varies with foraging ecology for different marine mammal species.

A number of approaches have been used to study the energetics of animals at sea. One approach, time budget analysis, sums the daily metabolic costs associated with various activities. Field observations of behavior are coupled with metabolic rate measurements made in captivity. Other methods rely on predictive relationships between physiological variables and metabolic rate. For example, metabolic costs can be assessed indirectly by measurement of changes in body mass and composition (Fedak and Anderson 1987), variations in heart rate or ventilation rate (Sumich 1983; Dolphin 1987a,b; Butler et al. 1992, 1995; Williams et al. 1993), or dilution of isotopically labeled water (Costa 1987, Anderson et al. 1993). Metabolic rates calculated from heart rate or ventilation rate assume that the level of oxygen consumed is directly related to each breath or heart beat. Estimates of metabolic rate from changes in body mass may be based on direct measurements or estimated changes in the relative proportion of fat and protein of an animal. Isotopic measurements of metabolic rate rely on changes in the concentration of two isotopes of water within body compartments. In this method, two isotopes of water are given to the animal orally or by injection. After equilibration, an initial blood sample is taken and the animal is released. After a period of 3 to 10 days, a final blood sample is taken. The metabolic rate is then determined from the change in concentration of the two isotopes in the blood. The first isotope, tritium- or deuterium-labeled water, decreases in the animal's blood as a function of the water exchange through the animal. The second isotope, oxygen-18-labeled water, declines as a function of the animal's water exchange and CO_2 production. The arithmetic difference between the disappearance rate of deuterium- or tritium-labeled water (water only) and oxygen-18-labeled water (water and CO_2) provides an estimate of the animal's rate of CO_2 production, and hence, metabolic rate (Lifson and McClintock 1966, Nagy 1980, Costa 1987).

Field metabolic rates (FMR) provide insight into the energetic strategies used by marine mammals. For example, to determine whether phocids and otariids use different strategies while foraging, we could compare the metabolic rates of each species actively foraging at sea. Comparisons of the metabolic costs associated with pup maintenance while onshore may be assessed by monitoring females while they suckle their pups on the beach. Not surprisingly, there are limited data on field metabolic rates to make such comparisons for marine mammals. The best data exist for pinnipeds and the bottlenose dolphin (Fig. 5-22) and indicate that foraging otariids and bottlenose dolphins expend energy at six times the predicted basal metabolic level (Costa and Gentry

1986; Costa et al. 1989b, 1995). In contrast, the metabolic rate of diving elephant seals is only two to three times the predicted basal rate (D. P. Costa, unpubl.). Metabolic rates of Weddell seals freely diving from a hole in the ice have been determined by indirect calorimetry at a rate of 1.5 to 2 times the predicted basal level (Kooyman et al. 1973, Castellini et al. 1992).

How might differences in field metabolic rate affect the foraging energetics and behavioral strategies of a marine mammal? As air breathers, foraging marine mammals are physiologically constrained by the amount of oxygen they have stored in their tissues and the rate at which this oxygen is used. Although animals can increase the duration of individual dives by using anaerobic metabolism, it is energetically more efficient to rely on aerobic metabolism (Kooyman et al. 1980, 1983; Kooyman 1989; Elsner, Chapter 3, this volume). The maximum time an animal can remain submerged without using anaerobic metabolic pathways is called the aerobic dive limit (ADL) as defined by:

$$\text{ADL} = \text{Total oxygen store (mL } O_2\text{) / Metabolic rate during dive (mL } O_2 \times \text{min}^{-1}\text{)}$$

The total oxygen store is represented by the oxygen reserves in the blood, muscle, and lung (Kooyman 1985). The bigger the ADL, the longer a marine mammal can carry out an uninterrupted search or the deeper it can dive. Thus, the foraging capability of an animal is related to the balance between its total oxygen store and its utilization rate as dictated by diving metabolic rate (see Elsner, Chapter 3, this volume).

To understand the relationship between diving metabolism and foraging behavior, the oxygen storage capability of marine mammals must be considered. As a group, phocids have large oxygen stores in comparison to other marine mammals. This suggests that phocids demonstrate the

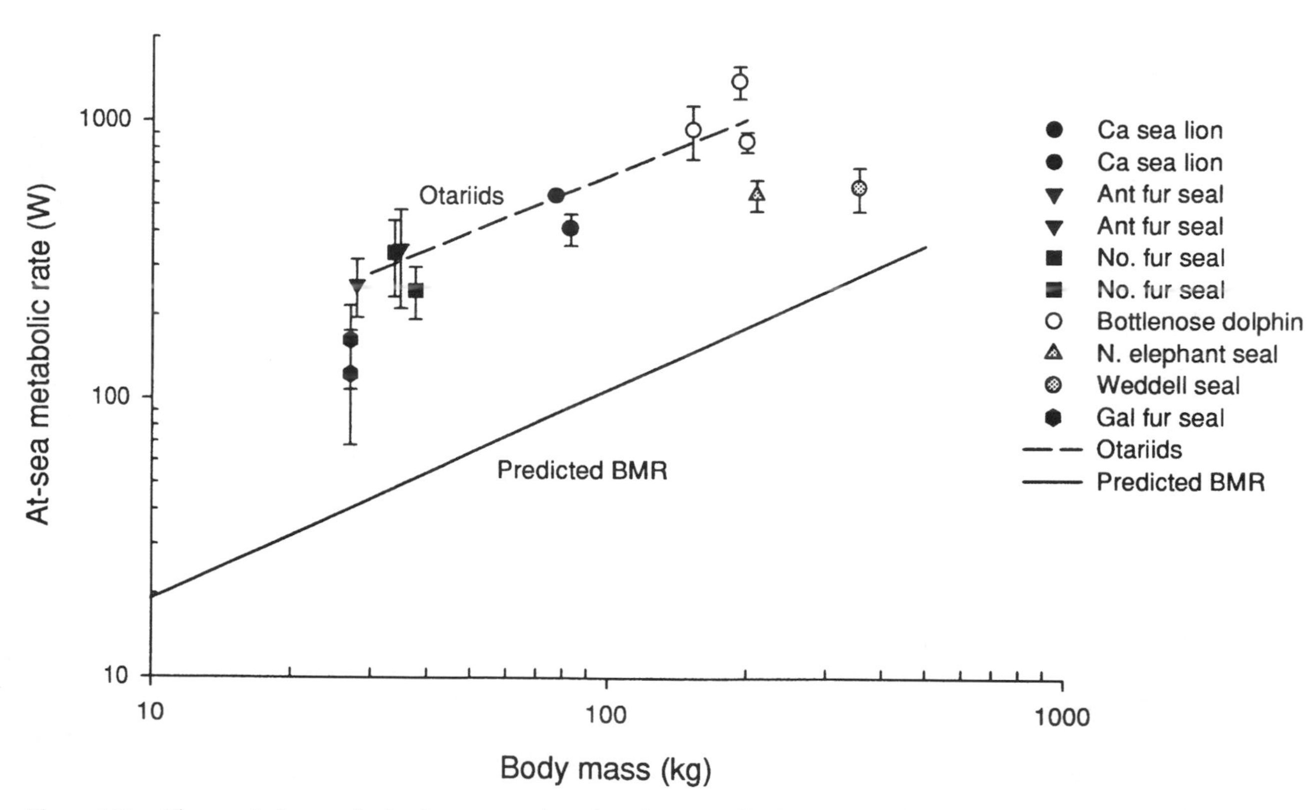

Figure 5-22. The metabolic rate of animals at sea was plotted as a function of body mass. Metabolic rates were determined using the oxygen-18 doubly labeled water method and are from northern fur seals (Costa and Gentry 1986), Antarctic fur seals (Costa et al. 1989a), Australian sea lions (Costa et al. 1989b), California sea lions (Costa et al. 1991), Galapagos fur seals (Trillmich 1990), bottlenose dolphins (Costa et al. 1995), and northern elephant seals (D. P. Costa, unpubl.). Weddell seals were measured using open circuit respirometry on seals diving from an ice hole (Castellini et al. 1992). The solid line represents the predicted basal metabolic rate (BMR) for a terrestrial animal of equal size (Kleiber 1975); the dashed line is the best fit linear regression for the otariids with the exception of the Galapagos fur seal ($r^2 = 0.53$). Error bars represent ± one standard deviation.

greatest diving ability. Small electronic data recorders, which can be attached to marine mammals, allow us to test this prediction by providing information about the diving patterns of marine mammals (Figs. 5-23 and 5-24). These devices record at programmed intervals the animal's depth and swim speed, water temperature, and other parameters. In most cases, data are stored on computer memory chips, and the unit must be recovered to retrieve the information. In some units, the information can be downloaded via a satellite link. However, current technology significantly limits the amount of information that can be obtained from a satellite uplink (Costa 1993b). Data on dive depth and duration obtained with such instruments are summarized in Figures 5-25 and 5-26; these data indicate that phocids and sperm whales are exceptional divers, often displaying long and deep dive patterns, whereas otariids and most cetaceans are shallow, short-duration divers.

The greater diving ability of phocids is not unexpected as they can store more oxygen per kilogram of body mass than otariids or most cetaceans (Kooyman 1989). However, the greater oxygen storage capacity of phocids can only account for a 50% increase in dive duration. From Figure 5-26 we can see that phocid seals can actually dive up to 10 times longer than most other marine mammals. The additional diving ability can be explained by their extremely low metabolism during diving (Castellini et al. 1992, Costa 1993a) compared to the high at-sea metabolic rate typical of otariids (see Fig. 5-22). Using the ADL equation, we can model the relative importance of metabolic rate and oxygen stores to diving ability of a phocid and an otariid (Fig. 5-27). In this model, the ADL for a phocid is calculated from the metabolic rates measured for Weddell seals (1.4 and 2 times basal metabolic rate) and an oxygen storage capacity of 60 mL O_2/kg. We calculated ADL based on an oxygen storage capacity of 40 mL O_2/kg (for otariids), 35 mL O_2/kg (for dolphins), and metabolic rates of 6 times basal metabolic rate determined from measurements of each species' sea metabolism. Although differences among dolphin, otariid, and phocid oxygen storage capacities have a significant effect on ADL, the differences in at-sea metabolism have a greater impact. Figure 5-27 also demonstrates that larger phocids have a greater aerobic dive limit than smaller ones. This is attributable in part to differences in scaling parameters for metabolic rate and total oxygen stores with body mass. Metabolic rate scales to body mass$^{0.75}$, whereas oxygen storage capacity scales to body mass$^{1.0}$ (same pattern in Fig. 5-15). This means that larger mammals have a lower mass-specific metabolism for a relatively constant proportion of oxygen storage capacity (Kooyman et al. 1983, Gentry et al. 1986, Kooyman 1989). All things being equal, large mammals should be able to dive longer and deeper than small ones based simply on body size. Thus, for phocid seals greater diving capability is due to the combined effects of (1) greater oxygen storage capacity, (2) significantly reduced metabolism during diving, and (3) large body size.

The effect of body size on ADL is not as great for otariids or dolphins as it is for phocids and may be attributable to differences in diving metabolism. For marine mammals with a high metabolic rate we find little change in ADL as body mass increases. However, larger body mass does result in a greater absolute demand for food. Small body size, although

Figure 5-23. A male northern elephant seal male rests on the beach with a satellite-linked transmitter attached to the top of his head and a time-depth recorder attached to the middle of his back. The satellite transmitter only transmits when the animal is at the surface and provides a fix of his location at sea. The time-depth recorder must be recovered to acquire the data.

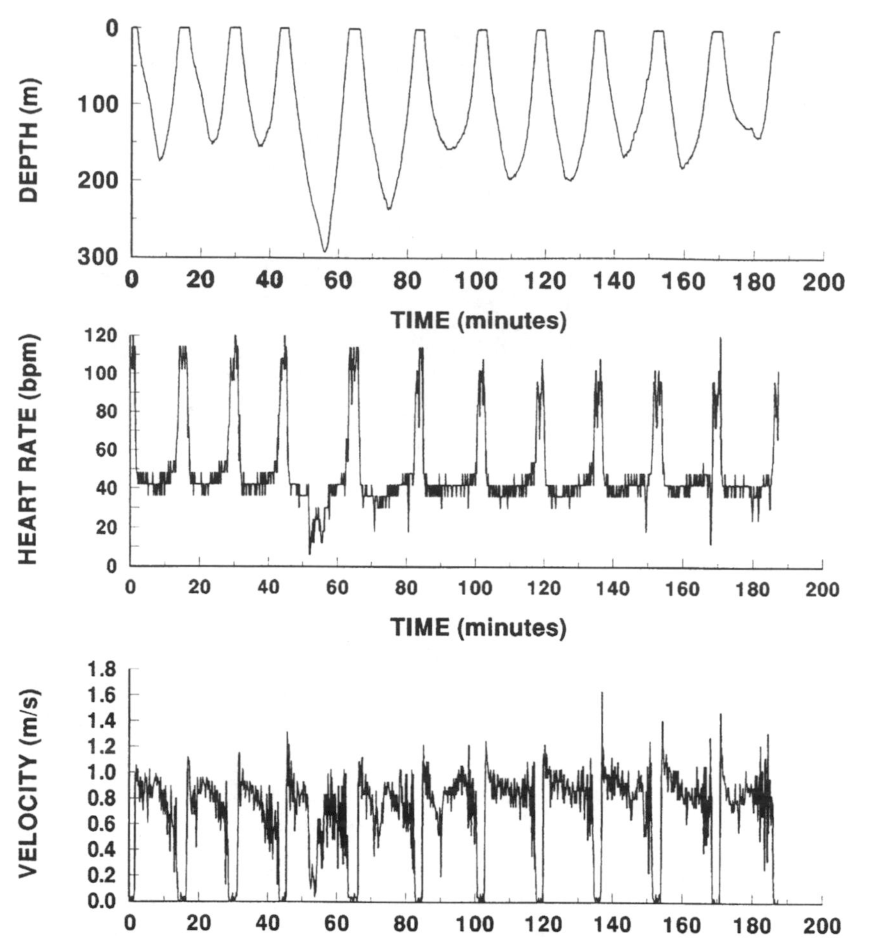

Figure 5-24. An example of the kind of information that can be obtained using archival data loggers. This record included data on time, depth, swim velocity, and heart rate that were obtained from a freely swimming juvenile northern elephant seal. (P. M. Webb, unpubl.)

resulting in a higher mass-specific metabolism, decreases absolute food energy requirements (Peters 1983, Millar and Hickling 1990). An advantage of lower absolute food requirements is that one can sustain oneself on smaller prey patches, which are likely to be more abundant. In terrestrial communities, high-quality food tends to be quite patchy and of lower overall abundance, providing an advantage for the smaller animal. These differences provide important insights into the evolution of optimal body size in marine mammals. If we assume that diving ability is an important component of the life history pattern of phocids, we would expect large body size to be favored because of the benefits of increased diving capability. In contrast, smaller size might be favored in otariids because they gain more from the reductions in absolute energy requirements than they get by minimal increases in dive duration resulting from greater mass.

There are other well-established differences in diving physiology and behavior among phocids, otariids, and small cetaceans. Phocids and monodontids have comparatively higher blood oxygen storage capacities owing to an increased proportion of red blood cells (hematocrit) (Lenfant et al. 1970). However, an elevation in hematocrit increases the blood viscosity and reduces the ability of the blood to optimally transport oxygen (Hedrick et al. 1986; Hedrick and Duffield 1991; Elsner, Chapter 3, this volume). Fortunately, maximum oxygen-carrying capacity is not an issue for phocids and monodontids that are able to maintain low metabolic rates while diving. In addition, behavioral differences among these groups result in differences in the costs of foraging (see Wells, Boness, and Rathbun, Chapter 8, and Bowen and Siniff, Chapter 9, this volume). For example, otariids and delphinids typically travel at the surface porpoising, whereas phocids tend to surface quietly and descend before swimming. The energetic implication of these behaviors has been addressed in the preceding section on activity costs. In general, otariids appear to use a more expensive foraging strategy that is optimal when resources are abundant. Phocids use a comparatively slow, energetically economical foraging pattern. Thus, the otariid life history pattern favors time minimization, whereas that of phocid favors energy minimization.

Differences in the Cost and Efficiency of Foraging

Different diving patterns and energy expenditures undoubtedly affect the ability of pinnipeds to obtain prey and the effi-

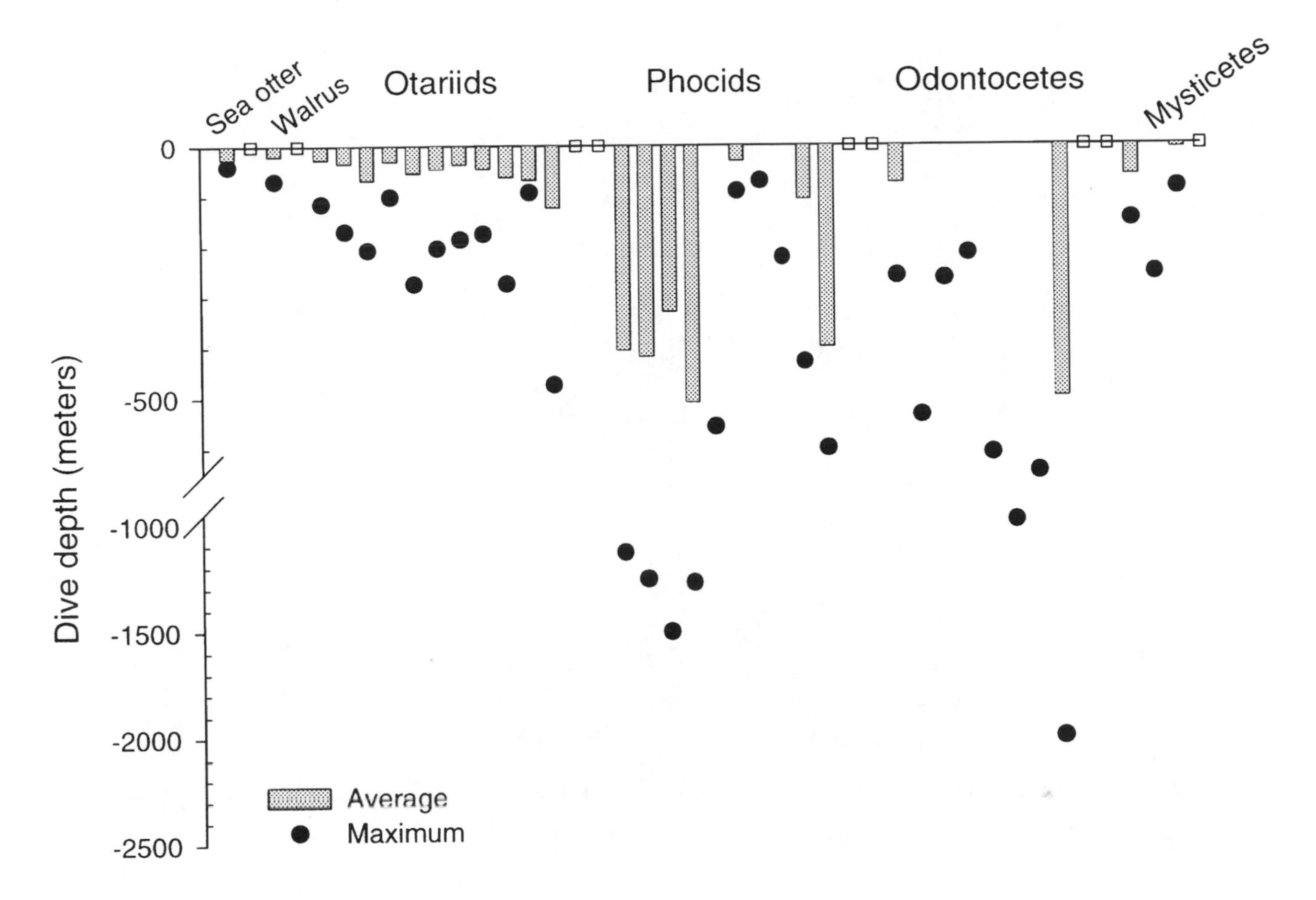

Figure 5-25. Maximum (dots) and where available mean (histogram) diving depth for 32 species of marine mammals. The squares at 0 depth represent breaks between the different taxonomic groups. The data are from left to right: sea otter; walrus: otariids: Galapagos fur seal, South American fur seal (*Arctocephalus australis*), northern fur seal, New Zealand fur seal (*Arctocephalus forsteri*), South African fur seal (*Arctocephalus pusillus pusillus*), Galapagos sea lion, South American sea lion (*Otaria byronia*), California sea lion, Australian sea lion, New Zealand sea lion; phocids: southern elephant seal male, female, northern elephant seal male, female, harbor seal, harp seal, gray seal, ringed seal, crabeater seal (*Lobodon carcinophagus*), Weddell seal; odontocetes: common dolphin (*Delphinus delphis*), bottlenose dolphin, killer whale, Pacific white-sided dolphin (*Lagenorhynchus obliquidens*), pilot whale (*Globicephala* spp.), narwhal (*Monodon monoceros*), beluga whale (*Delphinapterus leucas*), sperm whale; mysticetes: humpback whale (*Megaptera novaeangliae*), fin whale (*Balaenoptera physalus*), northern right whale (*Eubalaena glacialis*). Data are from Kooyman and Andersen 1969, Evans 1974, Kolb and Norris 1982, Gentry et al. 1986, Gales and Mattlin 1997, Ridgway 1986, Dolphin 1987b, Gentry et al. 1987, Le Boeuf et al. 1988, Costa et al. 1989b, Feldkamp et al. 1989, Riedman and Estes 1990, Hindell et al. 1991, Stewart and DeLong 1991, Thompson et al. 1991, Bengtson and Stewart 1992, Castellini et al. 1992, Wartzok et al. 1992, Lydersen and Kovacs 1993, Watkins et al. 1993, Wiig et al. 1993, Heide-Jorgensen and Dietz 1995, Werner and Campagna 1995, Winn et al. 1995.

ciency with which they acquire energy. Dive performance for a sea lion, a fur seal, and an elephant seal are presented in Table 5-3. Although elephant seals obtain more prey energy per dive than either northern fur seals or California sea lions, they take more time to do it, and acquire less energy per unit time. The ratio of energy acquired to energy expended is the same between the sea lion and the elephant seal, but is significantly greater for the fur seal. These data imply that a high metabolic rate may be advantageous in achieving high rates of prey consumption. There is a cost to this strategy, however, in terms of the animal's absolute energy requirements. Although the elephant seal is 10 times larger than the fur seal and four times larger than the sea lion, its absolute energy expenditure is almost identical to both. Despite large size, the absolute energy intake of the elephant seal is equivalent to that of the sea lion and only one quarter that of a fur seal. This

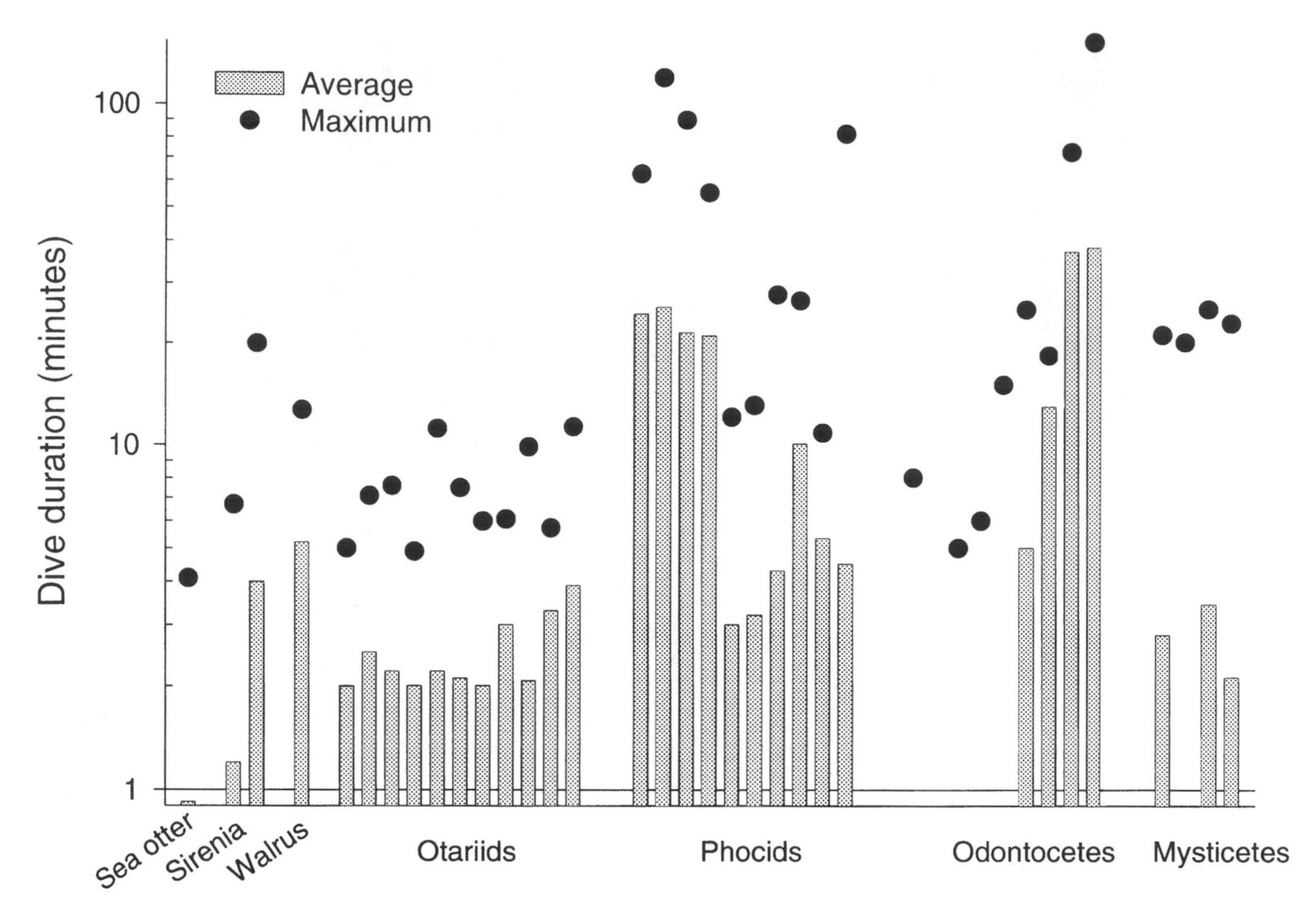

Figure 5-26. Maximum (dots) and where available mean (histogram) diving duration for sirenia: dugong, manatee; otariids: Galapagos fur seal, South American fur seal, northern fur seal, Antarctic fur seal, New Zealand fur seal, South African fur seal, Galapagos sea lion, southern sea lion, California sea lion, Australian sea lion, New Zealand sea lion; phocids: southern elephant seal male, female, northern elephant seal male, female, harbor seal, harp seal, gray seal, ringed seal, crabeater seal and Weddell seal; odontocetes: bottlenose dolphin, common dolphin, Pacific white-sided dolphin, pilot whale, narwhal, beluga whale, sperm whale, Arnoux's beaked whale (*Berardius arnuxii*); mysticetes: humpback whale, fin whale, bowhead whale, and northern right whale. Data are from same sources as Fig. 5-25 with addition of Watkins et al. 1984, Würsig et al. 1984, Caldwell and Caldwell 1985, Hobson and Martin 1996.

analysis is consistent with the idea that phocids expend less energy than otariids to obtain a similar amount of prey. Lower existence costs may enable phocids to subsist on a poorer or more dispersed prey resource than otariids. The dramatic performance of northern fur seals also suggests that in the right circumstances, such as upwelling regions, otariids may be able to better use resources when prey is plentiful.

These differences in the pinniped foraging energetics suggest that phocids have a conservative foraging mode that can net similar amounts of energy to those obtained by otariids, but at significantly lower relative costs. Furthermore, the breeding pattern of phocid seals allows them to occupy habitats where productivity is lower (Costa 1993a). In contrast, otariids have a foraging and reproductive pattern that is energetically more costly, but appears to be optimal in highly productive regions where prey resources are not limited. This argument is consistent with pinniped distributions and global productivity. Otariids are only found in locations where productivity is high, whereas phocids breed in both the highly productive regions of the world and in areas of poorer productivity, like the Hawaiian Islands, the Mediterranean, and historically in the Caribbean.

There are interesting consequences if higher metabolic rates enable greater food acquisition when conditions are right. Predatory lizards that use a more costly, high activity, widely foraging behavior expend more energy. However, they may acquire proportionately more energy than lizards

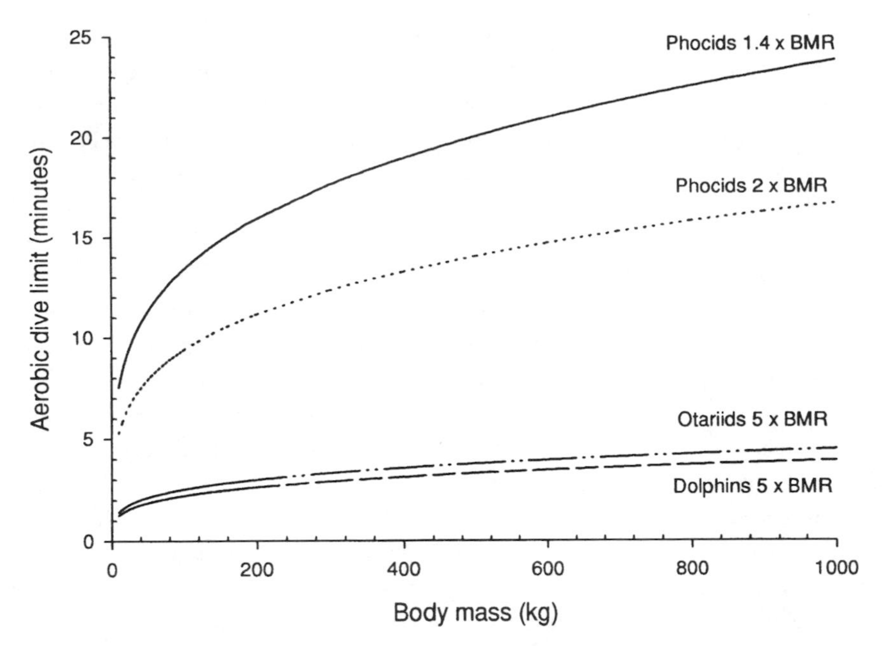

Figure 5-27. The variation in aerobic dive limit as a function of body mass calculated for a phocid seal operating at 1.4 and 2.0 times basal metabolic rate (BMR) and an otariid and a dolphin operating at 5 times BMR, respectively. Oxygen stores were assumed to be 60 mL O_2/kg for a phocid, 40 mL O_2/kg for an otariid, and 35 mL O_2/kg for a dolphin (Kooyman 1985).

Table 5-3. Dive Rate and Duration, Rate of Prey Energy Acquired, Energy Expenditure and Metabolic Rate of Two Otariids and a Phocid

Species/Diet	Mass (kg)	Dive Dive Rate (No. per day)	Dive duration (min)	Energy Acquired (kJ/min)	(kJ/dive)	Energy Expended (kJ/min)	(kJ/dive	Ratio of Acquired/ Expended	Metabolic Rate FMR/BMR
Northern elephant seal Squid	350	65	19.2	1770	92	403	21	4	1.3
California sea lion Fish	85	202	2.0	224	112	52	26	4	4.8
Northern fur seal Squid	37	38	2.1	853	406	38	18	23	6.0

Data on prey intake were estimated from water influx and metabolic rate of the sea lion and fur seals from oxygen-18 doubly-labeled water measurements (Costa and Gentry 1986, Costa 1988a, Costa et al. 1991). Metabolic rate of northern elephant seal was estimated from dive behavior (Le Boeuf et al. 1988).

FMR/BMR = field metabolic rate/basal metabolic rate.

using a more economical, sit-and-wait foraging behavior (Anderson and Karasov 1981, 1988; Karasov and Anderson 1984; Nagy et al. 1984). An important consequence is that the widely foraging predators are able to devote more energy to reproduction than the sit-and-wait predators. Consistent with these findings is the observation that mammals with high metabolic rates reproduce faster than similar-sized mammals with lower metabolic rates (McNab 1980, 1983, 1984, 1986a,b). Such a pattern holds true when marine mammals are included in the analysis (Schmitz and Lavigne 1984). These arguments support the hypothesis that otariids expend more energy foraging, but they get more for their effort than phocids. This requires that sufficient resources be available to support such an expensive life style, but this is likely in the upwelling environments that otariids typically inhabit (Repenning and Tedford 1977, King 1983).

The Energetics of Prey Choice

The amount of work, and therefore, energy expenditure, that an animal puts into locating prey varies as a function of the energy content, availability, and location of the individual prey items. Both size and proximate composition (fat, carbohydrate, protein, and water content) affect the energy content of prey. Larger prey items or those with higher fat contents contain more energy. Fat contains almost twice as much energy per unit mass as protein or carbohydrate, and it is stored with significantly less water in body tissues. Prey availability varies as a function of the absolute abundance of prey (amount of prey per unit of habitat) and its distribution in the environment. This distribution dictates predator efficiency. A predator is more efficient foraging on prey that is clumped than on prey that is evenly dispersed. Marine mammals appear to forage in areas where prey has been concentrated as a result of oceanographic processes including eddies, fronts, and bottom topography (Costa 1993b; Bowen and Siniff, Chapter 9, this volume).

Because marine mammals must periodically return to the surface to breathe, the location or depth of the prey will determine how hard the animal must work at foraging. Deeper prey require greater swimming distances. Some prey may simply be beyond the animal's diving ability.

Because they forage in near-shore waters and are easy to observe, sea otters are excellent subjects for assessing variations in foraging behavior and ecology. Sea otters consume their prey on the water surface where it is easy to identify from shore. They generally forage in water depths that are easily reached by scientists using scuba gear, enabling the abundance and distribution of prey to be measured.

Sea otters select prey based on a combination of factors—energy content, its local abundance or availability, the time it takes to acquire and process it, and previous experience of the animal. In areas that have only recently been occupied by sea otters, the diet is made up almost exclusively of preferred prey items like clams, abalone, or sea urchins (Table 5-4). In these environments, sea otters find large, energy-rich, abundant prey, which is easy to handle, consume, and digest. In such situations lower quality prey items (turban snails, sea stars, mussels, chitons) generally do not appear in the diet. These items may be abundant, but are energy-poor and difficult to eat and digest. Abalone, clams, sea urchins, and crabs grow slowly. Therefore, sea otter foraging pressure rapidly causes a reduction in their abundance. As the abundance and size of their preferred prey items decline, sea otters switch to less preferred but more accessible prey items like turban snails, kelp crabs, and in some cases even chitons and sea stars (Table 5-4). Some sea otters specialize on different types of prey items (Riedman and Estes 1990). These specialists are more efficient predators on the selected prey than nonspecialists because of the foraging tactics involved. Turban snails are small and easy to find but require extended processing time; many small snails must be captured and the shells broken to get a decent meal. Nonetheless, snails are abundant throughout the kelp forest and are easy to locate. Conversely, abalone are found deep in rock crevices and use their muscular foot to hold firmly to the substrate. It usually takes an otter several dives to obtain one abalone, and it often requires the use of a rock to break the abalone's hold. Otters feeding on clams must learn how to dig them out of the mud or sand and how to break open their thick shells.

Polar bears represent another example of optimal prey choice and its relation to the prey energy quality. Feeding predominately on ringed seals, polar bears eat the energy-rich blubber layer and leave behind the lean "core" of the carcass. Because of its high lipid content, the blubber has a per unit mass energy content almost 10 times greater than that of the lean tissue of the seal (Stirling and McEwan 1975). More important, as mentioned above, fat retains relatively little water (approximately 10%) when it is deposited as blubber. This compares with protein and carbohydrate, which are stored with 70% to 80% water by mass (Kleiber 1975). Thus, polar bears have learned to consume the most energy-dense part of the ringed seal and then move on to find another kill (Stirling and McEwan 1975).

Fur seals and sea lions provide insight into the factors that govern prey choice of pelagic marine mammals. Fur seals are tied to shore for breeding and rely on the availability of nearby prey resources to produce milk for their young. Female fur seals must optimize the time they spend feeding at sea to the time spent nursing the pup onshore (Costa et al. 1989b; Costa 1991a,b; Lunn and Boyd 1993). Movements of foraging fur seals indicate that they have ranges near the breeding colony. Because fur seal and sea lion mothers periodically feed at sea and return to their pups onshore, they offer a tractable system to examine the foraging ecology of open-ocean marine mammals.

Northern fur seal females exhibit three different foraging patterns: they feed near the bottom over the continental shelf

Table 5-4. Foraging Behavior of Sea Otters in Central California

	Prey Energy Content (kJ)		Number of Prey Consumed		Proportion of Diet (%)	
Prey Item	Newly Occupied	Established	Newly Occupied	Established	Newly Occupied	Established
Kelp crab	—	207	—	58	—	49
Rock crab	761	761	3	11	28	9
Turban snail	—	90	—	108	—	15
Abalone	2908	607	9	12	69	10
Red sea urchin	971	27	—	19	—	14

Data from Costa 1978.

where the depth seldom exceeds 200 m, over deep water off the continental shelf making only shallow dives, or they show a combination of deep and shallow dives (Loughlin et al. 1987). Deep-diving females apparently feed through the day and night on semidemersal juvenile walleye pollock that remain near the bottom (Sinclair et al. 1994). Females foraging over deep water beyond the continental shelf wait until gonatid squid (*Gonatopsis borealis* and *Berryteuthis magister*) and deep-sea smelt (*Leuroglossus schmidti*) move into shallow water before preying on them (Loughlin et al. 1987, Antonelis et al. 1993, Sinclair et al. 1994). The energetic difference between feeding on squid or pollock can be examined by comparing the total number of individual prey that would have to be captured per dive. Each individual pollock contains 10 times as much energy as each squid due to differences in both the energy density and size of individual prey (squid = 152 kJ, pollock = 1584 kJ) (Costa and Gentry 1986). Using estimates of the food requirements of foraging females and the total number of dives over a foraging trip, we find an average 695 kJ of prey energy obtained per dive when feeding on squid compared to 1500 kJ when feeding on pollock. A female fur seal would have to obtain five squid per dive compared to one pollock (Costa 1988b). Deep diving may only be economical when female fur seals feed on large prey that can supply a significant fraction of the energy requirement with each dive. Predation on small prey that require many individuals to be captured per dive is limited to shallow depths. However, one must be careful extrapolating these observations, as the terms shallow and deep are relative to the diving capability of the predator. Shallow to an elephant seal may be 300 m, whereas shallow to a northern fur seal is 30 m.

Some species feed exclusively on one type of prey. During the summer, Antarctic fur seals feed only on Antarctic krill. Nearly 75% of their night dives are shallower than 30 m. Daytime dives average 40 to 75 m (Croxall et al. 1985, Boyd and Croxall 1992). This pattern closely follows the vertical migration of krill, which remain below 50 m during the day and move near the surface at night. More than 40% of the krill are below 75 m depth at any time of day, but fur seal dives seldom exceed this depth. Although fur seals are physiologically capable of reaching this depth, it appears that it is too deep for them to efficiently feed on krill. Similar patterns have been observed for northern fur seals feeding on vertically migrating squid and other marine predators, such as macaroni (*Eudyptes chrysolophus*), chinstrap (*Pygoscelis antarctica*), and gentoo (*Pygoscelis papua*) penguins, feeding on krill (Croxall et al. 1988, Fraser et al. 1989).

What constraints account for these different foraging depths? Shallow dives use relatively little time in transit, leaving more time at the bottom of the dive to search for or pursue prey (Fig. 5-28). In contrast, deep dives require more time in transit, leaving less time to search for or pursue prey at the bottom of the dive. Thus, fewer prey can be obtained per long dive. If the same amount of time is spent per dive, and there is less time available to capture prey, it would be prudent to pursue prey of greater size and energy content. Likewise, if dives are of the same duration and prey is captured at a consistent rate per dive, more dives would be required when pursuing the prey of lower energy content. For short dives, increasing the number of sequential dives with a lower premium on energy return per dive would be the most economical strategy. For deep dives where transit time is long, a small number of long-duration dives with a high energy return per dive would be favored.

Faster swimming predators will use oxygen stores more rapidly and are limited to shallow dives or "spiked" dives with minimal bottom time (Fig. 5-28). In contrast, slow-swimming animals use oxygen less rapidly and make dives of longer duration. Presumably, the pursuit of many small prey

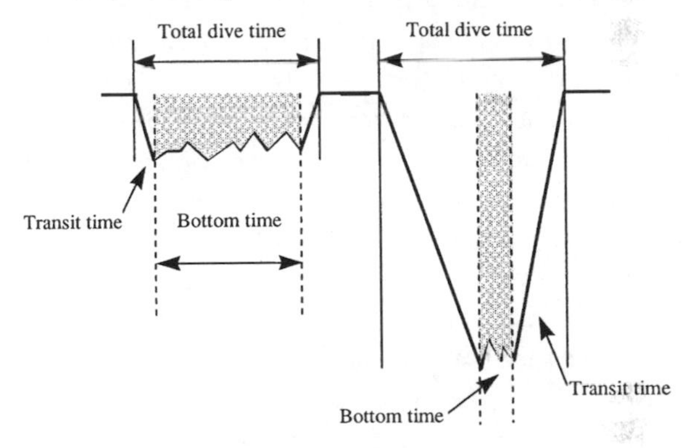

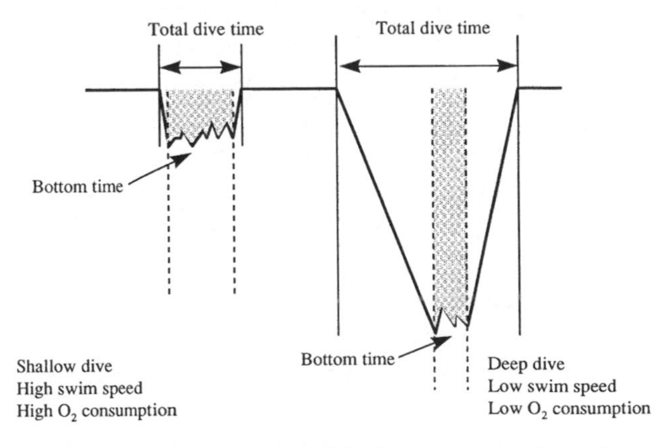

Figure 5-28. The top portion of the figure presents schematic representations of the diving pattern for deep- and shallow-diving predators when oxygen utilization is constant and total dive durations are equivalent. The bottom figure demonstrates patterns when the rate of oxygen utilization is greater for shallow dives than for deep dives. In this case, total dive duration is not constant.

requires fast and agile swimming with high rates of oxygen consumption. Such a high rate of oxygen consumption would constrain the predator to many short dives. However, for a shallow-diving predator this may not be a serious constraint as transit time to foraging depth is brief and most of the dive can be spent in pursuit and acquisition of prey. The opposite is true for deep dives. If a deep diver were to swim fast, most, if not all, of the oxygen stores would be used in transit to and from depth with little left over for the search and pursuit of prey. Such a situation probably results in the "spiked" dives observed in many pinnipeds (Gentry et al. 1986, Le Boeuf et al. 1988, Crocker et al. 1994). By swimming slowly and reducing oxygen consumption, the predator gains bottom time at depth but probably is limited to feeding on slow-moving prey or prey that is only encountered occasionally. As a result, the predator should select large prey of high energy content. Flat-bottomed dives observed for deep-diving northern elephant seals (Le Boeuf et al. 1988) and New Zealand sea lions (Gentry et al. 1987, Gales and Mattlin 1997) are examples of marine mammals that use this strategy. The predator may also limit the pursuit phase to slow methodical swimming until the prey has been spotted and then switch to a rapid, energetically costly form of high-speed swimming. This strategy could limit the dive to a single capture before oxygen stores are depleted. If deep divers can only capture a few individuals per dive, then they should pursue prey containing the highest energy content.

Variations in Foraging Energetics

The foraging success of marine mammals declines as food resources change in response to their own foraging activities (Estes et al. 1978, Hines and Pearse 1982, Kvitek et al. 1992) or as a result of changes in local oceanographic or climatic conditions. Pelagic marine mammals may respond to changes in local food availability by migrating or moving to areas where conditions are optimal (Schoenherr 1991, Kenney et al. 1995, Whitehead 1996). Other species, like bottlenose dolphins in Sarasota Bay (Wells et al. 1987) or sea otters, exist in specialized habitats or have limited home ranges. If prey becomes depleted within their home range, they have a limited ability to move.

The relationship between prey availability and reproductive success has been examined for a wide variety of pinniped species. Otariids are more susceptible to variations in nearby food resources because their breeding pattern is linked to continuous prey availability directly offshore (Costa 1993a); phocids are buffered against fluctuations in prey availability near the rookery because they accumulate the resources they need for lactation over the previous year (Trillmich et al. 1991). Successful reproduction by otariids requires that mothers use a foraging pattern that optimizes the amount of time spent at sea feeding with the amount of milk energy delivered to her pup waiting on the rookery. Studies of female otariids with dependent young show that as food resources decrease mothers can either increase the time spent at sea foraging or they can increase the intensity of their foraging effort (Trillmich 1990, Costa 1991b, Trillmich et al. 1991). However, simply increasing the time spent at sea increases the time between visits to the pup. Consequently, more of the ingested milk energy is spent on pup maintenance rather than directed to pup growth. Increases in trip duration that are associated with declining prey resources result in slower pup growth because less milk is delivered over the same time interval (Oftedal et al. 1987, Costa and Croxall 1988, Croxall et al. 1988, Costa et al. 1989b, Trillmich et al. 1991, Lunn and Boyd 1993).

A more optimal strategy when faced with reduced prey is for otariid mothers to adjust their foraging behavior to keep the same attendance pattern and provide their pups with the same rate of milk delivery. This can be done by modifying foraging behavior by (1) spending more time at sea, (2) decreasing the time spent resting or in transit, (3) switching to more abundant, deeper, shallower, or higher energy-dense prey, or (4) increasing the intensity of their foraging effort by diving deeper, faster, and with a shorter surface interval. All of these responses have been observed in otariid mothers. Northern fur seal females responded to changes in prey availability by apparently changing both the type of prey consumed (Sinclair et al. 1994) and the intensity of their foraging effort (Costa and Gentry 1986). These fur seal mothers had significantly different FMRs without changes in trip duration or pup mortality (Costa and Gentry 1986).

In response to more severe reductions in prey availability associated with the 1983 El Niño event, California sea lions responded by changing prey and by increasing both their at-sea energy expenditure and trip duration (Costa et al. 1991, DeLong et al. 1991, Feldkamp et al. 1991). While at sea, sea lion mothers spent a greater percentage of total trip time diving (40.4% in 1982; 66.2% in 1983), exhibited diving bouts of longer duration (2.9 hr in 1982; 4.6 hr in 1983), made dives of greater duration (1.9 min in 1982; 2.6 min in 1983), and possibly dived to deeper depths as well as spent less time swimming (50.6% in 1982; 29.2% in 1983) and resting (4.1% in 1982; 1.5% in 1983) than pre-El Niño animals. These changes in behavior were accompanied by increases in the rate of energy expenditure at sea, which indicates that sea lions worked harder when prey was scarce (Costa et al. 1991, Feldkamp et al. 1991).

Some species of fur seals may have a limited capacity to accommodate changes in prey availability because they are prey specialists or are already foraging near maximum levels. In the South Georgia area of the South Atlantic Ocean, the Antarctic fur seal feeds almost exclusively on Antarctic krill, resulting in tight coupling of the reproductive success of the fur seals with local krill availability. Antarctic fur seals feeding on krill possess one of the highest at-sea metabolic rates; they are shallow divers and spend considerably more of their time at sea diving than other otariids examined to date. By operating close to their metabolic maximum, Antarctic fur seals have a limited ability to increase their foraging intensity and are less able to respond to reductions in prey availability without increasing the time spent away from their pups. During periods of poor krill availability, females increase the time spent foraging, the amount of activity while at sea, and the depth of their dives. These behavioral changes are insufficient to maintain a normal attendance pattern (Boyd et al. 1994) because Antarctic fur seal females appear to be working near their maximal metabolic effort even in normal years (Costa et al. 1989b, Costa 1991a). In response to variations in local prey resources, other fur seals and sea lions may choose different prey or have greater flexibility in their time-activity budgets than do Antarctic fur seals.

Phocids are buffered from short-term fluctuations in prey availability owing to their unique reproductive pattern. In phocids, reproductive performance (maternal investment) during a given season reflects prey availability over the preceding year and represents the mother's foraging activities over a much larger spatial and temporal scale than is the case for otariids (Trillmich and Ono 1991, Stewart and Lavigne 1984, Costa 1993a). It follows that the weaning mass of a phocid pup is an indicator of its mother's foraging success over the previous year (Stewart and Lavigne 1984), whereas the subsequent postweaning survival of the pup is related to both its weaning mass (energy reserves provided by the mother) and the resources available to the pup after weaning. Variations in the success of individual pups born during a single season have been observed for many ice-breeding seals. Reproductive rates in Weddell seals at McMurdo Sound exhibit 4- to 6-year fluctuations (Testa and Siniff 1987) that appear to be in phase with that of El Niño events. Changes in the occurrence of leopard seals (*Hydrurga leptonyx*) on sub-Antarctic islands associated with the proximity of the pack-ice edge (Rounsevell and Eberhard 1980) also correlate with El Niño events. Similar 4- to 5-year fluctuations in cohorts are evident in the age structure of crabeater seals (*Lobodon carcinophagus*) in the Antarctic peninsula (Bengtson and Laws 1983). These findings support the concept that large-scale oceanographic processes affect prey availability and thus population processes for many pinnipeds (Croxall and Rothery 1991, Testa et al. 1991).

Summary

It is apparent from this discussion that both intrinsic and extrinsic factors affect the flow of energy in marine mammals. Each species must balance the ability to acquire energy with the many avenues of energy expenditure. Energetic costs may range from supporting basal metabolic processes to providing for subsequent generations. These internal factors are not isolated from the environment in which the animal lives. Water temperature, seasonal changes, and oceanographic variables, among others, will influence both the physiology and the behavior of marine mammals. The appropriateness of the response by an animal challenged by these extrinsic factors will be manifest as an energetic cost or benefit. Ultimately, these will dictate the survivorship of the individual marine mammal as well as of the species.

Acknowledgments

As is the case with a review of the literature, many colleagues have contributed either directly or indirectly to the ideas and concepts put forward here. The manuscript was considerably improved by the comments of S. Rommel, J. Reynolds, F. Fish, and two anonymous reviewers. We thank J. Reynolds and S. Rommel for asking us to write this chapter and especially for their patience in waiting for its completion. D. Waples' contributions were critical in the final stages of manuscript preparation. The writing of this chapter and some of the unpublished results were funded by grants from the Office of Naval Research N000149411013 and N00014-94-1-0455 to D. Costa, N00014-95-1-1023 and N00014-96-1-1023 to T. Williams, and the National Science Foundation OPP-9500072 and OCE 9018626 and ARPA funds for the Acoustic Thermography of the Ocean Climate Project (ATOC) to D. Costa.

Literature Cited

Acuña, H. O. 1995. Ecological and physiological factors that influence pup birth weight and postnatal growth of Juan Fernandez fur seals, *Arctocephalus philippii*. Ph.D. thesis. University of Florida, Gainesville, FL, 63 pp.

Anderson, R. A., and W. H. Karasov. 1981. Contrasts in energy intake and expenditure in sit-and-wait and widely foraging lizards. Oecologia 49:67–72.

Anderson, R. A., and W. H. Karasov. 1988. Energetics of the lizard *Cnemidophorus tigris* and life history consequences of foraging mode. Ecological Monographs 58:79–110.

Anderson, S. A., and M. A. Fedak. 1985. Grey seals males: Energetics and behavioural links between size and sexual success. Animal Behaviour 33:829–838.

Anderson, S. A., D. P. Costa, and M. A. Fedak. 1993. Bioenergetics. Pages 291–315 *in* R. Laws (ed.). Antarctic Pinnipeds: Research Methods and Techniques. Cambridge University Press, Cambridge, U.K.

Antonelis, G. A., E. H. Sinclair, R. R. Ream, and B. W. Robson. 1993. Inter-island variation in the diet of female northern fur seals (*Callorhinus ursinus*) in the Bering Sea (Abstract). Tenth Biennial Conference on the Biology of Marine Mammals. Galveston, TX, November 11–15.

Arnould, J. P. Y., and I. L. Boyd. 1995a. Temporal patterns of milk production in Antarctic fur seals (*Arctocephalus gazella*). Journal of Zoology (London) 237:1–12.

Arnould, J. P. Y., and I. L. Boyd. 1995b. Inter- and intra-annual variation in milk composition in Antarctic fur seals (*Arctocephalus gazella*). Physiological Zoology 68:1164–1180.

Arnould, J. P. Y., I. L. Boyd, and D. G. Socha. 1996a. Milk consumption and growth efficiency in Antarctic fur seal (*Arctocephalus gazella*) pups. Canadian Journal of Zoology 74:254–266.

Arnould, J. P. Y., I. L. Boyd, and J. R. Speakman. 1996b. Measuring the body composition of Antarctic fur seals (*Arctocephalus gazella*): Validation of hydrogen isotope dilution. Physiological Zoology 69:93–116.

Ashwell-Erickson, S., and R. Elsner. 1981. The energy cost of free-existence for Bering Sea harbor and spotted seals. Pages 869–899 *in* D.W. Hood and J.A. Calder (eds.). Eastern Bering Sea Shelf: Oceanography and Resources, Vol. 2. University of Washington Press, Seattle, WA.

Au, D., and D. Weihs. 1980. At high speeds dolphins save energy by leaping. Nature (London) 284:548–550.

Barbour, A. S. 1993. Heat increment of feeding in juvenile Northern elephant seals. M.S. thesis. University of California, Santa Cruz, 47 pp.

Barnes, L. G., D. P. Domning, and C. E. Ray. 1985. Status of studies on fossil mammals. Marine Mammal Science 1:15–53.

Bartholomew, G. A. 1970. A model for the evolution of pinniped polygyny. Evolution 24:546–559.

Bartholomew, G. A. 1977a. Energy Metabolism. Pages 57–110 *in* M. S. Gordon (ed.). Animal Physiology: Principles and Adaptations. Macmillan, New York, NY.

Bartholomew, G. A. 1977b. Body temperature and energy metabolism. Pages 364–449 *in* M. S. Gordon (ed.). Animal Physiology: Principles and Adaptations. Macmillan, New York, NY.

Bartsh, S. S., S. D. Johnston, and D. B. Siniff. 1992. Territorial behaviour and breeding frequency of male Weddell seals (*Leptonychotes weddellii*) in relation to age, size, and concentrations of serum testosterone and cortisol. Canadian Journal of Zoology 70:680–692.

Baudinette, R. V., and P. Gill. 1985. The energetics of "flying" and "paddling" in water: Locomotion in penguins and ducks. Journal of Comparative Physiology B 155:373–380.

Bengtson, J. L., and R. M. Laws. 1983. Trends in crabeater seal age at maturity: An insight into Antarctic marine interactions. Pages 669–675 *in* W. R. Siegfried, P. R. Condy, and R. M. Laws (eds.). Nutrient Cycles and Food Chains. Proceedings of the 4th SCAR symposium on Antarctic Biology. Springer-Verlag, Berlin.

Bengtson, J. L., and B. S. Stewart. 1992. Diving and haulout behavior of crabeater seals in the Weddell Sea, Antarctica, during March 1986. Polar Biology 12:635–644.

Berta, A., C. E. Ray, and A. R. Wyss. 1989. Skeleton of the oldest known pinniped, *Enaliarctos mealsi*. Science 244:60–62.

Best, P. B. 1982. Seasonal abundance, feeding, reproduction, age and growth in minke whales off Durban (with incidental observations from the Antarctic). Report of the International Whaling Commission 32:759–786.

Best, R. C. 1976. Ecological energetics of the polar bear (*Ursus maritimus*). M.S. thesis. University of Guelph, Guelph, Ontario, Canada, 136 pp.

Best, R. C., and V. M. F. da Silva. 1984. Preliminary analysis of reproductive parameters of the boutu, *Inia geoffrensis*, and the tucuxi, *Sotalia fluviatilis*, in the Amazon River system. Reports of the International Whaling Commission (Special Issue 6):361–369.

Bjorndal, K. A. 1979. Cellulose digestion and volatile fatty acid production in the green turtle *Chelonia mydas*. Comparative Biochemistry and Physiology A 63:127–133.

Blake, R. W. 1983. Energetics of leaping in dolphins and other aquatic animals. Journal of the Marine Biological Association of the United Kingdom 63:61–70.

Boness, D. J., and W. D. Bowen. 1996. The evolution of maternal care in pinnipeds. Bioscience 46:645–654.

Boness, D. J., W. D. Bowen, and O. T. Oftedal. 1994. Evidence of a maternal foraging cycle resembling that of otariid seals in a small phocid, the harbor seal. Behavioral Ecology and Sociobiology 34:95–104.

Bonner, W. N. 1984. Lactation strategies in pinnipeds: Problems for a marine mammalian group. Pages 253–272 *in* M. Peaker, R.G. Vernon, and C.H. Knight (eds.). Physiological Strategies in Lactation. Symposium of the Zoological Society of London Number 51, Academic Press, London.

Bowen, W. D. 1991. Behavioral ecology of pinniped neonates. Pages 66–127 *in* D. Renouf (ed.). The Behaviour of Pinnipeds. Chapman and Hall, London.

Bowen, W. D., D. J. Boness, and O. T. Oftedal. 1985. Birth to weaning in 4 days: Remarkable growth in the hooded seal, *Cystophora cristata*. Canadian Journal of Zoology 63:2841–2846.

Bowen, W. D., O. T. Oftedal, and D. J. Boness. 1992. Mass and energy transfer during lactation in a small phocid, the harbor seal (*Phoca vitulina*). Physiological Zoology 65:844–866.

Boyd, I. L., and J. P. Croxall. 1992. Diving behaviour of female Antarctic fur seals. Canadian Journal of Zoology 70:919–928.

Boyd, I. L., and C. D. Duck. 1991. Mass changes and metabolism in territorial male Antarctic fur seals (*Arctocephalus gazella*). Physiological Zoology 64:375–392.

Boyd, I. L., J. P. Y. Arnould, T. Burton, and J .P. Croxall. 1994. Foraging behaviour of Antarctic fur seals during periods of contrasting prey abundance. Journal of Animal Ecology 63:703–713.

Brodie, P. F. 1971. A reconsideration of aspects of growth, reproduction, and behavior of the beluga (*Delphinapterus leucas*), with references to the Cumberland Sound, Baffin Island, population. Journal of the Fisheries Research Board of Canada 28:1309–1318.

Bryden, M. M. 1968. Lactation and suckling in relation to early growth of the southern elephant seal. Australian Journal of Zoology 16:739–747.

Bryden, M. M., and G. S. Molyneux. 1978. Arteriovenous anastomoses in the skin of seals. II. The California sea lion *Zalophus californianus* and the northern fur seal *Callorhinus ursinus* (Pinnipedia: Otariidae). Anatomical Record 191:253–260.

Burn, D. M. 1986. The digestive strategy and efficiency of the west indian manatee, *Trichechus manatus*. Comparative Biochemistry and Physiology A 85:139–142.

Butler, P. J., A. J. Woakes, I. L. Boyd, and S. Kanatous. 1992. Relationship between heart rate and oxygen consumption during steady-state swimming in California sea lions. Journal of Experimental Biology 179:31–46.

Butler, P. J., R. M. Bevan, A. J. Woakes, J. P. Croxall, and I. L. Boyd. 1995. The use of data loggers to determine energetics and physiology of aquatic birds and mammals. Brazilian Journal of Medical and Biological Research 28:1307–1317.

Calder, W. A. 1984. Size, function and life history. Harvard University Press, Cambridge, MA.

Caldwell, D. K., and M. C. Caldwell. 1985. Manatees—*Trichechus manatus, Trichechus senegalensis* and *Trichechus inunguis*. Pages 33–66 *in* S. H. Ridgway and R. Harrison (eds.). Handbook of Marine Mammals, Vol. 3: The Sirenians and Baleen Whales. Academic Press, New York, NY.

Caldwell, D. K., and H. M. Fields. 1959. Surf-riding by Atlantic bottlenosed dolphins. Journal of Mammalogy 40:454–455.

Castellini, M. A., G. L. Kooyman, and P. J. Ponganis. 1992. Metabolic rates of freely diving Weddell seals: Correlations with oxygen stores, swim velocity and diving duration. Journal of Experimental Biology 165:181–194.

Christen, I. 1984. Growth and reproduction of killer whales, *Orcinus orca*, in Norwegian coastal waters. Reports of the International Whaling Commission (Special Issue 6):253–258.

Clarke, A. 1980. The biochemical composition of krill *Euphasia superba* from South Georgia. Journal of Experimental Marine Biology and Ecology 43:221–236.

Coltman, D. W. 1996. Phenotype and mating success of male harbour seals, *Phoca vitulina*, at Sable Island, Nova Scotia. Ph.D. thesis. Dalhousie University, Halifax, Nova Scotia, 185 pp.

Costa, D. P. 1978. The sea otter: Its interaction with man. Oceanus 21:24–30.

Costa, D. P. 1982. Energy nitrogen and electrolyte flux and sea water drinking in the sea otter, *Enhydra lutris*. Physiological Zoology 55:35–44.

Costa, D. P. 1987. Isotopic methods for quantifying material and energy intake of free-ranging marine mammals. Pages 43–66 *in* A. C. Huntley, D. P. Costa, G. A. J. Worthy, and M. A. Castellini (eds.). Approaches to Marine Mammal Energetics. Allen Press, Lawrence, KS.

Costa, D. P. 1988a. Assessment of the impact of the California sea lion and northern elephant seal on commercial fisheries. Pages 36–43 *in* California Sea Grant: Biennial Report of Completed Projects 1984–86. California Sea Grant College Program, University of California, La Jolla Publication R CSCP 024.

Costa, D. P. 1988b. Methods for studying the energetics of freely diving animals. Canadian Journal of Zoology 66:45–52.

Costa, D. P. 1991a. Reproductive and foraging energetics of pinnipeds: Implications for life history patterns. Pages 300–344 *in* D. Renouf (ed.). Behaviour of Pinnipeds. Chapman and Hall, London.

Costa, D. P. 1991b. Reproductive and foraging energetics of high latitude penguins, albatrosses and pinnipeds: Implications for life history patterns. American Zoologist 31:111–130.

Costa, D. P. 1993a. The relationship between reproductive and foraging energetics and the evolution of the Pinnipedia. Pages 293–314 *in* I. Boyd (ed.). Recent Advances in Marine Mammal Science. Symposium Zoological Society of London No. 66, Oxford University Press, Oxford.

Costa, D. P. 1993b. The secret life of marine mammals: New tools for the study of their biology and ecology. Oceanography 6:120–128.

Costa, D. P., and J. P. Croxall. 1988. The effect of prey availability on the foraging energetics of Antarctic fur seals, *Arctocephalus gazella* (Abstract). Scientific Committee on Antarctic Research Fifth Symposium on Antarctic Biology. Hobart, Tasmania. 29 August–3 September.

Costa, D. P., and R. L. Gentry. 1986. Free-ranging energetics of northern fur seals, *Callorhinus ursinus*. Pages 79–101 *in* R. L. Gentry and G. L. Kooyman (eds.). Fur Seals: Maternal Strategies on Land and at Sea. Princeton University Press, Princeton, NJ.

Costa, D. P., and G. L. Kooyman. 1982. Oxygen consumption thermoregulation, and the effect of fur oiling and washing on the sea otter, *Enhydra lutris*. Canadian Journal of Zoology 60:2761–2767.

Costa, D. P., and G. L. Kooyman. 1984. Contribution of specific dynamic action to heat balance and thermoregulation in the sea otter, *Enhydra lutris*. Physiological Zoology 57:199–203.

Costa, D. P., and F. Trillmich. 1988. Mass changes and metabolism during the perinatal fast: a comparison between Antarctic (*Arctocephalus gazella*) and Galapagos fur seals (*A. galapagoensis*). Physiological Zoology 61:160–169.

Costa, D. P., P. H. Thorson, J. G. Herpolsheimer, and J. P. Croxall. 1985. Reproductive bioenergetics of the Antarctic fur seal. Antarctic Journal of the United States 20:176–177.

Costa, D. P., B. J. Le Boeuf, C. L. Ortiz and A. C. Huntley. 1986. Energetics of lactation in the northern elephant seal, *Mirounga angustirostris*. Journal of Zoology 209:21–33.

Costa, D. P., S. D. Feldkamp, J. P. Schroeder, W. Friedl, and J. Haun. 1989a. Oxygen Consumption and Thermoregulation in Bottlenose Dolphins. Thermal Physiology Satellite Symposium IUPS, Tromso, Norway.

Costa, D. P., J. P. Croxall, and C. Duck. 1989b. Foraging energetics of Antarctic fur seals, *Arctocephalus gazella*, in relation to changes in prey availability. Ecology 70:596–606.

Costa, D. P., G. P. Antonelis, and R. DeLong. 1991. Effects of El Niño on the foraging energetics of the California sea lion. Pages 156–165 *in* F. Trillmich and K. Ono (eds.). Pinnipeds and El Niño: Responses to Environmental Stress. Ecological Studies 88. Springer-Verlag, Berlin.

Costa, D. P., G. A. J. Worthy, R. Wells, R. W. Read, D. Waples, M. D. Scott, and A. B. Irvine 1995. Free ranging energetics of the bottlenose dolphin (Abstract). Eleventh Biennial Conference on the Biology of Marine Mammals. Orlando, FL. December 14–18.

Crocker, D. E., B. J. Le Boeuf, Y. Naito, T. Asaga, and D. P. Costa. 1994. Swim speed and dive function in a female Northern elephant seal. Pages 328–342 *in* B. J. Le Boeuf and R. M. Laws (eds.). Elephant Seals: Population Biology, Ecology, Behavior, and Physiology. University of California Press, Berkeley, CA.

Croxall, J. P., and P. Rothery. 1991. Population regulation of seabirds: implications of their demography for conservation. Pages 272–296 *in* C. M. Perrins, J. D. Le Breton, and G. M. Hirons (eds.). Bird Population Studies: Relevance to Conservation and Management. Oxford University Press, London.

Croxall, J. P., I. Everson, G. L. Kooyman, C. Ricketts, and R. W. Davis. 1985. Fur seal diving behavior in relation to vertical distribution of krill. Journal of Animal Ecology 54:1–8.

Croxall, J. P., T. S. McCann, P. A. Prince, and P. Rothery. 1988. Reproductive performance of seabirds and seals at South Georgia and Signy Island, South Orkney Islands, 1976–87: implications for Southern Ocean monitoring studies. Pages 261–285 *in* D. Sahrhage (ed.). Antarctic Ocean and Resources Variability. Springer Verlag, Berlin.

Davis, R. W., and G. S. Fargion (eds.). 1996. Distribution and abundance of cetaceans in the north-central and western Gulf of Mexico: Final report. Vol. II: Technical Report. OCS Study MMS 96-0027. U.S. Department of the Interior, Minerals Management Service, Gulf of Mexico OCS Region, New Orleans, LA.

Davis, R. W., and T. M. Williams. 1992. Effect of water temperature on the swimming energetics of sea lions (Abstract). American Physiological Society Conference: Integrative Biology of Exercise. Colorado Springs, CO. September 23–26.

Davis, R. W., T. M. Williams, and G. L. Kooyman. 1985. Swimming metabolism of yearling and adult harbor seals, *Phoca vitulina*. Physiological Zoology 58:590–596.

DeLong, R. L., G. A. Antonelis, C. W. Oliver, B. S. Stewart, M. S. Lowry, and P. K. Yochem. 1991. Effects of the 1982–83 El Niño on several population parameters and diet of California sea lions on the California Channel Islands. Pages 166–172 *in* F. Trillmich and K. Ono (eds.). Pinnipeds and El Niño: Responses to Environmental Stress. Ecological Studies 88. Springer-Verlag, Berlin.

Derocher, A. E., D. Andriashek, and J. P. Y. Arnould. 1993. Aspects of milk composition and lactation in polar bears. Canadian Journal of Zoology 71:561–567.

Deutsch, C. J., M. P. Haley, and B. J. Le Boeuf. 1990. Reproductive effort of male northern elephant seals: Estimates from mass loss. Canadian Journal of Zoology 68:2580–2593.

Doidge, D. W., T. S. McCann, and J. P. Croxall. 1986. Attendance behavior of Antarctic fur seals. Pages 102–114 *in* R. L. Gentry and G. L. Kooyman (eds.). Fur Seals: Maternal Strategies on Land and at Sea. Princeton University Press, Princeton, N.J.

Dolphin, W. F. 1987a. Dive behavior and estimated energy expenditure of foraging humpback whales in southeast Alaska. Canadian Journal of Zoology 65:354–362.

Dolphin, W. F. 1987b. Ventilation and dive patterns of humpback whales, *Megaptera novaeangliae,* on their Alaskan feeding grounds. Canadian Journal of Zoology 65:83–90.

Domning, D. 1978. Sirenian evolution in the North Pacific Ocean. University California Publication Geological Science 118:1–176.

Drabek, C. M., and G. L. Kooyman. 1984. Histological development of the terminal airways in pinniped and sea otter lungs. Canadian Journal of Zoology 62:92–96.

Elsner, R. 1986. Limits to exercise performance: Some ideas from comparative studies. Acta Physiologica Scandinavia. 128(suppl. 556):45–51.

Elsner, R. 1987. The contribution of anaerobic metabolism to maximum exercise in seals. Pages 109–114 *in* A. C. Huntley, D. P. Costa, G. A. J. Worthy, and M. A. Castellini (eds.). Approaches to Marine Mammal Energetics. Society for Marine Mammalogy Special Publication No.1. Allen Press, Lawrence, KS.

Estes, J. A., N. S. Smith, and J. F. Palmisano. 1978. Sea otter predation and community organization in the Western Aleutian Islands, Alaska. Ecology 59: 822–833.

Evans, W. E. 1974. Radio-telemetric studies of two species of small odontocete cetaceans. Pages 385–394 *in* W. E. Schevill (ed.). The Whale Problem. Harvard University Press, Cambridge, MA.

Fadely, B. S., G. A. J. Worthy, and D. P. Costa. 1990. Assimilation efficiency of marine mammals determined using dietary manganese. Journal of Wildlife Management 54:246–251.

Fay, F. H. 1982. Ecology and Biology of the Pacific Walrus, *Odobenus rosmarus divergens illiger.* U.S. Department of the Interior Fish and Wildlife Service North American Fauna Series Number 74.

Fedak, M. A. 1986. Diving and exercise in seals: A benthic perspective. Pages 11–32 *in* A. Brubakk, J. W. Kanwisher, and G. Sundnes (eds.). Diving in Animals and Man. Kongsvold Symposium, Royal Norwegian Society of Sciences and Letters, Tapir Publishers, Tronndheim, Norway.

Fedak, M. A., and S. A. Anderson. 1982. The energetics of lactation: Accurate measurements from a large wild mammal, the gray seal (*Halichoerus grypus*). Journal of Zoology 198:473–479.

Fedak, M. A., and S. A. Anderson. 1987. Estimating the energy requirements of seals from weight changes. Pages 205–226 *in* A. C. Huntley, D. P. Costa, G. A. J. Worthy, and M. A. Castellini (eds.). Approaches to Marine Mammal Energetics. Society for Marine Mammalogy Special Publication No.1. Allen Press, Lawrence, KS.

Fedak, M. A., T. Arnbom, and I. L. Boyd. 1996. The relation between the size of southern elephant seal mothers, the growth of their pups, and the use of maternal energy, fat, and protein during lactation. Physiological Zoology 60:887–911.

Feldkamp, S. D. 1985. Swimming and diving in the California sea lion, *Zalophus californianus.* Ph.D. thesis. University of California, San Diego, 176 pp.

Feldkamp, S. D. 1987. Swimming in the California sea lion: Morphometrics, drag and energetics. Journal of Experimental Biology 131:117–135.

Feldkamp, S. D., D. P. Costa, and G. K. DeKrey. 1989. Energetics and behavioral effects of net entanglement on juvenile northern fur seals, *Callorhinus ursinus.* Fishery Bulletin 87:85–94.

Fish, F. E. 1983. Metabolic effects of swimming velocity and water temperature in the muskrat (*Ondatra zibethicus*). Comparative Biochemistry and Physiology A 75:397–400.

Fish, F. E. 1993. Influence of hydrodynamic design and propulsive mode on mammalian swimming energetics. Australian Journal of Zoology 42:79–101.

Fonnesbeck, P. V. 1968. Digestion of soluble and fibrous carbohydrates of forage by horses. Journal of Animal Science 27:1336–1344.

Fonnesbeck, P. V., R. K. Lydman, G. W. Vandercoot, and L. D. Symons. 1967. Digestibility of the proximate nutrients of forage by horses. Journal of Animal Science 26:1039–1045.

Fraser, W. R., R. L. Pitman, and D. G. Ainley. 1989. Seabird and fur seal responses to vertically migrating winter krill swarms in Antarctica. Polar Biology 10:37–41.

Gales, N. J., and R. H. Mattlin. 1997. Summer diving behavior of lactating New Zealand sea lions *Phocarctos hookeri.* Canadian Journal of Zoology 75:1696–1706.

Gallivan, G. J., and R. C. Best. 1980. Metabolism and respiration of the Amazonian manatee (*Trichechus inunguis*). Physiological Zoology 53:245–253.

Gallivan, G. J., and K. Ronald. 1979. Temperature regulation in freely diving harp seals (*Phoca groenlandica*). Canadian Journal of Zoology 57:2256–2263.

Gallivan, G. J., and K. Ronald. 1981. Apparent specific dynamic action in the harp seal (*Phoca groenlandica*). Comparative Biochemistry Physiology A 69:579–581.

Gallivan, G. J., R. C. Best, and J. W. Kanwisher. 1983. Temperature regulation in the Amazonian manatee *Trichechus inunguis.* Physiological Zoology 56:255–262.

Gentry, R. L., and J. R. Holt. 1986. Attendance behavior of northern fur seals. Pages 61–78 *in* R. L. Gentry and G. L. Kooyman (eds.). Fur Seals: Maternal Strategies on Land and at Sea. Princeton University Press, Princeton, NJ.

Gentry, R. L., D. P. Costa, J. P. Croxall, J. H. M. David, R. W. Davis, G. L. Kooyman, P. Majluf, T. S. McCann, and F. Trillmich. 1986. Synthesis and conclusions. Pages 220–264 *in* R. L. Gentry and G. L. Kooyman (eds.). Fur Seals: Maternal Strategies on Land and at Sea. Princeton University Press, Princeton, NJ.

Gentry, R. L., W. E. Roberts, and M. W. Cawthorn. 1987. Diving behavior of the Hooker's sea lion (Abstract). Seventh Biennial Conference on the Biology of Marine Mammals. Society of Marine Mammalogy, Miami, FL, December 5–9.

Gingerich, P. D., S. M. Raza, M. Arlf, M. Anwar, and X. Zhou. 1994. New whale from the Eocene of Pakistan and the origin of cetacean swimming. Nature 368:845–847.

Haley, M. P., C. J. Deutsch, and B. J. Le Boeuf. 1994. Size, dominance and copulatory success in male northern elephant seals, *Mirounga angustirostris.* Animal Behavior 48:1249–1260.

Hampton, I. F. G., and G. C. Whittow. 1976. Body temperature and heat exchange in the Hawaiian spinner dolphin (*Stenella longirostris*). Comparative Biochemistry Physiology A 55:195–197.

Hampton, I. F. G., G. C. Whittow, J. Szekerczes, and S. Rutherford. 1971. Heat transfer and body temperature in the Atlantic bottlenose dolphin (*Tursiops truncatus*). International Journal of Biochemistry and Biometerology 15:247–253.

Hart, J. S., and L. Irving. 1959. The energetics of harbor seals in air and in water with special consideration of seasonal changes. Canadian Journal of Zoology 37:447–457.

Hartman, D. S. 1979. Ecology and Behavior of the Manatee (*Trichechus manatus*) in Florida, Special publication no. 5, American Society of Mammalogists. Allen Press, Lawrence, KS.

Hedrick, M. S., and D. A. Duffield. 1991. Haematological and rheological characteristics of blood in seven marine mammal species: Physiological implications for diving behaviour. Journal of Zoology (London) 225:273–283.

Hedrick, M. S., D. A. Duffield, and L. H. Cornell. 1986. Blood viscosity and optimal hematocrit in a deep-diving mammal, the northern elephant seal (*Mirounga angustirostris*). Canadian Journal of Zoology 64:2081–2085.

Heide-Jorgensen, M. P., and R. Dietz. 1995. Some characteristics of narwhal, *Monodon monoceros*, diving behaviour in Baffin Bay. Canadian Journal of Zoology 73:2120–2132.

Hertel, H. 1966. Structure, Form, and Movement. Reinhold Publishing Corp., New York, NY.

Higgins, L. V., D. P. Costa, A. C. Huntley, and B. J. Le Boeuf. 1988. Behavioral and physiological measurements of maternal investment in the Steller sea lion, *Eumetopias jubatus*. Marine Mammal Science 4:44–58.

Hindell, M. A, D. J. Slip, and H. R. Burton. 1991. The diving behaviour of adult male and female southern elephant seals, *Mirounga leonina* (Pinnipedia:Phocidae). Australian Journal of Zoology 39:595–619.

Hines, A. H., and J. S Pearse. 1982. Abalones, shells and sea otters: Dynamics of prey populations in Central California. Ecology 63:1547–1560.

Hobson, R. P., and A. R. Martin. 1996. Behaviour and dive times of Arnoux's beaked whales, *Berardius arnuxii*, at narrow lead in fast ice. Canadian Journal of Zoology 74:388–393.

Hui, C. A. 1989. Surfacing behavior and ventilation in free-ranging dolphins. Journal of Mammalogy 70:833–835.

Hurley, J. 1996. Metabolic rate and heart rate during trained dives in adult California sea lions. Ph.D. thesis. University of California, Santa Cruz, 109 pages.

Irvine, A. B. 1983. Manatee metabolism and its influence on distribution in Florida. Biological Conservation 25:315–334.

Irving, L. 1969. Temperature regulation in marine mammals. Pages 147–174 *in* H. T. Anderson (ed.). Biology of Marine Mammals. Academic Press, New York, NY.

Irving, L. 1973. Aquatic mammals. Pages 47–96 *in* G. C. Whittow (ed.). Comparative Physiology of Thermoregulation, Vol. 3. Academic Press, New York, NY.

Irving, L., and J. S. Hart. 1957. Metabolism and insulation of seals as bare-skinned mammals in cold water. Canadian Journal of Zoology 35:497–511.

Irving, L., P. F. Scholander, and S. W. Grinnell. 1941. Respiration of the porpoise, *Tursiops truncatus*. Journal of Comparative Physiology 17:145–168.

Kanwisher, J., and S. H. Ridgway. 1983. The physiological ecology of whales and porpoises. Scientific American 248:110–120.

Kanwisher, J., and G. Sundnes. 1966. Thermal regulation in cetaceans. Pages 397–409 *in* K. S. Norris (ed.). Whales, Dolphins, and Porpoises. University of California Press, Berkeley, CA.

Karasov, W. H., and R. A. Anderson. 1984. Interhabitat differences in energy acquisition and expenditure in a lizard. Ecology 65:235–247.

Kasuya, T., and H. Marsh. 1984. Life history and reproductive biology of the short-finned pilot whale, *Globicephala macrorhynchus*, off the Pacific coast of Japan. Reports of the International Whaling Commission (Special Issue 6):259–310.

Keiver, K. M., K. Ronald, and F. W. H. Beamish. 1984. Metabolizable energy requirements for maintenance and fecal and urinary losses of juvenile harp seals (*Phoca groenlandica*). Canadian Journal of Zoology 62:769–776.

Kenney, R. D., H. E. Winn, and M. C. Macaulay. 1995. Cetaceans in the Great South Channel, 1979–1989: Right whale (*Eubalaena glacialis*). Continental Shelf Research 15:385–414.

King, J. 1983. Seals of the World. Oxford University Press, Oxford.

Kleiber, M. 1975. The Fire of Life. Krieger, New York, NY.

Kolb, P. M., and K. S. Norris. 1982. A harbor seal, *Phoca vitulina richardi*, taken from a sablefish trap. California Fish and Game 68:123–124.

Kooyman, G. L. 1981. Weddell seal: Consummate diver. Cambridge University Press, Cambridge, U.K.

Kooyman, G. L. 1985. Physiology without restraint in diving mammals. Marine Mammal Science 1:166–178.

Kooyman, G. L. 1989. Diverse Divers: Physiology and Behavior. Springer-Verlag, Berlin.

Kooyman, G. L., and H. T. Andersen. 1969. Deep-diving behavior. Pages 65–92 *in* H. T. Andersen (ed.). The Biology of Marine Mammals. Academic Press, New York, NY.

Kooyman, G. L., D. H. Kerem, W. B. Campbell, and J. J. Wright. 1973. Pulmonary gas exchange in freely diving Weddell seals. Respiratory Physiology 17:283–290.

Kooyman, G. L., E. A. Wahrenbrock, M. A. Castellini, R. W. Davis, and E. E. Sinnett. 1980. Aerobic and anaerobic metabolism during voluntary diving in Weddell seals: Evidence of preferred pathways from blood chemistry and behavior. Journal of Comparative Physiology B 138:335–346.

Kooyman, G. L., M. A. Castellini, R. W. Davis, and R. A. Maue. 1983. Aerobic diving limits of immature Weddell seals. Journal of Comparative Physiology B 151:171–174.

Koski, W. R., R. A. Davis, G. W. Miller, and D. E. Withrow. 1993. Pages 239–274 *in* J. J. Burns, J. J. Montague, and C. J. Cowles (eds.). The Bowhead Whale. Society for Marine Mammalogy Special Publication No.2. Allen Press, Lawrence, KS.

Kovacs, K. M., and D. M. Lavigne. 1986. Maternal investment and neonatal growth in phocid seals. Journal of Animal Ecology 55:1035–1051.

Kovacs, K. M., and D. M. Lavigne. 1992a. Maternal investment in otariid seals and walruses. Canadian Journal of Zoology 70:1953–1964.

Kovacs, K. M., and D. M. Lavigne. 1992b. Mass-transfer efficiency between hooded seals (*Cystophora cristata*) mothers and their pups in the Gulf of St. Lawrence. Canadian Journal of Zoology 70:1315–1320.

Kretzmann, M. B., D. P. Costa, L. V. Higgins, and D. J. Needham. 1991. Milk composition of Australian sea lion, *Neophoca cinerea*. Canadian Journal of Zoology 69:2556–2561.

Kretzmann, M. B., D. P. Costa, and B. J. Le Boeuf. 1993. Maternal energy investment in elephant seal pups: Evidence for sexual equality. American Naturalist 141:466–480.

Kriete, B. 1995. Bioenergetics of the killer whale, *Orcinus orca*. Ph.D. thesis. University of British Columbia, Vancouver, B.C., 138 pp.

Kruse, 1975. Swimming metabolism of California sea lions, *Zalophus californianus.* M.S. thesis. San Diego State University, San Diego, 53 pp.

Kvitek, R. G., J. S. Oliver, A. R. DeGange, and B. S. Anderson. 1992. Changes in Alaskan soft-bottom prey communities along a gradient in sea otter predation. Ecology 73:413–428.

Lang, T. G., and K. S. Norris. 1966. Swimming speed of a Pacific bottlenose dolphin. Science 151:588–590.

Lavigne, D. M., S. Innes, G. A. J. Worthy, K. M. Kovacs, O. J. Schmitz, and J. P. Hickie. 1986. Metabolic rates of seals and whales. Canadian Journal of Zoology 64:279–284.

Le Boeuf, B. J., D. P. Costa, A. C. Huntley, and S. D. Feldkamp. 1988. Continuous, deep diving in female northern elephant seals, *Mirounga angustirostris.* Canadian Journal of Zoology 66:446–458.

Le Boeuf, B. J., Y. Naito, T. Asaga, D. Crocker, and D. P. Costa. 1992. Swim speed in a female northern elephant seal: Metabolic and foraging implications. Canadian Journal of Zoology 70:786–795.

Lenfant, C., K. Johansen, and J. D. Torrance. 1970. Gas transport and oxygen storage capacity in some pinnipeds and the sea otter. Respiratory Physiology 9:277–286.

Liao, J. 1990. An investigation of the effect of water temperature on the metabolic rate of the California sea lion (*Zalophus californianus*). M.S. thesis. University of California, Santa Cruz, 55 pp.

Lifson, N., and R. McClintock. 1966. Theory of use of the turnover rates of body water for measuring energy and material balance. Journal of Theoretical Biology 12:46–74.

Lindstedt, S. L., and M. S. Boyce. 1985. Seasonality, fasting endurance and body size in mammals. American Naturalist 125:873–878.

Lockyer, C. 1977. Some Estimates of Growth in the Sei Whale, *Balaenoptera borealis.* (Special Issue 1). SC / SP / Doc 17. Reports of the International Whaling Commission, London, U.K.

Lockyer, C. H. 1978. A theoretical approach to the balance between growth and food consumption in fin and sei whales, with special reference to the female reproductive cycle. Report of the International Whaling Commission 28:243–249.

Lockyer, C. H. 1981a. Estimation of the energy costs of growth, maintenance and reproduction in the female minke whale, *Balaenoptera acutorostrata* from the southern hemisphere. Report of the International Whaling Commission 31:337–343.

Lockyer, C. H. 1981b. Growth and energy budgets of large baleen whales from the southern hemisphere. Pages 379–487 *in* Mammals in the Seas. FAO Fisheries Series No. 5, Vol. III. FAO, Rome.

Lockyer, C. H. 1984. Review of baleen whale (Mysticeti) reproduction and implications for management. Reports of the International Whaling Commission (Special Issue 6):27–50.

Lockyer, C. H. 1995. Investigations of aspects of the life history of the harbour porpoise, *Phocoena phocoena,* in British waters. Report of the International Whaling Commission (Special Issue 16):189–197.

Loughlin, T. R., J. L. Bengston, and R. L. Merrick. 1987. Characteristics of feeding trips of female northern fur seals. Canadian Journal of Zoology 65:2079–2084.

Lunn, N. J., and I. L. Boyd. 1993. Influence of maternal characteristics and environmental variation on reproduction in Antarctic fur seals. Symposia Zoological Society (London) 66:115–129.

Lydersen, C., and D. M. Kovacs. 1993. Diving behaviour of lactating harp seal, *Phoca groenlandica,* females from the Gulf of St. Lawrence, Canada. Animal Behaviour 46:1213–1221.

Markussen, N. H., M. Ryg, and N. A. Øritsland. 1994. The effect of feeding on the metabolic rate in harbour seals (*Phoca vitulina*). Journal of Comparative Physiology 164:89–93.

Marmontel, M. 1995. Age and reproduction in female Florida manatees. Pages 98–119 *in* T. J. O'Shea, B. B. Ackerman, and H. F. Percival (eds.). Population Biology of the Florida Manatee. Information and Technology Report 1. National Biological Service, U.S. Department of the Interior, Washington, D.C.

Marsh, H. 1995. The life history pattern of breeding, and population dynamics of the dugong. Pages 75–83 *in* T. J. O'Shea, B. B. Ackerman, and H. F. Percival (eds.). Population Biology of the Florida Manatee. Information and Technology Report 1. National Biological Service, U.S. Department of the Interior, Washington, D.C.

Marsh, H., A. V. Spain, and G. E. Heinsohn. 1978. Physiology of the dugong. Comparative Biochemistry and Physiology A 61:159–168.

Martin, A. R., and P. Rothery. 1993. Reproductive parameters of female long-finned pilot whales (*Globicephala melas*) around the Faroe Islands. Report of the International Whaling Commission (Special Issue 14):263–304.

Matsuura, D. T., and G. C. Whittow. 1973. Oxygen uptake of the California sea lion and harbor seal during exposure to heat. American Journal of Physiology 225:711–715.

Matsuura, D. T., and G. C. Whittow. 1975. Thermal insulation of the California sea lion during exposure to heat. Comparative Biochemistry Physiology A 51:27–30.

Maxwell, D. C. 1974. Marine primary productivity of the Galapagos archipelago. Ph.D. thesis. Ohio State University, Columbus.

McGinnis, S. M., G. C. Whittow, C. A. Ohata, and H. Huber. 1972. Body heat dissipation and conservation in two species of dolphins. Comparative Biochemistry Physiology A 43:417–423.

McNab, B. K. 1980. Food habits, energetics and the population biology of mammals. American Naturalist 116:106–124.

McNab, B. K. 1983. Ecological and behavioral consequences of adaptation to various food resources. Pages 664–697 *in* J. F. Eisenberg and D. G. Kleiman (eds.). Advances in the Study of Mammalian Behavior, Special Publication. American Society of Mammalogy. Allen Press, Lawrence, KS.

McNab, B. K. 1984. Basal Metabolic rate and the intrinsic rate of natural increase: An empirical and theoretical re-examination. Oecologia 64:423–424.

McNab, B. K. 1986a. The influence of food habitats on the energetics of eutherian mammals. Ecological Monographs 56:1–29.

McNab, B. K. 1986b. Food habits, energetics, and the reproduction of marsupials. Journal of Zoology (London) 208:595–614.

Millar, J. A. 1977. Adaptive features of mammalian reproduction. Evolution 31:370–386.

Millar, J. S., and G. J. Hickling. 1990. Fasting endurance and the evolution of mammalian body size. Functional Ecology 4:5–12.

Miller, K. L. 1978. Energetics of the Northern Fur Seal in Relation to Climate and Food Resources of the Bering Sea. Final Report to the U.S. Marine Mammal Commission, Contract MM5AC025. National Technical Information Service, Springfield, VA.

Miller, K., and L. Irving. 1975. Metabolism and temperature regulation in young harbor seals *Phoca vitulina richardi.* American Journal of Physiology 229:506–511.

Miller, K., M. Rosenmann, and P. Morrison. 1976. Oxygen uptake and temperature regulation of young harbor seals. (*Phoca vitulina richardii*) in water. Comparative Biochemistry and Physiology A 54:105–107.

Morrison, P. M., M. Rosenmann, and J. A. Estes. 1974. Metabolism and regulation in the sea otter. Physiological Zoology 47:218–299.

Murray, R. M., M. Marsh, G. E. Heinsohn, and A. V. Spain. 1977. The role of the midgut caecum and large intestine in the digestion of sea grasses by the dugong (Mammalia: Sirenia). Comparative Biochemistry and Physiology A 56:7–10.

Nadel, E. R., I. Holmer, U. Bergh, P. O. Astrand, and J. A. Stolwijk. 1974. Energy exchanges of swimming man. Journal of Applied Physiology 36:465–471.

Nagy, K. A. 1980. CO_2 production in animals: Analysis of potential errors in the doubly labeled water method. American Journal of Physiology 238:R466–R473.

Nagy, K. A., R. B. Huey, and A. F. Bennett. 1984. Field energetics and foraging mode of Kalahari lacertid lizards. Ecology 65:588–596.

Oftedal, O. T., D. J. Boness, and R. A. Tedman. 1987. The behavior, physiology, and anatomy of lactation in the Pinnipedia. Current Mammalogy 1:175–245.

Oftedal, O. T., D. J. Boness, and W. D. Bowen. 1988. The composition of hooded seal (*Cystophora cristata*) milk: An adaptation for postnatal fattening. Canadian Journal of Zoology 66:318–322.

Oftedal, O. T., W. D. Bowen, and D. J. Boness. 1996. Lactation performance and nutrient deposition in pups of the harp seal, *Phoca groenlandica,* on ice floes of Southeastern Labrador. Physiological Zoology 69: 635–657.

Orians, G. H., and N. E. Pearson. 1977. On the theory of central place foraging. Pages 153–177 *in* D. J. Horn, G. R. Stairs, and R. D. Mitchell (eds.). Analysis of Ecological Systems. Ohio State University Press, Columbus, OH.

Ortiz, C. L., D. P. Costa, and B. J. Le Boeuf. 1978. Water and energy flux in fasting weaned elephant seal pups (*Mirounga angustirostris*). Physiological Zoology 51:166–178.

Ortiz, C. L., B. J. Le Boeuf, and D. P. Costa. 1984. Milk intake of elephant seal pups: An index of parental investment. American Naturalist 124:416–422.

Parry, D. A. 1949. The structure of whale blubber and a discussion of its thermal properties. Quarterly Journal of Microscopy 90:13–26.

Parsons, J. L. 1977. Metabolic studies in ringed seals (*Phoca hispida*). M.S. thesis. University of Guelph, Canada, 82 pp.

Perez, M. A., and E. E. Mooney. 1986. Increased food and energy consumption of lactating northern fur seals, *Callorhinus ursinus.* Fishery Bulletin 84:371–381.

Perrin W. F., and J. R. Henderson. 1984. Growth and reproductive rates in two populations of spinner dolphins, *Stenella longirostris,* with different histories of exploitation. Reports of the International Whaling Commission (Special Issue 6):417–432.

Perrin, W. F., and A. A. Hohn. 1994. Pantropical spotted dolphin *Stenella attenuata.* Pages 71–98 *in* S. H. Ridgway and R. Harrison (eds.). Handbook of Marine Mammals, Vol. 5: The First Book of Dolphins. Academic Press, London.

Peters, R. H. 1983. The Ecological Implications of Body Size. Cambridge University Press, Cambridge, U.K.

Pierce, R. W. 1970. Design and operation of a metabolic chamber for marine mammals. Ph.D. thesis. University of California, Berkeley, CA, 82 pp.

Pilson, M. E. Q., and A. L. Kelly. 1962. Composition of milk from *Zalophus californianus,* the California sea lion. Science 135:104.

Pond, C. M. 1977. The significance of lactation in the evolution of mammals. Evolution 31:177–199.

Ponganis, P. J., E. P. Ponganis, K. V. Ponganis, G. L. Kooyman, R. L. Gentry, and F. Trillmich. 1990. Swimming velocities in otariids. Canadian Journal of Zoology 68:2105–2112.

Ramsay, M. A., and I. Stirling. 1988. Reproductive biology and ecology of female polar bears (*Ursus maritimus*). Journal of Zoology (London) 214:601–634.

Rathbun, G. B., J. P. Reid, R. K. Bonde, and J. A. Powell. 1995. Reproduction in free-ranging Florida manatees. Pages 135–156 *in* T. J. O'Shea, B. B. Ackerman, and H. F. Percival (eds.). Population Biology of the Florida Manatee. Information and Technology Report 1. National Biological Service. U.S. Department of the Interior, Washington, D.C.

Rea, L., and D. P. Costa. 1991. Changes in resting metabolic rate during long-term fasting in northern elephant seal pups (*Mirounga angustirostris*). Physiological Zoology 65:97–111.

Read, A. J. 1990. Age at sexual maturity and pregnancy rates of harbour porpoises *Phocoena phocoena* from the Bay of Fundy. Canadian Journal of Fishes and Aquatic Science 47:561–565.

Reiter, J., N. L. Stinson, and B. J. Le Boeuf. 1978. Northern elephant seal development: The transition from weaning to nutritional independence. Behavioral Ecology and Sociobiology 3:337–367.

Repenning, C. A., and R. H. Tedford. 1977. Otarioid Seals of the Neogene. U.S. Geological Survey Professional Papers 992, Washington, D.C.

Reyes, J. C., and K. Van Waerebeek. 1995. Aspects of the biology of Burmeister's porpoise from Peru. Report of the International Whaling Commission (Special Issue 16):349–364.

Reynolds, J. E., and S. A. Rommel. 1996. Structure and function of the gastrointestinal tract of the Florida manatee, *Trichechus manatus latirostris.* The Anatomical Record 245:539–558.

Rice, D. W., and A. A. Wolman. 1971. The life history and ecology of the gray whale *(Eschrichtius robustus).* Special Publication No. 3. The American Society of Mammalogists. Allen Press, Lawrence, KS.

Ridgway, S. H. 1986. Diving by cetaceans. *in* A.O. Brubakk, J. W. Kanwisher and G. Sundness (eds.). Diving in Animals and Man. Kongsvold Symposium, Royal Norwegian Society of Science and Letters, Tapir Publ., Trondheim, Norway.

Ridgway, S. H., and G. S. Patton. 1971. Dolphin thyroid: Some anatomical and physiological findings. Z. vergl. Physiologie 71:129–141.

Riedman, M. L., and J. A. Estes. 1990. The sea otter (*Enhydra lutris*): Behavior, ecology, and natural history. U.S. Fish and Wildlife Service, Biological Report 90 (14).

Ronald, K., K. M. Keiver, F. W. Beamish, and R. Frank. 1984. Energy requirements for maintenance and faecal and urinary losses of the grey seal (*Halichoerus grypus*). Canadian Journal of Zoology 62:1101–1105.

Rounsevell, D., and I. Eberhard. 1980. Leopard seals, *Hydrurga leptonyx,* at Macquarie island from 1949–1979. Australian Wildlife Research 7:403–415.

Schmidt-Nielsen, K. 1979. Animal physiology: Adaptations and environment. Cambridge University Press, Cambridge, U.K.

Schmitz, O. J., and D. M. Lavigne. 1984. Intrinsic rate of increase, body size, and specific metabolic rate in marine mammals. Oecologia 62:305–309.

Schoenherr, J. R. 1991. Blue whales on high concentrations of euphasiids around Monterey submarine canyon. Canadian Journal of Zoology 69:583–594.

Scholander, P. F. 1959. Wave-riding dolphins: How do they do it? Science 129:1085–1087.

Scholander, P. F., and L. Irving. 1941. Experimental investigations on the respiration and diving of the Florida manatee. Journal of Cellular and Comparative Physiology 17:67–78.

Scholander, P. F., and W. E. Schevill. 1955. Counter-current vascular heat exchange in the fins of whales. Journal of Applied Physiology 8:279–292.

Shane, S. H. 1990. Behavior and ecology of the bottlenose dolphin at Sanibel Island, Florida. Pages 245–265 *in* S. Leatherwood and R. R. Reeves (eds.). The Bottlenose Dolphin. Academic Press, New York, NY.

Sinclair, E. H., T. Loughlin, and W. Pearcy. 1994. Prey selection by northern fur seals (*Callorhinus ursinus*) in the Eastern Bering Sea. Fishery Bulletin 92:144–156.

South, F. E., R. H. Luecke, M. L. Zatzmann, and M. D. Shanklin. 1976. Air temperature and direct partitional calorimetry of the California sea lion (*Zalophus californianus*). Comparative Biochemistry and Physiology A 54:23–30.

Stewart, B. S., and R. L. DeLong. 1991. Diving patterns of northern elephant seal bulls. Marine Mammal Science 7:369–384.

Stewart, R. E. A., and D. M. Lavigne. 1984. Energy transfer and female condition in nursing harp seals *Phoca groenlandica*. Holarctic Ecology 7:182–194.

Stirling, I. 1975. Factors affecting the social behaviour in the Pinnipedia. Pages 205–212 *in* K. A. Ronald and W. A. Mansfield (eds.). The Biology of the Seal. Rapports et Procés-verbaux Des Réunions, Conseil International Pour L'Exploration de la Mer, Vol. 169.

Stirling, I. 1983. The evolution of mating systems in pinnipeds, Pages 489–527 *in* J. F. Eisenberg and D. G. Kleinman (eds.). Advances in the Study of Mammalian Behavior. American Society of Mammalogists. Allen Press, Lawrence, KS.

Stirling, I., and E. H. McEwan. 1975. The caloric value of ringed seals (*Phoca hispida*) in relation to polar bear (*Ursus maritimus*) ecology and hunting behavior. Canadian Journal of Zoology 53:1021–1027.

Sumich, J. L. 1983. Swimming velocities, breathing patterns and estimated costs of locomotion in migrating gray whales, *Eschrichtius robustus*. Canadian Journal of Zoology 61:647–652.

Tarasoff, F. J., and H. D. Fisher. 1970. Anatomy of the hind flippers of two species of seals with reference to thermoregulation. Canadian Journal of Zoology 48:821–829.

Taylor, C. R., G. M. O. Maloiy, E. R. Weibel, V. A. Langman, J. M. Z. Kamau, M. J. Seeherman, and N. C. Heglund. 1980. Design of the mammalian respiratory system. III. Scaling maximum aerobic capacity to body mass: Wild and domestic mammals. Respiration Physiology 44:25–37.

Taylor, C. R., R. H. Karas, E. R. Weibel, and H. Hoppeler. 1987. Adaptive variation in the mammalian respiratory system in relation to energetic demand. II. Reaching the limits to oxygen flow. Respiration Physiology 69:7–26.

Tedman, R., and B. Green. 1987. Water and sodium fluxes and lactational energetics in suckling pups of Weddell seals (*Leptonychotes weddellii*). Journal of Zoology (London) 212:29–42.

Testa, W., and D. B. Siniff. 1987. Population dynamics of Weddell seals (*Leptonychotes weddelli*) in McMurdo sound, Antarctica. Ecological Monographs 57:149–165.

Testa, W., S. E. B. Hill, and D. B. Siniff. 1989. Diving behavior and maternal investment in Weddell seals (*Leptonychotes weddelli*). Marine Mammal Science 5:399–405.

Testa, J. W., G. Oehlert, D. G. Ainley, J. L. Bengtson, D. B. Siniff, R. M. Laws, and D. Rounsevell. 1991. Temporal variability in Antarctic marine ecosystems: Periodic fluctuations in the phocid seals. Canadian Journal of Aquatic Science 48:631–639.

Thewissen, J. G. M., S. T. Hussain, and M. Arif. 1994. Fossil evidence for the origin of aquatic locomotion in Archaeocete whales. Science 263:210–212.

Thompson, D., P. S. Hammond, K. S. Nicholas, and M. A. Fedak. 1991. Movements and foraging behaviour of grey seals (*Halichoerus grypus*). Journal of Zoology (London) 224:223–232.

Toussaint, H. M. 1990. Differences in propelling efficiency between competitive and triathlon swimmers. Medicine and Science in Sports and Exercise 22:409–415.

Trillmich, F. 1990. The behavioral ecology of maternal effort in fur seal and sea lions. Behaviour 114:3–20.

Trillmich, F., and E. Lechner. 1986. Milk of the Galapagos fur seal and sea lion, with a comparison of the milk of eared seals (Otariidae). Journal of Zoology (London) 209:271–277.

Trillmich, F., and K. A. Ono. 1991. Pinnipeds and El Niño: Responses to Environmental Stress. Ecological Studies, Vol. 88. Springer-Verlag, Berlin.

Trillmich, F., G. L. Kooyman, P. Majluf, and M. Sanchez-Grinan. 1986. Attendance and diving behavior of South American fur seals during El Niño in 1983. Pages 153–167 *in* R. L. Gentry and G. L. Kooyman (eds.). Fur Seals: Maternal Strategies on Land and at Sea. Princeton University Press, Princeton, NJ.

Trillmich, F., K. Ono, D. P. Costa, R. L. DeLong, S. Feldkamp, J. Francis, R. L. Gentry, C. Heath, and B. J. Le Boeuf. 1991. Pages 247–270 *in* F. Trillmich and K. Ono (eds.). Pinnipeds and El Niño: Responses to Environmental Stress. Ecological Studies, Vol. 88. Springer-Verlag, Berlin.

Wartzok, D., R. Elsner, H. Stone, B. P. Kelly, and R. W. Davis. 1992. Under-ice movements and the sensory basis of hole finding by ringed and Weddell seals. Canadian Journal of Zoology 70:1712–1722.

Watkins, W. A., K. E. Moore, J. Sigurjonsson, D. Wartzok, and G. Notarbartolo di Sciara. 1984. Fin whale (*Balaenoptera physalus*) tracked by radio in the Irminger Sea. Rit Fiskideildar 1:1–14.

Watkins, W. A., M. A. Daher, D. M. Fristrup, and T. J. Howald. 1993. Sperm whales tagged with transponders and tracked underwater by sonar. Marine Mammal Science 9:55–67.

Wells, R. S., M. D. Scott, and A. B. Irvine. 1987. The social structure of free-ranging bottlenose dolphins. Pages 247–305 *in* H. H. Genoways (ed.). Current Mammalogy, Vol. 1. Plenum, New York, NY.

Werner, R., and C. Campagna. 1995. Diving behaviour of lactating southern sea lions (*Otaria flacescens*) in Patagonia. Canadian Journal of Zoology 73:1975–1982.

Whitehead, H. 1996. Variation in the feeding success of sperm whales: Temporal scale, spatial scale and relationship to migrations. Journal of Animal Ecology 65:429–438.

Whittow, G. C. 1987. Thermoregulatory adaptations in marine mammals: Interacting effects of exercise and body mass. A review. Marine Mammal Science 3:220–241.

Whittow, G. C., D. T. Matsuura, and Y. C. Lin. 1972. Temperature regulation in the California sea lion (*Zalophus californianus*). Physiological Zoology 45:68–77.

Wiig, Q., I. Gjertz, D. Griffiths, and C. Lydersen. 1993. Diving patterns of an Atlantic walrus *Odobenus rosmarus rosmarus* near Svalbard. Polar Biology 13:71–72.

Williams, T. M. 1983. Locomotion in the North American mink, a semi-aquatic mammal. I. Swimming energetics and body drag. Journal of Experimental Biology 103:155–168.

Williams, T. M. 1986. Thermoregulation of the North American mink during rest and activity in the aquatic environment. Physiological Zoology 59:293–305.

Williams, T. M. 1989. Swimming by sea otters: Adaptations for low energetic cost locomotion. Journal of Comparative Physiology A 164:815–824.

Williams, T. M. 1999. The evolution of cost efficient swimming in marine mammals: Limits to energetic optimization. The Royal Society Philosophical Transactions: Biological Sciences 354:193–201.

Williams, T. M., and G. L. Kooyman. 1985. Swimming performance and hydrodynamic characteristics of harbor seals *Phoca vitulina*. Physiological Zoology 58:576–589.

Williams, T. M., G. L. Kooyman, and D. A. Croll. 1991. The effect of submergence on heart rate and oxygen consumption of swimming seals and sea lions. Journal of Comparative Physiology B 160:637–644.

Williams, T. M., J. E. Haun, W. A. Friedl, R. W. Hall, and L. W. Bivens. 1992. Assessing the thermal limits of bottlenose dolphins: A cooperative study by trainers, scientists, and animals. Fall IMATA Soundings:16–17.

Williams, T. M., W. A. Friedl, and J. E. Haun. 1993. The physiology of bottlenose dolphins (*Tursiops truncatus*): Heart rate, metabolic rate and plasma lactate concentration during exercise. Journal of Experimental Biology 179:31–46.

Williams, T. M., B. Le Boeuf, R. Davis, D. Crocker, and R. Skrovan. 1996. Integrating behavior and energetics in diving marine mammals: New views using video technology. Fifth European Conference on Wildlife Telemetry. Strasbourg, France. August 25–30.

Winn, H. E., J. D. Goodyear, R. D. Kenney, and R. O. Petrong. 1995. Dive patterns of tagged right whales in the Great South Channel. Continental Shelf Research 15:593–611.

Woodcock, A. H., and A. F. McBride. 1951. Wave-riding dolphins. Journal of Experimental Biology 28:215–217.

Worthy, G. A. J., and E. F. Edwards. 1990. Morphometric and biochemical factors affecting heat loss in a small temperate cetacean (*Phocoena phocoena*) and a small tropical cetacean (*Stenella attenuata*). Physiological Zoology 63:432–442.

Worthy, G. A. J., and D. M. Lavigne. 1983a. Energetics of fasting and subsequent growth in weaned harp seal pups, *Phoca groenlandica*. Canadian Journal of Zoology 61:447–456.

Worthy, G. A. J., and D. M. Lavigne. 1983b. Changes in energy stores during postnatal development of the harp seal, *Phoca groenlandica*. Journal of Mammalogy 64:89–96.

Würsig, B., and M. Würsig. 1979. Behavior and ecology of the bottlenose dolphin, *Tursiops truncatus*, in the south Atlantic. Fishery Bulletin 77:399–412.

Würsig, B., E. M. Dorsey, M. A. Fraker, R. S. Payne, W. J. Richardson, and R. S. Wells. 1984. Behavior of bowhead whales, *Balaena mysticetus*, summering in the Beaufort Sea: Surfacing, respiration, and dive characteristics. Canadian Journal of Zoology 62:1919–1921.

Würsig, B., E. M. Dorsey, M. A. Fraker, R. S. Payne, and W. J. Richardson. 1985. Behavior of bowhead whales, *Balaena mysticetus*, summering in the Beaufort Sea: A description. Fishery Bulletin 83:357–377.

6

IAN L. BOYD, CHRISTINA LOCKYER, AND HELENE D. MARSH

Reproduction in Marine Mammals

Unlike many of the other chapters in this text, this one provides separate sections on carnivores (pinnipeds, sea otters, and polar bears), cetaceans, and sirenians; the authors and editors consider this the clearest way to thoroughly describe the topic. Perhaps most striking are the differences in information among the different groups. What factors permit so much detail on carnivore reproduction, yet limit it in the other groups? In general, why do scientists study certain systems and species more than others? Factors including availability of specimens, economic importance of species, legal restrictions, and implications for human health all influence what we study and know. Students entering careers in science should ponder such questions.

Although the available data differ among the groups considered here, a number of common threads persists. The fundamental structures and processes associated with reproduction are common to all marine mammals, although different groups have interesting variations on the common themes. Terms are used consistently among sections, and all sections consider at least some aspects of reproduction in the broader mammalian context. For more detail and comparison with domestic mammal reproductive biology, readers should consult books devoted to histology (e.g., Banks 1993, Dellman 1971), gross anatomy (e.g., Schummer and Nickel 1979, Evans 1993), and physiology (e.g., Swensen and Reece 1993, Cupps 1990).

Reproduction in Pinnipeds, Sea Otters, and Polar Bears [I. L. B.]

Pinnipeds, polar bears (*Ursus maritimus*), and sea otters (*Enhydra lutris*) include some of the largest members of the order Carnivora and they exhibit reproductive patterns that reflect both their common ancestry and their subsequent adaptations to aquatic or semiaquatic life. As would be expected in long-lived organisms, these species have low reproductive rates and, consequently, they are slow to recover from population reduction. Therefore, they have low resilience to environmental perturbations that increase mortality or reduce productivity. Documentation of reproductive life histories and the interaction between the environment and reproductive rate in these species can provide essential information to enable the dynamics of populations to be examined. Such research helps predict which environmental changes are likely to regulate populations and how regulatory processes are likely to operate.

The purpose of this section of the chapter is to review current knowledge of the reproductive strategies and systems of pinnipeds, sea otters, and polar bears. The ecological and evolutionary aspects of reproduction in these species are emphasized, with supporting information provided on the physiology and anatomy of reproduction. It is hoped that this will stimulate further thought and open new areas for investigation. Unlike some aspects of the biology of these groups, the study of their reproduction is not a fast-moving field. For some species, little is known about the pattern and physiological control of reproduction, reflecting both the difficulties of obtaining such information and the need for studies to focus on certain key features of reproduction, such as pregnancy rate or age at maturity (all measured at the level of populations). These are often components of population models and essential for understanding population dynamics. Increasingly, however, it is recognized that questions about the factors that influence reproductive success can only be addressed effectively by long-term studies at the level of individuals (e.g., northern elephant seals [*Mirounga angustirostris*], Reiter and Le Boeuf 1981; Antarctic fur seals [*Arctocephalus gazella*], Boyd et al. 1995). To obtain a clear picture of what controls the trajectory of populations, scientists must measure the variation in reproductive success among individuals.

Throughout this chapter a basic knowledge of mammalian reproductive systems is assumed; those with no prior knowledge of mammalian reproductive anatomy and physiology may also wish to consult a general text (e.g., Schummer and Nickel 1979). This chapter begins with general descriptions of the reproductive life histories of pinnipeds, sea otters, and polar bears and thereafter develops several themes, including environmental factors controlling seasonal reproductive cycles (specifically the effects of photoperiod and nutrition), the influence of age on reproduction, the functional anatomy of the reproductive system, the importance of delayed implantation, lactation patterns, and the evolution of reproduction in pinnipeds. Owing to the greater number of species involved, the reproductive biology literature for the pinnipeds is more extensive than that for either the polar bear or the sea otter. The relevant literature is scattered over at least four decades from the early anatomical studies by Harrison et al. (1952) and Laws (1956a) to the more recent experimental studies of reproductive endocrinology (e.g., Raeside and Ronald 1981, Boyd 1991a). Wherever possible, I have tried to compare and contrast the pinniped patterns with those of the sea otter and polar bear. I have also attempted to derive broad conclusions about the nature of reproduction in marine carnivores, but these conclusions are not rules and I do not claim to have covered all the angles. There will be exceptions, but the reader may wish to look closely at these apparent exceptions before rejecting the arguments put forward because they could provide important information on which to build a more informed understanding of the subject.

As with all animals, the question of when and how much to reproduce is central to all aspects of the ecology of individuals and populations (Smith 1976). This section shows that both photoperiod and nutrition are important environmental variables influencing the timing of reproductive cycles, and nutritional conditions also affect the age at sexual maturity and how many offspring can be produced. Among sea otters photoperiod appears to have little influence on reproductive timing, whereas among some of the pinnipeds, which have highly synchronous seasonal reproductive cycles, photoperiod probably strongly influences the timing of reproduction. Unlike nutrition, photoperiod has no direct effect on survival or reproductive success. Photoperiod is a convenient circannual clock to which pinnipeds have been able to adjust their reproductive cycles to ensure that offspring are produced at the most appropriate time of the year. Delayed implantation is a universal feature of the reproductive cycles of marine carnivores (although it is also a characteristic of many other carnivores) and the timing of implantation appears to be the key event in the reproductive cycle that is influenced by photoperiod and nutrition. Several lines of evidence support the view that the time when implantation occurs is a critical phase of the reproductive cycles for both sexes because it may represent the time at which the reproductive system is reactivated after a period of quiescence equivalent in length, and possibly physiologically related to, delayed implantation.

Reproductive Cycles

Here we define "reproductive cycle" as the normal minimum time period for all stages of reproduction in the female from ovulation through conception and pregnancy to birth and lactation followed by a short rest period. However, the reproductive interval (meaning the actual time between end of one cycle and start of another) may be extended or even shortened because of environmental, nutritional, and social circumstances.

Pinnipeds

The generalized view of the physiological processes controlling pinniped reproductive cycles derives from a range of studies of the anatomical and endocrine changes of several species. No one species has been studied comprehensively but, in those areas where information is available about the reproductive cycles of more than one species, there are sufficient consistency and overlap to suggest that a common pattern exists for pinnipeds.

Distinctive features of the pinniped reproductive system include delayed implantation in all species and highly seasonal, synchronized reproductive cycles in most species with the production of a single offspring at any single reproductive attempt. Many pinnipeds spend most of their lives in the water and studies of their ranging behavior have revealed that they can exploit large areas of the ocean from shallow coastal habitats to extreme depths beyond the continental margins. However, all pinnipeds have to return to land or ice to give birth and this probably carries enhanced risks of predation and disease compared to birth and nursing in the aquatic environment. The major features of pinniped reproduction (particularly delayed implantation and postpartum mating), many of which are also seen in other modern terrestrial carnivores, were probably essential prerequisites for carnivore radiation in the marine environment, but they may also have ultimately constrained pinnipeds to a life that is only partially aquatic.

The reproductive cycle in most species is annual and synchronous (Table 6-1), but there are exceptions to the rules. The Australian sea lion (*Neophoca cinerea*) has a reproductive cycle lasting approximately 18 months (Ling and Walker 1978), resulting in an aseasonal pattern of births, although there appears to be some synchrony of births and mating within colonies (Gales and Williamson 1989). Similarly, in the walrus (*Odobenus rosmarus*), which has a 15-month gestation, the reproductive cycle extends to 2 years (Fig. 6-1), although it remains seasonal (Fay 1981, Sease and Chapman 1988). Reproduction probably occurs annually in the Hawaiian monk seal (*Monachus schauinslandi*), but births and matings extend over approximately 6 months of the year; therefore, this species displays only mild synchrony of reproduction (Atkinson and Gilmartin 1992). This situation is not too surprising because the ability of tropical organisms to use photoperiod to tightly regulate timing is limited. There is no information about reproductive patterns in either the Mediterranean (*Monachus monachus*) or the Caribbean (*M. tropicalis*) monk seals, but we can probably assume they follow or followed a similar pattern to the Hawaiian species (Table 6-1).

The three distinct phases of the reproductive cycle of female pinnipeds are estrus, delayed implantation, and fetal growth (Fig. 6-1). Apart from the exceptions mentioned above, the duration of each of these phases is remarkably consistent across species. The greatest difference exists in the duration of estrus. Among the otariids, the postpartum estrus is normally 5 to 8 days, whereas among the phocids, the estrus cycle duration is linked to the duration of lactation (4 to 60 days; Oftedal et al. 1987); the significance of this dichotomy is discussed later in the chapter. Adjustments in the duration of estrus lead to equivalent adjustments in the duration of the period of delayed implantation but probably not in the duration of fetal growth. The few studies that have examined the timing of implantation have demonstrated that the duration of the fetal growth phase is remarkably consistent and unaffected by body size (Laws 1956a, Boyd 1984a, Stewart et al. 1989, Trites 1991, Temte and Temte 1993). This implies that the fetal growth rate of the largest pinniped, the southern elephant seal (*Mirounga leonina*), is approximately ten times that of the smallest, the Galapagos fur seal (*Arctocephalus galapagoensis*).

Among male pinnipeds, there is a profound annual cycle of testicular regression and recrudescence (Laws 1956b; Boyd 1982; Griffiths 1984a,b; Ashchepkova and Fedoseev 1988; Bester 1990; Noonan et al. 1991; Atkinson and Gilmartin 1992). The short breeding seasons observed in many species of pinniped may be imposed on males by the extreme seasonality and synchrony of estrus in females. Cyclicity of male reproductive potential has been measured using indices such as testicular mass, concentrations of circulating plasma testosterone, and the presence or absence of sperm in testes or epididymides (Fig. 6-2). Most of these indices conclude that there are some fertile males in the population for up to 6 months or more of each year. This is clearly shown in Figure 6-2 for the gray seal (*Halichoerus grypus*) where there is little difference in the duration of the testicular cycles of males from eastern Canada and those from the British coast, although the actual breeding seasons in these areas are 3 months apart. From a small population sample, Griffiths (1984b) showed mature sperm to be present in the epididymides of adult male southern elephant seals for at least 5 months of the year, extending to 1 to 2 months on each side of the extremes of the breeding season. However, he also showed that plasma testosterone concentrations were elevated only during the breeding season itself. This raises the possibility that elevated plasma testosterone concentrations may be a consequence rather than the cause of male sexual behavior in pinnipeds. Therefore, whereas males may be potent for up to half the year, females may be more restricted in the timing of their breeding by seasonal environmental considerations involving energetic constraints.

Pregnancy rates among adult females have been measured directly in remarkably few species of pinnipeds. In long-term studies of individually marked Weddell seals (*Leptonychotes weddellii*) annual pregnancy rate varied from 0.46 to 0.89 (Testa 1987). Northern elephant seals have a mean pregnancy rate of >0.90 (Reiter and Le Boeuf 1991) and for Antarctic fur seals the mean is 0.70 with an interannual range of 0.59 to 0.88 (Boyd et al. 1995). In none of these cases was there any relationship between pregnancy rate and population density. From cross-sectional samples of populations, Bowen et al. (1981) found that annual pregnancy rates of harp seals (*Phoca groenlandica*) varied from 0.81 to 0.99 and that pregnancy rate

Table 6-1. Characteristics of the Reproductive Cycles of Female Pinnipeds Showing Timing of Birth (B), Fertilization (F), and Implantation (I)

Northern Hemisphere	Month (J F M A M J J A S O N D)	Sources
Harbor seal		
Phoca vitulina concolor (Nova Scotia)	B+++F----------I	Boulva and McLaren 1979
P. vitulina richardsi (Vancouver Island)	B+++F-----------I	Bigg 1973, Bigg and Fisher 1974
P. vitulina vitulina (North Sea)	B+++F------------I	Thompson 1988, Reijinders 1990
Spotted seal (*P. largha*)	B++++F------------I	Burns 1981b, Thomas et al. 1980
Ringed seal (*P. hispida*)	B+++F---------I	Kelly 1988b, Smith 1987
Harp seal (*P. groenlandica*)	B+F----------I	Sergeant 1976, Stewart 1987, Stewart and Lavigne 1984, Stewart et al. 1989
Ribbon seal (*P. fasciata*)	B+++F----------I?	Burns 1970, 1981b, Kelly 1988c, Thomas et al. 1980
Bearded seal (*Erignathus barbatus*)	B++F---------I	Burns 1970, 1981a, Kelly 1988a, Potelov 1975; Thomas et al. 1980
Hooded seal (*Cystophora cristata*)	BF--------------I	Born 1982, Bowen et al. 1987, Sergeant 1976
Gray seal (*Halichoerus grypus*)		
Northeast Atlantic	----I B++F-------	Boyd 1982, 1983, 1984a, Coulson and Hickling 1964, 1985a, Coulson 1981
Northwest Atlantic	B+ +F---------------I	Mansfield 1958 (see footnote 1), Mansfield and Beck 1977
Monk seal (*Monachus sp.*)	Birth season lasts for 8 months; little other information	Kenyon 1981
Northern elephant seal (*Mirounga angustirostris*)	B+++F---------I?	Laws 1956b, Le Boeuf and Briggs 1977
Steller sea lion (*Eumetopias jubatus*)	B+F---------I	Hoover 1988, Pitcher and Calkins 1981
Northern fur seal (*Callorhinus ursinus*)	BF-------------I	Croxall and Gentry 1987, Daniel 1981, Gentry 1981, Gentry et al. 1986, Walker and Ling 1981

Continued on next page

Table 6-1 continued

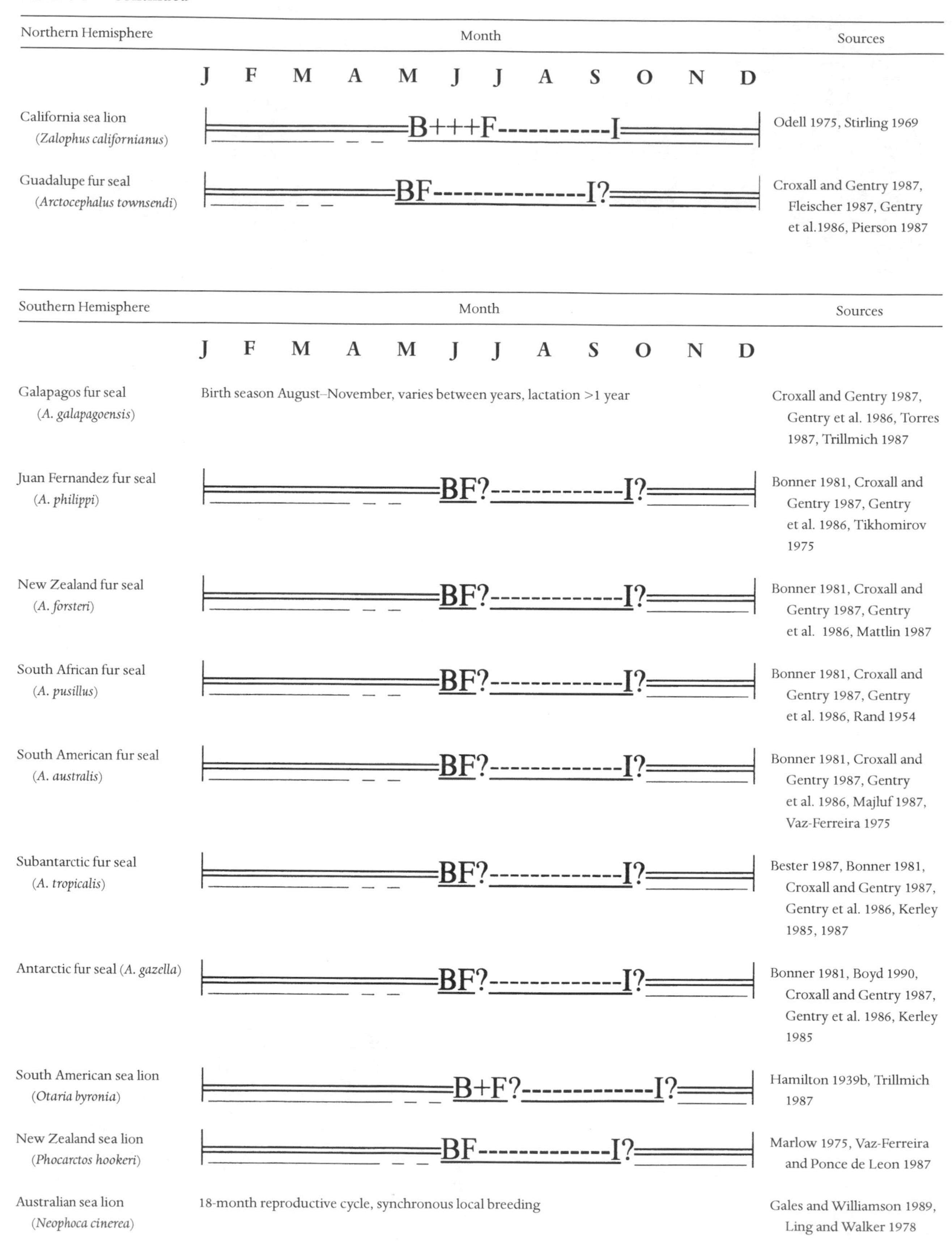

Northern Hemisphere	Month J F M A M J J A S O N D	Sources
California sea lion (*Zalophus californianus*)	═══════B+++F----------I═══════	Odell 1975, Stirling 1969
Guadalupe fur seal (*Arctocephalus townsendi*)	═══════BF---------------I?═══════	Croxall and Gentry 1987, Fleischer 1987, Gentry et al.1986, Pierson 1987

Southern Hemisphere	Month J F M A M J J A S O N D	Sources
Galapagos fur seal (*A. galapagoensis*)	Birth season August–November, varies between years, lactation >1 year	Croxall and Gentry 1987, Gentry et al. 1986, Torres 1987, Trillmich 1987
Juan Fernandez fur seal (*A. philippi*)	═══════BF?------------I?═══════	Bonner 1981, Croxall and Gentry 1987, Gentry et al. 1986, Tikhomirov 1975
New Zealand fur seal (*A. forsteri*)	═══════BF?------------I?═══════	Bonner 1981, Croxall and Gentry 1987, Gentry et al. 1986, Mattlin 1987
South African fur seal (*A. pusillus*)	═══════BF?------------I?═══════	Bonner 1981, Croxall and Gentry 1987, Gentry et al. 1986, Rand 1954
South American fur seal (*A. australis*)	═══════BF?------------I?═══════	Bonner 1981, Croxall and Gentry 1987, Gentry et al. 1986, Majluf 1987, Vaz-Ferreira 1975
Subantarctic fur seal (*A. tropicalis*)	═══════BF?------------I?═══════	Bester 1987, Bonner 1981, Croxall and Gentry 1987, Gentry et al. 1986, Kerley 1985, 1987
Antarctic fur seal (*A. gazella*)	═══════BF?------------I?═══════	Bonner 1981, Boyd 1990, Croxall and Gentry 1987, Gentry et al. 1986, Kerley 1985
South American sea lion (*Otaria byronia*)	═══════B+F?------------I?═══════	Hamilton 1939b, Trillmich 1987
New Zealand sea lion (*Phocarctos hookeri*)	═══════BF------------I?═══════	Marlow 1975, Vaz-Ferreira and Ponce de Leon 1987
Australian sea lion (*Neophoca cinerea*)	18-month reproductive cycle, synchronous local breeding	Gales and Williamson 1989, Ling and Walker 1978

Table 6-1 continued

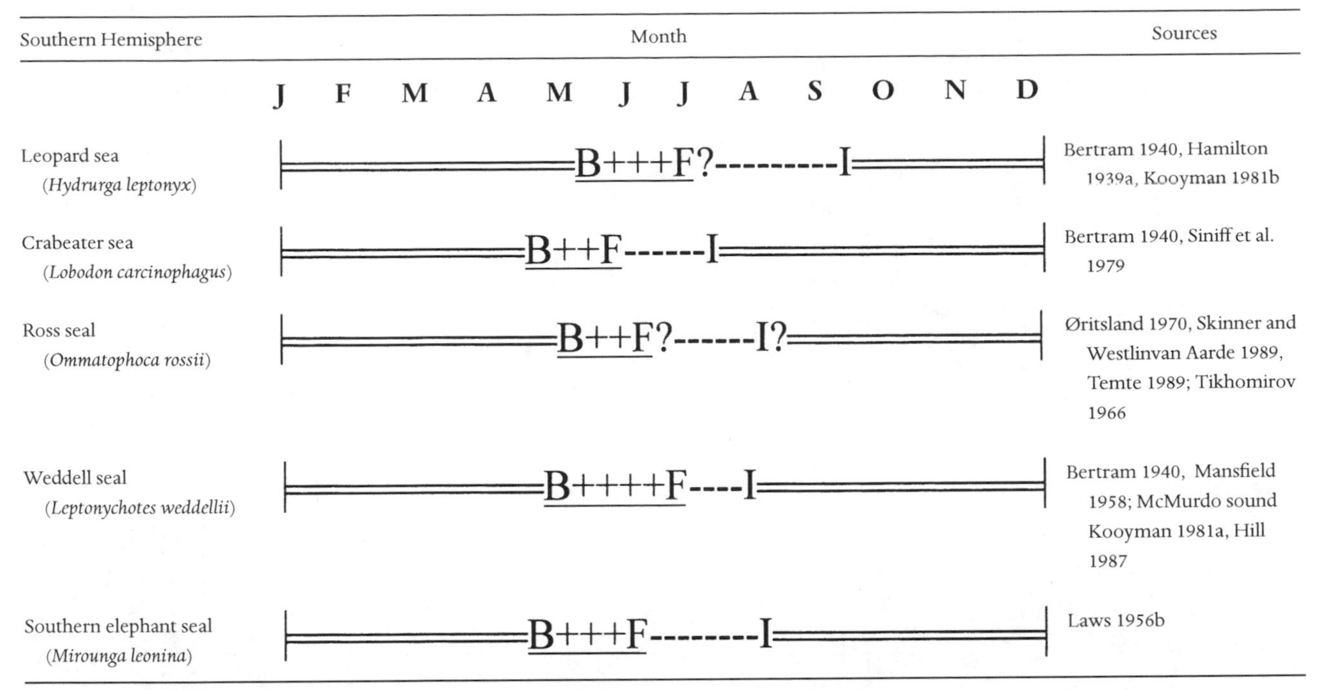

Southern Hemisphere	Month (J F M A M J J A S O N D)	Sources
Leopard sea (*Hydrurga leptonyx*)	═══B+++F?---------I═══	Bertram 1940, Hamilton 1939a, Kooyman 1981b
Crabeater sea (*Lobodon carcinophagus*)	═══B++F------I═══	Bertram 1940, Siniff et al. 1979
Ross seal (*Ommatophoca rossii*)	═══B++F?------I?═══	Øritsland 1970, Skinner and Westlinvan Aarde 1989, Temte 1989; Tikhomirov 1966
Weddell seal (*Leptonychotes weddellii*)	═══B++++F----I═══	Bertram 1940, Mansfield 1958; McMurdo sound Kooyman 1981a, Hill 1987
Southern elephant seal (*Mirounga leonina*)	═══B+++F--------I═══	Laws 1956b

Delayed implantation is shown as a dashed line, lactation as an underline, and active gestation as a double line. The broken line represents the period of embryonic diapause and the boldface solid line the period of active gestation. The crossed line indicates the period between birth and fertilization. The period of lactation is indicated by underlining. Note that these are highly generalized representations that are intended to depict the reproductive cycle of an average female. Each symbol represents approximately 1 week.

was lowest when the population density was highest, suggesting a density-dependent regulation. Similar effects have been observed for population samples of northern fur seals (*Callorhinus ursinus*) in which the interannual range in pregnancy rate was 0.68 to 0.99. In a rapidly increasing gray seal population, pregnancy rate was 0.89 (Boyd 1985) and for Arctic ringed seals (*Phoca hispida*) it was 0.48 to 0.91 (Smith 1987). Therefore, pregnancy rates between 0.5 and 0.9 appear to be normal for pinnipeds, but there can be substantial interannual variation, some of which may be a response to changes in population density and food availability.

Sea Otters

Unlike pinnipeds and polar bears, sea otters do not need to return to ice or land to give birth to young. In contrast to the other species of aquatic carnivores, sea otter reproduction is largely aseasonal, although there is some evidence for peaks of births at different times in different locations (Rotterman and Simon-Jackson 1988). Individual male sea otters appear to have no seasonal cycle of reproduction (Brosseau et al. 1975).

In common with other aquatic carnivores, sea otters have a period of delayed implantation (Sinha et al. 1966), although the duration is uncertain and perhaps highly variable. The estimated total duration of gestation measured from direct observations of individuals and from examination of reproductive tracts is 6 to 8 months (Rotterman and Simon-Jackson 1988, Jameson and Johnson 1993) and some of this variation may be attributable to flexibility in the duration of delayed implantation. Kenyon (1969) estimated that the period of active fetal growth lasts 4.5 to 5.5 months.

Sea otters can reproduce annually (Loughlin et al. 1981, Garshelis et al. 1984, Siniff and Ralls 1991, Jameson and Johnson 1993), but it is probably more normal for the birth interval to be more than 1 year because pups are normally nursed for 4 to 8 months and there is little evidence for concurrent pregnancy and lactation. Estimates for the birth rate among adult females vary from 0.5 to 0.9 (Rotterman and Simon-Jackson 1988, Siniff and Ralls 1991). Twins occur rarely. Overall, sea otter reproductive biology is not well documented but the information suggests that their reproductive pattern varies according to local environmental conditions, in terms of the timing of the peak of births, possibly the duration of delayed implantation, and the duration of lactation.

Polar Bears

Comparatively little is known about the reproductive pattern of polar bears. Like most aspects of the biology of the polar bear, the annual cycle of reproduction is dominated by the annual cycle of the sea ice (Ramsay and Stirling 1988). Because polar bears feed mainly on seals within the Arctic sea ice region (Stirling and McEwan 1975, Stirling and Archibald 1977, Smith 1980, Stirling and Øritsland 1995),

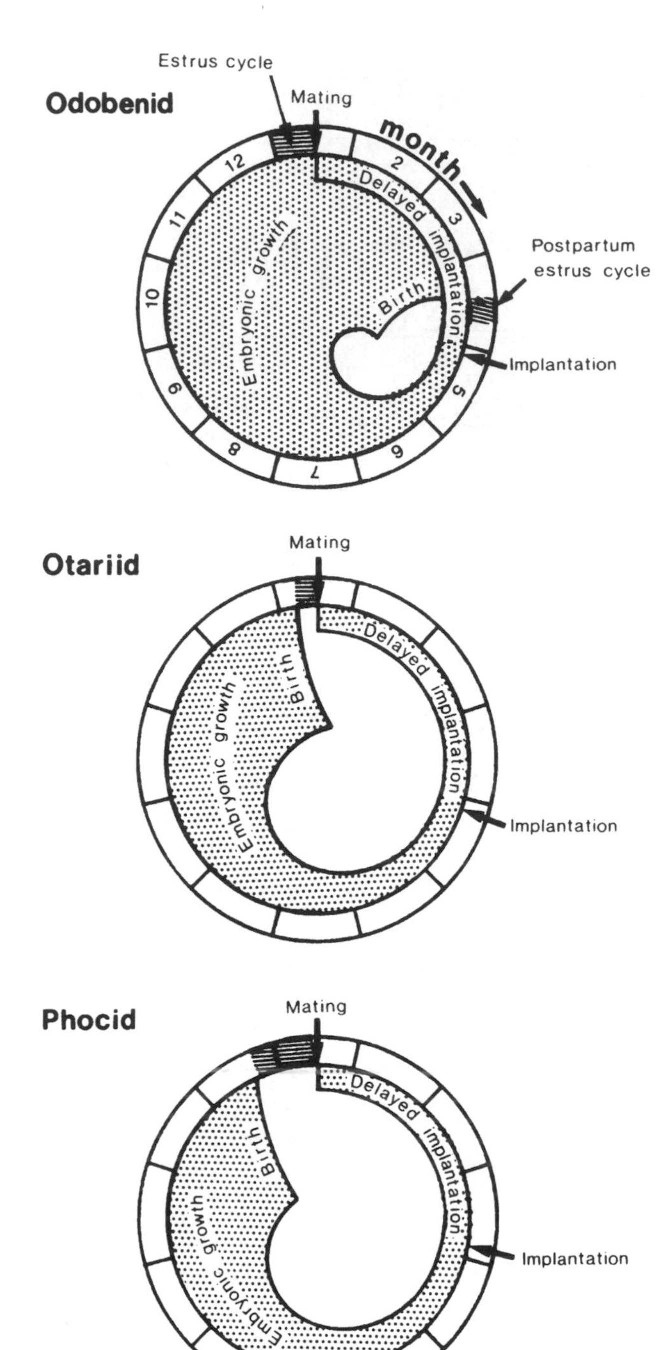

Figure 6-1. Three diagrams showing the reproductive cycles of the female walrus (odobenid) and female otariid and phocid pinnipeds. The shaded areas show the period of pregnancy.

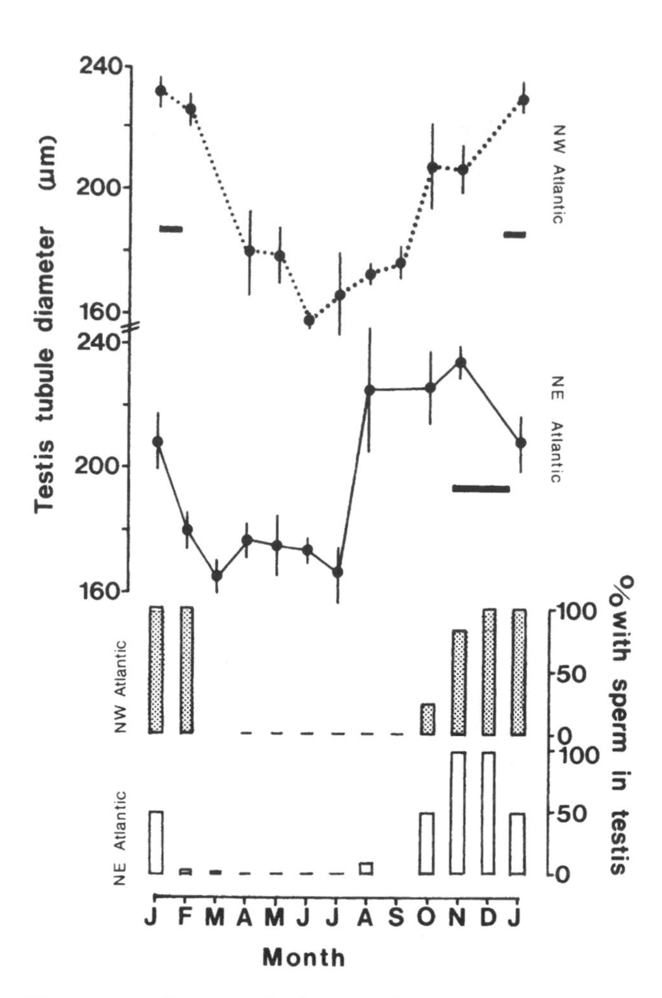

Figure 6-2. Changes in the diameter of testis tubules and the proportion of tubules containing sperm among adult gray seals from the northeast and northwest Atlantic populations. The breeding seasons in the two areas are shown by the black bars.

they are forced to fast ashore during the ice-free periods between late July and early November (Fig. 6-3) or they move north and summer on the drifting pack ice (Amstrup and DeMaster 1988, Ramsay and Stirling 1988). The breeding season occurs between March and May when females with yearling or, more often, 2-year-old cubs come into estrus. Like pinnipeds (and several other species of bears) they have a period of delayed implantation (Wimsatt 1963, Lønø 1972) that appears to last 4 to 5 months and ends simultaneously with females entering their wintering dens in late September and early October. Gestation is only 3 to 4 months, which is shorter than would be expected for an animal the size of a polar bear.

In contrast to the young of pinnipeds and sea otters at birth, polar bear cubs are poorly developed (altricial), which is probably a reflection of the lower premium on being able to accompany the mother during the early postnatal period or on the offspring becoming independent rapidly after birth. Ramsay and Dunbrack (1986) proposed that physiological constraints associated with the energetic costs of concurrent gestation and lactation while fasting have also led to the production of altricial young. The mean litter size and fecundity rate vary among locations; in the Hudson Bay area of Canada mean litter size was 1.9 (measured in spring), whereas at an equivalent stage in Alaska litters averaged 1.7 cubs (Amstrup and DeMaster 1988, Ramsay and Stirling 1988). The proba-

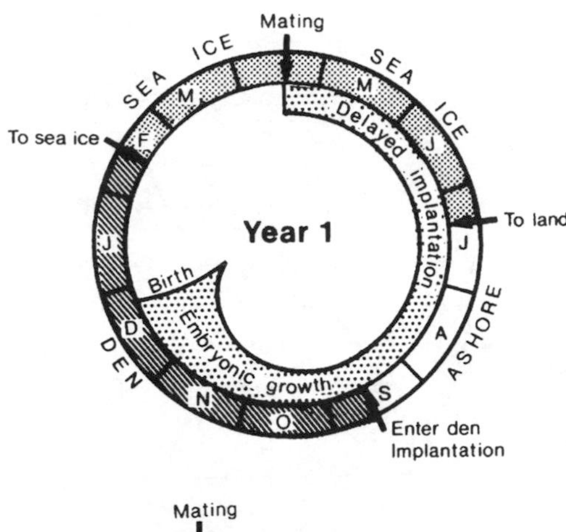

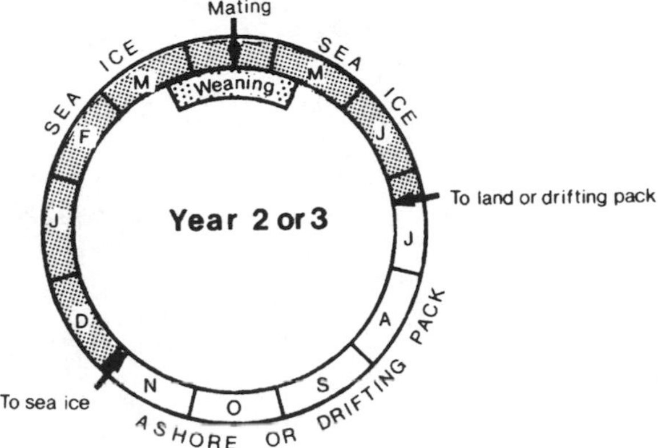

Figure 6-3. The annual cycle of reproduction in female polar bears, including years in which pregnancy occurs (year 1) and the following years (years 2 or 3) when females have cubs in attendance. The locations of individuals at various times of year are also shown. The cycle as shown is for the region of western Hudson Bay and the exact timing of events may vary with region. In more northerly regions, denning may not end until the middle of March.

bility that a sexually mature female may give birth in any year could be as low as 0.13 (Amstrup and DeMaster 1988), but is more likely to lie between 0.3 and 0.6 (Amstrup and DeMaster 1988). Therefore, the majority of sexually mature females do not breed in successive years probably because of the presence of suckling young that appear to inhibit the recurrence of estrus. The interval between births varies between averages of 3.1 and 3.6 years (Amstrup and DeMaster 1988, Derocher and Stirling 1995) with intervals as low as 2 years being rare (Ramsay and Stirling 1988, Derocher and Stirling 1995). Judging by the sensitivity of other Holarctic bear species to reduced food resources, large-scale reproductive failure may occur in polar bears due to environmental fluctuations (Ramsay and Stirling 1988). Polar bears potentially range over very large distances in search of food and they may be less sensitive to spatial fluctuations in food than many other species. Long-term declines in the reproductive rate of polar bears in Western Hudson Bay are related to declines in body mass and therefore, may be caused by changes in food availability (Derocher and Stirling 1995). Maternal condition may be an important factor contributing to lifetime reproductive success. Both maternal mass and litter mass tend to increase up to 14 to 16 years of age, which may be associated with gradually improving hunting skills. After age 16 years maternal mass and litter mass decline (Derocher and Stirling 1994). Body mass (particularly the predenning fat stores) seems to be particularly important to the reproductive success of females (Atkinson and Ramsay 1995).

The polar bear reproductive cycle appears to be well adapted to making maximum use of the spatial and temporal fluctuations that are likely to occur in prey distribution and abundance in relation to the unstable and ephemeral sea ice substrate where most hunting occurs. Delayed implantation would appear to ensure that the sequence of events leading to parturition can begin at a time when the priority for pregnant females is to establish a winter den and not to find a mate. The short period of gestation results in altricial young and, if the numbers of young produced are greater than can be sustained by the maternal resources, then reduction in the litter size can occur postpartum without significant cost to the mother and without the total loss of her reproductive attempt for that year. Cubs stay with mothers for up to 2 years and their presence appears to reduce the probability of the mother being mated. Under good feeding conditions where cubs have high growth rates and may rely less on milk for food, it may be possible for mothers to conceive at intervals of about 2 years. The reproductive system may be geared to respond to changing conditions to maximize reproductive rate using a variety of different time schedules.

Polar bears are particularly difficult animals to study but the use of mark–recapture methods and satellite telemetry (e.g., Messier et al. 1992, Atkinson and Ramsay 1995) is providing new insights into reproductive patterns as well as ecology.

Morphology, Functional Anatomy, and Physiology of the Pinniped Reproductive Tract

Anatomy of the Male Reproductive Tract

The anatomy of the reproductive tract of male pinnipeds is consistent with the general carnivore pattern (Fig. 6-4). The paired testes are scrotal in the otariids and para-abdominal (inguinal and deep to the blubber, but not contained with the abdomen) in the phocids and the walrus (Rommel et al. 1995). The position of the testes in the phocids may be partly associated with streamlining of the body. Among mammals, spermatogenesis normally requires reduced temperatures and the anatomy of blood flow to and from the testes sug-

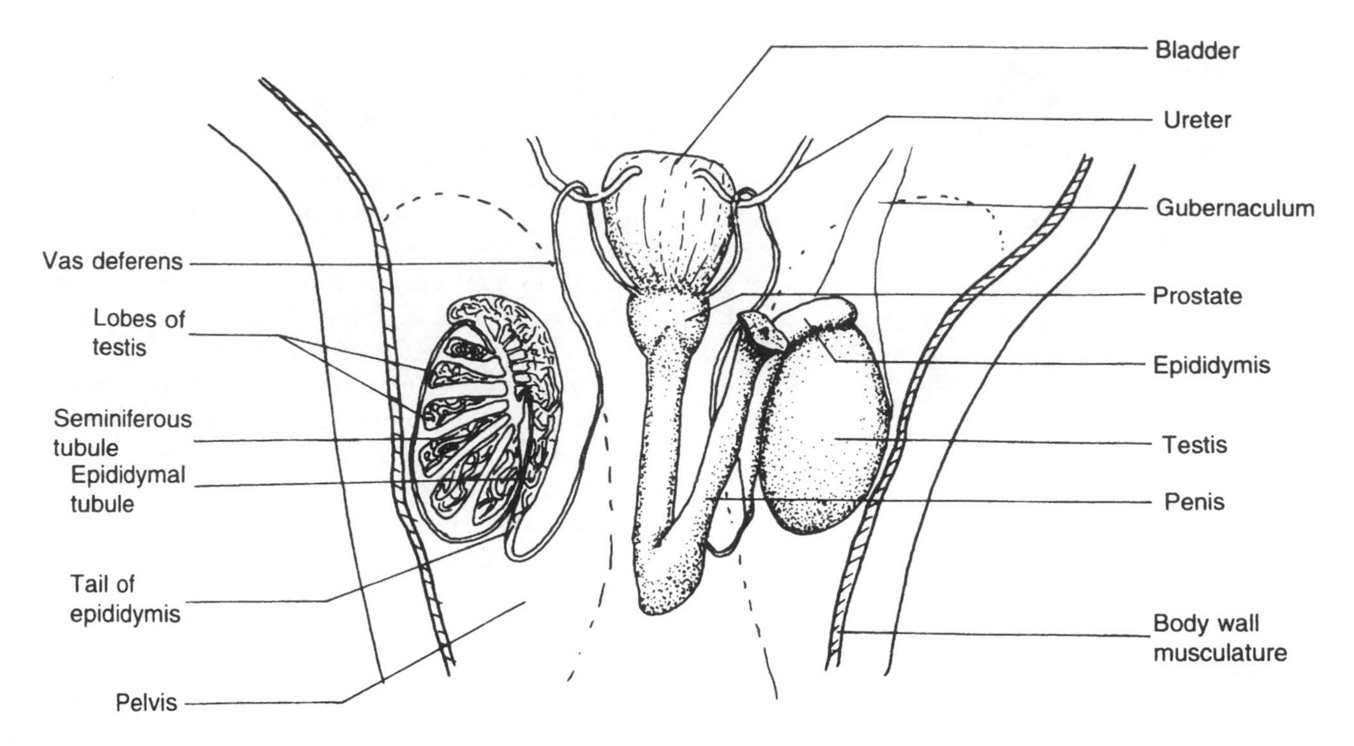

Figure 6-4. A diagram of the reproductive tract of an adult male otaxiid. The left side of the diagram shows a section through the testis and epididymis.

gests that heat exchange maintains the testes of phocids at several degrees below normal body temperature (Rommel et al. 1995). The only accessory gland of reproduction is the prostate, the size of which increases during the breeding season and which, if the pattern is similar to that of other mammals, responds to increased circulating concentrations of testosterone. Evidence from southern elephant seals appears to support this view (Griffiths 1984b).

Spermatogenesis is entirely consistent with the mammalian pattern (Laws 1956b). Spermatocytes within the seminiferous tubules develop into primary and secondary spermatids over a period of several weeks and are ultimately released into the lumen of the seminiferous tubules of the testis. Leydig cells located between the closely packed seminiferous tubules are probably the main source of testosterone. The sperm pass into the epididymis where final sperm maturation takes place and where spermatozoa are stored. Griffiths (1984a,b) showed profound annual cyclicity in both spermatogenesis and the epithelium of the epididymal tubules of male southern elephant seal. This is also illustrated in Figure 6-2 for gray seals in terms of the diameter of the seminiferous tubules in the testis; during periods of reproductive inactivity, when sperm are not being produced, there is involution of the seminiferous tubules, causing a reduction in both the dimensions and mass of the testes during the nonbreeding season. A similar process also causes reduction in the linear dimensions and mass of the epididymides.

The pinnipeds, sea otter, and polar bear all have an os penis or baculum. The size of bacula in these species has been shown to be consistent with other carnivore species in relation to body mass. The presence of a baculum and its size were explained mainly by sexual behavior at copulation; those species having a large baculum typically had copulation with a single prolonged intromission (Dixson 1995). The enlarged baculum may serve to strengthen the penis and protect the urethra during prolonged intromission.

Before ejaculation, sperm are transported along the vas deferens, which emerge from the tail of the epididymis and loop over the ureters before entering the prostate gland where they merge into the urethra at the base of the bladder (Fig. 6-4). The prostate gland secretes the vehicle in which the sperm are transported at ejaculation. This secretion also maintains the sperm at the correct pH and contains glucose, which can be used by the sperm as a metabolic fuel.

Anatomy of the Female Reproductive Tract

The female pinniped reproductive tract shows no unusual adaptations to aquatic life. It is generally consistent with the carnivore pattern, and the reproductive tracts of both sea otters (Sinha and Conway 1968) and polar bears are similar to those of pinnipeds. The uterus is bicornuate with both horns joining just above the cervix (Fig. 6-5). In immature individuals the uterus has a thin outer wall (myometrium) with little of the connective tissue thickening found in sexually mature individuals. In both mature and immature individuals, the inner wall (endometrium) is ridged longitudinally along the full length of each uterine horn. At the distal apex of each horn, the uterus tapers toward the uterotubal junction leading into the oviduct, which terminates in the fimbria, a funnel-like aperture to the oviduct that guides the egg to the oviduct after ovulation (Harrison et al. 1952). Uterine glands, which secrete mucus into the uterine lumen and appear to develop under the influence of estrogens, are particularly abundant in the endometrium around the time of mating (Harrison et al. 1952, Laws 1956b, Bigg and Fisher 1974, Ouelette and Ronald 1985). There is also an increase in the abundance of vaginal cornified epithelial cells at this time (Bigg and Fisher 1974) and relaxation of smooth muscle contraction of the cervix.

The bilateral ovaries of pinnipeds are of equal size and are attached to the uterus by the ovarian ligament (Fig. 6-5). Each ovary and its adjacent uterine horn share a common blood supply so that they are in close humoral contact. The surface of the ovaries is smooth but small subsurface crypts are present that are in direct contact with the surface. The significance of these crypts is unknown (Harrison et al. 1952). As in other mammals, oocytes, which form in the ovaries during early fetal development, lie dormant within distal regions of the ovarian cortex. After puberty, the oocytes continue their development at times that are partly related to the annual cycle of reproduction (see later sections) but the mechanisms that determine when each specific oocyte continues development are not understood for mammals in general. The oocytes continue their development surrounded by a narrow band of granulosa cells that are, in turn, surrounded by theca cells. This ball of cells with the oocyte at its center is known as a primordial follicle and these are present within pinniped ovaries in large numbers. The great majority of primordial follicles that begin to grow never reach the next stage of maturity, that of a secondary follicle. This mortality of follicles is often viewed as a selective process allowing only follicles containing a healthy oocyte to develop to maturity and to become a mature egg ready for fertilization. Secondary follicles develop when a fluid-filled vacuole develops and begins to expand within the primordial follicle. As this happens the oocyte retains a cluster of granulosa cells around it, the cumulus oophorus, but this cluster with the oocyte at its center is positioned around the edge of the expanding vacuole. Even at the secondary follicle stage many follicles fail to develop fully, become atretic, and eventually disappear completely. Large secondary follicles, some of which may have vacuoles more than 1 cm in diameter, which become atretic often give rise to a corpus atreticum that is recognized as a small, light-colored and elongated corpus of connective tissue in the ovarian cortex of the ovary. In the case of pinnipeds, only one follicle eventually matures to the stage that is ready to ovulate. As the largest secondary follicles expand, the external epithelium of

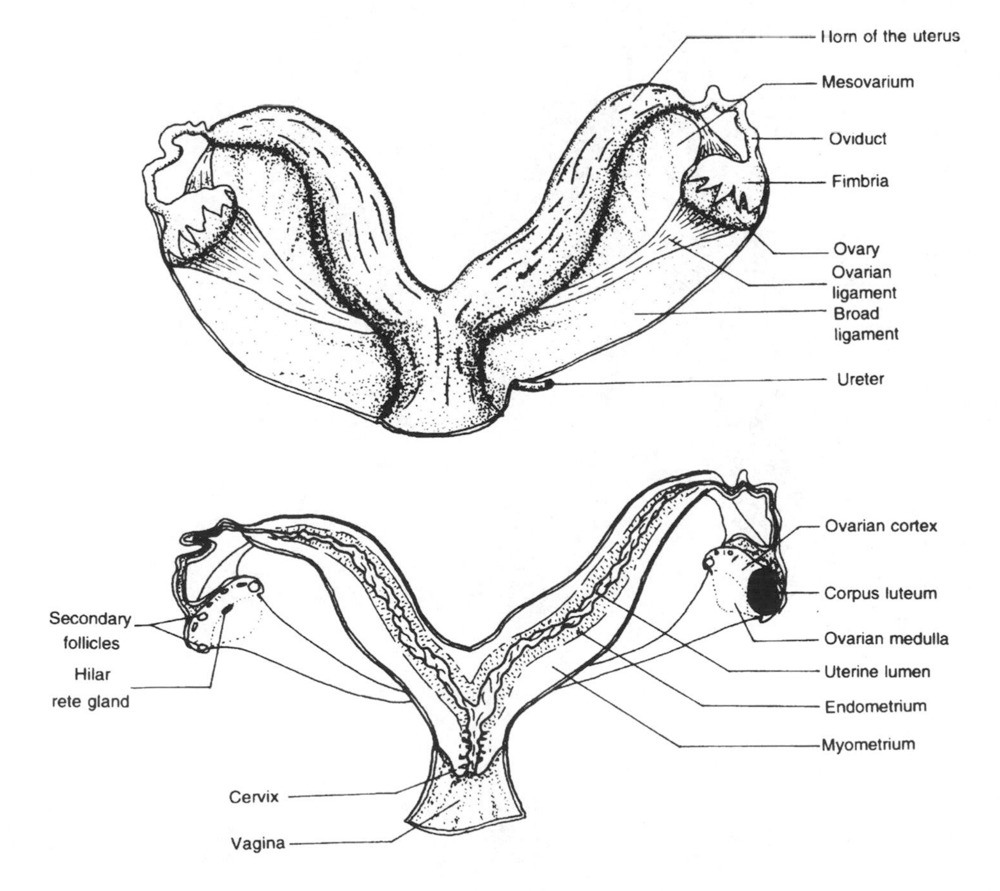

Figure 6-5. The reproductive tract of an adult, parous female pinniped. (Top) Diagram showing the intact reproductive tract. (Bottom) Diagram showing a section through the tract exposing the lumen of the uterus and longitudinal sections of the ovaries.

the ovary, the tunica albuginea, gradually thins and weakens. At the same time the cells surrounding the mature follicle, the theca interna, begin to secrete estrogen in large quantities and these reach the systemic circulation, causing induction of estrus behavior. Ovulation is marked by the rupture of the follicle, and the slightly positive pressure within the follicular fluid carries the oocyte, within the cumulus oophorus, out of the ovary and into the oviduct where fertilization normally takes place. Normally a small stigma forms on the surface of the ovary at the point where ovulation occurred.

After ovulation, the granulosa cells that lined the inside of the mature follicle begin to luteinize, which means that they "yellow," a reflection of the accumulation of lipid. These lipids are the precursors required for the synthesis of steroid hormones. The luteal cells expand and multiply to fill the empty and collapsed vacuole of the follicle. In pinnipeds this process is complete within 24 hr of ovulation and the new luteinized follicle has formed into a corpus luteum (Figs. 6-5 and 6-6). The corpus luteum secretes progesterone and, as I shall describe in subsequent sections, it probably plays a vital role in the control of pregnancy in pinnipeds.

An interesting but, as yet unexplained feature of ovarian anatomy in pinnipeds is the presence of hilar rete glands (Boyd 1984c; Fig. 6-5). These have been observed in both gray and ringed seal ovaries and are likely to also occur in other species. The function of the hilar rete gland is not understood, although it may be vestigial adrenocortical tissue. In the gray seal hilar rete glands are most abundant through delayed implantation.

Estrus

The various functional relationships between the uterus, the ovary, and the conceptus, which are integral to the highly regulated process leading to ovulation, are not well understood. Moreover, there is little information about how this process operates in nonpregnant individuals. From observations of Antarctic fur seals we know that nonpregnant females are mated at about the same time of year as pregnant females (Boyd et al. 1995). The general timing of mating and the release of follicle growth from inhibition are unlikely to depend entirely on release from inhibition of follicle growth due to parturition and, in particular, the loss of the placenta (Fig. 6-6). It is possible that

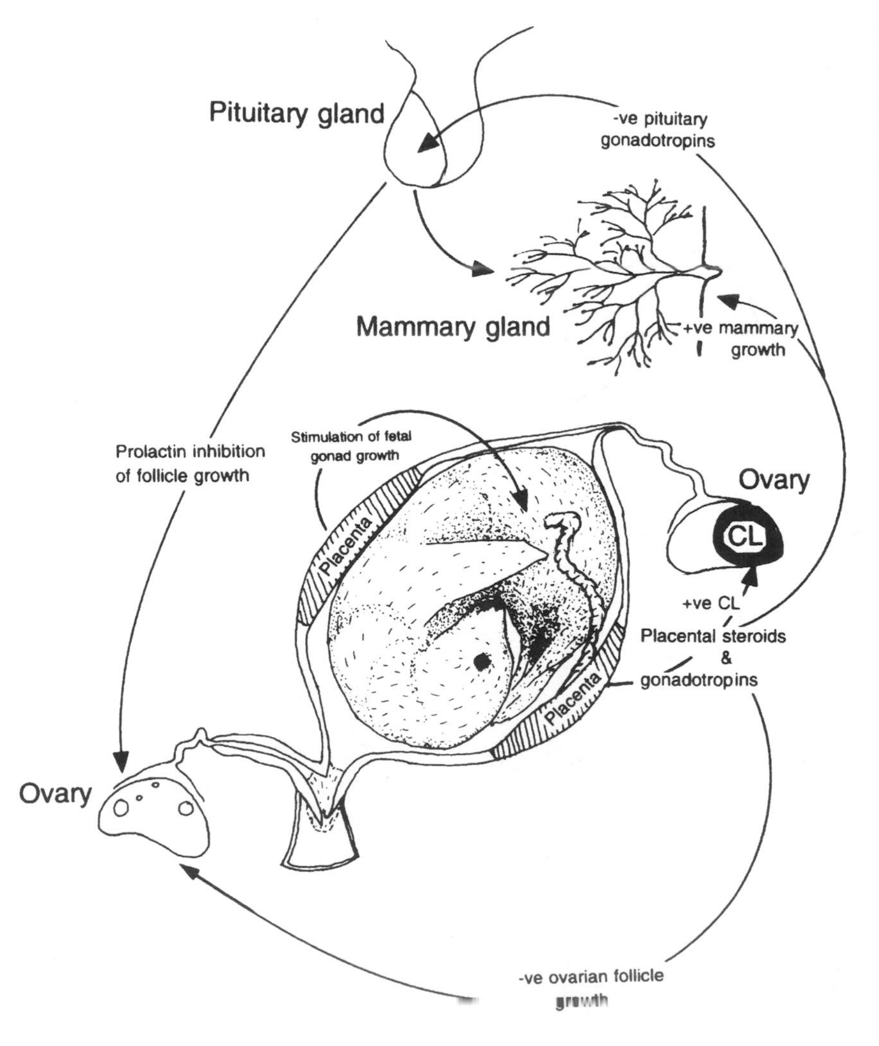

Figure 6-6. A diagrammatic view of the probable role of the placenta in regulating the reproductive system during active gestation. The corpus luteum (CL) is shown in the right-hand ovary.

pheromonal or behavioral cues could play a part in the induction of estrus, although there is no direct evidence for this from pinnipeds. As I shall discuss later in the chapter, the processes leading to ovulation in pregnant and nonpregnant individuals may have different physiological and evolutionary backgrounds.

As a general rule pinnipeds ovulate from alternate ovaries at each successive postpartum estrus. Inhibition of follicular growth in the ovary ipsilateral (same side) to the concurrent pregnancy may be achieved through local endocrine control from the adjacent pregnant horn of the uterus. A wave of follicle growth begins at parturition in the contralateral (opposite) ovary that leads to maturation of a single follicle (Laws 1956b, Craig 1964, Boyd 1983) (Fig. 6-7); this mature follicle is likely to have a volume equivalent to at least 50% of the contralateral inactive ovary. Increasing estrogens in the circulation during this postpartum phase of follicle growth (Boyd 1983, 1991a; Reijnders 1990) probably derives from the theca interna of the mature follicle (Harrison 1960) and is likely to be responsible for induction of behavioral estrus.

The duration of this follicular growth phase appears to be remarkably consistent among otariids but, among phocids, its duration varies with the duration of lactation. The advantages of having parturition and mating in close juxtaposition derive largely from the fact that pinnipeds breed on land or ice (Bartholomew 1970, Stirling 1975, Boness 1991, Le Boeuf 1991). In otariids, in which lactation lasts many months, there would be little to be gained from delaying mating until the end of lactation. Having mating concurrent with the season of births enables females to synchronize mating more easily than at the end of lactation and to extend lactation into the next reproductive cycle. However, this begs the question as to why phocids have adopted the alternative strategy because there would presumably be no additional cost to them if females were mated within several days of parturition as opposed to waiting the extra time until the end of lactation (Fig. 6-8). In some cases, there may even be a gain in mating early in lactation because females could then have the option to suckle their pups away from interference and aggression from males fighting over their position on the breeding grounds (Boness et al. 1995). A potential explanation—that phocids have modified the duration of the postpartum estrous cycle—is addressed in the section examining the evolution of pinniped reproductive patterns.

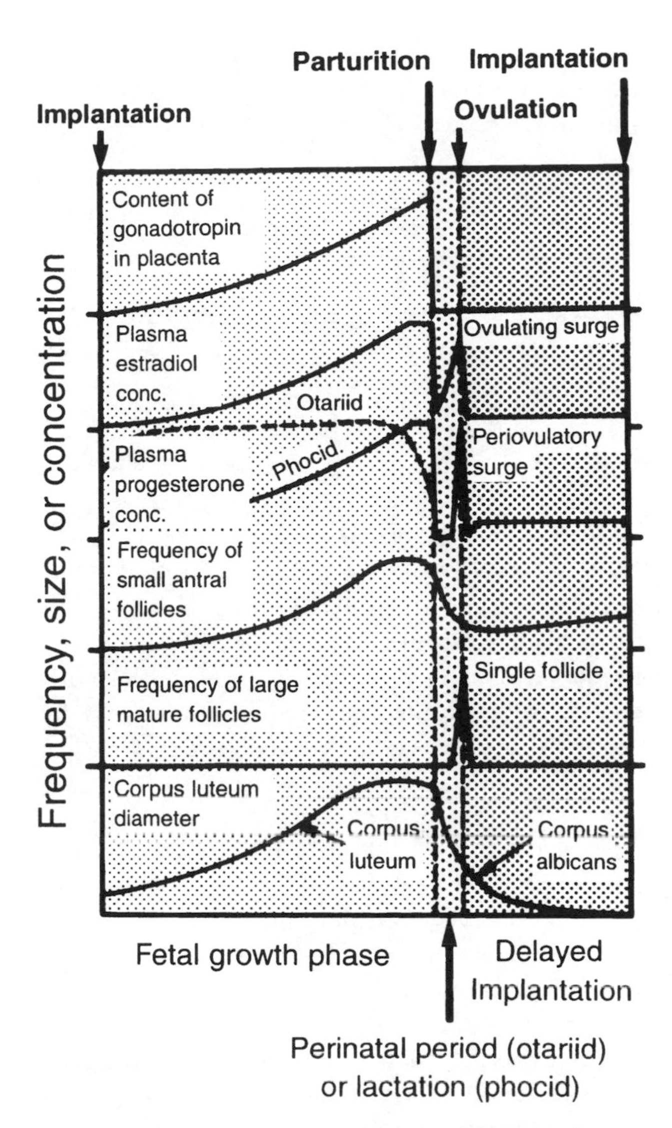

Figure 6-7. Diagrammatic representation of changes in the plasma concentrations of some reproductive hormones, diameter of the corpus luteum, and the frequency of ovarian follicles in the ovary not containing the corpus luteum of pregnancy during a reproductive cycle. For the purpose of the diagram, the perinatal period (otariids) and lactation (phocids) are considered to be equivalent. (The diagram is compiled from information from Laws 1956b; Craig 1964; Bigg and Fisher 1974, 1975; Daniel 1974, 1975; Raeside and Ronald 1981; Boyd 1982, 1983, 1984b, 1991a; Hobson and Boyd 1984; and Reijnders 1986, 1990.)

Delayed Implantation and Pseudopregnancy

Delayed implantation lasts 120 to 160 days in pinnipeds (Boshier 1981, Daniel 1981; Table 6-1), and this trait was probably inherited from carnivore ancestors, as it is also present in most ursids and mustelids. Delayed implantation is also referred to as embryonic diapause (Renfree and Calaby 1981), but because slow growth of the blastocyst may continue during delay in pinnipeds (Hewer and Backhouse 1968, Daniel 1971), it is more accurately described as delayed implantation.

Several researchers have speculated about the adaptive function of delayed implantation (e.g., Harrison 1964, Enders 1981); it has probably developed independently on several occasions in the course of mammalian evolution. One possible advantage is that it confers flexibility on the time

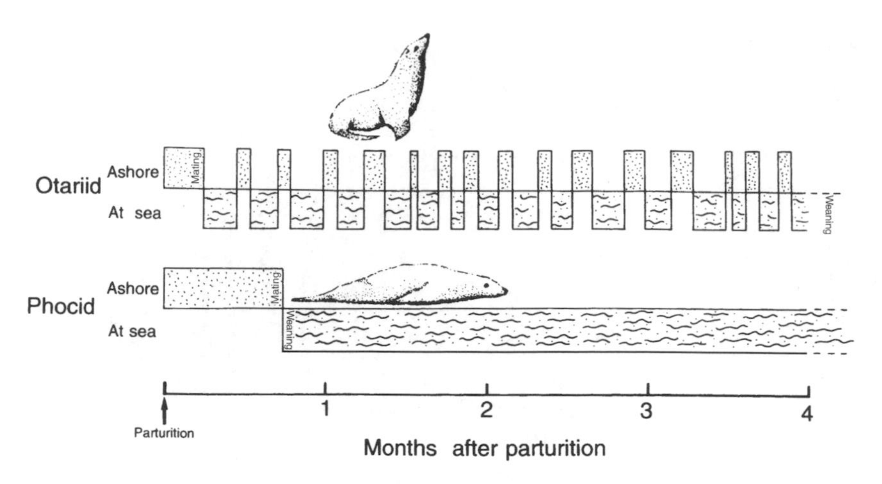

Figure 6-8. Contrasting lactation and reproductive strategies in otariid and phocid pinnipeds. Note the times of weaning relative to parturition. In this case, periods spent ashore represent periods when a mother attends and feeds her pup. Note also that this diagram is highly generalized; the exact durations of the different phases will vary among species and individuals.

schedule (phenology) of reproduction. This allows females to dissociate the time of mating from parturition. Without delayed implantation, gestation is normally of a fixed duration in mammals (Racey 1981); in species that do not have delayed implantation the time of mating largely determines the time of parturition. Although delayed implantation is facultative in some mammals (Renfree and Calaby 1981), in pinnipeds, as in ursids and mustelids, it appears to be obligate. However, without some flexibility in the duration of delayed implantation, female pinnipeds of most species would have little scope for altering the timing of their reproductive cycles. As will be described below, the time at which delayed implantation ends appears to depend on environmental factors, thus allowing individuals to adjust the timing of their breeding cycles to environmental change.

The physiology of delayed implantation is still largely unexplained. Daniel (1974, 1975) identified changes in proteins secreted by the uterus of northern fur seals that were associated with the preimplantation blastocyst, although the preimplantation uterus of harp seals shows no anatomical differences from the nongravid uterus (Ouellette and Ronald 1985). In phocids, there is little indication from observations of either plasma progesterone concentration or anatomy of the corpus luteum that implantation is associated with reactivation of the corpus luteum (Boyd 1982, 1984b; Reijnders 1990) as it is in both the mustelids (Martinet et al. 1981, Mead 1981) and the two otariids that have been studied (northern fur seal, Daniel 1981; Antarctic fur seal, Boyd 1991a). The contrasting pattern of progesterone secretion in phocids and otariids is illustrated in Figure 6-7.

There are few data from pinnipeds indicating which endocrine factors cause termination or maintenance of delayed implantation. Daniel (1974) found a peak of estrogen associated with termination of delay in northern fur seals, but this has not been demonstrated in other studies. Moreover, Daniel (1981) was unable to induce artificial implantation by treating females with ovarian steroids, which suggests that changes in the levels of these steroids are not the cause of implantation. Although the corpus luteum maintains a low level of activity throughout delayed implantation in pinnipeds (Figs. 6-7 and 6-8), we cannot be certain what function this serves. Renouf (pers. comm.) has shown that the corpora lutea of harp seals remain active beyond the normal period of delayed implantation even in the absence of pregnancy. Noonan and Ronald (1989) observed similar patterns in hooded seals (*Cystophora cristata*). The conclusion from these observations is that elevated progesterone throughout delayed implantation is independent of the presence of an embryo and only after implantation has occurred does the female "know," in a physiological sense, that she is pregnant. Before implantation she is in a state of pseudopregnancy and this implies that, in terms of their physiological control, delayed implantation and pseudopregnancy may be identical. In the event that a female is not pregnant, pseudopregnancy appears to extend beyond the normal time of implantation but eventually terminates as the corpus luteum collapses in the absence of hormones from the placenta to maintain its activity (Fig. 6-7). This lifetime of the corpus luteum probably represents the maximum window of time over which implantation can occur. These points are illustrated diagrammatically in Figure 6-9.

Gestation

Although the presence of a blastocyst probably has no role in corpus luteum maintenance, as in mink (Martinet et al. 1981), the corpus luteum may be required for blastocyst survival. The role of progesterone is equivocal during the fetal growth phase of gestation. From counts of secretory cells, as defined by Boshier (1977), progesterone secretion from the corpus luteum of gray seals is thought to decline in the last 2 months of gestation (Boyd 1982), although corpus luteum size increases along with plasma progesterone levels at this time. Evidence that progesterone is produced by the placenta throughout gestation (Harrison and Young 1966, Hob-

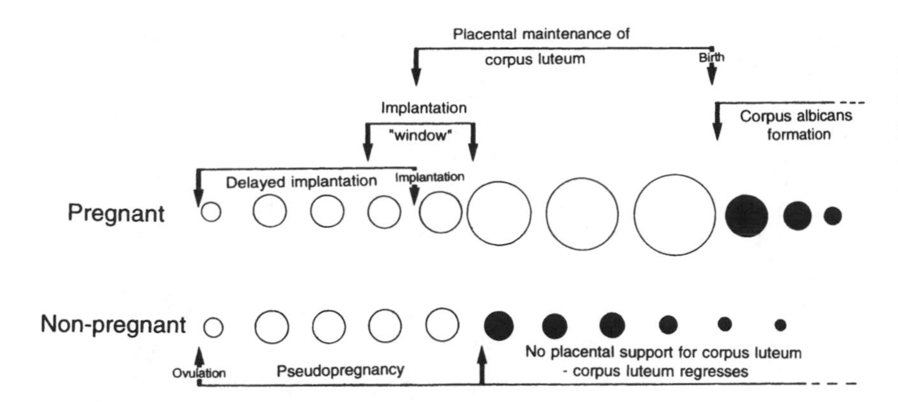

Figure 6-9. The development and regression of the corpus luteum in adult female pinnipeds. Each circle is representative of the size of the corpus luteum at various stages of the reproductive cycle from ovulation through implantation to birth. Open circles represent active corpora lutea, whereas closed circles represent regressing corpora lutea with declining or no steroidogenic activity. Such regressing corpora lutea are also known as corpora albicantia.

son and Boyd 1984, Boyd 1990a) suggests that the placenta assumes an early role in the endocrine maintenance of pregnancy (Figs. 6-6 to 6-8). After parturition (and subsequent loss of the placenta), the corpus luteum collapses rapidly to become a knot of light-colored connective tissue in the ovary known as a corpus albicans (Laws 1956b; Boyd 1983, 1984b; Bengtson and Laws 1985; Smith 1987) and this process is illustrated in Figure 6-9. In some species, such as the gray seal, the corpus albicans remains as a "scar" in the ovary for about 1 year before disappearing (Boyd 1984b), whereas in other species, such as the crabeater seal (*Lobodon carcinophagus*) (Bengtson and Laws 1985), the corpus albicans takes more than 3 years to regress completely. Bengtson and Laws (1985) used this delayed regression to estimate the reproductive histories of individuals.

The pinniped placenta is similar to that of other carnivores, including the sea otter (Harrison and Young 1966, Sinha and Mossman 1966). The shape of the placenta is described as zonary. It forms a broad band around the lumen of one uterine horn (see Fig. 6-6). The placentas of both gray seals and California sea lions (*Zalophus californianus*) also produce a chorionic gonadotropin (CG) that is immunoreactive with anti-human CG and has a similar submolecular structure to human CG (Hobson and Boyd 1984, Hobson and Wide 1986). Therefore, most evidence points to a major role for the fetoplacental unit in the maintenance and control of gestation, and this is underlined by the hypertrophy of the fetal gonads of pinnipeds during the final 2 months of gestation (Bonner 1955, Amoroso et al. 1965). As in some other mammals, this hypertrophy means that the ovaries of newborn females can be larger than those of most adult females (Fig. 6-10). In both horses and humans, the fetal gonads provide important precursors for estrogen synthesis by the placenta. This is supported by the observation that circulating estrogen levels increase late in gestation in parallel with the fetal gonadal hypertrophy (Raeside and Ronald 1981, Reijnders 1990; Fig. 6-7).

In the nonpregnant seal, uterine development and regression is most probably under the control of ovarian steroid

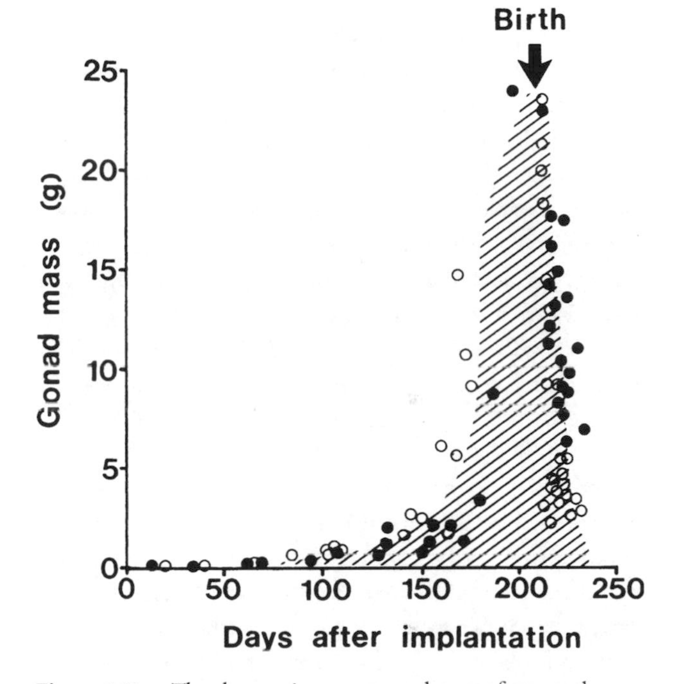

Figure 6-10. The changes in mean gonad mass of gray seal fetuses and neonates. Females are shown with open circles and males with closed circles.

hormones. The occurrence of a postpartum estrus is consistent for all pinnipeds; however, different patterns have emerged for phocids and otariids. Among otariids, estrus normally occurs 5 to 8 days postpartum and among phocids it normally occurs at the end of lactation, which can be 4 to 60 days postpartum depending on the species (Oftedal et al. 1987; see Fig. 6-1 and Table 6-1). Some ovarian follicular growth appears to occur in the nonpregnant ovary before birth in northern fur seals (Craig 1964) but no follicles develop beyond a certain critical size. At parturition the effects of fetal inhibition of ovulation, resulting most probably from placental hormones (Hobson and Boyd 1984), ceases. Ovarian follicular development extends through to the point where a single mature follicle is present in one ovary (Laws 1956b, Craig 1964, Boyd 1983). In postpartum individuals, ovarian follicle development is normally restricted to the ovary contralateral to the one that had the corpus luteum associated with the previous pregnancy. As mentioned above, pinnipeds ovulate from alternate ovaries at each successive postpartum estrus. Also as a general rule, pinnipeds appear to ovulate spontaneously (Harrison 1960) only once per year (Boyd 1991a), although recent evidence from a captive individual has cast doubt on this for Hawaiian monk seals; this individual was shown to be polyestrous (Atkinson et al. 1993).

Lactation

Although the phenology of the prepartum reproductive pattern of pinnipeds appears to have been highly conserved, the postpartum pattern has become adapted to a wide range of different environmental constraints. Lactation in pinnipeds has been comprehensively reviewed by Bonner (1984), Anderson and Fedak (1987a), and Oftedal et al. (1987). In general, lactation is a continuous process that is maintained by pituitary hormones, including prolactin and oxytocin. Mammary gland development normally begins in advance of parturition, and the high concentration of placental gonadotropins in female pinnipeds during the later stages of gestation (Hobson and Boyd 1984, Hobson and Wide 1986) may contribute to lactogenesis (mammary gland growth and development) (see Fig. 6-6). However, prolactin also contributes to lactogenesis; in Antarctic fur seals and southern elephant seals prolactin concentrations are greatest at or immediately before parturition and, in the fur seals, administration of a prolactin inhibitor causes involution of the mammary gland and cessation of lactation, at least during its early stages (Boyd 1990b, 1991a). As in other mammals, lactation is probably maintained over extended periods by the presence of the stimulus from the sucking offspring so that at weaning or when the pup / cub dies, lactation ends and mammary involution takes place rapidly.

Otariid lactation lasts from 4 months to more than 1 year, whereas lactation in phocid pinnipeds lasts only 4 to 60 days depending on species (Table 6-1). In terms of duration, lactation in sea otters and polar bears is more similar to the otariid than the phocid pinniped pattern (discussed above). Other than duration, the most profound contrast between otariid and phocid lactation is found in the behavioral pattern of the mothers. Whereas phocid mothers in general do not leave their pups for extended periods to feed, otariid mothers alternate between periods spent feeding at sea and periods ashore feeding their pups. These contrasting patterns are illustrated diagrammatically in Figure 6-8. Periods spent at sea by female otariids vary from overnight to more than 8 days depending on the species and individual and periods spent ashore during which pups are fed normally from a few hours to 2 to 3 days. However, there can be a large degree of variation in these foraging cycles between individuals within a species (Lunn et al. 1994). Among most of the phocids, periods of absence from the pups normally last only a few hours and, in many species, mothers will not leave the pup for the duration of lactation.

The behavioral pattern associated with lactation in otariids poses some interesting questions about the physiological control of lactation. Among the sea otters, polar bears, and phocids, lactation appears to conform to a pattern suggesting that it is a continuous process because mothers and young are together for almost the whole period of lactation. However, among the otariids lactation appears to be discontinuous because mothers spend up to several days feeding at sea between visits ashore to feed their pups (Oftedal et al. 1987; see Fig. 6-8). In a study of lactation in Antarctic fur seals, Arnould and Boyd (1998) examined the alternative hypotheses that (1) even when at sea, milk production continues at a slow rate and gradually fills the mammary gland, which is then emptied when the mother returns to land to feed her pup, or (2) milk production slows or stops when mothers are away from their pups and most milk is produced when mothers return to land and suckle their pups. The first of these would be expected based on the continuous nature of lactation in most mammals studied to date, but the results supported the second hypothesis. This indicates that at least one otariid has developed an unusual method of controlling lactation, which appears to be maintained for extended periods even without a sucking stimulus. Part of this process appears to involve the temporary cessation or slowing of the rate of milk production. Perhaps the otariid mammary gland is simply slow to regress as a result of the withdrawal of the sucking stimulus and responds rapidly once suckling is recommenced.

The approximate lactation durations in pinnipeds are described in Table 6-1, which contrasts durations of the otariid and phocid pinnipeds (for more details, see Oftedal et al. 1987). The hooded seal has one of the most extreme lactation patterns of all mammals and it illustrates the degree to which reduction in the duration of lactation has occurred in phocids. In this species, lactation lasts only 4 days during which the birth mass of the pup doubles (Bowen et al. 1987; Oftedal et al. 1989, 1993). For a mammal the size of a hooded seal, this is a remarkably short period of time (Oftedal 1984). This system of lactation may have evolved because of the instability of the pack ice habitat occupied by hooded seals during the breeding season; the faster a mother can deliver the milk energy to her pup the greater will be the chances of the pup surviving. Harp seals, which also breed in the unstable pack ice, have short (9 to 12 days) lactation durations (Stewart and Lavigne 1980). In these two species at least, there is a high priority given to weaning the pup early and before the pack ice becomes untenable for the pups.

The duration of lactation in phocid pinnipeds can be viewed as a compromise between the necessity to raise a pup to a viable mass at weaning and the requirement of the mother to spend as little time ashore as possible, either because she is vulnerable to predation or because she is restricted in her capacity to feed while attending her pup. In many species, mothers probably cannot both feed their pups, which are restricted to the land and ice, and themselves efficiently while lactating. At one extreme are species, such as the harp, hooded, gray, and elephant seals, that probably rarely feed during lactation. These species must begin lactation with sufficient body reserves to cover both the nutritional requirements of the growing pup and the metabolic overhead of the mother (Costa 1991; and see Costa and Williams, Chapter 5, this volume), that is, what it costs the mother to maintain herself during the period ashore. In contrast, species such as ringed and harbor seals (*Phoca vitulina*), in which the location of the pup allows mothers to feed during lactation, have relatively long lactation durations.

The milk produced by pinnipeds is normally highly concentrated with fat contents on the order of 40% of wet mass (Oftedal et al. 1987). This highly concentrated milk may be a consequence of spatial and temporal separation of foraging and suckling in most pinnipeds. Some otariids also appear to have the ability to adjust the fat content of their milk depending on the amount of energy they have available with which to produce milk; when females are able to deliver a lot of milk to the pup, both the fat content and the volume of the milk can increase (Arnould and Boyd 1995). It may be impractical to transport large volumes of milk or to produce large volumes of milk in isolation from feeding because of problems with the regulation of water balance. The milk composition of polar bears, although lower in total fat than for most pinnipeds, is also highly concentrated (Derocher et al. 1993) when compared with terrestrial carnivores.

Pinniped lactation has been used to test theories of differential investment in the sexes because, among some phocids, it is possible to measure all of the maternal resources that are given to the pup. In addition, several pinnipeds, such as elephant seals, provide examples of extreme polygyny and sexual dimorphism where we might expect mothers to provide more resources for male than female offspring because the fitness advantage accrued from producing a high-quality male offspring is potentially much greater than that from a female offspring. The underlying theory for this is complex and a more detailed description of the various arguments involved is presented in several reviews (e.g., Charnov 1982). Although Anderson and Fedak (1987b) found that female gray seals in Britain gave more resources to male offspring during lactation, this was not observed in gray seals from Canada (Bowen et al. 1992). In addition, no differences in male and female pup growth rates were found for southern or northern elephant seals, species in which differences would have been predicted (McCann et al. 1989, Kretzmann et al. 1993). In southern elephant seals, weaning mass (which is probably an important determinant of fitness) is most closely correlated with maternal mass at birth and pup sex is relatively unimportant (Arnbom et al. 1996, Fedak et al. 1996); the mass difference between the sexes of Antarctic fur seals at weaning is caused by different body composition, not different total body energy reserves (Arnould et al. 1996). Consequently, differences observed in weaning mass between the sexes do not necessarily reflect differences in parental investment. Therefore, despite being apparently good subjects for the study of differential maternal investment in pups of each sex, the evidence from pinnipeds in support of this theory is equivocal. See Trillmich (1996) for a comprehensive review of this subject in pinnipeds.

Variation in the sex ratio of mammals at birth has also been postulated as a mechanism by which mothers may alter the degree of investment they make in offspring of each sex. Once again the evidence for variation in the sex ratio at birth is equivocal (Clutton-Brock and Iason 1986). There is no evidence that the sex ratio of pinnipeds at birth varies significantly from 1:1, although Coulson and Hickling (1961) found that female gray seals tended to be born later than males, and this indicated the possibility of some form of systematic sex ratio variation. However, more recent equivalent studies of gray seals and other pinnipeds have not found similar changes in the sex ratio (e.g., McCann et al. 1989, Bowen et al. 1992).

Reproductive Seasonality

Endogenous Cycles and Metabolism

An important, and as yet largely ignored, aspect of the reproductive cycles of marine carnivores is the interaction with metabolism. Stewart and Lavigne (1984) presented a simple model to explain the choices available to female pinnipeds at each stage of the reproductive cycle based on energetic constraints. A prospective mother must match the future energetic demands of reproduction with the seasonal cycle of environmental productivity to ensure that sufficient resources are available at critical stages of reproduction. For example, it seems reasonable to assume that a mother with a diapause blastocyst in utero will not embark on further development of that conceptus if she is in poor condition and has no real prospect of raising her pup to weaning. The logic for this is that the energetic costs of producing a pup/cub to weaning are significant (Oftedal et al. 1987) and a mother will not risk her long-term survival and future reproductive investment for the sake of a single offspring. In Antarctic fur seals (Boyd et al. 1995) and northern elephant seals (Reiter and Le Boeuf 1991) there are measurable costs of reproduction in terms of maternal survival and reduced probability of reproducing in the year after a pregnancy. In contrast to polar bears it is also rare for pinnipeds to produce twins (Spotte 1982). Pinnipeds are precocial at birth, and even by the time of birth, mothers have invested significant resources in their offspring. Production of twins by pinnipeds would place greater energetic demands on mothers reducing the chances of twins surviving and reducing future potential reproductive success and the probability of mothers surviving. This contrasts with polar bears in which there is a relatively small prepartum investment in individual offspring and in which reduction in litter size postpartum may be a strategy to adjust the level of maternal investment in reproduction to the resources that are available. This suggests that the reproductive strategy adopted by polar bears (which differs little from a typical ursid pattern) can adapt more rapidly than that of pinnipeds to short-term fluctuations in environmental conditions.

To match the energetic requirements of reproduction with the seasonal environmental cycle of productivity, reproduction and metabolism appear to be under common physiological control. Seasonal changes in the metabolic rate of pinnipeds are probably closely linked to the central nervous and endocrine control of reproduction and may be involved in the mechanism by which mothers assess their condition at each stage of the reproductive cycle. There is an extensive literature from both ungulates and rodents that shows seasonal appetite and body mass cycles; and, these have also been shown to exist in harbor and harp seals (Renouf and Noseworthy 1991, Renouf et al. 1993). Although Renouf and Gales (1994) found only equivocal evidence of regular seasonal changes in metabolic rate in harp seals, they did find a relationship with reproductive condition. These studies have suggested that metabolic rate varies seasonally in pinnipeds and other studies have demonstrated well-defined cycles of adipose fat reserves (Boyd 1984a, Boyd et al. 1993). The mink (*Mustela vison*) is a close carnivore relative of the pinnipeds and, in particular, the sea otter. This species also has delayed implantation in its reproductive cycle, and Martinet et al. (1992) have shown that cycles of body mass and coat growth are under the same or similar central nervous control as reproduction. The reproductive system appears to be stimulated at certain times of year by day length changes; metabolic changes occur in response to an apparently endogenous circannual rhythm entrained to photoperiod. As discussed in the next section, photoperiod is undoubtedly important in the reproductive cycles of pinnipeds and probably acts principally on the timing of implantation in the reproductive cycle, but, using the example of the mink, it is probably also the main environmental cue for changes in metabolic rate.

The hormonal mechanisms involved in the common control of reproduction and metabolism are likely to involve the pineal gland and hypothalamus (Fig. 6-11). In most mammals studied to date the hormone melatonin is secreted from the pineal gland in the brain during the night phase of the diel light–dark cycle. Melatonin controls prolactin secretion from the anterior pituitary gland. The exact relationship between melatonin secretion and prolactin will depend on the time of year at which specific reproductive events occur. Prolactin normally inhibits reproductive activity (in both males and females) so we would expect it to be elevated during pe-

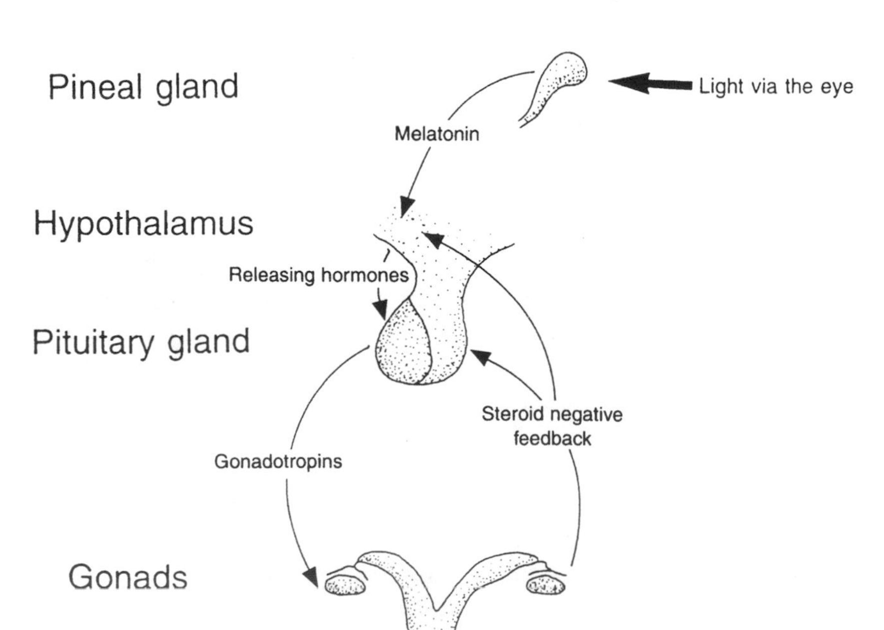

Figure 6-11. Diagram showing the interactions at various levels in the reproductive system. Central control is in the hypothalamus, but the activity of the hypothalamus is probably modified by melatonin from the pineal gland. The response of the hypothalamus will depend on, among other things, modifications of melatonin secretion through the effects of photoperiod.

riods of reproductive quiescence, such as the period of testicular regression after the breeding season in males. The little information about prolactin in pinnipeds is mainly restricted to its role in lactation and corpus luteum function in females (Boyd 1990b, 1991a), but we know that there are both diel and annual changes in melatonin in southern elephant seals that conform to the pattern expected of seasonally breeding mammals (Griffiths et al. 1979, Griffiths and Bryden 1981). Elden et al. (1971) also provided evidence of involvement of the pineal gland and melatonin in the timing of northern fur seal reproductive cycles.

There is both direct and indirect evidence that metabolism and reproduction are controlled by a common physiological mechanism. The advantages of this are obvious in that there will be a close correlation between the reproductive processes that lead to production of offspring and the processes that ensure this takes place at a time when there is sufficient food (or adipose reserves in the case of some species) to raise the pup/cub. How then does this suggest that proximate control of reproduction might operate in relation to environmental productivity?

Stewart and Lavigne (1984) suggest that there are five points in the reproductive cycle at which mothers may "decide" to proceed with or abandon the current reproductive attempt (note, however, that such decisions are not conscious but are driven purely by the physiological mechanisms that have evolved through natural selection). These are ovulation, conception, implantation, parturition, and weaning. If we assume that the probability of implanting, for example, is a declining function of the projected cost to a mother of continuing with her reproductive investment (Fig. 6-12), we can expect that implantation may not occur if a mother's condition declines below some threshold level for that time of year. In most pinnipeds, however, the costs of lactation represent by far the greatest energetic cost of rearing a pup and, by that logic, we may expect the greatest reduction in fecundity to occur between birth and weaning (Fig. 6-12). From the evidence that pinnipeds probably regulate their body condition within an annual cycle and that this condition cycle may be endogenous but environmentally entrained, it seems possible that pinnipeds have an expectation of the level of body reserves they should have at specific times of the year. When reserves decrease below this minimum expectation the response may be to abandon the current reproductive attempt.

The Role of Photoperiod

Reproduction in most, if not all, mammals is under the proximate control of the environment, and we have seen in the preceding section how one environmental factor, energy intake, could modify the outcome of a reproductive attempt. This is likely to affect reproduction by preventing females from reproducing if their body condition is insufficient to ensure potential success and subsequent survival to reproduce again. The major proximate control of reproductive cycles occurs through natural selection of particular phenotypes. For example, the large majority of pinnipeds from either the northern or the southern hemisphere give birth in the spring (Table 6-1). There is enough evidence from pinnipeds to show that the duration of active gestation (i.e., the period of fetal growth) is affected little by environmental conditions (Trites 1991). To achieve springtime births, pinnipeds must coordinate the timing of implantation with environmental changes, and this appears to have been achieved through the use of photoperiod. Similarly, photoperiod is most likely to be involved in the timing of mating and implantation in polar bears (Palmer et al. 1988). Females need to build up body reserves in advance of entering their winter den and this requires advanced warning. Mating must occur so far in advance of denning that its timing is likely to depend on an environmental cue that has a reasonably predictable relationship with the season when denning begins. Although in high latitudes it is possible that sea ice conditions, for example, could provide such a cue, even small local variations could lead to a wide range of denning times; one of the advantages of photoperiod is that its effects are not subject to local fluctuations, although photoperiod itself varies with latitude except at the equinoxes.

In a series of papers, Temte has examined data from zoo records of births in captive pinnipeds at different latitudes. He found that, whatever the latitude, blastocyst implantation normally takes place in response to a specific photoperiod, which is generally characteristic of the species (Spotte and Adams 1981; Temte 1985, 1991; Temte and Temte 1993). Many ecologists find it difficult to believe that day length and not food is likely to be the main proximate environmental cue for reproduction in animals like pinnipeds, but in large mammals, where mating takes place many months in advance of the time when food abundance may be critical, in most circumstances photoperiod is the only reliable environmental cue that can provide precise synchronization of reproduction with the seasonal cycle. As Temte et al. (1991) have discussed, even the timing of reproduction in harbor seals is likely to be controlled by photoperiod, despite a pattern of births that, at first sight, does not suggest this is the case. Bigg and Fisher (1975) and Temte (1994) demonstrated experimentally that the timing of estrus in harbor seals could be altered by manipulation of photoperiod and Temte (1991) showed that, even for captive harbor seals in zoos, where there was no food restriction, individuals had highly precise breeding times indicative of a photoperiodic cue. In fact, the timing of births in many wild pinniped populations has such

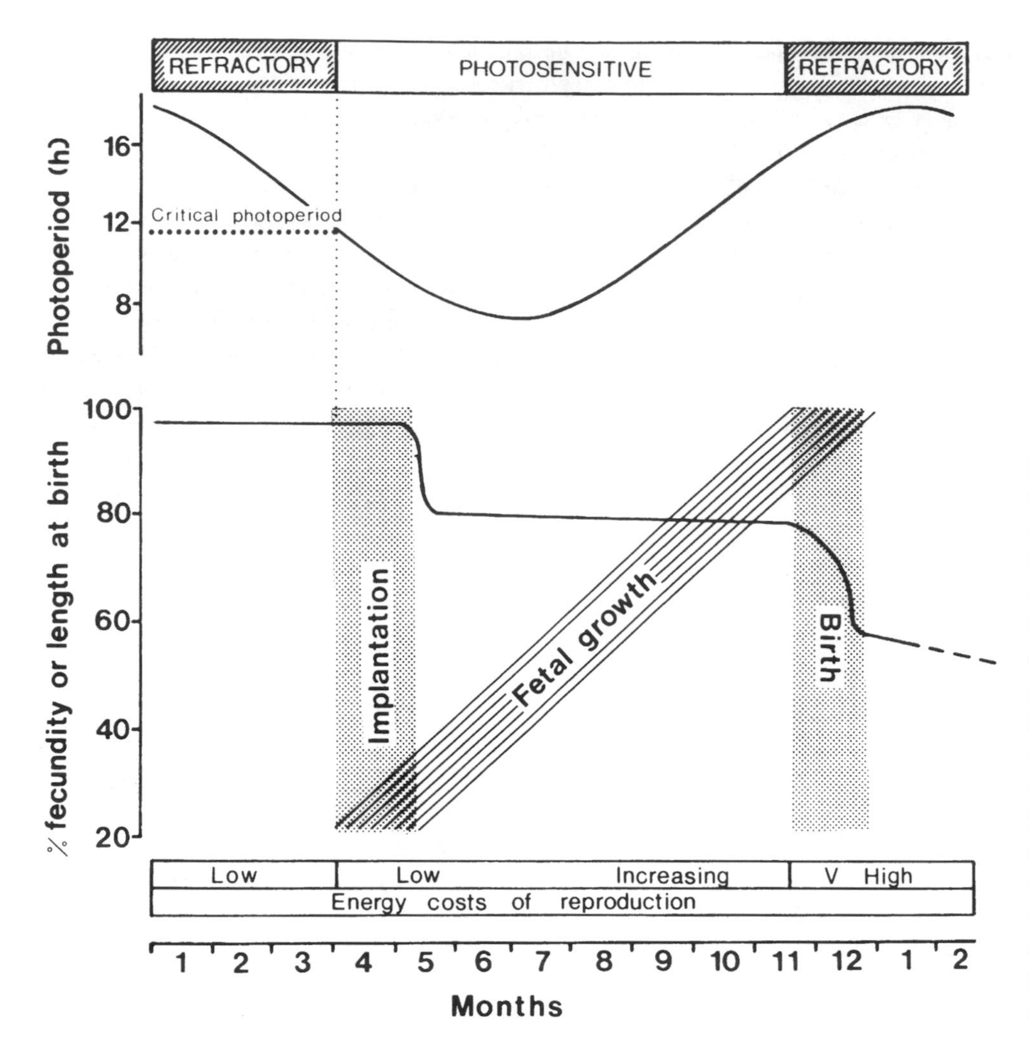

Figure 6-12. A diagrammatic representation of the theoretical change in fecundity of a female pinniped at various stages of her annual reproductive cycle in relation to the approximate energetic costs of reproduction at each stage. A hypothetical cycle of photoperiod is also shown to indicate when the critical photoperiod is likely to occur (close to the autumnal equinox in this case) and the resulting fundamental changes in the reproductive system (refractory—photosensitive) likely to be caused by the change in photoperiod.

high precision between years that probably the only environmental cue that varies on a circannual basis, and that could provide such precision, is the cycle of photoperiod.

Exactly how photoperiod leads to timing of reproduction is not fully understood, but the fundamental mechanisms may have developed early in vertebrate evolution because even distantly related groups, such as birds and fishes, have features of a common mechanism (Nicholls et al. 1988). We can assume, therefore, that pinnipeds also possess the main elements of this mechanism. However, it must be emphasized that in mammals such as pinnipeds, photoperiod (unlike the food supply) has no direct effect on survival or reproductive success; it is merely a convenient circannual clock to which pinnipeds can adjust their reproductive cycles to ensure that offspring are produced at the most advantageous time of year.

Among deep-diving pinnipeds, such as elephant seals, which spend significant portions of their lives at considerable depths (> 300 m) where light levels are below the threshold for photoperiodic stimulation, the question arises as to how photoperiod can be used by these animals to coordinate their seasonal reproductive cycles. The fundamental mechanism of photoperiodic time measurement is probably similar across the vertebrates and involves the entrainment by the diel light cycle of a circadian (approximately daily) cycle of photosensitivity (Elliott 1976, Follett 1982). Mammals do not have a system for totaling the hours of light or darkness experienced in any one day but they sample the dark–light cycle. Experience of even a single flash of light at the correct stage of the endogenous circadian cycle of photosensitivity can lead to entrainment of the seasonal reproductive cycle. Therefore, the deep-diving behavior of pinnipeds does not preclude the use of photoperiod as the environmental cue because a single surfacing during the photosensitive phase of the circadian cycle will be enough to ensure that photoperiodic time measurement can take place.

Using the mink (Martinet et al. 1992) as a general model for marine carnivores, seasonal breeding in pinnipeds could result from a single photoperiodic cue each year. Endogenous circannual rhythms that persist in the absence of seasonal cues appear to occur in long-lived mammals (Woodfill et al. [illegible]) and, given that the marine carnivores are among the longest lived of mammals, we might expect them to show a similar pattern. In general terms, there is an inactive (photorefractory) and an active (photosensitive) phase of seasonal reproductive cycles in mammals (Fig. 6-12), and the transition from the inactive to the active phase (or vice versa depending on the species) is often stimulated by a critical photoperiod (Fig. 6-12). The direction of change in photoperiod is also important. As can be seen from Table 6-1 (and diagrammatically in Fig. 6-12), implantation in most pinnipeds occurs during the phase of declining day length, with most occurring around the autumnal equinox when day length duration is 12 h in all parts of the globe. The critical photoperiod for northern fur seals (12.5 h; Temte 1985), California sea lions (11.5 h; Temte and Temte 1993), eastern Atlantic harbor seals (11.7 h; Temte 1994), and western Pacific harbor seals (14.3 h; Temte 1994) has been identified. The beginning of denning and implantation in polar bears also occurs around the autumnal equinox at least in the region of Hudson Bay (see Fig. 6-3). Species such as the migratory harp seal, which may be distributed over a wide latitudinal range at the time of implantation, when photoperiod probably has its effect on the timing of the reproductive cycle, could only synchronize pupping if the critical photoperiod were close to 12 hours, the day length at the equinox.

The critical photoperiod, which is the photoperiod used by animals to synchronize their reproductive cycles (Fig. 6-12), and the direction of change in photoperiod, are likely to be determined by natural selection. There is polymorphism within mammal species for different photoperiods, depending on location or breed (e.g., Belyaev and Klotchov 1965, Desjardins et al. 1986, Bronson 1988). Among pinnipeds, such polymorphism appears to be most profound in harbor and gray seals (Coulson 1981, Boyd 1991b, Temte et al. 1991, Temte 1994). Clines in pupping seasons have been identified in both these species within certain portions of their range and, for gray seals, there is a strong relationship between the minimum sea–surface temperature during the year and the timing of births (Boyd 1991b). Although Coulson (1981) developed an elaborate model to explain gray seal breeding seasons using temperature as the environmental cue, the relationship is almost certainly the result of selection for particular gray seal phenotypes with different responses to photoperiod in the various locations (e.g., northeastern Atlantic, northwestern Atlantic, and Baltic Sea) rather than a direct response to temperature. In this case, sea temperature probably reflects changes in the food supply to which gray seals have adapted their breeding cycles, using photoperiod as the environmental cue. Similar clinal variation or polymorphism may occur in the polar bear because denning occurs in September to October in the Hudson Bay region, whereas further north it is later in October to November.

In contrast, the apparent lack of seasonality and synchronization of reproduction in sea otters suggests that photoperiod may have little or no influence on their breeding cycles; individuals can breed at virtually any time providing the local environmental conditions are appropriate. This difference in the breeding cycles of sea otters and pinnipeds is probably caused by differences in the energetic and behavioral priorities of the two groups. Pinnipeds tend to range widely and generally have polygynous mating systems associated, in most cases, with colonial aggregation. Sea otters range over relatively small areas and have a system of overlapping male and female territories. Consequently, there is no necessity for sea otters to synchronize breeding so that males and females are in a predictable location at a specific time because potential mates and food occur in the same locations. In addition, the food of sea otters (usually sublittoral invertebrates) is relatively static and aseasonal in its occurrence. Therefore, sea otter reproduction probably does not rely on being synchronized to large-scale seasonal cycles of prey availability (although it may be linked to local environmental conditions), either for pregnant or lactating mothers or for their weaned or partly weaned young.

Both pinnipeds and polar bears have the ability to exploit widely dispersed prey over large areas and to capitalize on seasonally abundant prey; the complete separation of prey from the habitat (ice or land) that is essential for raising offspring, or the highly seasonal nature of prey availability, has almost certainly been a major factor in leading to the evolution of the highly synchronous and seasonal nature of reproduction. Such restrictions do not exist for sea otters, which are unique among the marine carnivores in not having to leave the water to give birth and which rely on a food source with little seasonal change in availability. Overall, this contrast between the sea otter and the other marine carnivores illustrates some of the possible reasons why seasonality of reproduction has evolved.

Secondary Influences of Nutrition on Reproductive Cycles

The exception of the sea otter points to other explanations for the timing of pinniped reproductive cycles, because photoperiod does not explain all of the variation in the timing of births in pinnipeds. The Hawaiian monk seal and the Australian sea lion are exceptions to the general pattern described because they have nonsynchronous and aseasonal reproduction, respectively. Part of the probable explanation for these exceptions can be found from detailed studies of pinnipeds that conform to the usual pattern of synchronous seasonal reproduction. For example, mean birth date for Antarctic fur seals can vary by as much as 10 days (Lunn and Boyd 1993; Fig. 6-13). At first this may seem to be a small degree of variance, but this variance can provide some valuable

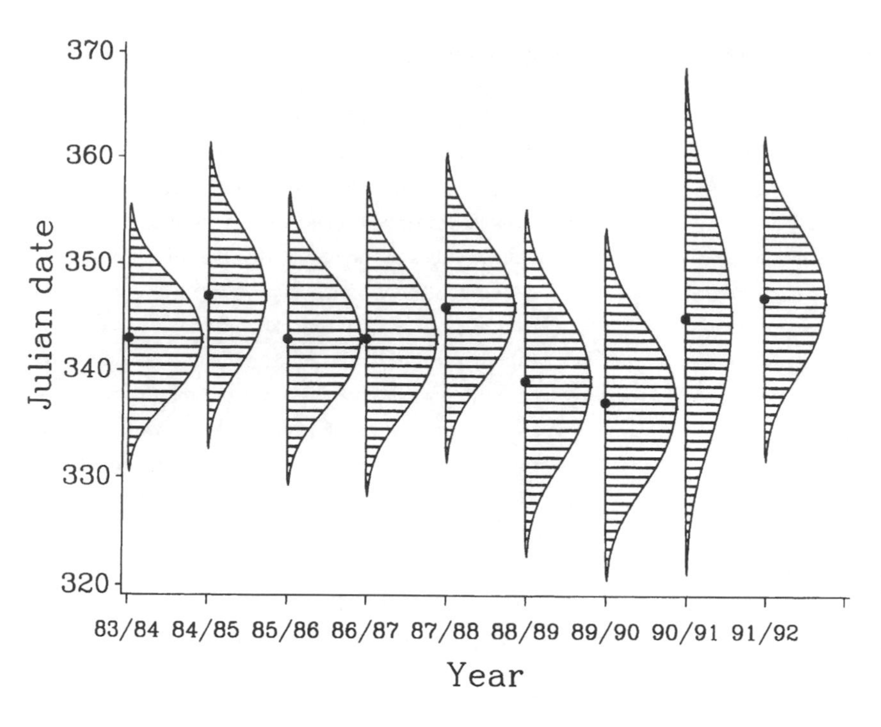

Figure 6-13. The distribution of births around the mean date (dot) for a subpopulation of Antarctic fur seals over 9 consecutive years. (Data from Lunn and Boyd 1993.)

insights into the biology of reproduction in this pinniped. The timing of parturition was correlated with several indicators of food availability in the previous year. Therefore, for example, those years when pups grew slowly and mothers spent long periods searching for food at sea because of low food availability, were followed by a later average birth date in the next year (Lunn and Boyd 1993). This was caused by an increase in the total length of gestation in individuals after years when food was scarce (Boyd 1996). Moreover, pregnancy rate was reduced after those years in which feeding conditions were poor. Even in this highly synchronized, seasonally breeding pinniped, we can find evidence that the food supply can influence the timing of reproduction.

There is, however, an important distinction between this effect and that of photoperiod. Whereas photoperiodic effects have presumably been selected to result in births at the most advantageous time of year, the effect referred to above is a damage-limitation response to suboptimal environmental conditions. The most appropriate explanation for the underlying mechanism is that female Antarctic fur seals have a threshold in their body condition below which implantation will not take place (see previous section for a discussion of the interaction of metabolism and reproduction), even after they have experienced the critical stimulatory photoperiod (see Fig. 6-12). It appears to have taken Antarctic fur seals longer to reach this critical condition in years when food was scarce than in more normal years (hence, the later mean date of parturition in the following season) and, overall, fewer females reached their critical threshold (hence, the lower pup production in the following season). This suggests a combination of two processes leading to implantation: (1) detection of the critical photoperiod followed by (2) attainment of the threshold body condition within a specific time frame (see Fig. 6-12). A similar pattern exists for gray seals in which individuals in poor condition were found to implant later than those in good condition (Boyd 1984a) and this also led to females in poor condition giving birth later than those in good condition (Boyd 1982). A similar pattern of late birth date with low body mass is present in northern elephant seals (Reiter et al. 1981). In Antarctic fur seals, there is evidence of consistent variation among individuals in the time of parturition so that individuals tend to give birth at the same time each year (Boyd 1996). This suggests that the timing of implantation also varies among individuals.

When the interval in which implantation can occur under a stimulatory photoperiod expands to several months, then the effect will be to reduce the synchrony with which breeding occurs in populations. Hawaiian monk seals, in which the breeding season is spread over 6 to 7 months (Atkinson and Gilmartin 1992), probably still need a photoperiodic cue to synchronize reproduction, but the combination of small annual changes in the photoperiod in the tropics compared with the higher latitudes and a greater dependence on fluctuating food availability may also mean that the time frame in which implantation can take place is relatively longer in this species. In this sense, the Hawaiian monk seal reproductive pattern is similar to that of a sea otter, which is completely aseasonal and may no longer make use of a photoperiodic cue. In the Australian sea lion (Ling and Walker 1978), it is probable that the influence of photoperiod has declined even more and that local environmental conditions have taken over as the main cue.

A Model for Pinniped Reproductive Cycles

In a previous section, it was suggested that a critical photoperiod was responsible for stimulating implantation and that pinniped reproductive systems alternate between an active and a quiescent state, perhaps under the influence of an endogenous annual rhythm entrained by photoperiod. I also suggest that the critical photoperiod for terminating delayed implantation is the environmental cue that leads to recrudescence of the reproductive system each year after its period of quiescence.

In mammalian reproductive systems, there is a hierarchy of control from the hypothalamus of the brain to the anterior pituitary gland and then to the gonads. The system needs to be operational at all of these levels before reproduction (ovulation or spermatogenesis) can proceed (see Fig. 6-11). For example, it is possible for reproduction to be activated at an hypothalamic level but, because of social, behavioral, or energetic factors, reproduction can be prevented by inhibition at the peripheral level of the gonads (see Fig. 6-6). To examine the fundamental cycle of reproduction in harbor seals, Gardiner (1994) tested the activity of the anterior pituitary gland at different times of year by stimulating it with exogenous doses of the gonadotropin releasing hormone (GnRH), which is normally produced by the hypothalamus. She discovered that there was indeed a cyclical response of the pituitary gland with the lowest response in the period after the normal breeding season. A reactivation of the pituitary response to GnRH occurred after the time of year at which implantation would normally occur.

Additional evidence for the underlying and fundamental cyclicity of the reproductive system of pinnipeds can be seen in the pattern of follicle growth in northern fur seal ovaries (Craig 1964). In the ovary contralateral to the current pregnancy, there is reactivation of follicular growth after implantation and, among females due to ovulate for the first time, follicular growth increased rapidly after the time of implantation and ended after mating. Similarly in male pinnipeds, testicular recrudescence begins soon after implantation normally occurs in females (Griffiths 1984b, Bester 1990, Noonan et al. 1991; see Fig. 6-2). Gentry (1981) observed a "false" estrus at the end of diapause in northern fur seals and there have been similar observations for harbor seals (Venables and Venables 1959) and the walrus (Fay 1981; see Fig. 6-1).

Moreover, molting in mammals is often associated with an increase in plasma prolactin, which coincides with the inactive (or refractory) phase of the reproductive cycle (Lynch and Wichman 1981, Duncan and Goldman 1984, Boyd 1985, Maurel et al. 1986, Nicholls et al. 1988). Most evidence suggests that the annual molt in pinnipeds begins during delayed implantation. Female harp seals begin to molt up to 2 months after parturition (Ronald and Healey 1981) and a similar pattern is seen in southern elephant seals (Laws 1956b, Carrick et al. 1962) and harbor seals (Boulva and McLaren 1979, Thompson and Rothery 1987). Renouf and Noseworthy (1991) identified a "breeding–molting" season in harbor seals by the different metabolic changes occurring during this time and this may be equivalent to the refractory phase illustrated in Figure 6-12. Ashwell-Erickson et al. (1986) showed that resting metabolic rate was depressed during molt in pinnipeds, possibly as an adaptation to fasting. However, the hormonal changes associated with molt could also play a significant role in the reactivation of the reproductive system and in the induction of implantation.

These observations—circumstantial though they may be—tend to conform to the general pattern observed in other seasonally breeding mammals in which there is a photoperiodic control of the reproductive cycle. The implication is that the inactive phase (also known as the photorefractory phase) of the cycle is equivalent to the period of pseudopregnancy or delayed implantation and that the switch to the active phase of the cycle is cued by photoperiod, and among pregnant females, this also leads to implantation. Among nonpregnant females it leads to involution of the corpus luteum and a gradual increase in follicular development. Among males similar cues lead to testicular regrowth.

Effects of Age on Reproduction

Puberty occurs at 2 to 7 years of age in pinnipeds, but the exact age at puberty depends on the species, sex, and environmental factors that affect growth rate. Delayed sexual maturity is also a feature of sea otter reproductive life histories as both sexes may not be sexually mature until as much as age 6 years (Garshelis 1983, in Rotterman and Simon-Jackson 1988; Jameson and Johnson 1993), although in some areas females have produced their first pup at about age 4 years (Rottermann and Simon-Jackson 1988). Sexual maturity in female polar bears can occur at age 4 but is normally age 5 to 6 years. Breeding success is lower in animals younger than 7 years old, suggesting that either experience or body size influence breeding success. Although samples of old animals are small, there is an indication of reproductive senescence in female polar bears beyond age 16 to 20 years (Ramsay and Stirling 1988, Derocher and Stirling 1994).

A major function of the time required to reach sexual maturity may be for individuals to grow to an adequate size, or gain sufficient experience, to allow them to maximize reproductive success. In an interspecific comparison, Harvey and Zammuto (1985) showed that, even allowing for differences in body size, the time required for maturation in mammals is a direct function of life expectancy. In general, it is difficult to apply this principle to the pinnipeds; maximum longevity can be a reflection of how intensively a particular species has been studied (the more animals that are examined the greater is the chance of finding a very old individual). A better measure of longevity would be the average annual adult mortality rate, but this has been measured directly in only a few species of pinnipeds.

The reasons for the delay in sexual maturity are most likely the result of the evolution of a life history that tends to maximize fitness. Although it is obviously impossible to observe the costs of reproduction in immature individuals, even in mature individuals there are significant fitness costs involved in reproduction. For example, in Antarctic fur seals 40% to 50% of the annual adult female mortality is caused by reproduction (i.e., an individual female is almost half as likely to survive after a year in which she reproduces than in one in which she does not reproduce) (Boyd et al. 1995). Moreover, 40% to 50% of reproductive failures among adult females are attributable to reproduction in the previous year (Boyd et al. 1995). Taken together this constitutes a substantial fitness cost of making a reproductive attempt. We can only assume that among juvenile individuals, this fitness cost will be even greater and will cancel out any fitness benefits from reproducing. Thus, physiological mechanisms have evolved to delay reproduction until the balance of fitness costs and benefits of reproduction in individuals is more favorable.

Absolute body size may be more significant for reproductive success in the males of some species than for females. Anderson and Fedak (1985) found that the largest male gray seals had the greatest apparent reproductive success. Sexual maturity, in terms of presence on the breeding grounds, is delayed even longer in male gray seals compared to females (Harwood and Prime 1978). A similar effect of body size on reproductive success was present for northern elephant seals (Deutsch et al. 1990). It is probable that delayed puberty in males allows sufficient growth to take place before males begin to compete within the breeding colonies. Large size not only potentially confers greater fighting ability but also permits males to fast longer and therefore, hold territory on the breeding grounds longer (Boyd and Duck 1991). The ability to hold territories for long periods by staying ashore and fasting is probably critical in determining male reproductive success in species that breed colonially.

Fitness advantages, measured in terms of the number of offspring that survive to reproduce, appear to exist for individuals maturing at particular ages. Female northern elephant seals and Antarctic fur seals that give birth for the first time at 3 years of age have lower subsequent survival than those that give birth for the first time when older (Huber 1987, Reiter and Le Boeuf 1991, Lunn and Boyd 1993, Lunn et al. 1994, Boyd et al. 1995). The gain from reproducing early is that those first-time breeders that begin breeding at an early age and survive, will have higher lifetime reproductive success. In both northern elephant seals and Antarctic fur seals this would appear to be supported by current observations (Reiter and Le Boeuf 1991, Boyd et al. 1995).

Changes in the average age at first reproduction have been used as an index of the reproductive potential of populations (DeMaster 1981). This has largely stemmed from the idea that pinnipeds, and some other mammals, reach puberty at a critical mass (Laws 1959a, 1977a; Widdowson 1981) and that, given favorable environmental conditions, growth rates will be high and the prepubertal period will be shortened. Thus, some studies have suggested that age at puberty declines with declining density or increasing food availability (Laws 1977a, Bengtson and Siniff 1981, Bengtson and Laws 1985). However, methods used to estimate age at puberty (especially those using growth layer groups in teeth) make the assumption that the adult survival rate is independent of age. This assumption is probably not upheld and leads to detection of spurious declines in age at maturity. In addition, some of the observed changes may have been an artifact of the distribution of errors around the ages obtained from teeth (Cooke and de la Mare 1984). Using data from ovaries, Bengtson and Laws (1985) still showed a decline in the age at sexual maturity in crabeater seals, which they related to shifts in food availability within the Southern Ocean. Shifts in the age at puberty have been demonstrated in harp seals in which the age varied between 4 and 6 years under various degrees of exploitation (Bowen et al. 1981). Age at puberty was low when harp seal population size was small; a similar effect has been observed for northern fur seals (Fowler 1990). This is broadly what one would expect if growth rate was sensitive to food availability and if these species had a broadly stable food source. In contrast, Testa and Siniff (1987) found no density-dependent changes in age at puberty in Weddell seals. Therefore, the interaction of growth and age at puberty remains uncertain.

In polygynous species, such as northern elephant seals (Reiter et al. 1981), there may also be an important distinction between physiological sexual maturity, which may depend on physical condition or growth, and behavioral maturity, which may depend on breeding experience. This is well known for male pinnipeds (e.g., southern elephant seals, Laws 1956b), but is less well established for females. In addition, the maturation process can continue after puberty. For example, in Antarctic fur seals, young females give birth later in the pupping season than middle-aged or old females;

they also give birth to smaller pups and they are more likely to fail to breed in the year following a pregnancy (Lunn et al. 1994). Overall, females of this species do not reach their peak reproductive performance until 7 to 8 years of age, whereas they produce their first pup at 3 to 4 years. Reproductive senescence may also occur in Antarctic and Northern fur seals (York and Hartley 1981, Trites 1991, Lunn et al. 1994) involving a steadily declining pregnancy rate after about 13 years of age. However, such apparent senescence can result from the presence of a fitness cost of reproduction, in terms of future survival (Boyd et al. 1995); those individuals that reproduce infrequently survive longest and therefore, there may be a decline in fecundity with age when considered at the level of the population rather than the individual.

The physiology of sexual maturation has not been addressed in pinnipeds and is poorly understood in mammals generally. In immature harbor seals, Gardiner (1994) found a seasonal cycle of pituitary activity that is similar to the cycle of reproduction in adults. Karsch and Foster (1981) have suggested that young mammals go through the same seasonal cycles as adults, with alternate activation and deactivation of the reproductive system in response to environmental stimuli (see Fig. 6-12). Juveniles may be highly sensitive to reproductive inhibition by negative feedback from gonadal steroids (see Fig. 6-11), with the result that the gonads never receive sufficient stimulation from gonadotropins for gametogenesis to proceed. Puberty may represent a decline in the high sensitivity to negative feedback from steroids.

Evolution of Pinniped Reproductive Cycles

Otariid lactation is characterized by mothers leaving pups ashore for variable periods while they search for food at sea but, between birth and departure for the first foraging trip, mothers remain ashore to feed and protect their newborn pup. This period of attendance ashore also represents the postpartum estrous cycle, and mothers are mated immediately before their first foraging trip to sea (see Fig. 6-8). Mothers arriving to give birth must also have all the energy reserves to sustain themselves and feed their pups throughout this period ashore. This is similar to the phocid lactation pattern (Oftedal et al. 1987) where, in general, mothers also remain with their pups throughout the postpartum estrus, are mated, and then leave for sea (see Fig. 6-8). The difference is, of course, that phocids bring greater body reserves ashore with them, extend (in most cases) the duration of the postpartum estrus cycle, and, unlike the otariids, they wean their pups around the time of mating (see Fig. 6-8). Therefore, it appears that to achieve a short lactation duration phocids have extended the duration of the postpartum estrus cycle. One view is that some small phocids have come full circle and, owing to constraints on the amount of energy they can carry as body fat reserves, they feed during lactation. Harbor seal mothers can perform feeding trips to sea, which emulate the pattern of foraging by otariids (Bowen et al. 1992) and was probably abandoned by the ancestral phocids (Fig. 6-14).

Although the duration of lactation in phocids will be ultimately determined by natural selection, the proximate mechanism controlling the duration of lactation may be the rate of follicular growth after parturition. This follows because estrogen, which controls estrous behavior in females, is produced by the mature preovulatory follicle in the ovaries. High rates of follicle growth and development would lead to a rapid return to estrus and a short lactation duration, as in hooded seals (Bowen et al. 1985), whereas low rates of follicle growth and development would lead to a later return to estrus and a long lactation, as in the ringed seal (McLaren 1958).

We can envisage the ancestral pinniped leading a semiaquatic existence in coastal habitats, and its reproductive cycle may have been similar to that of the sea otter, that is, asynchronous, aseasonal, and each reproductive cycle lasting at least 1 year. As the pinnipeds moved into more seasonal environments and began to exploit prey that were distant from the habitat suitable for parturition and the rearing of young or were highly seasonal in availability, there would have been a gradual spatial and temporal dissociation of foraging and mating activities. There may have been selective pressures for larger body size to increase fasting duration and, as a consequence, the duration of fetal growth would have increased as the size of the young at birth increased. In this case, the reproductive cycle may have extended to more than 1 year and may have been biennial. A variety of selective forces, including predation and mate selection, led to the development of colonial behavior to the level now seen in the walrus (Fay 1981; see Fig. 6-1), which, under this model, is an early divergence from the otariid/phocid lineage. The reproductive cycle of the walrus has many features in common with both the otariids and the phocids, such as an annual mating season and a postpartum estrus. The walrus also has delayed implantation (see Fig. 6-1) of a similar duration to other pinnipeds and a molting season that appears to have close temporal ties to the mating season and to implantation in the same way as we have seen with the other pinnipeds. The most profound difference between the walrus and other pinnipeds is that gestation in the walrus is 15 months (see Fig. 6-1) and therefore, breeding occurs at most biennially. As the early pinnipeds became better adapted to the marine environment and began to exploit food sources more efficiently, there may have been significant fitness advantages in breeding annually (individuals that breed annually will have a selective advantage over those that breed less frequently

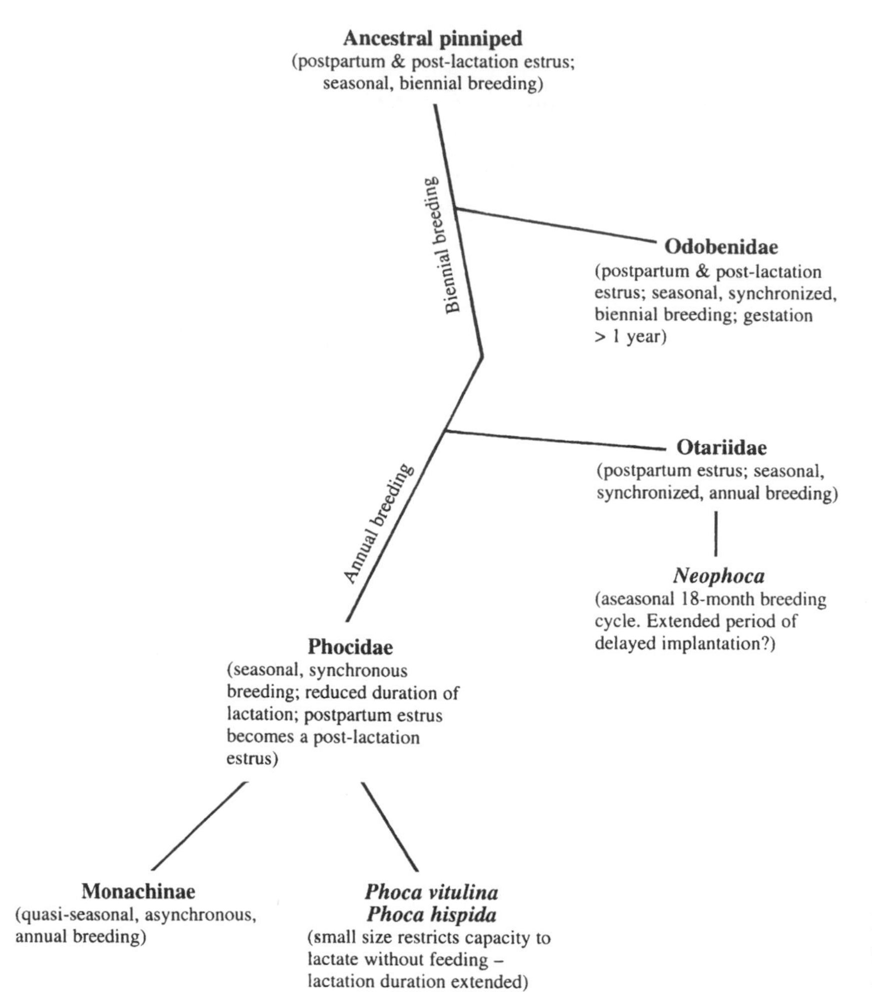

Figure 6-14. A probable sequence for the evolution of reproductive patterns in pinnipeds.

because, all other things being equal, they will produce more offspring). In the otariid/phocid lineage, all modern species (with the exception of the Australian sea lion) have gestation constrained so that the reproductive cycle lasts 365 days. This appears to have been achieved through selection for high fetal growth rates and has resulted in the unusual variation in fetal growth rates in pinnipeds where elephant seal fetal growth rates are about 10 times those of the Galapagos fur seal, the smallest pinniped. Fetal growth rates tend to be consistent between closely related mammal species and this variation in fetal growth rate is the best evidence that, at least in the larger pinnipeds, selection has reduced the length of the breeding cycle. Annual breeding, which has subsequently been abandoned by the Australian sea lion (Ling and Walker 1978), probably developed before the phocid/otariid divergence, which may account for the extreme similarity between the patterns of gestation in the two groups (Fig. 6-14).

Neither ancestral phocid nor otariid reproductive cycles would have had to change much to reach the point of modern pinnipeds, although the otariid pattern is probably the more primitive of the two. Both appear to have retained the postpartum estrus, albeit highly modified in phocids, and both have seasonal mating. The mating season coincides with the birth period. Therefore, postpartum matings and matings of nonpregnant females occur at the same time. However, the postpartum matings and matings of nonpregnant females would appear to have quite different physiological and evolutionary roots. In a temporal sense, pinnipeds have one mating season but this appears to be brought about by the overlaying of two different physiological processes, seasonal estrus and postpartum estrus. If it were not for delayed implantation, pinnipeds would probably have two mating periods each year, one postpartum estrus for individuals pregnant in the previous cycle and one seasonal estrus for previously nonpregnant individuals.

Artificial Control and Pathology of Reproduction

There is evidence that pollutants can disrupt the reproductive function of pinnipeds (see O'Shea, Chapter 10, this volume). Reijnders (1986) demonstrated reduced reproduction in captive harbor seals fed on fish from polluted waters and DeLong (1973) and Gilmartin et al. (1976) showed that abortion of California sea lions was associated with increased concentrations of pesticides in maternal and fetal tissues. There are also correlations between pesticide and polychlorinated biphenyl (PCB) concentrations and pathology of the reproductive tract in gray, harbor, and ringed seals from the Baltic Sea (Helle et al. 1976). The pathology took two forms, complete occlusion of the uterine lumen (see Fig. 6-5) at a point midway along one or both horns or a stenosis (narrowing) of the uterus at about the same point. This pathology is not understood and its linkage with environmental pollution is circumstantial. However, the effect of this condition on reproduction was so severe that pregnancy rates in ringed seals from the Gulf of Bothnia declined to as low as 25% (60% to 90% would be in the normal range) in the 1970s (Helle 1981), and this was likely to have been a major factor in causing declines in pinniped populations in the Baltic Sea.

There is an increasing interest in artificially manipulating reproduction in pinnipeds as a means to either reduce the number of individuals in a population without resorting to killing, or as a method for accelerating reproduction in species that are in danger of extinction. Immunocontraception is being tested experimentally on gray seals in Canada (Parsons et al. 1993). This method relies on a single immunization against the glycoproteins of the zona pellucida that remain around the blastocyst for much of delayed implantation. If successful, this could provide a method suitable for use on pinnipeds under special circumstances where reduction of the reproductive rate is deemed both desirable and practical as a long-term solution to problems of population management.

In pinnipeds less progress has been made toward manipulating reproduction to assist in the recovery of endangered species. Recent experiments have been carried out to reduce the sexual drive of male Hawaiian monk seals, which are mobbing and killing females, by administering drugs that are antagonistic to natural androgens. The manipulation of reproduction would normally need to be carried out on captive individuals and could involve techniques that are used regularly in domesticated breeds of mammals. These techniques include embryo transfer, artificial insemination, and even the transfer of blastocysts from endangered species, such as monk seals, to the primed uterus of a common species, such as harbor or gray seals, which could then act as surrogate mothers. Artificial insemination, using semen from wild males, could be used to maintain genetic variability within small captive groups of pinnipeds. These approaches to pinniped conservation may be practical given that pinnipeds appear to show a high degree of consistency in the reproductive pattern among species. However, our detailed knowledge of the endocrinology of pinniped reproduction is still insufficient to allow such an approach to be used in the near future. Perhaps these issues can provide valid objectives and the stimulus to expand the experimental approach to reproductive physiology research in pinnipeds.

Reproduction in Sirenians [H. D. M.]

It is probable that sirenians evolved from primitive terrestrial herbivores early in the Tertiary period. Like cetaceans, they spend their entire life in the water and do not return to land to give birth and suckle their young. Sirenians are believed to share a common origin with elephants and two groups of small African and/or middle eastern mammals called hyraxes and elephant shrews. The four groups are collectively referred to as the superorder Paenungulata. Although features of the external form of sirenians reflect adaptations to their life of swimming and diving, the phylogenetic history of dugongs (*Dugong dugon*) and manatees is reflected in many features of their reproductive anatomy, which in some respects is strikingly similar to that of elephants as outlined later.

Understanding the reproductive biology of sirenians is basic to the development of effective strategies for their conservation. We need to know when and where they breed to assess the likely effectiveness of establishing sanctuaries to protect breeding habitat. Knowledge of the mating system, the average ages at which females start and cease breeding, the litter size and the average interbirth interval, and how these are affected by changing environmental conditions are essential for estimating the levels of human-induced mortality that are likely to be sustainable.

This section reviews current information on sirenian reproduction and assesses the implications of this information for their conservation. I conclude that, because manatees and dugongs have a low lifetime fecundity, they are highly susceptible to human impacts. The long-term survival of sirenians is dependent on long-term conservation initiatives.

Methodology

The reproductive biology of sirenians is difficult to study. There are three basic approaches: (1) the analysis of specimens collected from carcasses, (2) monitoring the reproductive history of known free-ranging individuals during their life span (longitudinal studies), and (3) studying individuals in captivity. All these methods have their biases and it is important to understand these when interpreting the results. The optimum approach is probably to use all three methods. When they all come up with the same result there is great strength in the findings.

Even if a carcass study is based on large sample, the sample is unlikely to be random. For example, some conditions (e.g., twinning) may increase the chances of natural mortality and females with twins may be over-represented in the sample. If a sample results from a directed fishery, some life history stages may be targeted; for example, indigenous dugong hunters prefer breeding females to resting females or males because they are fatter (Roberts et al. 1996). The amount of information derived from a carcass depends on the season (animals decompose more quickly in summer) and the age of the animal (it is harder to determine the age of older manatees due to bone resorption, Marmontel 1995); in addition, a pregnancy becomes easier to detect as it progresses, and early pregnancies are likely to be missed (Marmontel 1995).

The sample of animals included in a long-term observational study is also not random. Animals are more likely to be included in the sample if they bear distinctive marks and if they are easy to observe and photograph. Thus, the sample is biased in favor of distinctively marked approachable animals that frequent aggregation sites in clear water.

Captive animals are also a biased sample of the free-ranging population. They have the advantage of being generally tractable to handling and available for recapture but the results are confounded by their captivity.

Our knowledge of reproduction in the four extant species of sirenians is uneven. The Florida manatee (*Trichechus manatus latirostris*), a distinct subspecies of the West Indian manatee (*Trichechus manatus*) (Domning and Hayek 1986) is the most comprehensively studied sirenian and our understanding of its reproductive biology is based on data from all three sources outlined above: (1) extensive long-term observational studies of the life history of up to 700 recognizable, free-ranging individuals (O'Shea and Hartley 1995, O'Shea and Langtimm 1995, Rathbun et al. 1995, Reid et al. 1995); (2) the necropsy of more than 2000 individuals between 1976 and 1991 through the southeastern United States Stranding Network (O'Shea and Ackerman 1995) including the examination of 67 males (Hernandez et al. 1995) and 275 females (Marmontel 1995) with estimated ages; and (3) observations of captive manatees including 28 calves born in captivity (Odell et al. 1995). In contrast, there are few data for the Antillean subspecies (*Trichechus manatus manatus*). Information on the reproductive biology of the dugong comes mostly from the analysis of the material from the carcasses of some 1400 dugongs (Bertram and Bertram 1973; Marsh 1980, 1986, 1995; Marsh et al. 1984a,b,c; Nietschmann 1984; Hudson 1986), including some 256 individuals whose ages have been estimated, plus some relatively short-term behavioral observations (Anderson and Birtles 1978, Preen 1989). Dugongs have never bred in captivity. Although Amazonian manatees (*Trichechus inunguis*) have been kept in captivity for many years (e.g., Best et al. 1982), data on the reproductive biology of free-ranging individuals are sparse. There is virtually no published information on reproduction in the West African manatee (*Trichechus senegalensis*) (Husar 1978).

Reproductive Anatomy

External Sexual Dimorphism

Sirenians exhibit little sexual dimorphism. Although the asymptotic length of females may be slightly larger than that of males in dugongs (Marsh 1980), the sex of sirenians cannot be inferred from body size. The distance between the genital opening and the anus is the most reliable index of gender. In females the genital opening is close to the anus, in males the genital opening is situated between the umbilical scar and the anus but closer to the former. Two axillary mammae are present in sirenians of both sexes. Their size varies with reproductive status.

Tusks (second incisors) erupt in male dugongs after puberty and can be regarded as a secondary sexual characteristic. However, not all dugongs with erupted tusks are male. Tusks also erupt in some old females, although the patterns of eruption and wear are different from those in males (see Marsh 1980). Manatees do not have tusks.

Reproductive Organs

The reproductive anatomy of male and female sirenians reflects both their paenungulate origins and their fusiform body shape. As in cetaceans, the testes of sirenians are permanently situated in the abdominal body cavity. In sirenians the testes occur immediately posterior to the kidneys. This position is characteristic of all paenungulates and is apparently not a secondary adaptation to the aquatic environment of sirenians. However, the gross morphology of the testes and excurrent ducts of dugongs and manatees resembles that of cetaceans much more closely than elephants and hyraxes (Marsh et al. 1984a). The resemblance reflects the similarities of the abdominal cavity of sirenians and cetaceans, which are dictated by the morphological adaptations of both groups to their aquatic environments (e.g., reduced pelvic girdle, streamlined shape, and absence of hind limbs).

There are striking similarities between the reproductive anatomies of female sirenians and African elephants (*Loxodonta africana*) (see Marsh et al. 1984a). These include the number and form of corpora lutea and corpora albicantia in the ovaries (Fig. 6-15), the zonary placenta, which leaves conspicuous scars (Fig. 6-16) in the uterus at former attachment

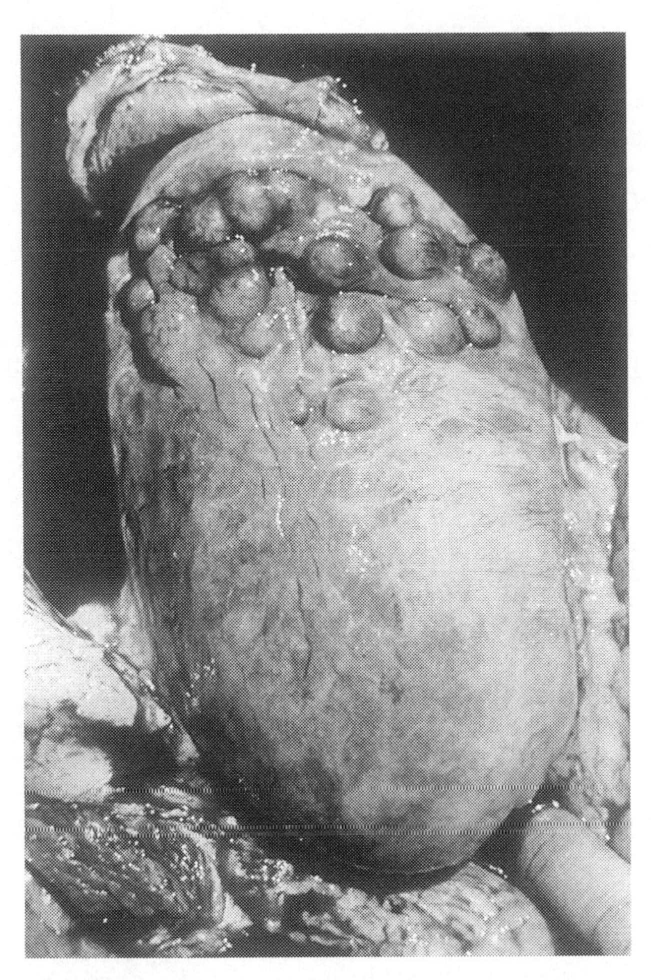

Figure 6-15. Unfixed ovary of a pregnant dugong with a 13.5-kg fetus. Numerous highly vascular corpora lutea can be seen at the cranial pole.

sites in dugongs (but not always in manatees) and elephants, the gross uterine morphology, and the cycle of histological changes in the uterine endometrium.

Reproductive Behavior

Sexual Behavior

Wells, Boness, and Rathbun (Chapter 8, this volume) discuss the sexual behavior of manatees and dugongs. Florida manatees form mating herds composed of a focal female in estrus surrounded by 5 to 22 males. Similar mating herds have also been reported for the West African manatee (see Powell 1996). The mating system of dugongs is not well understood and may vary geographically. Preen (1989) observed dugongs that were interacting in a manner superficially similar to manatee mating herds at the southern end of the dugong's range on the east coast of Australia and I have made similar observations off the east coast of Cape York, 13 degrees further north (unpubished data). In contrast, Anderson (1997) has suggested that some male dugongs establish leks in Shark Bay at the southern end of the dugong's range in Western Australia. Presumed males aggregate at the lek site, which is a small bay. Females apparently visit only for the purpose of copulation. Geographic plasticity such as reported for the dugong is also a feature of the mating systems of some other mammals including terrestrial ungulates (e.g., African impala, Jarman 1979).

Birth and Lactation

Florida manatees tend to seek the safety and seclusion of quiet backwaters to give birth (Hartman 1979, O'Shea and Hartley 1995, Reid et al. 1995); dugongs seek the shelter of

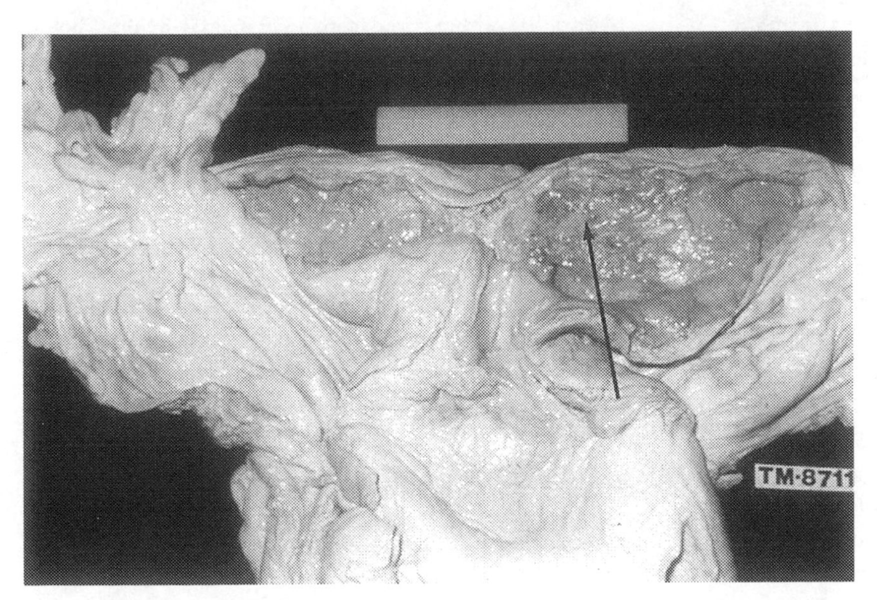

Figure 6-16. Reproductive tract of a lactating Florida manatee with one placental scar (arrowed) in a uterine horn. (Photograph by Miriam Marmontel.)

shallows, estuaries, and reef tops (see Marsh et al. 1984c). In both species this is presumably a strategy to reduce the risks of predation. O'Shea and Hartley (1995) suggest that calving in shallow water may also reduce the chances of female Florida manatees being harassed by males. Reducing the risk of such disturbance would presumably also reduce the risk of perinatal death.

Both manatees and dugongs nurse their young throughout a prolonged period of mother–calf dependence. Sirenian calves start eating aquatic plants soon after birth (Hartman 1979, Marsh et al. 1982). They are probably weaned between the ages of 1 and 2 years, as discussed below.

Length of Reproductive Cycle

Criteria of Sexual Maturity

Age at attainment of reproductive maturity is an important parameter in population biology. Knowledge of length at sexual maturity allows an evaluation of the potential reproductive status of animals for which the only available index of age is body length (Perrin and Reilly 1984). These parameters have been estimated in various ways for species for which large data sets are available, such as some cetaceans (as summarized by Perrin and Reilly 1984). The data sets for sirenians have been too small for such mathematical treatments and most researchers have simply given the size/age ranges within which maturity is reached. There are no data on the age or length at sexual maturity of Antillean, Amazonian, or West African manatees.

ASSESSMENT OF SEXUAL MATURITY IN FEMALES. The most usual criterion of female sexual maturity is first ovulation, evidenced by the presence of at least one corpus luteum or corpus albicans in the ovaries (Perrin and Reilly 1984). The corpus luteum is the endocrine gland that normally develops from the cellular components of an ovarian follicle after ovulation. Regressing or regressed corpora lutea are referred to as corpora albicantia. A follicle that has not ovulated may undergo histological changes similar to those exhibited by a corpus luteum. Such a follicle is known as a luteinized follicle. It can be difficult to differentiate a luteinized follicle from a corpus luteum unless the ovulation scar is visible. Ovulation scars are small in sirenians and difficult to see. They are much more obvious in cetaceans and luteinized follicles are generally easier to distinguish from corpora lutea in cetaceans (Marsh 1984) than in sirenians.

Using the presence of one or more corpora lutea/albicantia as the indicator of sexual maturity assumes that: (1) corpora lutea/albicantia can be reliably distinguished from unerupted luteinized large follicles, and (2) ovarian ovulatory scars remain visible indefinitely. At least the first of these criteria is certainly not satisfied in sirenians as explained above.

Both dugongs and manatees are polyovular (Marsh et al. 1984a, Marmontel 1995), producing a large and variable number of presumed corpora lutea (e.g., a mean of 36.3 per ovary per pregnancy in the Florida manatee, Marmontel 1995; see Fig. 6-15). These "corpora lutea" are also variable in size (e.g., ranging from 1–2 mm to 13–14 mm in the dugong; Marsh et al. 1984a). At least in the dugong, only some "corpora lutea" have visible stigmata confirming follicular rupture. It is likely that bodies classified as corpora lutea or corpora albicantia include luteinized follicles and their scars, respectively. Thus, it can be difficult to confirm from ovarian examination whether a nonpregnant, nulliparous female sirenian has ovulated.

Presumed corpora lutea or corpora albicantia have been observed in the ovaries of nulliparous Florida manatees and dugongs (Marsh et al. 1984a, Marmontel 1995), suggesting that they may undergo several estrous cycles before they first conceive. This finding is also supported by behavioral observations of Florida manatees (Bengtson 1981, Rathbun et al. 1995). Some females are attended by herds of males on several occasions separated by inactive periods of a few weeks.

The distinction between age of sexual maturity and age of first conception is probably of limited relevance to a population study. As summarized in Table 6-2, the usual criteria used to determine at necropsy whether or not a female sirenian is mature include at least one of the following: (1) the size, appearance, and vascularization of the ovaries and uterus (Figs. 6-15 and 6-16); (2) presumed corpora lutea and/or corpora albicantia in the ovaries (Fig. 6-15); (3) a fetus (Fig. 6-17); (4) at least one placental scar in the uterus (Fig. 6-16); and (5) colostrum or milk in the mammary glands. In addition, maturity may be inferred from body size as explained below.

External assessment of the maturity of a female sirenian is difficult unless: (1) she is accompanied by a suckling calf; (2) her mammary glands are elongated and active, and/or her abdomen is distended by a large fetus, or (3) her age and/or body length are known and outside the species-specific ranges within which females mature.

ATTAINMENT OF SEXUAL MATURITY IN FEMALES. The age of sexual maturity of female sirenians varies between individuals and species. It may also vary between populations, suggesting resource-dependent variation in the timing of reproduction. However, the evidence for this is equivocal.

Free-ranging Florida manatees apparently reach sexual maturity between the ages of 2.5 and 6 years. Marmontel (1995) and Marmontel et al. (1996) estimated the age of

Table 6-2. Characteristics of the Ovaries, Uteri, and Mammary Glands of Female Sirenians in Different Stages of Reproduction

Stage	Ovaries	Uterus	Mammary Gland
Immature, prepubescent	Smooth and flattened surface with numerous small follicles	Glands undeveloped, lumen small, blood supply undeveloped	Inactive
Immature, maturing	Several LGF	Early proliferative endometrium, size intermediate between juvenile and adult	Inactive
Sexually mature, nulliparous	CL and/or CA present, may have several LGF	Adult sized, no PS	Inactive
Parous	CA present, may have LGF or CL	At least one PS (dugong only)	Active or inactive
Ovulating	Luteal: LGF and CL	Secretory endometrium with or without PS. Not pregnant	Inactive or active[a]
Pregnant	CL present, may have CA	Embryo or fetus present, secretory endometrium	Usually inactive but may be active[a]
Lactating	Recent CA	Secretory endometrium, recent PS present	Active
Resting	CA but no LGF or CL	Involuting endometrial glands, PS present or absent	Inactive

Modified from Table 2 in Marmontel (1995) to be applicable to the dugong as well as the Florida manatee. Corresponding data are not available for Amazonian, Antillian, or West African manatees.

[a]Occasional pregnant and lactating Florida manatees (Marmontel 1995) and dugongs (Marsh 1989) have been recorded; see text for details.

CA = corpora albicantia; CL = corpora lutea; LGF = large graafian follicle; PS = placental scar.

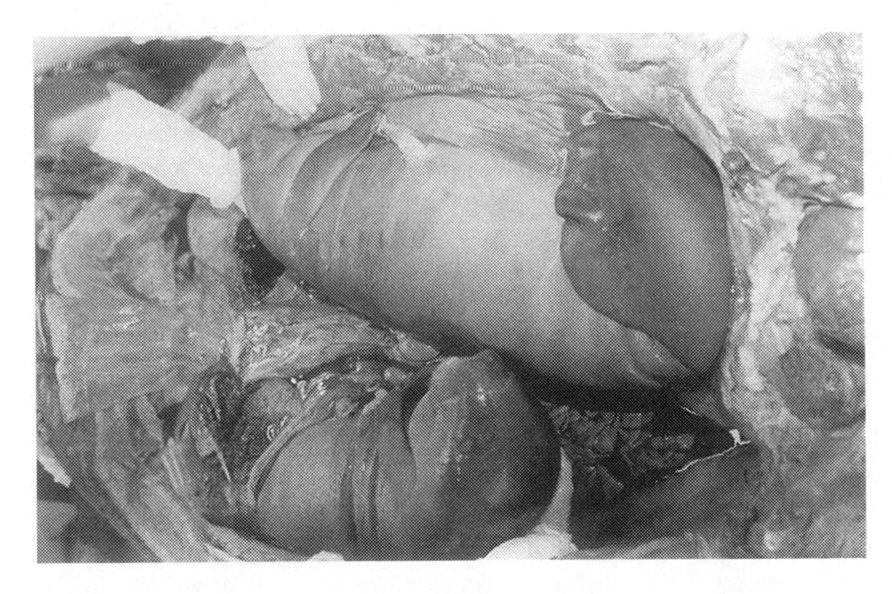

Figure 6-17. Twin, near-term fetuses found during the necropsy of a female Florida manatee. Both fetuses were female and measured 132 and 135 cm total length. (Photograph by Cathy Beck.)

necropsied specimens by counting growth-layer groups in the dome portion of the typanoperiotic bone complex. Marmontel (1995) found that all 143 females 5 years old or older for which reproductive information was available were mature. Some females attained sexual maturity at age 2.5 years. This age is consistent with data from long-term observational studies of free-ranging manatees on the west and east coasts of Florida. However, some individuals did not give birth until they reached 7 years old (Table 6-3). The youngest known-age Florida manatee to reproduce in captivity was a captured orphan that conceived when she was about 5.5 years old (Odell et al. 1995).

Table 6-3. The Ranges[a] of Ages (Years) of First Calving of Female Florida Manatees and Dugongs

Species/Location	Youngest Parous Female	Oldest Nulliparous Female	Evidence of Parity	Source of Data
Florida manatee				
Both coasts	4	[b]	Necropsy	Marmontel 1995
West coast	3.5	7	Longitudinal study	Rathbun et al. 1995
East coast	4	7	Longitudinal study	O'Shea and Hartley 1995; Reid et al. 1995
Dugong				
Townsville	10	9.5	Necropsy	Marsh et al. 1984c
Mornington Island	14.5	17.5	Necropsy	Marsh et al. 1984c
Daru	13	12[c]	Necropsy	Marsh 1986

[a]Ranges rather than means have been given here because of the small sample sizes.

[b]All female carcasses aged 5 years or older were sexually mature.

[c]One 18-year-old had recently given birth to her first calf.

The minimum age at which females first give birth should be about one year after the age of maturation if the gestation period lasts 12 to 14 months.

These estimates indicate that Florida manatees reach sexual maturity at younger ages than those suggested by earlier studies based on much smaller samples (Hartman 1979, Odell et al. 1981, Rathbun and Powell 1982). However, Marmontel (1995) reports that mature females, which have subsequently been aged as 3 years old, were collected in Florida as early as 1978, suggesting that the results reported in the earlier studies were due to small sample sizes rather than reductions in age at maturity over the intervening years.

Although less robust, the data for free-ranging dugongs suggest that they mature several years later than free-ranging Florida manatees (Table 6-3). All information is from carcass studies in which ages have been estimated from growth-layer groups in the tusks. However, the sample sizes are much smaller than those for Florida manatees and there are no longitudinal records for known individuals (Marsh 1980; Marsh et al. 1984a,c). The youngest female dugong with corpora lutea or corpora albicantia in her ovaries was a 9.5-year-old nulliparous female from Townsville, Australia. All 10-year-old or older females from Townsville were parous and had one or more placental scars in the uterus (Marsh et al. 1984a,c). However, data from two other locations (Mornington Island in north Queensland, Australia, and Daru in southern Papua New Guinea) suggest that the age at which females bear their first calves is even more variable than in Florida manatees; three females from Mornington Island were still nulliparous at 15 to 17 years (Table 6-3). The ovaries of two of these animals contained corpora lutea or corpora albicantia (Marsh et al. 1984c). One 18-year-old female from Daru had recently borne her first calf.

Female Florida manatees and dugongs reach sexual maturity over a range of sizes. In northern Australia dugong females less than 2.20 m long are almost certainly nonparous, whereas those larger than 2.50 m are likely to be parous (Marsh et al. 1984c). Nulliparous females with active ovaries have been recorded at lengths of between 2.35 and 2.50 m long. The smallest female with a placental scar was 2.34 m long (Marsh et al. 1984c).

Reflecting their generally larger body size, Florida manatee females are longer than dugongs when they mature. The smallest mature female recorded by Marmontel (1995) was a 2.54-m long 3-year-old. The average length of immature 3-year-old females was 2.54 m (n = 5; range, 2.42–2.72 m) and the average length of mature 3-year-old females was 2.80 m (n = 7; range, 2.54–3.02 m).

ASSESSMENT OF SEXUAL MATURITY IN MALES. It is relatively easy to establish whether or not a female sirenian is sexually mature; assessing males is more problematic. Maturity can be inferred from testis weight, for dugongs where a single testis weight of 30 g or more is diagnostic of maturity (Marsh et al. 1984b). Histological examination of a testis and/or epididymis may be required in the case of necropsy specimens. Even then it may be difficult to distinguish the testicular histology of pubertal males (those approaching first spermiogenesis) from that of mature males with testes that are reentering breeding condition (Marsh et al. 1984b, Hernandez et al. 1995) without data on the animal's age or body length.

External assessment of the reproductive maturity of a male sirenian is difficult unless (1) his age and/or body length are known and outside the range within which males mature for that species, or (2) he has erupted tusks (dugongs only, Marsh 1980). Participation in a mating herd is not diagnostic of sexual maturity. Rathbun et al. (1995) recorded young of the year and yearlings in a mating herd of manatees.

ATTAINMENT OF SEXUAL MATURITY IN MALES. As with females, the ages at which male sirenians become sexually mature vary between individuals and species and perhaps between populations (Table 6-4). The age of sexual maturity of males is similar to that of females in the same population. Male Florida manatees mature at a much earlier age than male dugongs.

Male Florida manatees and dugongs, like the females, reach sexual maturity at a range of sizes. In northern Australia, dugong males less than 2.2 m long are almost certainly immature, whereas those larger than 2.5 m are likely to be mature (Marsh et al. 1984c). This size range at maturity is similar to that of females. Tusk eruption appears to precede testicular competence in dugongs (Marsh et al. 1984b). The smallest male with erupted tusks was 2.26 m long and aged at more than 33 years; the largest male with unerupted tusks was 2.49 m long and aged 6 years (Marsh et al. 1984b). Young sexually mature males are usually smaller than older animals. They tend to have smaller testes and their tusks may not have erupted. This may diminish their ability to compete with other males for mates (Marsh et al. 1984c).

As with females, male Florida manatees are longer than dugongs when they mature. Hernandez et al. (1995) examined the testes of 67 Florida manatees. Animals with prepubescent or recrudescent testes ranged in body length from 237 to 295 cm. The most precocious male was a 237-cm long 2-year-old individual that already had some sperm in his testes. The smallest animal with fully spermatogenic testes was 252 cm long, close to the minimum body length of mature females (254 cm, see above).

Length of Gestation

Estimates of the gestation period of Florida manatees have not improved significantly since the data obtained by Hartman (1979), despite the births of 28 individuals in captivity (Odell et al. 1995). Estimates range from 12 to 14 months for captive Florida manatees (Cardeilhac et al. 1984, Qi Jingfen 1984, Odell et al. 1995). Rathbun et al. (1995) and Reid et al. (1995) obtained crude approximations of the gestation period from field observations of the time between individual females being observed in a mating herd and sighted with a small dependent calf. Most of these observations also suggested gestation periods within the 12- to 14-month range. Best (1982) assumed that the length of gestation in the Amazonian manatee to be about 1 year, apparently based on information from the West Indian manatee.

The best estimate of the gestation period for the dugong (13.9 months, Marsh 1995) is within the range for manatees. Marsh applied the method of Huggett and Widdas (1951) and Laws (1959b) to data on the body lengths and dates of death of 26 fetuses from Daru in southern Papua New Guinea. However, as a result of the small sample size and the diffusely seasonal breeding pattern of the dugong, the 95% confidence interval for this estimate is so imprecise as to be meaningless.

Litter Size and Size and Sex Ratio at Birth

Sirenians usually bear a single young. Twins have been confirmed for both free-ranging and captive Florida manatees on numerous occasions (e.g., Marmontel 1995, Odell et al. 1995, O'Shea and Hartley 1995, Rathbun et al. 1995; Fig. 6-17) and for the Antillean subspecies (Charnock-Wilson

Table 6-4. The Ranges[a] of Ages (Years) of Sexual Maturity of Male Florida Manatees and Dugongs

Location	Youngest Male with Mature Testes	Oldest Male with Immature or Recrudescent Testes	Oldest Male with Unerupted Tusks[b]	Source of Data
Florida manatee				
All Florida	2	11	n/a	Hernandez et al. 1995
Dugong				
Townsville	9	6[c]	10.5	Marsh et al. 1984c
Mornington Island	15	15.5	15.5	Marsh et al. 1984c
Daru	11	16	18	Marsh 1986

[a]Ranges rather than means have been given here because of the small sample sizes.

[b]Dugongs only.

[c]No males examined aged between 7 and 9 years inclusive.

1968). The records for the Florida manatee include twins of either or both sexes. There are also anecdotal reports of the occasional occurrence of twin fetuses in the dugong (Norris 1960, Jarman 1966, Thomas 1966, Bertram and Bertram 1968).

The carcass studies of the Florida manatee (Marmontel 1995) suggest that the incidence of twinning is 4%. This estimate is higher than the percentages observed in nursing manatee young in samples of free-ranging individuals from the east (1.79%, O'Shea and Hartley 1995) and west (1.4%, Rathbun et al. 1995) coasts. The discrepancy could be explained by (1) the estimates for free-ranging animals being lowered by the incidence of neonatal death of one or both twins; and (2) the estimates for carcasses being inflated by an increase in female mortality attributable to twinning. The true incidence is likely to lie somewhere between these estimates.

One fetus has been found in each of the 34 pregnant dugongs examined in recent years (Marsh 1995) and a litter size of one is reported for both the Amazonian manatee (Husar 1977, Best 1984, Timm et al. 1986) and the West African manatee (Beal 1939). Taken together, these results suggest that the mean litter size is close to one for all species of modern sirenian.

The size of sirenian neonates varies both among and within species (Table 6-5). The variation probably mirrors specific and individual differences in adult sizes. I suggest three reasons for the variation within a species: (1) the difficulty of distinguishing true neonates from large fetuses and young calves in carcass studies; (2) variation in the length of gestation; and (3) variation in neonatal size, which is independent of individual variation in the length of gestation and may reflect variation in the size of the mother. None of the data is sufficient for the method of estimating size at birth recommended by Perrin and Reilly (1984), and information for both the Florida manatee and the dugong has been expressed as a mean birth length. This tends to overestimate the actual size at birth (Perrin and Reilly 1984).

Reliable estimates of the sex ratio at birth are available only for the Florida manatee. Although they are not significantly different from 1:1 (O'Shea and Hartley 1995, Rathbun et al. 1995), the sex ratio of carcasses aged less than 1 year slightly favored males (Marmontel 1995). This may indicate a sex difference in neonatal mortality rather than a difference in the sex ratio at birth. The limited data for the dugong (Marsh et al. 1984c) also suggest a 1:1 sex ratio at birth.

Calf Dependency

Field observations of the time during which calves accompany their mothers have been obtained for the Florida manatee based on long-term observations of individual females that are (1) recognized at winter refugia on the basis of their scars, and/or (2) that give birth and wean young while being radio-tracked. Neither source of information is conducive to precise estimates. However, the trend is clear. Most calves (67%, O'Shea and Hartley 1995; 77%, Rathbun et al. 1995) are seen with their mothers during only one winter season; the remainder accompany their mothers for two winters. This pattern is independent of the sex of the calf, but may vary with the age of the mother; calves of young females seem more likely to be dependent for two winters than those of older females. Data from radio-tracked Florida manatees also indicate that some calves accompany their mothers for about 1 year, whereas other calves have been nursed for up to 24 months before weaning (Reid et al. 1995). Data on the length of lactation in dugongs are extremely sparse. A calf estimated to be 1.5-year-old (on the basis of dentinal layer counts) was caught in a shark net with its presumed mother who was still lactating. This suggests

Table 6-5. Sizes of Neonates and Adults of Various Species/Subspecies of Manatees and Dugongs

Species/Subspecies	Body Length (m, range [mean])[a]	Adult Length (m, mean)	Neonatal Body Weight (kg, range)	Adult Body Weight (kg, mean)	Sources of Data on Neonates
Florida manatee	0.80–1.60(1.22)	3	30–50	500	Odell 1982, O'Shea and Hartley 1995, Marmontel 1995, Rathbun et al. 1995
Amazonian manatee	0.85–1.05	2.8[b]	10–15	480[b]	Best 1982
West African manatee	1.00	[c]			Cadenat 1957
Dugong	1.00–1.30(1.15)	2.7	25–35	250–300	Marsh et al. 1984c, Marsh 1995

[a]If available.

[b]Measurements of large individuals not means.

[c]Not available but probably similar to Florida manatee.

The adult sizes are from Reynolds and Odell (1991).

that lactation can last at least 1.5 years (Marsh et al. 1984c), although dugongs start eating sea grass soon after birth (Marsh et al. 1982).

Pregnancy Rate and Calving Interval

The calving interval of the Florida manatee has been estimated by direct observation of known free-ranging individuals (O'Shea and Hartley 1995, Rathbun et al. 1995, Reid et al. 1995) and in captivity (Odell et al. 1995). These estimates (Table 6-6) suggest that on average, free-ranging manatees have a calf about every 2.5 years.

Marmontel (1995) and Marsh (1995) have also estimated the interbirth interval from carcasses of Florida manatees and dugongs, respectively. They estimated the interbirth interval as the reciprocal of the annual pregnancy rate. The annual pregnancy rate is the percentage of mature females that is pregnant (including those pregnant and lactating) divided by the length of gestation in years (Perrin and Reilly 1984). Calculation of the annual pregnancy rate assumes that (1) the length of the gestation period is known accurately; (2) all pregnancies are detected; (3) there are no biases due to seasonal birthing; and (4) the distribution of reproductive status in the sample is representative of the population.

None of these assumptions is likely to be true for existing studies of sirenians. Marmontel (1995) and Marsh (1995) attempted to address the first two assumptions by calculating calving intervals (1) using three estimates of the gestation period (12, 13, and 14 months); and (2) estimating the proportion pregnant based on both (a) a conservative scenario using confirmed pregnancies only, and (b) an optimistic scenario based on all pregnancies suggested by various indicators of reproductive condition. Marmontel also calculated calving intervals based on an intermediate scenario.

These data are summarized in Table 6-7, which indicates that the mean estimates of calving interval for Florida manatees tend to be between 2.5 and 5 years. Comparison of the data in Tables 6-6 and 6-7 suggests that the estimates based on all possible pregnancies are more reliable than those based on confirmed pregnancies only. Although the minimum interbirth interval for dugongs is not very different from that of manatees, the maximum estimates are longer, ranging up to 6.8 years.

Table 6-6. Interbirth Intervals of Free-Ranging and Captive Florida Manatees Based on Observations of Known Individuals

Location	No. of Females	No. of Births	Interbirth Interval (years mean ± SE; range)	References
Blue Springs	7	25	2.6 ± 0.17	O'Shea and Hartley 1995
Crystal River	33	99	2.48 ± 0.08 (1–5)	Rathbun et al. 1995
Atlantic coast	10	11	2.6 (2–4)	Reid et al. 1995
Captivity			(1.2–8.6)	Odell et al. 1995

Table 6-7. Estimates of Interbirth Interval (± SE) of Florida Manatees (Marmontel 1995) and Dugongs (Marsh 1995) Based on the Annual Pregnancy Rate and Three Possible Gestation Periods (12, 13, and 14 Months)

		Estimated Interbirth Interval (± SE) for Three Possible Gestation Periods		
Location	No. of Females	12 Months	13 Months	14 Months
Manatee				
Florida	212	4.5 ± 1.4	4.9 ± 1.6	5.3 ± 1.9
	286	**2.5 ± 0.7**	**2.7 ± 0.8**	**3.0 ± 1.0**
Dugong				
Numbulwar	86	3.1 ± 0.47	3.3 ± 0.55	3.6 ± 0.62
	86	**2.7 ± 0.38**	**2.9 ± 0.43**	**3.1 ± 0.49**
Townsville	18	4.5 ± 1.98	4.9 ± 2.26	5.3 ± 2.55
Daru	168	5.8 ± 0.98	6.3 ± 1.11	6.8 ± 1.25
	168	**4.9 ± 0.76**	**5.4 ± 0.86**	**5.8 ± 0.97**

Estimates are based on confirmed pregnancies and all possible pregnancies (**bold**).

A maximum of nine placental scars has been counted in a dugong uterus. Assuming that the scars persist, the calving interval can also be estimated by regressing the number of placental scars against age for parous dugongs. This calculation gave values in the same range as those in Table 6-7 (Marsh et al. 1984c). Placental scars do not persist in manatees and cannot be used to estimate the number of parous events (Marmontel 1995).

Marsh (1989) necropsied one lactating female dugong with a 41-cm-long fetus, indicating that ovulation is not always suppressed by lactation in this species. In contrast, in Florida manatees, there are only two confirmed cases of females that were simultaneously pregnant and lactating. Both females were carrying very large fetuses and Marmontel (1995) considered that the milk was probably for the new infants.

Evidence of Senescence

The age determination studies indicates that both dugongs and manatees are long-lived animals. The maximum estimated ages are 73 years for dugongs (Marsh 1995) and 59 years for manatees (Marmontel et al. 1996). Although there are reports of female manatees that have not given birth for long periods (O'Shea and Hartley 1995, Rathbun et al. 1995) and old male dugongs whose testicular histology suggested prolonged aspermatogenesis (Marsh et al. 1984c), there is no evidence of a definite postreproductive stage in the life cycle of female or male sirenians, similar to that recorded for some female cetacean species (Marsh and Kasuya 1986). Marmontel (1995) could find no evidence of age-related changes in the fecundity of Florida manatees older than 4 to 5 years, but warned that this might be an artifact of sample size. It may also reflect technical difficulties in estimating the ages of older individuals due to bone resorption. Although determining whether or not senescence in reproduction occurs in sirenians is of theoretical interest, the effect of this knowledge on our understanding of sirenian population dynamics is likely to be trivial (see Eberhardt and O'Shea 1995).

Timing of Reproductive Activity in Mature Sirenians

Perhaps because sirenians are limited to the subtropics and tropics, some early researchers assumed that they breed year-round (e.g., see Hartman 1979). However, there is increasing evidence that the reproductive activity of Florida and Amazonian manatees and dugongs is diffusely seasonal (Fig. 6-18) and that sirenians have the capacity to delay reproduction when their food supply is low.

Seasonality of Reproduction

FEMALES. Data from necropsies indicate that female Florida manatees (Marmontel 1995) and dugongs (Marsh et al. 1984c, Marsh 1995) do not breed continuously. Graafian follicles and/or corpora lutea were absent from the ovaries of 33% of 27 mature female manatees and 56% of 25 mature female dugongs whose ovaries were examined histologically and that were neither pregnant nor lactating. Although the reproductive activity of the females in a population is not highly coordinated, there is evidence of some seasonality. The ovaries of female dugongs in Northern Australia and southern Papua New Guinea are more likely to contain follicles or corpora lutea in the second half of the year (winter/spring) than in summer and autumn (Marsh et al. 1984c, Marsh 1995). Sterile cycles seem to be common (Marsh et al. 1984a) and may also occur in Florida manatees (Marmontel 1995, Rathbun et al. 1995).

Odell et al. (1995) note that captive Florida manatees participate in mating behavior throughout pregnancy, suggesting that females mate even when they are not fertile. This behavior may be an artifact of captivity but observations of mating herds in winter when at least some males are probably not spermatogenic suggest otherwise (J. E. Reynolds III, pers. comm. 1996). These observations make the temporal patterns observed in mating activity of free-ranging sirenians difficult to interpret in the absence of information on the levels of reproductive hormones in female participants.

Despite these methodological difficulties, observations suggest some seasonality of mating activity (Fig. 6-18). In Florida, manatee mating is concentrated in spring and summer, a pattern similar to that observed in dugongs in Queensland by Preen (1989) and H. Marsh (unpublished data), but inconsistent with the observations of Anderson and Birtles (1978) in the same region. Calving also appears to show some seasonality. Observations of free-ranging and captive animals and necropsy studies indicate that very few calves are born to free-ranging or captive Florida manatees in the winter months (Odell et al. 1995, O'Shea and Hartley 1995, Marmontel 1995, Rathbun et al. 1995, Reid et al. 1995). Although births have been recorded throughout the remainder of the year in Florida, calving peaks in spring and early summer (Fig. 6-18). The timing of births is similar for the dugong in northern Australia/southern Papua New Guinea (Fig. 6-18). Anecdotal information from this region suggests that the calving period may be longer in lower latitudes than at the southern limit of the dugong's range (Fig. 6-18), but insufficient data are available to test this statistically. Calving is also seasonal in the Amazonian manatee, although this species occurs in equatorial latitudes. In Brazil, calves are born between December and July, especially during the period from February to May when the water level in Amazonia is highest (Best 1982). In Ecuador, young are apparently born in January and June (Timm et al. 1986). Powell (1996) reports that African manatees mate

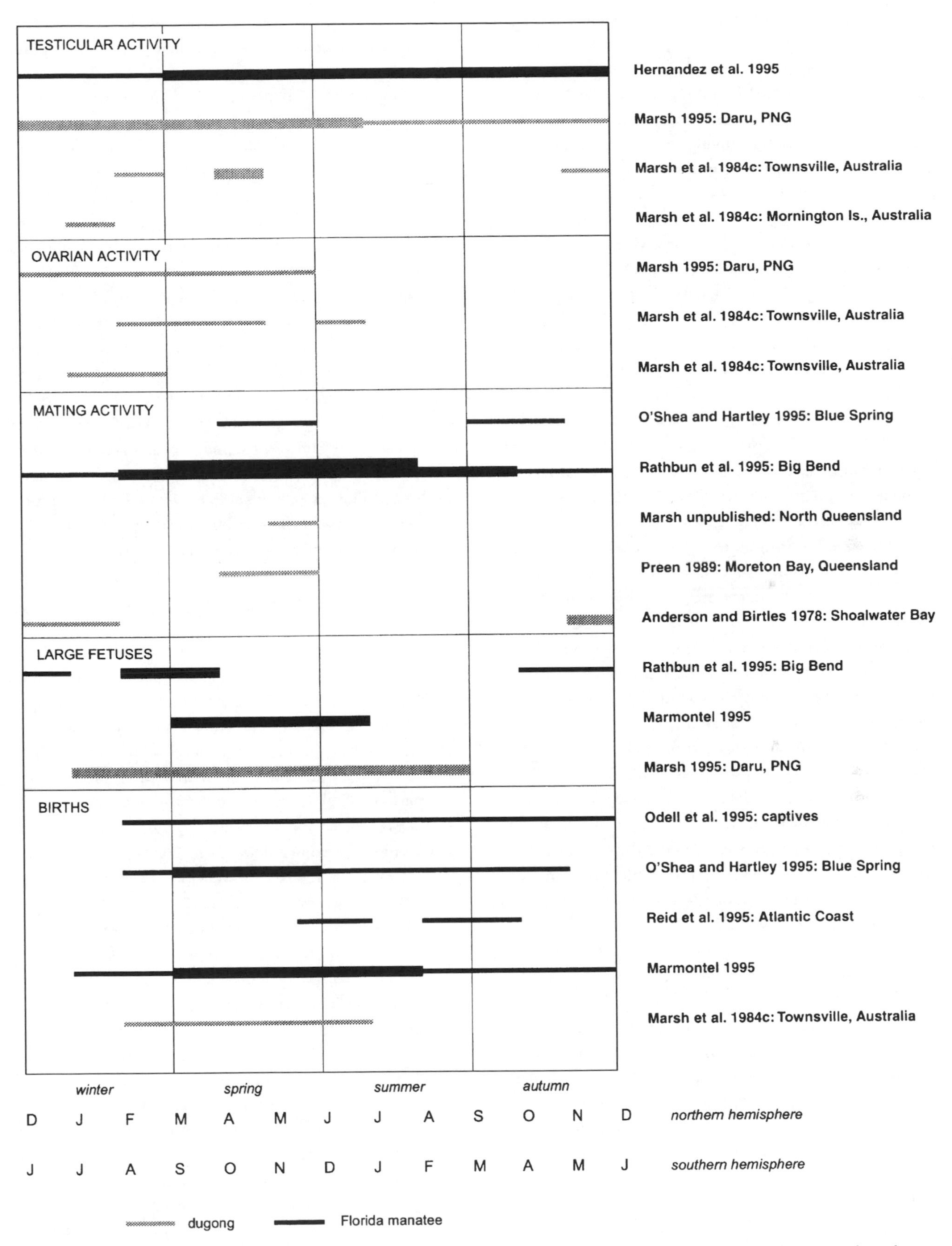

Figure 6-18. Timing of various indices of breeding in Florida manatees and dugongs relative to the seasons at the locations where the research was performed. Line thickness is an indication of the proportion of the population exhibiting the index.

during the rainy season between June and September when water levels are rising.

The timing of births in the Florida manatee reduces the likelihood that calves will be exposed to potentially dangerous low winter temperatures. However, this cannot be the explanation for seasonal calving in sirenians living in the tropics. In all species, calving peaks at the time of maximum plant productivity. Timing calving to coincide with plant productivity helps the mother meet the high energy demands of her offspring during late pregnancy and early lactation. Both free-ranging dugongs (Marsh et al. 1982) and free-ranging and captive Florida manatees (Hartman 1979, Odell 1982) start to eat plants a few weeks after birth; therefore, the calf can also supplement its milk intake with abundant new shoots that are low in fiber and high in soluble carbohydrates and nitrogen.

MALES. The reproductive status of mature male sirenians has been evaluated on the basis of histological examination of the testes and/or epididymides. This provides information on the reproductive status of a male at the time of his death but gives few insights on his reproductive history. It is clear from these studies that not all of the mature males in a population of Florida manatees or dugongs are producing spermatozoa continuously. For example, more than half of 41 pubertal and mature males from north Queensland and 141 mature male dugongs from Daru (Papua New Guinea) were infertile at the time of sampling. Spermatogenic activity is not highly coordinated within a population. For example, the testes of 16 mature male dugongs from Mornington Island in northern Australia were examined, all of which were killed in July. The testes of four males were in active spermatogenesis, four were approaching spermatogenesis, four intermediate, and four resting (Marsh et al. 1984a). However, it is difficult to confirm from such studies whether (1) individual males have periods of sexual inactivity followed by periods of sexual recrudescence and activity, or (2) the reproductive activity of some males is permanently suppressed (Marsh et al. 1984b, Hernandez et al. 1995). It is likely that both these explanations are true. Some dugong males had regressed testes (*sensu* Marsh et al. 1984b), suggesting that they had been sterile for long periods or even permanently (Marsh et al. 1984b). On the other hand, there is evidence of a seasonal pattern of gonadal activity that overlaps the one in females. Dugongs with active testes (fully spermatogenic or recrudescent testes *sensu* Marsh et al. 1984c) were a significantly higher proportion of the sample from Daru, Papua New Guinea between June and January than between February and May (Marsh 1995). Male manatees in Florida also show decreased spermatogenesis in winter (Hernandez et al. 1995), which is consistent with the frequency of mating herds (Rathbun et al. 1995).

Seasonality of testicular activity is more difficult to explain than seasonality of ovarian activity and calving as there is no evidence that sperm production per se is energetically expensive. However, the reproductive activity of male sirenians must be demanding of time and energy. Male manatees seem to travel further in search of females in estrus during warm months (Bengtson 1981), and if most females are not in estrus in winter, selection would favor reduction in testicular activity of males in winter to save energy. Geist (1974) points out that in the tropics the overall cost of social life for males with breeding seasons extending over several months should be higher than for those of temperate species where the mating season is shorter. He postulated that mechanisms may evolve to ration the mating activity of each male in species where the breeding season is prolonged over several months. The discontinuous breeding pattern of male sirenians would allow an individual to recuperate from his reproductive activities and increase his life expectancy. This would allow him to sire more offspring than if he were in rut continuously and died young (Marsh et al. 1984c).

Tendency to Delay Reproduction

Like most other long-lived species, sirenians are iteroparous, producing their offspring in a series of separate events. As discussed above, neither mature males nor mature females are continuously in breeding condition. This situation cannot be attributed solely to some seasonality of reproduction. There is considerable individual variation in both the prereproductive period and calving interval of both Florida manatees and dugongs (Tables 6-3, 6-4, 6-6, and 6-7) suggesting that sirenians postpone breeding under certain conditions.

The evidence for Florida manatees delaying reproduction is somewhat equivocal and hard to interpret. Necropsy data suggest that the pregnancy rate of Florida manatees increased between 1976 and 1991 (Marmontel 1995). The reason for such change is unknown. An increase in the annual pregnancy rate is expected in exploited populations as a density-dependent response. However, this seems an unlikely explanation for Marmontel's observations as the increases in numbers of manatees counted in aggregation areas in winter suggest some increase in the Florida population from the 1970s through the 1980s (O'Shea and Ackerman 1995). In contrast to Marmontel's results, Reynolds and Wilcox (1994) reported a decline in the percentage of calves in winter counts of manatees at some power plants during a 10-year period to winter 1991–1992, particularly at Atlantic coast aggregation sites.

The evidence for dugongs delaying reproduction is stronger. Marked temporal fluctuations have been documented in the following population parameters: the apparent pregnancy rate, the proportion of juveniles, and the incidence of testicular activity. These fluctuations seem to track major changes in their food supply. Anecdotal reports (Johannes and MacFarlane 1991) suggest that there was a major dieback of sea grasses in Torres Strait (between Australia and Papua New Guinea) in the mid-1970s. Nietschmann (1984) also reports that sea grasses were overgrazed in Torres Strait in 1976–1977. Nietschmann and Nietschmann (1981) observed that what the Torres Strait Islanders call *wati dangal* (lean dugongs with poor-tasting meat) were quite common in Torres Strait during this period. Hudson (1986) presents anecdotal evidence that none of 35 dugong females caught in a native fishery in the region between October 1976 and July 1977 was pregnant. Carcasses were collected over the succeeding 4 years (1978–1982). The proportion of mature females that was pregnant increased monotonically from 0.09 to 0.35 during this time (Marsh 1995). The difference among years was significant and was paralleled by a significant increase in the proportion of males with active testes between 1978 and 1981 (Marsh 1995), suggesting that a common factor (presumably the sea grass dieback and recovery) was affecting both female and male reproductive activity during this period.

The loss of more than 1000 km^2 of sea grass in Hervey Bay in southern Queensland in 1992 after two floods and a cyclone was associated with regional reduction in the dugong population from an estimated 2,206 ± 420 animals in 1988 to 600 ± 126, 21 months later (Preen and Marsh 1995). The proportion of the dugong population classified as calves during aerial surveys also declined from 22% in 1988 to 2.2% in 1993 and 1.5% in 1994 (Marsh et al. 1995, Marsh and Corkeron 1996), suggesting that the impacts of habitat loss on fecundity may last several years.

Best (1983) provides anecdotal evidence suggesting that Amazonian manatees may also be forced to delay reproduction during prolonged dry seasons when they may have to fast for up to almost 7 months. This leaves them in an emaciated condition similar to the *wati dangal* described above. Best (1983) cautions that large-scale deforestation in the Amazon may exacerbate the impacts of prolonged dry seasons on Amazonian manatees.

Implications for Conservation

These observations suggest considerable potential plasticity in the life history parameters of sirenians in response to food availability. If this is true, habitat conservation is critical. Food shortages probably cause sirenians to reproduce later and less often.

Sea grasses are prone to large-scale episodic diebacks (Poiner and Peterken 1995). Most losses, both natural and anthropogenic, are attributed to reduced light intensity due to sedimentation and increased epiphytism from nutrient enrichment. In some cases sediment instability, dredging, and poor catchment management interact to make the process more complex (Poiner and Peterken 1995). These problems are particularly acute for dugongs, which are essentially sea grass specialists (Marsh et al. 1982; but see Heinsohn and Spain 1974, and Preen 1995).

Florida manatees are opportunistic, generalist feeders on marine and freshwater plants (Reynolds and Odell 1991). In Florida, their habitat has increased dramatically due to the proliferation of exotic weeds and warm water from power plants. They reach sexual maturity earlier and have a higher calving rate than dugongs (see Tables 6-3, 6-4, 6-6, and 6-7). This means that the maximum rate of increase of a manatee population is likely to be higher than that for dugongs (Table 6-8) and it is significant that the fossil evidence suggests that trichecids may have displaced dugongids from the Caribbean–West Atlantic in the late Pliocene or early Pleistocene about 1.9 million years ago (Domning 1982, and pers. comm. 1996). These life history attributes as well as dietary specialization may have enhanced the outcome of this competitive displacement if it occurred.

Thus, manatee populations should be able to withstand anthropogenic impacts somewhat better than dugongs. There are indications that manatee numbers increased in Florida between the 1970s and 1990s (O'Shea and Ackerman 1995) despite a growth in the number of deaths attributable to human factors during this period (Reynolds 1995). In contrast, dugongs have declined along much of the urbanized

Table 6-8. Annual Percentage Rate of Increase of Stable Sirenian Populations Estimated Using a Stage-Based Leslie Matrix Population Model (Crouse et al. 1987; Somers 1994) for Various Combinations of Prereproductive Period and Interbirth Interval (Tables 6-2, 6-5, 6-6)

	Mean Interbirth Interval (years)		
Mean Age of First Birth (years)	2.5	3	5
4 (Florida manatee)	6.6	5.2	1.9
10 (Townsville dugong)	3.5	2.7	0.5

The assumed annual survival based on empirical data for the Florida manatee (Eberhardt and O'Shea 1995) is 0 to 2 years (0.822); 3 to 5 years (0.965). The model is truncated at age 50. Extending it beyond this age makes only a trivial difference.

Queensland coast since the mid-1980s (Marsh et al. 1995), although the human population density is much lower in Queensland than in Florida.

But perhaps these differences between Florida manatees and dugongs are relevant only in marginal situations. Marmontel (1993) used the computer program VORTEX to model the viability of the manatee in Florida under various sets of reproductive and mortality parameters. She determined that the population would be driven to extinction in the long term (1000 years) if adult mortality increased by 10% or if fecundity declined by 10%.

Given the inevitable increase in the human population, the mortality of dugongs and manatees will likely increase throughout most of their ranges. Their reproductive biology means that survivorship, particularly adult survivorship, must be high for populations to be maintained (Marmontel 1993, Eberhardt and O'Shea 1995, Marsh 1995). The prospects for the long-term survival of sirenians look bleak unless the momentum for their conservation is substantially increased.

Reproduction in Whales, Dolphins, and Porpoises [C. L.]

The cetaceans belong to two main categories: baleen whales and toothed whales. The former are all large whales, but the second category comprises a size range from the large sperm whale (*Physeter macrocephalus*), down to the small harbor porpoise (*Phocoena phocoena*), and a number of small freshwater river dolphins. The distinction between the two categories is important because the ecology of baleen whales, such as the southern blue (*Balaenoptera musculus*), fin (*Balaenopetera physalus*), and humpback (*Megaptera novaeangliae*) whales, incorporates a regular cycle of seasonal migration, often over long distances involving changes of latitude and feeding (Lockyer and Brown 1981). The entire life history, and in particular the reproduction, is geared to this cycle (Mackintosh 1965). The odontocetes represent a variety of life styles, some specialized to adapt to seasonal changes, but not as rigidly as the baleen whales (Gaskin 1982). Certain species will be referred to frequently either because they represent a good model (e.g., the fin whale for the baleen whales) or there is much known about them.

It is important to recognize the potential sources of information on reproduction in cetaceans. Animals have been largely studied from dissections of carcasses and subsequent morphological and histological samples (e.g., Laws 1961, Marsh and Kasuya 1984, Read and Hohn 1995). Yet again aspects of behavior and physiology have been documented from studies of animals both in captivity (e.g., Schroeder 1990, Robeck et al. 1994) and in the wild (e.g., Tyack and Whitehead 1983, Swartz 1986, Thomas 1986, McSweeney et al. 1989, Scott et al. 1990).

Reproductive Cycles

Mysticetes

There is slight sexual dimorphism in body size in the mysticetes, in which the female generally attains a body length about 5% larger than the male (Lockyer 1981a). Possibly the larger size of the female is to facilitate the carrying of a large fetus with the resulting large neonate. The fin whale has a 2-year reproductive cycle comprising a gestation period of about 11 months and a lactation period of about 6 to 7 months, followed by a period of anestrus (Laws 1961). The cycle starts in winter in low latitudes and warmer waters with ovulation and conception leading to pregnancy, during which the female migrates to summer feeding grounds in higher latitudes. The female returns to low latitudes to give birth in winter. The calf is weaned when the female migrates into high latitudinal waters for feeding during the summer of the second year (Mackintosh 1965, Lockyer 1984a). The generalized cycle shown in Figure 6-19 for the southern fin whale is applicable to other species of large baleen whales, although individual timings of conception, birth, and lactation may vary and are presented in Table 6-9. Blue, humpback, sei (*Balaenoptera borealis*), and gray whales (*Eschrichtius robustus*), all tend to follow a 2-year seasonal cycle of migration, feeding, and breeding. The minke whale (*Balaenoptera acutorostrata*), bowhead (*Balaena mysticetus*), and right whales (*Eubalaena glacialis, E. australis*), are exceptions with a different breeding interval. The minke whale may reproduce annually, whereas the balaenid whales may reproduce once only every 3 or 4 years (Evans 1987). Evidence for a seasonal reproductive cycle in male baleen whales comes from behavioral observations on coastally breeding species, such as the right whales that mate in winter, and from seasonal increases in testis weight and spermatogenic activity in humpback, gray, blue, and fin whales (Lockyer 1984a). In these species, male seasonal reproductive activity coincides with the period of estrus in the female. A protracted period of breeding for both sexes has the advantage of ensuring conception, should pregnancy fail or terminate in the first instance.

In fin whales, the summer period of feeding has been shown to be critical for energy storage in the form of fat to support lactation, which commences in the following winter. In fin whales, the seasonal cycle of migration, feeding, and subsequent fattening has a great influence on reproductive interval, ovulation rate, and fecundity, as well as (by in-

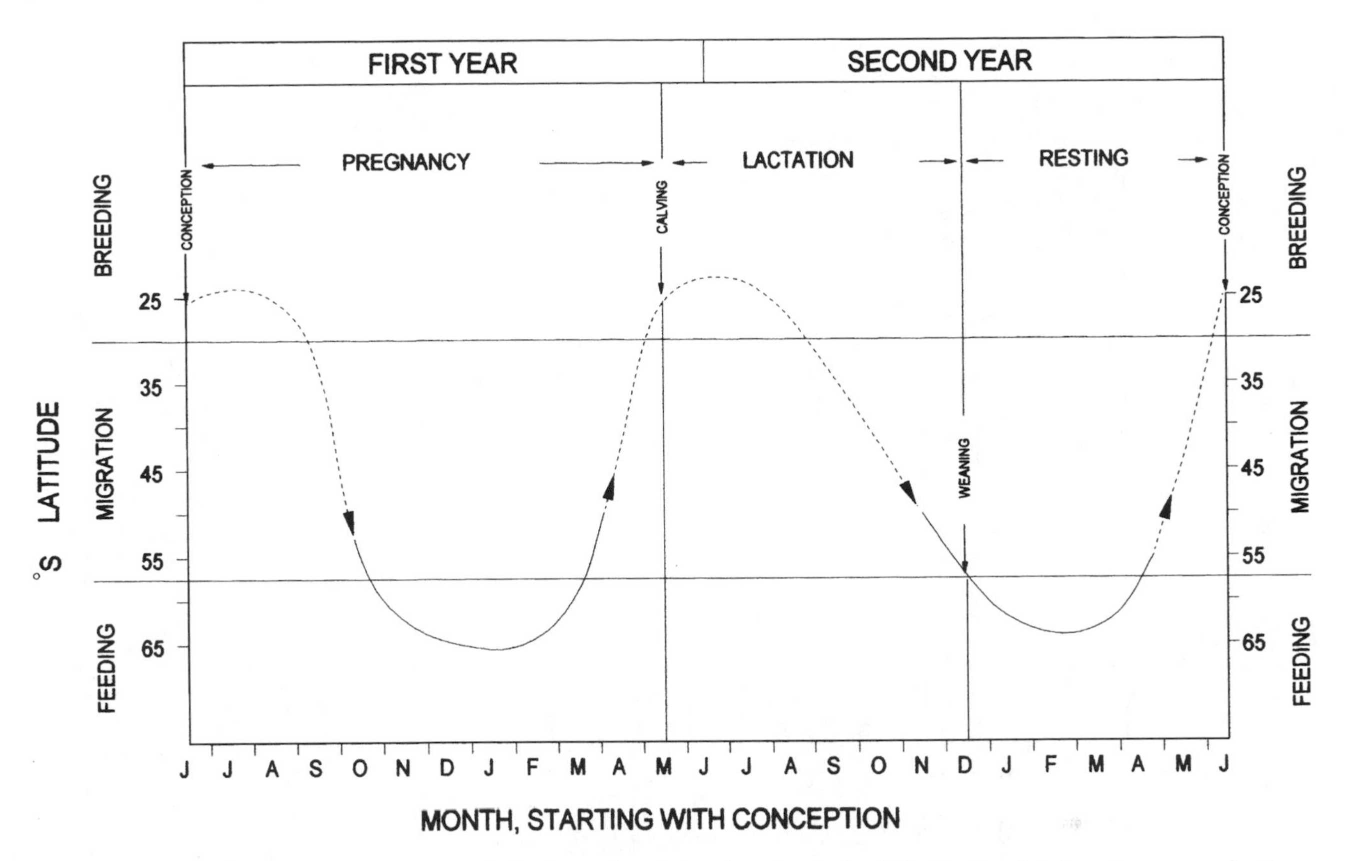

Figure 6-19. Schematic 2-year reproductive cycle for the southern hemisphere female fin whale. The fin whale has a 2-year reproductive cycle comprising a gestation period of about 11 months and a lactation period of about 6 to 7 months, followed by a period of anestrus. The cycle starts in winter in low latitudes and warmer waters with ovulation and conception leading to pregnancy, during which the female migrates to summer feeding grounds in higher latitudes. The female returns to low latitudes to give birth in winter. The calf is weaned when the female migrates into high latitudinal waters for feeding during the summer of the second year. (Courtesy of the International Whaling Commission; figure redrawn by S. A. Rommel with permission of the author.)

ference) survival of the calf. The summer feeding period in the Antarctic lasts about 4 months, but pregnant females are the first to arrive on the summer feeding grounds and they are among the last to depart, thus taking the opportunity to fatten as much as possible. Lockyer (1981a, 1987a) has calculated that the difference in body mass in the form of fat energy storage between the end of lactation and late pregnancy could be 50% to 75% of lactating body mass. The assumption is that much of this is available for milk production for the calf through catabolic breakdown of the fat.

The baleen whales are generally long lived, with ages being recorded from growth layer groups (GLGs) in ear plugs (Lockyer 1984b). In the balaenopterid species (Martin 1990), ages of up to 94 years have been recorded in the fin whale and potential longevity may be up to 70 years in sei whales. The smaller minke whale may live up to 60 years. Most rorquals have been long exploited, and thus their potential longevity may rarely be attained. Nevertheless, their long life span and female ability to produce offspring into old age means that some species, such as fin and minke whales, could experience as many as 40 pregnancies in a lifetime if ovulations are used as an indicator (Laws 1961, Lockyer 1987b).

Odontocetes

The odontocetes comprise a broad group of species ranging greatly in body form and size. The largest species is the sperm whale, which exhibits extreme sexual dimorphism with the adult males being as much as 1.6 times larger than the female (Best 1970, Lockyer 1981b). Such sexual dimorphism also exists in some smaller species, but is not so marked. Also, in the smaller marine species the larger size may be reached by either sex, for example, males are larger in killer whales (*Orcinus orca*), whereas females are larger in harbor porpoises. In river dolphins, which are all at the small end of the size scale, the female is generally larger except in one species, the Amazonian river dolphin (*Inia geoffrensis*) (Brownell 1984). Virtually all species have a breeding and calving season, but unlike the baleen whales, the season is often

Table 6-9. Reproductive and Associated Parameters for Baleen Whales

Species	Location	Peak Month(s) of Conception	Gestation Period (months)	Peak Months of Birth and Length of neonate (m)	Lactation Duration (months)	Length at Weaning (m)	Mean Lengths and Ages at Sexual Maturity: Length (m) Male	Length (m) Female	Age (yr)
Blue	Antarctic + South Africa	June/July	11	May; 7.0	7	12.8	22.6	24.0	5
Pygmy Blue	Antarctic	Two seasons: Main: Feb.–April Compl.: Oct.–Jan.?	< 12?	Main: March–April Compl.: Nov.–Jan.? 6.3	?	?	ca. 18.9	19.2	ca. 5–6
Fin	Antarctic (all areas) + South Africa	June/July	11	May; 6.4	7	11.5	19.0	20.0	10 → 6 or 7
	North Atlantic	December?	11?	November?	?6–7	?	16.8–17.6	17.7-19.1	11 → 8
	North Pacific	December	11–12	Nov./Dec.; 6.4	?	?	17.6–17.7	18.3–18.6	8–12
Sei	Antarctic (all areas) + South Africa	July	11–11.5	June; 4.5	6	8.0	13.6	14.0	11 → 8
	North Atlantic	Nov.–Feb.	> 10.75	Nov./Dec.; ?	?6	?	12.9	13.3	8
	North Pacific								
	(West)	Oct.–Nov.	> 10.5	Nov./Dec.; ?	7	?	12.8	13.3	10 → 6
	(East)	Oct.–Nov.	12–13	Nov.; 4.4	9	9.0	12.9	13.4	10
Minke	Antarctic + South Africa + Brazil	Aug.-Sept.	10	May/June; 2.8	4	4.5	7.2	8.0	14 → 6 or 7
	North Atlantic	February	10	Dec.; 2.6	< 6	ca. 4.5–5.5	6.9	7.3–7.45	7.3
	North Pacific	Feb./March	10	Dec./Jan.; 2.8	?6	?4.6	6.9	7.3	?
	Huanghai Sea	July –Sept.	10-11	May–July; 2.5-2.7	?	?	?	6.6 –7.0	?

Bryde	South Africa								
	(inshore)	Year-round	12	Year-round; 3.96	?6	ca. 7.1	12.0	12.5	ca. 10
	(offshore)	March	12	Feb./Mar.; 3.96	?6	ca. 7.1	(13.0)	(12.0)	8–11
	North Pacific (coastal Japan)	Mainly Dec. but protracted	?12	?Nov. but protracted 3.95	?	ca. 7.1	11.9	12.0	9
	North Pacific (pelagic)	Protracted	?	Protracted; > 3.95	?	?	11.9	12.0	10 and 8
	South Pacific (two stocks)	Protracted	?	Protracted; < 4.15	?	?	11.6–12.4	12.2–12.8	10
Humpback	Antarctic + Australia	July/Aug.	11.5	July/Aug.; 4.3	10.5–11	8.8	11.5	12.0	4 or 5(x 2)
	North Pacific								
	(West)	February	10–11	January; 4.1	10–11 or 12	ca. 8.0	11.6	12.0	11 (→ 2)?
	(East)		ca. 12						
Gray	Northeastern Pacific	December	13.75	Jan.–Feb.; 4.6	7	8.5	11.1	11.7	8
Bowhead	Arctic	May	ca. 12–13	April/May; 4.5	5–6	6.1	11.6	12.2–14.0	4
Pygmy Right	Southern hemisphere	?Probably extended	ca. 12	Probably extended? ca. 1.6-2.2	?5-6	ca. 3.2–3.6	?	?	?
Right	Southern hemisphere	Aug./Oct.	10	May/July; 5.5	?	?	13.0–16.0		?
	South Africa + South America			Aug.; 6.1					
	Northern hemisphere	?	10	?; 4.4-4.8?					

Adapted from Lockyer 1984a.

Age at sexual maturation derived from ear plug growth layers assuming one forms annually except in humpback whales where two may form annually. Where ?, ca., ( ) are indicated, the data are of limited reliability or based on small samples.

more protracted. Many species have a gestation period of less than 1 year, like the baleen whales. Lactation can be very variable, and rather than lasting a few months, can last for several years, as in sperm and pilot whales (*Globicephala* spp.). The function here can be social rather than nutritional in the later stages of lactation, and may be linked to schooling behavior. The reproductive interval can then be extended for 3 to 5 years, and is not necessarily associated with seasonal feeding habits or accumulation of body fat reserves. The period may be useful as a learning phase for the young in methods of cooperative feeding and foraging strategy where echolocation may be an important function (Brodie 1969).

The breeding season for the sperm whale appears to occur between October and December in the southern hemisphere and April and June in the northern hemisphere (Best et al. 1984). This information has been inferred from examination of ovarian activity studied from specimens collected from whaling operations. Best and Butterworth (1980) found evidence for synchrony in ovulation within schools of females. The synchronous estrus may be caused by the arrival of adult bulls, the effect of which would be advantageous in increasing the efficiency of fertilization, thereby minimizing the period that bulls must remain with the school and reducing the interference and disruption of the social organization within the school, especially to mother–infant bonds. Whitehead (1987) observed that bull sperm whales off the Galapagos Islands adopted a "searching" strategy moving between groups of females for a period of about 2 months, reaching a peak during April. Gestation lasts 14 to 15 months (Best et al. 1984, Whitehead et al. 1989), and true lactation lasts about 2 years (Best 1974), although suckling by the calf may occur up to age 7 to 8 years in females and 13 years in males (Best et al. 1984). This has been interpreted as a form of social behavior, and may not be significant nutritionally. Evidence from stomach contents suggests that solid food is taken by the calf at some stage before the end of its first year of life (Best et al. 1984).

Seasonality of mating and calving has been demonstrated for all delphinids that have been studied in detail (Perrin and Reilly 1984). However, depending on the duration of gestation and lactation, and the degree of seasonality, the reproductive interval may vary within a population because of individual circumstances. The length of gestation for small odontocetes varies from 10 to 16 months (Perrin and Reilly 1984), and recently killer whales have been reported to have a gestation of 17 months (Duffield et al. 1995). These times have been estimated from both observations on animals caught in fishery operations and captive live animals. Most dolphins have a gestation period of about 1 year. Pilot whales are now believed to have a gestation period closer to 1 year rather than the 15 to 16 months previously thought (Martin and Rothery 1993).

In river dolphins, most evidence points to a gestation period of less than 1 year (Brownell 1984, da Silva 1994), and in boutu (*Inia geoffrensis*), da Silva (1994) reported that peak births occurred right after the river flood period May to July in the Amazon, although the birth season was more protracted. Lactation probably lasts a year or longer in boutu, and most lactating females are observed during the dry season when the river levels are low and prey become concentrated (da Silva 1994).

Information for monodontids indicates a gestation period of about 14 to 15 months (Braham 1984), with conception in April to May and parturition between April and September. More recently, the gestation for Greenlandic belugas (*Delphinapterus leucas*) has been estimated to be at least 330 days (Heide-Jørgensen and Teilmann 1994), and mating probably takes place in May with implantation occurring in May to June. Birth takes place April to May, indicating that gestation may be less than 1 year, contrary to earlier findings. The proportion of pregnant females in the mature female population was estimated at 31%, suggesting that perhaps calf production is about once every 3 years. The period of lactation probably lasts 20 to 24 months, and Heide-Jørgensen and Teilmann (1994) found 10% of all mature females were simultaneously pregnant and lactating, representing about 16% of all lactating females.

The smaller cetaceans vary considerably in life span according to species. Age has been determined retrospectively from GLGs in teeth (Perrin and Myrick 1981). Bottlenose dolphins (*Tursiops truncatus*) may live for more than 50 years based on individuals monitored both in the wild and captivity and on tooth GLG studies (Hohn et al. 1989), and harbor porpoises may live up to 25 years (Lockyer 1995a,b). The Amazonian river dolphin may live in excess of 35 years (da Silva 1994), and pilot whales can live up to 60 years (Bloch et al. 1993, Kasuya and Tai 1993). Perhaps the longest lived odontocete is the sperm whale, which may live more than 70 years if age is based on tooth GLGs (Martin 1980, and pers. obs.). Clearly longer lived species can afford to invest more time in each offspring and still produce several calves during the lifetime. However, short-lived species, such as harbor porpoises, that generally live less than 25 years and perhaps only until age 15 years (Lockyer 1995b), must gear the reproductive cycle to enable the production of a calf annually whenever possible (Read and Hohn 1995). The gestation in the phocoenids is usually 10 to 11 months, and simultaneous pregnancy and lactation is observed (Gaskin et al. 1984). The reproductive cycle parameters for some odontocetes are shown in Table 6-10.

The timing of events in the reproductive cycle for all

Table 6-10. Reproductive and Associated Parameters for Some Odontocete Species

							Mean Lengths and Ages at Sexual Maturity		
			Gestation	Peak Months of	Lactation	Length at	Length (m)		
		Peak Month(s)	Period	Birth and Length	Duration	Weaning			
Species	Location	of Conception	(mo)	of Neonate (cm)	(years)	(cm)	♂	♀	Age (yr)
Sperm whale	Southern hemisphere	October–December	15–16	February–March; 400	>2	670–760	12.5–13.7	8.7	19(♂); 9(♀)
White whale (Beluga)	W. Greenland	May–June	>11	April–May; 160	?	?	3.9	3.45	6–7(♂); 4–7(♀)
Pilot whale (long-finned)	Northeast Atlantic	June (May–Sept.)	12	May–August; 177	3.67	?	4.94	3.75	14.3(♂); 8.7(♀)
Pilot whale (short-finned)	North Pacific northern Japan	August–January	14.5–16	December–January; 185	2–2.78	?	5.6	3.95	16.5(♂); 8.5(♀)
Killer whale	Northeast Pacific	?	17;(12–15)	October–March; 246	ca. 1.0	?	5.8	4.6–4.9	14–15 (♂+♀)
	Northeast Atlantic	October–December	17;(12–15)	December – ?; 208–220	<3.0?	?	5.8	4.6–4.9	15(♂); 8(♀)
False killer whale	Northeast Atlantic	?	15.5	?; 193	1.5	?	5.32	4.47	16(♂); 17(♀)
Bottlenose dolphin	Northwest Atlantic	Spring through June	12	March–June; 115	1.6	?	2.58	2.39	10–15(♂);ca. 10(♀)
Striped dolphin	Northwest Pacific	Two peaks: summer and winter	12	Summer and winter; 100.0	0.67–1.67	?	2.27	2.2	14(♂+♀)
Harbor porpoise	Bay of Fundy, Canada	June–Aug.	10.6	Mid-May; 78	0.67–0.75	ca. 115–118	?	1.43	3.15–3.44 (♂+♀)
	Northwest USA, California	July–Aug.	ca. 10	May–July; 80–86	?	?	1.3–1.49	1.42–1.52	4.6(♂+♀)
	British Isles	June–Aug.	10–11	June–July; ca. 70	?	?	>1.30	>1.40	3–4 (♂+♀)
	West Baltic	July–Aug.	11	June–July; 75	0.67	?	?	?	?
	Norway	?	ca. 10	May–July; 75 86	0.67	?	?	?	
Dall's porpoise	Coastal Japan	Mid-Aug.–late Oct., peaking Sept.	11.4	Aug.–Sept.; 99.7	0.5–3.5 (av. 2.07)	?	1.76–1.87	1.72–1.83	5–6(♂); 3–4(♀)
	North Pacific + Bering Sea	July and Aug.	11	Late Aug.; 100	ca. 2.0	?	?	?	?

Adapted from Gaskin et al. 1984, Perrin and Reilly 1984, Best et al. 1984, Donovan et al. 1993, Heide-Jørgensen and Teilmann 1994, Lockyer 1995a,b, Read 1990b,c, Kasuya 1995, Kasuya and Jones 1984, Hohn and Brownell 1990, Christensen 1984, Mead and Potter 1990, Duffield et al. 1995.

Where data are unavailable or incomplete, this is indicated by ?

cetaceans is clearly geared to optimize the seasonal changes in environmental conditions to benefit the ecology of the species and favor maximal survival of the young (see Bowen and Siniff, Chapter 8, this volume). Therefore, the cycle will differ slightly with each species according to that species' life style and ecology, and may even vary according to population (Urian et al. 1996).

Morphology and Histology

A general overview of morphology of the reproductive organs is provided by Pabst, Rommel, and McLellan (Chapter 2, this volume), and here it is intended only to describe specific cetacean features. The reproductive tract of cetaceans is typically mammalian. In individual females there are ovaries, fallopian tubes, uterus, vaginal opening to the vulva, and a hymen-equivalent possibly present in juveniles. In individual males there are testes, vasa deferentia, epididymides, and a penis, which normally is retracted within the body. Unlike pinnipeds, the male cetacean does not have a baculum in the penis. The penis is made erect by muscular contraction, is relatively elongate at the tip, and can be manipulated at will. The structure of the vagina in females is such that there is a pseudocervix (false cervix), an infolding of the vaginal wall (Harrison 1969). The uterus is bicornuate (twin horned), and the fetus usually develops in one of the horns. Proliferation of the endometrial lining of the uterus is stimulated as part of the seasonal reproductive cycle, and there is no menstrual blood loss as in some other mammalian species. The ovaries are both equally functional in baleen whales as far as ovulation is concerned. However, there is a tendency in odontocetes that only the left ovary ovulates initially, with right ovarian activity occurring later in life.

The ripe egg is released from the ovary with the subsequent development of the corpus luteum, which shrinks if fertilization does not take place, but continues to function if pregnancy occurs. In either event, the corpus luteum eventually changes into a smaller corpus albicans, which continues diminishing slowly until a permanent body associated with the ovulatory scar on the ovary surface remains. There is evidence that these bodies remain for the life of the animal. It appears that old corpora albicantia are not resorbed; this is rather unusual for mammals generally because usually these bodies disappear in time. Theoretically, cetacean ovaries should record the total ovarian cycles experienced, most of which represent pregnancies (Laws 1961, Ivashin 1984, Marsh and Kasuya 1984). Sometimes accessory corpora lutea occur in the pregnant ovary, and other bodies may also occur associated with failed ovulation, for example, corpora atretica, corpora aberrantia, and cystic follicles, as well as cysts of undetermined origin (Laws 1961; Perrin and Myrick 1981; Kasuya and Marsh 1984; Lockyer 1984a, 1987b; Lockyer and Smellie 1985). The regular seasonal ovulation in cetaceans, especially balaenopterid whales, permits another method for determining relative age after maturation. It is likely that cetaceans ovulate spontaneously, and Marsh (1985) reported such ovulation taking place in a solitary minke whale trapped in a lagoon off the Great Barrier Reef for 3 months before death.

The ovaries of the baleen whales may weigh several kilograms (Fig. 6-20), whereas those of a porpoise may only weigh a few grams (Lockyer 1984a). The corpus luteum of the blue whale in pregnancy may average 137 mm in diameter and weigh several hundred grams, whereas that of the minke whale may average 66 mm in diameter and weigh 156 g (Lockyer 1984a). The corpus luteum may shrink to less than 40% of its active size rapidly after birth, and then shrink to about 25% of size within a few years of birth in the minke whale (Lockyer 1987b). It is thought that the rate of shrinkage of the corpus albicans slows during pregnancy because of the effects of hormones. The ovaries may remain active throughout the lifetime of the female (discussed under Age and Reproductive Performance). However, there is evidence that in short-finned pilot whales (*Globicephala macrorhynchus*) there may be senescence of the ovaries in the oldest females of 50 years or so (Marsh and Kasuya 1984). Also, it should be pointed out that whereas both ovaries appear to be fully functional in baleen whales right from maturation (Lockyer 1987b) with little or no evidence of activity on one side more than another, the picture in many odontocetes is very different, with dominant ovulatory activity in the left ovary (Slijper 1949) and activity in the right ovary only starting after a certain number of corpora have been accumulated or there is disease rendering inactivity of the left ovary (e.g., short-finned pilot whales, Marsh and Kasuya 1984).

The reproductively mature males may only actively produce sperm, at least in quantity, seasonally. As discussed earlier, some species show a seasonality of breeding, and this is accompanied by seasonal growth, hypertrophy, and then atrophy of the testes.

The endometrial development is in concert with ovarian function (Lockyer and Smellie 1985), and proliferates during ovulation and pregnancy. During lactation the myometrium and endometrium shrink in thickness and glandular activity.

During pregnancy the developing embryo attaches via the placenta and umbilical cord (Fig. 6-20), both of which are completely lost during parturition. There is no subsequent

Figure 6-20. Fin whale fetus showing part of the umbilical cord and the two ovaries, one of which bears a large corpus luteum. (Sea Mammal Research Unit; formerly The Whale Research Unit, Institute of Oceanographic Sciences, U.K.)

bleeding or uterine shedding. Slijper (1949) provides detailed information about the umbilical cord and its size in relation to the mechanics of cetacean birth. There is no evidence of placental scars in the cetacean uterus, therefore these cannot be used as an indication of past pregnancies. The neonate is generally born tail first (Slijper 1949), after a period of uterine contractions, which are often very evident in captive dolphins, and expansion of the vulval opening. The female usually flexes her body during the birth process.

The young are suckled at two mammary slits positioned ventrally on either side of the midline just forward of the vulva, and each of which enclose a nipple normally retracted in the mammary slit fold. The mammary glands are elongate along the main body axis, and produce a thick creamy, sometimes yellowish milk that has a high fat content ranging from 14% to 53% (Lockyer 1984a, 1987a,c, 1993a).

Physiology of Reproduction

Hormone Cycles

Hormones and hormone cycles in pinnipeds, specifically, are discussed by Boyd earlier in this chapter. Much of what is known about hormone cycles in cetaceans has been discovered through captive breeding programs, where it has been possible to closely monitor the reproductive state by behavioral observation and blood and serum sampling for hormone assay. Hormone assays have also been performed on cetaceans taken in fisheries to assess pregnancy rates and breeding condition. In belugas a serum progesterone level higher than 3.00 ng/mL generally indicated pregnancy (Stewart 1994). In another study ofbelugas, blood serum and urine were tested for testosterone, progesterone, estrogen, and prolactin (Høier and Heide-Jørgensen 1994). In mature males, an average of 4.14 nmol/L testosterone were reported compared to 0.96 nmol/L in immatures. Pregnant females averaged 29.1 nmol/L progesterone, compared to 1.76 nmol/L in juveniles or nonpregnant individuals. Similarly, the estrogen levels were elevated in the former at 2.43 nmol/L compared to 0.37 nmol/L in the latter. Prolactin levels were high at 4.35 nmol/L in lactating females compared to 1.97 nmol/L in nonlactating females. Studies have also been reported on North Atlantic fin whales, using blood and urine collected from freshly dead animals taken by whalers (Kjeld et al. 1992). Recently, hormone studies have been undertaken on minke whales captured in the Antarctic, and comparisons made between ovarian information and blood hormone assays to ascertain reproductive status of females (Iga et al. 1996). Progesterone levels were 4.3 to 4.5 ng/mL in pregnant and ovulating females, and estradiol levels were 8.7 pg/mL in pregnant and 13.8 pg/mL in anestrous females. Juveniles had no detectable progesterone and their estradiol levels were about 7.4 pg/mL.

In captivity, radioimmunoassay is used to test levels of circulating progesterone and immunoreactive estrogens, the relative levels of which indicate ovulatory activity and likely pregnancy (Sawyer-Steffan et al. 1983, Schroeder 1990). Robeck et al. (1994) reported that levels of serum progesterone in bottlenose dolphins were generally more than 3000 pg/mL, decreasing to less than 1000 pg/mL within 1 month, indicating ovulation. Sustained levels above 3000 pg/mL were indicative of pregnancy. The baseline levels of progesterone and estrogen in noncycling female bottlenose dolphins averaged 286 pg/mL and 25 pg/mL, respectively. Robeck et al. (1994) reported evidence that bottlenose dolphins are spontaneous ovulators (Kirby and Ridgway 1984), with a 21- to 42-day estrous cycle. However, ovulatory cycles of variable length and seasonal reproductive activities may occur. Seasonal polyestrus and 1 to 2 years of anestrus followed by polyestrus may also occur. In a study of a captive colony of Hawaiian spinner dolphins (*Stenella longirostris*), a seasonal peak in testosterone levels in the blood of males was reported by Wells (1984).

Some cycles may be nonfertile in pubertal female bottlenose dolphins, and as reported later under Age and Reproductive Performance, pregnancy rate in newly mature females is lower than average. Even when conception occurs, the likelihood is greater that the calf will die as a result of poor parenting and difficulties during the birth (Bryden and Harrison 1986). Reduced fertility at this time may, therefore, not be disadvantageous to the population.

The effects of hormones on the development of the endometrium of cetaceans are marked and cyclical (Matthews 1948), and the increase in glandular presence, size, and activity in the stratum spongiosum of the mature, ovulating, and subsequently pregnant female compared to the immature and anestrous females are very noticeable (Lockyer and Smellie 1985). The overlying stratum compactum becomes increasingly engorged with blood and the surface proliferates with frond-like projections as pregnancy proceeds. The appearance of the endometrium in the lactating female is thin, and the glands have shrunk, and the stratum compactum has become smooth with a reduced blood capillary supply (Lockyer and Smellie 1985).

Lactation may be maintained in cetaceans for variable periods after parturition (up to several years in sperm whales), and as noted above, may serve dual functions of nutrition and social bonding for the calf. The hormonal aspects of lactation are especially interesting in relation to simultaneous pregnancy and lactation, the duration of which (lactation) may extend to several years in sperm whales, and the special cases of lactation in previously nonpregnant females with senescent ovaries. Kasuya and Marsh (1984) reported the continued lactation in short-finned pilot whales that were effectively past reproductive age. Two adult female bottlenose dolphins that were nonpregnant and nonlactating were able to commence lactation naturally after each had close contact with orphaned calves and were subsequently able to adopt a calf (Ridgway et al. 1995). The ensuing lactations were so successful that after several weeks the calves no longer required additional milk supplements.

Captive Breeding

There are now established bottlenose dolphin stud books (records of reproductive performance by both male and female captive animals) within North America and Europe. In response to both national and international regulations limiting the capture of bottlenose dolphins in the wild, there has been an upsurge in captive breeding programs, which also include artificial insemination to obtain new resources (Schroeder 1990, Schroeder and Keller 1990, Robeck et al. 1994). Species that have been successfully bred in captivity include bottlenose dolphin, killer whale, and Commerson's dolphin (*Cephalorhynchus commersoni*), to name the most successful.

Artificial insemination programs could facilitate breeding between dolphins geographically at some distance from each other, without the need to transfer animals. Such programs could also mean that the pedigree of any offspring is predictable. The program requires potentially receptive females to be monitored hormonally, and semen with spermatozoa must be collected from male candidates and stored frozen until required (Robeck et al. 1994). Semen has been successfully collected manually (Hill and Gilmartin 1977, Keller 1986) and also by electrical stimulation (Seager et al. 1981). Schroeder and Keller (1990) reported success in cryopreservation in liquid nitrogen of pelletized semen that had good sperm motility after thawing. Actual insemination techniques are detailed in Robeck et al. (1994), and the timing of semen introduction is controlled by results from endocrine monitoring and sonographic detection of ovulation in the female, as well as methods used in the inducing of ovulation (Robeck 1996). Such techniques may have greatest application in rare and endangered species such as some river dolphins, but have yet to be proved successful.

Species and Hybridization

There are several cases of hybrids being born in captivity, including crosses between bottlenose dolphins and Risso's dolphin (*Grampus griseus*) (Hirosaki et al. 1981), bottlenose dolphins and false killer whale (*Pseudorca crassidens*) (Nishiwaki and Tobayama 1982), and short-finned pilot whale and bottlenose dolphin (Lockyer 1993b). Most of these hybrids resulted in stillbirth or abortion, although there have been some survivors. This hybridization could be ascribed to the unnatural circumstances surrounding captivity, but more recently cases of hybridization in free-living balaenopterids have been documented genetically in Iceland (Arnason et al. 1991); two hybrids (a male and a female) had a blue whale mother and fin whale father, and a third (a male) had a fin whale mother and blue whale father. Furthermore, the female hybrid was reproductively active and in her second pregnancy as a result of mating with a blue whale, showing that in this case fertility was normal (Spilliaert et al. 1991). However, one of the males reached adult size and age but appeared to be sterile with undeveloped testes and no spermatogenesis. This kind of evidence suggests that speciation may not be as clear-cut as thought first by some taxonomists and geneticists.

Environmental Effects on Reproduction

Natural Ecological Factors

MYSTICETES. Off Iceland, the variation in food availability from year to year can influence the amount of fat storage in female fin whales; therefore, in years of great food abundance fat stores are very high, whereas in years of low abundance, the fat storage may be diminished. This can interfere with the 2-year reproductive cycle, so that in good years the female might produce a calf in consecutive years and in poor years the cycle could be prolonged to 3 years to enable fat reserves to build up; probably ovulation is suppressed if a certain threshold level of body weight or fat is not reached, a recognized strategy in terrestrial mammals (e.g., red deer [*Cervus elaphus*], Hamilton and Blaxter 1980; and reindeer [*Tarandus rangifer*], Leader-Williams and Ricketts 1982). The combined effects of sea temperature, plankton, and food production on body fat condition and fecundity have been investigated in fin whales off Iceland (Lockyer 1987a,c). There appeared to be a close correlation between these factors, some being directly and others indirectly linked to fecundity. The pattern of food abundance, body fat condition, and fecundity for female fin whales is shown in Figure 6-21. The age at sexual maturation and birth size have also been shown to vary with external environmental factors as well as exploitation (Lockyer 1990, Lockyer and Sigurjonsson 1992).

There is very little field evidence for reproductive effects in other baleen whales. It was believed at one time that the reduction of blue, fin, sei, and humpback whales in the southern oceans was anticipated to have the effect of releasing more food to the smaller competitive minke whale and allowing earlier maturation. A similar interpretation was also given for the earlier maturation of fin and sei whales with the dramatic reduction in numbers of their species. However, there is considerable controversy about this interpretation,

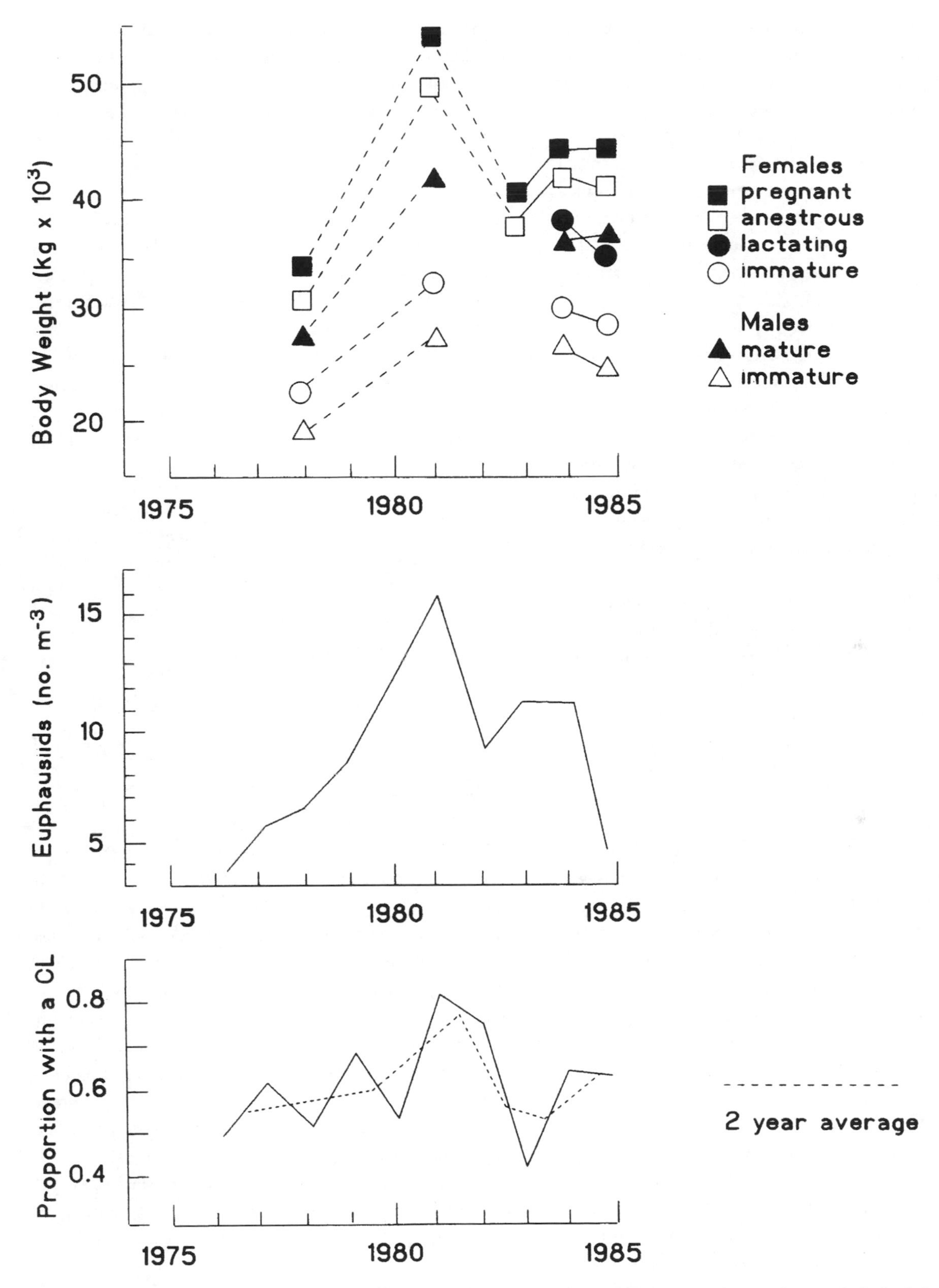

Figure 6-21. Yearly variations in female fin whale body condition (weight) for different reproductive classes, whale food abundance (euphausiid density), and potential whale fecundity (proportion of females with a corpus luteum) off southwestern Iceland. (Society for Marine Mammalogy; figure redrawn by S. A. Rommel with permission of the author.)

and a change in age at maturation may not equate directly with increase in pregnancy rate and recruitment to the population. This will be discussed more thoroughly under Significance of Reproduction in Management.

For humpback whales, estimated to have a 2- to 3-year reproductive cycle (Clapham and Mayo 1990), production of a calf each year is feasible. Weinrich et al. (1993) identified the same female in three consecutive years, and each year she was accompanied by a new calf in the southern Gulf of Maine. Subsequent monitoring showed that each calf survived to at least 2 years.

ODONTOCETES. There is some indirect evidence for environmental effects on reproduction in dolphins. The effects of the 1982 to 1983 El Niño off Peru diminished food available to mature female dusky dolphins (*Lagenorhynchus obscurus*). Pregnant and lactating females deposited poorly calcified dentinal GLGs in the teeth during the period of the El Niño, indicating stress in nutrition, whereas teeth of dolphins of other reproductive classes showed no such phenomena (Manzanilla 1989).

Similar interference in the GLG formation in teeth was observed in captive short-finned pilot whales from California. There was a link between the appearance of mineralization interference in the teeth and reproductive events in females such as ovulation and pregnancy, suggesting dietary change as the intermediary factor (Lockyer 1993b). Reproductive events, such as sexual maturation and parturition, have been correlated with the tooth GLG patterns of known-age free-ranging bottlenose dolphins in the Gulf of Mexico (Hohn et al. 1989). Such reproductive events were also correlated in teeth of *Stenella* (Klevezal and Myrick 1984). It is possible that severe dietary inadequacies might lead to similar phenomena as observed in fin whales, including lowered fecundity, reduced birth weight (Lockyer 1990), and perhaps lowered infant survival. Clearly this is a topic that merits some investigation because of the potentially significant effect factors such as dietary change may have on subsequent recruitment to a population.

Anthropogenic and Disease Factors

Anthropogenic and disease factors may include any foreign item introduced into the environment or a disease-causing agent that could affect reproduction. Also included are contaminants of all types (e.g., DDT, PCB, dioxins, chlordanes, toxophenes, and heavy metals [see O'Shea, Chapter 10, this volume]), and many recently described diseases caused by viruses (see also Geraci et al. [Twiss and Reeves 1999, Chapter 18]). There has been much interest in such factors especially in relation to reproduction, and not least are the potential effects on fecundity, fetal growth, contamination of milk, and subsequent calf survival. The possible teratological effects of contaminants in causing abnormalities of reproductive organs and fetuses is considered later under Pathology and Teratology. Many organochlorine compounds are toxic because they are able to mimic hormones. The following is a brief review of some studies on this huge topic.

Levels of organochlorines are generally lower in adult females than males for all cetacean species, because of lactational transfer to the calf and placental transfer to the fetus in utero (for example, see Addison and Brodie 1977; Reijnders 1980; Subramanian et al. 1987, 1988). The high transference level in milk is because of the concentration of organochlorines in lipophilic tissues such as blubber, which is generally used as an energy resource for milk production. The amount of transfer is related to the actual level in the mother, duration of lactation, milk quality in terms of fat content, and age of the mother (Stern et al. 1994).

MYSTICETES. Levels of contaminants are reportedly lower in the blubber of baleen whales than of odontocetes, with ranges in the former of 0.1 to 10 ppm wet weight for both DDT and PCBs (Wolman and Wilson 1970, Taruski et al. 1975, Aguilar and Jover 1982, Aguilar and Borrell 1988). Baleen whales mostly feed on euphausiids or planktonic organisms rather than fish and higher predators, and their consumption is thus at a lower trophic level than is the case for odontocetes. This information suggests that there are likely to be fewer problems that might be linked to mysticete reproduction associated with these contaminants.

ODONTOCETES. The diet of most odontocetes includes fish, squid, and generally higher trophic level consumers than the prey of mysticetes, therefore the actual intake of contaminants is likely to be higher in odontocetes than in baleen whales. In the Bay of Fundy, harbor porpoises have relatively high PCB levels (Gaskin et al. 1983), especially in immature animals and adult males, increasing with age to a maximum recorded level of 310 ppm in blubber. The adult females had relatively low levels, as anticipated from lactational mobilization of lipids. Most other tissues had low levels of 2 ppm as expected from their low lipid content. There were no reports indicating that the contaminants had affected reproductive health. Belugas in the St. Lawrence estuary have high levels of PCBs, and Martineau et al. (1987) have suggested that hormonal interference in reproduction may be the cause of low recruitment in this population.

Organochlorine pollutants are known to suppress the immune system, but the effect these may have on cetaceans in crisis (e.g., the striped dolphin population [*Stenella coeruleoalba*] in the Mediterranean in 1990 to 1991 when devastated by the morbillivirus epizootic), is uncertain (Borrell and

Aguilar 1992, Hall 1992, Aguilar and Borrell 1994). A similar virus, PDV (phocine distemper virus) in harbor seals, certainly caused spontaneous abortions apart from death of the adults infected (Heide-Jørgensen et al. 1992). An increase in abortions and stillbirths may occur in cetaceans. Many of the events observed in this field are largely speculative. Certainly new discoveries are being made constantly. A recent example is the description of a form of genital herpes in harbor porpoises in British waters (Ross et al. 1994a), which may interfere with calf production. More seriously, a cetacean form of brucellosis (known to cause abortions in cattle) has been described in harbor porpoises in British waters (Ross et al. 1994b). It is probably safe to say that many of these "new" discoveries are not really new, but have merely failed to be recognized in previous years through ignorance and lack of interest. The whole field of investigation of the interaction of health status of cetaceans, pollutant loads, and incidence of disease with reproductive success is one that needs to be examined carefully in the future. Contaminant loads may be significant in their effect or may be given undue importance and be misleading.

Energetics of Reproduction

Mysticetes

Feeding is a highly seasonal activity for most baleen whales, although food is taken year-round when available in sufficient density. To survive long periods of poor food availability, baleen whales fatten intensively during the summer feeding period (Rice and Wolman 1971; Lockyer 1981a,

Table 6-11. Predicted Energy Costs of the Reproductive Cycle of the Female Fin Whale off Iceland, and the Calculated Fat Energy Store from Observations on Carcasses of Different Reproductive Status

	Carcass lipid and energy store in female								
	Predicted data from girth and length measurements and lipid analysis				Observed data from individual weighings and lipid analysis, July–August				
	Body weight (t) (for 19.4 m length)		Calorific conversion[a] of carcass lipid store					Energy costs of reproduction	
Reproductive status	Week 0	Week 13	Week 0	Week 13	Body length (m)	Body weight (t)	Calorific conversion[a] of carcass lipid store	Pregnancy –11 months	Lactation –7 months
Anestrous	av. 41.5 throughout		77×10^6	102×10^6	19.2	39.0	79×10^6	Fetal growth to 1750 kg @ 2940 kcal/kg	Milk production of 72 kg/day @ 3320 kcal/kg at 90% gland efficiency
Pregnant	41.5	55.5	78×10^6	152×10^6	19.8	56.5	161×10^6		
Lactating	34.5	39.5	50×10^6	57×10^6	20.5	32.0	53×10^6		
	av. 37.0 throughout		av. 54×10^6 throughout						
								$= 5.1 \times 10^6$ plus, Heat of gestation – $Q_G = 4{,}400M^{1.2}$ $= 34.3 \times 10^6$ (Brody 1968)	$= 55.8 \times 10^6$
Average difference between week 13 pregnant and near-end lactating =								Total: 39.4×10^6 +	55.8×10^6
Reproductive costs	18.5 (range: 16–21)		98×10^6 (range: $95–102 \times 10^6$)			24.5	107×10^6	$= 95.2 \times 10^6$	

Adapted after Lockyer, 1987a.

[a]Lipid calorific value = 9450 kcal/kg.

1984a, 1987a,c), which is especially important for pregnant females preparing for lactation. The energetic demands of pregnancy have been shown to be considerably less than the overall demands of lactation (Lockyer 1987a,c); in northern fin whales, the costs of pregnancy are spread over a longer period than those of lactation, therefore the demands of lactation must be very intense on a daily basis. The thickest blubber has been described for pregnant females (Lockyer 1981a for fin and blue whales; Lockyer 1987a,c for fin whales; Rice and Wolman 1971 for gray whales). The fattening is not confined only to the blubber layer but is significant in the major locomotory muscles, viscera, and even bone (Lockyer 1981a, 1987a,c). Table 6-11 shows the estimated costs of reproduction in female fin whales from Icelandic waters, and also indicates the actual observed fat energy storage in an average sized (19.4 m) mature female. The implications of such energy storage not being met have already been discussed earlier under Environmental Effects on Reproduction; see also Costa and Williams (Chapter 5, this volume).

Odontocetes

As mentioned earlier, despite some seasonality in breeding and calving, lactation is often rather protracted so that the costs of lactation, even if high overall, are not as temporally focused as in baleen whales. Also, feeding is generally not as intensively seasonal as in the baleen whales. Therefore even

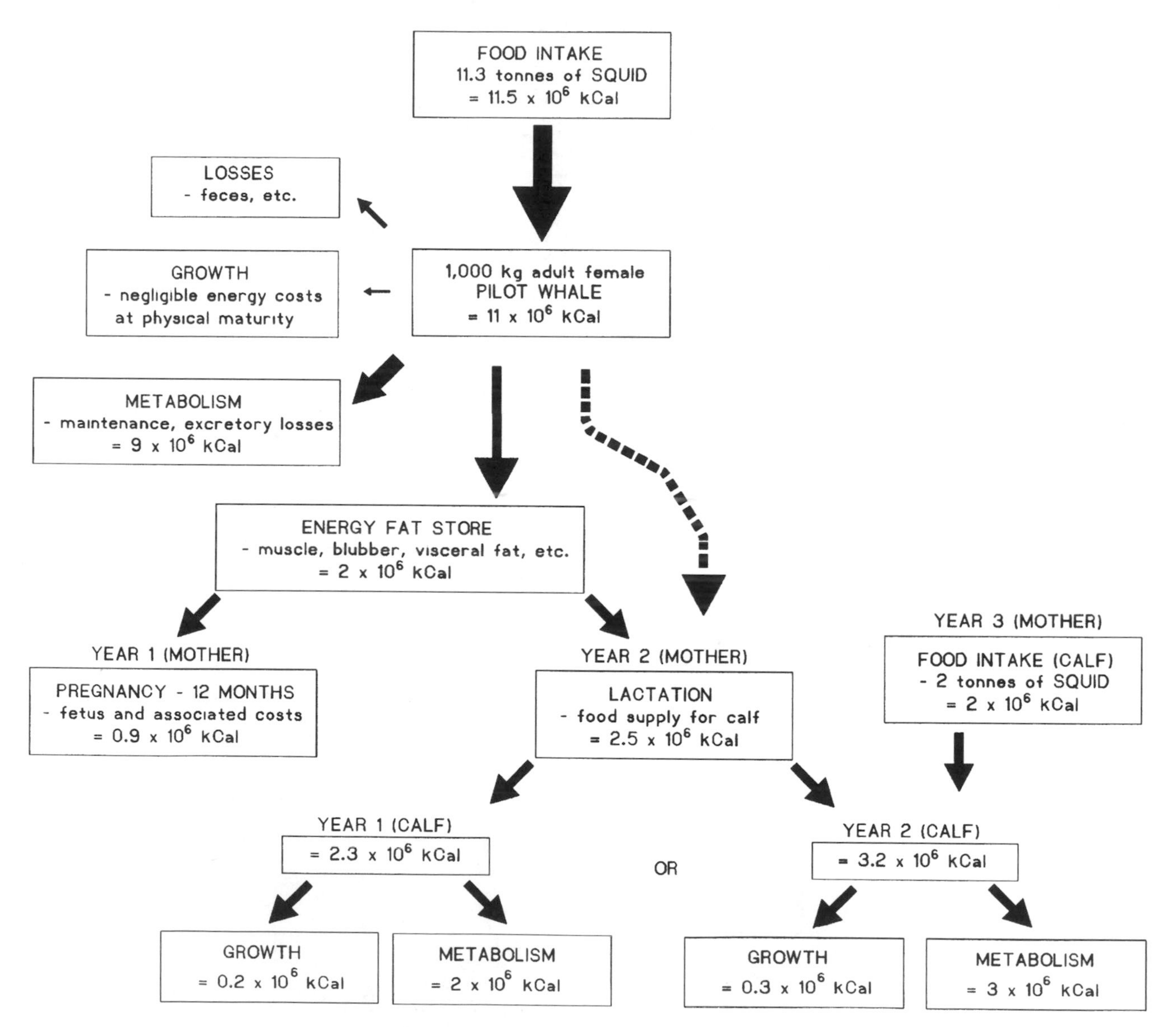

Figure 6-22. Schematic diagram of reproductive energy costs in a female long-finned pilot whale of average adult size (1000 kg) off the Faroe Islands. The scheme follows energy intake and utilization, and fat storage in year 1 used to support or supplement lactation in years 2 and 3, and the energy used for growth and maintenance of the calf. (Courtesy of International Whaling Commission; figure redrawn by S. A. Rommel with permission of the author.)

when seasonal fat storage occurs it is not as dramatic as in baleen whales. The energetic demands of reproduction are reflected in the measurements of girth and blubber thickness in adult females of different reproductive status (e.g., in harbor porpoises, Read 1990a), and some estimates of energy cost of reproduction have been made for this species (Yasui and Gaskin 1986). The energetic costs of pregnancy and lactation have been calculated for the sperm whale (Lockyer 1981b) and long-finned pilot whale (*Globicephala melaena*) (Lockyer 1993a). A schematic diagram of energy costs of reproduction in the long-finned pilot whale is presented in Figure 6-22. The stored fat in the female represents most of the energy required for lactation during the first year. Although pilot whale feeding is not seasonal as in mysticetes, off the Faroe Islands, birth coincides with movement of animals into the area for feeding (in August) with subsequent fattening in winter (Lockyer 1993a). This way, lactation is sustained by dietary intake through fall and winter, and then sustained by fat stores in spring.

In Faroese pilot whales, the energy demands of reproduction in fighting and being actively mobile (Bloch et al. 1993, Desportes et al. 1993, Bloch 1994) may be great on the males who have stored fat energy during winter. Most breeding appears to take place in mid-July, although Desportes et al. (1993) found two peaks of male reproductive activity. Body fat becomes rapidly depleted through spring and summer, first from the muscle and then from the blubber. By May, muscle lipid is negligible; by July, the weight of blubber is reduced; and by August, the lipid content of the blubber is less (Lockyer 1993a).

Fetal Growth

Fetal Sex Ratio

The fetal sex ratio appears to be close to unity in baleen whales (Lockyer 1984a). Although Kato and Shimadzu (1983) found a slight predominance of males in utero in large baleen whales, they estimated that at birth, the sex ratio was equal.

Desportes et al. (1994) reported a changing fetal sex ratio throughout gestation in long-finned pilot whales off the Faroe Islands, where they examined a large sample, year-round, from the fishery. Their results indicated that overall, sex ratio was significantly biased to females, and that the proportion of males declined as gestation advanced. They also observed that females more than 25 years old bore more females. From birth, selection appeared to act against females, and after 3 years, sexual parity was restored. However, their study demonstrated a significant uterine fetal mortality.

Fetal Growth Pattern

The growth pattern of the fetus has been described in formulae derived for a variety of mammalian species, mainly terrestrial (Huggett and Widdas 1951; Frazer and Huggett 1973, 1974; Sacher and Staffeldt 1974). The growth pattern comprises a nonlinear phase, t_0, which represents time in days from conception to the start of the linear phase, and which appears to be constant and related to the overall gestation period. In rorquals with a gestation period approaching 1 year, t_0 was predicted to be about 67 to 72 days (Huggett and Widdas 1951), or 73 to 74 days (Lockyer 1981a). In baleen whales it appears that an accelerated growth phase occurs in the last trimester (Laws 1959b; Lockyer 1981a, 1984a). The exponential growth phase coincides with the arrival on the feeding grounds of the large rorquals, and thus may be significant in terms of energy budget.

The growth pattern of odontocetes appears more steady (Lockyer 1981b, Gaskin et al. 1984, Perrin and Reilly 1984, Martin and Rothery 1993). Certainly the gestation period of some odontocetes lasts more than 12 months, whereas that of baleen whales is well within this period, and may demand accelerated growth to allow birth at a favorable time in relation to migration and feeding conditions.

One notable difference between cetaceans and pinnipeds is the apparent absence in the former of any delayed implantation of the blastocyst to a time when conditions are more favorable for ultimate survival of young. Clearly such a mechanism may be important to a marine mammal that only has a limited window of time for mating while onshore, but such considerations do not apply to cetaceans.

Normally one calf is produced from each pregnancy. The size of the calf relative to the mother is large and varies between about 29% of maternal length in large baleen whales to as much as 42% to 48% of maternal length in odontocete species as diverse as sperm whales and harbor porpoises. There is considerable overlap in body size between neonates and near-term fetuses in both long-finned pilot whales and harbor porpoises, for example. Once the calf is born, it must be able to swim, breathe, and dive. This ability is particularly important in schooling species where staying with other animals represents protection. After birth, the lipid content of fetal muscle, which has been increasing during gestation, decreases very rapidly in long-finned pilot whales (Lockyer 1993a), suggesting that this fat may be an important energy reserve that is drawn upon immediately after birth, perhaps before suckling is well established, thus playing an important role in calf survival.

Reproductive Behavior

Mating Behavior

Information about reproductive behavior in cetaceans is somewhat limited for many large or rare species because of their inaccessibility for observation in the wild (see Wells, Boness, and Rathbun, Chapter 8, this volume). Most information is available for species maintained in captivity that have bred successfully, such as bottlenose dolphins and killer whales.

From studies of captive cetaceans it is clear that postural displays by males and females are important in courtship, and that mating and intromission can take place belly to belly or from the side. Also, mating behavior can continue over several hours. The triggering of such behavior may be that the female is in estrus and communicates this to the male, but there is also the possibility of reflex ovulation perhaps in response to mating (Benirschke et al. 1980, Kirby and Ridgway 1984). Behavior described as male alliances or coalitions has been reported for bottlenose dolphins, where groups of males, often three, "capture" a female and accompany her for a period of time acting as consorts (Connor et al. 1996). The herding is often accompanied by aggressive postures and vocal threats, and appears to occur most frequently when females are likely to be in estrus. The female may attract the males when in estrus, and females may cycle repeatedly and thus favor success in terms of conception.

Several behaviors occur between males, between females, and between females and their male offspring. As mentioned earlier, the penis is made erect by muscular contraction, and can be manipulated at will. The organ may even be used in a tactile way to explore texture and the environment. Sexual behavior between mother and young may be part of a process of bonding or social training.

Such behaviors are not restricted to captive situations. Apparent misdirected sexual behavior by solitary adult male bottlenose dolphins in the wild have been reported (Lockyer 1978a, Lockyer and Morris 1986). Activities included attempted copulation with the underside of small boats, exploration of swimmers with the penis, and insertion of the penis into a pipe of flowing water. Clearly the erect penis is not necessarily an indication of sexual activity, and it could well be used as a sensory organ.

Among humpback whales and right whales (Payne 1988) there have been observations of copulations and attempted copulations with a female by a group of several males (Tyack and Whitehead 1983). In such instances the female may attempt to repel the males by turning upside down so that the ventral aspect of the body is at the water's surface, if she does not want their attentions or maybe is not in estrus. Many baleen whale species appear to have promiscuous mating systems (Evans 1987). The possible significance of this type of mating is discussed by Brownell and Ralls (1986) in relation to sperm competition, and they report a correlation between large testis size and penis length as well as a tendency for reduced intermale aggression with a disposition to multiple matings in certain baleen whale species.

In many odontocete species aggressive interactions may take place between adult males. Tooth marks and scars have been reported from sperm whales (Kato 1984), and interpreted as intrasexual fighting. Fighting also occurs between male long-finned pilot whales, as indicated by tooth marks and scars (Bloch 1994). This fighting activity would support the idea of intermale competition and explain the rapid drain of fat energy reserves in the adult males. DNA fingerprinting studies of long-finned pilot whale pod structure indicate that all members, including adult males, are closely related, and mating, when it occurs, is usually between different pods (Amos et al. 1993). More discussion on such interactive behavior is presented in Wells, Boness, and Rathbun (Chapter 8, this volume). Also, the potential significance of secondary sexual characteristics (e.g., the tusks in male narwhal, Martin 1990, and the large size of the bull sperm whale) in gaining dominance and access to females should be recognized.

Mating can occur with no reproductive purpose in short-finned pilot whales (Kasuya et al. 1993). In this study, the uterine fluid from females taken in a Japanese drive fishery was examined for presence of spermatozoa. The results of this study indicated that sperm could survive up to 3 to 4 days after ejaculation, that copulation rarely occurred during driving, and that sperm presence was not correlated with season, age, or reproductive status of the females. Pregnant females, those in anestrus, and even postreproductive females with senescent ovaries were found to have sperm present. The researchers opined that such mating "may enhance school stability and increase the reproductive success of female kin."

Mating can also take place between different species. (This aspect has been dealt with in more detail under Physiology of Reproduction.) Copulation resulting in pregnancy and even live births have been recorded in captive odontocetes such as hybrids from female bottlenose dolphin and male false killer whale (Nishiwaki and Tobayama 1982). This incident resulted in four pregnancies, three of which were either aborted or stillborn, while the fourth survived 277 days.

Mother–Calf Relationships

At the time of birth, females are often attended by other females. This has been observed in captive bottlenose dolphins where one female will act as an "aunt" (McBride and Kritzler 1951, Schroeder 1990) and help push the newborn to the surface if needed. Such behavior has also been observed at sea in a pod of sperm whales where the female giving birth was vertical in the water and surrounded by other whales lying in three ranks all facing the same direction (Gambell et al. 1973); the calf, later seen among a group of whales, was about 3.5 m in length with wrinkled appearance and furled flukes and dorsal fin. The calf was able to dive with the other whales for 7 min before returning to the surface.

In many odontocete species the mother–calf bond is very strong, to the extent that bottlenose dolphins have been observed to stay in attendance of their calves even after death. Connor and Smolker (1990) recorded a free-ranging female bottlenose dolphin in Shark Bay, Australia, staying with her dead calf for a prolonged period. In captivity, dolphins have been observed pushing dead offspring to the surface and staying in attendance, and similar instances in the wild have been recorded (Hubbs 1953, Moore 1955, Tayler and Saayman 1972). Accounts of mother–calf relationships are well documented for wild bottlenosed dolphins off Sarasota, where the local population has been monitored for more than two decades and the individual reproductive histories of several females and the fate of their young have been recorded (Wells 1991).

In captivity, it has been possible to observe the mother–calf relationship in detail, and in particular the frequency of suckling. The suckling bouts of bottlenose dolphin calves appear to have no pattern nor peak during the day, and the frequency of suckling decreases throughout the first 4 months of life (Cockcroft and Ross 1990). However, the bouts themselves become longer with age. The mother initially presents the mammaries from the side and later vertically as well. Suckling frequency and duration are similar at each mammary slit. The suckling period was observed to often be preceded by nudging/bumping of the mammary slit (Gurevich 1977, Cockcroft and Ross 1990). Perhaps this is a signal to the mother for "let-down" of the milk. After a few weeks, the period of suckling may last 3 to 9 sec and there may be four bouts per hour. Interesting differences in calf behavior were observed in two captive bottlenose mother–calf dolphin pairs where one calf was of a primiparous female and the other of a multiparous female. The former showed greater dependence on the mother (Reid et al. 1995).

Age and Reproductive Performance

In general, fertility and reproductive success are depressed in the newly mature female cetaceans (e.g., fin whales; Lockyer and Sigurjonsson 1991, 1992). However, after reaching a peak in younger animals, fertility maintains a plateau until it tends to decrease in later age (Best et al. 1984, Martin and Rothery 1993, Robeck et al. 1994). With advancing age, the interbirth interval and duration of lactation increase in long-finned pilot whales. Martin and Rothery (1993) propose that this may mean (1) higher survival of calves, (2) provision of milk to calves other than the mother's own, and (3) increased energetic investment in later calves with advancing age of the mother. At least two of these (1 and 3) have been demonstrated in terrestrial mammals (Clutton-Brock et al. 1982, Clutton-Brock 1984).

Perhaps one of the most interesting discoveries is that ovaries in short-finned pilot whales may become senescent in old age, with complete absence of any follicles (Marsh and Kasuya 1984, 1986). However, at the same time, old females with such ovaries may continue to lactate for several years, suckling not only their own previous offspring but other calves and juveniles in the pod (Marsh and Kasuya 1984). The function of this may be social rather than nutritional, and has not been observed in other species. The social organization of pilot whales may predispose this species to certain characteristics. The pilot whale exists in a matriarchal schooling system where there is considerable support from other females and the survival of young may be more favorable than for certain of the odontocetes or for the baleen whales where the mother is alone with her calf. In the former situation, the pressure to keep reproducing is allayed by potentially increased survival of the young, therefore the long-term investment in any offspring is worthwhile. For the more solitary species fecundities may depend more on reproducing regularly at shorter intervals over a longer life span.

In baleen whales, the probability of producing multiplets increases with age, and the incidence is as high as 2.3% for all age groups in sei whales (Kimura 1957). The record for multiple offspring may be sextuplets in a fin whale (Kimura 1957), and quintuplets in a fin whale (Laws 1961) that were supported by 13 corpora lutea. Multiplets have been reported from sperm whales where incidence may be about 0.4% to 0.55% (Best et al. 1984). The ultimate fate of such multiplets is uncertain and probably unfavorable. However, there is an historic record of a humpback whale nursing two calves (Scammon 1874).

Pathology and Teratology

Abnormalities of reproduction range from multiple and minor malformations of organs and offspring to serious life-threatening events that may also lead to sterility (Slijper 1949). Death in utero (Ichihara 1962) involves resorption of fetal tissue, necrosis, and even calcification of the products of pregnancy. Sometimes in the case of multiplets, some fetuses were dead, but the remaining ones were maintained by active corpora lutea. A death rate in utero of 0.14% was calculated by Ichihara (1962) for Antarctic fin whales, but he reported a higher rate in the region of 10 degrees E and 40 degrees E. Ivashin (1977) and Ivashin and Zinchenko (1982) reported nearly three times as many malformed fetuses in minke whales in Pridz Bay between longitudes 70 degrees E and 130 degrees E than elsewhere in the Antarctic; they speculated on the effect of some teratogenic factor in this region.

Deformities including malformations of the central nervous system, skeleton and skull, and visceral organs resulting in hernias, have been described for several species of cetaceans (Ivashin 1960). Monsters were reported from fin whales (Nishiwaki 1957, Ohsumi 1959) and minke whales (Ivashin and Zinchenko 1982). Siamese twins have been recorded in minke whales (Ivashin and Zinchenko 1982, Zinchenko and Ivashin 1987), sei whales (Kawamura 1969), and humpback whales (Zemsky and Budylenko 1970).

Extrauterine pregnancies sometimes include development of multiplets within the abdominal cavity (Ivashin 1960, Ivashin and Zinchenko 1982). The largest intraabdominal fetus observed was a 290-cm term-sized minke whale. The fate of such fetuses is clearly doomed, including perhaps that of the mother. However, in one instance reported for minke whales, whereas the abnormal fetuses were necrotized and calcified, there was a live normal fetus inside the uterus.

There have been cases of hermaphroditism where both sets of sex organs, male and female were present. Bannister (1963) recorded such a case in a fin whale. Such animals do not have functioning sex organs, and must be sterile.

Significance of Reproduction in Management

The important matters concerning reproduction in management are related to population production and recruitment, and how reproductive performance may be affected by pressures brought to bear by exploitation or other external factors. Therefore, knowledge of biological parameters (such as age and length at first ovulation, age and length at first pregnancy, age at sexual maturation in males, reproductive performance with age, general fecundity, length of gestation, timing of parturition, birth weight, pattern of lactation, and survival of young), are all necessary; and in cases where exploitation occurs, it is important to know the proportion of each sex and of reproductively mature whales taken in the catches. Many biological parameters have been shown to be density dependent (Fowler 1981, 1987, 1995), and also variable according to environmental changes that might affect food supply (Laws 1962; Lockyer 1987a, 1990). Fowler (1984) reported that cetacean populations are regulated by density-dependent changes in reproduction and survival, and these are expressed most often in the form of birth and juvenile survival in connection with food resources. These important factors are not exclusive, and social and behavioral factors are also important. I demonstrate the significance of reproduction in management below with reference to a few examples.

Perhaps the most publicized examples are related to baleen whales where cases of age at sexual maturation change in response to exploitation and the supposed subsequent reduced intra- and interspecific competition for food resources have been reported for fin, sei, and minke whales (Lockyer 1978b). The age at maturation has become younger in Antarctic populations (Lockyer 1972, 1974, 1979; Masaki 1978, 1979; Kato 1983, 1987; Ohsumi 1986a; Kato and Sakuramoto 1991). A similar trend was also observed in Icelandic fin whales, but subsequently reversed (Sigurjonsson 1990, 1995; Lockyer and Sigurjonsson 1991, 1992), most probably because of worsening feeding conditions. Trends in decreasing age at sexual maturation were also reported for North Pacific fin whales (Ohsumi 1986b), perhaps in response to exploitation. Fowler (1995) pointed out the need to consider, in addition to the environmental component, the often ignored genetic component in population response, which may be unique in each case.

Changes in pregnancy rate were also reported for Antarctic fin and sei whales, for which an increasing trend was noted from the time of severe overexploitation of southern whale stocks (Gambell 1973; Laws 1977a,b). This phenomenon was further investigated by Mizroch (1981a,b) and Mizroch and York (1982), and the findings were inconclusive as to whether a trend had really occurred once biases were taken into account (Lockyer 1984a). However, there has been more clear-cut evidence of correlation of reproductive performance with changes in food abundance off Iceland, where the pregnancy rate for fin whales appeared closely correlated with food resource and body fat condition (Lockyer 1986, 1987a,c; Vikingsson 1990, 1995).

Off Iceland, it was also reported that in seasons when feeding conditions were poor, the growth of the fetus was deleteriously affected with significantly lower body weight (but not length) (Lockyer 1990; Lockyer and Sigurjonsson 1991, 1992). This clearly has a potential effect on calf survival and population recruitment.

There has already been some discussion of the effects of age on reproductive performance, and clearly the age frequency distribution of the population, most often disturbed by exploitation, will affect population production.

In striped dolphins off Japan, reproductive rates have been observed to change in a density-dependent manner such that net reproduction increases as density decreases from carrying capacity of the population (Kasuya 1976, Kasuya and Miyazaki 1982). In the eastern tropical Pacific, heavily exploited and less exploited populations of spinner dolphins display some responses (such as decreases in age at maturation) predictable from density dependence (Smith 1983, Perrin and Henderson 1984, Perrin and Reilly 1984), although pregnancy rate and length of lactation were not correlated.

In the Bay of Fundy, there is evidence of changes in growth and reproductive parameters in harbor porpoises, with animals growing faster and maturing sexually at an earlier age; these changes might be mediated by increased prey availability to the porpoises, brought about by reduction in porpoise population size through incidental mortality in fisheries (Read and Gaskin 1990). Brodie (1995) proposes that density-dependent changes of harbor porpoises may be associated with fluctuations in herring abundance and reduction in predatory shark populations in the area.

Acknowledgments

I wish to thank Lloyd Lowry, Malcolm Ramsay, and an anonymous reviewer for their constructive comments on this chapter [I. L. B.].

I wish to thank the reviewers of this chapter for helpful suggestions and new references on captive breeding [C. L.].

I thank the reviewers of this chapter for helpful suggestions, Daryl Domning for his advice on sirenian evolution in the Caribbean, and Henry Nix and his staff at the Center for Research and Environmental Studies at the Australian National University for hosting the Special Studies Program that provided me with time to revise this manuscript [H. D. M.]

Literature Cited

Addison, R. F., and P. F. Brodie. 1977. Organochlorine residues in maternal blubber, milk and pup blubber from grey seals (*Halichoerus grypus*) from Sable Island, Nova Scotia. Journal of the Fisheries Research Board of Canada 34:937–941.

Aguilar, A., and A. Borrell. 1988. Age- and sex-related changes in organochlorine compound levels in fin whales (*Balaenoptera physalus*) from the Eastern North Atlantic. Marine Environmental Research 25:195–211.

Aguilar, A., and A. Borrell. 1994. Abnormally high polychlorinated biphenyl levels in striped dolphins (*Stenella coeruleoalba*) affected by the 1990–1992 Mediterranean epizootic. Science of the Total Environment 154:237–247.

Aguilar, A., and L. Jover. 1982. DDT and PCB residues in the fin whale, *Balaenoptera physalus*, of the North Atlantic. Report of the International Whaling Commission 32:299–301.

Amoroso, E. C., G. H. Bourne, R. J. Harrison, L. Harrison Mathews, I. Rowlands, and J. C. Sloper. 1965. Reproductive and endocrine organs of foetal, newborn and adult seals. Journal of Zoology (London) 147:430–486.

Amos, B., D. Bloch, G. Desportes, T. Majerus, D. R. Bancroft, J. A. Barrett, and G. A. Dover. 1993. A review of molecular evidence relating to social organisation and breeding system in the long-finned pilot whale. Report of the International Whaling Commission (Special Issue 14):209–217.

Amstrup, S. C., and D. P. DeMaster. 1988. Polar bear *Ursus maritimus*. Pages 39–56 *in* J. W. Lentfer (ed.). Selected Marine Mammals of Alaska. U.S. Marine Mammal Commission.

Anderson, P. K. 1997. Shark Bay dugongs in summer I: Lek mating. Behaviour 134:433–462.

Anderson, P. K., and A. Birtles. 1978. Behaviour and ecology of the dugong, *Dugong dugon* (Sirenia): Observations in Shoalwater and Cleveland bays, Queensland. Australian Wildlife Research 5:1–23.

Anderson, S. S., and M. A. Fedak. 1985. Grey seal males: Energetic and behavioural links between size and sexual success. Animal Behaviour 33:829–838.

Anderson, S. S., and M. A. Fedak. 1987a. The energetics of sexual success of grey seals and comparison with the costs of reproduction in other seals. Symposia of the Zoological Society (London) 57:319–341.

Anderson, S. S., and M. A. Fedak. 1987b. Grey seal, *Halichoerus grypus*, energetics: Females invest more in male offspring. Journal of Zoology (London) 211:667–679.

Arnason, U., R. Spilliaert, A. Palsdottir, and A. Arnason. 1991. Molecular identification of hybrids between the two largest whale species, the blue whale (*Balaenoptera musculus*) and the fin whale (*B. physalus*). Hereditas 115:183–189.

Arnbom, T. R., M. A. Fedak, and I. L. Boyd. 1997. Factors affecting maternal expenditure in southern elephant seals during lactation. Ecology 78:471–483.

Arnould, J. P. Y., and I. L. Boyd. 1995. Temporal patterns of milk production in Antarctic fur seals (*Arctocephalus gazella*). Journal of Zoology (London) 237:1–12.

Arnould, J. P. Y., Boyd, I. L., and D. G. Socha. 1996. Milk consumption and growth efficiency in Antarctic fur seal (*Arctocephalus gazella*) pups. Canadian Journal of Zoology 74:254–266.

Ashchepkova, L. I., and V. I. Fedoseev. 1988. Development of male germ cells in the fur seal (*Callorhinus ursinus*). Arkhiv Anatomii Gistologii I Embriologii 95:55–66.

Ashwell-Erickson, S., F. H. Fay, R. Elsner, and D. Wartzok. 1986. Metabolic and hormonal correlates of molting and regeneration of pelage in Alaskan Harbor and spotted seals (*Phoca vitulina* and *Phoca largha*). Canadian Journal of Zoology 64:1086–1094.

Atkinson, S., and W. G. Gilmartin. 1992. Seasonal testosterone pattern in Hawaiian monk seals (*Monachus schauinslandi*). Journal of Reproduction and Fertility 96:35–39.

Atkinson, S. N., and M. A. Ramsay. 1995. The effects of prolonged fasting on the body composition and reproductive success of female polar bears (*Ursus maritimus*). Functional Ecology 9:559–567.

Atkinson, S., J. Pietraszek, B. Becker, and T. Johanos. 1993. The estrous cycle of the Hawaiian monk seal and its relation to male seal aggression. Proceedings of the Tenth Biennial Conference on the Biology of Marine Mammals, Galveston, TX, November 11–15.

Banks, W. J. 1993. Applied Veterinary Histology, 3rd ed. Mosby Year Book, Baltimore, MD.

Bannister, J. L. 1963. An intersexual fin whale *Balaenoptera physalus* (L.) from South Georgia. Proceedings of the Zoological Society London 141:811–822.

Bartholomew, G. A. 1970. A model for the evolution of pinniped polygyny. Evolution 24:546–559.

Beal, W. P. 1939. The manatee as a food animal. Nigerian Field 8:124–126.

Belyaev, D. K., and D. V. Klotchkov. 1965. Increasing mink fertility by extra light. Krolikovodstvo I Zverovodstvo 8:2–5.

Bengston, J. L. 1981. Ecology of manatees in the St. Johns River, Florida. Ph.D. thesis, University of Minnesota, Minneapolis. 126 pp.

Bengtson, J. L., and R. M. Laws. 1985. Trends in crabeater seal age at maturity: An insight into Antarctic marine interactions. Pages 669–675 *in* W. R. Siegried, P. R. Condy and R. M. Laws (eds.). Antarctic Nutrient Cycles and Food Webs. Springer-Verlag, Berlin.

Bengtson, J. L., and D. B. Siniff. 1981. Reproductive aspects of female crabeater seals (*Lobodon carcinophagus*) along the Antarctic Peninsula. Canadian Journal of Zoology 59:92–102.

Benirschke, K., M. L. Johnson, and R. V. Benirschke. 1980. Is ovulation in dolphins, *Stenella longirostris* and *Stenella attenuata,* always copulation induced? Fisheries Bulletin 78:507–528.

Bertram, C. K. R., and G. C. L. Bertram. 1968. The Sirenia as aquatic meat producing herbivores. Symposia of the Zoological Society (London) 21:385–391.

Bertram, G. C. L. 1940. The biology of the Weddell seal and crabeater seal; with a study of comparative behaviour of the Pinnipedia. British Grahamland Expedition, 1934–37, Scientific Report No. 1.

Bertram, G. C. L., and C. K. R. Bertram. 1973. The modern sirenia: Their distribution and status. Biological Journal of the Linnean Society 5:297–338.

Best, P. B. 1970. The sperm whale (*Physeter catodon*) off the west coast of South Africa. 5. Age, growth and mortality. South Africa Division of Sea Fisheries Investigational Report 79:1–27.

Best, P. B. 1974. The biology of the sperm whale as it relates to stock management. Pages 257–293 *in* W. E. Schevill (ed.). The Whale Problem: A Status Report. Harvard University Press, Cambridge, MA.

Best, P. B., and D. S. Butterworth 1980. Timing of oestrus within sperm whale schools. Reports of the International Whaling Commission (Special Issue 2):137–140.

Best, P. B., P. A. S. Canham, and N. Macleod. 1984. Patterns of reproduction in sperm whales, *Physeter macrocephalus*. Reports of the International Whaling Commission (Special Issue 6):51–79.

Best, R. C. 1982. Seasonal breeding in the Amazonian manatee, *Trichechus inunguis* (Mammalia; Sirenia). Biotropica 14:76–78.

Best, R. C. 1983. Apparent dry-season fasting in Amazonian manatees, (Mammalia; Sirenia). Biotropica 15:61–64.

Best, R. C. 1984. The aquatic mammals and reptiles of the Amazon. Pages 371–412 *in* H. Sioli (ed.). The Amazon. Limnology and Landscape Ecology of a Mighty Tropical River and its Basin. Junk, Dordrecht.

Best, R. C., G. A. Robeiro, M. Yamakoshi, and V. M. F. da Silva. 1982. Artificial feeding for unweaned Amazonian manatees (*Trichechus inunguis*). International Zoo Yearbook 22:263–267.

Bester, M. N. 1987. Subantarctic fur seal, *Arctocephalus tropicalis,* at Gough Island. Pages 57–60 *in* J. P. Croxall and R. L. Gentry (eds.). Status, Biology, and Ecology of Fur Seals. National Oceanic and Atmospheric Administration Technical Report. National Marine Fisheries Service 51.

Bester, M. N. 1990. Reproduction in the male subantarctic fur seal *Arctocephalus tropicalis*. Journal of Zoology (London) 222:177–185.

Bigg, M. A. 1973. Adaptations in the breeding of the harbour seal, *Phoca vitulina*. Journal of Reproduction and Fertility (Suppl.) 19:131–142.

Bigg, M. A., and H. D. Fisher. 1974. The reproductive cycle of the female harbour seal off southeastern Vancouver Island. Pages 329–347 *in* R. J. Harrison (ed.). Functional Anatomy of Marine Mammals, Vol. 2. Academic Press, London.

Bigg, M. A., and H. D. Fisher. 1975. Effect of photoperiod on annual reproduction in female harbour seals. Rapports et Proces-Verbaux des Reunions Conseil Internationale pour l'Exploration Scientifique de la Mer 169:141–144.

Bloch, D. 1994. Intermale competition in schools of long-finned pilot whales as indicated by abundance of fighting marks. Doctoral thesis, Pilot whales in the N. Atlantic. Age, growth and social structure in Faroese grinds of the long-finned pilot whale, *Globicephala melas*. Lund University, Sweden.

Bloch, D., C. Lockyer, and M. Zachariassen. 1993. Age and growth parameters of the long-finned pilot whale off the Faroe Islands. Report of the International Whaling Commission (Special Issue 14):163–208.

Boness, D. J. 1991. Determinants of mating systems in the Otariidae (Pinnipedia). Pages 1–44 *in* D. Renouf (ed.). Behaviour of Pinnipeds. Chapman and Hall, London.

Boness, D. J., W. D. Bowen, and S. J. Iverson. 1995. Does male harassment of females contribute to reproductive synchrony in the grey seal by affecting maternal performance? Behavioral Ecology and Sociobiology 36:1–10.

Bonner, W. N. 1955. Reproductive organs of foetal and juvenile elephant seals. Nature (London) 179:982–983.

Bonner, W. N. 1981. Southern fur seals—*Arctocephalus*. Pages 161–208 *in* S. H. Ridgway and R. J. Harrison (eds.). Handbook of Marine Mammals, Vol. 1. Academic Press, London.

Bonner, W. N. 1984. Lactation strategies in seals: Problems for a marine mammalian group. Symposia of the Zoological Society (London) 51:253–272.

Born, E. W. 1982. Reproduction in the female hooded seal, *Crystophora cristata* Erxleben, at South Greenland. Journal of the Northwest Atlantic Fishery Science 3:57–62.

Borrell, A., and A. Aguilar. 1992. Pollution by PCBs in striped dolphins affected by the western Mediterranean epizootic. Pages 121–127 *in* X. Pastor and M. Simmonds (eds.). The Mediterranean Striped Dolphin Die-off. Proceedings of the Mediterranean striped dolphins mortality international workshop, 4–5 November 1991, Greenpeace.

Boshier, D. P. 1977. Observations on the corpus luteum of the grey seal (*Halichoerus grypus*) at the time of ovaimplantation. Pages 333–359 *in* R. J. Harrison (ed.). Functional Anatomy of Marine Mammals, Vol. 3. Academic Press, London.

Boshier, D. P. 1981. Structural changes in the corpus luteum and endometrium of seals before implantation. Journal of Reproduction and Fertility (Suppl.) 29:143–149.

Boulva, J., and I. A. McLaren. 1979. Biology of the harbour seal (*Phoca vitulina*) in eastern Canada. Bulletin of the Fisheries Resource Board Canada No. 200.

Bowen, W. D., C. K. Capstick, and D. E. Sergeant. 1981. Temporal changes in the reproductive potential of female harp seals (*Pagophilus froenlandicus*). Canadian Journal of Fisheries and Aquatic Sciences 38:495–503.

Bowen, W. D., O. T. Oftedal, and D. J. Boness. 1985. Birth to weaning in 4 days: Remarkable growth in the hooded seal, *Cystophore cristata*. Canadian Journal of Zoology 63:2841–2846.

Bowen, W. D., D. J. Boness, and O. T. Oftedal. 1987. Mass transfer from mother to pup and subsequent mass loss by the weaned pup in the hooded seal, *Cystophora cristata*. Canadian Journal of Zoology 65:1–8.

Bowen, W. D., W. T. Stobo, and S. J. Smith. 1992. Mass changes of grey seal *Halicheorus grypus* pups on Sable Island: Differential maternal investment reconsidered. Journal of Zoology (London) 227:607–622.

Boyd, I. L. 1982. Reproduction of grey seal with reference to factors influencing fertility. Ph.D. dissertation, Cambridge University, Cambridge, U.K. 167 pp.

Boyd, I. L. 1983. Luteal regression, follicle growth and the concentration of some plasma steroids during lactation in grey seals (*Halichoerus grypus*). Journal Reproduction and Fertility 69:157–164.

Boyd, I. L. 1984a. The relationship between body condition and the timing of implantation in pregnant grey seals (*Halichoerus grypus*). Journal of Zoology (London) 203:113–123.

Boyd, I. L. 1984b. Development and regression of the corpus luteum in grey seal (*Halichoerus grypus*) ovaries and its use in determining fertility rates. Canadian Journal of Zoology 62:1095–1100.

Boyd, I. L. 1984c. Occurrence of hilar rete glands in the ovaries of grey seals (*Halichoerus grypus*). Journal Zoology (London) 204:585–588.

Boyd, I. L. 1985. Pregnancy and ovulation rates in two stocks of grey seals on the British coast. Journal of Zoology (London) 205:265–272.

Boyd, I. L. 1990a. Mass and hormone content of the placentae of grey seals related to foetal sex. Journal of Mammalogy 71:101–103.

Boyd, I. L. 1990b. Role of prolactin in the maintenance of lactation and ovulation in Antarctic fur seals and southern elephant seals. Journal of Reproduction and Fertility Abstract Series No. 5. 29: 20.

Boyd, I. L. 1991a. Changes in plasma progesterone and prolactin concentrations during the annual cycle and the role of prolactin in the maintenance of lactation and luteal development in the Antarctic fur seal (*Arctocephalus gazella*). Journal of Reproduction and Fertility 91:637–647.

Boyd, I. L. 1991b. Environmental and physiological factors controlling the reproductive cycles of pinnipeds. Canadian Journal Zoology 69:1135–1148.

Boyd, I. L. 1996. Individual variation in the duration of pregnancy and birth date in Antarctic fur seals: The role of environment, age, and sex of fetus. Journal of Mammalogy 77:124–133.

Boyd, I. L., and C. D. Duck. 1991. Mass change and metabolism of territorial male Antarctic fur seals (*Arctocephalus gazella*). Physiological Zoology 64:375–392.

Boyd, I. L., and C. D. Duck. 1991. Mass change and metabolism of territorial male Antarctic fur seals (*Arctocephalus gazella*). Physiological Zoology 64:375–392.

Boyd, I. L., T. Arnbom, and M. A. Fedak. 1993. Water flux, body composition, and metabolic rate during molt in female southern elephant seals (*Mirounga leonina*). Physiological Zoology 66:43–60.

Boyd, I. L., J. P. Croxall, N. J. Lunn, and K. Reid. 1995. Population demography of Antarctic fur seals: The costs of reproduction and implications for life-histories. Journal of Animal Ecology 64:505–518.

Braham, H. 1984. Review of reproduction in the white whale, *Delphinapterus leucas*, narwhal, *Monodon monoceros*, and Irrawaddy dolphin, *Orcaella brevirostris*, with comments on stock assessment. Reports of the International Whaling Commission (Special Issue 6):81–89.

Brodie, P. F. 1969. Duration of lactation in Cetacea: An indicator of required learning? American Midland Naturalist 82:312–314.

Brodie, P. F. 1995. The Bay of Fundy / Gulf of Maine harbour porpoise (*Phocoena phocoena*): Some considerations regarding species interactions, energetics, density dependence and bycatch. Report of the International Whaling Commission (Special Issue 16):182–187.

Bronson, F. H. 1988. Mammalian reproductive strategies: Genes, photoperiod and latitude. Reproduction Nutrition Development 28:335–347.

Brosseau, C., A. M. Johnson, A. M. Johnson, and K. W. Kenyon. 1975. Breeding the sea otter at Tacoma Aquarium. International Zoo Yearbook 15:144–147.

Brownell, R. L. Jr. 1984. Review of reproduction in Platanistid dolphins. Reports of the International Whaling Commission (Special Issue 6):149–158.

Brownell, R. L. Jr., and K. Ralls. 1986. Potential for sperm competition in baleen whales. Reports of the International Whaling Commission (Special Issue 8):97–112.

Bryden, M. M., and R. J. Harrison 1986. Gonads and reproduction. Pages 149–159 *in* M. M. Bryden and R. Harrison (eds.). Research on Dolphins. Clarenden Press, Oxford.

Burns, J. J. 1970. Remarks on the distribution and natural history of pagophilic pinnipeds in the Bering and Chukchi Seas. Journal of Mammalogy 51:445–454.

Burns, J. J. 1981a. Bearded seal, *Erignathus barbatus* Erxleben, 1777. Pages 145–170 *in* S. H. Ridgway and R. J. Harrison (eds.). Handbook of Marine Mammals, Vol. 2. Academic Press, New York.

Burns, J. J. 1981b. Ribbon seal, *Phoca fasciata* Erxleben, 1777. Pages 89–109 *in* S. H. Ridgway and R. J. Harrison (eds.). Handbook of Marine Mammals, Vol. 2. Academic Press, New York.

Cadenat, J. 1957. Observations de cetaces, sireniens, chelonians et sauriens en 1955–1956. Bulletin de Institut Francais D'Afrique Noire 19:1358–1383.

Cardeilhac, P. T., J. R. White, and R. Francis-Floyd. 1984. Initial information on the reproductive biology of the Florida manatee. Paper presented at the Fifteenth Annual Conference and Workshop, International Association for Aquatic Animal Medicine, Tampa, FL, 30 April–2 May 1984.

Carrick, R., S. E. Csordas, S. E. Ingham, and K. Keith. 1962. Studies of the southern elephant seal, *Mirounga leonina* (L.). III. The annual cycle in relation to age and sex. CSIRO Wildlife Research 7:119–160.

Charnock-Wilson, J. 1968. The manatee in British Honduras. Oryx 9:293–294.

Charnov, E. L. 1982. The Theory of Sex Allocation. Princeton University Press, Princeton, NJ.

Christensen, I. 1984. Growth and reproduction of killer whales, *Orcinus orca*, in Norwegian coastal waters. Reports of the International Whaling Commission (Special Issue 6):253–258.

Clapham, P. J., and C. A. Mayo 1990. Reproduction of humpback whales (*Megaptera novaeangliae*) observed in the Gulf of Maine. Report of the International Whaling Commission (Special Issue 12):171–175.

Clutton-Brock, T. H. 1984. Reproductive effort and terminal investment in iteroparous animals. American Naturalist 123:212–229.

Clutton-Brock, T. H., and G. R. Iason. 1986. Sex ratio variation in mammals. Quarterly Review of Biology 61:339–374.

Clutton-Brock, T. H., F. E. Guinness, and S. D. Albon. 1982. Red Deer: Behaviour and Ecology of Two Sexes. University of Chicago Press, Chicago, IL.

Cockcroft, V. G., and G. J. B. Ross. 1990. Observations on the early development of a captive bottlenose dolphin calf. Pages 461–478 *in* S. Leatherwood and R. R. Reeves (eds.). The Bottlenose Dolphin. Academic Press, New York, NY.

Connor, R. C., and R. A. Smolker. 1990. Quantitative description of a rare behavioral event: a bottlenose dolphin's behavior toward her deceased offspring. Pages 355–360 *in* S. Leatherwood and R. R. Reeves (eds.). The Bottlenose Dolphin. Academic Press, New York, NY.

Connor, R. C., A. F. Richards, R. A. Smolker, and J. Mann. 1996. Patterns of female attractiveness in Indian Ocean bottlenose dolphins. Behaviour 133:37–69.

Cooke, J. G., and W. A. De La Mare. 1984. A note on the estimation of time trends in the age at sexual maturity in baleen whales from transition layer data, with reference to the North Atlantic fin whale. Report of the International Whaling Commission 34:701–709.

Costa, D. P. 1991. Reproductive and foraging energetics of pinnipeds: Implications for life history patterns. Pages 300–388 *in* D. Renouf (ed.). Behaviour of Pinnipeds. Chapman and Hall, London.

Coulson, J. C. 1981. A study of the factors influencing the timing of breeding in the grey seal *Halichoerus grypus*. Journal of Zoology (London) 194:553–571.

Coulson, J. C., and G. Hickling. 1964. The breeding biology of the grey seal, *Halichoerus grypus* (Fab.), on the Farne Islands, Northumberland. Journal of Animal Ecology 33:485–512.

Craig, A. M. 1964. Histology of reproduction and the estrus cycle in the female fur seal, *Callorhinus ursinus*. Journal of Fisheries Research Board of Canada 21:773–811.

Crouse, D. T., L. B. Crowder, and H. Caswell. 1987. A stage-based model for loggerhead sea turtles and implications for conservation. Ecology 68:1412–1423.

Croxall, J. P., and R. L. Gentry. 1987. The 1984 fur seal symposium: An introduction. Pages 1–4 *in* J. P. Croxall and R. L. Gentry (eds.). Status, Biology, and Ecology of Fur Seals. National Oceanic and Atmospheric Administration Technical Report. National Marine Fisheries Service 51.

Cupps, P. L. 1990. Reproduction in Domestic Animals, 4th ed. Academic Press, New York, NY.

Daniel, J. C. Jr. 1971. Growth of the pre-implantation embryo of the northern fur seal and its correlation with changes in uterine protein. Developmental Biology 26:316–322.

Daniel, J. C. Jr. 1974. Circulating levels of estradiol-17β during early pregnancy in the Alaska fur seal showing an estrogen surge preceding implantation. Journal of Reproduction and Fertility 37:425–428.

Daniel, J. C. Jr. 1975. Concentrations of circulating progesterone during early pregnancy in the northern fur seal, *Callorhinus ursinus*. Journal of the Fisheries Research Board of Canada 32:65–66.

Daniel, J. C. Jr. 1981. Delayed implantation in the northern fur seal (*Callorhinus ursinus*) and other pinnipeds. Journal of Reproduction and Fertility (Suppl.) 29:35–50.

Da Silva, V. M. F. 1994. Aspects of the biology of the Amazonian dolphins genus *Inia* and *Sotalia fluviatalis*. Ph.D. dissertation, University of Cambridge, U.K. 327 pp.

Dellman, H. D. 1971. Veterinary Histology; An Outline Text-Atlas. Lea and Febiger, Philadelphia, PA.

DeMaster, D. P. 1981. Estimating the average age of first brith in marine mammals. Canadian Journal of Fisheries and Aquatic Sciences 38:237–239.

Derocher, A. E., and I. Stirling. 1994. Age-specific reproductive performance of female polar bears (*Ursus maritimus*). Journal of Zoology (London) 234:527–536.

Derocher, A. E., and I. Stirling. 1995. Temporal variation in reproduction and body mass of polar bears in western Hudson Bay. Canadian Journal of Zoology 73:1675–1665.

Derocher, A. E., D. Andriashek, and J. P. Y. Arnould. 1993. Aspects of milk composition and lactation in polar bears. Canadian Journal of Zoology 71:561–567.

Desjardins, C., F. H. Bronson, and J. L. Blank. 1986. Genetic selection for reproductive photoresponsiveness in deer mice. Nature (London) 322:172–173.

Desportes, G., M. Saboureau, and A. Lacroix. 1993. Reproductive maturity and seasonality of male long-finned pilot whales, off the Faroe Islands. Report of the International Whaling Commission (Special Issue 14):234–262.

Desportes, G., L. W. Andersen, and D. Bloch. 1994. Variation in foetal and postnatal sex ratios in long-finned pilot whales. Ophelia 39:183–196.

Deutsch, C. J., M. P. Haley, and B. J. Le Boeuf 1990. Reproductive effort of male northern elephant seals—estimates from mass loss. Canadian Journal of Zoology 68:2580–2593.

Dixson, A. F. 1995. Baculum length and copulatory behaviour in carnivores and pinnipeds (Grand Order Ferae). Journal of Zoology (London) 235:67–76.

Domning, D. P. 1982. Evolution of manatees— a speculative history. Journal of Palaeontology 56:599–619.

Domning, D., and L. C. Hayek. 1986. Interspecific and intra-specific variation in manatees (Sirenia: *Trichechus*). Marine Mammal Science 2:87–144.

Donovan, G. P., C. H. Lockyer, and A. R. Martin (eds.). 1993. Biology of northern hemisphere pilot whales. Report of the International Whaling Commission (Special Issue 14): 479 pp.

Duffield, D. A., D. A. Odell, J. F. McBain, and B. Andrews. 1995. Killer whale (*Orcinus orca*) reproduction at sea world. Zoo Biology 14:417–430.

Duncan, M. J., and B. D. Goldman. 1984. Hormonal regulation of the annual pelage color cycle in the Djungarian hamster, *Phodopus sungorus*. I. Role of the gonads and the pituitary. Journal of Experimental Zoology 230:89–96.

Eberhardt, L. L., and T. J. O'Shea. 1995. Integration of manatee life-history data and population modeling. Pages 269–279 *in* T. J. O'Shea, B. B. Ackerman, and H. F. Percival (eds.). Population Biology of the Florida Manatee. U.S. Department of the Interior, National Biological Service, Information and Technology Report 1.

Elden, C. A., M. C. Keyes, and C. E. Marshall. 1971. Pineal body of the northern fur seal (*Callorhinus ursinus*): A model for studying the probable function of the mammalian pineal body. American Journal of Veterinary Research 32:639–647.

Elliott, J. A. 1976 Circadian rhythms and photoperiodic time measurement in mammals. Federation Proceedings, Federation of American Societies for Experimental Biology Journal 35:2339–2346.

Enders, A. C. 1981. Embryonic diapause—Perspectives. Journal of Reproduction and Fertility (Suppl.) 29:229–241.

Evans, H. E. 1993. Anatomy of the Dog. W.B. Saunders, Philadelphia, PA.

Evans, P. G. H. 1987. The Natural History of Whales and Dolphins. Christopher Helm, London.

Fay, F. H. 1981. Walrus—*Odobenus rosmarus*. Pages 1–24 *in* S. H. Ridgway and R. J. Harrison (eds.). Handbook of Marine Mammals, Vol. 2. Academic Press, New York, NY.

Fedak, M. A., T. Arnbom, and I. L. Boyd. 1996. The relation between the size of southern elephant seal mothers, the growth of their pups, and the use of maternal energy, fat, and protein during lactation. Physiological Zoology 69:887–911.

Fleischer, L. A. 1987. Guadalupe fur seal, *Arctocephalus townsendii*. Pages 43–48 *in* J. P. Croxall and R. L. Gentry (eds.). Status, Biology, and Ecology of Fur Seals. National Oceanic and Atmospheric Administration Technical Report National Marine Fisheries Service 51.

Follett, B. K. 1982. Physiology of photoperiodic time-measurement. Pages 268–275 *in* J. Aschoff, S. Daan, and G. Gross (eds.). Vertebrate Circadian Systems. Springer-Verlag, Berlin.

Fowler, C. W. 1981. Density dependence as related to life history strategy. Ecology 62:602–610.

Fowler, C. W. 1984. Density dependence in cetacean populations. Reports of the International Whaling Commission (Special Issue 6):373–379.

Fowler, C. W. 1987. A review of density dependence in populations of large mammals. Pages 410–441 *in* H. Genoways (ed.). Current Mammalogy. Plenum, New York, NY.

Fowler, C. W. 1990. Density-dependence in northern fur seals (*Callorhinus ursinus*). Marine Mammals Science 6:171–195.

Fowler, C. W. 1995. Population dynamics: Species traits and environmental influence. Pages 403–412 *in* A. S. Blix, L. Walloe, and O. Ulltang (eds.). Whales, Seals, Fish and Men. Elsevier Science, Amsterdam.

Frazer, J. F. D., and A. St. G. Huggett. 1973. Specific foetal growth rates of cetaceans. Journal of Zoology (London) 169:111–126.

Frazer, J. F. D., and A. St. G. Huggett. 1974. Species variations in the foetal growth rates of eutherian mammals. Journal of Zoology (London) 174:481–509.

Gales, N. J., and P. Williamson. 1989. Preliminary results of a long term study of the demography and the breeding biology of the Australian sea lion, *Neophoca cinerea*. *In* Proceedings of 8th Biennial Conference on the Biology of Marine Mammals, Pacific Grove, CA, 7–11 December 1989.

Gambell, R. 1973. Some effect of exploitation on reproduction in whales. Journal Reproduction and Fertility (Suppl.) 19:533–553.

Gambell, R., C. Lockyer, and G. J. B. Ross. 1973. Observations on the birth of a sperm whale calf. South African Journal of Science 69:147–148.

Gardiner, K. J. 1994. Some aspects of the reproductive endocrinology of harbour seals (*Phoca vitulina*) and grey seals (*Halichoerus grypus*). Ph.D. dissertation, University of Aberdeen, MD.

Garshelis, D. L., A. M. Johnson, and J. A. Garshelis. 1984. Social organization of sea otters in Prince William Sound Alaska. Canadian Journal of Zoology 62:2648–2658.

Gaskin, D. E. 1982. The Ecology of Whales and Dolphins. Heinemann, London.

Gaskin, D. E., R. Frank, and M. Holdrinet. 1983. Polychlorinated biphenyls in harbour porpoises *Phocoena phocoena* (L.) from the Bay of Fundy, Canada and adjacent waters, with some information on chlordane and hexachlorobenzene levels. Archives Environmental Contamination and Toxicology 12:211–219.

Gaskin, D. E., G. J. D. Smith, A. P. Watson, W. Y. Yasui, and D. B. Yurick 1984. Reproduction in porpoises (*Phocoenidae*): Implications for management. Reports of the International Whaling Commission (Special Issue 6):135–148.

Geist, V. 1974. On the relationship of ecology and behaviour in the evolution of ungulates: theoretical considerations. Pages 235–246 *in* V. Geist and F. Walther (eds.). The Behaviour of Ungulates and its Relation to Management. IUCN, Morges.

Gentry, R. L. 1981. Northern fur seal—*Callorhinus ursinus*. Pages 143–160 *in* S. H. Ridgway and R. J. Harrison (eds.). Handbook of Marine Mammals, Vol. 2. Academic Press, New York, NY.

Gentry, R. L., D. P. Costa, J. P. Croxall, J. H. M. David, R. W. Davis, G. L. Kooyman, P. Majluf, T. S. McCann, and F. Trillmich. 1986. Synthesis and conclusions. Pages 220–282 *in* R. L. Gentry and G. L. Kooyman (eds.). Fur Seals: Maternal Strategies on Land and at Sea. Princeton University Press, Princeton, NJ.

Gilmartin, W. G., R. L. DeLong, A. W. Smith, J. C. Sweeney, B. W. DeLappe, R. W. Riseborough, L. A. Griner, M. D. Dailey, and D. B. Peakall. 1976. Premature parturition in the California sea lion. Journal of Wildlife Disease 12:104–115.

Griffiths, D. J. 1984a. The annual cycle of the testis of the elephant seal (*Mirounga leonina*) at Macquarie Island. Journal of Zoology (London) 203:193–204.

Griffiths, D. J. 1984b. The annual cycle of the epididymis of the elephant seal (*Mirounga leonina*) at Macquarie Island. Journal of Zoology (London) 203:181–191.

Griffiths, D. J., and M. M. Bryden. 1981. The annual cycle of the pineal gland of the southern elephant seal (*Mirounga leonina*). Pages 57–66 *in* C. D. Matthews and R. F. Seamark (eds.). Pineal Function. Elsevier, Amsterdam.

Griffiths, D. J., R. F. Seamark, and M. M. Bryden. 1979. Summer and winter cycles in plasma melatonin levels in the elephant seal (*Mirounga leonina*). Australian Journal of Biological Sciences 32:581–586.

Gurevich, V. S. 1977. Post-natal behavior of an Atlantic bottlenosed dolphin calf (*Turiops truncatus,* Montagu) born at Sea World. Pages 168–184 *in* S. H. Ridgway and K. Benirschke (eds.). Breeding Dolphins: Present Status, Suggestions for the Future. U. S. Marine Mammal Commission Report MMC-76/07, Washington, D.C.

Hall, A. J. 1992. Disease causation and the striped dolphin mortality. Pages 111–118 *in* X. Pastor and M. Simmonds (eds.). The Mediterranean Striped Dolphin Die-off. Proceedings of the Mediterranean striped dolphins mortality international workshop, 4-5 November 1991. 190pp.

Hamilton, J. E. 1939a. The leopard seal *Hydrurga leptonyx* (de Blainville). Discovery Reports 18:239–264.

Hamilton, J. E. 1939b. A second report on the Southern sea lion, *Otario byronia* (de Blainville). Discovery Reports 19:121–164.

Hamilton, W. J., and K. L. Blaxter 1980. Reproduction in farmed red deer. 1. Hind and stag fertility. Journal of Agricultural Science 95:261–273.

Harrison, R. J. 1960. Reproduction and reproductive organs of the common seal (*Phoca vitulina*) in the Wash, East Anglia. Mammalia 24:372–385.

Harrison, R. J. 1964. Delayed implantation—A curious reproductive phenomenon. Proceedings of the Royal Institute of Great Britain 40:143–158.

Harrison, R. J. 1969. Reproduction and reproductive organs. Pages 253–348 *in* H. T. Andersen (ed.). The Biology of Marine Mammals. Academic Press, New York, NY.

Harrison, R. J., and B. A. Young. 1966. Functional characteristics of the seal placenta. Symposia of the Zoological Society (London) 15:47–68.

Harrison, R. J., L. H. Matthews, and J. M. Roberts. 1952. Reproduction in some pinnipedia. Transactions of the Zoological Society of London 27:437–531.

Hartman, D. S. 1979. Ecology and behavior of the manatee (*Trichechus manatus*) in Florida. Special Publication No. 5. The American Society of Mammalogists.

Harvey, P. H., and R. M. Zammuto. 1985. Patterns of mortality and age at first reproduction in natural populations of mammals. Nature (London) 315:319–320.

Harwood, J., and J. H. Prime. 1978. Some factors affecting the size of British grey seal populations. Journal of Applied Ecology 15:401–411.

Heide-Jørgensen, M. P., and J. Teilmann 1994. Growth, reproduction, age structure and feeding habits of white whales (*Delphinapterus leucas*) in West Greenland waters. Meddr Grønland, Bioscience 39:195–212.

Heide-Jørgensen, M. P., T. Härkönen, and P. Åberg. 1992. Long-term effects of epizootic in harbour seals in the Kattegat-Skagerrak and adjacent areas. Ambio 21:511–516.

Heinsohn, G. E., and A. V. Spain. 1974. Effects of a tropical cyclone on littoral and sub-littoral biotic communities and on a population of dugongs (*Dugong dugon*). Biological Conservation 2:143–152.

Helle, E. 1981. Reproductive trends and occurrence of organochlorines and heavy metals in the Baltic seal population. International Council for the Exploration of the Sea, C.M. E:37. 13pp.

Helle, E., M. Olsson, and S. Jensen. 1976. PCB levels correlated with pathological changes in seal uteri. Ambio 5:261–263.

Hernandez, P., J. E. Reynolds, H. Marsh, and M. Marmontel. 1995. Age and seasonality in spermatogenesis of Florida manatees. Pages 84–97 *in* T. J. O'Shea, B. B. Ackerman, and H. F. Percival (eds.). Population Biology of the Florida Manatee. U.S. Department of the Interior, National Biological Service, Information and Technology Report 1.

Hewer, H. R., and K. Backhouse. 1968. Embryology and foetal growth rate in the grey seal (*Halichoerus grypus*). Journal of Zoology (London) 155:507–533.

Hill, H. J., and W. G. Gilmartin 1977. Collection and storage of semen from dolphins. Pages 205–210 *in* S. H. Ridgway and K. Benirschke (eds.). Breeding Dolphins: Present Status, Suggestions for the Future. U. S. Marine Mammal Commission Report MMC-76/07, Washington, D.C.

Hill, S. E. B. 1987. Reproductive ecology of Weddell seals (*Leptonychotes weddelli*) in McMurdo Sound, Antarctica. Ph.D. thesis, University of Minnesota, Minneapolis, MN.

Hirosaki, Y., M. Honda, and T. Kinuta 1981. On the three hybrids between *Tursiops truncatus* and *Grampus griseus.* (1) Their parents and external measurements. (In Japanese.) Journal of the Japanese Association of Zoological Gardens and Aquariums 23:46–48.

Hobson, B. M., and I. L. Boyd. 1984. Concentrations of placental gonadotrophin, placental progesterone and maternal plasma progesterone during pregnancy in the grey seal (*Halichoerus grypus*). Journal of Reproduction and Fertility 72:521–528.

Hobson, B. M., and L. Wide 1986. Gonadotrophin in the term placenta of the dolphin (*Tursiops truncatus*), the California sea lion (*Zalophus californianus*), the grey seal (*Halichoerus grypus*) and man. Journal of Reproduction and Fertility 76:637–644.

Hohn, A. A., and R. L. Brownell 1990. Harbor porpoise in central Californian waters: Life history and incidental catches. Document SC/42/SM47, The Scientific Committee of the International Whaling Commission. 21pp.

Hohn, A. A., M. D. Scott, R. S. Wells, J. C. Sweeney, and A. B. Irvine. 1989. Growth layers in teeth from known-age, free-ranging bottlenose dolphins. Marine Mammal Science 5:315–342.

Høier, R., and M. P. Heide-Jørgensen. 1994. Steroid hormones and prolactin in white whales (*Delphinapterus leucas*) from West Greenland. Meddr Grønland, Bioscience 39:227–238.

Hoover, A. A. 1988. Steller sea lion, *Eumetopias jubatus.* Pages 159–193 *in* J. W. Lentfer (ed.). Selected Marine Mammals of Alaska. Marine Mammal Commission, Washington, D.C.

Hubbs, C. L. 1953. Dolphin protecting dead young. Journal of Mammalogy 34:498.

Huber, H. R. 1987. Natality and weaning success in relation to age of first reproduction in northern elephant seals. Canadian Journal of Zoology 65:1311–1316.

Hudson, B. E. T. 1986. The hunting of dugong at Daru, Papua New Guinea, during 1978–1982. Community management and education initiatives. Pages 77–94 *in* A. K. Haines, G. C. Williams, and D. Coates (eds.). Torres Strait Fisheries Seminar, Port Moresby, February 1985. Australian Government Publishing Service, Canberra.

Huggett, A. St. G., and W. F. Widdas. 1951. The relationship between mammalian foetal weight and conception age. Journal of Physiology 114:306–317.

Husar, S. 1977. *Trichechus inungius.* Mammalian Species 72:1–4.

Husar, S. 1978. *Trichechus senegalensis.* Mammalian Species 89:1–4.

Ichihara, T. 1962. Prenatal dead foetus of baleen whales. Scientific Reports of the Whales Research Institute 16:47–60.

Iga, K., Y. Fukui, and A. Miyamoto. 1996. Endocrinological observations of female minke whales (*Balaenoptera acutorostrata*). Marine Mammal Science 12:296–301.

Ivashin, M. V. 1960. On the multi fetation development abnormalities and embryo mortality in whales (in Russian). Zoologicheskii Zhurnal 39:755–758.

Ivashin, M. V. 1977. Some abnormalities in embryogenesis of minke whales, *Balaenoptera acutorostrata* (*Cetacean, Balaenoptera*) of the Indian Ocean area of the Antarctic. Zoologicheskii Zhurnal 41:1736–1739.

Ivashin, M. V. 1984. Characteristics of ovarian corpora in dolphins and whales as described by Soviet scientists. Reports of the International Whaling Commission (Special Issue 6):433–444.

Ivashin, M. V., and V. L. Zinchenko. 1982. Occurrences of pathological development of minke embryos (*Balaenoptera acutorostrata*) of the Southern Hemisphere. Document SC/34/Mi29, The Scientific Committee of the International Whaling Commission, Cambridge, U.K.

Jameson, R. J., and A. M. Johnson. 1993. Reproductive characteristics of female sea otters. Marine Mammal Science 9:156–167.

Jarman, M. V. 1979. Impala social behaviour, territory, hierarchy, mating and use of space. Advances in Ethology 21:1–93.

Jarman, P. J. 1966. The status of the dugong (*Dugong dugon* Müller): Kenya 1961. East African Wildlife Journal 4:82–88.

Johannes, R. E., and J. W. MacFarlane. 1991. Traditional fishing in the Torres Strait Islands. CSIRO Division of Fisheries, Hobart.

Karsch, F. J., and D. L. Foster. 1981. Environmental control of seasonal breeding: A common final mechanism governing seasonal breeding and sexual maturation. Pages 30–53 *in* D. Gilmore and B. Cook (eds.). Environmental Factors in Mammal Reproduction. Macmillan, London.

Kasuya, T. 1976. Reconsideration of life history parameters of the spotted and striped dolphins based on cemental layers. Scientific Reports of the Whales Research Institute 28:73–106.

Kasuya, T. 1995. Overview of cetacean life histories: An essay in their evolution. Pages 481–497 *in* A. S. Blix, L. Walloe, and O. Ulltang (eds.). Whales, Seals, Fish and Men. Elsevier Science, Amsterdam.

Kasuya, T., and L. L. Jones. 1984. Behavior and segregation of the Dall's porpoise in the northwestern north Pacific ocean. Scientific Reports of the Whales Research Institute 35:107–128.

Kasuya, T., and H. Marsh. 1984. Life history and reproductive biology of the short-finned pilot whale, *Globicephala macrorhynchus,* off the Pacific coast of Japan. Reports of the International Whaling Commission (Special Issue 6):259–310.

Kasuya, T., and N. Miyazaki. 1982. The stock of *Stenella coeruleoalba* off the Pacific coast of Japan. Food and Agriculture Organization of the United Nations Fisheries Series 5, Mammals in the Seas 4:21–37.

Kasuya, T., and S. Tai. 1993. Life history of short-finned pilot whale stocks off Japan and a description of the fishery. Report of the International Whaling Commission (Special Issue 14):439–473.

Kasuya, T., H. Marsh, and A. Amino. 1993. Non-reproductive mating in short-finned pilot whales. Report of the International Whaling Commission (Special Issue 14):425–437.

Kato, H. 1983. Some considerations on the decline in age at sexual maturity of the Antarctic minke whale. Report of the International Whaling Commission 33:393–399.

Kato, H. 1984. Observations of tooth scars on the head of male sperm whale, as an indication of intra-sexual fightings. Scientific Reports of the Whales Research Institute 35:39–46.

Kato, H. 1987. Density dependent changes in growth parameters of the southern minke whale. Scientific Reports of the Whales Research Institute 38:47–73.

Kato, H., and K. Sakuramoto. 1991. Age at sexual maturity of southern minke whales: A review and some additional analyses. Report of the International Whaling Commission 41:331–337.

Kato, H., and Y. Shimadzu. 1983. Foetal sex ratio of the Antarctic minke whale. Report of the International Whaling Commission 33:357–359.

Kawamura, A. 1969. Siamese twins in the sei whale, *Balaenoptera borealis* (Lesson). Nature (London) 221:490–491.

Keller, K. V. 1986. Training of Atlantic bottlenose dolphins (*Tursiops truncatus*) for artificial insemination. Proceedings of the 14th Annual International Marine Animal Trainer Association Conference 22–24.

Kelly, B. P. 1988a. Bearded seal, *Erignathis barbatus*. Pages 77–94 *in* J. W. Lentfer (ed.). Selected Marine Mammals of Alaska. Marine Mammal Commission, Washington, D.C.

Kelly, B. P. 1988b. Ringed seal, *Phoca hispida*. Pages 57–76 *in* J. W. Lentfer (ed.). Selected Marine Mammals of Alaska. Marine Mammal Commission, Washington, D.C.

Kelly, B. P. 1988c. Ribbon seal, *Phoca fasciata*. Pages 95–106 *in* J. W. Lentfer (ed.). Selected Marine Mammals of Alaska. Marine Mammal Commission, Washington, D.C.

Kenyon, K. W. 1969. The Sea Otter in the Eastern Pacific Ocean. North American Fauna 68, U.S. Fish & Wildlife Service, Washington, D.C. 352 pp.

Kenyon, K. W. 1981. Monk seals, *Monachus* Flemming, 1822. Pages 195–220 *in* S. H. Ridgway and R. J. Harrison (eds.). Handbook of Marine Mammals, Vol. 2. Academic Press, New York, NY.

Kerley, G. I. H. 1985. Pup growth in the fur seals *Arctocephalus tropicalis* and *A. gazella* on Marion Island. Journal of Zoology (London) 205:315–324.

Kerley, G. I. H. 1987. *Arctocephalus tropicalis* on the Prince Edward Islands. Pages 61–64 *in* J. P. Croxall and R. L. Gentry (eds.). Status, Biology, and Ecology of Fur Seals. National Oceanic and Atmospheric Administration Technical Report National Marine Fisheries Service 51.

Kimura, S. 1957. The twinning in southern fin whales. Scientific Reports of the Whales Research Institute 12:103–125.

Kirby, V. L., and S. H. Ridgway. 1984. Hormonal evidence for spontaneous ovulation in captive dolphins, *Tursiops truncatus* and *Delphinus delphis*. Reports of the International Whaling Commission (Special Issue 6):459–464.

Kjeld, M., J. Sigurjonsson and A. Arnason. 1992. Sex hormones concentrations in blood serum from the North Atlantic fin whales (*Balaenoptera physalus*). Journal of Endocrinology 134:405–413.

Klevezal, G. A., and A. C. Myrick Jr. 1984. Marks in tooth dentine of female dolphins (genus *Stenella*) as indicators of parturition. Journal of Mammalogy 65:103–110.

Kooyman, G. L. 1981a. Weddell seal, *Leptonychotes weddelli* Lesson, 1826. Pages 275–296 *in* S. H. Ridgway and R. J. Harrison (eds.). Handbook of Marine Mammals, Vol. 2. Academic Press, New York, NY.

Kooyman, G. L. 1981b. Leopard seal, *Hydrurga leptonyx* Blainville, 1820. Pages 261–274 *in* S. H. Ridgway and R. J. Harrison (eds.). Handbook of Marine Mammals, Vol. 2. Academic Press, New York, NY.

Kretzmann, M. B., D. P. Costa, and B. J. Le Boeuf. 1993. Maternal energy investment in elephant seal pups: evidence for sexual equality? American Naturalist 141:466–480.

Laws, R. M. 1956a. Growth and sexual maturity in aquatic mammals. Nature (London) 178:193–194.

Laws, R. M. 1956b. The elephant seal (*Mirounga leonina* Linn.). III. The physiology of reproduction. Scientific Report No. 15.

Laws, R. M. 1959a. Accelerated growth in seals with special reference to the Phocidae. Norsk Hvalfangst-Tidende 48:425–452.

Laws, R. M. 1959b. Foetal growth rates of whales with special reference to the fin whale, *Balaenopetra physalus* (L.). Discovery Reports 29:281–308.

Laws, R. M. 1961. Southern fin whales. Discovery Reports 31:327–486.

Laws, R. M. 1962. Some effects of whaling on the southern stocks of baleen whales. Pages 137–158 *in* L. D. LeCren and M. W. Holdgate (eds.). The Exploitation of Natural Animal Population. Blackwell Scientific Publications, Oxford.

Laws, R. M. 1977a. The significance of vertebrates in the Antarctic marine ecosystem. Pages 411–438 *in* G. A. Llano (ed.). Adaptations Within Antarctic Ecosystems, 3rd Symposium on Antarctic Ecology, Smithsonian Institution, Washington, D.C.

Laws, R. M. 1977b. Seals and whales of the southern ocean. Philosophical Transactions of the Royal Society of London 279:81–96.

Leader-Williams, N., and C. Ricketts. 1982. Seasonal and sexual patterns of growth and condition of reindeer introduced into South Georgia. Oikos 38:27–39.

Le Boeuf, B. J. 1991. Pinniped mating systems on land, ice and in water: Emphasis on the Phocidae. Pages 45–65 *in* D. Renouf (ed.). Behaviour of Pinnipeds. Chapman and Hall, London.

Le Boeuf, B. J., and K. T. Briggs. 1977. The cost of living in a seal harem. Mammalia 41:167–195.

Ling, J. K., and G. E. Walker. 1978. An 18-month breeding cycle in the Australian sea lion. Search (Sydney) 9:464–465.

Lockyer, C. 1972. The age at sexual maturity of the southern fin whale (*Balaenoptera physalus*) using annual layer counts in the ear plug. Journal of the Conseil Internationale pour l'Exploration Scientifique de la Mer 34:276–294.

Lockyer, C. 1974. Investigation of the ear plug of the southern sei whale, *Balaenoptera borealis*, as a valid means of determining age. Journal of the Conseil Internationale pour l'Exploration Scientifique de la Mer 36:71–81.

Lockyer, C. 1978a. The history and behaviour of a solitary wild, but sociable, bottlenose dolphin (*Turiops truncatus*) on the west coast of England and Wales. Journal of Natural History 12:513–528.

Lockyer, C. 1978b. A theoretical approach to the balance between growth and food consumption in fin and sei whales, with special reference to the female reproductive cycle. Report of the International Whaling Commission 28:243–250.

Lockyer, C. 1979. Changes in a growth parameter associated with exploitation of southern fin and sei whales. Report of the International Whaling Commission 29:191–196.

Lockyer, C. 1981a. Growth and energy budgets of large baleen whales from the Southern Hemisphere. Food and Agriculture Organization of the United Nations Fisheries Series 5, Mammals in the Seas 3:379–487.

Lockyer, C. 1981b. Estimates of growth and energy budget for the sperm whale. Food and Agriculture Organization Fisheries Series 5, Mammals in the Seas 3:489–504.

Lockyer, C. 1984a. Review of baleen whale (*Mysticeti*) reproduction and implications for management. Reports of the International Whaling Commission (Special Issue 6):27–50.

Lockyer, C. 1984b. Age determination by means of the earplug in baleen whales. Report of the International Whaling Commission 34:692–696.

Lockyer, C. 1986. Body fat condition in northeast Atlantic fin whales *Balaenoptera physalus,* and its relationship with reproduction and food resource. Canadian Journal of Fisheries and Aquatic Sciences 43:142–147.

Lockyer, C. 1987a. The relationship between body fat, food resource and reproductive energy costs in north Atlantic fin whales (*Balaenoptera physalus*). Symposia of the Zoological Society (London) 57:343–361.

Lockyer, C. 1987b. Observations of the ovary of the southern minke whale. Scientific Reports of the Whales Research Institute 38:75–89.

Lockyer, C. 1987c. Evaluation of the role of fat reserves in relation to the ecology of North Atlantic fin and sei whales. Pages 183–203 *in* A. C. Huntley, D. P. Costa, G. A. J. Worthy and M. A. Castellini (eds.). Approaches to Marine Mammal Energetics. Society for Marine Mammalogy, Special Publication No. 1. Lawrence, KS.

Lockyer, C. 1990. The importance of biological parameters in population assessments with special reference to fin whales from the N.E. Atlantic. North Atlantic Studies 2:22–31.

Lockyer, C. 1993a. Seasonal changes in body fat condition of northeast Atlantic pilot whales, and their biological significance. Report of the International Whaling Commission (Special Issue 14):323–350.

Lockyer, C. 1993b. A report on patterns of deposition of dentine and cement in teeth of pilot whales, genus *Globicephala.* Report of the International Whaling Commission (Special Issue 14):137–161.

Lockyer, C. 1995a. Aspects of the biology of the harbour porpoise, *Phocoena phocoena,* from British waters. Pages 443–457 *in* A. S. Blix L. Wallaoe, and O. Ulltang (eds.). Whales, Seals, Fish and Men. Elsevier Science, Amsterdam.

Lockyer, C. 1995b. Investigation of aspects of the life history of the harbour porpoise, *Phocoena phocoena,* in British waters. Report of the International Whaling Commission (Special Issue 15):189–197.

Lockyer, C., and S. G. Brown. 1981. The migration of whales. Pages 105–137 *in* D. J. Aidley (ed.). Animal Migration. Society for Experimental Biology, Seminar series 13. Cambridge University Press, Cambridge, U.K.

Lockyer, C., and R. J. Morris. 1986. The history and behaviour of a wild, sociable bottlenose dolphin (*Tursiops truncatus*) off the north coast of Cornwall. Aquatic Mammals 12:3–16.

Lockyer, C., and J. Sigurjonsson. 1991. The Icelandic fin whale (*Balaenoptera physalus*): Biological parameters and their trends over time. Document Sc / F91 / F8, The Scientific Committee of the International Whaling Commission, Reykjavik.

Lockyer, C., and J. Sigurjonsson. 1992. The Icelandic fin whale (*Balaenoptera physalus*): Biological parameters and their trends over time (Summary). Report of the International Whaling Commission 42:617–618.

Lockyer, C., and C. G. Smellie. 1985. Assessment of reproductive status of female fin and sei whales taken off Iceland, from a histological examination of the uterine mucosa. Report of the International Whaling Commission 35:343–348.

Lønø, O. 1972. Polar bear fetuses found in Svalbard. Porsk Polarinst Arbok 149:294–298.

Loughlin, T. R., J. A. Ames, and J. E. Vandevere. 1981. Annual reproduction, dependency period, and apparent gestation period in two California sea otters, *Enhydra lutris.* Fishery Bulletin 79:347–349.

Lunn, N. J., and I. L. Boyd. 1993. Effects of maternal age and condition on parturition and the perinatal period of Antarctic fur seals. Journal of Zoology (London) 229:55–67.

Lunn, N. J., I. L. Boyd, and J. P. Croxall. 1994. Reproductive performance of female Antarctic fur seals: The influence of age, breeding experience, environmental variation and individual quality. Journal of Animal Ecology 63:827–840.

Lynch, G. R., and H. A. Wichman. 1981. Reproduction and thermoregulation in *Peromyscus:* Effects of chronic short days. Physiological Behavior 26:201–205.

Mackintosh, N. A. 1965. The Stocks of Whales. Fishing News (Book), London, U.K. 232 pp.

Majluf, P. 1987. South American fur seal, *Arctocephalus australis,* in Peru. Pages 83–94 *in* J. P. Croxall and R. L. Gentry (eds.). Status, Biology, and Ecology of Fur Seals. National Oceanic and Atmospheric Administration Technical Report National Marine Fisheries Series 51.

Malpaux, B., J. R. Robinson, M. B. Brown and F. J. Karsch. 1988. Importance of changing photoperiod and melatonin secretory pattern in determining the length of the breeding season in the Suffolk ewe. Journal of Reproduction and Fertility 83:461–470.

Mansfield, A. W. 1958. The breeding behaviour and reproductive cycle of the Weddell seal (*Leptonychotes weddelli* Lesson). Falkland Island Dependencies Survey Scientific Reports No. 18.

Mansfield, A. W., and B. Beck. 1977. The grey seal in eastern Canada. Can. Dep. Fish. Environ. Fisheries and Marine Services Technical Report No. 704.

Marlow, B. J. 1975. The comparative behaviour of the Australian sea lions *Neophoca cinerea* and *Phocarctos hookeri* (Pinnipedia: Otariidae). Mammalia 39:159–230.

Marmontel, M. 1993. Age determination and population biology of the Florida manatee, *Trichechus manatus latirostris.* Ph.D. dissertation, University of Florida, Gainesville, FL. 408 pp.

Marmontel, M. 1995. Age and reproduction in female Florida manatees. Pages 98–119 *in* T. J. O'Shea, B. B. Ackerman, and H. F. Percival (eds.). Population Biology of the Florida Manatee. U.S. Department of the Interior, National Biological Service, Information and Technology Report 1.

Marmontel, M., T. J. O'Shea, H. Kochman, and S. Humphrey. 1996. Age determination in manatees using growth-layer-group counts in bone. Marine Mammal Science 12:54–88.

Marsh, H. 1980. Age determination of the dugong (*Dugong dugon*) (Muller) in northern Australia and its biological implications. Reports of the International Whaling Commission (Special Issue 3):181–201.

Marsh, H. 1984. Terminology of reproductive morphology and physiology. Reports of the International Whaling Commission (Special Issue 6):18–22.

Marsh, H. 1985. Observations on the ovaries of an isolated minke whale: Evidence for spontaneous sterile ovulation and structure of the resultant corpus. Scientific Reports of the Whales Research Institute 36:35–39.

Marsh, H. 1986. The status of the dugong in Torres Strait. Pages 53–76 *in* A. K. Haines, G. C. Williams, and D. Coates (eds.). Torres Strait Fisheries Seminar, Port Moresby. February 1985. Australian Government Publishing Service, Canberra.

Marsh, H. 1989. Mass stranding of dugongs by a tropical cyclone. Marine Mammal Science 5:75–84.

Marsh, H. 1995. The life history, pattern of breeding and population dynamics of the dugong. Pages 75–83 *in* T. J. O'Shea, B. B. Ackerman, and H. F. Percival (eds.). Population Biology of the Florida Manatee. U.S. Department of the Interior, National Biological Service, Information and Technology Report 1.

Marsh, H., and P. Corkeron. 1996. The status of the Dugong in the

Great Barrier Reef region. Pages 231–247 *In* D. Wachenfeld and J. Oliver (eds.). Proceedings of State of the Great Barrier Reef Workshop. Great Barrier Reef Marine Park Authority, Townsville.

Marsh, H., and T. Kasuya. 1984. Changes in the ovaries of the short-finned pilot whale, *Globicephala macrorhynchus,* with age and reproductive activity. Reports of the International Whaling Commission (Special Issue 6):331–335.

Marsh, H., and T. Kasuya. 1986. Evidence for reproductive senescence in female cetaceans. Reports of the International Whaling Commission (Special Issue 8):57–74.

Marsh, H., P. W. Channells, G. E. Heinsohn, and J. Morissey. 1982. Analysis of stomach contents of dugongs from Queensland. Australian Wildlife Research 9:55–67.

Marsh, H., G. E. Heinsohn, and P. W. Channells. 1984a. Changes in the ovaries and uterus of the dugong, *Dugong dugon* (Sirenia: Dugongidae) with age and reproductive activity. Australian Journal of Zoology 32:743–766.

Marsh, H., G. E. Heinsohn, and T. D. Glover. 1984b. Changes in the male reproductive organs of the dugong, *Dugong dugon* (Sirenia: Dugongidae) with age and reproductive activity. Australian Journal of Zoology 32:721–742.

Marsh, H., G. E. Heinsohn, and L. M. Marsh. 1984c. Breeding cycle, life history and population dynamics of the dugong, *Dugong dugon* (Sirenia: Dugongidae). Australian Journal of Zoology 32:767–785.

Marsh, H., P. Corkeron, I. R. Lawler, J. M. Lanyon, and A. R. Preen. 1995. The status of the dugong in the southern Great Barrier Reef Marine Park. Report to the Great Barrier Reef Marine Park Authority.

Martin, A. R. 1980. An examination of sperm whale age and length data from the 1949–78 Icelandic catch. Report of the International Whaling Commission 30:227–231.

Martin, A. R. 1990. Whales and Dolphins. Salamander Books Ltd, London, U.K. 192 pp.

Martin, A. R., and P. Rothery. 1993. Reproductive parameters of female long-finned pilot whales (*Globicephala melas*) around the Faroe Islands. Report of the International Whaling Commission (Special Issue 14):263–304.

Martineau, D., P. Béland, C. Desjardins, and A. Lagacé. 1987. Levels of organochlorine chemicals in tissues of beluga whales (*Delphinapterus leucas*) from the St Lawrence Estuary, Québec, Canada. Archives of Environmental Contamination and Toxicology 16:137–147.

Martinet, L., C. Allais, and D. Allain. 1981. The role of prolactin and LH in luteal function and blastocyst growth in mink (*Mustela vison*). Journal of Reproduction and Fertility (Suppl.) 29:119–130.

Martinet, L., M. Mondainmonval, and R. Monnerie. 1992. Endogenous circannual rhythms and photorefractoriness of testis activity, molt and prolactin concentrations in mink (*Mustela vison*). Journal of Reproduction and Fertility 95:325–338.

Masaki, Y. 1978. Yearly change in the biological parameters for the Antarctic sei whale. Report of the International Whaling Commission 28:421–430.

Masaki, Y. 1979. Yearly change of the biological parameters for the Antarctic minke whale. Report of the International Whaling Commission 29:375–396.

Matthews, L. H. 1948. Cyclic changes in the uterine mucosa of balaenopterid whales. Journal of Anatomy 82:207–232.

Mattlin, R. H. 1987. New Zealand fur seal, *Arctocephalus forsteri,* within the New Zealand region. Pages 49–52 *in* J. P. Croxall and R. L. Gentry (eds.). Status, Biology, and Ecology of Fur Seals. National Oceanic and Atmospheric Administration Technical Report National Marine Fisheries Service 51.

Maurel, D., C. Coutant, L. Boissin-Agasse, and J. Boissin. 1986. Seasonal moulting patterns in three fur bearing mammals: The European badger (*Meles meles* L.), the red fox (*Vulpes vulpes* L.), and the mink (*Mustela vison*). A morphological and histological study. Canadian Journal of Zoology 64:1757–1764.

McBride, A. F., and H. Kritzler. 1951. Observations on pregnancy, parturition and post-natal behavior in the bottlenose dolphin. Journal of Mammalogy 32:251–266.

McCann, T. S., M. A. Fedak, and J. Harwood. 1989. Parental investment in southern elephant seals, *Mirounga leonina.* Behavioral Ecology and Sociobiology 25:81–87.

McLaren, I. A. 1958. The biology of the ringed seal (*Phoca hispida* Schreder) in the eastern Canadian arctic. Bulletin of the Fisheries Research Board of Canada No. 118.

McSweeney, D. J., K. C. Chu, W. F. Dolphin, and L. N. Guinee. 1989. North Pacific humpback whale songs: A comparison of southeast Alaskan feeding ground songs with Hawaiian wintering ground songs. Marine Mammal Science 5:139–148.

Mead, J. G., and C. W. Potter. 1990. Natural history of bottlenose dolphins along the central Atlantic coast of the United States. Pages 165–195 *in* S. Leatherwood and R. R. Reeves (eds.). The Bottlenose Dolphin. Academic Press, New York, NY.

Mead, R. A. 1981. Delayed implantation in mustelids, with special emphasis on the spotted skunk. Journal of Reproduction and Fertility (Suppl.) 29:11–24.

Messier, F., M. K. Taylor, and M. A. Ramsay. 1992. Seasonal activity patterns of female polar bears (*Ursus maritimus*) in the Canadian Arctic as revealed by satellite telemetry. Journal of Zoology (London) 226:219–229.

Mizroch, S. A. 1981a. Analysis of some biological parameters of the Antarctic fin whale (*Balaenoptera physalus*). Report of the International Whaling Commission 31:425–434.

Mizroch, S. A. 1981b. Further notes on Southern Hemisphere baleen whale pregnancy rates. Report of the International Whaling Commission 31:629–634.

Mizroch, S. A., and A. E. York. 1982. Have Southern Hemisphere baleen whales pregnancy rates increased? Document SC/34/Ba4, The Scientific Committee of the International Whaling Commission, Cambridge, U.K.

Moore, J. C. 1955. Bottle-nosed dolphins support remains of young. Journal of Mammalogy 36:466–467.

Nicholls, T. J., A. R. Goldsmith, and A. S. Dawson. 1988. Photorefractoriness in birds and comparison with mammals. Physiological Review 68:133–176.

Nietschmann, B. 1984. Hunting and ecology of dugongs and green turtles in Torres Strait. National Geographic Society Research Reports 17:625–651.

Nietschmann, B., and J. Nietschmann. 1981. Good dugong, bad dugong; good turtle, bad turtle. Natural History 90:54–62.

Nishiwaki, M. 1957. One-eyed monster of fin whale. Scientific Reports of the Whales Research Institute 12:193–195.

Nishiwaki, M., and T. Tobayama. 1982. Morphological study of the hybrid between *Tursiops* and *Pseudorca.* Scientific Reports of the Whales Research Institute 34:109–121.

Noonan, L. M., and K. Ronald. 1989. Determination of estrone sulfate, progesterone and testosterone for hooded seals, *Cystophora cristata.* Proceedings of 8th Biennial Conference on the Biology of Marine Mammals, Pacific Grove, CA, 7–11 December 1989.

Noonan, L. M., K. Ronald, and J. Raeside. 1991. Plasma testosterone concentrations of captive male hooded seals (*Cystophora cristata*). Canadian Journal of Zoology 69:2279–2282.

Norris, C. E. 1960. The distribution of the dugong in Ceylon. Loris 8:296–300.

Odell, D. K. 1975. Breeding biology of the California sea lion (*Zalophus californianus*). Rapports et Proces-Verbaux des Reunions Conseil Internationale pour l'Exploration Scientifique de la Mer 169:374–378.

Odell, D. K. 1982. West Indian manatee, *Trichechus manatus*. Pages 828–837 *in* J. A. Chapman and G. A. Feldhamer (eds.). Wild Mammals of North America: Biology, Management and Economics. The John Hopkins University Press, Baltimore, MD.

Odell, D. K., D. J. Forrester, and E. D. Asper. 1981. A preliminary analysis of organ weights and sexual maturity in the West Indian manatee (*Trichechus manatus*). Pages 52–65 *in* R. L. Brownell Jr., and K. Ralls (eds.). The West Indian Manatee in Florida. Proceedings of a workshop held in Orlando, FL, 27–29 March 1978. Florida Department of Natural Resources, Tallahassee, FL.

Odell, D. K., G. D. Bossart, M. T. Lowe, and T. D. Hopkins. 1995. Reproduction of the West Indian manatee in captivity. Pages 192–193 *in* T. J. O'Shea, B. B. Ackerman, and H. F. Percival (eds.). Population Biology of the Florida Manatee. U.S. Department of the Interior, National Biological Service, Information and Technology Report 1.

Oftedal, O. T. 1984. Body size and reproductive correlates of milk energy output in lactating mammals. Acta Zoologica Fennica 171:183–186.

Oftedal, O. T., D. J. Boness, and R. A. Tedman. 1987. The behaviour, physiology and anatomy of lactation in the Pinnipedia. Current Mammalogy 1:175–245.

Oftedal, O. T., W. D. Bowen, E. M. Widdowson, and D. J. Boness. 1989. Effects of suckling and the postsuckling fast on weights of the body and internal organs of harp and hooded seal pups. Biology of the Neonate 56:283–300.

Oftedal, O. T., W. D. Bowen, and D. J. Boness. 1993. Energy transfer by lactating hooded seals and nutrient deposition in their pups during the four days from birth to weaning. Physiological Zoology 66:412–436.

Ohsumi, S. 1959. A deformed fin whale foetus. Scientific Reports of the Whales Research Institute 14:145–147.

Ohsumi, S. 1986a. Ear plug transition phase as an indicator of sexual maturity in female Antarctic minke whales. Scientific Reports of the Whales Research Institute 37:17–30.

Ohsumi, S. 1986b. Yearly change in age and body length at sexual maturity of a fin whale stock in the eastern North Pacific. Scientific Reports of the Whales Research Institute 37:1–16.

Øritsland, T. 1970. Biology and population dynamics of Antarctic seals. Pages 361–366 *in* M. W. Holdgate (ed.). Antarctic Ecology, Vol. 1. Academic Press, London.

O'Shea, T. J., and B. B. Ackerman. 1995. Population biology of the Florida manatee: An overview. Pages 280–287 *in* T. J. O'Shea, B. B. Ackerman, and H. F. Percival (eds.). Population Biology of the Florida Manatee. U.S. Department of the Interior, National Biological Service, Information and Technology Report 1.

O'Shea, T. J., and W. C. Hartley. 1995. Reproduction and early-age survival of manatees at Blue Spring, Upper St. Johns River, Florida. Pages 157–170 *in* T. J. O'Shea, B. B. Ackerman, and H. F. Percival (eds.). Population Biology of the Florida Manatee. U.S. Department of the Interior, National Biological Service, Information and Technology Report 1.

O'Shea, T. J., and C. A. Langtimm. 1995. Estimation of survival of adult Florida manatees in the Crystal River, at Blue Spring, and on the Atlantic Coast. Pages 194–222 *in* T. J. O'Shea, B. B. Ackerman, and H. F. Percival (eds.). Population Biology of the Florida Manatee. U.S. Department of the Interior, National Biological Service, Information and Technology Report 1.

O'Shea, T. J., B. B. Ackerman, and H. F. Percival (eds.). 1995. Population Biology of the Florida Manatee. U.S. Department of the Interior, National Biological Service, Information and Technology Report 1.

Ouellette, J., and K. Ronald. 1985. Histology of reproduction in harp and grey seals during pregnancy, postparturition, and estrus. Canadian Journal of Zoology 63:1778–1796.

Palmer, S. S., R. A. Nelson, M. A. Ramsay, I. Stirling, and J. M. Bahr. 1988. Annual changes in serum sex steroids in male and female black (*Ursus americanus*) and polar (*Ursus maritimus*) bears. Biological Reproduction 38:1044–1050.

Parsons, J. L., R. G. Brown, W. C. Kimmins, M. Mezei, B. Pohajdak, A. Bowen, and W. Stobo 1993. Immunocontraception of seals. Proceedings of the Tenth Biennial Conference on the Biology of Marine Mammals, Galveston, TX, 11–15 November 1993.

Payne, R. 1988. Behavior of Southern Right Whales (*Eubalaena australis*). University of Chicago Press, Chicago, IL.

Perrin, W. F. and J. R. Henderson. 1984. Growth and reproductive rates in two populations of spinner dolphins, *Stenella longirostris*, with different histories of exploitation. Reports of the International Whaling Commission (Special Issue 6):417–430.

Perrin, W. F., and A. C. Myrick Jr. (eds.). 1981. Report of the workshop. Age determination of toothed whales and sirenians. Reports of the International Whaling Commission (Special Issue 3):1–50.

Perrin, W. F. and S. B. Reilly. 1984. Reproductive parameters of dolphins and small whales of the family *Delphinidae*. Reports of the International Whaling Commission (Special Issue 6):97–133.

Pierson, M. O. 1987. Breeding behaviour of the Guadalupe fur seal, *Arctocephalus townsendii*. Pages 83–94 *in* J. P. Croxall and R. L. Gentry (eds.). Status, Biology, and Ecology of Fur Seals. National Oceanic and Atmospheric Administration Technical Report National Marine Fisheries Service 51.

Pitcher, K. W., and D. Calkins. 1981. Reproductive biology of Steller sea lion, *Eumetopias jubatus*, in the Gulf of Alaska. Journal of Mammalogy 62:599–605.

Poiner, I. R., and C. Peterken. 1995. Seagrasses. Pages 107–118 *in* L. P. Zann and P. Kailola (eds.). State of the Marine Environment Report for Australia. Technical Annexe 1. The Marine Environment. Great Barrier Reef Marine Park Authority, Townsville.

Potelov, V. A. 1975. Reproduction of the bearded seal (*Erignathus barbatus*) in the Barents Sea. Rapports et Proces-Verbaux des Reunions Conseil Internationale pour l'Exploration Scientifique de la Mer 169:554.

Powell, J. A. 1996. The distribution and biology of the West African manatee (*Trichechus senegalensis* Link, 1795). Report to the United Nations Environment Programme, Nairobi, Kenya, 61 pp.

Preen, A. 1989. Observations of mating behavior in dugongs (*Dugong dugon*). Marine Mammal Science 5:382–387.

Preen, A. 1995. Diet of dugongs: Are they omnivores? Journal of Mammalogy 76:163–171.

Preen, A., and H. Marsh. 1995. Response of dugongs to large-scale loss of seagrass from Hervey Bay, Queensland, Australia. Wildlife Research 22:507–519.

Qi Jingfen. 1984. Breeding of the West Indian manatee (*Trichechus manatus* Linn) in captivity. Acta Theriologica Sinica 4:27–33.

Racey, P. A. 1981. Environmental factors affecting gestation lengths in mammals. Pages 197–213 *in* D. Gilmore and B. Cook (eds.). Environmental Factors in Mammal Reproduction. Macmillan, London.

Raeside, J. I., and K. Ronald. 1981. Plasma concentrations of estrone, progesterone and corticosteroids during late pregnancy and after parturition in the harbour seal, *Phoca vitulina*. Journal of Reproduction and Fertility 61:135–139.

Ramsay, M. A., and R. L. Dunbrack. 1986. Physiological constraints on life-history phenomena: the example of small bear cubs at birth. American Naturalist 127:735–743.

Ramsay, M. A., and I. Stirling. 1988. Reproductive biology and ecology of female polar bears (*Ursus maritimus*). Journal of Zoology (London) 214:601–634.

Rand, R. W. 1954. Reproduction in the female Cape fur seal (*Arctocephalus pusillus*). Proceedings of the Zoological Society of London 124:717–739.

Rathbun, G. B., and J. Powell. 1982. Reproduction in the manatee (*Trichechus manatus*) (Abstract). Sixty-second annual meeting of the American Society of Mammalogists, Snowbird, UT.

Rathbun, G. B., J. P. Reid, R. K. Bonde, and J. A. Powell. 1995. Reproduction in free-ranging Florida manatees. Pages 135–156 *in* T. J. O'Shea, B. B. Ackerman, and H. F. Percival (eds.). Population Biology of the Florida Manatee. U.S. Department of the Interior, National Biological Service, Information and Technology Report 1.

Read, A. J. 1990a. Estimation of body condition in harbour porpoises, *Phocoena phocoena*. Canadian Journal of Zoology 68:1962–1966.

Read, A. J. 1990b. Age at sexual maturity and pregnancy rates of harbour porpoises *Phocoena phocoena* from the Bay of Fundy. Canadian Journal of Fisheries and Aquatic Sciences 47:561–565.

Read, A. J. 1990c. Reproductive seasonality in harbour porpoises, *Phocoena phocoena*, from the Bay of Fundy. Canadian Journal of Zoology 68:284–288.

Read, A. J., and D. E. Gaskin. 1990. Changes in the growth and reproduction of harbour porpoises, *Phocoena phocoena*, from the Bay of Fundy. Canadian Journal of Fisheries and Aquatic Sciences 47:2158–2163.

Read, A. J., and A. A. Hohn. 1995. Life in the fast lane: The life history of harbor porpoises from the Gulf of Maine. Marine Mammal Science 11:423–440.

Reid, J. P., R. K. Bonde, and T. J. O'Shea. 1995. Reproduction and mortality of radio-tagged and recognizable manatees on the Atlantic Coast of Florida. Pages 171–191 *in* T. J. O'Shea, B. B. Ackerman, and H. F. Percival (eds.). Population Biology of the Florida Manatee. U.S. Department of the Interior, National Biological Service, Information and Technology Report 1.

Reid, K., J. Mann, J. R. Weiner, and N. Hecker. 1995. Infant development in two aquarium bottlenose dolphins. Zoo Biology 14:135–147.

Reijnders, P. J. H. 1980. Organochlorine and heavy metal residues in harbour seals from the Wadden Sea and their possible effects on reproduction. Netherlands Journal of Sea Research 14:30–65.

Reijnders, P. J. 1986. Reproductive failure in harbour seals feeding on fish from polluted coastal waters. Nature (London) 324:456–457.

Reijnders, P. J. 1990. Progesterone and oestradiol—17β concentration profiles throughout the reproductive cycles in harbour seals (*Phoca vitulina*). Journal of Reproduction and Fertility 90:403–409.

Reiter, J. 1980. The pineal and its hormones in the control of reproduction in mammals. Endocrinology Review 1:109–131.

Reiter, J., and B. J. Le Boeuf. 1991. Life history consequences of variation in age at primiparity in northern elephant seals. Behavioral Ecology and Sociobiology 28:153–160.

Reiter, J., K. J. Panken, and B. J. Le Boeuf. 1981. Female competition and reproductive success in northern elephant seals. Animal Behaviour 29:670–687.

Renfree, M. B., and J. H. Calaby. 1981. Background to delayed implantation and embryonic diapause. Journal of Reproduction and Fertility (Suppl.) 29:1–9.

Renouf, D., and R. Gales. 1994. Seasonal variation in the metabolic rate of harp seals: unexpected energetic economy in the cold ocean. Canadian Journal of Zoology 72:1625–1632.

Renouf, D., and E. Noseworthy. 1991. Changes in food intake, mass, and fat accumulation in association with variations in thyroid hormone levels of harbour seals (*Phoca vitulina*). Canadian Journal of Zoology 69:2470–2479.

Renouf, D., R. Gales, and E. Noseworthy. 1993. Seasonal variation in energy intake and condition of harp seals: Is there a harp seal morph? Problems for bioenergetic modelling. Journal of Zoology (London) 230:513–528.

Reynolds, J. E. III. 1995. Florida manatee population biology: Research progress, infrastructure and applications of conservation and management. Pages 6–12 *in* T. J. O'Shea, B. B. Ackerman, and H. F. Percival (eds.). Population Biology of the Florida Manatee. U.S. Department of the Interior, National Biological Service, Information and Technology Report 1.

Reynolds, J. E. III, and D. K. Odell. 1991. Manatees and Dugongs. Facts on File, Inc., New York, NY.

Reynolds, J. E. III, and J. R. Wilcox. 1994. Observations of Florida manatees (*Trichechus manatus latirostris*) around selected power plants in winter. Marine Mammal Science 10:163–177.

Rice, D. W., and A. A. Wolman. 1971. The Life History and Ecology of the Gray Whale (*Eschrichtius robustus*). Special Publication 3. The American Society of Mammalogists, Lawrence, KS.

Ridgway, S., T. Kamolnik, M. Reddy, C. Curry, and R. Tarpley. 1995. Orphan-induced lactation in *Tursiops* and analysis of collected milk. Marine Mammal Science 11:172–182.

Robeck, T. R., B. E. Curry, J. F. McBain, and D. C. Kraemer. 1994. Reproductive biology of the bottlenose dolphin (*Tursiops truncatus*) and the potential application of advanced reproductive technologies. Journal of Zoo and Wildlife Medicine 25:321–336.

Roberts, A., N. Clomp, and J. Birkhead. 1996. Monitoring the marine and terrestrial harvest in an Aboriginal community in North Queensland. Pages 152–166 *in* M. Bomford and J. Caughley (eds.). Sustainable Use of Wildlife by Aboriginal Peoples and Torres Strait Islanders. Australian Government Publishing Service, Canberra.

Rommel, S. A., G. A. Early, K. A. Matassa, D. A. Pabst, and W. A. McLellan. 1995. Venous structures associated with thermoregulation of phocid seal reproductive organs. The Anatomical Record 243:390–402.

Ronald, K., and P. J. Healey. 1981. Harp seal—*Phoca groenlandica*. Pages 55–88 *in* S. H. Ridgway and R. J. Harrison (eds.). Handbook of Marine Mammals, Vol. 2. Academic Press, New York, NY.

Ross, H. M., R. J. Reid, F. E. Howie, and E. W. Gray. 1994a. Herpes virus infection of the genital tract in harbour porpoise *Phocoena phocoena*. European Research on Cetaceans 8:209.

Ross, H. M., G. Foster, R. J. Reid, K. L. Jahans, and A. P. MacMillan.

1994b. *Brucella* species infection in sea mammals. Veterinary Record 134:359.

Rotterman, L. M., and T. Simon-Jackson. 1988. Sea otter, *Enhydra lutris*. Pages 237–275 *in* J. W. Lentfer (ed.). Selected Marine Mammals of Alaska. U.S. Marine Mammal Commission, Washington, D.C.

Sacher, G. A., and E. F. Staffeldt. 1974. Relation of gestation time to brain weight for placental mammals: Implications for the theory of vertebrate growth. American Naturalist 108:593–615.

Sawyer-Steffan, J. E., V. L. Kirby, and W. G. Gilmartin. 1983. Progesterone and estrogens in the pregnant and non-pregnant dolphin, *Tursiops truncatus*, and the effects of induced ovulation. Biological Reproduction 28:897–901.

Scammon, C. M. (ed.). 1874. The marine mammals of the northwestern coast on North America described and illustrated: Together with an account of the American whale fishery. J. H. Karmony, San Francisco, CA. (Reprinted in 1968 by Dover Publications Inc., New York, NY).

Schroeder, J. P. 1990. Breeding bottlenose dolphins in captivity. Pages 435–446 *in* S. Leatherwood and R. R. Reeves (eds.). The Bottlenose Dolphin. Academic Press, New York, NY.

Schroeder, J. P., and K. V. Keller. 1990. Artificial insemination of bottlenose dolphins. Pages 447–460 *in* S. Leatherwood and R. R. Reeves (eds.). The Bottlenose Dolphin. Academic Press, New York, NY.

Schummer, A., and R. Nickel. 1979. The Viscera of the Domestic Mammals. Springer-Verlag, New York.

Scott, M. D., R. S. Wells, and A. B. Irvine. 1990. A long-term study of bottlenose dolphins on the west coast of Florida. Pages 235–244 *in* S. Leatherwood and R. R. Reeves (eds.). The Bottlenose Dolphin. Academic Press, New York, NY.

Seager, S., W. Gilmartin, L. Moore, C. Platz, and V. Kirby 1981. Semen collection (electroejaculation), evaluation and freezing in the Atlantic bottlenosed dolphin (*Tursiops truncatus*). Proceedings of the Annual Meeting of the American Association of Zoo Veterinarians.

Sease, J. L., and D. G. Chapman. 1988. The Pacific walrus, *Odobenus rosmarus divergens*. Pages 17–38 *in* J. W. Lentfer (ed.). Selected Marine Mammals of Alaska. Marine Mammal Commission, Washington, D.C.

Sergeant, D. E. 1976. History and present status of populations of harp and hooded seals. Biological Conservation 10:95–118.

Sigurjonsson, J. 1990. Whale stocks off Iceland—Assessment and methods. North Atlantic Studies 2:64–76.

Sigurjonsson, J. 1995. On the life history and autecology of North Atlantic rorquals. Pages 425–441 *in* A. S. Blix, L. Walloe, and O. Ulltang (eds.). Whales, Seals, Fish and Men. Elsevier Science, Amsterdam.

Sinha, A. A., and C. H. Conway. 1968. The ovary of the sea otter. Anatomical Record 160:795–806.

Sinha, A. A., and H. W. Mossman. 1966. Placentation of the sea otter. American Journal of Anatomy 119:521–554.

Sinha, A. A., C. H. Conway, and K. W. Kenyon. 1966. Reproduction in the female sea otter. Journal of Wildlife Management 30:121–130.

Siniff, D. B., and K. Ralls. 1991. Population status of California sea otters. Minerals Management Service, U.S. Department of the Interior, Los Angeles, CA, OCS Study MMS 88-0021.

Siniff, D. B., I. Stirling, J. L. Bengtson, and R. A. Reichle. 1979. Social and reproductive behaviour of crabeater seals (*Lobodon carcinophagus*) during the austral spring. Canadian Journal of Zoology 57:2243–2255.

Skinner, J. D., and L. M. Westlin-Van Aarde. 1989. Aspects of reproduction in female Ross seals (*Ommatophoca rossii*). Journal of Reproduction and Fertility 87:67–72.

Slijper, E. J. 1949. On some phenomena concerning pregnancy and parturition of the Cetacea. Bijdragon tot de Dierkunde 28:416–448.

Smith, C. C. 1976. When and how much to reproduce: The trade-off between power and efficiency. American Zoology 16:763–774.

Smith, T. D. 1983. Changes in size of three dolphin populations (*Stenella* sp.) in the eastern tropical Pacific. Fisheries Bulletin 81:1–14.

Smith, T. G. 1980. Polar bear predation of ringed and bearded seals in the land-fast sea ice habitat. Canadian Journal of Zoology 58:2201–2209.

Smith, T. G. 1987. The ringed seal, *Phoca hispida*, of the Canadian Western Arctic. Bulletin of the Fisheries Research Board of Canada No. 216.

Somers, I. 1994. Modelling loggerhead turtle populations. Pages 142–145 *in* R. James (compiler). Proceedings of Australian Marine Turtle Conservation Workshop. Queensland Department of Environment and Heritage, Brisbane, and Australian Nature Conservation Agency, Canberra.

Spilliaert, R., G. Vikingsson, U. Arnason, A. Palsdottir, J. Sigurjonsson, and A. Arnason. 1991. Species hybridization between a female blue whale (*Balaenoptera musculus*) and a male fin whale (*B. physalus*): Molecular and morphological documentation. Journal of Heredity 82:269–274.

Spotte, S. 1982. The incidence of twins in pinnipeds. Canadian Journal of Zoology 60:2226–2233.

Spotte, S., and G. Adams. 1981. Photoperiod and reproduction in captive female northern fur seals. Mammalogy Review 11:31–36.

Stern, G. A., D. C. G. Muir, M. D. Segstro, R. Dietz, and M. P. Heide-Jørgensen. 1994. PCB's and other organochlorine contaminants in white whales (*Delphinapterus leucas*) from West Greenland: Variations with age and sex. Meddr Grønland, Bioscience 39:245–259.

Stewart, R. E. A. 1987. Behavioural reproductive effort of nursing harp seals, *Phoca groenlandica*. Journal of Mammalogy 68:348–358.

Stewart, R. E. A. 1994. Progesterone levels and reproductive status of white whales (*Delphinapterus leucas*) from the Canadian Arctic. Meddr Grønland, Bioscience 39:139–243.

Stewart, R. E. A., and D. M. Lavigne. 1980. Neonatal growth of northwest Atlantic harp seals, *Pagophilus groenlandicus*. Journal of Mammalogy 61:670–680.

Stewart, R. E. A., and D. M. Lavigne. 1984. Energy transfer and female condition in nursing harp seals *Phoca groenlandica*. Holarctic Ecology 7:182–194.

Stewart, R. E. A., B. E. Stewart, D. M. Lavigne, and G. W. Miller. 1989. Foetal growth of northwest Atlantic harp seals, *Phoca groenlandica*. Canadian Journal of Zoology 67:2147–2157.

Stirling, I. 1969. Ecology of the Weddell seal in McMurdo Sound, Antarctica. Ecology 50:673–586.

Stirling, I. 1975. Factors affecting the evolution of social behaviour in Pinnipedia. Rapports et Proces-Verbaux des Reunions Conseil Internationale pour l'Exploration Scientifique de la Mer 169:205–212.

Stirling, I., and W. R. Archibald. 1977. Aspects of predation of seals by polar bears. Journal of the Fisheries Research Board of Canada 34:1126–1129.

Stirling, I., and E. H. McEwan. 1975. The caloric value of whole ringed seals (*Phoca hispida*) in relation to polar bear (*Ursus maritima*) ecology and hunting behavior. Canadian Journal of Zoology 53:1021–1027.

Stirling, I., and N. A. Øritsland. 1995. Relationships between estimates of ringed seal (*Phoca hispida*) and polar bear (*Ursus maritimus*) populations in the Canadian Arctic. Canadian Journal of Fisheries and Aquatic Sciences 52:2594–2612.

Subramanian, A., S. Tanabe, R. Tatsukawa, R. Saito, and N. Miyazaki. 1987. Reduction in the testosterone levels by PCB's and DDE in Dall's porpoise of North Western North Pacific. Marine Pollution Bulletin 18:643–646.

Subramanian, A., S. Tanabe, and R. Tatsukawa. 1988. Use of organochlorines as chemical tracers in determining some reproductive parameters in *dalli*-type Dall's porpoises *Phocoenoides dalli*. Marine Environmental Research 25:161–174.

Swartz, S. 1986. Gray whale, migratory, social and breeding behavior. Reports of the International Whaling Commission (Special Issue 8):207–229.

Swenson, M. J., and W. O. Reece (eds.). 1993. Duke's Physiology of Domestic Animals, 11th ed. Cornell University Press, Ithaca, NY.

Taruski, A. G., C. E. Olney, and H. E. Winn. 1975. Chlorinated hydrocarbons in cetaceans. Journal of the Fisheries Research Board of Canada 132:2205–2209.

Tayler, C. K., and G. S. Saayman. 1972. The social organisation and behaviour of dolphins (*Tursiops aduncas*) and baboons (*Papio ursinus*): Some comparisons and assessments. Annals of the Cape Provencial Museum 9:11–49.

Temte, J. L. 1985. Photoperiod and delayed implantation in the northern fur seal (*Callorhinus ursinus*). Journal of Reproduction and Fertility 73:127–131.

Temte, J. L. 1991. Precise birth timing in captive harbour seal (*Phoca vitulina*) and California sea lions (*Zalophus californianus*). Marine Mammal Science 7:145–156.

Temte, J. L. 1994. Photoperiod control of birth timing in the harbour seal (*Phoca vitulina*). Journal of Zoology (London) 233, 369–384.

Temte, J. L., and J. Temte. 1993. Photoperiod defines the phenology of birth in captive California sea lions. Marine Mammal Science 9:301–308.

Temte, J. L., M. A. Bigg, and O. Wiig. 1991. Clines revisited: The timing of pupping in the harbour seal (*Phoca vitulina*). Journal of Zoology (London) 224:617–632.

Testa, J. W. 1987. Long-term reproductive patterns and sighting bias in Weddell seals (*Leptonychotes weddelli*). Canadian Journal of Zoology 65:1091–1099.

Testa, J. W., and D. B. Siniff. 1987. Population dynamics of Weddell seals (*Leptonychotes weddelli*) in McMurdo Sound, Antarctica. Ecological Monograph 57:149–165.

Thomas, D. 1966. Natural history of dugongs in Rameswaram waters. Madras Journal of Fisheries 2:80–82.

Thomas, J., D. DeMaster, S. Stone, and D. Andriashek. 1980. Observations of a newborn Ross seal pup (*Ommatophoca rossi*) near the Antarctic Peninsula. Canadian Journal of Zoology 58:2156–2158.

Thomas, P. 1986. Methodology of behavioural studies of cetaceans: Right whale mother–infant behaviour. Reports of the International Whaling Commission (Special Issue 8):113–119.

Thompson, P. 1988. Timing of mating in the common seal (*Phoca vitulina*). Mammal Review 18:105–112.

Thompson, P., and P. Rothery. 1987. Age and sex differences in the timing of moult in the common seal, *Phoca vitulina*. Journal of Zoology (London) 212:597–603.

Tikhomirov, E. A. 1966. On the reproduction of seals belonging to the family Phocidae in the North Pacific. Zoologicheskii Zhurnal 45:275–281.

Tikhomirov, E. A. 1975. Biology of the ice forms of seals in the Pacific section of the Antarctic. Rapports et Proces-Verbaux des Reunions Conseil Internationale pour l'Exploration Scientifique de la Mer 169:409–412.

Timm, R. M., V. L. Albuja, and B. L. Clauson. 1986. Ecology, distribution, harvest and conservation of the Amazonian manatee, *Trichechus inunguis*, in Ecuador. Biotropica 18:150–156.

Torres, N. D. 1987. Juan Fernandez fur seal, *Arctocephalus townsendii*. Pages 37–42 *in* J. P. Croxall and R. L. Gentry (eds.). Status, Biology, and Ecology of Fur Seals. National Oceanic and Atmospheric Administration Technical Report National Marine Fisheries Service 51.

Trillmich, F. 1987. Galapagos fur seal, *Arctocephalus galapagoensis*. Pages 23–28 *in* J. P. Croxall and R. L. Gentry (eds.). Status, Biology, and Ecology of Fur Seals. National Oceanic and Atmospheric Administration Technical Report National Marine Fisheries Service 51.

Trillmich, F. 1996. Parental investment in pinnipeds. Advances in the Study of Behavior 25:533–577.

Trites, A. W. 1991. Foetal growth of northern fur seals: Life-history strategy and sources of variation. Canadian Journal of Zoology 69:2608–2617.

Twiss, J. R., and R. R. Reeves (eds.). 1999. Conservation and Biology of Marine Mammals. Smithsonian Institution Press, Washington, D.C.

Tyack, P., and H. Whitehead. 1983. Male competition in large groups of wintering humpback whales. Behaviour 83:132–154.

Urian, K. W., D. A. Duffield, A. J. Read, R. S. Wells, and E. D. Shell. 1996. Seasonality of reproduction in bottlenose dolphins, *Tursiops truncatus*. Journal of Mammalogy 77:394–403.

Vaz-Ferreira, R. 1975. Behaviour of the southern sea lion *Otaria flavescens* (Shaw) in the Uruguayan islands. Rapports et Proces-Verbaux des Reunions Conseil Internationale pour l'Exploration Scientifique de la Mer 169:219–227.

Vaz-Ferreira, R., and A. Ponce de Leon. 1987. South American fur seal, *Arctocephalus australis*, in Uruguay. Pages 29–32 *in* J. P. Croxall and R. L. Gentry (eds.). Status, Biology, and Ecology of Fur Seals. National Oceanic and Atmospheric Administration Technical Report National Marine Fisheries Service 51.

Venables, U. M., and L. S. V. Venables. 1959. Vernal coition of the seal *Phoca vitulina* in Shetland. Proceedings of the Zoological Society of London 132:665–669.

Vikingsson, G. A. 1990. Energetic studies on fin and sei whales caught off Iceland. Report of the International Whaling Commission 40:365–373.

Vikingsson, G. A. 1995. Body condition of fin whales during summer off Iceland. Pages 361–369 *in* A. S. Blix, L. Walloe, and O. Ulltang (eds.). Whales, Seals, Fish and Men. Elsevier Science, Amsterdam.

Walker, G. E., and J. K. Ling. 1981. New Zealand sea lion, *Phocarctos hookeri* (Gray, 1844). Pages 29–38 *in* S. H. Ridgway and R. J. Harrison (eds.). Handbook of Marine Mammals, Vol. 2. Academic Press, New York, NY.

Weinrich, M. T., J. Bove, and N. Miller. 1993. Return and survival of humpback whale (*Megaptera novaeangliae*) calves born to a single female in three consecutive years. Marine Mammal Science 9:325–328.

Wells, R. S. 1984. Reproductive behavior and hormonal correlates in Hawaiian spinner dolphins, *Stenella longirostris*. Reports of the International Whaling Commission (Special Issue 6):465–472.

Wells, R. S. 1991. The role of long-term study in understanding the social structure of a bottlenose dolphin community. Pages 199–225 *in* K. Pryor and K. S. Norris (eds.). Dolphin Societies: Discoveries and Puzzles. University of California Press, Berkeley, CA.

Whitehead, H. 1987. Social organization of sperm whales off the Galapagos: Implications for management and conservation. Report of the International Whaling Commission 37:195–199.

Whitehead, H., L. Weilgart, and S. Waters. 1989. Seasonality of sperm whales off the Galapagos Islands, Ecuador. Report of the International Whaling Commission 39:207–210.

Widdowson, E. M. 1981. The role of nutrition in mammalian reproduction. Pages 145–159 *in* D. Gilmore and B. Cook (eds.). Environmental Factors in Mammal Reproduction. Macmillan, London.

Wimsatt, W. A. 1963. Delayed implantation in the Ursidae, with particular reference to the black bear (*Ursus americanus*). Pages 49–86 *in* A. C. Enders (ed.). Delayed Implantation. University of Chicago Press, Chicago, IL.

Wolman, A. A., and A. J. Wilson. 1970. Occurrence of pesticides in whales. Pesticides Monitoring Journal 4:8–10.

Woodfill, C. J. I., N. L. Wayne, S. M. Moenter, and F. J. Karsch. 1994. Photoperiodic synchronization of a circannual reproductive rhythm in sheep—Identification of season-specific time cues. Biology of Reproduction 50:965–976.

Yasui, W. Y., and D. E. Gaskin. 1986. Energy budget of a small cetacean, the harbour porpoise, *Phocoena phocoena*. Ophelia 25:183–197.

York, A., and J. R. Hartley. 1981. Pup production following harvest of female northern fur seals. Canadian Journal of Fisheries and Aquatic Sciences 38:84–90.

Zemsky, V. A., and G. A. Budylenko. 1970. Siamese twins of humpback whale. Trudy AtlantNIRO 29:225–230.

Zinchenko, V. L., and M. V. Ivashin. 1987. Siamese twins of minke whales of the southern hemisphere. Scientific Reports of the Whales Research Institute 38:165–169.

7

PETER L. TYACK

Communication and Cognition

What are your associations with the words "marine mammal cognition and communication?" If you immediately think of big brains, high intelligence, and complex communication, then you are not alone. We humans have big brains and like to think of ourselves as an intelligent species. During the course of evolution, our hominid ancestors evolved larger brains for their body size, and this increase in brain size has traditionally been interpreted as an increase in intelligence. What then are we to make of the fact that the largest brain on the planet belongs to the sperm whale (*Physeter macrocephalus*)? The brain of an adult human weighs about 1 to 1.5 kg, but the brain of a sperm whale may weigh near 8 kg. John Lilly argued that an animal with the sperm whale's brain must have philosophical abilities that are "truly godlike" (Lilly 1975:220). However, sperm whales are much larger than humans, and the larger an animal is, the larger its brain tends to be. What about a dolphin with a body size closer to our own than a whale weighing 30 to 40 metric tons? An adult bottlenose dolphin, *Tursiops truncatus,* weighing 230 kg might have a brain weighing 2 kg. How does this compare to humans, where body sizes of 50 to 100 kg may be associated with brains weighing from 1 to 1.5 kg? Jerison (1978, 1986) has analyzed variation in body weight and brain weight among mammals and concludes that the ratio of brain weight to body weight in bottlenose dolphins is similar to that for humans.

Some of the most successful popular books on animal behavior argue that dolphins with bodies similar in size to those of humans and with brains as big as ours must be as intelligent. The books of John Lilly (e.g., 1967, 1975) launched a strong popular conception linking large brains, intelligence, and languagelike communication in dolphins and whales. Less publicity has attended the reaction of biologists to Lilly's claims. For example, Wilson (1975) rated dolphins no more intelligent than dogs. Bottlenose dolphins have performed well in animal language experiments (Herman et al. 1984, Herman 1986), but there is disagreement about whether their performance is any better than that of California sea lions (*Zalophus californianus*) (Schusterman and Krieger 1984, 1986; Schusterman and Gisiner 1986, 1988; Herman 1989), which have brain and body size ratios similar to those of most mammals (Worthy and Hickie 1986, see Pabst, Rommel, and McLellan, Chapter 2, this volume).

I personally do not believe that it is meaningful to attempt to fit different species along a linear scale of intelligence. There are hundreds of tests for intelligence within our own species, but we still have trouble defining human intelligence. How then can we ever hope to rank different levels of non-human intelligence? Many psychologists have moved away from attempts to define some pure "general intelligence" in humans and are instead defining multiple human intelligences (e.g., Gardner 1983). Non-human animals also have different clusters of sensory, cognitive, and motor abilities. These differences make it difficult to test behaviorally for some pure "general intelligence," if such a thing exists at all (see Wartzok and Ketten, Chapter 4, this volume).

The recognition that it may be more useful to assess different cognitive skills than one general "intelligence" makes it clear that it was naive to relate brain size to intelligence. To the extent that different parts of the brain are designed to process different kinds of information, it may be more productive to relate specific brain areas to specific cognitive or sensory abilities. For example, variations in spatial learning abilities have been correlated with variation in size of the hippocampus in birds (Krebs et al. 1986) and rodents (Schwegler and Crusio 1995). However, we do not yet know enough about information processing in the brain to predict different cognitive abilities from measuring the size of different brain areas.

We humans use a language that appears to be more complex than communication in other animals. Can we use complexity of communication to relate intelligence to large brains? There are serious problems with this approach. For example, the honeybee, which has a brain weighing on the order of milligrams, has a dance language (von Frisch 1967) that, to my mind, represents just as high an achievement of animal communication as anything demonstrated in wild marine mammals, no matter how large their brains. This may say more about the brilliance of honeybee biologists and the difficulties of studying marine mammals than about the full potential of these different species. We often learn more about cognitive skills in mammals when we ask animals to learn artificial communication systems than when we attempt to understand the full complexity of their own natural communication systems.

Why discuss communication and cognition in the same chapter? Most approaches to both communication and cognition emphasize *information*. The cognitive sciences differentiate themselves by their focus on information flow within an organism or between the environment and the organism. Most discussions of communication emphasize the form and manner in which information is transmitted between organisms. How information is received and perceived provides a critical intersection between communication and cognition; this is described by Wartzok and Ketten (Chapter 4, this volume). However, the people who study animal communication usually are trained in different disciplines than those who study animal cognition. Studies of animal communication have typically come from an ethological tradition emphasizing observation of animals in the wild, coupled with limited experimentation (e.g., Tinbergen 1951). Studies of animal cognition have emphasized training animals under carefully controlled artificial conditions (e.g., Roitblat 1987). This has meant that "animal language" studies, in which people train captive animals to communicate using artificial languages, are viewed as part of animal cognition. It is surprisingly difficult to compare these studies with those of natural communication systems in the same species. I hope that by combining both topics in the same chapter, the reader will more easily see where they may benefit in the future from closer integration.

The marine mammals are a diverse group, but there has been relatively little communication or cognition research on several taxa such as sea otters, polar bears, dugongs, and manatees. A systematic comparison of communication and cognition in all marine mammal species would be difficult, given the paucity of information on so many species. My approach will be to illustrate principles by going into details about a few species, selected because they have been well studied. Most of the relevant research has been conducted with the pinnipeds and cetaceans. My own marine mammal research has focused on the whales and dolphins, and I have a bias toward discussing these species.

This chapter starts with a discussion of different sensory modalities for communication in marine mammals and why the marine environment might favor the acoustic mode for long-range communication. The best understood examples of acoustic communication in marine mammals are then described and related to the problems posed by different forms of social organization in different species. The cognition section begins with a short description of animal "language" experiments. We use our own human language to learn about each other's mental experiences. The underlying assumption behind animal language training is that it may open a similar window on the minds of animals. As you read this section, think about how well these experiments meet this goal. The second part of the cognition section discusses cognitive abilities such as imitation, emphasizing evidence from untrained behaviors in captivity and the wild. The cognition section closes with a review of brain size in marine mammals and of proposed functions for these large brains.

Communication

As Wartzok and Ketten (Chapter 4, this volume) indicate, marine mammals have well-developed senses of touch, sight, and hearing. In most terrestrial environments, a visual display can be seen farther than a sound can be heard. Of all the ways to transmit information through the sea, however, sound is the best for communicating over distance. Whales may hear one another at ranges of tens of kilometers, but they see one another at ranges of no more than tens of meters. The unique suitability of acoustic signals for long-range communication in the sea does not limit the usefulness of other senses for shorter range communication, however. Chemical, tactile, and visual modes of communication are briefly reviewed in the following sections, and then I focus in more detail on acoustic communication.

Chemical Communication

Chemical communication is common among terrestrial mammals and many marine organisms but appears to be limited among marine mammals. Pinnipeds do use odor cues, for example, for mother–infant recognition, but these are primarily used in air (e.g., Terhune et al. 1979). The olfactory bulbs and nerves are used by terrestrial mammals for sensing airborne odors. These are reduced in mysticetes and absent in odontocetes (Breathnach 1960, Morgane and Jacobs 1972). Little is known about how marine mammals may sense waterborne chemicals (Kuznetsov 1974; Kuznetsov 1979, cited in Bullock and Gurevitch 1979; Nachtigall 1986), but there are some suggestions of use of pheromones (e.g., Norris and Dohl 1980). If marine mammals have only limited use of chemical communication, this may stem in part from the limited ranges of diffusion in water compared to the mobility of these animals.

Tactile Communication

Marine mammals use tactile sensation for a variety of purposes. The whiskers or vibrissae of seals are very sensitive to movement or vibration, and the walrus (*Odobenus rosmarus*) can even discriminate the shapes of objects using its mustachelike vibrissae (Kastelein and van Gaalen 1988, Kastelein et al. 1990). Walruses feed on the ocean bottom in murky water, and presumably they use their vibrissae to detect and select their prey. It is likely that touch is also important for communication at short range, but the details of tactile communication among marine mammals are not well understood. In many species, a mother and her young keep literally "in touch" by maintaining physical contact as they swim, and sea otter (*Enhydra lutris*) mothers carry their young at sea. Social interactions between manatees (*Trichechus* spp.), a relatively solitary species, are characterized by "mouthing, nuzzling, nudging, and embracing" (Hartman 1979). Muzzle-to-muzzle contact is also common among pinnipeds when they greet one another (Evans and Bastian 1969). Dolphins and whales may rub or caress one another with their flippers or other appendages. Among active schools of wild spinner dolphins (*Stenella longirostris*), some 30% of the members may engage in caressing at any one time (Johnson and Norris 1994). Gentle rubbing seems to play an important role in maintaining affiliative relationships in some dolphin species, perhaps analogous to social grooming in primates (Norris 1991, Samuels et al. 1989). Gentle touching by humans can provide positive reinforcement to captive bottlenose dolphins (Defran and Pryor 1980).

For many cetacean species, sexual contact appears to have a variety of social and communicative functions in addition to procreation. Sexual activity is often reported for all-male groups, and copulation is commonly observed between animals that are not sexually mature. Caldwell and Caldwell (1972a) report that all of the infant male bottlenose dolphins they observed in captivity attempted to mate with their mothers within a few weeks of birth. Nursing appears to stimulate the mother, who often initiates sexual activity with the calf. Nursing itself may become a ritualized display reinforcing the mother–calf bond (Brodie 1969). Although pilot and sperm whales start taking solid food by their second year of life, some individuals may suckle for more than a decade (Best 1979, Kasuya and Marsh 1984). If these teenage whales obtain most of their nutrition from their own foraging, suckling may take on a communicative or affiliative role.

Marine mammals engage in a variety of contact behaviors in aggressive interactions, but few studies have isolated a signal role as opposed to the physical displacement, pain, or harm the contact causes. This raises an important distinction for communication researchers. If I tell you to jump in a lake, I am sending a signal to you, but if I push you in, then any communicative signal pales in comparison to the physical effects of my act. Communication is defined as the transfer of information between two organisms. If you jump in the lake after I tell you to, you are responding to the information I sent. If you jump into the lake because I pushed you, you are responding to the physical effects of my action. Signals are likely to be used both to assess potential competitors and in dominance interactions, but more research is needed to clarify the role of communication in aggressive interactions.

Visual Communication

Vision is well developed among most marine mammals, and most species are reported to have some visual signals. Both aggressive and sexual interactions often involve visual signals at close range. Many aggressive visual signals in marine mammals follow patterns that are common among other mammals, including vigorous moving of the head toward another animal, prolonged staring at another animal, jerking the head, opening the mouth, or even making threats that resemble biting actions. Some behaviors appear to increase the apparent size of a male and may function as visual displays. For example, male humpback whales (*Megaptera novaeangliae*) competing for access to females may lunge with their jaws open, expanding the pleated area under the lower jaw with water. Visual signals that have been identified in submissive interactions among dolphins include flinching, looking away, and orienting the body away from another animal. Thrusting or presenting the genital region toward another animal may function as sexual visual signals. Direction of gaze is an important visual cue among primates, and human observers can often tell immediately when a dolphin or seal is making eye contact (Pryor 1991). Although the behavioral consequences have not been well studied, gaze cues may be important for marine mammals in clear water as well as in air. Many male seals and toothed whales have secondary sexual features that may function as weapons or armor during fights between males. Males of many species have enlarged teeth or tusks compared to conspecific females. For example, although female narwhals (*Monodon monoceros*) seldom have erupted teeth, adult males have one left tooth that is elongated into a tusk up to 3 m long. Other males may have manes of hair or thickened areas of callused skin. Although these traits may initially have evolved as weapons or armor, they may also function as visual signals either for potential male competitors or for potential female mates.

Many pinniped and cetacean species have distinctive pigmentation patterns that may even be individually distinctive in species such as the humpback whale (Katona et al. 1979). Whereas biologists find these pigmentation patterns to be very useful for species and individual identification, little is known about whether marine mammals use them as signals in their own social interactions. Most biologists have emphasized the role of pigment patterns as camouflage or disruptive coloration against visual predators (e.g., Madsen and Herman 1980). Variation in pigmentation and morphology is also correlated with age–sex classes among dolphins of the genus *Stenella* (Perrin et al. 1991). Large adult male *Stenella* often have a large postanal hump, which accentuates a threat posture involving a peculiar downward curve of the tail. Norris (1991) suggests that a dolphin making this threat looks similar to an attacking shark, and he suggests that the postanal hump in male *Stenella* mimics the claspers of an adult male shark. Norris even describes a threatening male *Stenella* swimming with sideways tail motions like those of a shark predator of these dolphins but very unlike the up-and-down tail motion typically used by dolphins for swimming. This intriguing idea that a threat signal mimics a predator is speculative and remains to be tested.

Exhaling to produce underwater bubbles creates a set of visual displays that are unique to aquatic animals. Some dolphins occasionally blow streams of bubbles that are highly synchronized with the production of a whistle vocalization (Caldwell and Caldwell 1972a). These bubble streams are a highly visible marker identifying who vocalized, but it is not known whether dolphins respond to this visual accompaniment of the acoustic signal. Humpback whales produce bubble streams in aggressive interactions. Large competitive groups of humpback whales are common during the winter breeding season. Most of these groups have a clear structure in which one adult male, called the "principal escort," apparently guards a central or "nuclear animal," usually a female (Tyack and Whitehead 1983). Principal escorts emit streams of bubbles typically in a line as long as 30 m. The bubble streams often are placed between a challenging male and the nuclear animal, perhaps as a visual screen.

With the exception of bubbles, which are unique to aquatic animals, the visual signaling of marine mammals is similar to that of their terrestrial relatives. Terrestrial animals also have visual agonistic displays that appear to be ritualized from fighting behavior; they use gaze cues; and they use pigmentation patterns for camouflage or for species identification or individual identification. However, there are differences in the range of vision in air versus water. As terrestrial animals, we think of vision as the sense of choice for detecting distant objects. In most aquatic environments, however, daytime vision is limited to a few tens of meters.

Acoustic Communication

The communication modality where marine mammals really stand out is the acoustic channel. Over the course of evolution, marine mammals have come to exploit almost the entire spectrum of sounds that humans have learned to use for exploration and communication under the sea. Large whales can produce loud sounds well below our hearing range that can be detected at ranges of hundreds of kilometers (Payne and Webb 1971, Spiesberger and Fristrup 1990). Dolphins and porpoises can hear sounds more than five times above our upper limit of hearing (Au 1993). To illustrate acoustic communication in marine mammals, I first define some basic acoustic terms and then describe what we know about echolocation and acoustic communication in several of the best studied species.

Because underwater sounds are foreign to most people, I explain acoustic terms using more familiar sounds. Musical tones have different pitches, so that middle C, for example, is lower than the C an octave above it. The physical feature of these sounds that causes the different pitches we perceive is called *frequency*. You can lower the frequency of a stringed instrument by increasing the length of a vibrating string that is at a constant tension. The longer the string, the slower the vibration, and the lower the frequency of sound produced. The frequency of a sound is defined by the number of cycles per second. The modern name for the unit of frequency is the *Hertz*, and 1000 Hertz are called a *kiloHertz*, abbreviated kHz. A musical tone has a corresponding frequency. When an instrument plays a note, it produces a sound that is centered around this frequency. In acoustic terms, it has a narrow frequency *bandwidth,* or is *narrow band.* In the modern orchestra, the A above middle C is 440 Hz. The A an octave above this is double the frequency, or 880 Hz. When an instrument plays a tone, such as 440 Hz, it also often produces sounds at various multiples of this frequency, such as 880 Hz. These higher frequencies are called *harmonics* of the fundamental *frequency* of 440 Hz. A sound that contains just one frequency or harmonically related frequencies is called a *tonal* sound. A melody consists of a series of different notes or discrete frequencies. A trombone or a siren can also make continuous changes in frequency. These changes in frequency are called *frequency modulation.* Not all sounds are *narrow band.* For example, when a bat hits a baseball, it makes a crack or click that gets loud very quickly, lasts for a very short time, and includes lots of frequencies. This click has a sudden onset or rapid *rise-time,* short *duration,* and broad bandwidth. The range of frequencies in a sound is called the frequency *spectrum* of the signal. Marine mammals make an enormous diversity of sounds ranging from simple clicks to complex series of clicks to frequency-modulated tonal sounds.

Underwater Acoustics and Patterns of Acoustic Communication

The structure of the sounds that a marine animal uses to solve a problem is influenced by the physics of sound in the ocean. Seawater is an excellent medium for sound propagation, and this opens up the opportunity for remarkable abilities of echolocation and long-range acoustic communication. In this section, I explore how marine mammals have evolved specializations to solve these problems.

Echolocation

Echolocation is usually defined as the ability to produce high frequency clicks and to detect echoes that bounce off distant objects. Echolocation is a good example of how difficult it is to make ironclad distinctions between communication and cognition. Echolocating animals produce signals that are similar to those of communicating animals, but the echolocation signal is not produced to transfer information to another animal. Rather it is used for the signaling animal to learn about its environment. The flow of information from the animal to the environment is like a communicative process, but the flow of information from the environment to the animal is like a cognitive process.

The only marine mammals known to have evolved a specialized ability of echolocation are the toothed whales. We know the most about echolocation in the smaller toothed whales that can be kept in captivity. Dolphins can echolocate objects at a greater distance than they can typically see them. An echolocating dolphin can detect a target about the size of a ping-pong ball almost a football field away (Murchison 1980). This target is so small that you could not even see it in air from that range. The increased potential range of echolocation compared to vision may make it particularly useful for detecting obstacles or prey. If a toothed whale were swimming rapidly in murky water or at night, it seldom could see an obstacle rapidly enough to avoid it, but echolocation would detect an obstacle far enough away to give even fast-swimming animals plenty of time to respond. Many marine mammals also feed at great depth or at night when there is little light. Some may visually detect luminescent prey nearby, but there are circumstances where vision has a more limited range than echolocation for detecting prey underwater. Most studies of dolphin echolocation have taken place under carefully controlled conditions with captive animals and artificial targets. We know very little about how wild dolphins use echolocation to solve tasks such as avoiding obstacles or detecting, selecting, and capturing prey.

Echolocation is not used only at great distances. Many echolocation tasks may function primarily at ranges of less than several meters. Free-swimming bottlenose dolphins trained to discriminate fish from inedible targets generally did not turn toward the fish until it was only 3 to 3.5 m away (Airapetyants and Konstantinov 1973). Dolphins often inspect objects with echolocation at ranges as close as a few centimeters away. Dolphins can use echolocation to discriminate the shape of targets—even targets with exactly the same shape, differing only in composition (e.g., Kamminga and van der Ree 1976).

Bats are the other mammals that are highly skilled at echolocation. Extensive neurobiological studies have shown that bats have sophisticated neural mechanisms to process echolocation sounds. The processing by some bats of distance to a target object involves remarkable temporal precision (Simmons 1973), and bats have sophisticated mechanisms to compare an echo to the particular pulse from which it came (Simmons et al. 1975). We know toothed whales use high frequencies for echolocation, but we know little about how they process sounds for echolocation.

The optimal frequency for echolocation depends on the size of the target. Objects or features that are much smaller than the wavelength of the impinging sound do not reflect the sound very efficiently. The wavelength, λ, of a sound equals the speed of sound, c, divided by the frequency, f, therefore: $\lambda = c/f$. The speed of sound in seawater is close to 1500 m/sec (Urick 1983). This suggests using sound frequencies on the order of 150 kHz or higher to detect targets of a size around 1 cm (i.e., if $\lambda \approx 0.01$ m, then $f = c/\lambda = 1500/0.01$

= 1500 × 100 = 150,000). Dolphins and porpoises have evolved specializations for producing and hearing such high frequency sound. Bottlenose dolphins hear best around 50 kHz, but they hear well above 100 kHz. Figure 7-1 shows the specialization for high frequency hearing of several toothed whale species compared with seals and fish.

Dolphins produce echolocation clicks with frequency emphases that match their auditory sensitivity relatively closely, as might be expected for the task of extracting faint echoes from noise. The echolocation clicks of bottlenose dolphins are very short (<100 μsec), with a rapid rise-time and a relatively broad bandwidth from several tens of kHz up to near 150 kHz (Fig. 7-2A; Au 1993). Several other toothed whales, such as porpoises of the genus *Phocoena* and delphinids of the genus *Cephalorhynchus,* produce more narrow band pulses in the 110- to 150-kHz range (Fig. 7-2B; Kamminga and Wiersma 1981). Au (1993) suggests that the smaller animals producing clicks illustrated in Figure 2B may simply be incapable of producing clicks as loud, short, and broadband as those shown in Figure 2A. However, *Cephalorhynchus* is more closely related to *Tursiops* than to *Phocoena,* and this suggests that *Cephalorhynchus* and *Phocoena* have independently converged on very similar morphology, behavior, and patterns of vocalization. Their echolocation clicks may also be an adaptation to the niche for which they seem to have converged (Watkins et al. 1977). When different species of bats produce clicks as different from one another as the clicks of *Tursiops* and *Phocoena,* this difference is associated with a different mode of sonar processing. Among bats, narrow-band longer duration clicks are associated with an echolocation system that relies on using Doppler shift to detect moving targets in a cluttered environment (Neuweiler et al. 1988). It is possible that *Cephalorhynchus* and *Phocoena* may process echolocation signals very differently from the well-studied bottlenose dolphins (Ketten 1997, Tyack 1997). We know next to nothing about how these species echolocate. Imagine how exciting it will be to study how they use echolocation and how they process the echoes from their clicks.

Dolphins can vary both the loudness and the frequency spectrum of their clicks. There is some evidence that bottlenose dolphins and beluga whales (*Delphinapterus leucas*) shift the frequency of their clicks to avoid noise if it is present in the normal frequency range (Au 1993). This would be

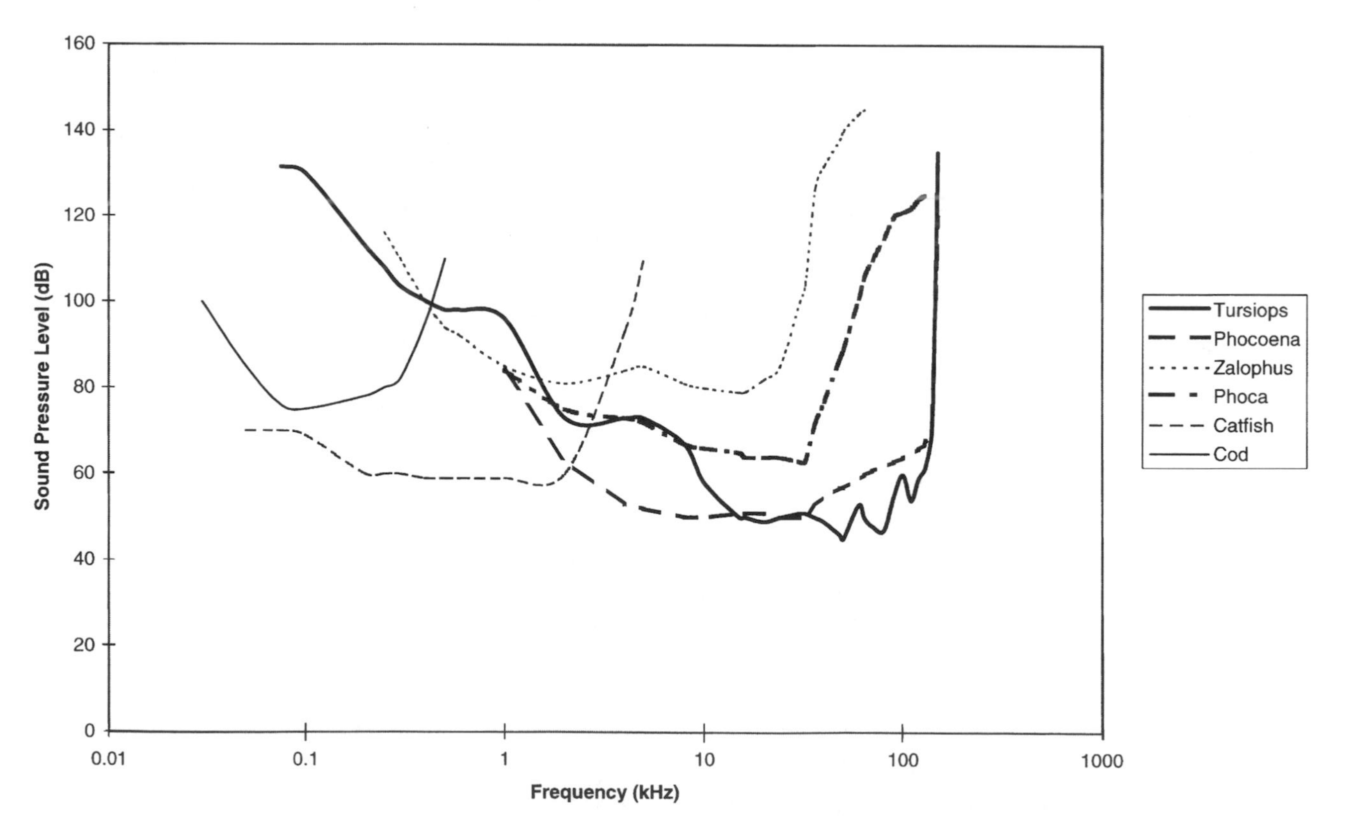

Figure 7-1. Audiograms of odontocete cetaceans specialized for high-frequency echolocation compared with several seal and marine fish species. (Audiograms: bottlenose dolphin [*Tursiops truncatus*], Johnson 1966; harbor porpoise [*Phocoena phocoena*], Andersen 1972; California sea lion [*Zalophus californianus*], Schusterman et al. 1972; harbor seal or common seal [*Phoca vitulina*], Møhl 1968; catfish, Poggendorf 1952; cod, Chapman and Hawkins 1973.)

A Tursiops truncatus

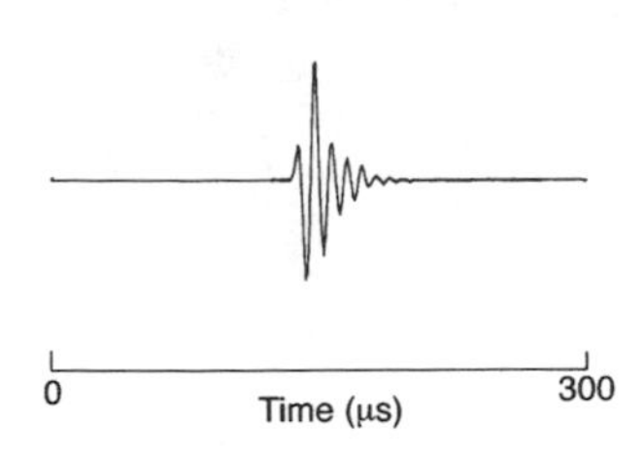

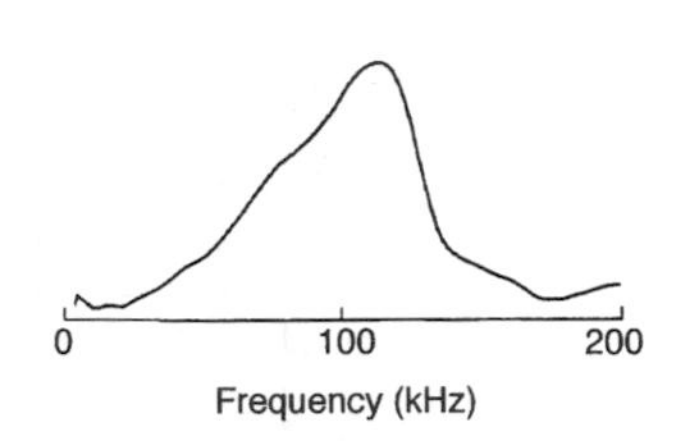

B Phocoena phocoena

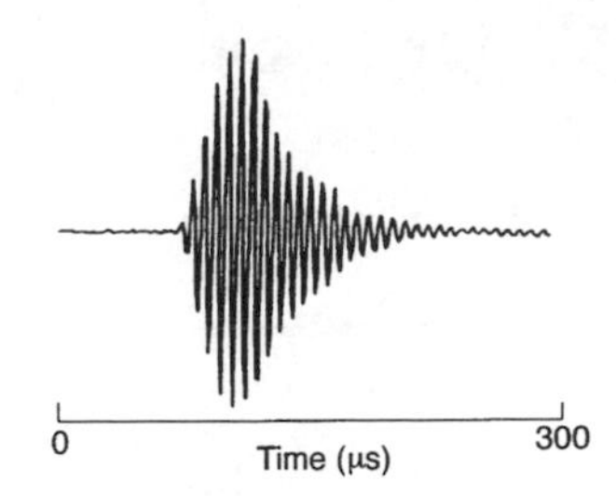

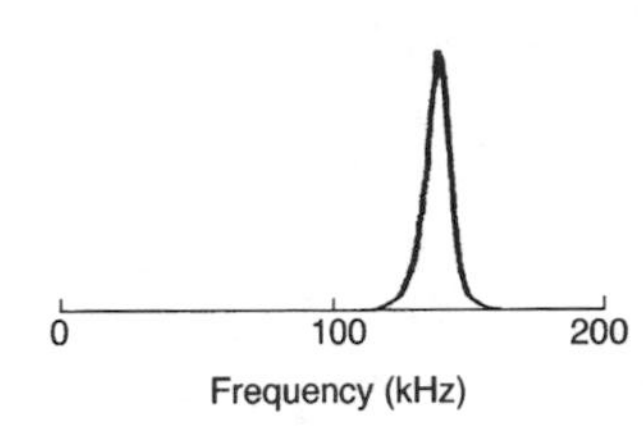

Figure 7-2. (A) Waveform and frequency spectrum of the echolocation click of a bottlenose dolphin (*Tursiops truncatus*). (From Au 1993.) (B) Waveform and frequency spectrum of the echolocation click of a harbor porpoise (*Phocoena phocoena*). (From Kamminga and Wiersma 1981.)

analogous to switching from a channel in a walkie-talkie with a lot of radio interference to one that is less noisy. Not much is known about whether or how dolphins modify their clicks depending on the echolocation problem on which they are working. For example, do dolphins adaptively modify their clicks to improve information in the echo from a particular target?

The clicks of bottlenose dolphins are very directional in the higher frequencies. Dolphins closing in on a target make both lateral and circular scanning motions, similar to shining a flashlight beam over an object. Bottlenose dolphins usually wait to hear the echo from a target before they produce the next click, and as they close in on the target, the interval between pulses usually decreases (Au 1993). This sounds to our ears as individual clicks blending into a buzz sound, but the dolphins are capable of much better temporal resolution in their hearing.

Low-frequency Sounds of Finback and Blue Whales

Not only have marine mammals evolved uses for sound well above our limit of hearing, but the largest of the baleen whales also produce sounds so low in frequency that they are an octave or more below the lowest sounds we can hear. Finback (*Balaenoptera physalus*) and blue (*B. musculus*) whales make sounds about as loud as a medium-sized ship, centered around 10 to 30 Hz (Fig. 7-3). Blue whale sounds last several tens of seconds (Cummings and Thompson 1971, Edds 1982), whereas those from finbacks are comprised of series of 1-sec pulses (Watkins et al. 1987). Identifying which species is producing a sound can be challenging. You can count on one hand the blue whales identified in published reports to have produced these calls. One good day at sea with blue whales could materially improve our data on this topic. This research area is at the early stages of exploration and discovery. Some people do not like problems where so little is known, but the most exciting experiences I have had as a marine mammalogist involved problems like this where a few days' observation made a significant contribution to our understanding.

The sounds of blue and finback whales have a variety of features suggesting that they are adapted for long-range communication (Payne and Webb 1971). These sounds have a simple structure that is often repeated over and over, increasing their detectability. During their breeding season, finback whales may produce a series of 20-Hz calls lasting typically many hours and up to longer than one day (Watkins et al. 1987). The low frequency of these calls also appears to be an adaptation for long-range communication. The higher the frequency of a sound, the more of its energy is dissipated into heat as it passes through water (Urick 1983, Tyack 1998). The frequency of these whale sounds is low enough that there is very little absorption of the sound energy, even over ranges of hundreds of kilometers. Sound energy is also lost when a signal interacts with the sea floor or surface, but there are a variety of ocean sound channels in which sound energy can be entrained, avoiding surface and bottom loss. In the shallow Arctic sound channel, the best propagation occurs in the 15- to 30-Hz frequency range (Urick 1983); many finback and blue whale signals concentrate energy in this band.

Calculations for sound propagation in the deep ocean suggest that the 20-Hz finback whale signals could be detected at ranges of hundreds of kilometers (Spiesberger and Fristrup 1990). Figure 7-4 shows many of the paths that such a sound would take over a range of 400 km. The sound rays shown in Figure 7-4 were calculated using the variation in speed of sound with depth shown on the left side of the

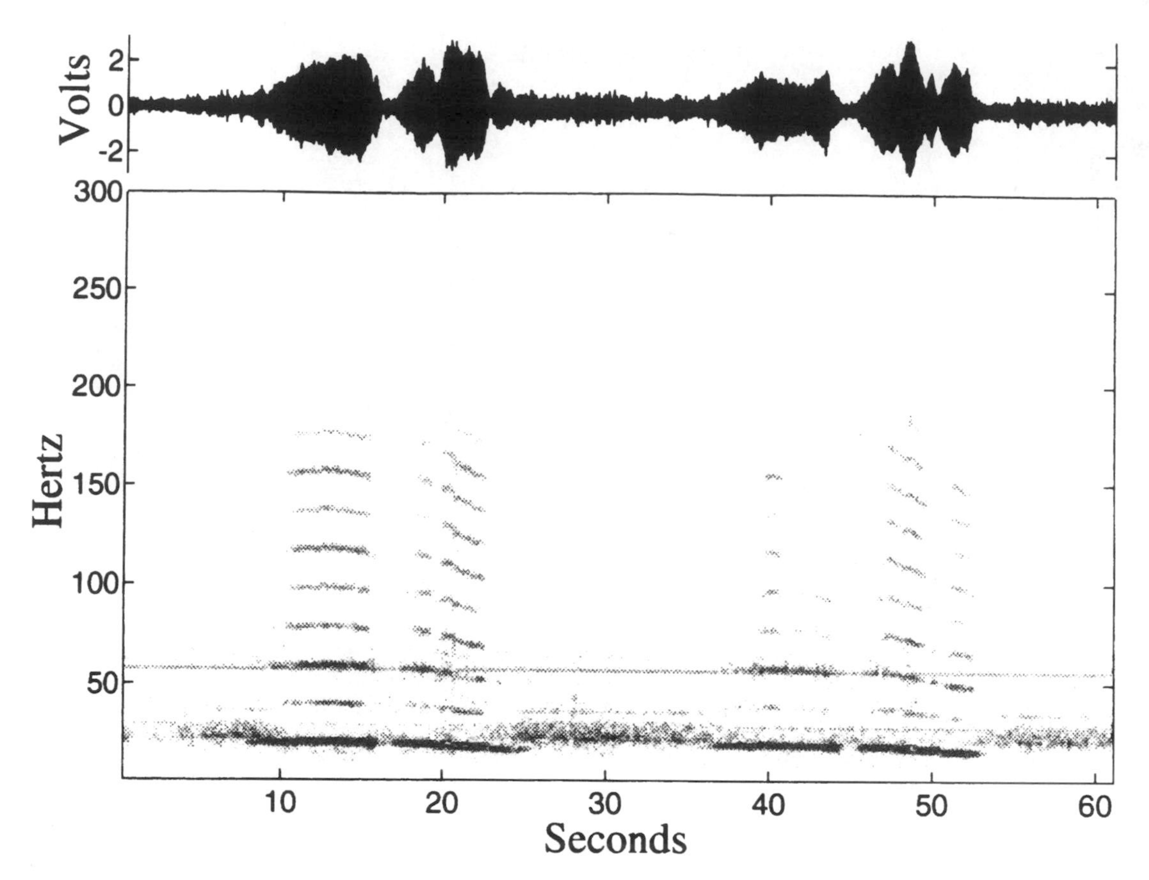

Figure 7-3. Low-frequency vocalization of a blue whale.

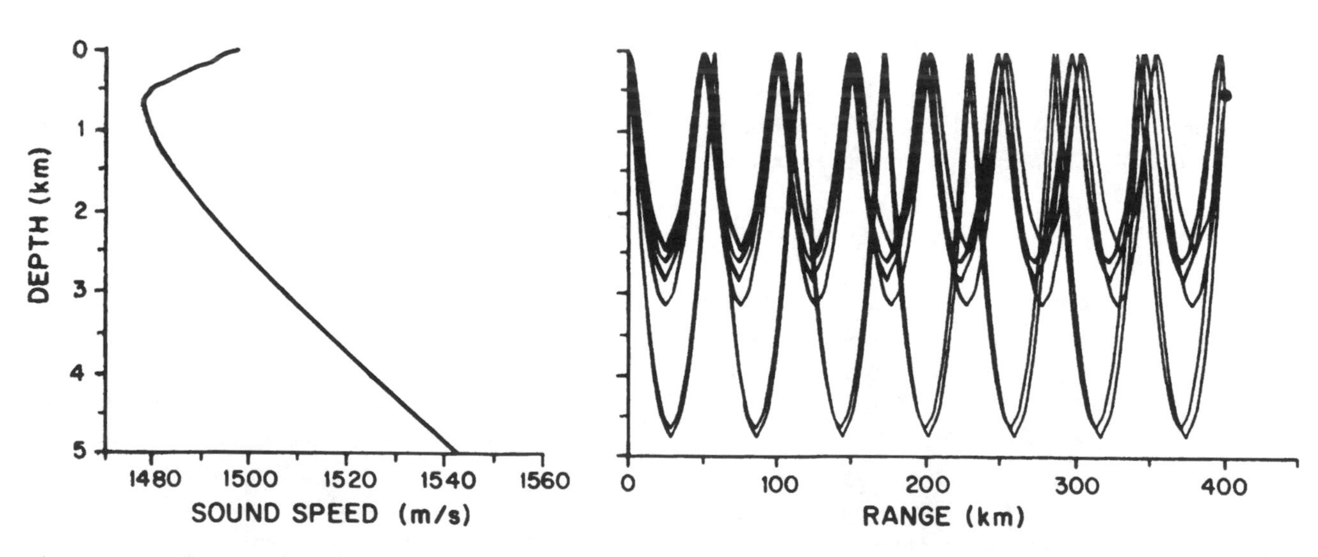

Figure 7-4. (Left) A typical profile of sound speed versus depth for temperate or tropical seas. (Right) Calculated ray paths for a finback 20-Hz pulse produced at a depth of 35 m and detected at a receiver 500 m deep and 400 km away. The ray paths illustrate the general patterns of propagation of this kind of signal in deep temperate or tropical seas. (From Spiesberger and Fristrup 1990.)

figure. The variation in sound speed in the ocean causes sound to refract, just as the variation in the speed of light between the air and the glass in an optical lens causes the light to refract. There is a minimum in sound speed in temperate and tropical waters near 1 km depth. Sound rays that start upward from this depth speed up and refract back downward. Downward-heading rays also speed up after they pass below the depth of the sound speed minimum, and they refract back upward. If you look at the density of rays in Figure 7-4, you can see that most of the sound energy concentrates near the depth of the sound speed minimum, also called the SOFAR channel (Urick 1983). An underwater microphone placed near this depth would be more likely to detect a distant sound than would one placed at random near the surface. However, the acoustic rays converge near the surface every 52 km, in what is known as a convergence zone. Although baleen whales are not known to dive as deep as the SOFAR channel, a whale swimming near the surface through a convergence zone might experience an increase in sound level of tenfold or more, perhaps even within a few tens of meters (Urick 1983). While finback calls carry long distances, it is hard to imagine that a whale would often find another whale after detecting it hundreds of kilometers away. Finback whales do swim 10 km/hr or more (Watkins et al. 1984), and their series of 20-Hz pulses may last for up to a day (Watkins et al. 1987). However, look at Figure 7-4 and think of a whale that seldom dives more than a hundred meters or so. Such a whale might hear a distant whale several convergence zones away, but to be able to find that whale, it would have to determine which direction to swim and then keep detecting it as it encountered another convergence zone every 50 km or so. Our ignorance of the diving patterns and of the low-frequency hearing sensitivity of whales makes it impossible to predict with confidence the ranges over which they can hear conspecific signals. Baleen whales are not thought to dive deep enough to enter the SOFAR channel, but we have little data on the dive patterns of whales. We do not know how deep they typically dive at sea, or whether they might dive especially deep to listen for distant signals.

The different rays shown in Figure 7-4 travel different distances and would vary in their time of arrival from 259 to 260 sec (about 4.3 min). Because the sound itself lasts about 1 sec, a whale would hear a complex superposition of arrivals. Acoustic oceanographers can often resolve these different rays or modes of travel, and can use this information to learn about the location of the source or about the ocean in between. We know nothing of whether whales perform similar processing. However, the high repeatability of these simple pulses clearly increases their detectability over long distances.

It is very difficult to record low-frequency signals from a boat because any motion of the sensor creates flow noise at these frequencies. However, the end of the Cold War opened the door for biologists to use a remarkable new tool (Costa 1993). For decades, the U.S. Navy placed underwater microphones, called hydrophones, deep in the ocean. These were cabled back to shore, and the signals were transmitted to central analysis rooms where sounds picked up from entire ocean basins were integrated. The operators of this sound surveillance system (the acronym is SOSUS) could listen for the sounds of ships over the entire North Atlantic or North Pacific. In the past few years, Christopher Clark of Cornell University has pioneered the use of this system to track whales over distances of hundreds of kilometers. Figure 7-5 shows a track of one whale, called Old Blue, tracked by Clark and Lt. Charles Gagnon of the U.S. Navy. Clark's group has recorded, mapped, and analyzed thousands of calls from baleen whales.

The SOSUS gives us the ability to follow vocalizing whales at ranges of hundreds of kilometers. The only other technique that allows us to track the whales over these great ranges involves tagging whales with a device that can telemeter their location (e.g., Watkins et al. 1984). Figure 7-6 shows how far a finback whale can move in just 1 week during the feeding season. Many whales also have annual migrations of thousands of kilometers, from polar feeding areas to tropical breeding areas. Several biologists have suggested that marine mammals might be able to sense echoes of low-frequency vocalizations from the sea floor to orient or navigate with respect to bathymetric features (Norris 1969, Payne and Webb 1971, Thompson et al. 1979). This would be quite different from the high frequency echolocation of toothed whales but might function, for example, as a depth sounder or to detect a distant island or continental shelf (Tyack 1997). It has also been suggested that bowhead whales (*Balaena mysticetus*) migrating in the Arctic could detect rough ice ahead by listening for echoes from their calls (Ellison et al. 1987). Development of more sophisticated tags that measure the depth of dives or that record sounds received by the whale may help us to determine whether the whales themselves can use sound to perceive information from the surface or bottom of the oceans in which they live and migrate.

Sexual Selection and the Evolution of Advertisement Displays

The introduction of this chapter describes communication as an exchange of information. Yet communication can also reflect an attempt by a signaler to manipulate the behavior of a recipient (Krebs and Dawkins 1984). For a human example, you can buy a newspaper to get information, but you cannot escape advertisements put on the page to get you to buy something else. Animals also produce advertisement dis-

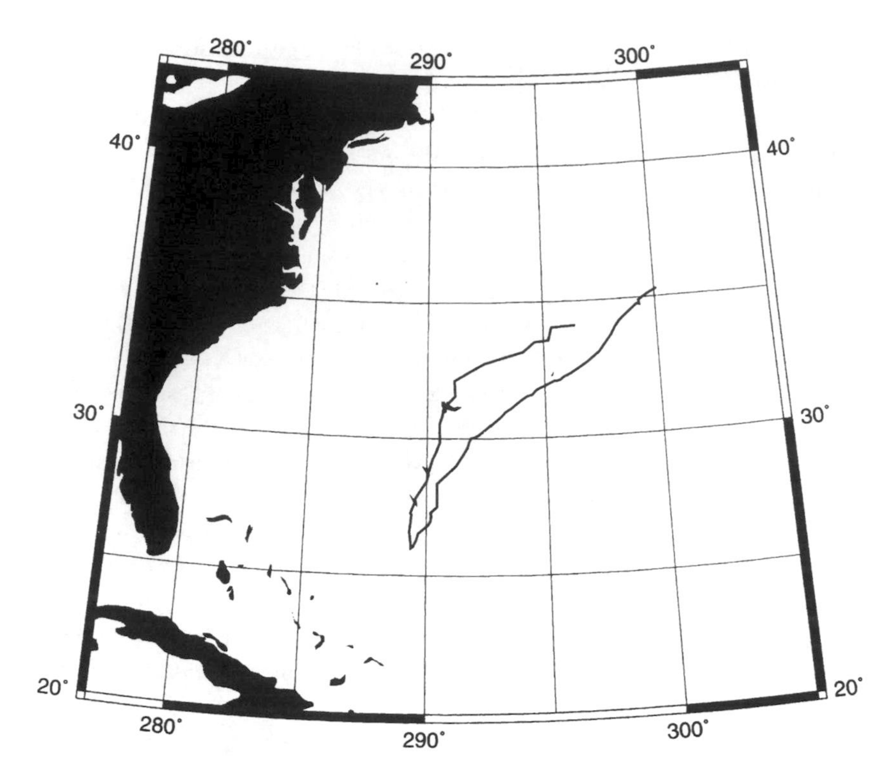

Figure 7-5. Track of vocalizing blue whale made using U.S. Navy's SOSUS arrays.

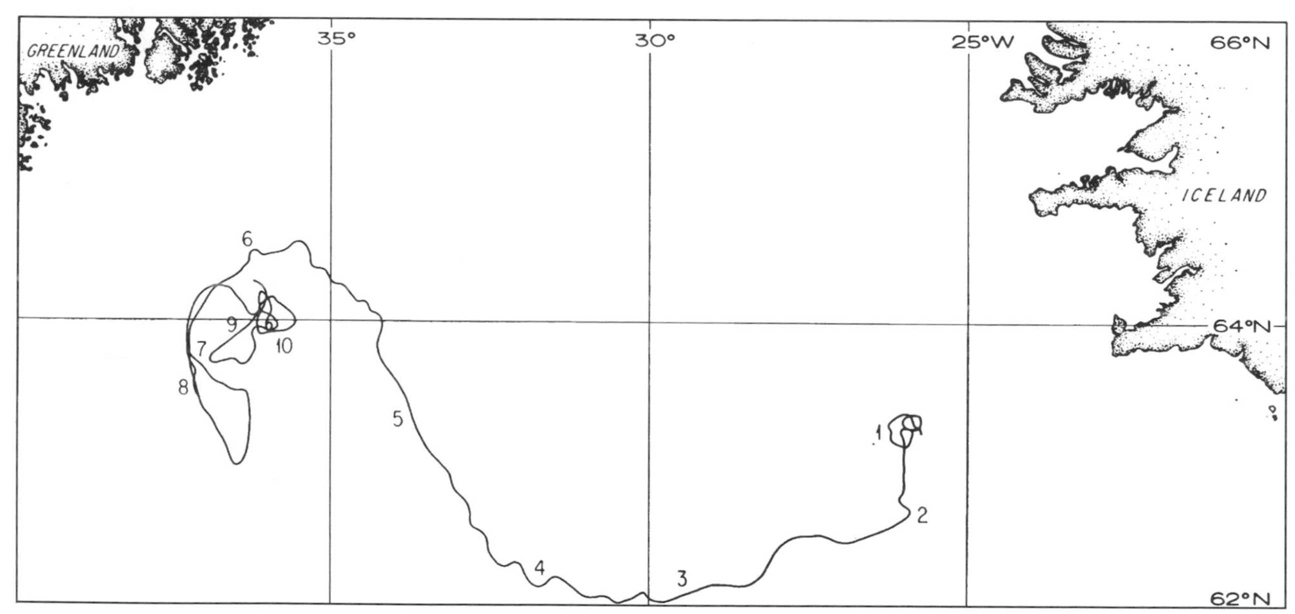

Figure 7-6. Track of a finback whale tagged with a radio and followed from a ship for one week. (From Watkins et al. 1984.)

plays, and their structure cannot be understood unless one considers how the signal is designed to manipulate choices of animals that hear the display. For example, communication is often described as an exchange of individual signals and immediate responses. However, one animal may produce a long series of advertisements to modify the outcome of one choice by a listening animal. The listening animal may only make its choice after hearing hundreds or thousands of advertisements from many different signalers. Advertisements are also typically very flashy and attention-getting. This makes them better known than many other kinds of animal signals. Examples include the songs of birds and whales. Songs are usually defined as acoustic displays in which a sequence of notes is repeated in a predictable pattern.

Songs of Humpback Whales

STRUCTURE. Perhaps the best known marine mammal vocalizations are the songs of the humpback whale (Payne and McVay 1971, Winn and Winn 1978). Unlike the low-frequency calls of finback and blue whales, which we can scarcely hear, the songs of humpbacks cover the frequency range of most human music: from about 30 Hz up to about 3 kHz. These songs sound so beautifully musical to our ears that they have been commercial bestsellers. Humpback whales sing continuously for hours, primarily during the breeding season. The song is made up of fewer than ten themes (Fig. 7-7), each of which is made up of phrases or series of sounds lasting about 15 sec. Phrases of one theme repeat a variable number of times before a new theme is heard. Humpbacks tend to sing themes in a particular order, and it often takes about 10 min before a singer comes back to the initial theme.

COMPARISONS WITH BIRDSONG. Many different animals produce long, complex songs. The best known are those of birds. Humpback songs might at first seem much more complex than birdsong, as the song of the whale lasts many minutes whereas the songs of most birds last only a few seconds. There are two reasons why these appearances may be misleading. First, what is called one song may differ between whales and birds. When the wren sings AAABBB, it is said to repeat song A three times and song B three times as it moves through its song repertoire. When the whale sings AAABBB, it is said to repeat three phrases from theme A and then three phrases from theme B as it completes its song. Thus, what is called a song in the bird may be more appropriately compared to a phrase from the song of the humpback whale. Second, if humpback song is speeded up about 14 times, it sounds to my ears remarkably like a bird such as the Bewick's wren (*Thryomanes bewickii*), which repeats one song type a few times before switching to another song type (Kroodsma 1982). Conversely, much more complexity is audible in each note of a slowed-down bird song than was apparent at the natural speed. The whale stretches its song over what seems a leisurely pace to a human, whereas the bird uses a greater frequency range but compresses its song over a duration so short that human ears miss some of the detail. We cannot be sure what level of nuance whales and birds pick up at the natural paces of their own signals. Signal-processing engineers often compare disparate signals by multiplying the duration and the frequency range to obtain what is called the time-bandwidth product. Changing the playback speeds makes bird and whale song sound similar; this suggests that the songs of some birds and whales may have comparable time-bandwidth products.

Kroodsma (1982) developed several different measures of the complexity of the song repertoires of Bewick's wrens, and these measures can be used to compare humpback song

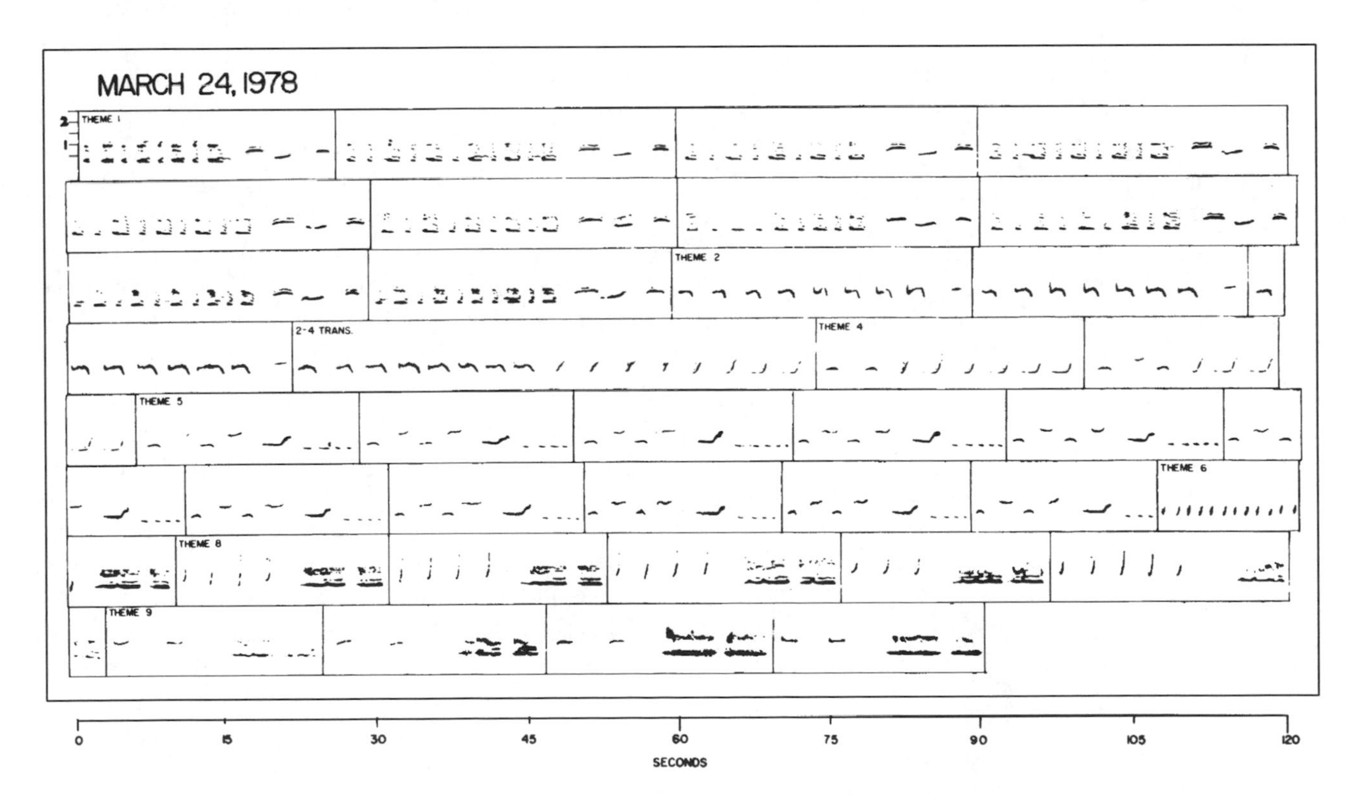

Figure 7-7. Spectrogram of the song of a lone humpback whale from the Hawaiian Islands. Each line represents 2 min, and this song took 14 min before repeating the initial theme. Each theme is made up of repeated phrases, and the boundary between phrases is marked by vertical lines. The first phrase of each theme indicates the number of the theme. (From Payne et al. 1983.)

to that of this avian songster (Table 7-1). Both the bird and whale songs are made up of sounds that may be repeated many times. The complexity of highly repetitive songs might be inflated compared to songs that contain sequences of different sounds. To correct for this, Table 7-1 includes a correction for durations and time-bandwidth products that counts just one example of each repeated sound.

Table 7-1 shows that the songs of humpback whales and Bewick's wren are roughly comparable by several indicators of song complexity. While the total song duration for the whale greatly exceeds that of the wren, the corrected song duration of the humpback (15–20 sec) is only slightly greater than that of the wren (9–15 sec). The corrected time–bandwidth products of whale and wren overlap.

Now let us compare the whale to a real champion avian songster: the long-billed marsh wren (*Telmatodytes palustris*). As mentioned above, transitions between *themes* in the songs of humpbacks should be compared to transitions between *songs* in the song repertoires of a songbird. Figure 7-8 illustrates these transitions for humpback whales and for the long-billed marsh wren. The large numbers in this figure represent themes for humpback whales and songs for the marsh wren. For the humpback themes, I have tallied almost 1200 transitions from one theme to another, using data from many different whales recorded during one singing season in the Hawaiian Islands (Tyack 1982). For the marsh wren, I reproduce a figure from Verner (1975), which tallies transitions between songs sung by one male during one long song bout. The marsh wren has more variable song transitions, but in general it does have preferred paths to switch between its 100 or so songs. Although the data from humpback song stems from many whales over an entire singing season, the transitions between themes of humpback song seems much simpler than transitions between songs of the wren, with more systematic transitions between fewer than ten themes.

Table 7-1. Comparison of Acoustic Complexity of Songs of Bewick's Wren and the Humpback Whale

Bewick's Wren[a]	Humpback Whale
Number of songs: 10–20	Number of themes: 5–8
Total phrases: 36–88	Total units: 15–20[b]
Song repertoire duration (sec): 27–40	Song duration (sec): 264–2100[c]
Corrected song repertoire duration (sec): 9–15	Corrected song duration (sec): 15–20 (average 1 sec unit duration × 15–20 sec)
Time–bandwidth product (kHz-sec): 87–147	Time–bandwidth product (kHz-sec): 528–4200 (assume 2 kHz bandwidth)
Corrected time–bandwidth product (kHz-sec): 31–72	Corrected time–bandwidth product (kHz-sec): 30–40

The songs of birds and whales may contain repeated sounds. The corrections for song duration and the time–bandwidth product only count repeated sounds once.

[a]From Kroodsma 1982.

[b]From Winn and Winn 1978; total number of different units and corrected durations depend strongly on subjective judgments of what constitutes the same unit. Here I have used Winn and Winn's (1978:113) definition and count of syllables that tends to lump rather than split. Splitters might double the number

[c]From Payne and Payne 1985

The real key to the complexity of humpback song is the way it changes over time. Until now we have compared the songs of humpbacks at any one time to song repertoires of birds. Some songbirds add songs to their repertoire over several years. Humpbacks do not show a similar slow increase in complexity of their repertoire with age; instead all individuals within a population sing songs that are very similar at any one time, and they all slowly change their entire songs more or less in synchrony over weeks and months. The song of each individual is much more like the songs of other individuals recorded at the same time than it is to itself recorded, say, a year later (Guinee et al. 1983). Within a population, song gradually evolves over time so that few elements of the song are preserved over many years. Sounds may change in duration, frequency, and timbre; they may disappear from the song entirely, and new sounds may appear in some other part of the song. Analysis of songs recorded off Bermuda over a period of more than two decades showed that once a particular song phrase disappeared, it never recurred (Payne and Payne 1985). Humpbacks live for decades, so the lifetime production of song is much more complex than suggested by the complexity of the song recorded at any one time.

In this section, I have emphasized comparisons of humpback song to birdsong, but some popular descriptions attempt to compare whale song to human language. For example, Sagan (1980) uses the length of the song to argue that a humpback song may contain the same amount of information as *The Iliad* or *The Odyssey* of Homer. This analogy is misleading for several reasons. The information content of the human text is calculated by the number of letters multiplied by the number of bits required to specify a letter. This requires much less information than would be required to record and reproduce the actual stream of speech. The information content of humpback song is calculated by how much data would be required to reproduce the sounds. This requires much more information than could be achieved by an efficient coding scheme, such as the alphabets used to represent speech or musical notation used to represent a symphony. No one knows how humpbacks represent or encode song. Humans use rhythmic structure and frequent repetition of similar sounds as an aid to remembering vocally transmitted material. Guinee and Payne (1988) suggest that the rhythmic structure and repetition in humpback song also function as mnemonic aids for the whales. Rhythm and rep-

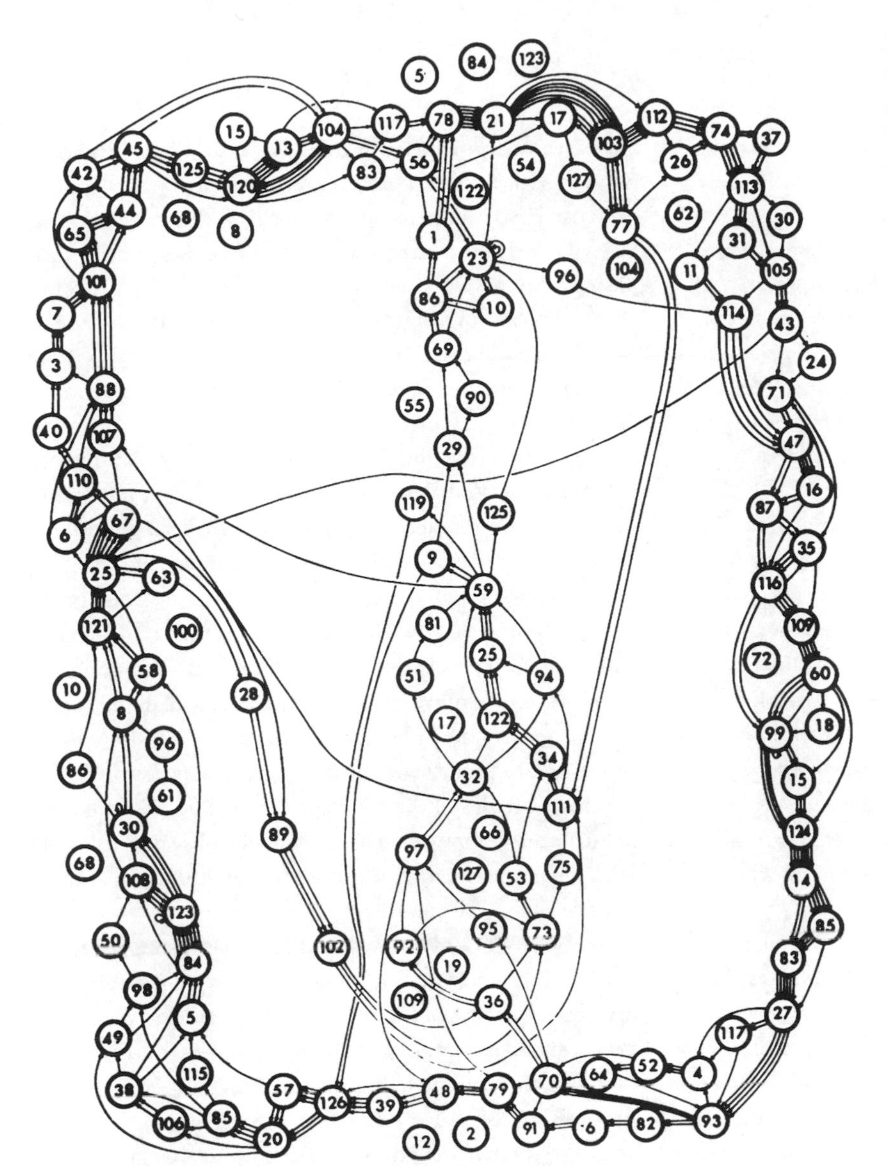

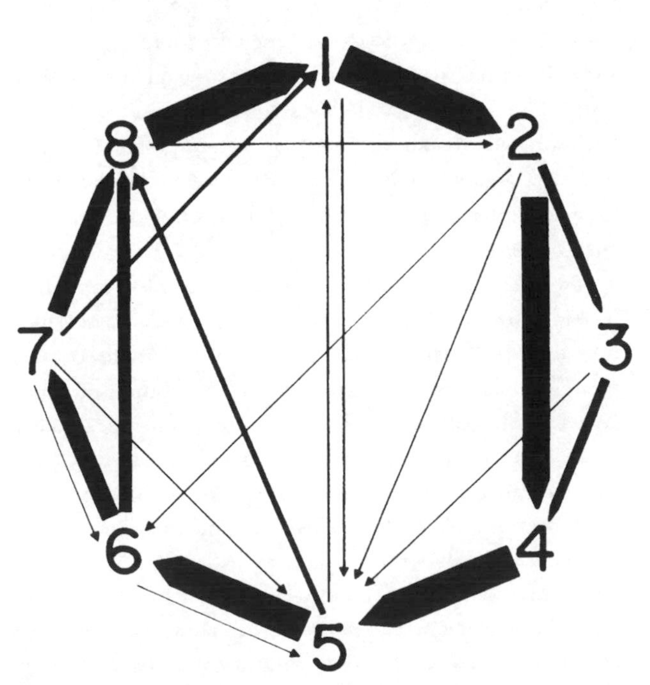

Figure 7-8. Transitions between songs in the repertoire of a long-billed marsh wren (A) and between themes in the song of a humpback whale (B). Both songs in the wren and themes in the whale tend to cycle in a particular order, but the wren cycles among many more songs than the whale cycles through themes. (Long-billed marsh wren, Verner 1975; humpback whale, Payne et al. 1983.)

etition of sounds appear both in music and in nonsense rhymes, not to mention many non-human signals; therefore these features need not be associated with linguistic processing. The Sagan (1980) analogy to human oral verse seems to imply that humpbacks use their songs to pass down an oral tradition with semantic content. There is no evidence that specific sounds within the song carry any such information. The process of song change is so rapid and so complete that it seems unlikely that each variation in the song reflects variation in what the song communicates to other whales.

FUNCTIONS OF HUMPBACK SONG. So much for the acoustic structure of humpback song. Can we say anything about why humpbacks sing? Humpbacks sing primarily during the winter breeding season, and nearly all of the singing humpbacks whose sex has been determined have been males (Baker et al. 1990, Glockner 1983, Palsbøll et al. 1992). Most singing humpbacks are alone, but they are highly motivated to join other whales (Tyack 1981, 1982). Aggressive behavior is often observed after a singer joins other males. Singers are less likely to join a female, but behavior associated with sexual activity is observed when a singer joins a female (Tyack 1981, 1982).

Charles Darwin (1871) coined the term sexual selection for traits such as reproductive advertisement displays that are concerned with increasing mating success. There are two ways sexual selection can work. It can increase the ability of an animal to compete with a conspecific of the same sex for an opportunity to mate with a member of the opposite sex (intrasexual selection) or it can increase the likelihood that an animal will be chosen by a potential mate (intersexual selection). A variety of results have encouraged biologists to suggest that humpback song plays a role in male–male competition: (1) song appears to maintain distance between singers (Tyack 1981, Helweg et al. 1992); (2) no known females were attracted to playbacks of song (Tyack 1983, Mobley et al. 1988); and (3) aggressive interactions (particularly between singers and known males) are much more commonly observed than sexual interactions (particularly between singers and known females) (Tyack 1981, 1982).

These behavioral observations are clearly consistent with the idea that song plays a role in mediating male–male interactions. Some acoustic features of humpback song are also consistent with intrasexual selection. Baker and Cunningham (1985) suggest that intrasexual selection tends to select for rapid song change and convergence in the songs of different male songbirds. Both rapid change and convergence between individuals are striking features of humpback song. Catchpole (1982) suggests that a small repertoire of syllables and repetition of bird song types (or repetition of phrases for humpbacks) are also typical of a male repulsion role for song.

However, just because humpback song appears to be used in male–male interactions does not mean that it is not also used by females to select a mate. Both intra- and intersexual selection often operate at the same time on the same display (for a discussion regarding bird song, see Catchpole 1982). The use of songs to mediate spacing between singers says nothing about whether females are also an important audience. Females are often more discriminating than males in responding to an advertisement display such as song. None of the song playbacks conducted with humpback whales duplicated all of the potentially relevant features of song, and this may account for some of the lack of response of females to playbacks. Furthermore, as Catchpole (1982) points out for songbirds, aggressive male–male interactions are much more obvious in many species than male–female interactions. Just because the responses of male humpbacks to song are seen more frequently than those of females does not mean that the subtler responses of females to singers are not biologically significant. The critical question here is whether females choose a male for mating based on the song. Because copulation has never been observed in humpback whales, the answer must wait observation of mating or genetic analysis of paternity.

In the absence of direct data on the use of song in female choice, there may be some indirect evidence from the acoustic structure of the song. Intersexual selection selects for increased complexity, beauty, or costliness of the display. Humans, at least, judge humpback song to be complex and beautiful. Catchpole (1982) also suggests that continuous singing, lack of matched countersinging between males, and lack of a singing response to song playback are also diagnostic of a female attraction role for song. Humpback whales often sing continuously for hours; they do not respond to the song of another male by matching the song; neither do they respond to song playbacks by singing themselves. Humpback song clearly has attributes of both intra- and intersexual selection and is more complex than would be expected for a signal used only for male–male interactions (Helweg et al. 1992).

Two different approaches predominate for modeling how intersexual selection may lead to the evolution of elaborate displays. The basic question is why do females choose particular features of males who contribute nothing but their genes to the offspring? The two approaches are called (1) "Fisher's hypothesis," or "the runaway process" versus (2) "good genes," or "the handicap hypothesis." The runaway process emphasizes positive feedback between the female preference and elaboration of the male display (Fisher 1958, Lande and Arnold 1981). Let us start by assuming that females have developed a preference for a longer tail or more complex acoustic display. This could arise because (1) the display was correlated with some valuable inherited trait; (2) the display could make the male more easy to find; or (3) females could simply have some bias to respond preferentially to a particular stimulus. Whatever the origin of the preference, the tendency for females with stronger preferences to mate with males with exaggerated displays means that genes for the preference will covary with genes for the exaggerated display. If females on average select males with exaggerated displays, then the next generation will have more sons with exaggerated displays and more daughters with the preference. This creates a positive feedback loop, potentially producing a runaway process leading to extreme and exaggerated development of secondary sexual characters in males and preferences in females.

The good genes approach emphasizes that male displays signal the male's genetic quality in some trait that is beneficial independent of the female's mating preference itself. Zahavi (1975) pointed out that you can think of extravagant displays like the peacock's tail as costly and stressful handicaps. The stress of the handicap may reveal the male's condition, which might otherwise be concealed, or the extent of the handicap may express the male's condition. For a humpback example, Chu and Harcourt (1986) suggest that humpback females may select singing males based on how long they can stay underwater and hold their breath. There are clear acoustic cues when a singer surfaces, and most singers surface once per song cycle. Chu and Harcourt (1986) argue that breath-holding ability may be a good indicator of a male's stamina and physical condition. The problem with this argument is that song duration changes as humpbacks slowly evolve every feature of their song. In the beginning of one year, the song may average 7 min, whereas 4 months later it will have doubled in length (Payne et al. 1983). The next season, it may start long and decrease in length. Each individual whale is more likely to sing songs of the current length than what they were singing a few months earlier or later. If humpbacks were using song to advertise their breath-holding ability, then each individual would be expected either always to sing as long as he was able, or to sing longest at that part of the breeding season when his chances of mating were highest. This does not seem consistent with the observations that whales at any one time sing songs of similar duration and that the songs change over time with no repeated seasonal pattern.

Female humpbacks may, however, be able to use song to monitor the outcome of competitive interactions of many males scattered over many tens of square kilometers. Singing males are frequently interrupted by other males. The singer usually stops singing when joined, providing a clear acoustic indicator of the joining. Shorter song bouts and higher joining rates for singers were observed by Tyack (1981) during the peak of ovulation as suggested by whaling data from other areas (Chittleborough 1958). During this period (in February) around 60% of the singers followed off Maui sang for less than 1 hr, whereas humpbacks often sang for many hours during other periods of the winter breeding season. Many of the interactions looked like one male displacing another, where either the singer or joiner would start singing relatively near where the two animals split, and the other whale would swim away. If most of the interactions that cause singers to stop involve this sort of male–male competition, and if successful males are interrupted less or for briefer periods of time, then a receptive female might be able to monitor many such interactions at a comfortable distance with minimal effort before she chooses a mate.

Even if female humpback whales use song to monitor interactions, this does not mean that they do not also have preferences for particular kinds of song. When I think admiringly of the incredible aesthetic design features of the peacock's tail or the humpback's song, I cannot help but believe that something more than handicaps are involved in the evolution of these traits. Many human artists are deeply moved by the artistic values of the ornaments and songs of non-human animals. As Darwin (1871) himself pointed out for other species, we clearly may have to thank the evolving aesthetic sensibilities of generations of female humpbacks for the musical features of the males' songs that have sold millions of recordings.

Reproductive Advertisement Displays in Other Marine Mammals

I have described humpback song in some detail because it is the best known of all advertisement displays in marine mammals. Yet songs are known from a variety of other marine mammals. Bowhead whales spend their winter breeding season in icy Arctic waters, where humans seldom venture. Their songs have been recorded in the spring as they migrate past Point Barrow, Alaska (Ljungblad et al. 1982). Bowhead songs are more simple than those of humpbacks, consisting of a few sounds that repeat in the same order for many song repetitions. As with humpback song, bowhead songs appear to change year after year. However, little is known about behavior concurrent with singing, and there are few reports of bowhead whales observed during their winter breeding season when they concentrate in the Bering Sea.

The long series of 20-Hz pulses produced by finback whales may also function as a reproductive advertisement display. The seasonal distribution of these 20-Hz series has been measured near Bermuda, and it matches the breeding season quite closely (Watkins et al. 1987). However, finback whales also appear to be more common in waters near the latitude of Bermuda during the winter breeding season.

Similar recordings in more polar waters will be required to test how frequently these whales produce 20-Hz series outside of the breeding season.

Some pinnipeds also repeat acoustically complex songs during the breeding season. Even casual listening in polar waters often reveals the strange songs of ice seals. Stirling (1973) sampled sounds of ringed seals (*Phoca hispida*) from winter to spring and found an increase in vocalizations correlated with increased agonistic behavior during the breeding season. The bearded seal (*Erignathus barbatus*) produces a downward-trending warbling song that sounds like an alien spaceship in some Grade B sci-fi movie (Cleator et al. 1989). The Alaskan Inuit call this seal the "singer" because part of the song can often be heard in air. The songs of bearded seals are heard frequently during the peak of the breeding season in May, but by July song is seldom heard around bearded seals. All 15 of the bearded seals collected after being identified as singers by Ray et al. (1969) were sexually mature adult males. Male walruses (*Odobenus rosmarus*) also perform ritualized visual and acoustic displays near herds of females during their breeding season (Fay et al. 1981, Sjare and Stirling 1993). Males inflate modified pharyngeal pouches to produce a metallic bell-like sound (Schevill et al. 1966). When walruses surface during these displays, they may make loud sounds in air, including knocks, whistles, and loud breaths. They then dive, producing distinctive sounds underwater, generally a series of sharp knocks followed by gong- or bell-like sounds. Usually several males attend each female herd, and it is not known whether females or other males are the most important audience for this display (Sjare and Stirling 1993).

Antarctic Weddell seals (*Leptonychotes weddellii*) also have extensive vocal repertoires, and males repeat underwater trills (rapid alternations of notes) during the breeding season. Males defend territories on traditional breeding colonies. These trills have been interpreted as territorial advertisement and defense calls (Thomas et al. 1983). Whether females may also use them in selecting a mate is unknown.

Correlations between Social Structure and Patterns of Acoustic Communication

Most communication signals evolve to solve specific problems in social behavior. In fact, communication and social behavior are just two different ways of expressing the same thing. This section traces correlations between the problems posed by the social lives of different species and the species' communication signals. Understanding the social functions of communication signals requires more detailed behavioral observations than are available for most marine mammal species. Some of the best-studied species include killer whales (*Orcinus orca*), bottlenose dolphins, and sperm whales (*Physeter macrocephalus*).

CALLS THAT IDENTIFY STABLE GROUPS IN KILLER WHALES. The most stable groups known in any mammal are those found in fish-eating killer whales in the coastal waters of the Pacific Northwest. Associations between individual killer whales have been tracked since before 1970, giving more than 25 years of longitudinal data. The only way a killer whale group, called a pod, changes is by birth, death, or rare fissions of very large groups (Bigg et al. 1987). Fish-eating killer whales are most unusual among mammals in that neither sex disperses from its natal group. Killer whales produce a variety of sounds, including clicks used in echolocation, tonal whistles, and pulsed calls, some of which form repeated discrete calls and others that are highly variable (Ford 1989). The discrete calls predominate when killer whales are traveling or foraging. Whistles and more variable pulsed calls are more common in groups engaged in social interaction. The whistles and variable-pulsed calls are difficult to subdivide, but the discrete calls form easy-to-categorize call types.

Each pod of killer whales has a group-specific repertoire of discrete call types. This pod-specific repertoire is stable for many years (Ford 1991). Each individual whale within a pod is thought to produce the entire call repertoire of that pod. Analysis of variation in call use within a pod suggests that some calls may be more common in resting groups, others more common in active groups. However, each discrete call in the pod's repertoire can be heard regardless of what the pod is doing. Different pods may share some discrete calls, but none share the same entire call repertoire. The entire repertoire of a pod's discrete calls can thus be thought of as a group-specific vocal repertoire. Different pods may have ranges that overlap and pods may even associate together for hours or days before diverging. Individual pods have clearly defined subpods and matrilineal groups that seldom split up, but these subpods may separate and converge. These group-specific call repertoires in killer whales are thought to indicate pod affiliation, maintain pod cohesion, and to coordinate activities of pod members.

MOTHER–OFFSPRING RECOGNITION IN MARINE MAMMALS. Killer whales have unusually stable groups. Other marine mammals have more fluid groupings, but there may be strong bonds between individuals within these groups. The mother–young bond is one of the most fundamental in mammals.

All mammalian young are born dependent on the mother. Newborn mammals need to suckle frequently and, in many species, depend on the mother for thermoregulation and protection from parasites and predators. Most mammals have a vocal system for regaining contact when mother and offspring are separated. These "isolation" or "distress" calls are produced by infants within days of birth and are particularly elicited by isolation. Most mammalian isolation calls are frequency-modulated tonal calls, are longer and louder than other infant calls, and become fixed in a stereotyped structure as the animal ages. Examples come from a variety of terrestrial taxa including primates (Newman 1985), felids (Buchwald and Shipley 1985), bats (Balcombe 1990), and ungulates (Nowak 1991). Once a mother and offspring have become separated, there is a risk that the mother might miss her own offspring and accept some other young animal as her own. Parents often devote considerable resources to their young, and this creates a risk that other animals might attempt to parasitize their parental care. These problems create a selective pressure for the evolution of mother–infant recognition mechanisms. In many of the species tested, mothers can recognize the calls of their young; in some, the young recognize similar calls from the mother. These infant isolation cries appear to represent a widespread and basic mammalian adaptation.

Colonially breeding seals often face difficult mother–young location and recognition problems. In many otariids, a mother leaves her young pup on land in a colony of hundreds to thousands of animals, feeds at sea for a day or more, and then must return to find and feed her pup. Among Galapagos fur seals (*Arctocephalus galapagoensis*) pups spend more time calling during their first day of life than later, and mothers learn to recognize the calls of their young within the first day of life (Trillmich 1981). Mothers give pup-contact calls as early as during birth. Later, mothers can signal with a pup-attraction call to a pup that is moving away. When a mother returns from feeding at sea, she comes up on the beach giving pup-attraction calls. Her own pup usually seems to recognize her call and approaches. If a pup approaches to suckle, the mother sniffs the pup for a final olfactory check. If it is not her offspring, she almost always rejects the pup, a rejection that can cause injury or occasionally death to the pup (Trillmich 1981). There is thus a strong incentive for both mother and pup to recognize each other correctly. Playback experiments of pup-attraction calls indicate that 10- to 12-day-old pups prefer their mother's call, and this recognition persists until they become independent at more than 2 years of age (Trillmich 1981). Other otariids, such as sea lions, show similar patterns of calling and recognition (Trillmich 1981, Schusterman et al. 1992).

Many phocid seals only suckle their young for a few weeks or less. They typically do not leave for long foraging trips but remain near the pup for the duration of suckling (see Costa and Williams, Chapter 5; Boyd, Lockyer, and Marsh, Chapter 6; and Wells, Boness, and Rathbun, Chapter 8, this volume). However, the female does frequently enter the water in between nursing bouts. Although nursing females are less gregarious than otariids, the phocid mothers still need to find their own pups and prevent suckling from others. Land-breeding phocid seals use a combination of geographic, acoustic, and olfactory cues for mother–infant recognition. Location cues are important among gray seals (*Halichoerus grypus*). Fogden (1971) reports that the pup remains where it was last suckled. In between suckling bouts, a mother usually returns to the water. When she next hauls out, the mother returns to where she last suckled. This means that a female can often use location cues to help narrow down the number of pups that might be hers. Among phocid seals that breed colonially on land, mothers do recognize calls of their pups (elephant seals [*Mirounga* spp.], Petrinovich 1974; gray seals, Fogden 1971).

Mother–pup recognition is more difficult for ice-breeding seals because the topography of the ice frequently changes. Terhune et al. (1979) suggest that female harp seals (*Phoca groenlandica*) have so few landmarks on the ice that they approach pups in a random manner, and therefore, must remain close to their pup for the duration of suckling. In spite of the less predictable location cues on the ice, Terhune et al. (1979) suggest that harp seal mothers may not identify the calls of their pups, but rely on a combination of visual and olfactory cues at close range. During most mother–pup approaches, neither the mother nor the pup vocalized, but the mother usually sniffed a pup upon approach.

Comparisons of recognition mechanisms in related species allow testing of the hypothesis that the cost and reliability of recognition mechanisms scale to the risk of misallocation of parental care. As in the study of animal songs, some of the strongest data come from birds. For example, barn swallows (*Hirundo rustica*) raise their young apart from other broods; therefore, location is a good predictor of kinship throughout the period of parental care. Although barn swallow chicks make a begging call, parents do not distinguish between the calls of their own and unrelated chicks (Medvin and Beecher 1986). Young cliff swallows (*Petrochelidon pyrrhonota*), on the other hand, intermingle within a colony while still being fed by their parents. Cliff swallow parents can discriminate the begging calls of their own offspring from those of other young (Stoddard and Beecher 1983). Cliff swallows have evolved both a more distinctive begging call in the young and more rapid discrimination of begging calls by adults (Loesche et al. 1991). Similar results suggest that colonial birds switch from location cues to identifying their own offspring at the time when the young from different broods intermix (Beer 1970, Miller and Emlen 1975). The differences in recognition systems of phocid and otariid seals seem to follow the predictions of this model. Mother–infant recognition systems in phocids do not seem to be as reliable as those of otariids (Reiter et al. 1978). Trillmich (1981) suggests that this difference may derive from the lower gregariousness, shorter duration of suckling, and increased predictability of location cues for phocid versus otariid seals.

The young of many dolphin and other odontocete species are born into groups comprised of many adult females with their young, and they rely on a mother–young bond that is even more prolonged than that of otariids. Many of these species have unusually extended parental care. For example, both sperm whales and pilot whales (*Globicephala macrorhynchus*) suckle their young for up to 13 to 15 years (Best 1979, Kasuya and Marsh 1984). Bottlenose dolphin calves typically remain with their mothers for 3 to 6 years (Wells et al. 1987). These dolphin calves are precocious in locomotor skills and swim out of sight of the mother within the first few weeks of life (Smolker et al. 1993). Calves this young often associate with animals other than the mother during these separations. This combination of early calf mobility with prolonged dependence would appear to select for a mother–offspring recognition system in bottlenose dolphins. Unless otherwise noted, the unmodified terms "dolphin," "mother," and "calf" refer in the following section to the bottlenose dolphin.

Dolphin mothers and young use frequency-modulated tonal whistles as signals for individual recognition. Observations of captive dolphins suggest that whistles function to maintain contact between mothers and young (McBride and Kritzler 1951). When a dolphin mother and her young calf are forcibly separated in the wild, they whistle at high rates (Sayigh et al. 1990); during voluntary separations, it is usually the calf that whistles to signal a reunion (Smolker et al. 1993). Caldwell and Caldwell (1965) demonstrated that each dolphin within a captive group produced an individually distinctive whistle or signature whistle. The Caldwells postulated that signature whistles function to broadcast individual identity. Experimental playbacks have demonstrated that mothers and offspring respond preferentially to each others' signature whistles even after calves become independent from their mothers (Sayigh et al. 1999).

In spite of the apparent premium on early development of mother–young recognition, there is great variability in the timing of signature whistle development in bottlenose dolphins. Caldwell and Caldwell (1979) studied whistle development in 14 calves born in captivity. They reported that calves whistle within days of birth, but that these early whistles are unstereotyped. Most of the calves in their study developed a stereotyped signature whistle by 1 to 3 months of age, but one calf had not yet developed a signature whistle when it was last recorded at 17 months of age. Preliminary longitudinal studies of whistle development in four free-ranging dolphin calves have also been conducted in waters near Sarasota, Florida, by Sayigh (1992). These results also indicate considerable individual variability in the timing of signature whistle development. Two calves developed signature whistles by 1 to 2 months of age; the third calf developed a signature whistle between 2 and 3.5 months of age, and the last calf did not develop a signature whistle until almost 2 years of age.

To understand mother–offspring recognition, it is more important to investigate when a mother (or calf) is first capable of discriminating her own offspring (or mother) from others, than to document changes in the signals. Adult dolphins have excellent auditory perception, and if a mother spends all her time near her calf, she may learn to discriminate a calf's unstereotyped whistles well before the calf develops a signature whistle. On the other hand, it may take an animal some time to learn to discriminate a signature signal that is already stereotyped. This illustrates a general point in communication research. Although the signal is the most obvious part of the communication process, how the recipient perceives and responds to the signal is just as important.

INDIVIDUAL RECOGNITION IN BOTTLENOSE DOLPHINS. Dolphins do not just use whistles for mother–infant recognition. Calves show no reduction in whistling as they wean and separate from their mother. Whereas adult males are not thought to provide any parental care, they are not known to whistle less than adult females. Bottlenose dolphins may take up to 2 years to develop an individually distinctive signature whistle, but once a signature whistle is developed, it remains stable for the rest of the animal's lifetime (Fig. 7-9; Caldwell et al. 1990; Sayigh et al. 1990, 1995). These results suggest that signature whistles may also function for individual recognition in contexts other than mother–offspring recognition.

Dolphins also rely on individual-specific social relationships throughout their lifespan. Bottlenose dolphins do not have stable groups as in resident killer whales, but rather live in a fission–fusion society in which group composition changes from hour-to-hour or even minute-by-minute (see Wells, Boness, and Rathbun, Chapter 8, this volume). Although dolphin groups are remarkably fluid, bonds between particular individuals within the groups may be very strong. Some wild individual bottlenose dolphins show stable patterns of association, even within the otherwise fluid patterns of grouping (Wells et al. 1987). As discussed, young dolphins in the wild often remain with their mothers for 3 to 6 years. After they leave their mothers, dolphins may both maintain close ties with members of their natal band as well as join with new individuals for periods of years. For example, some

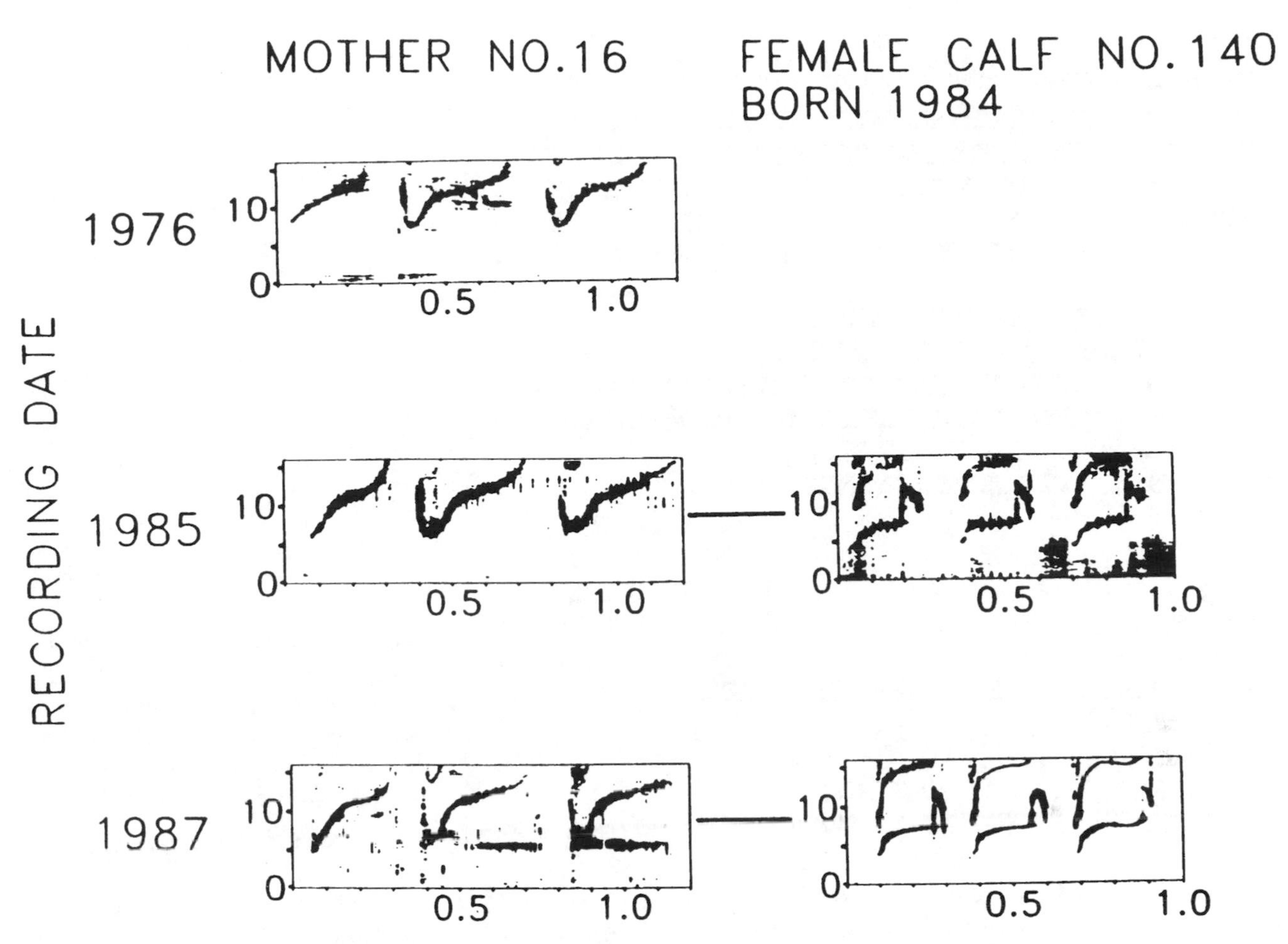

Figure 7-9. Spectrograms of signature whistles from one wild adult female bottlenose dolphin recorded over a period of 11 years and of her daughter at 1 and 3 years of age. Note the stability of both signature whistles. The x-axis indicates time in seconds and the y-axis indicates frequency in kHz. (From Fig. 7-2, Sayigh et al. 1990.)

pairs of adult males are almost always sighted together, even in different groups, for many years (Wells et al. 1987, Connor et al. 1992). Primiparous females may associate with their own mothers during the first few months of life of their own calves (Caldwell and Caldwell 1966). Figure 7-10 shows month-by-month changes in associations between individual wild dolphins sighted off a beach in Patagonia over a year and a half by Würsig (1978). Each row in this figure reflects sightings of one individual. Notice how some individuals always appeared together or left together, whereas other individuals started associating during the observation period. This combination of highly structured patterns of association between individuals, coupled with occasionally fluid patterns of social grouping, argues that individual-specific social relationships are an important element of bottlenose dolphin societies (Tyack 1986a).

Five of the dolphins shown in Figure 7-10 were resighted together 300 km away. Dolphins may range tens of kilometers a day. While swimming in turbid coastal waters, they often disperse out of sight of one another. It is difficult to imagine how dolphins that share a strong bond could remain together without an individually distinctive acoustic signal such as the signature whistle. Initial studies of signature whistles in adult dolphins, primarily isolated animals, suggested that more than 90% of an individual's whistle repertoire was made up of its signature whistle (for review, see Caldwell et al. 1990). However, the signature whistle hypothesis is not limited to the suggestion that dolphins just monotonously repeat the identical call to maintain contact, like a radio call sign with no other message. The acoustic features of the signature whistle vary as a function of behavioral context (Caldwell et al. 1990, Janik et al. 1994). Even after dolphins have developed the signature whistle, they also steadily increase their production of a repertoire of whistles that differ from the signature whistle (Caldwell et al. 1990).

When dolphins interact, they not only produce their own signature whistles but may also imitate the signature whistles of other individuals with whom they share strong bonds. In one study of two captive adult dolphins, Tyack (1986b) found that each imitated the signature whistle of the other at

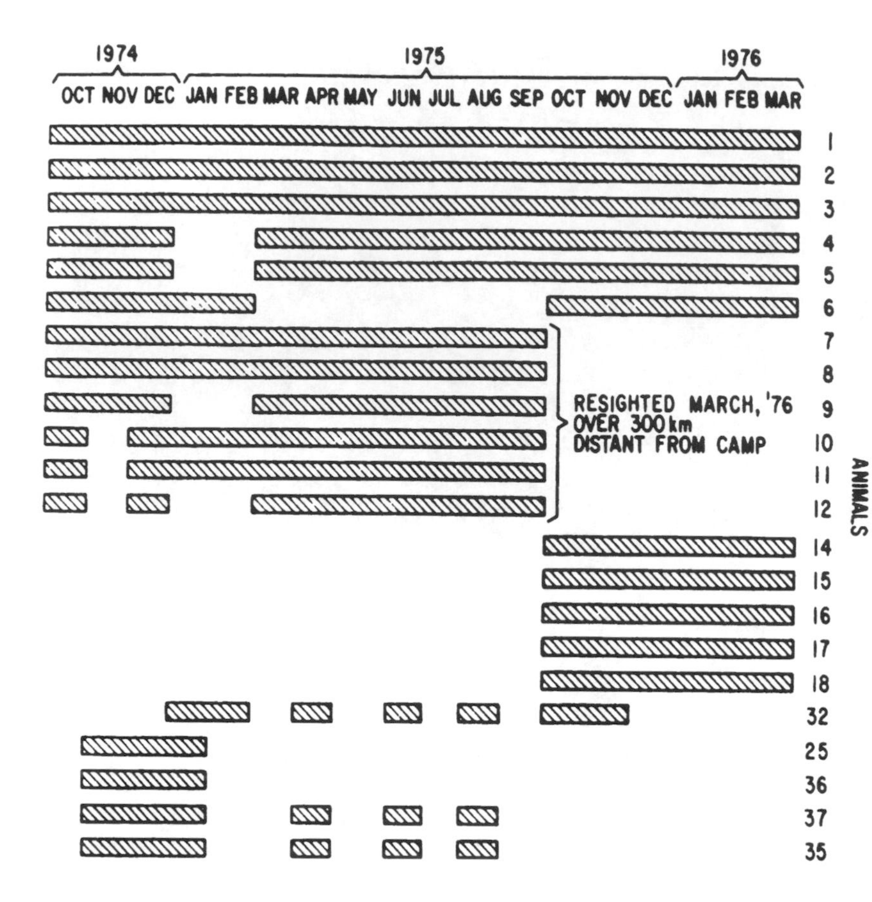

Figure 7-10. Monthly sighting patterns of individual bottlenose dolphins sighted off a beach in Patagonia. Each row represents an individual dolphin identified by natural markings on the dorsal fin. When a month is marked with a cross-hatched bar, it indicates that the individual was sighted from the beach during that month. (From Würsig 1978.)

rates of about 25% (i.e., 25% of all occurrences of each signature whistle were imitations). Rates of signature whistle imitation were near 1% between captive dolphins that were in separate pools but that could hear one another (Burdin et al. 1975, Gish 1979).

Imitation of signature whistles has also been observed between wild dolphins that share strong social bonds. For example, Figure 7-11 illustrates signature whistles of each member of a pair of males sighted together 75% of the time in Sarasota, Florida, along with imitations of the partner's whistle. Imitation of signature whistles has also been observed between wild dolphins that share strong social bonds. For example, Figure 11 illustrates signature whistles of each member of a pair of males sighted together 75% of the time in Sarasota, Florida, along with imitations of the partner's whistle. In a population of wild bottlenose dolphins in the Moray Firth, Scotland, Janik (1997) found whistle matching in 17% of all whistle interactions. An individual-specific response to imitation of signature whistles is illustrated in a case from the Sarasota population, where one adult female, Nicklo, imitated the signature whistle of an older female, Granny (Tyack 1993). Figure 7-12 shows the signature whistles of both dolphins, along with an imitation of Granny's whistle by Nicklo. The whistle imitation was recorded during a 28-min interval when Nicklo was held in a raft. Five other dolphins remained in the net corral: Granny, Nicklo's 3-year-old calf, another adult female with a 3-year-old, and an 11-year-old male. Nicklo started imitating Granny's whistle about halfway through her time in the raft. Figure 7-13 shows an example where a whistle by Granny (top) was synchronized within 3 sec of an imitation of Granny's whistle by Nicklo (bottom). Statistical analyses of the correlation of synchronous whistles showed that the only time whistles were synchronized occurred when Nicklo started to imitate Granny's whistle, and the only correlation was between Granny and Nicklo's imitation of Granny. Even Nicklo's own calf did not produce whistles synchronized to those of Nicklo. This suggests that one dolphin may imitate the signature whistle of another one to initiate an interaction with that particular individual. However, these animals were not able to interact directly, and more detailed study is required to determine whether dolphins imitate each other's signature whistles to call another individual.

MULTIPLE FUNCTIONS FOR THE CLICKS OF SPERM WHALES. Sperm whales have not been demonstrated to produce any vocalizations other than clicks. Although these clicks sound very simple to our ears compared to humpback song, sperm whales live in societies that would seem to require diverse kinds of communication. Sperm whales are born in relatively stable units of related females, but most individuals have varying degrees of association

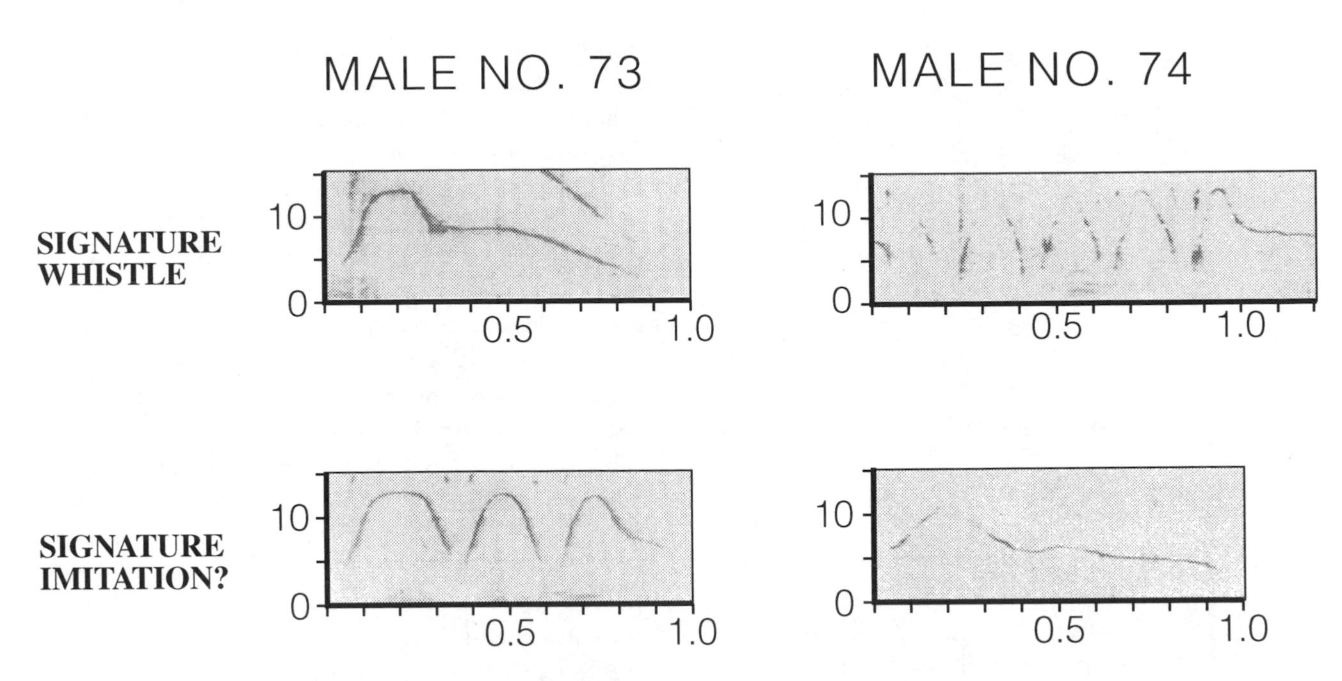

Figure 7-11. Spectrograms of signature whistles produced by two adult male dolphins from the wild population near Sarasota, Florida. These two animals were usually sighted together. Each male also repeated whistles that were similar to the signature of the partner, and these are interpreted as imitations of the partner's signature whistle. The x-axis indicates time in seconds and the y-axis indicates frequency in kHz. (From Fig. 4.16, Sayigh 1992.)

throughout their lifespan, and many rely on bonds with particular animals within their groups (Best 1979). For example, sperm whales appear to have social defenses against predators. Humans are among the most dangerous predators of sperm whales, and there are many reports from whalers of sperm whales coming to the aid of a harpooned comrade (e.g., Nishiwaki 1962, Caldwell and Caldwell 1966). Sperm whales also use a social defense from a killer whale attack (Arnbom et al. 1987). Unlike fish-eating killer whales, male sperm whales leave their natal group and may join mixed-sex juvenile groups or all-male groups as they mature. As subadult male sperm whales grow, they tend to be found in smaller groups of larger whales, until they finally become mostly solitary when they are sexually mature at about 20 to 25 years of age. During the breeding season these sexually mature males swim among the groups of adult females with which they mate. Stable units of related females tend to number about 10 whales, but two different female units often associate for about 10 days at a time (Whitehead and Kahn 1992). Sperm whales may rely on the stable units for protection of calves, whereas these stable units may join together in more fluid groups for benefits from socially coordinated feeding (Whitehead 1996).

Sperm whales feed on squid at depths of 400 to 600 m during dives that typically last 40 to 50 min (Papastavrou et al. 1989). Feeding and diving sperm whales typically produce long series of clicks at regular intervals of roughly 1 to 2/sec (Worthington and Schevill 1957). Echolocation has not been tested experimentally because sperm whales have never been maintained for long in captivity. Most researchers suggest that these regular clicks are used for echolocation, and also perhaps within a group to keep contact with one another while foraging. During the breeding season, large male sperm whales make especially loud and resonant clicks that sound to my ear like a firecracker exploding inside a metal can. Weilgart and Whitehead (1988) suggest that these resonant clicks may function as a threat display when males compete to accompany a breeding group of females. For example, a male may produce loud clicks when approaching a female group. If a male is already accompanying the group, he might click back. If some feature of these clicks, such as loudness or low frequencies, correlates with a male's competitive ability, then this information may help females to assess the male at a distance and may help a male to assess whether to challenge the other male or not. (Low frequency may correlate with size of the sound-producing organ, as in the musical example of the stringed instrument mentioned in the introductory section.) This dynamic would select for each male making as extreme a version of the display as possible. Figure 7-14 illustrates one of these loud clicks, showing how much longer the duration is than the regular clicks. A bottom echo from the click is also visible on the far right of the figure. If the bottom echo is so obvious to a hydrophone at the sea surface, then it is likely that a sperm whale diving

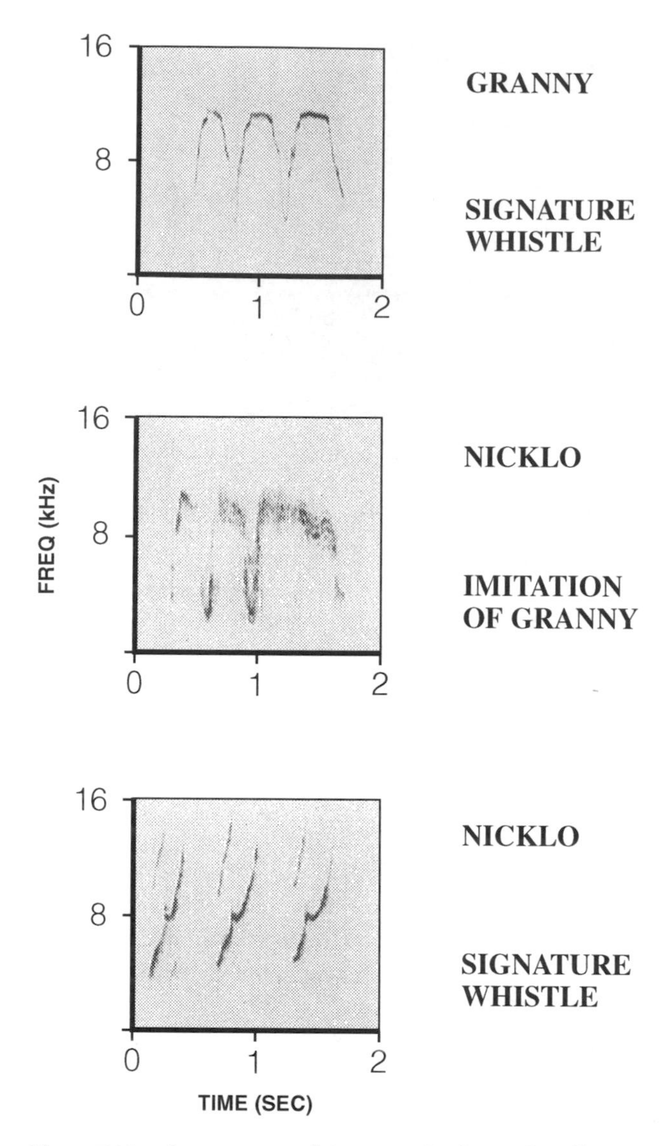

Figure 7-12. Spectrograms of signature whistles produced by two adult female dolphins from the wild population near Sarasota, Florida. The top spectrogram shows the signature whistle of Granny. The bottom spectrogram shows the signature whistle of Nicklo. The middle spectrogram shows an imitation of Granny's whistle produced by Nicklo.

near the bottom may be able to hear similar echoes from fainter regular clicks. Sperm whales dive at rates of 1 to 4 m/sec (Watkins et al. 1993). They have enormous momentum because they weigh up to 30 to 40 metric tons. It would take a diving whale some time to slow its descent. A depth-sounding sonar would be of obvious utility for early warning of the approaching sea floor for an animal that takes some time to slow its dive.

Sperm whales spend most of their time diving, but often in the late afternoon they spend a few hours at the surface, resting and socializing (Whitehead and Weilgart 1991). Especially when they are socializing, sperm whales also produce distinctive rhythmic patterns of clicks, called codas, often as exchanges between individual whales (Watkins and Schevill 1977). The limited data currently available on codas suggest a variety of potential functions, including individual, group, and regional identification. Although Watkins and Schevill (1977) described codas as individually distinctive, they also described an exchange in which each whale matched the coda of the other whale, an exchange surprisingly reminiscent of imitation of individually distinctive signature whistles in dolphins. Moore et al. (1993) described two shared coda patterns that comprised more than 50% of the codas from many individual whales within many different groups recorded over a large part of the southeast Caribbean. Weilgart and Whitehead (1993) described different shared coda patterns for sperm whales off the Galapagos, and Weilgart and Whitehead (1997) described geographical variation in the proportional usage of different codas. More work is needed to track coda usage of individual sperm whales, within stable units, and over large areas, but the current evidence suggests possible variety in usage for individual and regional identification that is consistent with the variety of problems posed by sperm whale societies.

In summary, there is a clear correlation between the communication patterns of marine mammals and their social organizations. Baleen whales and pinnipeds with large, somewhat anonymous breeding aggregations use reproductive advertisement displays to mediate male–male and male–female interactions on the breeding grounds. Killer whales with highly stable groups produce group-specific repertoires of stereotyped calls. Seals and dolphins with strong individual-specific bonds use a variety of different vocalizations for individual recognition. Sperm whales appear to use deceptively simple clicks to produce a diverse set of signals consistent with their diverse social groupings.

Cognition

As mentioned in the introductory section, cognition is usually defined as information processing within an animal. This is often contrasted with communication, which involves information transfer from one animal to another. The study of animal communication can start with measuring the signals exchanged between animals, but how can one study the internal signals associated with information processing within an animal? The direct neurophysiological approach involves invasive measurement of neural activity, but there has been very little neurobiological research conducted with marine mammals, in part because marine mammals are legally protected from invasive research. Three indirect approaches to studying animal cognition are discussed in this section. The first is "animal language" studies,

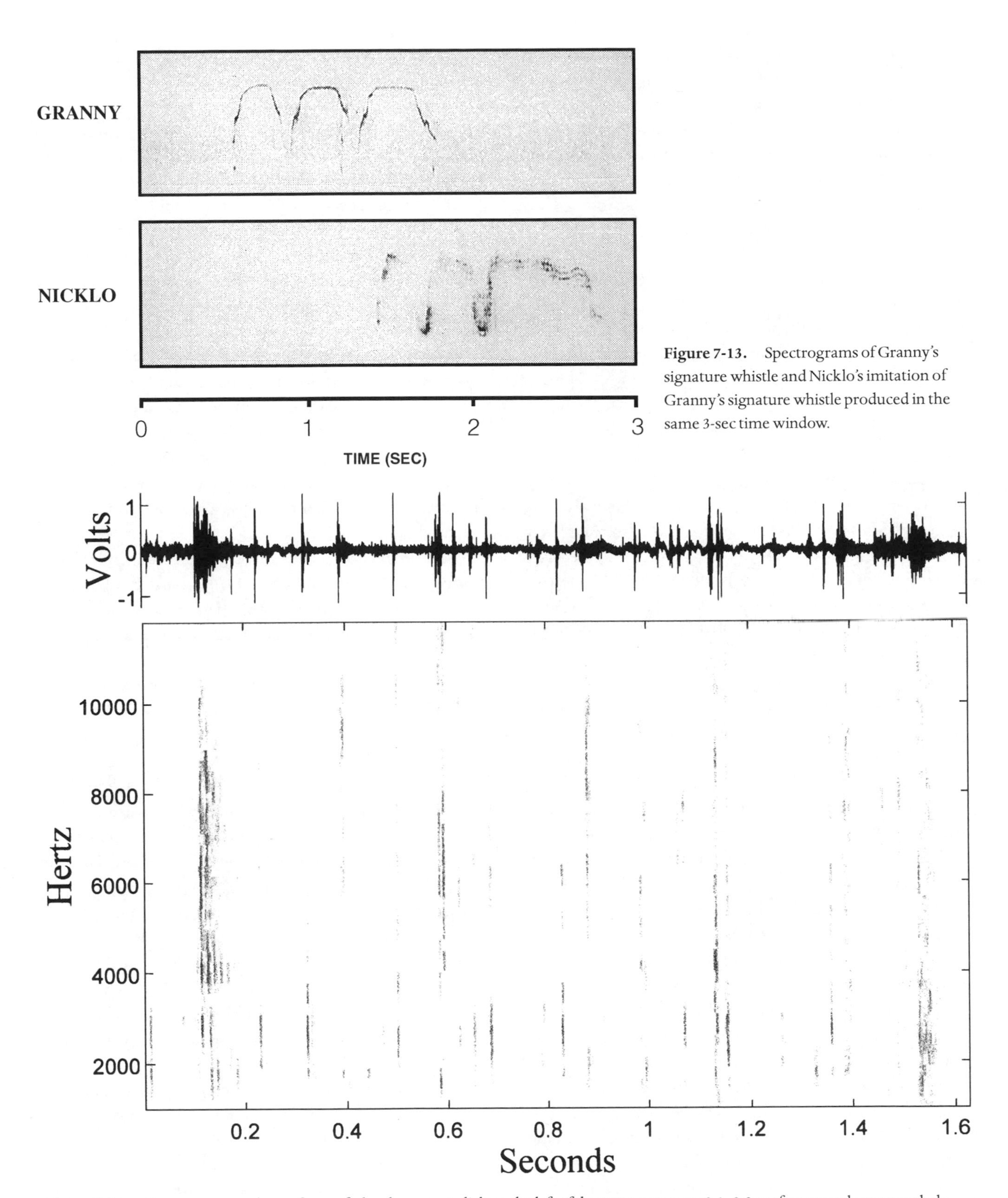

Figure 7-13. Spectrograms of Granny's signature whistle and Nicklo's imitation of Granny's signature whistle produced in the same 3-sec time window.

Figure 7-14. Spectrogram and waveform of a loud resonant click on the left of the spectrogram at 0.1–0.2 sec from a male sperm whale. Regular clicks typically last 2–30 msec, and many regular clicks show on the spectrogram as thin vertical lines from 0.3 to 1.4 sec. An echo of the loud click reflected from the sea floor is visible on the right side of the figure at 1.5–1.6 sec.

in which researchers train animals to use an artificial system of communication with some features of human language. The second reviews the importance of learning and imitation in the natural communication systems of marine mammals. The third reviews the question of why some marine mammals have such large brains.

Animal "Language" Studies

Language and Thought

How do humans keep one another informed of what they are thinking? We can use our own form of communication, language, to "share thoughts." Even when we are thinking of something nonlinguistic, say a mental image, we can report on the image using language. Language is not only used for communication. There are modes of thought that seem to be intimately related to language. When we are thinking about something, we often seem to be forming words and sentences within our heads, although we are completely silent. This suggests that the process of language development may influence how we think. Both the use of language to report thoughts and the potential for language to shape thinking have led people to search for an animal "language" to study animal cognition.

For almost a century, humans have attempted to train non-human animals to communicate with people using artificial human-made systems of communication in what are called animal language experiments. These experiments have attempted to find ways to allow animals to report on their thinking, to test the language competencies of animals, and to test whether language training might influence the abilities of animals to solve problems. The early animal language studies adopted the following approach: "We know the conditions under which human infants learn language. If we can just put a smart large-brained animal in a similar setting, perhaps it can also learn to speak." This approach assumes that language acquisition requires only very general intelligence and learning skills. Several psychologists have set out to test these ideas by raising baby chimpanzees in their own homes. A baby chimp named Gua was raised along with the infant son of a couple named Kellogg (Kellogg and Kellogg 1933). Although the boy acquired language at the usual rate, the chimp never did. Another chimp named Viki, raised by the Hayes family, learned to produce crude versions of the words "mama," "papa," "cup," and "up," but never progressed beyond this rudimentary stage (Hayes 1951).

Chimpanzees seem to have problems producing the sounds characteristic of human speech. Chimps are one of our closest evolutionary relatives, but not even they have the same vocal tract as we do, and they appear to be physically unable to produce some speech sounds (Lieberman 1984). There is also a large body of evidence suggesting that modification of vocalizations is very difficult for non-human terrestrial mammals, including our closest primate relatives (Newman and Symmes 1982, Janik and Slater 1997). If chimps cannot make speech sounds, then training them to speak may be no more promising than training them to flap their arms to fly. Do chimps have a basic cognitive problem with learning language or is it restricted to the vocal channel? These were the questions that Beatrice and Allen Gardner considered as they thought about raising a chimp. The Gardners knew that chimps were reputed to be skilled at imitating gestures so they decided to teach a gestural language used by humans with hearing impairment (Gardner and Gardner 1969). This was much more successful than the earlier attempts with speech, and there have been many similar studies training great apes with gestural languages. Chimpanzees, gorillas, and orangutans have all been trained with gestural languages.

Marine mammals have also been trained in animal language experiments. In the late 1950s, researcher John Lilly noticed that captive dolphins often emit sounds in air with the blowhole open. Under natural conditions, dolphins usually vocalize underwater with the blowhole closed, and the sounds are only faintly audible in air. Lilly became convinced that dolphins could mimic the speech sounds of talking humans (Lilly 1962). When he played the tapes of these purported imitations, few listeners were convinced. Lilly spent years attempting to shape the vocal responses of dolphins to human speech using both food and social rewards, but the imitations did not become more convincing to a human audience. In his own words, "Obviously the pronunciation of *Tursiops* is not very good" (Lilly 1975:346). The best he could do was document that dolphins match the number and duration of staccato bursts of speechlike sounds (Lilly 1965).

Frustrated at the lack of progress, Lilly went to considerable expense to design and build a facility in which humans and dolphins could live together for extended periods. A volunteer lived for 2.5 months with a subadult male dolphin in the specially built facility (Lilly 1967). In his popular writing, Lilly put as positive a spin on this project as he could, but it is difficult to identify positive results of this effort. At least the Hayes' were able to convince their audience that Viki did produce crude versions of four words.

In many ways Lilly's enterprise was even more naive than the similar attempts to raise infant chimps in the home. Among humans, it is much easier for young children to learn language than for adolescents. Even then, it takes years rather than weeks of exposure. For these reasons, the chimp experiments involved psychologists taking infant chimps in their homes and raising them for periods of years. In contrast, Lilly's volunteer spent a much shorter time living with a subadult dolphin. This shorter duration of 2.5 months with an older animal probably made it less likely that the exposure to humans would influence vocal development in the dolphin, but even this shorter duration was difficult for the volunteer. I am sure that the humans had their own difficulties living with a chimp, but imagine what it must be like for a volunteer to live in pools flooded with two feet of water all the time. For example, one "lesson" learned during this experiment was "Being able to sleep in a dry, comfortable bed each night would eliminate much of the discomfort in the program" (Lilly 1975:198). It obviously was not possible to create an environment where dolphins and humans lived together and both were equally at home.

Although Lilly's own research never yielded results anywhere near as promising as those of the gestural languages with apes, he published very popular books claiming that dolphins could talk. He made a series of unsubstantiated claims that strayed far from his own research. For example, "The sperm whale probably has 'religious' ambitions and successes quite beyond anything we know" (Lilly 1975:219). At the same time as these books stimulated popular interest, their unsubstantiated claims scared several generations of behavioral scientists away from a potentially fascinating group of mammals for comparative study.

There has also been controversy regarding the question: How languagelike is the performance of apes trained to use gestures? Some investigators report in glowing linguistic terms the production of sentences, the invention of novel compound words or even poetry in their charges (e.g., Patterson and Linden 1981). Other investigators maintain a more skeptical stance and question the data or interpretations of the claims for languagelike performance.

There have been difficulties interpreting studies in which animals are exposed to language as informally as human infants are. The flexibility and lack of control over the training have led critics to question whether the animals spontaneously generate meaningful sequences to which they had not been exposed (e.g., Terrace et al. 1979). Other investigators have responded to these criticisms by using more formal training with experimental controls to teach animals artificial languages. While these controversies were underway, Louis Herman of the University of Hawaii initiated animal language comprehension studies with bottlenose dolphins. Unlike the ape studies, the dolphins were trained to respond to commands but not to use these "commands" as communicative acts of their own. In these studies, Herman (1980, 1986) and Herman et al. (1984) trained dolphins to associate either objects or actions with human-made sounds or gestures using standard conditioning. Figure 7-15 shows some of the acoustic and gestural signals associated with the ob-

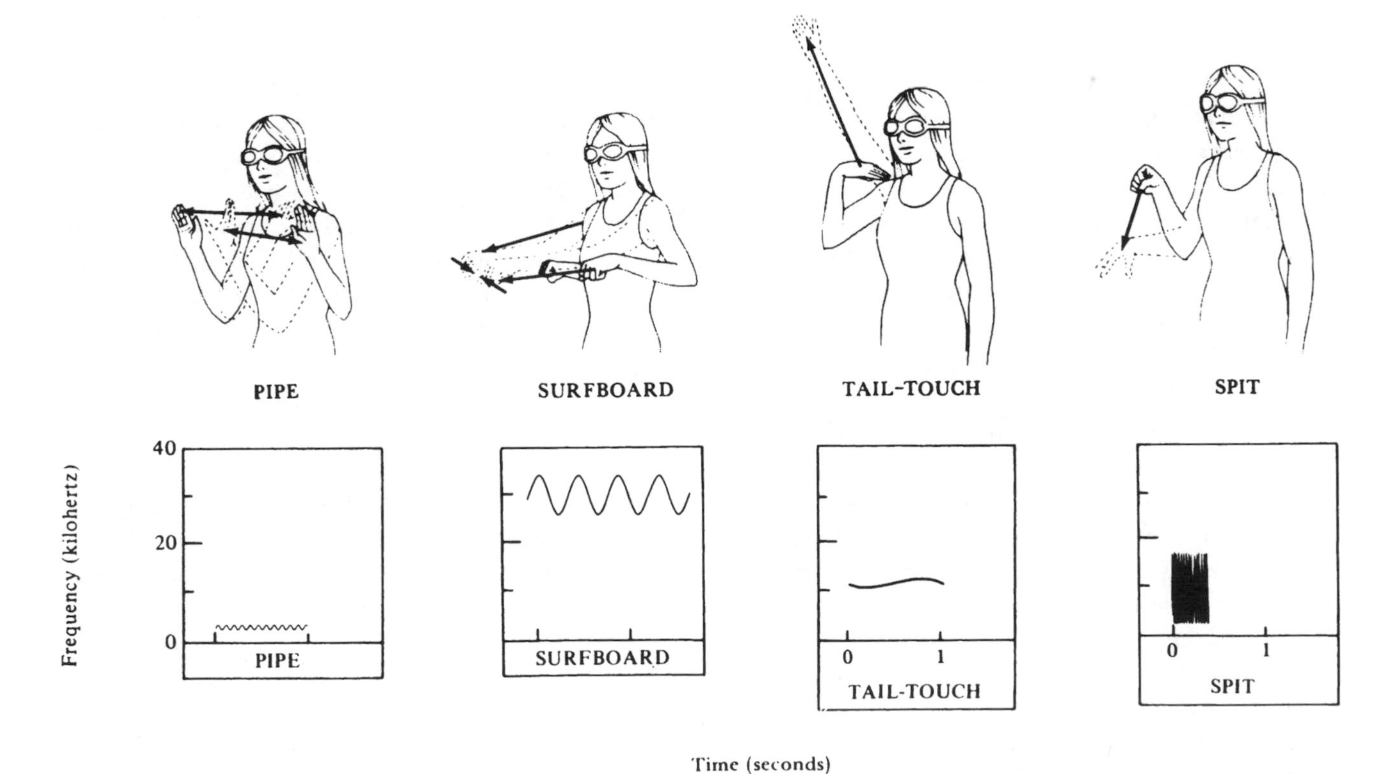

Figure 7-15. Line drawings of acoustic and gestural cues used by Herman's laboratory to train dolphins. (From Herman et al. 1984.)

jects *surfboard* and *pipe* and the actions *touch-with-tail* and *spit*. Herman started this training by presenting a cue when only one object was present. The dolphin would get a fish for touching the object with the tip of her jaw. Then two objects would be introduced, and the dolphin was only rewarded for touching the cued object. New actions were trained using a technique called "shaping." For example, dolphins were trained to put an object in the mouth by presenting the object along with a fish when the *mouth* cue was presented. The dolphin would open her mouth for the fish, and the trainer would put both the fish and the object in the mouth. In later sessions, the trainer would offer the object first and delay giving fish longer and longer. Ultimately, when the dolphin was shown the *mouth* cue, she would hold the requested object in her mouth in expectation of later reinforcement with the fish. Herman et al. (1989) report that dolphins perform with equal success to visual or auditory commands.

Although these experiments are called animal "language" experiments, it is difficult to compare the performance of the animals to the human use of language. One approach to making these comparisons more meaningful is to discuss specific features of language. Two of the most important features of human language are syntax and reference. Human languages are made up of words that can refer to objects, actions, almost anything. These words are structured into meaningful sentences using grammar or syntax. What do animal language studies tell us about syntactic or referential abilities of animals?

Syntax

Most commercial animal training involves training an animal to perform a complex series of behaviors in response to one simple command, but Herman trained the reverse, having animals make a relatively simple response to a complex series of action or object commands. This allowed Herman to study how well dolphins could generalize rules about command sequences designed with an artificial syntax. He trained one dolphin to interpret the string of commands OBJECT1–TAKE–OBJECT2 as if it meant "take object 1 to object 2." The other of his two dolphins was trained to interpret the string of commands OBJECT2–OBJECT1–TAKE as if it meant "take object 1 to object 2," reversing the order of cues compared with the requested actions. The first dolphin could have simply learned to go to objects and perform actions in the same sequence as the commands, but this was not possible for the second dolphin. After extensively training the dolphins with about half the objects that had been associated with cues, Herman et al. (1984) tested how the dolphins would interpret the same sequence involving new objects. For example, the dolphin might be trained on "take the Frisbee to the basket" and might then be asked for the first time "take the pipe to the surfboard." Only once in these tests did one of the dolphins reverse the requested order, taking object 2 to object 1. Because they had never been exposed to the particular three-command sequences used in the tests, the dolphins must have learned rules for interpreting the order of commands. This performance appears similar to the way English speakers might use word order to understand a sentence.

Dolphins are not the only marine mammal to have been trained in an artificial language. Schusterman and Krieger (1984) replicated many aspects of Herman's results using California sea lions (*Zalophus californianus*) instead of dolphins. Two sea lions learned 16 to 20 cues and also learned to interpret sequences of commands. The sequences of commands used with the sea lions took forms like modifier–object–action. For example, when given the sequence white pipe–flipper–touch, the sea lion was rewarded if it touched the white pipe with its flipper.

Although they used similar training paradigms, Herman and Schusterman have differing interpretations of their experiments. Herman tends to have a more linguistic interpretation, whereas Schusterman doubts that this is appropriate. Herman et al. (1984) and Herman (1986, 1989) describe their three-command sequences as sentences, and they describe the positions within these sentences as grammatical categories. OBJECT1 is defined as a direct object, OBJECT2 as an indirect object, and TAKE as a verb. Schusterman and Gisiner (1986, 1988, 1989), on the other hand, argue that it is inappropriate to describe these experiments in linguistic terms. They argue that although one can train an animal to associate an action with a command, this is not enough to justify calling the action command a verb.

What exactly does it mean to argue that the animals understand commands as grammatical categories? Many modern linguists emphasize that human infants do not seem to be exhaustively trained to use language. Given the enormous diversity of speech environments to which different children are exposed, it is amazing that children exposed to a language come up with such similar rules to generate and understand sentences. Furthermore, a child mastering the past tense, may say "runned" instead of "ran," applying a rule to generate a word the child never heard. Pinker (1989, 1994) argues that this ability of children to create more structure than is present in what they hear provides evidence that language acquisition in humans involves specific innate learning mechanisms preprogrammed to interpret grammar correctly. Animals that do not have the same innate predispositions may only be able to acquire languagelike performance through extensive training. Many linguists would agree with Schusterman that grammatical categories only make sense as part of a much richer syntactic structure, developed with much less formal training.

Reference

When humans name something with a word, they understand the word to act as a symbol for the thing that has just been named. But what about animals trained to associate a cue with an object? Does the animal understand that the cue refers to the object, or has it simply learned that it will be rewarded if it performs a response when shown a stimulus?

Animals do not need language training to learn associations between a cue and a response. For example, pigeons can be trained to press a red button if shown one object, say a square, and a green button if shown a circle (Carter and Eckerman 1975). This has been called symbolic matching, but does the pigeon think of the red button as a symbol for square, or has it just been trained to do a clever trick for food? Does Herman's dolphin think of a sideways hand-motion as a name for pipe? How can we discriminate the simple conditioning from real understanding of reference?

If an animal can use a signal as a symbol for an object, then it must be able to use the symbol more flexibly than the specific context in which it was trained. All the pigeon trained in symbolic matching has to do to perform correctly is to learn one simple and specific rule. When the pigeon is in a training box and is shown a square, it has to peck the red button to get food. Language-trained dolphins and sea lions were asked to be much more flexible in their responses to symbols. For example, when given a symbol for an object that was not present, the animal clearly acted as if it had a search image for the requested object. If asked to fetch a pipe, when none was present, the animal would search the pool for much longer than if a pipe had been in the pool. The sea lions would often balk at performing a requested action if the requested object was not there, and the dolphins were trained to press a special paddle if a requested object was not in the pool. These animals also had a relatively abstract understanding of requested actions. One of the sea lions was only trained in the water to touch objects with its tail. When for the first time it was asked to tail touch on land, it did so, although the motor pattern for the action was very different for walking than swimming.

To use a signal as a symbol for an object, an animal must also be able to associate properties of the object with the symbol. For example, if you are asked what color a banana is, the word banana allows you to remember that the object is yellow, although there is nothing yellow about the word banana. Some critics of ape language research have questioned whether chimps use signs as symbols or just produce a sign to get food when they see an object. Savage-Rumbaugh (1986) has pursued these questions about referential communication in her work with chimpanzees. She points out that referential communication depends critically on how chimps are trained. Three chimps in her laboratory were trained to use buttons on a keyboard to request or to label foods and objects such as tools. However, two of the chimps, Sherman and Austin, were trained specifically in skills related to symbolic usage of the buttons, whereas the other one, Lana, had been trained for other purposes such as interpreting strings of button pushes. Savage-Rumbaugh et al. (1980) designed an experiment to test whether these different training histories influenced the ability of each chimp to associate the properties of an object with its symbol. All three of the chimps were able to sort foods and tools into two bins. Therefore, they had clearly formed concepts of the categories of food and tool. Savage-Rumbaugh then introduced two new signs, one for food and another for tools, and they trained the chimps to label cake, orange, and bread with the food sign and key, money, and stick with the tool sign. The chimps were then asked to label new items such as banana as a food or magnet as a tool. Sherman and Austin did this readily, but Lana could not. Although she had learned to make individual associations between each object and the reinforced sign, she had not learned to associate the tool or food signs with the general properties of the objects they represented. Because all three chimps were trained in exactly the same way for this task, it seems that the earlier training influenced their tendency to treat signs as symbols. The abilities of Sherman and Austin to use these signs as symbols were amply demonstrated by the ease with which they could also label signs for objects that had not been used in the training as tools or food. There was nothing foodlike or toollike about the signs. This labeling required the chimps when seeing a symbol to be able to refer to properties of the referent.

Problems Comparing Animal Communication to Human Language

The basic paradigm of animal language studies involves training animals to use a set of signs created by humans in ways that appear similar to some features of human language. However, there are clear differences between how the animals learn to perform these experiments and how human infants learn language. What the animals have actually learned in animal language experiments depends on how they were trained. This is very different from how humans develop language. Human infants are not formally trained as they develop language in the first few years of life. Many parents would have a difficult time teaching grammar even if asked to do so. It is remarkable that different speakers develop such similar understandings of grammar and word meaning, given the unstructured and variable exposure to speech that they receive as infants. These problems reflect a dilemma in animal language studies. The more the study matches naturalistic language acquisition, as with the chimps raised at home, the more difficult it is to quantify results. On the other hand, the more the study uses formal training and testing sessions, the less the experiment looks like language. This dilemma is particularly problematic for artificial language experiments in which animals undergo extensive training to respond to commands but have no chance for real two-way communication (Locke 1993).

Debates about whether an animal has "true" language have generated more heat than light. We simply do not understand enough about how humans learn and process linguistic information, and not enough about how trained animals process artificial signs, to be able to determine where animal and human performances differ or overlap. It is more useful to isolate specific issues such as whether animals use signs as symbols or can learn syntactic rules. Some of these questions do not require training but can also be studied in the natural communication systems of animals. One of the most important cognitive issues here is how animals learn the signals and the rules for using these signals in natural communication.

Imitation and Social Learning

This section concentrates on the role of learning in the natural behavior of marine mammals. Animal communication does not need to involve complex cognitive processes. The classic ethological view emphasizes how evolution can shape both signaler and receiver (Tinbergen 1951). Natural selection may shape genetically fixed signals (*key* or *sign stimuli*). When a receiver detects the signal, the appropriate response may be released by an *innate-releasing mechanism.* Many such responses are *fixed action patterns* that are also shaped by natural selection. It can be difficult to distinguish between such highly adapted responses and those resulting from intelligence and learning. In Darwin's (1871:39) terms:

> . . . we may easily underrate the mental powers of the higher animals, and especially of man, when we compare their actions founded on the memory of past events, on foresight, reason, and imagination, with exactly similar actions instinctively performed by the lower animals; in this latter case the capacity of performing such actions having been gained, step by step, through the variability of the mental organs and natural selection, without any conscious intelligence on the part of the animal during each successive generation

How are we to discriminate between communication involving relatively hard-wired connections between species-specific signals and responses from more open systems susceptible to learning by experience?

Vocal Imitation

One place to start is with the signals themselves. The vocalizations of nonhuman terrestrial mammals, including our primate relatives, appear to be only slightly modified by experience and look like classic, genetically fixed species-specific signals (Janik and Slater 1997). Even such drastic treatments as deafening at birth produce only minor modifications of vocal development in some terrestrial mammals, such as slightly longer mews in deafened kittens (Buchwald and Shipley 1985). The importance of genetic factors compared to auditory experience is suggested by studies of primate hybrids, whose calls blend features from both parents but match those of neither (Newman and Symmes 1982, Brockelmann and Schilling 1984). Primates raised in isolation or with foster mothers of a different species still produce species-typical vocalizations (Winter et al. 1973, Owren et al. 1993). The hybrid and cross-fostering results are particularly striking as these animals are constantly exposed to, and must learn to respond to, vocalizations that differ from the ones they themselves produce.

Much more striking effects of auditory input are reported for vocalizations from a variety of marine mammal species. A captive harbor seal (*Phoca vitulina*) was reported to imitate human speech with a New England accent (Ralls et al. 1985). Captive beluga whales are also reported to imitate human speech well enough for caretakers to "perceive these sounds as emphatic human conversation" (Ridgway et al. 1985). One beluga named "Logosi" was reported to produce clear imitations of his name (Eaton 1979). Captive bottlenose dolphins of both sexes are highly skilled at imitating both human-made pulsed sounds and whistles (Caldwell and Caldwell 1972b, Herman 1980); in fact, they may imitate sounds spontaneously within a few seconds after the first exposure (Herman 1980) or after only a few exposures (Reiss and McCowan 1993). Dolphins can also be trained using food and social reinforcement to imitate human-made whistlelike sounds (Evans 1967, Richards et al. 1984, Sigurdson 1993). The only other non-human animals with such highly developed skills of vocal imitation are the most accomplished avian mimics such as parrots and mynahs (West and King 1990).

Why have marine mammals evolved such unusual abilities to modify their vocalizations based on what they hear? As discussed earlier in this chapter, some baleen whales, like many songbirds (Marler 1970, Catchpole and Slater 1995), appear to use vocal learning to increase the complexity, and perhaps the attractiveness, of their reproductive advertisement songs. The rapid change of humpback song coupled with convergence of singers within a population can only be achieved with well-developed abilities of vocal learning. Vocal imitation appears to function in dolphins both in the development of individually distinctive whistles and in maintaining individual-specific relationships among adults (Tyack and Sayigh 1997). Therefore, there is a diversity of evolutionary functions for vocal imitation among marine mammals. In general, vocal imitation allows the development of remarkably open systems of vocal communication, in which adults and young can learn new signals with new associations.

One of the ironies of animal language research with marine mammals is that Lilly actually was one of the first to discover vocal imitation in dolphins (Lilly 1965). At that early stage of animal language research, Lilly jumped to the conclusion that, if animals were capable of vocal imitation, they must also have most other skills that humans use for language. It also turned out to be more difficult to train imitation of speech signals than other sounds that are more like the natural sounds dolphins produce. The more recent animal language studies with marine mammals selected a very controlled experimental approach in which the subjects primarily responded to human commands in part because of Lilly's overinterpretation of very limited data. Yet Savage-Rumbaugh (1986) emphasizes the need for animal language training to incorporate two-way communication to develop more languagelike performance. No animal language studies with dolphins have fully exploited their imitative abilities.

Motor Imitation

Marine mammals also produce novel postures and movement patterns in imitation of other animals, a skill called motor imitation. Several different trainers at different marine zoological parks have described how a dolphin that had only observed show behaviors of other dolphins would be able to perform the entire show flawlessly (Caldwell and Caldwell 1972a, Pryor 1973). Other examples of playful imitation include bottlenose dolphins imitating distinctive swimming, self-grooming, and sleeping behaviors of fur seals and postures and swimming behavior of animals as diverse as fish, turtles, and penguins (Tayler and Saayman 1973). Perhaps the most striking example of playful imitation concerned a bottlenose dolphin calf that would watch humans who had gathered to watch her through an underwater viewing window. One day the calf swam off immediately after a human exhaled a large cloud of tobacco smoke. She went directly to her mother, suckled, swam back to the window, and squirted a mouthful of milk toward the smoker (Tayler and Saayman 1973).

Motor imitation by observational learning is more complex than first meets the eye. For an animal to imitate the act of another, it must be able to relate parts of the body of the performer to the appropriate parts of its own body. It must then somehow map its perception of the movements and configuration of these parts onto motor commands to perform similar movements. These abilities are quite mysterious. How did the young calf map the mouth of the smoker to her own mouth? How did she come up with the idea of using milk to simulate in water the motion of smoke in air? Motor imitation remains controversial both because of lack of repeated experimental demonstrations and because the cognitive processes leading to the skill are so poorly understood. However, the adaptive significance of this skill is obvious for marine mammals.

Many mammals appear to use social learning to learn how to select and handle prey (Zentall and Galef 1988). Many marine mammals are generalists, feeding on a variety of prey in diverse habitats. Marine mammals show remarkable flexibility in learning how to exploit feeding opportunities created by human fisheries. Seals, dolphins, and whales are often reported to learn how to take fish off hooks or remove them from nets. Even when they feed on the same kinds of prey, wild bottlenose dolphins from different areas typically use different foraging techniques. For example, dolphins may catch fish by chasing them down and catching them in the mouth, by striking them with their tails, or by washing them out of the water on a mud flat (see Wells, Boness, and Rathbun, Chapter 8, this volume). Lopez and Lopez (1985) suggest that young killer whales learn how to strand on a beach to capture seals by imitating adults. Norris and Prescott (1961) report that two captive-born bottlenose dolphins were clumsy at handling large prey items that were handled with ease by two wild-caught animals. This difference led them to suggest that this handling behavior was learned. Novel feeding behaviors have also been reported to spread through humpback whale populations with a pattern suggesting that young learn the feeding method by observation (Weinrich et al. 1992).

Observation of both wild and captive marine mammals clearly demonstrates the importance of learning and imitation in the development of foraging behavior and communicative displays. This is a promising area for future research—conducting carefully controlled tests of whether social learning, especially observational learning, is involved in the acquisition of these behaviors.

Brain Size in Cetaceans and Possible Functions

The review of cognition in the last two sections has relied on data from the behavior of animals. Different kinds of insights can derive from training captive animals and from observing wild ones. Cognition is more than just behavior, however. Cognition is typically defined as information processing within the organism, which means within the nervous system. Neurobiologists have made great progress in using neurophysiological methods to study how sensory systems convert external stimuli into neural signals, how these signals are analyzed within the brain, and how this may lead to structured patterns of muscular activity to generate a behavioral response. Neuroethological investigations of information processing in animals have had great success in explaining, for example, how bats echolocate and how birds learn their songs. Unfortunately, we know much less about how marine mammals actually process information in the brain. The development of noninvasive techniques to image activity within the nervous system of alert animals offers hope for the future, but for now we must rely on more gross comparisons of brain size and anatomy.

Worthy and Hickie (1986) compared the relationship between brain size and body size in 648 species of mammals, and they derived an equation to define this relationship for all mammals. They found that pinnipeds had brain-to-body weight ratios that were very similar to those predicted from the equation calculated from all mammals. Sirenians, mysticetes, and the sperm whale had smaller brains than would be predicted from their body size, using the equation from all mammals. O'Shea and Reep (1990) suggest that the small brain-to-body weight ratios of sirenians are correlated with their low-quality diet and low metabolic rate. Sirenians are specialized to eat tropical sea grasses, a food that is abundant but low in quality. Grazing on these plants does not appear to require complex foraging strategies or sensory adaptations, but does require a relatively large gut size to process larger quantities of low-quality food. O'Shea and Reep (1990) suggest that the foraging pattern of sirenians selected for an increase in body size without a corresponding increase in brain size.

The baleen whales and sperm whales are much larger than any other mammals. Big terrestrial mammals must support their body mass, but the weight of a large whale is supported by the water, and this may free them from some mechanical constraints. These differences in size and mechanical constraints on the bodies of big whales may cause problems in extrapolating a brain-to-body relationship derived from animals tending to weigh around 1 kg to animals weighing tens of thousands of times more. For example, the size of the head in these large whales seems less constrained than in their terrestrial counterparts; the whale head is much larger than is typical of mammals, between one-quarter to one-third of the entire body length (Clarke 1978, Lockyer 1982). Animals with large blubber stores, such as baleen whales, may also require a correction in body size to account for this relatively inert tissue.

Primates and most other odontocete cetaceans have brains that are larger than predicted from their body size (Worthy and Hickie 1986, Marino 1998). Humans have the highest brain-to-body weight ratios among primates. It has been known for centuries that some dolphins with bodies a little larger than those of an adult human have brains roughly the same size. Adult humans with bodies weighing from 36 to 95 kg have brains weighing from 1100 to 1540 g (Jerison 1973). An adult dolphin of the genus *Lagenorhynchus* weighing 100 kg might have a brain weighing more than 1250 g (Ridgway and Brownson 1984). Adult bottlenose dolphins weigh more than humans, with a typical range cited in Ridgway and Brownson (1984) of 130 to 200 kg. Their brains weigh on the order of 1600 g, comparable to that of a human.

Brains this large are extremely rare, in part because brain tissue is metabolically expensive (Parker 1990). The 1400-g brain of a human weighing 70 kg can account for about 20% of the basal metabolic rate (Sokoloff 1977). This investment in large brains may slow growth and development significantly. Holliday (1978) extrapolates from data on children 5 years old or older to estimate that a newborn human infant may devote 87% of its basal metabolic rate to the brain. Holliday (1978) estimates that the brain of a 1-year-old may still consume more than half of the basal metabolism, and data from 5-year-olds indicates that their brains still consume as much as 44% of basal metabolism. Large developing brains drain resources that could otherwise increase the number of offspring of the mother or reduce the time between generations (Parker 1990). Compared to most other mammals, odontocete cetaceans have both larger brain-to-body weight ratios and slow reproductive rates. Cetaceans rarely have more than one young at a time, and gestation periods are nearly 1 year. Some species do not reach sexual maturity until 8 to 10 years or more, and the interval between calves may be more than 5 years in some species. This focus on the costs of large brains immediately raises questions about why dolphins have invested so heavily in brain tissue. What are the potential benefits that might outweigh these formidable costs?

Several different functions have been suggested for the large brains of some cetaceans. Many researchers focus on the well-developed auditory system, especially the need for rapid auditory processing of high frequency echolocation pulses (Wood and Evans 1980, Ridgway 1986, Worthy and Hickie 1986). Many of the nuclei involved in processing auditory information are highly enlarged in dolphins compared to other mammals. The auditory cortex also appears to cover large areas in the dolphin (Bullock et al. 1968, Bullock and Ridgway 1972), but the question of how the dolphin cortex is organized to process auditory information is much less well understood than in echolocating bats (e.g., Suga 1977) or in owls specialized for passive acoustic localization (Knudsen 1982). The extraordinary echolocation abilities of the bat are achieved with such radically smaller brains that I find it hard to believe that echolocation per se requires such a large investment in neural tissue by dolphins. Brain size within marine mammals also does not appear to correlate well with what we know of echolocation abilities. For example, the harbor porpoise (*Phocoena phocoena*) and platanistid river dolphins appear to have well-developed echolocation systems, but have smaller brains than delphinids of the same size (Jerison 1978, Marino 1998).

As we have seen vocal learning is another vocal/auditory skill that is well developed in marine mammals, but rare in other non-human mammals. As was true for echolocation, however, small birds achieve comparable vocal learning abilities with much smaller brains, and what we know of the vocal learning abilities among marine mammals again does not appear to correlate well with brain-to-body size ratios. For example, baleen whales show impressive vocal learning capabilities, such as is seen in humpback song, but have low brain-to-body ratios (Jerison 1978, Worthy and Hickie 1986). Harbor seals are also capable of excellent vocal imitation (Ralls et al. 1985), but have brain-to-body ratios typical of most mammals (Worthy and Hickie 1986).

We are only beginning to understand the auditory processing abilities of marine mammals, and it is possible that they achieve processing qualitatively more complex than that of the bat or bird (Worthy and Hickie 1986). Not even results from the animal language research, however, provide strong evidence for processing abilities that would justify such a large investment in neural tissue compared to other mammals. Sea lions with brain-to-body ratios typical of most mammals have a performance in animal language experiments that is similar to that of dolphins with much higher investments in neural tissue.

These kinds of concerns have led some to argue that the cetacean brain retains some primitive and inefficient structure, which means that cetaceans need large brains to do the same things that land mammals achieve with much smaller brains (Glezer et al. 1988). Fossils suggest that cetaceans have had enlarged brains for many tens of millions of years, however, much longer than hominids had equivalent brain sizes (Jerison 1978). The high energetic demands of these large brains would have created a strong selective pressure for changes to improve the so-called primitive structure of cetacean brains to render them more efficient. Although cetacean brains do differ from those of most terrestrial mammals, I find it hard to believe that cetaceans have been constrained to retain for tens of millions of years inefficiencies that demand such a large metabolic cost.

I have just reviewed suggestions that cetaceans have large brains either to accomplish particularly complex auditory processing or to make up for an inefficient and primitive brain. A very different perspective comes from primatologists who emphasize the social functions of intelligence (e. g., Humphrey 1976). Think of the cognitive demands of a game like chess where players can use their intelligence to come up with new strategies to beat an opponent. Players must also be on guard against their opponents coming up with new strategies. Such a game could create positive feedback for the evolution of more and more sophistication. This is fundamentally different from the evolution of a specific sensory ability such as echolocation or a motor ability such as learning to manipulate an object. Once a neural system is constructed that can achieve, for example, a certain speed and precision of auditory processing, there need not be any selection for a more complex system. Competition between conspecifics for innovation of complex behavioral strategies could, on the other hand, create an arms race for any improvements in the cognitive abilities used to generate these strategies.

Now imagine a species in which the young are dependent for a large fraction of their lives on care from parents and possibly from other members of their group. Suppose that adults depend on one another for social defense from predators and for efficient foraging. At the same time, animals within these groups compete for access to food, mates, and other important resources. Individuals that lose in a one-on-one competition may form an alliance with other individuals to improve their competitive advantage. This kind of society puts a premium on animals that can recognize other individuals, understand how they interact with others, and remember the history of interactions. Any individual that can modify how she or he interacts with another depending on the history of their interaction has a significant advantage in obtaining benefits from cooperation and reducing risks and losses from competition. The young animals can spend years honing their social skills for the times when they become necessary.

This kind of picture has been drawn of the evolution of human social behavior and has been related to the evolution of human social intelligence, called Machiavellian intelligence by Byrne and Whiten (1988), in honor of an Italian Renaissance master of political advice. Descriptions of the complexities of political intrigue are not limited to human primates; complex dynamics of cooperation and competition in chimps at a zoo led Frans de Waal (1982) to title a book "Chimpanzee Politics." Are there marine mammal species where social politics of this sort might have a significant enough impact on reproductive success to justify the investment? We do not know much about such fine-grained interactions in societies of marine mammals, but many more general parallels are obvious. Some odontocete cetaceans have extended parental care and rely on conspecifics for social defense from predators and for efficient foraging. Some species with fluid social groupings provide particularly clear evidence for individual-specific social bonds. For example, bottlenose dolphins have a very fluid fission–fusion society in which group composition only remains stable over intervals of minutes to an hour or so. Yet within these fluid groupings different individuals usually remain together. Male bottlenose dolphins form stable alliances with one or two other males to defend females from other males, and groups of two alliances may temporarily cooperate to compete against a third (Connor et al. 1992). Evidence from social communication, such as the signature whistles of dolphins, suggests the importance of individual recognition, not just between parents and offspring but between animals of both sexes at each stage of life. Both the social structure and these communication signals suggest the importance of individual-specific social relationships of the sort implicated in the evolution of social intelligence.

Discussion of the social functions of intelligence emphasize particular kinds of societies involving balancing cooperation and competition within individual-specific social relationships (Cheney et al. 1986, Byrne and Whiten 1988). Species such as the bottlenose dolphin, for which there is the best evidence of these relationships, also have some of the highest brain-to-body weight ratios (Ridgway and Brownson 1984). There is little evidence for strong individual-specific social relationships in species with the smallest brain-to-body weight ratios such as the sirenians and baleen whales (O'Shea and Reep 1990). Most discussions of the social functions of intelligence have been limited to primates. Marine mammals are phylogenetically distant from primates and evolved in such a different environment that they will make a particularly interesting comparison (Marino 1996).

The remarkable diversity of social behavior, life history, ecology, and patterns of communication among marine mammals make them excellent subjects for studying the evolution of communication and cognition. Further neurobiological and behavioral studies offer great promise for comparative studies on the evolution of large-brained animals, what cognitive processes these brains support, and how they function in the natural behavior of marine mammals. Broad comparisons of communication and cognition across diverse mammalian taxa will help us to understand the evolution of these traits in many species including our own.

Literature Cited

Airapetyants, E. Sh., and A. I. Konstantinov. 1973. Echolocation in Animals. Israel Program of Scientific Translations, Jerusalem.

Andersen, S. H. 1972. Auditory sensitivity of the harbour porpoise, *Phocoena phocoena*. Investigations on Cetacea 2:255–259.

Arnbom, T., V. Papastavrou, L. S. Weilgart, and H. Whitehead. 1987. Sperm whales react to an attack by killer whales. Journal of Mammalogy 68:450–453.

Au, W. 1993. The Sonar of Dolphins. Springer Verlag, New York, NY.

Baker, C. S., R. H. Lambertsen, M. T. Weinrich, J. Calambokidis, G. Early, and S. J. O'Brien. 1990. Molecular genetic identification of the sex of humpback whales (*Megaptera novaeangliae*). Pages 105–111 *in* A. R. Hoelzel (ed.). Genetic Ecology of Whales and Dolphins. Report of the International Whaling Commission (Special Issue 13). Cambridge, U.K.

Baker, M. C., and M. A. Cunningham. 1985. The biology of bird song dialects. Behavioral and Brain Sciences 8:85–133.

Balcombe, J. P. 1990. Vocal recognition of pups by mother Mexican free-tailed bats *Tadarida brasiliensis mexicana*. Animal Behavior 39:960–966.

Beer, C. G. 1970. Individual recognition of voice in the social behavior of birds. Pages 27–74 *in* J. S. Rosenblatt, C. G. Beer, and R. A. Hinde (eds.). Advances in the Study of Behavior, Vol. 3. Academic Press, New York, NY.

Best, P. B. 1979. Social organization in sperm whales, *Physeter macrocephalus*. Pages 227–289 *in* H. E. Winn and B. L. Olla (eds.). Behavior of Marine Animals, Vol. 3: Cetaceans. Plenum, New York, NY.

Bigg, M. A., G. M. Ellis, J. K. B. Ford, and K. C. Balcomb. 1987. Killer Whales—A Study of Their Identification, Genealogy and Natural History in British Columbia and Washington State. Phantom Press, Nanaimo, B.C.

Breathnach, A. S. 1960. The cetacean central nervous system. Biological Review 35:187–230.

Brockelman, W. Y., and D. Schilling. 1984. Inheritance of stereotyped gibbon calls. Nature 312:634–636.

Brodie, P. 1969. Duration of lactation in cetacea: An indicator of required learning? American Midland Naturalist 82:312–314.

Buchwald, J. S., and C. Shipley. 1985. A comparative model of infant cry. Pages 279–305 *in* B. M. Lester and C. F. Z. Boukydis (eds.). Infant Crying. Plenum, New York, NY.

Bullock T. H., and S. H. Ridgway. 1972. Evoked potentials in the central auditory system of alert porpoises to their own and artificial sounds. Journal of Neurobiology 3:79–99.

Bullock, T. H., and V. S. Gurevitch. 1979. Soviet literature on the nervous system and psychobiology of cetacea. International Review of Neurobiology 21:47–127.

Bullock, T. H., A. D. Grinnell, E. Ikezono, K. Kameda, Y. Katsuki, M. Nomoto, O. Sato, N. Suga, and K. Yanagisawa. 1968. Electrophysiological studies of central auditory mechanisms in cetaceans. Zeitschrift für vergleichende Physiologie 74:372–387.

Burdin, V. I., A. M. Reznik, V. M. Shornyakov, and A. G. Chupakov. 1975. Communication signals of the Black Sea bottlenose dolphin. Soviet Physics-Acoustics 20:314–318.

Byrne, R., and A. Whiten. 1988. Machiavellian Intelligence. Clarendon Press, Oxford.

Caldwell, D. K., and M. C. Caldwell. 1972a. The World of the Bottlenose Dolphin. Lippincott, Philadelphia, PA.

Caldwell, M. C., and D. K. Caldwell. 1965. Individualized whistle contours in bottlenosed dolphins (*Tursiops truncatus*). Science 207:434–435.

Caldwell, M. C., and D. K. Caldwell. 1966. Epimeletic (care-giving) behavior in Cetacea. Pages 755–789 *in* K. S. Norris (ed.). Whales, Dolphins and Porpoises. University of California, Los Angeles, CA.

Caldwell, M. C., and D. K. Caldwell. 1972b. Vocal mimicry in the whistle mode by an Atlantic bottlenosed dolphin. Cetology 9:1–8.

Caldwell, M. C., and D. K. Caldwell. 1979. The whistle of the Atlantic bottlenosed dolphin (*Tursiops truncatus*)—ontogeny. Pages 369–401 *in* H. E. Winn and B. L. Olla (eds.). Behavior of Marine Animals, Vol. 3: Cetaceans. Plenum Press, New York, NY.

Caldwell, M. C., D. K. Caldwell, and P. L. Tyack. 1990. A review of the signature whistle hypothesis for the Atlantic bottlenose dolphin, *Tursiops truncatus*. Pages 199–234 *in* S. Leatherwood and R. Reeves (eds.). The Bottlenose Dolphin: Recent Progress in Research. Academic Press, San Diego, CA.

Carter, D. E., and D. A. Eckerman. 1975. Symbolic matching by pigeons: Rate of learning complex discriminations predicted from simple discriminations. Science 187:662–664.

Catchpole, C. K. 1982. The evolution of bird sounds in relation to mating and spacing behavior. Pages 297–319 *in* D. E. Kroodsma and E. H. Miller (eds.). Acoustic Communication in Birds, Vol. 1: Production, Perception and Design Features of Sounds. Academic Press, New York, NY.

Catchpole, C. K., and P. J. B. Slater. 1995. Bird Song. Cambridge University Press, Cambridge, U.K.

Chapman, C. J., and A. D. Hawkins. 1973. A field study of hearing in the cod, *Gadus morhua* L. Journal of Comparative Physiology 85:147–167.

Cheney, D. L., R. M. Seyfarth, and B. B. Smuts. 1986. Social relationships and social cognition in nonhuman primates. Science 234:1361–1366.

Chittleborough, R. G. 1958. The breeding cycle of the female humpback whale, *Megaptera nodosa* (Bonnaterre). Australian Journal of Marine Freshwater Resources 9:1–18.

Chu, K., and P. Harcourt. 1986. Behavioral correlations with aberrant patterns in humpback whale songs. Behavioral Ecology and Sociobiology 19:309–312.

Clarke, M. R. 1978. Structure and proportions of the spermaceti organ in the sperm whale. Journal of the Marine Biological Association of the U.K. 58:1–17.

Cleator, H. J., I. Stirling, and T. G. Smith. 1989. Underwater vocalizations of the bearded seal (*Erignathus barbatus*). Canadian Journal of Zoology 67:1900–1910.

Connor, R. C., R. A. Smolker, and A. F. Richards. 1992. Aggressive herding of females by coalitions of male bottlenose dolphins (*Tursiops* sp.). Pages 415–443 *in* A. H. Harcourt and F. B. M. de Waal (eds.). Coalitions and Alliances in Humans and Other Animals. Oxford University Press, Oxford.

Costa, D. P. 1993. The secret life of marine mammals. Oceanography 6:120–128.

Cummings, W., and P. O. Thompson. 1971. Underwater sounds from the blue whale, *Balaenoptera musculus*. Journal of the Acoustical Society of America 50:1193–1198.

Darwin, C. 1871. The descent of man and selection in relation to sex. J. Murray, London.

Defran, R. H., and K. Pryor. 1980. The behavior and training of cetaceans in captivity. Pages 319–362 *in* L. M. Herman (ed.). Cetacean Behavior: Mechanisms and Functions. Wiley-Interscience, New York, NY.

de Waal, F. 1982. Chimpanzee Politics. Harper & Row, New York, NY.

Eaton, R. L. 1979. A beluga whale imitates human speech. Carnivore 2:22–23.

Edds, P. 1982. Vocalizations of the blue whale, *Balaenoptera musculus* in the St. Lawrence River. Journal of Mammalogy 62:345–347.

Ellison, W. T., C. W. Clark, and G. C. Bishop. 1987. Potential use of surface reverberation by bowhead whales, *Balaena mysticetus*, in under-ice navigation: Preliminary considerations. Report of the International Whaling Commission 37:329–332.

Evans, W. E. 1967. Vocalization among marine mammals. Pages 159–186 *in* W. N. Tavolga (ed.). Marine Bioacoustics, Vol. 2. Pergamon Press, Oxford.

Evans, W. E., and J. Bastian. 1969. Marine mammal communication: Social and ecological factors. Pages 425–475 *in* H. T. Andersen (ed.). The Biology of Marine Mammals. Academic Press, New York, NY.

Fay, F. H., G. C. Ray, and A. A. Kibal'chich. 1981. Time and location of mating and associated behavior of the Pacific Walrus, *Odobenus rosmarus divergens* Illiger. Pages 89–99 *in* F. H. Fay and G. A. Fedoseev (eds.). Soviet-American Cooperative Research on Marine Mammals. NOAA Technical Report NMFS 12, Vol. 1, Pinnipeds. Washington, D.C.

Fisher, R. A. 1958. The Genetical Theory of Natural Selection. Dover, New York, NY.

Fogden, S. C. L. 1971. Mother–young behaviour at grey seal breeding beaches. Journal of Zoology (London) 164:61–92.

Ford, J. K. B. 1989. Acoustic behavior of resident killer whales (*Orcinus orca*) off Vancouver Island, British Columbia. Canadian Journal of Zoology 67:727–745.

Ford, J. K. B. 1991. Vocal traditions among resident killer whales (*Orcinus orca*) in coastal waters of British Columbia. Canadian Journal of Zoology 69:1454–1483.

Gardner, H. 1983. Frames of mind: The theory of multiple intelligences. Basic Books, New York, NY.

Gardner, R. A., and B. T. Gardner. 1969. Teaching sign language to a chimpanzee. Science 165:664–667.

Gish, S. L. 1979. A quantitative description of two-way acoustic communication between captive Atlantic bottlenosed dolphins (*Tursiops truncatus* Montagu). Ph.D. thesis, University of California, Santa Cruz, CA.

Glezer, I. I., M. S. Jacobs, and P. J. Morgane. 1988. Implications of the "initial brain" concept for brain evolution in Cetacea. Behavioral and Brain Sciences 11:75–116.

Glockner, D. A. 1983. Determining the sex of humpback whales (*Megaptera novaeangliae*) in their natural environment. Pages 447–464 *in* R. Payne (ed.). Communication and Behavior of Whales. Westview Press, Boulder, CO.

Guinee, L. N., and K. Payne. 1988. Rhyme-like repetitions in songs of humpback whales. Ethology 79:295–306.

Guinee, L., K. Chu, and E. M. Dorsey. 1983. Changes over time in the songs of known individual humpback whales (*Megaptera novaeangliae*). Pages 59–80 *in* R. Payne (ed.). Communication and Behavior of Whales. Westview Press, Boulder, CO.

Hartman, D. S. 1979. Ecology and behavior of the manatee (*Trichechus manatus*) in Florida. American Society of Mammology Special Publication 5. Pittsburgh, PA.

Hayes, C. 1951. The Ape in our House. Harper, New York, NY.

Helweg, D. A., A. S. Frankel, J. R. Mobley Jr., and L. M. Herman. 1992. Humpback whale song: Our current understanding. Pages 459–483 *in* J. A. Thomas, R. Kastelein, and A. Ya. Supin (eds.). Marine Mammal Sensory Systems. Plenum, New York, NY.

Herman, L. M. 1980. Cognitive characteristics of dolphins. Pages 363–429 *in* L. M. Herman (ed.). Cetacean Behavior: Mechanisms and Functions. Wiley-Interscience, New York, NY.

Herman, L. M. 1986. Cognition and language competencies of bottlenosed dolphins. Pages 221–252 *in* R. J. Schusterman, J. A. Thomas, and F. G. Wood (eds.). Dolphin Cognition and Behavior: A Comparative Approach. Lawrence Erlbaum, Hillsdale, NJ.

Herman, L. M. 1989. In which procrustean bed does the sea lion sleep tonight? Psychological Record 39:19–42.

Herman, L. M., D. G. Richards, and J. P. Wolz. 1984. Comprehension of sentences by bottlenosed dolphins. Cognition 16:129–219.

Herman, L. M., J. R. Hovancik, J. D. Gory, and G. L. Bradshaw. 1989. Generalization of visual matching by a bottlenosed dolphin (*Tursiops truncatus*): Evidence for invariance of cognitive performance with visual and auditory materials. Journal of Experimental Psychology: Animal Behavior Processes 15:124–136.

Holliday, M. A. 1978. Body composition and energy needs during growth. Pages 181–196 *in* F. Faulkner and N. Tanner (eds.). Human Growth, Vol. 2: Postnatal Growth. Plenum, New York, NY.

Humphrey, N. K. 1976. The social functions of intelligence. Pages 303–317 *in* P. P. G. Bateson and R. A. Hinde (eds.). Growing Points in Ethology. Cambridge University Press, Cambridge, U.K.

Janik, V. M. 1997. Whistle matching in wild bottlenose dolphins. Journal of the Acoustical Society of America 101:31–36.

Janik, V. M., and P. J. B. Slater. 1997. Vocal learning in mammals. Advances in the Study of Behavior 26:59–99.

Janik, V. M., G. Denhardt, and D. Todt. 1994. Signature whistle variations in a bottlenosed dolphin, *Tursiops truncatus*. Behavioral Ecology and Sociobiology 35:243–248.

Jerison, H. J. 1973. Evolution of the Brain and Intelligence. Academic Press, New York, NY.

Jerison, H. J. 1978. Brain and intelligence in whales. Pages 159–197 *in* S. Frost (ed.). Whales and Whaling, Vol. 2. Australian Government Publication Service, Canberra.

Jerison, H. J. 1986. The evolutionary biology of intelligence: Afterthoughts. Pages 447–463 *in* H. J. Jerison and I. Jerison (eds.). Intelligence and Evolutionary Biology. Springer-Verlag, Berlin.

Johnson, C. M., and K. S. Norris. 1994. Social behavior. Pages 243–286 *in* K. S. Norris, B. Würsig, R. S. Wells, and M. Würsig (eds.). The Hawaiian Spinner Dolphin. University of California Press, Berkeley, CA.

Johnson C. S. 1966. Auditory thresholds of the bottlenose porpoise (*Tursiops truncatus,* Montagu). U.S. Naval Ordnance Test Station, Technical Publication 4178:1–28.

Kamminga, C., and A. F. van der Ree. 1976. Discrimination of solid and hollow spheres by *Tursiops truncatus* (Montagu). Aquatic Mammals 4:1–9.

Kamminga, C., and H. Wiersma. 1981. Investigations on cetacean sonar II. Acoustical similarities and differences in odontocete sonar signals. Aquatic Mammals 8:41–62.

Kastelein, R. A., and M. A. van Gaalen. 1988. The sensitivity of the vibrissae of a Pacific walrus (*Odobenus rosmarus* divergens), Part 1. Aquatic Mammals 14:123–133.

Kastelein, R. A., S. Stevens, and P. Mosterd. 1990. The tactile sensitivity of the mystacial vibrissae of a Pacific walrus (*Odobenus rosmarus* divergens). Part 2: Masking. Aquatic Mammals 16:78–87.

Kasuya, T., and H. Marsh. 1984. Life history and reproductive biology of the short-finned pilot whale, *Globicephala macrorhynchus,* off the Pacific coast of Japan. Reports of the International Whaling Commission (Special Issue 6):259–310.

Katona, S. K., B. Baxter, O. Brazier, S. Kraus, J. Perkins, and H. Whitehead. 1979. Identification of humpback whales by fluke photographs. Pages 33–44 *in* H. E. Winn and B. L. Olla (eds.). The Behavior of Marine Animals, Vol. 3: Cetaceans. Plenum, New York, NY.

Kellogg, W. N., and L. A. Kellogg. 1933. The Ape and the Child. McGraw-Hill, New York, NY.

Ketten, D. R. 1997. Structure and function in whale ears. Bioacoustics 8:103–135.

Knudsen, E. I. 1982. Auditory and visual maps of space in the optic tectum of the owl. Journal of Neuroscience 2:1177–1194.

Krebs, J. R., and R. Dawkins. 1984. Animal signals: Mind reading and manipulation. Pages 380–402 *in* J. R. Krebs and N. B. Davies (eds.). Behavioural Ecology: An Evolutionary Approach, 2nd ed. Blackwell Scientific Publications, Oxford.

Krebs, J. R., D. F. Sherry, S. D. Healy, V. H. Perry, and A. L. Vaccarino. 1986. Hippocampal specialization of food-storing birds. Proceedings of the National Academy of Science USA 86:1388–1392.

Kroodsma, D. K. 1982. Song repertoires: Problems in their definition and use. Pages 125–146 *in* D. K. Kroodsma and E. H. Miller (eds.). Acoustic Communication in Birds, Vol. II. Academic Press, New York, NY.

Kuznetsov, V. B. 1974. A method of studying chemoreception in the Black Sea bottlenose dolphin (*Tursiops truncatus*). Pages 147–153 *in* V. Ye. Sokolov (ed.). Morphology, Physiology and Acoustics of Marine Mammals (in Russian). Science Press, Moscow.

Kuznetsov V. B. 1979. Chemoreception in dolphins of the Black Sea: Afalines (*Tursiops truncatus* Montagu), common dolphins (*Delphinus delphis* L.) and porpoises (*Phocoena phocoena* L.). Doklady Akademii Nauk SSSR (in Russian) 249:1498–1500.

Lande, R., and S. J. Arnold. 1981. Evolution of mating preference and sexual dimorphism. Journal of Theoretical Biology 117:651–664.

Lieberman, P. 1984. The Biology and Evolution of Language. Harvard University Press, Cambridge, MA.

Lilly, J. C. 1962. Vocal behavior of the bottlenose dolphin. Proceedings of the American Philosophical Society 106:520–529.

Lilly, J. C. 1965. Vocal mimicry in *Tursiops:* Ability to match numbers and durations of human vocal bursts. Science 147:300–301.

Lilly, J. C. 1967. The Mind of the Dolphin: A Nonhuman Intelligence. Doubleday, New York, NY.

Lilly, J. C. 1975. Lilly on Dolphins. Doubleday, New York, NY.

Ljungblad, D. K., P. O. Thompson, and S. E. Moore. 1982. Underwater sounds recorded from migrating bowhead whales, *Balaena mysticetus,* in 1979. Journal of the Acoustical Society of America 71:477–482.

Locke, J. L. 1993. The Child's Path to Spoken Language. Harvard University Press, Cambridge, MA.

Lockyer, C. 1982. Growth and energy budgets of large baleen whales from the Southern Hemisphere. Food and Agricultural Organization Fisheries Service 5:379–488.

Loesche, P., P. K. Stoddard, B. J. Higgins, and M. D. Beecher. 1991. Signature versus perceptual adaptations for individual vocal recognition in swallows. Behaviour 118:15–25.

Lopez, J. C., and D. Lopez. 1985. Killer whales (*Orcinus orca*) of Patagonia and their behavior of intentional stranding while hunting nearshore. Journal of Mammalogy 66:181–183.

Madsen, C. J., and L. M. Herman. 1980. Social and ecological correlates of cetacean vision and visual appearance. Pages 101–147 *in* L. M. Herman (ed.). Cetacean Behavior: Mechanisms and Functions. Wiley-Interscience, New York, NY.

Marino, L. 1996. What can dolphins tell us about primate evolution? Evolutionary Anthropology 5:73–110.

Marino, L. 1998. A comparison of encephalization between odontocete cetaceans and anthropoid primates. Brain Behavioral Evolution 51:230–238.

Marler, P. 1970. A comparative approach to vocal learning: Song development in white-crowned sparrows. Journal of Comparative and Physiological Psychology Monograph 71:1–25.

McBride, A. F., and H. Kritzler. 1951. Observations on pregnancy, parturition, and postnatal behavior in the bottlenose dolphin. Journal of Mammalogy 32:251–266.

Medvin, M. B., and M. D. Beecher. 1986. Parent–offspring recognition in the barn swallow (*Hirundo rustica*). Animal Behavior 34:1627–1639.

Miller, D. E., and J. T. Emlen. 1975. Individual chick recognition and family integrity in the ring-billed gull. Behaviour 52:124–144.

Mobley, J. R., L. M. Herman, and A. S. Frankel. 1988. Responses of wintering humpback whales (*Megaptera novaeangliae*) to playback of recordings of winter and summer vocalizations and of synthetic sound. Behavioral Ecology and Sociobiology 23:211–223.

Møhl, B. 1968. Auditory sensitivity of the common seal in air and in water. The Journal of Auditory Research 8:27–38.

Moore, K. E., W. A. Watkins, and P. L. Tyack. 1993. Pattern similarity in shared codas from sperm whales (*Physeter catodon*). Marine Mammal Science 9:1–9.

Morgane P. J., and J. S. Jacobs. 1972. Comparative anatomy of the cetacean central nervous system. Pages 117–224 *in* R. J. Harrison (ed.). Functional Anatomy of Marine Mammals, Vol. 1. Academic Press, New York, NY.

Murchison, A. E. 1980. Detection range and range resolution of echolocating bottlenose porpoise (*Tursiops truncatus*). Pages 43–70 *in* R. G. Busnel and J. F. Fish (eds.). Animal Sonar Systems. NATO ASI Series A 28:43–70. Plenum, New York, NY.

Nachtigall, P. E. 1986. Vision, audition, and chemoreception in dolphins and other marine mammals. Pages 79–113 *in* R. J. Schusterman, J. A. Thomas, and F. G. Wood (eds.). Dolphin Cognition and Behavior: A Comparative Approach. Lawrence Erlbaum, Hillsdale, NJ.

Neuweiler, G., A. Link, G. Marimuthu, and R. Rübsamen. 1988. Detection of prey in echocluttering environments. Pages 613–618 *in* P. E. Nachtigall and P. W. B. Moore (eds.). Animal Sonar: Processes and Performance. Plenum, New York, NY.

Newman, J. D. 1985. The infant cry of primates. Pages 307–323 *in* B. M. Lester and C. F. Z. Boudykis (eds.). Infant Crying. Plenum, New York, NY.

Newman, J. D., and D. Symmes. 1982. Inheritance and experience in the acquisition of primate acoustic behavior. Pages 259–278 *in* C. Snowdon, C. H. Brown, and M. Peterson (eds.). Primate Communication. Cambridge University Press, Cambridge, U.K.

Nishiwaki, M. 1962. Aerial photographs show sperm whales' interesting habits. Norsk Hvalfangstidende 51:395–398.

Norris, K. S. 1969. The echolocation of marine mammals. Pages 391–423 in H. T. Andersen (ed.). The Biology of Marine Mammals. Academic Press, New York, NY.

Norris, K. S. 1991. Dolphin Days. Avon, New York, NY.

Norris, K. S., and T. P. Dohl. 1980. The structure and functions of cetacean schools. Pages 211–261 *in* L. M. Herman (ed.). Cetacean Behavior: Mechanisms and Functions. Wiley-Interscience, New York, NY.

Norris, K. S., and J. H. Prescott. 1961. Observations on Pacific Cetaceans of Californian and Mexican Waters. University of California Press, Los Angeles, CA.

Nowak, R. 1991. Senses involved in discrimination of merino ewes at close contact and from a distance by their newborn lambs. Animal Behavior 42:357–366.

O'Shea, T. J., and R. L. Reep. 1990. Encephalization quotients and life-history traits in the sirenia. Journal of Mammology 71:534–543.

Owren M. J., J. A. Dieter, R. M. Seyfarth, D. L. Cheney. 1993. Vocalizations of rhesus (*Macaca mulatta*) and Japanese (*Macaca fuscata*) macaques cross-fostered between species show evidence of only limited modification. Developmental Psychobiology 26(7):389–406.

Palsbøll, P. J., A. Vader, I. Bakke, and M. R. El-Gewely. 1992. Determination of gender in cetaceans by the polymerase chain reaction. Canadian Journal of Zoology 70:2166–2170.

Papastavrou, V., S. C. Smith, and H. Whitehead. 1989. Diving behaviour of the sperm whale, *Physeter macrocephalus,* off the Galapagos Islands. Canadian Journal of Zoology 67:839–846.

Parker, S. T. 1990. Why big brains are so rare: Energy costs of intelligence and brain size in anthropoid primates. Pages 129–154 *in* S. T. Parker and K. R. Gibson (eds.). "Language" and Intelligence in Monkeys and Apes: Comparative Developmental Perspectives. Cambridge University Press, Cambridge, U.K.

Patterson, F. G., and E. Linden. 1981. The education of Koko. Holt, Rinehart and Winston, New York, NY.

Payne, K., and R. Payne. 1985. Large scale changes over 19 years in the songs of humpback whales in Bermuda. Zeitschrift für Tierpsychologie 68:89–114.

Payne, K. B., P. Tyack, and R. S. Payne. 1983. Progressive changes in the songs of humpback whales. Pages 9–59 *in* R. Payne (ed.). Communication and Behavior of Whales. Westview Press, Boulder, CO.

Payne, R. S., and S. McVay. 1971. Songs of humpback whales. Science 173:583–597.

Payne, R. S., and D. Webb. 1971. Orientation by means of long range acoustic signalling in baleen whales. Annals of the New York Academy of Science 188:110–141.

Perrin, W. F., P. A. Akin, and J. V. Kashiwada. 1991. Geographic variation in external morphology of the spinner dolphin *Stenella longirostris* in the eastern Pacific and implications for conservation. Fishery Bulletin 89:411–428.

Petrinovich, L. 1974. Individual recognition of pup vocalization by northern elephant seal mothers. Zeitschrift für Tierpsychologie 34:304–312.

Pinker, S. 1989. Learnability and cognition. MIT Press, Cambridge, MA.

Pinker, S. 1994. The Language Instinct. Morrow, New York, NY.

Poggendorf, D. 1952. The absolute threshold of hearing of the bullhead (*Amiurus nebulosus*) and contributions to the physics of the Weberian apparatus of the ostariophysi. Pages 147–181 translated and reprinted *in* W. N. Tavolga (ed.). Sound Reception in Fishes. 1976. Hutchinson and Ross, Dowden, Stroudsburg, PA.

Pryor, K. 1973. Behavior and learning in porpoises and whales. Naturwissenschaften 60:412–420.

Pryor, K. 1991. Non-acoustic communication in small cetaceans: Glance, touch, position, gesture, and bubbling. Pages 537–544 *in* J. A. Thomas and R. A. Kastelein (eds.). Sensory Abilities of Cetaceans. Plenum, New York, NY.

Ralls, K., P. Fiorelli, and S. Gish. 1985. Vocalizations and vocal mimicry in captive harbor seals, *Phoca vitulina*. Canadian Journal of Zoology 63:1050–1056.

Ray C., W. A. Watkins, and J. J. Burns. 1969. The underwater song of *Erignathus* (Bearded Seal). Zoologica 54:79–83 + 3 plates.

Reiss, D., and B. McCowan. 1993. Spontaneous vocal mimicry and production by bottlenose dolphins (*Tursiops truncatus*): Evidence for vocal learning. Journal of Comparative Psychology 107:301–312.

Reiter, J., N. L. Stinson, and B. J. Le Boeuf. 1978. Northern elephant seal development: the transition from weaning to nutritional dependence. Behavioral Ecology and Sociobiology 3:337–367.

Richards, D. G., J. P. Wolz, and L. M. Herman. 1984. Vocal mimicry of computer-generated sounds and vocal labeling of objects by a bottlenosed dolphin, *Tursiops truncatus*. Journal of Comparative Physiology 98:10–28.

Ridgway, S. 1986. Physiological observations of dolphin brains. Pages 31–59 *in* R. J. Schusterman, J. A. Thomas, and F. G. Wood (eds.). Dolphin Cognition and Behavior: A Comparative Approach. Lawrence Erlbaum, Hillsdale, PA.

Ridgway, S. H., and R. H. Brownson. 1984. Relative brain sizes and cortical surface areas of odontocetes. Acta Zoologica Fennica 172:149–152.

Ridgway, S. H., D. A. Carder, and M. M. Jeffries. 1985. Another "talking" male white whale. Page 67 *in* Abstracts, Sixth Biennial Conference on the Biology of Marine Mammals. Vancouver, BC, Canada.

Roitblat, H. L. 1987. Introduction to Comparative Cognition. Freeman, New York, NY.

Sagan, C. 1980. Cosmos. Random House, New York, NY.

Samuels, A., M. Sevenich, T. Gifford, T. Sullivan, and J. Sustman. 1989. Gentle rubbing among bottlenose dolphins. Page 58 *in* Abstracts, Eighth Biennial Conference on the Biology of Marine Mammals. Asilomar, CA.

Savage-Rumbaugh, E. S. 1986. Ape Language from Conditioned Response to Symbol. Columbia University Press, New York, NY.

Savage-Rumbaugh, E. S., D. Rumbaugh, S. T. Smith, and J. Lawson. 1980. Reference: The linguistic essential. Science 210:922–925.

Sayigh, L. S. 1992. Development and functions of signature whistles of free-ranging bottlenose dolphins, *Tursiops truncatus*. Ph.D. thesis, MIT/WHOI Joint Program, WHOI 92–37. Woods Hole, MA.

Sayigh, L. S., P. L. Tyack, R. S. Wells, and M. D. Scott. 1990. Signature whistles of free-ranging bottlenose dolphins, *Tursiops truncatus:* Stability and mother–offspring comparisons. Behavioral Ecology and Sociobiology 26:247–260.

Sayigh, L. S., P. L. Tyack, R. S. Wells, M. D. Scott, and A. B. Irvine. 1995. Sex difference in whistle production in free-ranging bottlenose dolphins, *Tursiops truncatus*. Behavioral Ecology and Sociobiology 36:171–177.

Sayigh, L. S., P. L. Tyack, R. S. Wells, M. D. Scott, and A. B. Irvine. 1999. Individual recognition in wild bottlenose dolphins: A field test using playback experiments. Animal Behavior 57:41–50.

Schevill, W. E., W. A. Watkins, and C. Ray. 1966. Analysis of underwater *Odobenus* calls with remarks on the development and function of the pharyngeal pouches. Zoologica 51:103–106 + 5 plates and phonograph disk.

Schusterman, R. J., and R. Gisiner. 1986. Animal language research: Marine mammals reenter the controversy. Pages 319–350 *in* H. J. Jerison and I. Jerison (eds.). Intelligence and Evolutionary Biology. Springer-Verlag, Berlin.

Schusterman, R. J., and R. Gisiner. 1988. Artificial language comprehension in dolphins and sea lions: The essential skills. The Psychological Record 38:311–348.

Schusterman, R. J., and R. Gisiner. 1989. Please parse the sentence: Animal cognition in the Procrustean bed of linguistics. The Psychological Record 39:3–18.

Schusterman, R. J., and K. Krieger. 1984. California sea lions are capable of semantic comprehension. The Psychological Record 34:3–23.

Schusterman, R. J., and K. Krieger. 1986. Artificial language comprehension and size transposition by a California sea lion (*Zalophus californianus*). Journal of Comparative Physiology 100:348–355.

Schusterman, R. J., R. F. Balliet, and J. Nixon. 1972. Underwater audiogram of the California sea lion by the conditioned vocalization technique. Journal of the Experimental Analysis of Behavior 17:339–350.

Schusterman, R. J., E. B. Hanggi, and R. Gisiner. 1992. Acoustic signalling in mother–pup reunions, interspecies bonding, and affiliation by kinship in California sea lions (*Zalophus californianus*). Pages 533–551 *in* J. Thomas, R. A. Kastelein, and A. Ya. Supin (eds.). Marine Mammal Sensory Systems. Plenum, New York, NY.

Schwegler, H., and W. E. Crusio. 1995. Correlations between radial-maze learning and structural variations of septum and hippocampus in rodents. Behavioural Brain Research 67:29–41.

Sigurdson, J. 1993. Whistles as a communication medium. Pages 153–173 *in* H. Roitblat, L. Herman, and P. Nachtigall (eds.). Language and Communication: Comparative Perspectives. Lawrence Erlbaum, Hillsdale, NJ.

Simmons, J. A. 1973. The resolution of target range by echolocating bats. Journal of the Acoustical Society of America 54:157–173.

Simmons, J. A., D. J. Howell, and N. Suga. 1975. Information content of bat sonar echoes. American Scientist 63:204–215.

Sjare, B., and I. Stirling. 1993. The breeding behavior and mating system of walruses. Page 10 *in* Abstracts, Tenth Biennial Conference on the Biology of Marine Mammals, Galveston, TX.

Smolker, R. A., J. Mann, and B. B. Smuts. 1993. Use of signature whistles during separation and reunions by wild bottlenose dolphin mothers and infants. Behavioral Ecology and Sociobiology 33:393–402.

Sokoloff, L. 1977. Circulation and energy metabolism of the brain. Pages 388–413 *in* G. J. Siegel (ed.). Basic Neurochemistry. Little Brown, Boston, MA.

Spiesberger, J. L., and K. M. Fristrup. 1990. Passive localization of calling animals and sensing of their acoustic environment using acoustic tomography. American Naturalist 135:107–153.

Stirling, I. 1973. Vocalization in the ringed seal (*Phoca hispida*). Journal of the Fisheries Research Board of Canada 30:1592–1594.

Stoddard, P. K., and M. D. Beecher. 1983. Parental recognition of offspring in the cliff swallow. Auk 100:795–799.

Suga, N. 1977. Amplitude spectrum representation in the Doppler shifted CF processing area of the auditory cortex of the mustache bat. Science 196:64–67.

Tayler, C. K., and G. S. Saayman. 1973. Imitative behavior by Indian Ocean bottlenose dolphins (*Tursiops aduncus*) in captivity. Behaviour 44:286–298.

Terhune, J. M., M. E. Terhune, and K. Ronald. 1979. Location and recognition of pups by adult female harp seals. Applied Animal Ethology 5:375–380.

Terrace, H. S., L. A. Pettito, R. J. Sanders, and T. G. Bever. 1979. Can an ape create a sentence? Science 206:891–902.

Thomas, J. A., K. C. Zinnel, and L. M. Ferm. 1983. Analysis of Weddell seal (*Leptonychotes weddelli*) vocalizations using underwater playbacks. Canadian Journal of Zoology 61:1448–1456.

Thompson, T. J., H. E. Winn, and P. J. Perkins. 1979. Mysticete sounds. Pages 402–431 *in* H. E. Winn and B. L. Olla (eds.). Behavior of Marine Animals, Vol 3: Cetaceans. Plenum, New York, NY.

Tinbergen, N. 1951. The Study of Instinct. Oxford University Press, Oxford.

Trillmich, F. 1981. Mutual mother–pup recognition in Galápagos fur seals and sea lions: cues used and functional significance. Behaviour 78:21–42.

Tyack, P. 1981. Interactions between singing Hawaiian humpback whales and conspecifics nearby. Behavioral Ecology and Sociobiology 8:105–116.

Tyack, P. 1982. Humpback whales respond to sounds of their neighbors. Ph.D. thesis, Rockefeller University, New York, NY.

Tyack, P. 1983. Differential response of humpback whales to playbacks of song or social sounds. Behavioral Ecology and Sociobiology 13:49–55.

Tyack, P. 1986a. Population biology, social behavior, and communication in whales and dolphins. Trends in Ecology and Evolution 1:144–150.

Tyack, P. 1986b. Whistle repertoires of two bottlenosed dolphins, *Tursiops truncatus:* Mimicry of signature whistles? Behavioral Ecology and Sociobiology 18:251–257.

Tyack, P. 1993. Why ethology is necessary for the comparative study of language and communication. Pages 115–152 *in* H. Roitblat, L. Herman, and P. Nachtigall (eds.). Language and Communication: Comparative Perspectives. Lawrence Erlbaum, Hillsdale, NJ.

Tyack, P. L. 1997. Studying how cetaceans use sound to explore their environment. Perspectives in Ethology 12:251–297.

Tyack, P. 1998. Acoustic communication under the sea. Pages 163–220 *in* S. L. Hopp, M. J. Owren, and C. S. Evans (eds.). Animal Acoustic Communication. Springer Verlag, Heidelberg.

Tyack, P. L., and L. S. Sayigh. 1987. Vocal learning in cetaceans. Pages 208–233 *in* C. T. Snowdon and M. Hausberger (eds.). Social Influences on Vocal Development. Cambridge University Press, Cambridge, U.K.

Tyack, P., and H. Whitehead. 1983. Male competition in large groups of wintering humpback whales. Behaviour 83:132–154.

Urick, R.J. 1983. Principles of Underwater Sound. McGraw-Hill, New York, NY.

Verner, J. 1975. Complex song repertoire of male long-billed marsh wrens in eastern Washington. The Living Bird 14:263–300.

von Frisch, K. 1967. The Dance Language and Orientation of Bees. Harvard University Press, Cambridge, MA.

Watkins, W. A., and W. E. Schevill. 1977. Sperm whale codas. Journal of the Acoustical Society of America 62:1485–1490.

Watkins, W. A., W. E. Schevill, and P. B. Best. 1977. Underwater sounds of *Cephalorhynchus heavisidii* (Mammalia: Cetacea). Journal of Mammology 58:316–320.

Watkins, W. A., K. E. Moore, J. Sigurjonsson, D. Wartzok, and G. Notarbartolo di Sciara. 1984. Fin whale (*Balaenoptera physalus*) tracked by radio in the Irminger Sea. Rit Fiskideildar 8:1–14.

Watkins, W. A., P. Tyack, K. E. Moore, and J. E. Bird. 1987. The 20-Hz signals of finback whales (*Balaenoptera physalus*). Journal of the Acoustical Society of America 82:1901–1912.

Watkins, W. A., M. A. Daher, K. M. Fristrup, T. J. Howald, and G. N. di Sciara. 1993. Sperm whales tagged with transponders and tracked underwater by sonar. Marine Mammal Science 9:55–67.

Weilgart, L., and H. Whitehead. 1988. Distinctive vocalizations from mature male sperm whales (*Physeter macrocephalus*). Canadian Journal of Zoology 66:1931–1937.

Weilgart, L., and H. Whitehead. 1993. Coda vocalizations in sperm whales (*Physeter macrocephalus*) off the Galapagos Islands. Canadian Journal of Zoology 71:744–752.

Weilgart, L. S., and H. Whitehead. 1997. Group-specific dialects and geographical variation in South Pacific sperm whales. Behavioral Ecology and Sociobiology 40:277–285.

Weinrich, M. T., M. R. Schilling, and C. R. Belt. 1992. Evidence for acquisition of a novel feeding behavior: Lobtail feeding in humpback whales. Animal Behavior 44:1059–1072.

Wells, R. S., M. D. Scott, and A. B. Irvine. 1987. The social structure of free-ranging bottlenose dolphins. Current Mammalogy 1:247–305.

West, M. J., and A. P. King. 1990. Mozart's starling. American Scientist 78:106–114.

Whitehead, H. 1996. Babysittong, dive synchrony, and indications of alloparental care in sperm whales. Behavioral Ecology and Sociobiology 38:237–244.

Whitehead, H., and B. Kahn. 1992. Temporal and geographical variation in the social structure of female sperm whales. Canadian Journal of Zoology 70:2145–2149.

Whitehead, H., and L. Weilgart. 1991. Patterns of visually observable behaviour and vocalizations in groups of female sperm whales. Behaviour 118:275–296.

Wilson, E. O. 1975. Sociobiology. Harvard University Press, Cambridge, MA.

Winn, H. E., and L. K. Winn. 1978. The song of the humpback whale *Megaptera novaeangliae* in the West Indies. Marine Biology 47:97–114.

Winter, P., P. Handley, D. Ploog, and D. Schott. 1973. Ontogeny of squirrel monkey calls under normal conditions and under acoustic isolation. Behaviour 47:230–239.

Wood, F. G., and W. E. Evans. 1980. Adaptiveness and ecology of echolocation in toothed whales. Pages 381–425 *in* R.-G. Busnel and J. Fish (eds.). Animal Sonar Systems. Plenum, New York, NY.

Worthington, L. V., and W. E. Schevill. 1957. Underwater sounds heard from sperm whales. Nature 180:291.

Worthy G. A. J., and J. P. Hickie. 1986. Relative brain size in marine mammals. American Naturalist 128:445–459.

Würsig, B. 1978. Occurrence and group organization of Atlantic bottlenose porpoises (*Tursiops truncatus*) in an Argentine Bay. The Biological Bulletin 154:348–359.

Zahavi, A. 1975. Mate selection—A selection for the handicap. Journal of Theoretical Biology 53:205–214.

Zentall, T. R., and B. G. Galef. 1988. Social Learning: Psychological and Biological Perspectives. Lawrence Erlbaum, Hillsdale, NJ.

8

RANDALL S. WELLS, DARYL J. BONESS, AND GALEN B. RATHBUN

Behavior

Behavior is an important component of the adaptive complex that allows animals to function (Bartholomew 1970). The behavior of animals results from interactions between the environment and the genetic make-up of individuals. Genetics, as expressed through characteristics such as morphology and physiology, both defines and is modified by behavior, which varies in response to environmental variables. How effective an organism is at applying its physical phenotypic features toward survival and the production of offspring is mediated in part by behavior. Yet, behavior, too, is influenced by selection pressures from the environment. As a demonstration of this, one might expect to find a similar, or greater, range of variation in behavior patterns as compared to the range in variation in morphology or physiology across closely related species. For example, the morphology or color pattern of an individual may have evolved concurrently with the behavior of the organism. One difference between behavior and other features is in the plasticity of behavior, especially in higher vertebrates such as marine mammals. Many physical features are a permanent part of the make-up of the individual, and can change only from generation to generation. The ability of behavior to vary within the life span of an individual, however, provides the individual with additional means to modify its life to respond to changing environmental characteristics or to optimize variability in other, less flexible, phenotypic features. Thus, behavioral flexibility may increase an individual's ability to survive and reproduce under conditions that approach the limits of more fixed phenotypic features.

The environment strongly influences the behavior of individuals, within the physical constraints of the animals' morphology and physiology. As such, ecological features may be identified as having a strong role in the development of the behavior patterns expressed by animals. Marine mammals are no exception to this pattern. As with other vertebrates, marine mammals must find sufficient food, survive a variety of environmental challenges, contend with predators, encounter and gain access to mates, and successfully rear offspring. Because these same factors affect terrestrial and marine mammals alike, there are many similarities in the behavior patterns of these animals.

Available information on the behavior of marine mammals is still very limited; many species have yet to be studied systematically. Yet, the integration of behavior (an organism's actions) and ecology (the interactions of an organism with its environment) is important for understanding the lives of marine mammals. This chapter uses the approach of behavioral ecology to identify and illustrate general patterns and possible functions of marine mammal behavior. We present these patterns as a point of departure for future in vestigations—to identify critical uncertainties and testable hypotheses for students of marine mammal science.

Some of the marine mammals, such as the pinnipeds and sea otters, have been better studied than the sirenians and cetaceans, because the semiaquatic nature of the former has provided increased accessibility for human observers. For many of the patterns we describe, there will be exceptions or gaps in our knowledge. It is important to recognize that the exceptions and unknowns provide much of the richness and range of variability that has helped marine mammals to survive for millions of years. Questions about the behavioral ecology of marine mammals continue to engage the interest of scientists and the public, and provide fruitful directions for future research.

Efforts to relate ecological variables to the behavior of terrestrial mammals have led to the development of matrices (tabular arangements of data on social structure relative to ecological factors) that may indicate trends in evolutionary frameworks of social organizations (e.g., Bartholomew 1970, Estes 1974, Geist 1974, Jarman 1974, Crook et al. 1976). Such matricies have facilitated the formulation of hypotheses. These researchers have noted a general correlation between ecological parameters and the type of social organization exhibited by a species. For example, species of African bovids that live in closed habitats tend to be solitary, small, and sedentary, whereas species in open habitats tend to be more gregarious, larger, wide-ranging, and cursorial (Estes 1974). Jarman (1974) identified correlations between antelope body size and food type and distribution, feeding patterns, group size, and methods of predator avoidance. We hope that the efforts of current and prospective marine mammal scientists soon will be able to advance such integrated schemes beyond the few preliminary attempts to date (e.g., Wells et al. 1980, Stirling 1983, Boness and Bowen 1996).

Many large mammals are long-lived. The variation and evolution of the behavior of long-lived mammals can be understood best if behavior is examined within the context of different stages of development. Ultimately, behavior must be related to reproductive success and fitness. Longitudinal studies of specific populations or social units are important in this regard. Much of what we have learned over the years about the behavior of chimpanzees (e.g., Goodall 1986), baboons (e.g., Altmann et al. 1988), vervet monkeys (e.g., Cheney et al. 1988), elephants (e.g., Douglas-Hamilton and Douglas-Hamilton 1975, Poole 1989), and lions (e.g., Packer et al. 1988), for example, have come from pioneering studies that have continued for decades. Such long-term studies of marine mammals have been few, but several have spanned at least two decades. Examples include longitudinal studies of northern elephant seals (e.g., Le Boeuf and Reiter 1988), southern right whales (e.g. Payne 1995), killer whales (e.g., Balcomb et al. 1982, Bigg et al. 1987), spinner dolphins (Norris et al. 1994), bottlenose dolphins (Scott et al. 1990a, Wells 1991a), and West Indian manatees (O'Shea and Hartley 1995, Rathbun et al. 1995).

One of the basic necessities for gathering information on the behavior of vertebrates is to reliably identify individuals (including determining their sex, age, genetic relationships, and reproductive status), and then be able to follow these individuals and gather behavioral and environmental information through time. With many terrestrial animals the techniques to identify individuals are well-established and relatively simple to implement. For example, birds are easily color-banded on their legs, small mammals can be ear tagged, and many large mammals (rhinoceroses, zebras, giraffes, lions, etc.) have individually distinctive pelage, skin, or horn patterns (Bookhout 1994). Many of these animals are relatively easy to observe with the aid of binoculars or a spotting scope. The behavior of marine mammals, however, has proven to be exceptionally difficult to study because of the difficulty in recognizing individuals; it is nearly impossible to follow many marine species in time and space, and in most cases we are unable to effectively see underwater. In addition, the sleek body forms of most marine mammals provide few straightforward attachment points for tags. The long and often frustrating process of developing new techniques and applying existing technology for the study of marine mammal behavior is well documented for the Sirenia.

In the early 1950s, Joseph Moore (1956) used unique deformities and scars on manatees to recognize individuals. He recorded these features and the behaviors of his manatees as they swam under a car bridge over the Miami River in Florida. In the late 1960s, Daniel Hartman (1979) studied manatees in the clear, spring-fed waters of Crystal River, Florida, and in the mid-1970s, John Reynolds (1981a) similarly worked on manatees in Blue Lagoon in Miami; both researchers further developed the technique by making sketches of the distinctive marks of known animals. The next development was the use of 35-mm underwater photography to document more accurately and in greater detail the subtle distinctions and changes through time of manatee scars (Powell and Rathbun 1984). Finally, in the 1990s, advanced technologies made possible by the boom in electronic and digital technology have allowed scar patterns to be saved, cataloged, and searched with the aid of laser disks and computers (Beck and Reid 1995). However, even with an efficient means of identifying individuals, it has proven difficult to follow manatees in the wild (Reid et al. 1991). This obstacle has been overcome by attaching radio transmitters to individual manatees. Each radio transmits a pulsed signal on a unique frequency that allows biologists to locate each transmitter (and animal, assuming the two are still attached). Directional antennas mounted on boats, cars, or airplanes, and triangulation methods are used to locate the radio-tags (Kenward 1987). The spatial information, plus any associated behavioral and ecological data, can then be cataloged and analyzed using geographical information systems (GIS). The Florida Department of Environmental Protection has developed an excellent GIS system that incorporates manatee spatial data (Reynolds and Haddad 1990).

The methods of attaching radio-tags to manatees, like the development of individual visual identification methods, have followed a slow but steady path to success. Early attempts to attach tags with belts and sutures were not successful (Hartman 1979, Irvine and Scott 1984). The first breakthrough came with the development of a peduncle (tail stock) belt by Diana Magor in Brazil and John Bengtson (1981) in Florida. However, radios mounted on peduncle belts were only effective in freshwater (salt water completely attenuates radio signals), which restricted the use of this tag to the St. Johns River. The next breakthrough came with the development of a floating radio-tag that was tethered to a peduncle belt (Rathbun et al. 1987). This attachment allowed biologists to radio-track manatees into salt water. Through the research efforts of Bruce Mate, manatees in Florida became the first marine mammals to be tracked successfully with satellite-monitored transmitters (Mate et al. 1986). After the successes in Florida, radio-tracking technology was transferred to dugongs in Australia (Marsh and Rathbun 1990). Now, radio-tracking sirenians has become a well-established technique (Reid and O'Shea 1989), allowing detailed information on habitat use, travel corridors, and seasonal movement patterns to be documented (Preen 1992; Reid et al. 1995, Deutsch et al. 1998).

Technology is always advancing and providing new and exciting methods of marking and following animals. Several years ago, Jeanette Thomas and her colleagues were among the first biologists to use passive integrated transponders (PITs) to permanently mark animals—sea otters in this case (Thomas et al. 1987). These electronic chips, the size of a rice grain, are now implanted subcutaneously (under the skin) in all sea otters that are captured. When recaptured, a special electronic wand is used to activate the transponder, which transmits back a unique alphanumeric (letter and number) code, thus providing biologists with an unequivocal identification of the individual—even if it is found on a beach and is nothing more than a badly decomposed carcass.

Marine mammals, perhaps more than any other group of vertebrates, have benefited greatly from technological advances in study methods. Many different types of identification, tracking, and data acquisition techniques have been developed, and many of these will be discussed where appropriate in this chapter. However, there are still numerous challenges awaiting future biologists because each species presents unique questions that will require new methods to be answered. Perhaps more than in any other group of vertebrates, marine mammalogists must be exceptionally innovative in developing and applying technology—and exceedingly patient to gather and interpret the results.

Some marine mammals are inherently easier to study and observe than others, and thus, more information has been gathered from these species compared to others. Pinnipeds, which all use the shore (or sea ice) during at least part of their life cycles, are probably the best known behaviorally, followed by the marine otters, which are restricted to near-shore areas where they are relatively easy to observe. The most difficult groups on which to gather behavioral data are the cetaceans and sirenians. Although electronic technology has enabled biologists to gather information on some of the most elusive marine mammals from the most remote areas of our planet, the application of innovative techniques has probably had the greatest impact on research of those species that present the least cumbersome logistical obstacles to biologist—the pinnipeds. Because some members of this group are comparatively well known, they are good subjects on which to begin our discussion of behavior. They illustrate nicely many of the concepts in behavioral ecology that apply, or are likely to apply, to the other groups of marine mammals. Therefore, we begin our review with the pinnipeds, progress through the sea otter and cetaceans, and end with the sirenians.

Pinnipeds [D. J. B.]

Pinnipeds appear to be relatively simple marine mammals behaviorally. They do not form highly structured social groups, do not have large numbers of predators or elaborate antipredator behavior, and reproduction does not generally involve complex communication and behavioral relationships. Their behavior has been strongly influenced by their evolutionary history as a semiaquatic marine mammal, whereby they feed at sea but reproduce on land or sea ice. Unlike the cetaceans and sirenians, and to a greater degree than the sea otters, pinnipeds are still strongly attached to using a solid substrate (land or sea ice; hereafter both habitats will be referred to simply as land or terrestrial) for some aspects of their life, particularly reproduction and molting. Among the first researchers to have clearly established the importance of an adaptive complex associated with marine feeding and terrestrial reproduction, molting, and resting were Nutting (1891), Bertram (1940), Peterson (1968), Bartholomew (1970), and Stirling (1975). The focus of these first efforts was the evolution of mating systems, but in recent years greater emphasis has been given to the adaptive complex and parental care and foraging behavior (Bonner 1984; Gentry and Kooyman 1986; Oftedal et al. 1987a; Costa 1991a, 1993; Boness and Bowen 1996).

Most pinnipeds exhibit an annual cycle of behavior that includes a very distinct and usually brief (relative to most other mammals) birthing and mating period (see Boyd, Lockyer, and Marsh, Chapter 6, this volume). It is during this time that females and males aggregate on oceanic islands, isolated mainland beaches, seasonal floe ice, or fast ice that may persist for years. The extent to which individual animals aggregate on land or ice outside of this breeding time is highly variable among species, although the different families show somewhat different trends. The phocid seals usually abandon breeding sites entirely to feed at sea. They return later to the same locations or other areas and form molting aggregations. The otariids do not show such clear abandonment of breeding sites. In most species male otariids disperse from breeding grounds after females have been mated, but females only make temporary foraging trips to sea and continue to return for various lengths of time to nurse their pups that are left behind on the breeding beaches. In some species the beaches are occupied by females and their offspring year-round, although numbers decline just before the beginning of a new breeding season (e.g., Miller 1975a, Kerley 1983, Trillmich et al. 1986, Heath et al. 1991, Higgins 1993). In other species, after 4 to 10 months the weaned juveniles and females also disperse, and the females return at the beginning of the next birthing season (e.g., Gentry 1970, Doidge et al. 1986, Gentry and Holt 1986, Francis et al. 1998).

For the most part, seals at sea are foraging. An exception to this involves those species that mate at sea. Like their terrestrially breeding counterparts, there is a distinct and brief mating period. Interestingly, a commonly espoused view is that seals are relatively solitary at sea, and yet unlike the case for cetaceans or sirenians, few researchers have attempted to follow seals once they depart from land (e.g., Loughlin et al. 1987, Hammond et al. 1993). Consequently, there is little direct observation of the behavior of pinnipeds during this part of their annual cycle. Our knowledge of foraging behavior and feeding habits comes primarily from studies that have used various techniques such as time–depth microprocessors (Kooyman et al. 1983, Hill 1986, DeLong et al. 1992), radio-telemetry, satellite-linked transmitters (Stewart et al. 1989, McConnell et al. 1992a), isotope dilution techniques (Costa 1987, Oftedal and Iverson 1987), scat analyses (Prime and Hammond 1987), and fatty acid (Iverson 1993, Iverson et al. 1998) and stable isotope (Wada et al. 1991) methods, all of which provide indirect views into the behavior of animals at sea.

Health and Maintenance Behaviors

Pinnipeds exhibit little in the way of maintenance behaviors. Probably the most apparent is behavior associated with thermoregulation. In contrast to what is most often seen as the typical problem of thermoregulation, many seals must actively seek to keep cool, not warm. As is true of all marine mammals, pinnipeds are physiologically and morphologically well equipped to handle the high thermal conductivity of an aquatic medium (Irving 1969; Whittow 1987; Trites 1990; see Pabst, Rommel, and McLellan, Chapter 2, and Elsner, Chapter 3, this volume). This has been accomplished through the evolution of small surface area-to-volume ratios, a thick subcutaneous layer of blubber, and in some species, the presence of an underfur protected by guard hairs (Ling 1970). These physical adaptations for conserving body heat in water often cause overheating problems on land, especially for the otariid and phocid species in subtropical and tropical climes. As there are limited physical adaptations for dissipating heat (see Matsura and Whittow 1974, Whittow et al. 1975, Riedman 1990), we see clear evidence of behavioral thermoregulation.

In many otariids during the breeding season, females shift their positions on the beach during the day so as to remain on the wetted surfaces associated with the tidal wash (Peterson and Bartholomew 1967, Stirling 1970, Gentry 1973, Odell 1974, Trillmich and Majluf 1981, Campagna and Le Boeuf 1988a, Heath 1989). In some instances, where temperatures are extreme or shorelines are not suitable places to locate, females enter the ocean or tide pools to keep cool (Walker and Ling 1981; Campagna and Le Boeuf 1988b; Heath 1989; Francis and Boness 1991; Carey 1991, 1992). For example, Francis and Boness (1991) showed that Juan Fernández fur seal (*Arctocephalus philippii*) females move to the water en masse when temperatures exceed 18°C (see Bonnot 1928, Odell 1974). Even otariid males, which defend land territories encompassing females, risk loss of potential copulations and even injury by passing through neighboring territories to move to and from the water during the peak temperatures of the day (Rand 1967, Francis and Boness 1991). At other locations where midday temperatures are high and beaches consist of large boulders, males and females avoid high solar radiation by seeking shelter in the shade of boulders (Miller 1975b, Trillmich 1984). In elegant studies in which the rookery topography was experimentally manipulated, Carey (1991, 1992) demonstrated the importance of tide pools and shade areas in determining the distribution of males and females on the rookery in the New Zealand fur seal (*Arctocephalus forsteri*).

There are relatively few phocids that exist in high temperature climates. The Hawaiian monk seal (*Monachus schauinslandi*) is the only extant species to be found in the subtropics. As with the otariids, both male and female monk seals enter the water when temperatures become extreme (34°C). However, monk seals also use intermediate behavioral means to thermoregulate. This may be important because monk seals are faced with significant predation pressure from tiger sharks (*Galeocerdo cuvier*). As with otariids, monk seals move up and down the beach slope from dry to wet sand in conjunction with the changing temperatures. Seals also alter their body postures to expose a previously moist part of the body to the air for evaporative cooling, and, more often than expected by chance, display the lighter underside to the sun whereby the seal might reduce solar heat gain (Whittow 1978). Monk seals and northern and southern elephant seals (*Mirounga angustirostris* and *M. leonina,* respectively) on sandy substrates dig in sand and often flip the sand on their backs (Whittow 1978, McCann 1983, Riedman 1990). This creates a cool, moist surface for the seal to lie on, and provides a further cooling effect from the moist sand on the back.

Activity budgets of seals reveal that individuals spend time self-grooming when on land (Stirling 1971, Crawley et al. 1978, Boness 1984, Anderson and Harwood 1985, Beentjes 1989). This behavior does not appear to have a social context as does mutual grooming as seen in many primates (e.g., Kurland 1977, Dunbar 1984, Andelman 1986) or other social animals (e.g., Armitage and Johns 1982, Rubenstein 1986). Nor is there evidence to suggest that grooming is a displacement activity associated with conflict (see Hinde 1970). With sea otters (*Enhydra lutris*), grooming is critical in keeping the pelage clean and the underfur dry, and thus accounts for a significant amount of time (up to 16% in the wild and 48% in captivity) (Kenyon 1969, Riedman and Estes 1990). Although the hair and underfur of fur seals is somewhat comparable to that of sea otters, the amount of time spent grooming by fur seals and other seals is minimal by comparison (e.g., 1.2% to 1.7% of time in New Zealand fur seals; Crawley et al. 1978). Furthermore, grooming activity occurs to a similar extent in some phocids and sea lions (e.g., 1.2% to 2.6% in Hooker's sea lion [*Phocarctos hookeri*]; Beentjes 1989). It is likely that even among the fur seals grooming serves more to keep the coat clean and the underfur dry. As Kenyon (1969) suggests, grooming in many animals may be a response to parasites such as lice. Many seals are infested with lice, and it may be that grooming is "scratching" in response to these parasites. In addition, the increased frequency of grooming in New Zealand fur seals during the summer might suggest that grooming could be a response to skin irritation during intense solar radiation (Johnstone and Davis 1987).

Foraging Patterns

Foraging behavior and feeding habits of marine mammals are shaped by the nature of the animals' physiological and physical adaptations to diving, their ecology (behavior, distribution, and abundance), and the ecology of their prey items. The physiological and physical adaptations are described by Elsner (Chapter 3, this volume) and will be mentioned here only briefly. Pinnipeds, in contrast to cetaceans, should be viewed as divers not surfacers; that is, they spend most of their time at the surface and dive to obtain their food (Kooyman 1989). Possible exception, as will be described later, are the two elephant seal species. Generally, the phocids are better adapted to diving deeper and for longer periods of time. Phocids can store more oxygen per body size (60 g of O_2/kg) than otariids (40 g of O_2/kg) because they have a larger blood volume and larger myoglobin content in the muscles (Kooyman 1985). The aerobic dive limit (ADL), defined as the maximum breath-hold that is possible without any increase in blood lactic acid, is dependent on a variety of factors, such as oxygen stores, swim velocity, and metabolic rate, and consequently, ADL relates to body size. Therefore, the larger phocid seals should be able to remain underwater for much longer periods than the smaller otariids. Calculated ADL for otariids are in the range of 3 to 6 min and for phocids 9 to 30 min (Kooyman 1989; Hindell et al. 1992; Lydersen et al. 1992; Costa 1993; Fedak and Thompson 1993; Ponganis et al. 1993; Elsner, Chapter 3, this volume).

Until very recently, foraging behavior could not be studied in any detail because of an inability to follow animals at sea. The development of the time–depth recorder (TDR) (Kooyman et al. 1976, 1983; Hill 1986) and the satellite-linked transmitter (Stewart et al. 1989; McConnell et al. 1992a,b; Bengtson et al. 1993) have provided the opportunity to infer foraging activity from diving behavior. In the past decade such studies have flourished, although the majority have focused on female otariids, primarily to understand the ecology and evolution of maternal strategies. Otariid females have an extended lactation period that is linked fundamentally to a maternal foraging cycle. In contrast, most phocid females fast during a short lactation period (see Costa and Williams, Chapter 5, this volume). These differing strategies are discussed in more detail later and are raised here only to explain why our knowledge of foraging habits is so skewed with respect to the different pinniped families and the different sexes. More recent studies of diving and foraging behavior are emerging for phocids, and include studies of both males and females (Hindell et al. 1991, Thompson et al. 1991, Le Boeuf et al. 1993, Lydersen and Kovacs 1993, Stewart and DeLong 1993, Boness et al. 1994, Le Boeuf 1994, Slip et al. 1994, Testa 1994, Folkow and Blix 1995, Coltman et al. 1998).

The use of diving patterns to infer foraging behavior is constantly being improved. In the early days of TDRs only a brief 2- to 3-day record was possible. Current TDRs and satellite-linked transmitters permit records that can span months and provide approximate location (McConnell et al. 1992a, Stewart and DeLong 1993, Hill 1994, Testa 1994). Furthermore, detailed analyses of individual dive types allow one to distinguish between what are likely feeding and traveling dives (Boyd and Croxall 1992, Le Boeuf et al. 1993, Asaga et al. 1994) (Fig. 8-1A). The validity of diving data as a reflection of foraging effort is being assessed by studies that combine these data with isotope dilution techniques to measure water turnover, which allows estimation of energy intake. Other studies combine TDR records with changes in body mass to produce a less precise measure of energy flow than do isotope dilution techniques (see Costa and Williams, Chapter 5, this volume).

Otariid females as a rule are relatively shallow foragers, that is, they typically dive to depths of less than 100 m, and the maximum recorded depths for most otariids is less than 250 m (Costa 1991a). This appears to be true for both fur seals and sea lions (see Gentry and Kooyman 1986, Boyd and Croxall 1992), although at this point there is little information on the latter (Kooyman and Trillmich 1986, Feldkamp et al. 1989, Antonelis et al. 1990). The duration of individual dives is correspondingly short in otariids compared to phocids. Typical dive durations are about 2 to 3 min. The pattern of diving in otariid females may also be described as bout diving (i.e., individuals dive for periods of time and then have periods where there is no diving) (Fig. 8-1B) (Kooyman 1989, Boyd et al. 1994). These dive bouts tend to occur mostly during the night or at dawn and dusk (times known as crepuscular) (Gentry and Kooyman 1986), but some species such as the California sea lion (*Zalophus californianus*) show a diurnal pattern of diving in which dives are restricted to daytime hours (Kooyman and Trillmich 1986).

This pattern of diving by otariid females, whereby they leave the breeding colony to make foraging trips, is consistent with what is known as central place foraging (Orians and Pearson 1979, Stephens and Krebs 1986). As noted by Gentry et al. (1986a), one of the predictions of central place foraging theory is that foraging trip duration should increase

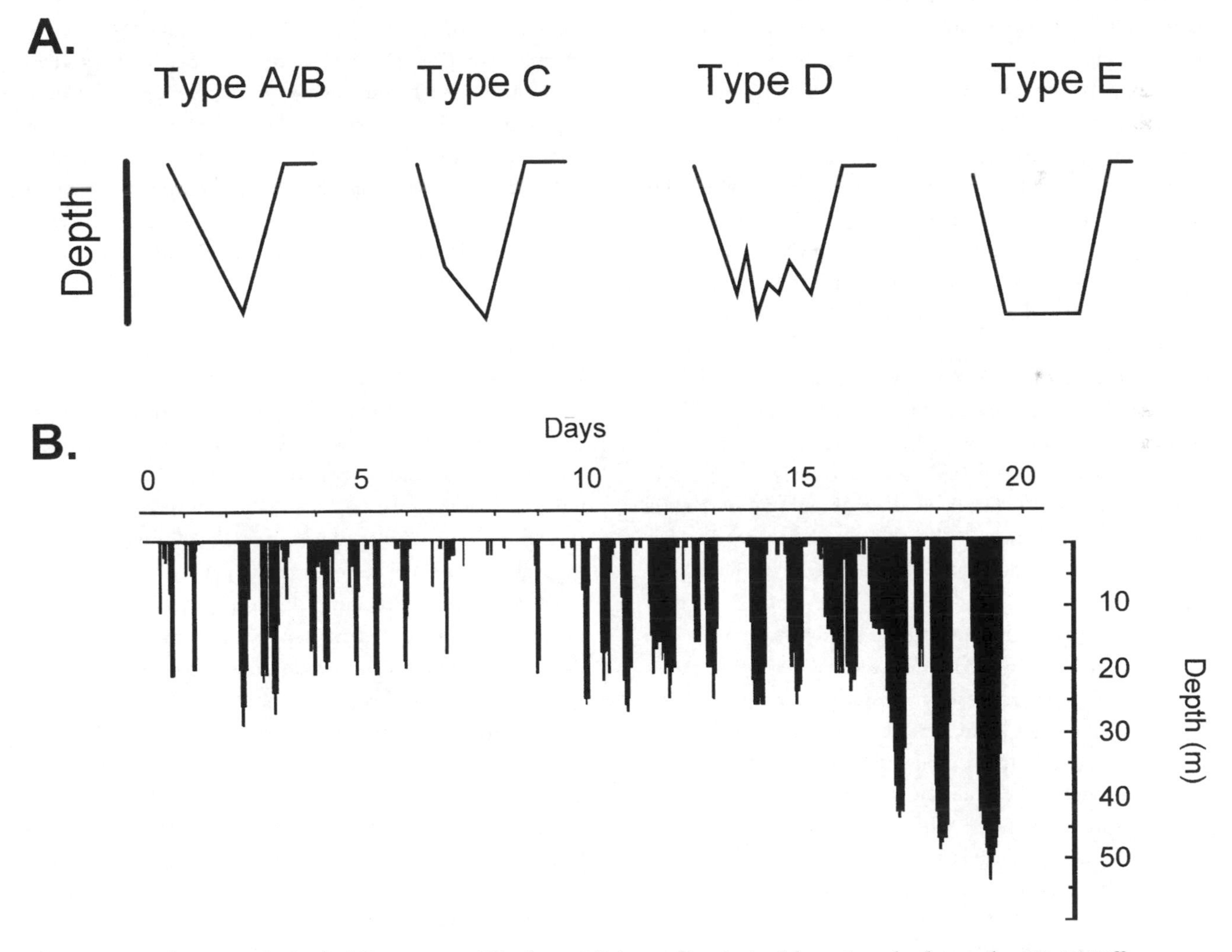

Figure 8-1. Schematics of individual dive patterns (A) and actual diving profiles obtained from time–depth recorders (B). (A) Different shaped dives from northern elephant seals. Dive types A/B are thought to be transit dives, type C "processing" dives (i.e., digestion and clearing of metabolites), and types D and E foraging dives. (B) A 20-day diving record from a female harbor seal shows a bout pattern of diving. Bout diving also occurs in otariids. (A is modified from Crocker et al. 1994 and B from Boness et al. 1994.)

as distance to foraging sites increases. This appears to be the case among otariids. For example, the average duration of a female Juan Fernández fur seal foraging trip is 12.2 days and individuals may travel more than 450 km to forage. In comparison, foraging trip duration of the northern fur seal (*Callorhinus ursinus*) ranges from 5.8 to 9.8 days (Gentry and Holt 1986) and foraging sites are from 140 to 400 km from the breeding colony (Loughlin et al. 1987, Antonelis et al. 1990, Goebel et al. 1991).

In at least one species, the northern fur seal, there appear to be different foraging patterns among individuals, based on dive depths. Some females are "shallow" divers (50 to 60 m), others are "deep" divers (170 to 180 m), and still others mix the two patterns, using shallow diving on one day and deep diving on another. The shallow divers are crepuscular and the depth of their dives changes over time, whereas the deep divers dive in bouts throughout the day and the depth to which they dive is constant (Fig. 8-2). The diet of this species is better known than in most others because of samples collected from the stomachs of seals shot at sea. These dive patterns correlate well with some of the known prey. For example, juvenile walleye pollack (*Theragra chalcogramma*) and Atka mackerel (*Pleurogrammus monopterygius*) move up and down the water column with the shift in the deep scattering layer (an assemblage of organisms that migrates vertically through the water column relative to light levels), which is at the surface at night (Fig. 8-2). It is likely that the shallow divers feed on these two species and follow their vertical movements.

The Antarctic fur seal (*Arctocephalus gazella*) is another fur seal for which we have accumulated data on diving patterns and diets. Its diving depths clearly track vertical positions of the krill on which it feeds almost exclusively (see Fig. 4 in Croxall et al. 1985). Most dives occur at night when the krill are near the surface and, when dives occur during the day, they tend to be deeper and to depths to which krill migrate during the daytime. Individual dives are V-shaped, suggesting seals pass through schools of their prey and return to the surface (Boyd and Croxall 1992). Presumably otariids feed on vertically migrating prey near the surface for energetic efficiency. Interestingly, other Antarctic aquatic predators that feed on krill (e.g., various penguins) exhibit a similar diving pattern to that of the Antarctic fur seal (Croxall et al. 1988, Fraser et al. 1989, Costa 1991b).

The distances traveled to forage and the length of foraging periods for those fur seals studied to date are, in part, constrained by the fact that the females are still lactating and tied to periodic visits to land to nurse their pups. We know little about foraging patterns in males, nonlactating females, and juveniles. The foraging periods and distances traveled for lactating otariids are highly variable and likely dependent on diets and local distribution of food resources in relation to the breeding colony site. For example, Galapagos fur seals (*Arctocephalus galapagoensis*), Galapagos sea lions (*Zalophus californianus wollebaeki*), and Steller sea lions (*Eumetopias jubatus*) travel very short distances and forage for less than a day at a time before returning to land (Gentry 1970; Trillmich 1986a,b; Higgins et al. 1988). In contrast, species such as the northern and Antarctic fur seals travel several hundred kilometers to feed and may be gone for nearly a week before returning to land (Doidge et al. 1986, Gentry and Holt 1986, Goldsworthy 1992). Juan Fernández fur seal females are extreme in this regard, remaining at sea for nearly 2 weeks on average and traveling almost 1000 km round trip before they return to the island where their pups are located (Francis et al. 1998).

The only phocid clearly exhibiting distinct foraging trips during lactation is the harbor seal (*Phoca vitulina*) (Boness et al. 1994). Like the lactating otariids, foraging of lactating harbor seals is constrained by the need for females to care for their pups on land. During the foraging trips harbor seal females also exhibit bout diving (Fig. 8-1B). Their foraging trips appear to be near the breeding colony because trip durations average only 7 hr (Boness et al. 1994). At the end of lactation, however, female harbor seals change their foraging patterns, beginning to dive deeper and continuously for extended periods (at least for several days before dive records ended because TDR memory was filled) (Boness et al. 1994). In a study of two harbor seal males and a female outside the breeding season, Thompson and Miller (1990) found that seals traveled as far as 25 and 46 km from haul-out sites on land and spent a maximum of 6 days at sea foraging. However, most foraging trips lasted less than 12 hr. There was a clear positive relationship between trip duration and distance to foraging location.

In most phocids, foraging is suspended during the breeding season, and thus the constraints imposed by the need to nurse young at the breeding colony are not a factor in shaping their foraging patterns. Our ability to make generalizations at this point is somewhat limited by the scarcity of data, but there are a few species for which we have a reasonable start to understanding their foraging patterns. The most extensive and dramatic data come from studies of northern and southern elephant seals (Le Boeuf et al. 1986, 1989, 1993; Boyd and Arnbom 1991; McConnell et al. 1992a; Stewart and DeLong 1993; Le Boeuf 1994; Slip et al. 1994). At the end of the breeding season, northern elephant seals, with depleted body fat stores from fasting (Costa et al. 1986, Deutsch et al. 1990), depart from the breeding site on extended foraging trips in which they dive nearly continuously for 120 and 70 days for males and females, respectively (Stewart and DeLong 1993). At the end of this period, the seals return to land

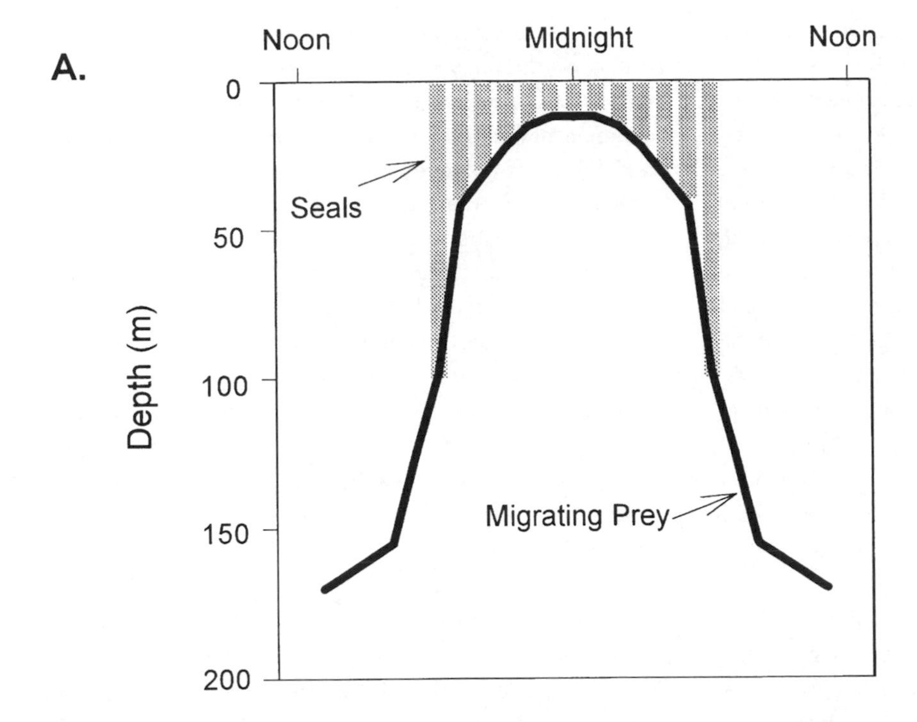

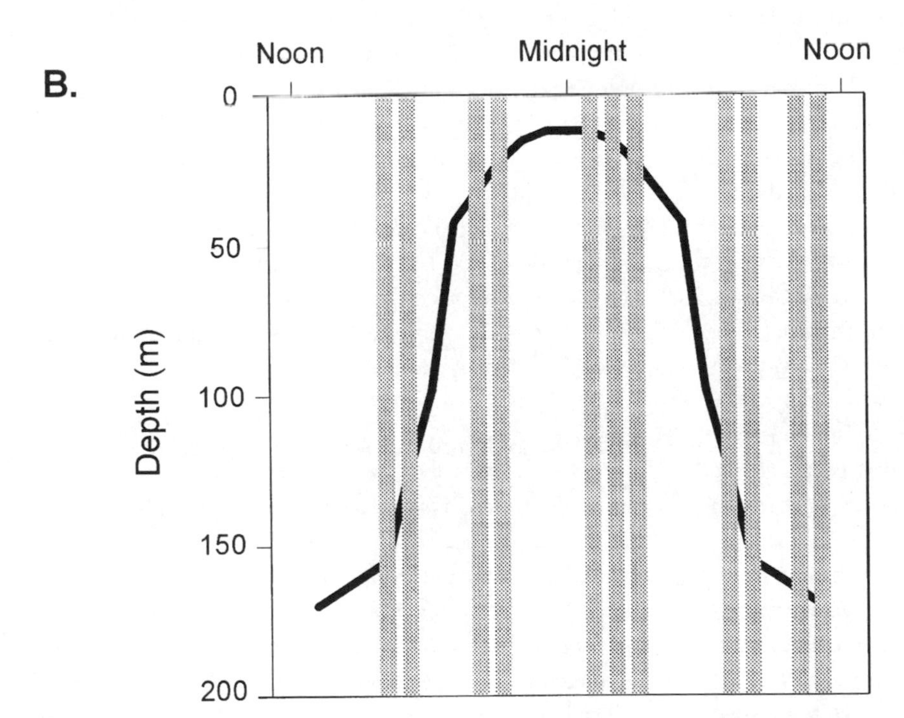

Figure 8-2. Schematic of shallow-diving (A) and deep-diving (B) northern fur seals in relation to vertically migrating prey. The gray bars represent the distribution of dive depths over time of day and the black curve represents the depth of the vertically migrating prey over time of day. Note that the shallow divers appear to be foraging on the migrating prey, adjusting their depth of diving according to the depth of the migrating prey. The deep divers, however, dive through the migrating prey and appear to feed on the bottom (benthic feeding), as evidenced by the consistent depth to which they go regardless of time of day. (Modified from Gentry et al. 1986b.)

to molt. An even longer continuous foraging trip occurs after the molt. During this period, males and females average 241 and 296 days at sea feeding, respectively; this period is terminated by a return to the breeding grounds. Therefore, northern elephant seals are at sea feeding virtually continuously for all but a few weeks of the year when they are either on land molting or breeding (Stewart and DeLong 1993). Similar patterns of continuous diving between breeding and molt appear to be true of the southern elephant seal as well, although fewer data are available for them (McConnell et al. 1992a).

Adult male and female elephant seals appear to show somewhat different diving patterns and foraging locations (Le Boeuf et al. 1993, Stewart and DeLong 1993). These may reflect differing foraging tactics and diets, yet analysis of scats and stomach contents suggest similar prey items (Condit and Le Boeuf 1984, Antonelis et al. 1987, Le Boeuf et al. 1993, Antonelis et al. 1994). For example, males dive less deep than females (about

360 m vs. 500 m), exhibit predominantly "flatbottom" rather than "wigglebottom" foraging dives (see Fig. 8-1A, type D dives), and travel much greater distances to primary foraging grounds (1500–3700 km vs. 1000–2500 km for males and females, respectively) (Le Boeuf et al. 1993, Stewart and DeLong 1993). The average duration of individual dives (21–23 min) does not differ between the sexes, however. Similar sexual differences in diving and foraging patterns have been reported in southern elephant seals that breed on Macquarie Island and then migrate to the Antarctic coast to forage (Hindell et al. 1991). As noted by Stewart and DeLong (1993), sexual segregation in foraging locations is not unique to pinnipeds—it is apparent in a number of cetacean species as well (see Gaskin 1982, Whitehead et al. 1992, Moore and Reeves 1993). Partitioning of food resources by the sexes has also been documented for numerous terrestrial vertebrates (Geist 1971, Hogstad 1978, Clutton-Brock et al. 1982, Vitt and Cooper 1985).

Seasonal differences in foraging patterns and diets have been noted in elephant seals (Stewart and DeLong 1993) as well as in other phocids (Plotz et al. 1991, Testa 1994). In the Weddell seal (*Leptonychotes weddellii*), shifts in dive depth and pattern occur with the seasons. Summer dives are primarily midwater (150–300 m) and distributed throughout the day, but in spring and fall, dives fluctuate diurnally between 100 m and 500 m (Testa 1994). The shift in time of diving and dive depth, as noted by Testa (1994), is likely associated with a shift from a diet of midwater fish and squid to krill as the primary food source (see Plotz et al. 1991).

Phocid seal diving and foraging behavior shows considerable individual variation in pattern and location (Thompson and Miller 1990, Thompson et al. 1991, McConnell et al. 1992b, Boness et al. 1994, Boyd et al. 1994). Most studies using TDRs and satellite-linked transmitters are based on relatively few animals; therefore, our understanding of foraging patterns in most species is still very elementary, especially as we have no direct data on prey for those species that go on long foraging trips. As indicated earlier, diet analysis from scats or stomach lavages of animals on land only reflect recent meals, not those during extended pelagic foraging. Furthermore, temporal and spatial differences in food resources for a given species will undoubtedly contribute to intraspecies differences in foraging patterns (e.g., Thompson et al. 1991; Boyd et al. 1994; Sinclair et al. 1994; Testa 1994; Bowen and Siniff, Chapter 9, this volume), therefore, there is a need for comparable studies of a given species at a variety of sites.

Walruses (*Odobenus rosmarus*) are mainly benthic (ocean floor) feeders, taking bivalve mollusks, other invertebrates, some fish, and occasionally other seals (Fay et al. 1984a, Lowry and Fay 1984, Fay 1985). Most feeding is done in areas over the continental shelf and thus dive depth rarely exceeds 100 m (Fay and Burns 1988). Female walruses feed throughout the winter breeding season and then remain with the receding northern edge of the pack ice. Males do not feed, or feed little, during the winter breeding period on ice, but in summer they use beaches and islets for resting and molting and disperse from these sites to forage (Fay 1985). In a study of male walruses in Greenland, individuals with satellite-monitored transmitters attached to their tusks traveled up to 80 km from their resting sites to forage. On the basis of quality satellite-determined locations, the swim speeds of these males averaged about 4 km/hr and ranged from about 0.5 to 15 km/hr (Born and Knutsen 1992).

Our understanding of walrus feeding behavior has been derived from a variety of sources, including anatomical features and observations of animals in both captivity and the wild. The mystacial vibrissae of the walrus are more developed and dense than those in any other pinniped, suggesting that the vibrissae are used in detecting prey buried in the substrate of the ocean floor (Fay 1982). Observations of the ocean floor in areas where walruses were known to be feeding further suggest that the walruses root in the bottom substrate with their snout, rather like a pig, to dig up their sedentary prey (Oliver et al. 1983). The soft fleshy parts of bivalve prey appear to be removed from the shells by suction (Fay 1985).

Most evidence of walruses feeding on seals is based on observation of stomach contents. However, there are a few direct observations of walruses feeding on seals (Lowry and Fay 1984), although no actual kills have been observed. The freshness of several carcasses, and lack of evidence of other sources of predation on the seal, has in part led to the conclusion that the behavior by walruses is predatory rather than scavenging on carrion.

Predation and Predator Avoidance Behavior

Pinnipeds on land or ice are particularly vulnerable to terrestrial predators because of limited mobility resulting from adaptations that make them efficient aquatic foragers. Many pinnipeds also have aquatic predators (Fig. 8-3). Virtually all species have been hunted by humans and to a large extent this pressure has been more extreme than that from any natural sources. It is very difficult to assess the impact human hunting has had on the behavior of pinnipeds, and as this activity has been relatively recent, we will not elaborate further. Examples of other predators of pinnipeds have been reviewed in several publications (Stirling 1983, Oftedal et al. 1987a, Riedman 1990), but quantitative estimates of predation level or dedicated studies of antipredator or predator avoidance behavior are rare. There are a few studies that indicate substantial levels of predation and these are usually on young seals (e.g., Kenyon 1973, Smith 1976, Fay 1982, Har-

Figure 8-3. The carcass of a harbor seal pup that washed ashore after being killed by a shark near Sable Island, Nova Scotia. (Photograph by Daryl J. Boness.)

court 1992, Hiruki et al. 1993). For example, predation of ringed seal (*Phoca hispida*) pups by Arctic fox (*Alopex lagopus*) on the fast-ice near Victoria Island averaged about 26% of the pups born, and may have been as high as 58% in some years (Smith 1976). These seals may also have suffered predation by polar bears (*Ursus maritimus*) (Stirling 1977, Smith et al. 1991). Less intense, but nevertheless significant, levels of predation of South American fur seal (*Arctocephalus australis*) pups have been reported in Peru, where up to 10% of the pups in a colony are killed by male southern sea lions (*Otaria byronia*) (Harcourt 1992).

Antipredator behavior (active defense), such as mobbing in birds and a few mammals (Altmann 1956, Kruuk 1964, Smythe 1970), or encircling young and facing off as in musk ox (*Ovibos moschatus*) (Tener 1965), Asian elephant (*Elephas maximus*), and water buffalo (*Bubalus bubalis*) (Eisenberg and Lockhart 1972), or in some cetaceans (see below) is essentially unknown in pinnipeds, with two possible exceptions. Trillmich (1996) has observed Galapagos sea lion females and juveniles and Galapagos fur seals mob sharks. Barlow (1972) and Eibl-Eibesfeldt (1984) described Galapagos sea lion males supposedly either chasing off sharks approaching sea lion pups or herding pups in the water away from sharks, and suggested that males typically protect pups in this manner. Miller (1974) argued that this was unlikely to be a typical behavior because there was little evolutionary basis for it as an adaptive response as the pups within a given male's territory during each breeding season are not likely to be the offspring of that male. At the time this argument was made, we knew little about site fidelity of male and female otariids, although the belief was that there was considerable variability in where females give birth from year to year. More recent data on several species of otariids and phocids suggest that high levels of site fidelity might be more common than previously thought (Croxall and Hiby 1983, Lunn and Boyd 1991, Pomeroy et al. 1994, Twiss et al. 1994). Thus, there may indeed be an adaptive antipredator basis to the behavior of males described by Barlow, yet there are no further descriptions of such behavior in this or other seal species to suggest it is common.

Most seals faced with the approach of potential predators on land or ice flee by entering the water. As dependent young are usually less efficient or maybe even incapable of fleeing, they are more prone to being taken than are adults (see Fay 1982). Formation of groups may, in part, be an adaptive response to interspecific predation, or in some species, conspecific infanticide (Terhune 1985, da Silva and Terhune 1988, Higgins and Tedman 1990, Campagna et al. 1992a, Harcourt 1992). Being a member of a group can reduce the probability of an individual being taken, through a "selfish herd" or "swamping" effect, as has been described for other animals (Hamilton 1971; Ims 1990; see also further discussion of this in the cetacean section). Another possible predator–avoidance advantage of groups is increased vigilance against predators (see Pulliam and Caraco 1984). In this regard, Terhune (1985) and da Silva and Terhune (1988) reported a reduction in scanning or vigilance behavior of individual harbor seals that were in large groups compared to those in small groups, but overall group vigilance increased in larger groups.

Other inferred adaptive behavioral responses to predation include reduced lactation length (Trillmich 1984, Bowen et al. 1985). This, however, is still speculative because it is difficult to demonstrate based on a comparative approach. For example, most northern hemisphere ice-breeding phocids, which are exposed to on-ice predation by polar bears and Arctic foxes, have short lactation periods, ranging from 4 to 21 days (Bonner 1984, Oftedal et al. 1987a, Bowen 1991). The two species that have longer lactation periods (about 60 days) are the ringed and Baikal seals (*Phoca siberica*), both of which breed in birth lairs (i.e., special dens that hide the young from easy polar bear predation) (Smith and Stirling 1975, Thomas et al. 1982, Furgal et al. 1996). There is, however, a confounded effect in that the latter species are also fast-ice rather than pack-ice species. Thus, the difference in lactation length could be related to stability of the ice. These possible explanations need not be mutually exclusive. Although there has been some suggestion that there are longer lactation periods in southern hemisphere ice-breeders because their principal predators are aquatic (leopard seals [*Hydrurga leptonyx*] and killer whales [*Orcinus orca*]), not ice-based (Oftedal et al. 1987a, Riedman 1990), recent data on crabeater seal (*Lobodon carcinophagus*) lactation length do not support this contention. Southern hemisphere crabeater seals, which breed on pack-ice, appear to have a much shorter lactation period (17 days; Shaughnessy and Kerry 1989) than previously estimated (28–42 days; Laws 1958, Siniff et al. 1979), and hence the length of their lactation period is in the range of that of northern hemisphere counterparts.

Use of habitat topography to avoid predation is clearly seen in the case of the ringed and Baikal seal birth lairs mentioned above. Other possible examples of this may be seen in Australian sea lions (*Neophoca cinerea*) (Higgins 1990) and Hawaiian monk seals (Westlake and Gilmartin 1990), although at this point the evidence for these species is circumstantial. Higgins reports that 19% of Australian sea lion pups are killed by conspecific males. Although the circumstances underlying this are not totally clear, such behavior was reported in this species by Marlow in 1972. In association with this conspecific "predation" (infanticide), young pups rest in crevices between rocks when they are not nursing. Although this behavior might be interpreted as thermoregulatory, as pups get older and presumably more capable of avoiding males, they no longer hide in crevices, despite the fact that thermal stress may not have been reduced with an increase in age. There are also several species in which seals breed in caves, a feature perhaps related to extensive hunting by humans (Davies 1949, Stirling 1983). As available data on the possible use of caves before intense hunting are inadequate, one must be careful in ascribing the use of caves as a response to human hunting. It may merely be a consequence of cave-dwelling animals surviving because they were not discovered or could not be hunted efficiently.

Observations of behavior of seals at sea are sparse, therefore we know little from direct observation about behavior associated with avoidance of aquatic predators. We have already mentioned the possibility that species with aquatic predators may be more prone to spending longer periods on land than species with lower predation pressure from aquatic predators. Another inferred effect of aquatic predation on behavior comes from an interesting study of foraging behavior in Galapagos fur seals. Trillmich and Mohren (1981) found a periodicity in the proportion of females ashore that is associated with the lunar cycle. In particular, when there is a full moon, 90% or more of marked females were ashore at night instead of foraging. Trillmich and Mohren suggested that this may, in part, be associated with avoiding increased risk of predation under the moonlit ocean surface.

Mating Patterns

The aspect of pinniped behavior that has been studied most intensively is reproductive behavior, both mating and pup-rearing patterns. On the basis of the number of mates, and the durability of relationships, mating patterns can be classified generally as monogamous (one mate for each individual), polygamous (multiple mates for one gender but not the other), or promiscuous (multiple mates for both genders). Two forms of polygamy can be described: polygyny, in which one male mates with multiple females during a breeding season, and polyandry, in which one female mates with multiple males. Studies of pinniped mating patterns have been influenced heavily by Bartholomew's (1970) early theoretical model of the evolution of polygyny. Various modifications and expansions of this model (e.g., Stirling 1975, 1983; Pierotti and Pierotti 1980; Le Boeuf 1991; Boness 1991; Boness et al. 1993) were based on more general theoretical works that emphasize the importance of social and ecological factors (Jarman 1974; Wilson 1975; Bradbury and Vehrencamp 1977; Emlen and Oring 1977; Rubenstein and Wrangham 1986; Clutton-Brock 1988, 1989; Davies 1991; Wiley 1991).

A general framework within which we can understand variation in mating patterns has evolved over the past couple of decades. Key elements of this framework include the following: (1) individual animals behave to maximize their reproductive success, either directly through the number of offspring produced and reared (Williams 1966, Wilson 1975, Clutton-Brock 1988) or indirectly through the success of relatives (inclusive fitness; Hamilton 1964); (2) the sexes should not be assumed to have the same goal of cooperating to care for young—generally females invest more in rearing offspring and males more in acquiring mates (Trivers 1972); (3) the potential for polygyny and the realization of that potential depends on the relative effort required by each sex to successfully rear offspring and the dispersion of females in space and time (Orians 1969, Jarman 1974, Emlen and Oring 1977, Bradbury and Vehrencamp 1977); (4) the spatial dispersion of females is a result of resource distribution, predation pressure, and costs and benefits of group living; the spatial pattern of males results from the dispersion pattern of females (Jarman 1974, Rubenstein and Wrangham 1986, Davies 1991); and (5) phylogenetic constraints or inertia (Simpson 1944) may predispose taxa to particular directions, although ecological factors can alter that predisposition.

The strategies by which males and females acquire mates and maximize reproductive success through mating are variable, but can be grouped into a few categories. Males tend to defend resources used by females (resource defense), follow or defend females directly (female defense: harems, multi- and single male groups), search for receptive females and move on to the next (scramble competition), sequentially defend single females through mating (sequential defense), or aggregate and attract mates (lekking). Females may simply accept males who attempt to mate with them, investigate and choose males based on direct or indirect benefits provided by males, encourage sperm competition by mating promiscuously with several males, or incite competition among the males and mate with the winner.

For virtually all pinnipeds studied to date the evidence indicates, or is highly suggestive of, a polygynous mating pattern (for recent reviews, see Boness 1991, Le Boeuf 1991,

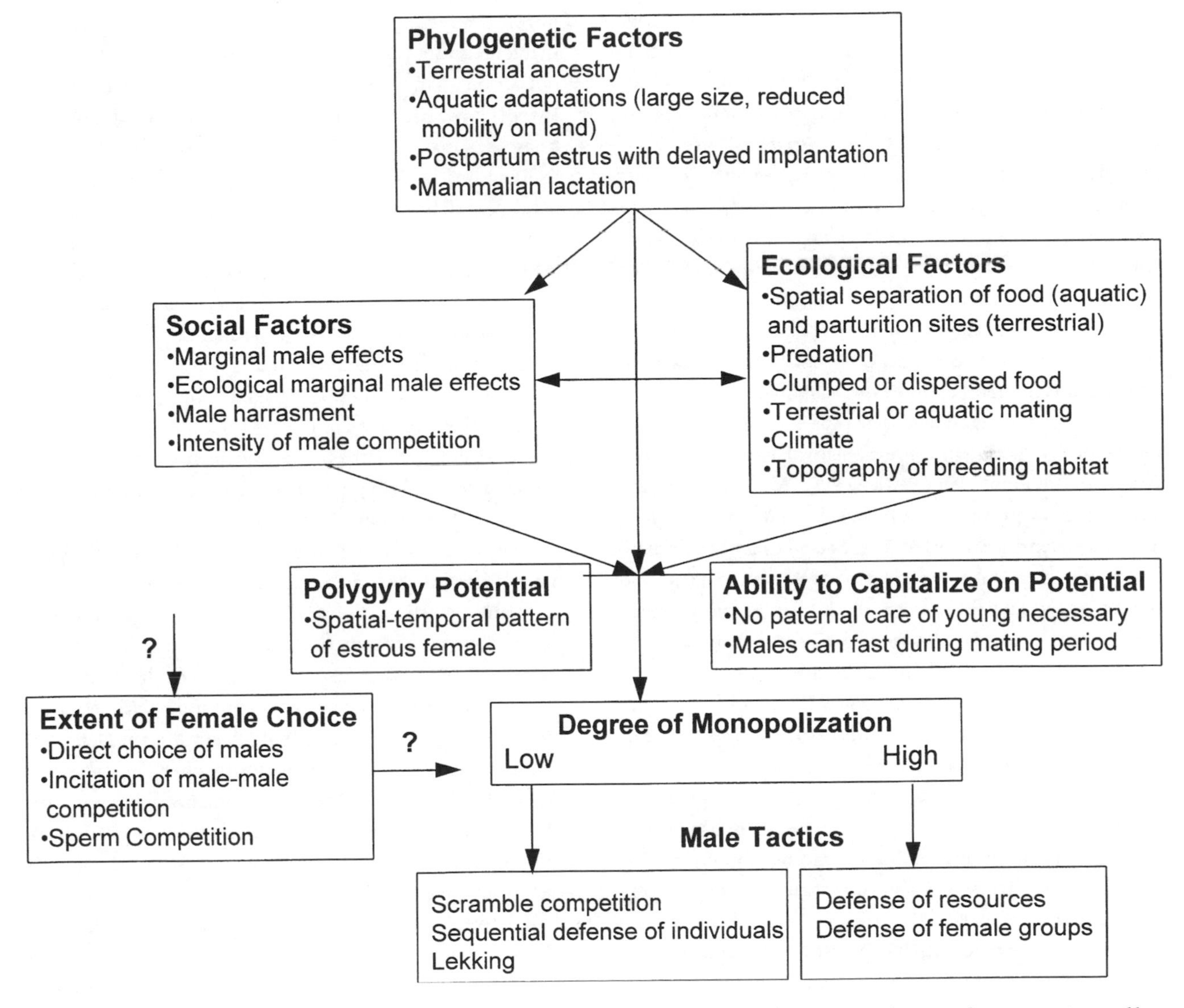

Figure 8-4. A schematic model of the major phylogenetic, ecological, and social factors that determine the type of mating tactics used by pinniped males. Note that little is known about mate choice by females and how it contributes to the tactics chosen by males. (Modified from Boness 1991.)

Boness et al. 1993). The degree of polygyny is extremely variable, and as we learn more about the strategies of male and female pinnipeds, it is apparent that there is considerable variability in the tactics used to acquire mates as well. Figure 8-4 illustrates the primary ecological, social, and phylogenetic factors and how they operate to produce polygynous mating systems in pinnipeds. Pinnipeds, as mammals with lactation and a terrestrial ancestry, but with various adaptations for marine feeding, are predisposed to polygynous systems. With mother's milk being the primary source of nutrients for dependent young (see section on parental care) and the lack of a role for males in protecting young from predators, males are not needed to successfully rear young and thus, are able to invest their efforts in acquiring more mates. The large body size of males permits substantial energy stores in the form of subcutaneous blubber, which in turn allow males to fast during the period when receptive females are available. Furthermore, the retention of terrestrial birthing and maternal care and seasonal variation in habitat likely played an important role in the evolution of both moderate spatial clustering and reproductive synchrony (Table 8-1). Some seal species exhibit extreme levels of spatial clustering and birth synchrony, which may be explained by a variety of factors including limited suitable breeding habitat (Le Boeuf and Condit 1983, Bowen et al. 1985), male harassment (Trillmich and Trillmich 1984, Campagna et al. 1992a, Boness et al. 1994), thermoregulatory needs (Campagna and Le Boeuf 1988a; Carey 1991, 1992), or predation (Harcourt 1992, Majluf 1992).

Pinnipeds exhibit all five basic types of polygynous mating tactics, although the prevalence of each type varies among the different families (Table 8-2). Among the otariids, all of which are primarily or exclusively land-mating species, resource defense (usually female parturition or thermoregulatory sites) by males is documented clearly in six species and suggested by data for six others. Studies of three otariid species, not necessarily exclusive of those mentioned above, suggest that males use female defense tactics to acquire mates (DeLong 1982, Campagna and Le Boeuf 1988b, Higgins 1990, Gales et al. 1994). With the exception of one of these studies, however, either the researchers themselves are not clear on how to interpret their data or there are conflicting interpretations. There is some evidence suggestive of lekking behavior in three species, although this interpretation is still controversial. No otariid species are known to exhibit scramble competition.

Among the phocids, all five types of mating tactics have been documented or postulated. There are only three land-mating phocid species and 15 that mate at sea. Unfortunately, the only species for which substantial behavioral data exist are those mating on land. All three of the land-mating species exhibit a form of female defense (Table 8-2). In those aquatically mating species, data on the behavior of males and females are limited, precluding clear interpretation of the primary strategy by which males acquire mates. Nonetheless, in a study of gray seal (*Halichoerus grypus*) males at Ramsey Island, Wales, where females rest in the water near where their pups are located on land, one or more males defend territories (defined as areas of exclusive use that are fixed in space and distinguished from locations from which neighboring males either retreat or do not enter without being attacked) off the beach and mate with females that are resting between nursing bouts (Davies 1949). Harbor seals at Miquelon Island appear to behave similarly (Perry 1993). In the latter study, however, only some paternities are accounted for by the males holding territories immediately off the beach, as discerned through DNA fingerprinting; therefore one cannot rule out the possibility that females are in some manner choosing males that reside outside the channel off the breeding grounds. In fact, at Sable Island, Nova Scotia, harbor seal males appear to take up positions 0.5 to 1.0 km off the beach (D. J. Boness, unpubl.) and are likely visited by females, resembling what amounts to the equivalent of a classic lek, as described in birds (see Bradbury and Gibson 1983, Wiley 1991). Boness et al. (1993) expanded on the ideas of others (Stirling 1983, Le Boeuf 1991) to suggest that the conditions for many aquatically mating phocids appear ideal for lekking (see Höglund and Alatalo 1995). There are no resources needed by females that males can defend, especially as the females have weaned their pups before becoming receptive, nor are the highly mobile females easily defended directly. On the other hand, because females are aggregated for breeding, nearby areas are essentially "hot spots" (see Beehler and Foster 1988, Wiley 1991, Balmford 1992) where males can gather and be readily available for females departing the breeding ground.

In the Hawaiian monk seal, Deutsch (1985) reported males visiting females on the breeding colony and "checking them out" by nosing them. With females that were near the end of lactation, males took up a position with them and followed them when they weaned their pup and departed. Deutsch suggested that this pattern of behavior was best viewed as scramble competition. Copulation, however, does not appear to occur for several weeks after the end of lactation; therefore this pattern might better be described as sequential female defense.

Relatively few studies have been done on walrus reproductive behavior; however, one comprehensive study (Sjare and Stirling 1996) suggests that, in the polynya (area of open water surrounded by stable land-fast ice) near Dundas Island, NWT, Atlantic walrus (*Odobenus rosmarus rosmarus*) males defended small groups of females. In contrast, in a lim-

Table 8-1. Spatial and Temporal Patterns of Females during the Mating Period for Various Species of Pinnipeds

Species	Common Name	Female Spatial Pattern, Movements, and Mating Location	Estimated (based on births) Period During Which There Are Estrous Females (days)
Otariidae			
Arctocephalus australis	South American fur seal	Clustered, daily thermoregulatory movements, mate on land	40
A. forsteri	New Zealand fur seal	Clustered, some thermoregulatory movements, mate on land	28
A. gazella	Antarctic fur seal	Clustered, no movements, mate on land	21
A. galapagoensis	Galapagos fur seal	Moderately clustered, may begin foraging before estrus, mate on land	70
A. philippii	Juan Fernandez fur seal	Clustered, daily thermoregulatory movements, mate on land and at sea	~30
A. pusillus	Cape fur seal	Clustered, may begin foraging before estrus, mate on land	30
A. pusillus doriferus	Australian fur seal	Clustered, some thermoregulatory movements, mate on land	26
A. tropicalis	Subantarctic fur seal	Clustered, some thermoregulatory movements, mate on land	33
A. townsendi	Guadalupe fur seal	Clustered, daily thermoregulatory movements, mate on land	42
Callorhinus ursinus	Northern fur seal	Clustered, some thermoregulatory movements, mate on land	27
Eumetopias jubatus	Steller sea lion	Clustered, may begin foraging before estrus, mate on land	20
Neophoca cinerea	Australian sea lion	Clustered, some thermoregulatory movements, mate on land	75
Otaria byronia	Southern sea lion	Clustered, daily thermoregulatory movements, mate on land	~45
Phocarctos hookeri	Hooker's sea lion	Clustered, some thermoregulatory movements, mate on land	18
Zalophus californianus	California sea lion	Clustered, daily thermoregulatory movements, mate on land	32
Phocidae			
Cystophora cristata	Hooded seal	Mildly clustered to dispersed on ice, mate at sea	15
Erignathus barbatus	Bearded seal	Dispersed on ice, mate at sea	24
Halichoerus grypus	Gray seal	Moderately clustered on land to mildly clustered on ice; may shift resting location, mate on land and at sea	36
Hydrurga leptonyx	Leopard seal	Dispersed on ice, mate at sea	~45
Leptonychotes weddellii	Weddell seal	Moderately clustered on ice, but mate at sea	40
Lobodon carcinophagus	Crabeater seal	Dispersed on ice, mate at sea	~25
Mirounga angustirostris	Northern elephant seal	Clustered, little movement, mate on land	36
M. leonina	Southern elephant seal	Clustered, little movement, mate on land	45
Monachus monachus	Mediterranean monk seal	Mildly clustered on land but mate at sea	~210
M. schauinslandi	Hawaiian monk seal	Mildly clustered on land but mate at sea	135
Ommatophoca rossi	Ross seal	Dispersed on ice but mate at sea	~45
Phoca caspica	Caspian seal		~12
P. fasciata	Ribbon seal	Dispersed on ice, mate at sea	~12
P. groenlandica	Harp seal	Evenly spaced on ice, mate at sea	15
P. hispida	Ringed seal	Dispersed on ice, mate at sea	30
P. largha	Spotted seal	Dispersed on ice, mate at sea	~60
P. sibirica	Baikal seal	Dispersed on ice, mate at sea	~30
P. vitulina	Harbor seal	Mildly clustered on land, but mate at sea	21
Odobenidae			
Odobenus rosmarus	Walrus	Clustered on ice, but mate at sea	60

Data are taken from Oftedal et al. 1987a, Riedman 1990, Boness 1991, and Boness et al. 1993.

ited study of Pacific walruses, Fay et al. (1984b) described a situation they interpreted as lekking behavior. Groups of males clustered near ice pans containing females with calves. The males interacted aggressively with each other and emitted "whistle" and "knock" displays, which may serve to attract females, although they may also be important in male–male relations. Although these studies illustrate what walrus males do, the conclusions drawn must be considered tentative at this stage. As with most of the aquatically mating seals, genetic assessment of paternity is needed to help resolve the limited behavioral observations.

In otariids that show resource defense, the usual resources defended are birth or thermoregulatory sites. The locations of breeding colonies are relatively constant from year to year, and males typically arrive and take up positions on territories encompassing these resources before females

Table 8-2. Types of Primary Male Tactics Exhibited for Various Species of Pinnipeds

Species	Common Name	Types of Primary Male Tactic Exhibited
Otariidae		
Arctocephalus australis	South American fur seal	Resource (birth and thermoregulatory sites) defense?; Lekking?
A. forsteri	New Zealand fur seal	Resource (birth and thermoregulatory sites) defense
A. gazella	Antarctic fur seal	Resource (birth sites) defense
A. galapagoensis	Galapagos fur seal	Resource (birth sites) defense?
A. philippii	Juan Fernandez fur seal	Resource (birth and thermoregulatory sites) defense
A. pusillus	Cape fur seal	Resource (birth sites) defense?
A. pusillus doriferus	Australian fur seal	Resource (birth and thermoregulatory sites) defense?
A. tropicalis	Subantarctic fur seal	Resource (birth sites) defense?
A. townsendi	Guadalupe fur seal	Resource (birth and thermoregulatory sites) defense
Callorhinus ursinus	Northern fur seal	Resource (birth sites) defense; Female defense
Eumetopias jubatus	Steller sea lion	Resource (birth and thermoregulatory sites) defense
Neophoca cinerea	Australian sea lion	Resource (birth sites) defense?; Female defense?
Otaria byronia	Southern sea lion	Female defense; Resource (birth and thermoregulatory sites) defense
Phocarctos hookeri	Hooker's sea lion	Resource (birth and thermoregulatory sites) defense?; Female defense?; Lekking?
Zalophus californianus	California sea lion	Lekking?; Resource (birth and thermoregulatory sites) defense?
Phocidae		
Cystophora cristata	Hooded seal	Sequential defense?; Scramble competition?
Erignathus barbatus	Bearded seal	Lekking?
Halichoerus grypus	Gray seal	Female defense
Hydrurga leptonyx	Leopard seal	?
Leptonychotes weddellii	Weddell seal	Resource defense?; Lekking?
Lobodon carcinophagus	Crabeater seal	Female defense?; Sequential defense?
Mirounga angustirostris	Northern elephant seal	Female defense
M. leonina	Southern elephant seal	Female defense
Monachus monachus	Mediterranean monk seal	?
M. schauinslandi	Hawaiian monk seal	Scramble competition?; Sequential defense?
Ommatophoca rossi	Ross seal	?
Phoca caspica	Caspian seal	?
P. fasciata	Ribbon seal	?
P. groenlandica	Harp seal	Evenly spaced on ice, mate at sea
P. hispida	Ringed seal	Resource defense (birth site)?; Scramble competition?
P. largha	Spotted seal	?
P. sibirica	Baikal seal	Resource defense (birth site)?; Scramble competition?
P. vitulina	Harbor seal	Resource (aquatic access route) defense?; Female defense?; Lekking?
Odobenidae		
Odobenus rosmarus	Walrus	Female defense; Lekking?

Data are taken from Boness et al. 1993.

begin to arrive. The extent to which individual males retain territories from year to year is not well known for most species, although we know that some males return to the same territory in at least 2 consecutive years, whereas others switch or are not seen again (e.g., Peterson 1968, Heath 1989, Lunn and Boyd 1991). Often, territories decrease in size from the beginning to the peak of the mating period, as more males become established. For example, Antarctic fur seal territories average about 60 m^2 at the beginning, but decline to about 22 m^2 during the peak (McCann 1980). In an excellent experimental study of New Zealand fur seals, designed to investigate the quality of territories with regard to relief of thermoregulatory stress, Carey (1991), during a 3-year period, added and removed pools and shade sites to areas of the beach. He found that males did not appear to be able to predict where females will settle based on shade or pool sites. Instead, males appeared to rely on previous experience of female distribution to select their territory sites.

Female otariids arrive at the traditional breeding sites after males have taken up positions on territories. There are few studies that have tried to assess what factors influence a female's choice of pupping location. However, in the study of New Zealand fur seals by Carey (1991), the evidence clearly showed that shaded areas, and to a lesser extent pools, influenced where females gave birth and resided on the beach. Female otariids may also choose a location because there are already other females there. In a group, especially where there is a cluster of territorial males, females may gain protection from harassment by roving males (Campagna et al. 1992a).

Site fidelity, defined as the tendency to return to the same location from year to year, is prevalent in at least one otariid, the Antarctic fur seal (Lunn and Boyd 1991). Although site fidelity may in part be a result of the quality of particular sites, as mentioned for thermoregulatory relief, Lunn and Boyd suggest that a constant location may be an important factor in mother–pup reunions, which occur every time a female returns from a foraging trip to nurse her pup (see Rearing Patterns section).

Female distribution and movements are important factors in determining male reproductive strategies, and at one time the common belief was that, for most otariids, females remained with their pups and did not move from the time they gave birth until they mated. This is being questioned more and more (e.g., Francis and Boness 1991, Boness et al. 1993, Majluf 1993) as new evidence is reported, or old evidence is revisited (see Table 2 in Boness et al. 1993). One of the most common reasons females move from their birth and care sites appears to be to obtain thermoregulatory reprieve, although in some species females may also begin foraging trips before they become receptive. The impact of female thermoregulatory movements on male behavior is seen in a recent study (Francis and Boness 1991). Some male Juan Fernández fur seal males defend entirely aquatic territories in response to groups of females rafting in predictable locations off the rookery (Fig. 8-5). The males holding aquatic territories are as successful in mating as males that defend land territories, which encompass birth and maternal care sites.

Among phocids exhibiting a female defense system, the behavior of males and females is variable, yielding somewhat different forms of female defense tactics. In the elephant seals, males arrive and establish dominance relationships before the females arrive (Carrick et al. 1962a; Le Boeuf and Peterson 1969; McCann 1980, 1981), as in the territory establishment in otariids. In contrast, gray seal males and females tend to arrive simultaneously at traditional breeding sites (Anderson et al. 1975, Boness and James 1979, Boness et al. 1995). For the multitude of ice-breeding species, we do not know about the establishment of breeding colonies.

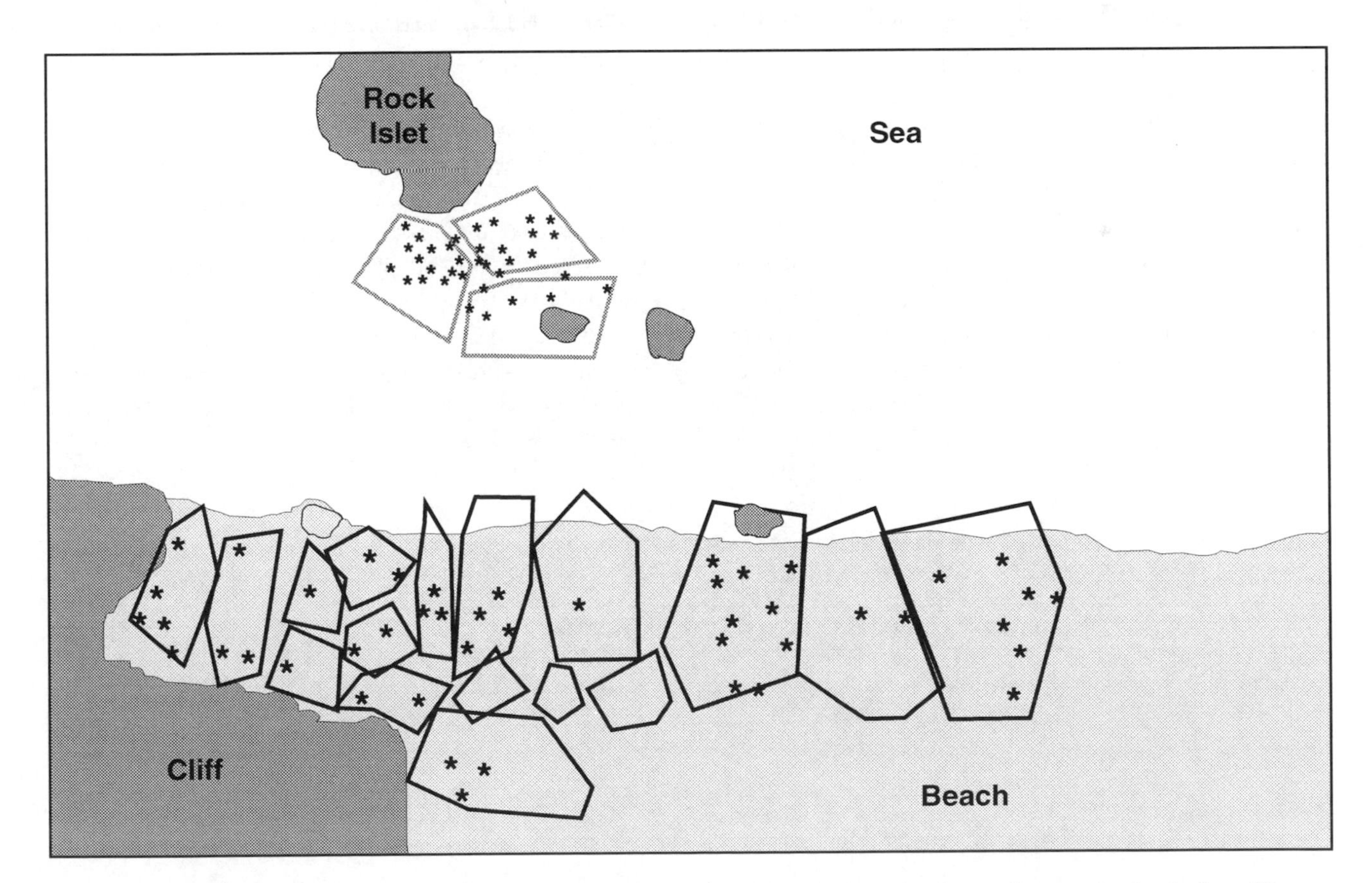

Figure 8-5. A schematic of the terrestrial and aquatic territoriality of Juan Fernández fur seal males in relation to the distribution of females. The asterisks on land represent the distribution of females with their pups. The asterisk in the water reflect where females go during the peak heat of the day to wallow and keep cool. (Modified from Francis and Boness 1991.)

Elephant seal males establish a dominance hierarchy through bloody fights and then maintain the relationships through vocal and visual threats (Le Boeuf 1972, 1974; Shipley et al. 1981). Contrary to what one might expect, the outcomes of fights are not predictable based on the body mass of males (Haley 1994). However, larger males do have greater stamina and can remain high in the hierarchy for longer periods than smaller males (Haley et al. 1994). Female elephant seals usually form one or more large groups, depending on the nature and size of the beach (Carrick et al. 1962a, Le Boeuf and Briggs 1977). Females compete for positions in the groups, with central positions conferring improved reproductive success (Reiter et al. 1981, McCann 1982). Both female aggressiveness and age appear to improve a female's success either directly or through the location or timing of breeding (Christenson and Le Boeuf 1978, Reiter et al. 1981). The dominant males maintain nearly exclusive access to females in groups and, by expending considerable energy to maintain their dominance for extended periods, achieve much higher mating successes (usually estimated by the number of females mated by a male) than do subordinate males (Deutsch et al. 1990, Haley et al. 1994).

Gray seal males and females behave quite differently. Females do not cluster as tightly as elephant seals, yet females are loosely clustered (see Boness and James 1979). Males cannot so easily be arranged in a hierarchy as can elephant seals, although there are clearly males that are more dominant and hold positions among the females, and others that roam about with less success in forming direct consortships with females (Anderson et al. 1975, Boness and James 1979, Anderson and Fedak 1985). Males maintain their associations with females primarily through visual threat displays (see Miller and Boness 1979), although sometimes threats escalate to fights when challengers do not back down. Interestingly, fights among gray seal males tend to be common later in the season in association with mating activity (Boness 1979), whereas in elephant seals fights are more common at the beginning when dominance status is being established (Le Boeuf 1972). Female gray seals also are more mobile in the breeding colony than are elephant seal females, often moving 5 to 10 m—a substantial distance given that males are on average 9 m apart. In an analysis of the response of males to this mobility of females, Boness and James (1979) showed that gray seal males were not tied to territories, but used their dominance to follow moving females to maximize being near females that were about to become receptive (Table 8-3). Length of tenure of gray seal males correlates highly with male success and is related to body mass (Boness and James 1979, Anderson and Fedak 1985).

Interannual site fidelity by gray seal females appears to be prevalent at some sites such as North Rona (Pomeroy et al. 1994). Male site fidelity, however, does not appear to be as strong as for females (Twiss et al. 1994). Interestingly, as noted by Pomeroy et al. (1994), female site fidelity in elephant seals may be linked to success in rearing pups, whereas success does not seem to be a factor in gray seals at North Rona. Generally, breeding site fidelity is likely to be higher in otariids than in phocids for several reasons, including the greater reliance on territories, a greater need for reprieve from overheating, and greater variation in beach topography and suitable sites at a given colony.

In the aquatically mating phocids, our knowledge of behavior of males and females is very limited. We do know that the spatial patterns of males and females are much more dispersed than in the land-mating phocids (Siniff et al. 1979, Boness et al. 1988, Godsell 1988, Boness 1990, Kovacs 1990, Van Parijis et al. 1997). In some species, such as the hooded seal (*Cystophora cristata*) and crabeater seal, males compete and the successful ones take up a position with a female and her pup on the ice, guarding her until she weans the pup, at

Table 8-3. The Movement of Male Gray Seals in Relation to the Movement and Sexual Status of Females Being Defended by a Male

Sexual Condition of Moving Female	Sexual Condition of Other Females within 5 m of the Focal Male		
	Some Receptive Females	Nonreceptive Females	No Other Females
Receptive	0.21* (19)	0.71^{ns} (14)	0.91* (11)
Not receptive	0.03* (38)	0.21* (78)	

* $P < 0.05$; $^{ns}P > 0.05$.

Values represent the proportion of times the male followed the moving female and, in parentheses, the number of female moves for which data were gathered. Modified from Boness and James 1979.

which time the male and female leave and presumably mate shortly thereafter (e.g., McRae and Kovacs 1994). However, at this point there has been no genetic confirmation that male consorts sire the females' offspring for any of these species. Furthermore, we do not know whether males sequentially take up with other females, although there is some circumstantial evidence of this for hooded seals (Boness et al. 1988, Kovacs 1990). In other aquatically mating phocids, males appear to position themselves in the water and wait for females to leave the beach after weaning pups (e.g., bearded seals [*Erignathus barbatus*]: Ray et al. 1969; harbor seals: Sullivan 1981, Perry 1993, Coltman et al. 1997; Weddell seals: Bartsh et al. 1992). The precise nature of competition and acquisition of mates in these species is unknown because of our inability to observe behavior other than that which occurs on land or ice (e.g., Walker and Bowen 1993). However, in some of these species males may use vocal or visual displays to attract females (Ray et al. 1969, Cleator et al. 1989, Hanggi and Schusterman 1994). Relatively new technologies, such as genetic techniques for paternity assessment, time–depth recorders, satellite-monitored transmitters, triangulation of positions using VHF (very high frequency) radio transmitters, hydrophone arrays (multiple underwater microphones arranged to locate output from sonic tags), and underwater video systems, are presently being used to advance our understanding. Moreover, technological advances such as paternity assessment from molecular genetic techniques are leading to modified views and greater understanding of mating systems in many animals, including pinnipeds (see Amos 1992, Pemberton et al. 1992, Boness et al. 1993). For example, a recent analysis of mating success among gray seal males on North Rona showed that in 36% of the cases where a male would have been judged by behavioral measures to be the father of an offspring, this was not the case (Amos et al. 1993b). In at least some of these cases, all of the established males were excluded as the father, leaving what was presumed to be a nonsuccessful peripheral male as the likely father.

What determines which particular male strategy is prevalent for a given species or population? One factor is suggested by the apparent absence of resource defense in land-mating phocids, but its predominance in otariids. Defense of territories requires frequent movement by males and maintenance of boundaries. Stereotypic boundary displays (e.g., Sandegren 1976), in which males move to mutual boundaries and face off, have been documented in all otariids studied (Fig. 8-6). High energetic costs of movement in phocids (crawling) compared to otariids (walking) may preclude defense of boundaries by phocid males. Consistent with this hypothesis is the fact that in at least three phocids that mate in the water, where movement should cost little, there is good evidence that males defend fixed territories (Hewer 1957, Kaufman et al. 1975, Perry 1993). Furthermore, in the two elephant seal species and the gray seal, there are conspicuous vocal and visual displays that help to maintain social order with minimal daily movement required (Bartholomew and Collias 1962, Miller and Boness 1979).

Intense male competition may also play a role in determining whether males exhibit a resource defense strategy or defend females directly. Numerous studies have suggested that territorial defense should break down when rates of challenges increase beyond some threshold (e.g., Wilson 1975, Hixon et al. 1983, Stamps and Buechner 1985). A possible example of this in pinnipeds may be seen in the South American sea lion (*Otaria byronia*), which exhibits resource defense at one colony and female defense at another (for a more thorough discussion, see Boness 1991).

The question of lekking in pinnipeds must still be considered hypothetical; preliminary data are intriguing. Clutton-Brock (1989) suggests that leks in ungulates tend to be associated with high local densities and where females range over large areas because effective female defense is not possible. Where lekking has been postulated in pinnipeds, conditions appear somewhat similar. For example, in many of the ice-breeding species, females are moderately aggregated on the ice but then leave it before they become receptive, as described above for harbor seals. These highly mobile females at sea would be difficult to defend and there are no clear resources that are defendable at that time. Males that cluster near the breeding colony and attempt to attract females could be more successful than males that choose to follow individual females until they are receptive and then seek to find another female. Among otariids, the suggestion of lekking is even less clear than in phocids. First, males and females as a whole are not spatially separated, whereby females visit a male area to mate. In the three species where lekking has been suggested (Heath 1989, Boness 1991, Majluf 1993), females still care for their pups in territories of males, but there are movements that expose females to other males and territories at or near the time of receptivity and provide an opportunity for females to choose. On the other hand, these movements may be thermoregulatory in nature; therefore it becomes necessary to separate whether mating simply follows the probability of being in a location either for maternal care or thermoregulation, or whether there is a propensity to mate more often than expected in a territory given these other needs for being there.

Female mating strategies in pinnipeds appear to be limited and constrained by maternal care needs. We have already noted that the question of direct female choice is not one that can be resolved at present, although there is some circumstan-

Figure 8-6. Two California sea lion males engaged in a boundary display. Note the rock between the two males. Such topographical features often appear to be used to help delineate boundaries. (Photograph by Daryl J. Boness.)

tial evidence for its occurrence in all three pinniped families. However, fairly convincing evidence of indirect choice exists in a couple of species. Two forms of indirect choice reported in other mammals and birds are incitation of male–male competition and sperm competition (see Ginsburg and Huck 1989, Petrie et al. 1992, Birkhead and Møller 1993). In the gray seal (Boness et al. 1982) and northern elephant seal (Cox and Le Boeuf 1977), females appear to indirectly choose males by inciting competition between them. This is accomplished by noisily resisting attempted copulations, which has the effect of inducing nearby males to challenge the offending male. The net effect of this activity by females is that if the male attempting to mount a female is a subordinate male, he is displaced by a dominant one, but if it is a dominant male the challenge by other males is usually warded off.

Another possible indirect choice mechanism of females is the encouragement of sperm competition through female promiscuity. Observed multiple matings of the land-breeding phocids are common, with females averaging between two to three known copulations. Although the common belief has been that otariid females only mate once, in a recent review of otariid studies Boness et al. (1993) found that in 9 of 15 otariid species up to 38% of the females in one study mated more than once, and often with a different male the second time. What we do not know yet is whether there is a pattern as to which male is successful. In the phocid studies, an assumption was made that the first copulation succeeded (Le Boeuf 1974, Boness and James 1979). However, as noted by Ginsburg and Huck (1989), order effects in multiple mating species are variable. In some species there is a primacy effect in which the first male is usually the one to sire the offspring, whereas in others it may be the last male that is more likely to be successful.

Rearing Patterns

As noted earlier, male seals do not appear to contribute to the care of offspring. Maternal care is relatively simple for most species; mothers do little beyond transferring milk to their offspring. In most phocids and otariids, mothers and pups do not go to sea together, therefore it is unlikely that females teach their young to forage. Walrus mothers, however, may play a greater maternal role given that the length of lactation is more than 1 year and calves travel with their mothers when they go to sea (Fay 1982). Maternal behavior in seals is thus concerned mainly with the provisioning of young with milk.

There have been three basic patterns of maternal behavior described in a number of review articles during the past decade: a female foraging cycle pattern, a fasting pattern, and simultaneous maternal foraging and nursing (Bonner 1984, Gentry and Kooyman 1986, Oftedal et al. 1987a, Costa 1991a). As with mating systems, there may be predispositions or a phylogenetic component to these maternal strategies (see Boness et al. 1994, Boness and Bowen 1996). For example, all otariids exhibit a foraging cycle strategy, most phocids exhibit a fasting strategy, and the single odobenid species, the walrus, is the only species in which simultaneous maternal foraging and nursing occur.

The major features of the foraging cycle maternal strategy are: (1) females acquire a moderate store of subcutaneous fat (blubber) before coming ashore to give birth; (2) after giving birth, females attend and nurse their young continuously for a few days, known as the perinatal period; and (3) at the end of the perinatal period, females begin alternating between making foraging trips to sea and making shore visits to nurse their young (Fig. 8-7). A commonality in the foraging strategy of otariid females is that lactation lasts for several months to more than a year. However, in the one phocid (the harbor seal) where this strategy has been unambiguously documented (Boness et al. 1994) the length of lactation is less than 1 month, that is, phocidlike. The fat content of milk of the harbor seal, which is the major energy source for dependent pups, is also still phocidlike. The otariids tend to have low-fat milk relative to the phocid species that fast, but the harbor seal has milk that is comparable to some of the fasting species, although it is at the low end of the phocid range (see Costa and Williams, Chapter 5, this volume).

The fasting maternal strategy of many phocids begins with females building up large stores of blubber before parturition. Once females haul out onto land or ice (either free-floating or land-fast) to give birth, they remain with their pups and nurse them daily until the end of lactation. Females in these species do not forage during lactation but fast, living off their blubber stores. Lactation is abruptly ended by the female leaving her pup on land or ice and heading off to sea to begin an extended period of foraging to replenish some of the depleted body fat (Fig. 8-7). Associated with the fasting behavior is a relatively short lactation period, extremely high fat content of milk, and rapid fattening of the pups (for more detail on the energetics and ecology associated with a fasting strategy, see Costa and Williams, Chapter 5; Boyd, Lockyer, and Marsh, Chapter 6; and Bowen and Siniff, Chapter 9, this volume). The most extreme example of a species exhibiting this pattern is the hooded seal—female hooded seals suckle for just under 4 days on average. They provide their pups with milk that is 61% fat about every 25 min, and pups gain 7.1 kg per day (82% of the gain is in the form of fat deposits) (Bowen et al. 1985, 1987; Oftedal et al. 1988; Perry and Stenson 1992).

The walrus is the only pinniped species in which a more typical terrestrial mammal pattern of behavior exists. Offspring care and maternal nutrient intake are not spatially or temporally partitioned; rather walrus calves travel with their mothers when the females leave ice floes to feed; walrus mothers and calves are able to nurse in the water (Fay 1982, Miller and Boness 1983). Associated with this is a long period of maternal care (2 to 3 years), although dependence solely on milk may last for only about 5 months (Fay 1982), after which time calves likely combine suckling with solid food intake.

Foraging cycles in otariids appear to be very adaptable. There is considerable variation among species, among populations within a species, and within the same population from year to year. This variation can be seen in Table 8-4, which presents values for the major components of the cycle (the perinatal period, foraging trips, and shore visits) for representative species. Although lactation length may vary by about fivefold among species, it does not vary within a species in the same manner as the other cycle components do. A major factor in the adaptability and variation in aspects of foraging cycle strategies in otariids, as described in Costa and Williams (Chapter 5, this volume) and Bowen and Siniff (Chapter 9, this volume), is the nutritional condition of females at parturition and the distribution, abundance, and type of food resources used during lactation.

It is only very recently that a foraging cycle pattern of behavior has been described in a phocid (Boness et al. 1994, Thompson et al. 1994), although possible feeding during lactation has been reported for some time (see Oftedal et al. 1987a, Costa 1991a, Lydersen and Kovacs 1993). We do not have enough data on the harbor seal, or any other phocid that might exhibit a foraging cycle (e.g., the ringed seal; Ryg et al. 1990), to know to what extent the behavior in these species is adaptable to local ecological conditions or environmental variation over time, as it is in otariids.

Variation in maternal behavior between fasting phocids is primarily in the length of lactation and in temporal patterns of suckling. Lactation length among phocids is short by otariid standards, but varies by a factor of 10, with the 4-day

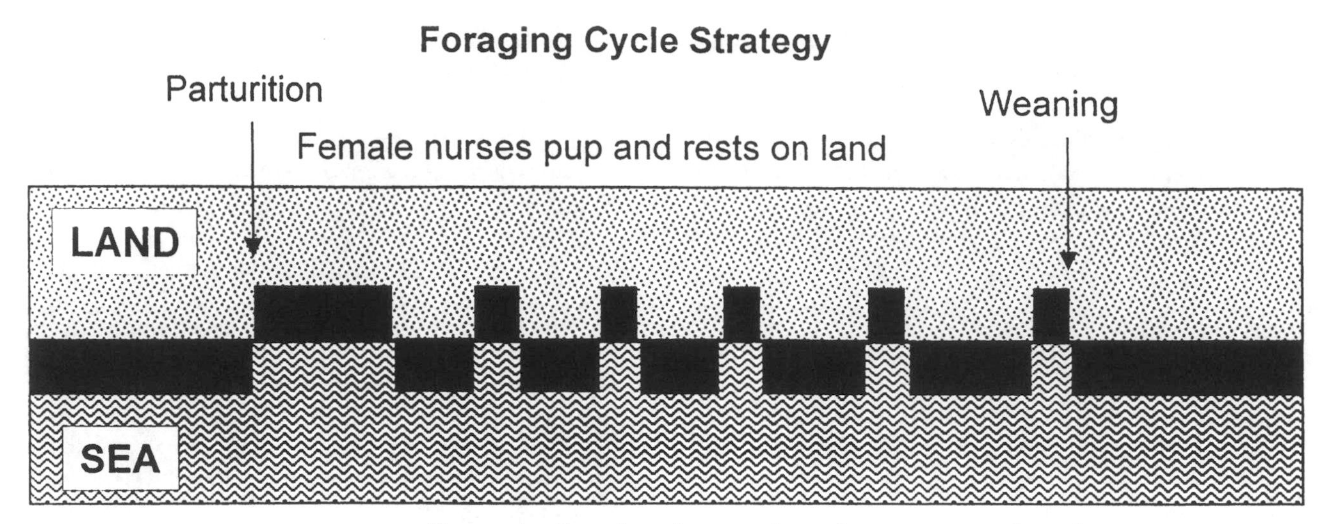

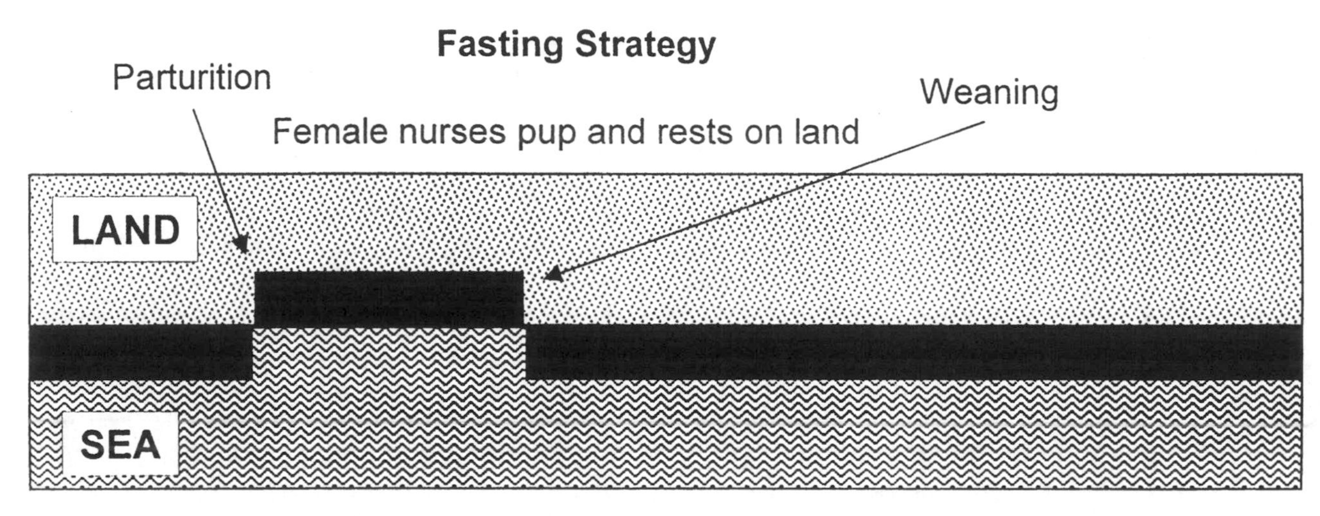

Figure 8-7. Schematics of the foraging cycle and fasting strategies of female pinnipeds. The black horizontal bars reflect the pattern and relative amounts of time spent on land and at sea foraging for each strategy during the period of maternal care (between parturition and weaning). See Table 8-4 for details on the variation in the temporal characteristics of the foraging cycles of otariids. (Modified from Bonner 1984.)

lactation period of the hooded seal being the shortest of any mammal, and some ice species such as the ringed seal or Baikal seal nursing their pups for about 60 days (see Table 3.1 in Bowen 1991). A variety of factors contribute to this variation, including habitat stability and predation, but these are discussed in more detail in Bowen and Siniff (Chapter 9, this volume) and will not be elaborated here. Although the temporal pattern of suckling is variable in phocids, with what it correlates is unclear. Correlations between milk fat and suckling bout duration or intersuckle interval are not simple, nor are relationships between length of lactation and parameters of suckling. There is a general developmental increase in bout duration as pups age, but in some species the interval between bouts declines, somewhat offsetting the increased bout duration (Oftedal et al. 1987a, Bowen 1991). In one species, the Weddell seal, both bout duration and the intersuckle interval decline as pups age. It may be that suckling behavior has a high individual component and simply does not represent milk and energy flow. Suckling may have a social component that has to do with maternal contact as well as the nutritive component. Non-nutritive suckling has been studied in considerable detail in domesticated animals (e.g., Koepke and Pribram 1971) and is also described in the cetacean section of this chapter.

Table 8-4. Variation in Temporal Characteristics of the Maternal Foraging and Attendance Behavior in Relation to Species Differences, Interannual Differences within a Colony, and Differences between Colonies within a Species

Species	Common Name	Perinatal Period (days)	Foraging Trip (days)	Shore Visits (days)	References
Arctocephalus australis	South American fur seal	—	4.6	1.3	Trillmich et al. 1986
A. galapagoensis	Galapagos fur seal	7.3	1.1	1.3	Trillmich 1986a
A. gazella	Antarctic fur seal	6.9	4.3	2.1	Doidge et al. 1986
		—	5.3	2.0	Lunn et al. 1993
		8.9	2.7	1.3	Goldsworthy 1992
		8.8	5.9	1.5	Goldsworthy 1995
A. philippii	Juan Fernandez fur seal	11.3	12.2	5.3	Francis et al. 1998
A. pusillus	Cape fur seal	4.3	3.0	2.4	David and Rand 1986
A. tropicalis	Subantarctic fur seal	13.8	2.0	1.7	Goldsworthy 1992
Callorhinus ursinus	Northern fur seal	—	7.7	1.5	Bartholomew and Hoel 1953
		8.0	9.4	2.0	Peterson 1968
		7.7	6.1	2.3	Gentry and Holt 1986 (Zapadni)
		8.3	8.2	2.2	Gentry and Holt 1986 (Kitovi)
		8.6	5.9	2.1	Gentry and Holt 1986 (East)
Eumetopias jubatus	Steller sea lion	6.7	1.2	0.9	Higgins et al. 1988
Zalophus californianus	California sea lion	7.9	1.4	1.5	Ono et al. 1987 (1982)
		6.1	2.1	1.6	Ono et al. 1987 (1983)
		8.0	2.5	1.4	Ono et al. 1987 (1984)
		7.7	1.6	1.7	Ono et al. 1987 (1985)

The foraging cycle pattern of behavior in the otariids may have placed strong selection on mother–pup recognition. Evidence of this is seen in strong rejections of attempts by otariid pups to obtain milk from females other than their mothers (Peterson and Bartholomew 1967, Trillmich 1981, Ono et al. 1987). Furthermore, numerous studies of otariids have described a distinct system for reuniting when mothers return from foraging at sea, based on a combination of vocalizations, returning to locations of last suckling and sniffing, (Bartholomew 1959, Peterson and Bartholomew 1967, Stirling 1971, Trillmich 1981, Roux and Jouventin 1987, Gisiner and Schusterman 1991, Insley 1992). Associated with this highly developed reunion and recognition system, otariid females rarely nurse pups other than their own (see Lunn 1992), and usually when they do, it appears to occur without the female's knowledge and nursing is terminated as soon as it is discovered.

In the fasting phocids and the walrus, females and pups are not usually separated for long periods or by great distances, although in some species, such as harp, gray, and harbor seals, females enter the water leaving their pups on ice or land (Fogden 1971, Kovacs 1987, Boness et al. 1994). Thus, vocal communication between mothers and pups may be less important for many phocids than for otariids. The close proximity of mothers and pups may have resulted in olfaction playing a more prominent role (for a discussion, see Bowen 1991). One phocid where vocalizations may be of some importance is the harbor seal (Perry and Renouf 1988). The harbor seal is much more aquatic during lactation than are most other phocids, and vocalizations may be used to maintain contact, although at this point the evidence suggesting individual recognition is sparse (see Renouf 1984). Studies of other phocid species provide some suggestion that vocal recognition of pups could occur, but again the evidence documenting it is weak (e.g., Petrinovich 1974, Insley 1992). In the Hawaiian monk seal, there is enough within-pup variation in calls to make it difficult for a female to distinguish her own young (Job et al. 1995); in fact, in a discriminant function analysis, pup vocalizations were only assigned to the correct pup about 14% of the time. Selection pressures for individual recognition in phocids may not have been strong.

Care (particularly nursing) of non-offspring in phocids is fairly common by comparison to otariids and many other mammals. For many species where it occurs, non-offspring nursing or parental care may be described as fostering, that is it lasts for an extended period of time and the female knowingly suckles the pup. In most birds and mammals where fostering is common, there appear to be direct benefits associated with its occurrence (Riedman 1982, Packer et al. 1992). On the other hand, in the otariids and some colonial birds and bats it seems to be a misdirected behavior that is costly. Hence fostering occurs infrequently, or there are non-related benefits of colonial behavior that outweigh the costs of fostering (McCracken 1984, Wilkinson 1992, Ferrer 1993). At present, we do not know enough about fostering in any of the phocids to be sure of whether it is primarily misdirected, or has direct or indirect benefits that outweigh its costs. Proximate factors, such as density (Fogden 1971, Riedman and Le Boeuf 1982, Boness et al. 1998) and storms, which cause separation of mothers and pups (Boness et al. 1992), contribute to its occurrence, and young females may be more prone to foster than older females (Riedman and Le Boeuf 1982), but how these factors might translate into costs and benefits needs further study. Boness (1990) has shown that the cost of fostering in Hawaiian monk seals is low, and suggests that the small population and high site fidelity may result in enough benefits derived from kin selection to favor fostering, or at least prevent it from being selected against. Recent genetic studies investigating the possibility of kin selection underlying fostering in gray and harbor seals suggest that it is unlikely that females acquire benefits by helping relatives (Perry et al. 1998, Schaeff et al., 1999).

Another possible benefit that might favor fostering behavior has been suggested in studies of the northern elephant seal. In this species, Riedman and Le Boeuf (1982) have found that most fostering is done by younger females, and they have suggested that these young females may gain maternal experience that would have otherwise been lost as fostering tended to occur after a female lost her pup. Fostering after losing a pup is the rule in harbor seals as well (Boness et al. 1992). What is currently lacking, however, are data showing that young females that foster gain an advantage in rearing future offspring over young females that lose their pups and do not foster. There is a clear need for more detailed and focused study of this unusual and interesting behavior, which in some phocids may be the most extreme instances among mammals.

Because many seal species are sexually dimorphic as adults and sexual dimorphism in some of these species is extreme, a large number of studies have focused on the question of differences in maternal investment and sex allocation (for review, see Kovacs and Lavigne 1986a, 1992; Trillmich 1996). The theoretical basis for differential investment and allocation in the sexes was brought to the forefront by Trivers (1972) and Trivers and Willard (1973), and has been expanded and expounded on for the past two decades (e.g., Maynard Smith 1980, Charnov 1982, Clutton-Brock 1991). Much of this work has to do primarily with the flow of energy from the mother to the pup, and behavior is the mechanism by which this happens. We do not propose to get into a lengthy discussion on differential transfer of energy or sex ratios here, therefore we restrict this discussion to differences in maternal behavior, such as maternal foraging, suckling, and length of lactation. Recognize, however, that a comprehensive treatment must incorporate an understanding of the energetics of lactation and reproductive success.

The first question is whether mothers behave differently in rearing their male and female offspring. Among the phocids there is very little evidence to suggest that this is the case. Reiter et al. (1978) found that northern elephant seal females nurse male pups for 1 day longer than female pups, but Kretzmann et al. (1993) failed to confirm this in a separate study and Campagna et al. (1992b) found no sex differences in lactation length and suckling rates in southern elephant seals. A similar equivocal picture is apparent in the gray seal. In one study, gray seal mothers spent a greater proportion of time nursing male offspring than female young, and male pups weaned at a significantly higher mass than females; however, lactation length did not differ (Kovacs and Lavigne 1986b, Kovacs 1987). In another study of gray seals (Bowen et al. 1992), lactation length and weaning mass did not differ between male and female pups, and that study noted an error in Kovacs and Lavigne's analysis that affected the outcome. Rosen and Renouf (1993) found that male harbor seal pups nursed for a greater proportion of haul-out time, but there were no differences in the duration of suckling bouts. These researchers attributed the sex differences to the behavior of pups rather than to the mothers; male pups initiated suckling bouts more often than did female pups, and the difference in proportion of time spent suckling was a function of a greater frequency of suckling by males.

The story may be somewhat different among the otariids, although recent findings point to a similar role for offspring controlling differences. Differences in suckling rates and foraging trip lengths between mothers of male and female pups have been reported in several species (Trillmich 1986b, Ono et al. 1987, Doidge and Croxall 1989, Goldsworthy 1992). The question that remains is to what extent these differences are driven by females attempting to influence the growth and development of pups of each sex differently, or simply females responding to pups with different demands. For example, Oftedal et al. (1987b) found that differences in milk intake between male and female California sea lion pups disappeared when the mass of pups was taken into account. Similarly, Lunn and Arnould (1997) failed to find evidence of differences in milk intake of male and female Antarctic fur seal pups. Unlike in the phocids, a common finding in otariids is a difference in birth mass between male and female pups (Kovacs and Lavigne 1992). In California sea lions, the maternal investment may occur prenatally, producing larger males at birth, but from then on the differences are attributable to the different demands of larger versus smaller pups and to sex-specific differences in the behavior and physiology of pups (Ono and Boness 1996). Somewhat similar conclusions are reached by Goldsworthy (1995) for Antarctic fur seals.

Migration and Social Organization

Many pinniped species migrate at various times during the year, although we know amazingly little about pelagic activity outside the reproductive period. These migrations are usually associated with seasonal changes in the environment, as in the case of the ice-breeding phocid species, the walrus, or subpolar otariids, or changes from breeding status to foraging or vice versa, as we described earlier for the postbreeding long-distance foraging trips of male and female elephant seals (e.g., Sergeant 1965, Doidge and Croxall 1985, Born and Knutsen 1992). There are other species, such as harbor and gray seals among the phocids and Australian sea lions among the otariids, that appear to disperse from breeding locations to forage, but do not actually undergo migrations (Thompson and Miller 1990; Gales et al. 1992; McConnell et al. 1992b; and see distinctions made by Bowen and Siniff, Chapter 9, this volume). In species like the California sea lion and northern elephant seal, males and females migrate to different locations to forage after the breeding season has ended (Mate 1973, Le Boeuf 1994). The extent to which there are sex differences in migratory patterns of seals generally is unknown.

The social organization of seals is most complex during the breeding season. In one of few studies of a nonbreeding group of mostly immature males and yearling Steller sea lions, but during the breeding season, Harestad and Fisher (1975) describe a spatial and social organization somewhat similar to the territorial system of males on the breeding grounds. When at sea, as best as we know at this point, seals appear to be relatively solitary. Unfortunately, studies that would give us insights into possible cooperation in feeding have not been done due to the logistic complexities of such work. Walruses appear to travel in groups (Fay 1982, Miller and Boness 1983), and when hauled out on land outside of the breeding season may be found in large groups in which aggressive interactions occur (Miller 1975c). Other nonbreeding groups occur in many species during an annual molting period. Usually these molting groups are extremely dense, but little interaction other than the occasional threat occurs (e.g., Carrick et al. 1962b). In other species, where seals haul out on land outside of the breeding or molting period, groups are formed but there is little indication that such groups are highly structured or stable. As noted earlier, some benefits may be gained by the increased vigilance afforded by grouping (for discussion, see Riedman 1990).

In summary, the behavior of pinnipeds has been influenced by their evolutionary history as well as selection pressures from both the marine and terrestrial environments. As the pinniped body shape and physiology have taken on those features making living in water not only possible but efficient, these features place certain constraints on the behavior of pinnipeds on land. Furthermore, a dependence both on marine food resources and on terrestrial conditions for giving birth and rearing offspring is probably the single most important factor that shapes behavioral patterns in these species. Unlike the other marine mammals that this chapter discusses, we know relatively little about the behavior of pinnipeds at sea. Most of our knowledge is focused on the foraging behavior of females in association with acquiring resources to care for their young. Little can be said about social patterns during periods at sea because direct observations of individuals during these periods are lacking. However, we know a great deal about the mating systems and maternal care patterns of nearly half of the pinniped species because these activities occur on land, where relatively long-term studies, following identifiable individuals, have been conducted. The remaining species that mate at sea are poorly studied and indeed, might be expected to exhibit somewhat different behavior patterns with respect to mating than do their land-mating relatives.

The quantity and quality of information available on the behavioral ecology of pinnipeds allows testing of fairly refined hypotheses, on the order of what is possible for terrestrial mammals. This state of research on and knowledge of pinnipeds can serve as a model to aim toward for those researchers who study the more elusive marine mammals that spend their entire life at sea. The following sections explore these marine mammals—the hints are tantalizing, but in many cases the conclusive, systematic data remain to be obtained.

Sea Otter [G. B. R.]

Sea otters (*Enhydra lutris*) represent a completely different evolutionary lineage than the pinnipeds just discussed, as are the cetaceans and sirenians, which are also unique. Because of their different evolutionary histories, it is not surprising that sea otters have solved many of the challenges of living in the marine environment completely differently than other marine mammals. In the following few pages, we will discuss some of the unique behaviors that sea otters have evolved, and compare these with those of other mammals, especially the pinnipeds that we have just discussed.

Perhaps one of the most striking behavioral differences in sea otters compared with other marine mammals is that sea otters spend most of their time at the surface of the ocean, a pattern that we just learned is also exhibited by most pinnipeds. However, unlike the seals, sea lions, and walrus, sea otters actually live *on* the surface of the ocean, where they usually float on their backs (Fig. 8-8)—a position that is rarely adopted by most other marine mammals. Even when traveling, sea otters tend to swim on or very near the surface of the water, again, often on their backs (Kenyon 1969, Packard and Ribic 1982). Living on the surface of the ocean has profound consequences for the behavioral ecology of sea otters, includ-

Figure 8-8. A raft of male sea otters. Sea otters often float on their backs, their flippers in the air to keep them dry. (Photograph by Richard A. Bucich.)

Figure 8-9. A sea otter grooming its face with its forepaws. Notice the fluffy appearance of the dry fur compared to the matted pelage of the foraging sea otter in Fig. 8-10. (Photograph by Richard A. Bucich.)

ing how they gather food, evade predators, use space, and interact with other sea otters. These factors are all intertwined and, together with their phylogeny (evolutionary history), help define and explain sea otter social structure or social organization, which we will discuss at the end of this section.

Health and Maintenance Behavior

Most marine mammals that occupy the temperate and polar seas have thick layers of insulating blubber to reduce heat loss to cold seawater. The sea otter, however, has little fat. Instead, it relies on dense fur for insulation. Sea otter hair, with densities in excess of 164,000 per square centimeter, is the thickest of any known mammal (Kenyon 1969, Williams et al. 1992). When sea otters dive in clear water a trail of air bubbles can be seen behind the swimming animal. These bubbles are created when water pressure compresses the fur, forcing some of the trapped air out, and thereby reducing the insulating quality of the hair. Sea otters must groom their pelage frequently to maintain its insulative properties. During daylight hours (they are also active at night, but are difficult to observe) they spend 5% to 16% of their time maintaining their fur. Grooming occurs most often after bouts of diving or swimming (Kenyon 1969, Estes et al. 1986, Riedman and Estes 1990), and entails rubbing and washing their pelage with their forepaws (Packard and Ribic 1982), which conditions the fur with body oils (Fig. 8-9). They also roll and tumble at the surface and blow bubbles into their fur, which replenishes the air lost while diving (Kenyon 1969, Packard and Ribic 1982). Because their fur is so important, it is not surprising that mothers observed in captivity spend most of their active time grooming their dependent pups (Hanson et al. 1993).

Diving and Foraging Behavior

Sea otters have relatively poor diving abilities when compared with most other marine mammals, largely because of the limitations imposed by their physiology (Costa and Kooyman 1982; see Elsner, Chapter 3, this volume), but also because of their reliance on fur for insulation. What would happen to the air trapped in their pelage if they were to dive deeply, or for long periods of time? Sea otter dive times are relatively short, usually less than 125 sec (Estes et al. 1981), although Ralls et al. (1995) recorded dives of up to 246 sec. Although their diving abilities are unremarkable, they are efficient predators of macroinvertebrates in shallow, nearshore communities. Because sea otters are restricted to these areas, and they bring their prey to the surface for ingestion (Kenyon 1969), their feeding biology is relatively easy to document and is fairly well known (see Bowen and Siniff, Chapter 9, this volume). Most of their foraging dives are less than 40 m deep (Kenyon 1969, Wild and Ames 1974, Riedman and Estes 1990), although a drowned sea otter was recovered from a crab trap set at about 100 m in the Aleutian Islands (Newby 1975). Foraging effort and success not only vary with location (depth and substrate type), but also with sex, reproductive status, and age class (Loughlin 1977, 1979; Garshelis et al. 1986; Ralls et al. 1995). In California, Loughlin (1977) determined that an average of nearly 75% of daytime foraging dives net prey, and several investigators have reported that sea otters spend at least 35% of their time foraging (for review, see Riedman and Estes 1990). Ralls and Siniff (1990) used radio-telemetry to monitor activity budgets in California and determined that juvenile females spend nearly 50% of their time foraging, which is greater than the other sex and age classes.

Favored prey, which are gathered mostly one at a time with their dexterous forepaws, include sea urchins, mollusks (clams, mussels, snails, abalones, and octopuses), crabs, and in some areas, fishes (for summaries, see Riedman and Estes 1990). Many food items must be removed from a shell or exoskeleton before ingestion. In addition to their bunodont (crushing) teeth, with low crowns and reduced cusps, sea otters break open their prey with the aid of tools. The evolution of tool use in mammals for feeding is rare (Alcock 1972), being most prominent in humans, the great apes, and the sea otter (Wilson 1975). When sea otters forage on prey that must be broken open, they select a stone (or even a discarded bottle, tin can, or another prey item) and either use it as an anvil on their chest upon which to pound the food item open (Fig. 8-10), or they use the tool as a hammer (Hall and Schaller 1964, Kenyon 1969, Houk and Geibel 1974, Riedman and Estes 1990). The same tool is often used during an entire foraging bout, being stored during dives in one of the axillae. Some individuals seem to prefer particular types of tools, such as stones or even other prey items like large clams (Riedman and Estes 1990).

Predation

Sea otters are killed by white sharks (*Carcharodon carcharias*) along the California coast (Ames et al. 1996) and killer whales (*Orcinus orca*) in Alaska (Kenyon 1969, Hatfield et al. 1998), but there are no reports of specialized avoidance behaviors

Figure 8-10. A tool-using sea otter, pounding a food item grasped in its forepaws on a stone anvil balanced on its chest. (Photograph by Richard A. Bucich.)

or alarm calls associated with predation (Kenyon 1969, Packard and Ribic 1982, McShane et al. 1995). A resting sea otter floating on its back on the surface of the ocean is probably unaware of most marine predators until it is too late. This apparent inability to effectively detect underwater predators has been used by biologists to develop an efficient capture method (Ames et al. 1983). A pair of divers using oxygen rebreathers (no exhalation bubbles are produced to alert the sea otters of an approaching diver) and pushing a purse seine net enclosed in an aluminum frame swim up beneath a resting sea otter. When the animal becomes aware of the "predator," its escape response is to dive—right into the "jaws" of the purse seine net. There is evidence that many sea otters are not actually eaten by the sharks that attack them, but are struck and abandoned. Apparently sharks forage optimally, like most other animals; it has been suggested that they prefer prey with abundant fatty tissue, such as young pinnipeds (Riedman and Estes 1990). It is possible that sharks strike sea otters, but reject them as food when they discover that they are not fat and blubbery, but rather lean and furry!

Sea otter fur takes on a different significance in Alaska, where nesting bald eagles are common. Sherrod et al. (1975) documented pup remains at several eagle nests on Amchitka Island. The eagles take neonates from the surface of the water, while their mothers dive nearby for food. Why don't the pups dive to escape the aerial predators? Young sea otters cannot dive because they lack the strength to overcome the buoyancy of their dense, air-filled pelage (Payne and Jameson 1984).

Mating Behaviors

Female sea otters usually come into estrus once a year, normally a few days after weaning or losing a pup (Jameson and Johnson 1993, Riedman et al. 1994). During estrus, which may last from 1 to 13 days (Garshelis et al. 1984, Riedman and Estes 1990), a female usually associates with only a single male. Precopulatory behaviors include synchronized activities, such as touching, nuzzling, and grooming. During mating the pair tumble and twirl at the surface of the water, and the male often grips the face of the female with his teeth, especially her nose (Kenyon 1969; Vandevere 1969, 1970). Copulation occurs in the water, and lasts from a few to 30 min (Kenyon 1969, Garshelis et al. 1984, Riedman and Estes 1990). In some cases, perhaps the result of young and inexperienced males, mating is especially violent and can even result in the death of the female (Staedler and Riedman 1993). Recently mated females often have deep, bloody wounds on their noses, which result in scar patterns that are individually identifiable (Foott 1971), but only until they mate again and gain new scars. At first consideration, the rough treatment of females by males during mating might seem maladaptive. However, many carnivores (especially felids and mustelids) include a "neck grip" during mating (Ewer 1973). This ritualized aggression is probably a very old, phylogenetic trait that possibly has an important role in stimulating ovulation and precopulatory behaviors (Ewer 1973). In the sea otter, why has the target of male biting been redirected from the neck to the nose and face region? Perhaps a female that suffers from the exposure of cold water as a result of damage to the insulative properties of her neck pelage also makes for a poor mother, and thus she may wean fewer pups.

After successfully copulating, the female leaves the male within a few days (Garshelis et al. 1984, Riedman and Estes 1990). Like many other mustelids, sea otters have delayed implantation (see Boyd, Lockyer, and Marsh, Chapter 6, this volume) that results in a variable gestation period that averages about 6 months, including an unimplanted period of 2 to 3 months (Jameson and Johnson 1993, Riedman et al. 1994, Monson and DeGange 1995).

Rearing Behaviors

Single, precocial neonates (the general rule in marine mammals) are normally born in the ocean, and are dependent on their mothers for about 5 months. However, pups sometimes remain with their mothers for up to 8 months (Jameson and Johnson 1993, Riedman et al. 1994, Monson and DeGange 1995). During dependency, pups are nursed and provisioned with prey by their mothers—fathers, as in most other mammals, have no role in rearing young. Vocal communication (above water) is frequent between a mother and her pup, and includes screams, cries, whines, hisses, growls, and barks. These sounds not only communicate different behavioral states, but perhaps individual identity (McShane et al. 1995). Learning about diving, foraging, tool use, and prey are all important features of the period of dependency (Sandegren et al. 1973, Payne and Jameson 1984, Riedman and Estes 1990). For example, individual adult sea otters often specialize on particular food items such as sea urchins, clams, or crabs. Prey specialization tends to be passed on from mother to pup, probably through learning during the period of dependency (Riedman and Estes 1990). The types of tools used by individual sea otters also appear to be learned by pups from their mothers (Riedman and Estes 1990).

In Alaska, some island populations of sea otters eat fish (Estes et al. 1982). Riedman and Estes (1990) speculate that fishing behavior is probably discovered by one or a few individual sea otters within a population when the sedentary macroinvertebrate prey decline because the carrying capacity of the sea otters has been reached (see Bowen and Siniff, Chapter 9, this volume). Through learning, especially from mother to pup, fishing becomes more widespread and eventually is established as a tradition within a population.

Spatial Organization

Sea otters generally occupy a well-defined, linear habitat bordered by shore on one side and deep water on the other. Even islands or chains of islands provide only narrow strips of habitat for sea otters. Perhaps the few exceptions to this are large enclosed bays, such as Prince William Sound in Alaska. Because of the long, narrow nature of sea otter habitat, and their relatively limited diving and swimming abilities, long-distance movements (measured in hundreds of kilometers) are unusual. In fact, most sea otters establish home ranges that are relatively small and stable (Ralls et al. 1996) and encompass suitable food resources and shelter from rough seas. Feeding and resting often occur in different parts of the home range (Garshelis et al. 1984). Depending on the configuration of the coast, home ranges typically include 1 to 17 km of coastline (Ribic 1982, Jameson 1989, Riedman and Estes 1990), but in Prince William Sound, Alaska, Monnett and Rotterman (1988) found females using larger areas.

In California, dense beds of giant kelp (*Macrocystis pyrifera*), and to a lesser degree bull kelp (*Nereocystis leutkeana*), are favored resting spots. Not only do kelp beds protect the sea otters from rough seas (the algae dissipate surface chop), they have learned to wrap themselves up in the fronds to keep from drifting away. In areas where large kelp beds are not a feature of the coast, such as some areas in Alaska, sheltered coves are more important features of home ranges. However, some sea otters (usually juvenile males) sometimes rest in open water up to several kilometers from shore (Ralls et al. 1996). But, even sea otters that establish offshore home ranges must remain in water shallow enough to enable them to forage effectively.

The importance of familiar home ranges to sea otters became evident during two experimental translocations in California, where mainland sea otters were either relocated along the mainland coast (Ralls et al. 1992), or were reintroduced to San Nicolas Island (Rathbun et al. 1990a). In both cases, sea otters attempted to return to their original home ranges. During the 4-year San Nicolas Island program, 139 sea otters were captured between Morro Bay and Monterey and flown about 200 km to the island and released. Despite the near-ideal habitat at the island, with an abundance of prey, most of the sea otters left. Many perished, probably because they headed home from the island in the wrong direction. There is little question that sea otters can swim long distances. Ralls et al. (1996) radio-tracked sea otters on the mainland, and found that occasionally they made sudden, short duration movements of 100 km outside their normal home ranges; juvenile males were especially prone to these excursions. However, in the case of San Nicolas Island, the sea otters dispersed into novel waters and apparently most were unable to navigate back home. What is amazing is that at least 31 did manage to return home, many to the exact area where they were captured! This home range fidelity is similar to that found in many terrestrial carnivores that rely on familiarity with an area to effectively gather food or capture prey. Nielsen and Brown (1988) have suggested that home range fidelity is one reason that the reintroduction of carnivores (as opposed to ungulate herbivores) is often difficult.

The home ranges of female sea otters generally overlap each other and are relatively stable through time. Males, however, use space in a more complicated and variable pattern through time. When females are in estrus, usually during the late summer and fall (more northern populations tend to breed later; Riedman and Estes 1990, Jameson and Johnson 1993), males establish home ranges among the females (Jameson 1989). The male areas often are continuous and are usually exclusive of other males—they seem to be defended territories (Garshelis et al. 1984, Jameson 1989). However, there is little actual aggression or fighting that occurs on the territories; most defense is accomplished by subtle postures (Loughlin 1977, Packard and Ribic 1982) and perhaps individual recognition based on a keen sense of smell (Riedman and Estes 1990). Vocalizations apparently are not used in territorial defense (McShane et al. 1995). The male territories overlap with parts of one or more female home ranges (female areas are not exclusive, and thus are not territories). Males with high-quality territories, which include optimal food resources and sheltered resting areas, attract and mate with most females (Garshelis et al. 1984). This is an example of resource defense polygyny, as already discussed for pinnipeds. Also, as in some pinnipeds, some male sea otters attempt to keep females from leaving their territories by blocking their exit, or even herding them (Garshelis et al. 1984, Riedman and Estes 1990). Male territories vary in size from 20 to 100 hectares, and in general they are about one and a half to two times smaller than female home ranges (Riedman and Estes 1990). Males often leave their territories for short periods of time, which results in their making more extensive and frequent movements than females (Ralls et al. 1996). This suggests that some males may practice a form of scramble competition by leaving their territories and searching for females in estrus. Although the modal sea otter mating system seems to be resource defense polygyny, there appears to be a certain amount of flexibility exhibited.

Sexual Segregation and Aggregations

In some areas during the winter and spring, when females are generally not in estrus (Jameson and Johnson 1993), many males abandon their territories and migrate to all-male areas (Jameson 1989). In California, when the sea otter population was rapidly expanding, male aggregations were typically found at the two ends of the distribution. However, in more stable populations, or in areas where essentially all available habitat is occupied, the all-male groups (rafts) are interspersed among the female home ranges, usually in exposed and presumably less optimal habitats (Riedman and Estes 1990). Male groups (see Fig. 8-8) are composed of adults that have left their territories and subadults that have left their maternal home ranges but are not old enough (possibly not large enough?) to hold territories. Male groups or rafts frequently include 40 to 80 individuals, but aggregations can exceed hundreds of animals in Alaska (Garshelis et al. 1984). The composition of the rafts is highly fluid, as individuals join and leave throughout the day. This is similar to the fission–fusion social structures of some cetaceans and the West Indian manatee (see below; Ghiglieri 1984).

Generally, the behavior of sea otters in rafts is synchronized, with periods of grooming and foraging separated by periods of inactivity. There are also frequent periods of intense social interactions, including play fighting (Packard and Ribic 1982, Garshelis et al. 1984). Dominance may be established between individuals during these interactions, which might help define the outcome of subsequent, less intense, encounters in territories. Rafts of females and dependent pups generally include fewer individuals than male groups, and there is usually less social interaction among the females (Garshelis et al. 1984, Riedman and Estes 1990). Garshelis et al. (1984) discuss the various functions that sea otter rafting may have, including establishment of a rudimentary rank order, predator avoidance, and social facilitation of foraging through the mechanism of reciprocal altruism (i.e., "good samaritan behavior"; see Trivers 1971, Wilson 1975). Any one of these functions does not explain all the factors associated with rafting. Biologists, in their attempts to understand natural systems, often try to fit their observations into single models when in fact a more sophisticated model, or several models, are needed.

Jameson (1989) and Riedman and Estes (1990) have speculated that the reason males migrate seasonally between territories and male raft areas relates to optimal foraging and access to estrous females. If food resources were not limited, males might set up and maintain year-round territories. However, with females occupying their home ranges year-round, food resources in these areas may be limited. The female reproductive cycle is diffusely seasonal (discussed above), and when there is little opportunity to mate, males may move from areas with limited prey populations to relatively food-rich regions, where there are few or no females. When females begin to come into estrus again, the adult males migrate back into the female areas and reestablish their territories. The breeding territories of males are often reoccupied by the same individuals year after year (Garshelis et al. 1984, Jameson 1989, Riedman and Estes 1990). This predictable type of movement between feeding and breeding areas is similar to that already discussed for some pinnipeds.

Social Organization

Estes (1989) has suggested that sea otter social organization is related to abundance and availability of prey (high densities of relatively easily obtained, sedentary macroinvertebrates) in near-shore environments. He reasons that with an abundant food supply, sea otters themselves are able to achieve high population densities. This, in turn, enables males to occupy home ranges (i.e., areas where daily activities are conducted) that include several females, and the areas are small enough to effectively defend as territories against other males. Adult and juvenile male sea otters that are excluded from an area because it is saturated with females and territorial males are relegated to habitats that females avoid. These males aggregate and form rafts.

A social structure similar to that of the sea otter, where males are polygynous and groups form that are sexually segregated ("bachelor herds" and "mixed or breeding herds"), is fairly common in terrestrial ungulates with medium to large body sizes. Examples include the eland, lechwe, gazelles, wildebeest, and waterbuck (Leuthold 1977). Like the sea otter, these ungulates have developed resource defense polygyny as a result of the dispersion of their food resources, but what makes the similarity especially interesting is that the two groups do not share a close evolutionary history (phylogeny). On the other hand, the social structure of sea otters is not commonly found in most terrestrial carnivores, which do have a close phylogeny with sea otters. Large home ranges are often required by terrestrial carnivores to obtain sufficient food, which is typically widely dispersed. Large areas usually cannot be effectively defended as territories. For example, males cannot easily find and mate with multiple females, or defend multiple females when they are dispersed on large home ranges. These conditions often result in a nearly solitary, weakly polygynous social organization (river otters, martins, bobcats, pumas, bears, etc.). In other cases, conditions are so adverse (prey is difficult to obtain, predation on young is heavy, etc.) that females require the assistance of other individuals to successfully raise young. In these cases, animals are still in low densities, but their social organizations are more structured, such as monogamy (i.e., foxes, jackals, and wolves) and extended family groups (some mongooses, cape hunting dog, etc.) (Eisenberg 1981).

We have stressed the importance of fur in understanding

many aspects of the behavioral ecology of sea otters. For example, on the one hand their dense pelage enables them to live in the marine environment, but on the other it helps to limit them to near-shore areas. Not surprisingly, there are many near-shore, temperate regions that support an abundance of macroinvertebrates, but where sea otters do not occur. Some of these areas, however, have been occupied by other species of otter that are similar in many respects to *Enhydra*. For example, in the eastern North Atlantic, river otters (*Lutra lutra*) occur in near-shore marine habitats, and along the coast of Chile the marine otter (*L. felina*) has filled the sea otter niche. Might otters be used to help us understand the roles of environment and phylogeny in the evolution of social organization? Given the apparent convergence of social structures by some terrestrial ungulates and the sea otter, as discussed, we might expect that the various closely related otters that occupy similar near-shore marine habitats to have very similar social organizations. Although biologists have only just begun to understand the behavioral ecology of *Lutra*, apparently the social organizations of the various marine forms are not that similar to that of *Enhydra* (Ostfeld et al. 1989, Kruuk and Moorehouse 1991). Perhaps the marine *Lutra* are a good example of one aspect of the influence of phylogenetic inertia (Simpson 1944) on social structure; the marine *Lutra* are relative newcomers to the near-shore community compared to *Enhydra*, and the genetically determined aspects of their behavior may prevent them from exhibiting a social organization like that of the sea otter, which undoubtedly has evolved over a greater period of time than has been available to the *Lutra* species. The role of phylogeny has been suggested as important in understanding the social behavior of some terrestrial mammals, such as elephant-shrews (Rathbun 1979) and zebras (Berger 1988), as well as pinnipeds, as we have already discussed.

Cetaceans [R. S. W.]

The advent of systematic study of the behavior of cetaceans has occurred within the past 50 years. Earlier information on whale behavior was provided by the log books and narrative accounts of commercial whalers (e.g., Scammon 1874). Interest in the behavior of smaller cetaceans, such as dolphins, increased dramatically as they began to be observed in captivity around the turn of the century (Townsend 1914). During the 1930s and 1940s colonies of dolphins, especially bottlenose dolphins (*Tursiops truncatus*) established at public display facilities (McBride 1940), provided unique opportunities for observing behavior. Descriptive studies of the behavior of captive small cetaceans flourished during the 1940s through the 1960s (e.g., McBride and Hebb 1948, McBride and Kritzler 1951, Lawrence and Schevill 1954, Tavolga and Essapian 1957, Essapian 1963, Tavolga 1966, Caldwell and Caldwell 1967). The results of these early observations motivated a number of researchers to also ask questions about the animals in their natural environment, away from the potential complications and influences of captivity.

A few seminal field studies of whales and dolphins were conducted in the 1950s and 1960s (e.g., Caldwell 1955, Brown and Norris 1956, Layne 1958, Norris and Prescott 1961, Caldwell et al. 1965, Brown et al. 1966, Evans and Bastian 1969, Caldwell and Caldwell 1972), but cetacean field studies expanded exponentially in number and scope beginning in the late 1960s and early 1970s with the advent of individual identification techniques, radio-tracking, and the development of comparative methods, quantitative measures, and statistical analysis techniques (e.g., Evans 1971; Evans et al. 1972; Irvine and Wells 1972; Tayler and Saayman 1972; Altmann 1974; Leatherwood 1975; Würsig and Würsig 1977, 1979, 1980; Katona et al. 1979; Norris and Dohl 1980a; Shane 1980; Watkins et al. 1981; Balcomb et al. 1982; Bigg 1982; Payne et al. 1983).

The first generation of cetacean behavioral scientists began their work before marine mammal science was well recognized as a field of study. Their efforts have established cetacean behavior as an important field of endeavor. As a result, our understanding of the lives of cetaceans has improved dramatically in the past few decades, but as will become clear through consideration of the overviews presented, many important areas of inquiry remain for the next generations of marine mammal scientists.

Before beginning the review of the behavior of cetaceans, several terms associated with ranging patterns, habitats, or social units that are used in this and the sirenian sections need to be defined:

Migration: a pattern of regular (often on an annual cycle) movements between at least two locations, typically separated by a long distance.

Home Range: a well-defined area of regular usage that typically provides for all or most of an animal's needs.

Core Area: within a home range, the region where most of an animal's activities are concentrated.

Territory: an area from which conspecific intruders are excluded.

Pelagic or Open Ocean: offshore deep-water habitat, including continental shelf waters, unbounded by shorelines.

Coastal: habitat along an open shoreline.

Inshore: enclosed bays and estuaries, often including associated coastal waters.

Riverine: river systems, including associated lakes and estuaries.

Population: a locally interbreeding group of organisms

occupying a clearly delimited space at the same or overlapping times.

Stock: a unit approximating a population, established for wildlife management purposes.

Community: a regional society of animals sharing ranges and social associates, but exhibiting genetic exchange with other similar units (populations are closed reproductive units).

Aggregation: a temporary collection of animals resulting predominantly from environmental rather than social factors.

Congregation: a temporary collection of animals resulting predominantly from social rather than environmental factors.

Group, School, or Herd: kinds of congregations exhibiting increased levels of organization; a cohesive collection of conspecifics in a limited area (typically, within several hundred meters), often engaged in similar activities and moving in the same general direction, maintained by social factors as a unit; groups may be stable over long periods of time or may change composition over periods ranging from minutes to weeks.

Subgroup: a distinct cluster of individuals within a group, often of similar sex, age, or reproductive condition.

Pod: a long-term group, of virtually unchanging composition.

Dominance Hierarchy: the physical domination of some members of a group by other members, in relatively orderly and and long-lasting patterns. Except for the highest and lowest ranking individuals, a given member dominates one or more of its companions and is dominated in turn by one or more of the others. The hierarchy is initiated and sustained by aggressive and agonistic behavior, albeit sometimes of a subtle and indirect nature (Wilson 1975).

Feeding Patterns

Food alone has defined the two divergent pathways of extant cetacean evolution, toward the mysticetes, or baleen whales, which use baleen to filter large quantities of small organisms from the water, and the odontocetes, or toothed whales, which capture individual prey items generally with the aid of teeth. Spatial and temporal distribution and abundance of food resources have shaped many behavioral adaptations in cetaceans.

The two suborders of cetaceans differ dramatically in their prey and feeding structures. These differences are reflected in their foraging and feeding behavior as well. The amount of time that a cetacean must spend in search of food defines the time that remains for other activities; in the absence of sufficient energy, all other activities cease. This basic fact has led to the development of patterns of activity, and the partitioning of time into identifiable cycles. In the case of many of the baleen whales, this has resulted in annual long-distance migrations between feeding and breeding grounds (see Bowen and Siniff, Chapter 9, this volume). In other cases, diurnal patterns of activity are dictated by the availability of specific kinds of prey at particular times of the day.

Different techniques are required for mysticetes to locate, capture, and filter larger quantities of small organisms (such as krill, copepods, amphipods, and schooling fish), than are needed for the more sizable individual prey items (such as fish, squid, or other marine mammals) taken by odontocetes. Different strategies for meeting energy requirements are involved for mysticetes that must secure all of the food for an entire year during a brief feeding season, as compared to the daily hunting of most odontocetes.

For many mysticetes, foraging involves an annual migration to a higher latitude region of abundant, concentrated food, from a lower latitude breeding area of limited food resources. Spring migrations to feeding areas after winter fasts on the breeding grounds are the norm for such baleen whales as the gray (*Eschrichtius robustus*), humpback (*Megaptera novaeangliae*), right (*Eubalaena australis*), and blue (*Balaenoptera musculus*), which swim thousands of kilometers to areas of seasonal abundance of food. Annual oceanographic and climatological cycles result in predictable regions of high summer productivity that not only support the immediate energy needs of the whales, but allow the animals to develop energy stores for the fast during their next migration and time on the breeding grounds. On the feeding grounds, mobility of the whales allows them to seek the most productive regions, or to select alternative prey species if necessary. This flexibility is exemplified by the appearance of large concentrations of blue whales along different parts of the California coastline since the mid-1980s (Calambokidis et al. 1990), and by occasional shifts in the regional abundance of humpbacks in the northwest Atlantic with changes in relative abundance of prey such as sand lance (*Ammodytes americanus*) and herring (*Clupea harengus*).

Mysticetes are batch feeders. The kind of prey taken by baleen whales is related in large part to the structure of their baleen plates, their ability to ingest and filter large volumes of water, and their excellent swimming ability. In general, baleen whales open their mouths and take in water and food, then close their mouths partially, forcing the water through the interwoven frayed edges of their baleen plates. The whales then scrape the trapped food off the filter with their tongues and swallow it. Different mysticetes are adapted to feeding primarily on benthic organisms, in the water column, or at the surface, although some flexibility for a given species to feed in more than one of these areas is evident. In some cases, as many as four different species of baleen whales have been observed feeding together, using different specialized techniques (Watkins and Schevill 1979, Tershy 1992).

Gray whales, with short, thick, coarse baleen, often feed on the bottom, obtaining amphipod crustaceans and other invertebrates in benthic community mats. They frequently surface with sediments clinging to the rostrum and streaming from the mouth (Nerini 1984, Würsig et al. 1986). They dive to the bottom, typically roll onto their right side (based on patterns of baleen wear; Kasuya and Rice 1970), and apply suction near the sea floor to ingest the mats. Sidescan sonar records and diver observations have found numerous fingerlike impressions, measuring on average 1 to 3 m long by 0.5 to 1.5 m wide, indicative of gray whale feeding in the Bering Sea sea floor in areas of high amphipod density (Nerini 1984). Nerini (1984) estimated that a single gray whale may consume about 36,821 kg/year of amphipods in this way. However, gray whales can also be generalists. Off Vancouver Island they feed on dense schools of small shrimplike mysids near the substrate (Murison et al. 1984). They are also capable of feeding on pelagic prey, such as fish, squid, or crab larvae, through surface skimming and engulfing. These behaviors provide gray whales with perhaps a greater diversity of feeding techniques, and thus greater dietary flexibility, than the other mysticetes (Nerini 1984).

The baleen of the rorqual whales, such as the blue, finback (*Balaenoptera physalus*), minke (*Balaenoptera acutorostrata*), or humpback, is intermediate in length and the coarseness of its frayed edges. The swiftness of rorquals may facilitate the pursuit of mobile prey, such as krill and small schooling fish like sand lance, herring, or capelin (*Mallotus villotus*). Thus, the rorquals tend to target prey in the water column and at the surface. Their expandable gular pleats (throat grooves or rorqual grooves; Fig. 8-11), along with a huge oral sac created when water pressure from swimming with the mouth open turns the tongue inside-out (Lambertsen 1983), allow for larger volumes of water to be engulfed in a single gulp than is possible for other mysticetes. Contraction of muscles that turn the tongue right-side-out again and others that contract the expanded throat region force the water through the baleen, leaving the food trapped on the baleen fringe filter, where it is collected by the tongue and swallowed.

The rorquals use a variety of behaviors to capture their prey. Lunge-feeding, in one of three forms, is commonly used by one or more whales feeding on prey near the surface (Jurasz and Jurasz 1979, Hain et al. 1982, Whitehead and Carlson 1988). For example, finback whales in the Gulf of California or blue whales in Monterey Bay, California, may surface at an oblique angle, with their mouths agape, capturing prey trapped by the water's surface. This is often done simultaneously with other whales close by. One complex variation, known as bubble feeding, has been developed by humpback whales. Bubble feeding involves individuals or several whales working together, apparently containing or concentrating prey by creating a cloud, column, or curtain of bubbles from their blowholes (Hain et al. 1982). When the bubbles form a ring or closing spiral, the pattern is termed a bubble net, with the escape of the prey blocked above by the water's surface, and from below by the mouth of the whale(s) (Jurasz and Jurasz 1979). The bubble net moves the fish into a tight school, and pushes them toward the surface. The participating whales then swim vertically through the water column and surface with their mouths

Figure 8-11. Humpback whale feeding offshore of southeast Alaska. The whale is closing its mouth. Note the tremendous expansion of the gular pleats (right) and water streaming through the baleen plates (left). (Photograph by Flip Nicklin, Minden Pictures.)

open, ingesting the concentrated prey. "Flick feeding" involves a humpback whale at the surface lashing its tail forward, concentrating krill, then diving rapidly forward into the prey with its mouth open (Jurasz and Jurasz 1979).

Humpback whales also feed within the top 120 m of the water column, on very dense patches of pelagic, shrimplike crustaceans known as euphausiids (Dolphin 1987a). This has been demonstrated through the use of sonar scans and remotely operated underwater cameras (Dolphin 1987a). Humpback whale dive depths are predictable based on ventilation patterns (breathing rates observed at the surface) and dive duration (Dolphin 1987b). Consideration of ventilation and dive variables suggests that dives in excess of 41 to 60 m may result in anaerobic metabolism and consequently, less efficient use of time and energy resources as compared to shallower dives (Dolphin 1987c, 1988). Midwater fish and euphausiid prey are mobile and distributed in patches in restricted regions of high productivity, such as upwellings. Thus, foraging whales may have to continue to expend energy traveling once they have arrived at the feeding terminus of the migration (Clapham 1993).

Individuals of the smallest of the rorquals, the minke whale, specialize in one of two feeding strategies: lunge feeding or bird-association feeding in which the whales exploit concentrations of fish fry below flocks of feeding gulls and diving birds such as alcids (Hoelzel et al. 1989). These individual strategies and locations of feeding activities are consistent over periods of up to at least 5 years. The bird-association feeders take advantage of a concentrated prey resource that is very ephemeral, whereas the lunge feeders make use of a patchy resource that can be found relatively predictably in particular areas. Lynas and Sylvestre (1988) characterized the feeding of minke whales in the St. Lawrence estuary as "line fishing," involving engulfing mouthfuls among prey scattered along a line, versus "patch fishing," a more intense and localized activity involving more concentrated prey.

Four sympatric species of rorquals were observed feeding simultaneously in the waters of the Canal de Ballenas, Mexico, exhibiting several variations in foraging and prey capture techniques (Tershy 1992). Bryde's (*Balaenoptera edeni*) and minke whales fed on fast schooling fish in small spatially and temporally predictable patches. Bryde's whales were relatively resident to the area and fed alone or in small aggregations. Finback and blue whales fed on slow and less nutritious euphausiids that were found in large spatially and temporally less predictable patches. Finbacks were relatively transient, and fed in larger aggregations in which there were coordinated groups of two to four individuals.

The bowhead (*Balaena mysticetus*) and right whales have longer baleen with a finer fringe than other mysticetes. Of these, the pygmy right whale (*Caperea marginata*) has the finest baleen fringe of all (Sekiguchi et al. 1992). There is a space between the right and left baleen rows at the front of the jaws of these species (Fig. 8-12). Taken together, these adaptations allow these whales to filter very small crustaceans such as copepods from the surface and water column. Northern right whales (Watkins and Schevill 1976) and bowheads (Würsig and Clark 1993) frequently engage in skim-feeding at the surface, wherein the whales move slowly through concentrations of prey, with their mouths open about 30 to 40 degrees. Effective skim feeding in the right whales is facilitated by the long baleen, arched upper jaw, and mobile lower lips that can control water flow. Bowheads have also been observed feeding while swimming on their sides, with the mouth open about 70 degrees, or ventrum up

Figure 8-12. Skim-feeding right whale. The space between the jaws facilitates water intake for skimming. Note the callosities on the top of the rostrum and along the lower jaw lines. (Photograph by Flip Nicklin, Minden Pictures.)

(Würsig et al. 1985, 1989; Würsig and Clark 1993). Where calanoid copepods are abundant, surface-feeding bowhead whales typically feed in loose groups of 2 to 25. Several whales, separated from each other by about 8 m, may move through prey patches in staggered, or V-shaped, echelon formations. These formations are dynamic, with participation changing at any time, and position changing when formations change direction. Similar formations occur in southern right whales (Würsig and Clark 1993). These formations may increase the efficiency of feeding for each of the participants by concentrating the prey; perhaps the barrier provided by each whale ahead limits the escape options for prey relative to the next whale in line (Würsig and Clark 1993).

Right and bowhead whale feeding is not limited to the surface, however. Most of the feeding of right whales appears to occur about 10 m or more below the surface (Watkins and Schevill 1976). Bowhead whales in shallow water (10 to 29 m) in the Beaufort Sea often surface with mud streaming from their heads, bellies, and sides, suggesting benthic feeding by skimming small prey from just above the substrate (Würsig et al. 1989, Würsig and Clark 1993). This evidence, and stomach contents indicative of benthic feeding, have only been reported for subadult bowheads (Würsig and Clark 1993).

In contrast to the batch-feeding mysticetes, most odontocetes capture single prey. The kind of prey taken by a given odontocete is related in part to the cetacean's body size, speed, maneuverability, and diving capabilities, as well as to the number, shape, position, and size of its teeth. Fish and squid tend to form the diets of most odontocetes, although a few species also eat other marine mammals. Many odontocetes use their teeth and jaws to grasp their prey, or in some cases to tear the prey, but not to chew. The tongues are used to maneuver the prey into an appropriate orientation for swallowing. It has also been suggested that a piston action of the gular (throat) muscles and tongue within the mouths of beaked whales may provide a means of sucking in prey (Heyning and Mead 1996). Odontocetes with few or no erupted teeth, or teeth only in the lower jaw, tend to eat soft-bodied prey such as squid. Odontocetes with numerous small teeth in the upper and lower jaws, tend to eat small prey, and those with large teeth in both jaws tend to eat large prey. The interdigitating arrangement of upper and lower teeth in some odontocetes facilitates the expulsion of water and retention of prey as the jaws close. Norris and Møhl (1983) hypothesized that some odontocetes may be able to emit sounds so intense and focused that their prey is debilitated and capture made easier; this possibility remains to be documented conclusively.

Another major difference between the mysticetes and the odontocetes is in the temporal and spatial aspects of their foraging. In part because they are smaller than mysticetes and cannot carry large quantities of food reserves, most odontocetes feed throughout the year, meaning that they are tied to regions where prey are available at all times, or they move with the prey. Therefore, foraging is an important component of each day's activities. Odontocetes also have the ability to use echolocation to locate prey; this has yet to be demonstrated for mysticetes.

Several general feeding patterns can be identified within the odontocetes relative to the different biological and ecological features. The largest of the odontocetes, including species such as the sperm whales and beaked whales, are deep divers that hunt squid. These cetaceans typically have few teeth—if erupted teeth are present, then they tend to be only in the lower jaw. They are capable of dives of more than 1 hr, and have been known to reach depths of thousands of meters (Heezen 1957). Northern bottlenose whales (*Hyperoodon ampullatus*), which feed on squid (typically, *Gonatus fabricii*) and are thought to be deep divers, dive for periods in excess of 1 hr, returning to the surface near the same location (Benjaminsen and Christensen 1979). Baird's beaked whales (*Berardius bairdii*) off Japan exhibit similar diving and foraging traits (Kasuya 1986).

Sperm whale (*Physeter macrocephalus*) females and juveniles often surface and dive in a coordinated manner, and presumably forage as groups. Foraging sperm whales off the Galápagos Islands travel in ranks about 550 m long, perpendicular to the direction of travel, with clusters of whales spaced out along the ranks (Whitehead 1989). These ranks move at about 4 km/hr, and maintain their headings for several hours. Feeding dives usually involve lifting of flukes above the water's surface to initiate a dive. Whales tracked with a depth sounder at feeding depths of approximately 440 m appear to swim about 40 m apart, with no particular positioning based on age or sex. The whales feed on histioteuthid squids, which are gregarious, and probably not strong swimmers. Whitehead (1989) suggests that rank foraging provides benefits to individuals by aiding in the gathering of information on prey densities, avoiding mutual interference, and possibly permitting capture of prey that has eluded other nearby whales.

In comparison to other sperm whales, bachelor males spend much of the year at higher latitudes foraging alone or in small groups. The segregation of sperm whales may be related to different sizes of squid eaten by the two sexes—the large squid eaten by the large males are found at greater depths and higher latitudes than the smaller squid eaten by the females and juveniles (Best 1979). Watkins and Schevill (1977) used a hydrophone array to track diving sperm whales in the western North Atlantic, and found these animals (probably large males based on the latitude) to be spread over greater horizontal and vertical distances (≥ 100 m apart) than were the females and young tracked by Whitehead (1989). The presumed males apparently regroup before they surface.

Variations in feeding behavior of sperm whales in apparent response to changes in environmental conditions have been reported from the Galapagos Islands. When histioteuthid squid availability declined during an El Niño warm water event, sperm whale dives became more shallow (to about 320 m vs. 420 m in previous years) and involved a greater variety of depths, the frequency of energetic aerial displays such as breaches declined, and directional changes became more frequent during foraging (Whitehead et al. 1989, Smith and Whitehead 1993). Whitehead (1996a) also examined sperm whale feeding success relative to environmental conditions by monitoring defecation rates, and suggested that the whales are able to cope with great variability in resource availability by migrating from areas of low food abundance. He also suggested that the presence of older, experienced whales in permanent social units may be important in directing such migrations, as has been proposed for other large, long-lived mammals such as elephants (Douglas-Hamilton and Douglas-Hamilton 1975). The older animals may, for example, retain memories of the locations of, and routes to, previous sites of resources.

Members of the family Monodontidae, the belugas (*Delphinapterus leucas*) and narwhals (*Monodon monoceros*), feed on fish and invertebrates in arctic waters. Their seasonal movements follow changes in ice coverage and movements of prey. Tracking using satellite-linked transmitters on 15 belugas indicated that during the summer, belugas spend at least some time each day in deep water, diving to the sea bed at up to 440 m depth to forage (Martin et al. 1993). The specific prey taken at such depths are unknown. Stomach content analyses indicate that narwhals may feed on fish on the bottom in water more than 500 m deep during the period of ice break-up, may nearly stop feeding during the open water season, and resume feeding after leaving the summering areas (Finley and Gibb 1982).

The largest dolphins, such as the killer whales, pilot whales, and false killer whales (*Pseudorca crassidens*), use large teeth in both upper and lower jaws to capture large prey at intermediate depths and near the surface. Although the most common diets of these species include schooling fish and squid, they also eat elasmobranch fishes and marine mammals (Martinez and Klinghammer 1970, Fertl et al. 1996). False killer whales and pilot whales have been observed attacking members of the genera *Stenella* and *Delphinus* near tuna purse seine nets (Perryman and Foster 1980). Brown et al. (1966) reported a captive pilot whale seizing and consuming the stillborn calf of a common dolphin tankmate. Recently, false killer whales and pilot whales have also been observed attacking sperm whales, but whether these attacks involved competition for prey, were kleptoparasitic (i.e., designed to cause the sperm whales to regurgitate prey that were then consumed by the smaller whales), or were direct predation (as suggested by observations of pieces of sperm whale flesh) was undetermined (Palacios and Mate 1996, Weller et al. 1996).

Killer whales feed on a variety of other cetaceans, including mysticetes, sperm whales, narwhals, dolphins and porpoises, and a variety of pinnipeds. Killer whales beach themselves at pinniped rookeries, on incoming tides, to capture pinnipeds at the water's edge (Lopez and Lopez 1985, Guinet 1991). In one report, killer whales spyhopped, or lifted their heads until their eyes were clear of the water, near an ice floe upon which a crabeater seal was lying, then created a wave that tipped the floe and threw the seal into the water (Smith et al. 1981). Smaller, single mammalian prey items may be captured individually, but schooling prey are often subject to a coordinated attack by an entire pod of killer whales. On one occasion a group of more than 300 narwhals, initially in full flight, was overtaken by a school of killer whales, which encircled and simultaneously attacked them (Steltner et al. 1984). Killer whales sometimes feed on small schools of Pacific white-sided dolphins (*Lagenorhynchus obliquidens*) in Alaskan waters (Dahlheim and Towell 1994). A pod of killer whales chased belugas into shallows at a creek mouth in an Alaskan river, capturing several and taking them into deeper water (Frost et al. 1992). With some of the smaller mammalian prey, the food handling process may be prolonged. Numerous descriptions exist of killer whales carrying, tossing, or beating smaller prey with their flukes, bodies, dorsal fins, or pectoral flippers (Felleman et al. 1991, Baird and Dill 1995). Sometimes they release prey, only to recapture it as it tries to escape.

With larger prey, such as baleen whales, members of killer whale pods work together (Tarpy 1979, Silber et al. 1990), and may approach silently. They may separate humpback whale calves from adults, to facilitate feeding (Flórez-González et al. 1994). Killer whales grab tails, lips, tongues, flippers, and blowholes in an effort to subdue the whales (Best 1982). Typically, a killer whale attack results in fatal wounds and the removal of large pieces of the prey, but the attack rarely kills the whale immediately, and the entire whale is not consumed.

The strong social bonds and extended periods of association of large dolphins have likely facilitated the development of coordinated and cooperative prey capture techniques, as appears to be the case for most cooperative terrestrial hunters, such as lions or wolves (Mech 1970, Packer and Ruttan 1988). Groups of the larger dolphins commonly spread in a wide line–abreast formation in an apparent food-search mode (Würsig 1986, Heimlich-Boran 1988, Felleman et al. 1991). Different social groups of killer whales near Vancouver Island have developed apparent "cultural" differences relative to their prey, and use different foraging strategies. "Resident" whales are present during the summer months, coinciding with the occurrence of large schools of salmon, their primary prey. The resident whales concentrate their feeding around three high-relief, kelp-covered reefs (Heimlich-Boran 1987). They may use percussive behaviors such as tail slaps to herd prey fish or to communicate with one another and coordinate movements (Felleman et al. 1991), but the coordination appears to be looser than for whales feeding on mammalian prey (Heimlich-Boran 1988, Hoelzel 1993). The residents apparently forage primarily at slack tides to optimize their ability to approach aggregations of salmon (Felleman et al. 1991), and use bathymetric features as barriers to aid in the collection of prey into higher densities (Heimlich-Boran 1988). "Transient" whales move sporadically through the shallower waters of the area throughout the year, and specialize on marine mammal prey, especially harbor seals, that frequent the shallows. They use a "sit-and-wait" approach, spending extended periods in bays where pinnipeds are present (Felleman et al. 1991). Foraging behaviors, including feeding, require on average 63% of the transients' time, but this proportion decreases with increasing group size (Baird and Dill 1995). The transients make fewer sounds than the residents, presumably in an effort to avoid alerting potential prey to their presence. Few interactions occur between residents and transients (Baird and Dill 1995).

Among the smaller odontocetes, biological and ecological factors also interact to define feeding patterns (Wells et al. 1980). Within species, feeding effort and diets vary by region, season, age, sex, and reproductive classes. For example, lactating female bottlenose dolphins (Cockcroft and Ross 1990a), common dolphins (*Delphinus delphis*) (Young and Cockcroft 1994), and pantropical spotted dolphins (*Stenella attenuata graffmani*) (Bernard and Hohn 1989) feed more, and on different size classes and species, than other dolphin classes, presumably reflecting different energy and hydration (osmoregulatory) requirements. In the Bay of Fundy, harbor porpoise calves feed on euphausiids, whereas adults feed primarily on fish (Smith and Read 1992). Bottlenose dolphins at three different study sites in Texas feed more in autumn, perhaps in response to presumed higher energy requirements due to declining water temperatures and seasonal deposition of additional subcutaneous blubber (Gruber 1981, Shane 1990a, Bräger 1993, Wells 1993c). Shane (1990a) reports significantly different amounts of time spent by bottlenose dolphins feeding in Texas (21%) versus Florida (36%), where water temperature changes are not as extreme as in Texas. Near Sarasota, Florida, about 90 km north of Shane's study site, bottlenose dolphins spend about 13% of their time feeding, with females spending less time feeding in winter than in summer (Waples 1995). D. Costa, G. Worthy, R. Wells, D. Waples, M. Scott, B. Irvine, and A. Read (unpubl. data; see Costa and Williams, Chapter 5, this volume), using doubly-labeled water techniques with free-ranging dolphins, found that Sarasota dolphin metabolic rates are lower in winter than in summer. These results suggest lower energy requirements when the water is colder, after completion of the autumn blubber deposition (Wells 1993c). Pacific coast bottlenose dolphins spend 19% of their time feeding (Hanson and Defran 1993).

In general, small odontocete feeding patterns are closely related to habitat type, and may be categorized grossly as open ocean, coastal, and riverine. Small open ocean dolphins and porpoises, lacking obstacles in their environment, are well adapted for high-speed swimming. They tend to be more streamlined, with smaller fins than the coastal or riverine species. The small pelagic cetaceans tend to have large numbers of small interdigitating teeth in both upper and lower jaws for capturing small schooling prey, which occur as very rich, but patchy and mobile, resources (Young and Cockcroft 1994). Many of the open ocean or open coast dolphins and porpoises swim in large schools; such schools are able to spread in broad ranks over large areas to search for food patches, and converge when prey is found (Würsig 1986, Hanson and Defran 1993).

Groups of small dolphins may be more ephemeral than killer whale pods, but they can also exhibit coordination to facilitate prey capture. Würsig and Würsig (1980) observed the formation of feeding aggregations of dusky dolphins (*Lagenorhyncus obscurus*) off Argentina. As many as 30 subgroups of 8 to 10 dolphins up to 8 km apart forage separately, but apparently maintain acoustic contact because they combine to form larger groups when schools of southern anchovies are found (Würsig 1986, Würsig et al. 1991). The combined groups engage in cooperative feeding by surrounding schools of southern anchovies, causing them to coalesce, and forcing them toward the barrier presented by the ocean's surface. Individual dolphins then pass through the tight ball of fish, capturing as many as five at a time, while other dolphins continue to hold the school together. Small groups of dolphins (six to eight) may feed in this manner for about 5 min, but larger combined groups may be able to increase the effectiveness of this feeding pattern. Additional dolphins can increase the size of the "fish ball" to include four or five fish schools within an area of 200 to 300 m in diameter, and prolong feeding opportunities for several hours (Würsig and Würsig 1980, Würsig 1986).

In a similar manner, common dolphins in the Gulf of

California may work together to surround and compress schools of fish, and drive them up to the surface where feeding occurs (Wells et al. 1981, Würsig 1986). After several minutes the feeding activity subsides, but several minutes later the dolphins begin dives to bring the reassembled fish school back to the surface. Often, the dolphins are joined by other predators, such as California sea lions and diving birds such as brown boobies (*Sula leucogaster*) or brown pelicans (*Pelecanus occidentalis*), that take advantage of the prey as it nears the surface. Coordinated feeding by large schools of dolphins on schooling fish and squid, sometimes in association with diving birds, also has been reported for common dolphins in the Atlantic (Major 1986), Atlantic spotted dolphins (*Stenella frontalis*) in the open waters of the Gulf of Mexico (Fertl and Würsig 1995) and near the Azores (Martin 1986), Fraser's dolphins (*Lagenodelphis hosei*) in the Caribbean (Watkins et al. 1994), and Pacific white-sided dolphins (Norris and Prescott 1961), among others. Simultaneous, noncooperative nocturnal feeding on schools of flying fish by spotted dolphins also occurs (Richard and Barbeau 1994).

The nighttime feeding of Hawaiian long-snouted spinner dolphins (*Stenella longirostris*) involves the synchronous dives of scattered groups (Norris et al. 1994). These dolphins move offshore in late afternoon in moderate to large schools, and spend the night over the 200- to 2000-m escarpment, feeding on mesopelagic (mid-water column, open ocean) fish such as myctophids (lantern fishes) and invertebrates associated with the deep scattering layer (DSL; so-named because of the echoes it produces on depth sounding equipment). The DSL is an assemblage of small fishes and invertebrates, such as squid, euphausiids, siphonophores, and sergestid shrimp, that migrates vertically from depths of 230 to 700 m toward the surface on a diurnal basis, in response to changing light levels (McConnaughey 1974). The different strata of the DSL move upward at night and merge, within diving range of the dolphins, and descend and reform separate strata during daylight. The nearly continuous echolocation heard in the vicinity of the dolphins at night suggests intensive foraging, but it has not been possible to determine whether cooperative efforts are used to concentrate the prey. At least it seems that the dolphins are taking advantage of a fairly dispersed but temporally abundant resource. In a very similar manner, dusky dolphins off New Zealand feed on prey associated with the DSL, but cooperative herding of prey is not as important as for conspecifics off Patagonia (Würsig et al. 1991). The dives of radio-tagged common dolphins in the eastern Pacific have also been correlated with vertical migration of the DSL (Evans 1974, 1975).

Coastal dolphins and porpoises inhabit an environment where physiographic features (bottom topography) can play an important role in the animals' movements and activities, and may eliminate the need for large numbers of schoolmates to aid in foraging. Speed-enhancing features are less evident. Morphological features that improve maneuverability around shorelines, reefs, and the sea floor, such as flexible necks and proportionally larger control surfaces (fins, flippers, flukes), are more common in strictly coastal species, or in the coastal forms of species such as bottlenose dolphins found both inshore and offshore. In some species, fewer larger teeth reflect an increase in prey size (e.g., bottlenose dolphins; Walker 1981, Van Waerbeek et al. 1990). Prey may occur in schools, or may be more predictable and more evenly distributed in association with sea floor features. School sizes of inshore, shallow water dolphins and porpoises tend to be smaller than in the open ocean habitat (Saayman and Tayler 1979, Wells et al. 1980). This may reflect, in part, the ability of individual cetaceans to locate and capture the more predictable food, their ability to take advantage of physical environmental features to limit the escape opportunities of the prey, and the absence of rich patches of food sufficient to support large schools.

Environmental features such as tides and currents may also influence the feeding patterns of coastal and inshore small odontocetes, as described for killer whales. Humpbacked dolphins (*Sousa chinensis*) in South Africa remain within 1 km of shore, where their small schools (< 25 individuals) generally disperse widely to forage and appear to capture prey individually; feeding activity increases on rising tides (Saayman and Tayler 1979). Similarly, bottlenose dolphins off Texas and Pine Island Sound, Florida, tend to feed while facing against a strong current (Shane 1990b). In contrast, Acevedo (1991) found that bottlenose dolphin feeding is most common during slack or ebb tides in Ensenada de La Paz, a coastal lagoon in Mexico. Hoese (1971) reported that bottlenose dolphins work tidal creeks at low tides, chasing fish out of the water onto mud banks. In other areas, no relationships between tides and bottlenose dolphin activities have been reported. These observations illustrate the plasticity of behavior in a species occupying a variety of habitats.

The predictability of coastal fish resources may facilitate foraging to some extent, but searching and hunting are still required. Echolocation may play a role in foraging, but passive listening may also be very important. For example, Barros and Odell (1990) hypothesized that the diets of bottlenose dolphins may emphasize certain species of fish that are noisy. Dolphins may be drawn to some fish because of the sounds they produce—the croaks, grunts, and splashes made by stridulatory mechanisms, hydrodynamic phenomena, leaps, and muscular contractions involving swim bladders. Thus, listening dolphins may be able to limit the use of their echolocation, and reduce the possibility of alerting prey to their presence.

Figure 8-13. "Fish whacking" by a bottlenose dolphin near Sarasota, Florida. The fish (in the air above the island) became airborne at the end of a chase, when it was struck by the dolphin's flukes. (Photograph by Randall S. Wells, Dolphin Biology Research Institute.)

Coastal small cetaceans use a variety of specialized prey capture techniques in addition to simply chasing down individuals. For example, bottlenose dolphins have developed specific feeding behaviors for certain habitats and situations, and have displayed a remarkable ability to adapt feeding patterns to changes in the environment. In shallow sea grass meadows bottlenose dolphins use "fish whacking" to stun and capture prey (Fig. 8-13; Wells et al. 1987, Shane 1990b). The dolphins chase individual fish through shallow waters and strike them with their flukes, often sending them flying. If not killed outright by the process, the fish are at least stunned.

Another bottlenose dolphin feeding technique associated with sea grass in Florida (R. Wells, pers. obs.) and Australia (R. Connor, pers. comm.), dubbed "kerplunking," involves lifting the flukes and caudal peduncle from the water, driving them through the surface creating a loud splash, and entraining a trail of bubbles. This has the effect of driving fish from the grasses and limiting their escape routes with the bubble barrier, apparently in much the same way as bubbles are used by humpback whales (discussed above). This behavior is engaged in by individuals or groups.

Bottlenose and Atlantic spotted dolphins on the Bahama Banks engage in "crater feeding" (Fig. 8-14; Herzing 1996). These dolphins use echolocation to scan the sandy sea floor, then burrow into the sand with their rostra to capture buried fish. At times, the dolphins may dig to the level of their scapulae. How the dolphins find the buried fish, and what kinds of fish are involved, remain unclear.

Figure 8-14. "Crater feeding" by a bottlenose dolphin on the Bahama Banks. After an intensive bout of echolocation, the dolphin worked its rostrum and head into the sand to capture a fish. (Photograph by Denise Herzing, Wild Dolphin Project.)

In Australia, bottlenose dolphins may use sponges to aid in prey capture (Smolker et al. 1997). In certain areas, primarily female dolphins carry sponges on their rostra while engaged in apparent foraging behavior. Whether the sponges are used to protect the dolphins' rostra, or are applied in some other way as tools to obtain fish is unclear. Although tool use by bottlenose dolphins has been described for captive dolphins (see Caldwell and Caldwell 1972), this is the first report of apparent tool use from the wild.

Dolphins have learned to take advantage of human fishing activities to supplement their own efforts. In many parts of the world, bottlenose dolphins have learned to approach shrimp boats and small purse seine boats fishing for bait fish, and take the bycatch either as it falls out of the net or as it is cleared from the net and discarded (e.g., Barros and Odell 1990, Corkeron et al. 1990, Fertl 1993, Wells 1993b). Humpbacked dolphins in Moreton Bay, Australia, gather to feed near trawlers (Corkeron 1990). Bottlenose dolphins in eastern Florida have learned to remove bait fish from crab traps. In Mauritania (Busnel 1973), the Black Sea (Bel'kovich et al. 1991), and Brazil (Pryor et al. 1990), bottlenose dolphins have even developed cooperative fishing patterns with wading fishermen, driving mullet into the fishermen's nets and presumably increasing their own effectiveness at prey capture (for additional examples, see Corkeron et al. 1990).

Coastal dolphins, like the more open water dolphins, may also work cooperatively to capture prey. Reports from the Black Sea (Morozov 1970, Bel'kovich et al. 1991), the coast of South Africa (Tayler and Saayman 1972), the Gulf of California (Würsig 1986), and elsewhere describe patterns of coordinated circling and driving of fish schools.

In tidal creeks in the southeastern United States, and elsewhere, bottlenose dolphins "strand feed" (Hoese 1971, Petricig 1995). Dolphins encountering fish schools rush at them, and the fish explode from the water. Some fish land on the muddy "strand" between the vegetation and the water, and the dolphins slide onto the banks to capture them (Fig. 8-15). In every observation to date in South Carolina, the dolphins beach on their right side, perhaps because bottlenose dolphins may be right-eyed in the near field (Petricig 1995). Strand feeding in South Carolina is limited to stable subgroups of resident dolphins. The fact that calves too young to eat fish follow older animals onto the banks suggests this is a learned behavior, which may be passed from generation to generation, another example of cultural transmission of knowledge within the cetaceans. Similarly, humpbacked dolphins off Mozambique work cooperatively and individually to drive fish into shallows and onto sandbanks, where they slide out of the water to capture the fish (Peddemors and Thompson 1994).

The riverine dolphins inhabit a complex habitat in which prey distribution may change dramatically from season to season with changes in rainfall. For example, boutu (*Inia geoffrensis*) inhabiting the Amazon River and its tributaries may move into the floodplain and forest during rainy season flooding (Layne 1958, Best and da Silva 1989), and return to river channels when water levels decline (Best and da Silva 1984). The greatest diversity of prey in the dolphins' stomachs occurs during the flood season, when the fish are reasonably abundant, but more dispersed and more difficult to catch (Best and da Silva 1989). Pelagic and diurnal prey predominate over benthic, littoral, or nocturnal fishes in the boutu's diet; schooling and nonschooling fish are taken in similar quantities (Best and da Silva 1989). These mostly solitary dolphins concentrate their feeding activities at river mouths and just below rapids, where currents may help to disorient fish schools (Best and da Silva 1989). Similarly, baiji (*Lipotes vexillifer*) concentrate their feeding activities near current eddies (Yuanyu et al. 1989). Fish in river systems offer a relatively predictable resource, but one that neither sup-

Figure 8-15. "Strand feeding" by bottlenose dolphins in South Carolina. (Photograph by Flip Nicklin, Minden Pictures.)

ports nor requires large numbers of dolphins for effective feeding. Thus, many of the river dolphins tend to feed individually or in small groups.

Responses to Predators and Other Threats

Most cetaceans face pressure from predators at some point during their lives. The principal predators of cetaceans include sharks, other cetaceans, polar bears, and as an evolutionarily more recent addition to the list, humans. Only responses to the non-human predators will be discussed here. Responses to potential predation include detection and avoidance, fleeing, using habitat features for cover, and active defense by individuals or coordinated groups. The need for protection from predation has likely been an important factor leading to the evolution of many group-type social systems in cetaceans, as have the benefits of group foraging.

The species of predator and frequency of predation vary with the size of the cetacean. Young calves of mysticetes and the larger odontocetes face predation from both sharks and killer whales, but attacks on older whales are mostly by killer whales. Even so, in some places the potential for predation on mysticetes is low (Dolphin 1987d, Clapham 1993). The smaller odontocetes, especially those smaller in size than their shark predators (< 3–4 m in body length), face the threat of predation from sharks or killer whales throughout their lives (Fig. 8-16). In some cases, as described, small cetaceans may also be eaten by delphinids such as false killer whales or pilot whales (Perryman and Foster 1980). The belugas and narwhals are also vulnerable to attacks by polar bears along the ice edge, in small openings in the ice, or when stranded in tidal pools (Smith 1985, Smith and Sjare 1990). The frequency of predation attempted on a given species may vary from one site to the next. For example, 22% of non-calf bottlenose dolphins in Sarasota Bay bear shark bite scars (Wells et al. 1987), compared to 37% in eastern Australia (Corkeron et al. 1987, Corkeron 1990) and 10.3% in South Africa (Cockcroft et al. 1989). The frequency of shark bite scars may also vary from species to species within the same region; 28% of South African humpbacked dolphins were scarred (Cockcroft 1991). Most river dolphins face little or no natural predation pressure, but remains of the more estuarine franciscana (*Pontoporia blainvillei*) have been found in 17% of the stomachs of seven-gilled sharks (*Notorhynchus cepedianus*) and 4% of hammerheads (*Sphryna* spp.) (Pinedo et al. 1989).

Detection of predators presumably involves vision, passive listening, or for the odontocetes, echolocation. The tendency of coordinated schools of Hawaiian long-snouted spinner dolphins to move in slow, synchronous circles over a light sandy sea floor may be an adaptation to facilitate visual detection of approaching dark-colored sharks when the dolphins are resting (Norris et al. 1994). During rest, the dolphins tend to rely on vision more than on echolocation. During periods of greater activity, echolocation occurs more frequently. The integration of sensory systems for detection of predators has been suggested as one function of cetacean social groups (Norris and Dohl 1980a,b; Norris et al. 1994). With sensory integration, information obtained from the sensory systems of one individual can be shared with the rest of the group, and may facilitate group responses such as flight, increased vigilance, or active defense.

Once potential predators have been detected, responses vary (e.g., Wood et al. 1970). Captive bottlenose dolphins vary in their responses to different species of sharks; as might be expected, those that pose a greater threat are avoided

Figure 8-16. Adult male spinner dolphin (*Stenella longirostris*) killed by a shark near Sarasota, Florida. Note body wound and removal of a fluke by sharks. (Photograph by Randall S. Wells.)

(Irvine et al. 1973). Humpbacked dolphins in South Africa have been reported to ignore, avoid, or chase large sharks in different instances (Saayman and Tayler 1979). Avoidance and flight in response to predators have been reported for many cetaceans. Common dolphins in South Africa "displayed precipitate flight in the open sea when pursued by three killer whales" (Saayman and Tayler 1979). Connor and Heithaus (1996) described the splitting and high-speed flight over 3 km of a group of bottlenose dolphins in Australia in response to approach by a great white shark. Migrating gray whales presented with recorded sounds of killer whales changed their paths dramatically to avoid the sound source, whereas recorded sounds of a variety of human industrial activities played back through the same system resulted in less pronounced course deviations (Cummings and Thompson 1971, Malme et al. 1983). Flight, often in silence, from killer whales has been reported for sperm whales (Arnbom et al. 1987), belugas, (Fish and Vania 1971, Cosens and Dueck 1991, Frost et al. 1992), narwhals (Steltner et al. 1984, Campbell et al. 1988, Cosens and Dueck 1991), humpbacked dolphins (Saayman and Tayler 1979), Dall's porpoises (*Phocoenoides dalli*) (Jefferson 1987), and dusky dolphins (Würsig and Würsig 1980, Würsig et al. 1991). These animals may move very close to shore or ice during flight. Such physical features of the environment may provide cover, or as Würsig and Würsig (1980) suggest for dusky dolphins, the turbulence and bubbles of the surf zone may reduce the effectiveness of the predators' visual and acoustic senses.

Members of the genus *Kogia* appear to use a unique method of protection from predation. Dwarf (*Kogia simus*) and pygmy (*Kogia breviceps*) sperm whales may discharge large clouds of dark red-brown feces for concealment from predators (e.g., great white sharks; Long 1991) or other threats (Scott and Cordaro 1987). Fluke movements within the cloud that result in rapid spreading of the cloud suggest that this behavior is more than simply a fright response without a specific function (R. S. Wells, pers. obs.).

Active defense may involve the use of a variety of weapons in physical contact. The primary defensive weapon available to all cetaceans is the fluke. Gray whale mothers respond aggressively to threats when their calves are disturbed, thrashing sideways or vertically with their flukes (Norris et al. 1977). Most odontocetes have the additional ability to use their teeth for biting, and all can use their rostra for ramming. Mysticetes and odontocetes have been observed slapping their flukes vertically or slashing them from side to side in response to an attack (Whitehead and Glass 1985), and dolphins may ram sharks or bite them (Wood et al. 1970). Although such defenses by an individual may be effective against a single predator, they are less effective against multiple predators attacking simultaneously.

Groups of cetaceans may improve their odds for survival in the face of predation by engaging in coordinated, cooperative defense. One of the best examples of such defense is the "marguerite formation" (resembling the petals of the marguerite flower; Fig. 8-17) by sperm whales, which has been described in response to harpooning (Nishiwaki 1962, Caldwell et al. 1966, Norris and Dohl 1980a), and attacks by false killer (Palacios and Mate 1996) and pilot whales (Weller et al. 1996). An injured or particularly vulnerable member of a group is surrounded by the remainder of the group, with their heads oriented toward the center of the circle, and their flukes oriented out. The formation can be assembled horizontally or vertically. Sperm whales also form tight groups facing attacking killer whales, with the most vulnerable young in the middle of the group (Arnbom et al. 1987). Such behavior is reminiscent of that of kinship groups of elephants (*Loxodonta africana*), where the individuals stand shoulder to shoulder with their weapons (in this case tusks) directed toward the threat, to provide a protective front (Douglas-Hamilton and Douglas-Hamilton 1975), or similar defensive formations by water buffalo (*Bubalus bubalis;* Eisenberg and Lockart 1972). Hawaiian long-snouted spinner dolphins also respond to threats by closing ranks, with mothers and young centrally located within the school (Johnson and Norris 1986, Norris et al. 1994).

Active and coordinated defense by smaller odontocetes has been reported. Common dolphins will work in a coordinated manner to defend the group from attack by multiple sharks (Springer 1967). In this case, many members of the dolphin school remained near the hull of Springer's research vessel, while other school members engaged in sorties to attack and drive off the sharks. Bottlenose dolphins in captivity were reported to work together to attack and kill a large sandbar shark (*Carcharhinus plumbeus*) (Essapian 1953).

Preliminary data suggest that free-ranging bottlenose dolphins in Sarasota Bay, Florida, may benefit from group defense from predation. Mortality patterns and the occurrence of shark bite scars have been examined relative to male group formation. Many, but not all, males form a lifelong pair bond with another male near the time of sexual maturity (Wells et al. 1987). Preliminary analyses indicate that paired males acquire more shark bite scars but live longer than do males that remain alone. One possible interpretation of these findings is that paired males may be more visible to sharks than are single males, but may be afforded more of an opportunity to recover from shark wounds due to the potential availability of protection by the partner. A wounded, unpaired male would presumably be more vulnerable to subsequent attacks by predators (R. S. Wells, unpubl.).

Group defense is used to varying degrees by different cetaceans and other aquatic animals, depending in large part

Figure 8-17. Artist's representation of the "marguerite" defensive formation used by sperm whales. (Drawing by Faith Keller.)

on the nature of the habitat. In the same three-dimensional aquatic environment, fish form schools for protection from predators because the groups provide cover for the individuals, their movements can confuse the predators ("confusion effect"), and the probability of selection of a given individual is reduced ("dilution effect"), although the probability of detection of the group may be increased ("encounter effect") (Inman and Krebs 1987). School formation offers similar advantages to cetaceans, and protection from predation is likely one of the primary evolutionary forces behind cetacean sociality. In addition to the benefits described, school formation can provide the presumed benefits of sensory integration and communication of information, and provide for a coordinated defense. In support of this hypothesis, the group size in the smaller odontocetes appears to be correlated with the risk of predation (Norris and Dohl 1980b, Wells et al. 1980, Scott 1991).

Open ocean dolphins may be at a higher risk of predation than inshore cetaceans. Offshore, the movements of large predators are not limited by physical features of the habitat (e.g., shorelines and channels); therefore, encounters with potential predators may be more frequent, as suggested by reports that sharks and dolphins were captured together in about 13% of eastern tropical Pacific Ocean tuna purse seine net sets (Au 1991). As might be expected based on grouping patterns of medium- to large-sized terrestrial mammals that are exposed to predation in open habitats (e.g., large herds of ungulates faced with lions on the African savanna or caribou faced with wolves on the Arctic tundra), open ocean dolphins form large groups, numbering hundreds to thousands of individuals. In coastal waters, sea floor features or land masses may interfere with a predator's ability to detect or approach small cetaceans. In contrast, schools of coastal small cetaceans tend to be smaller than those found offshore. In the most extreme case, river dolphins are virtually free from predation, and tend to be solitary or to form the smallest groups. Thus, for open ocean dolphins, group formation may provide some protection from predation, whereas habitat features may be more important in inshore waters (Wells et al. 1980). The sizes of groups are ultimately determined by the interplay of predator defense benefits, group foraging benefits, and limitations imposed by prey availability, among other factors.

Health and Maintenance Behaviors

Few cetacean behaviors have been described as being strictly for health or maintenance. The removal of sloughing skin is one maintenance behavior engaged in by a variety of both mysticetes and odontocetes. Normal continuous skin sloughing serves to limit the attachment of ectoparasites, and in many cases it is apparently accomplished simply from the forces of water passing over the swimming animal. In some cases, the animals facilitate the process by actively rubbing on inanimate objects or other animals. Belugas have an annual molt, one of the most extreme cases of skin sloughing of any cetacean. Each summer they spend from a few hours to 5 days in shallow estuaries, where freshwater and opportunities to rub on the substrate apparently facilitate the removal of the old skin (St. Aubin et al. 1990, Smith et al. 1992, Martin et al. 1993). Killer whales near Vancouver Island, Canada, frequent specific "rubbing beaches" where they rub and roll on a shallow sea floor of small smooth stones. The tactile tendencies of many of the cetaceans, especially the dolphins, may help to remove sloughing skin as they rub one another (Norris et al. 1994).

Parasite removal is one possible function of some cetacean leaps. Many photographs of spinning Hawaiian long-snouted spinner dolphins have shown multiple remoras attached to the dolphins (Fig. 8-18). Bottlenose dolphins in Florida sometimes leap repeatedly, orienting their reentry to impact on places where remoras are attached. In some cases, the same remora-bearing dolphins have engaged in the same kinds of leaps on consecutive days. The success of parasite removal through leaping is unknown. It seems curious that conspecific associates, including mothers, do not aid effectively in the removal of parasites such as remoras. One spinner dolphin calf had a remora attached above its right eye for several months (Norris et al. 1994); no attempts by the calf's mother to remove the remora were observed. In contrast, marks on small plastic cattle ear tags trailing from bottlenose dolphin dorsal fins suggest occasional attempts by other dolphins to remove the tags with their teeth (Scott et al. 1990a), but whether these were directed allogrooming efforts, as described for other mammals such as primates, or occurred incidental to other interactions is unknown.

Because of the voluntary nature of their respirations, cetaceans are unable to engage in a deep sleep in the same way as many other mammals. Instead, many cetaceans enter into periods of reduced activity each day. The animals slow their swimming speeds, and may remain in a limited area during rest. As an extreme example, boutu in captivity have been frequently reported to rest, sometimes upside down, on the bottoms of their pools (Caldwell et al. 1989); such "relaxed" behavior may reflect the fact that predation pressure on boutu in the wild is nearly nonexistent. South African humpback dolphins appear to rest while moving slowly in a compact group within a limited area (Saayman and Tayler 1979). Groups tend to become more synchronous in their surfacings and dives during rest.

Hawaiian long-snouted spinner dolphins enter shallow bays early in the morning, and demonstrate a steady decline in their activity level, including cessation of echolocation, as they begin a period of rest (Norris and Dohl 1980b, Norris et al. 1994). The spinner dolphin schools circle slowly and synchronously over the sandy sea floor, remaining below the surface for extended periods (Fig. 8-19). In the afternoon, a few individuals become more active, swimming more rapidly, with increasing aerial behavior. Activity levels change in an oscillatory manner, increasing as the school moves toward the mouth of the bay, subsiding as the school turns back into the bay. This pattern, termed zigzag swimming, appears to be a period of social facilitation, in which increasing numbers of individuals are recruited into a more active state. When the school as a whole has made the transition from resting to activity, it leaves the bay for a night of foraging. Norris et al. (1994) suggest that this pattern of increased activity is akin to the whistling choruses of African wild dogs, which are thought to prepare packs for the hunt (Estes and Goddard 1967).

Recent observations of captive Pacific white-sided dol-

Figure 8-18. Spinning Hawaiian spinner dolphin with three remoras attached. (Photograph by Randall S. Wells.)

Figure 8-19. Resting school of spinner dolphins in a shallow bay in Hawaii. The animals rest in a synchronous, slow-moving formation, typically remaining over a shallow, sandy bottom. (Photograph by Randall S. Wells.)

phins indicate that sleep is accomplished through the resting of one hemisphere of the brain at a time. This was demonstrated by dolphins sleeping in circling groups with only the eye toward the center of the group open, then switching eyes to maintain visual contact with group members as they changed position within the group (Goley 1996).

Cetaceans engage in epimeletic behavior—care- or attention-giving behavior for young, sick, or injured associates (Caldwell and Caldwell 1966, Connor and Norris 1982). Care directed toward young is termed nurturant, whereas care for individuals in distress is termed succorant. Succorant behavior includes: (1) *standing by,* when an individual remains with another in distress, but does not provide obvious aid (e.g., northern bottlenose whales remaining with injured companions; Benjaminsen and Christensen 1979), (2) *excitement,* which involves the interposition of an individual between a threat and a distressed associate, attacking a would-be captor, or pushing an injured animal away from a would-be captor, and (3) *supporting,* when one or more individuals maintain an injured associate at the water's surface (Caldwell and Caldwell 1966). In each of these cases, the actions of the fully functional individuals usually require that they remain in a dangerous situation longer than if there was no distress. At least four mysticete genera engage in nurturant behaviors, and at least three in succorant behaviors, including supporting associates (Caldwell and Caldwell 1966). At least eight genera of odontocetes have been reported to engage in nurturant behavior, and at least ten in succorant behavior, although only members of the family Delphinidae engage in supporting behavior (Caldwell and Caldwell 1966). Dolphin mothers may remain near or may support the remains of their dead young (e.g., Moore 1955, Connor and Smolker 1990, Palacios and Day 1995), but the tending of sick or dead individuals does not appear to be limited to kin relationships, or even to the same species (Connor and Norris 1982).

Mating Patterns

Mating patterns include both the specific behaviors involved in the mechanics of mating, as well as the systems that lead to fertile matings. Matings have been observed for a few of the baleen whales and for a number of the odontocetes, especially those maintained in oceanaria. Little information is available about the mating systems of most of these animals. The lack of information is attributable to a paucity of observations, as well as to confusion created by the actions of the animals themselves. In some species, changeable associations between males and females during the breeding season have been reported, but no copulations have been observed; in other species, copulations are observed frequently, but which of these are fertile matings and which have other social functions remain unclear. For many of the odontocetes, sexual behavior is a major component of nonreproductive social interactions (Saayman and Tayler 1979, Norris and Dohl 1980a, Wells 1984). The advent and application of techniques for hormonal analyses in conjunction with behavioral observations (e.g., study of captive Hawaiian spinner dolphins; Wells 1984) and genetic paternity testing (as used for wild bottlenose dolphins, Duffield and Wells 1991; and long-finned pilot whales [*Globicephala melaena*], Amos et al. 1991, 1993a) are beginning to provide new avenues for research into these questions.

Cetaceans exhibit a variety of precopulatory and copulatory behaviors. In preparation for mating, physical contact includes gentle rubbing of bodies against each other, stroking with the pectoral flippers, rubbing of the body along an outstretched pectoral flipper, stroking with the flukes, vigorous back-and-forth rubbing of the pectoral flippers ("pec-whetting"; Norris et al. 1994), rubbing of the genitalia, insertion of the rostrum into the genital slit (Fig. 8-20), "beak-to-genital propulsion," (Norris et al. 1994), and violent head butting (Caldwell and Caldwell 1972, Puente and Dewsbury 1976). Cetaceans live in a three-dimensional, essentially weightless environment that facilitates copulations from a variety of positions. The large size and curvature of the fibroelastic penis increases this flexibility, especially for mysticetes. Common positions for bottlenose dolphins include ventrum to ventrum with the pair oriented in the same direction, or the male may mount the female from a nonparallel position, with intromission lasting only a few seconds, and accompanied by vigorous pelvic thrusting (Saayman et al. 1973). Copulations involving humpbacked dolphins appear to be more stereotyped than for bottlenose dolphins, with the male displaying an erection and swimming inverted beneath a slowly swimming female, effecting intromission and gliding for 20 to 30 sec, without thrusting (Saayman and Tayler 1979). These episodes might be repeated at intervals over several hours.

Nonreproductive copulations may occur between a variety of partners. In bottlenose dolphins, copulations between subadults occur more commonly than between adults. Such interactions may be either hetero- or homosexual. Copulatory behavior is exhibited very early in life. Male bottlenose dolphins within a few weeks of birth may copulate with their mothers (Caldwell and Caldwell 1972). Similarly, copulatory patterns have been observed for all age classes of humpback dolphins in the wild (Saayman and Tayler 1979).

Of the three basic mating systems (see section on pinnipeds)—monogamy, polygamy, and promiscuity—the latter two may account for most cetacean matings. Monogamy, in which males and females mate with only one partner during a breeding season, has yet to be described for any cetacean. Promiscuity differs from the other two forms in that no durable mating relationships are formed; members of both sexes may mate with multiple partners, and matings occur more or less randomly. This description resembles association patterns seen for many cetaceans, but few examples of truly random matings exist in nature (Halliday 1980). Most cetaceans probably engage in polygamy in which an individual of one sex mates with several members of the opposite sex, and relationships are maintained for the duration of a breeding cycle. Identification of the precise form of polygamy, either polygyny (one male mates with multiple females), as described for many of the pinnipeds, or the more rarely seen polyandry (one female mates with multiple males), is confounded by confusion surrounding observations of matings.

Available evidence supports polygyny as the mating system for most mysticetes, but the specific form of the polygyny—resource defense, female defense, scramble competition, sequential defense, or lekking (see section on pinnipeds)—remains to be defined in most cases. Because of the extreme mobility of most cetaceans, resource defense polygyny is unlikely, but aspects of the other tactics are evident in observations from the field. For most mysticete whales, breeding occurs during a well-defined winter season, and follows a long migration to specific low latitude breeding grounds. This pattern is best exemplified by humpback, gray, and southern right whales, which congregate in near-shore waters that also serve as calving grounds (see Ranging Pat-

Figure 8-20. "Beak-to-genital propulsion" by Hawaiian spinner dolphins. (Photograph by Randall S. Wells.)

terns below). Presumed estrous females may be accompanied by several adult males, each competing for access to the female. Brownell and Ralls (1986) have suggested that sperm competition may be a factor in the mating systems of some of the baleen whales. Males with longer penises and larger testes relative to body weight would be expected to engage in efforts to dilute or displace the sperm of another male, whereas males of species with shorter penises and smaller ratios of testis to body weight would be expected to engage in aggressive interactions with other males. Right and bowhead whales have penises that are more than 14% of body length, whereas those of all other mysticetes are 7.5% to 11% of body length. Right, bowhead, and gray whales have larger than predicted testes (see Pabst, Rommel, and McLellan, Chapter 2, this volume) relative to body size (11.5% to 14.9%; Brownell and Ralls 1986), whereas all of the balaenopterids and the pygmy right whale have testes near or below the predicted size (Brownell and Ralls 1986). Thus, right, bowhead, and gray whales would be expected to engage in sperm competition (scramble competition), whereas the other mysticetes should engage in more direct male–male competition for females (female defense, sequential defense, or lekking).

The predictions of Brownell and Ralls (1986) appear to be borne out for many of those species for which behavioral data are available. In contrast to relatively limited agonistic interactions between male right, bowhead, and gray whales, battles between male balaenopterids may become quite violent, involving blocking, bubbling, tail lashing, ramming, and striking with the tail and caudal peduncle (Darling et al. 1983, Tyack and Whitehead 1983, Clapham et al. 1992). During the breeding season, humpback whales gather in large numbers over tropical banks where mature males are dispersed as individual singers, and they battle with one another for access to females (Tyack and Whitehead 1983). Competition takes place in groups, with multiple males vying for access to a single female (Tyack and Whitehead 1983, Baker and Herman 1984, Clapham et al. 1992). Individual males have been observed in more than one competitive group both during a single breeding season, and over multiple seasons (Clapham et al. 1992). These males inflict bleeding wounds on one another, but injuries recorded to date have not been serious (Clapham 1996). Clapham (1996) suggested that male humpback whales remain in close association with females over multiple days both to mate and to engage in mate guarding to prevent other males from mating with her (Fig. 8-21). Cooperative efforts by pairs of males to displace principal male escorts of females have been reported on occasion.

The humpback whale mating system, as a variation of male dominance polygyny, has been termed by Clapham (1996) as a floating lek based on the following facts: economic defense of either multiple females or resources is impossible, male parental care is absent, communal display behavior by males occurs, males engage in direct competition for estrous females, and the displaying males are mobile. In comparison to other lekking species, such as the dugong (*Dugong dugon*) (see section on sirenians) and American sage grouse (*Centrocercus urophasianus*), the lek is strictly a mating site that consists of a confined area in which a number of males attempt to attract females and compete with each other for possession of their very small territories or mating stations. Leks contain no ecological resources of use to the females. Typically, leks are occupied by male grouse throughout the breeding season, whereas females come only when they are ready to mate, and stay only long enough

Figure 8-21. On the Hawaiian breeding grounds, a humpback whale mother swims with her calf while a male escort (below) engages in mate guarding. (Photograph by Flip Nicklin, Minden Pictures.)

to select a mate and copulate. Conclusions regarding the mating system of humpback whales remain tentative, given the fact that copulations have yet to be observed (unlike right, gray, or bowhead whales), but genetic paternity tests indicate that female humpback whales mate with different males in different seasons. Biopsies from three female whales and three to five of each of their offspring showed different fathers (Clapham and Palsbøll 1997).

Bowhead whales deviate from the typical mysticete pattern in terms of the timing of breeding activity and the nature of their migration. Much socializing, including copulations, has been observed for western arctic bowhead whales during their spring migration from the Bering Sea to summer feeding grounds in the Beaufort Sea (Würsig and Clark 1993). This corresponds to the period of calving after a 12-month gestation period. Sexual activity often takes place in groups, where boisterous nudging, pushing, rolling, and rapid turns result in the creation of whitewater, and may involve more than one male attempting to copulate with a single female. Social activity declines through the summer (Würsig and Clark 1993).

Gray whale sexual behavior in calving lagoons tends to increase as tidal currents slacken. Extended penises, probing of the genital areas and caressing have been observed in as many as seven or eight duos or trios at a time within an area of 100-m diameter (Norris et al. 1983). These groups include both adults and juveniles. Courting whales move relatively frequently between breeding lagoons (Swartz 1986).

Male–male competition on breeding grounds occurs in many mysticetes, but what is the role of females in mate selection? Female choice by mysticetes may be exhibited in several ways. For example, female southern right whales and bowhead whales may invert themselves at the surface of the water, denying or increasing the difficulty of access for males (Würsig and Clark 1993). Female humpback whales have demonstrated aggressive behavior toward males in competitive groups (Clapham et al. 1992). The songs produced by male humpback whales on the breeding grounds may also play a role in female choice; for example, by attracting females and allowing them to compare males (Tyack 1981, Chu and Harcourt 1986). Less is known about the breeding patterns of the other mysticetes because of their more pelagic tendencies.

Whereas breeding may be seasonal for most odontocetes, it usually does not involve a long distance migration to a breeding ground as with mysticetes. The most notable exception to this pattern is the sperm whale. Adult male sperm whales may travel for mating purposes to equatorial waters from feeding areas at latitudes as high as 45 degrees. For most odontocetes, however, breeding and feeding activities occur concurrently in the same waters. Contact between male and female odontocetes may be relatively frequent because they tend to occupy overlapping ranges much of the year, or in some cases they may be members of the same schools.

Frequent encounters facilitate interactions when the animals become reproductively active. Females may be seasonally polyestrous, and may ovulate repeatedly within the period the males are active, as indicated from monitoring of levels of estrogen and progesterone in the blood (Kirby and Ridgway 1984). In many dolphins, the testes enlarge dramatically in response to a seasonal increase in testosterone production, providing males with the ability to produce large quantities of sperm in high concentrations for periods of several months (Ridgway and Green 1967; R. S. Wells, unpubl.). Concentrations of sperm of bottlenose dolphins have been measured at up to $54{,}569 \times 10^9$ sperm per ejaculate (Schroeder and Keller 1990). The large size of cetacean testes relative to other mammals (Kenagy and Trombulak 1986) and the high sperm concentrations are adaptations that allow frequent copulations and the possibility of "swamping" the sperm of another male that has recently copulated with the same female. Individual male common dolphins may mate with at least 25 females in 1.5 hr (Allen 1977).

Ephemeral mating relationships, in some cases with males from outside of stable female groups, have been documented or suggested for several odontocetes. Recent genetic studies have demonstrated that calves born into relatively stable female groups of pilot whales and bottlenose dolphins are not necessarily sired by males from within the same social groups. DNA analyses of long-finned pilot whale calves taken in the Faroe Island fishery where entire pods were sampled, indicated that males often mate successfully with several females, but the adult males in a pod at a given moment are only represented as sires of the current fetuses, and these males originate in other pods (Amos et al. 1991). This suggests a high level of gene flow as males move between stable matrilineal pods (Amos et al. 1991, 1993a). Preliminary DNA fingerprint analyses of bottlenose dolphin calves in Sarasota Bay indicated that about 32% of the calves were sired by nonresident males. Although some resident female bottlenose dolphins in Sarasota Bay mate with the same sire to produce more than one calf, more often different sires are responsible for subsequent calves from a given female (Duffield and Wells 1991).

Observational data provide additional suggestions of nondurable mating relationships for other odontocetes. The fact that males in killer whale pods appear to be sons of the related females in the pod, and therefore, are closely related to their pod mates, suggests that sires are probably not pod mates (Bigg et al. 1987). Sperm whale males are not long-term members of the cohesive schools of adult females and their young ("mixed schools") where breeding apparently occurs. The proportion of sexually mature males in association with mixed schools is low, and associations are brief. Males move between groups of females, spending about 8 hr with each, and sometimes revisiting the group over several days (Best 1979, Whitehead and Arnbom 1987, Whitehead 1993). Male sperm whales infrequently return to the same breeding grounds in different years (Whitehead 1993).

These initial genetic and observational findings, as well as a general pattern of school fluidity for many odontocetes, indicate that long-term monogamy is not the primary mating system for these species—polygyny or promiscuity seems to be the primary pattern. A pattern of mobile males traveling among different groups of females, briefly associating with females, appears consistent across the few diverse species for which data are available, but information from more species would help define the general odontocete patterns (Miyazaki and Nishiwaki 1978, Wells et al. 1980, Slooten 1994). The observed patterns of brief association may indicate mate guarding (female defense polygyny) in which females are monopolized long enough to ensure conception, while still allowing the males the possibility of inseminating multiple females during a single season (sequential defense polygyny).

In odontocetes, the importance of male–male competition for females is suggested by examples of sexual dimorphism favoring larger males, and by patterns of scarring of conspecific origin (tooth rakes). For example, male narwhals frequently bear scars and broken tips of tusks from encounters with other males (Fig. 8-22). In most beaked whales, teeth erupt through the gums of only adult males. Male beaked whales typically exhibit a higher incidence of scarring from the teeth of other males than do females (McCann 1974). Similarly, large male sperm whales often bear scars attributed to wounds inflicted by other large males (Best 1979, Kato 1984, Whitehead 1993). A breeding system model developed by Whitehead (1994) suggests that roving male sperm whales should delay competitive breeding until attaining such a size that only two to four larger males may be attending each receptive female (Fig. 8-23). The conclusions of this model also appear to hold for other large, sexually dimorphic mammals with roving breeding males, such as African elephants and polar bears (Whitehead 1994).

The use of aggression by male bottlenose dolphins to obtain mating opportunities, either directed toward females or through male–male competition, appears to vary from one location to the next. In Sarasota Bay, genetic paternity testing indicates that both single males and members of long-term male pair bonds sire offspring, but obvious aggression by males toward females appears to be infrequent (Wells et al. 1987, Duffield and Wells 1991, Duffield et al. 1994, Moors 1997). In Shark Bay, Western Australia, temporary consortships involving stable alliances of two to three male bottlenose dolphins and a female are established and maintained, often through aggressive herding (Connor et al. 1996). Male coalitions at both sites remain stable throughout the year, and over multiple years. During the breeding season in Shark Bay, male alliances work either individually or with closely associated alliances to capture females from other male alliances, or to defend females against such efforts (Connor et al. 1992a,b). The period of association between the males and a given female is variable (days to weeks), and may relate to the variable reproductive cycling of females, as documented in studies of captive bottlenose dolphins (Kirby and Ridgway 1984, Schroeder 1990). During this time, members of male pairs often flank females, with one male on each side. Such positioning may function to

Figure 8-22. Male narwhals "joust" with their tusks. The frequency and severity of these interactions are indicated by scarring and broken tusk tips embedded in the heads of males. (Photograph by Flip Nicklin, Minden Pictures.)

control the movements of the female, and it may also place the males in a hydrodynamically advantageous position that will provide energetic benefits to them (T. Williams, pers. comm.). In contrast, male bottlenose dolphins near the Moray Firth in Scotland apparently do not form alliances with other males, and may use a resident rather than a roving strategy by accompanying groups of females throughout the breeding season (Wilson 1995). The reasons for this difference are unknown, but may relate to the fact that the Scottish dolphin population appears to be isolated from other bottlenose dolphin populations.

In Sarasota Bay, consideration of the infrequent occurrence of aggressive interactions, lack of prolonged pair bonds between males and females, and high numbers of different males (up to 13) associating with receptive females during the breeding season led to a prediction of a promiscuous mating system (Wells et al. 1987), but subsequent observations of apparent mate guarding by male pairs at the time of conception suggest female defense or sequential defense polygyny may also occur (Moors 1997). Connor et al. (1996) suggest that consorting with females to monopolize them is not an entirely successful strategy. They also suggest that repeated brief ovulation events (assuming fertilization does not occur initially), rather than a single prolonged estrous period, may be part of a female strategy to avoid being monopolized by particular males, may reduce the risk of infanticide (as observed in captivity), or may allow females to mate with preferred males after monopolization by less desirable males.

Male coalitions of three or four Atlantic spotted dolphins sometimes chase and surround a female and eventually mate with her (Herzing 1996). In other cases, courtship involves genital inspection and echolocating or "buzzing" by a male. He then assumes an exaggerated S-posture during his approach beneath the female, inverts, and copulates. When females are unreceptive to approach, they direct aggression toward the courting males (Herzing 1996, 1998).

The relative importance of male–male competition versus female choice remains unclear, except as described above where males seem to control the movements of a female. Where males are larger than females, sexual dimorphism could be a product of sexual selection favoring large males for enhanced fighting abilities or because largeness in males may be an indicator of superior genes and thus preferentially selected by females (Trivers 1985). The same features that appear to provide advantages in combat between males may also be used by females to select mates, and may function to control or intimidate females. Large tusks erupt in male narwhals only, and wounds from tusks are most often inflicted on other males (Fig. 8-22; Silverman and Dunbar 1980, Best 1981a). Larger male size may be attractive to females as an indication of potential advantages to be passed on to offspring, but larger size may also result in faster swimming speeds, which may discourage a smaller and slower female from attempting to escape from a male. Larger size typically means that the male's propulsion structures and weapons—flukes, caudal peduncle, dorsal fin, rostrum, and teeth—are proportionally larger than those of the female. In Sarasota, male bottlenose dolphins do not become successful sires until they have achieved their full adult size, nearly 10 years after they have reached sexual maturity (Duffield and Wells 1991, Duffield et al. 1994). These males are significantly larger than the females, and their potential weapons are allometrically larger still (Tolley et al. 1995). Differences in sexual dimorphism between bottlenose dolphins in Sarasota and Shark Bays may explain the apparent differences in complexity and intensity of male coalition activities. The seemingly higher level of interaction between males in Australia may be compensation for a lack of obvious dimorphism—control of females may be accomplished through the involvement of a larger number of males, in the absence of a body size differential between the sexes.

The hypothesis that strong male-biased sexual size dimorphism in the larger dolphins, such as pilot whales and killer whales, is related to polygyny (for example, males need to be large to defend a group of breeding females from other competing breeding males), is in apparent conflict with observations of low permanent dispersal of males from their natal groups (Heimlich-Boran 1993). Heimlich-Boran (1993) suggests that instead these animals have a promiscuous mating system, and that the dimorphism is related to factors other than the mating system. He suggests that males return to their natal groups after mating because low densities of conspecifics may make a roving male's travel time between receptive females greater than the duration of estrus, because the costs of moving to an unfamiliar area and receiving a potentially unwelcome reception may be high, and because the benefits of cooperatively hunting with familiar pod mates may increase feeding efficiency. There is also increased inclusive fitness within a pod, where the male may contribute to the reproductive success of his close relatives through assistance in protection, foraging, or training, for example, and thus augment his own reproductive success.

Rearing Patterns

Parental care occurs over periods of at least several months to 1 year for most cetaceans, and in some cases it may last for many years. Primary responsibility for the rearing of young appears to be left to the mother, at least in all cetaceans studied to date. Maternal investment includes providing nourishment, protection, and learning opportunities to the calves. In some cases, such as killer whales and pilot whales, care of

Figure 8-23. Adult male sperm whale interacting with adult female and young at the breeding grounds off Dominica in the Caribbean Sea. Note the strong male-biased sexual size dimorphism and the male's erection. (Photograph by Flip Nicklin, Minden Pictures.)

Figure 8-24. Atlantic spotted dolphin (*Stenella frontalis*) nurses her calf. Note the lack of speckles on the younger dolphin; speckles are acquired with age. (Photograph by Denise Herzing, Wild Dolphin Project.)

young by individuals other than the mother (termed allomaternal care) is evident and is often provided by kin.

Few cetacean births have been observed in the wild, but in captive odontocetes most births are accomplished by the mother without the direct assistance of conspecifics; however, several noteworthy exceptions have been reported. Brown et al. (1966) reported on the birth in captivity of a stillborn common dolphin, in which the dorsal fin of the calf apparently prevented complete passage of the calf. A striped dolphin tankmate grabbed the flukes of the fetus, and withdrew it from the birth canal. Within several hours, the mother swam to a false killer whale tankmate, which gently grasped a remnant of the umbilical cord and removed the placenta after several attempts. The degree to which such behaviors would have occurred under normal conditions (i.e., among conspecifics in the wild), is difficult to assess.

Mother's milk provides the primary source of nutrition for cetacean calves for at least the first few months of life. Milk is obtained during brief suckling bouts from either of two nipples positioned on each side of the genital aperture. In humpback whales, nursing occurs at depths of 10 to 13 m, with the mother stationary and horizontal, and the calf vertical, head up to the mammary grooves (Glockner and Venus 1983). Bowhead whale mothers often remain motionless at the surface while calves nurse (Würsig and Clark 1993); a similar pattern has been reported for southern right whales (Thomas and Taber 1984). Delphinid suckling typically occurs as the mother glides, with either the mother or calf tilted to one side (Fig. 8-24). To initiate a nursing bout, a dolphin calf may nudge or bump the mother's mammary glands (Essapian 1953, Cockcroft and Ross 1990b), perhaps effecting the initiation of milk let down, as is the case in many terrestrial mammals. Contraction of abdominal wall muscles may result in the squirting of milk into the calf's mouth (Essapian 1953), although calves are also able to suck (J. E. Heyning, pers. comm.; R. S. Wells, pers. obs.). Initial nursing bouts for bottlenose dolphins tend to last only about 5 sec (but increase with the age of the calf), and may be repeated as many as 20 times within an hour (Cockcroft and Ross 1990b). Nursing comprises about 5% of the southern right whale's diurnal behavior, and the length of individual bouts increases with calf age (Thomas and Taber 1984).

The precise time of weaning is difficult to determine for many cetaceans because suckling may continue beyond the time when milk is a primary source of nutrition. Calves may accompany their mothers for periods beyond the cessation of nursing, and calves may suckle from females other than their own mother (Table 8-5). Observations of humpback whale calves feeding suggest that weaning may begin when they are 5 to 6 months old (Clapham and Mayo 1987), with an average duration of lactation of about 10.5 months (Chittleborough 1958). Finback whale calves become independent after the first year (Clapham and Seipt 1991). Odontocete calves may begin to chase small prey when they are only a few months old, but the ability to meet energetic needs through prey capture may be achieved at about 1 year of age, at least for bottlenose dolphins (Cockcroft and Ross 1990a). A stranded killer whale calf believed to be less than 2 months old had solid food in its stomach (Heyning 1988). A bottlenose dolphin calf in Sarasaota Bay, orphaned at about 16 months of age, survived on her own without having been adopted by another female, and gave birth to her own calf at 7 years of age (R. S. Wells, unpubl.). In contrast, female bottlenose dolphins in Sarasota Bay accompanied by calves up to 7 years of age have been found still lactating, suggesting continued suckling by the calf. Sperm whales may continue to suckle to an age of 13 years in some cases (Best 1979). Similarly, short-finned pilot whales (*Globicephala macrorhynchus*) up to 16 years of age were found to have milk in their stomachs, and females beyond the age of cessation of reproduction (about 39.5 years) were found to be lactating (Kasuya and Marsh 1984; Marsh and Kasuya 1984, 1991). In this species, it seems likely that nursing of older individuals has greater social value than nutritional value, and may help to maintain affiliative relationships. The recent discovery of orphan-induced lactation in nonlactating, nongravid captive bottlenose dolphins raises additional questions regarding the evolutionary origins and implications of suckling nonrelated offspring (Ridgway et al. 1995).

The strength of the mother–calf bond is apparent in the frequent reports of mothers supporting or remaining near dead dolphin calves. This has been reported frequently for bottlenose dolphins (e.g., Moore 1955, Connor and Smolker 1990), and also for Risso's dolphins (*Grampus griseus*) (Palacios and Day 1995) and other species. The degree of calf independence while with its mother varies from species to species. In bowhead whales, the mother–neonate association is very close in early summer, with the mother fetching the calf if it strays more than about 75 m (Würsig and Clark 1993). By late summer, the calf is left alone at the surface for periods of up to 30 min, while the mother apparently feeds, at times up to 1.6 km away (Würsig et al. 1989, Würsig and Clark 1993). Mother and calf southern right whales remain in close association during the early months. Soon after birth they begin a period of rapid travel that is hypothesized to develop the calf's swimming abilities, enhance the ability of the nonbuoyant calf to rise to the surface to breathe, and hinder detection by predators (Thomas and Taber 1984).

Atlantic spotted dolphin mothers produce their own individually specific signature whistles (see Tyack, Chapter 7, this volume) to call their calves (Herzing 1996). If this is unsuccessful, the mothers swim inverted toward their calves, direct echolocation clicks at the calves' genital regions, then touch their rostra to the calf's flank. Discipline for failure to respond to retrieval attempts involves pinning the calf down on the sandy bottom.

Protection of offspring is enhanced in many odontocetes by the formation of nursery groups, in which the sensory systems of a number of mothers may be integrated (Norris and Dohl 1980b). Clusters of mothers and young are common features in odontocete schools, and they may form discrete schools of their own, or distinct subgroups within a larger school of mixed sex and age classes. Nursery groups are evident in schools of many species of open ocean dolphins, including pantropical spotted dolphins (Pryor and Shallenberger 1991), Hawaiian spinner dolphins (Norris et al. 1994), and Risso's dolphins (Kruse 1989).

Many, but not all, bottlenose dolphins in Sarasota Bay raise their young within nursery groups (Wells et al. 1987; Wells 1991b, 1993a). The mothers tend to associate most closely with other mothers that have young of similar age. Associates are drawn from a pool of resident females representing multiple generations of several maternal lineages with a history of association during calf rearing. "Playpens," in which mothers form a protective perimeter around groups of interacting offspring, are common features of the nursery groups. Mothers raising their calves within larger, more stable groups are significantly more successful than females attempting to raise young outside such groups (Wells 1993a).

Raising young within a group of conspecifics also provides opportunities for allomaternal care. Assistance with the care of young has been documented for a variety of odontocetes, such as bottlenose dolphins (Shane 1990b) and spinner dolphins (Norris et al. 1994). Older sisters may "babysit" for younger bottlenose dolphin siblings (Wells et al. 1987). Similar behavior occurs within killer whale pods, with females teaching calves that are not their own the specialized techniques involved in feeding out of water (Guinet 1991). Sperm whale young are accompanied near the surface by a variety of group members while mothers presumably feed by diving beyond the capabilities of their calves (Whitehead 1996b). Such accompaniment may function to protect the vulnerable young from predation, and may have evolved as a form of kin selection in which non-mothers increase the survivorship of related young. In pilot whale schools, postreproductive females may still produce milk, presumably to nurse young other than their own (Marsh and Kasuya 1991). Similar allosuckling has also been suggested for sperm whales (Best et al. 1984, Gordon 1987).

Environmental features can also enhance survivorship of young. Many cetaceans, such as harbor porpoises and bottlenose dolphins, give birth during the summer, or as in the case of some of the mysticete whales, they migrate to low latitudes to give birth during winter. Warmer temperatures at the time of birth presumably reduce the thermal stresses on newborns, and birth before the warmest temperatures of the year may relieve thermal stresses on pregnant females.

Physiographic features are correlated with calf rearing for many cetaceans. Harbor porpoise mothers with calves in the Bay of Fundy appear to seek areas that are warm and stable, with high levels of secondary productivity (Smith and Gaskin 1983). Many baleen whales give birth or raise their young over relatively shallow banks and near shorelines. Most of the humpback whale mothers and calves off Maui, Hawaii, are found in waters less than 20 m deep (Glockner and Venus 1983). Gray whales give birth and provide initial care in sheltered, shallow lagoons. Bottlenose dolphins in Sarasota Bay tend to raise their young in shallow, highly productive nursery areas (Wells 1993b). Shallow areas also have the potential to reduce predation on young.

As described for pinnipeds, cetaceans exhibit several different rearing strategies. Typically, the birth of the next calf defines the duration of maternal investment in a given calf. At one extreme, females may invest a few months in raising each calf, but produce calves at frequent intervals. In contrast, others may invest many years in the raising of a given offspring. The time mothers and calves spend in close association presumably influences the amount of learning that can occur and enhances juvenile survivorship. Some cetacean mothers and calves separate within the first year or two of life. Harbor porpoises have an annual reproductive cycle in which the mother and calf separate within the first year (Read 1989, Read and Hohn 1995). Mysticete whale calves may accompany their mothers for 1 year, through a migratory roundtrip and then separate on the calving grounds. More than 91% of humpback whale calves observed by Clapham and Mayo (1987) separated from their mothers during their second winter. Some yearling southern right whales return to Patagonia with their mothers and separate there (Thomas and Taber 1984). Young striped dolphins join juvenile schools at 2.5 to 3 years of age (Miyazaki and Nishiwaki 1978). Bottlenose dolphins in Sarasota Bay typically remain with their mothers for 3 to 6 years (Wells 1991a,b). Young male sperm whales disperse from their natal groups when they are about 6 years old (Richard et al. 1996). In contrast, killer whale offspring apparently remain with their natal pod through their entire lives (Heimlich-Boran 1993). Lengthy periods of mother–calf association tend to be correlated with the development of more complex social organizations.

Selective pressures associated with production and rearing of offspring may have led to the evolution of the female-biased size dimorphism exhibited in several cetacean species (e.g., baleen whales, harbor porpoises), in contrast to male-biased size dimorphism, which likely evolved through sexual selection (described previously). Larger mothers may be better mothers, in that they may produce a larger baby, enable it to grow more rapidly by providing more or better milk, and be better at maternal care or protection (Ralls 1976). The pattern of dimorphism in baleen whales is probably influenced by the energetic demands on the mothers to nurse their young for extended periods while fasting on the calving grounds and during migration (see Costa and Williams, Chapter 5, this volume).

Ranging Patterns

Cetaceans live in most of the world's marine and estuarine habitats, as well as some freshwater systems. As a group, they exhibit a variety of patterns of habitat use, varying from year-round residency in a limited home range to annual long distance migrations. The ranging patterns of a species are closely tied to the ability of the habitat to meet the animals' needs. In some cases, this can be accomplished within a single bay or estuary system, whereas for other species it requires much of an ocean basin. The ranging patterns are defined in large measure by the influence of geographical and oceanographical features on physiological factors such as energetics and thermal considerations, as well as ecological factors, such as resource availability and predation pressure (see Bowen and Siniff, Chapter 9, this volume).

Long distance migrations have been documented for a number of mysticete whales, as well as adult male sperm whales. Important documentation of these patterns has been obtained through resightings or recaptures of identifi-

Table 8-5. Cetacean Life History Parameters

	Max. Adult Length (cm)		Dimorphism	Age at Sexual Maturity		Max. Age of Adults (yr)						
Species	Male	Female	Male / Female	Males	Females	Male	Female	Average Age at Weaning (mo)	Calving Interval (yrs)	Gestation (mo)	Birth Length (% of mother)	Length at Birth (cm)
Odontoceti												
Delphinidae												
Cephalorhynchus commersonii	130	134	0.97	5–6	5–6	17	18				0.56	75
Sotalia fluviatilis	146	145	1.01							10.0	0.52	75
Delphinus delphis	178	170	1.05	7.0	6.0	22	20	6	2.6	10.5	0.48	81
Lagenorhynchus obscurus	188	191	0.98					18		11.0		
L. obliquidens	190	192	0.99							10.0	0.46	88
Stenella longirostris	192	189	1.02	7.0	5.0	19	23	11–34	3.0	10.7	0.41	77
S. attenuata	201	187	1.07	12.0	9.0	40	46	20	2.5–3.9	11.5	0.44	83
S. coeruleoalba	227	220	1.03	9.0	9.0	29	28	15–18	3.3	12.0	0.45	100
Steno bredanensis	232	231	1.00			32	30					
Lagenodelphis hosei	236	235	1.00								0.43	100
Lagenorhynchus acutus	250	224	1.12			22	27			10.0	0.49	110
Orcaella brevirostris	250	250	1.00								0.34	85
Lagenorhynchus albirostris	260	259	1.00								0.48	125
Tursiops truncatus	261	251	1.04	8–12	5–10	41+	52	18–20	3.0–6.0	12.0	0.46	115
Lissodelphis borealis	263	217	1.21								0.41	90
Peponocephala electra	268	260	1.03								0.33	85
Feresa attenuata	270	270	1.00									
Sousa chinensis	280											100
Grampus griseus	380	380	1.00			13+	17+				0.39	150
Globicephala macrorhynchus	453	358	1.27	16.0	9.0	45.5	62.5	42	6.9	12.0	0.39	140
Pseudorca crassidens	532	447	1.19	11.0	11.0	20	22			15.5	0.43	193
Globicephala melaena	545	381	1.43	12.0	6.0	14	20+		3.3	15.0	0.46	177
Orcinus orca	945	566	1.67	16.0	10.0	35	34	12+	3.0–8.3	15.0	0.41	230
Phocoenidae												
Phocoena phocoena	160	180	0.89	4.0	4.6	12	24	8	1.0	11.0	0.44	80
Neophocaena phocaenoides	160	150	1.07					6–15		11.0	0.37	55

Phocoenoides dalli	220				7.9	6.8		24	3.0	11.0		100
Phocoena dioptrica	220											48
Platanistidae												
Pontoporia blainvillei	160	170	0.94	2.3	2.7	16	13	8–9	2.0	10.5	0.43	73
Platanista gangetica	210	250	0.84			28					0.30	75
Lipotes vexillifer	220	230	0.96							10.0	0.33	75
Inia geoffrensis	290					18+	28			10.0		79
Monodontidae												
Delphinapterus leucas	450	430	1.05	4–7	8–9	25–30	25–30	20–24		14.5	0.35	150
Monodon monoceros	500	450	1.11	5–8	11–13	50	50	20		14.5	0.36	160
Ziphiidae												
Ziphius cavirostris	700	754	0.93			36	30				0.36	270
Hyperoodon ampullatus	980	870	1.13	7–11	11.0	37	27	12	2.0	12.0	0.41	360
Berardius bairdii	1190	1280	0.93	8–10	8–10	71	39		3.0	17.0	0.35	450
Physeteridae												
Kogia simus	270	270	1.00						1.0		0.37	100
Kogia breviceps	300	300	1.00						1.0	11.0		
Physeter macrocephalus	1500	1090	1.38	10+	8–11			Extended	5.2–6.5	16.0	0.37	400
Mysticeti												
Balaenopteridae												
Balaenoptera acutorostrata		1070	< 1.0	5–8	6–8	< 50	< 50	3–6	1.0–1.2	10–11	0.25	270
Megaptera novaeangliae		1520	< 1.0	2–5	2–5	48	48	5	2.0	11–11.5	0.28	430
Balaenoptera edeni		1560	< 1.0	9–13	8–11			6	2+	12.0	0.26	400
B. borealis		1830	< 1.0	< 10	6–12	60	60	6–9	2–3	11–13	0.25	450
B. physalus		2600	< 1.0	6–7	6–12	90–100	90–100	6–7	2–3	10–12	0.25	640
B. musculus		3100	< 1.0	5–10	5–10	80–90	80–90	7	2–3	11.0	0.23	700
Balaenidae												
Caperea marginata	609	645	0.94					5–6		12.0	0.34	220
Eubalaena glacialis		1800	< 1.0	10.0	10.0			6–7	2.0	10.0		460
E. australis		1830	< 1.0							10.0	0.33	610
Balaena mysticetus		2000	< 1.0		4.0			5–6		12–13	0.23	450
Eschrichtiidae												
Eschrichtius robustus	1460	1500	0.97		5–11	> 40	> 40	7	2.0	13.7	0.31	460

Values from summaries presented in Perrin et al. 1984 and Ridgway and Harrison 1985.

able individuals. Much of what we know today about mysticete and sperm whale migrations has been provided by returns of "Discovery Tags." Numbered metal cylinders were shot into whales from whaling ships or from research vessels during dedicated cruises, starting during the 1932 to 1933 Antarctic whaling season (Brown 1962, 1978). Tags were recovered when whales were captured and rendered, providing information on two points within the animal's range. More than 20,000 of these tags were used, with generally low return rates (≤ 15%), depending on the fishery. For example, Ivashin (1981) estimated that only about 3% of more than 1400 marks applied to sperm whales during 1952 to 1979 were recovered. The moratorium on commercial whaling in 1985 effectively ended the use of this technique.

More recently individual identification from photographs of distinctive natural markings have provided a wealth of information on cetacean ranging patterns. Gray whales (Darling 1984, Jones and Swartz 1984, Swartz 1986) and blue whales (Calambokidis et al. 1990) have distinctive patterns of pigmentation in the region near the dorsal ridge or the fin that are especially useful. Right whales develop callosities, areas of toughened skin, on their heads that have allowed researchers to keep track of individuals over decades (Payne et al. 1983, Payne 1995). This technique has allowed researchers to monitor virtually every one of the approximately 300 northern right whales in the western North Atlantic (Kraus et al. 1986). Bowhead whales develop distinctive white pigmentation patches on their chins and caudal peduncles. The distinctive black and white patterns on the ventral surfaces of humpback whale flukes (Katona et al. 1979) provide excellent identifying marks (Fig. 8-25). Finback whales have distinctive pigmentation patterns involving their shoulder stripes, as well as distinctive dorsal fin shapes (Whitehead and Carlson 1988). Many minke whales have distinctive dorsal fin shapes and pigmentation patterns (Dorsey 1983). Sperm whales can be identified from the shapes of their flukes (Arnbom and Whitehead 1989, Dufault and Whitehead 1995a). Dorsal fin shape and features, along with pigment patterns in some cases, allow the photographic identification of killer whales (Fig. 8-26; Bigg et al. 1987), pilot whales (Heimlich-Boran 1993), Risso's dolphins (Kruse 1989), dusky dolphins (Würsig and Würsig 1980), Hawaiian spinner dolphins (Norris et al. 1994), humpback dolphins (Karczmarski 1997), tucuxi (*Sotalia fluriatilis*) (P. Flores, pers. comm.), bottlenose dolphins (e.g., Würsig and Würsig 1977, Wells et al. 1987), and other species (for reviews, see Scott et al. 1990b, Würsig and Jefferson 1990). Radio and satellite-linked tracking packages, with improved attachment technology, are decreasing in size and becoming increasingly useful with small cetaceans, allowing the tracking of movements around the clock, over long periods of time and over great distances (for reviews, see Scott et al. 1990b; Watkins et al. 1981, Irvine et al. 1982b, Swartz et al. 1987, Tanaka 1987, Würsig et al. 1991, Mate et al. 1995, Davis et al. 1996). The accumulated records of repeated sightings and extended tracking of large numbers of identifiable individuals within seasons and across years have greatly advanced our understanding of ranging patterns, social interactions, life history, and population biology.

Migrations

Gray whales in the northeastern Pacific Ocean travel southward along the shoreline from summer coastal feeding areas off Alaska and Canada to winter breeding and calving lagoons in Mexico, covering a distance of about 7600 km (see Bowen and Siniff, Chapter 9, this volume; Rice and Wolman

Figure 8-25. Distinctive ventral fluke patterns of humpback whales used in photographic identification of individuals. (Photograph by Randall S. Wells.)

Figure 8-26. Pioneering researchers Mike Bigg and Graeme Ellis photograph the individually identifiable killer whales of A-5 pod off Vancouver Island. The distinctive dorsal fin nicks and scars, and saddle patches, have allowed the researchers to trace the movements and behavior of these whales for decades. (Photograph by Flip Nicklin, Minden Pictures.)

1971; Swartz 1986). Pregnant females move first and penetrate deep into the lagoons where they and their newborns tend to avoid other whales. Resting females (nonpregnant), males, and juveniles of both sexes follow somewhat later, courting and mating along the migratory path, and aggregate in the calving lagoon entrances (Norris et al. 1983, Jones and Swartz 1984). Nonparturient adults and immatures leave the calving lagoons first, followed in about a month by cows and young of the year. The first group tends to migrate north from headland to headland, whereas the mothers and calves tend to follow the contour of the coastline (Poole 1984). Radio-tracking found no differences in migration rates between day and night (Swartz et al. 1987). Although most feeding by gray whales occurs at the northern terminus of their migration, there is evidence that some feeding may occur in the breeding lagoons as well (Norris et al. 1983). Migration occurs within about 5 km of shore where waters are typically less than 30 m deep. Braham (1984) suggests that this inshore migration may be influenced by the availability of sublittoral food resources, facilitating feeding along parts of the migratory pathway.

Individual identification has also shown that some gray whales return to the same waters at each end of their migrations year after year. Mothers and courting whales may return to the same calving lagoons each year, but may also circulate among lagoons (Jones and Swartz 1984, Swartz 1986). About 63% of the gray whales seen in the summering grounds off British Columbia have been resighted in the same area over periods of 2 to at least 8 years (Darling 1984).

The world's longest study of individually identifiable whales has been conducted with southern right whales off Patagonia, Argentina (Payne et al. 1983, Payne 1995). Southern right whales move between winter nursery grounds along the coasts of South America, South Africa, and Australia to summer feeding areas in Antarctic waters. Many of the southern right whales demonstrate strong site fidelity to the Patagonian nursery area, although they may be absent some years (Payne 1995). Reidentifications of southern right whales in distant locations, as much as 4424 km apart, fill gaps in our knowledge of their migration patterns. Such well-separated sightings of identifiable individuals have shown that females may use multiple nursery sites. In addition, these long-distance sightings link coastal nursery grounds and high latitude feeding areas, and provide indications that males may emigrate from coastal waters on occasion (Best et al. 1993).

The southern right whale findings are consistent with those for northern right whales. Northern right whales in the northwestern Atlantic migrate from wintering grounds off the Florida and Georgia coastlines to summer feeding grounds in the Gulf of Maine and associated offshore waters. A satellite-tagged male northern right whale was tracked more than 1500 km in 22 days as it visited known right whale aggregation areas in the summer feeding grounds (Mate 1990). A female and calf traveled nearly 4300 km in 6 months (Knowlton et al. 1992).

Bowhead whales are the only baleen whales to remain in near-freezing water through most of their lives, following the seasonal movements of the Arctic ice (Moore and Reeves 1993, Würsig and Clark 1993). Their spring migrations often take them through areas of heavy ice coverage where they take advantage of existing leads (long narrow open water spaces in ice-covered areas). They may break new ice up to 18 cm thick, or push up ice and breathe through thin cracks (Würsig and Clark 1993). How they locate these breathing holes or ice breaks is unknown, but George et al. (1989) have suggested that individuals may use their calls to assess acoustically the thickness of ice in their pathway. The migrating whales usually move singly or in small groups, with smaller whales migrating earlier. Breaching, milling, and socializing occur frequently during the spring migration of larger whales into the Beaufort Sea, suggesting the occurrence of mating aggregations. Feeding takes place during spring, summer, and fall; it is not known whether it continues into the winter, or whether the whales fast at this time. During summer, bowhead whales may concentrate their feeding activities in a single location for days to weeks (Moore and Reeves 1993, Würsig and Clark 1993). Much of their activity occurs along estuarine fronts (where water masses of different salinities meet) and regions of upwelling where zooplankton are likely to concentrate. In the fall migration, the whales intersperse traveling through waters 20 to 49 m deep with stops of several days for feeding. Some individually identifiable bowhead whales, especially subadults, occupy the same general summering areas in subsequent years (as summarized by Moore and Reeves 1993).

Humpback whales in all oceans move from winter low latitude, near-shore breeding and calving grounds to summer feeding areas in coastal waters at higher latitudes. The migrations in both directions involve a staggering of age and sex classes. Typically, newly pregnant females leave the winter breeding grounds first, followed by nonpregnant females, mature males, and immature whales. This order is roughly reversed for the return migration from feeding to calving grounds (Chittleborough 1958, Dawbin 1966). Intensified efforts at individual identification at a large number of research sites have allowed the definition of migration destinations for a number of individuals. In some cases, the same whales swim together in southeastern Alaska in summer and in the same interisland channel in Hawaii the following winter (Darling and Jurasz 1983). Some individuals have traveled for four to six consecutive seasons between Hawaii and southeast Alaska, suggesting traditional migratory routes are used (Darling and McSweeney 1985). Mothers frequently return to the wintering grounds with their calves from the previous year, and some older calves continue to return to the same waters after separation (Clapham and Mayo 1987; also see Weinrich et al. 1993). Darling and Jurasz (1983) report cases of whales being identified in Hawaii and on a different calving ground near Socorro Island, Mexico. Similarly, Darling and Cerchio (1993) have identified the same humpback whale in the waters of Japan during winter, and off Hawaii the following winter. Occasional feeding in low latitude waters has been observed (Baraff et al. 1991, Gendron and Urbán 1993).

The one-way distances of the humpback migrations vary. In the North Atlantic, they travel 4600 km from the Gulf of Maine to the West Indies, or about 5900 km between Iceland and the Caribbean (Martin et al. 1984). In the Pacific, migratory distances range from about 4600 km between southeast Alaska and Hawaii, to nearly 8000 km from the Antarctic Peninsula, across the equator to the Pacific coast of Colombia (Stone et al. 1990, Clapham 1996). The out-of-phase timing of northern and southern hemisphere seasons results in northern whales being in the high latitudes feeding during the summer, whereas the southern hemisphere whales are breeding in the low latitudes during austral winter, and vice versa. This timing has presumably reduced the exchange of whales between the hemispheres, but recent observations suggest that occasional transequatorial movements occur that could bring northern and southern whales together (Acevedo and Smultea 1995).

The movements of the more pelagic rorquals, such as blue whales and finbacks, are less well known, but appear to follow an annual migratory pattern similar to that of humpbacks. At least 45% of identifiable finback whales summering in Massachusetts Bay were resighted there in multiple years (Seipt et al. 1989), as was at least one finback summering in the St. Lawrence estuary (Edds and MacFarlane 1987). In the western North Atlantic, at least 25% of individually identified finback calves returned to their mothers' summer feeding grounds after separation, suggesting that at least in some cases fidelity to such areas is determined matrilineally (Clapham and Seipt 1991). North Atlantic finback whale movements of more than 1500 km have been determined with radio and satellite-linked transmitters (Watkins et al. 1996). Individual blue whales have been identified in multiple years moving between southern destinations in Pacific and Gulf of California waters off Baja California and the more northerly destinations of Monterey Bay and the Gulf of the Farallones, California (Calambokidis et al. 1990).

As described, bachelor sperm whales may travel to latitudes higher than 45 degrees to feed, but return to equatorial waters to breed. The migration of adult male sperm whales may be dictated by the extreme sexual dimorphism that requires males to meet greater energetic demands than females (Best 1979). The results of photographic identification surveys by Dufault and Whitehead (1995b) in the South Pacific suggested that it is unlikely that female and immature sperm whales undergo transoceanic movements, and their movements are limited to about 1000 km. Females and immatures tend to inhabit large areas characterized by high secondary productivity and steep underwater topography (Jaquet and Whitehead 1996), and to move out of areas of low productivity as necessary (Whitehead 1996a). The distribution of sightings of sperm whales near and in the Gulf Stream suggests that they also use features such as warm core rings (self-contained, circular warm water areas that have spun off main current systems), the edge of the stream, and the margin of the continental shelf, all of which are associated with high productivity (Waring et al. 1993).

What factors lead to the initiation of whale migration? The availability of food at the higher latitudes may be a principal driving factor. The whales must be able to obtain sufficient food during the feeding season to support them throughout the year, including during the energetically costly times of migration and breeding. Blooms of zooplankton, such as krill, are the primary prey of many of these whales, and the krill feed on phytoplankton that are dependent on sunlight. Thus, photoperiod, or the availability of sufficient sunlight to support the phytoplankton that support the whales' prey, may serve as one cue to initiate migration (Dawbin 1966). The lipid content of krill and capelin changes seasonally, resulting in summer increases in the energy content of these prey (Mårtensson et al. 1996). Thus, a related consideration may be the increased energetic costs of remaining in increasingly cold waters as feeding becomes less efficient outside of summer months.

In some of the more extreme cases the factors leading to seasonal movements seem more clear. The movement of bowhead whales, belugas, and narwhals into arctic waters is dependent on the seasonal retreat of the polar ice pack. These species use prey associated with the ice edge, or habitats that are seasonally unavailable to the whales because of ice coverage. During summer months, they move through such narrow openings as the Bering, Franklin, and Davis Straits into the Arctic Ocean, Beaufort Sea, and Baffin Bay, but they are forced to leave these waters for the remainder of the year. Water temperature may influence these movements as well, either as a direct result of thermoregulatory demands, or indirectly because of effects on prey. Stomach content analyses and observations suggest that narwhal movements into open water summering areas are related more to calving than to feeding, as little feeding is done at that time (Finley and Gibb 1982).

Fifteen belugas near Baffin Bay were tracked with satellite-linked transmitters during 1988 to 1992 (Martin et al. 1993) and showed extensive movements corresponding to the retreat of the ice, and in advance of the reforming sea ice. During this time, the belugas moved between estuaries and deeper waters where they made dives to 300 m or more, probably feeding near the sea floor. The reasons belugas travel thousands of kilometers for a brief stay in the summer grounds appear to be related to access to relatively warm freshwater outlets during their molt, and to the availability of relatively rich nearby food resources associated with the deep water seabed (Martin et al. 1993). Some of the belugas in Hudson Bay remain within the same estuary throughout the summer season, or return to the same estuary in at least 2 consecutive years (Caron and Smith 1990).

Seasonal changes in water temperature and other environmental factors, such as salinity, tidal amplitude, wind patterns, and nutrient upwelling in areas of high sea floor relief, are believed to be responsible for the shorter migrations or changes in distributions of other cetaceans as well, either directly or indirectly through effects on food resources (Norris 1967, Leatherwood et al. 1980, Gaskin and Watson 1985, Selzer and Payne 1988). For example, bottlenose dolphins in the coastal waters along the east coast of the United States range as far north as New Jersey in the summer, but they are rarely found north of North Carolina during the winter (for review, see Wang et al. 1994). Similarly, harbor porpoises along the same coast are found from the Gulf of Maine, northward during the summer, and some move as far south as the Carolinas in the winter. Whether these movements are primarily in direct response to thermal challenges, or occur because of shifts in prey distribution is not known.

Nonmigratory Ranging Patterns

The ranging patterns of open ocean cetaceans are influenced largely by oceanographic and physiographic features that influence the distribution of their prey. These species may range widely through ocean basins or along the continental shelf in search of rich patches of prey. For example, changes in common dolphin abundance in southern California waters are significantly correlated with the abundance and distribution of prey. The dolphins apparently move freely through the area, locate highly productive feeding areas, exploit them, and then move on in search of new feeding areas (Evans 1975).

Physiographic features appear to limit the range of the southernmost northern bottlenose whales in the western North Atlantic. A resident, year-round population of about 230 whales inhabits a core area of about 160 km^2 in the Gully, off Nova Scotia (Whitehead et al. 1997). The Gully is the most prominent canyon in the western North Atlantic. The movements of these individually identified whales are focused in waters more than 1000 m deep.

Pantropical spotted dolphins and spinner dolphins in the eastern tropical Pacific Ocean may range through the same habitats within a huge area of open ocean, but the distances they cover are quite different. Spinner dolphins may move a net distance of no more than 500 km/yr, whereas spotted dolphins may move more than 1800 km in the same period and swim an average of 63 km/day (Perrin et al. 1979).

Radio-tagged dusky dolphins travel on average about 19 km/day off Patagonia (Würsig 1982). They move faster and farther from shore, and over a greater area during the summer than in winter. Swimming speeds are greater in deeper water than in shallow water, and in the afternoon versus the morning, with faster swimming corresponding to bouts of surface feeding (Würsig and Würsig 1980). Dusk and night dives tend to last longer than daytime dives. Some of the dolphins are seen in the same area during both summer and winter, whereas others are not. The reasons for this incomplete migration are not known. Two tagged dusky dolphins were seen together 780 km away from their original sighting location along the Argentine coast after 8 years (Würsig and Bastida 1986).

Some small cetaceans may exhibit more than one ranging pattern from population to population, either migrating or shifting their distributions on a seasonal basis in apparent response to shifts in prey. For example, Dall's porpoises are typically found in waters more than 200 m deep (Morejohn 1979, Jefferson 1991). In the eastern North Pacific, they may be found throughout the north–south extent of the species' range during all months, but they may move inshore and offshore seasonally (Morejohn 1979). In Japan, seasonal migrations have been reported, with immature individuals having different routes than adults (Amano and Kuramochi 1992).

In regions where physiographic features make the locations of prey concentrations more predictable, cetacean movements may be more limited. For example, at least some spinner dolphins in Hawaiian waters are resident to specific island regions, where they take advantage of feeding opportunities afforded by the deep scattering layer formed in association with nutrient upwelling along a steep escarpment (Norris et al. 1994). Similarly, identifiable Risso's dolphins use the continental shelf waters near the Monterey Canyon, California, an area of periodic strong upwelling, over periods of several years (Kruse 1989). Radio-tracking has shown common dolphin movements to be concentrated in the vicinity of seamounts (Evans 1971, 1974, 1975). As a species, Atlantic spotted dolphins tend to be open ocean dolphins, but identifiable individuals are resident to a region that includes the shallow Bahama Banks and associated escarpment (Herzing 1998). Individually identifiable harbor porpoises occupy consistent "territories" and exhibit patrolling patterns during their summer occupancy of waters in New Brunswick (Gaskin and Watson 1985).

The tucuxi occurs in both marine and freshwater forms (Borobia et al. 1991). The freshwater form is somewhat smaller and is strongly influenced by seasonal fluctuations in river levels. During the low water period, fish are concentrated in river mouths, which are among the preferred habitats for the tucuxi. This period of concentrated prey is when most tucuxi calves are born in the Amazon River system (Best and da Silva 1984). The dolphins generally appear to prefer the more open riverine habitats and do not penetrate the flooded forests or swampy areas exploited by the boutu. The marine form of tucuxi is found primarily in brackish water and estuaries (Borobia et al. 1991).

The ranges of riverine dolphins are limited by the confines of the rivers and associated tributaries, lakes, flood plains, and estuaries. Most riverine dolphins spend their lives within these systems. Use of particular portions of the systems may vary seasonally with changes in environmental conditions, such as sometimes moving into the forests and swimming among the trees during rainy season flooding (Layne 1958, Pilleri 1969, Kasuya and Aminul Haque 1972, Best and da Silva 1989, Shrestha 1989). Boutu calves tend to be born during peak or declining water levels in the Amazon River system, when the adults' prey begin to be concentrated in permanent water bodies or river channels and therefore, are more readily available (Best and da Silva 1984). On the basis of changes in dolphin densities, Shrestha (1989) suggested that Ganges River dolphins (*Plantanista gangetica*) in Nepal collect in deep pools or main river channels during low water and move into smaller and shallower tributaries during the high water season. The movements of baiji in the Yangtze River are influenced by such features as current eddies, sandbars, and banks, with the range size influenced by seasonal water levels; baiji have been reported to move 200 km to remain near large eddies (Yuanyu et al. 1989).

Home Ranges

When cetaceans occur in coastal waters, home ranges can often be defined. These areas can take a variety of spatial and temporal forms, depending on the energetic requirements of the animals. Both seasonal and year-round home ranges have been identified for cetaceans. Minke whales occupy the same inland marine waters off Washington State over multiple years, with seven individuals identified as using three exclusive adjoining ranges within a 600-km^2 study area (Dorsey 1983). In some cases, both seasonal and year-round home ranges have been documented within the same species in the same region. For example, the "resident" pods of killer whales near Vancouver Island are present during the summer months of each year, but are not seen during the remainder of the year, whereas the "transient" pods are seen in Vancouver Island waters throughout the year, but may also travel as far as Alaska (the apparently contradictory nomenclature was established before the patterns were well defined) (Leatherwood et al. 1984, Bigg et al. 1987). Some identifiable humpbacked dolphins in South Africa use shallow waters (≤ 25 m deep) within 1 km of shore in Plettenberg Bay and Algoa Bay throughout the year, and over multiple years (but most were considered to be transient; Saayman and Tayler 1979, Karczmarski 1997). Overlapping, long-term home ranges have been suggested for *Inia* in the Amazon region, with long distance migrations considered unlikely (Best and da Silva 1989). Individuals of the marine form of tucuxi inhabit long-term home ranges in the vicinity of Florianópolis, Brazil (P. Flores, pers. comm.).

Home ranges have been well-documented for bottlenose dolphins from several areas (for review, see Shane et al. 1986). Along the central west coast of Florida about 100 resident bottlenose dolphins have been studied in the Sarasota area since 1970. The original animals, as well as three subsequent generations, which form a well-defined community, have used essentially the same year-round home range of about 125 km² for more than 28 years (Irvine and Wells 1972, Wells et al. 1980, Irvine et al. 1981, Scott et al. 1990a, Wells 1991a). The dolphins occupy the home range throughout the year, but emphasize different portions seasonally, apparently in response to changes in prey availability (Wells 1993b). Other studies suggest that at least some bottlenose dolphins may be long-term year-round or seasonal residents to coastal waters in Tampa Bay, Florida (Wells 1986, Wells et al. 1996b), Charlotte Harbor and Pine Island Sound, Florida (Shane 1990a,b; Wells et al. 1996a), Florida's east coast (Odell and Asper 1990), North Carolina (Wang et al. 1994), the Adriatic Sea (Bearzi 1999), Texas (Shane 1980, 1990b; Bräger 1993; Lynn 1995), southern California (Hansen 1990), Scotland (Wilson 1995), Argentina (Würsig and Würsig 1977, 1979; Würsig and Harris 1990), Ecuador (Felix 1994), and Western Australia (Connor and Smolker 1985). Bottlenose dolphins were observed repeatedly near Bahia Kino, in the Gulf of California, but their ranges extended beyond the estuary (Ballance 1990, 1992). Extensive use was made of the estuary for feeding when the animals were in the area. Caldwell and Caldwell (1972) described a pattern of seasonal home ranges connected by travelling corridors. In many cases where resident bottlenose dolphins have been observed, nonresident individuals have also been reported.

Although in many cases bottlenose dolphin community home ranges (the collective home ranges of a group of individuals in the same region) appear to be relatively stable under normal circumstances, changes can occur. Seasonal changes in habitat use by resident bottlenose dolphins in Ecuador were related to changes in salinity (Felix 1994). During 1982 to 1983, bottlenose dolphins in southern California shifted their range more than 500 km to the north to Monterey Bay during an El Niño warm water incursion, and identifiable individuals have continued to occupy the waters of Monterey Bay ever since (Wells et al. 1990).

Social Organization

Cetaceans exhibit a wide range of social patterns, from single individuals to groups of several thousand, and from ephemeral associations to lifelong bonds. In addition, patterns within a species can change from season to season, or from one developmental or reproductive stage to the next. Most cetaceans tend to be social at some point during their lives. Our success to date in studying their social patterns has been attributable in large part to the discovery of appropriate study sites where cetaceans can be found predictably under satisfactory viewing conditions, and the development of new techniques. Individual identification through time has allowed application of focal animal techniques of observation for quantifying behavioral patterns (Fig. 8-26; Altmann 1974, Whitehead 1995). Technological advances have allowed application of observation techniques to behavior of cetaceans in the open ocean (e.g., Whitehead and Arnbom 1987, Whitehead 1989), and underwater where the animals spend most of their lives. The integration of acoustic recording with underwater behavioral observations has been an important advancement in our attempt to comprehend the animals' interactions (e.g., Norris and Dohl 1980a, Norris et al. 1994, Ostman 1994, Herzing 1996). Genetic techniques have allowed determination of gender and genetic relationships (e.g., Duffield and Wells 1991, Amos et al. 1993a). Age determination has also provided a crucial clue for unraveling social organizations and interactions (e.g., Hohn et al. 1989).

Group Size

Cetacean group size, in addition to being related to feeding and predator avoidance, appears in some cases to be related to body size (Fig. 8-27). Large cetaceans, such as baleen whales, adult male sperm whales, and beaked whales, tend to be found in small groups or sometimes alone, forming somewhat larger groups for breeding purposes. Predation pressure on these larger cetaceans may be less than for the smaller animals, especially as they approach adult size. Group size for many of the smaller cetaceans appears to vary directly with the openness of the habitat in which they live. Open habitat is correlated with increased predation pressure and rich, but patchy resources. Open ocean, continental shelf, and open coastline dolphins and porpoises tend to live in larger groups, numbering sometimes into the hundreds or thousands, than do coastal inshore and river dolphins and porpoises, which usually occur in groups of 20 or fewer. Compared to open ocean species, inshore cetaceans may face reduced levels of predation and have access to somewhat more evenly distributed and predictable resources; consequently, they tend to be found in intermediate size groups. At one end of the continuum, river dolphins are essentially free from predation, use fairly predictable resources within a limited habitat, and tend to be found alone or in groups of less than 10 (Layne 1958; Best and da Silva 1984, 1989; Yuanyu et al. 1989).

Among the smaller odontocetes, the relationship between habitat openness and group size is not strictly linear. Differences exist within species. For example, bottlenose dolphin group size increases predictably from complex in-

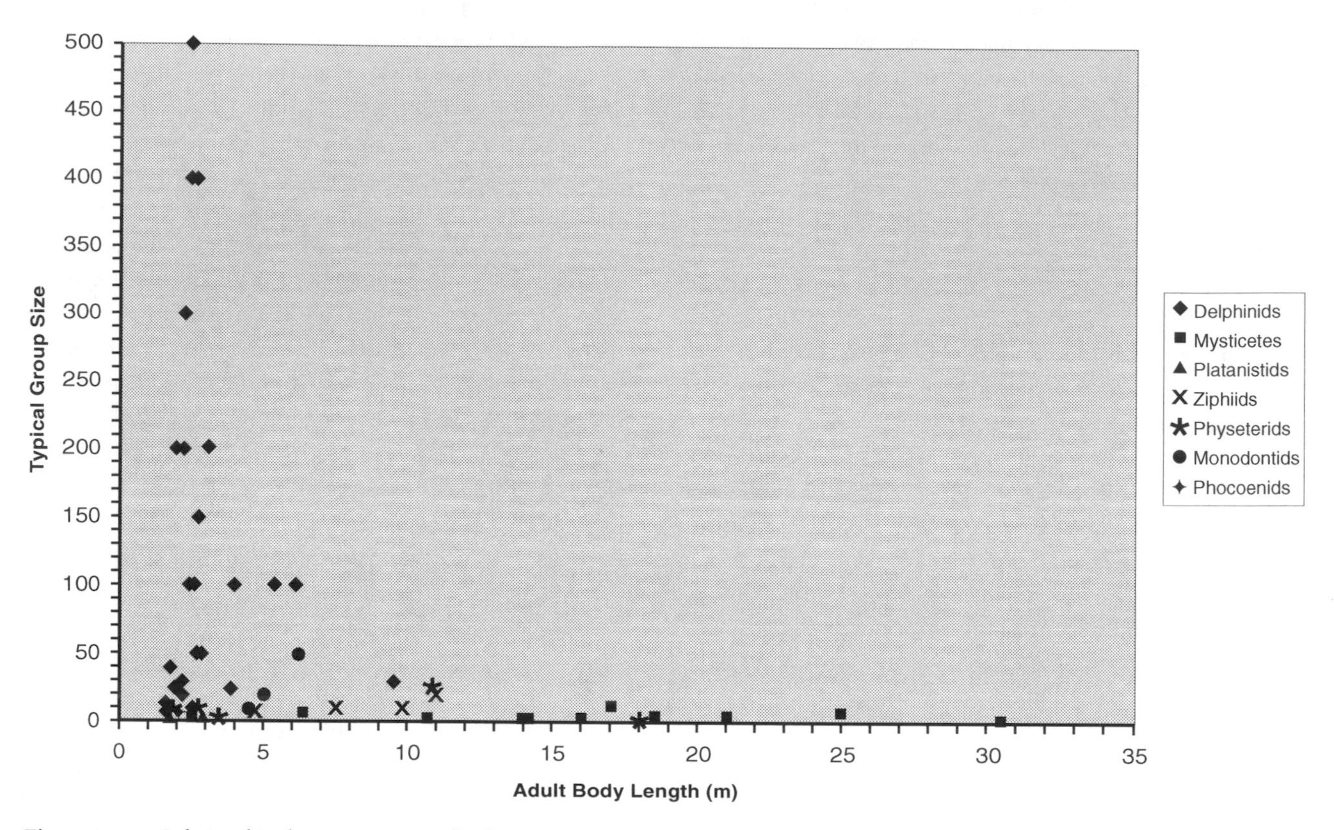

Figure 8-27. Relationships between cetacean body size and group size. Each point represents the approximate maximum adult body length for a cetacean species and the approximate upper limit for typical group size (not including large aggregations for feeding, breeding, etc.). (Estimates are from Leatherwood and Reeves [1983] and summary information in Wells et al. [1980].)

shore habitats, including small shallow bays and channels, to larger, deeper bays into open coastal waters (Wells et al. 1980), but the progression in group size does not necessarily continue linearly into deep pelagic waters (Kenney 1990, Scott and Chivers 1990). Average group size in open waters of the eastern Pacific is 57 animals (range, 1 to > 1000; Scott and Chivers 1990) and in the North Atlantic, 7 to 19 (Kenney 1990), as compared to 8 to 140 in open coastal waters, and typically less than 10 dolphins per group for small, shallow bays and channels (as summarized in Wells et al. 1980). This suggests that there may exist an optimal group size beyond which disadvantages of additional animals outweigh benefits of group living. Perhaps food resources are unable to support more than a certain number of animals. Scott and Chivers (1990) noted that decreases in group sizes for bottlenose dolphins in the eastern Pacific tended to coincide with El Niño warm water events, in which productivity was reduced, and the largest groups were on or near the continental slope where productivity was high. Walker (1981) demonstrated differences between stocks of bottlenose dolphins from one habitat to another. The differences in tooth number and size suggest prey differences, with the offshore animals often feeding on smaller schooling fish and squid. In turn, these prey were larger than those of some of the pelagic dolphins. The prey of the bottlenose dolphins may form smaller schools than do the prey of some of the pelagic dolphins, and therefore, each prey school may not be able to support as many dolphins.

Group Composition

Group size is just one measure of sociality. Other measures, such as stability of composition and complexity of organization, can be highly variable across species. Cetaceans can aggregate in response to environmental variables such as prey distribution, or they can congregate into more complex groups, with social organizations based on such factors as interindividual interactions, age, sex, genetic relationships, and reproductive conditions. Baleen whales on feeding grounds may be examples of resource-based aggregations. Hoek (1992) reported an unusual aggregation of harbor porpoises in the Gulf of St. Lawrence in apparent response to concentrations of surface schooling mackerel. Harbor porpoises typically swim singly or in very small groups, and have not been considered to have a complex social system. River dolphins, such as *Inia,* are predominantly solitary, with groups greater than two seldom encountered (Best and da Silva 1989). Average group size for the baiji, *Lipotes vexillifer,* is three to six individuals (range, 1 to 17; Kaiya and Yuemin 1989, Yuanyu et al. 1989), with the larger groups probably representing temporary feeding aggregations (Kaiya and Yuemin 1989).

Congregations of cetaceans, or collections of conspecifics at least in part in response to social factors, range from ephemeral interactions to lifelong affiliations with family members or other associates. For example, pantropical spotted dolphins (*Stenella attenuata*) in the eastern tropical Pacific Ocean may form very large schools in which there is no evidence of age or sex segregation (Perrin et al. 1976), whereas bottlenose dolphin schools may be organized largely on the basis of age and sex (Wells et al. 1987). In general, short-term interactions may involve a variety of combinations of age / sex classes, but longer term affiliations are built typically around similarities in age, gender, reproductive status, and kinship.

Large schools of open ocean dolphins, such as Hawaiian spinner, pantropical spotted, striped, or Risso's dolphins, typically change composition from day to day. Within these ephemeral groupings, however, some long-term associations may be found, functional division of labor may be seen (as in examples of the highly synchronized adult male cadres interposing themselves between threats and the rest of the school), and juvenile and nursery subgroups are evident (Miyazaki and Nishiwaki 1978, Kruse 1989, Pryor and Shallenberger 1991, Norris et al. 1994). Bottlenose dolphins tend to form small, fluid schools, but division of labor within these schools has been reported in the form of "scouting" (Caldwell et al. 1965), and in nursery groups (Wells et al. 1987). Regardless of the particular individual composition of a group, classes of subgroups may be identifiable across species.

The social organization of humpback whales on both feeding and breeding grounds can be characterized by small unstable groups (Whitehead et al. 1982; Clapham 1993, 1996), except in a few cases where associations continue through one year or across multiple years (Weinrich 1991). Most associations are in constant flux, and interactions among relatives are uncommon, occurring in less than 1% of associations in the Gulf of Maine (Clapham 1993). Clapham (1993) and others observed coordinated feeding groups and suggested that the upper limit of group size is determined by quantity of prey immediately available, and that group size varies from patch to patch. As described above, lack of predation obviates a need for large groups as a means for predator avoidance or for protection (Clapham 1996). Fluctuating, mobile distribution of patchy prey may also explain lack of territoriality or aggressive contests on feeding grounds. A similar pattern has been noted for finback whales, in which the number of whales in a group was closely related to horizontal size of prey schools on which whales were feeding, and associations tended to be brief (Whitehead and Carlson 1988).

Bowhead whale social associations appear to be brief, lasting only hours or days. Bowheads engage in much social activity during their spring migration, and sometimes during the fall migration (Würsig et al. 1993), including sexual behaviors, but their groups tend to be small and ephemeral (Würsig and Clark 1993). Feeding in areas of rich prey patches may involve coordinated efforts of up to 14 individuals in echelon formations, but these groupings are also temporary (Würsig and Clark 1993). Some sex and age segregation is seen during both the spring and fall migrations (Moore and Reeves 1993).

Kinship appears to be especially important in the structure of groups of some of the larger odontocetes, including killer whales, pilot whales, and sperm whales. The most stable groupings found to date are those of killer whales. Where they have been studied in detail for more than 25 years near Vancouver Island, Canada, killer whales live in stable extended family units called pods (Balcomb et al. 1982, Bigg 1982, Bigg et al. 1987). These pods are composed of mothers and their offspring, including sons and brothers, and it is believed that both sexes may spend their lives within their natal pods (Heimlich-Boran 1986). Each pod has developed its own pod-specific acoustic dialect (Ford and Fisher 1982, 1983; see Tyack, Chapter 7, this volume). More than 300 killer whales reside in the vicinity of Vancouver Island, divided into two nonoverlapping communities including 19 different resident pods, and one community of 30 transient pods that overlaps ranges of resident communities (Bigg et al. 1987). Transient pods appear to avoid residents (Baird and Dill 1995).

Genetic data from long-finned pilot whales (Amos et al. 1991, 1993a; Amos 1993), along with observational data from short-finned pilot whales (Heimlich-Boran 1993), suggest that pods are composed of related females with their female and male offspring, with one or more unrelated adult males in temporary attendance. Neither sex apparently disperses permanently from its natal group of 100 or so pod mates, although adult males may move among pods to mate and then return to their natal pods (Amos et al. 1993a).

A similar social structure exists for sperm whales and in at least some beaked whales. Temporal and spatial segregation by sex and age is apparent in sperm whale groups, with females and young remaining in lower latitudes throughout the year, and many adult males moving to higher latitudes for feeding, and returning to interact with female and young groups during the breeding season (Ohsumi 1966, 1971; Best

1979). Best (1979) reported that composition of schools of female sperm whales appears to be very stable. Sperm whales observed near the Galapagos Islands occur in stable groups of females and immatures (Fig. 8-28), accompanied by transient males that remain with female groups for a few hours (Whitehead and Arnbom 1987, Arnbom and Whitehead 1989). Genetic analyses indicate that female groups may be comprised of long-term associations of several different matrilines (Richard et al. 1996). Much like the killer whales, group-specific dialects in codas, or differences in patterned series of clicks produced by neighboring, potentially interacting female groups, have recently been identified (Weilgart and Whitehead 1997). Similar group stability was found off Sri Lanka, although small stable groups at times formed part of larger labile groups (Gordon 1987). Adult males were rarely seen outside of the breeding season. Northern bottlenose whales in the Gully, off Nova Scotia, also appear to form permanent female groups, with males moving between them (Whitehead 1990). Thus, in sperm, northern bottlenose, pilot, and killer whales it seems that females provide long-term stability to groups, whereas males moving among groups provide a mechanism for genetic exchange, exhibiting social patterns reminiscent of matriarchal societies of elephants (Weilgart et al. 1996).

Kin groups tend to be more stable than most other cetacean groupings, and they are associated with some of the most complex social behaviors seen among cetaceans. Coordinated attacks of killer whales on large whales, and cooperative defense demonstrated by sperm whales, are patterns that would presumably have benefited from long-term, stable social relationships, and their evolution may be easily explained theoretically through benefits provided both to individuals and to their kin. Heimlich-Boran (1993) has suggested that killer whales and pilot whales live in matriarchal "avunculate societies," in which males help to care for their female relations' young rather than their own. For this kind of society to be formed, four conditions need to be met. There must be (1) large benefits for remaining with kin, (2) low rates of dispersal of both sexes, (3) a mechanism for extreme inbreeding avoidance through out-group mating, and (4) low degrees of paternity certainty in which a male is more closely related to his sisters' offspring than he will be to the offspring of his mate. Heimlich-Boran (1993) suggested that killer whales and pilot whales benefit from group living through increased foraging efficiency and protection from predation, that prolonged maturation and intergenerational bonds indicate importance of learning and provide opportunities for cultural transmission of knowledge, and that benefits accrue from the long postreproductive life spans for females as a form of terminal investment. Female life spans may continue more than 20 years beyond the birth of the last calf (Kasuya and Marsh 1984; Marsh and Kasuya 1984, 1991). Males may function to defend pods, provide alloparental care, or aid in food detection and capture, resulting in benefits through inclusive fitness. The lack of dispersal by young whales of either gender is unusual for mammals. For example, male chimpanzees, gorillas, and hamadryas baboons are likely to remain within their natal groups, whereas females emigrate when reaching sexual maturity; macaques and other baboons exhibit the opposite pattern (Gouzoules and Gouzoules 1987).

Stable associations between individuals, including kin, are less evident for some smaller dolphin species than for the larger odontocetes previously described. Whereas some associations between small numbers of individuals may be consistent over periods of years, most are less frequent. For

Figure 8-28. Group of female and young sperm whales off the Galapagos Islands. (Photograph by Flip Nicklin, Minden Pictures.)

example, a small proportion of identifiable Hawaiian spinner dolphins has been observed together for several years, with males forming the strongest associations (Norris et al. 1994, Ostman 1994). Risso's dolphins in Monterey Bay, California, are found in large, labile groups, but some associations have been seen repeatedly over periods of 10 to 15 months (Kruse 1989). Unlike very stable long-term associations of larger odontocetes, societies of many dolphins can be described as being built around repeated, rather than constant, associations among individuals or closely affiliated groups. Although specific composition of a group may change from day to day, the same individuals may come into contact with one another frequently over periods of years, resulting in a more extended and more loosely defined society.

Social patterns of the bottlenose dolphins in Sarasota Bay have been observed for more than 28 years (Scott et al. 1990a, Wells 1991a). All 100 members of the resident community are never seen together in the same school, but they share a home range and encounter each other as they move through the range. Similar patterns have been noted in Tampa Bay (Wells et al. 1996a) and Charlotte Harbor (Wells et al. 1996b), immediately to the north and south of Sarasota. The amount of time the dolphins spend together depends on their gender, age, reproductive conditions, and genetic relationships (Wells et al. 1987, Wells 1991a). Females spend the most time with other females at a similar point in their reproductive lives, especially those with calves the same age as their own, and those that were raised with the same female associates. These nursery groups may include related females of several generations, and babysitting by older female offspring is not uncommon. Mothers and calves typically remain together for 3 to 6 years, after which time calves separate and often join juvenile groups of mixed gender. One of the most striking features of the society is the existence of pair-bonded adult males. Male pairs are formed at sexual maturity and may last through the lives of these males. Genetic studies indicate that males within these pairs are not related to one another, as is often the case for male lion pairs (Packer et al. 1991) or male chimpanzee coalitions (Goodall 1986), but the paired dolphin males tend to be of similar age, and often have been raised in the same nursery groups.

Interactions between dolphins sharing a common home range in Sarasota are more frequent than they are with animals in adjacent waters, but mixing with members of adjacent communities does occur (Wells 1986). The term community rather than population has been applied to local population units such as the 100 residents of Sarasota Bay because, although they form a relatively discrete behavioral unit, significant genetic exchange occurs with other such units (Duffield and Wells 1986, 1991). This exchange takes several forms. Males may travel outside of the home range and breed with females in other areas, or they may escort nonresident females through the Sarasota area. A few resident females may disappear for periods of time and conceive elsewhere.

The fluid, long-term, repeated rather than constant associations described for the bottlenose dolphins of Sarasota Bay are indicative of the fission–fusion societies described for chimpanzees (Goodall 1986). Many of the smaller open ocean dolphins, described previously, and sirenians (see next section) exhibit variations on this theme, as do bottlenose dolphins from a number of coastal sites (for review, see Shane et al. 1986). In Shark Bay, bottlenose dolphin group size and composition are unstable and groups contain both males and females, but long-term, consistent associations typically involve members of the same gender (Smolker et al. 1992). Males form consistent groups of two or three individuals that remain together for years, and preferentially associate with certain other male groups. Associations between females are less strong. Associations between males and females are inconsistent and are related to female reproductive condition (Smolker et al. 1992). Tagged bottlenose dolphins in North Carolina have been reported together and separately throughout their range (Read et al. 1996). Associations in Bahia Kino in the Gulf of California fall into three patterns: (1) those animals seen repeatedly in the same subgroups, (2) those seen only once, and (3) those seen repeatedly but in different subgroups (Ballance 1990). Low levels of association indicative of group fluidity have also been reported for bottlenose dolphins in Galveston Bay, Texas (Shane 1980, 1990a; Bräger et al. 1994), the Moray Firth, Scotland (Wilson 1995), the northern Adriatic Sea (Bearzi 1999), Pine Island Sound, Florida (Shane 1990a,b), southern California (Hansen 1990), and elsewhere. In Patagonia, 10 of 53 bottlenose dolphins seen regularly during a 1974 to 1976 study were again identified during 1984 to 1986, and five of these were seen with consistent associates from earlier sightings (Würsig and Würsig 1977, 1979; Würsig and Harris 1990).

Other small coastal cetaceans exhibit the pattern of fission–fusion societies described for bottlenose dolphins. In South Africa, humpback dolphins share home range waters, but their groups typically number seven individuals or fewer, and group composition is unstable and temporary (Saayman and Tayler 1979, Karczmarski 1997). Strong bonds between individuals other than mothers and calves are uncommon.

Social Development

Glimpses into the process of cetacean social development have come from two sources: "snapshot" views of age–sex composition of groups (e.g., sperm whales, Best 1979), and following individuals from birth through progressive life history milestones (e.g., bottlenose dolphins, Wells 1991a;

killer whales, Bigg et al. 1987; humpback whales, Clapham 1994; right whales, Payne 1995). As described above, bottlenose dolphins in Sarasota Bay spend their first 3 to 6 years with their mothers, often in nursery groups of regular female associates in which cohorts may be formed (Wells et al. 1987; Wells 1991a,b). Calves in Shark Bay (Smolker et al. 1992) and Scotland (Wilson 1995) typically remain with their mothers for at least 4 years. In Sarasota, with the birth of the next calf, the older calf may spend a period on its own in protected waters within the home range, or it may immediately join a group of juveniles and young mature individuals, ranging up to 15 years of age. Females interact within these groups until they give birth to their first calf, at which time they often return to their natal group to raise their calf.

The period after separation from mother appears to be a time in which bottlenose dolphins continue to learn skills that will serve them later in life, and they develop long-term social relationships. Juvenile groups engage in much more active socializing and play behavior than any other category except calves. Before primiparity, young females may spend time babysitting for their siblings. Males remain in juvenile groups until they reach sexual maturity at about 8 to 12 years of age, 3 years or more after females. During this time most develop a strong pair bond with another male that continues through life. Similar development of strong pair or triplet bonds between males has been reported from Shark Bay (Connor et al. 1992b), but not from Scotland (Wilson 1995). Mature males leave juvenile groups and begin traveling outside of the home range. They begin to travel from one female group to the next, but preliminary data indicate that they do not sire calves until they reach 20 years of age or more, after having achieved full adult size (Wells et al. 1987, Duffield and Wells 1991, Wells 1991a, Read et al. 1993, Tolley et al. 1995). Stability of these patterns within the Sarasota dolphin society is indicated by observations of four generations of individuals (Wells 1991a).

Sperm whales apparently follow a similar pattern of social development. Males leave nursery groups at about 4 to 6 years of age to form bachelor groups, whereas most females apparently remain within their natal groups (Best 1979, Richard et al. 1996). Some females may join juvenile groups, but return to the mixed school before attaining puberty. At about 21 to 28 years of age, males begin leaving bachelor groups. Many larger males move to higher latitudes in summer to feed and return to join female groups during the breeding season, although when larger males reach schoolmaster status (the largest, oldest males, > 25 years of age, presumably the primary breeding males), they may migrate less frequently (Best 1979).

Individual identification has allowed the examination of social development in baleen whale species as well. Humpback whales undergo maturational changes in social associations, becoming less solitary with age (Clapham 1994). Males typically increase their associations with adult females, and females increase their associations with adults of both sexes. Association patterns of young whales are nearly indistinguishable from those of adults when they reach 4 to 5 years of age (Clapham 1994).

Society Complexity and Social Interactions

The complexity of cetacean societies is determined at least in part by the species' life history. For example, most odontocetes share the following: a long life span, delayed maturity, diffusely seasonal breeding, production of a single offspring at a time, a high level of maternal investment, and low lifetime reproductive output (for review, see Perrin and Reilly 1984). Variations in life history strategies among the odontocetes (Table 8-5) are correlated with different levels of social system complexity. At one extreme, the small-bodied harbor porpoises attain sexual maturity early in life (4 years of age or younger), wean their calves within a year, produce a calf every year, and die by the relatively young age of 12 to 17 years (Read and Hohn 1995; Table 8-5). This fast-paced life is correlated with a fairly simple social system, lacking in long-term social relationships. At the other extreme, pilot whales, for example, attain much larger body sizes, reach sexual maturity much later in life, have a longer reproductive life span, invest more in each offspring through delayed weaning and multiyear calving intervals, retain the ability to nurse young even beyond their own reproductive years, and live to more than 60 years (Table 8-5). These life history parameters are correlated with a very stable, complex social system built on long-term, multigenerational bonds among related females. The heavy investment in each pilot whale offspring is in stark contrast to the high productivity but low investment strategy of the harbor porpoise and similar small odontocetes, reminiscent perhaps of the differences in strategies between small antelopes and elephants.

Cetacean societies are ordered by more than simply age, size, gender, or sexual condition. Complexities of social interactions range from those behaviors required for everyday school functioning, to the formation of individual relationships. Coordination of interactions between dolphins may involve subtle gestures, such as tilting or inverting the body, changing orientation of the head, rotating or extending flippers, and synchronizing movements. Apparent surfacing and orientation synchrony among nonmigratory bowhead whales spread over several kilometers has been observed, but it is not known whether this is a result of social interactions or simply a response to environmental cues (Würsig et al. 1985). Geometrical arrangements of dolphins within schools may improve school hydrodynamics and facilitate the functioning of sensory systems. Horizontal and vertical staggering creates "sensory windows," optimizing visual as well as acoustic transmission and reception capabilities (Johnson and Norris 1986). In some species, such as bowhead and right whales, such coordinated formations may increase feeding efficiency (Würsig and Clark 1993). Cetaceans with striking, high-contrast color patterns may use these to signal changes in orientation across a school (Evans and Bastian 1969, Norris et al. 1994).

Observations of cetaceans in captivity and, to a lesser extent, in the wild have shown the importance of affiliative and agonistic behaviors in development and maintenance of social relationships. Affiliative interactions may include such behaviors as changes in orientation to present the belly to another animal, stroking or caressing with fins or other parts of the body, sociosexual behaviors such as copulations or insertion of other body parts into the genital slits, close interanimal distances, or synchronous swimming and breathing (e.g., Wells 1984, Norris et al. 1994, Slooten 1994, Herzing 1996). The epimeletic behaviors already described could be considered extensions of affiliative behaviors.

Agonistic behaviors include striking with flukes, caudal peduncle, or dorsal fin, ramming with the rostrum or head, biting, tooth raking, or in the case of narwhals, jousting with tusks (see Fig. 8-22; Silverman 1979). Mothers may discipline their young by lifting them into the air with their rostra or flukes, or by holding them underwater (similar behavior has been demonstrated by captive killer whales toward humans) (Cockcroft and Ross 1990b). Between older and larger animals, such agonistic interactions can lead to serious injury, especially when adult males interact during the breeding season. This is evident in the agonistic herding and mate capture engaged in by male bottlenose dolphins in Shark Bay (Connor et al. 1992a,b), by battles between male dolphins in captivity in which individuals are sometimes killed (McBride and Hebb 1948), by the tusk wounds on narwhal males (see Fig. 8-22; Silverman and Dunbar 1980, Gerson and Hickie 1985), by tooth marks on male beaked whales (McCann 1974, Heyning 1984) and large male sperm whales (Kato 1984), by scrape marks on male southern right whales from the callosities of other whales (Payne and Dorsey 1983), and by scar patterns and violent interactions between male humpback whales on the breeding grounds (Darling et al. 1983, Baker and Herman 1984, Chu and Nieukirk 1988).

As in many animal groups, potentially dangerous physical contact may be preceded by threat displays. Among cetaceans, head-to-head displays, lobtailing, tail slaps to the head or other parts of the body, open mouth postures, jaw claps, chuffs (forceful exhalations), "S-posturing," and various kinds of chases, body charges, leaps, or body slaps may serve this purpose (Helweg et al. 1992, Norris et al. 1994, Slooten 1994, Herzing 1996). Bowhead whales use tail slaps and flipper slaps in aggressive contexts with conspecifics (Würsig and Clark 1993). Connor and Smolker (1996) report the use of "pops" by bottlenose dolphins as a threat vocalization to induce a female to remain with a male. "Squawks," "screams," and "barks" have been reported in agonistic contexts for Atlantic spotted dolphins (Herzing 1996).

Escalation of physical contact to dangerous levels appears to occur relatively infrequently among dolphins in captivity (Tavolga 1966, Ostman 1991), leading some researchers to conclude that dominance hierarchies may be an important component in the organization of day-to-day life in some cetacean societies. Typically, such hierarchies are ordered by size of the animals, with largest males dominant over others in the group, although older females may serve as a focus of much activity within groups (Tavolga 1966, Bateson 1974, Johnson and Norris 1986, Ostman 1991, Pryor and Shallenberger 1991, Samuels and Gifford 1997). This dominance may be manifested in positioning relative to other dolphins, in interanimal distances, with dominant animals swimming slightly in advance and above others, by threats, or by homosexual mounting (Evans and Bastian 1969, Ostman 1991, Pryor and Shallenberger 1991, Herzing 1996). In Moreton Bay, Australia, bottlenose dolphins feeding near trawlers apparently exhibit dominance through their positioning near boats. Adult males generally have priority of access to first choice of food, and this is sometimes reinforced through overt aggression (Corkeron et al. 1990). Best (1981a) suggests that displays of tusk length serve in dominance assessment for male narwhals. *Inia,* which are predominantly solitary in the wild, appear to be exceptions to this pattern. Aggression by males in captivity is frequently directed toward females and young, resulting in injuries and deaths; dominance hierarchies resulting from, or mediating such aggression, have not been identified (Best and da Silva 1989, Caldwell et al. 1989).

In summary, we find a progression of social complexity among cetaceans, from simple societies of river dolphins, most baleen whales, or relatively short-lived porpoises, through the somewhat fluid fission–fusion societies of many of the inshore, coastal, and open ocean dolphins, to stable social units of larger dolphins and toothed whales. The level of coordinated, cooperative behavior, including epimeletic behavior (Caldwell and Caldwell 1966) roughly correlates with this progression. Long-term associations tend to facilitate more complex coordinated activities. In the case of the inshore, coastal, and open ocean dolphins, such cooperation appears not to have a strong basis in kin relationships, and may be explained through the trading of altruistic acts, known as reciprocal altruism (Trivers 1971, Connor and Norris 1982). In larger animals, such as killer whales, pilot whales, and sperm whales, these activities tend to involve related individuals, and therefore may have resulted through kin selection. Regardless of the evolutionary pathway, strength of social relationships appears at times extraordinary. Mass strandings (e.g., Fehring and Wells 1976, Irvine et al. 1979, Mead et al. 1980) tend to involve those species (e.g., pilot whales, false killer whales, sperm whales, killer whales, spinner dolphins) that have demonstrated some of the strongest social bonds. The fact that often only a small proportion of stranded animals exhibit any obvious indication of illness or injury suggests that at least in some cases social factors may contribute to bringing the entire group onto the beach (Geraci and Lounsbury 1993). In some cases, loss or removal of sick or injured members has facilitated return of the remainder of the group to normal activities (Odell et al. 1980; R. S. Wells, pers. obs.) (Fig. 8-29).

Cetacean social complexity may also extend across species lines. Aggression between species has been observed, as have interspecific and intergeneric epimeletic behavior (Caldwell and Caldwell 1966). For example, bottlenose dolphins have exhibited aggression toward humpback dolphins (Saayman and Tayler 1979). In Moreton Bay, Australia, bottlenose dolphins appeared to be dominant over humpback dolphins because they maintained preferred positions for feeding around trawlers (Corkeron 1990). On the Bahama Banks, agonistic interactions between individuals or groups of male Atlantic spotted and bottlenose dolphins occur (Fig. 8-30; Herzing 1996). Captive groups of tucuxi make coordinated attacks on bottlenose dolphin tank mates (Terry 1984). Mixed species groups, such as spinner dolphins and pantropical spotted dolphins, in the eastern tropical Pacific Ocean, Risso's dolphins, Pacific white-sided dolphins, and northern right whale dolphins (*Lissodelphis borealis*) off the central California coast, and pilot whales with bottlenose dolphins in many locations (Leatherwood and Walker 1979, Norris and Dohl 1980b, Kenney 1990, Scott and Chivers 1990) are pervasive. Scott and Chivers (1990) reported bottlenose dolphins in association with at least 13 other species of cetaceans in the eastern tropical Pacific Ocean. The reasons for these associations remain unclear, but they likely represent yet another level of organization to meet the ecological pressures faced by these animals in the wild.

Although the body of available information on the behavioral ecology of cetaceans is somewhat limited, the importance of ecological factors such as food availability and predation pressure in shaping behavior in these animals is clear. The details of the interrelationships of genetics, ecology, and behavior, remain to be defined. Continued systematic scientific effort by a new generation of scientists, along with innovative field techniques to study the animals in their world, and as much as possible, on their terms, should go far toward placing our understanding of the behavior of the cetaceans on a par with that of other mammals in more familiar environments. Just such an approach, as described in the introductory paragraphs of this chapter, has led to our current understanding of the behavior of the sirenians.

Sirenians [G. B. R.]

The mammalian radiation into the marine environment has produced an impressive array of specialized forms that exploit a wide range of prey items (see Bowen and Siniff, Chapter 9, this volume). The sirenians, however, are unique in that they alone eat marine macrophytes as food. They share this unique niche with relatively few other marine vertebrates: marine iguana (*Amblyrhynchus cristatus*), green turtle (*Chelo-*

Figure 8-29. Part of a mass stranding of 175 to 200 short-finned pilot whales near Jacksonville, Florida, in February 1977 (Irvine et al. 1979). Two of the whales that were pushed offshore stranded again 6 days later in South Carolina. (Photograph by Randall S. Wells.)

Figure 8-30. Aggressive interactions between Atlantic spotted dolphins (smaller, darker) and two bottlenose dolphins on the Bahama Banks. (Photograph by Denise Herzing, Wild Dolphin Project.)

nia mydas), black brant (*Branta nigricans*), a few ducks, and several fishes. Although the fossil record suggests that the sirenians were much more diverse 30 million years ago during the late Oligocene period (Domning 1994), only five species have persisted into recent times (see Reynolds, Odell, and Rommel, Chapter 1, this volume): Steller's sea cow (*Hydrodamalis gigas*), which was extirpated in the 18th century (Whitmore and Gard 1977), and the dugong (*Dugong dugon*) are the only truly marine forms (Heinsohn et al. 1977), whereas the West African manatee (*Trichechus senegalensis*) and the West Indian manatee (*Trichechus manatus*) are found mainly in rivers, estuaries, bays, and sometimes near-shore marine environments associated with sources of freshwater (Powell and Rathbun 1984, Lefebvre et al. 1989). The Amazonian manatee (*Trichechus inunguis*) is restricted to the freshwaters of the Amazon River basin (Lefebvre et al. 1989). These species are morphologically and ecologically well defined and they share many similar behaviors. However, there are some significant differences in the behavioral ecology of dugongids and trichechids that make comparisons between these two families particularly interesting. Comparisons, however, are sometimes difficult because the research effort and knowledge available on each species varies greatly (Reynolds and Odell 1991).

Activity and Diving Behavior

The diet of sirenians, perhaps more than any other factor, influences much of their life history and behavior. The marine and freshwater macrophytes (some algae, sea grasses, and freshwater vascular plants) on which sirenians feed are limited in their distribution by a number of factors. For example, the light requirements of sea grasses have been determined to be fairly high, on the order of 10% to 20% of surface irradiance (Kenworthy and Haunert 1991, Dunton 1994). This limits the depths at which plants can grow; in many coastal rivers, bays, and estuaries the depth where macrophytes occur may be reduced to 1 m or less. Therefore, compared to many cetaceans and pinnipeds, sirenians are restricted to relatively shallow, near-shore waters by the distribution of their food—not unlike sea otters (discussed previously). West Indian manatees and dugongs probably rarely descend deeper than 20 m (Hartman 1979, Marsh et al. 1994), although dugongs are found in waters up to 70 m deep and nearly 60 km offshore (Marsh and Saalfeld 1989). Manatees and dugongs usually do not remain underwater for longer than 2 to 3 min when active, but when bottom resting West Indian manatees may remain submerged for up to 24 min (Reynolds 1981b). They typically remain at the surface to inhale for only a couple of seconds, and often do so very cryptically. Manatees and dugongs are active during day and night (Anderson and Birtles 1978, Hartman 1979, Best 1984, Rathbun et al. 1990b). Several factors influence this pattern, including human activities such as boating and diving (Buckingham 1990), tidal cycles (Anderson and Birtles 1978; Zoodsma 1991), and water temperatures (Bengtson 1981, Preen 1992). In Palau, where there is heavy hunting pressure on dugongs, Anderson (1981) suggests that dugongs feed mainly at night in shallow, near-shore waters to avoid poachers. The same may hold true for manatees in parts of Central America where they have been hunted (Reynolds and Odell 1991).

Predators

Another factor that probably is important in understanding sirenian behavior is that they have few natural predators, especially as adults. In South America, jaguars (*Panthera onca*) have been reported to prey on the Amazonian manatee (Marcoy 1875), whereas in Florida, alligators (*Alligator mississippiensis*) and sharks have rarely left wounds on animals (U.S. Geological Survey, Sirenia Project, unpubl.). Only in Australia is there good documentation of non-human predation on adult dugongs—by killer whales (Anderson and Prince 1985). It is also possible that sharks and crocodiles take dugongs (Anderson 1981). In general, few specialized antipredator behaviors such as those just discussed for some cetaceans have been observed in sirenians, probably because predation is uncommon in the relatively shallow near-shore coastal waters, rivers, and bays where their food grows. Indeed, none of the sirenians are particularly strong swimmers—they normally swim at speeds of 4 to 10 km/hr, but they are capable of short bursts of speed of up to 25 km/hr (Hartman 1979; Preen 1992). Manatees and dugongs also lack armament to protect themselves. Their secretive habits are perhaps their most effective protection. In addition, their relatively large size as adults, their very thick (ca. 5 cm) and tough skin, and dense, strong ribs are also effective deterrents to predation. However, neither their stealth nor robust anatomy are effective against humans (see Reynolds, Chapter 12, Twiss and Reeves 1999).

Foraging Behavior

The West Indian manatee (and presumably the very similar West African manatee) forages on a wide array of freshwater and marine plants (Best 1981b; Ledder 1986), sometimes including mangrove leaves (Best 1981b) and emergent terrestrial grasses and herbs growing on banks (Reynolds 1981b, Shane 1983b). Amazonian manatees appear to forage mainly on emergent plants, often those that form dense floating meadows composed of many species, especially grasses (Best 1981b, Montgomery et al. 1981). Dugongs, however, forage almost exclusively on marine angiosperms (i.e., sea grasses) that are well-rooted on the ocean bottom (Marsh et al. 1982; Lanyon 1991). Domning (1977, 1980) suggests that the growth forms of the plants normally eaten by each species of sirenian are related to their behavioral preferences for feeding position in the water column and their facial morphology. The dugong is nearly obligated to forage on the substrate because of its sharply down-turned rostrum. In fact, dugong calves often nurse in a horizontal, upside-down position because of their highly deflected nose (Anderson 1984). The Amazonian manatee, which has the least deflected rostrum, prefers to forage at the surface, whereas the West Indian manatee (and presumably the West African manatee, too) has an intermediate deflection to the rostrum, and has the widest foraging preference in the water column. For example, manatees in Florida are known to bottom-feed on acorns that have fallen into creeks from overhanging oak trees (O'Shea 1986). Having only a moderately deflected snout, West Indian manatee calves always nurse in a horizontal, upright position (Fig. 8-31). Although sirenians forage mainly on plants, dugongs and West Indian manatees sometimes feed on various sessile marine macroinvertebrates, such as ascidians, sea cucumbers, and sea pens (Anderson 1989, O'Shea et al. 1991, Preen 1995).

West Indian manatees often use their pectoral flippers to aid their prehensile upper lip and oral bristles (Marshall 1997) in stuffing floating or loose plant material into their mouths, and to rub their faces and mouth parts. They also sometimes use their flippers to "walk" along the bottom while feeding (Hartman 1979). Dugongs, however, rarely use their pectoral flippers to assist ingestion, nor have they been observed to walk on the bottom, although they do use their flippers

Figure 8-31. A West Indian manatee cow nurses her calf. (Photograph by Galen B. Rathbun, U.S. Geological Survey, Sirenia Project.)

as props while stationary on the substrate (Anderson and Birtles 1978; Anderson 1982; Preen 1992). Dugongs often leave clear evidence of their foraging activity in the form of meandering furrows in the sediments that are about 10 cm wide (Heinsohn et al. 1977, Anderson and Birtles 1978), or circular scars about 25 cm across (Aragones 1994). These excavations are where a single animal has used its highly specialized rostral disk to excavate and remove 50% to 70% of the sea grass biomass (including roots and rhizomes) from the substrate during one dive (Anderson and Birtles 1978; Preen 1992). In some marine habitats West Indian manatees also excavate the substrate to remove rooted plants. These sites are characterized by irregular, shallow craters where 50% to 90% of the plant biomass may be removed (Packard 1984, Lefebvre and Powell 1990).

Thermoregulation and Movements

The sirenian diet is associated with several important physiological and behavioral traits. Aquatic macrophytes, when compared to equivalent volumes of terrestrial plants, are poor sources of nutrients because they are composed of 90% to 95% water. Marine angiosperms, compared to some freshwater plants, have even lower concentrations of protein, carbohydrates, and some minerals (Birch 1975, Heinsohn et al. 1977). Relying on foods that are of relatively poor quality, it is not surprising that sirenians spend large amounts of their time foraging. West Indian manatees spend up to one-third of their daily time budgets feeding, and ingest on average about 7% of their body weight in plant matter (wet weight) in 24 hr (Bengtson 1981, Best 1981b, Etheridge et al. 1985).

Manatees (and probably dugongs) have a metabolic rate that is 20% to 30% lower than expected for a mammal of this body size and, because of thermal conductances that may be 30% higher than predicted weight-specific values, they are not able to maintain their body temperatures effectively outside of tropical and some subtropical waters (Gallivan and Best 1980, Irvine 1983). Their need for water temperatures above 19°C has a marked influence on their behaviors. For example, at some sites in Florida manatees rest at the surface, exposing their backs to the air (Shane 1983b). Presumably these basking animals absorb radiant energy from the sun.

More commonly, manatees in Florida migrate south to avoid cold air and water temperatures. During the winter months of November through March, they aggregate into herds of up to several hundred individuals at natural and artificial warm water sites (Shane 1984, Kinnaird 1985, Kochman et al. 1985, Reynolds and Wilcox 1994). During the summer months some individuals disperse in excess of 800 km, often into northern Florida and beyond (Rathbun et al. 1982; Reid and O'Shea 1989; Reid et al. 1991; Zoodsma 1991). On the basis of resightings of Florida manatees that have distinctive scars (Shane 1983a, Reid et al. 1991) and manatees tracked with satellite-monitored radio-tags (Reid and O'Shea 1989, Rathbun et al. 1990b), migrants swim about 30 to 40 km/day along near-shore travel corridors. While migrating they often stop at predictable sites along the way that are protected from open ocean waters and heavy boat traffic, and that have sources of fresh or warm water. However, not all manatees in Florida migrate long distances. Radio-tracking data for some individuals in southern Florida suggest that they remain in a particular region year-round, perhaps making shorter seasonal movements within the area (Reid et al. 1995). In the tropical waters of Puerto Rico and Mexico, there is no evidence that radio-tagged manatees make long seasonal migrations (J. Reid and B. Morales-Vela, pers. comm.).

Temperature-related, seasonal migrations have also been documented for dugongs. At the southern limit to their distribution at Shark Bay in western Australia, dugongs migrate between warm and cold water sites about 100 km apart (Anderson 1986). In Moreton Bay, eastern Australia, dugongs also respond to water temperatures, but there the movements are within a winter season and cover short distances, similar to manatee movements in southern Florida. In Moreton Bay, dugongs migrate daily up to 20 km between the relatively deep and warm oceanic waters off shore and the cooler, shallow waters inside the bay where they must feed (Preen 1992). Marsh and Rathbun (1990) radio-tracked dugongs in tropical eastern Australia and found that they are capable of swimming 140 km within 2 days in response to dropping air temperatures.

Resource Aggregations

The large groups of manatees that form seasonally in Florida are very different from the large social groups found in many cetaceans. The manatee aggregations are purely resource-based. The aggregation sites are either winter locations where natural springs and power or industrial plants discharge predictable sources of warm water (Shane 1984), or they are locations in marine habitats where springs, sewer plants, storm drains, or garden hoses discharge freshwater that the manatees drink (Shane 1983a). Winter aggregations around power plant outfalls throughout Florida can exceed 1000 individuals during a particularly cold winter day (J. Reynolds, pers. comm.). The maximum daily manatee count at any one outfall in Florida was nearly 330 (Reynolds and Wilcox 1994), but recent counts at each of two plants have exceeded 400 (J. Reynolds, pers. comm.). Dugongs also form large herds; one aggregation in the Arabian Gulf numbered 670 individuals (Preen 1989b). Preen (1992, 1995) has made a convincing argument that some of these herds are also resource-based, but in a more complicated way than those of manatees. He suggests that the grazing activity of a large herd of dugongs modifies the species composition and growth stages of sea grasses so that they become more nutritious for the dugongs. Without the modifying influence of a large herd of dugongs, the sea grass meadows probably could not sustain as large a population of grazers. Preen's (1995) idea of "cultivation grazing" is very similar to the positive feedback relationship that exists between savanna ungulates and their food plants on the Serengeti Plain of Tanzania (McNaughton 1984).

Unfortunately, little research has been published on dugong or manatee behavioral ecology in tropical waters (within the center of their distributions), therefore it is not known how widespread cultivation grazing or seasonal migrations might be. Preen (1992) believes that cultivation grazing, and thus some dugong aggregations, may be unique to the subtropics, just as warm water aggregations of manatees seem to be unique to subtropical regions of the United States (Lefebvre et al. 1989). Seasonal migrations, however, are not restricted to sirenian populations in the subtropics. Robin Best (1982, 1983) has shown how Amazonian manatees move seasonally between river channels during low water and flooded meadows and forests during high water in the Amazon River basin. In this case, the seasonal movements are driven by a wet and dry cycle, rather than cold and warm periods. There is also evidence that some dugong populations move or migrate in response to rough seas created by monsoonal weather patterns (Marsh et al. 1980).

During aerial surveys in northeastern Australia, nearly 70% of dugong "groups" ("subjectively distinct clumpings" that include single animals) were of lone animals (Marsh and Saalfeld 1989), and on another survey nearly 60% of sightings were of single animals (Marsh and Saalfeld 1990). During aerial surveys in western Australia about 90% of sightings were of lone individuals (Marsh et al. 1994). In Florida, similar figures are difficult to obtain because of the influences of cold season aggregations. However, in one summer/fall aerial survey about 60% of "groups" (individuals within ca. 100 m, or single animals) involved lone manatees separated from one another, and about 40% of sightings were of lone manatees (Irvine et al. 1982a). In both species, when groups are seen they are usually composed of fewer than 10 to 15 individuals and there is little indication that there is any coordinated herd structure. In the West Indian manatee, however, there are two types of small herds that are occasionally seen and exhibit well-defined social behaviors: cavorting herds and mating herds (Hartman 1979, Bengtson 1981).

Mating Patterns

The interactions between individuals in cavorting herds are characterized by intense pushing and shoving, and frequent nasogenital, anogenital, and genital–genital contacts, including intromission between males (Fig. 8-32). The "homosexual" behaviors (Hartman 1979) probably are best understood as bouts of play-fighting, and practice for participation in the sometimes intense competition for access to females in mating herds (discussed later). It is also possible that some form of rank or dominance is established between juvenile males in cavorting herds that later helps determine not only who participates in mating herds, but more important, which males actually mate (Rathbun and O'Shea 1984). Cavorting herds form at all times of the year, but they usually only last for a few hours. These herds have no definable structure; associations between individuals are only temporary as animals join and leave the group (Hartman 1979, Bengtson 1981, Reynolds 1981a).

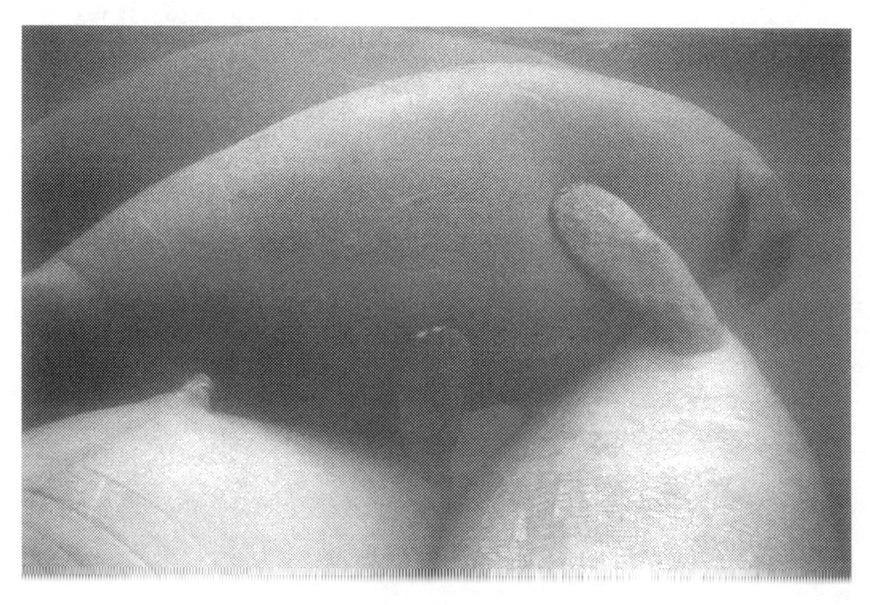

Figure 8-32. A group of cavorting male West Indian manatees, participating in "homosexual" behavior. (Photograph by James P. Reid, U.S. Geological Survey, Sirenia Project.)

Mating herds in the West Indian manatee are composed of a focal female that is in estrus and an average of 10 males (range, 5 to 22) that constantly and closely follow her (Rathbun et al. 1995). These herds can last up to 3 or 4 weeks, but the individual composition of a herd is fluid, with different males joining and leaving daily (Hartman 1979, Bengtson 1981, Rathbun et al. 1995). The cow is receptive to actual mating for only a short time, perhaps a day or two. During this period, the males engage in sometimes quite violent bouts of pushing and shoving as each individual tries to gain a favorable position to achieve intromission (Hartman 1979). O'Shea and Hartley (1995) have suggested that frenzied courting males may result in the deaths of calves still dependent on females in estrus. They further speculate that some neonatal mortality may be the result of infanticide, as described for several terrestrial mammals (Hausfater and Hrdy 1984). The frenzied persistence of males in mating herds is demonstrated by observations of their attempting to copulate with recently dead females (J. Reynolds, pers. comm.). Smaller males, perhaps subadults, often join mating herds but do not participate as actively as larger animals (Bengtson 1981). This system of mating closely fits the "scramble competition polygyny" model, as we first discussed for pinnipeds. Although it may resemble the description of mating herds in humpback whales (discussed previously), which have been described as floating leks, male humpback whales appear to display by "singing," and this may serve to attract females. There is no evidence to suggest that male manatees vocalize to attract females before mating herds develop. Although a receptive female manatee apparently copulates with multiple males (Hartman 1979), she probably exercises considerable choice in which males to accept by moving into shallow water where she has more control over the males. As with pinnipeds and cetaceans, we can only speculate whether sperm competition is a factor in the male and female mating strategies.

Mating herds are usually only seen during the warm season months (Rathbun et al. 1995). Bengtson (1981), based on radio-tracking data, found that during the summer males spend much of their active nonfeeding time patrolling their home ranges searching for females in estrus. The home ranges of several females are encompassed by the larger male areas. Bengtson (1981) also found that females that are in estrus travel more often than usual, at times even outside their normal home ranges. These greater female movements, along with the long estrous period and wide-ranging travels of males, probably allows a female to attract a large number of males into her mating herd, which results in a wider choice of mates. These tactics are similar to those practiced by some female pinnipeds on breeding beaches, as we discussed earlier in this chapter.

The mating system of dugongs is not well understood, principally because the gender of free-ranging dugongs is difficult to determine, and they are not easily observed. Indeed, the three most reliable descriptions of dugong sexual behavior are different, and unfortunately rather incomplete. In tropical eastern Australia a small group of dugongs performed stereotyped rushing behaviors at the surface of the water, which in one case was associated with a presumed male attempting to mate with a presumed female by rolling her onto her back in the water (Anderson and Birtles 1978). In subtropical eastern Australia, Preen (1989a) made several observations of small groups of dugongs that were intensely interacting, in a manner similar to manatee mating herds (Fig. 8-33). However, there were several differences, including that the dugong mating herds only lasted for a short time, perhaps no more than 1 day. There were also stereotyped displays associated with the presumed males, such as tail slapping on the surface of the water. Oddly, the rushing behavior described by Anderson and Birtles (1978) was not seen by Preen (1989a). In subtropical western Australia, Anderson and Barclay (1995) indicate that some male dugongs are territorial. Anderson (1998) has further suggested that the dugongs he studied establish breeding leks: aggregations of small, nonresource-based territories where males attract and mate with receptive females. This is similar to the lek mating system of some medium-sized terrestrial ungulates in Africa such as the Uganda kob (Jarman 1974, Leuthold 1977). Whether some pinnipeds and cetaceans also exhibit lekking behavior is still not clear, as we discussed earlier. Nor is it clear whether what has been called a floating lek for humpback whales is a modified lek or more like the scramble competition of manatees.

Obviously, Anderson's observations of dugongs from western Australia differ markedly from the two accounts from eastern Australia. Perhaps the mating system of the dugong is variable, taking on different forms depending on local environmental conditions. Some terrestrial ungulates, such as the impala in Africa, shift from one mating scheme to another in different habitats (Jarman 1979). Lott (1991) describes intraspecific variation in social and mating systems in a variety of vertebrate species. This type of flexibility apparently is not characteristic of all species. Earlier in this chapter we discussed phylogenetic inertia in terms of the behavioral ecology of pinnipeds, and the social structure of marine otters.

Although the seemingly benign disposition and gentle nature of sirenians, especially manatees, has often been commented on (Hartman 1979), it probably has been overemphasized, especially in popular accounts. Perhaps this is partly because sirenians lack any obvious armaments, such as horns, claws, or fangs. During the bouts of pushing and shoving that

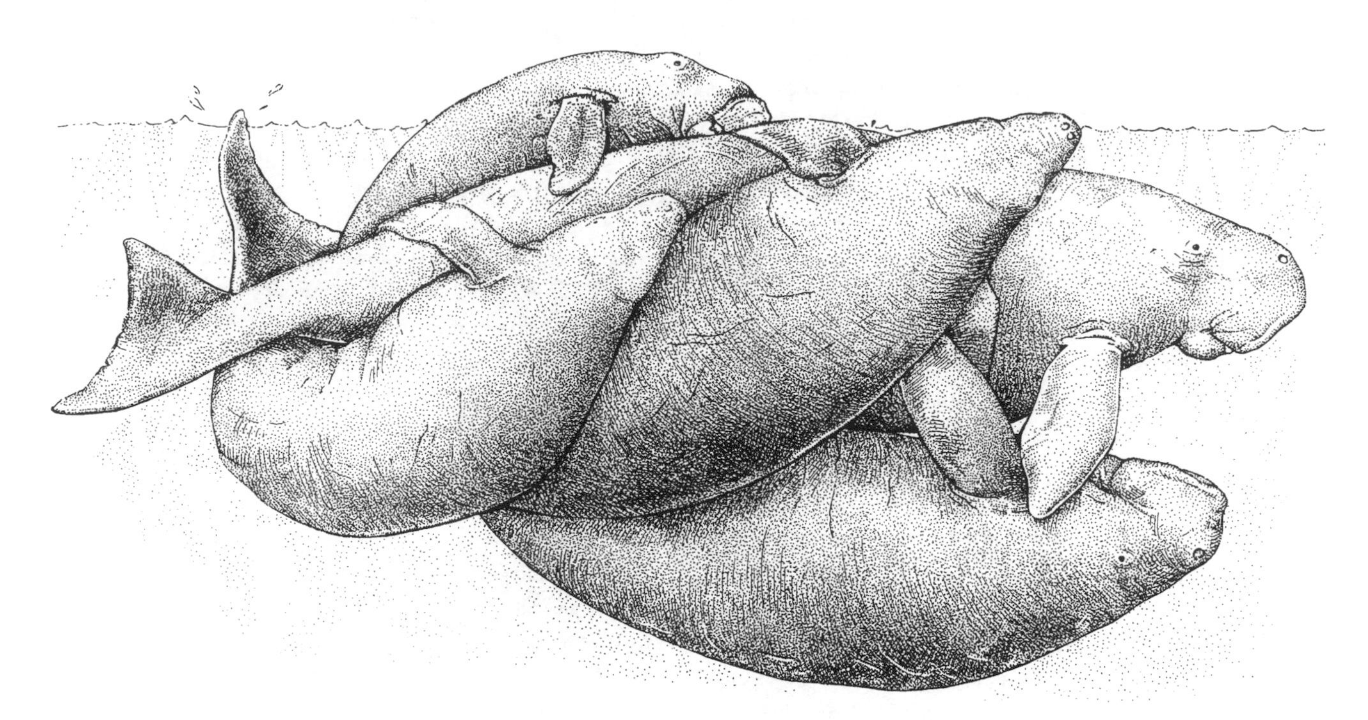

Figure 8-33. A hypothetical mating herd of dugongs, based on observations by Anthony R. Preen in eastern Australia. (Illustration by Lucy Smith.)

occur in manatee cavorting and mating herds, the physical contact between individuals can become quite intense, and in the dugong even violent because the males have short tusks (Marsh 1980). In female dugongs, the tusks are small and usually unerupted, but in males they are capable of inflicting wounds on other dugongs (Fig. 8-34). Anderson and Birtles (1978) observed parallel scars on the back of an adult female dugong, which they thought were the result of wounds inflicted by males during mating. Preen (1989a) actually observed presumed males raking their tusks across the backs of other dugongs in the sometimes vigorous pushing and shoving, or "fighting" associated with mating. Apparently dugong tusks are not used in the feeding process, therefore they may be important only in social contexts like the tusks of walruses and narwhals, which were discussed earlier in the sections on pinnipeds and cetaceans, respectively.

Rearing Patterns

The only close, easily identifiable social bond in sirenians seems to be between a mother and her calf (Hartman 1979; Preen 1992). During dependency, calves (in Florida 1% to 2% of births are twins) rarely stray farther than several meters from their mothers (Fig. 8-35). In the West Indian manatee, calves usually swim to one side or the other of the female behind the pectoral flippers (Reynolds 1981a), whereas in the dugong calves usually swim above the cow, sometimes even riding on her back (Anderson and Birtles 1978, Anderson 1982). In West Indian manatees, cows maintain nearly constant acoustic contact with their calves. The various chirps and squeaks often develop into duets between the pair (Hartman 1979, Reynolds 1981a, Steel 1982). The close cow–calf bond lasts until weaning, which in free-ranging dugongs and West Indian manatees usually occurs at 1 to 2 years of age (Marsh 1995, O'Shea and Hartley 1995), although nursing may continue for up to 3 or 4 years in manatees (Rathbun et al. 1995). As is the case in most other mammals, there is no paternal care of calves in any sirenian.

Although fostering is common in some pinnipeds, as we discussed earlier, it has been observed only occasionally in captive Florida manatees, and rarely in the wild. Rathbun et al. (1995) suggest that fostering in manatees may be an artifact of captivity and the relatively recent phenomenon of large resource-based aggregations.

The relatively long period of calf dependence in the West Indian manatee may be important in transmitting environmental information from adults to calves, which may result in the establishment of learned traditions (Bengtson 1981). For example, manatees in some of the better-studied winter aggregation sites in Florida have yearly return rates (nearly 90%), which include at least several generations of weaned calves (Rathbun et al. 1990b, O'Shea and Hartley 1995). Sim

Figure 8-34. The parallel scars on the back of this dugong are most likely caused by the tusks of males during intense mating activity, as illustrated in Fig. 8-33. (Photograph by Anthony R. Preen, James Cook University of North Queensland.)

Figure 8-35. Dugong calves rarely stray far from their mothers and typically travel above the cows. (Photograph by Anthony R. Preen, James Cook University of North Queensland.)

ilarly, some manatees use the same summer areas year after year (Shane 1983a; Reid et al. 1991; Zoodsma 1991). The locations of sources of freshwater where manatees can drink, including springs, dock-side garden hoses, and sewer outfalls (Shane 1983a), are probably also learned by calves from their mothers and become part of a population's tradition. However, manatees also show some flexibility in adapting to human-induced changes to their environment, as when power plant effluents are altered (Packard et al. 1989) or areas are established as manatee sanctuaries (Provancha and Provancha 1988; Buckingham 1990).

Communication

During the summer, when manatees in Florida are widely dispersed, they sometimes are found in small, ephemeral groups at well-defined, traditional locations that Bengtson (1981) and Zoodsma (1991) have called rendezvous sites. These sites are often located where several waterways (and manatee travel corridors) intersect. Two or more manatees that meet at rendezvous sites often engage in social interactions, including "kissing" (Hartman 1979), nuzzling, rubbing, and cavorting. Bengtson (1981) suggests that rendezvous sites are important in maintaining communication between individuals that use a common winter refuge, but are dispersed during the summer months.

Intraspecific communication has also been proposed as the function of communal rubbing sites used by manatees at winter aggregations. At clear water sites in Florida, where observation conditions are near ideal, manatees use specific rocks and stumps that protrude from the bottom as rubbing posts (Hartman 1979). Manatees traveling into or out of the warm water areas often visit these objects individually for a few minutes and rub specific parts of their bodies, such as their genitals, axillae, eyes, or chins. The frequent rubbing results in the objects becoming polished and visually very obvious. Rubbing posts are traditional, being used year after year, or until they are dislodged, when new sites are developed. Although these sites may be nothing more than spots where manatees scratch themselves, because the regions of the body that are most often rubbed are associated with excretions of various types, Rathbun and O'Shea (1984) have suggested that rubbing posts may function in chemical communication. Unfortunately, little research has been done on the ability of sirenians to communicate chemically through smell, taste, or some other mechanism. West Indian manatees apparently lack a vomeronasal apparatus (Mackay-Sim et al. 1985), therefore the function of rubbing posts, as well as the various greeting behaviors such as kissing, remains speculative.

The sounds produced by dugongs and West Indian manatees are well documented (Schevill and Watkins 1965; Anderson and Barclay 1995; see Tyack, Chapter 7, this volume). Neither species uses ultrasound, but rather a wide array of squeaks, trills, chirps, and barks in the 1- to 20-kHz range. These sounds are used to communicate to nearby individuals basic behavioral states such as fear, aggression, or affiliation (Steel 1982, Bengtson and Fitzgerald 1985, Anderson and Barclay 1995). In the West Indian manatee, cow–calf duets serve mainly as contact calls, and can become quite intense if a calf is disturbed, becomes separated from its mother, or wishes to nurse (Rathbun and O'Shea 1984).

Social Organization

Some marine mammals (mainly odontocetes) exhibit close, long-term, and stable social relationships between individuals (often kin), which result in well-defined, close-knit herd structures. In the West Indian manatee, and possibly other sirenians as well, herd structure and social organization do not appear to be well defined (Hartman 1979; Bengtson 1981). Individuals constantly join and leave groups, which results in great difficulty in defining any herd membership or structure. The seemingly amorphous social organization of all sirenians (Best 1984; Preen 1992), but especially the West Indian manatee (Hartman 1979), has resulted in terms such as essentially solitary, weakly social, and moderately social being applied (Hartman 1979, Reynolds 1981a).

This "solitary" life may mask the existence of a more complicated social relationship among individuals in a population, as suggested by West Indian manatee "rubbing posts," "kissing," "rendezvous" sites, and the long cow–calf bond. Indeed, manatee sociality in Florida resembles the fission–fusion social organization model developed for the chimpanzee (*Pan troglodytes*) by Ghiglieri (1971), and also described in the bottlenose dolphin (Würsig 1978, Wells et al. 1987, Smolker et al. 1992), the spider monkey (*Ateles paniscus*) in Peru by Symington (1990), and the African lion (*Panthera leo*) (Packer et al. 1990). In fission–fusion societies, individuals frequently join and leave herds or groups, resulting in all members of a local population knowing one another but not always associated in a single herd or group. These ephemeral groups are often territorial. If biologists could consistently follow individual manatees for a relatively long time, such as is often done relatively easily with terrestrial mammals, it is likely that long-lasting, but not particularly strong, social bonds would be identified (sirenians live for at least 50 years; Marsh 1980, Marmontel 1995). Because fission–fusion societies are often not highly structured (not to be confused with lack of complexity), they can be difficult to study. In addition, the ecological and social factors that are associated with fission–fusion social organizations are not well understood, partly because there are relatively few examples (Ghiglieri 1984). Ghiglieri (1984) and Symington (1990) suggest that fission–fusion systems develop where food resources are limited and patchy in space and time. For example, spider monkeys rely on food produced by widely spaced tropical forest trees that only periodically produce a limited supply of fruits. As a result, temporary groups of monkeys defend territories that include these fruiting trees.

We started our discussion of sirenian behavior with a statement that their herbivorous diet was perhaps the most important overall factor in understanding their life history and behavior. This is still undoubtedly the case for most populations of sirenians. On the basis of studies of fission–fusion social organizations in other groups (Ghiglieri 1984, Symington 1990), manatee feeding ecology in Florida should hold the secret to a better understanding of their social structure. Unfortunately, feeding is perhaps the least studied and understood aspect of manatee behavioral ecology. In addition, the often ubiquitous occurrence of aquatic food plants in Florida (including a large number of introduced freshwater aquatic weeds) does not fit the "patchy food" model of fission–fusion sociality. There is another critical resource for manatees in Florida that does have a temporal and spatial patchy occurrence—warm water. Thermal refuges are widely spaced (Reid et al. 1991), and sometimes are unpredictable (power plants sometimes fail; Packard et al. 1989). But warm water in Florida is not usually limited in quantity at any one site (as is fruit in a tree). In fact, we doubt that even if resources such as food or warm water did become limiting in a manner such as fruit in trees, manatees would become territorial, as predicted by the primate model of fission–fusion sociality (Ghigliere 1984, Symington 1990). Perhaps future biologists in Florida will be able to test our prediction, because manatee numbers may continue to increase in Florida (Garrott et al. 1995), thus creating limited resources in a density-dependent fashion. We believe, however, that the most likely outcome of this scenario will be that some other aspect of manatee behavior will shift to accommodate the situation because manatee evolutionary history has not equipped them with the behaviors necessary to become territorial. If this is indeed the case, then we propose that the current model of fission–fusion sociality needs to be modified to include a continuum of examples. At one end is the relatively simple form as found in the West Indian manatee, with its relatively weak social bonds and lack of territoriality. At the other end of the continuum would be a primate such as the spider monkey, with its more complicated social behaviors that include territoriality (Symington 1990). In between these two extremes are all other fission–fusion species (Ghigliere 1984), including other primates, several carnivores, and the bottlenose dolphins, as described earlier in this chapter. Where a species falls on this continuum will be determined by the complicated interactions between environmental and behavioral factors, filtered through the evolutionary history of the species. Unraveling this complex for manatees, dolphins, and other mammals will be up to future behavioral ecologists, blending the information from new field studies with ever-expanding theoretical insights.

Summary

Available information indicates that the behavior of marine mammals is influenced by their phylogenies and features of the marine environment in much the same way that the behavior of terrestrial mammals has evolved. Many of the environmental pressures are analogous across habitat types, and many of the behavioral solutions of marine and terrestrial mammals are similar. Coping with predators in an open habitat, either in the form of an ocean basin or an African savanna, tends to encourage formation of large groups, whereas more closed habitats such as forests, rivers, or shallow bays may reduce the threat of predation, and is correlated with smaller groups. The distribution of resources is an important driving feature in the behavioral ecology of all mammals. Locating rich patches of food, either in the form of fish schools or isolated fruit trees, may be facilitated through the cooperative efforts of group members, whereas nonschooling fish associated with sea grass meadows, or other evenly distributed or predictable food sources may neither require nor be able to support the foraging of groups. Highly coordinated cooperative hunting of large prey by stable family units occurs both in the marine and terrestrial ecosystems.

The importance of ecological variables, such as habitat structure, predation pressure, and resource (e.g., food, thermoregulatory sites, kelp beds to protect against energetically costly and dangerous drifting) abundance and distribution, is clear in determining the temporal and spatial patterns of receptive females in marine mammals, as is predicted by models of mating systems based primarily on terrestrial vertebrates. In turn, the modal pattern and variety of strategies by which marine mammals acquire mates can be reasonably predicted based on the spatiotemporal distribution of females, when combined with an understanding of the influence of social factors and constraints resulting from phylogenetic inertia. The importance of female choice, both direct (behavioral) and indirect (sperm competition), in mating systems is perhaps the least well understood in marine mammals. In this area, studies of terrestrial mammals are substantially ahead of those of marine mammals.

In short, although the medium in which marine mammals live has led to the evolution of considerable morphological and physiological features that are unique to marine mammals, their behavioral patterns are remarkably similar to those of terrestrial mammals, and appear to be derived from the same fundamental principles. The influence of the marine environment, however, does appear to produce some of the most extreme patterns of behavior found among mammals, such as a 4-day period of maternal care and lactation in the hooded seal, or a 3- to 4-month fast by lactating mysticete females. We believe that including marine mammals in analyses of evolution of behavior using a comparative approach is justifiable and, in fact, valuable because their behavior often increases the variance seen among mammals for a particular aspect of behavior.

Acknowledgments

Kim Bassos-Hull, Kristi Brockway, Tristen Moors, Edward Owen, Barbara Piel, Stephanie Smathers, and Caryn Weiss provided tremendous assistance with the gathering and organization of the literature used for the Cetacean section of this chapter; Michael Scott reviewed a draft of this section. Early drafts of the Sirenia section were reviewed by Chip Deutsch, Lynn Lefebvre, John Reynolds III, and Norman Scott Jr., and an early version of the sea otter material was reviewed by Brian Hatfield and Ron Jameson. The chapter benefited significantly from reviews by John Eisenberg, John Reynolds III, and Sentiel Rommel.

Literature Cited

Acevedo, A. 1991. Behaviour and movements of bottlenose dolphins, *Tursiops truncatus,* in the entrance to Ensenada de La Paz, Mexico. Aquatic Mammals 17:137–147.

Acevedo, A., and M. A. Smultea. 1995. First records of humpback whales including calves at Golfo Dulce and Isla del Coco, Costa

Rica, suggesting geographical overlap of northern and southern hemisphere populations. Marine Mammal Science 11:554–560.

Alcock, J. 1972. The evolution of the use of tools by feeding animals. Evolution 26:464–473.

Allen, J. F. 1977. Dolphin reproduction in oceanaria in Australasia and Indonesia. Pages 85–108 *in* S. H. Ridgway and K. Benirschke (eds.). Breeding Dolphins: Present Status, Suggestions for the Future. National Technical Information Service PB-273 673.

Altmann, J. 1974. Observational study of behavior: Sampling methods. Behaviour 49:227–267.

Altmann, J., G. Hausfater, and S. A. Altmann. 1988. Determinants of reproductive success in savanna baboons, *Papio cynocephalus*. Pages 403–418 *in* T. H. Clutton-Brock (ed.). Reproductive success. The University of Chicago Press, Chicago and London.

Altmann, S. A. 1956. Avian mobbing behavior and predator recognition. Condor 58:241–253.

Amano, M., and T. Kuramochi. 1992. Segregative migration of Dall's porpoise (*Phocoenoides dalli*) in the Sea of Japan and Sea of Okhotsk. Marine Mammal Science 8:143–151.

Ames, J. A., R. A. Hardy, F. E. Wendell, and J. J. Geibel. 1983. Sea Otter Mortality in California. Marine Resources Branch, California Department of Fish and Game, Monterey, CA.

Ames, J. A., J. J. Geibel, F. E. Wendell, and C. A. Pattison. 1996. White-shark-inflicted wounds of sea otters in California, 1968–1992. Pages 309–316 *in* A. P. Klimley and D. G. Ainley (eds.). Great White Sharks. Academic Press, New York, NY.

Amos, B., C. Schlötterer, and D. Tautz. 1993a. Social structure of pilot whales revealed by analytical DNA profiling. Science 260:670–672.

Amos, W. 1992. Analysis of polygamous systems using DNA fingerprinting. Symposia of Zoological Society (London) 64:151–165.

Amos, W. 1993. Use of molecular probes to analyse pilot whale pod structure: Two novel analytical approaches. Symposia of Zoological Society (London) 6:33–48.

Amos, W., J. Barrett, and G. A. Dover. 1991. Breeding system and social structure in the Faroese pilot whale as revealed by DNA fingerprinting. Pages 255–270 *in* A. R. Hoelzel (ed.). Genetic Ecology of Whales and Dolphins. Report of the International Whaling Commission (Special Issue 13), Cambridge, U.K.

Amos, W., S. Twiss, P. Pomeroy, and S. S. Anderson. 1993b. Male mating success and paternity in the gray seal, *Halichoerus grypus:* A study using DNA fingerprinting. Proceedings of the Royal Society of London 252:199–207.

Andelman, S. J. 1986. Ecological and social determinants of Cercopithecine mating patterns. Pages 201–216 *in* D. I. Rubenstein and R. W. Wrangham (eds.). Ecological Aspects of Social Evolution. Princeton University Press, Princeton, NJ.

Anderson, P. K. 1981. The behavior of the dugong (*Dugong dugon*) in relation to conservation and management. Bulletin of Marine Science 31:640–647.

Anderson, P. K. 1982. Studies of dugongs at Shark Bay, Western Australia. II. Surface and subsurface observations. Australia Wildlife Research 9:85–99.

Anderson, P. K. 1984. Suckling in *Dugong dugon*. Journal of Mammalogy 65:510–511.

Anderson, P. K. 1986. Dugongs of Shark Bay, Australia—Seasonal migration, water temperature, and forage. National Geographic Research 2:473–490.

Anderson, P. K. 1989. Deliberate foraging on macroinvertebrates by dugongs. National Geographic Research 5:4–6.

Anderson, P. K. 1997. Shark Bay dugongs in summer. I: Lek mating. Behaviour 134:433–462.

Anderson, P. K., and R. M. R. Barclay. 1995. Acoustic signals of solitary dugongs: physical characteristics and behavioral correlates. Journal of Mammalogy 76:1226–1237.

Anderson, P. K., and A. Birtles. 1978. Behaviour and ecology of the dugong, *Dugong dugon* (Sirenia): Observations in Shoalwater and Cleveland Bays, Queensland. Australian Wildlife Research 5:1–23.

Anderson, P. K., and R. I. T. Prince. 1985. Predation on dugongs: Attacks by killer whales. Journal of Mammalogy 66:554–556.

Anderson, S. S., and M. A. Fedak. 1985. Gray seal males: Energetic and behavioural links between size and sexual success. Animal Behaviour 33:829–838.

Anderson, S. S., and J. Harwood. 1985. Time budgets and topography: How energy reserves and terrain determine the breeding behavior of gray seals. Animal Behaviour 33:1343–1348.

Anderson, S. S., R. W. Burton, and C. F. Summers. 1975. Behaviour of gray seals (*Halichoerus grypus*) during a breeding season at North Rona. Journal of Zoology (London) 177:179–195.

Antonelis, G. A., M. S. Lowery, D. P. DeMaster, and C. H. Fiscus. 1987. Assessing northern elephant seal feeding habits by stomach lavage. Marine Mammal Science 3:308–322.

Antonelis, G. A., B. S. Stewart, and W. F. Perryman. 1990. Foraging characteristics of female northern fur seals (*Callorhinus ursinus*) and California sea lions (*Zalophus californianus*). Canadian Journal of Zoology 68:150–158.

Antonelis, G. A., M. S. Lowery, C H. Fiscus, B. S. Stewart, and R. L. DeLong. 1994. Diet of the northern elephant seal. Pages 211–223 *in* B. J. Le Boeuf and R. M. Laws (eds.). Elephant Seals. University of California Press, Berkeley, CA.

Aragones, L. V. 1994. Observations on dugongs at Calauit Island, Busuanga, Palawan, Philippines. Australian Wildlife Research 21:709–717.

Armitage, K. B., and D. W. Johns. 1982. Kinship, reproductive strategies and social dynamics of yellow-bellied marmots. Behavioral Ecology and Sociobiology 11:55–63.

Arnbom, T., and H. Whitehead. 1989. Observations on the composition and behaviour of groups of female sperm whales near the Galapagos Islands. Canadian Journal of Zoology 67:1–7.

Arnbom, T., V. Papastavrou, L. Weigart and H. Whitehead. 1987. Sperm whales react to an attack from killer whales. Journal of Mammalogy 68:450–453.

Asaga, T., B. J. Le Boeuf, and H. Sakurai. 1994. Functional analysis of dive types of female northern elephant seals. Pages 310–327 *in* B. J. Le Boeuf and R. M. Laws (eds.). Elephant Seals. University of California Press, Berkeley, CA.

Au, D. W. 1991. Polyspecific nature of tuna schools: Shark, dolphin, and seabird associates. Fishery Bulletin 89:343–354.

Baird, R. W., and L. M. Dill. 1995. Occurrence and behaviour of transient killer whales: Seasonal and pod-specific variability, foraging behaviour, and prey handling. Canadian Journal of Zoology 73:1300–1311.

Baker, C. S., and L. M. Herman. 1984. Aggressive behavior between humpback whales (*Megaptera novaeangliae*) wintering in Hawaiian waters. Canadian Journal of Zoology 62:1922–1937.

Balcomb, K. C., J. R. Boran, and S. L. Heimlich. 1982. Killer whales in Greater Puget Sound. Report of the International Whaling Commission 32:681–686.

Ballance, L. T. 1990. Residence patterns, group organization, and surfacing associations of bottlenose dolphins in Kino Bay, Gulf of California, Mexico. Pages 267–283 *in* S. Leatherwood and R. R. Reeves (eds.). The Bottlenose Dolphin. Academic Press, San Diego, CA.

Ballance, L. T. 1992. Habitat use patterns and ranges of the bottlenose dolphin in the Gulf of California, Mexico. Marine Mammal Science 8:262–274.

Balmford, A. 1992. Social dispersion and lekking in Uganda kob. Behaviour 120:177–191.

Baraff, L. S., P. J. Clapham, D. K. Mattila, and R. S. Bowman. 1991. Feeding behaviour of a humpback whale in low-latitude waters. Marine Mammal Science 7:197–202.

Barlow, G. W. 1972. A paternal role for bulls of the Galapagos sea lion. Evolution 26:307–310.

Barros, N. B., and D. K. Odell. 1990. Food habits of bottlenose dolphins in the southeastern United States. Pages 309–328 *in* S. Leatherwood and R. R. Reeves (eds.). The Bottlenose Dolphin. Academic Press, San Diego, CA.

Bartholomew, G. A. 1959. Mother–young relations and the maturation of pup behavior in the Alaskan fur seal. Animal Behaviour 7:163–171.

Bartholomew, G. A. 1970. A model for the evolution of pinniped polygyny. Evolution 24:546–559.

Bartholomew, G. A., and N. E. Collias. 1962. The role of vocalization in the social behaviour of the Northern Elephant seal. Animal Behaviour 10:7–14.

Bartholomew, G. A., and P. G. Hoel. 1953. Reproductive behavior of the Alaska fur seal, *Callorhinus ursinus*. Journal of Mammalogy 34:417–436.

Bartsh, S. S., S. D. Johnston, and D. B. Siniff. 1992. Territorial behavior and breeding frequency of male Weddell seals (*Leptonychotes weddelli*) in relation to age, size and concentrations of serum testosterone and cortisol. Canadian Journal of Zoology 70:680–692.

Bateson, G. 1974. Observations of a cetacean community. Pages 146–165 *in* J. McIntyre (ed.). Mind in the Waters. Scribner's, New York, NY.

Bearzi, G. 1999. Social ecology of bottlenose dolphins in the Kvarneric (Northern Adriatic Sea). Marine Mammal Science (in press).

Beck, C. A., and J. P. Reid. 1995. An automated photo-identification catalog for studies of the life history of the Florida manatee. Pages 120–134 *in* T. J. O'Shea, B. B. Ackerman, and H. F. Percival (eds.). Population Biology of the Florida Manatee. National Biological Service, Information and Technology Report 1, Washington, D.C.

Beehler, B. M., and M. Foster. 1988. Hotshots, hotspots and female preference in the organization of lek mating systems. American Naturalist 131:203–219.

Beentjes, M. P. 1989. Haul-out patterns, site fidelity and activity budgets of male Hooker's sea lions (*Phocartos hookeri*) on the New Zealand mainland. Marine Mammal Science 5:281–297.

Bel'kovich, V. M., E. E. Ivanova, O. V. Yefremenkova, L. B. Kozarovitsky, and S. P. Kharitonov. 1991. Searching and hunting behavior in the bottlenose dolphin (*Tursiops truncatus*) in the Black Sea. Pages 38–67 *in* K. Pryor and K. S. Norris (eds.). Dolphin Societies: Discoveries and Puzzles. University of California Press, Berkeley, CA.

Bengtson, J. L. 1981. Ecology of manatees (*Trichechus manatus*) in the St. Johns River, Florida. Ph.D. dissertation, University of Minnesota. 126 pages.

Bengtson, J. L., and S. M. Fitzgerald. 1985. Potential role of vocalizations in West Indian manatees. Journal of Mammalogy 66:816–819.

Bengtson, J. L., R. D. Hill, and S. E. Hill. 1993. Using satellite telemetry to study the ecology and behavior of Antarctic seals. Korean Journal of Polar Research 4:109–115.

Benjaminsen, T., and I. Christensen. 1979. The natural history of the bottlenose whale, *Hyperoodon ampullatus* (Forster). Pages 143–164 *in* H. E. Winn, and B. L. Olla (eds.). Behavior of Marine Animals, Vol. 3: Cetaceans. Plenum, New York, NY.

Berger, J. 1988. Social systems, resources, and phylogenetic inertia: an experimental test and its limitations. Pages 157–186 *in* C. N. Slobodchikoff (ed.). The Ecology of Social Behavior. Academic Press, New York, NY.

Bernard, H. J., and A. A. Hohn. 1989. Differences in feeding habits between pregnant and lactating spotted dolphins *Stenella attenuata*. Journal of Mammalogy 70:211–215.

Bertram, G. C. L. 1940. The biology of the Weddell and crabeater seals, with a study of the comparative behaviour of the Pinnipedia. British Graham Land Expeditions, 1934–1937, Scientific Report 1:1–139.

Best, P. B. 1979. Social organization in sperm whales, *Physeter macrocephalus*. Pages 227–289 *in* H. E. Winn and B. L. Olla (eds.). Behavior of Marine Animals, Vol. 3: Cetaceans. Plenum Press, New York, NY.

Best, P. B. 1982. Seasonal abundance, feeding, reproduction, age and growth in minke whales off Durban (with incidental observations from the Antarctic). Report of the International Whaling Commission 32:759–786.

Best, P. B., P. A. S. Canham, and N. Macleod. 1984. Patterns of reproduction in sperm whales, *Physeter macrocephalus*. Pages 51–79 *in* W. F. Perrin, R. L. Brownell Jr., and D. P. DeMaster (eds.). Reproduction in Whales, Dolphins, and Porpoises. Reports of the International Whaling Commission (Special Issue 6), Cambridge, U.K.

Best, P. B., R. Payne, V. Rowntree, J. T. Palazzo, and M. D. C. Both. 1993. Long-range movements of South Atlantic right whales *Eubalaena australis*. Marine Mammal Science 9:227–234.

Best, R. C. 1981a. The tusk of the narwhal (*Monodon monoceros* L.): Interpretation of its function (Mammalia: Cetacea). Canadian Journal of Zoology 59:2386–2393.

Best, R. C. 1981b. Foods and feeding habits of wild and captive Sirenia. Mammal Review 11:3–29.

Best, R. C. 1982. Seasonal breeding in the Amazonian manatee, *Trichechus inunguis* (Mammalia: Sirenia). Biotropica 14:76–78.

Best, R. C. 1983. Apparent dry-season fasting in Amazonian manatees (Mammalia: Sirenia). Biotropica 15:61–64.

Best, R. 1984. Managing the manatee—Conservation of Amazonian manatees. Pages 302–303 *in* D. Macdonald (ed.). The Encyclopedia of Mammals. Facts on File, Inc., New York, NY.

Best, R. C., and V. M. F. da Silva. 1984. Preliminary analysis of reproductive parameters of the boutu, *Inia geoffrensis*, and the tucuxi, *Sotalia fluviatilis*, in the Amazon River system. Pages 361–369 *in* W. F. Perrin, R. L. Brownell Jr., and D. P. DeMaster (eds.). Reproduction in Whales, Dolphins, and Porpoises. Reports of the International Whaling Commission (Special Issue 6), Cambridge, U.K.

Best, R. C., and V. M. F. da Silva. 1989. Biology, status, and conservation of *Inia geoffrensis* in the Amazon and Orinoco River basins. Pages 23–34 *in* W. F. Perrin, R. L. Brownell Jr., Z. Kaiya, and L. Jiankang (eds.). Biology and Conservation of the River Dolphins. Occasional Papers of the IUCN Species Survival Commission (SSC), Number 3, Gland, Switzerland.

Bigg, M. A. 1982. Assessment of killer whale (*Orcinus orca*) stocks off Vancouver Island, British Columbia. Report of the International Whaling Commission 32:655–666.

Bigg, M. A., G. M. Ellis, J. B. Ford, and K. C. Balcomb. 1987. Killer whales: A study of their identification, genealogy and natural history in British Columbia and Washington State. Phantom Press and Publishers, Nanaimo, B.C.

Birch, W. R. 1975. Some chemical and calorific properties of tropical marine angiosperms compared with those of other plants. Journal of Applied Ecology 12:201–212.

Birkhead, T., and A. Moller. 1993. Female control of paternity. Trends in Ecology and Evolution 8:100–104.

Boness, D. J. 1979. The Social System of the gray seal, *Halichoerus grypus* (Fab.), on Sable Island, Nova Scotia. Ph.D. dissertation, Dalhousie University, Halifax, Nova Scotia.

Boness, D. J. 1984. Activity budget of male gray seals, *Halichoerus grypus*. Journal of Mammalogy 65:291–297.

Boness, D. J. 1990. Fostering behavior in Hawaiian monk seals: Is there a reproductive cost? Behavioral Ecology and Sociobiology 27:113–122.

Boness, D. J. 1991. Determinants of mating systems in the Otariidae (Pinnipedia). Pages 1–44 *in* D. Renouf (ed.). Behaviour of Pinnipeds. Chapman and Hall, London.

Boness, D. J., and W. D. Bowen 1996. The evolution of maternal care in pinnipeds. BioScience 46:645–654.

Boness, D. J., and H. James. 1979. Reproductive behaviour of the gray seal (*Halichoerus grypus*) on Sable Island, Nova Scotia. Journal of Zoology (London) 188:477–500.

Boness, D. J., S. S. Anderson, and C. R. Cox. 1982. Functions of female aggression during the pupping and mating season of gray seals, *Halichoerus grypus* (Fabricius). Canadian Journal of Zoology 60:2270–2278.

Boness, D. J., W. D. Bowen, and O. T. Oftedal. 1988. Evidence of polygyny from spatial patterns of hooded seals (*Cystophora cristata*). Canadian Journal of Zoology 66:703–706.

Boness, D. J., W. D. Bowen, S. J. Iverson, and O. T. Oftedal. 1992. Influence of storms and maternal size on mother-pup separations and fostering in the harbor seal, *Phoca vitulina*. Canadian Journal of Zoology 70:1640–1644.

Boness, D. J., W. D. Bowen, and J. M. Francis. 1993. Implications of DNA fingerprinting for mating systems and reproductive strategies of pinnipeds. Symposia of the Zoological Society (London) 66:61–93.

Boness, D. J., W. D. Bowen, and O. T. Oftedal. 1994. Evidence of a maternal foraging cycle resembling that of otariid seals in a small phocid, the harbor seal. Behavioral Ecology and Sociobiology 34:95–104.

Boness, D. J., W. D. Bowen, and S. J. Iverson. 1995. Does male harassment of females contribute to reproductive synchrony in the gray seal by affecting maternal performance? Behavioral Ecology and Sociobiology 36:1–10.

Boness, D. J., M. P. Craig, L. Honigman, and S. Austin. 1998. Fostering behavior and the effect of female density in Hawaiian monk seals, *Monachus schauinslandi*. Journal of Mammalogy 79:1060–1069.

Bonner, W. N. 1984. Lactation strategies in pinnipeds: Problems for a marine mammalian group. Symposia of the Zoological Society (London) 51:253–272.

Bonnot, P. 1928. The sea lions of California. California Fish and Game 14:1–16.

Bookhout, T. A. (ed.). 1994. Research and management techniques for wildlife and habitats. The Wildlife Society, Bethesda, MD.

Born, E. W., and L. O. Knutsen. 1992. Satellite-linked radio-tracking of Atlantic walruses (*Odobenus rosmarus rosmarus*) in northeastern Greenland, 1989–1991. Zeitschrift für Säugetierkunde 57:275–287.

Borobia, M., S. Siciliana, L. Lodi, and W. Hoek. 1991. Distribution of the South American dolphin *Sotalia fluviatilis*. Canadian Journal of Zoology 69:1025–1039.

Bowen, W. D. 1991. Behavioural ecology of pinniped neonates. Pages 66–127 *in* D. Renouf (ed.). Behaviour of Pinnipeds. Chapman and Hall, London.

Bowen, W. D., O. T. Oftedal, and D. J. Boness. 1985. Birth to weaning in four days: Remarkable growth in the hooded seal, *Cystophora cristata*. Canadian Journal of Zoology 63:2841–2846.

Bowen, W. D., D. J. Boness, and O. T. Oftedal. 1987. Mass transfer from mother to pup and subsequent mass loss by the weaned pup in the hooded seal, *Cystophora cristata*. Canadian Journal of Zoology 65:1–8.

Bowen, W. D., W. T. Stobo, and S. J. Smith. 1992. Mass changes of gray seal *Halichoerus grypus* pups on Sable Island differential maternal investment reconsidered. Journal of Zoology (London) 227:607–622.

Boyd, I. L., and T. Arnbom. 1991. Diving behaviour in relation to water temperature in the southern elephant seal: Foraging implications. Polar Biology 11:259–266.

Boyd, I. L., and J. P. Croxall. 1992. Diving behavior of lactating fur seals. Canadian Journal of Zoology 70:919–928.

Boyd, I. L., J. P. Y. Arnould, T. Barton, and J. P. Croxall. 1994. Foraging behaviour of Antarctic fur seals during periods of contrasting prey abundance. Journal of Animal Ecology 63:703–713.

Bradbury, J. W., and R. Gibson. 1983. Leks and mate choice. Pages 109–138 *in* P. Bateson (ed.). Mate Choice. Cambridge University Press, Cambridge, U.K.

Bradbury, J. W., and S. L. Vehrencamp. 1977. Social organization and foraging in emballonurid bats. III. Mating systems. Behavioral Ecology and Sociobiology 2:1–17.

Bräger, S. 1993. Diurnal and seasonal behavior patterns of bottlenose dolphins (*Tursiops truncatus*). Marine Mammal Science 9:434–438.

Bräger, S., B. Würsig, A. Acevedo, and T. Henningsen. 1994. Association patterns of bottlenose dolphins (*Tursiops truncatus*) in Galveston Bay, Texas. Journal of Mammalogy 75:431–437.

Braham, H. W. 1984. Distribution and migration of gray whales in Alaska. Pages 249–266 *in* M. L. Jones, S. L. Swartz, and S. Leatherwood (eds.). The Gray Whale *Eschrichtius robustus*. Academic Press, San Diego, CA.

Brown, D. H., and K. S. Norris. 1956. Observations of captive and wild cetaceans. Journal of Mammalogy 37:311–326.

Brown, D. H., D. K. Caldwell, and M. C. Caldwell. 1966. Observations on the behavior of wild and captive false killer whales, with notes on associated behavior of other genera of captive delphinids. Los Angeles County Museum of Natural History Contributions in Science 95:1–32.

Brown, S. G. 1962. International co-operation in Antarctic whale marking from 1957–1960, and a review of the distribution of marked whales in the Antarctic. Norsk-Hvalfangst-tid 3:93–104.

Brown, S. G. 1978. Whale marking techniques. Pages 71–80 *in* B. Stonehouse (ed.). Animal Marking: Recognition Marking of Animals in Research. University Park Press, Baltimore, MD.

Brownell, R. L., and K. Ralls. 1986. Potential for sperm competition in baleen whales. Reports of the International Whaling Commission (Special Issue 8):97–112.

Buckingham, C. A. 1990. Manatee response to boating activity in a thermal refuge. M.S. thesis, University of Florida, Gainesville, FL. 83 pages.

Busnel, R. G. 1973. Symbiotic relationship between man and dolphins. Transcripts of the New York Academy of Science (Series 2) 35:112–131.

Calambokidis, J., G. H. Steiger, J. C. Cubbage, K. C. Balcomb, C. Ewald, S. L. Kruse, R. S. Wells, and R. Sears. 1990. Sightings and movements of blue whales off central California 1986–88 from photo-identification of individuals. Pages 343–348 *in* P. S. Hammond, S. A. Mizroch, and G. P. Donovan (eds.). Individual Recognition of Cetaceans: Use of Photo-identification and Other Techniques to Estimate Population Parameters. Report of the International Whaling Commission (Special Issue 12), Cambridge, U.K.

Caldwell, D. K. 1955. Evidence of home range of an Atlantic bottlenose dolphin. Journal of Mammalogy 36:304–305.

Caldwell, D. K., and M. C. Caldwell. 1972. The World of the Bottlenose Dolphin. Lippincott, New York, NY.

Caldwell, D. K., M. C. Caldwell, and D. W. Rice. 1966. Behavior of the sperm whale, *Physeter catodon* L. Pages 617–717 *in* K. S. Norris (ed.). Whales, Dolphins, and Porpoises. University of California Press, Berkeley, CA.

Caldwell, M. C., and D. K. Caldwell. 1966. Epimeletic (care-giving) behavior in Cetacea. Pages 755–789 *in* K. S. Norris (ed.). Whales, Dolphins, and Porpoises. University of California Press, Berkeley, CA.

Caldwell, M. C., and D. K. Caldwell. 1967. Dolphin community life. Los Angeles County Museum of Natural History Contributions in Science 5:12–15.

Caldwell, M. C., D. K. Caldwell, and J. B. Siebenaler. 1965. Observations on captive and wild Atlantic bottlenosed dolphins, *Tursiops truncatus,* in the northeastern Gulf of Mexico. Los Angeles County Museum of Natural History Contributions in Science 91:1–10.

Caldwell, M. C., D. K. Caldwell, and R. L. Brill. 1989. *Inia geoffrensis* in captivity in the United States. Pages 35–41 *in* W.F. Perrin, R.L. Brownell Jr., Z. Kaiya, and L. Jiankang (eds.). Biology and Conservation of the River Dolphins. Occasional Papers of the IUCN Species Survival Commission (SSC), Number 3, Gland, Switzerland.

Campbell, R. R., D. B. Yurick, and N. B. Snow. 1988. Predation on narwhals, *Monodon monoceros,* by killer whales, *Orcinus orca,* in the eastern Canadian arctic. Canadian Field-Naturalist 102:689–696.

Campagna, C., and B. J. Le Boeuf. 1988a. Reproductive behaviour of Southern sea lions. Behaviour 104:233–262.

Campagna, C., and B. J. Le Boeuf. 1988b. Thermoregulatory behaviour of southern sea lions and its effect on mating strategies. Behaviour 107:72–90.

Campagna, C., C. Bisioli, F. Quintana, F. Perez, and A. Vila. 1992a. Group breeding in sea lions: pups survive better in colonies. Animal Behaviour 43:541–548.

Campagna, C., B. J. Le Boeuf, M. Lewis, and C. Bisioli. 1992b. Equal investment in male and female offspring in southern elephant seals. Journal of Zoology (London) 226:551–561.

Carey, P. W. 1991. Resource-defense polygyny and male territory quality in the New Zealand fur seal. Ethology 88:63–79.

Carey, P. W. 1992. Agonistic behaviour in female New Zealand fur seals, *Arctocephalus forsteri.* Ethology 92:70–80.

Caron, L. M. J., and T. G. Smith. 1990. Philopatry and site tenacity of belugas, *Delphinapterus leucas,* hunted by the Inuit at the Nastapoka estuary, eastern Hudson Bay. Pages 69–79 *in* T. G. Smith, D. J. St. Aubin, and J. R. Geraci (eds.). Advances in Research on the Beluga Whale, *Delphinapterus leucas.* Canadian Bulletin of Fisheries and Aquatic Sciences 224:206.

Carrick, R., S. E. Csordas, and S. E. Ingham. 1962a. Studies on the southern elephant seal, *Mirounga leonina* (L.) IV. Breeding and development. CSIRO Wildlife Research 7:161–197.

Carrick, R., S. E. Csordas, S. E. Ingham, and K. Keith. 1962b. Studies on the southern elephant seal, *Mirounga leonina* (L.) III. The annual cycle in relation to age and sex. CSIRO Wildlife Research 7:119–160.

Charnov, E. L. 1982. Monographs in Population Biology, 18: Theory of Sex Allocation. Princeton University Press, Princeton, NJ.

Cheney, D. L., R. M. Seyfarth, S. J. Andelman, and P. C. Lee. 1988 Reproductive success in vervet monkeys. Pages 384–402 *in* T. H. Clutton-Brock (ed.). Reproductive Success. The University of Chicago Press, Chicago and London.

Chittleborough, R. G. 1958. The breeding cycle of the female humpback whale, *Megaptera nodosa* (Bonnaterre). Australian Journal of Marine and Freshwater Research 9:1–18.

Christenson, T. E., and B. J. Le Boeuf. 1978. Aggression in the female northern elephant seal, *Mirounga angustirostris.* Behaviour 64:158–172.

Chu, K., and P. Harcourt. 1986. Behavioral correlations with aberrant patterns in humpback whale songs. Behavioural Ecology and Sociobiology 19:309–312.

Chu, K., and S. Nieukirk. 1988. Dorsal fin scars as indicators of age, sex, and social status in humpback whales (*Megaptera novaeangliae*). Canadian Journal of Zoology 66:416–420.

Clapham, P. J. 1993. Social organization of humpback whales on a North Atlantic feeding ground. Symposia of the Zoological Society (London) 66:131–145.

Clapham, P. J. 1994. Maturational changes of association in male and female humpback whales, *Megaptera novaeangliae.* Journal of Zoology (London) 234:265–274.

Clapham, P. J. 1996. The social and reproductive biology of humpback whales: An ecological perspective. Mammal Review 26:27–49.

Clapham, P. J., and C. A. Mayo. 1987. Reproduction and recruitment of individually identified humpback whales, *Megaptera novaeangliae,* observed in Massachusetts Bay, 1979–1985. Canadian Journal of Zoology 65:2853–2863.

Clapham, P. J., and P. J. Palsbøll. 1997. Molecular analysis of paternity shows promiscuous mating in female humpback whales (*Megaptera novaeangliae* Borowski). Proceedings of the Royal Society of London, Part B 264:95–98.

Clapham, P. J., and I. E. Seipt. 1991. Resightings of independent fin whales, *Balaenoptera physalus,* on maternal summer ranges. Journal of Mammalogy 72:788–790.

Clapham, P. J., P. J. Palsbøll, D. K. Mattila, and O. Vásquez. 1992. Composition and dynamics of humpback whale competitive groups in the West Indies. Behaviour 122:182–194.

Cleator, H. J., I. Stirling, and T. G. Smith. 1989. Underwater vocalizations of the bearded seal (*Erignathus barbatus*). Canadian Journal of Zoology 67:1900–1910.

Clutton-Brock, T. H. (ed). 1988. Reproductive Success. University of Chicago Press, Chicago.

Clutton-Brock, T. H. 1989. Mammalian mating systems. Proceedings of the Royal Society of London B236:339–372.

Clutton-Brock, T. H. 1991. The Evolution of Parental Care. Princeton University Press, Princeton, NJ.

Clutton-Brock, T. H., F. E. Guiness, and S. D. Albon. 1982. Red Deer: The Behaviour and Ecology of Two Sexes. University of California Press, Chicago, IL.

Cockcroft, V. G. 1991. Incidence of shark bites on Indian Ocean humpbacked dolphins off Natal, South Africa. Marine Mammal Technical Report 3:277–282.

Cockcroft, V. G., and G. J. B. Ross. 1990a. Food and feeding of the Indian Ocean bottlenose dolphin off southern Natal, South Africa. Pages 295–308 *in* S. Leatherwood and R. R. Reeves (eds.). The Bottlenose Dolphin. Academic Press, San Diego, CA.

Cockcroft, V. G., and G. J. B. Ross. 1990b. Observations on the early development of a captive bottlenose dolphin calf. Pages 461–478 *in* S. Leatherwood and R. R. Reeves (eds.). The Bottlenose Dolphin. Academic Press, San Diego, CA.

Cockcroft, V. G., G. Cliff, and G. J. B. Ross. 1989. Shark predation on Indian Ocean bottlenose dolphins *Tursiops truncatus* off Natal, South Africa. South African Journal of Zoology 24:305–310.

Coltman, D. W., W. D. Bowen, D. J. Boness, and S. J. Iverson. 1997. Balancing foraging and reproduction in the male harbour seal, an aquatically mating pinniped. Animal Behaviour 54:663–678.

Condit, R., and B. J. Le Boeuf. 1984. Feeding habits and feeding grounds of the northern elephant seal. Journal of Mammalogy 65:281–290.

Connor, R. C., and M. R. Heithaus. 1996. Approach by great white shark elicits flight response in bottlenose dolphins. Marine Mammal Science 12:602–606.

Connor, R. C., and K. S. Norris. 1982. Are dolphins reciprocal altruists? American Naturalist 119:358–374.

Connor, R. C., and R. A. Smolker. 1985. Habituated dolphins (*Tursiops* sp.) in western Australia. Journal of Mammalogy 66:398–400.

Connor, R. C., and R. A. Smolker. 1990. Quantitative description of a rare behavioral event: A bottlenose dolphin's behavior toward her deceased offspring. Pages 355–360 *in* S. Leatherwood and R. R. Reeves (eds.). The Bottlenose Dolphin. Academic Press, San Diego, CA.

Connor, R. C., and R. A. Smolker. 1996. "Pop" goes the dolphin: A vocalization male bottlenose dolphins produce during consortships. Behaviour 133:643–662.

Connor, R. C., R. A. Smolker, and A. F. Richards. 1992a. Two levels of alliance formation among male bottlenose dolphins (*Tursiops* sp.). Proceedings of the National Academy of Science 89:987–990.

Connor, R.C., R. A. Smolker, and A. F. Richards. 1992b. Dolphin alliances and coalitions. Pages 415–443 *in* A. H. Harcourt and F. B. M. DeWaal (eds.). Coalitions and Alliances in Humans and Other Animals. Oxford University Press, U.K.

Connor, R. C., A. F. Richards, R. A. Smolker, and J. Mann. 1996. Patterns of female attractiveness in Indian Ocean bottlenose dolphins. Behaviour 133:37–69.

Corkeron, P. J. 1990. Aspects of the behavioral ecology of inshore dolphins *Tursiops truncatus* and *Sousa chinensis* in Moreton Bay, Australia. Pages 285–293 *in* S. Leatherwood and R. R. Reeves (eds.). The Bottlenose Dolphin. Academic Press, San Diego, CA.

Corkeron, P. J., R. J. Morris, and M. M. Bryden. 1987. Interactions between bottlenose dolphins and sharks in Moreton Bay, Queensland. Aquatic Mammals 13:109–113.

Corkeron, P. J., M. M. Bryden, and K. E. Hedstrom. 1990. Feeding by bottlenose dolphins in association with trawling operations in Moreton Bay, Australia. Pages 329–336 *in* S. Leatherwood and R. R. Reeves (eds.). The Bottlenose Dolphin. Academic Press, San Diego, CA.

Cosens, S. E., and L. P. Dueck. 1991. Group size and activity patterns of belugas (*Delphinapterus leucas*) and narwhals (*Monodon monoceros*) during spring migration in Lancaster Sound. Canadian Journal of Zoology 69:1630–1635.

Costa, D. P. 1987. Isotopic methods for quantifying material and energy intake of free-ranging marine mammals. Pages 43–66 *in* A. C. Huntley, D. P. Costa, G. A. J. Worthy, and M. A. Castellini (eds.). Marine Mammal Energetics. Society for Marine Mammalogy, Lawrence, KS.

Costa, D. P. 1991a. Reproductive and foraging energetics of pinnipeds: Implications for life history patterns. Pages 300–344 *in* D. Renouf (ed.). The Behaviour of Pinnipeds. Chapman and Hall, London.

Costa, D. P. 1991b. Reproductive and foraging energetics of high latitude penguins, albatrosses and pinnipeds: Implications for life history patterns. American Zoologist 31:111–130.

Costa, D. P. 1993. The relationship between reproductive and foraging energetics and the evolution of the Pinnipedia. Symposia of the Zoological Society (London) 66:293–314.

Costa, D. P., and G. L. Kooyman. 1982. Oxygen consumption, thermoregulation, and the effect of fur oiling and washing on the sea otter, *Enhydra lutris*. Canadian Journal of Zoology 60:2761–2767.

Costa, D. P., B. J. Le Boeuf, A. C. Huntley, and C. L. Ortiz. 1986. The energetics of lactation in the northern elephant seal, *Mirounga angustirostris*. Journal of Zoology (London) 209:21–33.

Cox, C. R., and B. J. Le Boeuf. 1977. Female incitation of male competition: A mechanism in sexual selection. American Naturalist 111:317–335.

Crawley, M. C., J. D. Stark, and P. S. Dodgshun. 1978. Activity budgets of New Zealand fur seals *Arctocephalus forsteri* during the breeding season. New Zealand Journal of Marine and Freshwater Research 11:777–788.

Crocker, D. E. , B. J. Le Boeuf, Y. Naito, T. Asaga, and D. P. Costa. 1994. Swim speed and dive function in a female northern elephant seal. Pages 328–342 *in* B. J. Le Boeuf and R. M. Laws (eds.). Elephant Seals. University of California Press, Berkeley.

Crook, J. H., J. E. Ellis, and J. D. Goss-Custard. 1976. Mammalian social systems: Structure and function. Animal Behaviour 24:261–274.

Croxall, J. P., and L. Hiby. 1983. Fecundity, survival and site fidelity in Weddell seals, *Leptonychotes weddelli*. Journal of Applied Ecology 20:19–32.

Croxall, J. P., I. Everson, G. L. Kooyman, C. Ricketts, and R. W. Davis. 1985. Fur seal diving behaviour in relation to vertical distribution of krill. Journal of Animal Ecology 54:1–8.

Croxall, J. P., R. W. Davis, and M. J. O'Connell. 1988. Diving patterns in relation to diet of gentoo and macaroni penguins at South Georgia. Condor 90:157–167.

Cummings, W. C., and P. O. Thompson. 1971. Gray whales, *Eschrichtius robustus*, avoid the underwater sounds of killer whales, *Orcinus orca*. Fishery Bulletin 69:525–530.

Dahlheim, M. E., and R. G. Towell. 1994. Occurrence and distribution of Pacific whitesided dolphins (*Lagenorhynchus obliquidens*) in southeastern Alaska, with notes on an attack by killer whales (*Orcinus orca*). Marine Mammal Science 10:458–464.

da Silva, J., and J. M. Terhune. 1988. Harbour seal grouping as an antipredator strategy. Animal Behaviour 36:1309–1316.

Darling, J. D. 1984. Gray whales off Vancouver Island, British Columbia. Pages 267–287 *in* M. L. Jones, S. L. Swartz, and S. Leatherwood (eds.). The Gray Whale *Eschrichtius robustus*. Academic Press, San Diego, CA.

Darling, J. D., and S. Cerchio. 1993. Movement of a humpback whale (*Megaptera novaeangliae*) between Japan and Hawaii. Marine Mammal Science 9:84–89.

Darling, J. D., and C. M. Jurasz. 1983. Migratory destinations of North Pacific humpback whales (*Megaptera novaeangliae*). Pages 359–368 *in*

R. Payne (ed.). Communication and Behavior of Whales. Westview Press, Boulder, CO.

Darling, J. D., and D. J. McSweeney. 1985. Observations on the migrations of North Pacific humpback whales (*Megaptera novaeangliae*). Canadian Journal of Zoology 63:308–314.

Darling, J. D., K. M. Gibson, and G. K. Silber. 1983. Observations on the abundance and behavior of humpback whales (*Megaptera novaeangliae*) of West Maui, Hawaii, 1977–79. Pages 201–222 *in* R. Payne (ed.). Communication and Behavior of Whales. Westview Press, Boulder, CO.

David, J. H. M., and R. W. Rand. 1986. Attendance behavior of South African fur seals. Pages 126–141 *in* R. L. Gentry and G. L. Kooyman (eds.). Fur Seals: Maternal Strategies on Land and at Sea. Princeton University Press, Princeton, NJ.

Davies, J. L. 1949. Observations of the gray seal (*Halichoerus grypus*) at Ramsey Island, Pembrokeshire. Proceedings of the Zoology Society of London 119:673–692.

Davies, N. B. 1991. Mating systems. Pages 263–300 *in* J. R. Krebs and N. B. Davies (eds.). Behavioural Ecology, 3rd ed. Blackwell Scientific, London.

Davis, R. W., G. A. J. Worthy, B. Würsig, S. K. Lynn, and F. I. Townsend. 1996. Diving behavior and at-sea movements of an Atlantic spotted dolphin in the Gulf of Mexico. Marine Mammal Science 12:569–581.

Dawbin, W. H. 1966. The seasonal migratory cycle of humpback whales. Pages 145–170 *in* K. S. Norris (ed.). Whales, Dolphins, and Porpoises. University of California Press, Berkeley, CA.

DeLong, R. L. 1982. Population biology of northern fur seals at San Miguel Island, California. Ph.D. thesis, University of California, Berkeley, CA.

DeLong, R. L., B. S. Stewart, and R. D. Hill. 1992. Documenting migrations of northern elephant seals using day length. Marine Mammal Science 8:155–159.

Deutsch, C. J. 1985. Male–male competition in the Hawaiian monk seal. Biennial Conference on the Biology of Marine Mammals (Abstract) 6:25.

Deutsch, C. J., M. P. Haley, and B. J. Le Boeuf. 1990. Reproductive effort of male northern elephant seals: Estimates from mass loss. Canadian Journal of Zoology 68:2580–2593.

Deutsch, C. J., R. K. Bonde, and J. P. Reid. 1998. Radio-tracking manatees from land and space: Tag design, implementation, and lessons learned from long-term study. Marine Technology Society Journal 32:18–29.

Doidge, D. W., and J. P. Croxall. 1985. Diet and energy budget of the Antarctic fur seal, *Arctocephalus gazella*, at South Georgia. Pages 543–550 *in* W. R. Siegfried, P. R. Condy, and R. M. Laws (eds.). Antarctic Nutrient Cycles and Food Webs. Springer-Verlag, Heidelberg.

Doidge, D. W., and J. P. Croxall. 1989. Factors affecting weaning weight in Antarctic fur seals, *Arctocephalus gazella* at South Georgia. Polar Biology 9:155–160.

Doidge, D. W., T. S. McCann, and J. P. Croxall. 1986. Attendance behavior of Antarctic fur seals. Pages 102–114 *in* R. L. Gentry and G. L. Kooyman (eds.). Fur Seals: Maternal Strategies on Land and at Sea. Princeton University Press, Princeton, NJ.

Dolphin, W. F. 1987a. Prey densities and foraging of humpback whales, *Megaptera novaeangliae*. Experientia 43:468–471.

Dolphin, W. F. 1987b. Ventilation and dive patterns of humpback whales, *Megaptera novaeangliae,* on their Alaskan feeding grounds. Canadian Journal of Zoology 65:83–90.

Dolphin, W. F. 1987c. Dive behavior and estimated energy expenditure of foraging humpback whales in southeast Alaska. Canadian Journal of Zoology 65:354–362.

Dolphin, W. F. 1987d. Observations of humpback whale, *Megaptera novaeangliae*—killer whale, *Orcinus orca,* interactions in Alaska: Comparison with terrestrial predator–prey relationships. Canadian Field-Naturalist 101:70–75.

Dolphin, W. F. 1988. Foraging dive patterns of humpback whales, *Megaptera novaeangliae,* in southeast Alaska: A cost–benefit analysis. Canadian Journal of Zoology 66:2432–2441.

Domning, D. P. 1977. An ecological model for late tertiary sirenian evolution in the North Pacific Ocean. Systematic Zoology 25:352–362.

Domning, D. P. 1980. Feeding position preference in manatees (*Trichechus*). Journal of Mammalogy 61:544–547.

Domning, D. P. 1994. A phylogenetic analysis of the Sirenia. Proceedings of the San Diego Society of Natural History 29:177–189.

Dorsey, E. M. 1983. Exclusive adjoining ranges in individually identified minke whales (*Balaenoptera acutorostrata*) in Washington state. Canadian Journal of Zoology 61:175–181.

Douglas-Hamilton, I., and O. Douglas-Hamilton. 1975. Among the Elephants. Viking Press, New York, NY.

Dufault, S., and H. Whitehead. 1995a. An assessment of changes with time in the marking patterns used for photoidentification of individual sperm whales, *Physeter macrocephalus*. Marine Mammal Science 11:335–343.

Dufault, S., and H. Whitehead. 1995b. The geographic stock structure of female and immature sperm whales in the South Pacific. Report of the International Whaling Commission 45:401–405.

Duffield, D. A., and R. S. Wells. 1986. Population structure of bottlenose dolphins: Genetic studies of bottlenose dolphins along the central west coast of Florida. Final contract report to the National Marine Fisheries Service, Southeast Fisheries Science Center, Miami, FL. Contract No. 45-WCNF-5-00366. 10 pp.

Duffield, D. A., and R. S. Wells. 1991. The combined application of chromosome, protein and molecular data for the investigation of social unit structure and dynamics in *Tursiops truncatus*. Pages 155–169 *in* A. R. Hoelzel (ed.). Genetic Ecology of Whales and Dolphins. Report of the International Whaling Commission (Special Issue 13), Cambridge, U.K.

Duffield, D. A., R. S. Wells, J. S. Lenox, and T. Moors. 1994. Analysis of paternity in a free-ranging bottlenose dolphin society by DNA fingerprinting and behavioral coefficients of association (Abstract). International Symposium of Marine Mammal Genetics, 23–24 September, La Jolla, CA.

Dunbar, R. I. M. 1984. Reproductive Decisions: An Economic Analysis of Gelada Baboon Social Strategies. Princeton University Press, Princeton, NJ.

Dunton, K. 1994. Seasonal growth and biomass of the subtropical seagrass (*Halodule wrightii*) in relation to continuous measurements of underwater irradiance. Marine Biology 120:479–489.

Edds, P. L., and J. A. F. MacFarlane. 1987. Occurrence and general behavior of balaenopterid cetaceans summering in the St. Lawrence Estuary, Canada. Canadian Journal of Zoology 65:1363–1376.

Eibl-Eibesfeldt, I. 1984. The Galapagos seals. Part 1. Natural history of the Galapagos sea lion (*Zalophus californianus* Wollebaek, Sivertsen). Pages 207–214 *in* R. Perry (ed.). Key Environments: Galapagos. Pergamon Press, Oxford.

Eisenberg, J. F. 1981. The Mammalian Radiations—An Analysis of Trends in Evolution, Adaptation, and Behavior. University of Chicago Press, Chicago, IL.

Eisenberg, J. F., and M. Lockhart. 1972. An Ecological Reconnaisance of Wilpattu National Park, Ceylon. Smithsonian Institution Press, Washington, D.C.

Emlen, S. T., and L. W. Oring. 1977. Ecology, sexual selection and the evolution of mating systems. Science 197:215–223.

Essapian, F. S. 1953. The birth and growth of a porpoise. Natural History November:392–399.

Essapian, F. S. 1963. Observations on abnormalities of parturition in captive bottlenosed dolphins, *Tursiops truncatus,* and concurrent behavior of other porpoises. Journal of Mammalogy 44:405–414.

Estes, J. A. 1989. Adaptations for aquatic living by carnivores. Pages 242–282 *in* J. L. Gittleman (ed.). Carnivore Behavior, Ecology, and Evolution. Cornell University Press, Ithaca, NY.

Estes, J. A., R. J. Jameson, and A. M. Johnson. 1981. Food selection and some foraging tactics of sea otters. Pages 606–641 *in* J. A. Chapman and D. Pursley (eds.). The Worldwide Furbearer Conference Proceedings, Frostburg, MD, 1980. Worldwide Furbearer Conference Inc., Frostburg, MD.

Estes, J. A., R. J. Jameson, and E. B. Rhode. 1982. Activity and prey selection in the sea otter: influence of population status on community structure. American Naturalist 120:242–258.

Estes, J. A., K. Underwood, and M. Karmann. 1986. Activity time budgets of sea otters in California. Journal of Wildlife Management 50:626–639.

Estes, R. D. 1974. Social organization of the African Bovidae. Pages 166–205 *in* V. Geist and F. Walther (eds.). The Behaviour of Ungulates and its Relation to Management. I.U.C.N. Publications, New Series No. 24, Morges, Switzerland.

Estes, R. D., and J. Goddard. 1967. Prey selection and hunting behavior of the African wild dog. Journal of Wildlife Management 31:52–70.

Etheridge, K., G. B. Rathbun, J. A. Powell, and H. I. Kochman. 1985. Consumption of aquatic plants by the West Indian manatee. Journal of Aquatic Plant Management 23:21–25.

Evans, W. E. 1971. Orientation behavior of delphinids: Radio telemetric studies. Annals of the New York Academy of Science 188:142–160.

Evans, W. E. 1974. Radio-telemetric studies of two species of small odontocete cetaceans. Pages 385–394 *in* W. E. Schevill (ed.). The Whale Problem. Harvard University Press, Cambridge, MA.

Evans, W. E. 1975. Distribution, differentiation of populations, and other aspects of the natural history of *Delphinus delphis* Linnaeus in the northeastern Pacific. Ph.D. dissertation, University of California, Los Angeles, CA. 147 pages.

Evans, W. E., and J. Bastian. 1969. Marine mammal communication: Social and ecological factors. Pages 425–476 *in* H. T. Andersen (ed.). The Biology of Marine Mammals. Academic Press, New York, NY.

Evans, W. E., J. D. Hall, A. B. Irvine, and J. S. Leatherwood. 1972. Methods for tagging small cetaceans. Fishery Bulletin 70:61–65.

Ewer, R. F. 1973. The Carnivores. Cornell University Press, Ithaca, NY.

Fay, F. H. 1982. North American fauna, no. 74: Ecology and biology of the Pacific walrus, *Odobenus rosmarus divergens* Illiger. U.S. Fish and Wildlife Service, Washington, D.C.

Fay, F. H. 1985. *Odobenus rosmarus.* Mammalian Species 238:1–7.

Fay, F. H., and J. J. Burns. 1988. Maximal feeding depths of walruses. Arctic 41:239–240.

Fay, F. H., Y. A. Bukhtiyarov, S. W. Stoker, and L. M. Shults. 1984a. Foods of the Pacific walrus in winter and spring in the Bering Sea. Pages 81–88 *in* F. H. Fay and G. A. Fedoseev (eds.). Soviet–American Cooperative Research on Marine Mammals, Vol. 1: Pinnipeds. NOAA Technical Report NMFS 12, Washington, D.C.

Fay, F. H., G. C. Ray, and A. A. Kibal'chick. 1984b. Time and location of mating and associated behaviour of the Pacific walrus, *Odobenus rosmarus divergens* Illiger. Pages 89–99 *in* F. H. Fay and G. A. Fedoseev (eds.). Soviet–American Cooperative Research on Marine Mammals, Vol 1: Pinnipeds. NOAA Technical Report NMFS 12, Washington, D.C.

Fedak, M. A., and D. Thompson. 1993. Behavioural and physiological options in diving seals. Symposia of the Zoological Society (London) 66:333–348.

Fehring, W. K., and R. S. Wells. 1976. A series of strandings by a single herd of pilot whales on the west coast of Florida. Journal of Mammalogy 57:191–194.

Feldkamp, S. D., R. L. DeLong, and G. A. Antonelis. 1989. Diving patterns of California sea lions. Canadian Journal of Zoology 67:872–883.

Felix, F. 1994. Ecology of the coastal bottlenose dolphin *Tursiops truncatus* in the Gulf of Guayaquil, Ecuador. Investigations on Cetacea 25:235–256.

Felleman, F. L., J. R. Heimlich-Boran, and R. W. Osborne. 1991. The feeding ecology of killer whales (*Orcinus orca*) in the Pacific northwest. Pages 113–147 *in* K. Pryor and K. S. Norris (eds.). Dolphin Societies: Discoveries and Puzzles. University of California Press, Berkeley, CA.

Ferrer, M. 1993. Natural adoption of fledglings by Spanish imperial eagles *Aquila adalberti.* Journal of Ornithology 134:335–337.

Fertl, D. 1993. Occurrence and behavior of bottlenose dolphins (*Tursiops truncatus*) in association with the shrimp fishery in Galveston Bay, Texas (Abstract). 10th Biennial Conference on the Biology of Marine Mammals, November 11–15, 1993, Galveston, TX.

Fertl, D., and B. Würsig. 1995. Coordinated feeding by Atlantic spotted dolphins (*Stenella frontalis*) in the Gulf of Mexico. Aquatic Mammals 21:3–5.

Fertl, D., A. Acevedo-Gutiérrez, and F. L. Darby. 1996. A report of killer whales (*Orcinus orca*) feeding on a carcharhinid shark in Costa Rica. Marine Mammal Science 12:606–611.

Finley, K. J., and E. J. Gibb. 1982. Summer diet of the narwhal (*Monodon monoceros*) in Pond Inlet, northern Baffin Island. Canadian Journal of Zoology 60:3353–3363.

Fish, J. F., and J. S. Vania. 1971. Killer whale, *Orcinus orca,* sounds repel white whales, *Delphinapterus leucas.* Fishery Bulletin 69:531–535.

Flórez-González, L., J. J. Capella, and H. C. Rosenbaum. 1994. Attack of killer whales (*Orcinus orca*) on humpback whales (*Megaptera novaeangliae*) on a South American Pacific breeding ground. Marine Mammal Science 10:218–222.

Fogden, S. C. L. 1971. Mother–young behaviour at gray seal breeding beaches. Journal of Zoology 164:61–92.

Folkow, L. P., and A. S. Blix. 1995. Distribution and diving behaviour of hooded seals. Pages 193–202 *in* A. S. Blix, L. Walløe, and Ø. Ulltang (eds.). Whales, Seals, Fish and Man. Elsevier, Amsterdam.

Foott, J. O. 1971. Nose scars in female sea otters. Journal of Mammalogy 51:621–622.

Ford, J. K. B., and H. D. Fisher. 1982. Killer whale (*Orcinus orca*) dialects as an indicator of stocks in British Columbia. Report of the International Whaling Commission 32:671–680.

Ford, J. K. B., and H. D. Fisher. 1983. Group-specific dialects of killer whales (*Orcinus orca*) in British Columbia. Pages 129–161 *in* R. S. Payne (ed.). Communication and Behavior of Whales. Westview Press, Boulder, CO.

Francis, J. M., and D. J. Boness. 1991. The effect of thermoregulatory behavior on the mating system of the Juan Fernández fur seal, *Arctocephalus philippii.* Behaviour 119:104–127.

Francis, J. M., D. J. Boness, and H. Ochoa-Acuña. 1998. A protracted foraging and attendance cycle in female Juan Fernández fur seals. Marine Mammal Science 14:552–574.

Fraser, W. R., R. L. Pitman, and D. G. Ainley. 1989. Seabird and fur seal responses to vertically migrating winter krill swarms in Antarctica. Polar Biology 10:37–41.

Frost, K. J., R. B. Russell, and L. F. Lowry. 1992. Killer whales, *Orcinus orca,* in the southeastern Bering Sea: Recent sightings and predation on other marine mammals. Marine Mammal Science 8:110–119.

Furgal, C. M., S. Innes, and K. M. Kovacs. 1996. Characteristics of ringed seal, *Phoca hispida,* subnivean structures and breeding habitat and their effects on predation. Canadian Journal of Zoology 74:858–874.

Gales, N. J., A. J. Cheal, G. J. Pobar, and P. Williamson. 1992. Breeding biology and movements of Australian sea-lions, *Neophoca cinerea,* off the west coast of Western Australia. Wildlife Research 19:447–456.

Gales, N. J., P. D. Shaughnessy, and T. E. Dennis. 1994. Distribution, abundance and breeding cycle of the Australian sea lion *Neophoca cinerea* (Mammalia: Pinnipedia). Journal of Zoology (London) 234:353–370.

Gallivan, G. J., and R. C. Best. 1980. Metabolism and respiration of the Amazonian manatee (*Trichechus inunguis*). Physiological Zoology 53:245–253.

Garrott, R. A., B. B. Ackerman, J. R. Cary, D. M. Heisey, J. E. Reynolds III, and J. R. Wilcox. 1995. Assessment of trends in size of manatee populations at several Florida aggregation sites. Pages 34–55 *in* T. J. O'Shea, B. B. Ackerman, and H. F. Percival (eds.). Population Biology of the Florida Manatee. National Biological Service, Information and Technology Report 1, Washington, D.C.

Garshelis, D. L., A. M. Johnson, and J. A. Garshelis. 1984. Social organization of sea otters in Prince William Sound, Alaska. Canadian Journal of Zoology 62:2648–2658.

Garshelis, D. L., J. A. Garshelis, and A. T. Kimker. 1986. Sea otter time budgets and prey relationships in Alaska. Journal of Wildlife Management 50:637–647.

Gaskin, D. 1982. The Ecology of Whales and Dolphins. Heinemann, London.

Gaskin, D. E., and A. P. Watson. 1985. The harbor porpoise, *Phocoena phocoena,* in Fish Harbour, New Brunswick, Canada: Occupancy, distribution, and movements. Fishery Bulletin 83:427–442.

Geist, V. 1971. Mountain Sheep: A Study in Behaior and Evolution. University of Chicago Press, Chicago, IL.

Geist, V. 1974. On the relationship of social evolution and ecology in ungulates. American Zoologist 14:205–220.

Gendron, D., and J. R. Urbán. 1993. Evidence of feeding by humpback whales (*Megaptera novaeangliae*) in the Baja California breeding ground, Mexico. Marine Mammal Science 9:76–81.

Gentry, R. L. 1970. Social behavior of the Steller sea lion. Ph.D. dissertation, University of California, Santa Cruz, CA.

Gentry, R. L. 1973. Thermoregulatory behavior of eared seals. Behaviour 46:73–93.

Gentry, R. L., and J. R. Holt. 1986. Attendance behavior of northern fur seals. Pages 41–60 *in* R. L. Gentry and G. L. Kooyman (eds.). Fur Seals: Maternal Strategies on Land and at Sea. Princeton University Press, Princeton, NJ.

Gentry, R. L., and G. L. Kooyman. 1986. Fur Seals: Maternal Strategies on Land and at Sea. Princeton University Press, Princeton, NJ.

Gentry, R. L., D. P. Costa, J. P. Croxall, J. H. M. David, R. W. Davis, G. L. Kooyman, P. Majluf, T. S. McCann, and F. Trillmich. 1986a. Synthesis and conclusions. Pages 220–264 *in* R. L. Gentry and G. L. Kooyman (eds.). Fur Seals: Maternal Strategies on Land and at Sea. Princeton University Press, Princeton, NJ.

Gentry, R. L., G. L. Kooyman, and M. E. Goebel. 1986b. Feeding and diving behavior of northern fur seals. Pages 61–78 *in* R. L. Gentry and G. L. Kooyman (eds.). Fur Seals: Maternal Strategies on Land and at Sea. Princeton University Press, Princeton, NJ.

George, J. C., C. Clark, G. M. Carroll, and W. T. Ellison. 1989. Observations on the ice-breaking and ice navigation behavior of migrating bowhead whales (*Balaena mysticetus*) near Point Barrow, Alaska, spring 1985. Arctic 42:24–30.

Geraci, J. R., and V. J. Lounsbury. 1993. Marine mammals ashore: A field guide for strandings. Texas A & M University Sea Grant College Program, Galveston, TX.

Gerson, H. B., and J. P. Hickie. 1985. Head scarring on male narwhals *Monodon monoceros:* Evidence for aggressive tusk use. Canadian Journal of Zoology 63:2083–2087.

Ghiglieri, M. P. 1971. The socioecology of chimpanzees in Kibale Forest, Uganda. Ph.D. dissertation, University of California, Davis, CA. 303 pages.

Ghiglieri, M. P. 1984. The Chimpanzees of Kibale Forest. Columbia University Press, New York, NY.

Ginsberg, J. R., and U. W. Huck. 1989. Sperm competition in mammals. Trends in Ecology and Evolution 4:74–79.

Gisiner, R., and R. J. Schusterman. 1991. California sea lion pups play an active role in reunions with their mothers. Animal Behaviour 41:364–367.

Glockner, D. A., and S. C. Venus. 1983. Identification, growth rate, and behavior of humpback whale (*Megaptera novaeangliae*) cows and calves in the waters off Maui, Hawaii, 1977–79. Pages 223–258 *in* R. Payne (ed.). Communication and Behavior of Whales. Westview Press, Boulder, CO.

Godsell, J. 1988. Herd formation and haul-out behaviour in harbour seals (*Phoca vitulina*). Journal of Zoology (London) 215:83–98.

Goebel, M. E., J. L. Bengtson, R. L. DeLong, R. L. Gentry, and T. R. Loughlin. 1991. Diving patterns and foraging locations of female northern fur seals. Fishery Bulletin 89:171–179.

Goldsworthy, S. D. 1992. Maternal care in three species of southern fur seal (*Arctocephalus* spp.). Ph.D. dissertation, Monash University, Australia.

Goldsworthy, S. D. 1995. Differential expenditure of maternal resources in Antarctic fur seals, *Arctocephalis gazella,* at Heard Island, southern Indian Ocean. Behavioral Ecology 6:218–228.

Goley, P. D. 1996. Two aspects of group organization in a captive Pacific whitesided dolphin school (*Lagenorhynchus obliquidens*): Sleep and dawn–dusk chorusing. Ph.D. dissertation, University of California, Santa Cruz, CA. 97 pages.

Goodall, J. 1986. The Chimpanzees of Gombe: Patterns of Behavior. The Belknap Press of Harvard University Press, Cambridge, MA.

Gordon, J. 1987. Sperm whale groups and social behaviour observed off Sri Lanka. Report of the International Whaling Commission 37:205–217.

Gouzoules, S., and H. Gouzoules. 1987. Kinship. Pages 299–305 *in* B. B. Smuts, D. L. Cheney, R. M. Sefarth, R. W. Wrangham, and T. T. Struhsaker (eds.). Primate Societies. The University of Chicago Press, Chicago, IL.

Gruber, J. A. 1981. Ecology of the Atlantic bottlenosed dolphin (*Tursiops truncatus*) in the Pass Cavallo area of Matagorda Bay, Texas. M.Sc. thesis, Texas A & M University, College Station, TX.

Guinet, C. 1991. Intentional stranding apprenticeship and social play in killer whales (*Orcinus orca*). Canadian Journal of Zoology 69:2712–2716.

Hain, J. H. W., G. R. Carter, S. D. Kraus, C. A. Mayo, and H. E. Winn. 1982. Feeding behaviour of the humpback whale, *Megaptera novaeangliae,* in the western North Atlantic. Fishery Bulletin 80:259–268.

Haley, M. P. 1994. Resource-holding power asymmetries, the prior residence effect, and reproductive payoffs in male northern elephant seal fights. Behavioral Ecology and Sociobiology 34:427–434.

Haley, M. P., C. J. Deutsch, and B. J. Le Boeuf. 1994. Size, dominance and copulatory success in male northern elephant seals, *Mirounga angustirostris.* Animal Behaviour 48: 1249–1260.

Hall, K. R. L., and G. B. Schaller. 1964. Tool-using behavior of the California sea otter. Journal of Mammalogy 45:287–298.

Halliday, T. 1980. Sexual Strategy. University of Chicago Press, Chicago, IL.

Hamilton, W. D. 1964. The genetical theory of social behavior I. Journal of Theoretical Biology 7:116.

Hamilton, W. D. 1971. Geometry of the selfish herd. Journal of Theoretical Biology 31:295–311.

Hammond, P. S., B. J. McConnell, and M. A. Fedak. 1993. Gray seals off the east coast of Britain: Distribution and movements at sea. Symposia of the Zoological Society (London) 66:211–224.

Hanggi, E. B., and R. J. Schusterman. 1994. Underwater acoustic displays and individual variation in male harbour seals, *Phoca vitulina.* Animal Behaviour 48:1275–1283.

Hansen, L. J. 1990. California coastal bottlenose dolphins. Pages 403–420 *in* S. Leatherwood and R. R. Reeves (eds.). The Bottlenose Dolphin. Academic Press, San Diego, CA.

Hanson, M. B., L. J. Bledsoe, B. C. Kirkevold, C. J. Casson, and J. W. Nightingale. 1993. Behavioral budgets of captive sea otter mother–pup pairs during pup development. Zoo Biology 12:459–477.

Hanson, M. T., and R. H. Defran. 1993. The behaviour and feeding ecology of the Pacific coast bottlenose dolphin, *Tursiops truncatus.* Aquatic Mammals 19:127–142.

Harcourt, R. 1992. Factors affecting early mortality in the South American fur seal (*Arctocephalus australis*) in Peru: Density-related and predation. Journal of Zoology (London) 226:259–270.

Harestad, A. S., and H. D. Fisher. 1975. Social behavior in a non-pupping colony of Steller sea lion (*Eumetopias jubatus*). Canadian Journal of Zoology 53:1596–1613.

Hartman, D. S. 1979. Ecology and behavior of the manatee (*Trichechus manatus*) in Florida. Special Publication No. 5, American Society of Mammalogists.

Hatfield, B. B., D. Marks, M. T. Tinker, K. Nolan, and J. Peirce. 1998. Attacks on sea otters by killer whales. Marine Mammal Science 14:888–894.

Hausfater, G., and S. B. Hrdy (eds.). 1984. Infanticide: Comparative and evolutionary perspectives. Aldine Publishing, NY.

Heath, C. B. 1989. The behavioral ecology of the California sea lion. Ph.D. thesis. University of California, Santa Cruz, CA.

Heath, C. B., K. A. Ono, D. J. Boness and, J. M. Francis. 1991. The influence of El Nino on female attendance patterns in the California sea lion. Pages 138–145 *in* F. Trillmich and K. A. Ono (eds.). Pinnipeds and El Niño: Responses to Environmental Stress. Springer Verlag, Heidelberg.

Heezen, B. C. 1957. Whales entangled in deep sea cables. Deep Sea Research 4:105–115.

Heimlich-Boran, J. R. 1987. Habitat use patterns and behavioral ecology of killer whales (*Orcinus orca*) in the Pacific northwest. M.Sc. thesis, Moss Landing Marine Laboratories and San Jose State University, CA. 59 pages.

Heimlich-Boran, J. R. 1988. Behavioral ecology of killer whales (*Orcinus orca*) in the Pacific northwest. Canadian Journal of Zoology 66:565–578.

Heimlich-Boran, J. R. 1993. Social organisation of the short-finned pilot whale, *Globicephala macrorhynchus,* with special reference to the comparative social ecology of delphinids. Ph.D. dissertation, University of Cambridge, U.K. 134 pages.

Heimlich-Boran, S. L. 1986. Cohesive relationships among Puget Sound killer whales. Pages 251–284 *in* B. C. Kirkevold and J. S. Lockard (eds.). Behavioral Biology of Killer Whales. A. R. Liss, New York, NY.

Heinsohn, G. E., J. Wake, H. Marsh, and A. V. Spain. 1977. The dugong (*Dugong dugon* (Müller)) in the seagrass system. Aquaculture 12:235–248.

Helweg, D. A., G. B. Bauer, and L. M. Herman. 1992. Observations of an S-shaped posture in humpback whales (*Megaptera novaeangliae*). Aquatic Mammals 18:74–78.

Herzing, D. L. 1996. Vocalizations and associated underwater behavior of free-ranging Atlantic spotted dolphins, *Stenella frontalis* and bottlenose dolphins, *Tursiops truncatus.* Aquatic Mammals 22:61–79.

Hewer, H. R. 1957. A Hebridean breeding colony of gray seals, *Halichoerus grypus* (Fab.), with comparative notes on the gray seals of Ramsey Island, Pembrokeshire. Proceedings of the Zoological Society of London 128:23–66.

Heyning, J. E. 1984. Functional morphology involved in intraspecific fighting of the beaked whale, *Mesoplodon carlhubbsi.* Canadian Journal of Zoology 62:1645–1654.

Heyning, J. E. 1988. Presence of solid food in a young calf killer whale (*Orcinus orca*). Marine Mammal Science 4:68–71.

Heyning, J. E., and J. G. Mead. 1996. Suction feeding in beaked whales: Morphological and observational evidence. Natural History Museum of Los Angeles County Contributions in Science 464:1–12.

Higgins, L. V. 1990. Reproductive behavior and maternal investment of Australian sea lions. Ph.D. thesis, University of California, Santa Cruz, CA.

Higgins, L. V. 1993. The nonannual, nonseasonal breeding cycle of the Australian sea lion, *Neophoca cinerea.* Journal of Mammalogy 74:270–274.

Higgins, L. V., and R. A. Tedman. 1990. Effect of attacks by male Australian sea lions, *Neophoca cinerea,* on mortality of pups. Journal of Mammalogy 71:617–619.

Higgins, L. V., D. P. Costa, A. C. Huntley, and B. J. Le Boeuf. 1988. Behavioral and physiological measurements of maternal investment in the Steller sea lion, *Eumetopias jubatus.* Marine Mammal Science 4:44–58.

Hill, D. R. 1994. Theory of geolocation by light levels. Pages 227–236 *in* B. J. Le Boeuf and R. M. Laws (eds.). Elephant Seals. University of California Press, Berkeley, CA.

Hill, R. D. 1986. Development of microcomputer monitor and blood sampler for freediving Weddell seals. Journal of Applied Physiology 61:1570–1576.

Hinde, R. A. 1970. Animal Behavior, 2nd ed. McGraw-Hill, New York, NY.

Hindell, M. A., D. J. Slip, and H. R. Burton. 1991. The diving behaviour of adult male and female southern elephant seals, *Mirounga leonina* (Pinnipedia Phocidae). Australian Journal of Zoology 39:595–619.

Hindell, M. A., D. J. Slip, H. R. Burton, and M. M. Bryden. 1992. Physiological implications of continuous, prolonged, and deep dives of the southern elephant seal (*Mirounga leonina*). Canadian Journal of Zoology 70:370–379.

Hiruki, L. M., I. Stirling, W. G. Gilmartin, T. C. Johanos, and B. L. Becker. 1993. Significance of wounding to female reproductive success in Hawaiian monk seals (*Monachus schauinslandi*) at Laysan Island. Canadian Journal of Zoology 71:469–474.

Hixon, M. A., F. L. Carpenter, and D. C. Patton. 1983. Territory area, flower density, and time budgeting in hummingbirds: An experimental and theoretical analysis. American Naturalist 122:366–391.

Hoek, W. 1992. An unusual aggregation of harbor porpoises (*Phocoena phocoena*). Marine Mammal Science 8:152–155.

Hoelzel, A. R. 1993. Foraging behaviour and social group dynamics in Puget Sound killer whales. Animal Behaviour 45:581–591.

Hoelzel, A. R., E. M. Dorsey, and S. J. Stern. 1989. The foraging specializations of individual minke whales. Animal Behaviour 38:786–794.

Hoese, H. D. 1971. Dolphin feeding out of water in a salt marsh. Journal of Mammalogy 52:222–223.

Höglund, J., and R. V. Alatalo. 1995. Leks. Princeton University Press, Princeton, NJ.

Hogstad, O. 1978. Sexual dimorphism in relation to winter foraging and territorial behaviour of the three-toed woodpecker *Picoides tridactylus* and three *Dendrocopos* species. Ibis 120:198–203.

Hohn, A.A., M.D. Scott, R.S. Wells, J.C. Sweeney, and A.B. Irvine. 1989. Growth layers in teeth from known-age, free-ranging bottlenose dolphins. Marine Mammal Science 5:315–342.

Houk, J. A., and J. J. Geibel. 1974. Observation of underwater tool use by the sea otter, *Enhydra lutris* Linnaeus. California Fish and Game 60:207–208.

Ims, R. A. 1990. On the adaptive value of reproductive synchrony as a predator swamping strategy. American Naturalist 136:485–498.

Inman, A. J., and J. Krebs. 1987. Predation and group living. Trends in Ecology and Evolution 2:31–32.

Insley, S. J. 1992. Mother–offspring separation and acoustic stereotypy: A comparison of call morphology in two species of pinnipeds. Behaviour 120:103–122.

Irvine, A. B. 1983. Manatee metabolism and its influence on distribution in Florida. Biological Conservation 25:315–334.

Irvine, A. B., and M. D. Scott. 1984. Development and use of marking techniques to study manatees in Florida. Florida Scientist 47:12–26.

Irvine, A. B., and R. S. Wells. 1972. Results of attempts to tag Atlantic bottlenose dolphins (*Tursiops truncatus*). Cetology 13:1–5.

Irvine, A. B., R. S. Wells, and P. W. Gilbert. 1973. Conditioning an Atlantic bottlenose dolphin, *Tursiops truncatus*, to repel various species of sharks. Journal of Mammalogy 54:503–505.

Irvine, A. B., M. D. Scott, R. S. Wells, and J. G. Mead. 1979. Stranding of the pilot whale, *Globicephala macrorhynchus*, in Florida and South Carolina. Fishery Bulletin 77:511–513.

Irvine, A. B., M. D. Scott, R. S. Wells, and J. H. Kaufmann. 1981. Movements and activities of the Atlantic bottlenose dolphin, *Tursiops truncatus*, near Sarasota, Florida. Fishery Bulletin 79:671–688.

Irvine, A. B., J. E. Caffin, and H. I. Kochman. 1982a. Aerial surveys for manatees and dolphins in western peninsular Florida. Fishery Bulletin 80:621–630.

Irvine, A. B., R. S. Wells, and M. D. Scott. 1982b. An evaluation of techniques for tagging small odontocete cetaceans. Fishery Bulletin 80:135–143.

Irving, L. 1969. Temperature regulation in marine mammals. Pages 147–174 *in* H. T. Andersen (ed.). The Biology of Marine Mammals. Academic Press, London.

Ivashin, M. V. 1981. Some results of the marking of sperm whales (*Physeter macrocephalus*) in the Southern Hemisphere under the Soviet marking programme. Report of the International Whaling Commission 31:707–718.

Iverson, S. J. 1993. Milk secretion in marine mammals in relation to foraging: Can fatty acids predict diet? Pages 263–292 *in* I. L. Boyd (ed.). Marine Mammals: Advances in Behavioural and Population Biology. Symposia of the Zoological Society of London, No. 66. Clarendon Press, Oxford.

Iverson, S. J., K. J. Frost, and L. F. Lowry. 1997. Fatty acid signatures reveal fine scale structure of foraging distribution of harbor seals and their prey in Prince William Sound, Alaska. Marine Ecology Progress Series 151:255–271.

Jaquet, N., and H. Whitehead. 1996. Scale-dependent correlation of sperm whale distribution with environmental features and productivity in the South Pacific. Marine Ecology Progress Series 135:1–9.

Jameson, R. J. 1989. Movements, home range, and territories of male sea otters off central California. Marine Mammal Science 5:159–172.

Jameson, R. J., and A. M. Johnson. 1993. Reproductive characteristics of female sea otters. Marine Mammal Science 9:156–167.

Jarman, M. V. 1979. Impala social behaviour: Territory, hierarchy, mating, and the use of space. Advances in Ethology 21:1–93.

Jarman, P. J. 1974. The social organization of antelope in relation to their ecology. Behaviour 48:215–267.

Jefferson, T. A. 1987. A study of the behavior of Dall's porpoise (*Phocoenoides dalli*) in the Johnstone Strait, British Columbia. Canadian Journal of Zoology 65:736–744.

Jefferson, T. A. 1991. Observations on the distribution and behaviour of Dall's porpoise (*Phocoenoides dalli*) in Monterey Bay, California. Aquatic Mammals 17:12–19.

Job, D. A., D. J. Boness, and J. M. Francis. 1995. Individual variation in nursing vocalizations of Hawaiian monk seal pups, *Monachus schauinslandi* (Phocidae, Pinnipedia), and lack of maternal recognition. Canadian Journal of Zoology 73:975–983.

Johnson, C. M., and K. S. Norris. 1986. Delphinid social organization and social behavior. Pages 335–346 *in* R. J. Schusterman, J. A. Thomas, and F. G. Wood (eds.). Dolphin Cognition and Behavior: A Comparative Approach. Lawrence Erlbaum Assoc., Hillsdale, NJ.

Johnstone, R. M., and L. S. Davis. 1987. Activity in a non-breeding colony of New Zealand fur seals, *Arctocephalus forsteri* (Note). New Zealand Journal of Marine and Freshwater Research 21:153–155.

Jones, M. L., and S. L. Swartz. 1984. Demography and phenology of gray whales and evaluation of whale-watching activities in Laguna San Ignacio, Baja California Sur, Mexico. Pages 309–374 *in* M. L. Jones, S. L. Swartz, and S. Leatherwood (eds.). The Gray Whale *Eschrichtius robustus*. Academic Press, San Diego, CA.

Jurasz, C. M., and V. P. Jurasz. 1979. Feeding modes of the humpback whale, *Megaptera novaeangliae*, in southeast Alaska. Scientific Report of the Whales Research Institute 31:69–83.

Kaiya, Z., and L. Yuemin. 1989. Status and aspects of the ecology and behavior of the baiji, *Lipotes vexillifer*, in the lower Yangtze River. Pages 86–91 *in* W. F. Perrin, R. L. Brownell Jr., Z. Kaiya, and L. Jiankang (eds.). Biology and Conservation of the River Dolphins. Occasional Papers of the IUCN Species Survival Commission (SSC), Number 3, Gland, Switzerland.

Karczmarski, L. 1997. Ecological studies of humpback dolphins *Sousa chinensis* in the Algoa Bay region, Eastern Cape, South Africa. Ph.D. dissertation, University of Port Elizabeth, South Africa. 201 pages.

Kasuya, T. 1986. Distribution and behavior of Baird's beaked whales off the Pacific coast of Japan. Scientific Report of the Whales Research Institute Tokyo 37:61–83.

Kasuya, T., and A. K. M. Aminul Haque. 1972. Some informations on distribution and seasonal movement of the Ganges dolphin. Scientific Report of the Whales Research Institute 24:109–116.

Kasuya, T., and H. Marsh. 1984. Life history and reproductive biology of the short-finned pilot whale, *Globicephala macrorhynchus,* off the Pacific coast of Japan. Report of the International Whaling Commission (Special Issue 6):259–310.

Kasuya, T., and D. W. Rice. 1970. Notes on baleen plates and on arrangement of parasitic barnacles of gray whales. Scientific Report of the Whales Research Institute Tokyo 22:39–43.

Kato, H. 1984. Observation of tooth scars on the head of male sperm whale, as an indication of intra-sexual fightings. Scientific Report of the Whales Research Institute 35:39–46.

Katona, S., B. Baxter, O. Brazier, S. Kraus, J. Perkins, and H. Whitehead. 1979. Identification of humpback whales by fluke photographs. Pages 33–44 *in* H. E. Winn and B. L. Olla (eds.). Behavior of Marine Animals, Vol. 3: Cetaceans. Plenum, New York, NY.

Kaufman, G. W., D. B. Siniff, and R. Reichle. 1975. Colony behavior of Weddell seals, *Leptonychotes weddelli,* at Hutton Cliffs, Antarctica. Rapports et Procès-verbaux des Réunions. Conseil International pour L'Exploration de la Mer 169:228–246.

Kenagy, G. J., and S. C. Trombulak. 1986. Size and function of mammalian testes in relation to body size. Journal of Mammalogy 67:1–22.

Kenney, R. D. 1990. Bottlenose dolphins off the northeastern United States. Pages 369–386 *in* S. Leatherwood and R. R. Reeves (eds.). The Bottlenose Dolphin. Academic Press, San Diego, CA.

Kenward, R. 1987. Wildlife Radio-Tagging—Equipment, Field Techniques and Data Analysis. Academic Press, San Diego, CA.

Kenworthy, W. J., and D. E. Haunert (eds.). 1991. The light requirements of seagrasses: proceedings of a workshop to examine the capability of water quality criteria, standards and monitoring programs to protect seagrasses. NOAA Technical Memorandum NMFS-SEFC-287, Beaufort, NC.

Kenyon, K. W. 1969. The sea otter in the eastern Pacific Ocean. North American Fauna 68:1–352.

Kenyon, K. W. 1973. Hawaiian monk seal *Monachus schauinslandi*. Pages 88–97 *in* Seals. Proceedings of a working meeting of seal specialists on threatened and depleted seals of the world. IUCN Publications New Series, Morges, Switzerland.

Kerley, G. I. H. 1983. Comparison of seasonal haul-out patterns of fur seals *Arctocephalus tropicali* and *A. gazella* on subantarctic Marion Island. South African Journal of Wildlife Research 13:71–77.

Kinnaird, M. F. 1985. Aerial census of manatees in northeastern Florida. Biological Conservation 32:59–79.

Kirby, V. L., and S. H. Ridgway. 1984. Hormonal evidence of spontaneous ovulation in captive dolphins, *Tursiops truncatus* and *Delphinus delphis*. Pages 459–464 *in* W. F. Perrin, R. L. Brownell Jr., and D. P. DeMaster (eds.). Reproduction in Whales, Dolphins, and Porpoises. Reports of the International Whaling Commission (Special Issue 6), Cambridge, U.K.

Knowlton, A. R., J. Sigurjónsson, J. N. Ciano, and S. D. Kraus. 1992. Long-distance movements of North Atlantic right whales (*Eubalaena glacialis*). Marine Mammal Science 8:397–405.

Kochman, H. I., G. B. Rathbun, and J. A. Powell. 1985. Temporal and spatial distribution of manatees in Kings Bay, Crystal River, Florida. Journal of Wildlife Management 49:921–924.

Koepke, J. E., and K. H. Pribram. 1971. Effect of milk on the maintenance of suckling behavior in kittens from birth to 6 mo. Journal of Comparative Physiology and Psychology 75:363–377.

Kooyman, G. L. 1985. Physiology without restraint in diving mammals. Marine Mammal Science 1:166–178.

Kooyman, G. L. 1989. Diverse Divers. Springer-Verlag, Berlin.

Kooyman, G. L., and F. Trillmich. 1986. Diving behavior of Galapagos fur seals. Pages 186–195 *in* R. L. Gentry and G. L. Kooyman (eds.). Fur Seals: Maternal Strategies on Land and at Sea. Princeton University Press, Princeton, NJ.

Kooyman, G. L., R. L. Gentry, and D. L. Urquhart. 1976. Northern fur seal diving behavior: A new approach to its study. Science 193:411–412.

Kooyman, G. L., J. O. Billups, and W. D. Farwell. 1983. Two recently developed recorders for monitoring diving activity of marine birds and mammals. Pages 197–214 *in* A. G. MacDonald and I. G. Priede (eds.). Experimental Biology at Sea. Academic Press, New York, NY.

Kovacs, K. M. 1987. Maternal behaviour and early behavioural ontogeny of gray seals (*Halichoerus grypus*) on the Isle of May, U.K. Journal of Zoology (London) 213:697–715.

Kovacs, K. M. 1990. Mating strategies in male hooded seals (*Cystophora cristata*)? Canadian Journal of Zoology 68:2499–2502.

Kovacs, K. M., and D. M. Lavigne. 1986a. Maternal investment and neonatal growth in phocid seals. Journal of Animal Ecology 55:1035–1051.

Kovacs, K. M., and D. M. Lavigne. 1986b. Growth of gray seal (*Halichoerus grypus*) neonates: differential maternal investment in the sexes. Canadian Journal of Zoology 64:1937–1943.

Kovacs, K. M., and D. M. Lavigne. 1992. Maternal investment in otariid seals and walruses. Canadian Journal of Zoology 70:1953–1964.

Kraus, S. D., K. E. Moore, C. A. Price, M. J. Crone, W. A. Watkins, H. E. Winn, and J. H. Prescott. 1986. The use of photographs to identify individual North Atlantic right whales (*Eubalaena glacialis*). Pages 145–151 *in* R. L. Brownell Jr., P. B. Best, and J. H. Prescott (eds.). Reports of the International Whaling Commission (Special Issue 10), Cambridge, U.K.

Kretzmann, M. B., D. P. Costa, and B. J. Le Boeuf. 1993. Maternal energy investment in elephant seal pups: evidence for sexual equality? American Naturalist 141:466–480.

Kruse, S. L. 1989. Aspects of the biology, ecology, and behavior of Risso's dolphins (*Grampus griseus*) off the California coast. M.Sc. thesis, University of California, Santa Cruz, CA. 120 pages.

Kruuk, H. 1964. Predators and anti-predator behaviour of the black-headed gull (*Laurus ridibundus*). Behaviour 11(Suppl.):1–129.

Kruuk, H., and A. Moorehouse. 1991. The spatial organization of otters (*Lutra lutra*) in Shetland. Journal of Zoology 224:41–57.

Kurland, J. A. 1977. Kin Selection in the Japanese Monkey. S. Karger, Basel.

Lambertsen, R. H. 1983. Internal mechanism of rorqual feeding. Journal of Mammalogy 64:76–88.

Lawrence, B., and W. E. Schevill. 1954. *Tursiops* as an experimental subject. Journal of Mammalogy 35:225–232.

Laws, R. M. 1958. Growth rates and ages of crabeater seals, *Lobodon carcinophagus*. Proceedings of the Zoological Society of London 130:275–288.

Lanyon, J. 1991. The nutritional ecology of the dugong (*Dugong dugon*) in tropical North Queensland. Ph.D. dissertation, Department of Ecology and Evolutionary Biology, Monash University, Victoria, Australia. 337 pages.

Layne, J. N. 1958. Observations on freshwater dolphns in the upper Amazon. Journal of Mammalogy 39:1–22.

Leatherwood, S. 1975. Some observations of feeding behavior of bottlenosed dolphins (*Tursiops truncatus*) in the northern Gulf of Mexico and (*Tursiops* cf. *T. gilli*) off southern California, Baja California, and Nayarit, Mexico. Marine Fisheries Review 37:10–16.

Leatherwood, S., and R. R. Reeves. 1983. The Sierra Club Handbook of Whales and Dolphins. Sierra Club Books, San Francisco, CA.

Leatherwood, S., and W. A. Walker. 1979. The northern right whale dolphin *Lissodelphis borealis* Peale in the eastern North Pacific. Pages 85–141 *in* H. E. Winn and B. L. Olla (eds.). Behavior of Marine Animals, Vol. 3: Cetaceans. Plenum, New York, NY.

Leatherwood, S., W. F. Perrin, V. L. Kirby, C. L. Hubbs, and M. Dahlheim. 1980. Distribution and movements of Risso's dolphin, *Grampus griseus*, in the eastern North Pacific. Fishery Bulletin 77:951–963.

Leatherwood, S., K. C. Balcomb, C. O. Matkin, and G. Ellis. 1984. Killer whales (*Orcinus orca*) of southern Alaska: Results of field research 1984. Preliminary report, Hubbs-Sea World Research Institute, San Diego, CA. Technical Report No. 84-175.

Le Boeuf, B. J. 1972. Sexual behaviour in the northern elephant seal, *Mirounga angustirostris*. Behaviour 41:1–25.

Le Boeuf, B. J. 1974. Male–male competition and reproductive success in elephant seals. American Zoologist 14:163–176.

Le Boeuf, B. J. 1991. Pinniped mating system on land, ice, and in the water emphasis on the Phocidae. Pages 45–65 *in* D. Renouf (ed.). Behaviour of Pinnipeds. Chapman & Hall, London.

Le Boeuf, B. J. 1994. Variation in diving pattern of northern elephant seals with age, sex, and reproductive condition. Pages 237–252 *in* B. J. Le Boeuf and R. M. Laws (eds.). Elephant Seals. University of California Press, Berkeley, CA.

Le Boeuf, B. J., and K. T. Briggs. 1977. The cost of living in a seal harem. Mammalia 41:167–195.

Le Boeuf, B. J., and R. S. Condit. 1983. The high cost of living on the beach. Pacific Discovery 36:12–14.

Le Boeuf, B. J., and R. S. Peterson. 1969. Social status and mating activity in elephant seals. Science 163:91–93.

Le Boeuf, B. J., and J. Reiter. 1988. Lifetime reproductive success in northern elephant seals. Pages 344–362 *in* T. H. Clutton-Brock (ed.). Reproductive Success. The University of Chicago Press, Chicago and London.

Le Boeuf, B. J., D. P. Costa, A. C. Huntley, G. L. Kooyman, and R. W. Davis. 1986. Pattern and depth of dives in northern elephant seals, *Mirounga angustirostris*. Journal of Zoology 208:1–7.

Le Boeuf, B. J., Y. Naito, A. C. Huntley, and T. Asaga. 1989. Prolonged, continuous, deep diving by northern elephant seals. Canadian Journal of Zoology 67:2514–2519.

Le Boeuf, B. J., D. E. Crocker, S. B. Blackwell, P. A. Morris, and P. H. Thorson. 1993. Sex differences in diving and foraging behaviour of northern elephant seals. Symposia of the Zoological Society (London) 66:149–178.

Ledder, D. A. 1986. Food habits of the West Indian manatee, *Trichechus manatus latirostris*, in South Florida. M.S. thesis, University of Miami, Miami, FL. 114 pages.

Lefebvre, L. W., and J. A. Powell. 1990. Manatee Grazing Impacts on Seagrasses in Hobe Sound and Jupiter Sound in Southeast Florida During the Winter of 1988–89. NTIS Report PB90-271883. National Technical Information Service, Springfield, VA.

Lefebvre, L. W., T. J. O'Shea, G. B. Rathbun, and R. C. Best. 1989. Distribution, status, and biogeography of the West Indian manatee. Pages 567–610 *in* C.A. Woods (ed.). Biogeography of the West Indies—Past, Present, and Future. Sandhill Crane Press, Gainesville, FL.

Leuthold, W. 1977. African Ungulates—A Comparative Review of their Ethology and Behaviour. Springer-Verlag, New York, NY.

Ling, J. K. 1970. Pelage and molting in wild mammals with special reference to aquatic forms. Quarterly Review of Biology 45:16–54.

Long, D. J. 1991. Apparent predation by a white shark *Carcharodon carcharias* on a pygmy sperm whale *Kogia breviceps*. Fishery Bulletin 89:538–540.

Lopez, J. C., and D. Lopez. 1985. Killer whales of Patagonia and their behavior of intentional stranding while hunting near shore. Journal of Mammalogy 66:181–183.

Lott, D. P. 1991. Intraspecific Variation in the Social Systems of Wild Vertebrates. Cambridge University Press, Cambridge, U.K.

Loughlin, T. R. 1977. Activity patterns, habitat partitioning, and grooming behavior of the sea otter, *Enhydra lutris*, in California. Ph.D. thesis, University of California, Los Angeles, CA. 110 pages.

Loughlin, T. R. 1979. Radio telemetric determination of the 24-hour feeding activities of sea otters, *Enhydra lutris*. Pages 717–724 *in* C. J. Amlaner and D. W. MacDonald (eds.). A Handbook on Biotelemetry and Radiotracking. Pergamon Press, Oxford and New York.

Loughlin, T. R., J. L. Bengtson, and R. L. Merrick. 1987. Characteristics of feeding trips of female northern fur seals. Canadian Journal of Zoology 65:2079–2084.

Lowry, L. F., and F. H. Fay. 1984. Seal eating by walruses in the Bering and Chuckchi Seas. Polar Biology 3:11–18.

Lunn, N. J. 1992. Fostering behaviour and milk stealing in Antarctic fur seals. Canadian Journal of Zoology 70:837–839.

Lunn, N. J., and J. P. Y. Arnould. 1997. Maternal investment in Antarctic fur seals: Evidence for equality of the sexes? Behavioral Ecology and Sociobiology 40:351–362.

Lunn, N. J., and I. L. Boyd. 1991. Pupping-site fidelity of Antarctic fur seals at Bird Island, South Georgia. Journal of Mammalogy 72:202–206.

Lunn, N. J., I. L. Boyd, T. Barton, and J. P. Croxall. 1993. Factors affecting the growth rate and mass at weaning of Antarctic fur seals at Bird Island, South Georgia. Journal of Mammalogy 74:908–919.

Lydersen C., and K. M. Kovacs. 1993. Diving behaviour of lactating harp seal, *Phoca groenlandica*, females from the Gulf of St. Lawrence, Canada. Animal Behaviour 46:1213–1221.

Lydersen, C., M. S. Ryg, M. O. Hammil, and P. J. O'Brien. 1992. Oxygen stores and aerobic dive limit of ringed seals (*Phoca hispida*). Canadian Journal of Zoology 70:458–461.

Lynas, E. M., and J. P. Sylvestre. 1988. Feeding techniques and foraging strategies of minke whales (*Balaenoptera acutorostrata*) in the St. Lawrence River estuary. Aquatic Mammals 14:21–32.

Lynn, S. K. 1995. Movements, site fidelity, and surfacing patterns of bottlenose dolphins on the central Texas coast. M.Sc. thesis, Texas A & M University, College Station, TX. 92 pages.

Mackay-Sim, A., D. Duvall, and B. M. Graves. 1985. The West Indian manatee (*Trichechus manatus*) lacks a vomeronasal organ. Brain, Behavior and Evolution 27:186–194.

Majluf, P. 1992. Timing of births and juvenile mortality in the South American fur seal in Peru. Journal of Zoology (London) 227:367–383.

Majluf, P. 1993. Lekking in the South American Fur Seal in Peru. XXII International Ethological Conference, Torremolinos, Spain, 1993.

Major, P. F. 1986. Notes on a predator–prey interaction between common dolphins (*Delphinus delphis*) and short-finned squid (*Illex illecebrosus*) in Lydonia Submarine Canyon, western North Atlantic Ocean. Journal of Mammalogy 67:769–770.

Marcoy, P. 1875. Travels in South America, Vol. II. Scribner, Armstrong, and Company, New York, NY.

Marlow, B. J. 1972. Pup abduction in the Australian sea-lion, *Neophoca cinerea*. Mammalia 36:161–165.

Marmontel, M. 1995. Age and reproduction estimates in female Florida manatees. Pages 98–119 *in* T. J. O'Shea, B. B. Ackerman, and H. F. Percival (eds.). Population Biology of the Florida Manatee. National Biological Service, Information and Technology Report 1, Washington, D.C.

Marsh, H. 1980. Age determination of the dugong (*Dugong dugon* (Müller)) in northern Australia and its biological implications. Reports of the International Whaling Commission (Special Issue 3):81–201.

Marsh, H. 1995. Life history and patterns of breeding and population dynamics of the dugong. Pages 75–83 *in* T. J. O'Shea, B. B. Ackerman, and H. F. Percival (eds.). Population Biology of the Florida Manatee. National Biological Service, Information and Technology Report 1, Washington, D.C.

Marsh, H., and T. Kasuya. 1984. Ovarian changes in the short-finned pilot whale, *Globicephala macrorhynchus*, off the Pacific coast of Japan. Reports of the International Whaling Commission (Special Issue 6):311–335.

Marsh, H., and T. Kasuya. 1991. An overview of the changes in the role of a female pilot whale with age. Pages 281–285 *in* K. Pryor and K. S. Norris (eds.). Dolphin Societies: Discoveries and Puzzles. University of California Press, Berkeley.

Marsh, H., and G. B. Rathbun. 1990. Development and application of conventional and satellite radio-tracking techniques for studying dugong movements and habitat usage. Australian Wildlife Research 17:83–100.

Marsh, H., and W. K. Saalfeld. 1989. Distribution and abundance of dugongs in the northern Great Barrier Reef Marine Park. Australian Wildlife Research 16:429–440.

Marsh, H., and W. K. Saalfeld. 1990. The distribution and abundance of dugongs in the Great Barrier Reef Marine Park south of Cape Bedford. Australian Wildlife Research 7:511–524.

Marsh, H., B. R. Gardner, and G. E. Heinsohn. 1980. Present-day hunting and distribution of dugongs in the Wellesley Islands (Queensland): Implications for conservation. Biological Conservation 19:255–267.

Marsh, H., P. W. Chanells, G. E. Heinsohn, and J. Morrissey. 1982. Analysis of stomach contents of dugongs from Queensland. Australian Wildlife Research 9:55–67.

Marsh, H., R. I. T. Prince, W. K. Saalfeld, and R. Shepherd. 1994. The distribution and abundance of the dugong in Shark Bay, Western Australia. Australian Wildlife Research 21:149–161.

Marshall, C. D. 1997. The Sirenian feeding apparatus: Functional morphology of feeding involving perioral bristles and associated structures. Ph.D. dissertation, University of Florida, Gainesville, FL.

Mårtensson, P.-E., A. R. Lager Gotaas, E. S. Nordøy, and A. S. Blix. 1996. Seasonal changes in energy density of prey of Northeast Altantic seals and whales. Marine Mammal Science 12:635–640.

Martin, A. R. 1986. Feeding association between dolphins and shearwaters around the Azore Islands. Canadian Journal of Zoology 64:1372–1374.

Martin, A. R., T. G. Smith, and O. P. Cox. 1993. Studying the behaviour and movements of high Arctic belugas with satellite telemetry. Symposia of the Zoological Society (London) 66:195–210.

Martinez, D. R., and E. Klinghammer. 1970. The behavior of the whale *Orcinus orca:* A review of the literature. Zeitschrift für Tierpsychologie 27:828–839.

Mate, B. R. 1973. Population kinetics and related ecology of the northern sea lion, *Eumetopias jubatus*. Ph.D. dissertation, University of Oregon.

Mate, B. R. 1990. Movements and dive patterns of a right whale monitored by satellite (Extended abstract). Fifth Conference on the Biology of the Bowhead Whale *Balaena mysticetus,* April 1–3, 1990, Anchorage, Alaska, pages 50–53. (Available from North Slope Borough, Department of Wildlife Management, P.O. Box 69, Barrow, AK 99723).

Mate, B., G. B. Rathbun, and J. P. Reid. 1986. An ARGOS-monitored radio-tag for manatees. Argos Newsletter 26:3–7.

Mate, B. R., K. A. Rossbach, S. L. Neukirk, R. S. Wells, A. B. Irvine, M. D. Scott, and A. J. Read. 1995. Satellite-monitored movements and dive behavior of a bottlenose dolphin (*Tursiops truncatus*) in Tampa Bay, Florida. Marine Mammal Science 11(4):452–463.

Matsura, D. T., and G. C. Whittow. 1974. Evaporative heat loss in the California sea lion and harbor seal. Comparative Biochemistry and Physiology A 48:9–20.

Maynard Smith, J. 1980. A new theory of sexual investment. Behavioral Ecology and Sociobiology 7:247–251.

McBride, A. F. 1940. Meet Mr. Porpoise. Natural History Magazine 45:16–29.

McBride, A. F., and D. O. Hebb. 1948. Behavior of the captive bottlenose dolphin, *Tursiops truncatus*. Journal of Comparative Physiology and Psychology 41:111–123.

McBride, A. F., and H. Kritzler. 1951. Observations on pregnancy, parturition, and postnatal behavior in the bottlenose dolphin. Journal of Mammalogy 32:251–266.

McCann, C. 1974. Body scarring on Cetacea-Odontocetes. Scientific Report of the Whales Research Institute Tokyo 26:145–155.

McCann, T. S. 1980. Territoriality and breeding behaviour of adult male Antarctic fur seal, *Arctocephalus gazella*. Journal of Zoology (London) 192:295–310.

McCann, T. S. 1981. Aggression and sexual activity of male Southern elephant seals, *Mirounga leonina*. Journal of Zoology (London) 195:295–310.

McCann, T. S. 1982. Aggressive and maternal activities of female southern elephant seals (*Mirounga leonina*). Animal Behavior 30:268–276.

McCann, T. S. 1983. Activity budgets of southern elephant seals, *Mirounga leonina,* during the breading season. Z Tierpsychol 61:111–126.

McConnaughey, B. H. 1974. Introduction to Marine Biology, C. V. Mosby, St. Louis, MO.

McConnell, B. J., C. Chambers, and M. A. Fedak. 1992a. Foraging ecology of southern elephant seals in relation to the bathymetry and productivity of the Southern Ocean. Antarctic Science 4:393–398.

McConnell, B. J., C. Chambers, K. S. Nicholas, and M. A. Fedak. 1992b. Satellite tracking of gray seals *Halichoerus grypus*. Journal of Zoology (London) 226:271–282.

McCracken, G. F. 1984. Communal nursing in Mexican free-tailed bat maternity colonies. Science 223:1090–1091.

McNaughton, S. J. 1984. Grazing lawns: Animals in herds, plant form, and coevolution. American Naturalist 124:863–886.

McRae, S. B., and K. M. Kovacs. 1994. Paternity exclusion by DNA fingerprinting, and mate guarding in the hooded seal *Cystophora cristata*. Molecular Ecology 3:101–107.

McShane, L. S., J. A. Estes, M. L. Riedman, and M. M. Staedler. 1995. Repertoire, structure and individual variation of vocalizations in the sea otter. Journal of Mammalagy 76:414–427.

Mead, J. G., D. K. Odell, R. S. Wells, and M. D. Scott. 1980. Observations on a mass stranding of spinner dolphins, *Stenella longirostris*, from the west coast of Florida. Fishery Bulletin 78:353–360.

Mech, L. D. 1970. The Wolf: The Ecology and Behavior of an Endangered Species. Natural History Press, Garden City, NY.

Miller, E. H. 1974. A paternal role in the Galapagos sea lions? Evolution 28:473–506.

Miller, E. H. 1975a. Annual cycle of fur seals, *Arctocephalus forsteri* (Lesson), on the Open Bay Islands, New Zealand. Pacific Science 29:139–152.

Miller, E. H. 1975b. Social and evolutionary implications of territoriality in adult male New Zealand fur seals, *Arctocephalus forsteri* (Lesson, 1828), during the breeding season. Rapports et Procès-verbaux des Réunions. Conseil International pour L'Exploration de la Mer 169:170–187.

Miller, E. H. 1975c. Walrus ethology. II. Herd structure and activity budgets of summering males. Canadian Journal of Zoology 54:704–715.

Miller, E. H., and D. J. Boness. 1979. Remarks on display functions of the snout of the gray seal, *Halichoerus grypus* (Fab.) with comparative notes. Canadian Journal of Zoology 57:140–148.

Miller, E. H., and D. J. Boness. 1983. Summer behaviour of the Atlantic walrus (*Odobenus rosmarus rosmarus*) on Coats Island, N. W. T. Zeitschrift für Säugetierkunde 48:298–313.

Miyazaki, N., and M. Nishiwaki. 1978. School structure of the striped dolphin off the Pacific coast of Japan. Scientific Report of the Whales Research Institute 30:65–115.

Monnett, C., and L. Rotterman. 1988. Movement patterns of adult female and weanling sea otters in Prince William Sound, Alaska. Pages 133–161 *in* D. B. Siniff and K. Ralls (eds.). Population Status of California Sea Otters. Final report to the Minerals Management Service, U.S. Department of the Interior 14-12-001-3003, Washington, D.C.

Monson, D. H., and A. R. DeGange. 1995. Reproduction, preweaning survival, and survival of adult sea otters at Kodiak Island, Alaska. Canadian Journal of Zoology 73:1161–1169.

Montgomery, G. G., R. C. Best, and M. Yamakoshi. 1981. A radio-tracking study of the Amazonian manatee *Trichechus inunguis* (Mammalia: Sirenia). Biotropica 13:81–85.

Moore, J. C. 1955. Bottlenosed dolphins support remains of young. Journal of Mammalogy 36:466–467.

Moore, J. C. 1956. Observations of manatees in aggregations. American Museum Novitates 1811:1–24.

Moore, S. E., and R. R. Reeves. 1993. Distribution and movement. Pages 313–386 *in* J. J. Burns, J. J. Montague, and C. J. Cowles (eds.). The Bowhead Whale. Special Publication No. 2. Society for Marine Mammalogy, Lawrence, KS.

Moors, T. L. 1997. Mating behavior of free-ranging female bottlenose dolphins, *Tursiops truncatus*. M.Sc. thesis, University of California, Santa Cruz, CA.

Morejohn, V. G. 1979. The natural history of Dall's porpoise in the North Pacific Ocean. Pages 45–83 *in* H. E. Winn and B. L. Olla (eds.). Behavior of Marine Animals, Vol. 3: Cetaceans. Plenum, New York, NY.

Morozov, D. A. 1970. The dolphins are hunting. Fisheries Management 6:16–17.

Murison, L. D., D. J. Murie, K. R. Morin, and J. da Silva Curiel. 1984. Foraging of the gray whale along the west coast of Vancouver Island. Pages 451–463 *in* M. L. Jones, S. L. Swartz, and S. Leatherwood (eds.). The Gray Whale *Eschrichtius robustus*. Academic Press, San Diego, CA.

Nerini, M. 1984. A review of gray whale feeding ecology. Pages 423–450 *in* M. L. Jones, S. L. Swartz, and S. Leatherwood (eds.). The Gray Whale *Eschrichtius robustus*. Academic Press, San Diego, CA.

Newby, T. C. 1975. A sea otter (*Enhydra lutris*) food dive record. The Murrelet 56:19.

Nielsen, L., and R. D. Brown (eds.). 1988. Translocation of Wild Animals. The Wisconsin Humane Society, Milwaukee, WI.

Nishiwaki, M. 1962. Aerial photographs showing sperm whales' interesting habits. Norsk Hvalfangst-Tidende 51:395–398.

Norris, K. S. 1967. Some observations on the migration and orientation of marine mammals. Pages 101–125 *in* R. M. Storm (ed.). Animal Orientation and Navigation. Oregon State University Press, Corvallis, OR.

Norris, K. S., and T. P. Dohl. 1980a. The behavior of the Hawaiian spinner porpoise, *Stenella longirostris*. Fishery Bulletin 77:821–849.

Norris, K. S., and T. P. Dohl. 1980b. The structure and function of cetacean schools. Pages 211–261 *in* L. M. Herman (ed.). Cetacean Behavior: Mechanisms and Functions. Wiley & Sons, New York, NY.

Norris, K. S., and B. Møhl. 1983. Can odontocetes debilitate prey with sound? American Naturalist 122:85–104.

Norris, K. S., and J. H. Prescott. 1961. Observations on Pacific cetaceans of Californian and Mexican waters. University of California Publications in Zoology 63:291–402.

Norris, K. S., R. M. Goodman, B. Villa-Ramirez, and L. Hobbs. 1977. Behavior of California gray whale, *Eschrichtius robustus*, in southern Baja California, Mexico. Fishery Bulletin 75:159–172.

Norris, K. B., B. Villa-Ramirez, G. Nichols, B. Würsig, and K. Miller. 1983. Lagoon entrance and other aggregations of gray whales (*Eschristius robustus*). Pages 259–293 *in* R. Payne (ed.). Communication and Behavior of Whales. Westview Press, Boulder, CO.

Norris, K. S., B. Würsig, R. S. Wells, and M. Würsig. 1994. The Hawaiian Spinner Dolphin. University of California Press, Berkeley, CA.

Nutting, C. C. 1891. Some of the causes and results of polygamy among the Pinnipedia. American Naturalist 25:103–112.

Odell, D. K. 1974. Behavioral thermoregulation in the California sea lion. Behavioral Biology 10:231–237.

Odell, D. K., and E. D. Asper. 1990. Distribution and movements of freeze-branded bottlenose dolphins in the Indian and Banana Rivers, Florida. Pages 515–540 *in* S. Leatherwood and R. R. Reeves (eds.). The Bottlenose Dolphin. Academic Press, San Diego, CA.

Odell, D. K., E. D. Asper, J. Baucom, and L. H. Cornell. 1980. A recurrent mass stranding of the false killer whale, *Pseudorca crassidens*, in Florida. Fishery Bulletin 78:171–177.

Oftedal, O. T., and S. J. Iverson. 1987. Hydrogen isotope methodology for measurement of milk intake and energetics of growth in suckling young. Pages 67–96 *in* A. C. Huntley, D. P. Costa, G. A. J. Worthy, and M. A. Castellini (eds.). Marine Mammal Energetics. Society for Marine Mammalogy, Lawrence, KS.

Oftedal, O. T., D. J. Boness, and R. A. Tedman. 1987a. The behavior, physiology, and anatomy of lactation in the Pinnipedia. Current Mammalogy 1:175–245.

Oftedal, O. T., S. J. Iverson, and D. J. Boness. 1987b. Milk and energy intakes of suckling California sea lion pups (*Zalophus californianus*) in relation to sex, growth and predicted maintenance requirements. Physiological Zoology 60:560–575.

Oftedal, O. T., D. J. Boness, and D. J. Bowen. 1988. The composition of hooded seal (*Cystophora cristata*) milk: An adaptation for postnatal fattening. Canadian Journal of Zoology 66:318–322.

Ohsumi, S. 1966. Sexual segregation of the sperm whale in the North Pacific. Scientific Reports of the Whales Research Institute (Tokyo) 20:1–16.

Ohsumi, S. 1971. Some investigations on the school structure of sperm whale. Scientific Reports of the Whales Research Institute (Tokyo) 23:1–25.

Oliver, J. S., P. N. Slattery, E. F. O'Connor, and L. F. Lowry. 1983. Walrus feeding in the Bering Sea: A benthic perspective. Fishery Bulletin 81:501–512.

Ono, K. A., and D. J. Boness. 1996. Sexual dimorphism in sea lion pups: Differential maternal investment, or sex-specific differences in energy allocation? Behavioral Ecology and Sociobiology 38:31–41.

Ono, K. A., D. J. Boness, and O. T. Oftedal. 1987. The effect of a natural environmental disturbance on maternal investment and pup behavior in the California sea lion. Behavioral Ecology and Sociobiology 21:109–118.

Orians, G. H. 1969. On the evolution of mating systems in birds and mammals. American Naturalist 103:589–603.

Orians, G. H., and N. E. Pearson. 1979. On the theory of central place foraging. Pages 153–177 *in* D. J. Horn, R. D. Mitchel, and T. F. Stairs (eds.). Analysis of Ecological Systems. Ohio State University Press, Columbus, OH.

O'Shea, T. J. 1986. Mast foraging by West Indian manatees (*Trichechus manatus*). Journal of Mammalogy 67:183–185.

O'Shea, T. J., and W. C. Hartley. 1995. Reproduction and early-age survival of manatees at Blue Spring, upper St. Johns River, Florida. Pages 157–170 *in* T. J. O'Shea, B. B. Ackerman, and H. F. Percival (eds.). Population Biology of the Florida Manatee. National Biological Service, Information and Technology Report 1, Washington, D.C.

O'Shea, T. J., G. B. Rathbun, R. K. Bonde, C. D. Buergelt, and D. K. Odell. 1991. An epizootic of Florida manatees associated with a dinoflagellate bloom. Marine Mammal Science 7:165–179.

Ostfeld, R. S., L. Ebensperger, L. L. Klosterman, and J. C. Castilla. 1989. Foraging, activity budget, and social behavior of the South American marine otter *Lutra felina* (Molina 1782). National Geographic Research 5:422–438.

Ostman, J. 1991. Changes in aggressive and sexual behavior between two male bottlenose dolphins (*Tursiops truncatus*) in a captive colony. Pages 305–317 *in* K. Pryor and K. S. Norris (eds.). Dolphin Societies: Discoveries and Puzzles. University of California Press, Berkeley, CA.

Ostman, J. S. O. 1994. Social organization and social behavior of Hawaiian spinner dolphins (*Stenella longirostris*) Ph.D. dissertation, University of California, Santa Cruz, CA, 114 pages.

Packard, J. M., and C. A. Ribic. 1982. Classification of the behavior of sea otters (*Enhydra lutris*). Canadian Journal of Zoology 60:1362–1373.

Packard, J. M. 1984. Impact of manatees *Trichechus manatus* on seagrass communities in eastern Florida. Acta Zoologica Fennica 172:21–22.

Packard, J. M., R. K. Frohlich, J. E. Reynolds III, and J. R. Wilcox. 1989. Manatee response to interruption of a thermal effluent. Journal of Wildlife Management 53:692–700.

Packer, C., and L. Ruttan. 1988. The evolution of cooperative hunting. American Naturalist 132:159–198.

Packer, C., L. Herbst, A. E. Pusey, J. D. Bygott, J. P. Hanby, S. J. Cairns, and M Borgerhoff Mulder. 1988. Reproductive success of lions. Pages 363–383 *in* T. H. Clutton-Brock (ed.). Reproductive Success. The University of Chicago Press, Chicago and London.

Packer, C., D. Scheel, and A. E. Pusey. 1990. Why lions form groups: Food is not enough. The American Naturalist 136:1–19.

Packer, C., D. A. Gilbert, A. E. Pusey, and S. J. O'Brien. 1991. A molecular genetic analysis of kinship and cooperation in African lions. Nature (London) 351:562–565.

Packer, C., S. Lewis, and A. Pusey. 1992. A comparative analysis of non-offspring nursing. Animal Behaviour 43:265–282.

Palacios, D. M., and D. Day. 1995. A Risso's dolphin (*Grampus griseus*) carrying a dead calf. Marine Mammal Science 11:593–594.

Palacios, D. M., and B. R. Mate. 1996. Attack by false killer whales (*Pseudorca crassidens*) on sperm whales (*Physeter macrocephalus*) in the Galápagos Islands. Marine Mammal Science 12:582–587.

Payne, R. 1995. Among Whales. Scribner, New York, NY.

Payne, R., and E. M. Dorsey. 1983. Sexual dimorphism and aggresive use of callosities in right whales (*Eubalaena australis*). Pages 295–329 *in* R. Payne (ed.). Communication and Behavior of Whales. Westview Press, Boulder, CO.

Payne, R., O. Brazier, E. M. Dorsey, J. S. Perkins, V. J. Rowntree, and A. Titus. 1983. External features in southern right whales (*Eubalaena australis*) and their use in identifying individuals. Pages 371–445 *in* R. Payne (ed.). Communication and Behavior of Whales. Westview Press, Boulder, CO.

Payne, S. F., and R. J. Jameson. 1984. Early behavioral development of the sea otter, *Enhydra lutris*. Journal of Mammalogy 65:527–531.

Peddemors, V. M., and G. Thompson. 1994. Beaching behaviour during shallow water feeding by humpback dolphins *Sousa plumbea*. Aquatic Mammals 20:65–67.

Pemberton, J. M., S. D. Albon, F. E. Guinness, T. H. Clutton-Brock, and G. A. Dover. 1992. Behavioral estimates of male mating success tested by DNA fingerprinting in a polygynous mammal. Behavioral Ecology 3:66–75.

Perrin, W. F., and S. B. Reilly. 1984. Reproductive parameters of dolphins and small whales of the family Delphinidae. Pages 97–133 *in* W. F. Perrin, R. L. Brownell Jr., and D. P. DeMaster (eds.). Reproduction in Whales, Dolphins, and Porpoises. Reports of the International Whaling Commission (Special Issue 6), Cambridge, U.K.

Perrin, W. F., J. M. Coe, and J. R. Zweifel. 1976. Growth and reproduction of the spotted porpoise. *Stenella attenuata,* in the offshore eastern tropical Pacific. Fishery Bulletin 74:229–269.

Perrin, W. F., W. E. Evans, and D. B. Holt. 1979. Movements of pelagic dolphins (*Stenella* spp.) in the eastern tropical Pacific as indicated by the results of tagging operations, 1969–1976. NOAA Technical Report NMFS SSRF-737, La Jolla, CA.

Perrin, W. F., R. L. Brownell Jr., and D. P. DeMaster (eds.). 1984. Reproduction in whales, dolphins, and porpoises. Reports of the International Whaling Commission (Special Issue 6), Cambridge, U.K.

Perry, E. A. 1993. Aquatic territory defence by male harbour seals (*Phoca vitulina*) at Miquelon: Relationship between active defence and male reproductive success. Ph.D. dissertation, Memorial University of Newfoundland, St. John's, Newfoundland.

Perry, E. A., and D. Renouf. 1988. Further studies of the role of harbour seal (*Phoca vitulina*) pup vocalizations in preventing separation of mother–pup pairs. Canadian Journal of Zoology 66:934–938.

Perry, E. A., and G. B. Stenson. 1992. Observations on nursing behaviour of hooded seals, *Cystophora cristata*. Behaviour 122:1–10.

Perry, E. A., D. J. Boness, and R. C. Fleischer. 1998. DNA fingerprinting evidence of nonfilial nursing in grey seals. Molecular Ecology 7:81–85.

Perryman, W. L., and T. C. Foster. 1980. Preliminary Report of Predation by Small Whales, Mainly the False Killer Whale, *Pseudorca crassidens,* on Dolphins (*Stenella* spp. and *Delphinus delphis*) in the Eastern Tropical Pacific. NOAA National Marine Fisheries Service, Southwest Fisheries Science Center Administrative Report LJ-80-05. 9 pages. (Available from the National Marine Fisheries Service, Southwest Fisheries Science Center, P.O. Box 271, La Jolla, CA 92038).

Peterson, R. S. 1968. Social behavior in pinnipeds, with particular reference to the northern fur seal. Pages 3–53 *in* R. J. Harrison, R. C. Hubbard, R. S. Peterson, D. E. Rice, and R. J. Schusterman (eds.). Behavior and Physiology of Pinnipeds. Appleton-Century Crofts, New York, NY.

Peterson, R. S., and G. A. Bartholomew. 1967. The natural history of the California seal lion. American Society of Mammalogy (Special Publications) 1:79.

Petricig, R. O. 1995. Bottlenose dolphins (*Tursiops truncatus*) in Bull Creek, South Carolina. Ph.D. dissertation, University of Rhode Island, Kingston, RI, 298 pages.

Petrie M., M. Hall, T. Halliday, H. Budgey, and C. Pierpont. 1992. Multiple mating in a lekking bird: Why do peahens mate with more than one male and with the same male more than once? Behavioral Ecology and Sociobiology 31:349–358.

Petrinovitch, L. 1974. Individual recognition of pup vocalization by northern elephant seal mothers. Zeitschrift für Tierpsychologie 34:308–312.

Pierotti, R., and D. Pierotti. 1980. Effects of cold climate on the evolution of pinniped breeding systems. Evolution 34:494–507.

Pilleri, G. 1969. On the behaviour of the Amazon dolphin, *Inia geoffrensis* in Beni (Bolivia). Revue Suisse de Zoologie 76:58–91.

Pinedo, M. C., R. Praderi, and R. L. Brownell Jr. 1989. Review of the biology and status of the franciscana, *Pontoporia blainvillei*. Pages 46–51 *in* W.F. Perrin, R.L. Brownell Jr., Z. Kaiya, and L. Jiankang (eds.). Biology and Conservation of the River Dolphins. Occasional Papers of the IUCN Species Survival Commission (SSC), Number 3, Gland, Switzerland.

Plotz, J., W. Ekau, and P. H. Reijnders. 1991. Diet of Weddell seals *Leptonychotes weddellii* at Vestkapp, eastern Weddell Sea (Antarctica), in relation to local food supply. Marine Mammal Science 7:136–144.

Pomeroy, P. P., S. S. Anderson, S. D. Twiss, and B. J. McConnell. 1994. Dispersion and site fidelity of breeding female gray seals (*Halichoerus grypus*) on North Rona, Scotland. Journal of Zoology (London) 233:429–447.

Ponganis, P. J., G. L. Kooyman, and M. A. Castellini. 1993. Determinants of the aerobic dive limit of Weddell seals: analysis of diving metabolic rates, postdive end tidal PO2's and blood and muscle oxygen stores. Physiological Zoology 66:732–749.

Poole, J. H. 1989. Mate guarding, reproductive success and female choice in African elephants. Animal Behavior 37:842–849.

Poole, M. M. 1984. Migration corridors of gray whales along the central California coast, 1980–1982. Pages 389–407 *in* M. L. Jones, S. L. Swartz, and S. Leatherwood (eds.). The Gray Whale *Eschrichtius robustus*. Academic Press, San Diego, CA.

Powell, J. A., and G. B. Rathbun. 1984. Distribution and abundance of manatees along the northern coast of the Gulf of Mexico. Northeast Gulf Science 7:1–28.

Preen, A. 1989a. Observations of mating behavior in dugongs (*Dugong dugon*). Marine Mammal Science 5:382–387.

Preen, A. 1989b. The Status and Conservation of Dugongs in the Arabian Region, Vol. I. Saudi Arabian Government, Meteorology and Environmental Protection. Administration, Coastal and Marine Management Series Report No. 10. 200 pages.

Preen, A. R. 1992. Interactions between dugongs and seagrasses in a subtropical environment. Ph.D. dissertation, Department of Zoology, James Cook University of North Queensland, Townsville, Queensland, Australia. 392 pages.

Preen, A. 1995. Impacts of dugong foraging on seagrass habitats: Observational and experimental evidence for cultivation grazing. Marine Ecology Progress Series 1224:201–213.

Preen, A. 1995. Diet of dugongs: Are they omnivores? Journal of Mammalogy 76:163–171.

Prime, J. H., and P. S. Hammond. 1987. Quantitative assessment of gray seal diet from fecal analysis. Pages 165–182 *in* A. C. Huntley, D. P. Costa, G. A. J. Worthy, and M. A. Castellini (eds.). Marine Mammal Energetics. Allen Press, Lawrence, KS.

Provancha, J. A., and M. J. Provancha. 1988. Long-term trends in abundance and distribution of manatees (*Trichechus manatus*) in the northern Banana River, Brevard County, Florida. Marine Mammal Science 4:323–338.

Pryor, K., and I. K. Shallenberger. 1991. Social structure in spotted dolphins (*Stenella attenuata*) in the tuna purse seine fishery in the eastern tropical Pacific. Pages 161–196 *in* K. Pryor and K. S. Norris (eds.). Dolphin Societies: Discoveries and Puzzles. University of California Press, Berkeley, CA.

Pryor, K., J. Lindbergh, S. Lindbergh, and R. Milano. 1990. A dolphin–human fishing cooperative in Brazil. Marine Mammal Science 6:77–82.

Puente, A. E., and D. A. Dewsbury. 1976. Courtship and copulatory behavior of bottlenosed dolphins (*Tursiops truncatus*). Cetology 21:1–9.

Pulliam, H. R., and T. Caraco. 1984. Living in groups: Is there an optimal group size? Pages 122–147 *in* J. R. Krebs and N. B. Davies (eds.). Behavioural Ecology, 2d. ed. Blackwell Scientific Publications, Oxford.

Ralls, K. 1976. Mammals in which females are larger than males. Quarterly Review of Biology 51:245–276.

Ralls, K., and D. B. Siniff. 1990. Time budgets and activity patterns in California sea otters. Journal of Wildlife Management 54:251–259.

Ralls, K., Doroff, A., and A. Mercure. 1992. Movements of sea otters relocated along the California coast. Marine Mammal Science 8:178–184.

Ralls, K., B. Hatfield, and D. B. Siniff. 1995. Foraging patterns of California sea otters as indicated by telemetry. Canadian Journal of Zoology 73:523–531.

Ralls, K., T. Eagle, and D. B. Siniff. 1996. Movement and spatial use patterns of California sea otters. Canadian Journal of Zoology 74:1841–1849.

Rand, R. W. 1967. The Cape fur seal (*Arctocephalus pusillus*). 3. General behaviour on land and at sea. Investigational Reports of the Division of Sea Fisheries of South Africa 60:1–39.

Rathbun, G. B. 1979. The social structure and ecology of elephant-shrews. Advances in Ethology 20:1–76.

Rathbun, G. B., and T. J. O'Shea. 1984. Manatee's simple social life. Pages 300–301 *in* D. Macdonald (ed.). The Encyclopedia of Mammals. Facts on File, Inc., New York, NY.

Rathbun, G. B., R. K. Bonde, and D. Clay. 1982. The status of the West Indian manatee on the Atlantic coast north of Florida. Pages 152–165 *in* R. R. Odom and J. W. Guthrie (eds.). Proceedings of the Symposium on Nongame and Endangered Wildlife. Georgia Department of Natural Resources, Game and Fish Division, Technical Bulletin WL5, Research Triangle, GA.

Rathbun, G. B., J. P. Reid, and J. B. Bourassa. 1987. Design and Construction of a Tethered, Floating Radio-tag Assembly for Manatees. National Technical Information Service, Springfield, VA. PB-161345 / AS.

Rathbun, G. B., R. J. Jameson, G. R. VanBlaricom, and R. L. Brownell Jr. 1990a. Reintroduction of sea otters to San Nicolas Island, California: Preliminary results for the first year. Pages 99–114 *in* P. J. Bryant and J. Remington (eds.). Memoirs of the Natural History Foundation of Orange County: Endangered Wildlife and Habitats in Southern California, Vol. 3. Natural History Foundation of Orange County, Newport Beach, CA.

Rathbun, G. B., J. P. Reid, and G. Carowan. 1990b. Distribution and movement patterns of manatees (*Trichechus manatus*) in northwestern peninsular Florida. Florida Marine Research Publications 48:1–33.

Rathbun, G. B., J. P. Reid, R. K. Bonde, and J. A. Powell. 1995. Reproduction in free-ranging Florida manatees. Pages 135–156 *in* T. J. O'Shea, B. B. Ackerman, and H. F. Percival (eds.). Population Biology of the Florida Manatee. National Biological Service, Information and Technology Report 1, Washington, D.C.

Ray, C., W. A. Watkins, and J. J. Burns. 1969. The underwater song of *Erignathus* (Bearded Seal). Zoologica 54:79–86.

Read, A. J. 1989. Incidental catches and life history of harbour porpoises Phocoena phocoena from the Bay of Fundy. Ph.D. dissertation, University of Guelph, Guelph, Ontario, Canada. 121 pp.

Read, A. J., and A. A. Hohn. 1995. Life in the fast lane: The life history of harbor porpoises from the Gulf of Maine. Marine Mammal Science 11:423–440.

Read, A. J., R. S. Wells, A. A. Hohn, and M. D. Scott. 1993. Patterns of growth in wild bottlenose dolphins, *Tursiops truncatus*. Journal of Zoology (London) 231:107–123.

Read, A. J., A. J. Westgate, K. W. Urian, R. S. Wells, B. M. Allen, and W. J. Carr. 1996. Monitoring movements and health status of bottlenose dolphins in Beaufort, NC, using radio telemetry. Final contract report to the National Marine Fisheries Service, Southeast Fisheries Science Center, Charleston, SC. Contract No. 40-GENF-500160.

Reid, J. P., and T. J. O'Shea. 1989. Three years operational use of satellite transmitters on Florida manatees: Tag improvements based on challenges from the field. Pages 217–232 *in* Proceedings of the 1989 North American Argos Users Conference and Exhibit. Service Argos, Landover, MD.

Reid, J. P., G. B. Rathbun, and J. R. Wilcox. 1991. Distribution patterns of individually identifiable West Indian manatees (*Trichechus manatus*) in Florida. Marine Mammal Science 7:180–190.

Reid, J. P., R. K. Bonde, and T. J. O'Shea. 1995. Reproduction and mortality of radio-tagged and recognizable manatees on the Atlantic Coast of Florida. Pages 171–191 *in* T. J. O'Shea, B. B. Ackerman, and H. F. Percival (eds.). Population Biology of the Florida Manatee. National Biological Service, Information and Technology Report 1, Washington, D.C.

Reiter, J., N. L. Stinson, and B. J. Le Boeuf. 1978. Northern elephant seal development: the transition from weaning to nutritional independence. Behavioral Ecology and Sociobiology 3:337–367.

Reiter, J., K. J. Panken, and B. J. Le Boeuf. 1981. Female competition and reproductive success in northern elephant seals. Animal Behaviour 29:670–687.

Renouf, D. 1984. The vocalization of the harbour seal pup (*Phoca vitulina*) and its role in the maintenance of contact with the mother. Journal of Zoology 202:583–590.

Reynolds, J. E. III. 1981a. Aspects of the social behaviour and herd structure of a semi-isolated colony of West Indian manatees, *Trichechus manatus*. Mammalia 45:431–451.

Reynolds, J. E. III. 1981b. Behavior patterns in the West Indian manatee, with emphasis on feeding and diving. Florida Scientist 44:233–242.

Reynolds, J. E. III, and K. D. Haddad (eds.). 1990. Report of the workshop on geographic information systems as an aid to managing habitat for West Indian manatees in Florida and Georgia. Florida Marine Research Publications 49:1–57.

Reynolds, J. E. III, and D. K. Odell. 1991. Manatees and Dugongs. Facts On File, Inc., New York, NY.

Reynolds, J. E. III, and J. R. Wilcox. 1994. Observations of Florida manatees (*Trichechus manatus latirostris*) around selected power plants in winter. Marine Mammal Science 10:163–177.

Ribic, C. A. 1982. Autumn movement and home range of sea otters in California. Journal of Wildlife Management 46:795–801.

Rice, D. W., and E. A. Wolman. 1971. The Life History and Ecology of the Gray Whale (*Eschrichtius robustus*). American Society of Mammologists Special Publication No. 3, Allen Press, Lawrence, KS.

Richard, K. R., and M. A. Barbeau. 1994. Observations of spotted dolphins feeding nocturnally on flying fish. Marine Mammal Science 10:473–477.

Richard, K. R., M. C. Dillon, H. Whitehead, and J. M. Wright. 1996. Patterns of kinship in groups of free-living sperm whales (*Physeter macrocephalus*) revealed by multiple molecular genetic analyses. Proceedings of the National Academy of Science 93:8792–8795.

Ridgway, S. H., and R. F. Green. 1967. Evidence for a sexual rhythm in male porpoises, *Lagenorhynchus obliquidens* and *Delphinus delphis bairdi*. Norsk Hvalfangst-Tidende 1:1–8.

Ridgway, S. H., and R. Harrison. 1985. Handbook of Marine Mammals, Vol. 3. Academic Press, San Diego, CA.

Ridgway, S. H., T. Kamolnick, M. Reddy, C. Curry, and R. J. Tarpley. 1995. Orphan-induced lactation in *Tursiops* and analysis of collected milk. Marine Mammal Science 11:172–182.

Riedman, M. L. 1982. The evolution of alloparental care and adoption in mammals and birds. Quarterly Review of Biology 57:405–435.

Riedman, M. 1990. The Pinnipeds. University of California Press, Berkeley, CA.

Riedman, M. L., and J. A. Estes. 1990. The Sea Otter (*Enhydra lutris*): Behavior, Ecology, and Natural History. U.S. Fish and Wildlife Service, Biological Report 90 (14), Washington, D.C.

Riedman, M. L., and B. J. Le Boeuf. 1982 Mother–pup separation and adoption in northern elephant seals. Behavioral Ecology and Sociobiology 11:203–215.

Riedman, M. L., J. A. Estes, M. M. Staedler, A. A. Giles, and D. R. Carlson. 1994. Breeding patterns and reproductive success of California sea otters. Journal of Wildlife Management 58:391–399.

Rosen, D. A. S., and D. Renouf. [illegible]. Sex differences in the nursing relationship between mothers and pups in the Atlantic harbour seal, *Phoca vitulina concolor.* Journal of Zoology (London) 231:291–299.

Roux, J. P., and P. Jouventin. 1987. Behavioral cues to individual recognition in the subantarctic fur seal, *Arctocephalus tropicalis.* Pages 95–102 *in* J. P. Croxall and R. L. Gentry (eds.). Status, Biology and Ecology of Fur Seals. U. S. Dept. of Commerce, NOAA Technical Report NMFS 51, Washington, D.C.

Rubenstein, D. I. 1986. Ecology and sociality in horses and zebras. Pages 282–302 *in* D. I. Rubenstein and R. W. Wrangham (eds.). Ecological Aspects of Social Evolution. Princeton University Press, Princeton, NJ.

Rubenstein, D. I., and R. W. Wrangham. 1986. Ecological Aspects of Social Evolution. Princeton University Press, Princeton, NJ.

Ryg, M., T. G. Smith, and N. A. Oritsland. 1990. Seasonal changes in body mass and body composition of ringed seals (*Phoca hispida*) on Svalbard. Canadian Journal of Zoology 68:470–475.

Saayman, G. S., and C. K. Tayler. 1979. The socioecology of humpback dolphins (*Sousa* sp.). Pages 165–226 *in* H. E. Winn and B. L. Olla (eds.). Behavior of Marine Animals, Vol. 3: Cetaceans. Plenum, New York, NY.

Saayman, G. S., C. K. Tayler, and D. Bower. 1973. Diurnal activity cycles in captive and free-ranging Indian Ocean bottlenose dolphins (*Tursiops aduncus* Ehrenburg). Behaviour 44:212–233.

St. Aubin, D. J., T. G. Smith, and J. R. Geraci. 1990. Seasonal epidermal molt in beluga whales, *Delphinapterus leucas.* Canadian Journal of Zoology 68:359–367.

Samuels, A., and T. Gifford. 1997. A quantitative assessment of dominance relations among bottlenose dolphins. Marine Mammal Science 13:70–99.

Sandegren, F. E. 1976. Courtship display, agonistic behavior and social dynamics in the steller sea lion (*Eumetopias jubatus*). Behaviour 57:159–172.

Sandegren, F. E., E. W. Chu, and J. E. Vandevere. 1973. Maternal behavior in the California sea otter. Journal of Mammalogy 54:668–679.

Scammon, C. M. 1874. The Marine Mammals of the Northwestern Coast of North America. John H. Carmany and Company, San Francisco, CA.

Schevill, W. E., and W. A. Watkins. 1965. Underwater calls of *Trichechus* (manatee). Nature 205:373–374.

Schaeff, C. M., D. J. Boness, and W. D. Bowen. 1999. Female distribution, genetic relatedness, and fostering behaviour in harbour seals, *Phoca vitulina.* Animal Behaviour 57:427–434.

Schroeder, J. P. 1990. Breeding bottlenose dolphins in captivity. Pages 435–446 *in* S. Leatherwood and R. R. Reeves (eds.). The Bottlenose Dolphin. Academic Press, San Diego, CA.

Schroeder, J. P., and K. V. Keller. 1990. Artificial insemination of bottlenose dolphins. Pages 447–460 *in* S. Leatherwood and R. R. Reeves (eds.). The Bottlenose Dolphin. Academic Press, San Diego, CA.

Scott, M. D. 1991. The size and structure of pelagic dolphin herds. Ph.D. dissertation, University of California, Los Angeles, CA. 165 pages.

Scott, M. D., and S. J. Chivers. 1990. Distribution and herd structure of bottlenose dolphins in the eastern tropical Pacific Ocean. Pages 387–402 *in* S. Leatherwood and R. R. Reeves (eds.). The Bottlenose Dolphin. Academic Press, San Diego, CA.

Scott, M. D., and J. G. Cordaro. 1987. Behavioral observations of the dwarf sperm whale, *Kogia simus.* Marine Mammal Science 3:353–354.

Scott, M. D., R. S. Wells, and A. B. Irvine. 1990a. A long-term study of bottlenose dolphins on the west coast of Florida. Pages 235–244 *in* S. Leatherwood and R. R. Reeves (eds.). The Bottlenose Dolphin. Academic Press, San Diego, CA.

Scott, M. D., R. S. Wells, A. B. Irvine, and B. R. Mate. 1990b. Tagging and marking studies on small cetaceans. Pages 489–514 *in* S. Leatherwood and R. R. Reeves (eds.). The Bottlenose Dolphin. Academic Press, San Diego, CA.

Seipt, I. E., P. J. Clapham, C. A. Mayo, and M.P. Hawvermale. 1989. Population characteristics of individually identified fin whales *Balaenoptera physalus* in Massachusetts Bay. Fishery Bulletin 88:271–278.

Sekiguchi, K., P. B. Best, and B. Z. Kaczmaruk. 1992. New infomation on the feeding habits and baleen morphology of the pygmy right whale *Caperea marginata.* Marine Mammal Science 8:288–293.

Selzer, L. A., and P. M. Payne. 1988. The distribution of white-sided (*Lagenorhynchus acutus*) and common dolphins (*Delphinus delphis*) vs. environmental features of the continental shelf of the northeastern United States. Marine Mammal Science 4:141–153.

Sergeant, D. E. 1965. Migration of harp seals *Pagophilus groenlandicus* (Erxleben) in the northwest Atlantic. Journal of Fisheries Research Board of Canada 22:433–463.

Shane, S. H. 1980. Occurrence, movements, and distribution of bottlenose dolphin, *Tursiops truncatus,* in southern Texas. Fishery Bulletin 78:593–601.

Shane, S. 1983a. Abundance, distribution, and movements of manatees (*Trichechus manatus*) in Brevard County, Florida. Bulletin of Marine Science 33:1–9.

Shane, S. 1983b. Manatees and power plants. Sea Frontiers 29:40–44.

Shane, S. H. 1984. Manatee use of power plant effluents in Brevard County, Florida. Florida Scientist 47:180–187.

Shane, S. H. 1990a. Comparison of bottlenose dolphin behavior in Texas and Florida, with a critique of methods for studying dolphin behavior. Pages 541–558 *in* S. Leatherwood and R. R. Reeves (eds.). The Bottlenose Dolphin. Academic Press, San Diego, CA.

Shane, S. H. 1990b. Behavior and ecology of the bottlenose dolphin at Sanibel Island, Florida. Pages 245–265 *in* S. Leatherwood and R. R. Reeves (eds.). The Bottlenose Dolphin. Academic Press, San Diego, CA.

Shane, S. H., R. S. Wells, and B. Würsig. 1986. Ecology, behavior, and social organization of the bottlenose dolphin: A review. Marine Mammal Science 2:34–63.

Shaughnessy, P. D., and K. R. Kerry. 1989. Crabeater seals *Lobodon carcinophagus* during the breeding season: Observations on five groups near Enderby Island, Antarctica. Marine Mammal Science 5:68–77.

Sherrod, S. K., J. A. Estes, and C. M. White. 1975. Depredation of sea otter pups by bald eagles at Amchitka Island, Alaska. Journal of Mammalogy 56:701–703.

Shipley, C., M. Hines, and J. S. Buchwald. 1981. Individual differences in threat calls of northern elephant seal bulls. Animal Behaviour 29:12–19.

Shrestha, T. K. 1989. Biology, status and conservation of the Ganges River dolpin, *Platanista gangetica,* in Nepal. Pages 70–76 *in* W. F. Perrin, R. L. Brownell Jr., Z. Kaiya, and L. Jiankang (eds.). Biology and Conservation of the River Dolphins. Occasional Papers of the IUCN Species Survival Commission (SSC), Number 3, Gland, Switzerland.

Silber, G. K., M. W. Newcomer, and H. M. Pérez-Cortés. 1990. Killer whales (*Orcinus orca*) attack and kill a Bryde's whale (*Balaenoptera edeni*). Canadian Journal of Zoology 68:1603–1606.

Silverman, H. B. 1979. Social organization and behaviour of the narwhal, *Monodon monoceros* L. in Lancaster Sound, Pond Inlet, and Tremblay Sound, Northwest Territories. M.Sc. thesis, McGill University, Montreal, Canada, 147 pages.

Silverman, H. B., and M. J. Dunbar. 1980. Aggressive tusk use by the narwhal (*Monodon monoceros* L.). Nature (London) 284:57–58.

Simpson, G. G. 1944. Tempo and Mode in Evolution. Columbia University Press, NY.

Sinclair, E., T. Loughlin, and W. Pearcy. 1994. Prey selection by northern fur seals (*Callorhinus ursinus*) in the eastern Bering Sea. Fishery Bulletin 92:144–156.

Siniff, D. B., I. Stirling, J. L. Bengtson, and R. A. Reichle. 1979. Social and reproductive behavior of crabeater seals (*Lobodon carcinophagus*) during the austral spring. Canadian Journal of Zoology 57:2243–2255.

Sjare, B., and I. Stirling. 1996. The breeding behavior of Atlantic walruses, *Odobenus rosmarus rosmarus,* in the Canadian High Arctic. Canadian Journal of Zoology 74:897–911.

Slip, D. J., M. A. Hindell, and H. R. Burton. 1994. Diving behavior of southern elephant seals from Macquarie Island. Pages 253–270 *in* B. J. Le Boeuf and R. M. Laws (eds.). Elephant Seals. University of California Press, Berkeley, CA.

Slooten, E. 1994. Behavior of Hector's dolphin: Classifying behavior by sequence analysis. Journal of Mammalogy 75:956–964.

Smith, G. J. D., and D. E. Gaskin. 1983. An environmental index for habitat utilization by female harbour porpoises with calves near Deer Island, Bay of Fundy. Ophelia 22:1–13.

Smith, R. J., and A. J. Read. 1992. Consumption of euphausiids by harbour porpoise (*Phocoena phocoena*) calves in the Bay of Fundy. Canadian Journal of Zoology 70:1629–1632.

Smith, S. C., and H. Whitehead. 1993. Variations in feeding success and behaviour of Galápagos sperm whales (*Physeter macrocephalus*) as they relate to oceanographic conditions. Canadian Journal of Zoology 71:1991–1996.

Smith, T. G. 1976. Predation of ringed seal pups (*Phoca hispida*) by the arctic fox (*Alopex lagopus*). Canadian Journal of Zoology 54:1610–1616.

Smith, T.G. 1985. Polar bears, *Ursus maritimus,* as predators of belugas, *Delphinapterus leucas.* Canadian Field Naturalist 99:71–75.

Smith, T.G., and B. Sjare. 1990. Predation of belugas and narwhals by polar bears in nearshore areas of the Canadian high arctic. Arctic 43:99–102.

Smith, T. G., and I. Stirling. 1975. The breeding habitat of the ringed seal (*Phoca hispida*). The birth lair and associated structures. Canadian Journal of Zoology 53:1297–1305.

Smith, T. G., D. B. Siniff, R. Reichle, and S. Stone. 1981. Coordinated behavior of killer whales, *Orcinus orca,* hunting a crabeater seal, *Lobodon carcinophagus.* Canadian Journal of Zoology 59:1185–1189.

Smith, T. G., M. O. Hammill, and G. Taugbol. 1991. A review of the developmental, behavioural and physiological adaptations of the ringed seal, *Phoca hispida,* to life in the Arctic winter. Arctic 44:124–131.

Smith, T. G., D. J. St. Aubin, and M. O. Hammill. 1992. Rubbing behaviour of belugas, *Delphinapterus leucas,* in a high arctic estuary. Canadian Journal of Zoology 70:2405–2409.

Smolker, R. A., A. F. Richards, R. C. Connor, and J. W. Pepper. 1992. Sex differences in patterns of association among Indian Ocean bottlenose dolphins. Behaviour 123:38–69.

Smolker, R. A., A.F. Richards, R.C. Connor, J. Mann, and P. Berggren. 1997. Sponge carrying by Indian Ocean bottlenose dolphins: Possible tool-use by a delphinid. Ethology 103:454–465.

Smythe, N. 1970. The adaptive value of the social organization of the coati (*Nasua narica*). Journal of Mammalogy 51:818–820.

Springer, S. 1967. Porpoises vs. sharks. Page 27 *in* Conference on the Shark–Porpoise Relationship. American Institute of Biological Sciences, Washington, D.C.

Staedler, M., and M. Riedman. 1993. Fatal mating injuries in female sea otters (*Enhydra lutris nereis*). Mammalia 57:135–139.

Stamps, J. A., and M. Buechner. 1985. The territorial defense hypothesis and the ecology of insular vertebrates. Quarterly Review of Biology 60:155–181.

Steel, C. 1982. Vocalization patterns and corresponding behavior of the West Indian manatee (*Trichechus manatus*). Ph.D. dissertation, Florida Institute of Technology, Melbourne, FL. 189 pages.

Steltner, H., S. Steltner, and D. E. Sergeant. 1984. Killer whales, *Orcinus orca,* prey on narwhals, *Monodon monoceros:* An eyewitness account. Canadian Field Naturalist 98:458–462.

Stephens, D. W., and J. R. Krebs. 1986. Foraging Theory. Princeton University Press, Princeton, NJ.

Stewart, B. S., and R. L. DeLong. 1993. Seasonal dispersion and habitat use of foraging northern elephant seals. Symposia of Zoological Society (London) 66:179–194.

Stewart, B. S., S. Leatherwood, P. K. Yochem, and M. P. Heide-Jorgensen. 1989. Harbor seal tracking and telemetry by satellite. Marine Mammal Science 5:361–376.

Stirling, I. 1970. Observations on the behavior of the New Zealand fur seal (*Arctocephalus forsteri*). Journal of Mammalogy 51:766–778.

Stirling, I. 1971. Studies on the behaviour of the South Australian fur seal, *Arctocephalus forsteri* (Lesson) II. Adult females and pups. Australian Journal of Zoology 19:267–273.

Stirling, I. 1975. Factors affecting the evolution of social behaviour in the Pinnipedia. Rapports et Procés-Verbaux des Réunions, Conseil International pour L'Exploration de la Mer 169:205–212.

Stirling, I. 1977. Adaptations of Weddell and ringed seals to exploit the polar fast ice habitat in the absence or presence of surface predators. *in* Proceedings of the Third SCAR Symposium on Antarctic Biology. Smithsonian Institution Press, Washington, D.C.

Stirling, I. 1983. The evolution of mating systems in pinnipeds. Pages 489–527 *in* J. F. Eisenberg and D. G. Kleiman (eds.). Recent Advances in the Study of Behavior. American Society of Mammalogists, Special Publications no. 7, Provo, UT.

Stone, G. S., L. Florez-Gonzalez, and S. Katona. 1990. Whale migration record. Nature (London) 346:705.

Sullivan, R. M. 1981. Aquatic displays and interactions in harbor seals (*Phoca vitulina*), with comments on mating systems. Journal of Mammalogy 62:825–831.

Swartz, S. L. 1986. Gray whale migratory, social and breeding behavior. Reports of the International Whaling Commission (Special Issue 8):207–229.

Swartz, S. L., M. L. Jones, J. Goodyear, D. E. Withrow, and R. V. Miller. 1987. Radio-telemetric studies of gray whale migration along the California coast: A preliminary comparison of day and night migration routes. Report of the International Whaling Commission 37:295–299.

Symington, M. M. 1990. Fission–fusion social organization in Ateles and Pan. International Journal of Primatology 11:47–61.

Tanaka, S. 1987. Satellite radio-tracking of bottlenose dolphins *Tursiops truncatus.* Nippon Suisan Gakkaishi 53:1327–1338.

Tarpy, C. 1979. Killer whale attack. National Geographic Magazine 155:542–545.

Tavolga, M. C. 1966. Behavior of the bottlenose dolphin (*Tursiops truncatus*): Social interactions in a captive colony. Pages 718–730 *in* K. S. Norris (ed.). Whales, Dolphins, and Porpoises. University of California Press, Berkeley and Los Angeles.

Tavolga, M. C., and F. S. Essapian. 1957. The behavior of the bottle-nosed dolphin (*Tursiops truncatus*): Mating, pregnancy, parturition, and mother–infant behavior. Zoologica 42:11–31.

Tayler, C. K., and G. S. Saayman. 1972. The social organization and behaviour of dolphins (*Tursiops aduncus*) and baboons (*Papio ursinus*): Some comparisons and assessments. Annals of the Cape Provence Museum of Natural History 9:11–49.

Tener, J. S. 1965. Muskoxen in Canada. Queen's Printer, Ottawa.

Terhune, J. 1985. Scanning behavior of harbor seals on haul-out sites. Journal of Mammalogy 66:392–395.

Terry, R. P. 1984. Intergeneric behavior between *Sotalia fluviatilis guianensis* and *Tursiops truncatus* in captivity. Z. Säugetierkunde 49:290–299.

Tershy, B. R. 1992. Body size, diet, habitat use, and social behavior of *Balaenoptera* whales in the Gulf of California. Journal of Mammalogy 73:477–486.

Testa, J. W. 1994. Over-winter movements and diving behavior of female Weddell seals (*Leptonychotes weddellii*) in the southwestern Ross Sea, Antarctica. Canadian Journal of Zoology 72:1700–1710.

Thomas, J. A., L. H. Cornell, B. E. Joseph, T. D. Williams, and S. Dreischman. 1987. An implanted transponder chip used as a tag for sea otters (*Enhydra lutris*). Marine Mammal Science 3:271–274.

Thomas, J. T., V. D. Pastukhov, R. Elsner, and E. Petrov. 1982. *Phoca sibirica*. Mammalian Species 188:1–6.

Thomas, P. O., and S. M. Taber. 1984. Mother–infant interaction and behavioral development in southern right whales, *Eubalaena australis*. Behaviour 88:42–60.

Thompson, D., P. S. Hammond, K. S. Nicholas, and M. A. Fedak. 1991. Movements, diving and foraging behaviour in gray seals (*Halichoerus grypus*). Journal of Zoology (London) 224:223–232.

Thompson, P. M., and D. Miller. 1990. Summer foraging activity and movements of radiotagged common seals (*Phoca vitulina* L.) in the Moray Firth, Scotland. Journal of Applied Ecology 27:492–501.

Thompson, P. M., D. Miller, R. Cooper, and P. Hammond. 1994. Changes in the distribution and activity of female harbour seals during the breeding season; Implications for their lactation strategy and mating patterns. Journal of Animal Ecology 63:24–30.

Tolley, K. A., A. J. Read, R. S. Wells, K. W. Urian, M. D. Scott, A. B. Irvine and, A. A. Hohn. 1995. Sexual dimorphism in wild bottlenose dolphins (*Tursiops truncatus*) from Sarasota, Florida. Journal of Mammalogy 76:1190–1198.

Townsend, C. H. 1914. The porpoise in captivity. Zoologica 1:289–299.

Trillmich, F. 1981. Mutual mother-pup recognition in Galapagos fur seals and sea lions: Cues and functional significance. Behavior 78:21–42.

Trillmich, F. 1984. The natural history of the Galapagos fur seal (*Arctocephalus galapagoensis*, Heller 1904). Pages 215–223 *in* R. Perry (ed.). Key Environment Series: Galapagos. Pergamon Press, Oxford.

Trillmich, F. 1986a. Attendance behavior of Galapagos fur seals. Pages 196–208 *in* R. L. Gentry and G. L. Kooyman (eds.). Fur Seals: Maternal Strategies on Land and at Sea. Princeton University Press, Princeton, NJ.

Trillmich, F. 1986b. Maternal investment and sex-allocation in the Galapagos fur seal, *Arctocephalus galapagoensis*. Behavioral Ecology and Sociobiology 19:157–164.

Trillmich, F. 1996. Parental investment in pinnipeds. Advances in the Study of Behavior 25:533–577.

Trillmich, F., and P. Majluf. 1981. First observations on colony structure, behavior, and vocal repertoire of the South American fur seal (*Arctocephalus australis*, Zimmerman, 1783) in Peru. Z SÑugetierkd 46:310–322.

Trillmich, F., and W. Mohren. 1981. Effects of lunar cycle on the Galapagos fur seal, *Arctocephalus galapagoensis*. Oecologia 48:85–92.

Trillmich, F., and K. G. K. Trillmich. 1984. The mating systems of pinnipeds and marine iguanas: convergent evolution of polygyny. Biological Journal Linn. Soc. 21:209–216.

Trillmich, F., G. L. Kooyman, P. Majluf, and M. Sanchez-Grinan. 1986. Attendance and diving behavior of South American fur seals during El Nino in 1983. Pages 153–167 *in* R. L. Gentry and G. L. Kooyman (eds.). Fur Seals: Maternal Strategies on Land and at Sea. Princeton University Press, Princeton, NJ.

Trites, A. 1990. Thermal budgets and climate spaces: The impact of weather on the survival of Galapagos *Arctocephalus galapagoensis* and Northern Fur Seal pups *Callorhinus ursinus* L. Functional Ecology 4:753–768.

Trivers, R. L. 1971. The evolution of reciprocal altruism. Quarterly Review of Biology 46:35–57.

Trivers, R. L. 1972. Parental investment and sexual selection. Pages 136–179 *in* B. Campbell (ed.). Sexual Selection and the Descent of Man, 1871–1971. Aldine, Chicago, IL.

Trivers, R. L. 1985. Social Evolution. Benjamin/Cummings, Menlo Park, CA.

Trivers, R. L., and D. E. Willard. 1973. Natural selection of parental ability to vary the sex ratio. Science 179:90–92.

Twiss, J. R., and R. R. Reeves. 1999. Conservation and Biology of Marine Mammals. Smithsonian Institution Press, Washington, D.C.

Twiss, S. D., P. P. Pomeroy, and S. S. Anderson. 1994. Dispersion and site fidelity of breeding male gray seals (*Halichoerus grypus*) on North Rona, Scotland. Journal of Zoology (London) 233:683–693.

Tyack, P. L. 1981. Interactions between singing Hawaiian humpback whales and conspecifics nearby. Behavioural Ecology and Sociobiology 8:105–116.

Tyack, P. L., and H. Whitehead. 1983. Male competition in large groups of wintering humpback whales. Behaviour 83:132–154.

Vandevere, J. E. 1969. Feeding behavior of the southern sea otter. Pages 87–94 *in* T. C. Poulter (ed.). Proceedings of the Sixth Annual Conference on Biological Sonar and Diving Mammals. Stanford Research Institute, Menlo Park, CA.

Vandevere, J. E. 1970. Reproduction in the southern sea otter. Pages 221–227 *in* Proceedings of the Seventh Annual Conference on Biological Sonar and Diving Mammals. Stanford Research Institute, Menlo Park, CA.

Van Parijis, S. M., P. M. Thompson, D. J. Tollit, and A. Makay. 1997. Distribution and activity of male harbour seals during the mating season. Animal Behaviour 54:35–43.

Van Waerebeek, K., J. C. Reyes, A. J. Read, and J. S. McKinnon. 1990. Preliminary observations of bottlenose dolphins from the Pacific coast of South America. Pages 143–154 *in* S. Leatherwood and R. R. Reeves (eds.). The Bottlenose Dolphin. Academic Press, San Diego, CA.

Vitt, J. L., and W. E. Cooper. 1985. The evolution of sexual dimorphism in the skink *Eumeces laticeps:* An example of sexual selection. Canadian Journal of Zoology 63:995–1002.

Wada, E., H. Mitzutani, and M. Minagawa. 1991. The use of stable isotopes for food web analysis. Critical Reviews in Food Science and Nutrition 30:361–371.

Walker, B. G., and W. D. Bowen. 1993. Behavioural differences among adult male harbour seals during the breeding season may provide evidence of reproductive strategies. Canadian Journal of Zoology 71:1585–1591.

Walker, G. E., and J. K. Ling. 1981. New Zealand sea lion—*Phocarctos hookeri.* Pages 25–38 *in* S. H. Ridgway and R. J. Harrison (eds.). Handbook of Marine Mammals, Vol. 1. Academic Press, London.

Walker, W. A. 1981. Geographical variation in morphology and biology of bottlenose dolphins (*Tursiops*) in the eastern North Pacific. NOAA, NMFS, SWFC Administrative Report LJ-81-03C, La Jolla, CA. 54 pages.

Wang, K. R., P. M. Payne, and V. G. Thayer (compilers). 1994. Coastal Stock(s) of Atlantic Bottlenose Dolphin: Status Review and Management. Proceedings and Recommendations from a Workshop held in Beaufort, NC, 13–14 September 1993. NOAA Technical Memorandum NMFS-OPR-4, Silver Spring, MD.

Waples, D. M. 1995. Activity budgets of free-ranging bottlenose dolphins (*Tursiops truncatus*) in Sarasota Bay, Florida. M.Sc. thesis, University of California, Santa Cruz, CA, 61 pages.

Waring, G. T., C. P. Fairfield, C. M. Ruhsam, and M. Sano. 1993. Sperm whales associated with Gulf Stream features off the north-eastern USA shelf. Fisheries Oceanography 2:101–105.

Watkins, W. A., and W. E. Schevill. 1976. Right whale feeding and baleen rattle. Journal of Mammalogy 57:58–66.

Watkins, W. A., and W. E. Schevill. 1977. Spatial distribution of *Physeter catodon* (sperm whales) underwater. Deep-Sea Research 24:693–699.

Watkins, W. A., and W. E. Schevill. 1979. Aerial observations of feeding behavior in four baleen whales: *Eubalaena glacialis, Balaenoptera borealis, Megaptera novaeangliae,* and *Balaenoptera physalus.* Journal of Mammalogy 60:155–163.

Watkins, W. A., K. E. Moore, D. Wartzok, and J. Johnson. 1981. Radio-tagging of finback (*Balaenoptera physalus*) and humpback (*Megaptera novaeangliae*) whales in Prince William Sound, Alaska. Deep-Sea Research 28:577–588.

Watkins, W. A., M. A. Daher, K. Fristrup, and G. Notarbartolo di Sciara. 1994. Fishing and acoustic behavior of Fraser's dolphin (*Lagenodelphis hosei*) near Dominica, Southeast Caribbean. Caribbean Journal of Science 30:76–82.

Watkins, W. A., J. Sigurjónsson, D. Wartzok, R. R. Maiefski, P. W. Howey, and M. A. Daher. 1996. Fin whale tracked by satellite off Iceland. Marine Mammal Science 12:564–569.

Weilgart, L., and H. Whitehead. 1997. Group-specific dialects and geographical variation in coda repertoire in South Pacific sperm whales. Behavioral Ecology and Sociobiology 40:277–285.

Weilgart, L., H. Whitehead, and K. Payne. 1996. A collossal convergence. American Scientist 84:278–287.

Weinrich, M. T. 1991. Stable social associations among humpback whales (*Megaptera novaeangliae*) in the southern Gulf of Maine. Canadian Journal of Zoology 69:3012–3018.

Weinrich, M. T., J. Bove, and N. Miller. 1993. Return and survival of humpback whale (*Megaptera novaeangliae*) calves born to a single female in three consecutive years. Marine Mammal Science 9:325–328.

Weller, D. W., B. Würsig, H. Whitehead, J. C. Norris, S. K. Lynn, R. W. Davis, N. Clauss, and P. Brown. 1996. Observations of an interaction between sperm whales and short-finned pilot whales in the Gulf of Mexico. Marine Mammal Science 12:588–593.

Wells, R. S. 1984. Reproductive behavior and hormonal correlates in Hawaiian spinner dolphins, *Stenella longirostris.* Pages 465–472 *in* W. F. Perrin, R. L. Brownell Jr., and D. P. DeMaster (eds.). Reproduction in Whales, Dolphins, and Porpoises. Reports of the International Whaling Commission (Special Issue 6), Cambridge, U.K.

Wells, R. S. 1986. Population structure of bottlenose dolphins: Behavioral studies of bottlenose dolphins along the central west coast of Florida. Contract Rept. to National Marine Fisheries Service, Southeast Fisheries Center, Miami, FL. Contract No. 45-WCNF-5-00366.

Wells, R. S. 1991a. The role of long-term study in understanding the social structure of a bottlenose dolphin community. Pages 199–225 *in* K. Pryor and K. S. Norris (eds.). Dolphin Societies: Discoveries and Puzzles. University of California Press, Berkeley, CA.

Wells, R. S. 1991b. Bringing up baby. Natural History Magazine August:56–62.

Wells, R. S. 1993a. Parental investment patterns of wild bottlenose dolphins. Pages 58–64 *in* N. F. Hecker (ed.). Proceedings of the 18th International Marine Animal Trainers Association Conference, November 4–9, 1990, Chicago, IL.

Wells, R. S. 1993b. The marine mammals of Sarasota Bay. Chapter 9, Pages 9.1–9.23 *in* P. Roat, C. Ciciccolella, H. Smith, and D. Tomasko (eds.). Sarasota Bay: 1992 Framework for Action. Sarasota Bay National Estuary Program. Sarasota Bay National Estuary Program, Sarasota, FL.

Wells, R. S. 1993c. Why all the blubbering? BISON Brookfield Zoo 7:12–17.

Wells, R. S., A. B. Irvine, and M. D. Scott. 1980. The social ecology of inshore odontocetes. Pages 263–317 *in* L. M. Herman (ed.). Cetacean Behavior: Mechanisms and Functions. Wiley & Sons, New York, NY.

Wells, R. S., B. G. Würsig, and K. S. Norris. 1981. A survey of the marine mammals of the upper Gulf of California, Mexico, with an assessment of the status of *Phocoena sinus.* Report No. PB81-168791. Marine Mammal Commission, Washington, D.C. 51 pp. (Available from the National Technical Information Service, Springfield, VA.)

Wells, R. S., M. D. Scott, and A. B. Irvine. 1987. The social structure of free-ranging bottlenose dolphins. Pages 247–305 *in* H. Genoways (ed.). Current Mammalogy, Vol. 1. Plenum, New York, NY.

Wells, R. S., L. J. Hansen, A. Baldridge, T. P. Dohl, D. L. Kelly, and R. H. Defran. 1990. Northward extension of the range of bottlenose dolphins along the California coast. Pages 421–431 *in* S. Leatherwood and R. R. Reeves (eds.). The Bottlenose Dolphin. Academic Press, San Diego, CA.

Wells, R. S., K. W. Urian, A. J. Read, M. K. Bassos, W. J. Carr, and M. D. Scott. 1996a. Low-level monitoring of bottlenose dolphins, *Tursiops truncatus,* in Tampa Bay, Florida: 1988–1993. NOAA Tech. Mem. NMFS-SEFSC-385, Miami, FL.

Wells, R. S., M. K. Bassos, K. W. Urian, W. J. Carr, and M. D. Scott. 1996b. Low-level monitoring of bottlenose dolphins, *Tursiops truncatus,* in Charlotte Harbor, Florida: 1990–1994. NOAA Tech. Mem. NMFS-SEFSC-384, Miami, FL.

Westlake, R. L., and W. G. Gilmartin. 1990. Hawaiian monk seal pupping locations in the Northwestern Hawaiian Islands. Pacific Science 44:366–383.

Whitehead, H. 1989. Formations of foraging sperm whales, *Physeter macrocephalus,* off the Galápagos Islands. Canadian Journal of Zoology 67:2131–2139.

Whitehead, H. 1990. Day of the Ziphiids. Whale and Dolphin Conservation Society Sonar 4:10–11.

Whitehead, H. 1993. The behaviour of mature male sperm whales on the Galapágos Islands breeding grounds. Canadian Journal of Zoology 71:689–699.

Whitehead, H. 1994. Delayed competitive breeding in roving males. Journal of Theoretical Biology 166:127–133.

Whitehead, H. 1995. Investigating structure and temporal scale in social organizations using identified individuals. Behavioral Ecology 6:199–208.

Whitehead, H. 1996a. Variation in the feeding success of sperm whales: Temporal scale, spatial scale and relationship to migrations. Journal of Animal Ecology 65:429–438.

Whitehead, H. 1996b. Babysitting, dive synchrony, and indications of alloparental care in sperm whales. Behavioral Ecology and Sociobiology 38:237–244.

Whitehead, H., and T. Arnbom. 1987. Social organization of sperm whales off the Galapagos Islands, February–April 1985. Canadian Journal of Zoology 65:913–919.

Whitehead, H., and C. Carlson. 1988. Social behaviour of feeding finback whales off Newfoundland: Comparisons with the sympatric humpback whale. Canadian Journal of Zoology 66:217–221.

Whitehead, H., and C. Glass. 1985. Orcas (killer whales) attack humpback whales. Journal of Mammalogy 66:183–185.

Whitehead, H., R. Silver, and P. Harcourt. 1982. The migration of humpback whales along the northeast coast of Newfoundland. Canadian Journal of Zoology 60:2173–2179.

Whitehead, H., V. Papastavrou, and S. C. Smith. 1989. Feeding success of sperm whales and sea-surface temperature off the Galápagos Islands. Marine Ecology Progress Series 53:201–203.

Whitehead, H., S. Brennan, and D. Grover. 1992. Distribution and behaviour of male sperm whales on the Scotian Shelf, Canada. Canadian Journal of Zoology 70:912–918.

Whitehead, H., A. Faucher, S. Gowans, and S. McCarrey. 1997. Status of the northern bottlenose whale, *Hyperoodon ampullatus,* in the Gully, Nova Scotia. The Canadian Field Naturalist 111:287–292.

Whitmore, F. C. Jr., and L. M. Gard Jr. 1977. Steller's sea cow (*Hydrodamalis gigas*) of Late Pleistocene age from Amchitka, Aleutian Islands, Alaska. U. S. Geological Survey Professional Paper 1036, Washington, D.C.

Whittow, G. C. 1978. Thermoregulatory behavior of Hawaiian monk seal, *Monachus schauinslandi.* Pacific Science 32:47–60.

Whittow, G. C. 1987. Thermoregulatory adaptations in marine mammals: Interacting effects of exercise and body mass. Marine Mammal Science 3:220–241.

Whittow, G. C., J. Szekerczes, E. Kridler, and D. L. Olsen. 1975. Skin structure of the Hawaiian monk seal (*Monachus schauinslandi*). Pacific Science 29:153–157.

Wild, P. W., and J. A. Ames. 1974. A report on the sea otter *Enhydra lutris,* L. Marine Resources Technical Report, California Department of Fish and Game 20:211–242.

Wiley, R. H. 1991. Lekking in birds and mammals: behavioral and evolutionary issues. Advances in the Study of Behavior 20:201–291.

Wilkinson, G. S. 1992. Communal nursing in the evening bat, *Nycticeius humaeralis.* Behavioral Ecology and Sociobiology 31:225–235.

Williams, G. C. 1966. Adaptation and Natural Selection. Princeton University Press, Princeton, NJ.

Williams, T. D., D. D. Allen, J. M. Groff, and R. L. Glass. 1992. An analysis of California sea otter (*Enhydra lutris*) pelage and integument. Marine Mammal Science 8:1–18.

Wilson, D. R .B. 1995. The ecology of bottlenose dolphins in the Moray Firth, Scotland: A population at the northern extreme of the species' range. Ph.D. dissertation, University of Aberdeen, Scotland, 201 pages.

Wilson, E. O. 1975. Sociobiology—The New Synthesis. The Belknap Press, Harvard University, Cambridge, MA.

Wood, F. G. Jr., D. K. Caldwell, and M. C. Caldwell. 1970. Behavioral interactions between porpoises and sharks. Investigations on Cetacea 2:264–277.

Würsig, B. 1978. Occurrence and group organization of Atlantic bottlenose porpoises (*Tursiops truncatus*) in an Argentine bay. The Biological Bulletin 154:348–359.

Würsig, B. 1982. Radio-tracking dusky porpoises in the South Atlantic. FAO Fisheries Serial 5:145–160.

Würsig, B. 1986. Delphinid foraging strategies. Pages 347–359 *in* R. J. Schusterman, J. A. Thomas and F. G. Wood (eds.). Dolphin Cognition and Behavior: A Comparative Approach. Lawrence Erlbaum Association, Hillsdale, NJ.

Würsig, B., and R. Bastida. 1986. Long-range movement and individual associations of two dusky dolphins (*Lagenorhynchus obscurus*) off Argentina. Journal of Mammalogy 67:773–774.

Würsig, B., and C. Clark. 1993. Behavior. Pages 157–199 *in* J. J. Burns, J. J. Montague, and C. J. Cowles (eds.). The Bowhead Whale. Special Publication No. 2. Society for Marine Mammalogy, Lawrence, KS.

Würsig, B., and G. Harris. 1990. Site and association fidelity in bottlenose dolphins off Argentina. Pages 361–365 *in* S. Leatherwood and R. R. Reeves (eds.). The Bottlenose Dolphin. Academic Press, San Diego, CA.

Würsig, B., and T. A. Jefferson. 1990. Methods of photo-identification for small cetaceans. Pages 43–52 *in* P. S. Hammond, S. A. Mizroch, and G. P. Donovan (eds.). Individual Recognition of Cetaceans: Use of Photo-identification and Other Techniques to Estimate Population Parameters. Report of the International Whaling Commission (Special Issue 12), Cambridge, U.K.

Würsig, B., and M. Würsig. 1977. The photographic determination of group size, composition, and stability of coastal porpoises (*Tursiops truncatus*). Science 198:755–756.

Würsig, B., and M. Würsig. 1979. Behavior and ecology of the bottlenose dolphin, *Tursiops truncatus,* in the south Atlantic. Fishery Bulletin 77:399–412.

Würsig, B., and M. Würsig. 1980. Behavior and ecology of the dusky dolphin, *Lagenorhynchus obscurus,* in the south Atlantic. Fishery Bulletin 77:871–890.

Würsig, B., E. M. Dorsey, M. A. Fraker, R. Payne, and W. J. Richardson. 1985. Behaviour of bowhead whales, *Balaena mysticetus,* summering in the Beaufort Sea: A description. Fishery Bulletin 83:357–377.

Würsig, B., R. S. Wells, and D. A. Croll. 1986. Behavior of gray whales summering near St. Lawrence Island, Bering Sea. Canadian Journal of Zoology 64:611–621.

Würsig, B., E. M. Dorsey, W. J. Richardson, and R. S. Wells. 1989. Feeding, aerial and play behaviour of the bowhead whale, *Balaena mysticetus,* summering in the Beaufort Sea. Aquatic Mammals 15:27–37.

Würsig, B., F. Cipriano, and M. Würsig. 1991. Dolphin movement patterns: Information from radio and theodolite tracking studies. Pages 79–111 *in* K. Pryor and K. S. Norris (eds.). Dolphin Societies: Discoveries and Puzzles. University of California Press, Berkeley, CA.

Würsig, B., J. Guerrero, and G. K. Silber. 1993. Social and sexual behavior of bowhead whales in fall in the western arctic: A re-examination of seasonal trends. Marine Mammal Science 9:103–110.

Young, D. D., and V. G. Cockcroft. 1994. Diet of common dolphins (*Delphinus delphis*) off the south-east coast of southern Africa: Opportunism or specialization? Journal of Zoology (London) 234:41–53.

Yuanyu, H., Qingzhong, Z., and Guocheng, Z. 1989. The habitat and behavior of *Lipotes vexillifer.* Pages 92–98 *in* W. F. Perrin, R. L. Brownell Jr., Z. Kaiya, and L. Jiankang (eds.). Biology and Conservation of the River Dolphins. Occasional Papers of the IUCN Species Survival Commission (SSC), Number 3, Gland, Switzerland.

Zoodsma, B. J. 1991. Distribution and behavioral ecology of manatees in southeastern Georgia. M.S. thesis, University of Florida, Gainesville, FL. 150 pages.

9

W. DONALD BOWEN AND DONALD B. SINIFF

Distribution, Population Biology, and Feeding Ecology of Marine Mammals

Ecology seeks to describe and explain the patterns of distribution and abundance of organisms. These patterns reflect the history of complex interactions with other organisms and with individuals of the same species and their environment. As such, there is little that cannot be discussed within this broad definition of ecology. However, partly because many relevant topics are dealt with in other chapters, and partly because of the inevitable limitations of space, we deal here mainly with the distribution, population biology, and feeding ecology of marine mammals. We finish with a discussion of the ecological role of marine mammals in aquatic ecosystems.

As discussed earlier (see Reynolds, Odell, and Rommel, Chapter 1, this volume), marine mammals represent diverse taxa grouped together because of a common reliance on aquatic or marine resources for their survival. Although clearly adapted to the aquatic environment, marine mammals share many characteristics with other large mammals, particularly in relation to population dynamics (Fowler and Smith 1981), energetics (Lavigne et al. 1986, Innes et al. 1987), and reproduction (Boyd 1991). Readers should consult specific chapters in this text dealing in detail with the latter topics. This permits those interested in marine mammals to draw ona large body of both empirical and theoretical literature to make inferences about species that are poorly known at present (McLaren 1990). Comparison of marine mammal biology against this broader landscape also highlights the features of marine mammals that are different, and thus, provides a powerful tool for understanding the adaptive nature of these differences.

The development of comparative methods based on well-resolved phylogenies and the rapid growth of DNA sequence data (e.g., Arnason et al. 1995, Milinkovitch 1995, Perry et al. 1995) provide molecular bases for testing hypotheses about the evolution of life history patterns and a means of distinguishing ecological adaptations from patterns that are present through common descent (Harvey and Purvis 1991, Brooks and McLennan 1991, Perry et al. 1995). In fact, a new approach, "genetic ecology," has gained prominence recently and provides important insight into relationships among individuals and groups through analysis of genetic variation and its consequences. Two excellent sources of information on genetic and molecular methods and applications are Hoelzel (1991) and Hoelzel and Dover (1991).

Distributional Ecology

An important part of ecology attempts to explain the restricted and generally patchy distribution of species. This is no less of a challenge with respect to marine mammals than it is for other taxa. Although marine mammals inhabit most marine environments: continental shelves (e.g., Phocidae, Otariidae, Delphinidae, Balaenopteridae, Balaenidae, the sea otter [*Enhydra lutris*]); deep ocean canyons (e.g., Ziphiidae); tropical seas (e.g., Delphinidae, Phocidae, Otariidae, Sirenia); the deep ocean (e.g., Physeteridae, Phocidae); and both Arctic (e.g., Odobenidae, Phocidae, Monodontidae, Balaenidae, the polar bear [*Ursus maritimus*]); and Antarctic polar seas (e.g., Delphinidae, Balaenopteridae, Balaenidae, Phocidae, Otariidae) (Katona and Whitehead 1988), they also inhabit some of the great rivers and their estuaries (e.g., the river dolphins and the Amazon manatee, *Trichechus inunguis*) and several species of phocid seals (*Phoca vitulina, P. hispida, P. caspica, P. sibirica*) also inhabit freshwater lakes.

General descriptions of the geographic distributions of pinnipeds and cetaceans are given in King (1983), Gaskin (1982), and Evans (1987), respectively. Geographic distributions of the sea otter and the polar bear are given in Riedman and Estes (1988) and Wiig et al. (1995), respectively. Reynolds and Odell (1991) provide an overview of sirenian distribution. However, in consulting these and other sources, it is important to remember that our knowledge of distribution and habitat use of different species varies considerably and in many cases may be strongly biased by the methods used to gather distributional information.

A number of factors affect the distribution of marine mammals. These may be grouped as follows: (1) habitat, including the type of substrate (e.g., ice), temperature, salinity, and bathymetry; (2) biological, including productivity, the distribution and abundance of predators, competitors, and prey; (3) demographical, including population size, age, sex, and reproductive status; (4) species adaptations, including morphological, physiological, and behavioral; and (5) human effects, including disturbance and pollutants (toxic materials, sound).

Although each of these factors may influence distribution, combinations of factors are generally responsible for the distribution patterns we observe. In this section, we first describe methods used to study the distribution of marine mammals, give examples of the distributional patterns of a number of species, and then discuss in more detail ways in which distributions are affected by both the structure of the habitat and predation. Others factors are discussed later in the chapter in connection with feeding ecology.

Methods Used to Study Distribution

Until relatively recently much of our knowledge about the distribution of marine mammals came from harvest records and from incidental observations made from the shore or ships. Although catch data provided considerable information on a number of cetacean (e.g., the baleen whales and sperm whale [*Physeter macrocephalus*], see Reports of the International Whaling Commission) and pinniped species (e.g., harp seals [*Phoca groenlandica*], reviewed in Sergeant 1991; northern fur seal [*Callorhinus ursinus*], Kajimura 1980), they often told us more about the distribution of hunting effort than the actual distribution of the population. In many cases catches were taken at only certain times of the year, thus limiting their value in determining seasonal patterns. In other cases (e.g., the harvest of harp and hooded seals [*Cystophora cristata*]), the movements of hunting vessels were limited by ice such that the entire distribution of the species could not be determined. Of course, many of the smaller or less abundant species were not harvested commercially and therefore, little was learned of their distribution in this manner. On the other hand, sea otters were nearly extirpated from the Pacific Rim by harvest because their distribution is limited to the near-shore, making them easily accessible in their entire distributional range (Riedman and Estes 1990).

Surveys on dedicated research vessels, ships of opportunity, and from aircraft have provided considerable new data on the distribution of many species (e.g., Cetacean and Turtle Assessment Program 1982; Seabirds at Sea Team, Northridge et al. 1995). Observations from the shore have been used to study the migration of gray whales (*Eschrichtius robustus*) along the west coast of North America (e.g., Herzing and Mate 1984). Each type of platform has advantages and disadvantages. Ships are slow and expensive, but they provide an opportunity to scan a particular location for a greater period of time than is usually possible from aircraft. Much larger areas can be covered rapidly by aircraft, but for only a relatively brief instant. In both cases, these methods provide only snapshots of dynamic phenomena.

Telemetry has played an important role in the development of our understanding of the use of both surface distribution (using very high frequency [VHF] radio-tags, satellite-linked radio-tags, and geolocation time–depth recorders [TDRs]) and underwater distribution (using TDRs and acoustic tags) of marine mammals (Fig. 9-1; Wells, Boness, and Rathbun, Chapter 8, this volume). VHF radio-telemetry has been used for some time to study the distribution of coastal species such as harbor seals, (Pitcher and McAllister 1981, Brown and Mate 1983, Thompson and Miller 1990), gray seals (*Halichoerus grypus*) (Hammond et al. 1993), gray whales (Mate and Harvey 1984), beluga whales (*Delphinapterus leucas*) (Frost et al. 1985), other cetaceans (Leatherwood and Evans 1979), sea otters (Ralls et al. 1996), sirenians (Rathbun et al. 1987), and polar bears (Derocher and Stirling 1990). Generally, this method is less useful on wide-ranging species because of both the limited range of receivers sta-

A

B

Figure 9-1. (A) Satellite-linked location transmitter glued to the fur of an adult female gray seal on Sable Island, Canada. (B) Geolocation time–depth recorder on an adult male harbor seal on Sable Island, Canada. This animal also has a very high frequency head-mounted radio transmitter used to locate the animal at finer spatial scales. (Photographs by W. D. Bowen.)

tioned on land and the high cost associated with aerial or shipboard tracking.

Acoustic transmitters have been used to track the underwater movements of gray seals (Hammond et al. 1993). Watkins et al. (1993) used transponders and sonar in underwater tracking of two sperm whales. Acoustic telemetry has the advantage of not requiring the animal to surface to obtain a location, but the disadvantage is that animals are often lost after only a short period of study because of the more limited range of such transmitters and the high costs associated with shipboard tracking.

More recently, satellite-linked telemetry, geolocation, and TDRs have revolutionized our ability to study the use of space by marine mammals (Fig. 9-1). Although attempts to use them during the early 1980s were of limited success, satellite transmitters (known as platform transmitter terminals or PTTs) have now been used to study a number of species. Fancy et al. (1988) provide a good introduction to the technical basis and practical limitations of satellite transmitters. The primary advantage of satellite telemetry is that the geographic scale of data collection is substantially increased and the study period is often extended compared to studies using traditional VHF or acoustic telemetry. The disadvantages of satellite telemetry include the high cost of instruments and satellite rental fees and the high variability in the accuracy of locations.

Unlike satellite telemetry, initially developed to study the use of space by terrestrial animals, TDRs were developed to study diving behavior (for review, see Gentry and Kooyman 1986). During the 1980s, TDR design was advanced considerably, such that TDRs are now capable of providing positional data using standard solar navigation equations (DeLong et al. 1992, Hill 1994). Unfortunately, they must be recovered to download the stored data and the locations are rather inaccurate (i.e., 60–100 km). Until recently TDRs could not be used on species that were difficult to recapture. Goodyear (1993) has developed a combination sonic/VHF tag that monitors the diving behavior and movements of marine mammals and does not require the animal to be recaptured. Initial application of the tag on right whales (*Eubalena glacialis*) and humpback whales (*Megaptera novaeangliae*) has provided encouraging results. Another approach has been to make a buoyant unit that contains both a data recorder and a VHF tag to permit the unit to be recovered after it has fallen off the animal (Ellis and Trites 1992).

Patterns of Seasonal Distribution

Like other vertebrates, many marine mammal species exhibit seasonal changes in distribution. These changes generally reflect the differing requirements for feeding and reproduction, the need to avoid predation, and response to changes (such as temperature) in the physical environment. As Gaskin (1982) noted for cetaceans "the study of the distributional ecology of these animals becomes largely a study of the distributional ecology of their prey species, and an analysis of the factors which may limit the ability of cetaceans to reproduce or rear calves to maturity." Although stated in relation to whales, this observation also applies to other marine mammals.

Seasonal movements can be either migratory or nonmigratory. It is clear that migratory and nonmigratory movements grade into one another such that the distinction is somewhat arbitrary (Baker 1978). Nevertheless, it is useful to distinguish two broad types of movement patterns: migration and dispersal. Migration is the periodic return movement of individuals from one place to another (Lockyer and Brown 1981). The spectacular long-distance movement between summer feeding areas and winter breeding grounds of the humpback whale is a good example of migration. Dispersal, on the other hand, applies to the movement of an individual from its birth place to its first or subsequent breeding site or group (Greenwood 1980, Shields 1987), although dispersal in marine mammals has also been used to describe a general movement of individuals over feeding areas. In this chapter, we use the term breeding dispersal to refer to the former and dispersal for the latter. Although remarkable long-distance annual migrations have been documented in a number of species of whales, seals, and in the polar bear, dispersal is generally less well documented because it may often involve smaller scale movements that may be difficult to detect. Not all individuals in a population necessarily behave in the same way and individual animals may show more than one pattern during their lives. The pattern of movement of an individual often may be related to its age, sex, and reproductive status, and may be affected by population size and prey abundance and distribution.

Pinnipeds

Perhaps the most important feature of pinniped natural history is that they must periodically abandon the sea (i.e., foraging areas) and return to a solid substrate (land or ice) to give birth, rear their offspring, and molt (Bartholomew 1970, Stirling 1975). Some species also haul out on land or ice at other times of the year to escape predators (Terhune 1985; Watts 1992, 1996), for thermoregulatory purposes (Watts 1992, 1996), to rest, or for reasons that are not understood. For most species, these requirements result in seasonal changes in distribution, as it is unusual for a single region to provide all of the resources needed by a species. In the case of pack ice-breeding species, such as harp and hooded seals and the walrus, *Odobenus rosmarus*, seasonal changes in ice cover virtually guarantee some change in the distribution of individuals.

Seasonal distribution varies greatly within and among pinniped families (Tables 9-1 and 9-2). For example, on the basis of current information, a greater proportion of phocid species (8 of 18, or 44%) are migratory than are otariids (2 of 14, or 14%). This difference may be partly explained by the variable quality of data available, meaning that recent and ongoing studies using satellite telemetry and geolocation TDRs should provide a more reliable assessment in the near future. However, it is also possible that this difference reflects a fundamental difference in the nature of the breeding habitat used by phocids and otariids (see Costa and Williams, Chapter 5, and Wells, Boness, and Rathbun, Chapter 8, this volume). All otariid species breed on land, whereas most phocids breed on ice, a strongly seasonal resource. Recent studies of mitochondrial DNA (mDNA) suggest that ice breeding is the ancestral character state of the Phocinae (i.e., northern phocids; Perry et al. 1995). In fact, 12 of 18 phocid species give birth and rear their offspring only on ice. In the gray seal, ice is used only by perhaps 25% of the world population (Mansfield and Beck 1977), and in the harbor seal, ice is used by part of the Alaskan population (Calambokidis et al. 1987). Of the 12 species that use only ice for birth and rearing offspring, six (50%) are migratory. If we include the distantly related walrus, in which ice breeding may represent convergent evolution, then 54% of ice-breeding species are migratory. In contrast, only 20% of the 20 land breeding species (i.e., 2 of 6 land-breeding phocids and 2 otariids) are known to be migratory.

Two species of North Atlantic phocid seals, the harp and the hooded seal, provide good examples of pinniped migration. Both species give birth, rear offspring, and molt on pack ice, the distribution of which varies seasonally, interannually, and over longer time scales (Vibe 1967).

The general distribution of the harp seal as a species of the subarctic pack ice has been known for some time from harvest records (Nansen 1925, Sergeant 1965). Harp seals gather in large concentrations to give birth in three areas: around Newfoundland, near Jan Mayen Island, and in the White Sea (Fig. 9-2). Evidence to date indicates that the Newfoundland population is isolated from the other two, but there is some degree of mixing between the Jan Mayen and White Sea populations (for review, see Sergeant 1991).

The Newfoundland population is the largest (Fig. 9-2) and the most highly migratory of the three harp seal populations. The following description is taken mainly from Sergeant (1991). In late September when new ice forms, harp seals begin their southward migration along the east and west coasts of Baffin Island and eastward through the Hudson Strait. On the basis of catches in coastal harvests, the first migrants reach northern Labrador in mid- to late-October and the

Table 9-1. Habitat Types and Seasonal Movement Patterns of Some Seals (Family Phocidae)

Species	Habitat: Breeding	Habitat: Molting	Habitat: Aquatic	Seasonal Movement	Source
Harp	Pack ice	Pack ice	Continental shelf	Migration	1
Hooded	Pack ice	Pack ice	Continental slope, oceanic	Migration	2
Harbor	Land, ice	Land, ice	Coastal	Dispersal	3
Gray	Land, ice	Land	Continental shelf	Dispersal	4
Bearded	Pack ice	Pack ice	Coastal–continental shelf	Migration	5
Ringed	Fast (lairs) and pack ice	Fast and pack ice	Continental shelf	Dispersal, migration	6
Ribbon	Pack ice	Pack ice	Continental shelf	Dispersal	7
Spotted	Pack ice	Pack ice	Continental shelf	Migration?	8
Baikal	Fast ice (lairs)	?	Freshwater	Migration	9
Caspian	Fast ice	Pack ice	Freshwater	Migration	10
Weddell	Fast ice, land	Pack ice?	Continental shelf	Migration?	11
Ross	Pack ice	Pack ice	Oceanic		12
Crabeater	Pack ice, fast ice	Pack ice	Continental shelf	Dispersal?	13
Leopard	Pack ice	Pack ice	Continental shelf	Migration	14
Southern elephant	Land, fast ice	Land	Oceanic, continental shelf	Migration	15
Northern elephant	Land	Land	Oceanic	Migration	16
Mediterranean monk	Land (caves)	Land	Coastal	Dispersal	17
Hawaiian monk	Land	Land	Oceanic	Dispersal?	18

1 = Sergeant 1991; 2 = Sergeant 1976, G. B. Stenson, pers. comm; 3 = Boulva and McLaren 1979, Bigg 1981; 4 = Stobo et al. 1990, Hammond et al. 1993, Hammond and Fedak 1994; 5 = Burns 1981a; 6 = Smith 1973, 1987; 7 = Burns 1981b; 8 = Naito and Nishiwaki 1975, Bigg 1981; 9 = Popov 1979; 10 = King 1983; 11 = Lindsey 1937, Stirling 1969; 12 = Ray 1981; 13 = Bertram 1940, Siniff et al. 1979; 14 = Rounsevell and Eberhard 1980; 15 = Slip et al. 1994; 16 = Stewart and DeLong 1995; 17 = Sergeant et al. 1978; 18 = Kenyon and Rice 1959.

Table 9-2. Habitats and Seasonal Movement Patterns of the Otariidae and Odobenidae Families

Species	Habitat		Seasonal Movement	Source
	Breeding	Aquatic		
Steller sea lion	Land	Continental shelf, pelagic	Dispersal	1
California sea lion	Land	Oceanic, coastal	Adult males migrate	2
Galapagos sea lion	Land	Oceanic	?	3
Southern sea lion	Land	Oceanic, coastal	?	4
Australian sea lion	Land	Coastal	Dispersal	5
Hooker's sea lion	Land	Coastal	?	6
Guadalupe fur seal	Land	?	?	7
Galapagos fur seal	Land	Oceanic?	?	8
Juan Fernandez fur seal	Land	Oceanic	Dispersal	9
S. American fur seal	Land	Coastal?	Dispersal	10
Subantarctic fur seal	Land	?	?	11
Antarctic fur seal	Land	Oceanic	Dispersal	12
South African fur seal	Land	Continental shelf	Dispersal	13
Australian fur seal	Land	Continental shelf	Dispersal	14
New Zealand fur seal	Land	Coastal ?	Non-migratory	15
Northern fur seal	Land	Oceanic, continental shelf	Migration	16
Walrus	Pack ice	Continental shelf	Migration	17

1 = Hoover 1988; 2 = King 1983; 3 and 4 = Vaz-Ferreira 1981; 5 = Walker and Ling 1981; 6 = Walker and Ling 1981; 7, 8, and 9 = Bonner 1981; 10 = Vaz-Ferreira 1979; 11 = Bonner 1981; 12 = Boyd et al. 1998; 13 = Shaughnessy 1982; 14 = Warneke and Shaughnessy 1985; 15 = Crawley and Warneke 1979; 16 = French et al. 1989; 17 = Fay 1981.

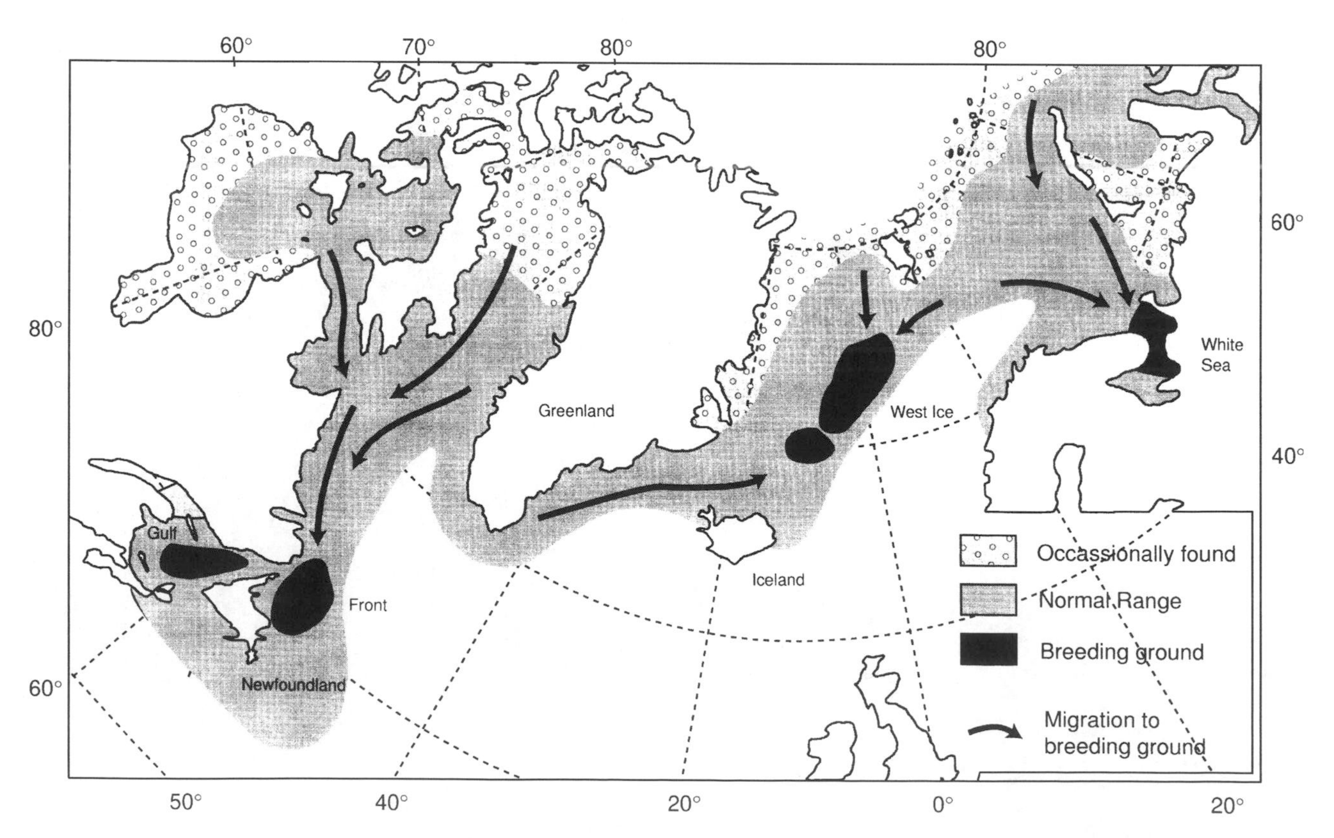

Figure 9-2. Overall geographic range and location of breeding areas of the three populations of harp seals. Arrows depict the direction of large-scale seasonal migrations back to breeding grounds. (After Sergeant 1991.)

Strait of Belle Isle (separating Newfoundland and Labrador) by mid-December. Here the migrating population splits, with roughly one-third entering the Gulf of St. Lawrence (referred as the Gulf population) and the remainder continuing down the east coast of Newfoundland (known as the Front). Adults lead this migration into the Gulf, but some juveniles are also present. At the Front, the situation is reversed with juveniles dominating coastal catches in December and January and adults arriving in January and February. Given that these data come from coastal harvests, this pattern may be biased if adults have a more offshore distribution. Not all juvenile harp seals make the southward migration. Some remain in the Arctic along the southwestern coast of Greenland. During January and February, adult harp seals appear to disperse widely throughout the Gulf and over the continental shelf off Newfoundland (Stenson and Kavanagh 1994) and fatten in preparation for reproduction. Recent data from satellite tags (G. Stenson and M. Hammill, pers. comm.) reveal extensive movement of adult harp seals covering much of the continental shelf within their winter range in the waters off Newfoundland.

The location, size, and number of harp seal whelping concentrations vary considerably both within and among years. Much of this variation seems to be the result of the combined effects of surface currents and wind, which can act to either concentrate or disperse the group. At the Front two large concentrations usually occur, each covering an area of up to 200 km^2. The density of female–pup pairs in these concentrations often ranges between 200 and 2000/km^2 (Fig. 9-3). During the early period of whelping, males are found in separate concentrations. After mating in the sea, adults disperse to feed, leaving weaned pups on the drifting ice floes.

In April, juveniles of both sexes and adult males form dense molting concentrations on the pack ice at the Front. Mean density (about 9700/km^2) in these concentrations is more than 10 times that found in whelping concentrations and can reach up to 21,000 seals/km^2. Adult females join these concentrations in late April. The proportion of time that harp seals spend on the ice during the molt is not known but is likely considerable. Individuals need to maximize skin temperature during the molt to promote rapid growth of hair (Feltz and Fay 1966). This can only be done by hauling out. Also both juveniles and adults lose considerable body mass over this period, indicating that relatively little feeding takes place during the molt (Chabot et al. 1995).

By the middle of May, most of the population follows the retreating ice edge north. Harp seals arrive along southwestern Greenland in June when capelin (*Mallotus villosus*), an

Figure 9-3. Aerial view of a harp seal whelping patch on the pack ice off Labrador, March 1979. (Photograph by W. D. Bowen.)

important food, concentrate to spawn (Sergeant 1991). All age groups appear to be represented at west Greenland, but juveniles and pups appear to move the farthest north. In the eastern Canadian Arctic, all age groups appear to be present, but pups and juveniles aged 1 and 2 years are underrepresented, presumably because most have gone to west Greenland. With the formation of new ice in September, harp seals begin their return migration along the east coast of Labrador. These large-scale movements represent an annual round trip of more than 4000 km.

Hooded seals (Fig. 9-4) give birth in March on the pack ice in four areas: in the southern Gulf of St. Lawrence, at the Front off Newfoundland, in the Davis Strait, and near Jan Mayen Island (Sergeant 1965). In this species, we see what seems to be an extreme adaptation to the unstable and temporary nature of pack ice. Hooded seal pups are weaned in just 4 days and almost immediately thereafter enter the sea to make their way to the edge of the pack ice (Bowen et al. 1985; see Costa and Williams, Chapter 5, and Boyd, Lockyer, and Marsh, Chapter 6, this volume). Unlike most other phocids, hooded seal pups also shed their long white birth coat, known as lanugo, in utero, therefore they are born with a short juvenile pelage. As the fetal coat of most phocids loses its insulation properties when wet, Oftedal et al. (1991) speculated that both the rapid growth of a blubber layer and in utero shedding are adaptations to the early entry of hooded seal pups into the cold sea.

In many ways, the hooded seal migration is similar to that of the harp seal, but hooded seals appear to be more closely associated with the edge of the continental shelf. Recent satellite telemetry data indicate that, after the breeding season, hooded seal adults feed along the continental slope off southern Newfoundland and the southern Grand Banks for about 20 days before migrating north across the Labrador Basin, arriving at west Greenland in June (Fig. 9-5; G. Stenson and M. Hammill, pers. comm.). Thus, unlike harp seals, hooded seals migrate before they molt. The reasons for this are unknown.

Recent studies using geolocation TDRs have shed considerable light on the distributional ecology of the two largest phocid species, the northern elephant seal (*Mirounga angustirostris*) and the southern elephant seal (*Mirounga leonina*). Northern elephant seals spend 8 to 10 months of each year at sea, and during that time they make two long-distance mi-

Figure 9-4. Adult hooded seal female and near-weaned pup (3 days old) on the pack ice off northeastern Newfoundland. (Photograph by W. D. Bowen.)

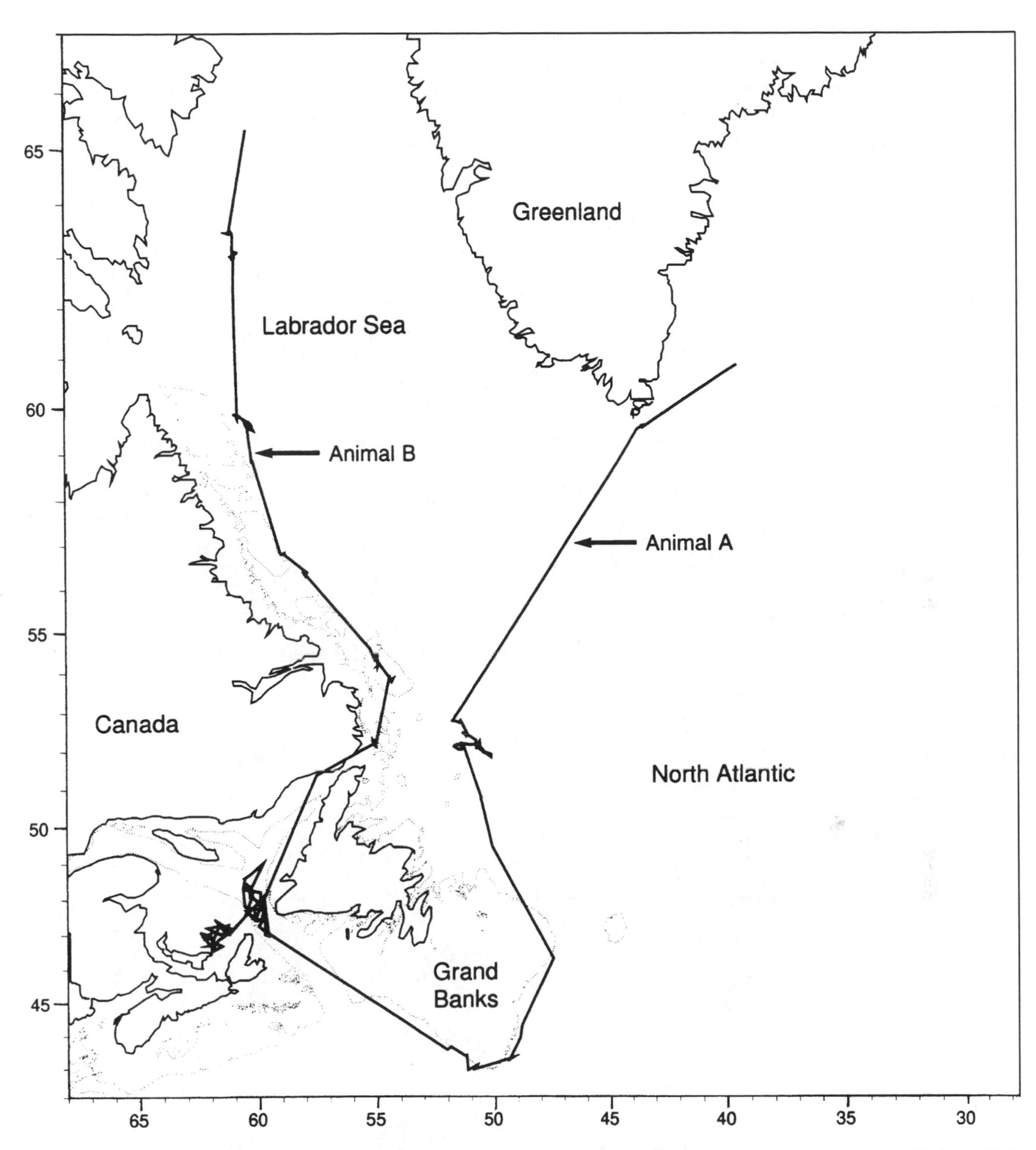

Figure 9-5. Adult hooded seal movements off eastern Canada as revealed by satellite-linked telemetry. Two animals tagged in the Gulf of St. Lawrence: one animal track (A) follows the southern Grand Banks and then across deep oceanic water to Greenland; one animal track (B) passes through the Strait of Belle Isle and north along the east coast of Canada to the Labrador Sea.

grations between breeding and molting sites on the southern California Channel Islands and pelagic foraging areas in the North Pacific (Stewart and DeLong 1993, 1995). At 18,000 to 21,000 km annually, this double migration is the longest reported for a mammal. However, as Stewart and DeLong (1995) point out, the application of new telemetry technology to study the migrations of whales and other seals may reveal annual movements that are similar to those observed in the northern elephant seal. In fact, as discussed later, recent studies suggest that the southern elephant seal exhibits a similar double migration.

The first migration of the northern elephant seal occurs

after the breeding season when adult females and adult males are at sea for an average of 73 days and 124 days, respectively (Fig. 9-6). During this postbreeding migration alone, adult females and males travel an average of 6289 km and 11,967 km, respectively, on a return trip from the breeding beaches in southern California to northern offshore foraging areas. The second migration occurs after the molt; females are at sea for approximately twice as long as males (an average of 234 days compared to 126 days) and cover an average distance of 12,264 km compared to an average of 9608 km by males (Stewart and DeLong 1995). During the course of a year, adult males and females are ashore simultaneously only for a relatively brief period (a few weeks) during the breeding season (Stewart and DeLong 1993).

Northern elephant seals begin to travel north immediately after the breeding and molting seasons; males and females use different foraging areas (Fig. 9-6), but both sexes use the California Current as a corridor to areas further north. Adult females remain south of 50°N latitude during both migrations but tend to forage in areas farther west during the postmolt period. There is preliminary evidence from three individuals that some females use similar migration routes and feeding areas between years. Males migrate farther north than females throughout the year with most traveling to the northern Gulf of Alaska and the eastern Aleutian Islands. These distribution patterns seem to correspond with three water masses in the North Pacific and with the distribution of cephalopods, which are the main prey of elephant seals. Females tend to feed near the Subarctic Current, between 40° and 50°N latitude, an area of cooler water to the north of warmer subtropical waters. Males aggregate along the offshore boundary of the Alaska Stream. These open

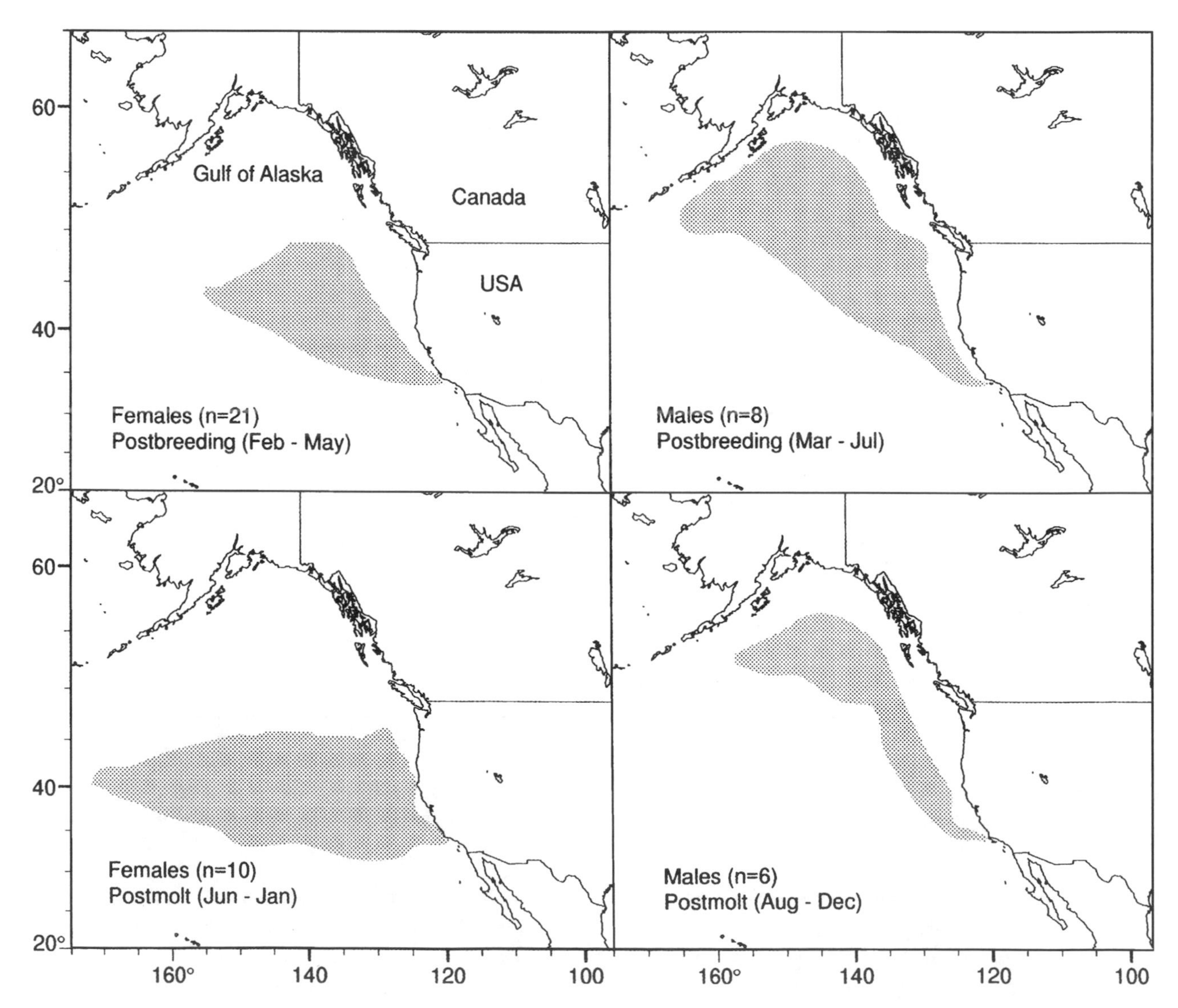

Figure 9-6. Migratory movements of male and female northern elephant seals in the eastern North Pacific as determined from geolocation time–depth recorders (from Stewart and DeLong 1995). Gray areas depict the areas used by the instrumented animals.

ocean frontal regions are known or thought to be areas of high biological productivity, including fish and cephalopod communities (Stewart and DeLong 1995).

Like the northern species, the annual cycle of adult southern elephant seals has separate aquatic periods after the breeding and molting seasons (Slip et al. 1994). Males and females studied at Macquarie Island appear to forage in somewhat different areas. Both sexes use cold Antarctic waters, but females also forage in the region around the Antarctic Polar Front during both the postbreeding and post-molting periods (Slip et al. 1994). In addition to their use of cold Antarctic water, some males also forage in the warmer subantarctic waters off Campbell Island near New Zealand after their molt. Studies of four satellite-tagged southern elephant seal females at South Georgia show a postbreeding migration toward the continental shelf margin of the Antarctic Peninsula (McConnell et al. 1992).

Unlike the species just discussed, harbor and gray seals are good examples of nonmigratory phocid species. Once again, however, our knowledge of the distributional ecology of these species is limited but accumulating rapidly, and hence, our conclusions about the nature of their seasonal movements may also change. Furthermore, both species have broad geographic ranges; therefore we might find differences among populations that reflect ecological variation.

The movement patterns of Atlantic harbor seals have recently been reviewed by Thompson (1993). In many parts of their range, harbor seals are seen in the same area throughout the year, leading to the conclusion that they are non-migratory. Studies using VHF telemetry confirm that they do not appear to forage more than 50 km from their haul-out sites (Stewart et al. 1989, Thompson and Miller 1990, Thompson et al. 1991). Larger scale movements (> 150–800 km) do occur in relation to a change in foraging sites (Thompson 1993) or juvenile dispersal (Thompson et al. 1994; W. D. Bowen, unpubl.). However, these greater movements appear to be unusual. For example, Thompson et al. (1994) found that 80% of recoveries of pups were less than 50 km from the site where they were tagged on Orkney, Scotland, and Härkönen (data cited in Thompson 1993) found that most pups from the Skaggerak remained within 20 km of their capture site until 3 or 4 years of age.

Seasonal changes in distribution are nonetheless commonly found in harbor seals. In some populations, certain sites are used for rearing young (Jefferies 1986, Kovacs et al. 1990, Olesiuk 1993), whereas other sites are mainly used during the molt (Thompson et al. 1989). Seasonal changes in movement patterns often appear to be related to changes in foraging areas. This is illustrated in the Moray Firth, Scotland (Thompson and Miller 1990, Thompson et al. 1996). Here, summer foraging typically occurs 20 to 40 km from haul-out sites, but during the winter, seals reduce their foraging to only 5 to 10 km from these sites when abundant prey (herring [*Clupea harengus*] and sprat [*Sprattus sprattus*]) move into inshore areas. Along the Pacific coast of North America, there are marked seasonal changes in the distribution of harbor seals at haul-out sites near estuaries during runs of salmonids (Brown and Mate 1983, Jefferies 1986, Olesiuk 1993).

The gray seal seems to exhibit seasonal dispersal from breeding and molting sites similar to that observed in the harbor seal but involving larger scale movements. In eastern Canada, our understanding of these movements comes mainly from more than 2500 recoveries of more than 50,000 gray seal pups that were flipper-tagged at breeding sites on Sable Island and in the southern Gulf of St. Lawrence between 1971 and 1990 (Stobo et al. 1990, Lavigueur and Hammill 1993). Most recoveries were from pups and juveniles (aged 1–3 years); therefore our understanding of the seasonal movements of adults is still somewhat scanty. Also, seasonal distribution patterns must be interpreted with some caution because recovery efforts varied geographically and seasonally in ways that cannot be quantified.

Gray seals in eastern Canada breed mainly on ice in the Gulf of St. Lawrence and on land at Sable Island and a few coastal islands off Cape Breton, Nova Scotia (Fig. 9-7). Gray seals of all ages use the waters near Sable Island and in the southern Gulf of St. Lawrence throughout the year, but large numbers of adults congregate in these areas to give birth and mate from December through mid-February (Fig. 9-8). Between the end of the breeding season and the molting period in May and June, recoveries of tagged adults suggest a general postbreeding dispersal over feeding areas on the Scotian Shelf and in the Gulf of St. Lawrence. It seems that only limited interchange between the Sable and Gulf components of the population occurs. Recoveries of juveniles and pups during this period suggest that younger gray seals are more wide ranging than adults, with recoveries coming not only from the Gulf of St. Lawrence and Scotian Shelf, but also from the Bay of Fundy, Gulf of Maine, and southern Newfoundland waters (Stobo et al. 1990, Lavigueur and Hammill 1993). Large numbers of juveniles and adults congregate on Sable Island and Anticosti Island to molt, but young gray seals are also recovered in large numbers along the coast of Nova Scotia and throughout the southern Gulf during this period. During the period from July to September, all age groups of gray seals appear to disperse more widely, such that there is greater mixing between these two populations. From October to December, most adult recoveries shift back to the eastern Scotian Shelf and southern Gulf of St. Lawrence; however, young animals remain broadly distributed through much of eastern Canadian waters.

The distribution of gray seals in waters surrounding the

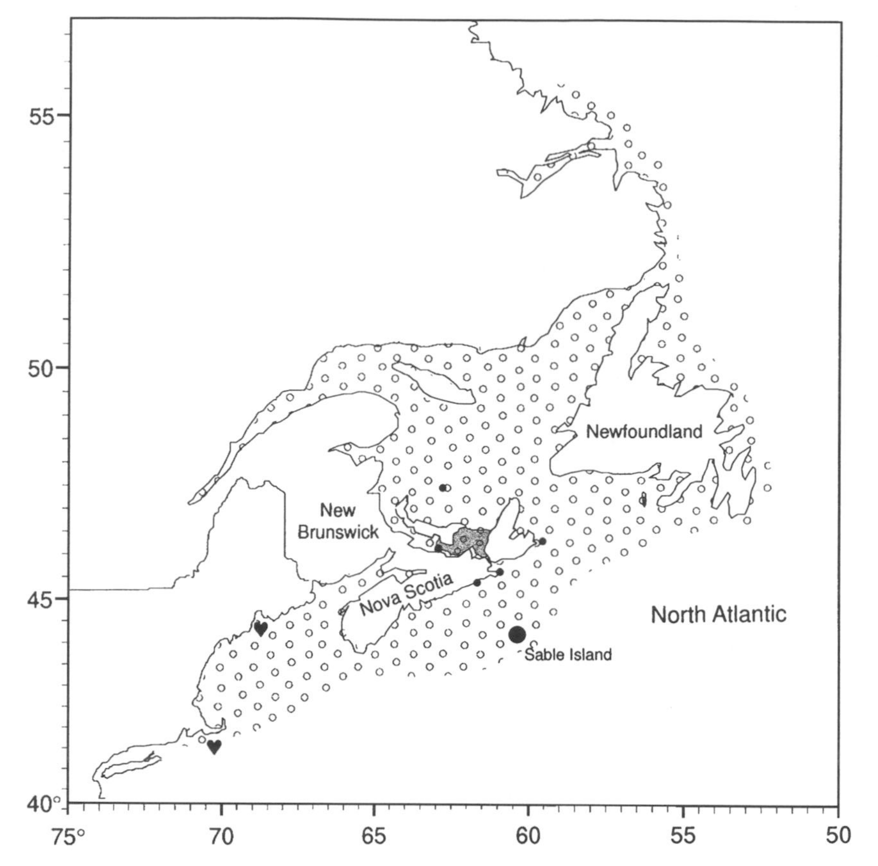

Figure 9-7. Distribution of gray seal breeding colonies in eastern North America. Gray shading indicates the area of ice used as pupping habitat in the southern Gulf of St. Lawrence. The large dot represents the largest colony located on Sable Island. The location of breeding colonies established since about 1987 are depicted by hearts.

Figure 9-8. Main colony of gray seals on Sable Island, Canada, during the breeding season in January. (Photograph by W. D. Bowen.)

United Kingdom is considerably more complex than in eastern Canada. Breeding occurs in the fall (from late September to early November), rather than in winter as in Canada, at eight major sites and at more than 40 colonies among those sites (J. Harwood, pers. comm.). Recoveries of pups from eight colonies around the United Kingdom 1 to 9 months after tagging indicate that the amount of geographic overlap among breeding sites varies considerably, but that, as in eastern Canada, pups disperse widely during the first year of life (Anonymous 1984). Researchers in the United Kingdom have used VHF tags, ultrasonic/VHF tags, and satellite telemetry to study details of seasonal movements of individual gray seals (for review, see Hammond et al. 1993). At the Isle of May on the east coast of Scotland, gray seal movements during the postbreeding period were widely distributed around the island extending some 70 km to the north, 45 km to the west, about 100 km to the south, and 200 km east. During the postmolt period from March to May, adult seals traveled more widely than juveniles. Throughout the summer, adults generally settled into a regular and often quite individual pattern of foraging. Similar patterns have also been found at the Farne Islands to the south, where juveniles spent more time near the islands but less time hauled out than adults during the postmolt period.

Satellite data from 23 gray seals captured at five sites on the east coast of the United Kingdom confirm that gray seals occasionally undertake long-distance movements (up to 2100 km). For example, one male made two trips of 26 and 35 days from the Farne Islands to Orkney Island, about 400 km away. However, gray seals captured at the Farne Islands spent most of their time within about 3 days' travel and 50 km of these haul-out sites (McConnell et al. 1994). These data further revealed that individuals often went to the same localized areas north of Orkney Island, which contain sandy sediment preferred by sand eels (*Ammodytes* spp.), a major prey of gray seals.

The Weddell seal (*Leptonychotes weddellii*) is an Antarctic pack ice phocid in which there is some evidence of migration. This species gives birth and breeds close to the Antarctic continent on sea ice that persists into the early austral summer. There is a tendency for this species to return to these areas after spending some time further north in pack ice regions. Not all individuals of a population display this behavior, and more data are needed to determine the extent to which the behavior develops (Testa 1994). The other species of Antarctic pack ice seals (crabeater [*Lobodon carcinophagus*], leopard [*Hydrurga leptonyx*], and Ross [*Ommatophoca* rossi]) appear not to migrate, although very little is known about their movement patterns.

The northern fur seal (*Callorhinus ursinus*) is the only otariid in which migratory behavior has been clearly described for both sexes. Northern fur seals at the Commander and Pribilof Islands undertake the longest migrations, moving as far south as 30° to 32°N, whereas those from Robben Island winter mainly in the Sea of Japan (Lander and Kajimura 1982). The seasonal distribution of animals from the Pribilof Islands is reasonably well understood based on at-sea research collections over many years (for reviews, see Kajimura 1980, Bigg 1982b). As in other species, the seasonal migration and local distribution patterns of northern fur seals are thought to have evolved in response to environmental variability and food supply. During the reproductive season (May to November), most of the northern fur seal population is in the eastern Bering Sea at the breeding colonies on St. Paul and St. George Islands. The timing of northern fur seal entry into the Bering Sea depends on age, sex, and reproductive status. Older mature males are the first to arrive at the Pribilof Islands beginning in May, and they hold territories for an average of 47 days (Peterson 1968). Younger bulls arrive with adult females in June. Some adult males winter in the Gulf of Alaska, indicating that they undertake a shorter migration than those of adult females (Bigg 1982b).

Pregnant females arrive in late June through July, and pupping is completed by early August. The oldest females arrive first, followed by a progression from older to younger animals of both sexes as the breeding season advances (Harry and Hartley 1981). Females mate about 5 days after giving birth and several days later leave the rookery on the first of a number of foraging trips to sea (Bartholomew and Hoel 1953). Loughlin et al. (1987) studied the characteristics of feeding trips in 40 female fur seals that had been equipped with head-mounted radio transmitters. In general, the feeding areas of these females were between 160 and 200 km from the rookery. Although most females foraged over the continental shelf, several did so in areas off the shelf where water depths exceeded 3000 m. With one exception, radio-tagged females were located west of the rookery near the continental shelf break. These results agree well with records of the location of vessels engaged in pelagic sealing of fur seals during the summers of 1883 through 1897 (Harry and Hartley 1981).

After nursing for an average of 118 days (Gentry and Holt 1986), adult females migrate from the Bering Sea through the Aleutian Islands and then southward to coastal areas along northern California. Immature females winter somewhat further north off Washington, and immature males even further north along the coast of British Columbia. Immature fur seals aged 3 years and older and adult females begin their northward migration in March (Wilke and Kenyon 1954). Yearlings and 2-year-olds appear to follow the winter migration routes used by older immatures, but few of these young animals seem to summer in the Bering Sea as these age classes are greatly under-represented at the rookeries on the Pribilof Islands (Lander and Kajimura 1982).

Cetacea

Gaskin (1982) provides an excellent analysis of the distributional ecology of cetaceans, and readers are encouraged to consult this source for additional detail. Cetaceans are entirely aquatic and therefore have undergone extreme anatomical adaptation compared to their terrestrial ancestors. The majority of species are marine, but estuarine and freshwater species do occur in tropical and subtropical habitats (e.g., the river dolphins). Of the two extant suborders of whales, the mysticetes exhibit the most notable long-distance seasonal migrations. Some odontocetes also undertake seasonal migration, but there does not appear to be a common pattern of seasonal movements comparable to that found in baleen whales (Lockyer and Brown 1981).

MYSTICETES. Nine of the 11 species of baleen whales undertake long-distance seasonal migrations. Eight species (blue [*Balaenoptera musculus*], fin [*Balaenoptera physalus*], sei [*Balaenoptera borealis*], minke [*Balaenoptera acutorostrata*], humpback, gray, northern right [*Eubalaena glacialis*], and southern right [*Eubalaena australis*]) migrate from their winter range in tropical or subtropical waters to summer feeding grounds in temperate and polar waters; the bowhead whale (*Balaena mysticetus*) migrates seasonally but never leaves polar waters. There is some evidence that Bryde's whales (*Balaenoptera edeni*) also migrate, although not as extensively (Lockyer and Brown 1981, Gaskin 1982). Although they remain in warm water throughout the year, large numbers of Bryde's whales are found in areas of high productivity brought about by upwellings near oceanic islands and continental slopes (Gaskin 1982). Not enough is known of the pygmy right whale (*Caperea marginata*) to draw firm conclusions about distribution or movements at this time.

The annual migration of baleen whales from low latitudes in winter to high latitudes in summer enables these filter feeders to exploit the high seasonal productivity of certain polar areas and yet mate and give birth in warm water areas of low productivity. It has been assumed that the movement to warm water areas during the annual migration has resulted in increased calf survival and, therefore, was selected. Interestingly, unlike the other mysticetes, the bowhead whale is a year-round resident of polar waters in both the North Atlantic and North Pacific. This, coupled with the fact that many small odontocetes are able to penetrate cold water, has led some researchers to conclude that the reason for the annual movement between warm and cold waters may not be as simple as believed (Gaskin 1982, Evans 1987).

Brodie (1975) argues that baleen whales migrate to low latitude, warmer waters during the winter because prey densities high enough to permit feeding at high latitudes are restricted to limited periods during the summer. Although prey density at low latitudes are insufficient to permit efficient feeding, inhabiting warmer waters at these latitudes enables individuals to fast until prey densities once again reach high levels at high latitude feeding areas. According to Brodie (1975), selection for larger body size among baleen whales is a way to adjust specific metabolic rate and permit efficient use of lipid stores during extended exclusion from feeding grounds. Thus, he argues that intraspecific variation in body size, such as that found in blue and fin whales where the Antarctic forms are larger than those in the northern hemisphere, is a response to differences in the duration of the feeding and fasting periods. Larger forms in the Antarctic probably feed over a much more restricted period of the summer than those in the northern hemisphere, where somewhat lower prey densities occur for longer periods of the year.

Our understanding of the seasonal movements of baleen whales is greatest for species such as humpback and gray whales that pass close to land where they can be observed over broad spatial scales. In the North Atlantic, humpback whales migrate between breeding areas in the West Indies and feeding areas from the mid-Atlantic states to Iceland (Fig. 9-9). More than 3600 North Atlantic humpback whales have been photographed in five regions where they feed from spring to fall and in four regions where mating and calving take place during the winter (Katona and Beard 1990). Photographic resightings of these individually identifiable whales have resulted in a better understanding of feeding group aggregations, migration patterns, and the size of the North Atlantic population. Only 26 of 643 humpbacks were resighted in more than one of five feeding ranges: Iceland, Greenland, Newfoundland, Gulf of St. Lawrence, and Gulf of Maine/Scotian Shelf (Fig. 9-9). Humpback whales from each of these feeding ranges migrate to breeding areas in the West Indies. Whales from all five feeding ranges interbreed at the Silver Bank area off the Dominican Republic. However, whales from Newfoundland had the highest representation at the Dominican Republic, Virgin Bank, and Bermuda, whereas those from Iceland and the Gulf of St. Lawrence were strongly represented at the breeding area off Puerto Rico. During the northward migration after breeding, humpbacks do not separate into relatively distinct feeding groups until they are north of Bermuda (Katona and Beard 1990).

Photo-identification data (see Wells, Boness, and Rathbun, Chapter 8, this volume) also have helped to clarify the migration patterns and population structure of humpback whales in the central and eastern North Pacific (Fig. 9-9).

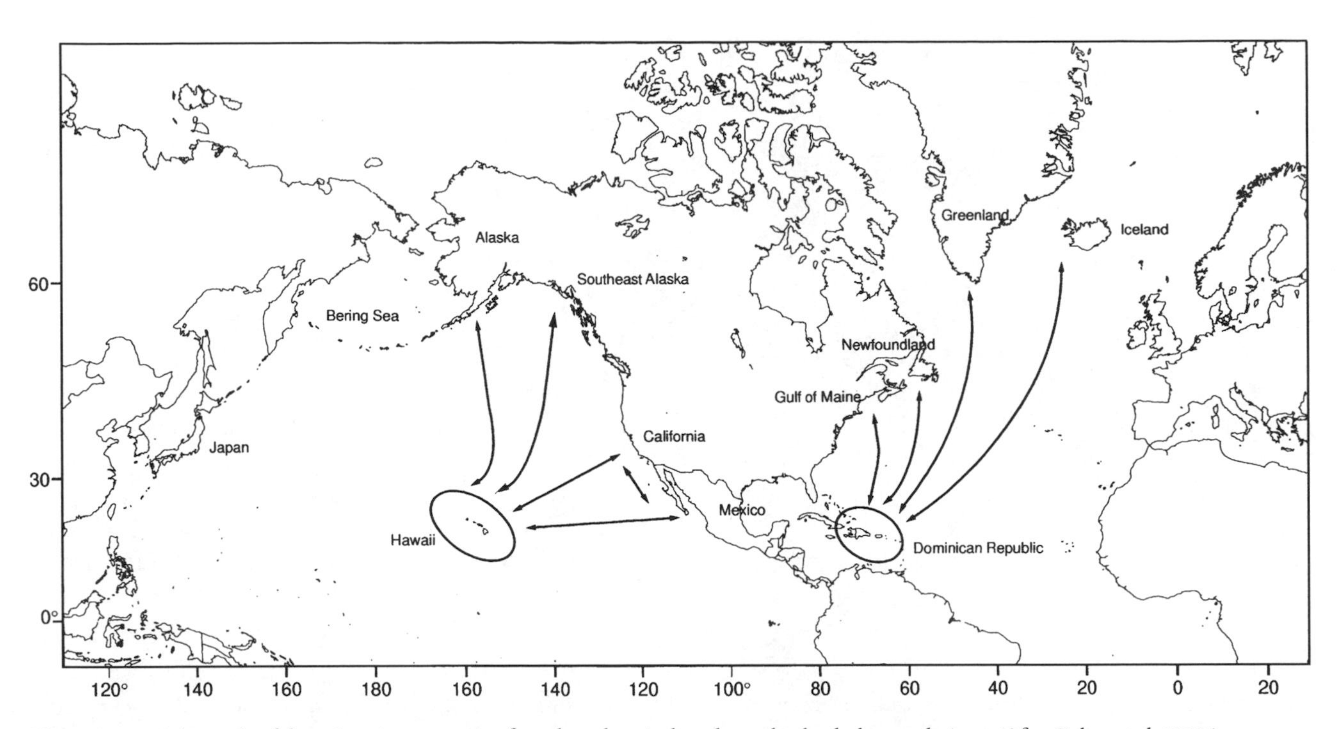

Figure 9-9. Schematic of the migration patterns of northern hemisphere humpback whale populations. (After Baker et al. 1990.)

Baker et al. (1986) summarized resightings between 1977 and 1983 from the two winter breeding areas (Hawaii and Mexico) and five summer feeding areas (the Farallon Islands, California; southeastern Alaska; Yakutat Bay, Alaska; Prince William Sound, Alaska; and the western Gulf of Alaska). They concluded that, as with the North Atlantic population, humpback whales in the central and eastern North Pacific occupy a number of geographically separate feeding ranges. Some exchange among winter breeding areas does occur, but more extensive intermingling appears to occur during summer feeding. More recent photo-identification data collected between 1986 and 1992 at nine feeding sites in the western North Pacific revealed a high degree of interchange among humpback whales seen off California, Oregon, and Washington (Calambokidis et al. 1996). However, a low rate of interchange occurred between whales seen off British Columbia and California, and none occurred between whales photographed in California and Alaska.

Lactating females are among the first to leave the feeding areas in the fall, followed by immature animals, mature males, nonpregnant females, and pregnant females. The order is roughly reversed in the late winter as the northward migration begins (see Clapham 1996).

Gray whales migrate annually between calving areas in the lagoons in Baja California, Mexico, and feeding grounds in the Arctic (Rice and Wolman 1971). Because of its the near-shore distribution, the seasonal movements are probably known better than for any other whale species. On the basis of sightings from ships and aerial surveys, Braham (1984) showed that the migration route of gray whales in Alaska is coastal, as they circumnavigate the Gulf of Alaska to Unimak Pass and into the Bering Sea (Fig. 9-10). The northbound migration from March to June appears to occur in two phases; the first includes most nonlactating adults and immatures, followed about a month later by lactating females and their young. The whales continue up the coast of Alaska, arriving at St. Lawrence Island by May as the ice recedes. By June, gray whales are common in the northern Bering Sea, but their distribution is dependent on the distribution of ice. Gray whales are widely dispersed throughout their northern foraging areas, including the Chukchi Sea, until late November, by which time the southward migration is well underway. Gray whales move south through the Unimak Pass in November and December, and then continue along the coast southward to Baja California. About 90% of the southbound gray whales pass along the Oregon coast between mid-December and late January (Herzing and Mate 1984). Gray whales arrive in the breeding lagoons beginning in mid-January, with peak numbers occurring about mid-February. Throughout most of the migration, gray whales stay within several kilometers of the coast where coastal and bottom topography over shallow waters may assist in orientation (Pike 1962).

ODONTOCETES. A number of odontocetes show strong seasonal changes in distribution. Wells et al. (1980) reviewed data on 24 species and concluded that migrations are generally seen in the more pelagic (i.e., offshore) species. How-

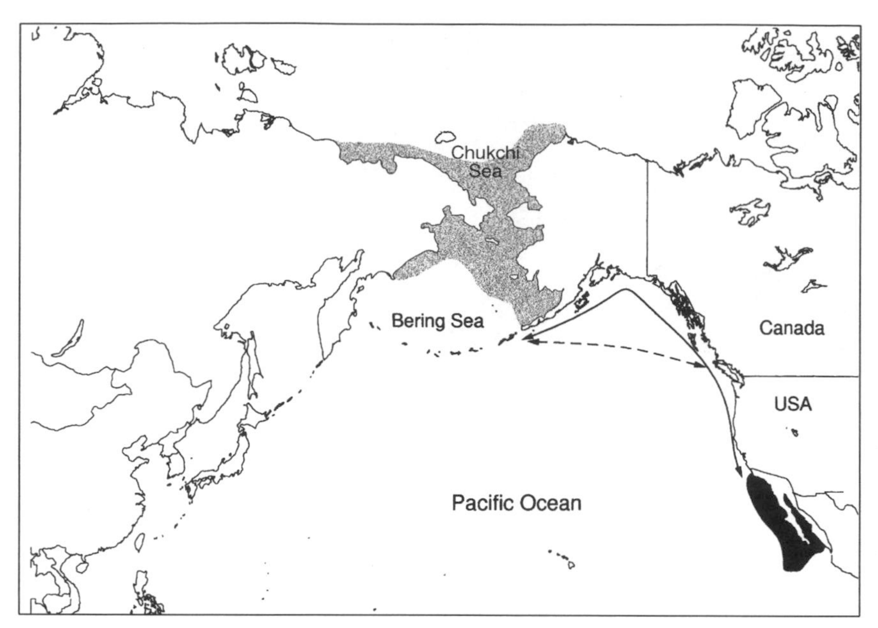

Figure 9-10. Migration patterns of the western Pacific population of gray whales. Light gray shading is the summer feeding area, whereas the dark gray area is the breeding ground. Dashed line indicates a less commonly traveled migration route. (After Evans 1987.)

ever, they noted several exceptions (the harbor porpoise [*Phocoena phocoena*], beluga whale, and narwhal [*Monodon monoceros*]) where temperature or ice strongly influence seasonal distributions.

On the basis of sightings from the Cetacean and Turtle Assessment Program surveys, Payne and Heinemann (1993) concluded that long-finned pilot whales (*Globicephala melaena*) moved off the continental shelf edge of the northeastern United States during the winter in response to changes in prey distribution. During the spring, pilot whales concentrated near the shelf edge/slope waters, moving up onto the shelf throughout the summer, particularly on Georges Bank. Waring et al. (1990) found the major concentrations of pilot whales during winter along the shelf edge from Cape Hatteras northward to Georges Bank, an area corresponding to the distribution of long-finned squid (*Loligo pealei*).

Several scales of movement patterns have been described from studies in the eastern tropical Pacific of species of *Stenella,* including daily movements of about 30 nautical miles (nm) per day (53 km/day) and movements over periods of months of 500 to 600 nm (about 1000 km) (Perrin 1975). During these longer periods, seasonal changes in distribution involved an inshore movement toward the tropical Americas in the fall and winter and an offshore movement during the spring and summer. Whether these movements might properly be described as migratory is perhaps a subjective judgment.

Selzer and Payne (1988) studied the seasonal distribution of Atlantic white-sided dolphins (*Lagenorhynchus acutus*) and common dolphins (*Delphinus delphis*) on the continental shelf of the northeastern United States. In spring, Atlantic white-sided dolphins were common in shallow coastal waters of the Gulf of Maine and through the Great South Channel along the western edge of Georges Bank, areas that coincide with the distribution of important prey, the sand lance (*Ammodytes americans*). Common dolphins used warmer waters along a broad band paralleling the continental slope to the south of Georges Bank. During fall, white-sided dolphins appeared to disperse northward into the deeper waters of the Gulf of Maine. Common dolphins moved onto the eastern side of Georges Bank during fall. Both in spring and fall, their distribution coincided with those of prey such as mackerel (*Scomber scombrus*), butterfish (*Peprilus triacanthus*), and squid (*Illex illecebrosus* and *Loligo pealei*).

The harbor porpoise is broadly distributed along the continental shelf on both sides of the North Atlantic (Donovan and Bjorge 1995). In the northwest Atlantic, several lines of evidence suggest the existence of three populations in the Gulf of Maine and Bay of Fundy, Canada, the Gulf of St. Lawrence, and off eastern Newfoundland and Labrador. In the Gulf of Maine and Bay of Fundy, large numbers of porpoise are present only during the summer months (Gaskin et al. 1975, Gaskin 1977). Most migrate south in the autumn through the Gulf of Maine to an unknown wintering area; however, a few harbor porpoises do remain in coastal waters off the southern Bay of Fundy throughout the winter. During winter in the North Sea, harbor porpoises appear to form two major groupings, one off Denmark and the other in deeper waters of the northeastern North Sea. These groupings disappear during the spring and summer as porpoises disperse over broader areas (Northridge et al. 1995).

In species with broad geographic distributions, such as the bottlenose dolphin (*Tursiops truncatus*), movement patterns vary from year-round residency to seasonal migrations (Shane et al. 1986). Along the west coast of Florida, dolphins frequently pass from embayments past barrier islands into the Gulf of Mexico during the fall through early spring, whereas, during the warmer summer months, they are generally found inside the bays (Scott et al. 1990). Mullet migrate into the Gulf to spawn during the fall, and it is likely that the dolphins shift their distribution to follow this principal prey. Scott et al. (1990) also speculate that the appearance of the bull sharks (*Carcharhinus leucas*) in the inshore waters of the Gulf may prompt bottlenose dolphins, and particularly females with young calves, to prefer the more protected waters of the bays where the risk of predation is reduced.

As indicated above, few odontocetes appear to show the kind of long distance seasonal migration that is found among the mysticetes. One odonotocete that migrates over long distances is the sperm whale, the species with the most extensive distribution of any marine mammal. The following account is taken largely from Best (1979). Sperm whales exhibit the greatest size dimorphism of the living cetaceans, with adult males, at 15.8 m and 43.5 tons, being 1.4 and 3.2 times the length and mass, respectively, of adult females. Sperm whales segregate seasonally by sex, but also by size among the male bachelor groups. In the southern hemisphere, mixed-sex schools and schools of small bachelors are believed to have similar distributions and migratory behavior. These groups move toward the equator in the fall and toward the subtropical oceanic convergence in the spring. Medium-sized bachelors travel toward the equator later in the year and leave equatorial waters earlier, whereas large males arrive in tropical waters last and leave the earliest for the higher latitude feeding grounds. There are few data on the distances traveled during the migration, but straight-line measurements suggest about 1550 km for males and 680 km for females. Using sightings of photo-identified sperm whales, Dufault and Whitehead (1995b) concluded that movements of female and immature whales from the Galapagos population were restricted to about 1000 km, the distance between the Galapagos Islands and the area off Ecuador and northern Peru where resightings occurred. Behavioral studies off the Galapagos Islands have revealed that female sperm whales form permanent units comprised of about a dozen individuals (Whitehead et al. 1991). Recent molecular genetic studies of this population indicated that these units consisted of matrilineal groups of related individuals (Richard et al. 1996). Males, however, disperse from these units when they are about 6 years old and join bachelor groups or remain alone. Maturing and mature males spend most of their time feeding at high latitudes (i.e., between 40° to 60°). Many younger males do not undertake the migration back to tropical waters as do mature males, concentrating on feeding in the more productive cool waters where there is no competition from females (Weilgart et al. 1996).

Polar Bears

The polar bear has a circumpolar distribution that is confined for the most part to areas above the Arctic Circle (DeMaster and Stirling 1981). Exceptions to this do occur in the northern Bering Sea and in the eastern Canadian Arctic and southern Greenland (Wiig et al. 1995). In all areas, bears undergo seasonal movements, which in some cases can be reasonably termed migrations, in response to changes in the amount and distribution of sea ice. In some populations, such as the one in western Hudson Bay, bears migrate onto land when the sea ice melts in summer. When on land they spend most of the time resting and feed little. Except for pregnant females, all bears return to the ice in November. Pregnant females remain on land where they give birth to cubs in late December or January. They move to the ice between mid-February and April. Activity levels and distance traveled by polar bears have also been studied in the Canadian Arctic archipelago. Here, Messier et al. (1992) found that bears were most active and traveled the greatest distances from May through July, coinciding with a period when seal pups are most available to predation (Archibald and Stirling 1977, Smith 1980) enabling bears to accumulate large fat reserves (Ramsay and Stirling 1988).

Sirenians

The distribution of extant sirenians is summarized by Reynolds and Odell (1991) and Reeves et al. (1992). All species occupy tropical or subtropical waters. The dugong (*Dugong dugon*) is the most abundant and widely distributed sirenian. It inhabits the waters of 43 different countries along the western Pacific and Indian Oceans, from Mozambique and Madagascar to regions north of the Philippines. Each of the manatee species occupies a more restricted range. Interestingly enough, early manatees may have outcompeted and eventually replaced dugongid sirenians in the Caribbean about 5 million years ago (Domning 1981).

Currently, the range of the West Indian manatee (*Trichechus manatus*) extends (discontinuously) from the southeastern United States through Central America and various Caribbean islands, to northeastern Brazil (Reynolds and Odell 1991). Lefebvre et al. (1989) provide a good overview of the species' status and distribution. The Amazonian manatee occupies freshwater lakes and rivers in the Amazon Basin; the species is found in four countries—Brazil, Colombia, Peru, and Ecuador (Reynolds and Odell 1991). Powell (1996) provides the most up-to-date account of the distribution and status of West African manatees

(*Trichechus senegalensis*), which are found in about 21 countries, from Angola to Senegal.

The sirenians occupy waters of different salinities. The dugong inhabits primarily marine areas, whereas the Amazonian manatee is confined to freshwater. The other manatees, West African and West Indian, are euryhaline (summarized by Reynolds and Odell 1991).

Seasonal changes in distribution and other biological attributes have recently been particularly well studied in Florida manatees (*Trichechus m. latirostris;* see O'Shea et al. 1995). Various field techniques, including photo-identification (Beck and Reid 1995), telemetry (e.g., see Reid et al. 1995), and aerial surveys (Reynolds and Wilcox 1994, Ackerman 1995, Garrott et al. 1995) have been used to establish long-term data bases. For this species, seasonal cold weather is the greatest determinant of distribution. In warm months, Florida manatees disperse as far north as Rhode Island and as far west as Texas, but during cold weather, virtually all animals retreat to Florida, where they often occupy natural and artificial warm water refugia (summarized in Reynolds and Odell 1991, Reynolds, Chapter 12, Twiss and Reeves 1999). Marsh and Rathbun (1990) found that cold weather induced radio-tagged dugongs to travel more than 100 km to warmer waters.

Amazon manatee distribution changes seasonally with the onset of wet and dry seasons (Best 1983). The same may hold true for West African (Powell 1996) and Antillean (*Trichechus m. manatus;* Lefebvre et al. 1989) manatees.

Factors Affecting Distribution

Habitat

Ice plays an important role in the seasonal distribution of many marine mammals, providing habitat for some while reducing access to habitat for others (Fay 1974). The presence of ice affects both the physical and biological oceanography of an area through attenuation of light, reduction of heat exchange, and mixing (Alexander 1981). The underside of ice supports a spring community dominated by diatoms, which serves as food for grazing zooplankton. Ice edges form another distinct environment in which there is a strong gradient in temperature and salinity, particularly in spring when rapid melting of sea ice occurs. Seasonal patterns of ice cover are associated with changes in marine productivity (e.g., Niebauer et al. 1981) and the distribution and availability of prey populations that, in turn, can affect the distribution of marine mammals. Evidence of this effect comes from the sharp peak in the abundance of marine birds and marine mammals that is often found in ice edge areas. Smith and Martin (1994) inferred that the movement of belugas into Peel Sound, in the Canadian high Arctic, was directly related to the persistence of a floe ice edge and the associated high food availability.

Fay (1974) suggested the following advantages associated with the use of seasonal ice by pinnipeds: isolation from terrestrial predators, unlimited space, proximity to food supply, transportation to new habitats, reduced risk of disease transmission, and shelter. Through its dynamic nature, ice can be an unstable and somewhat unpredictable habitat, as variation in ice conditions off eastern Canada in 1968 and 1969 illustrates so well (Fig. 9-11). This, and comparable variation in the Bering Sea (Burns and Gavin 1980), can have important effects on marine mammals. Sergeant (1991) found that lack of suitable ice in the Gulf of St. Lawrence, eastern Canada, in 1953, 1969, and 1981 reduced both births and subsequent survival of harp seal pups. Also in 1981, lack of ice resulted in premature weaning, interrupted neonatal molt, and high pup mortality among harp seals born off Labrador (W. D. Bowen, unpubl.).

Ice cover also plays an important role in the timing of births and the overall distribution of the breeding harp seals (Sergeant 1991). Pups are born in the Gulf in late February, usually about 5 days earlier than at the Front. The survival of harp seal pups during the first 2 weeks of life depends on the availability of stable habitat. At the Front, heavy ice provides this condition until late March. In the Gulf, however, the ice usually begins to disappear by mid-March, and thus, for pups to survive, they must be born earlier than at the Front. In years of little ice in the southern Gulf, females that would have whelped in the Gulf reproduce instead on the ice floes at the Front.

Perhaps the most obvious effect of ice on the distribution of marine mammals is simply its physical presence, which reduces access to areas that are used at other times of the year. Stirling et al. (1982) reported a 50% drop in the numbers of ringed seals and bearded seals (*Erignathus barbatus*) between 1974 and 1975 after heavy ice conditions during the winter of 1973 to 1974. Although the causes of the decline are uncertain, the heavy ice may have reduced food availability, resulting in increased winter mortality, reduced pup production, and emigration of seals.

Polar bears depend on sea ice as a platform from which they hunt seals (Stirling and Archibald 1977), as a place to mate (Ramsay and Stirling 1986), as a substrate on which to travel between denning areas and foraging areas, and in some areas for maternity denning itself (Lentfer 1975). The distribution of polar bears is strongly influenced not only by the availability of sea ice, but the characteristics of the ice that directly influence both the abundance and availability of

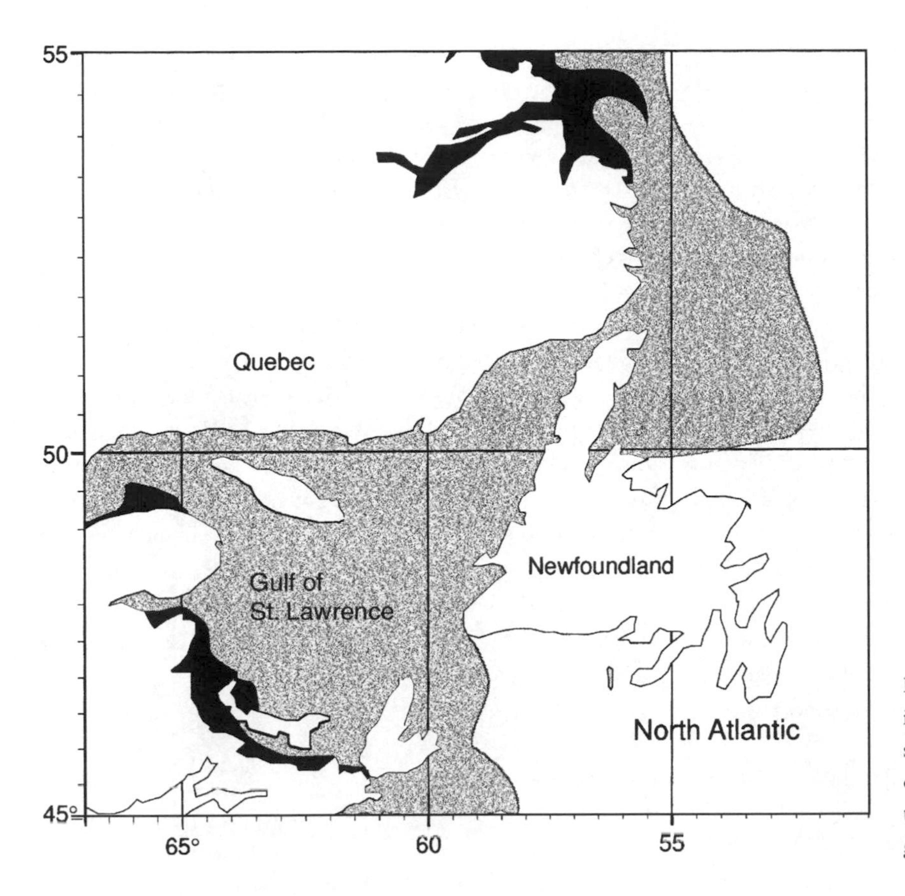

Figure 9-11. Variation on the extent of ice cover in eastern Canada in a year of extensive ice (1968, light gray shading) and a year of little ice (1969, dark shading) during the period of harp seal whelping. (After Sergeant 1991.)

seals (Stirling et al. 1993). Polar bears in the western Canadian Arctic generally showed a strong preference for the edge of ice floes and moving ice, habitats where bearded seals and nonbreeding ringed seals are generally abundant. Adult females with cubs of the year, however, generally avoided these less stable habitats in favor of the stable fast ice with drift areas suitable for ringed seal haul-out and birth lairs (Stirling et al. 1993).

Sea ice has a dominant influence on the ecology of the Southern Ocean, expanding to cover an area as large as 20 million km^2 in winter and contracting to about 4 million km^2 in summer (Laws 1984). The edge of the pack ice and the upwelling of nutrients along the Antarctic Divergence are areas of high biological productivity. The biological characteristics and our understanding of the dynamics of the Southern Ocean, particularly the areas covered annually by sea ice, have been slow in developing. In the 1960s and early 1970s, the Southern Ocean was thought to be the most productive region in the world. This view was a result of the high biomass levels of the primary producers (primarily diatoms) as well as at the upper trophic levels, including seabirds, seals, and whales. However, it now appears as more data become available, that the Southern Ocean's primary production is not as high as previously thought. Total annual primary production has been estimated to be less than 5% of the global total, although the Southern Ocean accounts for about 10% of the global ocean (Smith 1991). A recent book (El-Sayed 1994) summarizes studies from a large multinational research effort that took place in the 1980s and gives insight into functioning of this large ecosystem. The apparent dilemma of how the very large biomass that exists at the upper trophic levels is supported by relatively low primary production will certainly be the focus of future investigations.

The standing biomass of zooplankton in the Southern Ocean is significantly higher than that in either tropical or temperate oceans. Furthermore, a single species, Antarctic krill (*Euphausia superba*), probably accounts for half of the zooplankton biomass in the Southern Ocean. Increased zooplankton production, coupled with the retreat of ice in the summer, results in a great increase in krill availability to marine mammals.

Smith and Martin (1994) illustrate how several factors may interact to affect seasonal distribution. They examined the seasonal distribution of beluga whales over a 6-year period in the Canadian high Arctic. Access to summering areas around Somerset Island is limited by land-fast ice, which forms in winter and spring. Belugas begin to occupy their high Arctic summer range during July, and by early August, more than 4000 whales inhabit the shallow bays receiving the outflow of relatively warm freshwater rivers. The use of these warm water estuaries appears to promote the annual skin molt (St. Aubin et al. 1990). By late August to early September, belugas rapidly begin to migrate east to Lancaster Sound to feed on large schools of Arctic cod (*Boreogadus saida*).

Oceanographic conditions also appear to have a strong influence on the distribution of cetaceans. For example, although sperm whales inhabit the deep oceans throughout the globe, their distribution is linked to areas where upwelling of cool nutrient-rich water increases biological productivity (Volkov and Moroz 1977). It is likely that sperm whales congregate in these areas because of high concentrations of cephalopods (Smith and Whitehead 1993). Jaquet and Whitehead (1996) examined the spatial scales over which physical and biological oceanographic conditions influenced sperm whale distribution in the South Pacific. Sperm whales were encountered on 70 occasions in 264 segments of 80 nm. Although there was no correlation between whale density and oceanographic variables at fine spatial scales (80 nm), over large spatial scales (320–640 nm) female sperm whales were clustered in areas of high secondary productivity and high bathymetric relief.

Reilly and Thayer (1990) found that all sightings of blue whales in the eastern tropical Pacific occurred in relatively cool upwelling areas near the Galapagos Islands, off Baja California, and off Costa Rica. Although whaling data indicate that blue whales rarely fed in low latitudes, Reilly and Thayer speculate that feeding may be more important in these areas than previously thought.

Predation

Predation is an important selection pressure in the evolution of some pinnipeds (Stirling 1977) and no doubt this is also the case among the cetaceans, particularly the smaller species. Predation by polar bears and Arctic foxes undoubtedly plays a role in the selection by ringed seals in the Canadian Arctic of fast-ice breeding habitat that is suitable for the construction of birth lairs (Stirling 1977). Kingsley and Stirling (1991) argue that selection of haul-out sites by ringed seals also reflects the constant threat of polar bear predation. Hammill and Smith (1991) found that 24% of 562 ringed seal breathing holes examined during the spring showed evidence of attempted predation by polar bears. Pups appeared to be the target of this predation; in fact, up to 44% of annual pup production is estimated to have been consumed by polar bears before weaning. Walruses also prey on seals, and this seems to happen in the Bering Sea, particularly in years of reduced ice cover (Lowry and Fay 1984). This predatory interaction may partly explain the segregation of species that occurs during years of "normal" ice cover (Fay 1974). Risk of predation by transient killer whales may also affect the distribution and behavior of harbor seals (Baird and Dill 1995). Some transient killer whale pods spent much of their time foraging near harbor seal haul-out areas, particularly during the pupping season. Of 138 prey capture attempts by transient killer whales, 136 were successful, and young harbor seals comprised the majority of these kills. A number of small odontocetes are also subject to predation by killer whales and sharks (Wells et al. 1980).

In the Antarctic pack ice, the leopard seal is a major predator on many species of vertebrates (Siniff and Stone 1985) and is a significant predator on other seals, particularly the crabeater. It was observed during early explorations to the Antarctic that the crabeater seals often had long raking scars that were attributed to killer whale attacks. However, more recent data show that these scars were caused by leopard seal attacks, primarily on young of the year (Siniff and Bengtson 1977). Such predation is thought to be significant in many regions of the Antarctic as up to 80% of adult crabeaters bear scars caused by failed predation attempts by leopard seals.

Population Biology

Abundance

It is difficult to imagine how we could claim a reasonable understanding of the ecology of marine mammals without having good estimates of population numbers and trends. Yet, despite the interest in the ecology of marine mammals, the abundance of many species remains poorly known, and this lack of basic data presents a serious obstacle in the assessment of their population biology. Good estimates of abundance exist for only a small proportion of marine mammal species, and in fact, reliable estimates are usually available only for some populations within species. In general, the abundance of pinnipeds is better known than that of other marine mammals, and the abundance of species of commercial importance (past or present) is generally better known than for those species that have not been exploited. Evans (1987) lists as unknown the abundance of 50% of cetacean species. Even where information on numbers exists, the precision attached to the estimates varies greatly, owing to the difficulty in designing good census procedures or to the lack of effort to obtain good estimates. In many cases, we can only rank species as rare, uncommon, common, or abundant.

The abundance of marine mammals varies enormously, as illustrated by selected species listed in Table 9-3. The abundance estimates for the listed pinnipeds come from a review by Reijnders et al. (1993). The cetacean estimates are from Brownell et al. (1989) and Evans (1987), whereas those for the polar bear are from Wiig et al. (1995). These abundance estimates represent what might be thought of as mean values about which there are varying degrees of uncertainty de-

Table 9-3. Estimates of Worldwide Abundance of Selected Marine Mammals[a]

Species	Approximate Abundance	Time Period[b]
Pinnipeds		
Crabeater seal	12,000,000	Early 1980s
Harp seal	6,000,000	1988–1994
Gray seal	200,000	Early 1990s
Mediterranean monk seal	< 500	Late 1980s
Antarctic fur seal	1,500,000	1990–1991
Northern fur seal	1,200,000	Early 1990s
Guadalupe fur seal	6000	1987
California sea lion[b]	160,000	1970s–1980s
Hooker's sea lion	10,000	Early 1990s
Walrus	240,000	1980s
Cetaceans		
Sperm whale	2,000,000	1980s
Minke whale	825,000	1980s
Blue whale	14,000	1980s
Bowhead whale	8000	Late 1980s
Northern right whale	< 1000	1980s
Spinner dolphin (ETP)[c]	2,000,000	1980s
Indus river dolphin	500	1980s
White whale	40,000	1980s
Sirenians[d]		
Florida manatee	2500	1990s
Dugong	100,000	1980s
Polar bear[e]	21,000–28,000	Early 1990s
Sea otter[f]	100,000	1990s

[a]From Reijnders et al. 1993, Brownell et al. 1989, Evans 1987.
[b]Populations estimated at different times.
[c]Eastern tropical Pacific populations.
[d]Summarized by Reynolds and Odell 1991.
[e]Data from Wiig et al. 1995.
[f]Data from Bodkin et al. 1995.

pending on the species considered. We have not given estimates of precision because for the most part they are not available for the world population of these species, although particular populations may be well studied. For example, the abundance of the northwest Atlantic population of harp seals is known with good precision (Shelton et al. 1995), but the abundance of the other two populations is not.

It is important to recognize that commercial exploitation decimated many marine mammals species, in some cases to levels nearing extinction (e.g., elephant seals and right whales). During the past several decades, however, some have recovered or are continuing to recover. Thus, the present abundance of these species may not be a good guide to their preexploitation numbers. This is illustrated by reduction in the number of several baleen whales. Brownell et al. (1989) estimate that blue whale numbers were reduced from preexploitation levels of more than 200,000 to about 14,000 in the 1980s. Before exploitation, the population size of sei whales is thought to have been almost 550,000 animals. Today, only 120,000 sei whales are believed to inhabit our oceans. Southern right whales were reduced more than 90% from an estimated 100,000 to only about 3000 (Brownell et al. 1989).

Pinniped species range over four orders of magnitude in abundance, from the crabeater seals at about 12 million (probably the most abundant marine mammal in the world) to the Mediterranean monk seal at probably fewer than 500 individuals. Similar variation in species abundance is also found among cetaceans (Table 9-3). Although the magnitude of present variation has no doubt been increased by human activities, three orders of magnitude variation in abundance may well characterize the norm. This variation in abundance is relevant both with respect to present conservation efforts and to our understanding of the potential roles of marine mammals in the structure and function of marine ecosystems.

A closer look at the pinnipeds shows that phocids are generally more abundant than otariids (Table 9-4). Why this should be so is not entirely clear. During the past 50 years, both families have been commercially exploited and subjected to other anthropogenic factors that might have influenced abundance. More likely, the greater abundance of phocids is the result of their greater use of high productivity areas in temperate and polar waters than is the case in most otariid species. In fact, the three most abundant otariids, northern fur seal, Antarctic fur seal (*Arctocephalus gazella*), and South African fur seal (*Arctocephalus pusillus pusillus*), all forage seasonally in productive, high latitude ecosystems.

Aerial surveys have been used to estimate sizes of dugong populations (Marsh 1995) and trends in Florida manatee

Table 9-4. Distribution of Pinniped Species by Family and Abundance Class[a]

	Abundance Class			
Family (no. of species)	< 10,000	10,000–100,000	101,000–1,000,000	> 1,000,000
Phocidae (18)	2	1	12	3
Otariidae (14)	1	5	5	3
Odobenidae (1)			1	

[a]Abundance estimates from Reijnders et al. (1993).

population size (e.g., see Garrott et al. 1995, Lefebvre et al. 1995). Using both aerial survey data and other long-term data, Eberhardt and O'Shea (1995) developed a preliminary model for the latter species that suggested modest population growth in several parts of the range. The accuracy associated with estimates of sirenian population sizes is unknown. The dugong is almost certainly the most abundant species, with perhaps 100,000 individuals worldwide (for summary, see Reynolds and Odell 1991). All sirenians are still hunted at some level. One species that has suffered especially heavy documented losses due to commercial harvest is the Amazonian manatee, with perhaps 100,000 individuals being taken between 1935 and 1954 (Domning 1982).

Estimates of abundance have been made for all of the 15 polar bear populations that are currently recognized. However, the quality of these estimates varies considerably. Wiig et al. (1995) list estimates ranging from very poor to good, and in only five populations is there confidence in the estimates. On the basis of all available data, Wiig et al. (1995) estimate the worldwide population to be about 21,000 to 28,000 animals.

The exploitation and near-extinction of the sea otter is well documented (e.g., Kenyon 1969). However, by the 1980s sea otters had recolonized much of their historic range from the northeastern Gulf of Alaska, westward to the southern end of the Kuril Islands (Riedman and Estes 1988). Estimates of the worldwide population are rather imprecise as most populations are increasing and are widely distributed around the Pacific Rim. Estimates of more than 100,000 have been suggested, with sea otters occupying about 75% of their former range (Bodkin et al. 1995).

A number of countries and conservation organizations monitor information on the abundance of marine mammals. Of particular note are the Specialist Groups of the World Conservation Union, formerly called the International Union for Conservation of Nature and Natural Resources (IUCN), and the Scientific Committee of the International Whaling Commission (IWC). The U.S. Marine Mammal Commission also produces an annual report in which the marine mammals of special concern are discussed (e.g., Marine Mammal Commission 1995:9–88). In this recent report, the Commission lists 15 species of cetaceans, seven species of seals and sea lions, two species of otters, and three species of manatees and dugongs as either endangered or depleted. Most of the other species of marine mammals either have not been exploited or have recovered to the point where they are no longer listed in the endangered/threatened or depleted categories. Within U.S. coastal waters the major species of conservation concern are the right whale, the Florida manatee, the southern sea otter, the Hawaiian monk seal, and the Steller sea lion. Status and conservation of selected marine mammal species, including monk seals, manatees, and right whales, are discussed in Twiss and Reeves (1999). With the exception of the Steller sea lion, these species number, at most, a few thousand, and because of changes in the environment or conflicts with human-related resource development, they are in danger of declining to even lower population levels. Steller sea lions in the western part of Alaska and the Aleutian Islands are of concern because population numbers have diminished within the past 15 years from around 170,000 to 30,000 individuals. The reasons for this decline are the subject of some debate (National Resource Council 1996).

With respect to the cetacean species listed in the endangered or depleted category, many were heavily harvested during commercial whaling. The great whales have been slow to recover, partly because of their low reproductive rates; it takes many years for these populations to return to previous levels, but there is good evidence that some are recovering (Best 1993). A few of the small cetaceans, such as the river dolphins and the vaquita (*Phocoena sinus*), are in precarious positions because of habitat degradation and human exploitation. Their future looks bleak unless there are changes in human activities (e.g., Reeves et al. 1991, Shrestha 1993).

Other species of marine mammals have fully recovered and may be more abundant now than in the past. For example, in the southern hemisphere the minke whale was not heavily harvested until recently and, as other whale species declined, the minke whale population may have increased because more food was available due to the lack of competition by the heavily hunted baleen species (Laws 1985). Combined recent population estimates from some of the areas where minke whales are found suggest a worldwide population of about 935,000 (Gambell 1999). Recovery of the gray whale, now numbering more than 23,000 individuals, is such that it is no longer considered endangered and has been removed from the endangered species list. Historically, both the fur seals and elephant seals were harvested and driven to very low population levels. Beginning in the early 1900s, these species increased. In fact, the Antarctic fur seal now may occupy areas where it was previously absent and numbers may exceed historic abundance (D. Siniff, unpublished data). Southern elephant seals have reoccupied their previous range but have not yet reached preexploitation levels (D. Siniff, unpublished data), but northern elephant seals, once on the verge of extinction, now number about 50,000 individuals (for review, see Reijnders et al. 1993).

Estimating Abundance

Estimating the abundance of a marine mammal population is one of the most difficult tasks that marine mammal scien tists attempt. Many of the factors that limit our knowledge of the distribution of marine mammals contribute to the difficulty we have in determining their numbers. In general, this is true because marine mammals are visible only periodically when they come to the surface to breathe or when they haul out on ice or land to rest, molt, or reproduce. Thus, estimation methods must consider also the unseen portion of these populations. The many techniques that have been used to enumerate marine mammals all make assumptions that must be upheld. Where this is not possible, it is important to evaluate the impact of violation of assumptions on the abundance estimates. Three concepts associated with the estimation of abundance are precision, accuracy, and bias. Accuracy of an estimate is a measure of how close the estimate is to the true population, whereas precision is a measure of how close an estimate is to the expected value of the estimation model. Bias is the difference between the expected value of an estimation model and the true value.

Obtaining estimates of marine mammal abundance can be approached in three ways. One way is to obtain a census (i.e., total count) of the number of individuals in a population at a particular time and place. A census is rarely possible with marine mammals, but there are examples. The number of gray seal pups born on Sable Island, Canada, was determined by census between 1978 and 1990 (Stobo and Zwanenburg 1990). The number and genealogy of resident killer whales in the coastal waters of British Columbia and Washington were determined by counting individuals identified using natural markings (Bigg et al. 1990).

The second, and most common, way is to estimate abundance based on only a sample of the population, and the third is to obtain an index of population size (i.e., an estimate of some statistic that is correlated with the population size at a particular point in time; Caughley 1977). For example, an index that is used to indicate the size of the gray whale population is the number of whales counted annually as they migrate up the California coast. Not all the whales are counted, but it is assumed the same portion of the population is counted annually, and that this count reflects the size of the total gray whale population. If the count is declining or increasing, then it is assumed that the gray whale population is also following this trend. Population indices are often used in marine mammal studies simply because of the difficulty and expense of estimating total abundance. Perhaps the most common use of an index is as a measure of relative population trend. However, indices are not the same as population estimates and only certain population parameters can be obtained from such relative statistics.

Various methods are used to estimate the abundance of marine mammals. Among the more common methods are catch–effort analysis, mark–recapture, line and strip transect surveys, land-based visual surveys, and acoustic surveys.

Catch–effort methods are less commonly used today than in the past when commercial harvests were more common. Detailed discussions of these methods are found in Eberhardt et al. (1979), Seber (1982), Hammond (1986, 1990), Hiby and Jeffery (1987), Hiby and Hammond (1989), Krebs (1989), Skalski and Robson (1992), and Lefebrve et al. (1995).

Mark–recapture models are commonly used to estimate the abundance of pinniped (e.g., Siniff et al. 1977, Bowen and Sergeant 1983, York and Kozloff 1987) and baleen whale (for review, see Hammond 1986, 1990) populations. Permanent (i.e., hot brands) and a wide variety of semipermanent marks have been used including freeze-brands, flipper tags, and Discovery tags (for large cetaceans). Photographic records of natural markings are commonly used to identify cetaceans (for review, see Hammond et al. 1990) and have been used recently to estimate gray seal numbers (Hiby and Lovell 1990, Hiby 1994). One potential problem with natural markings is that they may change over time. Dufault and Whitehead (1995a) studied changes in natural marking on the flukes of sperm whales and found that 61% of 161 comparisons showed no change. The rate of change in fluke markings was low enough that recognition of individuals was unlikely to be affected, provided individuals were identified by more than one mark. However, the researchers also found that the gain rate of marks generally exceeded the loss rate, such that over time the accumulation of marks could make some individuals unrecognizable.

In many pinniped species, it is usually not possible to estimate directly the total number of individuals in a population. However, it is often possible to estimate the number of pups born annually. Mark–recapture techniques have been used extensively for this purpose in northern fur seals at the Pribilof Islands in the Bering Sea. Here pups were marked by shearing (i.e., removing) a spot of fur from selected individuals (York 1990). In some years, semipermanent tags were applied and estimates of annual survival were obtained for certain sex and age classes (York 1994). Mark–recapture techniques are particularly useful when populations return to the same locations, such that individuals can be permanently marked on more than one occasion and marked individuals can be censused over time. These multiple marking methods provide the opportunity to obtain not only abundance estimates but also estimates of survival probabilities. A good example of this approach involves the McMurdo Sound Weddell seal population (Siniff et al. 1977, Testa and Siniff, 1987, Testa et al. 1991). The Weddell seal is long lived (> 20 years), and long-term studies since 1968 have made it possible to estimate age-specific rates of survival and reproduction.

There are a number of statistical models that can be used

to estimate abundance and other population parameters from multiple mark and recapture studies, with recent advances being described by Lebreton et al. (1992) and Burnham (1989). Other publications supplement these papers in providing new methods for improving estimation techniques; for example, Buckland and Garthwaite (1991) provide guidance for improving the estimation of confidence limits for mark–recapture studies.

A number of assumptions are common to all mark–recapture methods, and departure from these assumptions can seriously bias the resulting estimates of abundance and survival. For example, the loss of marks is a common problem that can occur when semipermanent tags or natural marks are used to identify individuals. The bias resulting from tag loss can be removed if an estimate of the probability of losing a tag is available, for example, by using more than one identifying mark (e.g., Bowen and Sergeant 1983). Another problem in some mark–recapture studies is that not all recovered tags are reported. Bowen and Sergeant (1983) conducted sample surveys of fishing communities from which harp seal hunters operated to estimate the magnitude of this problem, and they found that nonreturn of tags was significant, approaching 20%. Differences in the probability of sighting different sex/age groups may bias results as equal probability of sighting is an assumption for each sampling period. However, recent statistical developments make it possible to obtain some measurement of the biases and to correct estimates accordingly. A computer program called RELEASE (Burnham et al. 1987) can test for heterogeneity among different groups in a data set, and once homogeneous groups have been identified, parameters can be estimated. The program SURGE (Lebreton et al. 1992) can be used to estimate survival probabilities by sex and age groups.

Transect surveys are another class of methods commonly used to estimate abundance. There are two basic approaches to such surveys: methods in which the transect has a fixed width (known as strip-transect surveys) and methods in which the transect width is theoretically infinite (known as line-transect surveys). In both of these approaches, an observer moves along a predetermined line on an observing platform (e.g., ship or aircraft) and sights individuals or groups. The perpendicular distance between the observer and the observed individual or group of marine mammals is measured or estimated.

In strip-transect surveys the probability of sighting an animal within the strip is designed to equal one, whereas in line-transect surveys, the frequency distribution of perpendicular distances is used to estimate the probability of sighting animals away from the line of travel. The estimation of g(0) (called g-naught), the probability of sighting an animal or group of animals on the track line, is often difficult, especially in species that spend only a small portion of their time at the surface. Recent software available for analysis of these data (Laake et al. 1994) uses the distribution of sighting distances to fit the most appropriate statistical model. Line-transect surveys are commonly used to estimate the abundance of cetaceans. Many examples of specific applications to cetaceans can be found in the annual reports or special issues (e.g., abundance of harbor porpoise, Palka 1995) of the International Whaling Commission. Marsh (1995) and Lefebvre et al. (1995) provide good reviews of the application of transect methods to estimate the abundance of dugongs and manatees.

Stratified, strip-transect, aerial photographic, and visual surveys are used to estimate pup production of harp seals (Stenson et al. 1993) and hooded seals (Bowen et al. 1987) on the pack ice of the northwest Atlantic. In both species, an added difficulty is that not all pups have been born at the time of the survey, whereas others may have been weaned and left the ice. To deal with this problem, a model of the distribution of births is used to correct or scale the aerial survey estimate to an estimate of total production. A similar approach is used to estimate gray seal pup production in the United Kingdom (Ward et al. 1987).

Although abundance estimation may appear as a relatively straightforward task, good surveys are difficult to achieve in practice and always require careful planning, particularly with respect to factors that may bias the results (Caughley 1974, Leatherwood et al. 1978, Myers and Bowen 1989, Lefebvre et al. 1995). In planning to estimate the abundance of a population, it is important to consider the precision of the estimate, as this will affect the number of surveys that are needed to detect a trend in population size and the magnitude of change that can be detected. Imprecise estimates have low statistical power to detect trends in abundance. The ability of a statistical procedure to distinguish a significant difference from a null hypothesis of no change is called the power of that procedure. Power analysis permits an investigator to estimate the probability of detecting a trend if one is present (Gerrodette 1987). Edwards and Perkins (1993) provide an example of the use of power analysis in evaluating linear trends in dolphin abundance in the eastern tropical Pacific.

Population Structure

Like other organisms, marine mammal species are generally comprised of more than one population. For our purposes, a population is defined as an interbreeding group of individuals of the same species that occurs within a particular geographic area. Populations may undergo genetic exchange with neighboring populations, but there is a greater likelihood of matings between individuals of the same population than among populations. Thus, genetic characteristics of populations are likely to differ over the species range (Slatkin 1987). The term metapopulation has been applied to a population that encompasses the entire range under study, with each subunit being considered a population. Population structure is of considerable interest. It potentially affects many aspects of marine mammal life histories; in turn, these aspects have implications for conservation and management. Although we cannot do justice to this important and complex topic, it is useful to briefly outline some of the methods that are used to investigate population structure.

The most frequently used methods include tagging studies and the analysis of morphological, protein, and DNA variation. Tagging studies, using both semipermanent (including VHF and satellite radio-tags noted earlier) and natural markings (see Hammond et al. 1990), have contributed to our understanding of the population structure of many species. Recently stable isotope ratios and fatty acid profiles have also been used to distinguish between populations of freshwater and marine harbor seals (Smith et al. 1996). Until the past two decades, morphological variation was probably the primary tool used to study population structure. For example, Yablokov and Sergeant (1963) analyzed cranial measurements and concluded that the White Sea and Jan Mayen populations of harp seals did not differ greatly, but both could be readily distinguished from the Newfoundland population.

During the past 20 years, the analysis of enzyme variation using gel electrophoresis has provided considerable data on genetic structure within and among populations. Testa (1986) and Hoelzel and Dover (1989) list 8 pinnipeds and 13 cetaceans, respectively, where more than 15 loci have been studied. More recently, variation in DNA itself has been used to analyze population structure (for review, see Hoelzel and Dover 1989; Hoelzel 1993, 1994). The DNA analyses often reveal population differences where previous techniques, such as morphometric measurements and gel electrophoresis, have not. Mitochondrial DNA sequences and genomic DNA typing are extremely useful in revealing population structure. However, as Hoelzel (1994) notes, the interpretation of molecular genetic data still relies heavily on a good understanding of the life history and behavior of the species under study.

It is not possible to provide a comprehensive review of recent molecular studies; readers are encouraged to consult Hoelzel (1994). However, several studies illustrate the value of DNA analysis in understanding population structure. For example, Paetkau et al. (1995) studied genetic variation in 126 polar bears from four populations using microsatellite markers. They found evidence of restricted gene flow among these populations despite known long-distance movements of individuals. In fact, 60% of the individuals sampled could be assigned to their population based on their genotype alone. In a series of papers, Baker et al. (1990, 1994) investigated the population structure of humpback whales at the global scale and found evidence of genetic segregation among subpopulations of humpback whales in the Pacific as well as differences between the Atlantic and Pacific populations (Fig. 9-12). They hypothesized that this segregation resulted from maternally directed fidelity to specific feeding areas. Palsboll et al. (1995) recently studied a larger sample of humpback whales from five feeding sites in the North Atlantic and found evidence of segregation between the Iceland and western Atlantic feeding areas that they too attributed to maternally directed philopatry. Allen et al. (1995) studied genetic variation in two gray seal colonies separated by about 500 km and also found evidence of restricted gene flow between sites. However, given the large number of breeding sites around the United Kingdom, considerably more work is needed before we have a complete understanding of gray seal population structure. Stanley et al. (1996) studied mitochondrial DNA in the harbor seals from 24 locations around the world. They found that populations in the Atlantic and Pacific and the east and west coast populations in these two oceans showed significant genetic differentiation. They interpreted their results to suggest that harbor seals are regionally philopatric on scales of several hundred kilometers.

Molecular genetic studies have also provided insight into the social structure of marine mammals, which, as we have seen in the case of humpback whales, may strongly influence population structure. Pilot whales live in large groups, but the stability and structure of these groups were unknown until recently. Studies of Faroese pilot whales, using highly variable microsatellite sequences, show that although mature males were present in the pods, all males within a pod could be excluded from paternity in 33 of 34 mother–fetus pairs (Amos et al. 1993). The researchers concluded that adult males do not mate within natal pods. Matings, and therefore, gene exchange, must occur when pods meet. Thus, the spatial distribution of pods may be expected to have a strong influence on gene flow within and among populations in this species. Similar mating systems have been documented in molecular studies of African lions (Packer et al. 1991) and sperm whales (Richard et al. 1996).

Population Dynamics

The analysis of how and why populations vary in abundance over time is the subject of population dynamics. Although not directed toward marine mammals, Pielou (1974) and

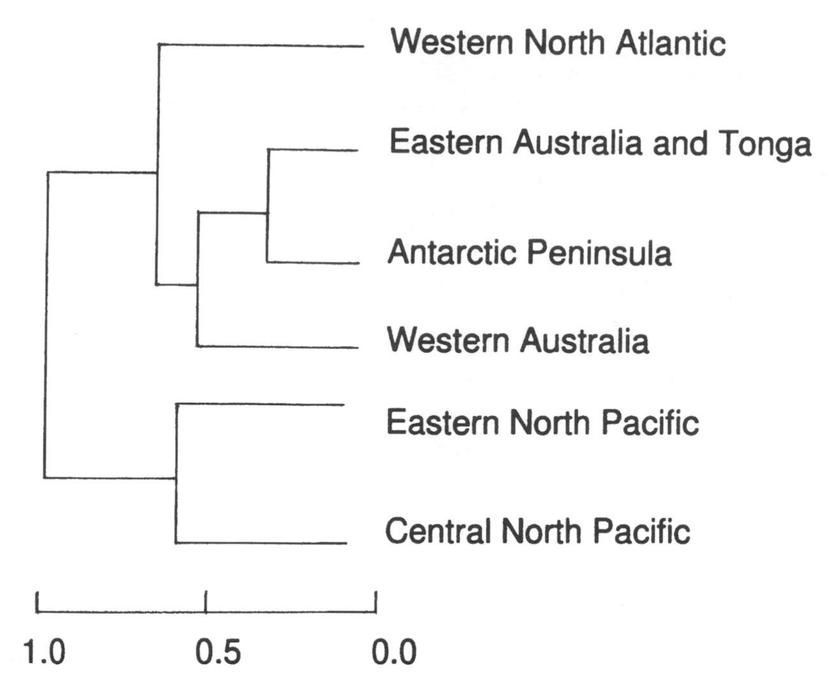

Figure 9-12. Genetic relationships among six subpopulations or stocks of humpback whales based on clustering of genetic distances among mitochondrial DNA (mtDNA) haplotypes. (After Baker et al. 1994.)

Caughley (1977) provide clear and reasonably comprehensive introductions to this topic. A good example of the analysis of the population dynamics of a marine mammal involves the study by Olesiuk et al. (1990b) of killer whales off British Columbia. Fowler and Smith (1981) include a number of good studies of the dynamics of large mammal populations, including several chapters on marine mammals. McLaren and Smith (1985) provide a good overview of the population ecology of seals.

Knowing the age of an animal is important in many types of research, but nowhere is it more important than in studies of population dynamics. Fortunately, there are a number of reliable methods that can be used to determine the age of most marine mammals. Incremental growth of cementum, dentine, and bone has been used to determine the age of many mammals (for reviews, see Sergeant 1959, Laws 1962, Jonsgard 1969, Perrin and Myrick 1980). It was not until about 1950 that the value of dental layers for determining age of marine mammals was recognized (Scheffer 1950, Laws 1952, Nishiwaki and Yagi 1953). Since then methods have been developed that help determine the age of most marine mammals. The use of dental growth layers groups (GLGs) in teeth has been validated, with reference to individuals of known age, in the northern fur seal (Scheffer 1950, Anas 1970), harp seal (Bowen et al. 1983), gray seal (Mansfield 1991), bottlenose dolphin (*Tursiops truncatus*) (Hohn et al. 1989), and Hawaiian long-snouted spinner dolphin (*Stenella longirostris*) (Myrick et al. 1984). The results of a workshop on age determination in odontocetes, with particular reference to phocoenids, are found in Bjorge and Donovan (1995). The use of GLGs has been extended to the use of incisors, which can be easily removed from living animals, enabling age to be determined for animals used in long-term studies (Arnbom et al. 1992, Bernt et al. 1996).

In baleen whales, ear plugs (i.e., a waxy structure found in the external auditory meatus) have been used to determine the age of individuals with varying degrees of success (for review, see Lockyer 1984). In manatees, GLGs found in the periotic bone provide a reliable means of age determination (Marmontel et al. 1996).

There is a subjective component to age determination with the result that errors do occur. These errors can lead to bias in the estimation of age-specific birth and death rates or add variance in estimates. Considerable effort has been directed to ensure consistency in age determination and to correct for bias resulting from inconsistent readings (Doubleday and Bowen 1980, Perrin and Myrick 1980, Lawson et al. 1992). Campana et al. (1995) provide some useful graphical and statistical methods for determining the consistency of age determination.

Marine mammals tend to be large-bodied, long-lived species in which reproduction is delayed and litter size is typically one. Therefore, despite being long lived, lifetime reproductive output of females is generally low. This type of life history, characteristic of k-selected species, makes marine mammals particularly vulnerable to overharvest because (1) the population level where maximum recruitment takes place for a given life history pattern is not well defined (Fowler and Baker 1991, Fowler 1994, de la Mare 1994), and (2) the potential rate of increase is relatively low. The rate of increase in a population can be expressed as a finite rate where λ = the ratio of numbers in 2 successive years or an exponential rate (r), where $\lambda = e^r$. Caughley (1977) distinguished among several exponential rates of increase: the intrinsic or Malthusian rate (r_m) which might be thought of as the maximum rate of increase under optimum conditions; the survival–fecundity rate (r_s) resulting from a particular combination of vital rates; and the "observed" rate (r_o), which is an average over some period of time.

Using currently accepted population parameters (International Whaling Commission 1982), the rate of increase of sperm whales is estimated at about 0.01 per year. The rate of increase of bowhead whales off Alaska is also low at about 0.03 per year (Zeh et al. 1991). However, the potential rate of population increase is not always low in marine mammals. Reilly and Barlow (1986) simulated r_s over a feasible range of dolphin population parameters (age at first birth, calving interval, and survival) and, although the maximum finite rates of about 0.09 were possible, they felt that rates of 0.02 to 0.05 apply to most dolphin species. Best (1993) reviewed "observed" estimates of the rate of increase for severely depleted populations of baleen whales; mean estimates varied widely from 0.031 to 0.144.

In pinnipeds, there is good evidence that some populations can increase at rates about 0.13 to 0.15 per year. Annual censuses of the number of gray seal pups born on Sable Island, Canada, between 1976 and 1990 show an observed exponential rate of increase of 0.13 (Mohn and Bowen 1996; Fig. 9-13). Similarly, harbor seals in British Columbia increased at a rate of 0.125 over a period of about 20 years (Olesiuk et al. 1990c). A similar, although somewhat higher, rate of increase (0.168) was measured in the Antarctic fur seal (Payne 1977).

For the dugong, Marsh (1995) suggested that the maximum asymptotic annual rate of increase could reach 0.063 for an unharvested population experiencing the most optimistic life history parameters. Not surprisingly, adult survivorship is the most critical factor affecting sirenian population trends (Marsh 1995, Marmontel et al. 1997).

The effect of these different rates is easily seen when they are presented as the doubling time of the population (t_d = [ln 2] / r; Pielou 1974). Doubling times for populations with r = 0.02 and 0.15 are 34.6 and 4.6 years, respectively!

Studies of the mechanisms that cause populations to change over time require estimates of age-specific survival and reproductive rates. These data are obtained from either cross-sectional or longitudinal studies. In longitudinal studies, also referred to as cohort studies, individual animals are followed over time to obtain data on their survival and reproductive performance (e.g., northern elephant seals [Sydeman and Nur 1994], Weddell seals [Testa and Siniff 1987], killer whales [*Orcinus orca*] [Olesiuk et al. 1990b]). In contrast, in cross-sectional studies, a sample of individuals of various ages is collected from a commercial harvest, from individuals sampled for scientific purposes, or from other means such as incidental catches in fisheries or strandings (e.g., harp seals [Roff and Bowen 1983, 1986]; northern fur seals [Trites and Larkin 1989, Trites and York 1993]; polar bears [Derocher and Stirling 1994]). Female fecundity is determined by examining ovaries and uteri for evidence of pregnancy (see Boyd, Lockyer, and Marsh, Chapter 6, this volume). Survival is sometimes estimated as the natural log-

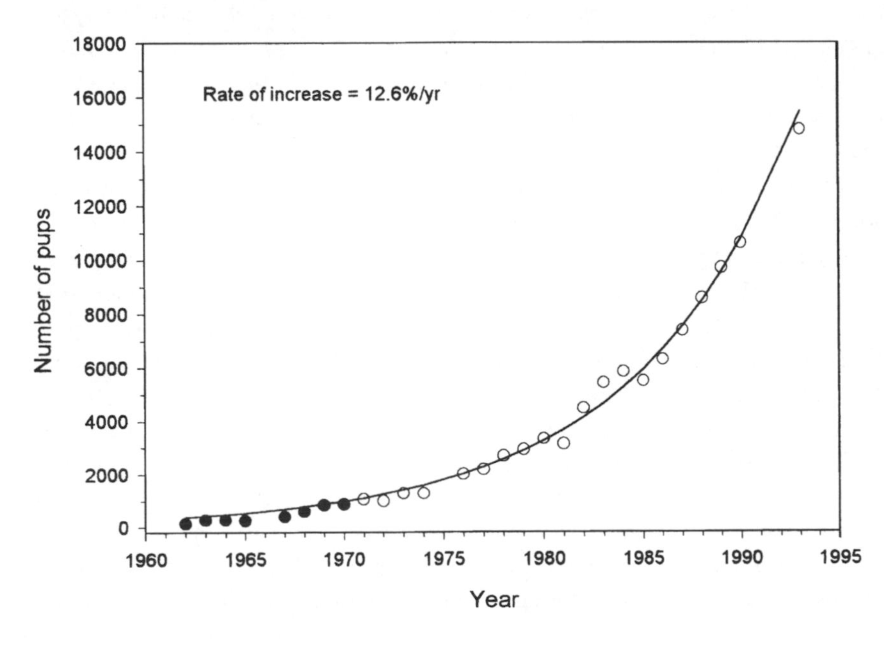

Figure 9-13. Counts of the number of gray seal pups born on Sable Island, Canada, between 1962 and 1993. The fitted line is an exponential curve with an exponent of 0.13 (after Mohn and Bowen 1996) using only the data from 1976 to 1990. Note the excellent fit even for years not included in the regression model.

arithm of numbers against age under the assumption of a stationary population (see Caughley 1977). Until recently, only cross-sectional data were available for most marine mammals that had been studied. However, longitudinal studies are increasingly common, and contribute greatly to our understanding of population dynamics.

Age-specific fecundities are often known in marine mammals, but age- and sex-specific survival rates are less commonly available from free-ranging populations. It is possible to estimate the survival rate from multiple mark–recapture studies (Weddell seals [Testa and Siniff 1987], polar bears [Derocher and Stirling 1995], humpback whales [Buckland 1990], manatees [O'Shea and Langtimm 1995]) or from a cross-sectional (gray seals [Harwood and Prime 1978]) or longitudinal (killer whales [Olesiuk et al. 1990c]) sample of aged animals. Where data either have not or cannot be obtained, Barlow and Boveng (1991) suggest that rates of births and deaths can be cautiously modeled after rates that have been measured in other long-lived mammals.

Matrix models developed by Leslie (1945, 1948) or the "Lotka equation" as developed by Cole (1954) are often used to study how a population behaves over time, given a certain age structure and age-specific reproductive schedule. Eberhardt and Siniff (1977) used these theoretical structures to study the effects of changes in such parameters as age of first reproduction, juvenile survival, and reproductive rates on population growth rates. These models assumed stable age structures and constant reproductive schedules. As these conditions are rarely met in natural populations, caution should be used in interpreting the results from such models. Many studies have used such models to investigate how the dynamics of marine mammal populations might be influenced by different types of management or naturally occurring environmental events (e.g., Fowler 1981, 1990; Boveng et al. 1988; Eberhardt 1990; Ragen 1995; Marmontel et al. 1997).

The practical importance of understanding marine mammal population dynamics can be appreciated with reference to the U.S. Marine Mammal Protection Act of 1972, which established that a marine mammal population should not be allowed to decline below a level that was designated the optimum sustainable population (OSP) level. This level has been interpreted as a range of numbers between the level where the population produces its greatest annual increment (i.e., the maximum net productivity level, MNPL) and the carrying capacity (K) of the environment. This legislative designation has been the focus of many studies that have sought to interpret the meaning of this idea (Eberhardt and Siniff 1977, Fowler 1981, Gerrodette and DeMaster 1990). Initially the U.S. Marine Mammal Commission produced a document setting forth certain guidelines that could be used to determine whether a population was within the OSP level. Eberhardt and Siniff (1977) carried out demographic analyses to evaluate the consequences of the value of different parameters on the rate of increase of marine mammals. DeMaster et al. (1982) examined the problem of defining when a population was at OSP and developed what has become known as the dynamic response analysis, which has since been examined in detail through modeling and simulation efforts (e.g., Gerrodette and DeMaster 1990). Ragen (1995) provided an analysis of the problems surrounding determination of maximum net productivity level in a very extensive analysis of northern fur seal data. Using a simulation model, population parameters were varied and the effects on the range of MNPL values were determined. This analysis suggested that when uncertainty exists about the estimates of population parameters, and when conditions exist where population equilibrium seems unlikely because of changes in potential density-dependent processes, MNPL levels are likely to vary.

This result is perhaps not very surprising; a basic assumption of the Marine Mammal Protection Act is that populations of marine mammals increase until they reach some equilibrium with their environment. This was the basic assumption under which nearly all fisheries and marine mammal harvests have been managed. However, as we have obtained more information about how populations, resources, and environments change over time, the idea of a specific equilibrium population size seems unlikely. DeMaster et al. (1982) expressed a similar opinion and suggested that we should consider populations of marine mammals as resource dependent rather than density dependent when carrying capacities are likely to vary annually depending on the food resources available. Furthermore, the idea of stable equilibrium seems even more unlikely when one considers the effects of environmental change at both global and local scales. All of these factors make the idea of equilibrium populations with fixed and predictable MNPL levels an uncertain criterion on which to base long-term management strategies.

Having examined the logic of the optimum sustainable population concept, a question arises: Is there a better way to define the relationship between a population and its resources? It is probably fair to say that while the underlying population concepts are useful in interpreting population status, qualitative measures also need to be considered. Perhaps the approach outlined within the Convention on the Conservation of Antarctic Marine Living Resources (CCAMLR) should receive more attention. Research under this Convention has centered on measures relating marine mammal populations to changes in the ecosystem. Measures of time spent foraging, length of trips to sea for female fur seals during lactation, pup mass gains, and other such measures attempt to relate current conditions for food gathering to ecosystem state (Fowler and Siniff 1992).

Currently none of the indices can be tied quantitatively to changes in population parameters. Obviously, the problem of interpreting population status in relation to resources is difficult and multidimensional. However, it seems clear that management policies derived from relationships among population parameters and the concept of equilibrium level need to be tempered with the realization that the feedback mechanisms that determine rates of population change are most likely to be resource dependent for many species of marine mammals (DeMaster et al. 1982).

Diets and Foraging Behavior

The types and amounts of food eaten by marine mammals and the means by which they obtain these foods are of both practical and theoretical interest. Unfortunately, the quality of information about marine mammal diets varies considerably. In many cases, we have an idea of the range of foods that are eaten but are considerably less certain about the relative contribution of different foods to the energy requirements of individuals or populations, and how diets change with age, sex, and prey abundance. Inevitably our understanding of the diets and foraging ecology of marine mammals is tied to our knowledge of their distributional ecology. As we have seen, new methods are rapidly extending our understanding of the distribution of marine mammals, and as we will see below, this extends to the methods used to study diet.

We briefly review the methods that are used to determine the composition of diets and new technologies that promise to advance our understanding of the foraging ecology of marine mammals, the kinds of foods eaten by different groups of marine mammals, and the factors that affect the composition of diets. Foraging behavior and energetics are dealt with in Wells, Boness, and Rathbun, Chapter 8, and Costa and Williams, Chapter 5, in this volume.

Methods Used to Study the Foraging Ecology of Marine Mammals

In many species of mammals, it is feasible to observe directly the various aspects of foraging (i.e., searching, capturing, and eating) by individuals. Divers have observed foraging gray whales (Darling 1977) and feeding can be inferred from surface observations of behavior (e.g., Simila and Ugarte 1993, Würsig and Clark 1993, Baird and Dill 1995), but with few exceptions (e.g., sea otters [Kvitek et al. 1993] and manatees [Hartman 1979]), direct observation of feeding is difficult in marine mammals because (1) many species inhabit remote areas for most of the year and access to these areas is often difficult, and (2) feeding generally occurs at depths where observation is not possible. Therefore for the most part, indirect methods are used to study the foraging ecology of marine mammals.

Prey Structures that Resist Digestion

The most common methods used to determine the diet of marine mammals rely on the identification of prey structures that are resistant to digestion and can be collected from stomachs, intestines, or feces. In piscivorous (i.e., fish-eating) species, sagittal otoliths (Fig. 9-14) are most commonly used for this purpose, but other structures such as bones, scales, and lenses also provide a means of prey identification (Fitch and Brownell 1968; for reviews, see Pierce and Boyle 1991, Pierce et al. 1993). Where cephalopods are eaten, beaks can be used for prey identification (e.g., Clarke 1986). Other invertebrates can be identified from the remains of exoskeletons or shells. Many marine mammal and fisheries laboratories have reference collections of these materials (e.g., for reference collection of otoliths from northeast Atlantic fishes, see Härkönen 1986).

In addition to the kind of prey eaten, the recovery of certain hard parts, particularly otoliths and cephalopod beaks, can be used to estimate the size and sometimes the age of the prey consumed. This is possible because of the close relationship between otolith size (mass, length, and width have been used) and fish length and body mass. Examples of predictive equations can be found in most papers on marine mammal diets published within the past decade (e.g., Frost and Lowry 1980, Recchia and Read 1989, Hammond and Prime 1990, Bowen and Harrison 1994). Information on prey size is important because scientists and managers are usually interested in the biomass, and ultimately the energy intake, associated with the consumption of prey.

Both stomach contents and fecal samples have been used as a source of prey hard parts in diet studies of pinnipeds. However, feces are being increasingly used for this purpose because they are often less expensive to collect. A high proportion of samples contain identifiable prey, and estimates of diet from fecal data are less affected by differential rates of digestion than are estimates from stomach samples (Hammond and Prime 1990). Regurgitated otoliths and squid beaks have been used to evaluate the diet of California sea lions (*Zalophus californianus*) (Ainley et al. 1982). Gastric lavage has been used as an alternative to lethal sampling to obtain stomach contents in harbor seals (Fig. 9-15; Bowen et al. 1989) and northern (Antonelis et al. 1987) and southern elephant seals (Green and Burton 1993). The feeding habits of Hawaiian monk seals (*Monachus schauinslandi*) are routinely studied by examining spews (Gilmartin 1993).

Despite their widespread use in diet studies, the use of

Figure 9-14. Examples of the sagittal otoliths of several species of teleost fish. From left to right: cod (*Gadus morhua*), redfish (*Sebastes fasciatus*), and silver hake (*Merluccius bilinearis*) (Courtesy of S. Campana.)

hard parts to determine the species composition and size of prey is subject to a number of biases that may seriously limit the value of such information (Murie and Lavigne 1985, Jobling and Breiby 1986, Dellinger and Trillmich 1988, Harvey 1989, Pierce and Boyle 1991, Pierce et al. 1993). One problem is that otoliths are present only if the head of the fish is consumed. Although there is evidence that heads of some prey species may not always be consumed, the bias resulting from this behavior is not well understood. Another potential difficulty in drawing inferences from such data is that stomach and fecal contents only provide a representative estimate of the diet near the point of collection (usually a haul-out site in the case of pinnipeds) such that offshore diets are not sampled (Bowen and Harrison 1994). This may not pose a problem in coastal species, such as the harbor seal or bottlenose dolphin, but could result in a significant bias in wide-ranging, offshore species such as elephant seals, harp seals, hooded seals, Juan Fernández fur seals (*Arctocephalus townsendi*), and northern fur seals. To overcome this problem, offshore collections of stomachs have been undertaken in both pinnipeds (e.g., northern fur seals [Kajimura 1984], harp seals [Lawson et al. 1994, Nilssen et al. 1995]) and cetaceans (e.g., baleen and sperm whales collected during pelagic hunting [Kawamura 1980, Kawakami 1980, Haug et al. 1995]).

Another serious difficulty with the use of hard parts is that they erode during digestion, such that the size of prey consumed may be underestimated and in some cases identification is not possible. A further complication is that the degree of erosion is species specific and a function of prey size within species. Otoliths from some species (e.g., Atlantic salmon [*Salmo salar*]) are quickly digested and thus are rarely found in stomach or fecal contents (Boyle et al. 1990). Differential rates of digestion among species may seriously bias stomach content analyses in favor of species with large and robust hard parts.

There is generally little that can be done about the potential biases caused by nonrepresentative sample collections and the consumption of species without hard parts. However, a number of studies have tried to correct for the reduction in the size of otoliths that occurs during digestion by conducting feeding experiments with captive seals (Murie and Lavigne 1985, Bigg and Fawcett 1985, da Silva and Neilson 1985, Dellinger and Trillmich 1988, Harvey 1989, Hammond and Prime 1990). Although such efforts are valuable, undoubtedly it is difficult to duplicate the feeding conditions that seals encounter in the wild. Meal size, feeding frequency, proximate composition of the diet, and activity are likely to affect passage time of food through the gut, and thus the degree to which otoliths and other hard parts may be digested (e. g., Marcus et al. 1998).

Various statistics are used to indicate the importance of different prey in the diet. The most common, but least informative, is the frequency of occurrence, defined simply as the percentage of samples in which a given prey was found. The relative frequency of occurrence expresses each prey type as a percentage of the total number of prey types found in a given number of samples and has the advantage of totaling 100%. Both of these measures suffer from the underlying assumption that equal amounts of different prey are eaten. In species where there is a large range in the size of prey eaten, frequency measures overestimate the importance of the smaller prey and underestimate the importance of larger prey. Attempts to overcome this shortcoming have led to the development of the modified relative frequency (Bigg and Perez 1985) and the split-plot frequency of occurrence (Olesiuk et al. 1990a), which take the volume of prey in the sample into account.

Figure 9-15. Contents (mainly sand lance [*Ammodytes dubius*]) of a gastric lavage sample obtained from an adult harbor seal, Sable Island, Canada. (Photograph by W. D. Bowen.)

Perhaps the most useful measure, but also the most difficult to achieve because of the biases discussed, is the percentage wet weight of each prey item in the diet. This information, coupled with data on the energy content of prey, can be used to estimate the contribution of prey species to the energy requirements of a population (see Costa and Williams, Chapter 5, this volume). Researchers differ on the extent to which they are willing to make the additional assumptions required for this kind of presentation of the data. No general conclusions are possible because the types and variety of prey eaten by marine mammals are so diverse. Nevertheless the reader should be aware that the various ways of expressing marine mammal diets may not be directly comparable.

Serology

Serological methods also have some potential to identify the prey of marine mammals (for review, see Pierce et al. 1993). Proteins are degraded during digestion but some survive well enough to react strongly with antisera to specific prey. Boyle et al. (1990) and Pierce et al. (1990) raised antisera to investigate the incidence of Atlantic salmon in fecal and stomach content samples of harbor seals, gray seals, and bottlenose dolphins. The antiserum was not completely specific to Atlantic salmon, as reactions were observed with other salmonids, but not with species from other families. The antiserum produced was not of sufficient strength and specificity to detect reliably salmon in fecal samples, but the results on stomach contents were more promising. Antiserum raised to sand lance (*Ammodytes marinus*) was successfully used on fecal samples of harbor seals from the Moray Firth (Pierce et al. 1990).

Antisera may be frozen (–20°C) for prolonged periods without significant loss of activity; however, fish muscle proteins deteriorate during frozen storage. Clearly the greatest advantage of serology is that the identification of prey is not dependent on the recovery of otoliths or other hard parts. However, there are disadvantages. First, separate antisera must be produced for each potential prey species of interest. This is time consuming and expensive. Second, the method is essentially qualitative at this time. Quantitative methods for using antisera do exist, but extensive calibration using captive feeding experiments is required (Pierce et al. 1993).

Fatty Acids

It has been recognized for some time (e.g., Klem 1935, Ackman and Eaton 1966, Hooper et al. 1973) that the fatty acid composition of prey influences the fatty acid composition of the lipids of baleen whales. Lipids in marine organisms are characterized by their diversity and high levels of long chain and polyunsaturated fatty acids that originate in various unicellular phytoplankton and seaweeds (Ackman 1980). Unlike other nutrients, such as proteins that are readily broken down during digestion, dietary fatty acids pass intact into the circulation and those of carbon chain-length greater than 14 are often deposited in animal tissue with little modification.

Iverson (1988, 1993) suggested that the pattern of tissue

fatty acids might be used to determine changes in, and the components of, the diets of marine mammals, and began to systematically evaluate the potential of this method. She proposed two approaches to the use of fatty acids in diet studies (Iverson 1993). In the first, rare fatty acids might be used as a tracer. For example, jellyfish were indicated in the diet of the ocean sunfish (*Mola mola*) based on the occurrence of a *trans*-6-hexadecenoic acid, a rare fatty acid (Hooper et al. 1973). The second and more common approach characterizes the pattern of prey fatty acids as a signature (Fig. 9-16) and compares reference signatures to the unidentified signature in a marine mammal sample (Iverson 1993). A study of the transfer of milk fatty acids from hooded seal mothers to their pups illustrates this approach (Iverson et al. 1995). Here milk is the prey and pup blubber is the tissue under analysis. At birth the fatty acid composition of pup blubber reflects the biosynthesis of fatty acids by the fetus and placental transfer. However, after only 4 days of nursing, the fatty acid composition of the pup is remarkably similar to that of the milk (i.e., prey) consumed (Fig. 9-17).

Although several multivariate statistical techniques, including discriminant function analysis, may be valuable in fatty acid signature analysis, the large number of fatty acids per sample (> 60) requires either large sample sizes to be used or variable selection before analysis. Unfortunately, such selection can be rather subjective and may eliminate important fatty acids from further consideration. To deal with this problem, Iverson et al. (1997) and Smith et al. (1997) have used classification trees with encouraging results. Classification trees recursively partition the animals from which the fatty acids were obtained into two or more groups based on a series of dichotomous splits of the fatty acids. The entire set of fatty acids is used in this procedure, but only a subset is chosen to classify samples into as homogeneous groups as possible. Tree-based models do not require assumptions about the distribution of the data, use computer power to test all possible combinations, and are visually easy to present and understand.

Fatty acids are identified and quantified using temperature-programmed gas liquid chromatography (e.g., Iverson et al. 1995). Three tissues (i.e., blood, blubber, and milk), each capable of providing information about foraging on a different temporal scale, can be used for fatty acid analysis. Chylomicrons (lipoproteins that transport dietary fatty acids) can determine the composition of a meal, after their recovery using density-gradient ultracentrifugation from blood samples up to 12 hr after that meal (H. Heras and S. J. Iverson, pers. comm.). Fatty acids stored in blubber, on the

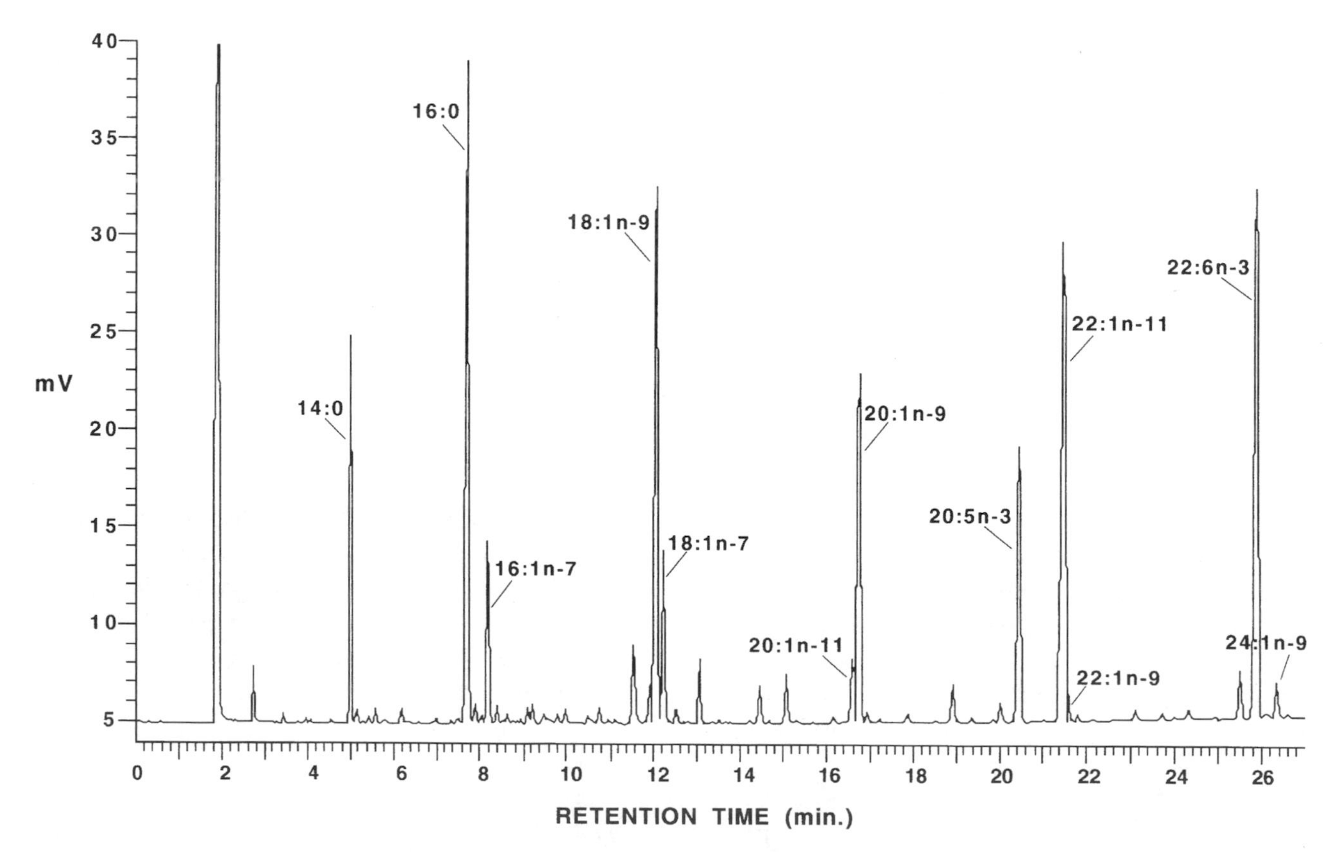

Figure 9-16. Chromatogram of fatty acids from whole capelin (*Mallotus villosus*), showing the characteristic signature of this species. Major fatty acids are labeled. (Courtesy S. J. Iverson.)

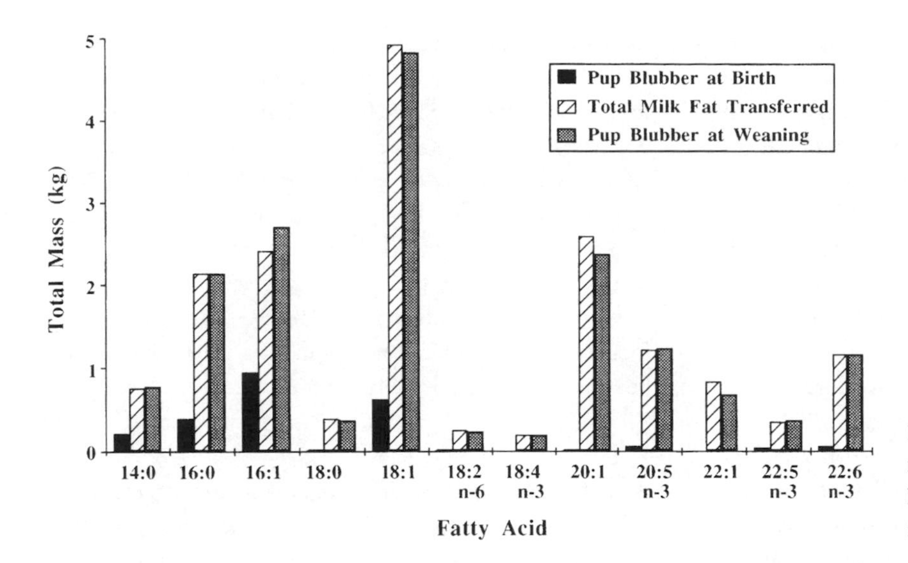

Figure 9-17. Changes over lactation in the fatty acid composition of the blubber of hooded seal pups compared to that ingested in milk. (After Iverson et al. 1995.)

other hand, represent the integration of feeding over periods of weeks to perhaps months depending on the rate and degree of lipid deposition that has occurred. Fatty acids in milk provide a short-term view of the diet if the mother is feeding (e.g., a lactating fur seal) or a longer term view of previous feeding if the mother is fasting and thus, mobilizing lipids from body stores (e.g., a lactating phocid).

There are several advantages to fatty acid signature analysis. First, samples can be obtained using relatively noninvasive techniques (i.e., blood and biopsy samples). Therefore, it is possible to conduct longitudinal studies on individuals and to obtain data from rare or endangered populations that might not otherwise be possible. Blubber samples provide a longer term integration of the diet and are less affected by the geographic location of sampling. In addition, the method is not dependent on the recovery of hard parts; therefore soft-bodied prey may be detected.

As with any method, there are disadvantages associated with fatty acid analysis. One is that not all prey signatures may be unique. Another is that signatures of the prey may vary significantly in relation to age, reproductive condition, and geographic location. Accurate identification of prey components in the diet requires that a comprehensive library of reference prey signatures be assembled. The method often provides only qualitative information on diet composition, although quantitative estimation is possible. Finally, metabolic alteration of the signature may occur and thus a clear understanding of the lipid metabolism of marine mammals is required for the confident use of the method (Iverson 1993).

Stable Isotopes

The carbon isotope ratio, $^{13}C/^{12}C$ (denoted $\delta^{13}C$), and the nitrogen isotope ratio, $^{15}N/^{14}N$ (denoted $\delta^{15}N$), of various animal tissues can be useful in diet studies because they reflect the foods that have been assimilated by the predator (DeNiro and Epstein 1978, 1981; Minagawa and Wada 1984; Peterson and Fry 1987). Like fatty acids, stable isotope ratios provide a longer record of the diet (e.g., half-life for turnover in muscle is 27 days; Tieszen et al. 1983) than stomach or fecal contents and are not dependent on the recovery of hard parts.

The $\delta^{15}N$ composition of an individual animal is typically about 3% to 4% greater than that of its diet (Minagawa and Wada 1984, Fry 1988). This enrichment occurs at each trophic level within a food web; therefore $\delta^{15}N$ values provide a good indication of the trophic level at which the predator feeds (Wada et al. 1987, Fry 1988). On the other hand, $\delta^{13}C$ values do not show the same kind of predictable trophic enrichment (Fry 1988, Hobson and Welch 1992), but do seem to vary geographically such that the location of feeding can often be deduced (e.g., Schell et al. 1989).

The use of naturally occurring stable isotopes of carbon and nitrogen has recently been applied to the study of the foraging ecology of several marine mammals (Schell et al. 1989, Ramsay and Hobson 1991, Hobson and Welch 1992, Rau et al. 1992, Ostrom et al. 1993, Ames et al. 1996). Generally, muscle tissue has been used in stable isotope studies, but skin, vibrissae, baleen, and blubber may also be used for this purpose. These latter tissues have the advantage of being relatively more easily obtained by nonlethal and noninvasive means. Although stable isotope δ values may provide information on trophic level of predation and the broad-scale geographic source of the diet, they do not usually permit the identification of the species of prey being eaten.

Diet

The ability to forage successfully is clearly essential for the survival of individuals, placing foraging ability under strong selective pressure. There is a large body of theoretical and empirical literature on the evolution of foraging strategies of vertebrates (e.g., Stephens and Krebs 1986). To date studies on marine mammals have played little role in either the development or testing of theory; the reasons are not difficult to imagine. Most of the information needed to parameterize foraging models (e.g., capture success rate, prey-handling time, search time, prey density) is not available, although with recent and future advances in technology, we can expect this to change. For example, using data derived from telemetry studies, Thompson et al. (1993) developed several models that predict how a seal should behave during foraging within the constraints placed on the animal by diving physiology. Nevertheless, with the exception of the detailed studies on the sea otter (e.g., Estes and Duggins 1995 and references therein), we are rarely in a position to test hypotheses about why marine mammals forage as they do. However, good descriptions of the diets of marine mammals are a prerequisite to this deeper understanding.

It is not feasible to review in this chapter the large literature on the diets of marine mammals. Riedman (1990) provides further references on the diets of pinnipeds, whereas Gaskin (1982), Evans (1987), and Goodall and Galeazzi (1985) do so for the cetaceans. Here we list selected studies that provide the reader with a window to the literature on diets of the better-studied pinniped (Tables 9-5 and 9-6) and cetacean (Tables 9-7 and 9-8) species. Stirling and Archibald (1977), Smith (1985), and Hammill and Smith (1991) present data on the diets of polar bears. Several papers in the monograph edited by VanBlaricom and Estes (1988) and a paper (Kvitek et al. 1993) discuss the diet and foraging behavior of sea otters. The foraging behavior and food habits of sirenians have been well documented by a number of researchers including Best (1983), Hartman (1979), Lefebvre and Powell (1990), Preen (1995), and Reynolds (1981); in addition, Wells, Boness, and Rathbun, Chapter 8 (this volume) describe sirenian feeding behavior and ecology. In the following sections, we briefly review some general features of marine mammal diets and then focus on examples that serve to illustrate factors that seem to be important in determining the composition of diets.

Pinnipeds

At first glance, pinniped diets are characterized by the large number of species that are eaten. For example, more than 100 taxa of crustaceans, cephalopods, and teleosts have been identified from 6457 stomach content samples of harp seals in the northwest Atlantic collected during the past 40 years (for review, see Wallace and Lavigne 1992). More than 40 taxa of fish and invertebrates were found in 682 gray seal stomachs examined from eastern Canada (Benoit and Bowen 1990a) and more than 48 taxa were identified from 2841 fecal samples collected from harbor seals in British Columbia (Olesiuk et al. 1990a). Lowry et al. (1990) found 52 prey types in 1476 fecal samples collected from California sea lions. More than 60 taxa, mainly fish, were found in the stomachs of northern fur seals sampled at sea from 1958 to 1974 (Kajimura 1985).

Table 9-5. Selected Studies of Phocid Seal Diets in which Fifty or More Stomachs or Feces Contained Food Remains

Species	No.	Location	Main Prey	Source
Gray seal	365	Eastern Canada	Sandlance, flatfishes, Atlantic cod	1
	1401	United Kingdom	Sandlance	2
Harbor seal	250	Eastern Canada	Herring, Atlantic cod, pollock, squid	3
	2841	Western Canada	Pacific hake, Pacific herring	5
	314	Sweden	Atlantic cod, sole, herring, sandlance	5
Harp seal	1334	Northwest Atlantic	Arctic cod, herring , capelin	6
	339	White Sea/East ice	Capelin, sand lance, herring	7
Hooded seal	853	Greenland	Greenland halibut, redfish, Gadidae	8
Ringed seal	973	Bering Sea	Saffron cod, Arctic cod, shrimps	9
Ribbon seal	61	Bering Sea	Pollock, eelpout, Saffron cod	10
Bearded seal	397	Bering Sea	Shrimp, crab, clam	11
N. elephant seal	193	California	Cephalopods, Pacific whiting	12
S. elephant seal	56/71	Heard/Macquarie Islands	Squids, pelagic and benthic fishes	13
	76	Heard Island	Squids, pelagic fishes	14
Leopard seal	143	Southern Ocean	Krill, cephalopods, penguins, seals	15

1 = Bowen and Harrison 1994; 2 = Hammond and Prime 1990; 3 = Bowen and Harrison 1996; 4 = Olesiuk et al. 1990a; 5 = Harkonen 1987; 6 = Lawson and Stenson 1995; 7 = Nilssen et al. 1995; 8 = Kapel 1982; 9 = Lowry et al. 1980b; 10 = Frost and Lowry 1980; 11 = Lowry et al. 1980; 12 = Antonelis et al. 1994b; 13 = Green and Burton 1993; 14 = Slip 1995; 15 = Siniff and Stone 1985.

Table 9-6. Selected Studies of Otariid Seal Diets in which Fifty or More Stomachs or Feces Contained Food Remains

Species	No.	Location	Main Prey	Source
Northern fur seal	7373	North Pacific	Anchovy, herring, capelin, sand lance	1
South African fur seal	697	Benquela	Anchovy, hakes, squid	2
Antarctic fur seal	50	South Georgia	Krill, cephalopods, fish	3
Sub-Antarctic fur seal	132	Gough Island	Squids	4
Australian fur seal	353	Tasmania	Squids	5
South American fur seal	100	Peru	Sardine, southern anchovy, jack mackerel	6
Juan Fernandez fur seal	370	Alejando	Myctophid fishes, squid	7
New Zealand fur seal	64	New Zealand	Octopus, squid, barracuda	8
Steller sea lion	153	Gulf of Alaska	Pollock, herring, squids	9
California sea lion	1476	California	Northern anchovy, Pacific whiting, squid	10

1 = Kajimura 1984, Perez and Bigg 1986; 2 = David 1987; 3 = Doidge and Croxall 1985; 4 = Bester and Laycock 1985; 5 = Gales et al. 1994; 6 = Majluf 1987; 7 = Acuna and Francis 1995; 8 = Street 1964; 9 = Pitcher 1981; 10 = Lowry et al. 1990.

Table 9-7. Selected Studies of Mysticete Diets

Species	No.	Location	Main Prey	Source
Blue whale	467	North Pacific	Euphausiids	1
	529	Southern Ocean	Euphausiids (*Euphausia*)	
Fin whale	19,511	North Pacific	Euphausiids, copepods, fish, squid	1
	16,261	Southern Ocean	Euphausiids	
Sei whale	12,048	North Pacific	Copepods, euphausiids, fish, squid	1
	10,037	Southern Ocean	Euphausiids, copepods, amphipods	
Minke whale	87	Northeast Atlantic	Atlantic herring, capelin, sand eel, krill	2
	88	Southern Ocean	Euphausiids	1
Humpback whale	308	North Pacific	Fish (herring, capelin, saffron cod), euphausiids	1
		South Africa	Euphausiids	3
Gray whale	324	Bering Sea	Benthic amphipods, decapods	4
	[a]	Puget Sound, U.S.A.	Ghost shrimp (*Callianassa*)	5
Bowhead whale	35	Beaufort and Chukchi Seas	Copepods (*Calanus*), euphausiids	6
Right whale	100	Bay of Fundy, Canada	Copepods (*Calanus*)	7

[a]Based on sampling feeding pits.

1 = Review by Kawamura 1980; 2 = Haug et al. 1995; 3 = Best 1967; 4 = reviewed in Nerini 1984; 5 = Weitkamp et al. 1992; 6 = Lowry 1993; 7 = Murison and Gaskin 1989.

In virtually all cases, however, careful inspection reveals that relatively few species (usually less than five and often two to three) account for most of the energy ingested by pinnipeds in any one season or geographic location. For example, of the 24 taxa eaten by gray seals on the eastern Scotian Shelf, Canada, two to four species accounted for more than 80% of the biomass of prey eaten in each of eight samples collected between June 1991 and January 1993 (Bowen and Harrison 1994). Similar results have been found in gray seals in the United Kingdom, but Hammond and Prime (1990) also show that there is likely to be seasonal and geographic variation in the number of species that comprise the diet. For example, at Donna Nook, six species accounted for 80% of the biomass eaten by gray seals in some months, whereas only two species comprised that percentage of the diet in other months. Only 5 of 52 species accounted for more than 80% of the relative frequency of prey in the diets of California sea lions, although both the species and their relative contribution to the diet differed before, during, and after the El Niño of the early 1980s (Lowry et al. 1990).

Although fish and cephalopods are the main prey taxa eaten by pinnipeds, several species consume substantial quantities of crustaceans. Krill (mainly *Euphausia superba)* comprises an important component of the diet of three

Table 9-8. Selected Studies of Odontocete Diets

Species	No.	Location	Main Prey	Source
Sperm whale	1627	Japan	Squids	1
	885	Bering Sea	Squids, fish	
Pilot whale	668	Faroe Islands	Cephalopods (*Gonatus*)	2
Killer whale				
Resident	139	British Columbia	Salmon (*Oncorhynchus*)	3
Transient		British Columbia	Harbor seals	4
Narwhal	73	Eastern Arctic	Arctic cod, Greenland halibut, polar cod	5
Dusky dolphin	136	Peru	Fish (anchoveta, hake), squid	6
Bottlenose dolphin	127	South Africa	Fish (*Pomadasys olivaceum*); cephalopods (*Sepia*)	7
Common dolphin	297	South Africa	Fish (pilchard, *Pomadasys olivaceum*), squids (*Loligo*)	8
Striped dolphin	27	Mediterranean	Squids (3 species), fish	9
	27	Japan	Fish (myctophids)	
Spotted dolphin	29	Eastern tropical Pacific	Cephalopods	10
Commerson's dolphin	45	Tierra del Fuego	Fish (silversides), mysids, squid (*Loligo*)	11
Harbor porpoise	127	Bay of Fundy, Canada	Atlantic herring, silver hake, Atlantic cod	12
	179	Scandinavian	Atlantic herring, benthic fish (cod, saithe, whiting)	
Dall's porpoise	73	Sea of Okhotsk	Japanese pilchard, gonatid squid (*Berryteuthis*), walleye pollock	13

1 = Reviewed in Kawakami 1980; 2 = Desportes and Mouritsen 1993; 3 = Nichol and Shackleton 1996; 4 = Baird and Dill 1995; 5 = Finley and Gibb 1982; 6 = McKinnon 1993; 7 = Cockcroft and Ross 1990, Barros and Odell 1990; 8 = Young and Cockcroft 1994; 9 = Blanco et al. 1995, Miyazaki et al. 1973; 10 = Perrin et al. 1973; 11 = Bastida and Lichtschein 1988; 12 = Recchia and Read 1989, Aarefjord et al. 1995; 13 = Walker 1996.

Antarctic species, although only tentative quantitative conclusions are possible given the limited quality of present data. Krill accounts for about 37% and 50%, respectively, of the diets of the crabeater seal and the leopard seal (Laws 1984). The importance of krill is further suggested by the similarity in maxillary dentition of the two species, which appears to be an adaptation to some form of filter feeding. Overall, about one-third of the Antarctic fur seal (*Arctocephalus gazella*) diet is krill, which is almost exclusively taken by lactating females (Doidge and Croxall 1985). Crustaceans also appear to be important prey of harp seals, particularly pups and immature animals, in the northwest Atlantic (Lawson et al. 1994) and the White Sea (Nilssen et al. 1995).

A feature shared by fish-eating pinnipeds is the size of prey that seems to contribute most to energy intake. Bearing in mind the biasing effect that digestion has on the estimates of prey size, studies on a number of phocids, including gray seals (Prime and Hammond 1990, Bowen et al. 1993, Bowen and Harrison 1994, Hammond et al. 1994), harbor seals (Harvey 1988, Bowen and Harrison 1996), ribbon seals (*Phoca fasciata*) (Frost and Lowry 1980), and harp seals (Beck et al. 1993) have shown that most fish eaten fall within the 10- to 35-cm length range (Fig. 9-18). There are fewer data on otariids, but the pattern seems to hold in California sea lions (Ainley et al. 1982), northern fur seals (Sinclair et al. 1994), and Steller sea lions (*Eumetopias jubatus*) (Frost and Lowry 1986). Larger fish are taken and in some cases contribute significantly to the overall energy budget (e.g., Frost and Lowry 1986, Hammond et al. 1994). Adults often take larger prey than immatures, and larger species tend to prey on somewhat larger prey (Frost and Lowry 1986). The tendency to consume relatively smaller sizes within a prey species is also found among squid-eating (teuthophagous) pinnipeds, such as the southern elephant seals studied both at South Georgia (Rodhouse et al. 1992) and Heard Island (Slip 1995).

Five species of pinnipeds (four otariids and one phocid) prey on other pinnipeds. This appears to be common in the southern sea lion where males prey on South American fur seal pups and females (Majluf 1987). Both adult male and female Steller sea lions consume ringed seal, spotted seal, harbor seal, bearded seal, and northern fur seal pups (for review, see Hoover 1988). Although consumption of these species appears to be infrequent, Steller sea lions may take 3% to 6% of the northern fur seal pups born on St. George Island each year (Gentry and Johnson 1981). Walruses feed primarily on benthic invertebrates, but they also prey on young ringed seals, spotted seals, and bearded seals (for review, see Lowry and Fay 1984). In fact, the stomach contents of walruses taken in the Chukchi Sea in summer indicate that seals occur in about 10% of the walrus stomachs. Seals also account for about 3% to 25% of the diet of leopard seals; the variation is apparently caused by a seasonal shift in diet as young crabeater seals become more difficult to capture (Oritsland 1977).

Cetaceans

As we have seen, determining the diets of pinnipeds is difficult. However, additional problems are encountered when studying the diets of cetaceans. Fecal samples are rarely available and thus most data have come from stomach contents of

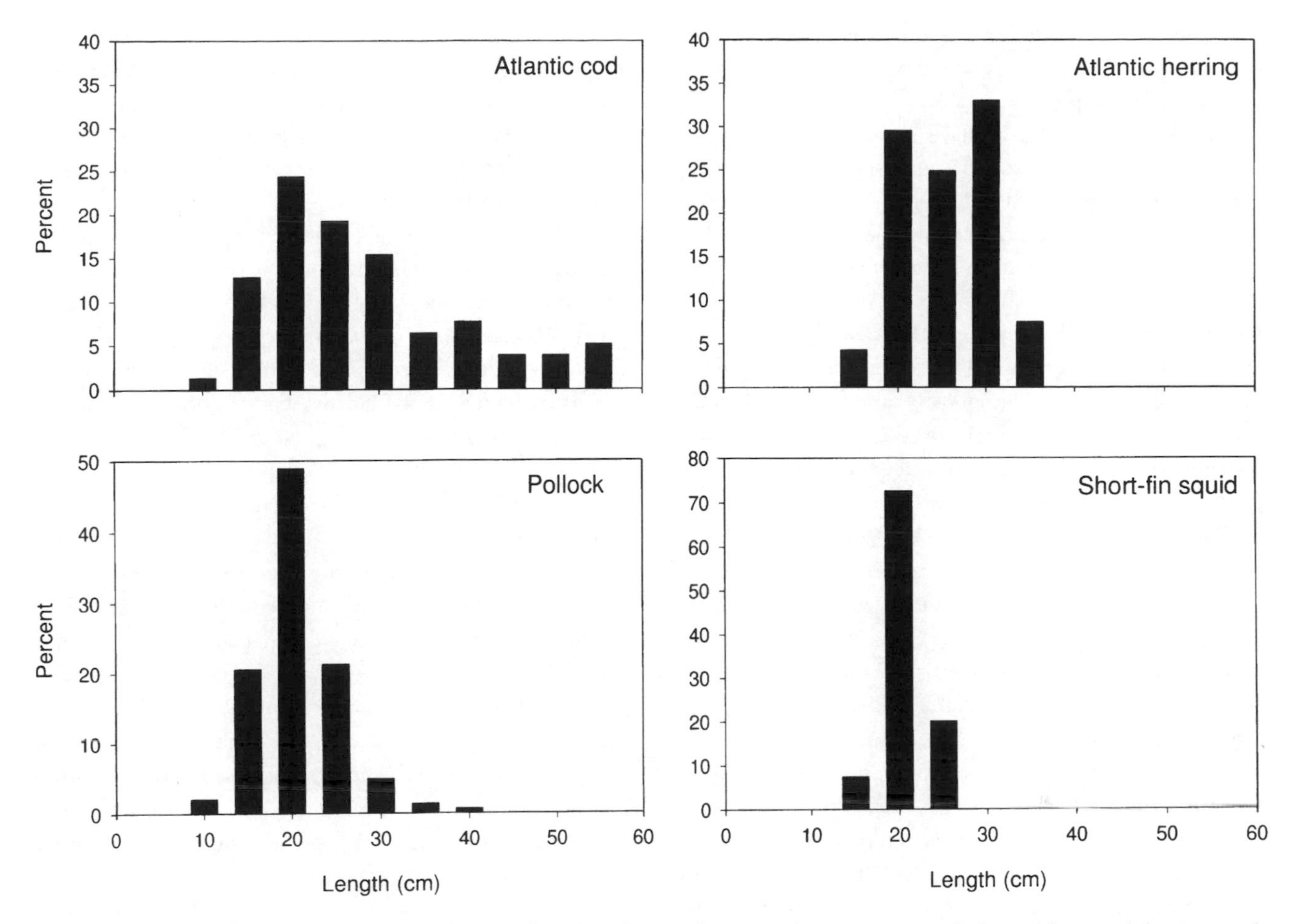

Figure 9-18. Examples of the estimated distribution of prey lengths eaten by gray seals in eastern Canada derived from otoliths recovered in feces.

animals taken during commercial whaling. This is particularly true in baleen whales, pilot whales, and sperm whales. Entangled and stranded cetaceans are also an important source of information on diets. Gaskin (1982) and Evans (1987) both provide reviews of the diets and feeding behavior of whales. Kawamura (1980, 1994) reviews the feeding ecology of baleen whales. Readers are encouraged to consult these sources and Wells, Boness, and Rathbun, Chapter 8, this volume.

The major anatomical adaptations of the feeding apparatus in cetaceans have had important consequences for their diets and foraging behavior. Gaskin (1982) noted that the straining system of the baleen whales is among the most bizarre adaptations found in mammals. Among baleen whale species, the shape, arrangement, and fine structure of the baleen plates reflect both the types of prey eaten and the methods used to capture food (Pivorunas 1979). The baleen whales feed primarily on planktonic crustaceans, such as copepods, euphausiids, and amphipods. They feed by engulfing prey (i.e., rorquals, the gulpers) or skimming the surface (i.e., right whales, the skimmers). In rorquals, ventral throat grooves and a specialized tongue allow tremendous distension of the throat area during feeding (Lambertsen 1983, Orton and Brodie 1987). The engulfed water is expelled from the buccal cavity by the upward and backward motion of the tongue, such that food is moved posteriorly toward the esophagus. Species with coarse baleen (e.g., blue whale) feed on densely swarming zooplankton. Some species (e.g., the humpback whale) may also consume fish, such as capelin and herring. The gray whale filters amphipods and other crustaceans from bottom mud by pumping sediment and prey through its baleen plates. The skimmers (e.g., right whales and bowhead whales) have long baleen plates and large lower jaws that allow more continuous straining of prey, such as copepods, from surface waters.

In contrast, the toothed whales feed almost entirely on individual fish and squids. Given the high degree of seasonal variation in zooplankton biomass and production (e.g., Cushing 1975), feeding in baleen whales is highly seasonal and localized, whereas that of toothed whales is more widely distributed and less seasonal. Therefore, it seems likely that the availability of food has played a major role in the evolution of whale distributions and seasonal movement patterns (Lockyer and Brown 1981, Gaskin 1982; also see earlier discussion in this chapter).

The right and bowhead whales are zooplankton grazers. They have long (often exceeding 3 m), finely fringed baleen plates (200 to 400 on each side of the jaw), which together form a fine-toothed comb that effectively separates and retains small zooplankton from the sea water engulfed while skimming the surface. The diet of the North Atlantic right whale consists mainly of the copepod *Calanus finmarchicus*. In the North Pacific, several species of *Calanus* are important foods of right whales. Considerably larger prey, mainly *Euphausia superba* (55 to 65 mm), are taken along with *Calanus* by right whales in the Southern Ocean. Copepods and euphausiids are the principal foods of bowheads sampled at communities along the northern Alaska coast in the Beaufort Sea, but some epibenthic organisms, such as mysids and gammarid amphipods, are also eaten (for review, see Lowry 1993).

In the rorquals, the baleen plates are much shorter (20 to 90 cm) and only a relatively small area in the central and posterior portion of the baleen is actually used to filter food. Sei whales have short baleen plates with the finest fringes. When skimming the surface like the right whales, they feed mainly on copepods, but they also exhibit "gulping" feeding that is more typical of the rorquals. The diet of sei whales varies by region and season, as might be expected in such a widely distributed species (Table 9-7). For example, euphausiids are most important in the areas off Australia and in the western South Pacific, whereas amphipods are commonly eaten in the Indian Ocean. In both the North Atlantic and North Pacific, calanoid copepods are the most important foods, but squids and fish are seasonally or locally important foods. Evidence of latitudinal effects on the diet of sei whales has also been found in the Antarctic with pelagic amphipods dominating the diet at low latitudes, calanoid copepods at intermediate latitudes, and euphausiids at high latitudes (Kawamura 1970).

The blue whale has fine-fringed baleen plates that are the longest of the rorquals. It exhibits marked diet specialization (stenophagy), feeding mainly on euphausiids throughout most of its range. In the Southern Ocean, blue whales feed almost exclusively on *Euphausia superba*. In the North Atlantic, only the euphausiids *Thysanoessa inermis, Temora longicornis,* and *Meganyctiphanes norvegica* have been reported in the diet, and in the North Pacific, *T. inermis, T. longipes,* and *Nematoscelis megalops* are commonly eaten species (Kawamura 1980).

The diet of fin whales at high latitudes is well known in all the major oceans (see Gaskin 1982). Fin whales have coarser baleen than other rorquals, enabling them to feed not only on larger krill and copepods, but on a variety of small schooling fish, such as capelin and herring. In the Southern Ocean, particularly beyond the Antarctic Convergence, *Euphausia superba* is the primary food of the fin whale. However, in the northern hemisphere the diet is considerably more diverse. In the North Pacific, herring, Japanese sardine (*Sardinella melanosticta*), capelin, and pollock (*Theragra chalcogramma*) are frequently eaten by fin whales taken near coastal areas, whereas euphausiids are generally the dominant foods further offshore. Squid (*Ommastrephes sloani pacificus*) is also eaten by southward migrating fin whales near the Kuril Islands in the fall. In the North Atlantic during summer, fin whales feed mainly on herring and mackerel that have come to feed on concentrations of pelagic crustaceans, *Meganyctiphanes norvegica* and *Thysanoessa inermis* (Hjort and Ruud 1929).

With their short and narrow baleen plates, both humpback and minke whales feed more heavily on fishes than do other baleen whales. Minke whales fed mainly on fish in all five regions of the Norwegian and Barents Seas (Haug et al. 1995). Capelin were most important in two regions, herring in two other regions, and a mixture of herring, sand eel (*Ammodytes*), and cod in the fifth region. Only at high northern latitudes do pelagic crustaceans such as *Meganyctiphanes norvegica* form a significant portion of the diet (Ohsumi 1979). In the southern hemisphere, *Euphausia superba* is the most important food of both the minke (Ohsumi 1979) and the humpback whale (Dawbin 1956). The diet of humpback whales in other areas, such as the Bering and Chukchi Seas, varies considerably, but prey such as herring, capelin, and the euphausiids *(Thysanoessa spinifera* and *T. inermis)* are commonly eaten (Kawamura 1980).

The Bryde's whale is another fish-eating species with rather short, coarse baleen, but its diet is poorly known. Best (1967, 1974) described two allopatric forms of this species off South Africa. The smaller inshore form, with relatively narrow baleen, fed mainly on anchovies (*Engraulis capensis*), pilchards (*Sardinops ocellata*), and other fish, whereas the larger offshore form, with wider baleen, fed on euphausiids. In the Coral Sea and South Pacific and Indian Oceans, Kawamura and Satake (1976) reported that Bryde's whales fed mainly on euphausiids, whereas, the anchovy (*Engraulis australis*) was eaten off southwestern Australia (Chittleborough 1959). In the Gulf of California, Bryde's whales preyed on schooling fish such as Pacific sardine (*Sardinops sagax*) and thread herring (*Opisthonema*) in 87% of 88 feeding events (Tershy 1992).

The gray whale specializes on benthic prey (Table 9-7). Its reduced number (130–180) of short and relatively stiff baleen plates is presumably adapted to filtering of food, mainly amphipod crustaceans, from bottom sediments. Studies conducted on feeding gray whales in the Bering and Chukchi Seas during summer show that gammaridian amphipods, especially *Ampelisca macrocephala,* are the major prey, often comprising 90% of the food remains in stomachs (for review, see Nerini 1984). The Ampeliscidae are tube builders that occur in dense concentrations of up to 900 g/m^2 in the upper several centimeters of bottom sediments. Little feeding appears to occur in gray whale winter range, the Baja California lagoons, but what does occur may be mainly on pelagic foods. The diet of migrating gray whales is also poorly known but is thought to include sessile polychaetes and tubeworms (Darling 1977).

Unlike the heterodont condition of mammals, most toothed whales (odontocetes) have a long row of uniformly shaped teeth (homodont) designed for grasping and holding fast-moving prey such as fish or squid. Piscivorous delphinid genera, including *Stenella, Delphinus,* and *Lagenorhynchus,* are typical of this group with 20 to 65 pairs of sharp conical teeth on each side of the head. Although primarily fish eating (ichthyophagous), these species also consume significant amounts of squids (Table 9-8). Fitch and Brownell (1968) found that lantern fish (Myctophidae) accounted for 90%, by number, of the contents of 13 stomachs of five species of delphinids in the North Pacific. Off the coast of Japan, striped dolphins fed mainly on myctophids but also consumed significant quantities of a semipelagic shrimp (Miyazaki et al. 1973). Spotted and spinner dolphins in the eastern tropical Pacific ate large numbers of the ommastrephid squid *(Dosidicus gigas*), but onychoteuthid and enoploteuthid squids were also consumed (Perrin et al. 1973). Young and Cockcroft (1994) studied the diet of 297 common dolphins off the southeast coast of Africa captured in shark nets between 1974 and 1992; pilchard (*Sardinops ocellatus*) accounted for 48% of the reconstructed mass of food eaten, whereas three other fish species accounted for another 30% of the diet. Squids, mainly *Loligo* species, accounted for about 11% of the diet. Another schooling pelagic fish (anchoveta, *Engraulis ringens*) made up almost 84% of the mass of food eaten by dusky dolphins (*Lagenorhynchus obscurus*) studied in the coastal waters of central Peru (McKinnon 1993). Twenty-five prey taxa, including a mysid shrimp, several pelagic fishes, and the squid (*Loligo gahi*), were found in 45 Commerson's dolphin (*Cephalorhynchus commersonii*) stomachs examined from individuals entangled in gillnets off Tierra del Fuego (Bastida and Lichtschein 1988).

The diet of another delphinid, the bottlenose dolphin, has been studied in several parts of its range. Cockcroft and Ross (1990) studied the stomach contents of 127 bottlenose dolphins off Southern Natal, South Africa. Although more than 72 species of prey were identified, two to three species accounted for 75% to 94% of the mass of food eaten. Two of these three species were fish (one benthic and the other a schooling pelagic species) and the third was a cephalopod, the cuttlefish (*Sepia officinalis*). Bottlenose dolphins in southeastern U.S. waters fed mainly on sciaenid groundfishes in three areas, and on a mixture of fish, shrimp, and cephalopods in the fourth area (Barros and Odell 1990).

Porpoises (Phocoenidae) generally have 13 to 28 pairs of spade-shaped teeth, which are set in each jaw at angles that provide a shearing action during feeding. Several studies have been conducted on the diet of harbor porpoise in the Bay of Fundy. Smith and Gaskin (1974) found that herring, Atlantic cod, and mackerel accounted for 78% of otoliths recovered from 54 stomachs collected during 1969 to 1972. Recchia and Read (1989) examined the stomach contents of 127 harbor porpoises over three summers, 1985 to 1987. Herring was the most important food and accounted for 80% of the estimated total energy intake, followed by silver hake and Atlantic cod, which together accounted for another 17% of the diet. Herring was a variable, but important prey of harbor porpoises at seven locations in Scandinavia, followed by benthic fishes such as Atlantic cod, saithe, and whiting (Aarefjord et al. 1995). Dall's porpoise (*Phocoenoides dalli*) also feeds heavily on pelagic fish. Walker (1996) studied the diet of 73 Dall's porpoises taken in the southern Sea of Okhotsk during the summer of 1988. Japanese pilchard (*Sardinops sagax melanosticutus*) accounted for about 81% of the estimated energy intake, followed by a squid (*Berryteuthis magister*) at 15%, and walleye pollock at 4%.

In species such as the pilot whales (*Globicephala*), which feed primarily on squids (see Desportes and Mouritsen 1993), there is a reduction in the number of teeth (7 to 12 pairs per tooth row), but an increase in their size. This is also true of another squid specialist, the sperm whale, where 20 to 25 teeth are found in the lower jaw and up to 10 highly curved and nonfunctional teeth are found in the upper jaw. Both the number and size of teeth is further reduced in females of some species. This reduction in dentition is most extreme in many beaked whales (Ziphiidae) in which only a single pair of teeth protrude from the lower jaw, usually only in males. In beaked whales, teeth have been replaced by a ribbed palate. Although relatively little is known of most ziphiid diets, squid appears to be an important food in all species (Gaskin 1982).

More is known about the diet of the sperm whale than of many other odondocetes because of its previous commercial importance (see references in Clarke 1980, Kawakami 1980, Gaskin 1982, Evans 1987, Clarke et al. 1993). Oceanic populations of sperm whales feed mainly on meso- and bathypelagic cephalopods, whereas at more near-shore locations fish comprise a much greater proportion of the diet. Medium-sized squids (standard length from 0.9–1.5 m) probably make up most of the diet of sperm whales in most populations. Near the Azores, Clarke et al. (1993) found that three families of cephalopods contributed about 80% of the estimated mass of food eaten. Most (77.5%) of the prey species have luminous organs and 82% are neutrally buoyant, leading Clarke et al. (1993) to speculate that sperm whales obtain about 77% of their food by swimming through luminous shoals of slow-swimming squids and only about 23% by chasing larger, fast-swimming species. Near Iceland, sperm whales feed mainly on fish, including lumpfish (*Cyclopterus lumpus*), redfishes (*Sebastes* sp.), and monkfish (*Lophius piscatorius*) rather than cephalopods. Fish also appear to be an important source of food for sperm whales off the Gulf of Alaska and British Columbia (Kawakami 1980).

The killer whale is the only consistently meat eating (sarcophagous) odontocete. Killer whales feed on a variety of prey, including 20 species of cetaceans, 14 species of pinnipeds, the sea otter, and the dugong. Fish, seabirds, and cephalopods are also eaten (Table 9-8; Jefferson et al. 1991). Despite the large number of reports of killer whale predation, there are few studies in which the composition of the diet has been determined. Two forms of killer whales are generally recognized in several parts of their range. In the eastern North Pacific, these two forms, called "transients" and "residents," can be distinguished by differences in behavior, morphology, mitochondrial DNA, and diet (Bigg 1982a, Bigg et al. 1987, Baird and Stacey 1988, Stevens et al. 1989). Resident killer whales are primarily piscivorous, with salmon (*Oncorhynchus* spp.) being important prey (Ford et al. 1994, Nichol and Shackleton 1995). The transient form feeds mainly on marine mammals, especially harbor seals (Bigg et al. 1987). Changes in seasonal distribution of transient killer whales off southern Vancouver Island appeared to be related in late summer to increased predation on young harbor seals (Baird and Dill 1995), which appear to meet the whales' energy requirements at that time. In northern Norway, the seasonal distribution of killer whales is closely associated with spring spawning herring (Christensen 1988). Simila and Ugarte (1993) described a feeding behavior, known as the carousel, that is used by killer whales, perhaps to concentrate and push schools of herring to the surface where they are more easily captured.

Seals, especially ringed seals, and to a lesser extent bearded seals, are the main food of the polar bear (Stirling and Archibald 1977, Smith 1980, Hammill and Smith 1991). During periods of hyperphagia in the spring, polar bears often consume only the blubber layer of their prey. Although polar bears spend 4 months or more on land, evidence from stable carbon isotope analysis indicates that bears consume little food from terrestrial ecosystems in the eastern Canadian Arctic (Ramsay and Hobson 1991). The mean isotope ratios for polar bears were close to those of ringed seals but were significantly different from isotope ratios in four berries collected in areas frequented by bears during the summer.

Sea otters forage in rocky substrate and soft bottom communities in the near-shore. They return to the surface to consume their catch, therefore observations of diet have been relatively easy to obtain. Observations in California have indicated a great deal of individual variation in diet and foraging patterns (Estes et al. 1981). Some investigations (Riedman and Estes 1990, Ralls et al. 1996) have indicated that individuals may specialize on particular food items. Some food items that seem to be particularly favored include sea urchins (*Strongylocentrotus* spp.), many genera of bivalves, and abalone (*Haliotus* spp). As these species become less abundant, many other species of sessile animals that occupy the near-shore areas are consumed. These include various crabs, starfish, sea cucumbers, annelid worms, and chitons. Fish also are taken by sea otters in some locations; for example, kelp fish species seem to play a central role in the diet of sea otters at Amchitka Island, Alaska (Estes et al. 1981).

Wells, Boness, and Rathbun, Chapter 8, this volume, provide a thorough review of sirenian feeding behavior and ecology; we refer interested readers to that chapter. It is especially interesting, however, that the various species of sirenians tend to feed to some extent on different species of plants due to the degree of rostral deflection of the skull (Domning 1980); for example, dugongs appear to be obligate bottom feeders as a consequence of their sharply downturned rostrum, whereas West Indian manatees, with their moderately downturned rostrum, feed equally well on the bottom, at the surface, or in the water column. Thus, despite their general similarities as aquatic herbivores, the effects of sirenian cropping are likely to differ among the various species.

Factors Affecting Marine Mammal Diets

A number of factors undoubtedly influence the foraging behavior and thus the composition of the diet of an individual. Among these, demographic factors (age, sex, and reproductive status), anatomical and physiological constraints of predators, the risk of predation, competitive interactions, and the distribution and abundance of potential prey are likely most important. In this section we examine evidence of the impact of such factors on the diets of marine mammals. Although we touch on foraging behavior, this is dealt with in greater detail by Wells, Boness, and Rathbun, Chapter 8, this volume.

Demographic Factors

AGE. Most marine mammals are relatively long-lived species and thus individuals may have many years to gain foraging experience. We should not be surprised to find that diets often change with age. Slip (1995) studied the diets of 32 juvenile and 44 adult southern elephant seals at Heard Island by means of lavage. Juveniles ate proportionally more fish than adults, and the species composition of squid consumed by juveniles differed significantly from that taken by adults. Slip (1995) speculated that these differences might reflect the more limited ability of juveniles to dive and capture swift and deep prey. Green and Burton (1993) also reported evidence of age-related differences in the diets of southern elephant seals at Heard and Macquarie Islands. Lowry et al. (1980a) found that the contribution of clams, crabs, and sculpins to the diet of bearded seals increased with age, from pups through animals 3 years old or more, whereas the contribution of shrimps, isopods, and saffron cod decreased with age. Harp seal pups consumed greater quantities of capelin, euphausiids, and hyperid amphipods than did older seals at inshore areas around Newfoundland (Lawson et al. 1994, Lawson and Stenson 1995). Gray seal pups ate more silver hake (*Merluccius bilinearis*) and fewer squid than did older seals on the eastern Scotian Shelf, but otherwise few differences were evident (Bowen et al. 1993). Lowry et al. (1980b) and Bradstreet and Finley (1983) found that ringed seal pups and immature animals relied more heavily on crustaceans than did adults in which the cod *Boreogadus* was of greater importance. Ringed seals during their first year of life also took smaller cod than did older ringed seals. Bowen and Harrison (1996) found that harbor seal pups ate relatively more pelagic prey (herring and squid) and less benthic prey than older animals. All these examples imply that the diet of younger seals is affected by their ability to capture prey. Pelagic prey may be more easily located (perhaps backlit against the surface) by inexperienced pups than concealed benthic prey such as flatfishes. The diet of pups may also reflect physiological limitations placed on their diving behavior by small body size, and hence the amount of oxygen that can be stored, compared to larger and older individuals (see Costa and Williams, Chapter 5, and Pabst, Rommel, and McLellan, Chapter 2, this volume).

Age effects on diet have also been reported in cetaceans. The stomachs of bottlenose dolphin calves contained significantly fewer prey taxa per stomach (mean 2.9) than did older animals (mean range 6.0–8.4; Cockcroft and Ross 1990). Although eaten by all age classes, one fish (*Pomadasys olivaceum*) was the single most important prey of calves. In contrast, cuttlefish were relatively uncommon in the diet of calves but commonly eaten by older dolphins. Bottlenose dolphin calves also ate significantly smaller prey than older animals (Cockcroft and Ross 1990).

Stirling and Latour (1978) found that polar bear cubs of the year and yearlings were less proficient hunters than adults. However, by the time they reach 2 years of age, young bears were as proficient as adults although they spent significantly less time hunting than older bears.

SEX. Few species of marine mammals have been studied sufficiently to enable firm conclusions to be reached with respect to sex differences in diet. There are several reasons for this. First, sample size is often small, such that statistical power is low. Second, the increasing use of feces in diet studies of pinnipeds means that the sex and age of individuals represented in some studies is unknown. Differences in the diets of males and females are most likely to occur when the foraging areas used by the two sexes differ. This may be more common in dimorphic species where differences in body size may lead to differences in habitat use. However, even where the same geographic area is used by both males and females, differences in the distribution of diving depths may lead to differences in diet.

As we have seen earlier, there is good evidence that the foraging ranges of males and females differ among size dimorphic pinnipeds (e.g., elephant seals) and cetaceans (e.g., the sperm whale). In the case of the northern elephant seal, unfortunately there are no data from offshore feeding areas to address sex differences in the diet of this species. Available data come from lavage samples collected at San Miguel Island, California, which represent recent feeding along the California coast (Antonelis et al. 1994b). These data show little evidence of sex differences. Slip (1995) found no significant difference in the diets of male and female southern elephant seals, another highly dimorphic species, apart from the tendency of males to take larger prey. However, only six adult males were sampled. In contrast, both Rodhouse et al. (1992) and Green and Burton (1993) found evidence of sex differences in the diets of adult southern elephant seals. The large sample of northern fur seal stomach content data collected at sea during the period 1958 to 1974 unfortunately contained few males and therefore, the question of sex difference in diet cannot be addressed (Kajimura 1984).

Among species in which males and females are similar in body size, the diets of males and females show few differences. Lowry et al. (1980a) concluded that the differences in the diets of male and female bearded seals were slight, if any, and Antonelis et al. (1994a) reached the same conclusion. Lowry et al. (1980b) and Bradstreet and Finley (1983) found no significant differences in the diets of male and female ringed seals. Kvitek et al. (1993) found no evidence of differences in the species composition of the diets of male and female sea otters in southeast Alaska. McKinnon (1993) studied the diet of the dusky dolphin (*Lagenorhynchus obscurus*) in the coastal waters of central Peru and found no evidence for sex differences. Male and female bottlenose dolphins in the southeastern

United States (Barros and Odell 1990) and off southern Natal, South Africa, (Cockcroft and Ross 1990) generally fed on the same prey, although the diet of lactating females did differ from that of adult males (Cockcroft and Ross 1990).

In several small odontocetes, reproductive condition of females appears to influence the composition of the diet. Young and Cockcroft (1994) reported that adult male and nonlactating female common dolphins consumed few cephalopods (about 4% to 5% by mass) compared to pregnant and lactating females (17% and 22% by mass). Males tended to concentrate on a single fish prey (pilchard), whereas the diet of females was more diverse. Bernard and Hohn (1989) compared the diet of 11 pregnant and 12 lactating female spotted dolphins that had been taken in the offshore tuna fishery over a 15-year period. Pregnant dolphins consumed significantly more squid and fewer fish than did lactating females. Conversely, lactating dolphins ate greater amounts of fish than pregnant females. They speculated that lactating females may benefit from the higher energy density and availability of fish compared to squid as a way of satisfying the increased energy demands of nursing. Clearly more data are needed to confirm these hypotheses.

Seasonal and Geographic Effects

Given the extensive data on seasonal changes in the distribution of marine mammals and their prey, it should come as no surprise that diets often exhibit strong seasonal and geographic differences. After all, the distributional ecology of marine mammals, as in other vertebrates, is strongly influenced by the availability of suitable food. Pelagic collections of northern fur seals, taken throughout the western North Pacific between 1958 and 1974, provide a good example of large-scale regional and seasonal variations in diet (Kajimura 1984, Perez and Bigg 1986). In addition to regional and seasonal differences, squids were the main food of northern fur seals in offshore areas, whereas fish dominated the diet at more coastal locations. These differences appeared to exist throughout the species range off western North America (Perez and Bigg 1986). Seasonal or regional variation in the composition of prey eaten is also evident among a number of phocid species, including gray seals (Benoit and Bowen 1990a,b; Hammond and Prime 1990; Bowen et al. 1993; Bowen and Harrison 1994; Hammond et al. 1994), harbor seals (Härkönen 1987, Olesiuk et al. 1990a, Pierce et al. 1991, Bowen and Harrison 1996), harp seals (Lawson and Stenson 1995, Nilssen et al. 1995), ringed seals (Lowry et al. 1980b), and bearded seals (Lowry et al. 1980a). Marked geographical variation in the composition of diet has also been clearly documented in bottlenose dolphins in four areas off the southeastern United States (Barros and Odell 1990), in inshore and offshore collections of harbor porpoises in the Bay of Fundy (Recchia and Read 1989), and at seven locations throughout Scandinavia (Aarefjord et al. 1995). Strong geographic influences are clearly exhibited among the mysticetes (for review, see Gaskin 1982). For example, like other mysticetes in the Southern Ocean, minke whales feed primarily on euphausiids; however, in the North Atlantic minke whales feed mainly on pelagic and demersal fishes.

Although seasonal and geographic variation in diets is common in marine mammals, the reasons for this variation are poorly understood. Most researchers attribute such variation to differences in the prey assemblages available at different locations or to changes in the availability of prey related to their life history (Olesiuk et al. 1990a). These explanations are often reasonable, but they are usually based on rather weak correlations. Bowen and Harrison (1996) concluded that many of the differences in the diets of harbor seals in two habitats in eastern Canada could be accounted for by differences in the distribution and abundance of species such as alewife (*Alosa pseudoharengus*), winter flounder (*Pseudopleuronectes americanus*), hake (*Urophysis* spp.), and capelin. Olesiuk et al. (1990a) suggested that hake were generally absent from the diet of harbor seals during the March to April spawning period when hake were further offshore and in deeper water. However, when hake formed schools in shallower water from May through November, they became a major food of harbor seals. Similarly the occurrence of herring and salmonids in the diet appeared to be explained by changes in the availability of these prey (Olesiuk et al. 1990a). Herring and mackerel were seasonal foods of gray seals when they were concentrated during spawning or in overwintering aggregations (Bowen et al. 1993).

Prey Abundance

Although changes in prey abundance are often thought to be an important cause of diet variation, the abundance of marine mammal prey is usually only poorly known. Furthermore, most estimates of prey abundance are from fisheries surveys. The spatial and temporal scales of these data are coarse compared to what marine mammals may experience during foraging. This is an area of research where greater collaboration between marine mammalogists and fisheries ecologists is badly needed.

Nevertheless, there have been several attempts to relate prey abundance to marine mammal diets. Bailey and Ainley (1982) reported that the decline in California sea lion predation on Pacific hake (*Merluccius productus*) was related to the decline in the abundance of 2- to 4-year-old hake, the most common ages eaten. Bowen and Harrison (1994) compared the diet of gray seals in the vicinity of Sable Island with estimates of the abundance of prey species from groundfish bottom trawl surveys conducted in the same area. Although there were exceptions, generally these comparisons suggested that the more abundant and widespread species were more frequently eaten by gray seals. Sinclair et al. (1994) found that interannual variation in the importance of walleye pollock (*Theragra chalcogramma*) in the diet of northern fur seals was positively related to year-class size of pollock (Fig. 9-19). Thompson et al. (1996) found that between-year differences in the diets of harbor seals in the Moray Firth, Scotland, were closely related to interannual differences in the abundance and distribution of herring and sprat (*Sprattus sprattus*). Furthermore, differences in the abundance of these clupeid species resulted in changes in the body condition of seals. Recent data from eastern Canada also point to changes in prey abundance as the cause of interannual differences in the diet of harbor seals (Bowen and Harrison 1996). Capelin was not identified in the diet from 1988 to 1990, but accounted for about 9% of the diet by wet weight in 1992. This increase in biomass was associated with a marked increase in the local abundance of capelin from about one to nine capelin per tow in 1985 to 1990 to more than 90 capelin per tow in 1992 (Frank et al. 1994). In other cases (e.g., small Atlantic cod and pollock), however, interannual changes in the diet of harbor seals were not readily explained by estimates of prey abundance (Bowen and Harrison 1996). However, as the researchers pointed out, it is not known whether estimates of prey abundance from offshore fisheries research surveys correspond to trends in inshore abundance in areas where harbor seals likely spend most of their time foraging.

There is evidence that prey distribution and abundance are also important factors in determining the diet composition of cetaceans. This is particularly the case in baleen whales where fronts between major water masses, eddies, and upwellings are typically areas of high biological productivity where zooplankton biomass is highly concentrated (Gaskin 1982). These areas attract pelagic fish predators such as herring and mackerel that are also important marine mammal foods. During the summer in the Bay of Fundy, right whales appear to feed mainly on copepods. Murison and Gaskin (1989) found that mean copepod biomass was significantly greater in areas where right whales were present and diving (therefore, presumably feeding) than in areas where they were absent. Kann and Wishner (1995) found that right whales tended to concentrate near the leading edge of a low salinity plume where the peak abundances of copepods were located. Increased inshore abundance of humpback, fin, and minke whales along the northeast coast of Newfoundland was negatively related to offshore year-class strength of cape-lin, their major food in the area (White-

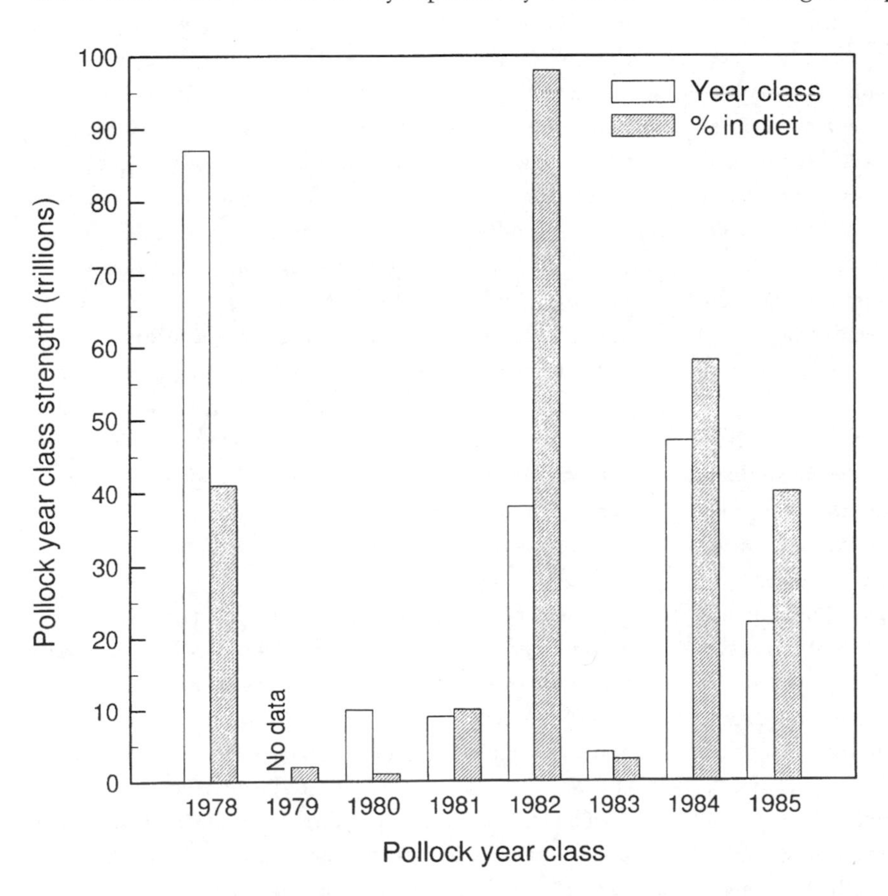

Figure 9-19. Estimates of walleye pollock year-class strength from 1978 to 1985 and the relative abundance of each year class in the diet of northern fur seals. (After Sinclair et al. 1994.)

head and Carscadden 1985). Both humpback and fin whales moved offshore in years of high offshore abundance of 2- and 3-year-old capelin, whereas minke whales seemed to respond more strongly to the abundance of 1- to 2-year-old capelin. Piatt et al. (1989) found that the abundance of these three species of baleen whales was strongly correlated with the abundance of capelin both within and among years at Witless Bay, Newfoundland. In fact, capelin abundance alone explained 63% of the interannual variation in total baleen whale abundance. Haug et al. (1995) compared the diets of minke whales with acoustic and trawl estimates of the abundance of prey near catcher ships in several areas of the Barents Sea. In the southern coastal areas, there was variation in the importance of the 1992 year-class of herring in the minke whale diet, which corresponded to the west to east abundance of this year-class. In the north, however, minke whales did not feed heavily on the large quantities of small cod and pelagic amphipods that dominated the trawls; capelin, the dominant prey in these areas, was found only sporadically in the trawls. This example serves to highlight the difficulty of estimating the amounts of food that are available to marine mammals. Nevertheless, Ichii (1990) demonstrated that the krill concentrations targeted by the commercial fishery corresponded closely with the distribution of feeding minke whales in the Southern Ocean.

Prey abundance can affect not only the composition of the diet but also whether feeding occurs. It has been suggested that baleen whales must forage on high-density prey concentrations to satisfy the metabolic demands associated with their large body size (Brodie et al. 1978, Kenny et al. 1986). Using data from three species, but mainly from humpback whales, Piatt and Methven (1992) found that few whales were observed until capelin schools exceeded a threshold density of five schools per linear kilometer surveyed. Also they found evidence that the threshold feeding density of capelin varied among years in relation to the overall abundance of capelin.

Fluctuations in the relative abundance of prey can have significant effects on the abundance and composition of cetacean assemblages. Payne et al. (1990) found a strong negative correlation between the abundance of sand lance and copepods (*Calanus finmarchicus*) over the period 1982 to 1988; both are major foods of baleen whales in the Gulf of Maine. This negative correlation was associated with negative correlations between planktivorous and piscivorous species of baleen whales feeding there. Right and sei whales were abundant only during 1986, the year copepod abundance peaked and sand lance abundance was at its lowest level. Given the documented importance of copepods in the diet of these species (e.g., Watkins and Schevill 1979), the strong positive correlation should be expected. In contrast, the piscivorous humpback and fin whales were most common in the area in years when sand lance was abundant.

Role of Marine Mammals in Marine Ecosystems

Marine mammals are major consumers of production at all trophic levels from primary production (i.e., sirenians) to predatory fish and even other marine mammals as in the case of killer whales, some pinnipeds, and polar bears. Because of their large body size and abundance, they have a major influence on the structure and functioning of some marine communities (Estes 1979, Ray 1981). The long-standing and growing debate surrounding ecological interactions between marine mammal and fisheries is rooted in the belief that marine mammals can have significant effects on prey populations that are of commercial interest to humans. An understanding of the role of marine mammals in marine ecosystems is important because it provides a context within which to evaluate the potential impact of marine mammal predation on prey populations and community structure, and the impact of variation in prey populations and environmental change on the dynamics of marine mammals. In turn, this knowledge fosters the rational discussion of the conservation and management of marine mammals.

However, before we can consider the evidence for the role of marine mammals in marine ecosystems, we must be clear about what is meant by the term role. In an ecological sense, role implies something about the functional significance of a species or other taxon. Katona and Whitehead (1988) suggest that we might consider this question by "asking whether the extinction of all cetaceans would create any noticeable difference in ecosystem function." Of course, we could pose the same question for other groups of marine mammals, and we could substitute large variation in distribution and abundance of species for extinction. Implicit here is the trophodynamic function of marine mammals as consumers and the effect of this consumption on species interactions and community structure. However, marine mammals may also play an important role in shaping the behavior and life history traits of prey species, in nutrient recycling, and in modifying benthic habitats (Katona and Whitehead 1988).

In general, there are few empirical data on "the roles of marine mammals in marine communities beyond poorly supported trophic arguments concerning marine mammal–fishery conflicts" (Estes 1979). And, with few exceptions, our understanding of the role of marine mammals in marine ecosystems has changed little during the past several decades. This may seem puzzling given the importance that society attaches both to the harvesting of marine living resources and to marine mammals as among the most visible components (i.e., as charismatic megafauna) of marine ecosystems. One major reason for this lack of understanding is that, in attempting to determine the function of marine mammals, we are faced with the intersection of three inexact sciences: marine ecology, marine mammal biology, and fisheries (see Katona and Whitehead 1988). In each of these complex fields, research is expensive, experiments are rarely possible, interactions occur at quite different spatial and temporal scales making measurement of system properties difficult, and there is an inherent indeterminacy in the behavior of complex systems that makes simplifying deterministic explanations problematic. This means that understanding comes slowly and mostly through the correlation of observed events.

Nevertheless, there are several systems where experiments have revealed rather clear evidence about the function of marine mammal predation. In other systems, studies conducted during large-scale changes in the abundance of predator and prey have provided insight as to the ecological significance of marine mammals. The term significance is used here to imply that marine mammals are known or thought to be important components of an ecosystem without specifying exactly what their function might be. In these latter cases, our inferences are not strong because we are dealing with the correlation of an uncontrolled event (say, the rapid overexploitation of a whale or seal population) with other changes in the ecosystem. Depending on the nature of the uncontrolled event, such methods are referred to as observational, analytical, or intervention studies (Eberhardt and Thomas 1991), as compared to manipulative, controlled experiments. Nevertheless, it is important to recognize that there is more than one way to study ecology and that the use of different approaches ought to be encouraged.

Insight into the role of marine mammals in ecosystems might be gained from studies of terrestrial carnivores (e.g., Skogland 1991) or predation in freshwater ecosystems (Kerfoot and Sih 1987). Although there is no doubt benefit in this comparative approach, it is important to bear in mind that there are differences in the structure and function of aquatic and terrestrial ecosystems that could bear on the function of predation (Steele 1985, 1991). For example, oceanic systems are more closely coupled to their physical environment (e.g., trends in temperature) on time scales of years and decades than are terrestrial systems. They are also inherently more unpredictable than terrestrial systems. Thus, marine organisms may have evolved different mechanisms than terrestrial animals to deal with variation in the physical environment over short time scales. Steele (1985) hypothesized that, particularly at higher trophic levels (i.e., piscivorous fish and marine mammals), we might expect the internal dynamics and structure of marine and terrestrial systems to differ in response to the differing character of their physical environments. Does this mean that marine mammal predation functions differently in the marine systems than does mammalian predation on land? At this time we simply do not know.

The energy flow through the cetacean component of marine ecosystems has been viewed as evidence of their ecological significance (Katona and Whitehead 1988). Kanwisher and Ridgway (1983) suggested that cetaceans as a group, and sperm whales alone, might consume a greater quantity of prey than all human fisheries combined. Although this is an impressive claim, it really does not help us evaluate the ecological significance of this consumption. Consumption by a marine mammal population or assemblage of species usually is a very large figure that alone suggests significance, but marine mammal predation can be properly evaluated only when compared to other sources of mortality. Bax (1991) compared estimates of finfish consumption by fish, marine mammals, birds, and commercial fisheries in six ecosystems. In each of these, fish predation on fish far exceeded predation from other sources. In three of these ecosystems, marine mammals took the second largest component of fish production. On the basis of available analyses, marine mammal predation on fish appears to be particularly large off the northeastern United States (including Georges Bank), where about 36% is attributable to whales (Overholtz et al. 1991), and in the Barents and Norwegian Seas where 45% is attributable to whales (Bax 1991). Recent work by Kenney et al. (1996) suggests that marine mammals (mainly whales) are even more important consumers of fish and squid biomass off the northeastern United States than was previously estimated by Overholtz et al. (1991).

Although considerably smaller in body size, pinnipeds are likely to be ecologically significant components of some marine ecosystems as well. For example, it is difficult to imagine that 12 to 15 million crabeater seals are not important in the structure and function of the Antarctic ecosystem (Laws 1984, 1985). In the northern hemisphere, the combined harp seal populations now number some 8 million individuals, and the northwest Atlantic population alone may have consumed about 7 million tons of prey in 1994 (Anonymous 1995). This and similar estimates of total annual food consumption for other species (e.g., gray seals; Mohn and Bowen 1996) are derived from estimates of age-specific population size, body mass at age, average daily energy expenditure, and the energy content of the diet (see Costa and Williams, Chapter 5, this volume). Each of these estimates can be subject to considerable uncertainty.

The above calculations refer to marine mammal food consumption over large geographic areas and numbers of prey species. At smaller spatial scales and for single prey species, there is evidence that marine mammals can affect prey populations. Fay et al. (1977) reported that walrus predation had apparently caused severe depletion of benthic bivalves, which resulted in a decrease in the proportion of bivalves and an increase in the proportion of fishes and ther prey in the diet. There is evidence that sea otters can have an impact on the commercial yields of some shellfish, such as abalone (*Haliotis* spp.; Riedman and Estes 1990), although the degree of impact is often questioned (Foster and Schiel 1988).

Kanwisher and Ridgway (1983) also speculated about the ecological function of some cetaceans in the cycling of nutrients by feeding at depth and then defecating in the euphotic zone. Katona and Whitehead (1988) attempted to evaluate this hypothesis by studying sperm whales off the Galapagos Islands. Using estimates of whale population density, rate of food consumption, and the nitrogen content of oceanic squid, the researchers were able to show that sperm whales could likely contribute only about 0.04% of the nitrogen content of primary production west of the Galapagos. Although apparently not a significant source of nutrients in this system, whale-induced nutrient cycling may at times be important at smaller spatial and temporal scales (Katona and Whitehead 1988).

Katona and Whitehead (1988) also address the role of cetaceans in the coevolution of predator and prey and note that sonar and the range of prey sizes included in filter-feeding baleen whales have likely had important effects on the behavior of prey. One example of this may be found in mysticetes. Individual prey such as herring and krill theoretically minimize the risk of predation by seabirds, seals, and larger fish by forming tight schools. However, when attacked by bulk feeders such as baleen whales, individuals should do better by scattering. There is evidence that capelin respond in this way when being attacked by humpback whales (Katona and Whitehead 1988).

An intriguing role of a marine mammal is in the physical restructuring of the benthos that results from feeding by gray whales, walruses, and dugongs. Nerini (1984) estimates that gray whales turn over between 9% and 27% of the benthic substrate in the northern Bering Sea annually. This disturbance helps to maintain early colonizing species at higher abundance than would otherwise be the case and may be important in providing habitat for the larvae of their primary amphipod prey. Walruses may structure the benthic fauna by selectively feeding on older individuals of a few species of bivalve molluscs (Vibe 1950, Fay and Stocker 1982). Ingestion and defecation by walruses may result in substantial redistribution of sediment, which may favor colonization of some species but not others (Fay et al. 1977, Oliver et al. 1983). However, the relationship between walruses and benthic invertebrates is not well understood. Preen (1995) suggests that dugongs are "cultivation feeders," whose feeding activities modify growth of sea grasses.

Estes (1979) and Ray (1981) used the concept of r and K selection as a basis for theoretical arguments about the ecological function of marine mammals in ecosystems. In general, K-selected species (i.e., long-lived species with low reproductive rates and high, relatively uniform abundance) are predicted to have the greatest effect in structuring their ecosystems (Ray 1981). The argument here is that these species function in some analogous sense as the "forests of the oceans" in tying up large amounts of nutrients and being able to buffer short-term fluctuations in resource availability. The cascading changes in community structure hypothesized to have occurred in both the Southern Ocean (Laws 1985) and the Bering Sea (National Research Council 1996) as a result of exploitation may well be illustrations of the effect of marine mammal "clear cutting" on marine ecosystems.

Large-scale, intensive harvesting of fish and whales occurred during the 1950s, 1960s, and early 1970s in the Bering Sea and Gulf of Alaska. The stocks of large whales, flatfishes, and slope rockfishes were severely reduced during this period. Mostly as a result of both environmental changes and this intensive exploitation of the ecosystem, the eastern Bering Sea fish assemblage became dominated by one species, the walleye pollock (*Theragra chalcogramma*). King crab and herring populations were also reduced severely, and although some finfish populations grew rapidly, a number of forage species appear to have declined in the 1970s and 1980s. In the face of these changes, both harbor seals and Steller sea lions have declined dramatically during the past several decades. These declines appear to be related to a lack of food, with juvenile pinnipeds likely affected most severely. The most likely explanation of these events is that a combination of environmental change and human exploitation of predators (both whales and fish) resulted in changes to the ecosystems that have been detrimental to pinnipeds (National Research Council 1996).

Case Studies

One species whose role in structuring the near-shore community is well documented is the sea otter (Estes and Palmisano 1974, Simenstad et al. 1978, Estes and Duggins 1995). Sea otters are predators on invertebrates, including sea urchins, clams, and crabs, that occupy the near-shore community. They provide a good example of a keystone predator, that is, a predator that has a disproportionately large effect on its community (Paine 1969). Sea urchins graze on kelp, a major component of these near-shore environments. The ecological interactions among sea otters, urchins, and kelp communities have been the area of study by several investigators, particularly Estes and his colleagues, for many years. In general, these studies have demonstrated that when sea otters are present, kelp forests are allowed to develop because otter predation on urchins reduces grazing on kelp. Much of the reduction in grazing is because otters tend to forage initially on the very large urchins, which are the most effective grazers of kelp. When otters invade a near-shore community, the large urchins are the first to decline in abundance, until over time most of the urchins that are present are small and not very effective at reducing kelp biomass. Thus, in general, the kelp populations are allowed to increase in abundance as sea otters become established. This allows other populations, such as the fish that occupy kelp cover, to increase in abundance and hence the community becomes more diverse. This general pattern is somewhat complicated by events such as storms that damage the kelp communities or by physical features such as currents or temperature changes that may influence reproductive patterns of both kelp and sea urchins. However, long-term trends produce the general picture described above (Estes and Duggins 1995).

The clearest example of the ecological role of pinniped predation comes from a study of lakes. But even here, the evidence is only indirect. Power and Gregoire (1978) compared the fish communities in nine northern Quebec lakes with no seal populations with Lower Seal Lake, which supported a population of harbor seals. Although the size of this seal population was unknown, the authors observed seals in 1 year and observations of seals were reported to the authors in the following year. Surveys of the fish communities in these lakes showed that Lower Seal Lake differed from the other lakes in the relative abundance of lake trout (*Salvelinus fontinalis*) and brook trout (*S. namaycush*). In Lower Seal Lake, brook trout was the dominant species, whereas lake trout was dominant in all other lakes. Power and Gregoire (1978) also noted that changes in life history characteristics of lake trout in Lower Seal Lake were consistent with effects associated with heavy exploitation. These lake trout were on average smaller, younger, grew more rapidly, and matured earlier than lake trout in neighboring lakes. Power and Gregoire (1978) concluded that seal predation was responsible for both the changes in community structure and life history traits of fish species in Lower Seal Lake. They observed that seal predation had apparently targeted species, such as lake trout, that aggregated in lakes during spawning. Brook trout became the dominant species in Lower Seal Lake because its spawning sites were widely dispersed in many small streams, and juveniles spent the first 2 to 3 years of life in streams where they were reasonably protected from seal predation. Although the conclusions of Power and Gregoire (1978) are compelling on the surface, they are based on strong inference rather than direct empirical evidence. For example, we do not know whether there were sufficient seals in the lake to account for the effects observed. Experimental introductions of harbor seals into other lakes in this area offer exciting opportunities for learning more about the role of seal predation on fish communities (McLaren and Smith 1985).

We can also gain some insight into the ecological significance of marine mammals from events that took place in the Southern Ocean earlier in this century. Commercial exploitation of marine mammals brought some species of seals and whales near extinction in the 19th and early 20th centuries (Laws 1977). This overexploitation amounted to nothing less than an enormous uncontrolled "experiment" (Laws 1985) or, more correctly, an "observational study" in the sense of Eberhardt and Thomas (1991). What have we learned about the function of marine mammals in this cold water ecosystem as a result of this perturbation? Before we can address this issue, we must first reiterate some of the main features of this ecosystem, as reviewed by Laws (1985).

Sea ice is a dominant influence on the ecology of the Southern Ocean, expanding during the winter to cover an area as large as 20 million km^2 and contracting to about 4 million km^2 in summer. The edge of the pack ice is a region of high biological productivity and serves as the habitat for large populations of seabirds and marine mammals. An area of high productivity is also associated with the upwelling of nutrients along the Antarctic Divergence. The standing stock biomass of zooplankton in the Southern Ocean is significantly higher than in either tropical or temperate oceans. Furthermore, a single species, the Antarctic krill, probably accounts for half of 105 mg/m^3 of zooplankton biomass. Increased zooplankton production, coupled with the retreat of ice in the summer, results in a great increase in krill available to marine mammals (Laws 1985). Squid and nototheniid fishes are abundant, but in contrast to other oceans, dense shoals of pelagic fish are not.

Six species of pinnipeds are found in the Southern Ocean. Four seal species (crabeater, leopard, Weddell, and Ross) are associated with either pack or fast ice throughout the year, whereas the southern elephant seal and fur seals are land breeders and are rarely found in areas of pack ice. Both the fur seals and southern elephant seals were brought to near extinction in the 19th century, but their numbers have recovered substantially. Crabeater seals depend almost entirely on krill. Antarctic fur seals and leopard seals feed mainly on krill at certain times of the year. Elephant seals and Ross seals consume mainly cephalopods, but fish are also important at times. Weddell seals eat mainly fish, with cephalopods taken in some areas.

Six mysticete species (blue, fin, sei, minke, humpback,

and southern right) and one subspecies (pygmy blue) inhabit the Southern Ocean, along with the sperm whale and 11 species of small toothed whales that penetrate south of the Antarctic Convergence. Krill is the major food of the baleen whales. Sperm whales feed mainly on deep sea squids, whereas the other odontocetes feed on a mixture of squids and fishes. The abundance of baleen whales and sperm whales declined by probably more than 50% between 1904 and 1973 as a result of intense exploitation. Because the largest species were taken first, the cetacean biomass declined from an estimated 45 million mt to only 9 million mt over this same period.

Laws (1985) estimated that this enormous reduction in the biomass of large whales may have released some 150 million mt of krill annually to predators (the remaining whales, seals, seabirds, and fish). We can gain a sense of the magnitude of this ecological perturbation by comparing this biomass of krill with the 101 million mt landed in 1993 by all the world's fisheries (FAO 1995). Although many of the estimates of population food consumption are necessarily tentative, given the quality of data, Laws (1985) calculated that most of the krill was redistributed to seals and birds, both of which have become considerably more abundant during the past three decades (Fig. 9-20). Krill-eating species, such as the crabeater seal and the Antarctic fur seal, have also become considerably more abundant. In fact the crabeater seal, which probably numbers about 12 million, is the most abundant pinniped in the world. Chinstrap (*Pygoscelis antarctica*), Adélie (*P. adeliae*), and macaroni penguins (*Eudyptes chrysolophus*), which together account for about 90% of the Antarctic avian biomass, have also increased in numbers. The present minke whale population is perhaps twice that which existed before whaling, possibly a response to the great reduction in blue whale numbers. Laws (1985) further points out that the increase in king penguins may have been caused by the increase in krill-feeding squids, which in turn became more abundant because of decreases in the number of sperm whales.

Therefore, it seems clear that the relative abundances of krill consumers and other prey in the Southern Ocean are quite different today than they were before the exploitation of the large whales. Although we cannot be certain, it seems reasonable that these population increases are attributable to the increase in food availability brought about by the overexploitation of the large whales (Laws 1985). If this is true then the recovery of these species of whales will be possible only to the extent that they are able to outcompete other species that have benefited during their period of reduced numbers.

Conclusion

We have much to learn about the ecological roles of marine mammals in marine ecosystems, but it seems from our cur-

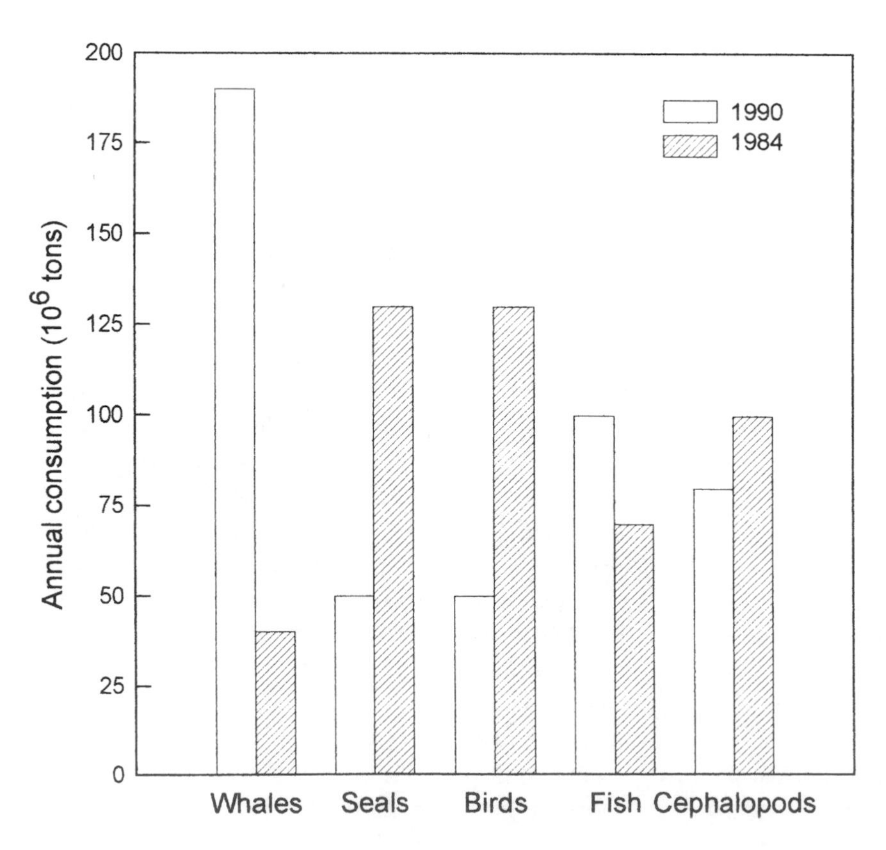

Figure 9-20. Speculated changes in the consumption of Antarctic krill by major groups of predators. (After Laws 1985.)

rent knowledge that the abundance and distribution of marine mammals do have important consequences on the structure and function of some systems. Developing a better understanding of the role of marine mammals in marine ecosystems is one of the greatest challenges facing those interested in marine mammal ecology. However, this understanding will likely come slowly and will be the product of long-term research. In our view, the search for this understanding will be most effective if it is based on broad, interdisciplinary studies that include modellers, oceanographers, and fisheries and marine mammal ecologists.

Literature Cited

Aarefjord, H., A. J. Bjorge, C. C. Kinze, and I. Lindstedt. 1995. Diet of the harbour porpoise (*Phocoena phocoena*) in Scandinavian waters. Pages 211–222 *in* A. Bjorge and G. P. Donovan (eds.). Biology of the Phocoenids. Reports of the International Whaling Commission, Cambridge, U.K.

Ackerman, B. B. 1995. Aerial surveys of manatees: A summary, progress report. Pages 13–34 *in* T. J. O'Shea, B. B. Ackerman, and H. F. Percival (eds.). Population Biology of the Florida Manatee. U.S. Department of the Interior, National Biological Service Information and Technology Report I, Washington, D.C.

Ackman, R. G. 1980. Fish lipids, part 1. Pages 86–103 *in* J. J. Connell (ed.). Advances in Fish Science and Technology. Fishing News Books, Ltd. Surrey, U.K.

Ackman, R. G., and C. A. Eaton. 1966. Lipids of the fin whale (*Balaenoptera physalus*) from north Atlantic waters. III. Occurrence of eicosenoic and docosenoic fatty acids in the zooplankter *Meganyctiphanes norvegica* (M. Sars) and their effect on whale oil composition. Canadian Journal of Biochemistry 44:1561–1566.

Acuna, H. O., and J. M. Francis. 1995. Spring and summer prey of the Juan Fernandez fur seal, *Artcocephalus philippii*. Canadian Journal of Zoology 73:1444–1452.

Ainley, D. G., H. R. Huber, and K. M. Bailey. 1982. Population fluctuations of California sea lions and the Pacific whiting fishery off Central California. Fishery Bulletin 80:253–258.

Alexander, V. 1981. Ice-biota interactions: An overview. Pages 757–761 *in* D. W. Hood and J. A. Calder (eds.). The Eastern Bering Sea Shelf: Oceanography and Resources. University of Washington Press, Seattle, WA.

Allen, P. J., W. Amos, P. P. Pomeroy, and S. D. Twiss. 1995. Microsatellite variation in gray seals (*Halichoerus grypus*) shows evidence of genetic differentiation between two British breeding colonies. Molecular Ecology 4:653–662.

Ames, A. L., E. S. Van Veet, and W. M. Sackett. 1996. The use of stable carbon isotope analysis for determining the dietary habits of the Florida manatee, *Trichechus manatus latirostris*. Marine Mammal Science 12:555–563.

Amos, B., C. Schlotteerer, and D. Tautz. 1993. Social structure of pilot whales revealed by analytical DNA profiling. Science 260:670–672.

Anas, R. E. 1970. Accuracy in assigning ages to fur seals. Journal of Wildlife Management 34:844–852.

Anonymous. 1984. The impact of gray and common seals on North Sea resources. Natural Environment Research Council, Sea Mammal Research Unit, Contract no. ENV 665 UK(H)-final report, Cambridge, U.K.

Anonymous. 1995. Report on the Status of Harp Seals in the Northwest Atlantic. Department of Fisheries and Oceans, Canada Stock Status Report 95 / 7.

Antonelis, G. A., M. Lowry, D. P. Demaster, and C. H. Fiscus. 1987. Assessing northern elephant seal feeding habits by stomach lavage. Marine Mammal Science 3:308–322.

Antonelis, G. A., S. R. Melin, and Y. A. Bukhtiyarov. 1994a. Early spring feeding habits of bearded seals (*Erignathus barbatus*) in the central Bering Sea, 1981. Arctic 47:74–79.

Antonelis, G. A., M. S. Lowry, C. Fiscus, B. S. Stewart, and R. L. DeLong. 1994b. Diet of the northern elephant seal. Pages 211–223 *in* B. J. LeBoeuf and R. M. Laws (eds.). Elephant Seals: Population Ecology, Behavior, and Physiology. University of California Press, Berkeley, CA.

Archibald, W. R., and I. Stirling. 1977. Aspects of predation of seals by polar bears. Journal of the Fisheries Research Board of Canada 34:1126–1129.

Arnason, U., K. Bodin, A. Gullberg, C. Ledje, and S. Mouchaty. 1995. A molecular view of pinniped relationships with particular emphasis on the true seals. Journal of Molecular Evolution 40:78–85.

Arnbom, T. A., N. J. Lunn, I. L. Boyd, and T. Barton. 1992. Aging live Antarctic fur seals and southern elephant seals. Marine Mammal Science 8:37–43.

Bailey, K. M., and D. G. Ainley. 1982. The dynamics of California sea lion predation on Pacific hake. Fisheries Research 1:163–176.

Baird, R. W., and L. M. Dill. 1995. Occurrence and behaviour of transient killer whales: Seasonal and pod-specific variability, foraging behaviour, and prey handling. Canadian Journal of Zoology 73:1300–1311.

Baird, R. W., and P. J. Stacey. 1988. Variation in saddle patch pigmentation in populations of killer whales (*Orcinus orca*) from British Columbia, Alaska, and Washington State. Canadian Journal of Zoology 66:2582–2585.

Baker, C. S., L. M. Herman, A. Perry, W. E. Lawton, J. M. Straley, A. A. Wolman, G. D. Kaufman, H. E. Winn, J. D. Hall, J. M. Reinke, and J. Ostman. 1986. Migratory movement and population structure of humpback whales (*Megaptera novaeangliae*) in the central and eastern North Pacific. Marine Ecology Progress Series 31:105–119.

Baker, C. S., S. R. Palumbi, R. H. Lambertsen, M. T. Weinrich, J. Calambokidis, and S. J. O'Brien. 1990. Influence of seasonal migration on geographic distribution of mitochondrial DNA haplotypes in humpback whales. Nature 344:238–240.

Baker, C. S., R. W. Slade, J. L. Bannister, R. B. Abernethy, M. T. Weinrich, J. Lien, J. Urban, P. Corkeron, J. Calambokidis, O. Vasquez, and S. R. Palumbi. 1994. Hierarchical structure of mitochondrial DNA gene flow among humpback whales, *Megaptera novaeangliae*, worldwide. Molecular Ecology 3:313–327.

Baker, R. R. 1978. The Evolutionary Ecology of Animal Migration. Hodder & Stoughton, London.

Barlow, J., and P. Boveng. 1991. Modeling age-specific mortality for marine mammal populations. Marine Mammal Science 7:50–65.

Barros, N. B., and D. K. Odell. 1990. Food habits of bottlenose dolphins in the Southeastern United States. Pages 309–328 *in* S. Leatherwood and R. R. Reeves (eds.). The Bottlenose Dolphin. Academic Press, San Diego, CA.

Bartholomew, G. A. 1970. A model for the evolution of pinniped polygyny. Evolution 24:546–559.

Bartholomew, G. A., and P. G. Hoel. 1953. Reproductive behavior of the Alaska fur seal, *Callorhinus ursinus*. Journal of Mammalogy 34:417–436.

Bastida, R., and V. Lichtschein. 1988. Food habits of *Cephalorhynchus commersonii* off Tierra del Fuego. Pages 143–160 *in* R. L. Brownell Jr., and G. P. Donovan (eds.). Biology of the Genus *Cephalorhynchus*. International Whaling Commission (Special Issue 9). Cambridge, U.K.

Bax, N. J. 1991. A comparison of the fish biomass flow to fish, fisheries, and mammals in six marine ecosystems. ICES Marine Science Symposia 193:217–224.

Beck, C. A., and J. P. Reid. 1995. An automated photo-identification catalog for studies of the life history of the Florida manatee. Pages 120–134 *in* T. J. O'Shea, B. B. Ackerman, and H. F. Percival (eds.). Population Biology of the Florida Manatee. U.S. Department of the Interior, National Biological Service Information and Technology Report I, Washington, D.C.

Beck, G. G., M. O. Hammill, and T. G. Smith. 1993. Seasonal variation in the diet of harp seals (*Phoca groenlandica*) from the Gulf of St. Lawrence and Western Hudson Strait. Canadian Journal of Fisheries and Aquatic Sciences 50:1363–1371.

Benoit, D., and W. D. Bowen. 1990a. Seasonal and geographic variation in the diet of gray seals (*Halicheorus grypus*) in eastern Canada. Pages 215–226 *in* W. D. Bowen (ed.). Population Biology of Sealworm (*Pseudoterranova decipiens*) in Relation to its Intermediate and Seal Hosts. Canadian Bulletin, Journal of Fisheries and Aquatic Sciences 222, Ottawa, Ontario.

Benoit, D., and W. D. Bowen. 1990b. Summer diet of gray seals (*Halicheorus grypus*) at Anticosti Island, Gulf of St. Lawrence, Canada. Pages 227–242 *in* W. D. Bowen (ed.). Population Biology of Sealworm (*Pseudoterranova decipiens*) in Relation to its Intermediate and Seal Hosts. Canadian Bulletin, Journal of Fisheries and Aquatic Sciences 222, Ottawa, Ontario, Canada.

Bernard, H. J., and A. A. Hohn. 1989. Differences in feeding habits between pregnant and lactating spotted dolphins (*Stenella attenuata*). Journal of Mammalogy 70:211–215.

Bernt, K. E., M. O. Hammill, and K. M. Kovacs. 1996. Age determination of gray seals (*Halichoerus grypus*) using incisors. Marine Mammal Science 12:476–482.

Bertram, G. C. L. 1940. The biology of the Weddell and crabeater seals: With a study of the comparative behaviour of pinnipedia, 1934–37. Scientific Reports of the British Graham Land Expedition, 1934–1937 1:1–139.

Best, P. B. 1967. Distribution and feeding habits of baleen whales off the Cape Province. Investigational Reports of the Division of Sea Fisheries, South Africa 57:1–44.

Best, P. B. 1974. Two allopatric forms of Bryde's whale off South Africa. Reports of the International Whaling Commission (Special Issue 1):10–38.

Best, P. B. 1979. Social organization of sperm whales, *Physeter macrocephalus*. Pages 227–289 *in* H. E. Winn and B. L. Olla (eds.). Behavior of Marine Animals. Current Perspectives in Research, Vol. 3: Cetaceans. Plenum, New York, NY.

Best, P. B. 1993. Increase rates in severely depleted stocks of baleen whales. ICES Journal of Marine Science 50:169–186.

Best, R. C. 1983. Apparent dry-season fasting in Amazon manatees (Mammalia: Sirenia). Biotropica 15:61–64.

Bester, M. N., and P. A. Laycock. 1985. Cephalopod prey of the sub-Antarctic fur seal, *Arctocephalus tropicalis*, at Gough Island. Pages 551–554 *in* W. R. Siegfried, P. R. Condy, and R. M. Laws (eds.). Antarctic Nutrient Cycles and Food Webs. Springer-Verlag, Berlin.

Bigg, M. A. 1981. Harbour seal—*Phoca vitulina* (Linnaeus, 1758) and *Phoca largha* (Pallas, 1811). Pages 1–27 *in* S. H. Ridgway and R. J. Harrison (eds.). Handbook of Marine Mammals. Academic Press, London.

Bigg, M. A. 1982a. An assessment of killer whale (*Orcinus orca*) stocks off Vancouver Island, British Columbia. Report of the International Whaling Commission 32:655–666.

Bigg, M. A. 1982b. Migration of northern fur seals in the eastern North Pacific and eastern Bering Sea: An analysis using effort and population composition data. Proceedings of the Annual Meeting, North Pacific Fur Seal Commission, Ottawa, Ontario, Canada.

Bigg, M. A., and I. F. Fawcett. 1985. Two biases in diet determination of northern fur seals (*Callorhinus ursinus*). Pages 284–291 *in* J. R. Beddington, R. J. H. Beverton, and D. M. Lavigne (eds.). Marine Mammals and Fisheries. George Allen & Unwin, London.

Bigg, M. A., and M. A. Perez. 1985. Modified volume: A frequency–volume method to assess marine mammal food habits. Pages 277–283 *in* J. R. Beddington, R. J. H. Beverton, and D. M. Lavigne (eds.). Marine Mammals and Fisheries. George Allen & Unwin, London.

Bigg, M. A., G. M. Ellis, J. K. B. Ford, and K. C. Balcomb. 1987. Killer Whales: A Study of Their Identification, Genealogy, and Natural History in British Columbia and Washington State. Phantom Press, Nanaimo, B.C.

Bigg, M. A., P. F. Olesiuk, G. M. Ellis, J. K. B. Ford, and K. C. Balcomb III. 1990. Social organization and genealogy of resident killer whales (*Orcinus orca*) in the coastal waters of British Columbia and Washington State. Report of the International Whaling Commission (Special Issue 12):383–405.

Bjorge, A., and G. P. Donovan. 1995. Biology of the Phocoenids, Special Issue 16. International Whaling Commission, Cambridge, U.K.

Blanco, C., J. Aznar, and J. A. Raga. 1995. Cephalopods in the diet of the striped dolphin *Stenella coeruleoalba* from the western Mediterranean during an epizootic in 1990. Journal of Zoology (London) 237:151–158.

Bodkin, J. L., R. J. Jameson, and J. A. Estes. 1995. Sea otters in the North Pacific Ocean. Pages 353–356 *in* E. T. LaRoe III, G. S. Farris, C. E. Puckett, P. D. Doran, and M. J. Mac (eds.). Our Living Resources: A Report to the Nation on the Distribution, Abundance and Health of U.S. Plants, Animals and Ecosystems. Department of the Interior, National Biological Service, Washington, D.C.

Bonner, W. N. 1981. Southern fur seals *Arctocephalus* (Geoffroy Sainte-Hilaire and Cuvier, 1826). Pages 161–208 *in* S. H. Ridgway and R. J. Harrison (eds.). Handbook of Marine Mammals, Vol. 1: The Walrus, Sea Lions, Fur Seals and Sea Otter. Academic Press, London.

Boulva, J., and I. A. McLaren. 1979. Biology of the harbour seal, *Phoca vitulina*, in eastern Canada. Bulletin of the Fisheries Research Board of Canada 200:1–24.

Boveng, P., D. P. DeMaster, and B. S. Stewart. 1988. Dynamic response analysis III. A consistency filter and application to four northern elephant seal colonies. Marine Mammal Science 4:210–222.

Bowen, W. D., and G. Harrison. 1994. Offshore diet of gray seals *Halichoerus grypus* near Sable Island, Canada. Marine Ecology Progress Series 112:1–11.

Bowen, W. D., and G. D. Harrison. 1996. Comparison of harbour seal diets in two habitats in Atlantic Canada. Canadian Journal of Zoology 74:125–135.

Bowen, W. D., and D. E. Sergeant. 1983. Mark-recapture estimates of harp seal pup (*Phoca groenlandica*) production in the northwest

Atlantic. Canadian Journal of Fisheries and Aquatic Sciences 40:728–742.

Bowen, W. D., D. E. Sergeant, and T. Oritsland. 1983. Validation of age determination in the harp seal, *Phoca groenlandica,* using dentinal annuli. Canadian Journal of Fisheries and Aquatic Sciences 40:1430–1441.

Bowen, W. D., O. T. Oftedal, and D. J. Boness. 1985. Birth to weaning in 4 days: Remarkable growth in the hooded seal, *Cystophora cristata.* Canadian Journal of Zoology 63:2841–2846.

Bowen, W. D., R. A. Myers, and K. Hay. 1987. Abundance estimation of a dispersed, dynamic population: hooded seals (*Cystophora cristata*) in the Northwest Atlantic. Canadian Journal of Fisheries and Aquatic Sciences 44:282–295.

Bowen, W. D., O. T. Oftedal, and D. J. Boness. 1989. Variation in the efficiency of mass transfer in harbour seals, *Phoca vitulina,* over the course of lactation (Abstract). Eighth Biennial Conference, Biology of Marine Mammals Pacific, Grove. CA.

Bowen, W. D., J. W. Lawson, and B. Beck. 1993. Seasonal and geographic variation in the species composition and size of prey consumed by gray seals (*Halichoerus grypus*) on the Scotian shelf. Canadian Journal of Fisheries and Aquatic Sciences 50:1768–1778.

Boyd, I. L. 1991. Environmental and physiological factors controlling the reproductive cycles of pinnipeds. Canadian Journal of Zoology 69:1135–1148.

Boyd, I. L., D. J. McCafferty, K. Reid, R. Taylor, and T. R. Walker. 1998. Dispersal of male and female Antarctic fur seals (*Arctocephalus gazella*). Canadian Journal of Fisheries and Aquatic Sciences 55:845–852.

Boyle, P. R., G. J. Pierce, and J. S. W. Diack. 1990. Sources of evidence for salmon in the diet of seals. Fisheries Research 10:137–150.

Bradstreet, M. S. W., and K. J. Finley. 1983. Diet of ringed seals (*Phoca hispida*) in the Canadian high Arctic. LGL Limited, Toronto, Ontario, Canada.

Braham, H. W. 1984. Distribution and migration of gray whales in Alaska. Pages 249–266 *in* M. L. Jones, S. L. Swartz, and S. Leatherwood (eds.). The Gray Whale *Eschrichtius robustus.* Academic Press, Orlando, FL.

Brodie, P. F. 1975. Cetacean energetics: An overview of intraspecific size variation. Ecology 56:152–161.

Brodie, P. F., D. D. Sameoto, and R. W. Sheldon. 1978. Population densities of euphausiids off Nova Scotia as indicated by net samples, whale stomach contents, and sonar. Limnology and Oceanography 23:1264–1267.

Brooks, D. R., and D. A. Mclennan. 1991. Phylogeny, Ecology, and Behavior. University of Chicago Press, Chicago, IL.

Brown, R. F., and B. R. Mate. 1983. Abundance, movements and feeding habits of harbor seals, *Phoca vitulina,* in Netarts and Tillamook Bays, Oregon. Fishery Bulletin 81:291–301.

Brownell, R. L. Jr., K. Ralls, and W. F. Perrin. 1989. The plight of the "forgotten" whales. Oceanus 32:5–11.

Buckland, S. T. 1990. Estimation of survival rates from sightings of individually identifiable whales. Report of the International Whaling Commission (Special Issue 12):149–153.

Buckland, S. T., and P. H. Garthwaite. 1991. Quantifying precision of mark–recapture estimate using bootstrap and related methods. Biometrics 47:255–268.

Burnham, K. P. 1989. Numerical survival rate estimation for capture–recapture models using SAS PRO NLIN. Pages 416–435 *in* L. McConald, B. Manly, J. Lockwood, and J. Logan (eds.). Estimation and Analysis of Insect Populations. Springer-Verlag, New York, NY.

Burnham, K. P., D. R. Anderson, G. C. White, C. Brownie, and K. H. Pollock. 1987. Design and analysis methods for fish survival experiments based on release-recapture. American Fisheries Society Monogram 5:1–437.

Burns, J. J. 1981a. The bearded seal (*Erignathus barbatus,* Erxleben, 1777). Pages 145–170 *in* S. H. Ridgway and R. J. Harrison (eds.). Handbook of Marine Mammals. Academic Press, London.

Burns, J. J. 1981b. Ribbon seal—*Phoca fasciata.* Pages 89–109 *in* S. H. Ridgway and R. J. Harrison (eds.). Handbook of Marine Mammals. Academic Press, London.

Burns, J. J., and A. Gavin. 1980. Recent records of hooded seals, *Cystophora cristata* Erxleben, from the western Beaufort Sea. Arctic 33:326–329.

Calambokidis, J., B. L. Taylor, S. D. Carter, G. H. Steiger, P. K. Dawson, and L. D. Antrim. 1987. Distribution and haul-out behavior of harbor seals in Glacier Bay, Alaska. Canadian Journal of Zoology 65:1391–1396.

Calambokidis, J., G. H. Steiger, J. R. Evenson, K. R. Flynn, K. C. Balcomb, D. E. Claridge, P. Bloedel, J. M. Straley, C. S. Baker, O. Vonziegesar, M. E. Dahlheim, J. M. Waite, J. D. Darling, G. Ellis, and G.A. Green. 1996. Interchange and isolation of humpback whales off California and other North Pacific feeding grounds. Marine Mammal Science 12:215–226.

Campana, S. E., M. C. Annand, and J. I. Mcmillan. 1995. Graphical and statistical methods for determining the consistency of age determination. Transactions of American Fishery Society 124:131–138.

Caughley, G. R. 1974. Bias in aerial survey. Journal of Wildlife Management 38:921–933.

Caughley, G. R. 1977. Analysis of vertebrate populations. John Wiley, London.

Cetacean and Turtle Assessment Program (CETAP). 1982. A Characterization of Marine Mammals and Turtles in the Mid- and North Atlantic Areas of the U.S. Outer Continental Shelf. University of Rhode Island. U.S. Bureau of Land Management, Contract No. AA551-CTB-48, Washington, D.C.

Chabot, D., G. B. Stenson, and N. B. Cadigan. 1995. Short- and Long-term Fluctuations in the Size and Condition of Harp Seals (*Phoca groenlandica*) in the Northwest Atlantic. NAFO CR Doc. 95 / 42., Serial No. N2551, Dartmouth, Nova Scotia.

Chittleborough, R. G. 1959. *Balaenoptera brydei* Olsen on the west coast of Australia. Norsk Hvalfangsttid. 48:62–66.

Christensen, I. 1988. Distribution, movements and abundance of killer whales (*Orca orcinus*) in Norwegian coastal waters. Rit. Fiskideildar 11:79–88.

Clapham, P. J. 1996. The social and reproductive biology of humpback whales: An ecological perspective. Mammal Review 26:27–49.

Clarke, M. R. 1980. Cephalopods in the diet of sperm whales in the Southern Hemisphere and their bearing on sperm whale biology. Discovery Reports 37:1–324.

Clarke, M. R. 1986. Cephalopods in the diet of odontocetes. Pages 281–321 *in* M. M. Bryden and R. Harrison (eds.). Research on Dolphins. Clarendon Press, Oxford.

Clarke, M. R., H. R. Martins, and P. Pascoe. 1993. The diet of sperm whale (*Physeter macrocephalus*) off the Azores. Philosophical Transactions of the Royal Society of London Series, Biological Sciences 339:67–82.

Cockcroft, V. G., and G. J. B. Ross. 1990. Food and feeding in the Indian ocean bottlenose dolphin off Southern Natal, South Africa. Pages 295–308 *in* S. Leatherwood and R. R. Reeves (eds.). The Bottlenose Dolphin. Academic Press, San Diego, CA.

Cole, L. C. 1954. The population consequences of life history phenomena. Quarterly Review of Biology 29:103–137.

Crawley, M. C., and R. Warneke. 1979. New Zealand fur seal. Pages 45–48 *in* Mammals of the Seas. Food and Agricultural Organization Fishery Service, Rome.

Cushing, D.H. 1975. Marine Ecology and Fisheries. Cambridge University Press, Cambridge, U.K.

Darling, J. 1977. Population biology and behaviour of the gray whale (*Eschrichutis robustus*) in Pacific Rim National Park, British Columbia. M.Sc. thesis. University of Victoria, Victoria, B.C.

da Silva, J., and J. D. Neilson. 1985. Limitations of using otoliths recovered in scats to estimate prey consumption in seals. Canadian Journal of Fisheries and Aquatic Sciences 42:1439–1442.

David, J. H. M. 1987. Diet of the South African (Cape) fur seal (1974–1985) and an assessment of competition with fisheries in southern Africa. South African Journal of Marine Science 5:693–713.

Dawbin, W. H. 1956. The migrations of the humpback whales which pass the New Zealand coast. Transactions of the Royal Society of New Zeland 84:147–196.

de la Mare, W. K. 1994. Some analyses of the dynamics of reduced mammal populations. Report of the International Whaling Commission 44:459–466.

Dellinger, T., and F. Trillmich. 1988. Estimating diet composition from scat analysis in otariid seals (Otariidae): Is it reliable? Canadian Journal of Zoology 66:1865–1870.

Delong, R. L., B. S. Stewart, and R. D. Hill. 1992. Documenting migrations of northern elephant seals using day length. Marine Mammal Science 8:155–159.

Demaster, D. P., and I. Stirling. 1981. *Ursus maritimus*. Mammal Species 145:1–7.

Demaster, D. P., D. J. Miller, D. Goodman, R. DeLong, and B. S. Stewart. 1982. Assessment of California sea lion fishery interactions. Transactions of the North American Wildlife National Research Conference 47:254–277.

Deniro, M. J., and S. Epstein. 1978. Influence of diet on the distribution of carbon isotopes in animals. Geochim. Cosmochim. Acta 42:495–506.

Deniro, M. J., and S. Epstein. 1981. Influence of diet on the distribution of nitrogen isotopes in animals. Geochim. Cosmochim. Acta 45:341–351.

Derocher, A. E., and I. Stirling. 1990. Distribution of polar bears (*Ursus maritimes*) during the ice-free period in western Hudson Bay. Canadian Journal of Zoology 68:1395–1403.

Derocher, A. E., and I. Stirling. 1994. Age-specific reproductive performance of female polar bears (*Ursus maritimus*). Journal of Zoology (London) 234:527–536.

Derocher, A. E., and I. Stirling. 1995. Mark-recapture estimation of polar bear population size and survival in western Hudson Bay. Journal of Wildlife Management 59:215–221.

Desportes, G., and R. Mouritsen. 1993. Preliminary results on the diet of long-finned pilot whales off the Faroe Islands. Pages 305–324 *in* G. P. Donovan, C. H. Lockyer, and A. R. Martin (eds.). Biology of Northern Hemisphere Pilot Whales. Report of the International Whaling Commission (Special Issue 14), Cambridge, U.K.

Doidge, D. W., and J. P. Croxall. 1985. Diet and energy budget of the Antarctic fur seal *Arctocephalus gazella* at South Georgia. Pages 543–550 *in* W. R. Siegfried, P. R. Condy, and R. M. Laws (eds.). Antarctic Nutrient Cycles and Food Webs. Springer-Verlag, Berlin.

Domning, D. P. 1980. Feeding position preference in manatees (*Trichechus*). Journal of Mammalogy 61:544–547.

Domning, D. P. 1981. Sea cows and sea grasses. Paleobiology 7:417–420.

Domning, D. P. 1982. Commercial exploitation of manatees *Trichechus* in Brazil c. 1785–1973. Biological Conservation 22:101–126.

Donovan, G. P., and A. Bjorge. 1995. Harbour porpoises in the North Atlantic: edited extract from the Report of the IWC Scientific Committee, Dublin 1995. Pages 3–25 *in* A. Bjorge and G. P. Donovan (eds.). Biology of the Phocoenids. International Whaling Commission, Cambridge, U.K.

Doubleday, W. G., and W. D. Bowen. 1980. Inconsistencies in Reading the Age of Harp Seal (*Pagophilus groenlandicus*) Teeth, Their Consequences, and Means of Reducing Resulting Biases. Northwest Atlantic Fisheries Organization, Serial No. N247, Ottawa, Ontario, Canada.

Dufault, S., and H. Whitehead. 1995a. An assessment of changes with time in the marking patterns used for photoidentification of individual sperm whales, *Physeter macrocephalus*. Marine Mammal Science 11:335–343.

Dufault, S., and H. Whitehead. 1995b. The geographic stock structure of female and immature sperm whales in the South Pacific. Report of the International Whaling Commission 45:401–405.

Eberhardt, L. L. 1990. A fur seal population model based on age structure data. Canadian Journal of Fisheries and Aquatic Sciences 47:122–127.

Eberhardt, L. L., and T. J. O'Shea. 1995. Integration of manatee life-history data and population modeling. Pages 269–279 *in* T. J. O'Shea, B. B. Ackerman, and H. F. Percival (eds.). Population Biology of the Florida Manatee. U.S. Department of the Interior, National Biological Service Information and Technology Report I, Washington, D.C.

Eberhardt, L. L., and D. B. Siniff. 1977. Population dynamics and marine mammal management policies. Journal of the Fisheries Research Board of Canada 34:183–190.

Eberhardt, L. L., and J. M. Thomas. 1991. Designing environmental field studies. Ecological Monographs 61:53–73.

Eberhardt, L. L., D. G. Chapman, and J. R. Gilbert. 1979. A review of marine mammal census methods. Wildlife Monographs 63:1–46.

Edwards, E. F., and P. C. Perkins. 1993. Power to detect linear trends in dolphin abundance: Estimates from tuna-vessel observer data, 1975–89. Fishery Bulletin 90:625–631.

Ellis, G. M., and A. W. Trites. 1992. The RAM-packs came back: a method for attaching and recovering pinniped data recorders. Aquatic Mammals 18:61–64.

El-Sayed, S. Z. 1994. Southern Ocean Ecology:The Biomass Perspective. Cambridge University Press, Cambridge, U.K.

Estes, J. A. 1979. Exploitation of marine mammals: r-selection of K-strategists? Journal of Fisheries Research Board of Canada 36:1009–1017.

Estes, J. A., and D. O. Duggins. 1995. Sea otters and kelp forests in Alaska: Generality and variation in a community ecological paradigm. Ecological Monographs 65:75–100.

Estes, J. A., and J. F. Palmisano. 1974. Sea otters: Their role in structuring nearshore communities. Science 185:1058–1060.

Estes, J. A., R. J. Jameson, A. M. Johnson. 1981. Food selection and some foraging tactics of sea otters. Pages 606–641 in J. A. Chapman and D. Pursley (eds.). Proceedings of the Worldwide Furbearer Conference, August 3–11, 1980, Frostburg, MD. University of Maryland Press, College Park, MD.

Evans, P. G. H. 1987. The Natural History of Whales and Dolphins. Facts on File, Inc., New York, NY.

Fancy, S. G., L. F. Pank, D. C. Douglas, C. H. Curby, G. W. Garner, S. C. Amstrup, and W. L. Regelin. 1988. Satellite Telemetry: A New Tool for Wildlife Research and Management. U. S. Dept. Interior, Fish and Wildlife Service, Resource Publication 172, Washington, D.C.

Fay, F. H. 1974. The role of ice on the ecology of mammals of the Bering Sea. Pages 383–399 *in* D. W. Hood and J. A. Calder (eds.). Oceanography of the Bering Sea. Institute of Marine Science, University Alaska, Fairbanks, AK.

Fay, F. H. 1981. Walrus—*Odobenus rosmarus*. Pages 1–24 *in* S. H. Ridgway and R. J. Harrison (eds.). Handbook of Marine Mammals. Academic Press, London.

Fay, F. H., and S. W. Stocker. 1982. Reproductive success and feeding habits of walruses taken in 1982 spring harvest, with comparisons from previous years. Final Report, Eskimo Walrus Commission, Nome, AK.

Fay, F. H., H. M. Feder, and S. W. Stoker. 1977. An Estimation of the Impact of the Pacific Walrus Population on its Food Resources in the Bering Sea. Final Report U.S. Marine Mammal Commission, PB-273-505, National Technical Informational Service, Springfield, VA.

FAO. 1995. Fishery Statistics Commodities. FAO Yearbook, Volume 77, 1993. FAO, United Nations, Rome.

Feltz, E. T., and F. H. Fay. 1966. Thermal requirements *in vitro* of epidermal cells from seals. Cryobiology 3:261–264.

Finley, K. J., and E. J. Gibb. 1982. Summer diet of the narwhal (*Monodon monoceros*) in Pond Inlet, northern Baffin Island. Canadian Journal of Zoology 60:3353–3363.

Fitch, J. E., and R. L. Brownell Jr. 1968. Fish otoliths in cetacean stomachs and their importance in interpreting feeding habits. Journal of Fisheries Research Board of Canada 25:2561–2575.

Ford, J. K. B., G. M. Ellis, and K.C. Balcomb. 1994. The Natural History and Genealogy of *Orcinus orca* in British Columbia and Washington State. UBC Press, Vancouver, B.C.

Foster, M. S., and D. R. Schiel. 1988. Kelp communities and sea otters: Keystone species or just another brick in the wall? Pages 92–108 *in* G. R. VanBlaricom and J. A. Estes (eds.). The Community Ecology of Sea Otters. Springer-Verlag, Berlin.

Fowler, C. W. 1981. Density dependence as related to life history strategy. Ecology 62:602–610.

Fowler, C. W. 1990. Density dependence in northern fur seals (*Callorhinus ursinus*). Marine Mammal Science 6:171–195.

Fowler, C. W. 1994. Further consideration of nonlinearity in density dependence among large mammals. Report of the International Whaling Commission 44:385–391.

Fowler, C. W., and J. D. Baker. 1991. A review of animal population dynamics at extremely reduced populations levels. Report of the International Whaling Commission 41:545–554.

Fowler, C. W., and D. B. Siniff. 1992. Determining population status and the use of biological indices in the management of marine mammals. Pages 1025–1037 *in* D. R. McCullough and R. H. Barrett (eds.). Wildlife 2001: Populations. Elsevier Applied Science, New York, NY.

Fowler, C. W., and T. D. Smith. 1981. Dynamics of large mammal populations. Wiley & Sons, New York, NY.

Frank, K. T., J. Simon, and J. E. Carscadden. 1994. Recent excursions of capelin (*Mallotus villosus*) to the Scotian shelf and Flemish Cap during anomalous hydrographic conditions. Northwest Atlantic Fisheries Organization, Scientific Council Studies Doc. 94/68, Serial No. N2446, Dartmouth, Nova Scotia.

Frankel, O. H., and M. E. Soule. 1981. Conservation and Evolution. Cambridge University Press, Cambridge, U.K.

French, D. P., M. Reed, J. Calambokidis, and J. C. Cubbage. 1989. A simulation model of seasonal migration and daily movements of the northern fur seal. Ecological Modelling 48:193–219.

Frost, K. J., and L. F. Lowry. 1980. Feeding of ribbon seal (*Phoca fasciata*) in the Bering Sea in spring. Canadian Journal of Zoology 58:1601–1607.

Frost, K. J., and L. F. Lowry. 1986. Sizes of walleye pollock, *Theragra chalcogramma,* consumed by marine mammals in the Bering Sea. Fishery Bulletin 84:192–197.

Frost, K. J., L. F. Lowry, and R. R. Nelson. 1985. Radiotagging studies of belukha whales (*Delphinapterus leucas*) in Bristol Bay, Alaska. Marine Mammal Science 1:191–202.

Fry, B. 1988. Food web structure on Georges Bank from stable C, N, and S isotopic compositions. Limnology and Oceanography 33:1182–1190.

Gales, N. J., P. D. Shaughnessy, and T. E. Dennis. 1994. Distribution, abundance and breeding cycle of the Australian sea lion *Neophoca cinerea* (Mammalia: Pinnipedia). Journal of Zoology (London) 234:353–370.

Gambell, R. 1999. The International Whaling Commission and the contemporary whaling debate. *in* J. R. Twiss Jr., and R. R. Reeves (eds.). Conservation and Management of Marine Mammals. Smithsonian Institution Press, Washington, D.C.

Garrott, R. A., B. B. Ackerman, J. R. Cary, D. M. Heisey, J. E. Reynolds, and J. R. Wilcox. 1995. Assessment of trends in sizes of manatee populations at several Florida aggregation sites. Pages 34–55 *in* T. J. O'Shea, B. B. Ackerman, and H. F. Percival (eds.). Population Biology of the Florida Manatee. U.S. Department of the Interior, Washington, D.C.

Gaskin, D. E. 1977. Harbour porpoise *Phocoena phocoena* (L.) in the western approaches to the Bay of Fundy 1969–75. Report of the International Whaling Commission 27:487–492.

Gaskin, D. E. 1982. The Ecology of Whales and Dolphins. Heinemann Educational Books, London.

Gaskin, D. E., G. J. D. Smith, and A. P. Watson. 1975. Preliminary study of movements of harbor porpoises (*Phocoena phocoena*) in the Bay of Fundy using radiotelemetry. Canadian Journal of Zoology 53:1466–1471.

Gentry, R. L., and J. R. Holt. 1986. Attendance behaviour of northern fur seals. Pages 41–60 *in* R. L. Gentry and G. L. Kooyman (eds.). Fur Seals: Maternal Strategies on Land and at Sea. Princeton University Press, Princeton, NJ.

Gentry, R. L., and J. H. Johnson. 1981. Predation by Sea Lions on Fur Seal Neonates in Alaska. U.S. Fish and Wildlife Service Report 28:23 pp.

Gentry, R. L., and G. L. Kooyman. 1986. Fur Seals: Maternal Strategies on Land and at Sea. Princeton University Press, Princeton, NJ.

Gerrodette, T. 1987. A power analysis for detecting trends. Ecology 68:1364–1372.

Gerrodette, T., and D. P. Demaster. 1990. Quantitative determination of optimum sustainable population level. Marine Mammal Science 6:1–16.

Gilmartin, W. G. 1993. Hawaiian Monk Seal Work Plan Fiscal Years 1994–96. National Marine Fisheries Service Administrative Report H-93-16.

Goodall, R. N. P., and A. R. Galeazzi. 1985. A review of the food habits of the small cetaceans of the Antarctic and Sub-Antarctic. Pages 566–572 *in* W. R. Seigfried, P. R. Condy, and R. M. Laws (eds.). Antarctic Nutrient Cycles and Food Webs. Springer-Verlag, Berlin.

Goodyear, J. D. 1993. A sonic/radio tag for monitoring dive depths and underwater movements of whales. Journal of Wildlife Management 57:503–513.

Green, K., and H. R. Burton. 1993. Comparison of the stomach contents of southern elephant seals, *Mirounga leonina*, at Macquarie and Heard Islands. Marine Mammal Science 9:10–22.

Greenwood, P. J. 1980. Mating systems, philopatry and dispersal in birds and mammals. Animal Behavior 28:1140–1162.

Hammill, M. O., and T. G. Smith. 1991. The role of predation in the ecology of the ringed seal in Barrow Strait, Northwest Territories. Marine Mammal Science 7:123–135.

Hammond, P. S. 1986. Estimating the size of naturally marked whale populations using mark–recapture techniques. Pages 253–282 *in* G. P. Donovan (ed.). Behaviour of Whales in Relation to Management. International Whaling Commission, Cambridge, U.K.

Hammond, P. S. 1990. Capturing whales on film—Estimating cetacean population parameters from individual recognition data. Mammal Review 20:17–22.

Hammond, P. S., and M. A. Fedak. 1994. Gray seals in the North Sea and their interactions with fisheries. Sea Mammal Research Unit, Cambridge, U.K. Final Report to Ministry of Agriculture, Fisheries, and Food. Contract MF 0503, Cambridge, U.K.

Hammond, P. S., and J. H. Prime. 1990. The diet of British gray seals (*Halichoerus grypus*). Pages 243–254 *in* W. D. Bowen (ed.). Population Biology of Sealworm (*Pseudoterranova decipiens*) in Relation to its Intermediate and Seal Hosts. Canadian Bulletin, Journal of Fisheries and Aquatic Sciences, Ottawa, Ontario, Canada.

Hammond, P. S., S. A. Mizroch, and G. P. Donovan. 1990. Individual recognition of cetaceans: Use of photo-identification and other techniques to estimate population parameters. Special Issue 12. International Whaling Commission, Cambridge, U.K.

Hammond, P. S., B. J. McConnell, and M. A. Fedak. 1993. Gray seals off the east coast of Britain: distribution and movements at sea. Symposia of the Zoological Society (London) 66:211–224.

Hammond, P. S., A. J. Hall, and J. H. Prime. 1994. The diet of gray seals around Orkney and other island and mainland sites in north-eastern Scotland. Journal of Animal Ecology 31:340–350.

Härkönen, T. 1986. Guide to the otoliths of the bony fishes of the Northeast Atlantic. Danbiu ApS., Biological Consultants, Hellerup, Denmark.

Härkönen, T. J. 1987. Seasonal and regional variations in the feeding habits of the harbour seal, *Phoca vitulina*, in the Skagerrak and the Kattegat. Journal of Zoology (London) 213:535–543.

Harry, G. Y., and J. R. Hartley. 1981. Northern fur seals in the Bering Sea. Pages 847–867 *in* D. W. Hood and J. A. Calder (eds.). The Eastern Bering Sea Shelf: Oceanography and Resources. University of Washington Press, Seattle, WA.

Hartman, D. S. 1979. Ecology and behavior of the manatee (*Trichechus manatus*) in Florida. Special Publication 5. American Society of Mammalogists, Lawrence, KS.

Harvey, J. T. 1988. Population dynamics, annual food consumption, movements, and dive behaviors of harbor seals, *Phoca vitulina richardsi*, in Oregon. Ph.D. thesis, Oregon State University, Corvallis, OR.

Harvey, J. T. 1989. Assessment of errors associated with harbour seal (*Phoca vitulina*) fecal sampling. Journal of Zoology (London) 219:101–111.

Harvey, P. H., and A. Purvis. 1991. Comparative methods for explaining adaptations. Nature 351:619–624.

Harwood, J., and J. H. Prime. 1978. Some factors affecting the size of British gray seal populations. Journal of Applied Ecology 15:401–411.

Haug, T., H. Gjosaeter, U. Lindstrom, and K. T. Nilssen. 1995. Diet and food availability for north-east Atlantic minke whales (*Balaenoptera acutorostrata*), during the summer of 1992. ICES Journal of Marine Science 52:77–86.

Herzing, D. L., and B. Mate. 1984. Gray whale migrations along the Oregon coast, 1978–1981. Pages 289–307 *in* M. L. Jones, S. L. Swartz, and S. Leatherwood (eds.). The Gray Whale, *Eschrichtius robustus*. Academic Press, Orlando, FL.

Hiby, A. R. 1994. Abundance estimates for gray seals in summer based on photo-identification data. Pages 5–22 *in* P. S. Hammond and M. A. Fedak (eds.). Gray Seals in the North Sea and their Interactions with Fisheries. Final report to Ministry of Agriculture, Fisheries and Food. Contract MF 0503, Cambridge, U.K.

Hiby, A. R., and P. S. Hammond. 1989. Survey techniques for estimating abundance of cetaceans. Reports of the International Whaling Commission (Special Issue 11):47–80.

Hiby, A. R., and J. S. Jeffery. 1987. Census techniques for small populations, with special reference to the Mediterranean monk seal. Symposia of the Zoological Society (London) 58:193–210.

Hiby, A. R., and P. Lovell. 1990. Computer aided matching of natural markings: A prototype system for gray seals. Report of the International Whaling Commission (Special Issue 12):57–61.

Hill, R. D. 1994. Theory of geolocation by light levels. Pages 227–236 *in* B. J. Le Boeuf and R. M. Laws (eds.). Elephant Seals: Population Ecology, Behavior, and Physiology. University of California Press, Berkeley, CA.

Hjort, J., and J. T. Rudd. 1929. Whaling and fishing in the North Atlantic. Rapports et Proces-verbaux des Réunions, Conseil Permanent International pour L'Exploration de la Mer 56:5–123.

Hobson, K. A., and H. E. Welch. 1992. Determination of trophic relationships within a high Arctic marine food web using del13C and del15N analysis. Marine Ecology Progress Series 84:9–18.

Hoelzel, A. R. (ed.). 1991. Genetic Ecology of Whales and Dolphins. Report of the International Whaling Commission (Special Issue 13), Cambridge, U.K.

Hoelzel, A. R. 1993. Genetic ecology of marine mammals. Symposia of the Zoological Society (London) 66:15–32.

Hoelzel, A. R. 1994. Genetics and ecology of whales and dolphins. Annual Review of Ecology and Systematics 25:377–399.

Hoelzel, A. R., and G. A. Dover. 1989. Molecular techniques for examining genetic variation and stock identity in cetacean species. Reports of the International Whaling Commission (Special Issue 11):81–120.

Hoelzel, A. R., and G. A. Dover. 1991. Molecular Genetic Ecology. Oxford University Press, New York, NY.

Hohn, A. A., M. D. Scott, R. S. Wells, J. C. Sweeney, and A. B. Irvine. 1989. Growth layers in the teeth from known-age, free-ranging bottlenose dolphins. Marine Mammal Science 5:315–342.

Hooper, S. N., M. Paradis, and R. G. Ackman. 1973. Distribution of trans-6-hexadecenoic acid, 7-methyl-7-hexadecenoic acid and common fatty acids in lipids of the ocean sunfish *Mola mola*. Lipids 8:509–516.

Hoover, A. A. 1988. Steller sea lion, *Eumetopias jubatus*. Pages 161–193 *in* J. W. Lentfer (ed.). Selected Marine Mammals of Alaska. Species Accounts with Research and Management Recommendations. Marine Mammal Commission, Washington, D.C.

Ichii, T. 1990. Distribution of Antarctic krill concentrations exploited by Japanese krill trawlers and minke whales. Proc. NIPR Symp Polar Biology 3:36–56.

Innes, S., D. M. Lavigne, W. M. Earle, and K. M. Kovacs. 1987. Feeding rates of seals and whales. Journal Animal Ecology 56:115–130.

International Whaling Commission 1982. Report of the sub-committee on sperm whales. Report of the International Whaling Commission 32:68–86.

Iverson, S. J. 1988. Composition, intake and gastric digestion of milk lipids in pinnipeds. Ph.D. thesis, University of Maryland, Washington, D.C.

Iverson, S. J. 1993. Milk secretion in marine mammals in relation to foraging: Can milk fatty acids predict diet? Symposia of the Zoological Society (London) 66:263–291.

Iverson, S. J., O. T. Oftedal, W. D. Bowen, D. J. Boness, and J. Sampugna. 1995. Prenatal and postnatal transfer of fatty acids from mother to pup in the hooded seal. Journal of Comparative Physiology 165:1–12.

Iverson, S. J., J. P. Y. Arnould, and I. L. Boyd. 1997. Milk fatty acids signatures indicate both major and minor shifts in diet of lactating Anarctic fur seals. Canadian Journal of Zoology 75:188–197.

Jaquet, N., and H. Whitehead. 1996. Scale-dependent correlation of sperm whale distribution with environmental features and productivitiy in the South Pacific. Marine Ecology Progress Series 135:1–9.

Jefferies, S. J. 1986. Seasonal Movements and Population Trends of Harbour Seals (*Phoca vitulina richardsi*) in the Columbia River and Adjacent Waters of Washington and Oregon: 1976–1982. Report to U.S. Marine Mammal Commission, Contract MM2079357-5, Washington, D.C.

Jefferson, T. A., P. J. Stacey, and R. W. Baird. 1991. A review of killer whale interactions with other marine mammals: Predation to co-existence. Mammal Review 21:151–180.

Jobling, M., and A. Breiby. 1986. The use and abuse of fish otoliths in studies of feeding habits of marine piscivores. Sarsia 71:265–274.

Jonsgard, A. 1969. Age determination in marine mammals. Pages 1–30 *in* H. A. Anderson (ed.). The Biology of Marine Mammals. Academic Press, New York, NY.

Kajimura, H. 1980. Distribution and migration of northern fur seals (*Callorhinus ursinus*) in the eastern Pacific. Pages 4–43 *in* H. Kajimura, R. H. Lander, M. A. Perez, A. E. York, and M. A. Bigg (eds.). Further Analysis of Pelagic Fur Seal Data Collected by the United States and Canada During 1958–1974. Proceedings of the 23rd Annual Meeting of the North Pacific Fur Seal Commission, Moscow, U.S.S.R.

Kajimura, H. 1984. Opportunistic Feeding of the Northern Fur Seal, *Callorhinus ursinus,* in the Eastern North Pacific Ocean and Eastern Bering Sea. NOAA Technical Report NMFS SSRF-779.

Kajimura, H. 1985. Opportunistic feeding by the northern fur seal (*Callorhinus ursinus*). Pages 301–318 *in* J. R. Beddington, R. J. H. Beverton, and D. M. Lavigne (eds.). Marine Mammals and Fisheries. Allen & Unwin, Cambridge, MA.

Kann, L. M., and K. Wishner. 1995. Spatial and temporal patterns of zooplankton on baleen whale feeding grounds in the southern Gulf of Maine. Journal of Plankton Research 17:235–262.

Kanwisher, J. W., and S. H. Ridgway. 1983. The physiological ecology of whales and porpoises. Scientific American 248:110–120.

Kapel, F. O. 1982. Studies on the hooded seal, *Cystophora cristata,* in Greenland, 1970–80. NAFO Scientific Council Studies 3:67–75.

Katona, S. K., and J. A. Beard. 1990. Population size, migrations and feeding aggregations of the humpback whale (*Megaptera novaeangliae*) in the western North Atlantic Ocean. Pages 295–306 *in* P. S. Hammond, S. A. Mizroch, and G. P. Donovan (eds.). Individual Recognition of Cetaceans: Use of Photo-identification and Other Techniques to Estimate Population Parameters. International Whaling Commission, Special Issue 12, Cambridge, U.K.

Katona, S., and H. Whitehead. 1988. Are Cetacea ecologically important? Oceanographic Marine Biological Annual Review 26:553–568.

Kawakami, T. 1980. A review of sperm whale food. Scientific Report of the Whale Research Institute 32:199–218.

Kawamura, A. 1970. Food of sei whales taken by Japanese whaling expeditions in the Antarctic season 1967/68. Scientific Report of the Whale Research Institute 22:127–152.

Kawamura, A. 1980. A review of food of Balaenopterid whales. Scientific Report of the Whale Research Institute 32:155–197.

Kawamura, A. 1994. A review of baleen whale feeding in the Southern Ocean. Report of the International Whaling Commission 44:261–271.

Kawamura, A., and Y. Satake. 1976. Preliminary report on the geographical distribution of the Bryde's whale in the North Pacific with special reference to the structure of the filtering apparatus. Scientific Report of the Whale Research Institute 28:1–35.

Kenny, R. D., M. A. M. Hyman, R. E. Owen, G. P. Scott, and H. E. Winn. 1986. Estimation of prey densities required by western North Atlantic right whales. Marine Mammal Science 2:1–13.

Kenney, R. D., G. P. Scott, T. J. Thompson, and H. E. Winn. 1996. Estimates of prey consumption and trophic impacts of cetaceans in the U.S.A. northeast continental shelf ecosystem. Journal of the Northwest Atlantic Fishery Science 22:155–171.

Kenyon, K. W. 1969. The sea otter in the eastern Pacific Ocean. North American Fauna 68:1–352.

Kenyon, K. W., and D. W. Rice. 1959. Life history of the Hawaiian monk seal. Pacific Science 13:215–252.

Kerfoot, W.C., and A. Sih. 1987. Predation: Direct and Indirect Impacts on Aquatic Communities. University Press of New England, Hanover, NH.

King, J. E. 1983. Seals of the World. Comstock, Ithaca, NY.

Kingsley, M. C. S., and I. Stirling. 1991. Haul-out behaviour of ringed and bearded seals in relation to defense against surface predators. Canadian Journal of Zoology 69:1857–1861.

Klem, A. 1935. Studies in the biochemistry of whale oils. Hvalradets Skr. Nr. 11:49–108.

Kovacs, K. M., K. M. Jonas, and S. E. Welke. 1990. Sex and age segregation by *Phoca vitulina concolor* at haul-out sites during the breeding season in the Passamaquoddy Bay region, New Brunswick. Marine Mammal Science 6:204–214.

Krebs, C. J. 1989. Ecological Methodology. Harper Collins, New York, NY.

Kvitek, R. G., C. E. Bowlby, and M. Staedler. 1993. Diet and foraging behavior of sea otters in Southeast Alaska. Marine Mammal Science 9:168–181.

Laake, J. L., S. T. Buckland, D. R. Andersen, and K. P. Burnham. 1994. Distance Sampling, Abundance Estimation of Biological Populations: Distance User's Guide. V2.1. Colorado Cooperative Fish and Wildlife Research, Colorado State University, Fort Collins, CO.

Lambertsen, R. H. 1983. Internal mechanism of rorqual feeding. Journal of Mammalogy 64:76–88.

Lander, R. H., and H. Kajimura. 1982. Status of the northern fur seals. Pages 319–345 *in* Mammals in the Seas. FAO Advisory Committee on Marine Resources Research Working Party on Marine Mammals. FAO Fisheries Service No. 5 4:319–345, Rome.

Lavigne, D. M., S. Innes, G. A. J. Worthy, K. M. Kovacs, O. J. Schmitz, and J. P. Hickie. 1986. Metabolic rates of seals and whales. Canadian Journal of Zoology 64:279–284.

Lavigueur, L., and M. O. Hammill. 1993. Distribution and seasonal movements of gray seals, *Halichoerus grypus*, born in the Gulf of St. Lawrence and Eastern Nova Scotia. Canadian Field-Naturalist 107:329–340.

Laws, R. M. 1952. A new method of age determination for mammals. Nature 169:972–973.

Laws, R. M. 1962. Age determination of pinnipeds with special reference to growth layers in the teeth. Zeitshrift fur Saugetierkunde 27:129–146.

Laws, R. M. 1977. The significance of vertebrates in the Antarctic marine ecosystem. Pages 411–438 *in* G. A. Llano (ed.). Adaptation Within Antarctic Ecosystems. Symposium on Antarctic Biology, 3rd ed. Smithsonian Institution Press, Washington, D.C.

Laws, R. M. 1984. Seals. Pages 621–716 *in* R. M. Laws (ed.). Antarctic Ecology. Academic Press, London, UK.

Laws, R. M. 1985. The ecology of the southern ocean. American Scientist 73:26–40.

Lawson, J. W., and G. B. Stenson. 1995. Historic variation in the diet of harp seals (*Phoca groenlandica*) in the northwest Atlantic. Pages 261–269 *in* A. S. Blix, L. Walloe, and O. Ultang (eds.). Whales, Seals, Fish, and Man. Elsevier Science, Amsterdam.

Lawson, J. W., G. D. Harrison, and W. D. Bowen. 1992. Factors affecting accuracy of age determination in the harp seal *Phoca groenlandica*. Marine Mammal Science 8:169–171.

Lawson, J. W., G. B. Stenson, and D. G. Mckinnon. 1994. Diet of harp seals (*Phoca groenlandica*) in Divisions 2J and 3KL during 1991–93. NAFO Scientific Council Studies 21:143–154.

Leatherwood, S., and W. E. Evans. 1979. Some recent uses and potentials of radiotelemetry in field studies of cetaceans. Pages 1–31 *in* H. E. Winn, and B.L. Olla (eds.). Behavior of Marine Animals. Current Perspectives in Research, Vol. 3: Cetaceans. Plenum, New York, NY.

Leatherwood, S., J. R. Gilbert, and D. G. Chapman. 1978. An evaluation of some techniques for aerial censuses of bottlenose dolphins. Journal of Wildlife Management 42:239–250.

Lebreton, L. D., K. P. Burnham, J. Clobert, and D. R. Anderson. 1992. Modeling survival and testing biological hypotheses using marked animals: A unified approach with case studies. Ecological Monographs 62:69–118.

Lefebvre, L. W., and J. A. Powell. 1990. Manatee Grazing Impacts on Seagrasses in Hobe Sound and Jupiter Sound in Southeast Florida During the Winter of 1988–89. NTIS Report PB90-271883. National Technical Information Service, Springfield, VA.

Lefebvre, L. W., T. J. O'Shea, G. B. Rathbun, and R. C. Best. 1989. Distribution, status, and biogeography of the West Indian manatee. Pages 567–610 *in* C. A. Woods (ed.). Biogeography of the West Indies. Sandhill Crane Press, Gainesville, FL.

Lefebvre, L. W., B. B. Ackerman, K. M. Poiter, and K. H. Pollock. 1995. Aerial survey as a technique for estimating trends in manatee population size—Problems and prospects. Pages 63–74 *in* T. J. O'Shea, B. B. Ackerman, and H. F. Percival (eds.). Population Biology of the Florida Manatee. U.S. Department of the Interior, Washington, D.C.

Lentfer, J. W. 1975. Polar bear denning on drifting sea ice. Journal of Mammalogy 56:716.

Leslie, P. H. 1945. On the use of matrices in certain population mathematics. Biometrika 33:183–212.

Leslie, P. H. 1948. Some further notes on the use of matrices in population mathematics. Biometrika 35:213–268.

Lindsey, A. A. 1937. The Weddell seal in the Bay of Whales, Antarctica. Journal of Mammalogy 18:127–144.

Lockyer, C. 1984. Age determination by means of the earplug in baleen whales. Report of the International Whaling Commission 34:692–696.

Lockyer, C. H., and S. G. Brown. 1981. The migration of whales. Pages 105–138 *in* D. J. Aidley (ed.). Animal Migration. Cambridge University Press, Cambridge, U.K.

Loughlin, T. R., J. L. Bengtson, and R. L. Merrick. 1987. Characteristics of feeding trips of northern fur seals. Canadian Journal of Zoology 65:2079–2084.

Lowry, L. F. 1993. Foods and feeding ecology. Pages 201–238 *in* J. J. Burns, J. J. Montague, and C. J. Cowles (eds.). The Bowhead Whale. Special Publication Number 2, Society of Mammalogists, Lawrence, KS.

Lowry, L. F., and F. H. Fay. 1984. Seal eating by walruses in the Bering and Chukchi Seas. Polar Biology 3:11–18.

Lowry, L. F., K. J. Frost, and J. J. Burns. 1980a. Feeding of bearded seals in the Bering and Chukchi Seas and trophic interaction with Pacific walruses. Arctic 33:330–342.

Lowry, L. F., K. J. Frost, and J. J. Burns. 1980b. Variability in the diet of ringed seals, *Phoca hispida*, in Alaska. Canadian Journal of Fisheries and Aquatic Sciences 37:2254–2261.

Lowry, M. S., C. W. Oliver, and C. Macky. 1990. Food habits of California sea lions, *Zalophus californianus*, at San Clemente Island, California, 1981–86. Fishery Bulletin (U.S.) 88:509–521.

Majluf, P. 1987. South American fur seal, *Artocephalus australis*, in Peru. Pages 33–35 *in* J. P. Croxall, R. L. Gentry (eds.). Status, Biology, and Ecology of Fur Seals. NOAA Technical Report NMFS 51, Springfield, VA.

Mansfield, A. W. 1991. Accuracy of age determination in the gray seal *Halichoerus grypus* of eastern Canada. Marine Mammal Science 7:44–49.

Mansfield, A. W., and B. Beck. 1977. The gray seal in eastern Canada. Fisheries and Marine Service. Technical Report 704:1–81.

Marcus, J., W. D. Bowen, and J. D. Eddington. 1998. Effects of meal size on otolith recovery from fecal samples of gray and harbor seal pups. Marine Mammal Science 14:789–802.

Marine Mammal Commission. 1995. Annual Report to Congress. 235 pp.

Marmontel, M. M., T. J. O'Shea, H. I. Kochman, and S. R. Humphrey. 1996. Age determination in manatees using growth-layer counts in bone. Marine Mammal Science 12:54–88.

Marmontel, M., S. R. Humphrey, and T. J. O'Shea. 1997. Population viability analysis of the Florida manatee (*Trichechus manatus latirostris*). Conservation Biology 11:467–481.

Marsh, H. 1995. The life history, pattern of breeding, and population dynamics of the dugong. Pages 75–83 *in* T. J. O'Shea, B. B. Ackerman, and H. F. Percival (eds.). Population Biology of the Florida Manatee. U.S. Department of the Interior, National Biological Service Information and Technology Report I, Washington, D.C.

Marsh, H., and G. B. Rathbun. 1990. Development and application of conventional and satellite radio-tracking techniques for studying dugong movements and habitat use. Australian Wildlife Research 17:83–100.

Mate, B., and J. T. Harvey. 1984. Ocean movements of radio-tagged gray whales. Pages 577–589 *in* Jones, M. L., S. L. Swartz, and S. Leatherwood (eds.). The Gray Whale, *Eschrichtius robustus*. Academic Press, Orlando, FL.

McConnell, B. J., C. Chambers, and M. A. Fedak. 1992. Foraging ecology of southern elephant seals in relation to the bathymetry and productivity of the Southern Ocean. Antarctic Science 4:393–398.

McConnell, B. J., M. A. Fedak, P. Lovell, and P. S. Hammond. 1994. The movements and foraging behaviour of gray seals. Pages 88–148 *in* P. S. Hammond and M. A. Fedak (eds.). Gray Seals in the North Sea and Their Interactions with Fisheries. SeaMammal Research Unit, Natural Environment research Council, Cambridge, U.K.

McKinnon, J. 1993. Feeding habits of the dusky dolphin, *Lagenorhynchus obscurus,* in the coastal waters of central Peru. Fishery Bulletin 92:569–578.

McLaren, I. A. 1990. Pinnipeds and oil: Ecological perspectives. Pages 55–101 *in* J. R. Geraci and D. J. St. Aubin (eds.). Sea Mammals and Oil: Confronting the Risks. Academic Press, San Diego, CA.

McLaren, I. A., and T. G. Smith. 1985. Population ecology of seals: Retrospective and Prospective views. Marine Mammal Science 1:54–83.

Messier, F., M. K. Taylor, and M. A. Ramsay. 1992. Seasonal activity patterns of female polar bears (*Ursus maritimus*) in the Canadian Arctic as revealed by satellite telemetry. Journal of Zoology (London) 226:219–229.

Milinkovitch, M. C. 1995. Molecular phylogeny of cetaceans prompts revision of morphological transformations. TREE 10:328–334.

Minagawa, M., and E. Wada. 1984. Stepwise enrichment of 15N along food chains. Further evidence and the relation between delta15N and animal age. Geochim. Cosmochim. Acta 48:1135–1140.

Miyazaki, N., T. Kusaka, and M. Nishiwali. 1973. Food of *Stenella caeruleoalba*. Scientific Report of the Whale Research Institute 14:265–275.

Miyazaki, N., T. Kasuya, M. Nishiwaki, and W. H. Dawbin. 1974. Distribution and migration of two species of *Stenella* in the Pacific coast of Japan. Scientific Report of the Whale Research Institute 26:227–243.

Mohn, R., and W. D. Bowen. 1996. Gray seal predation on the Eastern Scotian Shelf: effects on the dynamics and potential yield of Atlantic cod. Canadian Journal of Fisheries and Aquatic Sciences 53:2722–2738.

Murie, D. J., and D. M. Lavigne. 1985. Digestion and retention of Atlantic herring otoliths in the stomachs of gray seals. Pages 292–299 *in* J. R. Beddington, R. J. H. Beverton, and D. M. Lavigne (eds.). Marine Mammals and Fisheries. George Allen & Unwin, London, UK.

Murison, L. D., and D. E. Gaskin. 1989. The distribution of right whales and zooplankton in the Bay of Fundy, Canada. Canadian Journal of Zoology 67:1411–1420.

Myers, R. A., and W. D. Bowen. 1989. Estimating bias in aerial surveys of harp seal pup production. Journal of Wildlife Management 53:361–372.

Myrick, A. C. Jr., E. W. Shallenberger, I. Kang, and D. B. Mackay. 1984. Calibration of dental layers in seven captive Hawaiian spinner dolphins, *Stenella longirostris,* based on tetracycline labeling. Fishery Bulletin 82:207–225.

Naito, Y., and M. Nishiwaki. 1975. Ecology and morphology of *Phoca vitulina largha* and *Phoca kurilensis* in the southern Sea of Okhotsk and northeast of Hokkaido. Rapports Procès-Verbaux des Réunions, Conseil International pour l'Exploration de la Mer 169:379–386.

Nansen, F. 1925. Hunting and Adventure in the Arctic. Duffield and Co., New York, NY.

National Resource Council. 1996. The Bering Sea Ecosystem. National Research Council, Washington, D.C.

Nerini, M. 1984. A review of gray whale feeding ecology. Pages 423–450 *in* M. L. Jones, S. L. Swartz, and S. Leatherwood (eds.). The Gray Whale, *Eschrichtius robustus*. Academic Press, Orlando, FL.

Nichol, L. M., and D. M. Shackleton. 1995. Seasonal movements and foraging behaviour of northern resident killer whales (*Orcinus orca*) in relation to the inshore distribution of salmon (*Oncorhynchus* spp.) in British Columbia. Canadian Journal of Zoology 74:983–991.

Niebauer, H. J., V. Alexander, and R. T. Cooney. 1981. Primary production at the Eastern Bering Sea ice edge: the physical and biological regimes. Pages 763–772 *in* D. W. Hood and J. A. Calder (eds.). The Eastern Bering Sea Shelf: Oceanography and Resources. University of Washington Press, Seattle, WA.

Nilssen, K. T., T. Haug, V. Potelov, V. A. Stasenkov, and Y. K. Timoshenko. 1995. Food habits of harp seals (*Phoca groenlandica*) during lactation and moult in March–May in the southern Barents Sea and White Sea. ICES Journal of Marine Science 52:33–41.

Nishiwaki, M., and T. Yagi. 1953. On the age and growth of teeth in a dolphin—*Prodelphinus caeruleo-albus*. Scientific Reports of the Whales Research Institute (Tokyo) 8:133–146.

Northridge, S. P., M. L. Tasker, A. Webb, and J. M. Williams. 1995. Distribution and relative abundance of harbour porpoises (*Phocoena phocoena* L.), white-beaked dolphins (*Lagenorhynchus albirostris* Gray), and minke whales (*Balaenoptera acutorostrata* Lacepede) around British Isles. ICES Journal of Marine Science 52:55–66.

Oftedal, O. T., W. D. Bowen, E. M. Widdowson, and D. J. Boness. 1991. The prenatal molt and its ecological significance in hooded and harbour seals. Canadian Journal of Zoology 69:2489–2493.

Ohsumi, S. 1979. Feeding habits of the minke whale in the Antarctic. Report of the International Whaling Commission 29:473–476.

Olesiuk, P. F. 1993. Annual prey consumption by harbour seals (*Phoca vitulina*) in the Strait of Georgia, British Columbia. Fishery Bulletin 91:491–515.

Olesiuk, P. F., M. A. Bigg, G. M. Ellis, S. J. Crockford, and R. J. Wigen. 1990a. An assessment of the feeding habits of harbour seals (*Phoca vitulina*) in the Strait of Georgia, British Columbia, based on scat analysis. Canadian Technical Report of Fisheries and Aquatic Science No. 1730, Nanaimo, British Columbia, 135 p.

Olesiuk, P. F., M. A. Bigg, and G. M. Ellis. 1990b. Life history and population dynamics of resident killer whales (*Orcinus orca*) in the coastal waters of British Columbia and Washington State. Report of the International Whaling Commission (Special Issue 12):209–243.

Olesiuk, P. F., M. A. Bigg, and G. M. Ellis. 1990c. Recent trends in the abundance of harbour seals, *Phoca vitulina,* in British Columbia. Canadian Journal of Fisheries and Aquatic Sciences 47:992–1003.

Oliver, J. S., P. N. Slattery, E. F. O'Conner, and L. F. Lowry. 1983. Walrus, *Odobenus rosmarus,* feeding in the Bering Sea: A benthic perspective. Fishery Bulletin 81:501–512.

Oritsland, T. 1977. Food consumption of seals in the Antarctic pack ice. Pages 749–768 *in* G. A. Llano (ed.). Adaptations within Antarctic Ecosystems. Smithsonian Institution Press, Washington, D.C.

Orton, L. S., and P. F. Brodie. 1987. Engulfing mechanics of fin whales. Canadian Journal of Zoology 65:2898–2907.

O'Shea, T. J., and C. A. Langtrimm. 1995. Estimation of survival of adult Florida manatees in the Crystal River, at Blue Spring, and on the Atlantic coast. Pages 194–222 *in* T. J. O'Shea, B. B. Ackerman, and H. F. Percival (eds.). Population Biology of Florida Manatee.

U.S. Department of the Interior, National Biological Service Information and Technology Report I, Washington, D.C.

O'Shea, T. J., B. B. Ackerman, and H. F. Percival (eds.). 1995. Population Biology of the Florida Manatee. U.S. Department of the Interior, National Biological Service Information and Technology Report I, Washington, D.C.

Ostrom, P. H., J. Lien, and S. A. Macko. 1993. Evaluation of the diet of Sowerby's beaked whale, *Mesoplodon bidens,* based on isotopic comparisons among northwestern Atlantic cetaceans. Canadian Journal of Zoology 71:858–861.

Overholtz, W. J., S. A. Murawski, and K. L. Foster. 1991. Impact of predatory fish, marine mammals, and seabirds on the pelagic fish ecosystem of the northeastern USA. ICES Marine Science Symposium 193:198–208.

Packer, C., D. A. Gilbert, A. E. Pussey, and S. J. O'Brien. 1991. A molecular genetic analysis of kinship and cooperation in African lions. Nature 351:562–565.

Paetkau, D., W. Calvert, I. Stirling, and C. Strobeck. 1995. Microsatellite analysis of population structure in Canadian polar bears. Molecular Ecology 4:347–354.

Paine, R. T. 1969. A note on trophic complexity and species diversity. American Naturalist 103:91–93.

Palka, D. 1995. Abundance estimate of the Gulf of Maine harbor porpoise. Pages 27–50 *in* A. Bjorge and G. P. Donovan (eds.). Biology of the Phocoenids. International Whaling Commission, Special Issue 16, Cambridge, U.K.

Palsboll, P. J., P. J. Clapham, D. K. Mattila, F. Larsen, R. Sears, H. R. Seigismund, J. Sigurjonsson, O. Vasquez, and P. Arctander. 1995. Distribution of mtDNA haplotypes in North Atlantic humpback whales: The influence of behaviour on population structure. Marine Ecology Progress Series 116:1–10.

Payne, M. R. 1977. Growth of a fur seal population. Philosophical Transactions of the Royal Society of London Series, Biological Sciences 279:67–79.

Payne, P. M., and D. W. Heinenmann. 1993. The distribution of pilot whales (*Globicephala* spp.) in shelf/shelf-edge and slope waters of the northeastern United States. International Whaling Commission (Special Issue) 14:51–68.

Payne, P. M., D. N. Wiley, S. B. Young, S. Pittman, P. J. Clapman, and J. W. Jossi. 1990. Recent fluctuations in the abundance of baleen whales in the southern Gulf of Maine in relation to changes in selected prey. Fishery Bulletin 88:687–696.

Perez, M. A., and M. A. Bigg. 1986. Diet of Northern fur seals (*Callorhinus ursinus*) off western North America. Fishery Bulletin 84:957–971.

Perrin, W. F. 1975. Distribution and differentiation of populations of dolphins of the genus *Stenella* in the eastern tropical Pacific. Journal of Fisheries Research Board of Canada 32:1059–1067.

Perrin, W. F., and A. C. Myrick Jr. 1980. Age determination of toothed whales and sirenians. Reports of the International Whaling Commission (Special Issue 3), Cambridge, U.K.

Perrin, W. F., R. R. Warner, C. H. Ficus, and D. B. Holts. 1973. Stomach contents of porpoise, *Stenella* sp., and yellowfin tuna, *Thunnus albacares,* in mixed species aggregations. Fishery Bulletin 70:1077–1092.

Perry, E. A., S. M. Carr, S. E. Bartlett, and W. S. Davidson. 1995. A phylogentic perspective on the evolution of reproductive behavior in pagophilic seals of the northwest Atlantic as indicated by mitochondrial DNA sequences. Journal of Mammalogy 76:22–31.

Peterson, B. J., and B. Fry. 1987. Stable isotopes in ecosystem studies. Annual Review of Ecology and Systematics 181:293–320.

Peterson, R. S. 1968. Social behaviour in pinnipeds with particular reference to the northern fur seal. Pages 3–53 *in* R. J. Harrison, R. C. Hubbard, R. S. Peterson, D. E. Rice, and R. J. Schusterman (eds.). Behaviour and Physiology of Pinnipeds. Appleton-Century-Crofts, New York, NY.

Piatt, J. F., and D. A. Methven. 1992. Threshold foraging behavior of baleen whales. Marine Ecology Progress Series 84:205–210.

Piatt, J. F., D. A. Methven, A. E. Burger, R. L. Mclagan, V. Mercer, and E. Creelman. 1989. Baleen whales and their prey in a coastal environment. Canadian Journal of Zoology 67:1523–1530.

Pielou, E. C. 1974. Population and Community Ecology. Cordon and Breach, New York, NY.

Pierce, G. J., and P. R. Boyle. 1991. A review of methods for diet analysis in piscivorous marine mammals. Oceanographic Marine Biology, Annual Review 29:409–486.

Pierce, G. J., J. S. W. Diack, and P. R. Boyle. 1990. Application of serological methods to identification of fish prey in diets of seals and dolphins. Journal of Experimental Marine Biology and Ecology 137:123–140.

Pierce, G. J., P. M. Thompson, A. Miller, J. S. W. Diack, D. Miller, and P. R. Boyle. 1991. Seasonal variation in the diet of common seals (*Phoca vitulina*) in the Moray Firth area of Scotland. Journal of the Zoological Society of London 223:641–652.

Pierce, G. J., P. R. Boyle, J. Watt, and M. Grisley. 1993. Recent advances in diet analysis of marine mammals. Symposia of the Zoological Society (London) 66:241–261.

Pike, G. 1962. Migration and feeding of the California gray whale, *Eschrichtius robustus.* Journal of the Fisheries Research Board of Canada 19:815–838.

Pitcher, K. W. 1981. Prey of the Steller sea lion, *Eumetopias jubatus,* in the Gulf of Alaska. Fisheries Bulletin 79:467–472.

Pitcher, K. W., and D. C. Mcallister. 1981. Movements and haulout behaviour of radio-tagged harbor seals, *Phoca vitulina.* Canadian Field-Naturalist 95:292–297.

Pivorunas, A. 1979. The feeding mechanisms of baleen whales. American Scientist 67:432–440.

Popov, L. 1979. Baikal seal. FAO Fisheries Series No. 5. Pages 72–73 *in* Mammals of the Sea, Vol. 2. Rome.

Powell, J. A. 1996. The distribution and biology of the West Indian manatee (*Trichechus senegalensis* Link, 1795). Report to the United Nations Environment Programme, Nairobi, Kenya.

Power, G., and J. Gregoire. 1978. Predation by freshwater seals on the fish community of Lower Seal Lake, Quebec. Journal of the Fisheries Research Board of Canada 35:844–850.

Preen, A. 1995. Impacts of dugong foraging on seagrass habitats: Observational and experimental evidence for cultivation grazing. Marine Ecology Progress Series 124:201–213.

Prime, J. H., and P. S. Hammond. 1990. The diet of gray seals from the South-Western North Sea assessed from analyses of hard parts found in feces. Journal of Applied Ecology 27:435–447.

Ragen, T. J. 1995. Maximum net productivity level estimation for the northern fur seal (*Callorhinus ursinus*) population of St. Paul Island, Alaska. Marine Mammal Science 11:275–300.

Ralls, K., T. C. Eagle, and D. Siniff. 1996. Movement and spatial use patterns of California sea otters. Canadian Journal of Zoology 74:1841–1849.

Ramsay, M. A., and K. A. Hobson. 1991. Polar bears make little use of terrestrial food webs: evidence from stable-carbon isotope analysis. Oecologia 86:598–600.

Ramsay, M. A., and I. Stirling. 1986. On the mating system of polar bears. Canadian Journal of Zoology 64:2142–2151.

Ramsay, M. A., and I. Stirling. 1988. Reproductive biology and ecology of female polar bears (*Ursus maritimus*). Journal of Zoology (London) 214:601–634.

Rathbun, G. B., J. P. Reid, and J. B. Bourassa. 1987. Design and construction of a tethered, floating radio tag assembly for manatees. Document PB 87-161345 / A5. National Technical Information Service, Springfield, VA.

Rau, G. H., D. G. Ainley, J. L. Bengtson, J. J. Torres, and T. L. Hopkins. 1992. 15N / 14N and 13C / 12C in Weddell Sea birds, seals, and fish: Implications for diet and trophic structure. Marine Ecology Progress Series 84:1–8.

Ray, G. C. 1981. The role of large organisms. Pages 397–413 *in* A. R. Longhurst (ed.). Analysis of Marine Ecosystems. Academic Press, London, UK.

Recchia, C. A., and A. J. Read. 1989. Stomach contents of harbour porpoises, *Phocoena phocoena* (L.), from the Bay of Fundy. Canadian Journal of Zoology 67:2140–2146.

Reeves, R. R., A. A. Chaudhry, and U. Khalid. 1991. Competing for water on the Indus Plain: Is there a future for Pakistan's river dolphins? Environmental Conservation 18:341–350.

Reeves, R. R., B.S. Stewart, and S. Leatherwood. 1992. The Sierra Club Handbook of Seals and Sirenians. Sierra Club Books, San Francisco, CA.

Reid, J. P., R. K. Bonde, and T. J. O'Shea. 1995. Reproduction and mortality of radio-tagged and recognizable manatees on the Atlantic coast of Florida. Pages 171–191 *in* T. J. O'Shea, B. B. Ackerman and H. F. Percival (eds.). Population Biology of the Florida Manatee. U.S. Department of the Interior, National Biological Service Information and Technology Report I, Washington, D.C.

Reijnders, P., S. Brasseur, J. Van Der Torn, Van Der Wolf, I. Boyd, J. Harwood, D. Lavigne, and L. Lowry. 1993. Status Survey and Conservation Action Plan: Seals, Fur Seals, Sea Lions, and Walrus. IUCN Gland, Switzerland.

Reilly, S. B., and J. Barlow. 1986. Rates of increase in dolphin popula tion size. Fishery Bulletin 84:527–533.

Reilly, S. B., and V. G. Thayer. 1990. Blue whale (*Balaenoptera musculus*) distribution in the eastern tropical Pacific. Marine Mammal Science 6:265–277.

Reynolds, J. E. III. 1981. Behavior patterns in the West Indian manatee, with emphasis on feeding and diving. Florida Scientist 44:233–242

Reynolds, J. E. III, and D. K. Odell. 1991. Manatees and Dugongs. Facts on File, Inc., New York, NY

Reynolds, J. E. III, and J. R. Wilcox. 1994. Observations of Florida manatees (*Trichechus manatus latirostris*) around selected power plants in winter. Marine Mammal Science 10:163–177.

Rice, D. W., and A. A. Wolman. 1971. The life history and ecology of the gray whale, *Eschrichtius robustus*. Special Publication, American Society of Mammalogists 3:1–142.

Richard, K. R., M. C. Dillon, H. Whitehead, and J. M. Wright. 1996. Patterns of kinship in groups of free-living sperm whales (*Physeter macrocephalus*) revealed by multiple moleculae genetic analyses. Proceedings of the National Academy of Science 93:8792–8795.

Riedman, M. L. 1990. The Pinnipeds—Seals, Sea Lions, and Walruses. University of California Press, Berkeley, CA.

Riedman, M. L., and J. A. Estes. 1988. A review of the history, distribution and foraging ecology of sea otters. Pages 4–21 *in* G. R. VanBlaricom and J. A. Estes (eds.). The Community Ecology of Sea Otters. Springer-Verlag, Heidelberg, Germany.

Riedman, M. L., and J. A. Estes. 1990. The sea otter (*Enhydra lutris*): Behavior, ecology, and natural history. U.S. Fish and Wildlife Service, Biological Report 90(14).

Rodhouse, P. G., T. R. Arnbom, M. A. Fedak, J. Yeatman, and A. W. A. Murray. 1992. Cephalopod prey of the southern elephant seal, *Mirounga leonina* L. Canadian Journal of Zoology 70:1007–1015.

Roff, D., and W. D. Bowen. 1983. The population dynamics and management of northwest Atlantic harp seals (*Phoca groenlandica*). Canadian Journal of Fisheries and Aquatic Sciences 40:919–932.

Roff, D., and W. D. Bowen. 1986. Analysis of population trends in northwest Atlantic harp seals (*Phoca groenlandica*) from 1967–1983. Canadian Journal of Fisheries and Aquatic Sciences 43:553–564.

Rounsevell, D., and I. Eberhard. 1980. Leopard seals, *Hydrurga leptonyx* (Pinnipedia), at Macquarie Island from 1949 to 1979. Australian Wildlife Research 7:403–415.

Scheffer, V. B. 1950. Growth layers on the teeth of pinnipeds as an indication of age. Science 112:309–311.

Schell, D. M., S. M. Saupe, and N. Haubenstock. 1989. Bowhead whale (*Balaena mysticetus*) growth and feeding as estimated by del13C techniques. Marine Biology 103:433–443.

Scott, M. D., R. S. Wells, and A. B. Irvine. 1990. A long-term study of bottlenose dolphins on the west coast of Florida. Pages 235–265 *in* S. Leatherwood and R. R. Reeves (eds.). The Bottlenose Dolphin. Academic Press, San Diego, CA.

Seber, G. A. F. 1982. The Estimation of Animal Abundance and Related Parameters. C. Griffin, London.

Selzer, L. A., and P. M. Payne. 1988. The distribution of white-sided (*Lagenorhynchus acutus*) and common dolphins vs. environmental features of the continental shelf of the northeastern United States. Marine Mammal Science 4:141–153.

Sergeant, D. E. 1959. Age determination of odontocete whales from dentinal growth layers. Norwegian Whaling Gazette 6:273–288.

Sergeant, D. E. 1965. Exploitation and conservation of harp and hood seals. Polar Record 12:541–551.

Sergeant, D. E. 1976. History and present status of harp and hooded seals. Biological Conservation 10:95–118.

Sergeant, D. E. 1991. Harp seals, man, and ice. Canadian Special Publication of Fisheries and Aquatic Sciences 114:153.

Sergeant, D. E., K. Ronald, J. Boulva, and F. Berkes. 1978. The recent status of *Monachus monachus* the Mediterranean monk seal. Pages 31–55 *in* K. Ronald and R. Duguy (eds.). The Mediterranean Monk Seal, United Nations Environment Programme, Technical Series, Vol. 1. Pergammon Press, Oxford, England.

Shane, S. H., R. S. Wells, and B. Würsig. 1986. Ecology, behavior, and social organization of the bottlenose dolphin: A review. Marine Mammal Science 2:34–63.

Shaughnessy, P. D. 1982. The status of seals in South Africa and Namibia. Pages 383–410 *in* Mammals in the Seas. FAO Fisheries Series. FAO, Rome.

Shelton, P. A., G. B. Stenson, B. Sjare, and W. G. Warren. 1995. Model estimates of harp seal numbers at age for the Northwest Atlantic. Department of Fisheries and Oceans, Atlantic fisheries Research Doc. 95 / 21, Dartmouth, Nova Scotia.

Shields, W. M. 1987. Dispersal and mating systems: Investigating their causal connections. Pages 3–24 *in* B. D. Chepko-Sade and Z. T. Halpin (eds.). Mammalian Dispersal Patterns: The Effects of Social Structure on Population Genetics. University of Chicago Press, Chicago, IL.

Shrestha, T. K. 1993. Ecology, status appraisal, conservation and management of Gangetic dolphin *Platanista gangetica* in the Koshi River of Nepal. Journal of Freshwater Biology 5:93–105.

Simenstad, C. A., J. A. Estes, and K. W. Kenyon. 1978. Aleuts, sea otters, and alternate stable-state communities. Science 200:403–411.

Simila, T., and F. Ugarte. 1993. Surface and underwater observations of cooperatively feeding killer whales in northern Norway. Canadian Journal of Zoology 71:1494–1499.

Sinclair, E., T. Loughlin, and W. Pearcy. 1994. Prey selection by northern fur seals (*Callorhinus ursinus*) in the eastern Bering Sea. Fishery Bulletin 92:144–156.

Siniff, D. B., and J. L. Bengtson. 1977. Observations and hypotheses concerning the interactions among crabeater seals, leopard seals and killer whales. Journal of Mammalogy 58:414–416.

Siniff, D. B., and S. Stone. 1985. The role of the leopard seal in the trophodynamics of the Antarctic marine ecosystem. Pages 555–560 *In* W. R. Siegfried, P. R. Condy, and R. M. Laws (eds.). Nutrient Cycles and Food Chains. Proceedings of the 4th SCAR Symposium on Antarctic Biology. Springer-Verlag, Berlin.

Siniff, D. B., D. P. Demaster, R. J. Hofman, and L. L. Eberhardt. 1977. An analysis of the dynamics of a Weddell seal population. Ecological Monographs 47:319–335.

Siniff, D. B., I. Stirling, J. L. Bengtson, and R. A. Reichle. 1979. Social and reproductive behavior of crabeater seals (*Lobodon carcinophagus*) during the austral spring. Canadian Journal of Zoology 57:2243–2255.

Skalski, J. R., and D. S. Robson. 1992. Techniques for wildlife investigations. Design and analysis of capture data. Academic Press, San Diego, CA.

Skogland, T. 1991. What are the effects of predators on large ungulate populations? Oikos 61:401–411.

Slatkin, M. 1987. Gene flow and the geographic structure of natural populations. Science 236:787–792.

Slip, D. J. 1995. The diet of southern elephant seals (*Mirounga leonina*) from Heard Island. Canadian Journal of Zoology 73:1519–1528.

Slip, D. J., M. A. Hindell, and H. R. Burton. 1994. Diving behavior of southern elephant seals from Macquarie Island: An overview. Pages 253–270 *in* B. J. Le Boeuf and R. M. Laws (eds.). Elephant Seals: Population Ecology, Behavior, and Physiology. University of California Press, Berkeley, CA.

Smith, G. J. D., and D. E. Gaskin. 1974. The diet of harbor porpoises (*Phocoena phocoena* (L.)) in coastal waters of Eastern Canada, with special reference to the Bay of Fundy. Canadian Journal of Zoology 52:777–782.

Smith, R. J., K. A. Hobson, H. N. Koopman, and D. M. Lavigne. 1996. Distinguishing between populations of fresh-water and salt-water harbour seals (*Phoca vitulina*) using stable-isotope ratios and fatty acid profiles. Canadian Journal of Fisheries and Aquatic Sciences 53:272–279.

Smith, S. C., and H. Whitehead. 1993. Variations in the feeding success and behaviour of Galapagos sperm whales (*Physeter macrocephalus*) as they relate to oceanographic conditions. Canadian Journal of Zoology 71:1991–1996.

Smith, S. J., S. J. Iverson, and W. D. Bowen. 1997. Fatty acid signatures and classification trees: New tools for investigating the foraging ecology of seals. Canadian Journal of Fisheries and Aquatic Sciences 54:1377–1386.

Smith, T. G. 1973. Population dynamics of the ringed seal in the Canadian eastern Arctic. Fisheries Research Board, Canadian Bulletin 181:1–55.

Smith, T. G. 1980. Polar bear predation of ringed and bearded seals in the landfast sea ice habitat. Canadian Journal of Zoology 58:2201–2209.

Smith, T. G. 1985. Polar bears, *Ursus maritimus,* as predators of belugas, *Delphinapterus leucas.* Canadian Field-Naturalist 99:71–75.

Smith, T. G. 1987. The ringed seal, *Phoca hispida,* of the Canadian Western Arctic. Canadian Bulletin, Fisheries and Aquatic Sciences, Ottowa, Ontario.

Smith, T. G., and A. R. Martin. 1994. Distribution and movement of belugas, *Delphinapterus leucas,* in the Canadian High Arctic. Canadian Journal of Fisheries and Aquatic Sciences 51:1653–1663.

Smith, W. O. 1991. Nutrient distributions and new production in polar regions: Parallels and contrasts between the Arctic and Antarctic. Mar. Chem., 35:245–257 (cited by Thompson, S.K. 1992. Sampling. Wiley & Sons, New York, NY).

St. Aubin, D. J., T. G. Smith, and J. R. Geraci. 1990. Seasonal epidermal molt in beluga whales, *Dephinapterus leucas.* Canadian Journal of Zoology 68:359–367.

Stanley, H. F., S. Casey, J. M. Carnahan, S. Goodman, J. Harwood, and R. K. Wayne. 1996. Worldwide patterns of mitochondrial DNA differentiation in the harbor seal (*Phoca vitulina*). Molecular Biology and Evolution 13:368–382.

Steele, J. H. 1985. A comparison of terrestrial and marine ecological systems. Nature 313:355–358.

Steele, J. H. 1991. Marine functional diversity. BioScience 41:470–474.

Stenson, G. B., and D. J. Kavanagh. 1994. Distribution of harp and hooded seals in offshore waters of Newfoundland. NAFO Scientific Council Studies 21:121–142.

Stenson, G. B., R. A. Myers, M. O. Hammill, I. Ni, W. G. Warren, and M. C. S. Kingsley. 1993. Pup production of harp seals, *Phoca groenlandica,* in the Northwest Atlantic. Canadian Journal of Fisheries and Aquatic Sciences 50:2429–2439.

Stephens, D. W., and J. R. Krebs. 1986. Foraging Theory. Princeton University Press, Princeton, NJ.

Stevens, T. A., D. A. Duffield, E. D. Asper, K. G. Hewlett, A. Bolz, L. J. Gaga, and G. D. Bossart. 1989. Preliminary findings of restriction fragment differences in mitochondrial DNA among killer whales (*Orcinus orca*). Canadian Journal of Zoology 67:2592–2595.

Stewart, B. S., and R. L. DeLong. 1993. Seasonal dispersal and habitat use of foraging northern elephant seals. Symposia of the Zoological Society (London) 66:179–194.

Stewart, B. S., and R. L. DeLong. 1995. Double migrations of the northern elephant seal, *Mirounga angustirostris.* Journal of Mammalogy 76:196–205.

Stewart, B. S., S. Leatherwood, P. K. Yochem, and M. P. Heide-Jorgensen. 1989. Harbor seal tracking and telemetry by satellite. Marine Mammal Science 5:361–375.

Stirling, I. 1969. Ecology of the Weddell seal in McMurdo Sound, Antarctica. Ecology 50:573–586.

Stirling, I. 1975. Factors affecting the evolution of social behaviour in Pinnipedia. Rapports Procès-Verbaux des Réunions, Conseil pour l'Exploration de la Mer 169:205–212.

Stirling, I. 1977. Adaptations of Weddell and ringed seals to exploit the polar fast ice habitat in the absence or presence of surface predators. Pages 741–748 *in* G. A. Llano (ed.). Adaptations within Antarctic Ecosystems. Smithsonian Institution Press, Washington, D.C.

Stirling, I., and W. R. Archibald. 1977. Aspects of predation of seals by polar bears. Journal of Fisheries Research Board of Canada 34:1126–1129.

Stirling, I., and P. B. Latour. 1978. Comparative hunting abilities of polar bear cubs of different ages. Canadian Journal of Zoology 56:1768–1772.

Stirling, I., M. Kingsley, and W. Calvert. 1982. The distribution and abundance of seals in the eastern Beaufort Sea, 1974–79. Occasional Paper Number 47, Canadian Wildlife Service 23, Ottawa, Ontario.

Stirling, I., D., Andriashek, and W. Calvert. 1993. Habitat preferences of polar bears in the western Canadian Arctic in late winter and spring. Arctic 29:13–24.

Stobo, W. T., and K. C. T. Zwanenburg. 1990. Gray seal (*Halichoerus grypus*) pup production on Sable Island and estimates of recent production in the Northwest Atlantic. Pages 171–184 *in* W. D. Bowen (ed.). Population Biology of Sealworm (*Pseudoterranova decipiens*) in Relation to its Intermediate and Seal Hosts. Canadian Bulletin, Journal of Fisheries and Aquatic Sciences 222, Ottawa, Ontario.

Stobo, W. T., B. Beck, and J. K. Horne. 1990. Seasonal movements of gray seals (*Halichoerus grypus*) in the Northwest Atlantic. Pages 199–213 *in* W. D. Bowen (ed.). Population Biology of Sealworm (*Pseudoterranova decipiens*) in Relation to its Intermediate and Seal Hosts. Canadian Bulletin, Journal of Fisheries and Aquatic Sciences 222, Ottawa, Ontario.

Street, R. J. 1964. Feeding Habits of the New Zealand Fur Seal. New Zealand Marine Department of Fisheries, Technical Report 9:20, Wellington, New Zealand.

Sydeman, W. J., and N. Nur. 1994. Life history strategies of female northern elephant seals. Pages 137–153 *in* B. J. Le Boeuf and R. M. Laws (eds.). Elephant Seals: Population Ecology, Behavior and Physiology. University of California Press, Berkeley and Los Angeles, CA.

Terhune, J. M. 1985. Scanning behaviour of harbour seals on haul-out sites. Journal of Mammalogy 66:392–395.

Tershy, B. R. 1992. Body size, diet, habitat use, and social behaviour of Balaenoptera whales in the Gulf of California. Journal of Mammalogy 73:477–486.

Testa, J. W. 1986. Electromorph variation in Weddell seals (*Leptonychotes weddellii*). Journal of Mammalogy 67:606–610.

Testa, J. W. 1994. Over winter movements and diving behavior of female Weddell seals (*Leptonychotes weddellii*) in the southwestern Ross Sea, Antarctica. Canadian Journal of Zoology 72:1700–1710.

Testa, W., and D. B. Siniff. 1987. Population dynamics of Weddell seals (*Leptonychotes weddelli*) in McMurdo Sound, Antarctica. Ecological Monographs 57:149–165.

Testa, W. J., G. Oehlert, D. G. Ainley, J. L. Bengtson, D. B. Siniff, R. M. Laws, and D. Rounsevell. 1991. Temporal variability in Antarctic marine ecosystems: Periodic fluctuations in the phocid seals. Canadian Journal of Fisheries and Aquatic Sciences 48:631–639.

Thompson, D., P. S. Hammond, K. S. Nicholas, and M. A. Fedak. 1991. Movements, diving and foraging behaviour of gray seals (*Halichoerus grypus*). Journal of Zoology (London) 224:223–232.

Thompson, D., A. R. Hiby, and M. A. Fedak. 1993. How fast should I swim? Behavioural implications of diving physiology. Symposia of the Zoological Society (London) 66:349–368.

Thompson, P. M. 1993. Harbour seal movement patterns. Symposia of the Zoological Society (London) 66:225–239.

Thompson, P. M., and D. Miller. 1990. Summer foraging activity and movements of radio-tagged common seals (*Phoca vitulina* L.) in the Moray Firth, Scotland. Journal of Animal Ecology 27:492–501.

Thompson, P. M., M. A. Fedak, B. J. Mcconnell, and K. S. Nicholas. 1989. Seasonal and sex-related variation in the activity patterns of common seals (*Phoca vitulina*). Journal of Applied Ecology 26:521–535.

Thompson, P. M., K. M. Kovacs, and B. J. Mcconnell. 1994. Natal dispersion of harbour seals (*Phoca vitulina*) from breeding sites in Orkney, Scotland. Journal of Zoology (London) 234:668–673.

Thompson, P. M., D. J. Tollit, S. P. R. Greenstreet, A. Mackay, and H. M. Corpe. 1996. Between-year variations in the diet and behaviour of harbour seals, *Phoca vitulina* in the Moray Firth; causes and consequences. Pages 44–52 *in* S. P. R. Greenstreet and M. L. Tasker (eds.). Aquatic Predators and Their Prey. Blackwell Scientific Publications, Oxford, England.

Tieszen, L. L., T. W. Boutton, K. G. Tesdahl, and N. A. Slade. 1983. Fractionation and turnover of stable carbon isotopes in animal tissues: Implications for del 13 C analysis of diet. Oceologica 57:32–37.

Trites, A. W., and P. A. Larkin. 1989. The decline and fall of the Pribilof fur seal (*Callorhinus ursinus*): A simulation study. Canadian Journal of Fisheries and Aquatic Sciences 46:1437–1445.

Trites, A. W., and A. E. York. 1993. Unexpected changes in reproductive rates and mean age at first birth during the decline of the Pribilof northern fur seal (*Callorhinus ursinus*). Canadian Journal of Fisheries and Aquatic Sciences 50:858–864.

Twiss, J. R., and R. R. Reeves. 1999. Conservation and Biology of Marine Mammals. Smithsonian Institution Press, Washington, D.C.

VanBlaircom, G. R., and J. A. Estes. 1988. The community ecology of sea otters. Springer-Verlag, Berlin.

Vaz-Ferreira, R. 1979. South American fur seal. Pages 34–36 *in* Mammals in the Seas. FAO, Fisheries Series, Rome.

Vaz-Ferreira, R. 1981. South American sea lion, *Otaria Xavescens* (Shaw 1800). Pages 39–65 *in* S. H. Ridgway and R. J. Harrison (eds.). Handbook of Marine Mammals. Academic Press, London.

Vibe, C. 1950. The marine mammals and the marine fauna in the Thule district (northwest Greenland) with observations on ice conditions in 1939–41. Meddelelser om Groenland 150:1–115.

Vibe, C. 1967. Arctic animals in relation to climate fluctuations. Meddelelser Om Groenland 170 Nr. 5. Copenhagen.

Volkov, A. F., and I. F. Moroz. 1977. Oceanological conditions of the distribution of cetacea in the eastern tropical part of the Pacific Ocean. International Whaling Commission Report 27:186–188.

Wada, E., M. Terazaki, Y. Kabaya, and T. Nemoto. 1987. 15N and 13C abundances in the Antarctic Ocean with emphasis on the biogeochemical structure of the food web. Deep Sea Research 34:829–841.

Walker, G. E., and J. K. Ling. 1981. New Zealand sea lion *Phocarctos hookeri* (Gray 1844). Pages 25–38 *in* S. H. Ridgway and R. J. Harison (eds.). Handbook of Marine Mammals. Academic Press, London.

Walker, W. A. 1996. Summer feeding habits of Dall's porpoise, *Phocoenoides dalli*, in the southern Sea of Okhotsk. Marine Mammal Science 12:167–181.

Wallace, S. D., and D. M. Lavigne. 1992. A review of stomach contents of harp seals (*Phoca groenlandica*) from the Northwest Atlantic. Technical Report No. 92-03. International Marine Mammal Association Inc, Guelph, Ontario.

Ward, A. J., D. Thompson, and A. R. Hiby. 1987. Census techniques for gray seal populations. Symposia of the Zoological Society (London) 58:225–245.

Waring, G. T., P. Gerrior, P. M. Payne, B. L. Parry, and J. R. Nicolas. 1990. Incidental take of marine mammals in foreign fishing activities off the Northeast United States. Fishery Bulletin 88:347–360.

Warneke, R. M., and P. D. Shaughnessy. 1985. *Arctocephalus pusillus*. The South African and Australian fur seal: Taxonomy, evolution, biogeography, and life history. Pages 53–77 *in* J. K. Ling and M. M. Bryden (eds.). Studies of Sea Mammals in Southern Latitudes. South Australian Museum, Adelaide.

Watkins, W. A., and W. E. Schevill. 1979. Aerial observations of feeding behavior in four baleen whales: *Eubalaena glacialis, Balaenoptera borealis, Megaptera novaeangliae, Balaenoptera physalus*. Journal of Mammalogy 60:155–163.

Watkins, W. A., M. A. Daher, K. M. Fristrup, and T. J. Howald. 1993. Sperm whales tagged with transponders and tracked underwater by sonar. Marine Mammal Science 9:55–67.

Watts, P. 1992. Thermal constraints on hauling out by harbour seals (*Phoca vitulina*). Canadian Journal of Zoology 70:553–560.

Watts, P. 1996. The diel hauling-out cycle of harbour seals in an open marine environment: correlates and constraints. Journal of Zoology (London) 240:175–200.

Weilgart, L., H. Whitehead, and K. Payne. 1996. A colossal convergence. American Scientist 84:278–287.

Weitkamp, L. A., R. C. Wissmar, and C. A. Simenstad. 1992. Gray whale foraging on ghost shrimp (*Callianassa californiensis*) in littoral sand flats of Puget Sound, U.S.A. Canadian Journal of Zoology 70:2275–2280.

Wells, R. S., A. B. Irvine, and M. D. Scott. 1980. The social ecology of inshore odontocetes. Pages 263–317 *in* L. M. Herman (ed.). Cetacean Behavior: Mechanisms and Functions. Wiley & Sons, New York, NY.

Whitehead, H., and J. E. Carscadden. 1985. Predicting inshore whale abundance—Whales and capelin off the Newfoundland coast. Canadian Journal of Fisheries and Aquatic Sciences 42:976–981.

Whitehead, H., S. Waters, and T. Lyrholm. 1991. Social organization in female sperm whales and their offspring: Constant companions and causal acquaintances. Behavioral Ecology and Sociobiology 29:385–389.

Wiig, O., E. Born, and E. W. Garner. 1995. Polar bears: Proceedings of the eleventh working meeting of the IUCN / SSC polar bear specialist group. IUCN, Cambridge, U.K.

Wilke, F., and K. Kenyon. 1954. Migration and food of the northern fur seal. Transactions of the 19th North American Wildlife Conference 19:430–440.

Würsig, B., and C. Clark. 1993. Behavior. Pages 157–199 *in* J. Burns, J. J. Montague, and C. J. Cowles (eds.). The Bowhead Whale. Special Publication No. 2. Society of Marine Mammalogy, Lawrence, KS.

Yablakov, A. V., and D. E. Sergeant. 1963. Cranial variation in the harp seal [Original in Russian]. Zoologicheskii Zhurnal 42:1857–1865 (Fisheries Research Board Canadian Translation Service. No. 485).

York, A. E. 1990. Trends in numbers of pups born on St. Paul and St. George Islands 1973–88. Pages 31–37 *in* Kajimura H (ed.). Fur Seal Investigations, 1987 and 1988. NOAA Technical Memorandum NMFS F / NWC-180 U.S. Department of Commerce, Washington, D.C.

York, A. E. 1994. The population dynamics of Northern sea lions, 1975–1985. Marine Mammal Science 10:38–51.

York, A. E., and P. Kozloff. 1987. On the estimation of numbers of northern fur seal, *Callorhinus ursinus*, pups on St. Paul Island. Fishery Bulletin 85:367–375.

Young, D. D., and V. G. Cockcroft. 1994. Diet of common dolphins (*Delphinus delphis*) off the south-east coast of southern Africa: Opportunism or specialization? Journal of Zoology (London) 234:41–53.

Zeh, J. E., J. C. George, A. E. Raftery, and G. M. Carroll. 1991. Rate of increase, 1978–1988, of bowhead whales, *Balaena mysticetus*, estimated from ice-based census data. Marine Mammal Science 7:105–122.

10

THOMAS J. O'SHEA

Environmental Contaminants and Marine Mammals

Environmental contaminants have gained increasing attention as potential threats to marine mammals. In this chapter I present an overview of the current state of knowledge on marine mammals and the two classes of environmental contaminants that have received the most attention: organochlorine compounds and toxic elements. A brief introduction to other classes of contaminants and toxins is included. The scope of the chapter does not include marine mammals and oil, which has been treated in depth in other sources (Geraci and St. Aubin 1990, Loughlin 1994). The approach is to provide a general introduction and overview with examples of case studies. However, appendices with detailed tabulations of specific information are also provided as a key to the primary literature. This will allow the interested reader to pursue these topics in greater depth. General reference works should also be consulted (e.g., Peterle 1991, Hoffman et al. 1994, Klaassen 1996). Although numerous investigations have been conducted in this field, the state of knowledge about effects of many contaminants on marine mammals remains very incomplete. In addition, results of studies have occasionally been interpreted to be of significance in ways that are beyond the bounds that can actually be supported by existing data. Nevertheless, prudent interpretations of some of the findings in this field have implications that are profound and call for continued research into the future.

Organochlorine Compounds

One of humankind's major technological advances has been the capability to synthesize and manufacture chemicals on a massive scale for applications in industry, agriculture, and health. Remarkable changes in society have been enabled by these developments, but, like many technological advances, seemingly unpredictable environmental consequences have also occurred. Pollution of the planet by highly persistent organochlorine compounds is a classic example.

The organochlorines are a diverse group of industrial and agricultural compounds synthesized for various properties, including chemical stability. Many of the organochlorines are highly fat soluble (lipophilic) but have low water solubility (hydrophobic), and differentially accumulate in lipids of animals. Millions of tons of these chemicals have been produced and released into the environment, mostly during the latter half of this century (Hoffman et al. 1994). Designed for chemical stability, some of the organochlorines are extremely persistent in the environment and resistant to metabolic degradation, thereby increasing in concentrations through food webs. Because the ultimate sinks for many of these persistent compounds are the oceans of the world, where organochlorines are rapidly adsorbed to organic matter and taken up by plankton, marine mammals have been an end point in the food web accumulation of these compounds. Numerous studies have documented the presence of organochlorines in marine mammal tissues (Appendices 1 to 4). References cited in the appendices to this chapter indicate that organochlorines have been reported in tissues of at least 23 species of pinnipeds, 44 species of odontocetes, and 11 species of baleen whales, as well as sea otters, sirenians, and polar bears. Analyses have been conducted on samples from more than 7000 individuals (Appendices 1 to 4). Many of these analyses have focused on the organochlorines, although other halogenated compounds such as the polybrominated biphenyls have occasionally also been reported. Many of the organochlorines reported in marine mammal tissues were originally introduced to the environment as pesticides targeting the nervous systems of insects. However, the reader should bear in mind that many insecticides now in use, particularly in developed nations, are not organochlorines but organophosphates, carbamates, and other forms that are not persistent in tissues and do not appreciably accumulate in marine food chains. Thus, the common use of the generic term pesticides for organochlorine insecticide residues is misleading. Similarly, the term pesticide also encompasses the numerous synthetic herbicides (some of which include halogens in their structure), which are designed to affect plant physiological systems and have never been determined to be a serious contaminant issue for marine mammals.

Expression of Results of Organochlorine Residue Analyses

The reporting of concentrations of organochlorines in tissues can be bewildering to those people who are unfamiliar with the field. Most studies report contaminant residues on the basis of mass of chemical per unit mass of tissue. The latter, however, may be expressed on the basis of fresh weight (or "wet weight") of the tissue sample, on the basis of weight of the sample with water removed ("dry weight"), or on the basis of the extractable lipid components ("lipid weight"), which for all practical purposes will contain all of the organochlorine contaminants. The most typical expression of concentrations in the literature are given as parts per million (ppm) wet weight, which on an unit of mass basis may also be expressed as μg/g, or mg/kg. Lower concentrations may be expressed as parts per billion (ppb) or by the units ng/g or μg/kg. It is important to be certain of the units in comparing findings among various studies. Publications that most thoroughly document results give values of concentrations of contaminants, but will further provide percent lipid (which can be quite variable even in blubber) and percent water, thereby permitting the reader to recalculate values for comparison with other findings. Many factors can cause variation in organochlorine concentrations in marine mammals and these are discussed in other sections of this chapter. A few studies attempt to estimate total amounts of organochlorines in bodies and organ systems of marine mammals, expressing amounts in units of mass and total body burdens. Patterns of variation in total body burdens of organochlorines can differ from patterns in concentrations in blubber.

Concentrations of contaminants in tissues are also described using different summary statistics. Often results are expressed as means, ranges, and standard deviations or standard errors. However, in many samples the actual distribution of contaminant concentration data points is non-normal, and what is often found is a skewed array with many relatively low values and a few very high values. Arithmetic means or parametric statistical tests of hypotheses on these data can be inappropriate and misleading. Instead investigators often transform the data to a logarithmic scale, and compute a geometric mean and confidence intervals to help adjust for the non-normality of distributions. As with units of expression, it is important to be certain of the statistical basis of the summary data provided when interpreting results of various studies. Thorough presentations of results often include tables with original data in addition to summary statistics. This allows readers to make detailed comparisons and to perform their own statistical tests of hypotheses.

Major Compounds

The following section provides an overview of the characteristics of the various organochlorines frequently reported from marine mammals (Appendices 1 to 4). In subsequent sections I explore patterns of variation in organochlorine concentrations in marine mammal tissues, review evidence for impacts of these contaminants on marine mammal health and population dynamics, and summarize recent studies of organochlorine metabolism and biochemical toxicity in marine mammals. Over the years the analytical methodology used by chemistry laboratories to quantify organochlorine residues in tissues has become increasingly sophisticated. Several methods have been used, but modern, thorough studies typically use high-resolution capillary gas chromatography with electron capture detection, combined with confirmation by mass spectrometry, subsequent to use of standardized extraction and clean-up procedures using highest purity grade solvents. High-performance liquid chromatography is sometimes used in the process of isolating certain individual compounds (for example, see Varanasi et al. 1992, Wells and Echarri 1992). Quality assurance procedures also must be followed and well documented to ensure scientific credibility. These can entail determination and reporting of error in calibrated or "spiked" samples, multiple analyses of the same sample within the same laboratory, interlaboratory comparison studies, and use of standard reference materials. Systematic procedures for collecting tissues to ensure avoidance of spurious contamination must also be followed (see Geraci and Lounsbury 1993). Steps toward developing standardized protocols for analysis of marine mammal tissues have been taken (Lillestolen et al. 1993). However, an internationally standardized set of analytical methods and quality assurance procedures has not yet been developed and formally agreed on for determination of contaminants in marine mammals, and therefore methodology can vary among studies.

DDT and Metabolites

DDT (2,2-*bis*-(*p*-chlorophenyl)-1,1,1-trichloroethane or dichlorodiphenyltrichloroethane) and its metabolites (Fig. 10-1) are the best known and most widely reported organochlorines found in marine mammal tissues and as such, deserve treatment from a historic perspective. They are arguably one of the most notorious groups of compounds to be recognized as environmental contaminants, and indeed provided significant impetus to the tremendous surge in societal awareness of the potential dangers of persistent environmental contaminants. These compounds were at the forefront in sparking growth in both the science of wildlife toxicology and government regulation of contaminants (the latter, however, was not without much debate). DDT and metabolites were first reported in marine mammal tissues in the 1960s, at about the time it became evident from field studies that they were associated with eggshell thinning and declines in populations of some species of birds (Ratcliffe 1967, Hickey and Anderson 1968, Stickel 1973). DDT was also implicated in direct mortality of some species of wildlife. Contamination of the global marine ecosystem by organochlorines was exemplified by the discovery of residues of DDT and metabolites in Antarctic seals, far from significant sources of direct exposure, reported in the journal *Nature* more than three decades ago by investigators at the Patuxent Wildlife Research Center and Johns Hopkins University (Sladen et al. 1966). Much of the controversy surrounding the regulation of DDT, however, was on the possible human health impact, primarily carcinogenic potential, rather than impacts on wildlife or possible impairment of reproduction. The evidence for its carcinogenicity continues to cause debate and, although use of DDT has been banned in some nations (particularly the developed countries of the northern hemisphere), the chemical is still used in many parts of the world. Transport from these areas to some of the more "pristine" reaches of the globe has been well documented (for example, see Simonich and Hites 1995).

Today DDT has notoriety among marine mammalogists chiefly as a contaminant and environmental "villain." However, as pointed out by Metcalf (1973), this chemical ". . . has had an influence on human ecology perhaps unmatched by any other synthetic substance. Through its effectiveness in the conquest of malaria, typhus, and other insect-borne diseases it has played a decisive role in the population explosion.

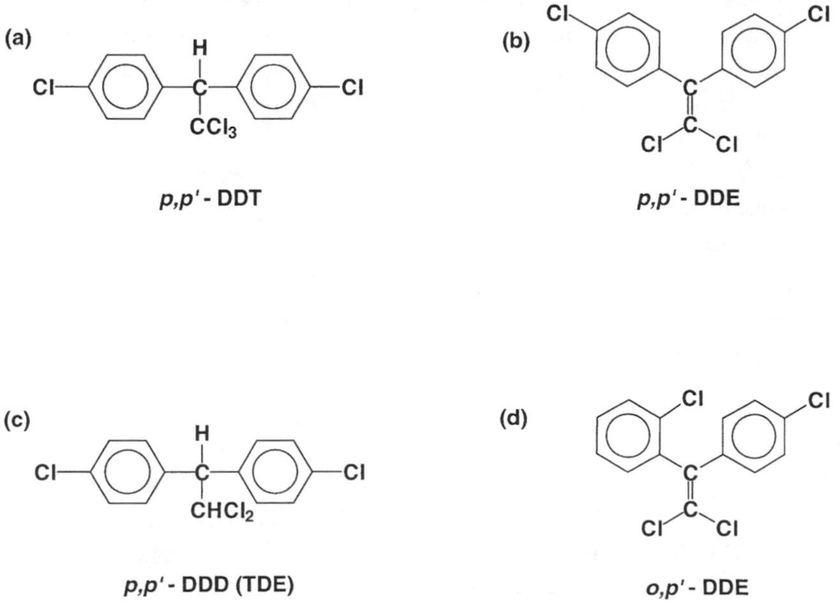

(a) *p,p′*-DDT (b) *p,p′*-DDE (c) *p,p′*-DDD (TDE) (d) *o,p′*-DDE

Figure 10-1. Chemical structures of (a) *p,p′*-DDT and the principal metabolites of this compound found in marine mammal tissues, (b) *p,p′*-DDE, and (c) *p,p′*-DDD or TDE. *p,p′*-DDE is usually the most widespread and abundant metabolite found in marine mammal blubber. The isomer *o,p′*-DDE (d) is reported less often but has greater estrogenic activity.

It has also become the classic example of an environmental micropollutant." DDT (and, to a lesser extent, its metabolites) is primarily neurotoxic but can also cause pathological changes to the liver and reproductive system in laboratory mammals.

The synthesis of DDT was first reported in 1874 by the German chemist Othmar Zeidler. However, the insecticidal properties of the chemical were not discovered until it was resynthesized in 1939 by Paul Müller. The magnitude of the technological impact of this discovery on humankind can be gauged by the awarding of a Nobel Prize to Müller in 1948 for "discovery of the strong action of DDT against a wide variety of arthropods" (Metcalf 1973). In India alone, the World Health Organization credited DDT spraying with a reduction in malaria from 100 million cases in 1 year in the 1930s to 150,000 per year by 1966 (Metcalf 1973). DDT was sprayed not only for control of disease vectors, but against crop and garden insects, and over entire landscapes of forested ecosystems in efforts aimed at eradication of timber pests. During World War II, soldiers were dusted with DDT to control body lice (which as vectors of typhus played a larger role than military tactics in turning the tide of many previous wars). The compound was also sprayed to protect troops from malaria and other diseases. As pointed out by Metcalf (1973), one can only speculate how history might have been changed had the insecticidal properties of DDT been recognized at the time of its discovery by Zeidler, rather than after it had rested on the shelf for some 65 years.

Unfortunately, history has a pattern whereby strides forward in technology are often accompanied by large scale setbacks to the natural functioning of ecosystems (Ehrenfeld 1981). Among such unforeseen ecosystem ramifications is the widespread occurrence of DDT and its metabolites in tissues of biota throughout the world, including marine mammals. Extremely high concentrations of these compounds, higher than found in tissues of most terrestrial mammals, have been reported in odontocete cetaceans and pinnipeds in some parts of the world. The presence of metabolites of DDT in marine mammals throughout the globe is fact. The significance of these metabolites for marine mammal populations is subject to difficulty in interpretation, as explored in further sections.

Most analyses of tissues of marine mammals for DDT and other organochlorines have focused on the blubber, which because of its high lipid content is the major storage compartment for these contaminants. The breakdown product DDE (2,2-*bis*-(*p*-chlorophenyl)-1,1-dichloroethylene or **d**ichloro**d**iphenyldichloro**e**thylene) is the most abundant metabolite found in blubber. DDE concentrations are usually far higher than of DDT or of TDE (2,2-*bis*-(*p*-chlorophenyl)-1,1-dichloroethane or **t**etrachloro**d**iphenyl**e**thane), also referred to as DDD (**d**ichloro**d**iphenyl**d**ichloroethane) (Fig. 10-1). DDE is much less toxic (in terms of lethality to laboratory animals) than DDT. A small portion of the technical grade mixtures of DDT also contain the *o,p'* isomers (Fig. 10-1), which are not frequently reported in environmental samples. The *o,p'* isomers have also been reported at lower concentrations in some studies of marine mammal tissues. Some studies have also reported additional metabolites of DDT, including methyl sulfone compounds (Appendices 1 to 4). It is common for publications on concentrations of DDT and metabolites in tissues to provide an arithmetic summation of these measurements in each tissue sample. This sum of the individual components is typically referred to as "total DDT" and denoted by notations such as ΣDDT, DDTR, or DDTs. Some of the earlier studies (generally before the 1970s) on concentrations of DDT and metabolites in marine mammals were carried out before the recognition of polychlorinated biphenyls (PCBs) as environmental contaminants (and therefore their separation during chemical analysis), making accuracy in quantification of values reported less reliable.

The dynamics of storage and metabolism of DDE and related compounds in marine mammal tissues are complex, and although studied in some detail as described in further sections, knowledge of these dynamics and their possible ramifications to health and population stability of marine mammals remains incomplete. These compounds can be expected to be found at some level in most biota on Earth, and certainly in every living marine mammal. However, great variation exists in concentrations of these and other organochlorines in tissues of marine mammals, making it difficult to describe "typical" or background levels of contamination. As a rough reference point, during the 1960s and early 1970s mean concentrations of ΣDDT in human adipose tissue in the United States ranged from 5 to 10 ppm, and concentrations in human adipose tissues in other nations ranged from 2 ppm (Australia, Netherlands) to 28 ppm (India) (presumed lipid weight basis; Matsumura 1985; see also Jensen 1983). More current values for ΣDDT in human adipose tissue obtained in Mexico during the late 1980s to early 1990s range from 1.0 to 90.0 ppm (lipid basis) in fat (Waliszewski et al. 1996), and recent determinations of average ΣDDT in mother's milk in African countries range from 3.0 to 20 ppm (lipid basis) (Ejobi et al. 1996). Extreme cases of ΣDDT contamination of marine mammals have resulted in concentrations of 1000 to 2000 ppm or more in blubber. However, typical concentrations range much less than 100 ppm, with many samples at 10 ppm or less, particularly in the baleen whales or other species from the open oceans or high latitudes. Residue concentrations of total DDT in milk of marine mammals have been rel domly reported (Appendices 1 to 4), but are typically less than or equivalent to those found in some contemporary African mothers.

The Cyclodienes

Aldrin, dieldrin (Fig. 10-2), and endrin are cyclodienes synthesized by a chemical process referred to as the Diels-Alder reaction (hence the etymology). They are insecticides that are generally much more acutely toxic than DDT; they also were used in large quantities before restrictions in some countries during the 1970s. The cyclodienes are highly neurotoxic, with likely different mechanisms of action than DDT at the cellular and biochemical levels, and can cause reproductive defects in laboratory mammals at high dosage levels. Dieldrin is an insecticide in its own right but is also a metabolite of aldrin, which breaks down in the environment much more quickly than dieldrin. Dieldrin is commonly found in blubber of marine mammals, whereas the less persistent aldrin and more toxic endrin are seldom reported (Appendices 1 to 4). Concentrations of these chemicals in marine mammal tissues are generally lower than those of DDT and its metabolites or the PCBs. However, dieldrin in particular has been found in marine mammals throughout the world. According to Matsumura (1985:58) dieldrin "is one of the most persistent chemicals ever known." Some publications refer to dieldrin by the alternate chemical abbreviation of HEOD.

The technical grade of the cyclodiene insecticide chlordane is a mixture of compounds and isomers, about 60% of which is chlordane. Chlordane occurs as a mixture of *cis*- and *trans*-isomers (Fig. 10-2). Other persistent components of the mixture include heptachlor, nonachlor, and oxychlordane. Heptachlor is also prepared from chlordane as an insecticide in its own right and generally is of higher acute toxicity. The more toxic heptachlor epoxide (Fig. 10-2) is the principal metabolite of heptachlor found in marine mammal tissues; isomers of chlordane, oxychlordane, and nonachlors have also been reported (Appendices 1 to 4). These compounds have been found in marine mammals throughout the global marine ecosystem, and include complex mixtures and metabolites from seemingly pristine reaches such as the Arctic and Antarctic (Norstrom and Muir 1994), but at concentrations that can generally be considered low.

Toxaphene

The literature on organochlorines in marine mammal tissues occasionally reports the presence of toxaphene (Appendices 1 to 4). Technical grade toxaphene released to the environment is not a compound but a mixture of compounds and isomers used as an agricultural pesticide, particularly on

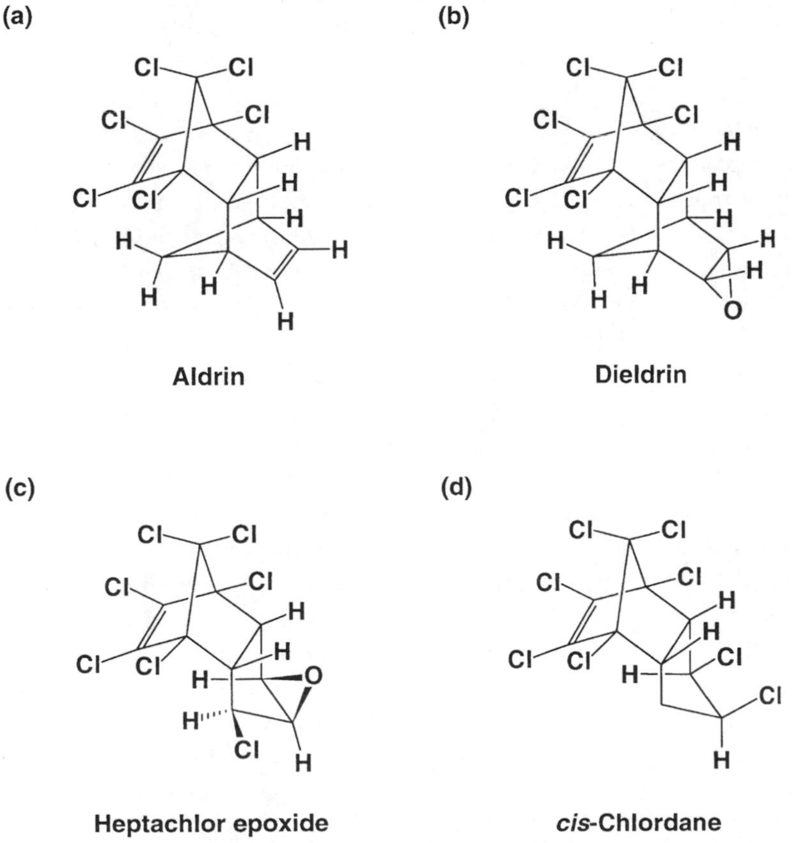

(a) Aldrin

(b) Dieldrin

(c) Heptachlor epoxide

(d) *cis*-Chlordane

Figure 10-2. Cage diagrams of chemical structures of representative cyclodiene compounds. The cyclodienes are organochlorine insecticides or metabolites.

cotton. It consists of many chlorinated terpenes (also referred to as chlorinated camphenes, polychlorocamphenes [PCCs], and as chlorinated norbornane derivatives). The theoretical number of isomers of toxaphene compounds and metabolites is in the thousands, with estimates based on empirical analytical chemistry numbering in the hundreds. Some congeners can be masked by PCBs during chemical analysis.

Toxaphene was widely used before restrictions imposed in some nations, including a ban on insecticidal use in the United States in 1982. Restrictions were based primarily on suspected carcinogenicity and mutagenicity. Toxaphene is intermediate in acute toxicity between dieldrin and DDT. Individual components vary widely in toxicity. It was the most widely used insecticide in the United States before restrictions and continues to be used in many nations. Like other organochlorines, it can be spread through the environment by atmospheric transport and is found in low concentrations in tissues of marine mammals in remote areas such as the Arctic (Zhu and Norstrom 1993). Cumulative world use during the period 1946 to 1974 was more than 409,000 metric tons (mt). Saleh (1991) provides an excellent literature review on these and other aspects of the chemistry, toxicology, and environmental kinetics of toxaphene.

Polychlorinated Biphenyls

The polychlorinated biphenyls (PCBs, also referred to as chlorobiphenyls or CBs) are a mixture of compounds (Figs. 10-3 and 10-4). They are produced by the chlorination of biphenyls and theoretically can include a mixture of 209 different isomers and congeners, although a smaller number have actually been reported as contaminants in biological samples. They have had a wide variety of industrial applications, including use as dielectric fluids in electrical transformers; in heat transfer systems, capacitors, and plastics; as inks, additives, and lubricants; in carbonless copy paper; and in hydraulics. Publications on PCBs in marine mammals sometimes refer to them by trade names such as the Aroclors (produced in the United States) in which the last two digits refer to the weight percentage chlorine of the mixture, e.g., Aroclor 1254 is 54% chlorine by weight; other trade names of mixtures produced in different countries and sometimes referred to in the marine mammal literature include the Kanechlors (Japan), Phenoclors (France), and Clophens (Germany). These commercial mixtures contained large numbers of individual PCB congeners (e.g., Aroclor 1254 typically contains some 50 to 70 PCB compounds). The total amount of PCBs produced since 1929 has been estimated at 1.5 million mt (de Voogt and Brinkman 1989). Individual PCB congeners may also be synthesized for specialized uses. PCBs were first discovered as environmental contaminants

(a)

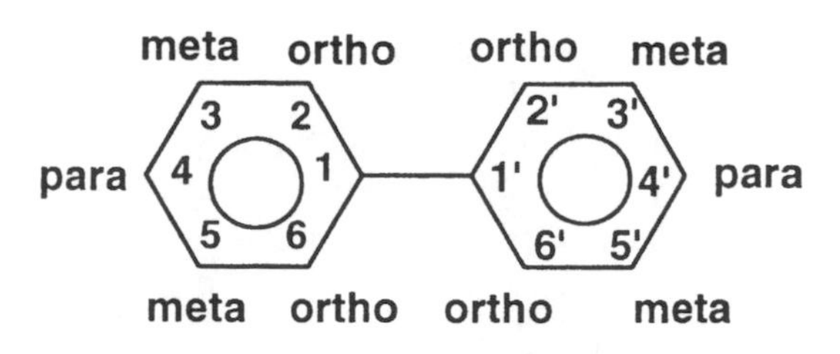

(b)

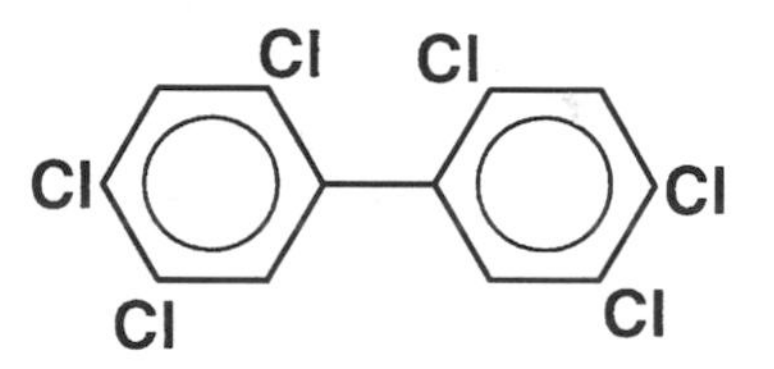

Figure 10-3. (a) Generalized structural formula for polychlorinated biphenyls (PCBs) with numbering of the carbon atoms in rings and potential positions of chlorine and vicinal H atoms. An example (b) is the metabolically resistant PCB 153 (2,2′,4,4′,5,5′-hexachlorobiphenyl).

in the late 1960s and subsequently were determined to be widespread throughout the biota of the globe.

During the 1970s and 1980s the manufacturing of PCBs was terminated in most industrialized nations with the recognition of their widespread distribution as ecological contaminants and the potential for detrimental health effects. This recognition was heightened by mass human exposure due to accidental mixing of PCBs with rice oils ("Yusho disease") in Japan in 1968 and Taiwan in 1979, resulting in a number of related ailments (although subsequent research suggests some of the resulting effects may have been attributable to unusually high concentrations of polychlorinated dibenzodioxins [PCDDs] and polychlorinated dibenzofurans [PCDFs] (Fig. 10-5) in the mixtures as well). Despite the marked cessation in their production and sale, most PCBs are still contained in systems for which they were originally designed (such as transformers and other machinery) and have not yet reached the environment. As these systems leak, degrade, and are disposed of,

(a) Non-ortho

PCB 77 PCB 126 PCB 169

(b) Mono-ortho

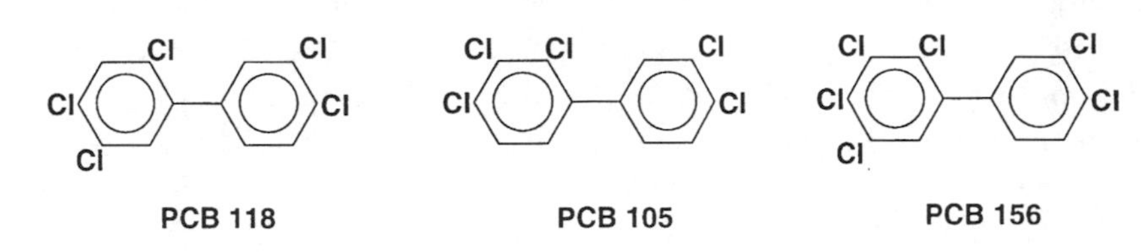

(c) Di-ortho

PCB 128 PCB 138

Figure 10-4. Examples of coplanar PCBs with chemical names and IUPAC numbers. Non-ortho: 3,3′,4,4′-tetrachlorobiphenyl (PCB 77); 3,3′,4,4′,5-pentachlorobiphenyl (PCB 126); 3,3′,4,4′,5,5′-hexachlorobiphenyl (PCB 169). Mono-ortho: 2,3′,4,4′,5-pentachlorobiphenyl (PCB 118); 2,3,3′,4,4′-pentachlorobiphenyl (PCB 105); 2,3,3′,4,4′,5-hexachlorobiphenyl (PCB 156). Di-ortho: 2,2′,3,3′,4,4′-hexachlorobiphenyl (PCB 128); 2,2′,3,4,4′,5′-hexachlorobiphenyl (PCB 138).

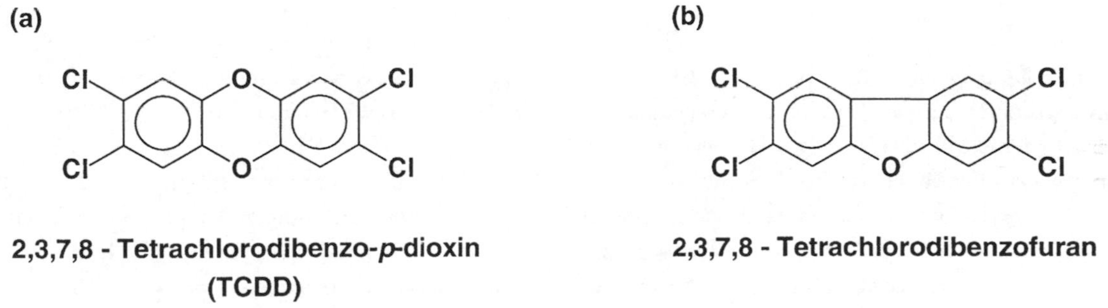

(a) 2,3,7,8 - Tetrachlorodibenzo-*p*-dioxin (TCDD)

(b) 2,3,7,8 - Tetrachlorodibenzofuran

Figure 10-5. Examples of polychlorinated dibenzodioxins (PCDDs) and polychlorinated dibenzofurans (PCDFs). The dioxin (2,3,7,8-tetrachlorodibenzo-*p*-dioxin or TCDD) is a potent enzyme inducer and one of the most toxic compounds known. Although not generally reported at elevated concentrations in marine mammals, it is the basis for calculation of toxic equivalencies. Some PCBs are considered to be approximate isostereomers of TCDD.

the quantities released to the environment will increase. The ultimate repository for these compounds will be the oceans. PCB concentrations in the environment, and in the tissues of marine mammals in particular, are projected to increase for many years to come (Tanabe 1988). Like other organochlorines, PCBs are lipophilic and therefore, are readily sequestered in marine food webs. However, there is wide variance among individual congeners in their toxicity and persistence in various metabolic systems, as detailed elsewhere in this chapter in discussions of PCB metabolism.

The analytical methodology used to quantify the presence of PCBs in marine mammal tissues has matured during the past two decades. Research into the presence of organochlorines in tissues before the 1970s generally did not include determination of PCBs; this caused some inaccuracies in the expression of amounts of other compounds. Subsequently, the practice was to compare the amounts of PCBs present in the sample with a standard mixture such as Aroclor 1254 or 1260. However, choice of standard and analytical methodology affects the estimated concentrations, with results from the same individuals showing concentration estimates with three- to fourfold differences using different procedures and Aroclor standards. Modern analytical procedures now can provide concentrations of individual congeners, which is of great importance because of wide differences among congeners in relative toxicity, and total PCBs can be estimated as the sum of congener concentrations. Because of the wide number of individual congeners, a systematic numeric system of nomenclature for individual PCBs has been adopted in addition to standard names based on structural formulae (Figs. 10-3 and 10-4). These are referred to in the literature as IUPAC, Ballschmiter, or PCB numbers (for example, 2,3,3′,4,4′,5-hexachlorobiphenyl (Fig. 10-4b) is simply referred to as PCB 156) and are listed in detail by Ballschmiter and Zell (1980) and Ballschmiter et al. (1989). For point of reference, PCBs generally range from 1 to 10 ppm (lipid basis) in human adipose tissue and 1 to 2 ppm (lipid basis) in human milk, typically quantified in comparison with an Aroclor 1254 or 1260 standard (Ballschmiter et al. 1989). Additional discussion on the patterns of occurrence and potential effects of PCBs in marine mammals are treated elsewhere in this chapter.

Other Organochlorines

PCDFs, PCDDs, polychloroquaterphenyls (PCQs), and polychlorinated napthalenes (PCNs) have been detected in commercial PCB mixtures where they have been formed in low quantities as by-products of the PCB manufacturing process. PCDDs and PCDFs are also formed as impurities in a variety of other industrial and combustion processes (Safe 1986, 1991). The PCDD compound 2,3,7,8-tetrachlorodibenzo-*p*-dioxin (TCDD; Fig. 10-5) is among the most toxic compounds known and is the most intensively studied. It was an impurity present in the defoliant Agent Orange used during the Vietnam War. A few reports indicate the presence of these compounds in marine mammal tissues (Appendices 1 to 4), although with few exceptions (Jarman et al. 1996) investigations have not been extensive.

Isomers of hexachlorocyclohexane (HCH), sometimes also referred to as benzene hexachloride (BHC), often appear in low concentrations in blubber of marine mammals (Appendices 1 to 4). It is an insecticide, the pure γ isomer of which is also named lindane. Lindane is neurotoxic, and some isomers of HCH have caused tumors in liver cells of laboratory mammals. Ratios and metabolism of various isomers of HCH were quantified and discussed for 10 odontocete species by Tanabe et al. (1996). Other organochlorine insecticide residues reported infrequently in marine mammals include the structurally related compounds kepone and mirex, and endosulfan (which is also used as an ascaricide). Hexachlorobenzene (HCB) is a fungicide that appears as a widespread environmental contaminant and has been detected in a variety of marine mammals (Appendices 1 to 4). Some related compounds with different numbers of chlorines are sometimes reported as total chlorobenzenes (CBz). The methyl sulfone metabolites of DDE and PCBs, which are toxic themselves, have also been quantified in marine mammal tissues (Appendices 1 to 4; Bergman et al. 1994). One recent study has also revealed the presence of hexachlorinated anthracenes in seals from the Baltic Sea (Koistinen 1990). In addition to the organochlorines, other organohalogens reported in marine mammal tissues include the polybrominated biphenyls (PBBs), manufactured as flame retardants, octachlorostyrene (OCS), and polybrominated diphenyl ethers (Kuehl et al. 1991, Kuehl and Haebler 1995). The substance *tris* (4-chlorophenyl) methanol (TCP) has been found in several species of pinnipeds and cetaceans, and may stem from synthetic dyes or DDT (Jarman et al. 1992, de Boer et al. 1996, Muir et al. 1996a).

Distribution and Kinetics of Organochlorines in Body Compartments of Marine Mammals

In marine mammals, blubber is the major repository for organochlorines, and most of the emphasis on determination of residue concentrations has focused on this tissue (Appendices 1 to 4). Tanabe et al. (1981) demonstrated that in an adult male striped dolphin (*Stenella coeruleoalba*) most of the body weight is contributed by muscle (56%), followed by blubber (17%) and bone (12%). Striped dolphin blubber, however, contained 95% of the total body burden of $\sum$DDT and PCBs and more than 90% of the body burden of HCHs and HCB. This is because of the higher lipid composition of blubber (about 75% of blubber weight was lipid and 91% of all lipids in the body were found in blubber, whereas muscle tissue is comprised of 1% to 4% lipid). The total amounts of organochlorines found in bone and other organs were negligible, leading these researchers to suggest that combined amounts in blubber and muscle (with its larger total mass) could be used to represent the total amounts of these contaminants in individual animals. Most organs had similar proportions of various organochlorines, except the brain and liver. Lower relative concentrations of DDT were found in the liver (implying active metabolism). The brain had lower concentrations of organochlorines on a lipid weight basis, as well as apparent differences in amounts of lower chlorinated PCBs and isomers of HCH. Lower relative con centrations of organochlorines in brains have also been reported for sperm whales (*Physeter macrocephalus*) (Aguilar 1983), harp seals (*Phoca groenlandica*) (Frank et al. 1973), small cetaceans (O'Shea et al. 1980), and a variety of other species (Aguilar 1985; see Appendices 1 to 4 for additional references).

Very few studies have attempted to estimate the absolute body load of organochlorines in marine mammals other than in the striped dolphin. One exception, however, is the detailed study of fin whales (*Balaenoptera physalus*) conducted by Aguilar and Borrell (1994b), who calculated tissue masses for five complete whales by dissection and determined organochlorine concentrations in samples of blubber, muscle, bone, liver, and kidney for 26 individuals, extrapolating results to a total sample of 169 whales. (For extrapolations, they estimated total body loads based on organochlorine concentrations in blubber samples and empirical relationships established in the smaller groups. Anatomical studies of reproductive tracts provided details on sexual maturity, and growth layer counts in ear plugs established age estimates.) Using this information, they were able to estimate quantities of organochlorines transferred to young by females during reproduction by differences in estimated total body loads of males and females of the same age. Body loads averaged 6.5 g for $\sum$DDT and 7.8 g for PCBs in the sample of 26, with estimates as high as 23.5 g $\sum$DDT and 20.1 g PCBs in males. These amounts are one to two orders of magnitude higher than estimates for some small cetaceans, which is attributable to the huge differences in body size and amounts of lipid.

The general framework for understanding the kinetics of organochlorines in marine mammals is similar to that established for other homeotherms, in which the blood is considered a central "compartment" through which various organs maintain a dynamic relationship with contaminant exposure (Aguilar 1985). As noted for striped dolphins, the organochlorine residues found in other organs are generally distributed according to their fat content, with concentrations expressed on a lipid weight basis being highly similar among all organs. The brain tends to have lower concentrations of organochlorines than would be predicted on the basis of its fat content. This is not thought to be attributable to a blood–brain barrier, which is unlikely to be effective for liposoluble organochlorines, but rather to the kinds of lipids present in the brain. The brain has a higher amount of phospholipid in comparison with other organs, and organochlorines tend to have a lower affinity for phospholipids than for other lipids. In general, organs with higher amounts of the less polar triglycerides and nonesterified fatty acids have higher organochlorine concentrations (Aguilar 1985). Characteristics of compounds present will vary among organs as well, with highly polar forms occurring at higher concentrations in phospholipids, and greater proportions of degradation products present in the liver because of its active metabolic function.

Although most of the organochlorines in the bodies of marine mammals can be found in blubber, the dynamics of this relationship are complex. Blubber is not metabolically inert. Many marine mammal species undergo extensive annual and seasonal changes in the amounts of fat stored in blubber. These changes correlate with cycles in breeding and lactation, and migrations off feeding grounds. Residue concentrations in blubber can be diluted with rapid expansion of the lipid component during seasonal fattening periods or growth. In contrast, marine mammals sampled as stranded carcasses may have depleted lipid reserves from disease or starvation, with consequent elevations in organochlorine residue concentrations in blubber. Although in marine mammals the rates at which organochlorines are either passed into the blood with lipid mobilization or are concentrated in the remaining fat are poorly known, it is likely that both processes take place (Aguilar 1985, 1987). It has also been suggested that organochlorine concentrations in blubber of different whale species with similar prey may be inversely related to body size, owing to both overall metabolic rate and the size of the lipid compartment available for dilution of residues (Aguilar 1989, cited in Borrell 1993).

A number of variables affecting organochlorine content of blubber can be controlled during sampling. Blubber fat content can vary by topographic location on the body and by structural stratification within areas. Fatty acids are more highly saturated in inner deposits, and changes to the inner deposits may contribute more greatly to variations in blubber thickness. These differences require consistency in field sampling protocols (see Geraci and Lounsbury 1993). Lipid composition of blubber can be dissimilar to other fat-rich areas such as visceral fat deposits and the melon area of cetaceans. In addition, lipid composition of the blubber itself can vary among species. Time since death may also alter residue concentrations in tissues. Borrell and Aguilar (1990) repeatedly sampled blubber of a stranded striped dolphin carcass for organochlorines over a 55-day period and found changes in concentrations (both declines and increases, depending on the compound), beginning 2 weeks after initial sampling.

Patterns of Organochlorine Concentrations in Marine Mammals

Age, Sex, and Reproductive Status

Age, sex, and reproductive status have significant effects on organochlorine residue concentrations in blubber of marine

Table 10-1. Summary of Selected Investigations on the Relationships among Sex, Age, and Organochlorine Concentrations in Blubber of Marine Mammals

Group and Species	Region	General Findings	References
Pinnipeds			
Callorhinus ursinus	Japan	PCBs and ΣDDT in females increased to 6 years of age, then decreased sharply, increasing again after age 20. No trend observed in HCHs.	Tanabe et al. 1994b
Eumetopias jubatus	Alaska	Concentrations of ΣDDT, PCBs, and chlordanes were higher in males than females. Age-related increases occurred in males; females increased in organochlorines until age 5, then decreased until about age 20–25, when increases coincident with likely reproductive senescence occurred.	Lee et al. 1996
Halichoerus grypus	Nova Scotia	Higher concentrations of PCBs in mature males; concentrations in mature females 30% of males; newborn to 3 weeks no sex difference, concentrations 10% of adult males.	Addison et al. 1988
H. grypus	North Sea (England)	Total organochlorine concentrations higher in mature males than immature males, and higher in males than females. Significant variation in total organochlorine concentrations with both age and blubber thickness.	Donkin et al. 1981
Odobenus rosmarus	Arctic (Canada)	Concentrations of multiple organochlorines higher in males than females.	Muir et al. 1995
O. rosmarus	Arctic (Greenland)	Concentrations of ΣDDT and PCBs increase with age in males, but not females. PCB concentrations decrease with age in females and are significantly higher in immature than mature females. No correlations existed between age and ΣDDT in females.	Born et al. 1981
Phoca groenlandica	North Atlantic (Canada)	ΣDDT, PCBs, and dieldrin increase with age in young females, reaching a plateau at sexual maturity.	Frank et al. 1973
P. groenlandica	Greenland Sea	Concentrations of PCBs, DDE, and HCH in blubber increased with age in males. PCDDs and PCDFs did not.	Oehme et al. 1995b
P. groenlandica	Gulf of St. Lawrence (Canada)	Concentrations of ΣDDT, DDE, and PCBs increased with age, but TDE, DDT, and dieldrin did not.	Addison et al. 1973
P. groenlandica	Gulf of St. Lawrence and Hudson Strait (Canada)	Concentrations of ΣDDT, PCBs, and other organochlorines higher in males than females.	Beck et al. 1994
P. groenlandica	Northwest Atlantic and Arctic oceans (Canada, Greenland)	ΣDDT, PCBs, heptachlor epoxide, dieldrin concentrations generally higher in adult males than in adult females, increase with age in males but not females. Variations in patterns occurred among areas.	Ronald et al. 1984a
P. hispida	Baltic Sea	ΣDDT and PCB residue concentrations significantly higher in males in comparison with reproductive females; concentrations increase with age in males but not in females.	Helle et al. 1976a,b
P. hispida	Arctic (Canada)	Concentrations of DDE, DDT, and PCBs in blubber lipids increased with age in males, but not in females, and were inversely proportional to blubber thickness. Concentrations in males were higher than in females.	Addison and Smith 1974; Addison et al. 1986a
P. hispida	Norway	Higher concentrations of PCBs and DDE in males than females.	Daelemans et al. 1993
P. hispida	Lake Saimaa (Finland)	Concentrations of ΣDDT and PCBs increase with age in males but not females.	Helle et al. 1983
P. hispida	Arctic Ocean (Canada)	PCB congeners, ΣDDT, and chlordane concentrations increased with age in males, but not in females. Concentrations were significantly higher in males. Age and sex differences also occurred among groups in PCB congeners and homologs.	Muir et al. 1988b
P. vitulina	North Sea (U.K.)	No significant variation in concentrations of ΣDDT, PCBs, dieldrin with sex; increase with age for PCBs only.	Hall et al. 1992
P. vitulina	Northern Ireland	Higher concentrations of DDE in males than females; no differences between sexes in other organochlorines.	Mitchell and Kennedy 1992
P. vitulina	North Sea (Germany, Iceland)	Higher concentrations of PCBs in males, but no tests of statistical significance.	Heidmann et al. 1992
Odontocete cetaceans			
Berardius bairdii	Western North Pacific (Japan)	DDE and PCB residue concentrations higher in males than females.	Subramanian et al. 1988a
Cephalorhynchus commersonii	Southern Indian Ocean (Kergulean Islands)	Males with higher concentrations of ΣDDT and PCBs than females. Concentrations decline with age in females, increase in males.	Abarnou et al. 1986
Delphinapterus leucas	St. Lawrence estuary (Canada)	PCBs, ΣDDT concentrations higher in adult males than females, increase with age in both sexes	Martineau et al. 1987

Table 10-1 continued

Group and Species	Region	General Findings	References
D. leucas	St. Lawrence estuary (Canada)	PCBs, ΣDDT, mirex, others positively correlated with age of adult females but not adult males. Perhaps due to shifts in diets by older males, lack of reproduction and age structure in females.	Muir et al. 1996a
D. leucas	Canada (several regions)	PCB concentrations higher in males than females, variable results with other organochlorines. Negative correlations between age and concentrations of some organochlorines in females and positive correlations in males in some stocks.	Muir et al. 1990
Delphinus delphis	Indian Ocean (South Africa)	Increased concentrations of PCBs and ΣDDT with age in males, rapid decline in females at the age of sexual maturity. Mature males with significantly higher concentrations than all age and sex classes except nonreproductive females.	Cockcroft et al. 1990
Globicephala macrorhynchus	Western North Pacific (Japan)	DDE and PCB residue concentrations increase with age in females to about 10 years, then decrease sharply with reproductive maturity, remain low until reproductive senescence.	Tanabe et al. 1987b
G. melaena	Faroe Islands	Higher ΣDDT and PCB concentrations in mature males than mature females. No differences between sexes in immature animals. ΣDDT and PCB concentrations decline significantly with age in mature females, but show no trend with age in mature males.	Borrell 1993; Borrell and Aguilar 1993; Borrell et al. 1995
Lagenorhynchus acutus	Faroe Islands	Higher ΣDDT and PCB concentrations in males.	Borrell 1993
L. albirostris	Northwestern Atlantic (Newfoundland)	Concentrations of ΣDDT, PCBs, chlordane, and PCB congeners increase with age in males but not in females.	Muir et al. 1988a
Phocoena phocoena	Faroe Islands	Higher ΣDDT and PCB concentrations in males.	Borrell 1993
P. phocoena	Northwestern Atlantic (Bay of Fundy)	Males with higher concentrations of ΣDDT, PCBs, and dieldrin than females. Concentrations in reproductive females lower than in immature and resting females. Concentrations increase with age in males, but decrease in females.	Gaskin et al. 1982, 1983
P. phocoena	North Sea (Scotland)	Significant increase in ΣDDT, PCBs, chlordanes, and dieldrin with age in males but not females. No changes in HCB concentrations.	Wells et al. 1994
P. phocoena	North Sea (Netherlands)	Higher PCB concentrations in blubber of males, increasing with age.	van Scheppingen et al. 1996
P. phocoena	Northeastern Atlantic	Concentrations of multiple organochlorines increased with age in blubber of males, except HCB, HCHs.	Kleivane et al. 1995
P. phocoena	Great Britain	Higher organochlorine concentrations with age.	Kuiken et al. 1994
Phocoenoides dalli	Bering Sea and Northwestern Pacific	Higher concentrations of PCBs and DDE in males than females; concentrations in females decrease with increasing age after 2 years until age 6 or 7. Relative concentrations of certain PCB congeners shift with age in females but not males.	Subramanian et al. 1988b
Physeter macrocephalus	North Atlantic (Spain)	Males with lower concentrations of PCBs and ΣDDT than females, perhaps due to differences in feeding habits and habitat.	Aguilar 1983
P. macrocephalus	North Sea	Positive correlations between body length and concentrations of HCB, dieldrin, TDE, *o,p'*-DDT, chlordanes, several PCB congeners.	Law et al. 1996
Stenella coeruleoalba	Mediterranean Sea	DDE and PCB concentrations higher in males than females.	Aguilar and Borrell 1994a
Tursiops truncatus	Indian Ocean (South Africa)	Mature males with significantly higher concentrations of PCB and ΣDDT than females or immatures; concentrations lowest in reproductively active females. Concentrations increase with age in males.	Cockcroft et al. 1989
T. truncatus	Western North Atlantic (U.S.A.)	DDE, PCB, and *t*-nonachlor concentrations higher in blubber and liver of adult males than adult females.	Geraci 1989
T. truncatus	Atlantic Coast (U.S.A.)	Higher organochlorine concentrations in males.	Kuehl et al. 1991
T. truncatus	Gulf of Mexico (U.S.A.)	Higher PCB and ΣDDT concentrations in males, other organochlorines variable. ΣDDT increased with age in males, but not in females.	Salata et al. 1995
T. truncatus	Gulf of Mexico (U.S.A.)	Higher organochlorines in adult males than females, immatures.	Kuehl and Haebler 1995
Mysticete cetaceans			
Balaenoptera acutorostrata	Antarctic	DDE and PCB concentrations increase with age in males but not females, and are generally higher in mature males than in mature females.	Tanabe et al. 1986

Continued on next page

Table 10-1 continued

Group and Species	Region	General Findings	References
B. borealis	Eastern North Atlantic	Higher organochlorine concentrations in males.	Borrell 1993
B. physalus	Eastern North Atlantic	No differences in concentrations of PCBs or ΣDDT in immature males and females. Concentrations increased with age and body size in males but decreased in females, tending to reach a plateau in both sexes. Patterns of total body loads were also similar. Relative abundance of degraded compounds increased with age in males, decreased in females. Amounts transferred with reproduction estimated.	Aguilar and Borrell 1988, 1994b
B. physalus	Eastern North Atlantic	Higher organochlorine concentrations in males.	Borrell 1993
Eschrichtius robustus	Pacific Coast North America	Higher concentrations of mirex, DDE, and ΣDDT in males.	Varanasi et al. 1994

DDE = 2,2,-*bis*-(*p*-chlorophenyl)-1,1-dichloroethylene; DDT = 2,2,-*bis*-(*p*-chlorophenyl)-1,1-trichloroethane; ΣDDT = arithmetic summation of concentrations of isomers and metabolites of DDT; HCB = hexachlorobenzene; HCH = hexachlorocyclohexane; PCBs = polychlorinated biphenyls; PCDDs = polychlorinated dibenzo-*p*-dioxins; PCDFs = polychlorinated dibenzofurans; TDE = 2,2,-*bis*-(*p*-chlorophenyl)-1,1-dichloroethane.

mammals (Table 10-1). The general pattern observed extends to most species examined thus far and is based primarily on studies of DDT and metabolites and PCBs. Immature males and females often show no differences in organochlorine concentrations in blubber. Organochlorine concentrations then increase with age in males but decrease with age in mature females. Mature animals often show significant differences between sexes (males with higher concentrations) and in ratios among compounds. These differences are attributable to transfer of organochlorines from females to young during gestation and lactation, and the lack of these avenues for excretion in males. Organochlorines are accumulated by the fetus transplacentally, but reproductive transfer in the lipid-rich milk is especially pronounced. Addison and Stobo (1993) reported that 98% of the organochlorine burden of gray seal pups at weaning is obtained through maternal milk; 30% of the total body burden of ΣDDT and 15% of PCBs in adult females are transferred to a single lactating young (Addison and Brodie 1977). Calculations indicate that a primiparous female Steller sea lion transfers about 80% of its PCBs and ΣDDT to its first-born pup through lactation (Lee et al. 1996). Among cetaceans, it has been estimated that more than 90% of the PCBs of adult female striped dolphins is transferred to the calf through lactation (Tanabe 1988); about 80% of the PCBs and ΣDDT of a female bottlenose dolphin is transferred to the first-born calf (Cockcroft et al. 1989); and young adult pilot whales transfer about 68% of the ΣDDT and 100% of the PCBs through lactation to their calves (Borrell et al. 1995). Therefore, these contaminants are physically passed from generation to generation. The proportion of the body burden of organochlorines transferred to young by females may be less in mysticetes than in odontocetes (Tanabe et al. 1986; Aguilar and Borrell 1988, 1994b). Female fin whales were estimated to transfer about 26% of the total body load of ΣDDT and 14% of PCBs to the first-born calf, with amounts transferred decreasing substantially with subsequent reproduction (Aguilar and Borrell 1994b). These percentages for fin whales are lower than some of the estimates of reproductive transfer in other marine mammals (Table 10-2), but the lower proportions may be attributable to shorter lactation periods in fin whales. Total amounts transferred to young "are not expected to have significant toxicological effects on the population" (Aguilar and Borrell 1994b), a conclusion that should not necessarily be extended to other species because fin whales are among the least contaminated cetaceans.

Differences between sexes in proportions and ratios of DDT metabolites and PCB congeners in North Atlantic fin whales and sei whales (*Balaenoptera borealis*) may be attributable to corresponding modifications in enzymatic activity of the liver produced by differences in organochlorine burdens within sexes, as well as by reproductive transfer of certain compounds by females to offspring (Aguilar and Borrell 1988, Borrell 1993). The repeated findings of differences in organochlorine residue concentrations among age and sex categories (Table 10-1) point out both the need to account for these factors in reporting research results and the difficulties inherent in interpretation of data based on small sample sizes. The only notable exception to this pattern occurs in belugas (*Delphinapterus leucas*) from the St. Lawrence River estuary (Canada), in which correlations with age and certain organochlorines exist in females but not males (Muir et al. 1996a). This may be attributable to an unusual age distribution of the sample population, lack of reproduction in older

Table 10-2. Studies Pertinent to Reproductive Transfer of Organochlorine Compounds between Females and Offspring in Marine Mammals

Group and Species	General Findings	References
Pinnipeds		
Arctocephalus gazella *Callorhinus ursinus* *Mirounga angustirostris* *Neophoca cinerea* *Zalophus californianus*	Determined organochlorine concentrations in milk, including specific PCB congeners. DDE and PCBs found in milk of pinnipeds in all areas, including Antarctica, but one to two orders of magnitude higher in northern hemisphere samples.	Bacon et al. 1992
Callorhinus ursinus	Determined HCH concentrations and ratios in milk, and in tissues of neonates.	Mössner et al. 1992
Eumetopias jubatus	Estimated lactational transfer based on changes in organochlorine concentrations in blubber of females at age of reproductive maturity in contrast to males. Calculations suggest that about 80% of PCBs and ΣDDT are transferred to the first-born pup.	Lee et al. 1996
Halichoerus grypus	Determined concentrations of DDE, DDT, ΣDDT, PCBs, and PCB congeners in maternal milk, blubber, blood and pup blubber and blood. Demonstrated selectivity in transfer of some organochlorines from blood to milk.	Addison and Brodie 1987
H. grypus	Determined organochlorine concentrations (ΣDDT, PCBs, and trans-nonachlor, HCB, α-HCH) in blubber of neonates, and changes in concentrations to weaning and 1 year of age. Total burdens increased sharply from birth to weaning, then did not change, except for α-HCH, which declined significantly.	Addison and Stobo 1993
H. grypus	Determined concentrations of ΣDDT, PCBs, and dieldrin in mothers and fetuses. Estimated that the near-term fetus contained less than 1% of the organochlorine burden of the mother, obtained through transplacental transfer.	Donkin et al. 1981
H. grypus	Determined concentration of ΣDDT and PCBs in maternal blubber and milk and pup blubber. Estimated that about 30% of ΣDDT and 15% of PCBs in females are transferred to pup through annual lactation.	Addison and Brodie 1977
Phoca hispida	Concentrations of ΣDDT and PCBs present in blubber of neonates, increased rapidly during suckling period.	Helle et al. 1983
Odontocete cetaceans		
Delphinapterus leucas	DDE, ΣDDT, PCBs, and PCB congeners reported in milk sample from one female.	Massé et al. 1986
Delphinus delphis	Marked decline in PCB and ΣDDT concentrations in blubber of females after age of sexual maturity; calculations of amounts transferred to first-born calves.	Cockcroft et al. 1990
Globicephala melaena	ΣDDT transfers to the fetus more readily than PCBs, but transfer rates are comparable in lactation. Reproductive transfer from females declines with age, with young females transferring three times that of older females by lactation and six times by gestation.	Borrell et al. 1995
Phocoena phocoena	Higher dieldrin and ΣDDT concentrations in immature and resting female blubber than in reproductive females.	Gaskin et al. 1971
P. phocoena	Determined presence of dieldrin, chlordane compounds, HCB, PCBs and PCB congeners in fetal blubber, demonstrating transplacental transfer.	Wells et al. 1994
P. phocoena	Determined presence of DDE, DDT, TDE, o,p'-TDE, PCBs, dieldrin, HCB, HCHs in tissues of a fetus, demonstrating transplacental transfer.	Duinker and Hillebrand 1979
Phocoenoides dalli	Decline in PCB and DDE concentrations in female blubber at sexual maturity. Greater transfer of lower chlorinated biphenyls in lactation than gestation.	Subramanian et al. 1988b
Physeter macrocephalus	PCB and ΣDDT concentrations reported in milk samples.	Aguilar 1983
Stenella coeruleoalba	Reported concentrations of ΣDDT, PCBs, and HCHs in milk.	Kawai et al. 1988
S. coeruleoalba	Determined organochlorine burdens in a pregnant female and near-term fetus, and estimated transplacental transfer rates from 4.0% to 9.7% for PCBs, ΣDDT, HCHs, and HCB.	Tanabe et al. 1982
Tursiops truncatus	Significant decline in PCB and ΣDDT concentrations in blubber of females after age of sexual maturity. Calculations that 80% of body load of female is passed to first-born calf.	Cockcroft et al. 1989
T. truncatus	Determined concentrations of DDE, PCBs, and other organochlorines in milk of captive females. Highest concentrations in 34-year old first lactation, lowest in youngest female.	Ridgway and Reddy 1995
Mysticete cetaceans		
Balaenoptera physalus	Estimated total body loads of ΣDDT and PCBs in males and females, and determined age-related trends in each sex. Differences at similar ages were considered due to reproductive transfer. About 26% of total body load of ΣDDT and 14% of PCBs transferred to first calf, with decreasing amounts transferred in subsequent reproduction.	Aguilar and Borrell 1994b
Other		
Ursus maritimus	Determined changes in concentrations of ΣDDT, PCBs, chlordane, HCHs, and CBz in adipose tissue and milk of females and cubs, and mean daily ingestion rates through lactation.	Polischuk et al. 1995

See Table 10-1 for summaries of additional studies on changes in organochlorine concentrations in females at reproductive maturity.

Cbz = chlorobenzenes; DDE = 2,2,-bis-(p-chlorophenyl)-1,1-dichloroethylene; DDT = 2,2,-bis-(p-chlorophenyl)-1,1-trichloroethane; ΣDDT = arithmetic summation of concentrations of isomers and metabolites of DDT; HCB = hexachlorobenzene; HCH = hexachlorocyclohexane; PCBs = polychlorinated biphenyls; TDE = 2,2,-bis-(p-chlorophenyl)-1,1-dichloroethane.

females, and age-related shifts in food habits (Muir et al. 1996a).

Regional, Ecological, and Temporal Variation

In addition to variation attributable to age and sex, major differences in observed concentrations of organochlorine residues in marine mammals occur according to geographic location and feeding ecology. Marine mammals in areas with generally high organochlorine inputs, such as near-shore waters close to coastal industrial and agricultural centers, tend to have higher concentrations of organochlorines in tissues. Similarly, species that tend to feed higher in marine food webs also tend to have increased concentrations in tissues. Baleen whales tend to have much lower organochlorine residue concentrations in blubber than other marine mammals, attributable to their generally more pelagic habits (resulting in reduced exposure to near-shore contamination) and low position in the food chain (O'Shea and Brownell 1994). Members of inshore populations of piscivorous species tend to have greatest concentrations of organochlorines in tissues, particularly where subject to industrial contamination. Extremely high concentrations of ∑DDT and PCBs have been reported in odontocete cetaceans from the eastern North Pacific off southern California, the mid-Atlantic coast of the United States, the Mediterranean Sea, and the St. Lawrence River estuary. Very high contamination of pinnipeds has been reported from the Baltic Sea and coastal southern California. Odontocetes and pinnipeds from such areas can have concentrations of PCBs and ∑DDT that are two to three orders of magnitude higher than in those from the open ocean or coastal reaches of more pristine latitudes. Greater numbers of kinds of organochlorines are also usually detected in these contaminated populations. However, even in remote reaches of the Arctic, a wide array of organochlorines has been demonstrated in marine mammal tissues (Muir et al. 1990, 1992a; Norstrom and Muir 1994). Marine mammals in the northern oceans tend to be more contaminated with organochlorines than those in southern oceans, but this trend may shift as usage of some organochlorines declines in the north and increases in developing nations of the southern tropics (O'Shea and Brownell 1994, Tanabe et al. 1994a, Oehme et al. 1995a). Southern oceans currently have less contamination than northern oceans because of less industrialization and less intensive agriculture, but contamination with DDT has already increased; in addition, much of the organochlorine deposition in tropical countries is removed to the atmosphere by high temperatures and heavy rainfall, with subsequent global redistribution (Hidaka et al. 1983; Tanabe et al. 1983, 1984b; Kawano et al. 1986, 1988; Tanabe 1988). However, little attention has focused on residue surveys in southern hemisphere marine mammals. More than 90% of the individuals sampled in the literature summarized in Appendices 1 to 4 have been in the northern hemisphere (more than 75% of these from European and Canadian waters).

Organochlorine contamination of marine mammals on a broad regional basis has been intensively studied in the north polar regions (particularly the Canadian Arctic and subarctic; see Appendix), where marine food chains include plankton, fish, seals, polar bears (*Ursus maritimus*), and humans. Excellent reviews on this topic have been provided by Muir et al. (1992a) and Norstrom and Muir (1994). Higher levels of more volatile organochlorines (such as HCHs and components of chlordane and toxaphene) relative to PCBs and ∑DDT are found in marine mammals of high latitudes, and these more volatile organochlorines are relatively more uniformly distributed across these regions. This is attributable to atmospheric transport from lower, warmer latitudes and subsequent condensation in cool regions—a process referred to as the global distillation effect (Simonich and Hites 1995). Whereas compounds such as the PCBs and the DDT group are the predominant organochlorine residues in tissues of marine mammals in temperate and tropical zones, concentrations of HCHs and components of chlordane and toxaphene are found at nearly equivalent concentrations in marine mammals of the polar regions. Although relative proportions change with increasing latitude, actual concentrations are low. PCBs in Arctic ringed seals (*Phoca hispida*) occur in concentrations some 10 to 50 times lower than in Scandinavian and Baltic ringed seals, and PCBs and ∑DDT concentrations are 25 to 30 times lower in Arctic belugas than in belugas from the St. Lawrence River (this difference is not apparent in the more volatile HCH and chlordane components in beluga tissues) (Muir et al. 1992a). Concentrations of organochlorines in marine mammals of the Arctic are, as a rule, higher than in those from the Antarctic. PCDDs are found at greater concentrations in marine mammals of the high Arctic than in the subarctic, possibly because of deposition by the particulate-laden "Arctic haze" transported over the pole from sources in Europe and Asia, "sometimes as thick as smog in major cities" (Norstrom et al. 1990). PCDDs and PCDFs do not biomagnify in seals or polar bears (Norstrom et al. 1990, de Wit et al. 1992). Polar bears have a capacity to metabolize certain organochlorines that is unique among marine mammals, and perhaps among all animals (Norstrom et al. 1988, Norstrom and Muir 1994). Biomagnification of ∑DDT from seal prey does not occur, and chlordane components are also much more readily metabolized by polar bears.

Variation in population structure and dynamics, particularly age at first reproduction, can also result in seeming differences in contamination among populations, species, and regions because of varying patterns of organochlorine accumulation with age and sex as noted previously (also see Wells et al. 1994). In addition, Aguilar et al. (1993) noted that after adjustment for age, females from different pods of long-finned pilot whales (*Globicephala melaena*) from the fishery of the Faroe Islands could be distinguished by aspects of their organochlorine residue concentrations, particularly by ratios of *p, p'*-DDE/ΣDDT. These researchers suggested that observed differences in organochlorine profiles likely corresponded to ecological segregation in use of food resources. Contaminant concentrations and ratios have also been used as possible indicators of populations that may segregate geographically (Aguilar 1987, Kleivane et al. 1995). Populations of walrus (*Odobenus rosmarus*) in the Canadian Arctic differ in concentrations and characteristics of organochlorines in blubber, and these differences suggest significant seal eating by walrus in some areas (Muir et al. 1995).

A few studies have examined changes in contamination of marine mammals with organochlorines over time. For example, Aguilar (1984) investigated changes in the ratio of DDE to ΣDDT in several species of marine mammals from the North Atlantic over the period 1964 to 1981. These ratios showed a significant increase with time, indicating continual transformation of DDT to DDE in marine systems and also indicating that such ratios may be used to assess the recency of chronology of DDT inputs to marine ecosystems. Tanabe et al. (1994b) examined trends in organochlorine contaminants in northern fur seals (*Callorhinus ursinus*) from Japan from 1971 to 1988. ΣDDT increased until the mid-1970s, then declined. The proportion of ΣDDT composed of DDT dropped off sharply after the early 1970s. PCBs also increased until the mid-1970s, then dropped off to a constant level during the 1980s. HCHs showed a slight decline. In the Baltic Sea, ΣDDT and PCBs decreased considerably from the 1969 to 1973 period to the 1980s in ringed seals, whereas ΣDDT (but not PCBs) declined similarly in gray seals (*Halichoerus grypus*). The differences in trends have been attributed to differences in feeding ecology between the two species (Blomkvist et al. 1992). In male harp seals from the St. Lawrence River estuary, concentrations of PCBs decreased, but ΣDDT did not decline with time (1982–1989); however, the percentage of ΣDDT composed of DDE increased, indicating continued decreases in input of the more readily metabolized DDT (Beck et al. 1994). Seasonal changes were also apparent, with winter and summer areas differing in exposure through the food. However, interpretations of temporal differences must be approached with caution because of potential differences in sample attributes and in analytical procedures. Prevalence of chlordane and toxaphene components in Hudson Bay beluga whales increased from the 1960s to the 1980s, consistent with the hypothesis of increased atmospheric deposition from the lower latitudes (Muir et al. 1990). St. Lawrence River belugas showed declines in ΣDDT, PCBs, and some other organochlorines in males (but not females) sampled in 1993 to 1994 in comparison with the early and mid-1980s (Muir et al. 1996b).

Impacts of Organochlorines on Marine Mammal Populations

Numerous studies have suggested that exposure to organochlorines could have impacts on marine mammal populations. Such impacts would most likely be manifested through mortality or reproductive impairment. To date only a few studies have been carried out that firmly support this possibility, but findings have confounding factors that complicate establishment of possible cause-and-effect relationships or prevent pinpointing specific compounds. However, mounting evidence suggests that organochlorines may be detrimental to marine mammal populations. Three potential ways in which organochlorine effects may manifest themselves are direct mortality, reproductive impairment, and increased susceptibility to disease. Studies that relate to these possibilities and the general link to population declines are examined below.

Direct Mortality

There is no evidence for direct mortality of marine mammals caused by organochlorine compounds. This topic was reviewed for the baleen whales by O'Shea and Brownell (1994), who noted that the brain is the only tissue in mammals and birds in which concentrations of organochlorines can be considered diagnostic of cause of death. Organochlorine concentrations reported in brains of baleen whales were far lower than diagnostic levels in other mammals. Concentrations reported in brains of small cetaceans, pinnipeds, or other marine mammals noted in studies summarized in Appendices 1 to 4 also do not approach those consistent with lethality in other species. However, such levels have not been accurately established specifically for marine mammals, brain tissue is not routinely obtained from marine mammals for organochlorine analysis, and sample sizes worldwide are generally small.

Reproductive Impairment

Organochlorines (largely PCBs) have been *experimentally* verified to be responsible for impaired reproduction only in the harbor seal (*Phoca vitulina*) (Reijnders 1986), although it has been suggested that even in the relatively controlled experimental studies of harbor seals, confounding effects may have occurred (Addison 1989). The harbor seal feeding experiment was conducted to follow up field surveys that revealed high concentrations of organochlorines in a declining population in the Wadden Sea (Reijnders 1980). Control and experimental groups of 12 female harbor seals each were fed diets low in organochlorines (mackerel from the eastern North Atlantic) or high in organochlorines (other fish species from the Dutch Wadden Sea), primarily PCBs and DDE, over a 2-year period. Blood samples were periodically monitored for progesterone and estradiol-17β, and breeding males were introduced to both groups. Reproductive success was significantly lower in females fed fish from the Wadden Sea, and failure was thought to occur at the implantation stage of pregnancy.

Association of elevated organochlorine concentrations with impaired reproduction (but not direct experimental evidence) has been noted in three other pinniped species from two other contaminated coastal areas: ringed and gray seals from the Baltic Sea (Helle et al. 1976a, 1976b) and California sea lions (*Zalophus californianus*) (DeLong et al. 1973). However, interpretation of these data is not simple. Female marine mammals with impaired reproduction for reasons other than contaminant exposure will also have higher organochlorine concentrations in tissues because they are not able to excrete these chemicals through lactation. Hence, in contrast to experimental feeding studies (Reijnders 1986), higher organochlorine concentrations found in field surveys of tissues of females with impaired versus nonimpaired reproduction does not constitute unequivocal evidence of cause-and-effect relationships. This was apparent in investigations of stillbirths and premature pupping in California sea lions. Initial studies associated this phenomenon with high organochlorine residues (DeLong et al. 1973), but later investigations showed that disease agents (*Leptospirosis*) and other factors may have been at least partly responsible (Smith et al. 1974, Gilmartin et al. 1976, Martin et al. 1976). Furthermore, although contaminated with organochlorines, California sea lion populations have not declined but have generally increased in recent decades (Le Boeuf and Bonnell 1980, Lowry et al. 1992, O'Shea and Brownell 1998). In the case of seals from the Baltic, Helle et al. (1976a,b) found higher concentrations of organochlorines in female ringed seals with uterine occlusions or stenosis, but Blomkvist et al. (1992) later reported no relation between organochlorine concentrations and uterine pathology in female gray seals from the Baltic. Reijnders (1984) and Addison (1989) provide further review of the evidence for effects of contaminants on pinniped reproduction. Without additional experimental studies such as those with harbor seals, links between organochlorine exposure and reproductive anomalies are unlikely to be entirely conclusive.

Evidence for impaired reproduction in cetaceans attributable to organochlorine exposure is very limited. Martineau et al. (1987) suggested that elevated PCBs in belugas in the St. Lawrence River affected their reproduction. Béland et al. (1991) provided additional data on a small sample of stranded adult female belugas from the St. Lawrence that included observations of reproductive pathology, but the nonspecific nature of the lesions and the representativeness of the sample are not clear enough to draw firm conclusions of cause-and-effect relationships. Addison (1989) pointed out that residue concentrations may not be markedly higher in St. Lawrence River belugas than in other populations when other factors are considered. For example, body condition may have influenced results, and other forms of habitat deterioration may also have affected this population (Addison 1989). Indirect evidence for potential impacts of organochlorines on cetacean reproduction is provided by biochemical lesions. In Dall's porpoises (*Phocoenoides dalli*), Subramanian et al. (1987) obtained a weak correlation between testosterone concentrations in blood and DDE concentrations in blubber of 12 males collected in May to June 1984 in the northwestern North Pacific. No significant correlations existed among testosterone levels and other organochlorines. This finding is of interest but should be followed with expanded studies that partition variation among other possible confounding factors such as nutritional status and age.

No evidence exists for organochlorine impacts on reproduction in sirenians, where contamination is low because of their position in the food chain (O'Shea et al. 1984). There are no published studies demonstrating impacts of organochlorines on reproduction in sea otters (*Enhydra lutris*), but this is a topic of interest for future study because of the high susceptibility of some other mustelids to reproductive impairment by PCBs (for review, see O'Shea and Brownell 1994). Although they are at the top of the Arctic food chain, it is "unlikely that organochlorines are currently having a significant effect on the polar bear reproduction," due both to exposure levels and the remarkable detoxification capabilities of this species (Norstrom et al. 1988).

Susceptibility to Disease

Bergman and Olsson (1985) examined 19 gray seals and 10 ringed seals from the polluted Baltic Sea and reported the occurrence of uterine stenoses and occlusions, benign uterine tumors, adrenocortical hyperplasia, hyperkeratosis, nail deformations, and other lesions. The nature of the pathology suggested the existence of a disease complex involving organochlorine interference with the endocrine system, resulting in hyperadrenocorticism. PCBs were especially suspect, in part because effects on adrenal function have been demonstrated in laboratory studies of other mammals (Fuller and Hobson 1986). In addition, evidence that Aroclor 1254 alters the synthesis of steroids after in vitro exposure of adrenals from gray seals has been reported (Freeman and Sangalang 1977). Changes in symmetry of the skull and frequencies of bone lesions, indicating possible developmental and hyperadrenocortical effects of organochlorines, have also been noted in museum specimens of gray seals collected from the Baltic Sea after 1960, when pollution was substantial, in comparison with specimens collected in previous years (Zakharov and Yablokov 1990, Bergman et al. 1992, Olsson et al. 1994). Similar, but less pronounced, increases in skull lesions in recent years in comparison with historic specimens have also been reported for Baltic Sea harbor seals (Mortensen et al. 1992), and both species have been found to have chromosomal aberrations possibly related to contaminant exposure (Hongell 1996). To further investigate the hypothesis that the adrenal cortex of marine mammals is enlarged as a result of exposure to organochlorines, Kuiken et al. (1993) examined 28 harbor porpoises (*Phocoena phocoena*) stranded singly in various areas of Great Britain. They used quantitative methods of histopathology and chemistry with detailed statistical analyses to investigate this hypothesized relationship. Adrenal hyperplasia was found, but not in association with concentrations of seven organochlorine pesticides and metabolites or 25 PCB congeners. It was generally related to chronic causes of death such as disease and starvation, and was thought to be a general indicator of stress rather than organochlorine exposure.

Because of the findings of immunosuppression by organochlorines in studies of laboratory animals, a number of investigators have considered that organochlorine exposure may have played a role in the recent mass die-offs of marine mammals caused by morbilliviruses. This is an area of much active research, conflicting results, and some contention. Hall et al. (1992) tested the relationship between organochlorine concentrations in blubber of harbor seals that died in a morbillivirus (phocine distemper virus, PDV) outbreak in Great Britain in 1988 and those that survived (sampled alive by capture and biopsy in 1989). Statistical analyses were adjusted for age class, sex, and location. Higher concentrations of organochlorines were observed in the seals that succumbed to the epizootic. However, as pointed out by Hall et al. (1992), seasonal differences in blubber thickness could also have contributed to the lower concentrations of organochlorines in survivors, and age-related effects may have been obscured. These investigators decided that "data are not sufficient to conclude that there was a direct link between mortality from PDV infection and OC [organochlorine] contamination."

Kuiken et al. (1994) used another approach to test the hypothesis that exposure to organochlorines causes immunosuppression in marine mammals. They examined 94 harbor porpoise carcasses found stranded in Great Britain from 1989 to 1992 and analyzed the blubber for a variety of organochlorines, including 25 individual PCB congeners. Each case was classified as having died from physical trauma (principally accidental deaths in fisheries) or from infectious or parasitic disease. (Death attributable to trauma was considered to be likely independent of organochlorine concentrations, whereas marine mammals found dead due to disease could have relatively higher organochlorine concentrations if the latter produced immunosuppression that rendered them more susceptible to disease.) Diseased porpoises had higher PCB concentrations in blubber than those found dead from trauma. However, body condition was also poorer in diseased porpoises, and the region of origin had a major effect on PCB concentrations in blubber. When the analyses were adjusted for region by limiting the data set to 69 individuals from a more well-defined area, there were no significant differences in concentrations of organochlorines between the diseased and physical trauma groups. Similarly, Blomkvist et al. (1992) found no significant differences in PCB and DDT concentrations in juvenile harbor seals from Sweden collected during and before a PDV outbreak in 1988. Schumacher et al. (1993) examined thyroid glands of harbor seals that died during a PDV epizootic as well as harbor seals and harbor porpoises that did not. They distinguished histological changes consistent with impacts of PCBs on laboratory animals but could not demonstrate conclusively that PCBs were responsible for the conditions observed in the seals. Jenssen et al. (1994) did not find a correlation between PCBs and thyroid hormone levels in blood of gray seal pups.

Kendall et al. (1992) examined the relationships among concentrations of organochlorines in blubber and plasma thymulin concentrations of harbor seals and gray seals during and after an epizootic of PDV at coastal Scotland and Northern Ireland. (Studies in other laboratories indicated that thymulin concentrations are low in mammals with immunodeficiency. Thymulin is produced in the thymus, influences development of T cells, and can be stimulated by adrenocorticotrophic hormone. The thymus is sensitive to some contaminants, such as dioxin.) Thymulin levels were negatively correlated with titers to morbillivirus in gray seals, but this relationship was not affected by organochlorine concentrations. Similarly, no relationships were detected between thymulin levels and organochlorine concentrations in blubber of harbor seals. Kendall et al. (1992) suggested the immunosuppressive effect of the morbillivirus itself may have obscured relationships between organochlorine concentrations and thymulin levels, and that additional study should further examine these relationships based on organochlorine concentrations in blood. Significant relationships between thymulin levels and time since exposure combined with levels of PCB 153 and PCB 180 in multiple regression analysis suggested such an effect, although these investigators stated that "these results should not be interpreted as implying that seals with high OC levels were therefore more vulnerable to mortality from PDV." This caution is reinforced by experimental dosing studies. Harder et al. (1992) exposed six harbor seals to dietary PCBs, followed by dosing with cell-cultured phocine distemper virus. Four control seals that had minimal PCB exposure were also dosed with PDV. All seals developed severe clinical signs of PDV, and both PCB exposed and unexposed groups suffered mortality and showed no differences in antibody production. Therefore, PCB dosing in this experiment had no influence on susceptibility to the morbillivirus, although concentrations of PCBs reached in the exposed animals were at levels exceeded by those measured in some field samples.

In contrast, Ross et al. (1995) fed two groups of 11 young harbor seals relatively uncontaminated Atlantic herring or herring from the Baltic Sea for 2 years. The latter group had higher concentrations of PCBs, dioxins, and furans. The seals fed Baltic Sea herring had lower in vivo immunological responses to ovalbumin injection. The researchers suggested that these findings support the contention that organochlorines played a role in the European PDV outbreaks through immunosuppression, but noted that "it is difficult to extrapolate from the immunological responses using ovalbumin as an antigen to a seal's ability to mount a specific immune response against a pathogen in the natural environment."

Considerable analytical efforts have gone into investigations of the potential connection between organochlorine contamination and the striped dolphin morbillivirus epizootic in the Mediterranean Sea during the early 1990s. Kannan et al. (1993c) and Aguilar and Borrell (1994a) observed concentrations of PCBs (particularly the coplanar PCB congeners) in blubber of some of these individuals that were much higher than reported in other marine mammals elsewhere in the world (reaching as much as 1000 ppm on a lipid weight basis). They speculate that PCBs and DDT may have played a role in immunosuppression and susceptibility to morbillivirus. Blubber from bow-riding striped dolphins sampled by biopsy dart (for a review of this technique, see Aguilar and Borrell 1994c) in the same region in years before and after the epizootic had PCB concentrations that were significantly lower than in blubber of those found dead during the epizootic. PCB concentrations in liver were higher than in blubber in those found dead during the epizootic, suggesting mobilization of lipids and release of PCBs to the bloodstream. Aguilar and Borrell (1994a) noted three possible hypotheses to explain the potential relationship between increased PCB concentrations and susceptibility to the morbillivirus epizootic: (1) PCBs could have caused immune system depression, perhaps through effects on the thymus. Individuals with higher PCBs in tissues may therefore have been more susceptible to the disease and suffered higher mortality; (2) liver lesions that were common in striped dolphins that succumbed to the infection may have been caused by mobilized PCBs, and interacted with the disease to cause higher mortality in individuals with higher PCB burdens; and (3) the liver lesions could have been caused by a previous condition, resulting in higher PCB concentrations in tissues of affected dolphins because of a reduced capacity of the liver to metabolize PCBs.

A very detailed field study was carried out by Hall et al. (1997) to address possible relationships between PCB exposure and susceptibility to disease in gray seal pups. During a 3-year period, wild gray seal mother–pup pairs were captured and sampled for individual PCB congener analysis in milk and for determination of various hematological and blood chemistry properties. Pups were also challenged with morbillivirus vaccines and stimulated with mitogens to determine immunocompetence. No relationships were found between the prevalence of infection in pups (as an indicator of possible immunosuppression) and cumulative exposure to PCBs in mother's milk. Pups born to females with high levels of PCBs in the milk did not show any biochemical, hematological, or immunological abnormalities in comparisons to pups suckling from females with lower concentrations. In a different approach, attempting to test the hypothesis that organochlorines suppress the immune system of dolphins, Lahvis et al. (1995) determined concentrations of PCBs, DDT, DDE, and *o,p'*-DDE in blood samples of five free-ranging male bottlenose dolphins. They compared these concentrations with in vitro mitogen-induced proliferation responses of lymphocyte cultures from these same individuals. Linear regression analyses indicated correlations between reduced immune responses and higher organochlorine concentrations. This was considered consistent with the hypothesis and with findings in studies of laboratory animals, although sample sizes were very small and spurious effects could possibly exist (for example, age of dolphins in this study also appeared to be correlated with organochlorine concentrations).

Controlled experimental studies related to immune function and organochlorine exposure were carried out on harbor seals in the Netherlands (de Swart et al. 1993, 1994, 1996; Ross et al. 1996). Patterned after similar studies investigating reproductive effects (discussed previously), two groups of captive seals were captured as newly weaned pups in a relatively uncontaminated area of Scotland and held for a 1-year acclimation period. They were then matched by weight and sex and fed either herring from the Atlantic (relatively low in organochlorine contamination) or more highly contaminated herring from the Baltic Sea over a 2.5-year period. Daily intakes of organochlorines were estimated based on chemical analysis of samples (including determination of specific PCB congeners) and calculations of toxic equivalents (TEQ, see definition in section on metabolism in this chapter). The Baltic herring diet was about five times as high in PCBs and $\sum$DDT and 10 times higher in toxic equivalents. As in previous studies (Brouwer et al. 1989), vitamin A levels were lower in the group of seals fed this diet. In addition, counts of total white blood cells and granulocytes (but not lymphocytes or monocytes) were higher. Immune function differed after stimulation with certain mitogens: natural killer cell activity and lymphocyte function assays were significantly lower in seals with greater organochlorine exposure (de Swart et al. 1994). Responses were inversely correlated with TEQ summations (predominantly PCBs) in blubber biopsies. Cellular rather than humoral immunity appeared affected, consistent with expectations from laboratory animal studies. Seals were then subjected to a 15-day fasting period at the end of the study. Although concentrations of organochlorines in blood were increased during mobilization of lipids as energy reserves, differences were not observed in either group after the 15-day period. These studies support the possibility that organochlorine exposure could have led to an increased susceptibility to the morbillivirus infections seen in wild populations of marine mammals. As pointed out by Kennedy (1995), however, immunosuppression also occurs as a direct result of morbillivirus damage to lymphoid tissues, resulting in the numerous secondary infections observed in recent morbillivirus-caused epizootics. Morbilliviruses are known for their very high virulence and have historically produced very high mortality in immunologically naive populations of terrestrial mammals, even before the widespread synthesis of organochlorines by humans. It is likely that morbillivirus infections alone were the primary cause of high mortality in recent marine mammal die-offs, but the possibility of organochlorines contributing to susceptibility cannot be fully excluded.

The St. Lawrence River estuary in Quebec, Canada, is highly contaminated with a wide variety of pollutants, including organochlorines, toxic elements, and polyaromatic hydrocarbons. Although it is not possible to separate the individual effects of these contaminants on disease processes, Béland et al. (1993) and Martineau et al. (1994) reported a high prevalence of tumors, digestive tract and mammary gland lesions, and other abnormalities (including true hermaphroditism) in belugas found dead and examined at necropsy. Reproductive abnormalities were also observed, along with high levels of numerous contaminants in comparison with tissues from other populations. Much interest centers around the role of organochlorines in the disease processes observed in these animals (Béland et al. 1993, Martineau et al. 1994). A high prevalence of carcinomas has also been reported in California sea lions, with a hypothesized link to possible contaminant exposure (Gulland et al. 1996).

Links between Organochlorine Levels and Marine Mammal Population Declines

It is often assumed that sublethal toxic effects of persistent contaminants will ultimately alter population size, survival, recruitment, and species composition of mammalian communities. However, as pointed out by McBee and Bickham (1990), changes in such parameters attributable to sublethal effects of contaminants have rarely been demonstrated in any wild mammalian populations, terrestrial or marine (see also Heinz 1989). Unfortunately, field studies are hampered by many confounding variables, obscuring the nature of various associations and possible cause-and-effect relationships. Marine mammal populations with high exposure to organochlorines are also likely to have been subjected to numerous other forms of human-induced stress, such as other contaminants, noise pollution and disturbance, habitat deterioration, or changes in food quantity and quality. The cause-and-effect relationships between contaminants and population declines of marine mammals may be further obscured by overhunting. In the cases of seals in the Baltic and belugas in the St. Lawrence, previous overhunting is well documented and was probably the primary cause of population declines, with contaminants thought to play a role in preventing population growth after hunting ceased (Helle 1980, Reijnders 1985, Sergeant and Hoek 1988). Very recently, more detailed assessments of trends in St. Lawrence River belugas showed that the population has not declined since the early 1970s, and that hunting in the 1970s was the major factor impacting population growth potential (Kingsley 1998). During recent years, in some parts of the Baltic, seal populations have increased or remained stable as concentrations of some organochlorines in blubber have decreased (Olsson et al. 1992). Sea lions in California, in contrast, have had elevated organochlorine concentrations, but populations continue to increase (O'Shea and Brownell 1998).

Although evidence for the impacts of organochlorines on reproduction in marine mammals is limited, it is supported by studies in which some of these compounds have been determined to affect reproduction in laboratory animals. In particular, dietary PCBs can profoundly impair reproduction of females in some mammals. The susceptibility of mammalian species to PCBs is variable, however, even among closely related taxa. Although the literature on contaminants in marine mammals often generalizes from studies of other mammals that show dramatic effects (such as mink), some species have very low sensitivity. O'Shea and Brownell (1994) reviewed the literature on reproductive effects of dietary exposure to PCBs in carnivores (mustelids), primates, lagomorphs, bats, and rodents and noted considerable differences in sensitivity of reproduction to PCB exposure, even among species in the same genus. This wide variance in sensitivity makes generalization to marine mammal species difficult. It is likely that concentrations of PCBs in food of the baleen whales and sirenians are typically lower than those demonstrated to cause effects in most other mammals. However, this may not be the case for piscivorous species of cetaceans and pinnipeds that feed near shore in contaminated regions.

Metabolism, Biotransformation, and Biochemical Toxicity

The metabolism, biotransformation, and excretion of organochlorines involves processes that convert these hydrophobic compounds to more polar metabolites. Details of the precise metabolic pathways are complex and incomplete, but models of these processes implicate certain organochlorines as inducers of enzymes that could also lead to endocrine imbalances, a critical biochemical link to reproductive impairment. A general overview of this rapidly advancing field is provided in this section. I encourage the reader to consult other sources, including texts or reviews in biochemical toxicology or biotransformation (e.g., Klaassen 1996), for more in-depth treatments. Reviews of the metabolism of organochlorines in relation to enzyme induction and possible mechanisms of toxicity have been prepared by a number of investigators, including Borlakoglu and Haegele (1991), Goldstein and Safe (1989), Hodgson et al. (1980), Matsumura (1985), Paasivirta (1991), Peakall (1992), Rice and O'Keefe (1995), and Safe (1984, 1990, 1994). De Voogt et al. (1990), Boon et al. (1992, 1994), and Reijnders (1994) have specifically reviewed this topic for marine mammals. This overview condenses information provided in these papers and references therein. This is a field that is growing rapidly.

Mixed Function Oxidase Induction

Research into the biochemical pathways of organochlorine metabolism in laboratory mammals has shown that initial steps take place on membranes of the endoplasmic reticulum of the microsomes of liver cells (hepatocytes). This is the site where the cytochrome P-450-dependent monoxygenase enzyme systems function, and an understanding of their role is critical for interpreting recent literature on contaminants in marine mammals. Cytochrome P-450 is actually a family of hemoproteins that gives a characteristic absorption spectrum of 450 nm, hence the derivation of the name. There are many types of cytochrome P-450 monoxygenases associated with the endoplasmic reticulum of the liver, any one or more of which may be induced by a particular foreign compound (xenobiotic). Each form in turn catalyzes the oxidative metabolism of a relatively specific group of lipophilic substrates. Referred to as the mixed function oxidase (MFO) system, these biochemical pathways initially evolved to allow animals to detoxify poisonous natural compounds (such as plant defensive chemicals) at a higher rate of metabolism, thus removing them from the body more quickly. [It has been suggested that natural selection may have resulted in carnivores being less able to metabolize xenobiotics than herbivores because of the wider number of plant toxins normally ingested by the latter (Peterle 1991).] The MFO system is also capable of enhancing the metabolism of anthropogenic chemicals such as the organochlorines, but during some metabolic transformations can actually render some compounds to forms that are more toxic. Many hundreds of xenobiotics from numerous sources are now known to cause induction of enzymes of the MFO system. All of these MFO-inducing xenobiotics are organic and lipophilic, and can only be excreted after metabolic conversion to polar products. Typically, the initial step involves oxygenation in the MFO system, which is characterized by the requirement of the reduced form of nicotinamide-adenine dinucleotide phosphate (NADPH), microsomes, and oxygen for degradation of the foreign compound. Induction of MFO activity is a true induction, resulting in synthesis of new enzymes (rather than activation of previously synthesized enzymes). In some cases the metabolites produced from synthetic organochlorines can have greater toxicity than the parent compounds.

Cytochrome P-450 is the component that binds with oxygen and the substrates, thereby converting them into compounds that are more polar and subject to enhanced excretion. The significance of the presence of many cytochrome P450s is that each can exhibit a different substrate preference, providing the system with an overall ability to oxidize many different xenobiotics. The number of known genes that encode these various cytochrome P-450 enzymes is large and continues to grow. Nebert et al. (1991) described more than 150 forms of cytochrome P-450, each belonging to one of 27 gene families. A gene family may respond to a characteristic group of xenobiotics as well as endogenous compounds. The literature includes an evolving nomenclature for these families. Boon et al. (1992) provide a summary as follows: the gene is indicated by an italicized root symbol CYP (cytochrome P-450), followed by an Arabic numeral indicating the family, a letter designating the subfamily, and an Arabic numeral representing the individual gene within the subfamily (e.g., CYP1A1).

Biochemists have tentatively grouped inducers of MFO

activity into two primary classes on the basis of drug metabolism studies. One class is characterized by induction by phenobarbital (PB) and includes a wide variety of anthropogenic compounds. The other class is induced by 3-methylcholanthrene (3-MC). PB-type inducers induce the CYP2B subfamily and result in a proliferation of the smooth endoplasmic reticulum and induction of cytochrome P-450, as well as a number of other oxidative activities, some of which have been measured in marine mammals (Table 10-3). There is little or no increase in arylhydrocarbonhydroxylase (AHH) activity. The 3-MC class of compounds, in contrast, does not cause a marked increase in the smooth endoplasmic reticulum, and induces cytochrome P-448. The 3-MC category of compounds induces oxidative activity involving AHH and *O*-deethylation of 7-ethoxycoumarin. The 3-MC class induces the CYP1A subfamily and includes the most potent enzyme inducer known, TCDD (Fig. 10-5). In addition to these two primary classes of enzyme inducers, some xenobiotic compounds induce both kinds of activities and are referred to as mixed-type inducers.

The Ah Receptor, PCBs, and the Toxic Equivalency Concept

Xenobiotics can function as enzyme inhibitors, inducers, or substrates, and the many interactions between different compounds and enzymes can be extremely complex. The mechanism of induction of cytochrome P-450 is incompletely known. Some of the observations can be accounted for by the Ah receptor model (Nebert and Gonzalez 1987, Safe 1994), in which highly hydrophobic xenobiotics enter cells through typical uptake processes and then possibly compete successfully with a hypothetical normal cellular ligand for a receptor protein, designated the arylhydrocarbon or Ah receptor. According to this model (which is very generalized and is rapidly being revised as new work develops), this results in the formation of an inducer–receptor complex in the cytosol. This complex is then translocated to an unknown site, leading to the transcription of specific mRNAs that, after translation on ribosomes, ultimately induce forms of cytochrome P-450 that are incorporated into the endoplasmic reticulum. This model may explain the enhanced metabolism of pollutants and the pattern of metabolites formed from xenobiotics present in the cytosol. However, metabolism may lead to the formation of reactive intermediates that are capable of binding to critical molecules in the cytosol or the nucleus, thereby initiating toxicity in the cell. A wide variety of toxic responses in laboratory animals have been correlated with P-450 induction. Most of the supporting information for the model has come from laboratory rats and inbred strains of laboratory mice, where it has been shown that strict molecular structure is required to bind successfully to the Ah receptor (Nebert and Gonzales 1987, Safe, 1984). The Ah receptor is thought to play a role in controlling portions of what biochemists refer to as Phase I and Phase II drug metabolism, which coordinate the oxidative metabolism and conjugation of xenobiotics. Phase I drug metabolism reactions include induction of aldrin epoxidation, AHH, ethoxycoumarin-*O*-deethylation (ECOD), and ethoxyresorufin-*O*-deethylation (EROD). Induction of enzymes involved in Phase II drug metabolism reactions include epoxide hydrolases, glucuronyl transferases, and glutathione-*S*-transferases. Patterns of enzyme induction can be used to distinguish among the various classes of inducers, and have been measured in some marine mammal species (Table 10-3; discussed later).

The toxic equivalency concept is based on the likely structure–activity relationships (SAR) of xenobiotics with receptors. TCDD has a planar aromatic structure that fits closely to the Ah receptors. Certain PCB congeners have a structure that can also fit this Ah or "dioxin receptor." Their toxicity depends on the location of the chlorine atoms. For example, when only one chlorine atom at the 4 position of PCB 126 (Fig. 10-4) is changed to 5′ (PCB 127), AHH induction decreases to 1/50,000 of that of PCB 126 (Paasivirta 1991). The measurement of AHH or EROD induction has been used as a screening method for possible TCDD-like toxicity of PCBs, with degree of induction showing correlation with other toxic effects in laboratory mammals. This has resulted in the "toxic equivalency factor (TEF)" concept. The concept has been most extensively applied to PCBs where it was first developed for marine mammals by Tanabe and colleagues, particularly with reference to cetaceans (see below). After careful scrutiny, a number of caveats to the use of TEFs have been recently raised (Reijnders 1994, Safe 1994). Nevertheless, a body of literature has developed that uses the TEF approach in studies of contaminants in marine mammals, requiring a brief overview.

A TEF can be calculated for a specific PCB congener based on the potency of a biological response in relation to the response to TCDD, the most toxic of the PCDDs and PCDFs (Safe 1990, Ahlborg et al. 1994). The coplanar and mono-orthoplanar PCBs (Fig. 10-4) give the most similar responses to TCDD, but activities are generally much less pronounced than in PCDDs and PCDFs. TEFs are calculated as the ratio of the magnitude of a biological response to the PCB to the same response to TCDD. TEFs are commonly based on the ability to induce components of the MFO system (an Ah receptor-mediated response) but can also be expressed in terms of immunotoxicity, carcinogenicity, or other responses. The PCB congener with the highest TEF is PCB 126 (0.1 TEF).

The TEF concept has been extended to include an overall estimate of the TCDD-like toxicity of the entire complex

Table 10-3. Summaries of Studies Reporting Enzyme Induction in Marine Mammals

Group and Species	General Findings	References
Pinnipeds		
Cystophora cristata	Hepatic cytochrome P-450 activities measured in 18 adults and 7 pups, and in response to phenobarbitol treatment in 1 pup. Enzyme activities determined for NADPH (cytochrome P-450), NADH (cytochrome b_5), EROD, PROD, ECOD, MCOD, E_2-OHase, UDP-GT. No contaminant concentrations determined.	Goksøyr et al. 1992
Halichoerus grypus	Hepatic cytochrome P-450 levels increased with age and were correlated with concentrations of PCBs in blubber.	Addison et al. 1988
H. grypus	EROD activity determined in 8 adults and 12 pups. Activity in liver samples was higher than in kidney, and higher in adults than pups. EROD activity did not differ by sex or age in adults. Concentrations of contaminants not reported.	Addison and Brodie 1984
Phoca groenlandica	Hepatic cytochrome P-450 activities measured in 10 adult females and 11 pups, and in response to phenobarbital treatment in 1 pup. Enzyme activities determined for NADPH (cytochrome P-450), NADH (cytochrome b_5), EROD, PROD, ECOD, MCOD, E_2-OHase, UDP-GT. No contaminant concentrations determined.	Goksøyr et al. 1992
P. vitulina	Hepatic EROD and BPOH activities similar to other mammals, whereas cytochrome P-450 and b_5 concentrations were slightly lower. MFO activities in newborns lower than in adult females. No data provided on contaminant concentrations.	Addison et al. 1986b
P. vitulina	Determined in vitro hepatic microsomal metabolism of PCB 127 and EROD activity in one seal. EROD activity was comparable to the rat, metabolite production was comparable to a harbor porpoise.	Murk et al. 1994
Odontocete cetaceans		
Delphinapterus leucas	Extensive biochemical and molecular characterization of hepatic microsomal enzymes from 8 males and 5 females from the Canadian Arctic. Determined activities of cytochrome P450, CYP1A, cytochrome b_5, EROD, PROD, E_2-OH, AHH. CYP1A activity verified and correlated with PCB concentrations in tissues. Activities consistently higher in males. Immunochemical similarities of other P-450 forms related to CYP2B and CYP2E1 observed.	White et al. 1994
Globicephala macrorhynchus	Hepatic cytochrome P-450 activities determined for 2 fetuses, and 33 immature and mature individuals. Enzyme activities determined for NADPH-cytochrome c reductase, ALDE, AH, AHH, EROD. No significant difference in monooxygenase activities between sexes or among age groups. Concentrations of contaminants not reported.	Watanabe et al. 1989
Orcinus orca	Hepatic cytochrome P-450 activities determined for 3 individuals. Enzyme activities determined for NADPH-cytochrome c reductase, ALDE, AH, AHH, EROD. Concentrations of contaminants not reported.	Watanabe et al. 1989
Phocoena phocoena	Determined in vitro hepatic microsomal metabolism of PCB 127 and EROD activity in one porpoise. EROD activity was less than in a harbor seal, but PCB metabolite formation was comparable.	Murk et al. 1994
Stenella coeruleoalba	BPMO activity in biopsy skin samples of 7 individuals determined in relation to organochlorine concentrations in blubber. Enzyme activities and organochlorine concentrations were much higher than in fin whales from the same area.	Fossi et al. 1992
S. coeruleoalba	Hepatic cytochrome P-450 activities determined for 5 individuals. Enzyme activities determined for NADPH-cytochrome c reductase, BPOH, ALDE, AH, EROD. Concentrations of contaminants not reported.	Watanabe et al. 1989
Mysticete cetaceans		
Balaenoptera acutorostrata	Determined concentrations of hepatic cytochrome P-450, cytochrome b_5, EROD, ECOD, NADPH-cytochrome P-450 reductase, AHH, APDM in four females, one male, and two fetuses. Intrahepatic differences observed only in EROD. Concentrations of contaminants not reported.	Goksøyr et al. 1985, 1986
B. acutorostrata	Determined concentrations of hepatic and renal cytochrome P-450, cytochrome b_5, NADPH-cytochrome P-450 reductase, EROD, ECOD, biph.-4OH, AHH, E-2OH, UDP-GT, GSH-T in four fetuses and 10 adults.	Goksøyr et al. 1988, 1989
B. physalus	BPMO activity in biopsy skin samples of 9 individuals determined in relation to organochlorine concentrations in blubber. Enzyme activities and organochlorine concentrations were much lower than in striped dolphins from the same area.	Fossi et al. 1992

Table 10-3 continued

Group and Species	General Findings	References
Other		
Ursus maritimus	Detailed analysis of CYP1A, CYP2B protein content, EROD, PROD, BROD in hepatic microsomes together with determination of concentrations of congeners of PCBs, PCDDs, PCDFs, other organochlorines. High correlations between CYP1A activity and EROD, PROD, TEQs, PCBs, PCDDs, PCDFs. Low BROD correlations. CYP2B correlations highest with chlordanes, o-PCBs.	Letcher et al. 1996

Source: See also Fossi and Marsili 1997.

AH = aniline hydroxylase; AHH = aryl (benzo[a]pyrene) hydrocarbon hydroxylase; ALDE = aldrin epoxidase; APDM =aminopyrine *N*-demethylase; biph.-40H = biphenyl 4-hydroxylase; BPMO = benzo[a]pyrene monooxygenase; BOPH = benzo[a]pyrene hydroxylase; CYP1A = subfamily of monoxygenases (see text); CYP2B = subfamily of monoxygenases (see text); CYP2E = subfamily of monoxygenases (see text); E_2-OH = estradiol 2-hydroxylase; ECOD = 7-ethoxycoumarin *o*-deethylase; EROD = 7-ethoxyresorufin *o*-deethylase; GSH-T = glutathione *S*-transferase; MFO = mixed function oxygenase; NADH = nicotinamide adenine dinucleotide (reduced); NADPH = nicotinamide adenine dinucleotide phosphate (reduced); PCBs = polychlorinated biphenyls; PROD = pentoxyresorufin *o*-depentylase; UDP-GT = uridine diphosphate glucuronyl transferase.

mixture of PCBs, PCDDs, and PCDFs present in environmental and tissue samples. This is commonly referred to as the total "toxic equivalents" or TEQ of a mixture, calculated as the sum of the concentration of each PCB, PCDD, or PCDF times its TEF for the entire measured sample. Although individual PCBs have much lower TEFs than the dioxins and dibenzofurans, they are nearly always present at much higher concentrations, and the TEQ of a mixture is thought to better reflect the overall potential toxic impact. However, a number of investigators have pointed out that toxic responses may not be additive (although this may be the case for Ah receptor-mediated toxicity) but can be non-additive with antagonistic effects. This is clearly seen in some response systems such as immunotoxicity, where the measured responses can be considerably less than predicted on the basis of the sum of TEFs (Safe 1994). Furthermore, as pointed out by Reijnders (1994), even when TEQs are calculated on the basis of Ah receptor-mediated responses, differences in induction of P-450-based enzymes exist among species and organs, and the toxicities of PCB metabolites are not included in calculations. This is important because some phenolic metabolites of PCBs have greater potencies than the parent compounds.

Metabolism and Possible Toxicity of PCB Congeners

The quantification of PCBs in marine mammal tissues was initially limited to comparisons with standard mixtures such as Aroclor 1254. However, Tanabe and coworkers at Ehime University in Japan (see references below) made groundbreaking advances in isomer-specific quantification of PCBs in cetacean tissues. Their initial findings on residue concentrations led to some interesting hypotheses suggesting possible links among PCB congener profiles, MFO activity, and potential for toxicity. Although there are in theory 209 possible isomers and congeners of PCBs (only about 120 are known from industrial mixtures), the investigations focused on a subset of the 20 coplanar PCBs with non-*ortho* chlorine substitution in the biphenyl rings, in particular those with four or more chlorine atoms in the *para* and *meta* positions (Fig. 10-4). These were chosen because of structural similarity to TCDD and their similar toxic responses (but requiring higher doses), that include induction of MFO activity.

Tanabe et al. (1987a, 1988) revealed the presence of trace amounts of coplanar PCB 77, PCB 126, and PCB 169 (Fig. 10-4) in small sample sizes of fish, cetaceans (finless porpoise [*Neophocaena phocaenoides*], Dall's porpoise, Baird's beaked whale [*Berardius bairdii*], Pacific white-sided dolphins [*Lagenorhynchus obliquidens*], killer whales [*Orcinus orca*]), humans, dogs, and cats. Samples in remote marine areas contained coplanar PCBs, and their concentrations in animal tissues were highly correlated with total PCB concentrations (although three to five orders of magnitude lower). Considered together with the proportions of coplanar PCBs determined in commercial PCB preparations, these findings constituted evidence that coplanar PCBs were widely distributed as contaminants of the global environment as a direct result of general industrial PCB pollution. The initial study (Tanabe et al. 1987a) revealed that the amounts of these coplanar PCBs in the environmental samples, although low, were much higher than those of PCDDs and PCDFs, both in terms of absolute concentrations and on the basis of toxic equivalency as related to enzyme induction in laboratory animals (discussed previously). On the basis of the pattern of relative concentrations of these coplanar PCBs in commercial mixtures and biota, these investigators further suggested that cetaceans may have a lower capacity to metabolize these congeners. (However, this suggestion was based on small sample sizes of a limited number of other taxa.) Tanabe et al. (1987a:158) further cautioned that, although coplanar PCBs may have a greater potential for harm than dioxins or dibenzofurans when viewed from a toxic equivalency standpoint, "the toxic effects of trace levels of coplanar PCBs, PCDDs, PCDFs on a long-term basis to humans and environmental animals are neither fully understood nor clearly demonstrated."

Additional investigations on this topic ensued. On the basis of calculations derived from the residue concentrations of coplanar PCBs (Fig. 10-4) in cetaceans relative to those in potential food organisms and a small number of other mammalian species, it was hypothesized that cetaceans lacked the capacity to metabolize these congeners, and, as a correlate, were likely to have no PB-type MFO systems and a relatively small capacity for induction of MC-type enzymes (Tanabe 1988, Tanabe et al. 1988). It was also suggested that low enzyme induction capacity could cause reproductive toxicity by these chemicals (Tanabe 1988), but this speculation was not based on direct measurements of enzyme induction capacity or strong evidence for reproductive impairment in cetaceans. Furthermore, the relationship between MFO activity and reproductive effects in mammals currently remains an area of considerable debate and uncertainty (Stone 1994).

Kannan et al. (1989) expanded on the earlier work and determined concentrations of mono- and di-*ortho* analogs of the coplanar PCBs (Fig. 10-4) as well as these non-*ortho* chlorine substituted congeners in the same or a similar series of samples. Investigations into the occurrence of these analogs were called for by their potentially comparable toxicity with the non-*ortho* chlorine substituted PCBs, as suggested by their structure and enzyme-inducing capability in laboratory mammals. The mono- and di-*ortho* coplanar PCBs were also found in all samples at low concentrations but in greater amounts than dioxins or dibenzofurans. Interestingly, however, the concentrations of all coplanar PCBs relative to total PCBs did not vary between humans and cetaceans, an observation inconsistent with the hypothesis that cetaceans lack an ability to metabolize these compounds (Kannan et al. 1989). Data on concentrations of specific PCB congeners in harbor porpoises from near the Netherlands, and other cetaceans, agreed with earlier findings of metabolism of congeners with vicinal H atoms in the *ortho, meta* position and a maximum of one *ortho*-Cl, but also suggested an ability for cetaceans to metabolize congeners with vicinal H atoms in *meta* and *para* positions, although at a lesser capacity than in seals (Duinker et al. 1989). PCB 153 is especially resistant to metabolism in the harbor porpoise (van Scheppingen et al. 1996).

The hypothesis that cetaceans lack an ability to metabolize coplanar PCBs was further tested by Watanabe et al. (1989) by the direct determination of MFO activity in liver microsomal samples from 33 short-finned pilot whales (*Globicephala macrorhynchus*) (and two fetuses), five striped dolphins, one killer whale, and one laboratory rat. MFO activity of both MC and PB types was measured (Table 10-3) and found in all cetaceans, with some significant differences among species. Levels of cytochrome P-450 in cetacean liver microsomes were comparable to those from other mammals. Activities of EROD and AHH in cetaceans were comparable to those in the rat, whereas activities of aldrin epoxidase (ALDE) and AH were lower (Watanabe et al. 1989). Although both MC and PB-type induction was demonstrated to occur, results were interpreted to support the hypothesis that PB-type induction (low ALDE and AH activity) is low in cetaceans and could account for the pattern of coplanar congener accumulation in tissues demonstrated by other studies. NADPH cytochrome c reductase, cytochrome P-450, and MFO activities were generally lower in the fetal pilot whales than in mature or immature individuals, but did not differ between sex or age groups (Kannan et al. 1989). This study was a milestone in providing measurements of MFO in cetaceans. However, it tempers earlier conclusions about a complete inability of cetaceans to metabolize certain coplanar PCBs because of a lack of enzyme induction capability. The work also underscores the need to better understand the variability in MFO activity levels in wild populations, exacerbated by very little data on relationships between organochlorine concentrations in tissues or food and MFO activity levels in marine mammals (Table 10-3). Additional studies, such as those by Letcher et al. (1996), in which MFO activities, CYP protein contents, and organochlorine concentrations in polar bear livers were determined and correlations established, will help fill these information gaps. The recent establishment of cell culture lines and verification of the existence of an Ah receptor in bottlenose dolphins should also allow more thorough investigation of various aspects of P-450 induction in the future (Carvan et al. 1994).

As progress in understanding details of PCB metabolism in cetaceans was advancing based on analytical chemistry studies in Japan, researchers in Europe also began congener-specific determinations in marine mammals, emphasizing seals. Boon et al. (1987) determined concentrations of individual congeners in the diet, blood, and feces of captive female harbor seals fed organochlorine-contaminated, as well as less-contaminated control fish in experimental studies in the Netherlands. The patterns of PCBs within fish and seal blood samples were nearly constant, but differed substantially between the two; differences in congener composition between seal groups were related solely to diet (Storr-Hansen et al. 1995). Molecular structure of congeners that persisted and accumulated in seal blood differed from that of congeners that were metabolized. Congeners that were untabolized by the seals were those with 3 to 6 chlorine atoms and penta- and hexa-chlorinated biphenyls with vicinal H atoms at *meta–para* positions, or vicinal H atoms at *ortho–meta* positions with single *ortho*-chlorine atoms. Congeners with 5 to 10 chlorine atoms and with either 2 or 3 *ortho*-chlorines were not metabolized. Boon et al. (1987) noted that congener persistence was likely related to enzymatic metabolism as it is influenced by molecular structure. Congeners with vicinal H atoms at *meta–para* positions were metabolized in "globular configurations" (rings perpendicular); such metabolism is likely to be carried out by the P-450 enzymes. Congeners with *ortho–meta* vicinal H atoms reach a planar configuration that is more likely to be metabolized by P-448 enzyme systems. Structurally, the mono-*ortho* chlorine-containing congeners can reach both globular and planar configurations, and are known to be mixed-type MFO inducers (Boon et al. 1987). The capacity of harbor seals to metabolize these two classes of PCB congeners is shared with some small cetaceans and the polar bear, but may be diminished in ringed seals (Boon et al. 1989). Ratios of the more easily metabolized PCB 52 to the resistant PCB 153 also are similar among porpoises and dolphins from the North Sea of Scotland but are much lower in harbor seals from the same region (Wells et al. 1994).

The pattern of congeners found in polar bears is especially unique. Many of the congeners prominent in their seal prey are absent or occur at relatively low concentrations in polar bear tissues. Identities of specific congeners indicate an ability of polar bears to metabolize PCBs with nonchlorinated *para* positions, adjacent nonchlorinated *ortho–meta* positions, or both *ortho* positions chlorinated in one ring (Norstrom et al. 1988). Differences in metabolism of congeners among marine mammal species, as well as differences between marine mammals and laboratory species, point "to the dangers of extrapolation between species" (Boon et al. 1992:152). Many recent studies have included congener-specific determinations in marine mammals and will help further elucidate patterns of PCB metabolism (Appendices 1 to 4). Such recent work with belugas of the St. Lawrence River estuary, for example, implies high activity of both CYP1A and CYP2B enzyme systems, perhaps through heightened induction from heavy PCB exposure (Muir et al. 1996a).

The use of enzyme induction information is growing in the study of contaminants in marine mammals. These are sensitive bioassays, but they do not firmly establish the existence of harmful impacts to individual animal health or population status. Much like the simple presence of contaminants in tissues, consequences can only be inferred based on results of laboratory studies of other species, which unfortunately sometimes show significant variation in responses. In addition to the caveats and research needs noted above, Peakall (1992:99) recently pointed out that for wildlife species in general, "it has not been established whether induction of cytochrome-mediated enzyme activities leads either directly or indirectly to toxicity or whether enzyme induction and toxicity are independent aspects of the responses." Clearly, this is an area of much needed additional, coordinated research.

Marine Mammals and the "Gender Benders"

Much recent attention has focused on the role of organochlorine contaminants as disrupters of the endocrine system in humans and wildlife. The existence of chemicals with endocrine-disrupting capacity ("gender benders") is well known, as exemplified by the drug diethylstilbestrol (DES), used to prevent miscarriages in women in the 1950s and 1960s. Laboratory animals and humans exposed to DES prenatally can exhibit a number of anomalies of the reproductive system as adults. Similar changes have been seen in wildlife species in areas very heavily contaminated with organochlorines, including altered hormone levels and dysfunctional manifestations of reproductive behavior (Colborn et al. 1993, Raloff 1994). Endocrine disrupters can include any xenobiotics that interfere with the normal function of hormone receptor proteins (which mediate the effects of endogenous hormones on gene activation). These chemicals can either mimic the hormone (Fig. 10-6) by binding to the receptor and activating a hormonelike response, or can act as an antagonist by binding to the receptor protein, making it unavailable to be activated by the natural hormone. In the case of the estrogen receptor (ER), the former action is termed estrogenic, whereas the latter is antiestrogenic.

Organochlorines exhibit a wide and mixed array of estrogenic activity in studies of laboratory animals. For example, *p,p′*-DDE is not estrogenic, whereas *o,p′*-DDT is estrogenic; Ah receptor agonists, however, such as TCDDs and PCDFs, are antiestrogenic. The presence of complex mixtures of both estrogenic and antiestrogenic organochlorines in marine mammals makes it difficult to predict ultimate biological effects. The organochlorines may be present at lower concentrations than natural hormones and may have much lower activities; hormonelike natural chemicals (such as those found in the human diet) often have much stronger endocrine effects than organochlorines in laboratory tests. The question of whether organochlorines at typical background concentrations exert any meaningful endocrine effects at all is a topic of much debate, particularly as applied to humans (Colborn and Clement 1992, Colborn et al. 1993, Stone 1994, Safe 1995). There has been little intensive study of this topic in marine mammals although the experimental

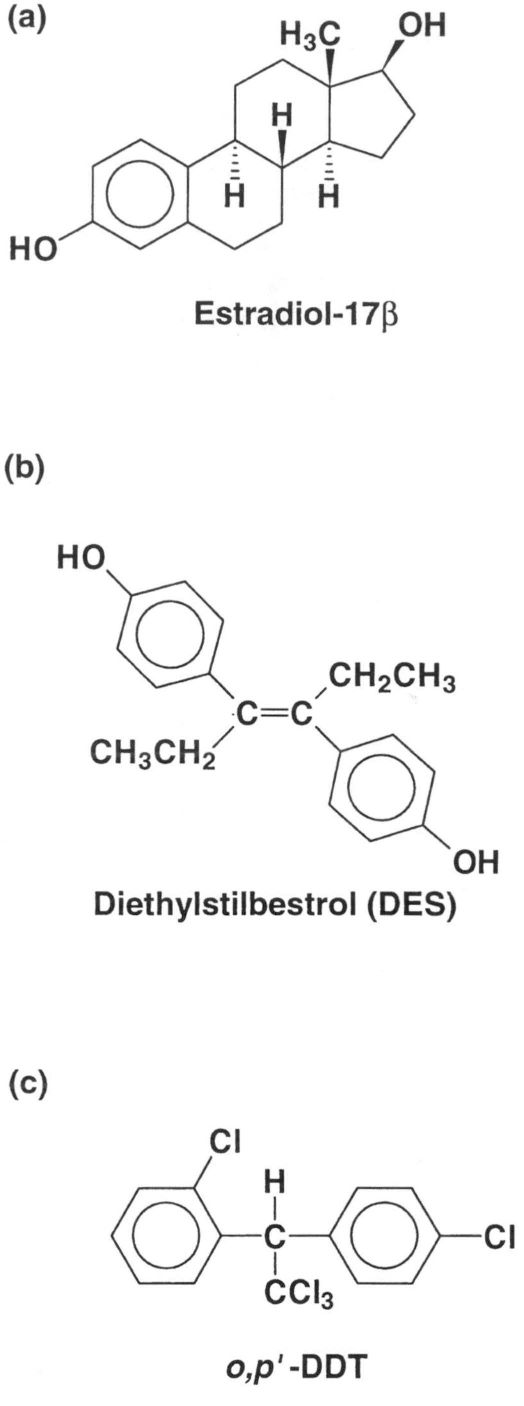

Figure 10-6. Structural similarities of some estrogenic compounds. Estradiol-17β is a "natural" ovarian estrogen, diethylstilbestrol (DES) is a synthetic pharmaceutical, and *o,p'*-DDT is an isomer of the insecticide DDT. Controversy surrounds the importance of synthetic organochlorines as estrogenic or antiestrogenic disrupters of endocrine metabolism at typical background levels.

studies of harbor seals support such a possibility. Not only do captive seals fed diets high in organochlorines fail to reproduce normally, they also show biochemical lesions (lowered retinol and thyroid hormones) compatible with an endocrine disruption hypothesis (Brouwer et al. 1989).

Toxic Elements

A long-standing postulate of toxicology, first attributed to the Renaissance scientist Paracelsus in the 1500s, is that "all substances are poisons; there is none which is not a poison. The right dose differentiates a poison from a remedy" (Gallo 1996). No better example of this can be found than in the study of toxic elements. Unlike organochlorines, elements are naturally occurring substances, many of which are essential to normal metabolic function and sustenance of life. However, at certain exposure levels, these elements can have toxic effects, and some elements have no known essential role in normal biological processes. About 80 of the elements on the periodic chart can be considered metals. Goyer (1996) provides an overview of toxicology of these metallic elements. A few elements have been a major focus in studies of marine mammals, primarily because of their known danger and toxicity to humans and other animals. Significant attention has thus been given to investigating cadmium, lead, and mercury in tissues of marine mammals. These metals are treated in depth in this chapter and an overview of some of the other elements is also provided. Original studies have reported results of analyses of up to 40 trace elements in tissues of nearly 6000 individuals in more than 60 species of marine mammals. Many of these studies are summarized in Appendices 5 to 8. The methods used to quantify residues of toxic elements in tissues can vary among studies. These should include systematic procedures for tissue collection and storage that prevent spurious contamination; up-to-date methods of sample preparation, including tissue homogenization and digestion; and the use of quality assurance procedures such as comparisons with certified standard reference materials, use of procedural blanks, multiple analyses of the same samples within the same laboratories, and interlaboratory calibration and comparison studies. Zeisler et al. (1993) provide an example of excellent quality assurance procedures for determination of elemental concentrations in marine mammal tissues. Actual chemical quantification procedures can vary with the laboratory, elements chosen for analysis, desired sensitivity, and overall objectives. Such techniques may involve atomic absorption spectrometry (flame, flameless, or cold vapor may be appropriate depending on the element), instrumental neutron activation analysis with gamma ray spectrometry (e.g., Mackey et al. 1996), inductively coupled argon plasma emission/mass spectrometry, [illegible] cence, prompt gamma activation analysis, and differential pulse and square wave stripping voltammetry (e.g., Zeisler et al. 1993).

Cadmium

Cadmium is a naturally occurring, nonessential element. It can become an environmental contaminant as a by-product of various industrial processes, such as mining and smelting, petroleum production, and manufacturing of a number of products. It has been used in items as diverse as storage batteries, paints, and nuclear reactors, and is also released into the environment through motor vehicle exhausts and the breakdown of automobile tires. Cadmium concentrations can increase in sediments of aquatic systems subject to industrial effluents. In mammals, toxic effects of cadmium exposure are mitigated by protective binding with endogenously produced metallothionein, a low molecular weight protein. Cadmium typically reaches its highest concentrations in the kidneys, although greater amounts can be found in the liver in cases of unusually excessive exposure. Humans chronically exposed to cadmium ingestion from polluted mine waste water used for crop irrigation near Toyama Bay in Japan suffered itai-itai ("ouch-ouch") disease. This syndrome, which included leg and back pain, skeletal deformities, and susceptibility to bone fractures, was attributable to loss of calcium through cadmium-damaged kidneys (Goyer 1996). A variety of other toxic effects has also been produced in laboratory mammals subjected to excessive cadmium exposure, including disorders of the circulatory system, nervous system, and reproductive system (particularly in males), as well as renal system pathology. However, in some cases relatively high amounts are required to produce these effects, some of which may fail to occur in the presence of adequate dietary zinc and selenium. Cadmium is only slowly excreted from the body and has a half-life of 30 years in humans.

Cadmium concentrations have been determined in organs of numerous marine mammal species (Appendices 5 to 8). As in other mammals, almost without exception the highest cadmium concentrations in organs of marine mammals occur in kidneys, with lesser concentrations in liver, followed by relatively lower amounts in muscle and most other organs and tissues, including bone. The relatively high accumulation of cadmium in kidneys has been demonstrated for at least 28 species of marine mammals in all major taxonomic groups (see references in Appendices 5 to 8). Great variation in cadmium concentrations in kidneys of marine mammals has been noted among individuals, with much of the variation attributable to increased accumulation with age (Table 10-4). Unusually high concentrations of hundreds of parts per million of cadmium have been reported in kidneys of a variety of species (Table 10-5).

Despite seemingly high concentrations in kidneys of some individuals (linked to age in many), I am unaware of any published observations demonstrating cadmium-induced pathology in marine mammals. Some of the high cadmium concentrations reported in marine mammals are undoubtedly a result of naturally high cadmium concentrations in prey species rather than anthropogenic contamination. High cadmium concentrations in organs of squid, for example, are a known source of elevated cadmium in marine mammals (Hamanaka et al. 1977; McClurg 1984; Leonzio et al. 1992; Szefer et al. 1993, 1994; Malcolm et al. 1994; Caurant and Amiard-Triquet 1995). Certain molluscs have also been suggested as a dietary source of cadmium in walruses (Miles and Hills 1994).

Marine mammals appear to share the protective action of metallothionein proteins against cadmium toxicity, first reported in terrestrial mammals. Metallothioneins are induced by the presence of divalent cations such as Hg^{++}, Cd^{++}, Cu^{++}, and Zn^{++} and have a high affinity for binding such cations. Forms of metallothioneins have been reported in an array of organisms ranging from blue-green algae to mammals. Metallothionein has been isolated from tissues of gray seals and northern fur seals (Olafson and Thompson 1974), California sea lions (Lee et al. 1977), ribbon seals (*Phoca fasciata*) (Mochizuki et al. 1985), harbor seals (Mochizuki et al. 1985, Tohyama et al. 1986), and narwhals (*Monodon monoceros*) (Wagemann et al. 1984). Other metal-binding proteins have been isolated from tissues of sperm whales and California sea lions (Ridlington et al. 1981). Most of the cadmium in the cytosol of sea lion liver and kidney is in association with metallothionein (Lee et al. 1977). In the harbor seal, metallothionein levels in livers and kidneys increased positively with age and with cadmium and zinc concentrations, indicating sequestration of these metals by this protein. The thiol groups of metallothionein proteins are thought to act as soft Lewis bases that bind readily with soft (easily polarizable) Lewis acids such as Zn^{++}, Cd^{++}, and Hg^{++} (Perttilä et al. 1986). In addition to zinc, some studies have shown positive correlations between cadmium and selenium concentrations in organs of marine mammals, indicating participation of these elements in formation of the biochemical complexes that reduce toxicity.

There appear to be no notable differences between the sexes in cadmium concentrations in tissues of marine mammals. Little cadmium seems to be transferred to the fetus (Honda and Tatsukawa 1983, Roberts et al. 1976, Wagemann et al. 1988), although minor amounts of such transfer have been documented (Meador et al. 1993). In striped dolphins, cadmium in bones increased primarily in the late suckling

Table 10-4. Summary of Studies Reporting Relationships between Age, Sex, and Metal Concentrations in Marine Mammals

Group and Species	Organ	Relation with Age and Sex	References
Pinnipeds			
Arctocephalus gazella	L	Cd increase with age. No relationship with age and concentrations of Rb, Mg, Sr, Mo, Pb, Cu, Zn, Cr, Hg.	Malcolm et al. 1994
A. pusillus	Multiple organs	Hg increase with age in liver, spleen, brain, and hair.	Bacher 1985
Callorhinus ursinus	Bone, K, L, M	Cd increase with age in liver, muscle, bone, but not kidney. Hg increase with age in liver. No relationships with age and Pb, Ni, Zn in kidney, liver, muscle, or bone.	Goldblatt and Anthony 1983
C. ursinus	K, L, M	Age positively correlated with Hg in muscle, liver, kidney, Fe in muscle and liver, Cd in kidney. Negative correlation with age and Mn in muscle, kidney, Cu in kidney.	Noda et al. 1995
C. ursinus	K, L, M	Concentrations of Cd, Pb in liver and kidney did not correlate with age. Hg in liver correlated with age. Hg in kidney and muscle did not.	Anas 1974a
Erignathus barbatus	L, M	Hg, Se increase with age.	Smith and Armstrong 1975, 1978
Eumetopias jubatus	K, L, M, O	Cd increase with age in kidney, liver.	Hamanaka et al. 1982
Halichoerus grypus	K, L	Positive correlation with Se, Hg, and age.	Perttilä et al. 1986
H. grypus	B, K, L	Positive correlations between age and Cd, Cr, V, Hg, and Se in either kidney, liver, or both. Negative correlation of Co with age. No correlations with age for Al, As, Pb. Relationships with other elements also determined	Frank et al. 1992
H. grypus	Teeth	Age and Cd, Cr, Pb show negative correlation, Zn positive correlation, Cu no correlation.	Heppleston and French 1973
H. grypus	Br, K, L	Cd increase with age in kidneys, Hg increased with age in livers. Hg in brain did not increase with age.	Heppleston and French 1973
Odobenus rosmarus	K, L	Cd, As increase with age in liver and kidney, Zn in kidneys. No correlations with age and Pb, Hg, or Se. No metals varied with sex except higher Se in livers of females and higher As in livers and kidneys of males.	Warburton and Seagars 1993
O. rosmarus	K, L, M	Hg increase with age in liver.	Born et al. 1981
Phoca groenlandica	Br, K, L, M	Cd, Hg, Se increase with age in kidney, liver, muscle, dependent on location. No consistent relation with age and Cu.	Ronald et al. 1984b
P. groenlandica	K, L, M, O	Cu and Zn higher in pups than mothers, higher proportion MeHg in livers of pups. MeHg correlated with age in muscle only.	Wagemann et al. 1988
P. groenlandica	L, M	Hg increase with age class.	Botta et al. 1983
P. hispida	K, L, M	Cd, Hg increase with age in kidney; Cu decrease with age in liver, kidney, muscle. As increase with age in liver; Pb independent of age; other relations also examined.	Wagemann 1989
P. hispida	K, L	No correlation with age and Se or Hg.	Perttilä et al. 1986
P. hispida	Hair	No differences in concentrations of Cd, Cr, Hg, Ni, Pb in hair of still-born, pups, adults, except Ni higher in still-borns than pups, yearlings; Hg lower in subadults to age 2 than adults or pups.	Hyvärinen and Sipilä 1984
P. hispida	K, L, M	Cd in kidney and Ag, Hg in liver increased with age; Cu in three tissues and Zn in muscle decreased with age. No age effects noted with Pb, Se.	Wagemann 1989
P. hispida	L, M	Hg, Se increase with age.	Smith and Armstrong 1975, 1978
P. vitulina	K, L	Cd, Hg in liver and kidney increase with body length.	Tohyama et al. 1986
P. vitulina	K, L, M, O	Cd increase with age in kidney, liver. No increase in Pb with age. Hg increase with age in liver, rate of increase varies with location.	Roberts et al. 1976
P. vitulina	B, K, L	Positive correlations between age and Cd, Cr, V, Hg, and Se in either kidney, liver, or both. Negative correlation of Co with age. No correlations with age for Al, As, Pb. Relationships with other elements also determined.	Frank et al 1992
P. vitulina	K, L	Cd increase with age in kidneys and Pb and Hg in livers of males but not females. No correlation with age and Pb in kidney, As or Se in liver.	Miles et al. 1992
P. vitulina	Br, K, L	Apparent increases with age in Cd and Pb in liver and kidney, Hg in liver. No age-related patterns in Zn, Cu concentrations in kidney, liver, or brain.	Drescher et al. 1977
P. vitulina	Hair, skin	Pb, Cd increase with age in hair of males, not females. Cd higher in males. No differences or effects noted for Hg.	Wenzel et al. 1993
P. vitulina	Br, K, L	Cd increase with age in kidneys, Hg increase with age in livers. Hg in brain did not increase with age.	Heppleston and French 1973
P. vitulina	L	Hg increase with age in livers.	Koeman et al. 1972
P. vitulina	Br, K, L	Hg, Se increase with age in liver.	Reijnders 1980

Table 10-4 continued

Group and Species	Organ	Relation with Age and Sex	References
P. vitulina	K, L	Hg increase with body length in liver, kidney; Se increase in liver.	Himeno et al. 1989
Zalophus californianus	Br, K, L, M, O	No apparent Hg increase with age.	Buhler et al. 1975
Odontocete cetaceans			
Delphinapterus leucas	L	Hg increase in liver with age.	Béland et al. 1991
D. leucas	K, L, M	Hg increase with age in liver and kidney, Cd increase with age in kidney, Cu decrease with age in liver, muscle, and kidney.	Wagemann et al. 1990
D. leucas	K, L, M	Positive correlations with age and Cd, Se, Hg in liver and Hg in kidneys.	Hansen et al. 1990
D. leucas	L	Ag, Hg, Se increase with age.	Becker et al. 1995
Globicephala melaena	L	Hg, Pb, Se increase with body length.	Meador et al. 1993
G. melaena	B, K, L, M	Hg increase with age in all organs, depending on location. Cd increase with age or length in kidneys, liver. No correlation with age and Cu, Zn, Pb. Other relationships noted.	Muir et al. 1988a
G. melaena	K, L, O	Hg, Se, Pb increase with body length in livers. Only Pb in brain differed by sex.	Meador et al. 1993
G. melaena	B, K, L, M	Correlations with body size investigated for Ag, Cd, Cu, Hg, Se, Zn. Hg in muscle, Se and Hg in liver, and Cd, Se and Zn in kidneys were positively correlated with size. No differences between sexes in total Hg, which was lower in immatures.	Julshamn et al. 1987
G. melaena	K, L, M, O	Cd, Hg, Se, in liver and kidney correlated with age and length. Other patterns depended on pod of origin. Metals in liver higher in females than males within age groups.	Caurant et al. 1993
G. melaena	L	Ag, Hg, Se increase with age.	Becker et al. 1995
G. melaena	L	Ag, Hg, Se increase with body length; eight other elements show no correlation with length.	Mackey et al. 1995
Lagenorhynchus albirostris	B, K, L, M	Cd increase with length in liver, kidney, with age in muscle. No correlation with age or length for Hg, Se, Cu, Zn, Pb.	Muir et al. 1988a
Monodon monoceros	B, K, L, M	Complex relationships among effects of size and sex on concentrations of Hg, Se, Cd, Cu, Zn, and Pb in four tissues. Only 7 of 26 relationships involved sex differences, with no obvious pattern. Cd increase with size in kidneys, but decrease in livers; Hg increase with size in livers of males but decrease in females, increased with size in kidneys of both sexes. Pb decrease with length in livers. Other relationships noted.	Wagemann et al. 1983
M. monoceros	K, L, M	Positive correlations with Cd and age in livers of males and kidneys of females, Hg and Se in all organs, Zn in kidneys of females.	Hansen et al. 1990
Phocoena phocoena	K, L, O	Hg and Cd, but not Cu or Zn increase with body length in livers and kidneys.	Falconer et al. 1983
P. phocoena	Br, K, L, M	Hg increase with age.	Gaskin et al. 1979
P. phocoena	K, L	Increase in total Hg, Se with age. No differences between sexes within age classes.	Teigen et al. 1993
P. phocoena	K, L, M	Total Hg increase with body length, with MeHg decreasing as a proportion of total. No differences between sexes.	Joiris et al. 1991
Physeter macrocephalus	M	Hg lower in smaller, nonreproductive females than reproductive females. No correlation between Hg and body length in females, negative correlation in males.	Cannella and Kitchener 1992
Stenella attenuata	K, L, M, O	Cd increase with age in multiple organs. Hg increase with age, higher in females.	André et al 1990, 1991b
S. coeruleoalba	K, L, M	Increases in concentration with age for Fe, Pb, Ni, Cd, Hg in muscle, Pb, Ni, Cd, Hg in liver, Hg in kidney; Mn, Zn and Cu decreased with age in liver, Mn and Cu decreased with age in kidney. Complex changes seen at weaning and between calves and older animals.	Honda et al. 1983
S. coeruleoalba	Bone	Complex relations with various metals and age and developmental state. Total Hg increased with age.	Honda et al. 1986a
S. coeruleoalba	K, L, M, O	Hg increased with total length in liver and muscle. No difference between sexes.	Andre et al. 1991a, b
S. coeruleoalba	K, L, M, O	Cd concentrations increase markedly from birth to 1.5 yr in most organs, remaining constant to age 15, then increasing with age in kidney, liver, muscle. Zn concentrations in various organs did not change with age in adults; Cu decrease with age in liver, kidney. Pb, Ni increase with age in muscle, liver, Other changes noted.	Honda and Tatsukawa 1983, Honda et al. 1983
S. coeruleoalba	Bone	Complex changes from fetal to old age noted in several elements. Cd, Hg, Pb, Zn increase with age in mature animals; MeHg constant. Pb in liver highest in mature males.	Honda et al. 1986a
S. coeruleoalba	Muscle	Hg in muscle increase with age.	Arima and Nagakura 1979

Continued on next page

Table 10-4 continued

Group and Species	Organ	Relation with Age and Sex	References
S. coeruleoalba	Br, K, L, M	Hg increase with length.	Leonzio et al. 1992
Tursiops truncatus	K, L, M	Cd concentrations in kidneys positively correlated with total length. Cu in livers highest in neonates, decreasing with age.	Wood and Van Vleet 1996
Mysticete cetaceans			
Balaenoptera acutorostrata	L	Correlations between age and Cd, Fe, and Hg in livers. No relationships with Mn, Zn, Cu, Pb, Ni, Co. Only Fe differed between sexes, higher in females.	Honda et al. 1987
B. acutorostrata	K, L, M	Correlations between age and Cd in muscle, Hg, Se in liver.	Hansen et al. 1990
B. physalus	K, L, M	Total Hg and organic Hg in liver, total Hg in muscle increase with age. No differences between sexes.	Sanpera et al. 1993
B. physalus	K, L, M	Complex relationships that vary with location and sex for Cd, Cu, Zn in three tissues.	Sanpera et al. 1996
Polar bears, sea otters, and sirenians			
Dugong dugon	K, L, M	Complex relationships and interactions. Cu, Mn in liver and kidney negatively correlated with age. Zn, Cd increase with age in liver and kidney; Fe increase with age in liver and muscle, Ag with age in kidney, Co in liver. No relationships with age and Ag in liver, Fe and Co in kidney, Zn, Cu, Mn in muscle.	Denton et al. 1980
Trichechus manatus	K, L	Size and Cu in livers varied negatively, Cd in kidneys increase, whereas Fe in livers and Pb in livers and kidneys showed no relationship. Metal concentrations did not vary by sex.	O'Shea et al. 1984
Ursus maritimus	L	Effect of age and location on Cd, Hg, and Se concentrations. No effect of age on Ag, As, Ca, Cu, Fe, K, Mg, Mn, Na, P, Sr, Zn.	Norstrom et al. 1986
U. maritimus	L	Cd, Hg, Se increase with age in livers; K, Mn, Mg, and P decrease with age; Ag, Ca, Na no trend with age.	Braune et al. 1991
U. maritimus	Hair	Total Hg in hair not related to age or sex.	Born et al. 1991
U. maritimus	L, M	Hg in livers of adults higher than young in some areas. No differences by sex.	Lentfer and Galster 1987

B = blubber; Br = brain; K = kidney; L = liver; M = muscle; MeHg = methyl mercury; O = other.

stage, probably because of a higher absorption efficiency from milk and rapid bone growth phases in comparison to later ages (Honda et al. 1986a). Premature and normal pups of ringed seals do not differ in cadmium concentrations in hair (Hyvärinen and Sipilä 1984). Broad regional trends in cadmium concentrations in livers of polar bears have been noted across the Arctic (Norstrom et al. 1986).

Lead

Lead is a nonessential element that is well known for its toxicity in mammals. It is a major airborne contaminant. Some primary sources of introduction of lead to the environment are from production of storage batteries, automotive exhaust, combustion of fuel with lead additives, smelting, pigments in paints, water pipes (particularly in older plumbing systems), and lead arsenate used as an insecticide. Pathology of lead poisoning has been well documented for domestic animals and humans, with evidence of the latter extending back to the ancient Romans. Modern causes of lead poisoning include ingestion of lead paint fragments by children and consumption of lead-tainted illegal moonshine whiskey. A variety of sublethal behavioral effects of lead are known from laboratory studies, and chronic exposure can cause pathological disorders of the nervous system, gastrointestinal tract, renal system (correlated histopathologically by the presence of lead-based intranuclear renal inclusion bodies), and immunotoxicity. Lead exposure also interferes with the production of hemoglobin and red blood cells. The inhibition of enzymes (delta aminolevulinic acid dehydratase) in the biosynthesis of heme provides a well-known biochemical marker of lead exposure (Peakall 1992).

Lead has generally not been found in marine mammal tissues at levels that are cause for concern. Most reports are of concentrations in soft tissues such as liver, kidney, and muscle (Appendices 5 to 8). The liver and kidneys of marine mammals tend to have higher lead residue concentrations than muscle, blubber, or other soft tissues. There is no clear pattern across studies that examine the relative amounts of lead in liver versus kidney of marine mammals. In most studies, concentrations in these tissues are less than 1 ppm (wet weight) and in nearly all cases are within the normal ranges

Table 10-5. Examples of Unusually High Concentrations (ppm) of Cadmium, Mercury, and Lead Reported in Soft Tissue of Marine Mammals

Group and Species	Cadmium	Mercury	Lead	References
Pinnipeds				
Arctocephalus gazella	684 (K, D)			Malcolm et al. 1994
Callorhinus ursinus	568 (K, D)		1.8 (K, W) 14.8 (L, D)	Anas 1974a, Goldblatt and Anthony 1983
Erignathus barbatus		420 (L, W)		Smith and Armstrong 1975
Halichoerus grypus		1097 (L, W)	7.0 (L, W)	Law et al. 1991, Simmonds et al. 1993
Odobenus rosmarus	458 (K, D)		11.6 (K, D)	Warburton and Seagars 1993
Phoca hispida	608 (K, D)			Wagemann 1989
P. vitulina		>700 (L, W)	2.3 (L, W)	Duinker et al. 1979, Reijnders 1980, Roberts et al. 1976
Zalophus californianus	569 (K, D)	240 (L, W)	3.0 (K, W)	Braham 1973, Buhler et al. 1975, Martin et al. 1976
Odontocete cetaceans				
Delphinus delphis			3.5 (L, W)	Kuehl et al. 1994
Delphinapterus leucas	275 (K, D)	756 (L, D)	2.1 (L, D)	Wagemann et al. 1990
Globicephala melanea	425 (K, D)	626 (L, D)		Meador et al. 1993
Kogia breviceps	412 (K, W)			Marcovecchio et al. 1990
Monodon monoceros	800 (K, D)			Wagemann et al. 1983
Phocoena phocoena			4.3 (L, W)	Law et al. 1991
Pseudorca crassidens	106 (K, W)	728 (L, W)		Baird et al. 1989
Stenella coeruleoalba		1544 (K, W)	12.4 (K, D)	André et al. 1991a, Leonzio et al. 1992
Tursiops truncatus		13,150 (L, D)		Leonzio et al. 1992
Mysticete cetaceans				
Balaenoptera acutorostrata	115 (K, D)		2.6 (L, D)	Honda et al. 1987
B. physalus	209 (K, D)			Sanpera et al. 1996
Other				
Dugong dugon	308 (K, D)			Denton et al. 1980
Trichechus manatus	190 (K, D)		7.1 (K, D)	O'Shea et al. 1984
Ursus maritimus			1.6 (L, W)	Norheim et al. 1992

K = kidney; L = liver; D = dry weight basis; W = wet weight basis.

seen in other mammals, including humans. Some of the higher concentrations reported in soft tissues of various species are provided in Table 10-5. Comparatively high concentrations reported in livers of a small number of harbor porpoises and one white-beaked dolphin (*Lagenorhynchus albirostris*) from Danish waters (to 5.3 ppm and 4.5 ppm wet weight, respectively) more than 20 years ago (Andersen and Rebsdorff 1976) have not been replicated. Similarly, early reports of high lead concentrations in livers of gray seals and harbor seals (to 17 and 12 ppm wet weight, respectively) around Great Britain (Holden 1975) have not been repeated in more recent sampling (Law et al. 1991).

Most studies of concentrations of lead in soft tissues of marine mammals show no consistent trend with age or sex (see references in Table 10-4). However, a few studies have correlated concentrations of lead in soft tissues with age: higher concentrations in liver and kidney of older harbor seals were reported by Drescher et al. (1977); decreases in lead concentrations with age were found in teeth of gray seals (Heppleston and French 1973); lead concentrations in livers of pilot whales increased with body length (Meador et al. 1993) but not age (Muir et al. 1988a); and lead in livers of harbor seals increased with age in males, but not in females or in kidneys of either sex (Miles et al. 1992). The most detailed study to show a trend in lead concentrations in soft tissues with age was conducted in striped dolphins. Lead concentrations in muscle, liver, and kidney increased with age until about 1 year, leveled off up to 18 years, then increased in older age and were highest in the oldest individuals (Honda et al. 1983).

Bone is the prominent depot for long-term storage of lead in vertebrates, and the highest concentration of lead known in tissues of marine mammals is 61.6 ppm (wet weight) in bone of a young bottlenose dolphin stranded on the edge of Spencer Gulf, Australia, an area known for emissions from a lead smelter (Kemper et al. 1994). A maximum of 62.8 ppm (dry weight) reported for the humerus of a California sea lion was considered comparable to values in normal human bone (Braham 1973). Bones of most other marine mammals examined have had far less lead and, although generally negligible concentrations are reported in marine mammal bone, the number of studies on this topic has been limited. Amounts present are usually higher than those in soft tissues (Braham 1973, Roberts et al. 1976, Goldblatt and Anthony 1983, Kemper et al. 1994). Honda et al. (1984) noted that, although bone of striped dolphin represented only 4% of the body weight, nearly 13% of the body burden of lead was in the skeleton, with certain bones having higher concentrations than others. Lead in female striped dolphin bone occurred at lower concentrations than in males, and lead accumulated most rapidly during the suckling period (Honda et al. 1986a). There are no published studies indicating histopathological damage or biochemical lesions caused by lead exposure in marine mammals. One account of a possible case of lead poisoning in an unspecified marine mammal (but without documentation of lead concentrations in tissues) was suggested by Britt and Howard (1983). Smith et al. (1990, 1992) used isotopic ratios to show that, although lead concentrations in teeth of sea otters have not changed between modern and preindustrial eras, the source of lead has shifted from naturally derived lead to anthropogenic aerosol-dominated forms.

Mercury

Long recognized for its poisonous effects, mercury is one of the few nonessential elements that shows appreciable biomagnification in marine food webs and has a relatively low threshold for toxicity. Despite the latter, marine mammals have evolved biochemical mechanisms to tolerate seemingly high exposure to mercury in the food chain. Mercury enters the environment as a contaminant resulting from a number of processes, including mining, combustion of fossil fuels, manufacturing of paper, and chlor-alkali plants. It is intentionally used as a fungicide, particularly as a seed dressing. Much mercury in the environment is also released through natural processes, and certain regions have higher mercury levels from geological sources. Mercury can also be highly volatile. Natural mercury brought to the sea surface from cold upwellings in the equatorial Pacific may volatize to the atmosphere in quantities approximately equal to all global anthropogenic emissions (Kim and Fitzgerald 1986). Atmospheric mercury can return to the Earth as fallout in particulate matter. Both aerobic and anaerobic microorganisms in sediments and soils convert various organic and inorganic forms of mercury to dimethyl (CH_3-Hg-CH_3) or methyl forms. The highly toxic methyl mercury ion (CH_3-Hg^+ or MeHg) is soluble in water, taken up by organisms, and can biomagnify several orders of magnitude in the food chain.

Mercury is best known for its neurotoxic effects and was once used in the production of felt hats, which inspired the character of the "Mad Hatter" in Lewis Carroll's *Alice in Wonderland*. It is also nephrotoxic, immunotoxic, mutagenic, crosses the placenta, and is found in milk. In recent history there have been notable cases of human mercury poisoning. In the 1950s and 1960s, people in the area of Minamata Bay and Nigata Bay in Japan suffered an epidemic of paralysis and mortality that was traced to seafood contaminated with mercury from effluents of chlor-alkali plants. In these instances, mercury concentrations in fish were up to 2000 times that of mercury concentrations in the surrounding water. Mass poisonings also occurred in recent history in Iraq, where 459 deaths were attributed to ingesting bread in which flour was made from mercury-treated seed. In comparison to other forms of mercury, methyl mercury is readily absorbed by the gastrointestinal tract. Mercury accumulates in the kidneys, liver, and brain (particularly the methyl form), but the organ distribution and toxicity also varies with the chemical form. In marine mammals, most chemical analyses have determined the concentrations of total mercury in tissues rather than the more toxic methyl mercury fraction (Appendices 5 to 8). A few studies report "organic mercury" (not strictly equivalent to methyl mercury) as a fraction of the total mercury.

One of the most outstanding cases of interactions between toxic elements is the apparent protective effect of selenium against mercury toxicity. Laboratory studies of a number of organisms show that various toxic effects of mercury were prevented or reduced in severity by simultaneous or prior exposure to selenium (Cuvin-Aralar and Furness 1991). In marine mammals, tissue concentrations of mercury that would indicate toxicity in other species are often exceeded with no evidence of harm, but these concentrations are typically accompanied by increased selenium in the liver in a 1:1 molar ratio (Koeman et al. 1973, 1975). The interactions and molecular level complexities of this protective effect are not well understood (Lee et al. 1977, van de Ven et al. 1979, Cuvin-Aralar and Furness 1991), but may include redistribution of mercury away from sensitive organs (such as the kidney) to muscle and other tissue. In the absence of selenium, mercury is bound to metallothionein proteins (see section on cadmium), which detoxify mercury, but also may cause long-term mercury retention. Selenium apparently diverts binding of mercury away from metallothionein to higher molecular weight proteins. Very little of the mercury in sea lion livers, for example, was bound to metallothionein (Lee et al. 1977), and no significant correlation exists between metallothionein and mercury in livers in harbor seals (Tohyama et al. 1986). Some of the proposed mechanisms for the protective effect of selenium against mercury have been reviewed by Cuvin-Aralar and Furness (1991), and include redistribution of mercury, competition for binding sites, formation of a mercury–selenium complex (described in cetacean livers by Martoja and Viale 1977), conversion of toxic forms of mercury to more benign forms, and prevention of oxidative damage.

Mercury determinations in tissues of marine mammals typically focus on the liver, although other tissues and organs sometimes examined include kidney, muscle, blubber, and hair (Appendices 5 to 8). Concentrations of total mercury are usually higher in liver than in kidney, muscle, or other tissues and have often been shown to increase significantly in liver with age (although the proportion that is methyl mercury decreases with age) (Table 10-4). Some of the concentrations of total mercury found in livers of marine mammals are far in excess of those that would be toxic to other mammals, but lethal effects have generally not been observed. This is apparently because of the metabolic capacity of marine mammals to protectively guard against mercury toxicity. Although most of the mercury in fish prey is in the highly toxic methylated form, the proportion of total mercury in marine mammal livers that is actually methylated is usually very low, with the methyl mercury fraction highest in other tissues like muscle, which are less active as sites of metabolic detoxification. Examples of extraordinarily high concentrations of total mercury in livers of marine mammals without evidence of accompanying toxicity include cases where mercury reached hundreds and sometimes thousands of parts per million (Table 10-5). In humans poisoned at Minamata Bay, in contrast, mercury in livers ranged from 22 to 70 ppm, but unlike in marine mammals, nearly all was methyl mercury (Britt and Howard 1983). Areas where mercury of geologic origin is naturally high, such as the Mediterranean Sea, produce very high mercury concentrations in marine mammals (André et al. 1991a, 1991b; Leonzio et al. 1992; Kemper et al. 1994). Marine mammals that typically feed lower in the food chain, such as baleen whales and sirenians, have very low mercury concentrations in liver in comparison with piscivorous species (Denton and Breck 1981; O'Shea et al. 1984; Byrne et al. 1985; Honda et al. 1986b, 1987; Dietz et al. 1990; Sanpera et al. 1993). Unlike organochlorines in blubber, which are usually highest in adult males, mercury concentrations in livers of marine mammals generally show no differences between sexes or are higher in females (Table 10-4). Concentrations of mercury in tissues of some marine mammals could pose human health risks in areas where people consume organs from these animals (for example, see Botta et al. 1983, Andersen et al. 1987, Simmonds et al. 1994), but the potential risk relates to the extent to which mercury is methylated (Eaton et al. 1980).

Mercury–selenium correlations have been determined in tissues of numerous marine mammals, and results are consistent with a role for selenium in protection against mercury toxicity (Koeman et al. 1973, 1975; Cuvin-Aralar and Furness 1991). Such positive correlations have been noted, for example, in muscle, bone, livers, kidneys, and brains of striped dolphins (Arima and Nagakura 1979, Honda et al. 1986a, Leonzio et al. 1992); in kidneys, livers, and muscle of belugas (Wagemann et al. 1990); in livers and kidneys of harbor, harp, gray, ringed, and bearded seals (*Erignathus barbatus*) (Smith and Armstrong 1978, van de Ven et al. 1979, Reijnders 1980, Ronald et al. 1984b, Perttilä et al. 1986, Wagemann et al. 1988, Frank et al. 1992); in livers of bottlenose dolphins (Kuehl et al. 1994); in livers and kidneys of pilot whales (Julshamn et al. 1987, Muir et al. 1988a, Caurant et al. 1993, Meador et al. 1993); in livers and kidneys of white-beaked dolphins (Muir et al. 1988a); in livers of polar bears (Norstrom et al. 1986, Braune et al. 1991, Norheim et al. 1992); in livers and kidneys of walrus (Taylor et al. 1989, Warburton and Seagars 1993); and in livers and kidneys of narwhals (Wagemann et al. 1983). Many of these studies also verify the approximately 1:1 molar ratio of mercury to selenium in liver, suggestive of a protective effect of selenium (Koeman et al. 1975). The mercury-to-selenium ratios in fish (as prey species) differ dramatically from those seen in marine mammals (Koeman et al. 1973, 1975; Kari and Kauranen 1978). Numerous other studies have conducted analyses on relations between mercury and various other elements in organs of marine mammals (Appendices 5 to 8). The gross distribution of mercury within different parts of the liver has been determined for a few species and appears homogenous (Nielsen and Dietz 1990, Stein et al. 1992).

A detailed example of investigations on the distribution of mercury, methyl mercury, and selenium in marine mammal body compartments was conducted on striped dolphins (Itano et al. 1984a,b,c). Similar patterns revealed by these analyses have also been observed in other marine mammal species (see references in Appendices 5 to 8). Samples of 15 organs and tissues were analyzed from 55 individuals. Highest total mercury was found in the liver, but liver had the lowest proportion of methyl mercury. Selenium was also highest in the liver. Total mercury concentrations in muscle and liver increased with age, but leveled off at age 20 to 25 years (total body burdens reached a constant level at age 16 years). No differences in total mercury concentrations were apparent in tissues or whole bodies of males and females. Selenium and mercury concentrations were significantly correlated in nearly all tissues and organs. In the striped dolphins, 90% of the entire body burden of methyl mercury was located in muscle. In fetuses and calves of striped dolphins, most of the mercury was methylated, and in the term fetuses was about 1% of that in the pregnant female. Total mercury (most of which was methylated) in milk was much lower than that in any other tissue. Transfer of mercury across the placenta, relatively low concentrations in milk, and high proportions of methyl mercury in pup liver have also been reported in pinnipeds (Wagemann et al. 1988).

Unlike investigations on other metals in marine mammals, a few experimental studies have been conducted on toxicology of mercury in seals. Tillander et al. (1972) dosed a captive female ringed seal with radioactively labeled methyl mercury and determined a two-phased excretion rate. One component involved rapid excretion of about 55% of the mercury with a half-time of about 3 weeks, whereas the remainder was excreted with a half-time of 500 days. Ramprashad and Ronald (1977) administered methyl mercuric chloride at two dosage levels (0.25 and 25 mg/kg) to four harp seals and determined an effect of mercury exposure on sensory epithelium in the cochlea. Ronald et al. (1975, reported in Holden 1978) found that at the higher dosage level death occurred in 20 to 26 days due to renal failure in these seals, which showed increases in mercury concentrations in the brain and liver, as well as hepatitis and renal failure. Freeman et al. (1975) reported that in harp seals given 0.25 mg/kg methyl mercuric chloride for 61 days, more than 70% of the mercury in the liver was inorganic, demonstrating substantial demethylation. Muscle tissue had higher proportions of methyl mercury. In vitro synthesis of steroids in gonads and adrenals of harp and gray seals due to mercury exposure has also been reported (Freeman et al. 1975, Freeman and Sangalang 1977). Van de Ven et al. (1979) administered methyl mercuric chloride to captive gray seals and found that both mercury and selenium increased in livers and kidneys, but only mercury increased in other tissues. In vitro tests for several enzymatic demethylation mechanisms were negative, and no indication of demethylation processes by the microflora of the gut was found. Administration of methyl mercury also did not stimulate the P-450 enzyme system. Results suggested that the role of selenium in ameliorating mercury toxicity primarily involves tissue distribution and accumulation. However, Himeno et al. (1989) reported higher in vitro demethylation activity in liver and kidneys of harbor seals in comparison with laboratory rats and mice. Few studies of mercury contamination in marine mammals have included associated investigations of histopathology. Rawson et al. (1993, 1995) noted mercury-associated pigment granules and liver disease in bottlenose dolphins with relatively high concentrations of mercury in comparison to those without such conditions.

Other Elements

Various studies have provided data on concentrations of up to 40 trace elements in tissues of marine mammals (Appendices 5 to 8). Most of these elements are currently of lesser concern as toxic contaminants in marine mammals than cadmium, lead, or mercury, chiefly because of their relatively low concentrations, their necessity as essential dietary elements, or because of an absence of information suggesting harmful effects at reported levels of exposure. Arsenic has been reported in numerous species of marine mammals, but at levels not considered toxic. Copper is an essential element that typically decreases with age in livers of marine mammals (Table 10-4). It has not been implicated as a potential threat in marine mammals except for a specific localized case where it was applied as an aquatic herbicide in a winter feeding ground of Florida manatees (O'Shea et al. 1984). In this case, elevated hepatic copper concentrations were found to correspond with geographic patterns of copper herbicide use (after adjusting for age), with maximum concentrations in liver equivalent to those associated with toxic effects in some sensitive terrestrial mammals. Unusually high concentrations of silver have been documented in the livers of Alaska beluga whales, positively correlated with selenium and age, but the toxicological significance of these findings is unknown (Becker et al. 1995). Similarly, livers of some Alaskan marine mammals appear to have elevated vanadium concentrations relative to marine mammals from other areas (Mackey et al. 1996). Most of the work on selenium in marine mammals has focused on its protective association with mercury, but this essential element can be toxic to other mammals in its own right. Very little is known about interactions among other elements in marine mammals, but imbalances, particularly involving bromine, have been suggested as playing a role in premature parturition in California sea lions (Martin et al. 1976). Certainly, much remains to be learned about the roles of potentially toxic elements in the health of marine mammals. Recent work, for example, has revealed the presence of organotins (butyltins) in blubber of eight species of marine mammals, with higher concentrations in individuals from coastal areas (Iwata et al. 1994). Unlike organochlorines, the butyltins have a greater affinity for tissues with a higher protein-binding capacity, such as liver and hair, than lipid-rich tissues (Kannan et al. 1996, Kim et al. 1996). Butyltins have been used in a variety of applications, including marine antifouling paints, pesticides, and wood preservatives. The possible toxic significance of their recently discovered (Iwata et al. 1994, Kim et al. 1996, Kannan et al. 1998) accumulation in tissues of marine mammals is undetermined.

Other Contaminants and Toxins

Air Pollutants

Marine mammals inhabiting urbanized coastal areas are subject to contamination from various air pollutants at levels similar to those experienced by large fractions of the human population. However, little research has been conducted on potential impacts of air pollution on marine mammals. Rawson et al. (1991) noted the presence of carbon deposits in macrophages in mediastinal lymph nodes and lung tissue of Atlantic bottlenose dolphins from the west coast of Florida. Such deposits are also typical in humans and domestic animals in urban areas, and stem from inhaled carbon particles that enter the alveoli and are ingested by macrophages that pass to the mediastinal lymph nodes of the pulmonary lymph systems. No pathology was associated with these deposits in the limited number of cases observed in Florida. Mercuric selenide (tiemannite) granules in tissues of bottlenose dolphins and pilot whales from this area may possibly result from inhalation of atmospheric mercury (Rawson et al. 1995).

Aromatic and Polycyclic Aromatic Hydrocarbons

The aromatic and polycyclic aromatic hydrocarbons (AHs and PAHs) are often topics of investigation in marine pollution studies, but have not been an extensive focus of inquiry in marine mammals. These compounds can stem from numerous natural and anthropogenic sources, but as contaminants they are chiefly associated with components of petroleum. AHs have carbon atoms arranged in ring structures (1 to 6 rings with 6 carbon atoms per ring) and PAHs are AHs with up to 7 fused carbon rings that can have substitutions attached (Albers 1995). They include some well-known carcinogens such as benzo[*a*]pyrene. The PAHs and AHs do not show great biomagnification in food chains and are readily metabolized by many organisms. Hellou et al. (1991) found relatively low concentrations of PAHs in muscle tissue of 28 harp seals from the northwest Atlantic, and no accumulation with age. Low concentrations were also reported in muscle samples from smaller numbers of a wider range of species from the same region, including harbor, harp, hooded (*Cystophora cristata*), and ringed seals, and single beluga, sperm whale, and minke whales, common and white-sided dolphins (Hellou et al. 1990). Law and Whinnett (1992) also reported similarly low concentrations of PAHs in muscle tissue of 26 harbor porpoises from the coast of the United Kingdom. PAHs have been hypothesized to be responsible for tumors in belugas of the St. Lawrence River estuary through the formation of DNA adducts (e.g., Martineau et al. 1988), but this view is not fully accepted (Geraci et al. 1987), and DNA adducts have also been reported at similar levels from beluga whale livers in remote locations with negligible PAH contamination (Ray et al. 1992).

Dinoflagellate Toxins

Irruptions of toxin-producing marine dinoflagellates of various species occur throughout the world. Although not considered anthropogenic contamination, a seeming increase in irruptions worldwide is suspected to be associated with nutrient enrichment and other human activities (Anderson 1994). Some dinoflagellate blooms have been associated with mortality of fish, birds, and mammals, as well as human mortality and illness (particularly from ingesting shellfish and other seafood that have concentrated dinoflagellate toxins). The characteristics of poisoning vary with species of dinoflagellate, mode of exposure, and chemical characteristics of the toxins produced. The resulting toxic syndromes (known best from human exposure) fall into four categories: neurotoxic shellfish poisoning, paralytic shellfish poisoning (PSP), diarrhetic shellfish poisoning, and ciguatera poisoning (Baden 1983, Steidinger and Baden 1984). Most symptoms are neurological or gastrointestinal. Mortality of bottlenose dolphins and manatees has been associated with blooms of the dinoflagellate *Ptychodiscus brevis* (also referred to as *Gymnodinium breve*) in Florida (Gunter et al. 1948, Layne 1965, O'Shea et al. 1991), which were also held responsible for the widely publicized manatee die-off of 1996 (Bossart et al. 1998). In the latter case, manatees may have been exposed to lethal amounts of brevetoxin through ingestion of contaminated food or water or perhaps through inhalation of aerosols released during lysis of cells by wind and wave action at the water surface (severe lesions were present in the lungs of many manatees). Deaths attributable to toxins produced by other dinoflagellate species have also been observed in humpback whales (Geraci et al. 1989) and sea otters (De Gange and Vacca 1989). In the case of the humpback whales, the agent was saxitoxin produced by the dinoflagellate *Alexandrium tamarense,* a substance that can accumulate in food chains and had built up in mackerel ingested by the humpbacks. Experimental feeding studies have demonstrated that sea otters will selectively avoid the most toxic portions of prey clams with high quantities of PSP toxins (Kvitek et al. 1991).

Radionuclides

Ionizing radiation can produce a broad spectrum of injuries to mammals at the molecular, cellular, organ, and organismal levels, including effects on behavior, growth and development, and mutagenicity and carcinogenicity (for a review, see Eisler 1994). Anthropogenic radionuclides that contaminate today's ecosystems come primarily from fallout from nuclear weapons testing (which peaked 30 to 50 years ago), the Chernobyl accident in 1986, nuclear reactor operations, nuclear fuel processing and disposal, and applications in medicine, industry, agriculture, and research. Only a few studies have examined marine mammals to determine the extent of their contamination by radionuclides, and none has reported any associated effects. Anderson et al. (1990) examined milk and tissues of gray seals collected in 1987 from the North Sea and North Atlantic for cesium-137 (^{137}Cs) and the actinides plutonium and americium. Levels of ^{137}Cs were low in milk and tissues and about 70% was attributed to the nuclear reprocessing industry in England, with the remainder from the Chernobyl accident. Actinides were barely detectable. Anderson et al. (1990) concluded that there was no evidence for extensive concentration of radionuclides from fish prey and that the radiation doses received by seals were below the limit for human radiation workers, but likely higher than limits set for the general public. Calmet et al. (1992) collected muscle and liver tissues from spotted, spinner, and common dolphins from the eastern Pacific and determined ^{137}Cs concentrations as well as ^{40}K and ^{210}Pb. These investigators concluded that the concentration factor from seawater was about the same as in fish, and that the radiation doses measured are unlikely to have effects on dolphin populations. Other marine mammals in which radionuclides have been measured include fin whales (Osterberg et al. 1964, Samuels et al. 1970), harp seals (Samuels et al. 1970), sperm whales, spotted seals, and bearded seals (Holtzman 1969, cited in Eisler 1994).

Conclusions and Future Research Needs

Interpreting the significance of the presence of contaminants in marine mammal tissues is a difficult and sometimes controversial area. However, the reader should bear in mind that the published literature summarized in the appendices to this chapter includes results of organochlorine and toxic element residue surveys from more than 13,000 samples of marine mammals. Laboratory analyses for determination of contaminants are expensive; these results represent a cumulative investment by society of many millions of dollars. An enormous amount of information has been learned from these surveys about fundamental patterns of variation in contaminant residues in marine mammals. Some of the organochlorines, for example, are known to accumulate differentially in the most lipid-rich tissues, and to increase with age in males but not in reproductive females who largely shunt these substances to their young through lactation. DDE and PCBs are the most commonly reported organochlorines. PCBs vary widely in individual persistence and resistance to metabolism. Global transport of organochlorines has been verified, and these substances show up in the bodies of marine mammals, albeit at low concentrations, in some of the most remote areas of the world's oceans. In more inshore species in some of the most heavily contaminated environments, an alphabet soup of different organochlorines can be detected, some at incredibly high concentrations, particularly in species highest in the food web.

The study of organochlorines is complex, however, and analyses that focus on residues alone, although expensive, leave much to be desired for interpretation of significance to the health and dynamics of exposed populations. In addition, many studies have small sample sizes. Despite a considerable investment in research, no marine mammal deaths in the wild have conclusively been shown to be a direct result of organochlorine or toxic element exposure. Indirect effects have been difficult to pinpoint. Reproductive impairment and gross lesions *associated with* organochlorine contaminants have been convincingly demonstrated in a few highly polluted areas, and experimental cause-and-effect studies have been carried out in only one species. There is mixed evidence for linkages with increased susceptibility to disease, and the complex study of biochemical and physiological effects in marine mammals is in its infancy. Although there is a lack of absolute scientific certainty in linking the presence of specific organochlorine contaminants to detrimental impacts on marine mammal populations, the body of indirect, circumstantial evidence reviewed in this chapter continues to grow. Organochlorine pollution of the seas will continue, particularly from PCBs, and efforts to abate contamination without waiting for more "smoking guns" can only benefit marine mammal conservation in the long run.

The study of toxic elements has included determination of residue concentrations from thousands of individuals in numerous species. This work has revealed patterns of variation that include relationships among metals, such as the protective linkage between selenium and mercury, organ-specific sites of accumulation for certain metals, and age-dependent accumulation (cadmium in kidneys) or loss (copper in livers) of some elements. Some elements can be found in very high concentrations in marine mammal tissues because of high, naturally occurring geologic sources (for example, mercury in the Mediterranean Sea) or relatively high amounts in prey (for example, cadmium in squid). However, most of the mercury found in marine mammal livers is not in the toxic methylated form, and although mercury may sometimes occur at concentrations in livers of marine mammals that exceed toxic amounts in other species, it appears to be detoxified and no harmful effects have been conclusively demonstrated. Similarly, higher concentrations of cadmium in tissues of marine mammals are protectively bound by metallothionein proteins, and no associated pathological effects of cadmium are known from these animals. Very little is known about sources of variation, interactions and effects of most of the other potentially toxic elements found in tissues of marine mammals. Unlike organochlorines, most research has not gone beyond the study of residue concentrations, and little detail is known about relationships among metals and marine mammal physiology, disease, or population dynamics. Knowledge of other contaminants and toxins in marine mammals is even less comprehensive although, unlike anthropogenic chemicals, dinoflagellate biotoxins have been more firmly established as sources of direct mortality and morbidity of marine mammals.

To have a maximum impact on knowledge, future research on the presence of contaminant residues in marine mammals needs to emphasize obtaining sufficiently large sample sizes to partition variation in residue concentrations among such potentially significant sources as age, sex, location, reproductive status, feeding habits, and nutritional status. In addition, to allow for adequate interpretation, contaminant studies should include corollary research such as gross pathology and histopathology, analysis of reproductive tracts, and evidence of biochemical lesions such as MFO activity, circulating levels of hormones, and other biomarkers. These data should be collected from the same individuals for which contaminant concentrations are determined. Knowledge of overall habitat quality and the status of the population (increasing or declining, harvested or unharvested) from which samples are obtained is also useful for interpretation of contaminant data. These requirements are a tall order and would have greatest chances of success in establishing associations between contaminant exposure and biological effects if carried out simultaneously in areas of contrasting contamination. Because of sample size requirements, advantage should be taken of obtaining fresh samples from by-catches of fisheries or from harvested populations. As an alternative approach, utilization of biopsy sampling techniques (Aguilar and Borrell 1994c) would be of particular value if combined with studies that emphasize longitudinal record keeping on reproductive histories of some of the same individuals (see Wells, Boness, and Rathbun, Chapter 8, this volume) or where detailed observational follow-up is possible, as in the case of pinniped breeding colonies. These suggestions notwithstanding, impacts of contaminants on marine mammals would most conclusively be determined through carefully controlled captive feeding experiments. Thus far, expense and ethical–political considerations have prevented such research in all but a very few cases.

Acknowledgments

I thank John Reynolds for his patience during the development and writing of this chapter, and Joe Geraci for his discussions and insight. Pam Smith and Brenda Coen worked many hours in word processing, and Liz Lucke and Mary Jane Dodson located many of the references. Very helpful comments were made on the manuscript by David Grove, Melissa Etheridge, Barnett Rattner, John Reynolds, Butch Rommel, Edward Van Vleet, and anonymous referees. David Grove provided improved diagrams of chemical structures.

Literature Cited

Abarnou, A., D. Robineau, and P. Michel. 1986. Contamination par les organochlorés des dauphins de Commerson des Îles Kergulen. Oceanologica Acta 9:19–29.

Addison, R. F. 1989. Organochlorines and marine mammal reproduction. Canadian Journal of Fisheries and Aquatic Sciences 46:360–368.

Addison, R. F., and P. F. Brodie. 1973. Occurrence of DDT residues in beluga whales (*Delphinapterus leucas*) from the MacKenzie Delta, N.W.T. Journal of the Fisheries Research Board of Canada 30:1733–1736.

Addison, R. F., and P. F. Brodie. 1977. Organochlorine residues in maternal blubber, milk, and pup blubber from gray seals (*Halichoerus grypus*) from Sable Island, Nova Scotia. Journal of the Fisheries Research Board of Canada 34:937–941.

Addison, R. F., and P. F. Brodie. 1984. Characterization of ethoxyresorufin-O-deethylase in gray seal *Halichoerus grypus*. Comparative Biochemistry and Physiology C 79:261–263.

Addison, R. F., and P. F. Brodie. 1987. Transfer of organochlorine residues from blubber through the circulatory system to milk in the lactating gray seal *Halichoerus grypus*. Canadian Journal of Fisheries and Aquatic Sciences 44:782–786.

Addison, R. F., and T. G. Smith. 1974. Organochlorine residue levels in Arctic ringed seals: variation with age and sex. Oikos 25:335–337.

Addison, R. F., and W. T. Stobo. 1993. Organochlorine residue concentrations and burdens in gray seal (*Halichoerus grypus*) blubber during the first year of life. Journal of Zoology (London) 230:443–450.

Addison, R. F., M. E. Zinck, and R. G. Ackman. 1972. Residues of organochlorine pesticides and polychlorinated biphenyls in some commercially produced Canadian marine oils. Journal of the Fisheries Research Board of Canada 29:349–355.

Addison, R. F., S. R. Kerr, J. Dale, and D. E. Sergeant. 1973. Variation of organochlorine residue levels with age in Gulf of St. Lawrence harp seals (*Pagophilus groenlandicus*). Journal of the Fisheries Research Board of Canada 30:595–600.

Addison, R. F., P. F. Brodie, M. E. Zinck, and D. E. Sergeant. 1984. DDT has declined more than PCBs in eastern Canadian seals during the 1970s. Environmental Science and Technology 18:935–937.

Addison, R. F., M. E. Zinck, and T. G. Smith. 1986a. PCBs have declined more than DDT-group residues in Arctic ringed seals (*Phoca hispida*) between 1972 and 1981. Environmental Science and Technology 20:253–256.

Addison, R. F., P. F. Brodie, A. Edwards, and M. C. Sadler. 1986b. Mixed function oxidase activity in the harbour seal (*Phoca vitulina*) from Sable Is., N.S. Comparative Biochemistry and Physiology C 85:121–124.

Addison, R. F., K. W. Renton, A. J. Edwards, and P. F. Brodie. 1988. Polychlorinated biphenyls and mixed-function oxidase in gray seals (*Halichoerus grypus*). Marine Environmental Research 24:111–112.

Aguilar, A. 1983. Organochlorine pollution in sperm whales, *Physeter macrocephalus*, from the temperate waters of the eastern North Atlantic. Marine Pollution Bulletin 14:349–352.

Aguilar, A. 1984. Relationship of DDE / DDT in marine mammals to the chronology of DDT input into the ecosystem. Canadian Journal of Fisheries and Aquatic Sciences 41:840–844.

Aguilar, A. 1985. Compartmentation and reliability of sampling procedures in organochlorine pollution surveys of cetaceans. Residue Reviews 95:91–114.

Aguilar, A. 1987. Using organochlorine pollutants to discriminate marine mammal populations: A review and critique of the methods. Marine Mammal Science 3:242–262.

Aguilar, A. 1989. Organochlorine pollutants and cetaceans: a perspective. Pages 10–11 *in* P. G. H. Evans and C. Smeenk (eds.). European Research on Cetaceans 3. Proceedings of the third annual conference of the European Cetacean Society. Leiden, The Netherlands.

Aguilar, A., and A. Borrell. 1988. Age- and sex-related changes in organochlorine compound levels in fin whales (*Balaenoptera physalus*) from the eastern North Atlantic. Marine Environmental Research 25:195–211.

Aguilar, A., and A. Borrell. 1994a. Abnormally high polychlorinated biphenyl levels in striped dolphins (*Stenella coeruleoalba*) affected by the 1990–1992 Mediterranean epizootic. The Science of the Total Environment 154:237–247.

Aguilar, A., and A. Borrell. 1994b. Reproductive transfer and variation of body load of organochlorine pollutants with age in fin whales (*Balaenoptera physalus*). Archives of Environmental Contamination and Toxicology 27:546–554.

Aguilar, A., and A. Borrell. 1994c. Assessment of organochlorine pollutants in cetaceans by means of skin and hypodermic biopsies. Pages 245–267 *in* M. C. Fossi and C. Leonzio (eds.). Nondestructive Biomarkers in Vertebrates. Lewis Publishers, Boca Raton, FL.

Aguilar, A., and L. Jover. 1982. DDT and PCB residues in the fin whale, *Balaenoptera physalus,* of the North Atlantic. Report of the International Whaling Commission 32:299–301.

Aguilar, A., L. Jover, and J. Nadal. 1982. A note on the organochlorine contamination in a Blainville's beaked whale, *Mesoplodon densirostris* (de Blainville, 1817) from the Mediterranean Sea. Publicaciones del Departamento de Zoologia Universidad de Barcelona 7:85–90.

Aguilar, A., L. Jover, and A. Borrell. 1993. Heterogeneities in organochlorine profiles of Faroese long-finned pilot whales: Indication of segregation between pods? Report of the International Whaling Commission (Special Issue 14):359–367.

Ahlborg, U. G., G. C. Becking, L. S. Birnbaum, A. Brouwer, H. J. G. M. Derks, M. Feeley, G. Golor, A. Hanberg, J. C. Larsen, A. K. D. Liem, S. H. Safe, C. Schlatter, F. Waern, M. Younes, and E. Yrjänheikki. 1994. Toxic equivalency factors for dioxin-like PCBs. Chemosphere 28:1049–1067.

Albers, P. H. 1995. Petroleum and individual polycyclic aromatic hydrocarbons. Pages 330–355 *in* D. J. Hoffman, B. A. Rattner, G. A. Burton Jr., and J. Cairns Jr. (eds.). Handbook of Ecotoxicology. Lewis Publishers, Boca Raton, FL.

Alzieu, C. and R. Duguy. 1979. Teneurs en composés organochlorés chez les cétacés et pinnipèdes fréquentant les côtes françaises. Oceanologica Acta 2:107–120.

Ames, A. L. and E. S. Van Vleet. 1996. Organochlorine residues in the Florida manatee, *Trichechus manatus latirostris.* Marine Pollution Bulletin 32:374–377.

Anas, R. E. 1971. Organochlorine pesticides in northern fur seals, California sea lions, and birds, 1968–69. Pages 32–36 *in* Fur Seal Investigations, 1969. National Marine Fisheries Service, U.S. Department of Commerce, Special Scientific Report—Fisheries No. 628, Washington, D.C.

Anas, R. E. 1973. Mercury in fur seals. Pages 91–96 *in* D. R. Buhler (ed.). Proceedings of the workshop on mercury in the western environment. Oregon State University Continuing Education Publications, Corvallis, OR.

Anas, R. E. 1974a. Heavy metals in the northern fur seal, *Callorhinus ursinus,* and harbor seals, *Phoca vitulina richardi.* Fishery Bulletin 72:133–137.

Anas, R. E. 1974b. DDT plus PCB's in blubber of harbor seals. Pesticides Monitoring Journal 8:12–14.

Anas, R. E., and A. J. Wilson. 1970. Organochlorine pesticides in fur seals. Pesticides Monitoring Journal 3:198–200.

Anas, R. E., and D. D. Worlund. 1975. Comparison between two methods of subsampling blubber of northern fur seals for total DDT plus PCB's. Pesticides Monitoring Journal 8:261–262.

Andersen, A., K. Julshamn, O. Ringdal, and J. Mørkøre. 1987. Trace elements intake in the Faroe Islands II. Intake of mercury and other elements by consumption of pilot whales (*Globicephalus meleanus*). The Science of the Total Environment 65:63–68.

Andersen, S. H., and A. Rebsdorff. 1976. Polychlorinated hydrocarbons and heavy metals in harbour porpoise (*Phocoena phocoena*) and whitebeaked dolphin (*Lagenorhynchus albirostris*) from Danish waters. Aquatic Mammals 4:14–20.

Anderson, D. M. 1994. Red tides. Scientific American 271:62–68.

Anderson, G. R. V. 1991. Australia. Progress report on cetacean research, May 1989 to May 1990. Report of the International Whaling Commission 41:223–229.

Anderson, S. S., F. R. Livens, and D. L. Singleton. 1990. Radionuclides in gray seals. Marine Pollution Bulletin 21:343–345.

André, J. M., J. C. Amiard, C. Amiard-Triquet, A. Boudou, and F. Ribeyre. 1990. Cadmium contamination of tissues and organs of delphinids species (*Stenella attenuata*)—Influence of biological and ecological factors. Ecotoxicology and Environmental Safety 20:290–306.

André, J. M., A. Boudou, F. Ribeyre, and M. Bernhard. 1991a. Comparative study of mercury accumulation in dolphins (*Stenella coeruleoalba*) from French Atlantic and Mediterranean coasts. The Science of the Total Environment 104:191–209.

André, J. M., A. Boudou, and F. Ribeyre. 1991b. Mercury accumulation in Delphinidae. Water, Air, and Soil Pollution 56:187–201.

Arima, S., and K. Nagakura. 1979. Mercury and selenium content of *Odontoceti.* Bulletin of the Japanese Society of Scientific Fisheries 45:623–626.

Aucamp, P. J., J. L. Henry, and G. H. Stander. 1971. Pesticide residues in South African marine mammals. Marine Pollution Bulletin 2:190–191.

Augier, H., W. K. Park, and C. Ronneau. 1993. Mercury contamination of the striped dolphin *Stenella coeruleoalba* Meyen from the French Mediterranean coasts. Marine Pollution Bulletin 26:306–311.

Bacher, G. J. 1985. Mercury concentrations in the Australian fur seal *Arctocephalus pusillus* from SE Australian waters. Bulletin of Environmental Contamination and Toxicology 35:490–495.

Bacon, C. E., W. M. Jarman, and D. P. Costa. 1992. Organochlorine and polychlorinated biphenyl levels in pinniped milk from the Arctic, the Antarctic, California and Australia. Chemosphere 24:779–791.

Baden, D. G. 1983. Marine food-borne dinoflagellate toxins. International Review of Cytology 82:99–150.

Baird, R. W., K. M. Langelier, and P. J. Stacey. 1989. First records of false killer whales, *Pseudorca crassidens,* in Canada. Canadian Field-Naturalist 103:368–371.

Baker, A. N., 1978. The status of Hector's dolphin, *Cephalorhynchus hectori* (Van Beneden), in New Zealand waters. Report of the International Whaling Commission 28:331–334.

Ballschmiter, K., and M. Zell. 1980. Analysis of polychlorinated biphenyls (PCB) by glass capillary gas chromatography. Composition of technical Aroclor and Clophen-PCB mixtures. Zeitschrift für Analytische Chemie 302:20–31.

Ballschmiter, K., C. Rappe, and H. R. Buser. 1989. Chemical properties, analytical methods and environmental levels of PCB, PCTs, PCNs and PBBs. Pages 47–67 *in* R. D. Kimbrough and A. A. Jensen (eds.). Halogenated Biphenyls, Terphenyls, Napthalenes, Dibenzodioxins and Related Products, 2nd ed. Elsevier, Amsterdam.

Beck, G., T. G. Smith, and R. F. Addison. 1994. Organochlorine residues in harp seals, *Phoca groenlandica,* from the Gulf of St. Lawrence and Hudson Strait: An evaluation of contaminant concentrations and burdens. Canadian Journal of Zoology 72:174–182.

Beck, H., E. M. Breuer, A. Dross, and W. Mathar. 1990. Residues of PCDDs, PCDFs, PCBs and other organochlorine compounds in harbour seals and harbour porpoise. Chemosphere 20:1027–1034.

Becker, P. R., E. A. Mackey, R. Demiralp, R. Suydam, G. Early, B. J. Koster, and S. A. Wise. 1995. Relationship of silver with selenium and mercury in the liver of two species of toothed whales (odontocetes). Marine Pollution Bulletin 30:262–271.

Béland, P., S. DeGuise, and R. Plante. 1991. Toxicology and Pathology of St. Lawrence Marine Mammals. Final report, Wildlife Toxicology Fund, St. Lawrence National Institute of Ecotoxicology. World Wildlife Fund, Toronto.

Béland, P., S. DeGuise, C. Girard, A. Lagacé, D. Martineau, R. Michaud, D. C. G. Muir, R. J. Norstrom, É. Pelletier, S. Ray, and L. R. Shugart. 1993. Toxic compounds and health and reproductive effects in St. Lawrence beluga whales. Journal of Great Lakes Research 19:766–775.

Bergman, Å., and M. Olsson. 1985. Pathology of Baltic gray seal and ringed seal females with special reference to adrenocortical hyperplasia: Is environmental pollution the cause of a widely distributed disease syndrome? Finnish Game Research 44:47–62.

Bergman, Å., M. Olsson, and S. Reiland. 1992. Skull-bone lesions in the Baltic gray seal (*Halichoerus grypus*). Ambio 21:517–519.

Bergman, Å., R. J. Norstrom, K. Haraguchi, H. Kuroki, and P. Béland. 1994. PCB and DDE methyl sulfones in mammals from Canada and Sweden. Environmental Toxicology and Chemistry 13:121–128.

Bernhoft, A., and J. U. Skaare. 1994. Levels of selected individual polychlorinated biphenyls in different tissues of harbour seals (*Phoca vitulina*) from the southern coast of Norway. Environmental Pollution 86:99–107.

Blomkvist, G., A. Roos, S. Jensen, A. Bignert, and M. Olsson. 1992. Concentrations of sDDT and PCB in seals from Swedish and Scottish waters. Ambio 21:539–545.

Boon, J. P., P. J. H. Reijnders, J. Dols, P. Wensvoort, and M. T. J. Hillebrand. 1987. The kinetics of individual polychlorinated biphenyl congeners in female harbour seals (*Phoca vitulina*), with evidence for structure-related metabolism. Aquatic Toxicology 10:307–324.

Boon, J. P., F. Eijgenraam, J. M. Everaarts, and J. C. Duinker. 1989. A structure–activity relationship (SAR) approach towards metabolism of PCBs in marine animals from different trophic levels. Marine Environmental Research 27:159–176.

Boon, J. P., E. Van Arnhem, S. Jansen, N. Kannan, G. Petrick, D. Schulz, J. C. Duinker, P. J. H. Reijnders, and A. Goksøyr. 1992. The toxicokinetics of PCBs in marine mammals with special reference to possible interactions of individual congeners with the cytochrome P450-dependent monooxygenase system: An overview. Pages 119–159 *in* C. H. Walker and D. R. Livingstone (eds.). Persistent Pollutants in Marine Ecosystems. Pergamon Press, Oxford.

Boon, J. P., I. Oostingh, J. van der Meer, and M. T. J. Hillebrand. 1994. A model for the bioaccumulation of chlorobiphenyl congeners in marine mammals. European Journal of Pharmacology, Environmental Toxicology and Pharmacology Section 270 (2–3):237–251.

Borlakoglu, J. T., and K. D. Haegele. 1991. Comparative aspects on the bioaccumulation, metabolism and toxicity with PCBs. Comparative Biochemistry and Physiology C 100:327–338.

Born, E. W., I. Kraul, and T. Kristensen. 1981. Mercury, DDT, and PCB in the Atlantic walrus (*Odobenus rosmarus rosmarus*) from the Thule District, North Greenland. Arctic 34:255–260.

Born, E. W., A. Renzoni, and R. Dietz. 1991. Total mercury in hair of polar bears (*Ursus maritimus*) from Greenland and Svalbard. Polar Research 9:113–120.

Borrell, A. 1993. PCBs and DDTs in blubber of cetaceans from the northeastern North Atlantic. Marine Pollution Bulletin 26:146–151.

Borrell, A., and A. Aguilar. 1987. Variations in DDE percentage correlated to total DDT burden in the blubber of fin and sei whales. Marine Pollution Bulletin 18:70–74.

Borrell, A., and A. Aguilar. 1990. Loss of organochlorine compounds in the tissues of a decomposing stranded dolphin. Bulletin of Environmental Contamination and Toxicology 45:46–53.

Borrell, A., and A. Aguilar. 1993. DDT and PCB pollution in blubber and muscle of long-finned pilot whales from the Faroe Islands. Report of the International Whaling Commission (Special Issue 14):351–358.

Borrell, A., D. Bloch, and G. Desportes. 1995. Age trends and reproductive transfer of organochlorine compounds in long-finned pilot whales from the Faroe Islands. Environmental Pollution 88:283–292.

Bossart, G. D., D. G. Baden, R. Y. Ewing, B. Roberts, and S. D. Wright. 1998. Brevetoxicosis in manatees (*Trichechus manatus latirostris*) from the 1996 epizootic: Gross, histologic, and immunohistochemical features. Toxicologic Pathology 26:276–282.

Botta, J. R., E. Arsenault, and H. A. Ryan. 1983. Total mercury content of meat and liver from inshore Newfoundland-caught harp seal (*Phoca groenlandica*). Bulletin of Environmental Contamination and Toxicology 30:28–32.

Bowes, G. W., and C. J. Jonkel. 1975. Presence and distribution of polychlorinated biphenyls (PCB) in arctic and subarctic marine food chains. Journal of the Fisheries Research Board of Canada 32:2111–2123.

Bowes, G. W., and J. A. Lewis. 1974. Extraction of polychlorinated biphenyls: Evaluation of a column technique applied to polar bear and seal tissue. Journal of the Association of Official Analytical Chemists 57:138–144.

Bratton, G. R., C. B. Spainhour, W. Flory, M. Reed, and K. Jayko. 1993. Presence and potential effects of contaminants. Pages 701–744 *in* J. J. Burns, J. J. Montague, and C. J. Cowles (eds.). The Bowhead Whale. Special Publ. 2, Society for Marine Mammalogy, Lawrence, KS.

Braham, H. W. 1973. Lead in the California sea lion (*Zalophus californianus*). Environmental Pollution 5:253–258.

Braune, B. M., R. J. Norstrom, M. P. Wong, B. T. Collins, and J. Lee. 1991. Geographical distribution of metals in livers of polar bears from the Northwest Territories, Canada. The Science of the Total Environment 100:283–299.

Brewerton, H. V. 1969. DDT in fats of Antarctic animals. New Zealand Journal of Science 12:194–199.

Britt, J. O., and E. B. Howard. 1983. Tissue residues of selected environmental contaminants in marine mammals. Pages 79–94 *in* E. B. Howard (ed.). Pathobiology of Marine Mammal Diseases, Vol. II. CRC Press, Boca Raton, FL.

Brouwer, A., P. H. J. Reijnders, and J. H. Koeman. 1989. Polychlorinated biphenyl (PCB)-contaminated fish induces vitamin A and thyroid hormone deficiency in the common seal (*Phoca vitulina*). Aquatic Toxicology 15:99–106.

Buckland, S. J., D. J. Hannah, J. A. Taucher, E. Slooten, and S. Dawson. 1990. Polychlorinated dibenzo-*p*-dioxins and dibenzofurans in New Zealand's Hector's dolphin. Chemosphere 20:1035–1042.

Buhler, D. R., R. R. Claeys, and B. R. Mate. 1975. Heavy metal and chlorinated hydrocarbon residues in California sea lions (*Zalophus californianus californianus*). Journal of the Fisheries Research Board of Canada 32:2391–2397.

Byrne, C., R. Balasubramanian, E. B. Overton, and T. F. Albert. 1985. Concentrations of trace metals in the bowhead whale. Marine Pollution Bulletin 16:497–498.

Calmet, D., D. Woodhead, and J. M. André. 1992. ^{210}Pb, ^{137}Cs and ^{40}K in three species of porpoises caught in the Eastern Tropical Pacific Ocean. Journal of Environmental Radioactivity 15:153–169.

Cannella, E. G., and D. J. Kitchener. 1992. Differences in mercury levels in female sperm whale, *Physeter macrocephalus* (Cetacea: Odontoceti). Australian Mammalogy 15:121–123.

Carvan, M. J. III, M. Santostefano, S. Safe, and D. Busbee. 1994. Characterization of a bottlenose dolphin (*Tursiops truncatus*) kidney epithelial cell line. Marine Mammal Science 10:52–69.

Caurant, F., and C. Amiard-Triquet. 1995. Cadmium contamination in pilot whales *Globicephala melas:* Source and potential hazard to the species. Marine Pollution Bulletin 30:207–210.

Caurant, F., C. Amiard-Triquet, and J.-C. Amiard. 1993. Factors influencing the accumulation of metals in pilot whales (*Globicephala melas*) off the Faroe Islands. Report of the International Whaling Commission (Special Issue 14):369–390.

Cebrian Menchero, D., E. Georgakopoulos-Gregoriades, N. Kalogeropoulos, and R. Psyllidou-Giouranovits. 1994. Organochlorine levels in a Mediterranean monk seal (*Monachus monachus*). Marine Pollution Bulletin 28:181–183.

Clausen, J., and O. Berg. 1975. The content of polychlorinated hydrocarbons in Arctic ecosystems. Pure and Applied Chemistry 42:223–232.

Clausen, J., L. Braestrup, and O. Berg. 1974. The content of polychlorinated hydrocarbons in Arctic mammals. Bulletin of Environmental Contamination and Toxicology 12:529–534.

Cockcroft, V. G., A. C. De Kock, D. A. Lord, and G. J. B. Ross. 1989. Organochlorines in bottlenose dolphins, *Tursiops truncatus,* from the east coast of South Africa. South African Journal of Marine Science 8:207–217.

Cockcroft, V. G., A. C. De Kock, G. J. B. Ross, and D. A. Lord. 1990. Organochlorines in common dolphins caught in shark nets during the Natal "sardine run." South African Journal of Zoology 25:144–148.

Colborn, T., and C. Clement (eds.). 1992. Chemically induced alterations in sexual and functional development: The wildlife/human connection. Princeton Scientific Publishing, Princeton, NJ.

Colborn, T., F. S. vom Saal, and A. M. Soto. 1993. Developmental effects of endocrine-disrupting chemicals in wildlife and humans. Environmental Health Perspectives 101:378–384.

Cook, H. W., and B. E. Baker. 1969. Seal milk. I. Harp seal (*Pagophilus groenlandicus*) milk: composition and pesticide residue content. Canadian Journal of Zoology 47:1129–1132.

Cuvin-Aralar, L. A., and R. W. Furness. 1991. Mercury and selenium interaction: A review. Ecotoxicology and Environmental Safety 21:348–364.

Daelemans, F. F., F. Mehlum, C. Lydersen, and P. J. C. Schepens. 1993. Mono-ortho and non-ortho substituted PCBs in Arctic ringed seal (*Phoca hispida*) from the Svalbard area: analysis and determination of their toxic threat. Chemosphere 27:429–437.

de Boer, J., and P. G. Wester. 1993. Determination of toxaphene in human milk from Nicaragua and in fish and marine mammals from the northeastern Atlantic and the North Sea. Chemosphere 27:1879–1890.

de Boer, J., P. G. Wester, E. H. G. Evers, and U. A. T. Brinkman. 1996. Determination of *tris* (4-chlorophenyl) methanol and *tris* (4-chlorophenyl) methane in fish, marine mammals, and sediment. Environmental Pollution 93:29–47.

De Gange, A. R., and M. M. Vacca. 1989. Sea otter mortality at Kodiak Island, Alaska, during summer 1987. Journal of Mammalogy 70:836–838.

de Kock, A. C., P. B. Best, V. Cockcroft, and C. Bosma. 1994. Persistent organochlorine residues in small cetaceans from the east and west coasts of southern Africa. The Science of the Total Environment 154:153–162.

de Moreno, J. E. A., M. S. Gerpe, V. J. Moreno, and C. Vodopivez. 1997. Heavy metals in Antarctic organisms. Polar Biology 17:131–140.

de Swart, R. L., R. M. G. Kluten, C. J. Huizing, E. J. Vedder, P. J. H. Reijnders, I. K. G. Visser, F. G. C. M. UytdeHaag, and A. D. M. E. Osterhaus. 1993. Mitogen and antigen induced B cell and T cell responses of peripheral blood mononuclear cells from the harbour seal (*Phoca vitulina*). Veterinary Immunology and Immunopathology 37:217–230.

de Swart, R. L., P. S. Ross, E. J. Vedder, H. H. Timmerman, S. H. Heisterkamp, H. van Loveren, J. G. Vos, P. J. H. Reijnders, and A. D. M. E. Osterhaus. 1994. Impairment of immunological functions in harbour seals (*Phoca vitulina*) feeding on fish from polluted coastal waters. Ambio 23:155–159.

de Swart, R. L., P. S. Ross, J. G. Vos, and A. D. M. E. Osterhaus. 1996. Impaired immunity in harbour seals (*Phoca vitulina*) exposed to bioaccumulated environmental contaminants: review of a long-term feeding study. Environmental Health Perspectives 104 (Suppl. 4):823–828.

de Voogt, P., and U. A. T. Brinkman. 1989. Production, properties and usage of polychlorinated biphenyls. Pages 3–45 *in* R. Kimbrough and S. Jensen (eds.). Halogenated Biphenyls, Terphenyls, Napthalenes, Dibenzodioxins and Related Products. Elsevier Science Publishers, B. V. (Biomedical Division), Amsterdam.

de Voogt, P., D. E. Wells, L. Reutergårdh, and U. A. Th. Brinkman. 1990. Biological activity, determination and occurrence of planar, mono- and di-ortho PCB. International Journal of Analytical Chemistry 40:1–46.

de Wit, C., B. Jansson, S. Bergek, M. Hjelt, C. Rappe, M. Olsson, and Ö. Andersson. 1992. Polychlorinated dibenzo-*p*-dioxin and polychlorinated dibenzofuran levels and patterns in fish and fish-eating wildlife in the Baltic Sea, 1992. Chemosphere 25:185–188.

DeLong, R. L., W. G. Gilmartin, and J. G. Simpson. 1973. Premature births in California sea lions: Association with high organochlorine pollutant residue levels. Science 181:1168–1170.

Denton, G. R. W., and W. G. Breck. 1981. Mercury in tropical marine organisms from North Queensland. Marine Pollution Bulletin 12:116–121.

Denton, G. R. W., H. Marsh, G. E. Heinsohn, and C. Burdon-Jones. 1980. The unusual metal status of the dugong (*Dugong dugon*). Marine Biology 57:201–219.

Dietz, R., C. O. Nielsen, M. M. Hansen, and C. T. Hansen. 1990. Organic mercury in Greenland birds and mammals. The Science of the Total Environment 95:41–51.

Donkin, P., S. V. Mann, and I. E. Hamilton. 1981. Polychlorinated biphenyl, DDT and dieldrin residues in gray seal (*Halichoerus *grypus*) males, females and mother-foetus pairs sampled at the Farne Islands, England, during the breeding season. The Science of the Total Environment 19:121–142.

Drescher, H. E. 1978. Über den fund einer ringelrobbe, *Phoca hispida,* an der Nordseeküste von Schleswig-Holstein. Zoologische Anzeiger 200:141–144.

Drescher, H. E., U. Harms, and E. Huschenbeth. 1977. Organochlorines and heavy metals in the harbour seal *Phoca vitulina* from the German North Sea Coast. Marine Biology 41:99–106.

Dudok van Heel, W. H. 1972. Raised levels of mercury and chlorinated hydrocarbons in newly captured *Tursiops truncatus* from Florida waters. Aquatic Mammals 1:24–36.

Duinker, J. C., and M. T. J. Hillebrand. 1979. Mobilization of organochlorines from female lipid tissue and transplacental transfer to fetus in a harbour porpoise (*Phocoena phocoena*) in a contaminated area. Bulletin of Environmental Contamination and Toxicology 23:728–732.

Duinker, J. C., M. T. J. Hillebrand, and R. F. Nolting. 1979. Organochlorines and metals in harbour seals (Dutch Wadden Sea). Marine Pollution Bulletin 10:360–364.

Duinker, J. C., A. H. Knap, K. C. Binkley, G. H. Van Dam, A. Darrel-Rew, and M. T. J. Hillebrand. 1988. Method to represent the qualitative and quantitative characteristics of PCB mixtures: Marine mammal tissues and commercial mixtures as examples. Marine Pollution Bulletin 19:74–79.

Duinker, J. C., M. T. J. Hillebrand, T. Zeinstra, and J. P. Boon. 1989. Individual chlorinated biphenyls and pesticides in tissues of some cetacean species from the North Sea and the Atlantic Ocean: Tissue distribution and biotransformation. Aquatic Mammals 15:95–124.

Eaton, R. D. P., and J. P. Farant. 1982. The polar bear as a biological indicator of the environmental mercury burden. Arctic 35:422–425.

Eaton, R. D. P., D. C. Secord, and P. Hewitt. 1980. An experimental assessment of the toxic potential of mercury in ringed seal liver for adult laboratory cats. Toxicology and Applied Pharmacology 55:514–521.

Ehrenfeld, D. W. 1981. The Arrogance of Humanism. Oxford University Press, New York, NY.

Eisler, R. 1994. Radiation Hazards to Fish, Wildlife, and Invertebrates: A Synoptic Review. Biological Report 26. National Biological Service, Washington, D.C.

Ejobi, F., L. W. Kanja, M. N. Kyule, P. Müller, J. Krüger, and A. A. R. Latigo. 1996. Organochlorine pesticide residues in mothers' milk in Uganda. Bulletin of Environmental Contamination and Toxicology 56:873–880.

Evans, R. D., P. Richner, and P. M. Outridge. 1995. Micro-spatial variations of heavy metals in the teeth of walrus as determined by laser ablation ICP-MS: The potential for reconstructing a history of metal exposure. Archives of Environmental Contamination and Toxicology 28:55–60.

Falandysz, J., N. Yamashita, S. Tanabe, R. Tatsukawa, L. Rucińska, and K. Skóra. 1994. Congener-specific data on polychlorinated biphenyls in tissues of common porpoise from Puck Bay, Baltic Sea. Archives of Environmental Contamination and Toxicology 26:267–272.

Falconer, C. R., I. M. Davies, and G. Topping. 1983. Trace metals in the common porpoise, *Phocoena phocoena.* Marine Environmental Research 8:119–127.

Ford, C. A., D. C. G. Muir, R. J. Norstrom, M. Simon, and M. J. Mulvihill. 1993. Development of a semi-automated method for non-ortho PCBs: Application to Canadian Arctic marine mammal tissues. Chemosphere 26:1981–1991.

Forrester, D. J., F. H. White, J. C. Woodard, and N. P. Thompson. 1975. Intussusception in a Florida manatee. Journal of Wildlife Diseases 11:566–568.

Forrester, D. J., D. K. Odell, N. P. Thompson, and J. R. White. 1980. Morphometrics, parasites, and chlorinated hydrocarbon residues of pygmy killer whales from Florida. Journal of Mammalogy 61:356–360.

Fossi, M. C. and L. Marsili. 1997. The use of non-destructive biomarkers in the study of marine mammals. Biomarkers 2:205–216.

Fossi, M. C., L. Marsili, C. Leonzio, G. Notarbartolo di Sciara, M. Zanardelli, and S. Focardi. 1992. The use of non-destructive biomarker in Mediterranean cetaceans: Preliminary data on MFO activity in skin biopsy. Marine Pollution Bulletin 24:459–461.

Frank, A., V. Galgan, A. Roos, M. Olsson, L. R. Peterson, and A. Bignert. 1992. Metal concentrations in seals from Swedish waters. Ambio 21:529–538.

Frank, R., K. Ronald, and H. E. Braun. 1973. Organochlorine residues in harp seals (*Pagophilus groenlandicus*) caught in eastern Canadian waters. Journal of the Fisheries Research Board of Canada 30:1053–1063.

Freeman, H. C., and D. A. Horne. 1973. Mercury in Canadian seals. Bulletin of Environmental Contamination and Toxicology 10:172–180.

Freeman, H. C., and G. B. Sanglang. 1977. A study of the effects of methyl mercury, cadmium, arsenic, selenium, and a PCB (Aroclor 1254) on adrenal and testicular steroidogeneses *in vitro,* by the gray seal *Halichoerus grypus.* Archives of Environmental Contamination and Toxicology 5:369–383.

Freeman, H. C., G. Sanglang, J. F. Uthe, and K. Ronald. 1975. Steroidogenesis *in vitro* in the harp seal (*Pagophilus groenlandicus*) without and with methyl mercury treatment *in vivo.* Environmental Physiology and Biochemistry 5:428–439.

Fujise, Y., K. Honda, R. Tatsukawa, and S. Mishima. 1988. Tissue distribution of heavy metals in Dall's porpoise in the northwestern Pacific. Marine Pollution Bulletin 19:226–230.

Fuller, G. B., and W. C. Hobson. 1986. Effect of PCBs on reproduction in mammals. Pages 101–125 *in* J. S. Waid (ed.). PCBs and the Environment, Vol. II. CRC Press, Boca Raton, FL.

Gallo, M. A. 1996. History and scope of toxicology. Pages 3–12 *in* C. D. Klaassen (ed.). Casarett and Doull's Toxicology, the Basic Science of Poisons, 5th ed. McGraw-Hill, Health Professions Division, New York, NY.

Gaskin, D. E., M. Holdrinet, and R. Frank. 1971. Organochlorine pesticide residues in harbour porpoises from the Bay of Fundy region. Nature 233:499–500.

Gaskin, D. E., K. Ishida, and R. Frank. 1972. Mercury in harbour porpoises (*Phocoena phocoena*) from the Bay of Fundy region. Journal of the Fisheries Research Board of Canada 29:1644–1646.

Gaskin, D. E., R. Frank, M. Holdrinet, K. Ishida, C. J. Walton, and M. Smith. 1973. Mercury, DDT, and PCB in harbour seals (*Phoca vitulina*) from the Bay of Fundy and Gulf of Maine. Journal of the Fisheries Research Board of Canada 30:471–475.

Gaskin, D. E., G. J. D. Smith, P. W. Arnold, M. V. Louisy, R. Frank, M. Holdrinet, and J. W. McWade. 1974. Mercury, DDT, dieldrin, and PCB in two species of Odontoceti (Cetacea) from St. Lucia, Lesser Antilles. Journal of the Fisheries Research Board of Canada 31:1235–1239.

Gaskin, D. E., K. I. Stonefield, P. Suda, and R. Frank. 1979. Changes in mercury levels in harbor porpoises from the Bay of Fundy, Canada, and adjacent waters during 1969–1977. Archives of Environmental Contamination and Toxicology 8:733–762.

Gaskin, D. E., M. Holdrinet, and R. Frank. 1982. DDT residues in blubber of harbour porpoise, *Phocoena* (L.), from eastern Canadian waters during the five-year period 1969–1973. Pages 135–143 *in* Mammals in the Seas. FAO Fisheries Series No. 5, Vol. 4, Rome, Italy.

Gaskin, D. E., R. Frank, and M. Holdrinet. 1983. Polychlorinated biphenyls in harbor porpoises *Phocoena phocoena* (L.) from the Bay of Fundy, Canada and adjacent waters, with some information on chlordane and hexachlorobenzene levels. Archives of Environmental Contamination and Toxicology 12:211–219.

George, J. L., and D. E. H. Frear. 1966. Pesticides in the Antarctica. Pages 155–167 *in* N. W. Moore (ed.). Pesticides in the Environment and their Effects on Wildlife. Journal of Applied Ecology 3 (supplement).

Geraci, J. R. 1989. Clinical investigation of the 1987–88 mass mortality of bottlenose dolphins along the U.S. central and south Atlantic coast. Final report to National Marine Fisheries Service, U.S. Navy Office of Naval Research, and U.S. Marine Mammal Commission, Washington, D.C.

Geraci, J. R., and V. J. Lounsbury. 1993. Marine Mammals Ashore: A Field Guide for Strandings. Texas A&M Sea Grant Publications, Galveston, TX.

Geraci, J. R., and D. J. St. Aubin (eds.). 1990. Sea Mammals and Oil: Confronting the Risks. Academic Press, San Diego, CA.

Geraci, J. R., N. C. Palmer, and D. J. St. Aubin. 1987. Tumors in cetaceans: Analysis and new findings. Canadian Journal of Fisheries and Aquatic Sciences 44:1289–1300.

Geraci, J. R., D. M. Anderson, R. J. Timperi, D. J. St. Aubin, G. A. Early, J. H. Prescott, and C. A. Mayo. 1989. Humpback whales (*Megaptera novaeangliae*) fatally poisoned by dinoflagellate toxin. Canadian Journal of Fisheries and Aquatic Sciences 46:1895–1898.

Gilmartin, W. G., R. L. DeLong, A. W. Smith, J. C. Sweeney, B. W. De Lappe, R. W. Risebrough, L. A. Griner, M. D. Dailey, and D. B. Peakall. 1976. Premature parturition in the California sea lion. Journal of Wildlife Diseases 12:104–115.

Goksøyr, A., J. Tarlebö, J. E. Solbakken, and J. Klungsöyr. 1985. Characteristics of the hepatic microsomal cytochrome P-450 system of the minke whale (*Balaenoptera acutorostrata*). Marine Environmental Research 17:113–116.

Goksøyr, A., J. E. Solbakken, J. Tarlebo, and J. Klungsoyr. 1986. Initial characterization of the hepatic microsomal cytochrome P-450-system of the piked whale (minke) *Balaenoptera acutorostrata*. Marine Environmental Research 19:185–203.

Goksøyr, A., T. Andersson, L. Förlin, J. Stenersen, E. A. Snowberger, B. R. Woodin, and J. J. Stegeman. 1988. Xenobiotic and steroid metabolism in adult and foetal piked (minke) whales, *Balaenoptera acutorostrata*. Marine Environmental Research 24:9–13.

Goksøyr, A., T. Andersson, L. Förlin, E. A. Snowberger, B. R. Woodin, and J. J. Stegeman. 1989. Cytochrome P-450 monooxygenase activity and immunochemical properties of adult and foetal piked (minke) whales, *Balaenoptera acutorostrata*. Pages 698–701 *in* I. Schuster (ed.). Cytochrome P-450: Biochemistry and Biophysics. Taylor and Francis, New York, NY.

Goksøyr, A., J. Beyer, H. E. Larsen, T. Andersson, and L. Förlin. 1992. Cytochrome P450 in seals: monooxygenase activities, immunochemical cross-reactions and response to phenobarbital treatment. Marine Environmental Research 34:113–116.

Goldblatt, C. J., and R. G. Anthony. 1983. Heavy metals in northern fur seals (*Callorhinus ursinus*) from the Pribilof Islands, Alaska. Journal of Environmental Quality 12:478–482.

Goldstein, J. A., and S. Safe. 1989. Mechanisms of action and structure–activity relationships for the chlorinated dibenzo-*p*-dioxins and related compounds. Pages 239–293 *in* R. D. Kimbrough and A. A. Jensen (eds.). Halogenated Biphenyls, Terphenyls, Napthalenes, Dibenzodioxins and Related Products. Elsevier, Amsterdam.

Goyer, R. A. 1996. Toxic effects of metals. Pages 691–736 *in* C. D. Klaassen (ed.). Casarett and Doull's Toxicology, the Basic Science of Poisons, 5th ed. McGraw-Hill, Health Professions Division, New York, NY.

Granby, K., and C. C. Kinze. 1991. Organochlorines in Danish and west Greenland harbour porpoises. Marine Pollution Bulletin 22:458–462.

Guitart, R., X. Guerrero, A. M. Silvestre, J. M. Gutiérrez, and R. Mateo. 1996. Organochlorine residues in tissues of striped dolphins affected by the 1990 Mediterranean epizootic: Relationships with the fatty acid composition. Archives of Environmental Contamination and Toxicology 30:79–83.

Gulland, F. M. D., J. G. Trupkiewicz, T. R. Spraker, and L. J. Lowenstine. 1996. Metastatic carcinoma of probable transitional cell origin in 66 free-living California sea lions (*Zalophus californianus*), 1979 to 1994. Journal of Wildlife Diseases 32:250–258.

Gunter G., R. H. Williams, C. C. Davis, and F. G. Walton Smith. 1948. Catastrophic mass mortality of marine animals and coincident phytoplankton bloom on the west coast of Florida, November 1946 to August 1947. Ecological Monographs 18:309–324.

Hall, A. J., R. J. Law, D. E. Wells, J. Harwood, H. M. Ross, S. Kennedy, C. R. Allchin, L. A. Campbell, and P. P. Pomeroy. 1992. Organochlorine levels in common seals (*Phoca vitulina*) which were victims and survivors of the 1988 phocine distemper epizootic. The Science of the Total Environment 115:145–162.

Hall, A., P. Pomeroy, N. Green, K. Jones, and J. Harwood. 1997. Infection, haematology and biochemistry in gray seal pups exposed to chlorinated biphenyls. Marine Environmental Research 43:81–98.

Hall, J. D., W. G. Gilmartin, and J. L. Mattsson. 1971. Investigation of a pacific pilot whale stranding on San Clemente Island. Journal of Wildlife Diseases 7:324–327.

Hamanaka, T., and S. Mishima. 1981. Cadmium and zinc concentrations in marine organisms in the northern North Pacific Ocean. Pages 191–206 *in* Special Volume. Research Institute of North Pacific Fisheries, Hokkaido University, Hokkaido, Japan.

Hamanaka, T., H. Kato, and T. Tsujita. 1977. Cadmium and zinc in ribbon seal, *Histriophoca fasciata*, in the Okhotsk Sea. Pages 547–561 *in* Special Volume. Research Institute of North Pacific Fisheries, Hokkaido University, Hokkaido, Japan.

Hamanaka, T., T. Itoo, and S. Mishima. 1982. Age-related change and distribution of cadmium and zinc concentrations in the Steller sea lion (*Eumetopias jubata*) from the coast of Hokkaido, Japan. Marine Pollution Bulletin 13:57–61.

Hansen, C. T., C. O. Nielsen, R. Dietz, and M. M. Hansen. 1990. Zinc, cadmium, mercury and selenium in minke whales, belugas and narwhals from West Greenland. Polar Biology 10:529–539.

Haraguchi, K., M. Athanasiadou, A. Bergman, L. Hovander, and S. Jensen. 1992. PCB and PCB methyl sulfones in selected groups of seals from Swedish waters. Ambio 21:546–549.

Harder, T. C., T. Willhaus, W. Leibold, and B. Liess. 1992. Investigations on course and outcome of phocine distemper virus infection in harbor seals (*Phoca vitulina*) exposed to polychlorinated biphenyls. Journal of Veterinary Medicine B 39:19–31.

Harms, U., H. E. Drescher, and E. Huschenbeth. 1977/1978. Further data on heavy metals and organochlorines in marine mammals from German coastal waters. Meeresforschung 26:153–161.

Heidmann, W. A., G. Staats de Yanés, A. Büthe, and H. Rüssel-Sinn. 1992. Correlation between concentration and composition of PCB mixtures in seals (*Phoca vitulina*). Chemosphere 24:1111–1117.

Heinz, G. H. 1989. How lethal are sublethal effects? Environmental Toxicology and Chemistry 8:463–464.

Helle, E. 1980. Age structure and sex ratio of the ringed seal *Phoca (Pusa) hispida* Schreber population in the Bothnian Bay, northern Baltic Sea. Zeitschrift für Saugetierkunde 45:310–317.

Helle, E., M. Olsson, and S. Jensen. 1976a. DDT and PCB levels and reproduction in ringed seal from the Bothnian Bay. Ambio 5:188–189.

Helle, E., M. Olsson, and S. Jensen. 1976b. PCB levels correlated with pathological changes in seal uteri. Ambio 5:261–263.

Helle, E., H. Hyvärinen, H. Pyysalo, and K. Wickström. 1983. Levels of organochlorine compounds in an inland seal population in eastern Finland. Marine Pollution Bulletin 14:256–260.

Hellou, J., G. Stenson, I-H. Ni, and J. F. Payne. 1990. Polycyclic aromatic hydrocarbons in muscle tissue of marine mammals from the Northwest Atlantic. Marine Pollution Bulletin 21:469–473.

Hellou, J., C. Upshall, I.H. Ni, J. F. Payne, and Y. S. Huang. 1991. Polycyclic aromatic hydrocarbons in harp seals (*Phoca groenlandica*) from the Northwest Atlantic. Archives of Environmental Contamination and Toxicology 21:135–140.

Helminen, M., E. Karppanen, and I. Koivisto. 1968. Saimaan norpan elohopeapitoisuudesta 1967. Suomen Eläinlääkärilehti Finsk Veterinärtidskrift 74:87–89.

Henriksson, K., E. Karppanen, and M. Helminen. 1969. Kvicksilverhalter hos insjö-och havssälar. Nordisk Hygienisk Tidskrift 50:54–59.

Henry, J., and P. B. Best. 1983. Organochlorine residues in whales landed at Durban, South Africa. Marine Pollution Bulletin 14:223–227.

Heppleston, P. B. 1973. Organochlorines in British gray seals. Marine Pollution Bulletin 4:44–45.

Heppleston, P. B., and M. C. French. 1973. Mercury and other metals in British seals. Nature 243:302–304.

Hickey, J. J., and D. W. Anderson. 1968. Chlorinated hydrocarbons and egg-shell changes in raptorial and fish-eating birds. Science 162:271–273.

Hidaka, H., S. Tanabe, and R. Tatsukawa. 1983. DDT compounds and PCB isomers and congeners in Weddell seals and their fate in the Antarctic marine ecosystem. Agricultural and Biological Chemistry 47:2009–2017.

Himeno, S., C. Watanabe, T. Hongo, T. Suzuki, A. Naganuma, and N. Imura. 1989. Body size and organ accumulation of mercury and selenium in young harbor seals (*Phoca vitulina*). Bulletin of Environmental Contamination and Toxicology 42:503–509.

Hodgson, E., A. P. Kulkarni, D. L. Fabacher, and K. M. Robacker. 1980. Induction of hepatic drug metabolizing enzymes in mammals by pesticides: A review. Journal of Environmental Science and Health B 15:723–754.

Hoffman, D. J., B. A. Rattner, G. A. Burton Jr., and J. Cairns Jr. (eds.). 1994. Handbook of Ecotoxicology. Lewis Publishers, Boca Raton, FL.

Holden, A. V. 1970. Monitoring organochlorine contamination of the marine environment by the analysis of residues in seals. Pages 266–272 *in* M. Ruivo (ed.). Marine Pollution and Sea Life. Fishing News (Books) LtD., Surrey, England.

Holden, A. V. 1975. The accumulation of oceanic contaminants in marine mammals. Pages 353–361 *in* K. Ronald and A. W. Mansfield (eds.). Biology of the Seal. Conseil International Pour L'Exploration de la Mer. Rapports et Procès-Verbaux des Réunions Vol. 169, Charlottenlund Slot-Denmark.

Holden, A. V. 1978. Pollutants and seals: A review. Mammal Review 8:53–66.

Holden, A. V., and K. Marsden. 1967. Organochlorine pesticides in seals and porpoises. Nature 216:1274–1276.

Holtzman, R. B. 1969. Concentrations of the naturally occurring radionuclides ^{226}Ra, ^{210}Pb and ^{210}Po in aquatic fauna. Pages 535–546 *in* D. J. Nelson and F. C. Evans (eds.). Symposium on Radioecology. Proceedings of the second national symposium. The Clearinghouse for Federal Scientific and Technical Information, National Bureau of Standards, Springfield, VA.

Honda, K., and R. Tatsukawa. 1983. Distribution of cadmium and zinc in tissues and organs, and their age-related changes in striped dolphins, *Stenella coeruleoalba*. Archives of Environmental Contamination and Toxicology 12:543–550.

Honda, K., R. Tatsukawa, K. Itano, N. Miyazaki, and T. Fujiyama. 1983. Heavy metal concentrations in muscle, liver and kidney tissue of striped dolphin, *Stenella coeruleoalba*, and their variations with body length, weight, age and sex. Agricultural and Biological Chemistry 47:1219–1228.

Honda, K., Y. Fujise, K. Itano, and R. Tatsukawa. 1984. Composition of chemical components in bone of striped dolphin, *Stenella coeruleoalba:* Distribution characteristics of heavy metals in various bones. Agricultural and Biological Chemistry 48:677–683.

Honda, K., Y. Fujise, R. Tatsukawa, K. Itano, and N. Miyazaki. 1986a. Age-related accumulation of heavy metals in bone of the striped dolphin, *Stenella coeruleoalba*. Marine Environmental Research 20:143–160.

Honda, K., Y. Yamamoto, and R. Tatsukawa. 1986b. Heavy metal accumulation in the liver of Antarctic minke whale (*Balaenoptera acutorostrata*). Memoirs of the National Institute for Polar Research (Special Issue) 44:182–184.

Honda, K., Y. Yamamoto, H. Kato, and R. Tatsukawa. 1987. Heavy metal accumulations and their recent changes in southern minke whales *Balaenoptera acutorostrata*. Archives of Environmental Contamination and Toxicology 16:209–216.

Hongell, K. 1996. Chromosome survey of seals in the Baltic Sea in 1988–1992. Archives of Environmental Contamination and Toxicology 31:399–403.

Hyvärinen, H., and T. Sipilä. 1984. Heavy metals and high pup mortality in the Saimaa ringed seal population in eastern Finland. Marine Pollution Bulletin 15:335–337.

Itano, K., S. Kawai, N. Miyazaki, R. Tatsukawa, and T. Fujiyama. 1984a. Mercury and selenium levels in striped dolphins caught off the Pacific Coast of Japan. Agricultural and Biological Chemistry 48:1109–1116.

Itano, K., S. Kawai, N. Miyazaki, R. Tatsukawa, and T. Fujiyama. 1984b. Body burdens and distribution of mercury and selenium in striped dolphins. Agricultural and Biological Chemistry 48:1117–1121.

Itano, K., S. Sawai, N. Miyazaki, R. Tatsukawa, and T. Fujiyama. 1984c. Mercury and selenium levels at the fetal and suckling stages of striped dolphin, *Stenella coeruleoalba*. Agricultural and Biological Chemistry 48:1691–1698.

Itano, K., S. Kawai, and R. Tatsukawa. 1985a. Properties of mercury and selenium in salt-insoluble fraction of muscles in striped dolphin. Bulletin of the Japanese Society of Scientific Fisheries 51:1129–1131.

Itano, K., S. Kawai, and R. Tatsukawa. 1985b. Distribution of mercury and selenium in muscle of striped dolphins. Agricultural and Biological Chemistry 49:515–517.

Iwata, H., S. Tanabe, N. Miyazaki, and R. Tatsukawa. 1994. Detection of butyltin compound residues in the blubber of marine mammals. Marine Pollution Bulletin 28:607–612.

Jansson, B., S. Jensen, M. Olsson, L. Renberg, G. Sundström, and R. Vaz. 1975. Identification by GC-MS of phenolic metabolites of PCB and p,p'-DDE isolated from Baltic guillemot and seal. Ambio 4:93–97.

Jansson, B., R. Vaz, G. Blomkvist, S. Jensen, and M. Olsson. 1979. Chlorinated terpenes and chlordane components found in fish, guillemot and seal from Swedish waters. Chemosphere 4:181–190.

Jarman, W. M., M. Simon, R. J. Norstrom, S. A. Bruns, C. A. Bacon, B. R. T. Simaret, and R. W. Risebrough. 1992. Global distribution of *tris*-(4-chlorophenyl) methanol in high trophic level birds and mammals. Environmental Science and Technology 26:1770–1774.

Jarman, W. M., R. J. Norstrom, D. C. G. Muir, B. Rosenberg, M. Simon, and R. W. Baird. 1996. Levels of organochlorine compounds, including PCDDs and PCDFs, in the blubber of cetaceans from the west coast of North America. Marine Pollution Bulletin 32:426–436.

Jenness, R., and D. K. Odell. 1978. Composition of milk of the pygmy sperm whale (*Kogia breviceps*). Comparative Biochemistry and Physiology A 61:383–386.

Jensen, A. A. 1983. Chemical contaminants in human milk. Residue Reviews 89:1–128.

Jensen, S., A. G. Johnels, M. Olsson, and G. Otterlind. 1969. DDT and PCB in marine animals from Swedish waters. Nature 224:247–250.

Jenssen, B. M., J. U. Skaare, M. Ekker, D. Vongraven, and M. Silverstone. 1994. Blood sampling as a non-destructive method for monitoring levels and effects of organochlorines (PCB and DDT) in seals. Chemosphere 28:3–10.

Joiris, C. R., L. Holsbeek, J. M. Bouquegneau, and M. Bossicart. 1991. Mercury contamination of the harbour porpoise (*Phocoena phocoena*) and other cetaceans from the North Sea and the Kattegat. Water, Air, and Soil Pollution 56:283–293.

Jones, A. M., Y. Jones, and W. D. P. Stewart. 1972. Mercury in marine organisms of the Tay region. Nature 238:164–165.

Jones, D., K. Ronald, D. M. Lavigne, R. Frank, M. Holdrinet, and J. F. Uthe. 1976. Organochlorine and mercury residues in the harp seal (*Pagophilus groenlandicus*). The Science of the Total Environment 5:181–195.

Julshamn, K., A. Andersen, O. Ringdal, and J. Mørkøre. 1987. Trace elements intake in the Faroe Islands I. Element levels in edible parts of pilot whales (*Globicephalus meleanus*). The Science of the Total Environment 65:53–62.

Kallenborn, R., M. Oehme, W. Vetter, and H. Parlar. 1994. Enantiomer selective separation of toxaphene congeners isolated from seal blubber and obtained by synthesis. Chemosphere 28:89–98.

Kannan, K., R. K. Sinha, S. Tanabe, H. Ichihashi, and R. Tatsukawa. 1993a. Heavy metals and organochlorine residues in Ganges River dolphins from India. Marine Pollution Bulletin 26:159–162.

Kannan, K., J. Falandysz, S. Tanabe, and R. Tatsukawa. 1993b. Persistent organochlorines in harbour porpoises from Puck Bay, Poland. Marine Pollution Bulletin 26:162–165.

Kannan, K., S. Tanabe, A. Borrell, A. Aguilar, S. Focardi, and R. Tatsukawa. 1993c. Isomer-specific analysis and toxic evaluation of polychlorinated biphenyls in striped dolphins affected by an epizootic in the western Mediterranean Sea. Archives of Environmental Contamination and Toxicology 25:227–233.

Kannan, K., S. Corsolini, S. Focardi, S. Tanabe, and R. Tatsukawa. 1996. Accumulation pattern of butyltin compounds in dolphin, tuna, and shark collected from Italian coastal waters. Archives of Environmental Contamination and Toxicology 31:19–23.

Kannan, K., K. S. Guruge, N. J. Thomas, S. Tanabe, and J. P. Giesy. 1998. Butyltin residues in southern sea otters (*Enhydra lutris nereis*) found dead along California coastal waters. Environmental Science and Technology 32:1169–1175.

Kannan, N., S. Tanabe, M. Ono, and R. Tatsukawa. 1989. Critical evaluation of polychlorinated biphenyl toxicity in terrestrial and marine mammals: Increasing impact of non-ortho and mono-ortho coplanar polychlorinated biphenyls from land to ocean. Archives of Environmental Contamination and Toxicology 18:850–857.

Kapel, F. O. 1983. Denmark (Greenland) progress report on cetacean research June 1981 to May 1982. Report of the International Whaling Commission 33:203–207.

Kari, T., and P. Kauranen. 1978. Mercury and selenium contents of seals from fresh and brackish waters in Finland. Bulletin of Environmental Contamination and Toxicology 19:273–280.

Kawai, S., M. Fukushima, N. Miyazaki, and R. Tatsukawa. 1988. Relationship between lipid composition and organochlorine levels in the tissues of striped dolphin. Marine Pollution Bulletin 19:129–133.

Kawano, M., T. Inoue, H. Hidaka, and R. Tatsukawa. 1984. Chlordane compounds residues in Weddell seals (*Leptonychotes weddelli*) from the Antarctic. Chemosphere 13:95–100.

Kawano, M., T. Inoue, H. Hidaka, and R. Tatsukawa. 1986. Chlordane residues in krill, fish and Weddell seal from the Antarctic. Toxicology and Environmental Chemistry 11:137–145.

Kawano, M., T. Inoue, T. Wada, H. Hidaka, and R. Tatsukawa. 1988. Bioconcentration and residue patterns of chlordane compounds in marine animals: Invertebrates, fish, mammals, and seabirds. Environmental Science and Technology 22:792–797.

Kemper, C., P. Gibbs, D. Obendorf, S. Marvanek, and C. Lenghaus. 1994. A review of heavy metal and organochlorine levels in marine mammals in Australia. The Science of the Total Environment 154:129–139.

Kendall, M. D., B. Safieh, J. Harwood, and P. P. Pomeroy. 1992. Plasma thymulin concentrations, the thymus and organochlorine contaminant levels in seals infected with phocine distemper virus. The Science of the Total Environment 115:133–144.

Kennedy, S. 1995. Morbillivirus Epizootics in Aquatic Mammals. Report of the Workshop on Chemical Pollution and Cetaceans, SC/M95/P15. Report of the International Whaling Commission, Cambridge, U.K.

Kerkhoff, M., and J. de Boer. 1982. Identification of chlordane compounds in harbour seals from the coastal waters of the Netherlands. Chemosphere 11:841–845.

Kerkhoff, M., J. de Boer, and J. Geerdes. 1981. Heptachlor epoxide in marine mammals. The Science of the Total Environment 19:41–50.

Kim, G. B., J. S. Lee, S. Tanabe, H. Iwata, R. Tatsukawa, and K. Shimazaki. 1996. Specific accumulation and distribution of butyltin compounds in various organs and tissues of the Steller sea lion (*Eumetopias jubatus*): Comparison with organochlorine accumulation pattern. Marine Pollution Bulletin 32:558–563.

Kim, J. P., and W. F. Fitzgerald. 1986. Sea-air partitioning of mercury in the equatorial Pacific Ocean. Science 231:1131–1133.

Kim, K. C., R. C. Chu, and G. P. Barron. 1974. Mercury in tissues and lice of northern fur seals. Bulletin of Environmental Contamination and Toxicology 11:281–284.

Kingsley, M. C. S. 1998. Population index estimates for the St. Lawrence belugas, 1973–1995. Marine Mammal Science 14:508–530.

Klaassen, C. D. (ed.). 1996. Casarett and Doull's Toxicology, the Basic Science of Poisons, 5th ed. McGraw-Hill, Health Professions Division, New York, NY.

Kleivane, L., J. U. Skaare, A. Bjørge, E. de Ruiter, and P. J. H. Reijnders. 1995. Organochlorine pesticide residue and PCBs in harbour porpoise (*Phocoena phocoena*) incidentally caught in Scandinavian waters. Environmental Pollution 89:137–146.

Knap, A. H., and T. D. Jickells. 1983. Trace metals and organochlorines in the Goosebeak whale. Marine Pollution Bulletin 14:271–274.

Koeman, J. H., and H. van Genderen. 1966. Some preliminary notes on residues of chlorinated hydrocarbon insecticides in birds and mammals in the Netherlands. Pages 99–106 *in* N. W. Moore (ed.). Pesticides in the Environment and their Effects on Wildlife. Journal of Applied Ecology 3 (Supplement).

Koeman, J. H., W. H. M. Peters, C. J. Smit, P. S. Tjioe, and J. J. M. Goeij. 1972. Persistent chemicals in marine mammals. TNO-Nieuws 27:570–578.

Koeman, J. H., W. H. M. Peeters, and C. H. M Koudstaal-Hol. 1973. Mercury–selenium correlations in marine mammals. Nature 245:385–386.

Koeman, J. H., W. S. M. van de Ven, J. J. M. de Goeij, P. S. Tjioe, and J. L. van Haaften. 1975. Mercury and selenium in marine mammals and birds. The Science of the Total Environment 3:279–287.

Koistinen, J. 1990. Residues of planar polychloroaromatic compounds in Baltic fish and seal. Chemosphere 20:1043–1048.

Kucklick, J. R., L.L. McConnell, T. F. Bidleman, G. P. Ivanov, and M. D. Walla. 1993. Toxaphene contamination in Lake Baikal's water and food web. Chemosphere 27:2017–2026.

Kucklick, J. R., T. F. Bidleman, L. L. McConnell, M. D. Walla, and G. Ivanov. 1994. Organochlorines in the water and biota of Lake Baikal, Siberia. Environmental Science and Technology 28:31–37.

Kuehl, D., and R. Haebler. 1995. Organochlorine, organobromine, metal, and selenium residues in bottlenose dolphins (*Tursiops truncatus*) collected during an unusual mortality event in the Gulf of Mexico, 1990. Archives of Environmental Contamination and Toxicology 28:494–499.

Kuehl, D. W., R. Haebler, and C. Potter. 1991. Chemical residues in dolphins from the U. S. Atlantic Coast including Atlantic bottlenose obtained during the 1987/88 mass mortality. Chemosphere 22:1071–1084.

Kuehl, D. W., R. Haebler, and C. Potter. 1994. Coplanar PCB and metal residues in dolphins from the U.S. Atlantic coast including Atlantic bottlenose obtained during the 1987/1988 mass mortality. Chemosphere 23:1245–1253.

Kuiken, T., U. Höfle, P. M. Bennett, C. R. Allchin, J. K. Kirkwood, J. R. Baker, E. C. Appleby, C. H. Lockyer, M. J. Walton, and M. C. Sheldrick. 1993. Adrenocortical hyperplasia, disease and chlorinated hydrocarbons in the harbour porpoise (*Phocoena phocoena*). Marine Pollution Bulletin 26:440–446.

Kuiken, T., P. M. Bennett, C. R. Allchin, J. K. Kirkwood, J. R. Baker, C. H. Lockyer, M. J. Walton, and M. C. Sheldrick. 1994. PCB's, cause of death and body condition in harbour porpoises (*Phocoena phocoena*) from British waters. Aquatic Toxicology 28:13–28.

Kurtz, D. A., and K. C. Kim. 1976. Chlorinated hydrocarbon and PCB residues in tissues and lice of northern fur seals, 1972. Pesticides Monitoring Journal 10:79–83.

Kvitek, R. G., A. R. De Gange, and M. K. Beitler. 1991. Paralytic shellfish poisoning toxins mediate feeding behavior of sea otters. Limnology and Oceanography 36:393–404.

Lahvis, G. P., R. S. Wells, D. W. Kuehl, J. L. Stewart, H. L. Rhinehart, and C. S. Via. 1995. Decreased lymphocyte responses in free-ranging bottlenose dolphins (*Tursiops truncatus*) are associated with increased concentrations of PCBs and DDT in peripheral blood. Environmental Health Perspectives 103 (Suppl. 4):62–67.

Lake, C. A., J. L. Lake, R. Haebler, R. McKinney, W. S. Boothman, and S. S. Sadove. 1995. Contaminant levels in harbour seals from the northeastern United States. Archives of Environmental Contamination and Toxicology 29:128–134.

Law, R. J., and J. A. Whinnett. 1992. Polycyclic aromatic hydrocarbons in muscle tissue of harbour porpoises (*Phocoena phocoena*) from UK waters. Marine Pollution Bulletin 24:550–553.

Law, R. J., C. R. Allchin, and J. Harwood. 1989. Concentrations of organochlorine compounds in the blubber of seals from eastern and north-eastern England, 1988. Marine Pollution Bulletin 20:110–115.

Law, R. J., C. F. Fileman, A. D. Hopkins, J. R. Baker, J. Harwood, D. B. Jackson, S. Kennedy, A. R. Martin, and R. J. Morris. 1991. Concentrations of trace metals in the livers of marine mammals (seals, porpoises and dolphins) from waters around the British Isles. Marine Pollution Bulletin 22:183–191.

Law, R. J., B. R. Jones, J. R. Baker, S. Kennedy, R. Milne, and R. J. Morris. 1992. Trace metals in the liver of marine mammals from the Welsh Coast and the Irish Sea. Marine Pollution Bulletin 24:296–304.

Law, R. J., C. R. Allchin, and R. J. Morris. 1995. Uptake of organochlorines (chlorobiphenyls, dieldrin; total PCB & DDT) in bottlenose dolphins (*Tursiops truncatus*) from Cardigan Bay, West Wales. Chemosphere 30:547–560.

Law, R. J., R. L. Stringer, C. R. Allchin, and B. R. Jones. 1996. Metals and organochlorines in sperm whales (*Physeter macrocephalus*) stranded around the North Sea during the 1994/1995 winter. Marine Pollution Bulletin 32:72–77.

Layne, J. N. 1965. Observations on marine mammals in Florida waters. Bulletin of the Florida State Museum 9:131–181.

Le Bouef, B. J., and M. L. Bonnell. 1971. DDT in California sea lions. Nature 234:108–109.

Le Bouef, B. J., and M. L. Bonnell. 1980. Pinnipeds of the California islands: Abundance and distribution. Pages 475–493 *in* D. M. Power (ed.). The California Islands: Proceedings of a Multidisciplinary Symposium. Santa Barbara Museum of Natural History, Santa Barbara, CA.

Lee, J. S., S. Tanabe, H. Umino, R. Tatsukawa, T. R. Loughlin, and D. C. Calkins. 1996. Persistent organochlorines in Steller sea lion (*Eumetopias jubatus*) from the bulk of Alaska and the Bering Sea, 1976–1981. Marine Pollution Bulletin 32:535–544.

Lee, S. S., B. R. Mate, K. T. von der Trenck, R. A. Rimerman, and D. R. Buhler. 1977. Metallothionein and the subcellular localization of mercury and cadmium in the California sea lion. Comparative Biochemistry and Physiology C 57:45–53.

Lentfer, J. W., and W. A. Galster. 1987. Mercury in polar bears from Alaska. Journal of Wildlife Diseases 23:338–341.

Leonzio, C., S. Focardi, and C. Fossi. 1992. Heavy metals and selenium in stranded dolphins of the Northern Tyrhenian (NW Mediterranean). The Science of the Total Environment 119:77–84.

Letcher, R. J., R. J. Norstrom, and A. Bergman. 1995. Geographical distribution and identification of methyl sulfone PCB and DDE metabolites in pooled polar bear (*Ursus maritimus*) adipose tissue from western hemisphere Arctic and subarctic regions. The Science of the Total Environment 160/161:409–420.

Letcher, R. J., R. J. Norstrom, S. Lin, M. A. Ramsay, and S. M. Bandiera. 1996. Immunoquantitation and microsomal monoxygenase activities of hepatic cytochromes P4501A and P4502B and chlorinated hydrocarbon contaminant levels in polar bear (*Ursus maritimus*). Toxicology and Applied Pharmacology 137:127–140.

Lieberg-Clark, P., C. E. Bacon, S. A. Burns, W. M. Jarman, and B. J. Le Boeuf. 1995. DDT in California sea lions: A follow-up study after twenty years. Marine Pollution Bulletin 30:744–745.

Lillestolen, T. I., N. Foster, and S. A. Wise. 1993. Development of the national marine mammal tissue bank. The Science of the Total Environment 139/140:97–107.

Lock, J. W., and S. R. B. Solly. 1976. Organochlorines in New Zealand birds and mammals. 1. Pesticides. New Zealand Journal of Science 19:43–51.

Loganathan, B. G., S. Tanabe, H. Tanaka, S. Watanabe, N. Miyazaki, M. Amano, and R. Tatsukawa. 1990. Comparison of organochlorine residue levels in the striped dolphin from western North Pacific, 1978–79 and 1986. Marine Pollution Bulletin 21:435–439.

Loughlin, T. R. (ed.). 1994. Marine Mammals and the *Exxon Valdez*. Academic Press, San Diego, CA.

Lowry, M. S., P. Boveng, R. J. DeLong, C. W. Oliver, B. S. Stewart, H. DeAnda, and J. Barlow. 1992. Status of the California Sea Lion (*Zalophus californianus californianus*) Population in 1992. National Marine Fisheries Service Administrative Report LJ-92-32, Southwest Fisheries Science Center, La Jolla, CA.

Luckas, B., W. Vetter, P. Fischer, G. Heidemann, and J. Plötz. 1990. Characteristic chlorinated hydrocarbon patterns in the blubber of seals from different marine regions. Chemosphere 21:13–19.

Mackey, E. A., R. Demiralp, P. R. Becker, R. R. Greenberg, B. J. Koster, and S. A. Wise. 1995. Trace element concentrations in cetacean liver tissues archived in the national marine mammal tissue bank. The Science of the Total Environment 175:25–41.

Mackey, E. A., P. R. Becker, R. Demiralp, R. R. Greenberg, B. J. Koster, and S. A. Wise. 1996. Bioaccumulation of vanadium and other trace metals in livers of Alaskan cetaceans and pinnipeds. Archives of Environmental Contamination and Toxicology 30:503–512.

Malcolm, H. M., I. L. Boyd, D. Osborn, M. C. French, and P. Freestone. 1994. Trace metals in Antarctic fur seal (*Arctocephalus gazella*) livers from Bird Island, South Georgia. Marine Pollution Bulletin 28:375–380.

Marcovecchio, J. E., V. J. Moreno, R. O. Bastida, M. S. Gerpe, and D. H. Rodríguez. 1990. Tissue distribution of heavy metals in small cetaceans from the southwestern Atlantic Ocean. Marine Pollution Bulletin 21:299–304.

Marcovecchio, J. E., M. S. Gerpe, R. O. Bastida, D. H. Rodriguez, and S. G. Morón. 1994. Environmental contamination and marine mammals in coastal waters from Argentina: An overview. The Science of the Total Environment 154:141–151.

Marsili, L., and S. Focardi. 1996. Organochlorine levels in subcutaneous blubber biopsies of fin whales (*Balaenoptera physalus*) and striped dolphins (*Stenella coeruleoalba*) from the Mediterranean Sea. Environmental Pollution 91:1–9.

Martin, J. H., P. D. Elliott, V. C. Anderlini, D. Girvin, S. A. Jacobs, R. W. Risebrough, R. L. Delong, and W. G. Gilmartin. 1976. Mercury–selenium–bromine imbalance in premature parturient California sea lions. Marine Biology 35:91–104.

Martineau, D., P. Béland, C. Desjardins, and A. Lagacé. 1987. Levels of organochlorine chemicals in tissues of beluga whales (*Delphinapterus leucas*) from the St. Lawrence estuary, Quebec, Canada. Archives of Environmental Contamination and Toxicology 16:137–147.

Martineau, D., A. Lagacé, P. Béland, R. Higgins, D. Armstrong, and L. R. Shugart. 1988. Pathology of stranded beluga whales (*Delphinapterus leucas*) from the St. Lawrence estuary, Quebec, Canada. Journal of Comparative Pathology 98:287–311.

Martineau, D., S. De Guise, M. Fournier, L. Shugart, C. Girard, A. Lagacé, and P. Béland. 1994. Pathology and toxicology of beluga whales from the St. Lawrence estuary, Quebec, Canada. Past, present and future. The Science of the Total Environment 154:201–215.

Martoja, R., and D. Viale. 1977. Accumulation de granules de séléiure mercurique dans le foie d'Odontocètes (Mammifères, Cétacés): Un mécanisme possible de détoxication du méthyl-mercure par le sélénium. C. R. Acad. Sc. Paris 285(D):109–112.

Massé, R., D. Martineau, L. Tremblay, and P. Béland. 1986. Concentrations and chromatographic profile of DDT metabolites and polychlorobiphenyl (PCB) residues in stranded beluga whales (*Delphinapterus leucas*) from the St. Lawrence estuary, Canada. Archives of Environmental Contamination and Toxicology 15:567–579.

Matsumura, F. 1985. Toxicology of Insecticides, 2nd ed. Plenum, New York, NY.

McBee, K., and J. W. Bickham. 1990. Mammals as bioindicators of environmental toxicity. Pages 37–88 *in* H. H. Genoways (ed.). Current Mammalogy. Plenum, New York, NY.

McClurg, T. P. 1984. Trace metals and chlorinated hydrocarbons in Ross seals from Antarctica. Marine Pollution Bulletin 15:384–389.

Meador, J. P., U. Varanasi, P. A. Robisch, and S.-L. Chan. 1993. Toxic metals in pilot whales (*Globicephala melanea*) from strandings in 1986 and 1990 on Cape Cod, Massachusetts. Canadian Journal of Fisheries and Aquatic Sciences 50:2698–2706.

Metcalf, R. L. 1973. A century of DDT. Journal of Agricultural and Food Chemistry 21:511–519.

Miles, A. K., and S. Hills. 1994. Metals in diet of Bering Sea walrus: *Mya* sp. as a possible transmitter of elevated cadmium and other metals. Marine Pollution Bulletin 28:456–458.

Miles, A. K., D. G. Calkins, and N. C. Coon. 1992. Toxic elements and organochlorines in harbor seals (*Phoca vitulina richardsi*), Kodiak, Alaska, USA. Bulletin of Environmental Contamination and Toxicology 48:727–732.

Mishima, M., N. Yamagata, and T. Torii. 1977. Concentrations of trace metals in tissues of several animals around Syowa Station. Antarctic Record (Japan) 58:145–153.

Mitchell, S. H., and S. Kennedy. 1992. Tissue concentrations of organochlorine compounds in common seals from the coast of Northern Ireland. The Science of The Total Environment 115:163–177.

Miyazaki, N., K. Itano, M. Fukushima, S.-I. Kawai, and K. Honda. 1979. Metals and organochlorine compounds in the muscle of dugong from Sulawesi Island. Scientific Reports of the Whales Research Institute (Japan) 31:125–128.

Miyazaki, N., I. Nakamura, S. Tanabe, and R. Tatsukawa. 1987. A stranding of *Mesoplodon stejnegeri* in the Maizuru Bay, Sea of Japan. Scientific Reports of the Whales Research Institute (Tokyo) 38:91–105.

Mochizuki, Y., K. T. Suzuki, H. Sunaga, T. Kobayashi, and R. Doi. 1985. Separation and characterization of metallothionein in two species of seals by high performance liquid chromatography-atomic absorption spectrophotometry. Comparative Biochemistry and Physiology C 82:249–254.

Morris, R. J., R. J. Law, C. R. Allchin, C. A. Kelly, and C. F. Fileman. 1989. Metals and organochlorines in dolphins and porpoises of Cardigan Bay, West Wales. Marine Pollution Bulletin 20:512–523.

Mortensen, P., A. Bergman, A. Bignert, H.-J. Hansen, T. Harkonen, and M. Olsson. 1992. Prevalence of skull lesions in harbor seals (*Phoca vitulina*) in Swedish and Danish museum collections: 1835–1988. Ambio 21:520–524.

Mössner, S., T. R. Spraker, P. R. Becker, and K. Ballschmiter. 1992. Ratios of enantiomers of alpha-HCH and determination of alpha-, beta-, and gamma-HCH isomers in brain and other tissues of neonatal northern fur seals (*Callorhinus ursinus*). Chemosphere 24:1171–1180.

Muir, D. C. G., R. Wagemann, N. P. Grift, R. J. Norstrom, M. Simon, and J. Lien. 1988a. Organochlorine chemical and heavy metal contaminants in white-beaked dolphins (*Lagenorhynchus albirostris*) and pilot whales (*Globicephala melaena*) from the coast of Newfoundland, Canada. Archives of Environmental Contamination and Toxicology 17:613–629.

Muir, D. C. G., R. J. Norstrom, and M. Simon. 1988b. Organochlorine contaminants in Arctic marine food chains: Accumulation of specific polychlorinated biphenyls and chlordane-related compounds. Environmental Science and Technology 22:1071–1079.

Muir, D. C. G., C. A. Ford, R. E. A. Stewart, T. G. Smith, R. F. Addison, M. E. Zinck, and P. Béland. 1990. Organochlorine contaminants in belugas, *Delphinapterus leucas*, from Canadian waters. Pages 165–190 *in* T. G. Smith, D. J. St. Aubin, and J. R. Geraci (eds.). Advances in Research on the Beluga Whale, *Delphinapterus leucas*. Canadian Bulletin of Fisheries and Aquatic Sciences 224.

Muir, D. C. G., R. Wagemann, B. T. Hargrave, D. J. Thomas, D. B. Peakall, and R. J. Norstrom. 1992a. Arctic marine ecosystem contamination. The Science of the Total Environment 122:75–134.

Muir, D. C. G., C. A. Ford, N. P. Grift, and R. E. A. Stewart. 1992b. Organochlorine contaminants in narwhal (*Monodon monoceros*) from the Canadian Arctic. Environmental Pollution 75:307–316.

Muir, D. C. G., M. D. Segstro, K. A. Hobson, C. A. Ford, R. E. A. Stewart, and S. Olpinkski. 1995. Can seal eating explain elevated levels of PCBs and organochlorine pesticides in walrus blubber from eastern Hudson Bay (Canada)? Environmental Pollution 90:335–348.

Muir, D. C. G., C. A. Ford, B. Rosenberg, R. J. Norstrom, M. Simon, and P. Béland. 1996a. Persistent organochlorines in beluga whales (*Delphinapterus leucas*) from the St. Lawrence River estuary. I. Concentrations and patterns of specific PCBs, chlorinated pesticides and polychlorinated dibenzo-*p*-dioxins and dibenzofurans. Environmental Pollution 93:219–234.

Muir, D. C. G., K. Koczanski, B. Rosenberg, and P. Béland. 1996b. Persistent organochlorines in beluga whales (*Delphinaterus leucas*) from the St. Lawrence River Estuary. II. Temporal trends, 1982–1994. Environmental Pollution 93:235–245.

Munday, B. L. 1985. Mercury levels in the musculature of stranded whales in Tasmania. Tasmanian Fisheries Research 27:11–13.

Murk, A., D. Morse, J. Boon, and A. Brouwer. 1994. In vitro metabolism of 3, 3′, 4, 4′-tetrachlorobiphenyl in relation to ethoxyresorufin-O-deethylase activity in liver microsomes of some wildlife species and the rat. European Journal of Pharmacology, Environmental Toxicology and Pharmacology Section 270:253–261.

Nagakura, K., S. Arima, M. Kurihara, T. Koga, and T. Fujita. 1974. Mercury content of whales. Bulletin of Tokai Registry of Fisheries Research Laboratory 78:41–46.

Nebert, D. W., and F. J. Gonzalez. 1987. P450 genes: Structure, evolution, and regulation. Annual Review of Biochemistry 56:945–993.

Nebert, D. W., D. R. Nelson, M. J. Coon, R. W. Estabrook, R. Feyereisen, Y. Fujii-Kuriyama, F. J. Gonzalez, F. P. Guengerich, I. C. Gunsalus, E. F. Johnson, J. C. Loper, R. Sato, M. R. Waterman, and D. J. Waxman. 1991. The P450 superfamily: Update on new sequences, gene mapping and recommended nomenclature. DNA and Cell Biology 10:1–14.

Newman, J. W., J. Vedder, W. M Jarman, and R. R. Chang. 1994. A method for the determination of environmental contaminants in living marine mammals using microscale samples of blubber and blood. Chemosphere 28:1795–1805.

Nielsen, C. O., and R. Dietz. 1990. Distributional pattern of zinc, cadmium, mercury, and selenium in livers of hooded seal (*Cystophora cristata*). Biological Trace Element Research 24:61–71.

Noda, K., H. Ichihashi, T. R. Loughlin, N. Baba, M. Kiyota, and R. Tatsukawa. 1995. Distribution of heavy metals in muscle, liver and kidney of northern fur seal (*Callorhinus ursinus*) caught off Sanriku, Japan and from the Pribilof Islands, Alaska. Environmental Pollution 90:51–59.

Norheim, G., J. Utne Skaare, and O. Wiig. 1992. Some heavy metals, essential elements, and chlorinated hydrocarbons in polar bear (*Ursus maritimus*) at Svalbard. Environmental Pollution 77:51–57.

Norstrom, R. J., and D. C. G. Muir. 1994. Chlorinated hydrocarbon contaminants in arctic marine mammals. The Science of the Total Environment 154:107–128.

Norstrom, R. J., R. E. Schweinsberg, and B. T. Collins. 1986. Heavy metals and essential elements in livers of the polar bear (*Ursus maritimus*) in the Canadian Arctic. The Science of the Total Environment 48:195–212.

Norstrom, R. J., M. Simon, D. C. G. Muir, and R. E. Schweinsburg. 1988. Organochlorine contaminants in arctic marine food chains: Identification, geographical distribution and temporal trends in polar bears. Environmental Science and Technology 22:1063–1071.

Norstrom, R. J., M. Simon, and D. C. G. Muir. 1990. Polychlorinated dibenzo-*p*-dioxins and dibenzofurans in marine mammals in the Canadian north. Environmental Pollution 66:1–19.

Norstrom, R. J., D. C. G. Muir, C. A. Ford, M. Simon, C. R. Macdonald, and P. Béland. 1992. Indications of P450 monoxygenase activities in beluga (*Delphinapterus leucas*) and narwhal (*Monodon monoceros*) from patterns of PCB, PCDD and PCDF accumulation. Marine Environmental Research 34:267–272.

Oehme, M., P. Fürst, C. Krüger, H. A. Meemken, and W. Groebel. 1988. Presence of polychlorinated dibenzo-*p*-dioxins, dibenzofurans and pesticides in Arctic seal from Spitzbergen. Chemosphere 18:1291–1300.

Oehme, M., M. Ryg, P. Fürst, H. A. Meemken, and W. Groebel. 1990. Re-evaluation of concentration levels of polychlorinated dibenzo-*p*-dioxins and dibenzofurans in arctic seal from Spitzbergen. Chemosphere 21:519–523.

Oehme, M., A. Biseth, M. Schlabach, and Ø. Wiig. 1995a. Concentrations of polychlorinated dibenzo-*p*-dioxins, dibenzofurans and non-Ortho substituted biphenyls in polar bear milk from Svalbard (Norway). Environmental Pollution 90:401–407.

Oehme, M., M. Schabach, K. Hummert, B. Luckas, and E. S. Nordoy. 1995b. Determination of levels of polychlorinated dibenzo-*p*-dioxins, dibenzofurans, biphenyls and pesticides in harp seals from the Greenland Sea. The Science of the Total Environment 162:75–91.

Ofstad, E. B., and K. Martinsen. 1983. Persistent organochlorine compounds in seals from Norwegian coastal waters. Ambio 12:262–264.

Olafson, R. W., and J. A. J. Thompson. 1974. Isolation of heavy metal binding proteins from marine vertebrates. Marine Biology 28:83–86.

Olsson, M., A. G. Johnels, and R. Vaz. 1975. DDT and PCB levels in seals from Swedish waters. The occurrence of aborted seal pups. Pages 43–53 *in* Proceedings of the Symposium on the Seal in the Baltic. Swedish National Environment Protection Board SNV. PM 591, Lidingö, Sweden.

Olsson. M., B. Karlsson, and E. Ahnland. 1992. Seals and seal protection: Summary and comments. Ambio 21:606.

Olsson, M., B. Karlsson, and E. Ahnland. 1994. Diseases and environmental contaminants in seals from the Baltic and the Swedish west coast. The Science of the Total Environment 154:217–227.

Ono, M., N. Kannan, T. Wakimoto, and R. Tatsukawa. 1987. Dibenzofurans a greater global pollutant than dioxins? Evidence from analyses of open ocean killer whale. Marine Pollution Bulletin 18:640–643.

O'Shea, T. J., and R. L. Brownell Jr. 1994. Organochlorine and metal contaminants in baleen whales: A review and evaluation of conservation implications. The Science of the Total Environment 154:179–200.

O'Shea, T. J., and R. L. Brownell Jr. 1998. California sea lion (*Zalophus californianus*) populations and ∑DDT contamination. Marine Pollution Bulletin 36:159–164.

O'Shea, T. J., R. L. Brownell Jr., D. R. Clark Jr., W. A. Walker, M. L. Gay, and T. G. Lamont. 1980. Organochlorine pollutants in small cetaceans from the Pacific and South Atlantic Oceans, November 1968–June 1976. Pesticides Monitoring Journal 14:35–46.

O'Shea, T. J., J. F. Moore, and H. I. Kochman. 1984. Contaminant concentrations in manatees in Florida. Journal of Wildlife Management 48:741–748.

O'Shea, T. J., G. B. Rathbun, R. K. Bonde, C. D. Buergelt, and D. K. Odell. 1991. An epizootic of Florida manatees associated with a dinoflagellate bloom. Marine Mammal Science 7:165–179.

Osterberg, C., W. Pearcy, and N. Kujala. 1964. Gamma emitters in the fin whale. Nature 204:1006–1007.

Paasivirta, J. 1991. Chemical Ecotoxicology. Lewis Publishers, Chelsea, MI.

Paasivirta, J., and T. Rantio. 1991. Chloroterpenes and other organochlorines in Baltic, Finnish and Arctic wildlife. Chemosphere 22:47–55

Pantoja, S., L. Pastene, J. Becerra, M. Silva, and V. A. Gallardo. 1984. DDTs in balaenopterids (Cetacea) from the Chilean coast. Marine Pollution Bulletin 15:451.

Pantoja, S., L. Pastene, J. Becerra, M. Silva, and V. A. Gallardo. 1985. Lindane, aldrin and dieldrin in some Chilean cetacea. Marine Pollution Bulletin 16:255.

Peakall, D. 1992. Animal Biomarkers as Pollution Indicators. Chapman and Hall, London.

Perttilä, M., O. Stenman, H. Pyysalo, and K. Wickström. 1986. Heavy metals and organochlorine compounds in seals in the Gulf of Finland. Marine Environmental Research 18:43–59.

Peterle, T. J. 1991. Wildlife Toxicology. Van Nostrand Reinhold, New York, NY.

Polischuk, S. C., R. J. Letcher, R. J. Norstrom, and M. A. Ramsay. 1995. Preliminary results of fasting on the kinetics of organochlorines in polar bears (*Ursus maritimus*). The Science of the Total Environment 160/161:465–472.

Raloff, J. 1994. The gender benders: Are environmental "hormones" emasculating wildlife? Science News 145:24–27.

Ramprashad, F., and K. Ronald. 1977. A surface preparation study on the effect of methyl mercury on the sensory hair cell population in the cochlea of the harp seal (*Pagophilus groenlandicus* Erxleben, 1777). Canadian Journal of Zoology 55:223–230.

Ratcliffe, D. A. 1967. Decrease in eggshell weight in certain birds of prey. Nature 215:208–210.

Rawson, A. J., H. F. Anderson, G. W. Patton, and T. Beecher. 1991. Anthracosis in the Atlantic bottlenose dolphin (*Tursiops truncatus*). Marine Mammal Science 7:413–416.

Rawson, A. J., G. W. Patton, S. Hofmann, G. G. Peitra, and L. Johns. 1993. Liver abnormalities associated with chronic mercury accumulation in stranded Atlantic bottlenose dolphins. Ecotoxicology and Environmental Safety 25:41–47.

Rawson, A. J., J. P. Bradley, A. Teetsov, S. B. Rice, E. M. Haller, and G. W. Patton. 1995. A role for airborne particulates in high mercury levels of some cetaceans. Ecotoxicology and Environmental Safety 30:309–314.

Ray, S., B. P. Dunn, J. F. Payne, L. Fancey, R. Helbig, and P. Béland. 1992. Aromatic DNA-carcinogen adducts in beluga whales from the Canadian Arctic and Gulf of St Lawrence. Marine Pollution Bulletin 22:392–396.

Reijnders, P. J. H. 1980. Organochlorine and heavy metal residues in harbour seals from the Wadden Sea and their possible effects on reproduction. Netherlands Journal of Sea Research 14:30–65.

Reijnders, P. J. H. 1984. Man-induced environmental factors in relation to fertility changes in pinnipeds. Environmental Conservation 11:61–65.

Reijnders, P. J. H. 1985. On the extinction of the southern Dutch harbour seal population. Biological Conservation 31:75–84.

Reijnders, P. J. H. 1986. Reproductive failure in common seals feeding on fish from polluted coastal waters. Nature 324:456–457.

Reijnders, P. J. H. 1994. Toxicokinetics of chlorobiphenyls and associated physiological responses in marine mammals, with particular reference to their potential for ecotoxicological risk assessment. The Science of the Total Environment 154:229–236.

Renberg, L., G. Sundström, and L. Reutergardh. 1978. Polychlorinated terphenyls (PCT) in Swedish white-tailed eagles and in gray seals: A preliminary study. Chemosphere 6:477–482.

Renzoni, A., and R. J. Norstrom. 1990. Mercury in the hairs of polar bears *Ursus maritimus*. Polar Record 26:326–328.

Rice, C. P., and P. O'Keefe. 1995. Sources, pathways, and effects of PCBs, dioxins, and dibenzofurans. Pages 424–468 *in* D. J. Hoffman, B. A. Rattner, G. A. Burton Jr., and J. Cairns Jr. (eds.). Handbook of Ecotoxicology. Lewis Publishers, Boca Raton, FL.

Richard, C. A., and E. J. Skoch. 1986. Comparison of heavy metal concentrations between specific tissue sites in the northern fur seal. Proceedings of the Annual International Association for Aquatic Animal Medicine Conference and Workshop 17:94–103.

Ridgway, S., and M. Reddy. 1995. Residue levels of several organochlorines in *Tursiops truncatus* milk collected at varied stages of lactation. Marine Pollution Bulletin 30:609–614.

Ridlington, J. W., D. C. Chapman, D. E. Goeger, and P. D. Whanger. 1981. Metallothionein and Cu-chelatin: Characterization of metal-binding proteins from tissue of four marine mammals. Comparative Biochemistry and Physiology 70B:93–104.

Roberts, T. M., P. B. Heppleston, and R. D. Roberts. 1976. Distribution of heavy metals in tissues of the common seal. Marine Pollution Bulletin 7:124–126

Robinson, J., A. Richardson, A. N. Crabtree, J. C. Coulson, and G. R. Potts. 1967. Organochlorine residues in marine organisms. Nature 214:1307–1311.

Ronald, K., J. F. Uthe, and H. Freeman. 1975. Effects of Methyl Mercury on the Harp Seal. International Council for the Exploration of the Sea. C.M. 1975/N:9. Mimeo. 15 pp.

Ronald, K., R. J. Frank, J. L. Dougan, R. Frank, and H. E. Braun. 1984a. Pollutants in harp seals (*Phoca groenlandica*). I. Organochlorines. The Science of the Total Environment 38:133–152.

Ronald, K., R. J. Frank, J. Dougan, R. Frank, and H. E. Braun. 1984b. Pollutants in harp seals (*Phoca groenlandica*). II. Heavy metals and selenium. The Science of the Total Environment 38:153–166.

Rosewell, K. T., D. C. G. Muir, and B. E. Baker. 1979. Organochlorine residues in harp seal (*Pagophilus groenlandicus*) tissues, Gulf of St. Lawrence, 1971, 1973. Pesticides Monitoring Journal 12:189–192.

Ross, P. S., R. L. De Swart, P. J. H. Reijnders, H. V. Loveren, J. G. Vos, and A. D. M. E. Osterhaus. 1995. Contaminant-related suppression of delayed-type hypersensitivity and antibody responses in harbor seals fed herring from the Baltic Sea. Environmental Health Perspectives 103:162–167.

Ross, P., R. De Swart, R. Addison, H. Van Loveren, J. Vos, and A. Osterhaus. 1996. Contaminant-induced immunotoxicity in harbour seals: Wildlife at risk? Toxicology 112:157–169.

Safe, S. 1984. Polychlorinated biphenyls (PCBs) and polybrominated biphenyls (PBBs): Biochemistry, toxicology and mechanism of action. CRC Critical Reviews in Toxicology 13:319–393.

Safe, S. 1986. Comparative toxicology and mechanism of action of polychlorinated dibenzo-*p*-dioxins and dibenzofurans. Annual Review of Pharmacology and Toxicology 26:371–399.

Safe, S. 1990. Polychlorinated biphenyls (PCBs), dibenzo-*p*-dioxins (PCDDs), dibenzofurans (PCDFs), and related compounds: Environmental and mechanistic considerations which support the development of toxic equivalency factors (TEFs). Critical Reviews in Toxicology 21:51–88.

Safe, S. 1991. Polychlorinated dibenzo-*p*-dioxins and related compounds: Sources, environmental distribution, and risk assessment. Environmental Carcinogenesis and Ecotoxicology Reviews C9:261–302.

Safe, S. H. 1994. Polychlorinated biphenyls (PCBs): Environmental impact, biochemical and toxic responses, and implications for risk assessment. Critical Reviews in Toxicology 24:87–149.

Safe, S. H. 1995. Environmental and dietary estrogens and human health: Is there a problem? Environmental Health Perspectives 103:346–351.

Salata, G. G., T. L. Wade, J. L. Sericano, J. W. Davis, and J. M. Brooks. 1995. Analysis of Gulf of Mexico bottlenose dolphins for organochlorine pesticides and PCBs. Environmental Pollution 88:167–175.

Saleh, M. A. 1991. Toxaphene: Chemistry, biochemistry, toxicity and environmental fate. Reviews of Environmental Contamination and Toxicology 118:1–85.

Samuels, E. R., M. Cawthorn, B. H. Lauer, and B. E. Baker. 1970. Strontium-90 and Cesium-137 in tissues of fin whales (*Balaenoptera physalus*) and harp seals (*Pagophilus groenlandicus*). Canadian Journal of Zoology 48:267–269.

Sanpera, C., R. Capelli, V. Minganti, and L. Jover. 1993. Total and organic mercury in North Atlantic fin whales: Distribution pattern and biological related changes. Marine Pollution Bulletin 26:135–139.

Sanpera, C., M. González, and L. Jover. 1996. Heavy metals in two populations of North Atlantic fin whales (*Balaenoptera physalus*). Environmental Pollution 91:299–307.

Saschenbrecker, P. W. 1973. Levels of DDT and PCB compounds in North Atlantic fin-back whales. Canadian Journal of Comparative Medicine 37:203–206.

Schafer, H. A., R. W. Gossett, C. F. Ward, and A. M. Westcott. 1984. Chlorinated hydrocarbons in marine mammals. Pages 109–114 *in* W. Bascom (ed.). Southern California Coastal Water Research Project Biennial Report, 1983–1984. Southern California Coastal Water Research Project, Long Beach, CA.

Schantz, M. M., B. J. Koster, S. A. Wise, and P. R. Becker. 1993. Determination of PCBs and chlorinated hydrocarbons in marine mammal tissues. The Science of the Total Environment 139/140:323–345.

Schintu, M., F. Jean-Caurant, and J.-C. Amiard. 1992. Organomercury determination in biological reference materials: Application to a study on mercury speciation in marine mammals off the Faröe Islands. Ecotoxicology and Environmental Safety 24:95–101.

Schumacher, U., S. Zahler, H.-P. Horney, G. Heidemann, K. Skirnisson, and U. Welsch. 1993. Histological investigations on the thyroid glands of marine mammals (*Phoca vitulina, Phocoena phocoena*) and the possible implications of marine pollution. Journal of Wildlife Diseases 29:103–108.

Schweigert, F. J., and W. T. Stobo. 1994. Transfer of fat-soluble vitamins and PCBs from mother to pups in gray seals (*Halichoerus grypus*). Comparative Biochemistry and Physiology C: Pharmacology, Toxicology Endocrinology 109:111–117.

Serat, W. F., M. K. Lee, A. J. Van Loon, D. C. Mengle, J. Ferguson, J. M. Burks, and T. R. Bender. 1977. DDT and DDE in the blood and diet of Eskimo children from Hooper Bay, Alaska. Pesticides Monitoring Journal 11:1–4.

Sergeant, D. E., and F. A. J. Armstrong. 1973. Mercury in seals from eastern Canada. Journal of the Fisheries Research Board of Canada 30:843–846.

Sergeant, D. E., and W. Hoek. 1988. An update of the status of white whales, *Delphinapterus leucas,* in the Saint Lawrence estuary, Canada. Biological Conservation 45:287–302.

Shaw, S. B. 1971. Chlorinated hydrocarbon pesticides in California sea otters and harbor seals. California Fish and Game 57:290–294.

Simmonds, M. P., P. A. Johnston, and M. C. French. 1993. Organochlorine and mercury contamination in United Kingdom seals. Veterinary Record 132:291–295.

Simmonds, M. P., P. A. Johnston, M. C. French, R. Reeve, and J. Hutchinson. 1994. Organochlorines and mercury in pilot whale blubber consumed by Faroe Islanders. The Science of the Total Environment 149:97–111.

Simonich, S. L., and R. A. Hites. 1995. Global distribution of persistent organochlorine compounds. Science 269:1851–1854.

Sladen, W. J. L., C. M. Menzie, and W. L. Reichel. 1966. DDT residues in Adelie penguins and a crabeater seal from Antarctica. Nature 210:670–673.

Smillie, R. H., and J. S. Waid. 1987. Polychlorinated biphenyls and organochlorine pesticides in the Australian fur seal, *Arctocephalus pusillus doriferus*. Bulletin of Environmental Contamination and Toxicology 39:358–364.

Smith, A. W., C. M. Prato, W. G. Gilmartin, R. J. Brown and M. C. Keyes. 1974. A preliminary report on potentially pathogenic microbiological agents recently isolated from pinnipeds. Journal of Wildlife Diseases 10:54–59.

Smith, D. R., S. Niemeyer, J. A. Estes, and A. R. Flegal. 1990. Stable lead isotopes evidence anthropogenic contamination in Alaskan sea otters. Environmental Science and Technology 24:1517–1521.

Smith, D. R., S. Niemeyer, and A. R. Flegal. 1992. Lead sources to California sea otters: Industrial inputs circumvent natural lead biodepletion mechanisms. Environmental Research 57:163–174.

Smith, T. G., and F. A. J. Armstrong. 1975. Mercury in seals, terrestrial carnivores, and principal food items of the Inuit, from Holman, N.W.T. Journal of the Fisheries Research Board of Canada 32:795–801.

Smith, T. G., and F. A. J. Armstrong. 1978. Mercury and selenium in ringed and bearded seal tissues from Arctic Canada. Arctic 31:75–84.

Solly, S. R. B., and V. Shanks. 1976. Organochlorine residues in New Zealand birds and mammals. 2. Polychlorinated biphenyls. New Zealand Journal of Science 19:53–55.

Steidinger, K. A., and D. G. Baden. 1984. Toxic marine dinoflagellates. Pages 201–261 *in* D. L. Spector (ed.). Dinoflagellates. Academic Press, New York, NY.

Stein, J. E., K. L. Tilbury, D. W. Brown, C. A. Wigren, J. P. Meador, P. A. Robisch, S.-L. Chan, and U. Varanasi. 1992. Intraorgan distribution of chemical contaminants in tissues of harbor porpoises (*Phocoena phocoena*) from the Northwest Atlantic. U.S. Department of Commerce, National Oceanic and Atmospheric Administration, Technical Memorandum NMFS NWFSC-3, Seattle, WA.

Steinhagen-Schneider, G. 1986. Cadmium and copper levels in seals, penquins and skuas from the Weddell Sea in 1982/1983. Polar Biology 5:139–143.

Stickel, L. F. 1973. Pesticide residues in birds and mammals. Pages in 254–312 *in* C. A. Edwards (ed.). Environmental Pollution by Pesticides. Plenum, New York, NY.

Stickney, R. R., H. L. Windom, D. B. White and F. Taylor. 1972. Mercury concentration in various tissues of the bottlenose dolphin (*Tursiops truncatus*). Proceedings of the Annual Conference of Southeastern Game and Fish Commissions 26:634–636.

Stone, R. 1994. Environmental estrogens stir debate. Science 265:308–310.

Stoneburner, D. L. 1978. Heavy metals in tissues of stranded short-finned pilot whales. The Science of the Total Environment 9:293–297.

Storr-Hansen, E., and H. Spliid. 1993a. Coplanar polychlorinated biphenyl congener levels and patterns and the identification of separate populations of harbor seals (*Phoca vitulina*) in Denmark. Archives of Environmental Contamination and Toxicology 24:44–58.

Storr-Hansen, E., and H. Spliid. 1993b. Distribution patterns of polychlorinated biphenyl congeners in harbor seal (*Phoca vitulina*) tissues: Statistical analysis. Archives of Environmental Contamination and Toxicology 25:328–345.

Storr-Hansen, E., H. Spliid, and J. P. Boon. 1995. Patterns of chlorinated biphenyl congeners in harbor seals (*Phoca vitulina*) and in their food: Statistical analysis. Archives of Environmental Contamination and Toxicology 28:48–54.

Subramanian, A., S. Tanabe, R. Tatsukawa, S. Saito, and N. Miyazaki. 1987. Reduction in the testosterone levels by PCBs and DDE in Dall's porpoises of northwestern North Pacific. Marine Pollution Bulletin 18:643–646.

Subramanian, A., S. Tanabe, and R. Tatsukawa. 1988a. Estimating some biological parameters of Baird's beaked whales using PCBs and DDE as tracers. Marine Pollution Bulletin 19:284–287.

Subramanian, A., S. Tanabe, and R. Tatsukawa. 1988b. Use of organochlorines as chemical tracers in determining some reproductive parameters in *Dalli*-type Dall's porpoise *Phocoenoides dalli*. Marine Environmental Research 25:161–174.

Suzuki, T., and T. Miyoshi. 1973. Changes of mercury content in fish and whale muscles by rinsing. Bulletin of the Japanese Society of Scientific Fisheries 39:917.

Szefer, P., W. Czarnowski, J. Pempkowiak, and E. Holm. 1993. Mercury and major essential elements in seals, penquins, and other representative fauna of the Antarctic. Archives of Environmental Contamination and Toxicology 25:422–427.

Szefer, P., K. Szefer, J. Pempkowiak, B. Skwarzec, R. Bojanowski, and E. Holm. 1994. Distribution and coassociations of selected metals in seals of the Antarctic. Environmental Pollution 83:341–349.

Takei, G. H., and G. H. Leong. 1981. Macro-analytical methods used to analyze tissues of the Hawaiian monk seal, *Monachus schauinslandi,* for organochlorine pesticides, polychlorobiphenyls, and pentachlorophenol. Bulletin of Environmental Contamination and Toxicology 27:489–498.

Tanabe, S. 1988. PCB problems in the future: Foresight from current knowledge. Environmental Pollution 50:5–28.

Tanabe, S., R. Tatsukawa, H. Tanaka, K. Maruyama, N. Miyazaki, and T. Fujiyama. 1981. Distribution and total burdens of chlorinated hydrocarbons in bodies of striped dolphins (*Stenella coeruleoalba*). Agricultural and Biological Chemistry 45:2569–2578.

Tanabe, S., R. Tatsukawa, K. Maruyama, and N. Miyazaki. 1982. Transplacental transfer of PCBs and chlorinated hydrocarbon pesticides from the pregnant striped dolphin (*Stenella coeruleoalba*) to her fetus. Agricultural and Biological Chemistry 46:1249–1254.

Tanabe, S., T. Mori, R. Tatsukawa, and N. Miyazaki. 1983. Global pollution of marine mammals by PCBs, DDTs and HCHs (BHCs). Chemosphere 12:1269–1275.

Tanabe, S., T. Mori, and R. Tatsukawa. 1984a. Bioaccumulation of DDTs and PCBs in the southern minke whale (*Balaenoptera acutorostrata*). Memoirs of the National Institute of Polar Research (Special Issue) 32:140–150.

Tanabe, S., H. Tanaka, and R. Tatsukawa. 1984b. Polychlorobiphenyls, ΣDDT, and hexachlorocyclohexane isomers in the western North Pacific ecosystem. Archives of Environmental Contamination and Toxicology 13:731–738.

Tanabe, S., S. Miura, and R. Tatsukawa. 1986. Variations of organochlorine residues with age and sex in Antarctic minke whale. Memoirs of the National Institute for Polar Research (Special Issue) 44:174–181.

Tanabe, S., N. Kannan, A. Subramanian, S. Watanabe, and R. Tatsukawa. 1987a. Highly toxic coplanar PCBs: Occurrence, source, persistency and toxic implications to wildlife and humans. Environmental Pollution 47:147–163.

Tanabe, S., B. G. Loganathan, A. Subramanian, and R. Tatsukawa. 1987b. Organochlorine residues in short-finned pilot whale: Possible use as tracers of biological parameters. Marine Pollution Bulletin 18:561–563.

Tanabe, S., S. Watanabe, H. Kan, and R. Tatsukawa. 1988. Capacity and mode of PCB metabolism in small cetaceans. Marine Mammal Science 4:103–124.

Tanabe, S., A. Subramanian, A. Ramesh, P. L. Kumaran, N. Miyazaki, and R. Tatsukawa. 1993. Persistent organochlorine residues in dolphins from the Bay of Bengal, South India. Marine Pollution Bulletin 26:311–316.

Tanabe, S., H. Iwata, and R. Tatsukawa. 1994a. Global contamination by persistent organochlorines and their ecotoxicological impact on marine mammals. The Science of the Total Environment 154:163–177.

Tanabe, S., J. Sung, D. Choi, N. Baba, M. Kiyota, K. Yoshida, and R. Tatsukawa. 1994b. Persistent organochlorine residues in northern fur seal from the Pacific coast of Japan since 1971. Environmental Pollution 85:305–314.

Tanabe, S., P. Kumaran, H. Iwata, R. Tatsukawa, and N. Miyazaki. 1996. Enantiomeric ratios of α-hexachlorocyclohexane in blubber of small cetaceans. Marine Pollution Bulletin 32:27–31.

Taruski, A. G., C. E. Olney, and H. E. Winn. 1975. Chlorinated hydrocarbons in cetaceans. Journal of the Fisheries Research Board of Canada 32:2205–2209.

Taylor, D. L., S. Schliebe, and H. Metsker. 1989. Contaminants in blubber, liver and kidney tissue of Pacific walruses. Marine Pollution Bulletin 20:465–468.

Teigen, S. W., J. U. Skaare, A. Bjørge, E. Degre, and G. Sand. 1993. Mercury and selenium in harbor porpoise (*Phocoena phocoena*) in Norwegian waters. Environmental Toxicology and Chemistry 12:1251–1259.

ten Noever de Brauw, M. C, C. Van Ingen, and J. H. Koeman. 1973. Mirex in seals. The Science of the Total Environment 2:196–198.

Theobald, J. 1973. D.D.T. levels in the sea lion. Journal of Zoo Animal Medicine 4:23.

Tillander, M., J. K. Miettinen, and I. Koivisto. 1972. Excretion rate of methyl mercury in the seal (*Pusa hispida*). Pages 303–305 *in* M. Ruivo (ed.). Marine Pollution and Sea Life. Fishing News (Books) Ltd., Surrey, England.

Tohyama, C., S.-I. Himeno, C. Watanabe, T. Suzuki, and M. Morita. 1986. The relationship of the increased level of metallothionein with heavy metal levels in the tissue of the harbor seal (*Phoca vitulina*). Ecotoxicology and Environmental Safety 12:85–94.

van Scheppingen, W. B., A. J. I. M. Verhoeven, P. Mulder, M. J. Addink, and C. Smeenk. 1996. Polychlorinated biphenyls, dibenzo-*p*-dioxins, and dibenzofurans in harbor porpoises (*Phocoena phocoena*) stranded on the Dutch coast between 1990 and 1993. Archives of Environmental Contamination and Toxicology 30:492–502.

van de Ven, W. S. M., J. H. Koeman, and A. Svenson. 1979. Mercury and selenium in wild and experimental seals. Chemosphere 8:539–555.

van der Zande, T., and E. de Ruiter. 1983. The quantification of technical mixtures of PCBs by microwave plasma detection and the analysis of PCBs in the blubber lipid from harbour seals (*Phoca vitulina*). The Science of the Total Environment 27:133–147.

Varanasi, U., J. E. Stein, W. L. Reichert, K. L. Tilbury, M. M. Krahn, and S.-L. Chan. 1992. Chlorinated and aromatic hydrocarbons in bottom sediments, fish and marine mammals in U.S. coastal waters: Laboratory and field studies of metabolism and accumulation. Pages 83–15 *in* C. H. Walker and D. R. Livingstone (eds.). Persistent Pollutants in Marine Ecosystems. Pergamon Press, New York, NY.

Varanasi, U., J. E. Stein, K. L. Tilbury, J. P. Meador, C. A. Wigren, R. C. Clark, and S.-L. Chan. 1993a. Chemical contaminants in gray whales (*Eschrichtius robustus*) stranded in Alaska, Washington, and California, U.S.A. U.S. Department of Commerce, National Oceanic and Atmospheric Administration, National Marine Fisheries Service. NOAA Technical Memorandum NMFS-NWFSC-11, Seattle, WA.

Varanasi, U., J. E. Stein, K. L. Tilbury, D. W. Brown, J. P. Meador, M. M. Krahn, and S.-L. Chan. 1993b. Contaminant monitoring for NMFS marine mammal health and stranding response program. Pages 1–15 *in* Coastal Zone 93 Proceedings, The Eighth Symposium on Coastal and Ocean Management, New Orleans, LA.

Varanasi, U., J. E. Stein, K. L. Tilbury, J. P. Meador, C. A. Sloan, R. C. Clark, and S.-L. Chan. 1994. Chemical contaminants in gray whales (*Eschrichtius robustus*) stranded along the west coast of North America. The Science of the Total Environment 145:29–53.

Vetter, W., B. Luckas, and M. Oehme. 1992. Isolation and purification of the two main toxaphene congeners in marine organisms. Chemosphere 25:1643–1652.

Viale, D. 1981. Lung pathology in stranded cetaceans on the Mediterranean coasts. Aquatic Mammals 8:96–100.

Viale, D. 1994. Cetaceans as indicators of a progressive degradation of Mediterranean water quality. International Journal of Environmental Studies 45:183–198.

Vicente, N., and D. Chabert. 1978. Recherches de polluants chimiques dans le tissu graisseux d'un dauphin échoué sur la côte méditerranéenne. Oceanologica Acta 1:331–334

Wagemann, R. 1989. Comparison of heavy metals in two groups of ringed seals (*Phoca hispida*) from the Canadian Arctic. Canadian Journal of Fisheries and Aquatic Sciences 46:1558–1563.

Wagemann, R., N. B. Snow, A. Lutz, and D. P. Scott. 1983. Heavy metals in tissues and organs of the narwhal (*Monodon monoceros*). Canadian Journal of Fisheries and Aquatic Sciences 40:206–214.

Wagemann, R., R. Hunt, and J. F. Klauerkamp. 1984. Subcellular distribution of heavy metals in liver and kidney of a narwhal whale (*Monodon monoceros*): An evaluation for the presence of metallothionein. Comparative Biochemistry and Physiology C 78:301–307.

Wagemann, R., R. E. A. Stewart, W. L. Lockhart, B. E. Stewart, and M. Povoledo. 1988. Trace metals and methyl mercury: Associations and transfer in harp seal (*Phoca groenlandica*) mothers and pups. Marine Mammal Science 4:339–355.

Wagemann, R., R. E. A. Stewart, P. Béland, and C. Desjardins. 1990. Heavy metals and selenium in tissues of beluga whales, *Delphinapterus leucas*, from the Canadian Arctic and the St. Lawrence estuary. Canadian Bulletin of Fisheries and Aquatic Sciences 224:191–206.

Waliszewski, S. M., V. T. S. Pardio, J. N. P. Chantiri, R. M. R. Infanzón, and J. Rivera. 1996. Organochlorine pesticide residues in adipose tissue of Mexicans. The Science of the Total Environment 181:125–131.

Walker, W., R. W. Risebrough, W. M. Jarman, B. W. de Lappe, J. A. Tefft, and R. L. DeLong. 1989. Identification of tris(chlorophenyl)-methanol in blubber of harbor seals from Puget Sound. Chemosphere 18:1799–1804.

Warburton, J., and D. J. Seagars. 1993. Heavy Metal Concentrations in Liver and Kidney Tissues of Pacific Walrus: Continuation of a Baseline Study. U.S. Fish and Wildlife Service Technical Report R7/MMM/93-1, Anchorage, AK.

Watanabe, S., T. Shimada, S. Nakamura, N. Nishiyama, N. Yamashita, S. Tanabe, and R. Tatsukawa. 1989. Specific profile of liver microsomal cytochrome P-450 in dolphin and whales. Marine Environmental Research 27:51–65.

Wells, D. E., and I. Echarri. 1992. Determination of individual chlorobiphenyls (CBs), including non-ortho, and mono-ortho chloro substituted CBs in marine mammals from Scottish waters. International Journal of Environmental Analytical Chemistry 47:75–97.

Wells, D., L. A. Campbell, H. M. Ross, P. M. Thompson, and C. H. Lockyer. 1994. Organochlorine residues in harbour porpoise and bottlenose dolphins stranded on the coast of Scotland, 1988–1991. The Science of the Total Environment 151:77–99.

Wenzel, C., D. Adelung, H. Kruse, and O. Wassermann. 1993. Trace metal accumulation in hair and skin of the harbour seal, *Phoca vitulina*. Marine Pollution Bulletin 26:152–155.

White, R. D., M. E. Hahn, W. L. Lockhart, and J. J. Stegeman. 1994. Catalytic and immunochemical characterization of hepatic microsomal cytochromes P450 in beluga whale (*Delphinapterus leucas*). Toxicology and Applied Pharmacology 126:45–57.

Windom, H. L., and D. R. Kendall. 1979. Accumulation and biotransformation of mercury in coastal and marine biota. Pages 303–323 *in* J. O. Nriagu (ed.). The Biogeochemistry of Mercury in the Environment. Elsevier / North-Holland Biomedical Press, Amsterdam.

Wolman, A. A., and A. J. Wilson. 1970. Occurrence of pesticides in whales. Pesticides Monitoring Journal 4:8–10.

Wood, C. M., and E. S. Van Vleet. 1996. Copper, cadmium and zinc in liver, kidney and muscle tissues of bottlenose dolphins (*Tursiops truncatus*) stranded in Florida. Marine Pollution Bulletin 32:886–889.

Woodley, T., M. Brown, S. Kraus, and D. Gaskin. 1991. Organochlorine levels in North Atlantic Right Whale (*Eubalaena glacialis*) blubber. Archives of Environmental Contamination and Toxicology 21:141–145.

Yamamoto, Y., K. Honda, H. Hidaka, and R. Tatsukawa. 1987. Tissue distribution of heavy metals in Weddell seals (*Leptonychotes weddelli*). Marine Pollution Bulletin 18:164–169.

Yediler, A., A. Panou, and P. Schramel. 1993. Heavy metals in hair samples of the Mediterranean monk seal (*Monachus monachus*). Marine Pollution Bulletin 26:156–159.

Zakharov, V. M., and A. V. Yablokov. 1990. Skull asymmetry in the Baltic gray seal: Effects of environmental pollution. Ambio 19:266–269.

Zeisler, R. R. Demiralp, B. J. Koster, P. R. Becker, M. Burow, P. Ostapczuk, and S. A. Wise. 1993. Determination of inorganic constituents in marine mammal tissues. The Science of the Total Environment 139 / 140:365–386.

Zhu, J., and R. J. Nostrom. 1993. Identification of polychlorocamphenes (PCCs) in the polar bear (*Ursus maritimus*) food chain. Chemosphere 27:1923–1936.

Zhu, J., R. J. Norstrom, D.C.G. Muir, L. A. Ferron, J.-P. Weber, and E. Dewailly. 1995. Persistent chlorinated cyclodiene compounds in ringed seal blubber, polar bear fat, and human plasma from northern Québec, Canada: Identification and concentrations of photoheptachlor. Environmental Science and Technology 29:267–271.

Appendices

The literature on environmental contaminants in marine mammals spans numerous disciplines, making access difficult to those beginning work in this field. Sources include journals in the fields of chemistry, biochemistry, environmental sciences, physiology, toxicology, wildlife biology, marine mammalogy,and other areas of study. For ease of entry into this literature, I have tabulated information from most of the published references in this field that have appeared during the past 30 years (through most of 1996). Tables can be used to find information on various species, regions of the world, and types of contaminants, organs, and numbers of individuals investigated. I hope these tabulations prove useful in planning future studies as well as in making comparisons with completed work. Original sources should be obtained for more detailed observations and interpretation. Species names are as used in the original publications and may differ from those provided in Chapter 1.

Appendix 10-1. Summary of Selected Organochlorine Residue Surveys in Pinnipeds

Species	Region	Sample Period	Tissues Sampled	Number of Individuals Sampled	Compounds Reported	References
Arctocephalus forsteri	New Zealand (captive)	—	L	5	PCBs, DDE, DDT, TDE, ΣDDT	Solly and Shanks 1976; Lock and Solly 1976
A. gazella	Antarctica	1984–85	Mi	3	PCB congeners, DDE, *o,p*′-DDE, DDT, *o,p*′-DDT, TDE, *o,p*′-TDE, chlordanes, heptachlor epoxide, dieldrin	Bacon et al. 1992
A. gazella	Antarctica	1987	B	11	PCB congeners, PCDDs, PCDFs	Oehme et al. 1995b
A. pusillus	Southeastern Australia	—	B, L, M, O	11	DDT, TDE, DDE, ΣDDT, PCB, HCB, HCH	Smillie and Waid 1987
Callorhinus ursinus	Alaska	1981	Mi	7	PCB congeners, DDE, *o,p*′-DDE, DDT, *o,p*′-DDT, TDE, *o,p*′-TDE, chlordanes, heptachlor epoxide, dieldrin	Bacon et al. 1992
C. ursinus	Alaska	1990	B, L	2	ΣDDT, PCBs	Varanasi et al. 1992, 1993b
C. ursinus	Alaska (Pribilof Islands)	1990	B, Br, L, Lu, Mi	4	HCHs	Mössner et al. 1992
C. ursinus	Bering Sea (Alaska)	1987	B, K, L, M	2	DDT, *o,p*′-DDT, TDE, DDE, *o,p*′-DDE, ΣDDT, PCBs, PCB congeners, HCB, HCHs, heptachlor epoxide, chlordane, *t*-nonachlor, dieldrin	Schantz et al. 1993

Appendix 10-1 continued

Species	Region	Sample Period	Tissues Sampled	Number of Individuals Sampled	Compounds Reported	References
C. ursinus	Eastern North Pacific (Alaska, Washington)	1968–69	Br, L	30	DDE, TDE, DDT, dieldrin	Anas and Wilson 1970
C. ursinus	Eastern North Pacific (Southern California)	1979–80	L	4	DDE, TDE, DDT, ΣDDT, PCBs	Britt and Howard 1983
C. ursinus	Eastern North Pacific (Washington)	1972	B	12	ΣDDT plus PCBs combined	Anas and Worlund 1975
C. ursinus	Japan	1971–88	B	105	DDE, DDT, TDE, ΣDDT, PCBs, PCB congeners, HCHs	Tanabe et al. 1994b
C. ursinus	North Pacific (Pribilof Islands; Washington)	1968–69	B, Br, L, M, Mi	37	DDE, TDE, DDT, dieldrin	Anas 1971
C. ursinus	Pribilof Islands (Alaska)	1972	B, O	7	DDE, TDE, DDT, *o,p′*-DDT, *o,p′*-TDE, ΣDDT, dieldrin, PCBs	Kurtz and Kim 1976
Cystophora cristata	Greenland	1972	B	5	DDE, PCBs, heptachlor epoxide, aldrin, lindane	Clausen et al. 1974; Clausen and Berg 1975
C. cristata	Gulf of St. Lawrence (Canada)	1960s	B	1	DDE, TDE, DDT, PCBs, dieldrin	Holden 1970
Erignathus barbatus	Arctic (Canada)	1971–72	Fat	2	DDE, TDE, DDT, PCBs	Bowes and Jonkel 1975; Bowes and Lewis 1974
E. barbatus	Greenland	1972	B	5	DDE, PCBs, heptachlor, heptachlor epoxide, aldrin, lindane	Clausen et al. 1974; Clausen and Berg 1975
Eumetopias jubatus	Japan	1990	B	4	DDE, DDT, TDE, ΣDDT, PCBs, PCB congeners, HCHs	Tanabe et al. 1994b
E. jubatus	Alaska	1985–90	B, L	8	ΣDDT, PCBs	Varanasi et al. 1992, 1993b
E. jubatus	Alaska, Bering Sea	1976–81	B, L	43	DDE, TDE, DDT, ΣDDT, HCHs, HCB, chlordanes, nonachlors, oxychlordane, PCBs	Lee et al. 1996
E. jubatus	Japan	1994	B, K, L, M, O	1	ΣDDT, PCBs, HCHs, chlordanes	Kim et al. 1996
Halichoerus grypus	Baltic Sea	1969–73	B	60	ΣDDT, PCBs	Olsson et al. 1975
H. grypus	Baltic Sea	1991	B, L	1	ΣDDT, PCBs, methyl sufones of PCB congeners and DDE	Bergman et al. 1994
H. grypus	Baltic Sea (Finland)	1981–87	B	7	PCB congeners, PCDFs, PCDDs, PCNs, hexachlorinated anthracenes	Koistinen 1990
H. grypus	Baltic Sea (Finland)	1985–89	B	1	ΣDDT, HCH, lindane, HCB, chlordane, toxaphene, PCN, PCBs, PCB congeners, dibenzofurans	Paasivirta and Rantio 1991
H. grypus	Baltic Sea (Gulf of Finland)	1976–82	B, K, L, M	9	DDT, TDE, DDE, ΣDDT, PCBs, chlordanes, trans-nonachlor, oxychlordane	Perttilä et al. 1986
H. grypus	Baltic Sea (Sweden)	1968	B, L, Mi	8	DDT, ΣDDT, PCBs	Jensen et al. 1969
H. grypus	Baltic Sea (Sweden)	1979–87	B	—	PCDDs, PCDFs	deWit et al. 1992
H. grypus	Baltic Sea (Sweden)	1976	Fat	3	PCBs, PCTs	Renberg et al. 1978
H. grypus	Baltic Sea (Sweden)	1974–77	B	5	ΣDDT, PCBs, chlordanes, HCHs, toxaphene	Jansson et al. 1979
H. grypus	Baltic Sea (Sweden)	—	B, K, L, other	5	DDE, PCBs, phenolic metabolites	Jansson et al. 1975
H. grypus	Baltic Sea (Sweden)	1979–90	B	37	ΣDDT, PCBs, PCB congeners, methyl sulfones of PCBs and DDE	Blomkvist et al. 1992; Haraguchi et al. 1992
H. grypus	Britain	—	B, Br, K, L, M, O	25	ΣDDT, PCBs, dieldrin	Heppleston 1973
H. grypus	Eastern North Atlantic (Norway)	1977–79	B, L	12	ΣDDT, PCBs, HCB, HCHs	Ofstad and Martinsen 1983

Continued on next page

Appendix 10-1 continued

Species	Region	Sample Period	Tissues Sampled	Number of Individuals Sampled	Compounds Reported	References
H. grypus	England	1960's	B, L	13	DDE, TDE, DDT, PCBs, dieldrin	Holden 1970
H. grypus	England	1988–89	B, K, L	8	DDE, TDE, DDT, dieldrin, HCH, HCB, PCBs	Simmonds et al. 1993
H. grypus	Gulf of St. Lawrence (Canada)	1960's	B	7	DDE, TDE, DDT, PCBs, dieldrin	Holden 1970
H. grypus	North Atlantic (Wales)	1988	B, K, L, M	2	HCB, HCHs, dieldrin, DDE, DDT, TDE, PCBs, PCB congeners	Morris et al. 1989
H. grypus	North Sea (England)	1965	L	1	DDE, dieldrin	Robinson et al. 1967
H. grypus	North Sea (England)	1972	B, L	189	DDE, TDE, DDT, ΣDDT, PCBs, dieldrin, total organohalogens	Donkin et al. 1981
H. grypus	North Sea (England)	1988	B	3	DDE, TDE, DDT, PCBs, PCB congeners, dieldrin, HCHs, HCB	Law et al. 1989
H. grypus	North Sea (Scotland)	1988	B	7	ΣDDT, PCBs	Blomkvist et al. 1992
H. grypus	Northeast Atlantic (France)	—	B, K, L, M	4	ΣDDT, PCBs	Alzieu and Duguy 1979
H. grypus	Northwest Atlantic (Nova Scotia)	1987	B, Mi	25	PCBs	Schweigert and Stobo 1994
H. grypus	Norway	1991	Blood	17	DDE, DDT, ΣDDT, PCBs, PCB congeners	Jenssen et al. 1994
H. grypus	Nova Scotia	—	B	38	PCBs	Addison et al. 1988
H. grypus	Scotland	1960s	B	31	DDE, TDE, DDT, PCBs, dieldrin	Holden 1970
H. grypus	Scotland	1965–67	B, Br, K, L, M, O	—	DDE, TDE, DDT, ΣDDT, dieldrin	Holden and Marsden 1967
H. grypus	Western North Atlantic (Canada)	1967	B	—	DDE, TDE, DDT, dieldrin	Holden and Marsden 1967
H. grypus	Western North Atlantic (Nova Scotia)	1982	B	16	DDE, TDE, DDT, ΣDDT, PCBs	Addison et al. 1984
H. grypus	Western North Atlantic (Nova Scotia)	1984–85	B, O	30	DDE, DDT, ΣDDT, PCBs, PCB congeners	Addison and Brodie 1987
Hydrurga leptonyx	Australia	—	B	1	DDE, DDT, PCBs, oxychlordane, heptachlor epoxide, HCB	Kemper et al. 1994
H. leptonyx	New Zealand (captive)	—	L	1	PCBs	Solly and Shanks 1976
Leptonychotes weddeli	Antarctic	1981	B	1	ΣDDT, PCBs, HCHs	Tanabe et al. 1983
L. weddeli	Antarctic	1981	B	3	ΣDDT, PCBs, *cis*-chlordane, *cis*-nonachlor, *t*-nonachlor, oxychlordane	Kawano et al. 1984
L. weddeli	Antarctica	1980–82	B, Br, K, L, M, O	5	ΣDDT, PCBs, PCB congeners, *cis*-chlordane, *t*-chlordane, *cis*-nonachlor, *t*-nonachlor, oxychlordane	Kawano et al. 1986, 1988; Hidaka et al. 1983
L. weddellii	Antarctica	—	B	1	toxaphene	Vetter et al. 1992
L weddellii	Antarctica (McMurdo Sound)	1965–67	Fat	20	DDE, DDT	Brewerton 1969
L. weddellii	Antarctica (Ross Island)	1964	Br, fat, K, L, M, O	16	DDE, DDT	George and Frear 1966
L. weddellii	Antarctic (Weddell Sea)	—	B	4	HCBs, HCHs, DDE, DDT, TDE, ΣDDT, PCBs, PCB congeners, toxaphene, chlordane	Luckas et al. 1990
L. weddellii	Ross Sea Antarctica	1965–67	Fat	20	DDT, DDE	Brewerton 1969
Lobodon carcinophagus	Ross Sea (Antartica)	1964	B, L	1	DDE, TDE, DDT, ΣDDT	Sladen et al. 1966

Appendix 10-1 continued

Species	Region	Sample Period	Tissues Sampled	Number of Individuals Sampled	Compounds Reported	References
Mirounga angustirostris	California	—	B, blood	—	PCB congeners, HCB, mirex, *t*-nonachlor, oxychlordane, heptachlor epoxide	Newman et al. 1994
M. angustirostris	California	1974–81	L	7	DDE, TDE, DDT, ΣDDT, PCBs	Britt and Howard 1983
M. angustirostris	California	1983–84	B, K, L, M	4	ΣDDT, PCBs	Schafer et al. 1984
M. angustirostris	California	1986–87	Mi	4	PCB congeners, DDE, *o,p′*-DDE, DDT, *o,p′*-DDT, TDE, *o,p′*-TDE, chlordanes, heptachlor epoxide, dieldrin	Bacon et al. 1992
Monachus monachus	Mediterranean (Greece)	1990	B, Br, K, L, M, O	1	PCBs, ΣDDT, HCHs	Cebrian Menchero et al. 1994
M. schauinslandi	Hawaii	—	B, K, L, Lu, O	1	DDE, PCBs, PCP	Takei and Leong 1981
Neophoca cinerea	Australia	1987	Mi	5	PCB congeners, DDE, *o,p′*-DDE, DDT, *o,p′*-DDT, TDE, *o,p′*-TDE, chlordanes, heptachlor epoxide, dieldrin	Bacon et al. 1992
Odobenus rosmarus	Arctic (Greenland)	1975, 1977	B	28	ΣDDT, DDE, DDT, PCBs	Born et al. 1981
O. rosmarus	Arctic (Baffin Bay, Canada)	—	B	4	PCBs, PCB congener ratios	Norstrom et al. 1992
O. rosmarus	Arctic (Canada)	1985–92	B	53	DDE, ΣDDT, PCBs, PCB congeners, dieldrin, mirex, oxychlordane, chlordanes, toxaphene, HCHs, CBz, PCDDs, PCDFs	Muir et al. 1995
O. rosmarus	Bering Sea	1981–84	B	53	Dieldrin, oxychlordane, 12 others analyzed but not found	Taylor et al. 1989
Ommatophoca rossi	Antarctica	1981–82	B	20	DDT, DDE, TDE, PCBs, dieldrin	McClurg 1984
Phoca groenlandica	Arctic (Greenland Sea)	1991	B, Br	11	DDE, TDE, DDT, ΣDDT, HCHs, HCB, PCBs, PCB congeners, PCDDs, PCDFs	Oehme et al. 1995b
P. groenlandica	Gulf of St. Lawrence (Canada)	1968	Mi	1	DDE, DDT	Cook and Baker 1969
P. groenlandica	Gulf of St. Lawrence (Canada)	1971	B	18	DDE, TDE, DDT, ΣDDT, PCBs, dieldrin	Addison et al. 1973
P. groenlandica	Gulf of St. Lawrence (Canada)	1971–73	B, Br, K, L, M, O	31	DDE, TDE, DDT, ΣDDT, PCBs, dieldrin, HCBs	Rosewell et al. 1979
P. groenlandica	Gulf of St. Lawrence (Canada)	1972–73	B, Br, K, L	20	DDE, TDE, DDT, ΣDDT, dieldrin, PCBs	Jones et al. 1976
P. groenlandica	Gulf of St. Lawrence (Canada)	1982	B	22	DDE, TDE, DDT, ΣDDT, PCBs	Addison et al. 1984
P. groenlandica	Gulf of St. Lawrence and Hudson Strait (Canada)	1988–89	B	50	ΣDDT, *p,p′*-DDE, PCBs, PCB congeners	Beck et al. 1994
P. groenlandica	North Atlantic (Canada)	1969–71	B, Br, L, M	78	ΣDDT, PCBs, dieldrin	Frank et al. 1973
P. groenlandica	Northwest Atlantic and Arctic oceans (Canada, Greenland)	1976–78	B, Br, K, L, M, O	248	ΣDDT, PCBs, dieldrin, chlordane, heptachlor epoxide, HCB	Ronald et al. 1984a
P. groenlandica	Western North Atlantic	1954–62	commercial oils	—	ΣDDT, DDE, TDE, DDT, dieldrin, PCBs	Addison et al. 1972
P. hispida	Arctic (Canada)	—	B	8	PCBs, PCB congeners	Ford et al. 1993
P. hispida	Arctic (Canada)	1960s	B	3	DDE, TDE, DDT, PCBs, dieldrin	Holden 1970

Continued on next page

Appendix 10-1 continued

Species	Region	Sample Period	Tissues Sampled	Number of Individuals Sampled	Compounds Reported	References
P. hispida	Arctic (Canada)	1970–72	Fat, L, M	11	DDE, TDE, DDT, *o,p*-DDT, PCBs	Bowes and Jonkel 1975; Bowes and Lewis 1974
P. hispida	Arctic (Canada)	1972	B	28	DDE, DDT, ΣDDT, PCBs	Addison and Smith 1974
P. hispida	Arctic (Canada)	1989–91	B	11	DDE, ΣDDT, PCBs, PCB congeners, dieldrin, mirex, oxychlordane, chlordanes, toxaphene, HCHs, HCBz, PCDDs, PCDFs	Muir et al. 1995
P. hispida	Arctic (Canada)	1992	B	1	PCBs, ΣDDT, HCHs, chlordane, toxaphene, PCB congeners	Zhu et al. 1995; Zhu and Norstrom 1993
P. hispida	Arctic (Norway)	—	B	11	HCBs, HCHs, DDE, DDT, TDE, ΣDDT, PCBs, PCB congeners	Luckas et al. 1990
P. hispida	Arctic (Norway)	1960s	B	2	DDE, TDE, DDT, PCBs, dieldrin	Holden 1970
P. hispida	Arctic (Baffin Bay, Canada)	—	B	4	PCBs, PCB congener ratios, PCDDs, PCDFs	Norstrom et al. 1992
P. hispida	Arctic (Canada)	1983–84, 1975–76, 1972	B, L	67	DDE, ΣDDT, PCBs, PCB congeners, HCHs, chlordanes, toxaphene, dieldrin, CBZ	Muir et al. 1988b
P. hispida	Arctic (Canada)	1983–84, 1985–86	B	78	PCDDs, PCDFs, PCBs, HCB	Norstrom et al. 1990
P. hispida	Baltic Sea	1969–73	B	33	ΣDDT, PCBs	Olsson et al. 1975
P. hispida	Baltic Sea	1969–74	B	73	ΣDDT, PCBs	Helle et al. 1976a; Olsson et al. 1975
P. hispida	Baltic Sea	1975	B	109	ΣDDT, PCBs	Helle et al. 1976b
P. hispida	Baltic Sea (Finland)	1981–87	B	7	PCB congeners, PCDFs, PCDDs, PCNs, hexachlorinated anthracenes	Koistinen 1990
P. hispida	Baltic Sea (Finland)	1985–89	B	1	ΣDDT, HCH, lindane, HCB, chlordane, toxaphene, PCN, PCBs, PCB congeners, dibenzofurans	Paasivirta and Rantio 1991
P. hispida	Baltic Sea (Gulf of Bothnia, Sweden)	1968	B	2	DDT, ΣDDT, PCBs	Jensen et al. 1969
P. hispida	Baltic Sea (Sweden)	1960s	B	1	DDE, TDE, DDT, PCBs, dieldrin	Holden 1970
P. hispida	Baltic Sea (Sweden)	1979–87	B	—	PCDDs, PCDFs	deWit et al. 1992
P. hispida	Baltic Sea (Sweden)	1980–88	B	17	ΣDDT, PCBs, PCB congeners, methyl sulfones of PCBs and DDE	Blomkvist et al. 1992; Haraguchi et al. 1992
P. hispida	Chukchi Sea, Norton Sound (Alaska)	1988	B, K, L	4	DDT, *o,p′*-DDT, TDE, DDE, *o,p′*-DDE, ΣDDT, PCBs, PCB congeners, HCB, HCHs, heptachlor epoxide, chlordane, *t*-nonachlor, dieldrin	Schantz et al. 1993
P. hispida	Greenland	1972	B	5	DDE, PCBs, heptachlor, heptachlor epoxide, aldrin, lindane	Clausen et al. 1974; Clausen and Berg 1975
P. hispida	Lake Saimaa (Finland)	1977–81	B, K, L, M, O	14	DDE, TDE, DDT, ΣDDT, PCBs, chlordane, chlorophenols	Helle et al. 1983
P. hispida	North Sea (Germany)	—	B	1	DDE, DDT, TDE, ΣDDT, PCBs, lindane, dieldrin	Harms et al. 1977/78
P. hispida	North Sea (Germany)	1975	B	1	ΣDDT, PCBs, dieldrin, lindane	Drescher 1978
P. hispida	Northern Quebec, Canada	1989–90	B	41	PCBs, chlordanes, nonachlors, oxychlordane, heptachlor epoxide, photoheptachlor	Zhu et al. 1995
P. hispida	Norway	1986	B	7	PCDDs, PCDFs, HCB, HCHs, DDE, TDE, DDT, ΣDDT, PCBs	Oehme et al. 1988, 1990
P. hispida	Norway	[illegible]	B, L, K	13	PCBs, DDE, PCB congeners	Daelemans et al. 1993

Appendix 10-1 continued

Species	Region	Sample Period	Tissues Sampled	Number of Individuals Sampled	Compounds Reported	References
P. hispida	Western Arctic (Canada)	1981	B	31	DDE, DDT, ΣDDT, PCBs	Addison et al. 1986a
P. hispida	Western North Atlantic (Canada)	1967	B	—	DDE, TDE, DDT, dieldrin	Holden and Marsden 1967
P. largha	Alaska	1975	O	—	DDE, DDT, *o,p′*-DDT	Serat et al. 1977
P. largha	Japan	1991	B	4	DDE, DDT, TDE, ΣDDT, PCBs, PCB congeners, HCHs	Tanabe et al. 1994b
P. siberica	Lake Baikal	—	B	1	ΣDDT, PCBs, chlordane, HCHs, toxaphene	Kucklick et al. 1993
P. siberica	Lake Baikal	—	B	1	DDE, TDE, DDT, *o,p′*-DDT, ΣDDT, PCBs, PCB congeners, HCB, HCHs, heptachlor, chlordanes, *t*-nonachlor, toxaphene	Kucklick et al. 1994
P. vitulina	Alaska	1989–90	B, L	9	ΣDDT, PCBs	Varanasi et al. 1992, 1993b
P. vitulina	Baltic Sea (Sweden)	—	B	9	HCB, HCHs, DDE, TDE, DDT, ΣDDT, PCBs, PCB congeners	Luckas et al. 1990
P. vitulina	Baltic Sea (Sweden)	1979–89	B	55	ΣDDT, PCBs, PCB congeners, methyl sulfones of PCBs and DDE	Blomkvist et al. 1992; Haraguchi et al. 1992
P. vitulina	Baltic Sea (Sweden)	1979–87	B	—	PCDDs, PCDFs	de Wit et al. 1992
P. vitulina	Britain	—	B, Br, K, L, M, O	5	ΣDDT, PCBs, dieldrin	Heppleston 1973
P. vitulina	California	1970	B, Br, L, M, O	2	DDE, TDE, DDT	Shaw 1971
P. vitulina	Cook Inlet (Alaska)	1976–78	B	23	DDE, DDT, oxychlordane, PCBs, others analyzed but not detected	Miles et al. 1992
P. vitulina	Eastern North Atlantic (Norway)	1977–80	B, K, L, M, O	10	ΣDDT, PCBs, HCB, HCHs	Ofstad and Martinsen 1983
P. vitulina	Eastern North Pacific (Southern California)	1976–80	L	4	DDE, TDE, DDT, ΣDDT, PCBs	Britt and Howard 1983
P. vitulina	Eastern North Pacific (U.S.A.)	1971	B	13	ΣDDT + PCBs	Anas 1974b
P. vitulina	England	1960s	B	15	DDE, TDE, DDT, PCBs, dieldrin	Holden 1970
P. vitulina	England	1988	B, K, L	15	DDE, TDE, DDT, dieldrin, HCH, HCB, PCBs	Simmonds et al. 1993
P. vitulina	Iceland	1988–89	B	10	PCBs	Heidmann et al. 1992
P. vitulina	Netherlands	—	B	1	PCBs, mirex	ten Noever de Brauw et al. 1973
P. vitulina	Netherlands (captive)	1981–83	Blood, feces	—	PCB congeners	Storr-Hansen et al. 1995
P. vitulina	North Atlantic (U.S.A.)	1980, 1990–92	B, L	15	DDE, mirex, HCB, ∝-chlordane, *trans*-nonachlor, PCBs, PCB congeners	Lake et al. 1995
P. vitulina	North Atlantic (Iceland)	—	B	7	HCB, HCHs, DDE, TDE, DDT, ΣDDT, PCBs, PCB congeners	Luckas et al. 1990
P. vitulina	North Atlantic (Bay of Fundy, Gulf of Maine)	1971	B, Br, L, M	12	DDE, TDE, DDT, *o,p′*-DDT, ΣDDT, dieldrin, PCBs	Gaskin et al. 1973
P. vitulina	North Sea (Denmark)	1988–91	B, Br, Li, M, K, O	5	DDE, HCB, PCBs, PCB congeners	Storr-Hansen and Spliid 1993b
P. vitulina	North Sea (Denmark)	1988	B	21	DDE, HCB, PCBs, PCB congeners	Storr-Hansen and Spliid 1993a
P. vitulina	North Sea (England)	1988	B	10	DDE, TDE, DDT, PCBs, PCB congeners, dieldrin, HCHs, HCB	Law et al. 1989
P. vitulina	North Sea (Germany)	—	B, Br, K, L	70	DDE, DDT, TDE, ΣDDT, PCBs, lindane, dieldrin	Drescher et al. 1977; Harms et al. 1977/78

Continued on next page

Appendix 10-1 continued

Species	Region	Sample Period	Tissues Sampled	Number of Individuals Sampled	Compounds Reported	References
P. vitulina	North Sea (Germany)	—	B	11	HCB, HCHs, DDE, TDE, DDT, ΣDDT, PCBs, PCB congeners	Luckas et al. 1990
P. vitulina	North Sea (Germany)	1988	B	1	toxaphene	Vetter et al. 1992
P. vitulina	North Sea (Germany)	1988	B	1	toxaphene	Kallenborn et al. 1994
P. vitulina	North Sea (Germany)	1988	B	5	DDE, TDE, DDT, HCHs, HCB, PCB congeners, PCDDs, PCDFs	Beck et al. 1990
P. vitulina	North Sea (Germany)	1988–90	B	27	PCBs	Heidmann et al. 1992
P. vitulina	North Sea (Germany, Denmark)	1975–76	B, K, L	16	DDE, TDE, DDT, ΣDDT, PCBs, dieldrin, HCHs, heptachlor epoxide, others analyzed but not detected	Reijnders 1980
P. vitulina	North Sea (Netherlands)	—	B, Br, K, L, O	8	DDE, TDE, DDT, *o,p′*-TDE, HCHs, PCBs, dieldrin, mirex	Duinker et al. 1979
P. vitulina	North Sea (Netherlands)	1960s	Fat, L	3	DDE, TDE, DDT, dieldrin	Koeman and van Genderen 1966
P. vitulina	North Sea (Netherlands)	1970–71	B	8	DDE, TDE, DDT, PCBs, dieldrin, HCB	Koeman et al. 1972
P. vitulina	North Sea (Netherlands)	1972–81	B	175	PCBs	Van der Zande and De Ruiter 1983
P. vitulina	North Sea (Netherlands)	1975–76	B, K, L	14	DDE, TDE, DDT, ΣDDT, PCBs, dieldrin, HCHs, heptachlor epoxide, others analyzed but not detected	Reijnders 1980
P. vitulina	North Sea (Netherlands)	1978	B, Br, L	3	ΣDDT, PCBs, dieldrin, heptachlor epoxide, oxychlordane, *trans*-nonachlor	Kerkhoff et al. 1981; Kerkhoff and de Boer 1982
P. vitulina	North Sea (Norway)	1988	B, Br, K, L	10	PCB congeners	Bernhoft and Skaare 1994
P. vitulina	North Sea (U.K.)	1988–89	B	89	PCB congeners, DDE, TDE, DDT, ΣDDT, dieldrin	Hall et al. 1992
P. vitulina	Northern Ireland	1988	B, L, K	55	DDE, TDE, *o,p′*-DDT, DDT, ΣDDT, PCBs, HCB, HCHs, heptachlor, heptachlor epoxide, chlordanes, dieldrin	Mitchell and Kennedy 1992
P. vitulina	Puget Sound, U.S.A	1972–82	B	17	DDE, DDT, PCBs, heptachlor epoxide, HCHs, nonachlors, oxychlordane, chlordane	Walker et al. 1989
P. vitulina	Scotland	1960's	B	21	DDE, TDE, DDT, PCBs, dieldrin	Holden 1970
P. vitulina	Scotland	1965–66	B	—	DDE, TDE, DDT, dieldrin	Holden and Marsden 1967
P. vitulina	Scotland	1988	B	9	DDE, PCBs	Simmonds et al 1993
P. vitulina	Scotland	1988	B	1	PCB congeners	Wells and Echarri 1992
P. vitulina	Sweden	1969–73	B	8	ΣDDT, PCBs	Olsson et al. 1975
Zalophus californianus	California	—	Fat, L	1	DDE	Theobald 1973
Z. californianus	California	1969	B, Br, L, M	11	DDE, TDE, DDT, dieldrin	Anas 1971
Z. californianus	California	1988	Mi	1	PCB congeners, DDE, *o,p′*-DDE, DDT, *o,p′*-DDT, TDE, *o,p′*-TDE, chlordanes, heptachlor epoxide, dieldrin	Bacon et al. 1992
Z. californianus	California	—	L	—	DDE	Hall et al. 1971
Z. californianus	California	1970	B, Br, L	20	ΣDDT, PCBs	DeLong et al. 1973
Z. californianus	California	1970–81	L	69	DDE, TDE, DDT, ΣDDT, PCBs	Britt and Howard 1983
Z. californianus	California	1970	B, Br, L, M	25	DDE, TDE, DDT, ΣDDT	LeBouef and Bonnell 1971

Appendix 10-1 continued

Species	Region	Sample Period	Tissues Sampled	Number of Individuals Sampled	Compounds Reported	References
Z. californianus	California	1988–92	B	7	DDE, DDT	Lieberg-Clark et al. 1995
Z. californianus	Oregon	1970–73	Br, L, M, fat	19	DDE, TDE, DDT, ΣDDT, PCBs	Buhler et al. 1975

B = blubber; Br = brain; K = kidney; L = liver; Lu = lung; M = muscle; Mi = milk; O = other. Dashes appear where values were not available in original source.

CBZ = chlorobenzenes; DDE = 2,2.-*bis* (p-chlorophenyl)-1, 1-dichloroethylene; DDT = 2,2,-*bis*-(*p*-chlorophenyl)-1, 1-trichloroethane; ΣDDT = arithmetic summation of concentrations of isomers and metabolites of DDT; HCB = hexachlorobenzene; HCH = hexachlorocyclohexane; PCBs = polychlorinated biphenyls; PCDDs = polychlorinated dibenzo-*p*-dioxins; PCDFs = polychlorinated dibenzofurans; PCNs = polychlorinated napthalenes; PCP = pentachlorophenol; PCTs = polychlorinated terphenyls; TDE = 2,2,-*bis*-(*p*-chlorophenyl)-1, 1-dichloreoethane.

Appendix 10-2. Summary of Selected Organochlorine Residue Surveys in Odontocete Cetaceans

Species	Region	Sample Period	Tissues Sampled	Number of Individuals Sampled	Compounds Reported	References
Berardius bairdii	Western North Pacific (Japan)	1985	B	3	PCBs, PCB congeners, PCDFs, PCDDs	Kannan et al. 1989
B. bairdii	Western North Pacific (Japan)	1985	B	37	DDE, PCBs	Subramanian et al. 1988a
B. bairdii	Western North Pacific (Japan)	1985–89	B	3	HCHs	Tanabe et al. 1996
Cephalorhynchus commersonii	Southern Indian Ocean (Kerguelen Island)	1983	Melon	11	DDE, DDT, ΣDDT, PCBs, HCH, ratios, PCB congeners	Abarnou et al. 1986
C. heavisidii	Southwestern Indian Ocean (South Africa)	1977–87	B	9	ΣDDT, PCBs, HCB	de Kock et al. 1994
C. hectori	New Zealand	—	B	1	ΣDDT, PCBs, lindane	Baker 1978
C. hectori	New Zealand	1985–87	B	6	PCDDs, PCDFs	Buckland et al. 1990
Delphinapterus leucas	Arctic (Canada)	—	B	6	ΣPCBs, PCB congeners	Ford et al. 1993
D. leucas	Arctic Ocean (Canada)	1972	B, L, M	14	ΣDDT, DDE, TDE, DDT, *o,p*′-DDT, PCBs analyzed but not detected	Addison and Brodie 1973
D. leucas	Arctic Ocean (Canada)	1983–84	B	29	PCDDs, PCDFs, PCBs, HCB	Norstrom et al. 1990
D. leucas	Baltic Sea (Germany)	—	B, L	1	DDE, TDE, DDT, ΣDDT, PCBs, lindane, dieldrin	Harms et al. 1977/78
D. leucas	Canada (six locations)	1966–87	B, L, M	88	DDE, ΣDDT, PCBs, PCB congeners, ratios, HCH, toxaphene, dieldrin, mirex, chlordane, *trans*-nonachlor, HCB	Muir et al. 1990
D. leucas	Chukchi Sea (Alaska)	1989–90	B	4	DDT, *o,p*′-DDT, TDE, DDE, *o,p*′-DDE, ΣDDT, PCB congeners, HCB, HCHs, heptachlor epoxide, chlordane, t-nonachlor, dieldrin	Schantz et al. 1993
D. leucas	Hudson Bay (Canada)	1966	Oils	—	DDE, TDE, DDT, ΣDDT, dieldrin, PCBs	Addison et al. 1972
D. leucas	St. Lawrence River (Canada)	1982–85	B, K, L, Mi, O	26	ΣDDT, DDT, TDE, DDE, PCBs, PCB congeners	Martineau et al. 1987; Massé et al. 1986
D. leucas	St. Lawrence River (Canada)	—	B	10	PCBs, PCB congener ratios, PCDDs, PCDFs	Norstrom et al. 1992
D. leucas	St. Lawrence River (Canada)	1988	B, L	1	ΣDDT, PCBs, methyl sulfones of PCB congeners and DDE	Bergman et al. 1994

Continued on next page

Appendix 10-2 continued

Species	Region	Sample Period	Tissues Sampled	Number of Individuals Sampled	Compounds Reported	References
D. leucas	St. Lawrence River (Canada)	1993–94	B	16	DDE, TDE, DDT, ΣDDT, *o,p'*-DDE, *o,p'*-TDE, *o,p'*-DDT, dieldrin, mirex, toxaphene, CBz, HCHs, chlordanes, nonachlors, oxychlordane, heptachlor epoxide, TCP-methane, OCS, PCBs, PCB congeners	Muir et al. 1996b
D. leucas	St. Lawrence River (Canada), and Eastern Newfoundland	1987–90	B, L	38	ΣDDT, DDE, TDE, DDT, dieldrin, toxaphene, t-nonachlor, oxychlordane, mirex, TCP-methane, PCDD congeners, PCDF congeners, PCB congeners, CBz, HCHs	Muir et al. 1996a
Delphinus delphis	Atlantic coast (U.S.A.)	1986–88	B	4	DDE, dieldrin, HCB, mirex, chlordanes, lindane, heptachlor epoxide, PCB congeners, PCDD, PCDFs, PBBs	Kuehl et al. 1991
D. delphis	California	1974–76	B, Br, M	13	DDE, TDE, DDT, *o,p'*-DDE, *o,p'*-TDE, *o,p'*-DDT, PCBs, dieldrin, HCB, heptachlor epoxide, *trans*-nonachlor, oxychlordane, *cis*-chlordane	O'Shea et al. 1980
D. delphis	Eastern North Atlantic (Spain)	1979	B	1	DDE, TDE, DDT, *o,p'*-DDT, ΣDDT	Aguilar 1984
D. delphis	Western North Pacific	1987	B	2	HCHs	Tanabe et al. 1996
D. delphis	Eastern North Pacific (California)	1978–84	B, Br, K, L, M	11	ΣDDT, PCBs	Schafer et al. 1984
D. delphis	France	—	B, K, L, M, O	25	ΣDDT, DDE, DDT, TDE, PCBs	Alzieu and Duguy 1979
D. delphis	Indian Ocean (South Africa)	1980–86	B, L	97	PCBs, ΣDDT	Cockcroft et al. 1990
D. delphis	Japan	1968–75	B, M	2	DDE, TDE, DDT, *o,p'*-DDE, *o,p'*-TDE, *o,p'*-DDT, PCBs, dieldrin, HCB, toxaphene, heptachlor epoxide, *trans*-nonachlor, endrin, mirex	O'Shea et al. 1980
D. delphis	Mediterranean Sea (France)	1977	B	1	DDE, TDE, DDT, HCH, PCBs	Vicente and Chabert 1978
D. delphis	New Zealand (captive)	—	L	3	PCBs, DDE, TDE, DDT, ΣDDT	Solly and Shanks 1976; Lock and Solly 1976
D. delphis	North Sea (England)	1965	B, L	1	DDE, dieldrin	Robinson et al. 1967
D. delphis	North Sea (Netherlands)	1979	B, Br, K, L, M, O	1	DDE, TDE, DDT, dieldrin, HCB, HCHs, PCBs, PCB congeners	Duinker et al. 1989
D. delphis	North Atlantic (U.S.A.)	1971–75	B	1	PCBs, chlordane, dieldrin, DDE, TDE, DDT, ΣDDT	Taruski et al. 1975
D. delphis	Southeastern Atlantic (South Africa)	1980–87	B	—	ΣDDT, PCBs	de Kock et al. 1994
D. delphis	Southwestern Indian Ocean (South Africa)	1984–87	B	17	ΣDDT, PCBs, HCB	de Kock et al. 1994
D. delphis	Victoria, Australia	—	B	1	DDT, PCBs	Kemper et al. 1994
D. delphis	Western North Atlantic	—	B	3	PCB congeners	Kuehl et al. 1994
Feresa attenuata	Florida (Gulf and Atlantic)	1975–78	B, Br, K, L, M, O	3	ΣDDT, DDE, TDE, DDT, dieldrin, PCBs	Forrester et al. 1980
Globicephala macrorhynchus	California	1974–76	B, M	2	DDE, TDE, DDT, *o,p'*-DDE, *o,p'*-TDE, *o,p'*-DDT, PCBs	O'Shea et al. 1980

Appendix 10-2 continued

Species	Region	Sample Period	Tissues Sampled	Number of Individuals Sampled	Compounds Reported	References
G. macrorhynchus	Caribbean Sea (St. Lucia)	1972	B, K, L, M	5	DDE, TDE, DDT, ΣDDT, PCBs, dieldrin	Gaskin et al. 1974
G. macrorhynchus	Japan	1968–75	B	6	DDE, TDE, DDT, *o,p′*-DDE, *o,p′*-TDE, *o,p′*-DDT, PCBs, dieldrin, HCB, toxaphene	O'Shea et al. 1980
G. macrorhynchus	Eastern North Pacific (California)	1971	L	5	DDE	Hall et al. 1971
G. macrorhynchus	Western North Atlantic (U.S.A.)	—	B, L	5	ΣDDT, PCBs	Varanasi et al. 1992
G. macrorhynchus	Western North Pacific (Japan)	1985	B	29	DDE, PCBs	Tanabe et al. 1987b
G. melaena	France	—	B, K, L, M	16	ΣDDT, DDE, DDT, TDE, PCBs	Alzieu and Duguy 1979
G. melaena	North Atlantic	1962–67	Oils	—	DDE, TDE, DDT, ΣDDT, dieldrin, PCBs	Addison et al. 1972
G. melaena	North Atlantic (U.S.A.)	1971–75	B	2	PCBs, chlordane, dieldrin, DDE, TDE, DDT, ΣDDT	Taruski et al. 1975
G. melaena	Northwestern Atlantic (Newfoundland)	1980	B	41	DDE, ΣDDT, PCBs, PCB congeners, dieldrin, toxaphene, HCB, HCHs, t-nonachlor, chlordane, mirex	Muir et al. 1988a
G. melaena	Western North Atlantic	—	B, L	11	DDE, PCBs, *trans*-nonachlor	Geraci 1989
G. melanea	Western North Atlantic (U.S.A.)	1986–90	B, L	23	ΣDDT, PCBs, chlordanes	Varanasi et al. 1993b
G. melaena	Australia	—	B	9	DDE, DDT, TDE, ΣDDT, PCBs, dieldrin, HCB, HCH	Kemper et al. 1994
G. melaena	North Atlantic (Faroe Islands)	1986–88	B	130	DDE, TDE, DDT, *o,p′*-DDT, ΣDDT, PCBs	Borrell et al. 1995
G. melaena	North Atlantic (Faroe Islands)	1987	B, M	211	DDE, TDE, *o,p′*-DDT, DDT, ΣDDT, PCBs	Borrell 1993; Borrell and Aguilar 1993
G. melaena	North Atlantic (Faroe Islands)	1986	B, O	50	DDE, PCBs, dieldrin, HCH, heptachlor epoxide	Simmonds et al. 1994
Grampus griseus	British Columbia	1988	B	1	DDE, TDE, DDT, ΣDDT, CBz, HCB, HCHs, chlordanes, nonachlors, oxychlordane, heptachlor epoxide, photoheptachlor, mirex, toxaphene, OCS, non-ortho PCB congeners, PCBs, PCDD congeners, PCDF congeners	Jarman et al. 1996
G. griseus	France	—	B, K, L, M, O	6	ΣDDT, DDT, DDE, TDE, PCBs	Alzieu and Duguy 1979
G. griseus	Eastern North Atlantic (Spain)	1978	B	1	DDE, TDE, DDT, *o,p′*-DDT, ΣDDT	Aguilar 1984
G. griseus	Eastern South Atlantic (South Africa)	1970	B, L	1	ΣDDT, dieldrin	Aucamp et al. 1971
G. griseus	Southwestern Indian Ocean (South Africa)	1984	B	2	ΣDDT, PCBs, HCB	de Kock et al. 1994
Hyperoodon spp.	North Atlantic	1962	Oils	—	DDE, TDE, DDT, ΣDDT, dieldrin, PCBs	Addison et al. 1972
H. ampullatus	North Sea (Germany)	—	B	1	DDE, TDE, DDT, ΣDDT, PCBs, lindane, dieldrin	Harms et al. 1977/78
Kogia breviceps	Australia	—	B	1	DDT, dieldrin	Kemper et al. 1994
K. breviceps	Southwestern Indian Ocean (South Africa)	1978–87	B	5	ΣDDT, PCBs, HCB	de Kock et al. 1994

Continued on next page

Appendix 10-2 continued

Species	Region	Sample Period	Tissues Sampled	Number of Individuals Sampled	Compounds Reported	References
K. breviceps	Western Atlantic (Florida)	1974	Mi	1	DDE, TDE, DDT, *o,p′*-DDE, *o,p′*-DDT, ΣDDT, PCBs, dieldrin	Jenness and Odell 1978
K. simus	Southwestern Indian Ocean (South Africa)	1976–84	B	4	ΣDDT, PCBs, HCB	de Kock et al. 1994
Lagenodelphis hosei	Eastern Tropical Pacific	—	B, M	1	DDE, TDE, DDT, *o,p′*-DDE, *o,p′*-TDE, *o,p′*-DDT, PCBs	O'Shea et al. 1980
L. hosei	Japan	1991	B	4	HCHs	Tanabe et al. 1996
Lagenorhynchus acutus	Faroe Islands	1987	B	13	DDE, TDE, *o,p′*-DDT, DDT, ΣDDT, PCBs	Borrell 1993
L. acutus	Atlantic Coast (U.S.A.)	1989	B	3	DDE, dieldrin, HCB, mirex, chlordanes, lindane, heptachlor epoxide, PCB congeners, PCDD, PCDFs, PBBs	Kuehl et al. 1991
L. acutus	North Atlantic (Nova Scotia)	1971–75	B	1	PCBs, chlordane, dieldrin, DDE, TDE, DDT, ΣDDT	Taruski et al. 1975
L. acutus	Western North Atlantic	—	B	2	PCB congeners	Kuehl et al. 1994
L. albirostris	North Sea	—	B	1	Toxaphene	de Boer and Wester 1993
L. albirostris	North Sea (Denmark)	1972	B, Br, L, M, O	1	DDE, TDE, DDT, ΣDDT, PCBs, dieldrin	Andersen and Rebsdorff 1976
L. albirostris	North Sea (Netherlands)	1977	B, Br, M	3	DDE, TDE, DDT, dieldrin, HCB, HCHs, PCBs, PCB congeners	Duinker et al. 1989
L. albirostris	North Sea (Scotland)	1977	B	5	ΣDDT, PCBs, dieldrin, heptachlor epoxide	Kerkhoff et al. 1981
L. albirostris	Northwestern Atlantic (Newfoundland)	1982	B	27	DDE, ΣDDT, PCBs, PCB congeners, dieldrin, toxaphene, HCB, HCHs, *t*-nonachlor, chlordane, mirex	Muir et al. 1988a
L. obliquidens	North Pacific	1991	B	3	HCHs	Tanabe et al. 1996
L. obliquidens	North Pacific (Japan)	1981	B	5	ΣDDT, PCBs, HCHs	Tanabe et al. 1983
L. obscurus	Southwestern Indian Ocean (South Africa)	1977–87	B	12	ΣDDT, PCBs, HCB	de Kock et al. 1994
L. obscurus	Western South Pacific	1980	B	1	ΣDDT, PCBs, HCHs	Tanabe et al. 1983
Lissodelphis borealis	North Pacific	1991	B	2	HCHs	Tanabe et al. 1996
Mesoplodon bidens	France	—	L, O	1	ΣDDT, PCBs	Alzieu and Duguy 1979
M. densirostris	Mediterranean (Spain)	1980	B, Br, M	1	DDE, *o,p′*-DDT, TDE, DDT, ΣDDT, PCBs	Aguilar et al. 1982
M. densirostris	North Atlantic (U.S.A.)	1971–75	B	2	PCBs, chlordane, dieldrin, DDE, TDE, DDT, ΣDDT	Taruski et al. 1975
M. densirostris	Southwestern Indian Ocean (South Africa)	1984–86	B	6	ΣDDT, PCBs, HCB	de Kock et al. 1994
M. layardi	Southwestern Indian Ocean (South Africa)	1978–85	B	2	ΣDDT, PCBs, HCB	de Kock et al. 1994
M. layardi	Australia	1989	B	1	ΣDDT, DDE, DDT, TDE, PCBs	Anderson 1991
M. mirus	Southwestern Indian Ocean (South Africa)	1986	B	2	ΣDDT, PCBs, HCB	de Kock et al. 1994
M. stejnegeri	Japan	1984	B, K, L, M	1	DDE, PCBs	Miyazaki et al. 1987
M. sp.	Australia	—	B	2	DDE, DDT, PCBs, dieldrin, oxychlordane, HCB	Kemper et al. 1994
Monodon monoceros	Arctic (Baffin Bay, Canada)	—	B	17	PCBs, PCB congener ratios	Norstrom et al. 1992
M. monoceros	Arctic (Canada)	—	B	17	ΣPCBs, PCB congeners	Ford et al. 1993

Appendix 10-2 continued

Species	Region	Sample Period	Tissues Sampled	Number of Individuals Sampled	Compounds Reported	References
M. monoceros	Arctic (Canada)	1982–83	B, L	21	Chlorobenzenes, HCH, chlordanes, DDE, ΣDDT, mirex, dieldrin, toxaphene, PCBs, PCB congeners	Muir et al. 1992b
Neophocoena phocoenoides	Seto-Inland Sea (Japan)	1985	B	1	PCBs, PCB congeners, PCDFs, PCDDs	Kannan et al. 1989
N. phocoenoides	Western North Pacific (Japan)	1968–75	B, Br, M	6	DDE, TDE, DDT, *o,p′*-DDE, *o,p′*-TDE, *o,p′*-DDT, PCBs, dieldrin, HCB, toxaphene, heptachlor epoxide, *trans*-nonachlor, oxychlordane, mirex	O'Shea et al. 1980
Orcinus orca	British Columbia, Washington	1986–89	B	6	DDE, TDE, DDT, ΣDDT, CBz, HCB, HCHs, chlordanes, nonachlors, oxychlordane, heptachlor epoxide, photoheptachlor, mirex, toxaphene, OCS, PCBs, non-ortho PCB congeners, PCDD congeners, PCDF congeners	Jarman et al. 1996
O. orca	Western North Pacific (Japan)	1986	B	3	PCBs, PCDDs, PCDFs, PCB congeners	Ono et al. 1987; Kannan et al. 1989
O. orca	Victoria, Australia	—	B	1	ΣDDT, dieldrin, heptachlor, endrin, HCB	Kemper et al. 1994
Peponocephala electra	Japan	1982	B	5	ΣDDT, PCBs, HCHs	Tanabe et al. 1983
P. electra	Japan	1982	B	3	HCHs	Tanabe et al. 1996
Phocoena phocoena	Baltic Sea (Germany)	—	B, L	2	DDE, TDE, DDT, ΣDDT, PCBs, lindane, dieldrin	Harms et al. 1977/78
P. phocoena	Baltic Sea (Poland)	1989–90	B, L, M	3	PCB congeners	Falandysz et al. 1994
P. phocoena	Baltic Sea (Poland)	1989–90	B, L, M	3	PCBs, ΣDDT, HCHs, HCB, aldrin, dieldrin, heptachlor, heptachlor epoxide, chlordanes	Kannan et al. 1993b
P. phocoena	British Columbia	1987–89	B	7	DDE, TDE, DDT, ΣDDT, CBz, HCB, HCHs, chlordanes, nonachlors, oxychlordane, heptachlor epoxide, photoheptachlor, mirex, toxaphene, octachlorostyrene, PCBs, non-ortho PCB congeners, PCDD congeners, PCDF congeners	Jarman et al. 1996
P. phocoena	California	1974–76	B, Br, M	1	DDE, TDE, DDT, *o,p′*-DDE, *o,p′*-TDE, *o,p′*-DDT, PCBs, dieldrin, HCB, heptachlor epoxide, trans-nonachlor	O'Shea et al. 1980
P. phocoena	California	1987–88	B	3	DDE, TDE, DDT, ΣDDT, CBz, HCB, HCHs, chlordanes, nonachlors, oxychlordane, heptachlor epoxide, photoheptachlor, mirex, toxaphene, octachlorostyrene, PCBs, PCDD congeners, PCDF congeners	Jarman et al. 1996
P. phocoena	Denmark (North Sea, Baltic), West Greenland	1986–88	B	27	HCHs, HCB, DDE, TDE, DDT, ΣDDT, PCBs, PCB congeners	Granby and Kinze 1991
P. phocoena	Eastern North Pacific (Washington)	1992	B, L	3	ΣDDT, PCBs, chlordanes	Varanasi et al. 1993b

Continued on next page

Appendix 10-2 continued

Species	Region	Sample Period	Tissues Sampled	Number of Individuals Sampled	Compounds Reported	References
P. phocoena	Faroe Islands	1987–88	B	6	DDE, TDE, *o,p′*-DDT, DDT, ΣDDT, PCBs	Borrell 1993
P. phocoena	France	—	B, L, O	3	PCBs, DDE, ΣDDT	Alzieu and Duguy 1979
P. phocoena	Great Britain	1990–91	B	28	DDE, TDE, DDT, PCBs, HCB, HCHs, dieldrin	Kuiken et al. 1993
P. phocoena	Great Britain	1989–92	B	94	DDE, TDE, DDT, dieldrin, HCB, HCH, PCBs, PCB congeners	Kuiken et al. 1994
P. phocoena	Greenland	1972	B	2	DDE, PCBs, heptachlor epoxide, aldrin, lindane	Clausen et al. 1974; Clausen and Berg 1975
P. phocoena	Greenland	1989	B, L	4	PCB congeners, PCDD congeners, PCDF congeners	van Scheppingen et al. 1996
P. phocoena	North Atlantic (Bay of Fundy)	1969–70	B, L	36	DDE, TDE, DDT, *o,p′*-DDT, ΣDDT, dieldrin	Gaskin et al. 1971
P. phocoena	North Atlantic (Wales)	1988	B, M, L, O	4	HCB, HCHs, dieldrin, DDE, DDT, TDE, PCBs, PCB congeners	Morris et al. 1989
P. phocoena	North Atlantic (U.S.A.)	1971–75	B	1	PCBs, chlordane, dieldrin, DDE, TDE, DDT, ΣDDT	Taruski et al. 1975
P. phocoena	Northeast North Atlantic (Atlantic, North Sea, Kattegat)	1987–1991	B	34	DDE, TDE, DDT, *o,p′*-DDE, *o,p′*-DDT, dieldrin, endrin, HCHs, oxychlordane, *trans*-nonachlor, heptachlor epoxide, HCB, PCB congeners	Kleivane et al. 1995
P. phocoena	North Sea	—	B	1	Toxaphene	de Boer and Wester 1993
P. phocoena	North Sea	1970–71	B	7	DDE, TDE, DDT, PCBs, dieldrin, HCB	Koeman et al. 1972
P. phocoena	North Sea (Denmark)	1972–73	B, Br, K, L, M, O	7	DDE, TDE, DDT, ΣDDT, PCBs, dieldrin	Andersen and Rebsdorff 1976
P. phocoena	North Sea (Germany)	—	B, L	1	DDE, TDE, DDT, ΣDDT, PCBs, lindane, dieldrin	Harms et al. 1977/78
P. phocoena	North Sea (Germany)	1988	B	1	DDE, TDE, DDT, HCHs, HCB, PCB congeners, PCDDs, PCDFs	Beck et al. 1990
P. phocoena	North Sea (Scotland)	1988–91	B	48	HCB, ΣDDT, PCBs, PCB congeners, endrin, dieldrin, heptachlor epoxide, chlordanes, *t*-nonachlor, oxychlordane	Wells et al. 1994
P. phocoena	North Sea (Netherlands)	1977–79	B, Br, K, L, O M,	11	DDE, TDE, DDT, dieldrin, HCB, HCHs, PCBs, PCB congeners	Duinker et al. 1989
P. phocoena	North Sea (Netherlands)	1978	B, K, L, M, O	4	PCBs, PCB congeners	Duinker et al. 1988
P. phocoena	North Sea (Netherlands)	1979	B, K, L	1	DDE, TDE, DDT, o,p′–TDE, PCBs, dieldrin, HCB, HCHs	Duinker and Hillebrand 1979
P. phocoena	North Sea (Netherlands)	1990–93	B, K, L	22	PCBs, PCB congeners, PCDD congeners, PCDF congeners	van Scheppingen et al. 1996
P. phocoena	Northwestern Atlantic (Bay of Fundy, Rhode Island, Maine, Newfoundland, Prince Edward Island)	1971–77	B, Br, K, L, M, Mi, O	107	PCBs, HCB, chlordanes	Gaskin et al. 1983
P. phocoena	Northwestern Atlantic (Bay of Fundy, Rhode Island, Maine, Nova Scotia, Newfoundland, Prince Edward Island)	1969–73	B	115	ΣDDT	Gaskin et al. 1982

Appendix 10-2 continued

Species	Region	Sample Period	Tissues Sampled	Number of Individuals Sampled	Compounds Reported	References
P. phocoena	Scotland	1967	B, Br, K, L, M, O	4	DDE, TDE, DDT, ΣDDT, dieldrin	Holden and Marsden 1967
P. phocoena	Scotland	1990	B	1	PCB congeners	Wells and Echarri 1992
P. phocoena	Western North Atlantic (U.S.A.)	1991	B, L	3	ΣDDT, PCBs, chlordanes	Varanasi et al. 1993b
P. phocoena	Western North Atlantic (U.S.A.)	—	B, L	9	DDE, PCBs, *trans*-nonachlor	Geraci 1989
Phocoenoides dalli	Japan	1983	B	4	DDE, DDT, TDE, ΣDDT, PCBs, PCB congeners, HCHs	Tanabe et al. 1994b
P. dalli	Northern North Pacific	1980–82	B	3	DDE, TDE, DDT, HCHs, chlordanes	Kawano et al. 1988
P. dalli	Northern North Pacific	1980, 1985	B	5	PCBs, PCB congeners	Kannan et al. 1989
P. dalli	Bering Sea and Northwestern Pacific	—	B	42	PCBs, DDE, PCB congeners	Subramanian et al. 1988b
P. dalli	British Columbia	1987–88	B	3	DDE, TDE, DDT, ΣDDT, CBz, HCB, HCHs, chlordanes, nonachlors, oxychlordane, heptachlor epoxide, photoheptachlor, mirex, toxaphene, OCS, non-ortho PCB congeners, PCBs, PCDD congeners, PCDF congeners	Jarman et al. 1996
P. dalli	North Pacific (Bering Sea, North Pacific, Japan)	1980–82	B	6	ΣDDT, PCBs, HCHs	Tanabe et al. 1983
P. dalli	North Pacific (Bering Sea, North Pacific, Japan)	1985–89	B	9	HCHs	Tanabe et al. 1996
P. dalli	Northwestern Pacific	1984	B	12	PCBs, DDE	Subramanian et al. 1987
P. dalli	California	1974–76	B, M	1	DDE, TDE, DDT, *o,p'*-DDE, *o,p'*-TDE, *o,p'*-DDT, PCBs, dieldrin, HCB, *trans*-nonachlor	O'Shea et al. 1980
P. dalli	Western North Pacific (Japan)	1968–75	B, Br, M	1	DDE, TDE, DDT, *o,p'*-DDE, *o,p'*-TDE, *o,p'*-DDT, PCBs, dieldrin, HCB, *trans*-nonachlor	O'Shea et al. 1980
Physeter catodon	Antarctic	1962–66	Oils	—	DDE, TDE, DDT, ΣDDT, dieldrin, PCBs	Addison et al. 1972
P. catodon	Eastern North Pacific (California)	1983	B, M	1	ΣDDT, PCBs	Schafer et al. 1984
P. catodon	Eastern North Pacific (California)	1968	B, Br, L	6	DDE, TDE, DDT, dieldrin, others analyzed but not detected	Wolman and Wilson 1970
P. catodon	Lesser Antilles	1971–75	B	2	PCBs, chlordane, dieldrin, DDE, TDE, DDT, ΣDDT	Taruski et al. 1975
P. catodon	North Atlantic	1967	Oils	—	DDE, TDE, DDT, ΣDDT, dieldrin, PCBs	Addison et al. 1972
P. macrocephalus	Australia	1989	B	2	DDE, TDE, DDT, PCBs	Anderson 1991
P. macrocephalus	France	1976	B, L, M, O	2	ΣDDT, PCBs	Alzieu and Duguy 1979
P. macrocephalus	Iceland	1982	B	10	DDE, TDE, *o,p'*-DDT, DDT, ΣDDT, PCBs	Borrell 1993
P. macrocephalus	North Atlantic (Spain)	1979-80	B, Br, K, L, M, O	14	DDE, TDE, DDT, *o,p'*-DDT, ΣDDT, PCBs	Aguilar 1983
P. macrocephalus	North Sea	1994–95	B	7	DDE, TDE, DDT, *o,p'*-DDE, *o,p'*-DDT, *o,p'*-TDE, dieldrin, HCB, HCHs, *trans*-chlordane, *cis*-chlordane, *trans*-nonachlor, PCBs, PCB congeners	Law et al. 1996

Continued on next page

Appendix 10-2 continued

Species	Region	Sample Period	Tissues Sampled	Number of Individuals Sampled	Compounds Reported	References
P. macrocephalus	North Sea (Netherlands)	1979	B, Br, M, O	1	DDE, TDE, DDT, dieldrin, HCB, HCHs, PCBs, PCB congeners	Duinker et al. 1989
P. macrocephalus	Scotland	1990	B	1	PCB congeners	Wells and Echarri 1992
P. macrocephalus	South Africa	1994	B	12	DDE, TDE, DDT, ΣDDT, PCBs, dieldrin	Henry and Best 1983
P. macrocephalus	Southwestern Indian Ocean (South Africa)	1986	B	1	ΣDDT, PCBs, HCB	de Kock et al. 1994
P. macrocephalus	Victoria, Australia	—	B	1	ΣDDT, PCBs, dieldrin, endrin, HCB, BHC	Kemper et al. 1994
Platanista gangetica	India (Ganges River)	1988–92	B, K, L, M	4	PCBs, ΣDDT, HCHs, HCB, aldrin, dieldrin, heptachlor, heptachlor epoxide, chlordanes	Kannan et al. 1993a
Pontoporia blainvillei	South Atlantic (Uruguay)	1974	B, Br, M	8	DDE, TDE, DDT, *o,p'*-DDE, *o,p'*-TDE, *o,p'*-DDT, PCBs, dieldrin, HCB, toxaphene, heptachlor expoxide, *trans*-nonachlor, oxychlordane, cis-chlordane	O'Shea et al. 1980
Pseudorca crassidens	British Columbia	1987–89	B	2	DDE, TDE, DDT, ΣDDT, CBz, HCB, HCHs, chlordanes, nonachlors, oxychlordane, heptachlor epoxide, photoheptachlor, mirex, toxaphene, OCS, non-ortho PCB congeners, PCBs, PCDD congeners, PCDF congeners	Jarman et al. 1996
P. crassidens	Eastern North Pacific (Canada)	1988	B	1	ΣDDT, PCBs, methyl sulfones of PCB congeners and DDE	Bergman et al. 1994
P. crassidens	North Pacific (Canada)	1987	B, Br, L	1	PCBs, DDE, TDE	Baird et al. 1989
Sotalia fluviatilis	Colombia	1977 (captive)	B, Br, K, L, M, O	2	DDE, TDE, DDT, dieldrin, HCB, HCHs, PCBs, PCB congeners	Duinker et al. 1989
S. fluviatilis	—	1970–71	B	1	DDE, TDE, DDT, PCBs, dieldrin	Koeman et al. 1972
Sousa chinensis	India (Bay of Bengal)	1990–91	B	3	HCHs, HCB, DDE, DDT, TDE, *o,p'*-DDT, ΣDDT, PCBs	Tanabe et al. 1993
S. chinensis	India (Bay of Bengal)	1992	B	2	HCHs	Tanabe et al. 1996
Stenella attenuata	Australia	—	B	1	DDE, TDE, DDT, ΣDDT, PCBs, heptachlor, oxychlordanes, HCB	Kemper et al. 1994
S. coeruleoalba	Eastern North Atlantic (Spain)	—	B	1	DDE, TDE, DDT, *o,p'*-DDT, PCBs	Borrell and Aguilar 1990
S. coeruleoalba	Eastern Tropical Pacific	1973–76	B, Br, M	14	DDE, TDE, DDT, *o,p'*-DDE, *o,p'*-TDE, *o,p'*-DDT, PCBs, dieldrin, toxaphene, heptachlor expoxide, *trans*-nonachlor, endrin, *cis*-chlordane	O'Shea et al. 1980
S. coeruleoalba	France	—	B, K, L, M, O	27	ΣDDT, DDE, DDT, TDE, PCBs	Alzieu and Duguy 1979
S. coeruleoalba	Japan	1968–75	B, Br, M	5	DDE, TDE, DDT, *o,p'*-DDE, *o,p'*-TDE, *o,p'*-DDT, PCBs, dieldrin, HCB, toxaphene, *trans*-nonachlor	O'Shea et al. 1980
S. coeruleoalba	Japan	1978–79, 1986	B	16	PCBs, ΣDDT, HCHs, HCB	Loganathan et al. 1990
S. coeruleoalba	Japan	1992	B	4	HCHs	Tanabe et al. 1996
S. coeruleoalba	Mediterranean	1991	B	7	PCBs, ΣDDT	Fossi et al. 1992
S. coeruleoalba	Mediterranean	1990	B	10	ΣDDT, PCBs, PCB congeners	Kannan et al. 1993c
S. coeruleoalba	Mediterranean	1987–91	B	181	PCBs	Aguilar and Borrell 1994a

Appendix 10-2 continued

Species	Region	Sample Period	Tissues Sampled	Number of Individuals Sampled	Compounds Reported	References
S. coeruleoalba	Mediterranean	1990–93	B	89	ΣDDT, PCBs, PCB congeners	Marsili and Focardi 1996
S. coeruleoalba	Mediterranean	1990	Br, K, L, Lu, M, O	10	ΣDDT, PCBs	Guitart et al. 1996
S. coeruleoalba	North Atlantic (U.S.A.)	1971–75	B	2	PCBs, chlordane, dieldrin, DDE, TDE, DDT, ΣDDT	Taruski et al. 1975
S. coeruleoalba	North Atlantic (Wales)	1988	B, M, O	1	HCB, HCHs, dieldrin, DDE, DDT, TDE, PCBs, PCB congeners	Morris et al. 1989
S. coeruleoalba	North Pacific (Japan)	1978	B, Br, K, L, M, O	6	HCB, PCBs, ΣDDT, HCHs, DDE, DDT, TDE	Tanabe et al. 1981
S. coeruleoalba	North Pacific (Japan)	1978–79	B, L, M, K, O	2	PCBs, ΣDDT, HCHs	Kawai et al. 1988
S. coeruleoalba	North Pacific (Japan)	1978	B	4	ΣDDT, PCBs, HCHs	Tanabe et al. 1983
S. coeruleoalba	North Pacific (Japan)	1978–79	B, Br, K, L, M, O	2	ΣDDT, PCBs, HCHs	Kawai et al. 1988
S. coeruleoalba	Southwestern Indian Ocean (South Africa)	1984–86	B	2	ΣDDT, PCBs, HCB	de Kock et al. 1994
S. coeruleoalba	Western North Pacific	1978	O	4	ΣDDT, PCBs, HCHs, PCB congeners, DDE, TDE, DDT	Tanabe et al. 1984b
S. coeruleoalba	Western North Pacific (Japan)	1978	B, Br, K, L, M, O	6	DDE, TDE, DDT, ΣDDT, PCBs, HCHs, HCB	Tanabe et al. 1981
S. longirostris	Caribbean Sea (St. Lucia)	1972	B, K, L, M	2	DDE, TDE, DDT, ΣDDT, PCBs, dieldrin	Gaskin et al. 1974
S. longirostris	Eastern Tropical Pacific	1980–82	B	2	HCHs	Tanabe et al. 1996
S. longirostris	India (Bay of Bengal)	1990–91	B	5	HCHs, HCB, DDE, DDT, TDE, *o,p′*-DDT, ΣDDT, PCBs	Tanabe et al. 1993, 1996
Steno bredanensis	Hawaii	1976	B, Br, M	7	DDE, TDE, DDT, *o,p′*-TDE, *o,p′*-DDT, PCBs, dieldrin, *trans*-nonachlor	O'Shea et al. 1980
Tursiops truncatus	—	1970–71	B	2	DDE, TDE, DDT, PCBs, dieldrin, HCB	Koeman et al. 1972
T. truncatus	Atlantic Coast (U.S.A.)	1987–89	B, Br, K, L	14	DDE, dieldrin, HCB, mirex, chlordanes, lindane, heptachlor epoxide, PCB congeners, PCDD, PCDFs, PBBs	Kuehl et al. 1991
T. truncatus	Atlantic Coast (U.S.A.)	1987–89	B	9	PCB congeners	Kuehl et al. 1994
T. truncatus	Australia	—	B	6	DDE, DDT, ΣDDT, PCBs, dieldrin, heptachlor, lindane	Kemper et al. 1994
T. truncatus	California	1974–76	B, M	2	DDE, TDE, DDT, *o,p′*-DDE, *o,p′*-TDE, *o,p′*-DDT, PCBs, dieldrin, HCB, heptachlor epoxide, *trans*-nonachlor, oxychlordane, *cis*-nonachlor, *cis*-chlordane	O'Shea et al. 1980
T. truncatus	Eastern North Pacific (California)	1980–84	B, Br, L, K, M	7	ΣDDT, PCBs	Schafer et al. 1984
T. truncatus	France	—	K, L, M, O	5	ΣDDT, DDT, DDE, TDE	Alzieu and Duguy 1979
T. truncatus	Gulf of Mexico (Florida)	1969–70 (captive)	B	4	DDE, TDE, DDT, PCBs, dieldrin	Dudok van Heel 1972
T. truncatus	Gulf of Mexico (Texas, Florida)	—	B	33	DDE, TDE, DDT, *o,p′*-DDE, *o,p′*-TDE, *o,p′*-DDT, ΣDDT, dieldrin, endrin, aldrin, mirex, HCB, HCHs, heptachlor, heptachlor epoxide, oxychlordane, nonachlors, chlordanes, PCBs, PCB congeners	Salata et al. 1995

Continued on next page

Appendix 10-2 continued

Species	Region	Sample Period	Tissues Sampled	Number of Individuals Sampled	Compounds Reported	References
T. truncatus	Gulf of Mexico (U.S.A.)	1990	B	26	DDE, DDT, dieldrin, mirex, HCB, *cis*-chlordane, oxychlordane, heptachlor epoxide, PCBs, PBBs, PBDPEs, OCS	Kuehl and Haebler 1995
T. truncatus	India (Bay of Bengal)		B	4	HCHs, HCB, DDE, DDT, TDE, *o,p'*-DDT, ΣDDT, PCBs	Tanabe et al. 1993
T. truncatus	Indian Ocean (South Africa)	1980–87	B	105	PCBs, ΣDDT, dieldrin	Cockcroft et al. 1989
T. truncatus	North Atlantic (Wales)	1988	B, M, O	1	HCB, HCHs, dieldrin, DDE, DDT, TDE, PCBs, PCB congeners	Morris et al. 1989
T. truncatus	North Atlantic (Wales)	1989–91	B	3	DDE, TDE, DDT, dieldrin, HCB, HCHs, PCBs, PCB congeners	Law et al. 1995
T. truncatus	North Sea (Scotland)	1988–91	B	6	ΣDDT, HCB, PCBs, PCB congeners, dieldrin, endrin, heptachor epoxide, chlordanes, *t*-nonachlor, oxychlordane	Wells et al. 1994
T. truncatus	Northern Gulf of Mexico (Texas)	1990	B, L	20	ΣDDT, PCBs, PCB congeners	Varanasi et al. 1992
T. truncatus	Northern Gulf of Mexico (U.S.A.)	1990–92	B, L	65	ΣDDT, PCBs, chlordanes	Varanasi et al. 1993b
T. truncatus	Scotland	—	B	1	PCB congeners	Wells and Echarri 1992
T. truncatus	Southeastern Atlantic (South Africa)	1976–87	B	—	ΣDDT, PCBs	de Kock et al. 1994
T. truncatus	Southwestern Indian Ocean (South Africa)	1976–87	B	6	ΣDDT, PCBs, HCB	de Kock et al. 1994
T. truncatus	Western North Atlantic (Florida)	1965 (captive)	B, Br, M	1	DDE, TDE, DDT, dieldrin, HCB, HCHs, PCBs, PCB congeners	Duinker et al. 1989
T. truncatus	Western North Atlantic (U.S.A.)	1987–88	B, L	56	DDE, PCBs, PCB congeners, *trans*-nonachlor	Geraci 1989
T. truncatus	captives	1992–93	Mi	5	DDE, TDE, DDT, dieldrin, HCB, heptachlor epoxide	Ridgway and Reddy 1995
Ziphius cavirostris	France	—	M, O	2	ΣDDT, PCBs	Alzieu and Duguy 1979
Z. cavirostris	North Atlantic (Bermuda Island)	1981	B, K, L, M, O	4	PCBs, PCB congeners	Duinker et al. 1988

B = blubber; Br = brain; K = kidney; L = liver; Lu = lung; M = muscle; Mi = milk; O = other. Dashes appear where values were not available in original source.

CBZ = chlorobenzenes; DDE = 2,2.-*bis* (p-chlorophenyl)-1, 1-dichloroethylene; DDT = 2,2,-*bis*-(*p*-chlorophenyl)-1, 1-trichloroethane; ΣDDT = arithmetic summation of concentrations of isomers and metabolites of DDT; HCB = hexachlorocyclohexane; OCS = octachlorostyrene; PBDPEs = polybrominated diphenyl ethers; PCBs; polychlorinated biphenyls; PCDDs = polychlorinated dibenzo-*p*-dioxins; PCDFs = polychlorinated dibenzofurans; TCP = *tris*(4-chlorophenyl)methane / methanol; TDE = 2,2,-*bis* (*p*-chlorophenyl)-1, 1-dichloroethane.

Appendix 10-3. Summary of Selected Organochlorine Residue Surveys in Baleen Whales

Species	Region	Sample Period	Tissues Sampled	Number of Individuals Sampled	Compounds Reported	References
Balaena mysticetus	Alaska	1992	B	2	ΣDDT, PCBs, chlordanes	Varanasi et al. 1993b
B. mysticetus	Arctic	1986–88	B, O	6	TDE, dieldrin, heptachlor epoxide, BHC, endrin, endosulfan sulfate, kepone	Bratton et al. 1993
Balaenoptera acutorostrata	Antarctic	1980–81	L	30	ΣDDT, PCBs	Tanabe et al. 1984a
B. acutorostrata	Antarctic	1984–85	B	37	DDE, PCBs	Tanabe et al. 1986
B. acutorostrata	Arctic	1985	L	1	PCBs	Goksøyr et al. 1988

Appendix 10-3 continued

Species	Region	Sample Period	Tissues Sampled	Number of Individuals Sampled	Compounds Reported	References
B. acutorostrata	Eastern North Pacific (California)	1977	B, Br, L, M	1	ΣDDT, PCBs	Schafer et al. 1984
B. acutorostrata	Mediterranean (France)	1977	K, L, O	1	PCBs	Alzieu and Duguy 1979
B. acutorostrata	South Africa	1974	B	29	DDE, TDE, DDT, ΣDDT, PCBs, dieldrin	Henry and Best 1983
B. acutorostrata	Southwestern Indian Ocean (South Africa)	1984	B	1	ΣDDT, PCBs, HCB	de Kock et al. 1994
B. acutorostrata	Eastern North Pacific (Washington)	1989–90	B	2	ΣDDT, PCBs, chlordanes	Varanasi et al. 1993b
B. acutorostrata	Canada (St. Lawrence River)	1988–90	B	5	ΣDDT, PCBs, mirex	Béland et al. 1991
B. borealis	Antarctica	1950	Oils	—	DDE, TDE, DDT, ΣDDT, dieldrin, PCBs	Addison et al. 1972
B. borealis	Iceland	1982–85	B	40	DDE, TDE, *o,p′*-DDT, DDT, ΣDDT, PCBs	Borrell 1993
B. borealis	North Atlantic (Iceland)	1982	B	23	DDE, ΣDDT	Borrell and Aguilar 1987
B. borealis	South Africa	1974	B	1	DDE, TDE, DDT, ΣDDT, PCBs, dieldrin	Henry and Best 1983
B. edeni	Eastern South Pacific (Chile)	1983	B, L	2	DDE, TDE, DDT	Pantoja et al. 1984, 1985
B. musculus	Antarctica	1950	Oils	—	DDE, TDE, DDT, ΣDDT, dieldrin, PCBs	Addison et al. 1972
B. musculus	Eastern Tropical Pacific	1980	L	1	DDE, TDE, DDT, ΣDDT, PCBs	Britt and Howard 1983
B. musculus	Canada (St. Lawrence River)	1988–90	B	2	ΣDDT, PCBs, mirex	Béland et al. 1991
B. physalus	Eastern North Atlantic (Spain)	1980	M, O	68	DDE, TDE, DDT, *o,p′*-DDT, ΣDDT, PCBs	Aguilar and Jover 1982
B. physalus	Eastern North Atlantic (Spain)	1980	B	1	DDE, TDE, DDT, *o,p′*-DDT, ΣDDT	Aguilar 1984
B. physalus	Eastern North Atlantic	1982–84	B	166	DDE, TDE, *o,p′*-DDT, DDT, ΣDDT, PCBs, PCB congeners	Aguilar and Borrell 1988
B. physalus	Eastern North Atlantic (Spain)	1982–84	B	137	DDE, ΣDDT	Borrell and Aguilar 1987
B. physalus	Eastern South Pacific (Chile)	1983	B, L	1	DDE, TDE, DDT, dieldrin, aldrin, lindane	Pantoja et al. 1984, 1985
B. physalus	France (Atlantic, Mediterranean)	1973, 1976	B, K, L, O	2	ΣDDT, PCBs	Alzieu and Duguy 1979
B. physalus	Iceland	1982–86	B	51	DDE, TDE, *o,p′*-DDT, DDT, ΣDDT, PCBs	Borrell 1993
B. physalus	Mediterranean (France)	—	B, Br, L	1	ΣDDT, PCBs	Viale 1981
B. physalus	Mediterranean	—	B, Br, L, K	1	PCBs, ΣDDT	Viale 1981
B. physalus	Mediterranean	1991	B	9	ΣDDT, PCBs	Fossi et al. 1992
B. physalus	Mediterranean	1990–93	B	68	ΣDDT, PCBs, PCB congeners	Marsili and Focardi 1996
B. physalus	North Atlantic	1967–70	Oils	—	DDE, TDE, DDT, ΣDDT, PCBs, dieldrin	Addison et al. 1972
B. physalus	South Africa	1974	B	6	DDE, TDE, DDT, ΣDDT, PCBs, dieldrin	Henry and Best 1983
B. physalus	Greenland	—	B	6	ΣDDT, PCBs, dieldrin	Holden 1975
B. physalus	Canada (St. Lawrence River)	1988–90	B	2	ΣDDT, PCBs, mirex	Béland et al. 1991
B. physalus	Western North Atlantic (Canada)	1970–71	B	12	DDE, TDE, DDT, *o,p′*-DDT, PCBs	Saschenbrecker 1973
Caperea marginata	Australia	1987	B	3	DDE, TDE, DDT, ΣDDT, dieldrin, HCB	Kemper et al. 1994
C. marginata	Southwestern Indian Ocean (South Africa)	1987	B	1	ΣDDT, PCBs, HCB	de Kock et al. 1994
Eschrichtius robustus	Eastern North Pacific (California)	1968–69	B, Br, L	23	DDE, TDE, DDT, dieldrin, others analyzed but not detected	Wolman and Wilson 1970
E. robustus	Eastern North Pacific (California)	1976	B, Br, L, M	1	ΣDDT, PCBs	Schafer et al. 1984

Continued on next page

Appendix 10-3 continued

Species	Region	Sample Period	Tissues Sampled	Number of Individuals Sampled	Compounds Reported	References
E. robustus	Eastern North Pacific (British Columbia)	1987–88	B	2	PCDD congeners, PCDF congeners	Jarman et al. 1996
E. robustus	Eastern North Pacific (U.S.A.)	1988–91	B, Br, L	22	DDE, TDE, DDT, *o,p*′-DDE, *o,p*′-TDE, *o,p*′-DDT, PCBs, PCB congeners, heptachlor, heptachlor epoxide, α-chlordane, *t*-nonachlor, HCB, BHC, dieldrin, mirex, aldrin	Varanasi et al. 1993a, Varanasi et al. 1994
Eubalaena australis	Southwestern Indian Ocean (South Africa)	1984	B	2	ΣDDT, PCBs, HCB	de Kock et al. 1994
E. glacialis	Western North Atlantic	1988–89	B	35	DDE, TDE, DDT, ΣDDT, PCBs, dieldrin, heptachlor epoxide, chlordane, HCB	Woodley et al. 1991
Megaptera novaeangliae	North Atlantic (Nova Scotia)	1971–75	B	1	PCBs, chlordane, dieldrin, DDE, TDE, DDT, ΣDDT	Taruski et al. 1975
M. novaeangliae	North Atlantic (U.S.A.)	1971–75	B	1	PCBs, chlordane, dieldrin, DDE, TDE, DDT, ΣDDT	Taruski et al. 1975
M. novaeangliae	Western Atlantic (Lesser Antilles)	1971–75	B	2	PCBs, chlordane, dieldrin, DDE, TDE, DDT, ΣDDT	Taruski et al. 1975
M. novaeangliae	Western North Atlantic (U.S.A.)	—	B	8	DDE, PCBs, *trans*-nonachlor	Geraci 1989

B = blubber; Br = brain; K = kidney; L = liver; M = muscle; O = other. Dashes appear where values were not available in original source.

BHC = benzene hexachloride; DDE = 2,2,-*bis*-(*p*-chlorophenyl)-1, 1-dichloroethylene; DDT = 2,2,-*bis*-(*p*-chlorophenyl)-1,1-trichloroethane; ΣDDT = arithmetic summation of concentrations of isomers and metabolites of DDT; HCB = hexachlorobenzene; PCBs = polychlorinated biphenyls; PCDDs = polychlorinated dibenzo-*p*-dioxins; PCDFs = polychlorinated dibenzofurans; TDE = 2,2,-*bis*-(*p*-chlorophenyl)-1,1-dichloroethane.

Appendix 10-4. Summary of Selected Organochlorine Residue Surveys in Sea Otters, Sirenians, and Polar Bears

Species	Region	Sample Period	Tissues Sampled	Number of Individuals Sampled	Compounds Reported	References
Enhydra lutris	California	1969–70	A, Br, K, L, O	10	DDE, TDE, DDT	Shaw 1971
Trichechus manatus	Florida	1974	B, Br, L, M, O	1	DDE, PCBs, dieldrin	Forrester et al. 1975
T. manatus	Florida	1977–81	B	26	ΣDDT, dieldrin, PCBs	O'Shea et al. 1984
T. manatus	Florida	1982	B, Br	4	DDE	O'Shea et al. 1991
T. manatus	Florida	1990–93	B, K, L	19	*o,p*′-TDE, *o,p*′-DDT, HCB, lindane	Ames and Van Vleet 1996
Dugong dugon	Sulawesi (Indonesia)	1975	M	2	PCB, DDT, BHC	Miyazaki et al. 1979
Ursus maritimus	Arctic (Canada)	1992–94	L	16	PCB congeners, PCDD congeners, PCDF congeners, ΣDDT, methylsulfone PCBs, chlordane	Letcher et al. 1996
U. maritimus	Arctic (12 regions)	1989–91	A	94 (pooled)	DDE, PCBs, PCB congeners, methyl sulfones of PCBs and DDE	Letcher et al. 1995
U. maritimus	Arctic/subarctic (Canada)	1969–84	A, L	141	DDE, TDE, DDT, ΣDDT, HCBs, HCHs, dieldrin, chlordane-related compounds, PCBs, PCB congeners	Norstrom et al. 1988
U. maritimus	Arctic Ocean (Canada)	1982–84	A, L	113	PCDDs, PCDFs, PCBs, HCB	Norstrom et al. 1990
U. maritimus	Arctic Ocean (Canada)	1982, 1984	A	20	ΣDDT, PCB congeners, heptachlor epoxide, chlordane isomers, HCHs, dieldrin, PCBs	Muir et al. 1988b
U. maritimus	Arctic Ocean (Norway)	1978–89	A, L	24	DDT, *o,p*′-DDT, DDE, PCBs, PCB congeners, HCB, HCH, oxychlordane, heptachlor, heptachlor epoxide, dieldrin	Norheim et al. 1992

Appendix 10-4 continued

Species	Region	Sample Period	Tissues Sampled	Number of Individuals Sampled	Compounds Reported	References
U. maritimus	Arctic (Norway)	1990–91	Mi	6	PCB congeners, PCDDs, PCDFs	Oehme et al. 1995a
U. maritimus	Baffin Bay (Canada)	—	A	6	PCBs, PCDDs, PCDFs, PCB congeners	Norstrom et al. 1992
U. maritimus	Canada	1968–72	A, Br, L, M, Mi	40	DDE, DDT, PCBs	Bowes and Jonkel 1975; Bowes and Lewis 1974
U. maritimus	Canada (Manitoba)	1992–93	A, Mi	7	ΣDDT, PCBs, chlordanes, HCHs, HCBs	Polischuk et al. 1995
U. maritimus	Greenland	1972	A	1	DDE, PCBs, hepachlor epoxide, aldrin	Clausen et al. 1974; Clausen and Berg 1975
U. maritimus	Hudson Bay (Canada)	1985	A, L	2	ΣDDT, PCBs, methyl sulfones of PCB congeners and DDE	Bergman et al. 1994
U. maritimus	Northern Quebec (Canada)	1989–90	A	52	PCBs, chlordanes, nonachlors, oxy-chlordane, heptachlor epoxide, photoheptachlor	Zhu et al. 1995
U. maritimus	Repulse Bay (Canada)	1989	A	1	ΣDDT, PCBs, HCHs, chlordane, toxaphene, PCB congeners	Zhu et al. 1995; Zhu and Norstrom 1993

A = adipose tissue; B = blubber; Br = brain; K = kidney; L = liver; M = muscle; Mi = milk; O = other.

BHC = benzene hexachloride; DDE = 2,2,-*bis*(*p*-chlorophenyl)-1, 1-dichloroethylene; DDT = 2,2,-*bis*(*p*-chlorophenyl)-1,1-trichloroethane; ΣDDT = arithmetic summation of concentrations of isomers and metabolites of DDT; HCB = hexachlorobenzene; HCH = hexachlorocyclohexane; PCBs = polychlorinated biphenyls; PCDDs = polychlorinated dibenzo-*p*-dioxins; PCDFs = polychlorinated dibenzofurans; TDE = 2,2,-*bis*-(*p*-chlorophenyl)-1,1-dichloroethane.

Appendix 10-5. Summary of Selected Surveys of Metals and Trace Element Concentrations in Tissues of Pinnipeds

Species	Region	Sample Period	Tissues Sampled	Max. No. of Individuals Sampled	Metals and Trace Elements Reported	References
Arctocephalus australis	Western South Atlantic (Argentina)	—	B, K, L, M	8	Cd, Cu, Hg, Zn	Marcovecchio et al. 1994
A. gazella	Antarctic (South Georgia)	1987	L	11	Ba, Cd, Ce, Co, Cu, Cr, Hg, I, La, Mg, Mo, Pb, Rb, Sn, Sr, Zn	Malcolm et al. 1994
A. gazella	Antarctic	1987–89	B, K, M, L	4	Cd, Cu, Hg, Zn	de Moreno et al. 1997
A. pusillus	Australia (Bass Strait)	—	Br, K, L, M, O	16	Hg	Bacher 1985
A. spp.	Australia	—	K, L, M, O	16	Cd, Hg, Pb	Kemper et al. 1994
Callorhinus ursinus	Alaska	1972	Bl, Mi, hair	7	Hg	Kim et al. 1974
C. ursinus	Bering Sea	1984	K, L, M, O	—	Cd, Cr, Cu, Ni, Pb, Se, Zn	Richard and Skoch 1986
C. ursinus	Bering Sea	1987	K, L, M	2	40 trace elements, MeHg	Zeisler et al. 1993
C. ursinus	Japan, Pribilof Islands	1990–92	K, L, M	67	Cd, Cu, Fe, Hg, Mn, Zn	Noda et al. 1995
C. ursinus	Pribilof Islands	1970	Br, L, M	39	As, Cd, Hg, Pb	Anas 1973, 1974a
C. ursinus	Pribilof Islands	1975	K, L, M, bone	39	Cd, Hg, Ni, Pb, Zn	Goldblatt and Anthony 1983
C. ursinus	Eastern North Pacific (Washington)	1970–71	Br, K, L, M	39	As, Cd, Hg, Pb	Anas 1973, 1974a
Cystophora cristata	Greenland	1984	L	3	Cd, Hg, Se, Zn	Nielsen and Dietz 1990
C. cristata	Greenland	1984–87	K, L, M	4	Hg, organic Hg	Dietz et al. 1990
C. cristata	Gulf of St. Lawrence	1971	B, L, M, hair	3	Hg	Sergeant and Armstrong 1973

Continued on next page

Appendix 10-5 continued

Species	Region	Sample Period	Tissues Sampled	Max. No. of Individuals Sampled	Metals and Trace Elements Reported	References
Erignathus barbatus	Alaska	—	L	3	Ag, Cd, Hg, Se, V	Mackey et al. 1996
E. barbatus	Arctic (Canada)	1973	L, M	6	Hg, MeHg	Smith and Armstrong 1975
E. barbatus	Arctic Ocean (Canada)	1973–76	L, M	64	Hg, MeHg, Se	Smith and Armstrong 1978
E. barbatus	Greenland	1984–87	K, L, M	3	Hg, organic Hg	Dietz et al. 1990
E. barbatus	Northwest Atlantic (Canada)	—	Claws	9	Hg, MeHg	Freeman and Horne 1973
Eumetopias jubatus	Japan	1976–77	Br, K, L, M, O	22	Cd, Zn	Hamanaka et al. 1982
E. jubatus	Japan	1994–95	B, K, L, M, O	7	Butyltin compounds	Kim et al. 1996
Halichoerus grypus	Baltic Sea (Sweden)	1979–90	B, K, L	19	Al, As, Ca, Cd, Co, Cr, Cu, Fe, Hg, Mg, Mn, Ni, Pb, Se, V, W, Zn	Frank et al. 1992
H. grypus	England	1988–89	B, K, L	8	Hg	Simmonds et al. 1993
H. grypus	Great Britain	1968–72	Br, K, L, teeth	73	Cd, Cr, Cu, Hg, Pb, Zn	Heppleston and French 1973
H. grypus	Great Britain	1977	L, bile	15	Hg, MeHg, Se	van de Ven et al. 1979
H. grypus	Gulf of Finland	1976–82	B, K, L, M	9	Cd, Cu, Hg, Pb, Se, Zn	Pertillä et al. 1986
H. grypus	North Sea (Germany)	—	L	1	Cd, Cu, Hg, Pb, Zn	Harms et al. 1977/78
H. grypus	North Sea (Scotland)	—	B, K, L, O	1	Hg	Jones et al. 1972
H. grypus	North Sea (U.K.)	—	B, Br, K, L, M, O	13	Cd, Cu, Hg, Pb, Zn	Holden 1975
H. grypus	Nova Scotia	1971	B, K, L, M	11	Hg	Sergeant and Armstrong 1973
H. grypus	Nova Scotia	1972	B, Br, K, L, M, O	6	Hg, MeHg	Freeman and Horne 1973
H. grypus	St. Lawrence River	1983–90	L	5	Cd, Hg	Béland et al. 1991
H. grypus	U.K.	1988–89	L	14	Cd, Cr, Cu, Hg, Ni, Pb, Zn	Law et al. 1991, 1992
H. grypus	Wales	1988	B, K, L, M	2	Cd, Cr, Cu, Hg, Ni, Pb, Zn	Morris et al. 1989
Histriophoca fasciata	Okhotsk Sea	1975	L, M, O	28	Cd, Zn	Hamanaka et al. 1977
Hydrurga leptonyx	Antarctic	1989	K, L, M, O	3	Ca, Hg, K, Mg, Na	Szefer et al. 1993
H. leptonyx	Antarctic	1989	K, L, M, O	3	Ag, Cd, Co, Cr, Cu, Fe, Mn, Ni, Pb, Zn	Szefer et al. 1994
H. leptonyx	Australia	—	B, K, L, M	2	Cd, Hg, Pb	Kemper et al. 1994
Leptonychotes weddellii	Antarctic	—	K, O	1	Cd, Cu, Fe, Mg, Pb, Zn	Mishima et al. 1977
L. weddellii	Antarctic	—	B, Bl, K, L, M, Bone, O	3	Cd, Cu, Fe, Hg, Mn, Ni, Pb, Zn	Yamamoto et al. 1987
L. weddellii	Antarctic	1983	K, L, M	8	Cd, Cu	Steinhagen-Schneider 1986
L. weddellii	Antarctic	1989	K, L, M	2	Ca, Hg, K, Mg, Na	Szefer et al. 1993
L. weddellii	Antarctic	1989	K, L, M	2	Ag, Cd, Co, Cr, Cu, Fe, Mn, Ni, Pb, Zn	Szefer et al. 1994
Lobodon carcinophagus	Antarctic	1983	K, L, M	9	Cd, Cu	Steinhagen-Schneider 1986
L. carcinophagus	Antarctic	1989	K, L, M	27	Ca, Hg, K, Mg, Na	Szefer et al. 1993
L. carcinophagus	Antarctic	1989	K, L, M	27	Ag, Cd, Co, Cr, Cu, Fe, Mn, Ni, Pb, Zn	Szefer et al. 1994
Mirounga leonina	Antarctic	1987–89	M, B, O	2	Cd, Cu, Hg, Zn	de Moreno et al. 1997
Monachus monachus	Greece	1986–91	Hair	—	Cd, Cu, Hg, Pb, Zn	Yediler et al. 1993
Neophoca cinerea	Australia	—	K, L	5	Cd	Kemper et al. 1994

Appendix 10-5 continued

Species	Region	Sample Period	Tissues Sampled	Max. No. of Individuals Sampled	Metals and Trace Elements Reported	References
Odobenus rosmarus	Alaska	1986–89	K, L	56	Ag, Al, As, B, Ba, Be, Cd, Cr, Cu, Fe, Hg, Mg, Mn, Mo, Ni, Pb, Sb, Se, Sn, Sr, Tl, V, Zn	Warburton and Seagars 1993
O. rosmarus	Bering Sea	1981–84	K, L	65	As, Cd, Hg, Pb, Se, Zn	Taylor et al. 1989
O. rosmarus	Greenland	1975–77	K, L, M	69	Hg, MeHg	Born et al. 1981
O. rosmarus	Arctic Ocean (Canada)	1988	Teeth	12	Cu, Pb, Sr, Zn	Evans et al. 1995
Ommatophoca rossi	Antarctic	1981–82	L	20	Cd, Co, Cr, Cu, Fe, Hg, Mn, Ni, Pb, Zn	McClurg 1984
Otaria flavescens	Western South Atlantic (Argentina)	—	B, K, L, M	7	Cd, Hg	Marcovecchio et al. 1994
Phoca groenlandica	Greenland	1984–87	K, L, M	3	Hg, organic Hg	Dietz et al. 1990
P. groenlandica	Arctic (Greenland, Canada)	1976–78	Bl, Br, K, L, M	43	Cd, Cu, Hg, Pb, Se	Ronald et al. 1984b
P. groenlandica	Newfoundland	1980	L, M	30	Hg	Botta et al. 1983
P. groenlandica	Canada (Gulf of St. Lawrence)	1973	B, Bl, Br, K, L	20	Hg, MeHg	Jones et al. 1976
P. groenlandica	Gulf of St. Lawrence	—	B, Br, K, L, M, O	10	Hg, MeHg	Freeman and Horne 1973
P. groenlandica	Gulf of St. Lawrence	1984	K, L, M, Mi	40	Cd, Cu, Hg, MeHg, Se, Zn	Wagemann et al. 1988
P. groenlandica	Northwest Atlantic (Canada)	1976–78	Bl, Br, K, L, M	205	Cd, Cu, Hg, Pb, Se	Ronald et al. 1984b
P. groenlandica	St. Lawrence River	1971	B, L, M	20	Hg	Sergeant and Armstrong 1973
P. groenlandica	St. Lawrence River	1983–90	L	1	Cd, Hg	Béland et al. 1991
P. hispida	Alaska	—	L	13	Ag, Cd, Hg, Se, V	Mackey et al. 1996
P. hispida	Alaska (Chukchi Sea, Nome)	1988–89	K, L	4	40 trace elements, MeHg	Zeisler et al. 1993
P. hispida	Arctic (Canada)	1972	L, M	80	Hg, MeHg	Smith and Armstrong 1975
P. hispida	Arctic (Canada)	1983	K, L, M	28	Ag, Cd, Cu, Hg, Pb, Se, Zn	Wagemann 1989
P. hispida	Arctic Ocean (Canada)	1972–77	L, M	390	Hg, MeHg, Se	Smith and Armstrong 1978
P. hispida	Baltic Sea (Sweden)	1979–90	B, K, L	17	Al, As, Ca, Cd, Co, Cr, Cu, Fe, Hg, Mg, Mn, Ni, Pb, Se, V, W, Zn	Frank et al. 1992
P. hispida	Finland	—	Hair	32	Cd, Cr, Hg, Ni, Pb	Hyvärinen and Sipilä 1984
P. hispida	Finland	1967–68	L, M, O	25	Hg	Henriksson et al. 1969
P. hispida	Finland (Lake Saimaa)	1967	K, L, M	7	Hg	Helminen et al. 1968
P. hispida	Finland (Saimaa, Bothnian Bay)	1974–75	B, K, L, M	15	Hg, MeHg, Se	Kari and Kauranen 1978
P. hispida	Greenland	1984–87	K, L, M	12	Hg, organic Hg	Dietz et al. 1990
P. hispida	Gulf of Finland	1976–82	B, K, L, M	11	Cd, Cu, Hg, Pb, Se, Zn	Pertillä et al. 1986
P. hispida	Netherlands (captive)	—	B, Br, K, L	1	Hg, MeHg, Se	van de Ven et al. 1979
P. hispida	North Sea (Germany)	—	L	1	Cd, Cu, Hg, Pb, Zn	Harms et al. 1977/78
P. hispida	North Sea (Germany)	1975	L	1	Cd, Cu, Hg, Pb, Zn	Drescher 1978
P. hispida	Northwest Atlantic (Canada)	—	Claws	14	Hg, MeHg	Freeman and Horne 1973
P. hispida	Okhotsk Sea	1975	O	1	Cd, Zn	Hamanaka et al. 1977
P. largha	Japan	1992	B	1	Butyltin compounds	Iwata et al. 1994
P. vitulina	Alaska	1976–78	K, L	23	As, Cd, Hg, Pb, Se	Miles et al. 1992
P. vitulina	Baltic Sea (Sweden)	1979–90	B, K, L	14	Al, As, Ca, Cd, Co, Cr, Cu, Fe, Hg, Mg, Mn, Ni, Pb, Se, V, W, Zn	Frank et al. 1992
P. vitulina	Eastern North Pacific (U.S.A.)	1970–71	L	13	Hg	Anas 1974a
P. vitulina	England	1988–89	B, K, L	14	Hg	Simmonds et al. 1993

Continued on next page

Appendix 10-5 continued

Species	Region	Sample Period	Tissues Sampled	Max. No. of Individuals Sampled	Metals and Trace Elements Reported	References
P. vitulina	Germany	1988	Hair, skin	47	Cd, Hg, Pb	Wenzel et al. 1993
P. vitulina	Great Britain	1968–72	Br, K, L, teeth	31	Cd, Cr, Cu, Hg, Pb, Zn	Heppleston and French 1973
P. vitulina	Japan	1984	K, L	15	Cd, Cu, Hg, Zn	Tohyama et al. 1986
P. vitulina	Netherlands	1974–75	B, Br, K, L, M, O	7	Hg, MeHg, Se	van de Ven et al. 1979
P. vitulina	North Sea (England)	1969–70	Br, K, L, M, O	9	Cd, Hg, Pb	Roberts et al. 1976
P. vitulina	North Sea (Germany)	—	K, L, M	7	Cd, Cu, Hg, Pb, Zn	Harms et al. 1977/78
P. vitulina	North Sea (Germany)	1974–76	Br, K, L	63	Cd, Cu, Hg, Pb, Zn	Drescher et al. 1977
P. vitulina	North Sea (Netherlands)	—	B, Br, K, L, O	9	Cd, Cr, Cu, Fe, Mn, Pb, Zn	Duinker et al. 1979
P. vitulina	North Sea (Netherlands)	1970–71	Br, L	11	As, Cd, Hg, Se, Zn	Koeman et al. 1972
P. vitulina	North Sea (Netherlands)	1975–76	Br, K, L	14	Br, Hg, MeHg, Se	Reijnders 1980
P. vitulina	North Sea (Schleswig-Holstein and Denmark)	1975–76	Br, K, L	16	Br, Hg, MeHg, Se	Reijnders 1980
P. vitulina	North Sea (Sweden)	1979–90	B, K, L	38	Al, As, Ca, Cd, Co, Cr, Cu, Fe, Hg, Mg, Mn, Ni, Pb, Se, V, W, Zn	Frank et al. 1992
P. vitulina	North Sea (U.K.)	—	B, Br, K, L, M, O	15	Cd, Cu, Hg, Pb, Zn	Holden 1975
P. vitulina	Nova Scotia	1971	B, L, M, hair	8	Hg	Sergeant and Armstrong 1973
P. vitulina	Nova Scotia	1972	B, Br, K, L, M, O	1	Hg	Freeman and Horne 1973
P. vitulina	Okhotsk Sea	—	K, L	15	Hg, MeHg, Se	Himeno et al. 1989
P. vitulina	Okhotsk Sea	1975	L, M, O	4	Cd, Zn	Hamanaka et al. 1977
P. vitulina	Scotland	1969–70	Br, K, L, M, O	11	Cd, Hg, Pb	Roberts et al. 1976
P. vitulina	Scotland	1988	B	9	Hg	Simmonds et al. 1993
P. vitulina	St. Lawrence River	1983–90	L	10	Cd, Hg	Béland et al. 1991
P. vitulina	U.K.	1988–89	L	28	Cd, Cr, Cu, Hg, Ni, Pb, Zn	Law et al. 1991, 1992
P. vitulina	Western North Atlantic (Bay of Fundy)	1971	B, Br, L, M	12	Hg, MeHg	Gaskin et al. 1973
P. vitulina	Western North Atlantic (U.S.A.)	1980, 1991	L	7	Hg	Lake et al. 1995
Zalophus californianus	California	1972	K, L	40	Ag, Br, Ca, Cd, Cu, Fe, Hg, K, Mg, Mn, Na, Se, Zn	Martin et al. 1976
Z. californianus	California	1971–72	Br, bone, K, L, M, O	6	Pb	Braham 1973
Z. californianus	Captive	—	L	1	Hg	Theobald 1973
Z. californianus	Oregon	1970s	K, L	7	Cd, Hg	Lee et al. 1977
Z. californianus	Oregon	1970–73	Multiple organs	20	Cd, Hg, MeHg	Buhler et al. 1975

Dashes appear where values were not available in original source.

B = bubber; Bl = blood; Br = brain; K = kidney; L = liver; M = muscle; MeHg = methyl mercury; Mi = milk; O = other.

Appendix 10-6. Summary of Selected Surveys of Metals and Trace Elements in Tissues of Odontocete Cetaceans

Species	Region	Sample Period	Tissues Sampled	Max. No. of Individuals Sampled	Metals and Trace Elements Reported	References
Delphinapterus leucas	Alaska (Chukchi Sea)	1989–90	L	6	40 trace elements, MeHg	Zeisler et al. 1993
D. leucas	Alaska	1989–92	L	15	Ag, Hg, Se	Becker et al. 1995
D. leucas	Alaska	—	L	15	Ag, Cd, Hg, Se, V	Mackey et al. 1996

Appendix 10-6 continued

Species	Region	Sample Period	Tissues Sampled	Max. No. of Individuals Sampled	Metals and Trace Elements Reported	References
D. leucas	Baltic Sea (Germany)	—	K, L, M	1	Cd, Cu, Hg, Pb, Zn	Harms et al. 1977/78
D. leucas	Greenland	1984–86	K, L, M	43	Cd, Hg, Se, Zn	Hansen et al. 1990
D. leucas	Greenland	1984–87	K, L, M	6	Hg, Organic Hg	Dietz et al. 1990
D. leucas	Canadian Arctic and St. Lawrence	1981–87	K, L, M	144	Cd, Cu, Hg, Pb, Se, Zn	Wagemann et al. 1990
D. leucas	St. Lawrence River	1988–90	L	35	Cd, Hg	Béland et al. 1991
Delphinus delphis	Australia	—	B, K, L, M, O	32	Cd, Hg, Pb	Kemper et al. 1994
D. delphis	Gulf of Guinea	1975	Lung	1	Cd, Cr, Fe, Hg, Pb, Ti	Viale 1981
D. delphis	Irish Sea	1990–91	L	8	Cd, Cr, Cu, Hg, Ni, Pb, Zn	Law et al. 1992
D. delphis	Tropical Atlantic	1975	B, K, L, M	1	Hg, Se	Martoja and Viale 1977
D. delphis	Mediterranean Sea	1973	B, L, M	1	Hg, Se	Martoja and Viale 1977
D. delphis	Mediterranean Sea (France)	1973	Lung	1	Cd, Cr, Fe, Hg, Pb, Ti	Viale 1981
D. delphis	Mediterranean Sea (France)	1977	B	1	Cd, Cu, Pb	Vicente and Chabert 1978
D. delphis	New Zealand	1970–71	L	2	As, Cd, Hg, Se, Zn	Koeman et al. 1972, 1975
D. delphis	North Pacific	1987	B	1	Butyltin compounds	Iwata et al. 1994
D. delphis	North Sea (Belgium)	1986	L, M	1	Hg, MeHg	Joiris et al. 1991
D. delphis	Wales	1990	L	1	Cd, Cr, Cu, Hg, Ni, Pb, Zn	Law et al., 1991, 1992
D. delphis	Western North Atlantic (U.S.A.)	1987–89	L	3	Cd, Cr, Hg, Mn, Pb, Se	Kuehl et al. 1994
D. delphis	Australia	1989	K, L, M	8	Cd, Cu, Hg, Pb, Se, Zn	Anderson 1991
Globicephala macrorhynchus	California	1971	L	6	Hg	Hall et al. 1971
G. macrorhynchus	Caribbean Sea	1972	K, L, M	5	Hg, MeHg	Gaskin et al. 1974
G. macrorhynchus	Japan	1975	M	12	Hg, Se	Arima and Nagakura 1979
G. macrorhynchus	Western North Atlantic (Georgia)	—	B, L, M	1	Hg, MeHg	Windom and Kendall 1979
G. macrorhynchus	Western North Atlantic (southeastern U.S.A.)	1977	B, L, K	4	Cd, Hg, Se	Stoneburner 1978
G. melaena	Newfoundland	1980–82	B, K, L, M	41	As, Cd, Cu, Hg, Pb, Se, Zn	Muir et al. 1988a
G. melaena	North Atlantic (Faroe Islands)	1977–78	B, K, L, M	30	As, Cd, Cu, Hg, MeHg, Se, Zn	Julshamn et al. 1987
G. melaena	North Atlantic (Faroe Islands)	1986	B, O	53	Hg	Simmonds et al. 1994
G. melaena	Northwestern Atlantic (Massachusetts)	1986–90	B, L, K, O	17	As, Cd, Cu, Hg, MeHg, Pb, Se	Meador et al. 1993
G. melaena	St. Lawrence River	1983–90	L	2	Cd, Hg	Béland et al. 1991
G. melaena	Tasmania	1981–83	M	2	Hg	Munday 1985
G. melaena	Northwestern Atlantic (Massachusetts)	1990–91	L	8	Ag, Hg, Se	Becker et al. 1995
G. melaena	Northwestern Atlantic (Massachusetts)	1990–91	L	8	20 elements	Mackey et al. 1995
G. melaena	North Atlantic (Faroe Islands)	1986–88	K, L, M, Mi, O	131	As, Cd, Cu, Hg, Se, Zn	Caurant et al. 1993
G. melaena	North Atlantic (Faroe Islands)	1986–88	Blood, urine	40	Cd	Caurant and Amiard-Triquet 1995
G. melaena	North Atlantic (Faroe Islands)	1987	L	14	Hg, organic Hg	Schintu et al. 1992
Globicephala sp.	Australia	1989	K, L	4	Cd, Cu, Hg, Pb, Se, Zn	Anderson 1991
Globicephala sp.	Australia	—	K, L, M, O	12	Cd, Hg, Pb	Kemper et al. 1994
Hyperoodon ampullatus	North Sea (Germany)	—	L, M	1	Cd, Cu, Hg, Pb, Zn	Harms et al. 1977/78

Continued on next page

Appendix 10-6 continued

Species	Region	Sample Period	Tissues Sampled	Max. No. of Individuals Sampled	Metals and Trace Elements Reported	References
H. ampullatus	St. Lawrence River	1983–90	L	1	Cd, Hg	Béland et al. 1991
H. planifrons	Australia	—	Bone	3	Pb	Kemper et al. 1994
Kogia breviceps	Australia	—	B, L, O	5	Cd, Hg, Pb	Kemper et al. 1994
K. breviceps	Western South Atlantic (Argentina)	—	B, K, L, M, O	1	Cd, Cu, Hg, Zn	Marcovecchio et al. 1990, 1994
Lagenorhynchus obscurus	New Zealand	1970–71	L	1	As, Cd, Hg, Se, Zn	Koeman et al. 1972, 1975
L. acutus	Ireland	1989	L	1	Cd, Cr, Cu, Hg, Ni, Pb, Zn	Law et al. 1991
L. acutus	Western North Atlantic (U.S.A.)	1987–89	L	2	Cd, Cr, Hg, Mn, Pb, Se	Kuehl et al. 1994
L. acutus	Western North Atlantic (Massachussetts)	1993	L	4	20 elements	Mackey et al. 1995
L. albirostris	Irish Sea	1989	L	1	Cd, Cr, Cu, Hg, Ni, Pb, Zn	Law et al. 1991, 1992
L. albirostris	Newfoundland	1980–82	K, L, M	27	Cd, Cu, Hg, Pb, Se, Zn	Muir et al. 1988a
L. albirostris	North Sea (Denmark)	1972	B, L, M, O	1	Cu, Hg, Pb, Zn	Andersen and Rebsdorff 1976
Mesoplodon ginkgodens	Japan	1993	B	1	Butyltin compounds	Iwata et al. 1994
Mesoplodon. sp.	Australia	—	B, K, L, M, O	18	Cd, Hg, Pb	Kemper et al. 1994
Monodon monoceros	Arctic Ocean (Canada)	1978–79	B, K, L, M	60	As, Cd, Cu, Hg, Pb, Se, Zn	Wagemann et al. 1983
M. monoceros	Greenland	1984–86	K, L, M	98	Cd, Hg, Se, Zn	Hansen et al. 1990
M. monoceros	Greenland	1984–87	K, L, M	6	Hg, organic Hg	Dietz et al. 1990
Neophocoena phocoenoides	Japan	1973	M	1	Hg, MeHg, Se	Arima and Nagakura 1979
N. phocoenoides	Japan	1981–92	B	3	Butyltin compounds	Iwata et al. 1994
N. phocoenoides	South China Sea	1990	B	1	Butyltin compounds	Iwata et al. 1994
Orcinus orca	Australia	—	K, L	1	Cd, Hg, Pb	Kemper et al. 1994
O. orca	Japan	1986	B	1	Butyltin compounds	Iwata et al. 1994
Phocoena phocoena	Baltic, North Sea (Germany)	—	K, L, M	3	Cd, Cu, Hg, Pb, Zn	Harms et al. 1977/78
P. phocoena	Irish Sea	1988–90	L	36	Cd, Cr, Cu, Hg, Ni, Pb, Zn	Law et al. 1992
P. phocoena	Western North Atlantic (Bay of Fundy)	1969–77	Br, K, L, M, O	146	Hg, MeHg	Gaskin et al. 1972, 1979
P. phocoena	Western North Atlantic (northeastern U.S.A.)	1991	Br, K, L, O	3	As, Cd, Cu, Hg, Pb, Se	Stein et al. 1992
P. phocoena	Western North Atlantic (northeastern U.S.A.)	1990–92	L	6	20 elements	Mackey et al. 1995
P. phocoena	North Sea (Belgium, Denmark)	1987–90	K, L, M	17	Hg, MeHg	Joiris et al. 1991
P. phocoena	North Sea (Denmark)	1972–73	B, L, M, O	4	Cu, Hg, Pb, Zn	Andersen and Rebsdorff 1976
P. phocoena	North Sea (Netherlands)	1970–71	Br, L	6	As, Cd, Hg, Se, Zn	Koeman et al. 1972, 1975
P. phocoena	North Sea (Scotland)	1974	Br, K, L, O	26	Cd, Co, Cr, Cu, Hg, MeHg, Ni, Pb, Zn	Falconer et al. 1983
P. phocoena	Norway	1989–90	K, L	92	Hg, Se	Teigen et al. 1993
P. phocoena	Baltic Sea (Poland)	1989–93	K, L, M	4	Ag, Cd, Cr, Cu, Mn, Ni, Pb, Zn	Szefer et al. 1994
P. phocoena	St. Lawrence River	1983–90	L	9	Cd, Hg	Béland et al. 1991
P. phocoena	Wales	1988	B, L, M, O	4	Cd, Cr, Cu, Hg, Ni, Pb, Zn	Morris et al. 1989
P. phocoena	U.K.	1988–90	L	20	Cd, Cr, Cu, Hg, Ni, Pb, Zn	Law et al. 1991, 1992
Phoceonoides dalli	North Pacific	1987	B	2	Butyltin compounds	Iwata et al. 1994

Appendix 10-6 continued

Species	Region	Sample Period	Tissues Sampled	Max. No. of Individuals Sampled	Metals and Trace Elements Reported	References
P. dalli	Northwestern Pacific	NR	Multiple organs	2	Cd, Cu, Fe, Hg, Mn, Ni, Pb, Zn	Fujise et al. 1988
P. dalli	Western North Pacific	—	B, Br, K, L, M, O	3	Cd, Cu, Fe, Hg, Mn, Ni, Pb, Zn	Fujise et al. 1988
P. dalli	Western North Pacific	1978	L, M, O	2	Cd, Zn	Hamanaka and Mishima 1981
Physeter catodon	—	—	M	1	Hg	Suzuki and Miyoshi 1973
P. catodon	Antarctic	1972	M	6	Hg, MeHg	Nagakura et al. 1974
P. catodon	North Pacific	1972	M	7	Hg, MeHg	Nagakura et al. 1974
P. macrocephalus	Australia	—	B, K, L, M	3	Cd, Hg, Pb	Kemper et al. 1994
P. macrocephalus	Mediterranean Sea	1974	Lung	1	Cd, Cr, Fe, Hg, Pb, Ti	Viale 1981
P. macrocephalus	North Sea (Netherlands)	1995	L	1	As, Cd, Cr, Cu, Hg, Ni, Pb, Se, Zn	Law et al. 1996
P. macrocephalus	North Sea (Belgium)	1989	L, M	1	Hg, MeHg	Joiris et al. 1991
P. macrocephalus	Southern Australia	1976	M	414	Hg	Cannella and Kitchener 1992
P. macrocephalus	Australia	1989	M	1	Hg	Anderson 1991
Platanista gangetica	India	1988–92	K, L, M	4	Cd, Cu, Fe, Mn, Ni, Pb, Zn	Kannan et al. 1993a
Pontoporia blainvillei	Western South Atlantic (Argentina)	—	B, K, L, M, O	7	Cd, Cu, Hg, Zn	Marcovecchio et al. 1990, 1994
Pseudorca crassidens	Australia	—	B, K, L, M	38	Cd, Hg, Pb	Kemper et al. 1994
P. crassidens	Eastern North Pacific (British Columbia)	1987	Br, K, L	1	As, Ca, Cd, Cu, Fe, Hg, Mg, Mn, Pb, Se, Zn	Baird et al. 1989
Sotalia guianenses	Surinam	1970 71	L	2	As, Cd, Hg, Se, Zn	Koeman et al. 1972, 1975
Stenella attenuata	Australia	—	B, L, M	2	Hg, Pb	Kemper et al. 1994
S. attenuata	Japan	1972	M	2	Hg, MeHg, Se	Arima and Nagakura 1979
S. attenuata	Eastern Tropical Pacific	1977–83	Multiple organs	27	Cd	André et al. 1990
S. attenuata	Eastern Tropical Pacific	1977–83	Multiple organs	44	Hg	André et al. 1991b
S. coeruleoalba	Eastern North Atlantic (France)	1972–80	K, L, M, O	8	Hg	André et al. 1991a
S. coeruleoalba	Irish Sea	1990–91	L	3	Cd, Cr, Cu, Hg, Ni, Pb, Zn	Law et al. 1992
S. coeruleoalba	Japan	1973–74	M	11	Hg, MeHg, Se	Arima and Nagakura 1979
S. coeruleoalba	Japan	1978–80	Bone	40	Ca, Cd, Cu, Fe, Hg, MeHg, Mn, Ni, Pb, Se, Zn	Honda et al. 1986a
S. coeruleoalba	Japan	1977–79	Multiple organs	55	Hg, MeHg, Se	Itano et al. 1984a,b,c
S. coeruleoalba	Japan	1977–80	M, L, K	59	Cd, Cu, Fe, Hg, Mn, Ni, Pb, Zn	Honda et al. 1983
S. coeruleoalba	Japan	1977–80	Multiple organs	76	Cd, Zn	Honda and Tatsukawa 1983
S. coeruleoalba	Japan	1979	Bone	1	Ca, Cd, Cu, Fe, Hg, Mn, Ni, Pb, Se, Zn	Honda et al. 1984
S. coeruleoalba	Japan	1980	M	5	Hg, MeHg, Se	Itano et al. 1985a,b
S. coeruleoalba	Mediterranean (Italy)	1987–89	Br, K, L, M	23	Cd, Hg, Pb, Se, Zn	Leonzio et al. 1992
S. coeruleoalba	Mediterranean (France)	1972–80	K, L, M, O	27	Hg	André et al. 1991a
S. coeruleoalba	Mediterranean (France)	1973	Lung	1	Cr, Hg	Viale 1981

Continued on next page

Appendix 10-6 continued

Species	Region	Sample Period	Tissues Sampled	Max. No. of Individuals Sampled	Metals and Trace Elements Reported	References
S. coeruleoalba	Mediterranean (France)	1988–90	B, Br, K, L, M, O	13	Hg	Augier et al. 1993
S. coeruleoalba	Mediterranean Sea	1973–92	Multiple organs	6	Cd, Hg, Pb, Se	Viale 1994
S. coeruleoalba	Wales	1988	B, M, O	1	Cd, Cr, Cu, Hg, Ni, Pb, Zn	Morris et al. 1989
S. coeruleoalba	Wales	1990	L	2	Cd, Cr, Cu, Hg, Ni, Pb, Zn	Law et al. 1991, 1992
S. longirostris	Bay of Bengal	1990	B	1	Butyltin compounds	Iwata et al. 1994
Stenella sp.	Australia	—	B, K, L, M	3	Cd	Kemper et al. 1994
Stenella sp.	Caribbean Sea	1972	K, L, M	2	Hg, MeHg	Gaskin et al. 1974
Tursiops gephyreus	Western South Atlantic (Argentina)	—	K, L, M, O	2	Cd, Cu, Hg, Zn	Marchovecchio et al. 1990, 1994
T. gilli	Japan	1973	M	1	Hg, MeHg, Se	Arima and Nagakura 1979
T. truncatus	Australia	—	B, K, L, M, O	24	Cd, Hg, Pb	Kemper et al. 1994
T. truncatus	Belgium (captive)	1989	K, L, M	2	Hg, MeHg	Joiris et al. 1991
T. truncatus	Florida (Gulf and Atlantic coasts)	1990–94	K, L, M	39	Cd, Cu, Zn	Wood and Van Vleet 1996
T. truncatus	Gulf of Mexico (Florida)	1969–71 (captive)	Blood, L	11	As, Cd, Hg, Sb, Se, Zn	Dudok van Heel 1972
T. truncatus	Gulf of Mexico (Florida)	1987–91	L	12	Hg	Rawson et al. 1993
T. truncatus	Gulf of Mexico	1990	L	26	Cd, Cr, Hg, Mn, Pb, Se	Kuehl and Haebler 1995
T. truncatus	Irish Sea	1989	L	2	Cd, Cr, Cu, Hg, Ni, Pb, Zn	Law et al. 1991, 1992
T. truncatus	Mediterranean Sea	1975–86	Multiple organs	3	Ca, Cd, Cr, Cu, Fe, Hg, Pb, Ti, V, Zn	Viale 1994
T. truncatus	Mediterranean Sea	1973–75	B, K, L, M	2	Hg, MeHg, Se	Martoja and Viale 1977
T. truncatus	Mediterranean Sea	1973–75	Lung	3	Cd, Cr, Fe, Hg, Pb, Ti	Viale 1981
T. truncatus	Mediterranean Sea (Corsica)	1975–86	B, K, L, M, O	3	Ca, Cd, Cr, Cu, Fe, Hg, Pb, Ti, V	Viale 1994
T. truncatus	Mediterranean Sea (Italy)	1987–89	Br, K, L, M	6	Cd, Hg, Pb, Se, Zn	Leonzio et al. 1992
T. truncatus	Mediterranean Sea (Italy)	1992	B, L	3	Butyltin compounds	Kannan et al. 1996
T. truncatus	Netherlands (captive)	1970–71	L	3	As, Cd, Hg, Se, Zn	Koeman et al. 1972, 1975
T. truncatus	Wales	1988	B, M, O	1	Cd, Cr, Cu, Hg, Ni, Pb, Zn	Morris et al. 1989
T. truncatus	Western North Atlantic (Georgia, USA)	1971	Br, K, L, M, O	1	Hg	Stickney et al. 1972
T. truncatus	Western North Atlantic (U.S.A.)	1987–89	L	9	Cd, Cr, Hg, Mn, Pb, Se	Kuehl et al. 1994
Ziphius cavirostris	Australia	—	Bone	1	Pb	Kemper et al. 1994
Z. cavirostris	Bermuda	1981	B, K, L, M, O	4	Cd, Cu, Fe, Mn, Ni, Pb, V, Zn	Knap and Jickells 1983
Z. cavirostris	Mediterranean Sea	1974	B, K, L, M	1	Hg, MeHg, Se	Martoja and Viale 1977
Z. cavirostris	Mediterranean Sea (France)	1974	Lung	1	Cd, Cr, Fe, Hg, Pb, Ti	Viale 1981
Z. cavirostris	Western South Atlantic (Argentina)	—	B, K, L, M	1	Cd, Hg	Marcovecchio et al. 1994

Dashes appear where values were not available in original source.

B = blubber; Br = brain; K = kidney; L = liver; M = muscle; MeHg = methyl mercury; Mi = milk; NR = not reported; O = other.

Appendix 10-7. Summary of Selected Surveys of Metals and Trace Elements in Tissues of Baleen Whales

Species	Region	Sample Period	Tissues Sampled	Max. No. of Individuals Sampled	Metals and Trace Elements Reported	References
Balaena mysticetus	Alaska	—	L	3	Ag, Cd, Hg, Se, V	Mackey et al. 1996
B. mysticetus	Arctic Ocean (Alaska)	1979–88	B, K, L, M	12	Ag, As, Ba, Be, Cd, Cr, Cu, Hg, Fe, Ni, Pb, Sb, Se, Ti, Zn	Bratton et al. 1993; Byrne et al. 1985
Balaenoptera acutorostrata	Antarctic	1980–85	L	135	Cd, Co, Cu, Fe, Hg, Mn, Ni, Pb, Zn	Honda et al. 1986b, 1987
B. acutorostrata	Antarctic	1985	B	1	Butyltin compounds	Iwata et al. 1994
B. acutorostrata	Greenland	1980	K, L, M	24	Cd, Hg, Se, Zn	Hansen et al. 1990
B. acutorostrata	Greenland	1980	K, L, M	22	Cd, Cu, Hg, Pb, Se, Zn	Kapel 1983
B. acutorostrata	Greenland	1984–87	K, L, M	3	Hg, organic Hg	Dietz et al. 1990
B. acutorostrata	St. Lawrence River	1983–90	L	9	Cd, Hg	Béland et al. 1991
B. acutorostrata	Wales	1991	L	1	Cd, Cr, Cu, Hg, Ni, Pb, Zn	Law et al. 1992
Balaenoptera sp.	Australia	—	Bone	5	Pb	Kemper et al. 1994
B. borealis	Tasman Sea	1972	M	9	Hg, MeHg	Nagakura et al. 1974
B. physalus	Southern Ocean	1947–48	M	8	Hg, MeHg	Nagakura et al. 1974
B. physalus	Eastern North Atlantic	1983–86	K, L, M	36	Hg, organic Hg	Sanpera et al. 1993
B. physalus	North Atlantic (Iceland, Spain)	1983–86	K, L, M	118	Cd, Cu, Zn	Sanpera et al. 1996
B. physalus	Mediterranean	1975	Lung	1	Cd, Cr, Fe, Hg, Pb, Ti	Viale 1981
B. physalus	St. Lawrence River	1983–90	L	1	Cd, Hg	Béland et al. 1991
Caperea marginata	Australia	—	B, L, M, O	8	Cd, Hg, Pb	Kemper et al. 1994
C. marginata	Tasmania	1980s	M	1	Hg	Munday 1985
Eschrichtius robustus	Eastern North Pacific (U.S.A.)	1988–91	Br, K, L, O	11	Ag, Al, As, Ba, Cd, Cr, Cu, Fe, Hg, Mn, Ni, Pb, Se, Sn, Sr, Zn	Varanasi et al. 1994

B = blubber; Br = brain; K = kidney; L = liver; M = muscle; MeHg = methyl mercury; Mi = milk; O = other.

Appendix 10-8. Summary of Selected Surveys of Metals and Trace Element Concentrations in Tissues of Sea Otters, Sirenians, and Polar Bears

Species	Region	Sample Period	Tissues Sampled	Max. No. of Individuals Sampled	Metals and Trace Elements Reported	References
Enhydra lutris	California	1987–89, ancient	Teeth	13	Pb	Smith et al. 1990
E. lutris	North Pacific (Amchitka Island)	1986–87, ancient	Teeth	10	Pb	Smith et al. 1990
Dugong dugon	Australia	—	K, L, M	2	Hg	Denton and Breck 1981
D. dugon	Australia	1974–78	Br, K, L, M	43	Ag, Cd, Co, Cr, Cu, Fe, Mn, Ni, Pb, Zn	Denton et al. 1980
D. dugon	Sulawesi (Indonesia)	1975	M	2	Hg, MeHg, Se	Miyazaki et al. 1979
Trichechus manatus	Florida	1977–81	K, L, M	54	Cd, Cu, Fe, Hg, Pb, Se	O'Shea et al. 1984
T. manatus	Florida	1982	Br, K, L	8	Cd, Cu, Hg, Pb	O'Shea et al. 1991
Ursus maritimus	Arctic (Greenland, Svalbard)	1978–89	Hair, K, L, M	97	Hg	Born et al. 1991
U. maritimus	Canada	1910–80	Hair	146	Hg	Eaton and Farant 1982
U. maritimus	Alaska	1972	L, M	~62	Hg	Lentfer and Galster 1987
U. maritimus	Arctic	1976–88	Hair	141	Hg	Renzoni and Norstrom 1990
U. maritimus	Canada (Northwest Territories)	1982–84	L	124	Ag, As, Ba, Be, Ca, Cd, Cu, Fe, Hg, K, Mg, Mn, Mo, Na, P, Se, Sr, Ti, V, Zn, Zr	Braune et al. 1991; Norstrom et al. 1986
U. maritimus	Greenland	1984–87	K, L, M	4	Hg, organic Hg	Dietz et al. 1990
U. maritimus	Svalbard	1978–89	K, L	22	As, Cd, Cu, Hg, Pb, Se, Zn	Norheim et al. 1992

B = blubber; Br = brain; K = kidney; L = liver; M = muscle; MeHg = methyl mercury; Mi = milk.

Contributors

Daryl J. Boness heads the Department of Zoological Research, National Zoological Park, Smithsonian Institution, in Washington, D.C. His research focuses on the reproductive behavior of pinnipeds, in particular how ecological and social factors affect the mating and parental care strategies of males and females. Studies carried out to date have involved 11 of the 31 species of pinnipeds in all parts of the world. He received his doctorate from Dalhousie University, Halifax, Nova Scotia, in 1979, and has served on the U.S. Marine Mammal Commission's Committee of Scientific Advisors since 1993.

W. Donald Bowen is adjunct professor in biology at Dalhousie University, Halifax, Nova Scotia, and head of the Marine Fish Division at the Bedford Institute of Oceanography, Dartmouth, Nova Scotia. His research has been focused on the reproductive and foraging ecology and population dynamics of gray seals and harbor seals and, more recently, on the ecology of northern right whales. He earned his B.S. and M.S. degrees in zoology from the University of Guelph, Guelph, Ontario, in 1971 and 1973, and received his doctorate in zoology from the University of British Columbia, Vancouver, in 1978.

Ian L. Boyd is the head of the marine ecosystem management programme at the British Antarctic Survey, Cambridge, England, where he has developed a research program on Southern Ocean top food chain predators including Antarctic fur seals and southern elephant seals. His research interests are focused on the behavior and ecology of foraging in marine predators and the effects of environmental variability on their populations. He has spent ten summer seasons carrying out studies in the Antarctic, and was awarded the W. S. Bruce Medal from the Royal Society of Edinburgh for his contributions to polar research and the Scientific Medal from the Zoological Society of London for his research on the foraging ecology of marine predators. He obtained a B.S. degree from Aberdeen University, Aberdeen, Scotland, in 1979 and his doctorate from Cambridge University in 1982. He was also awarded a D.Sc. degree from Aberdeen University in 1995.

Daniel P. Costa is professor and vice chair of biology at the University of California at Santa Cruz. While on leave of absence, he developed a marine mammal research program for the Office of Naval Research in Washington, D.C., first initiating and then supervising a program that funded basic research on the effects of low-frequency sound on marine mammals. Since returning to California, he has directed the marine mammal research program for the California Acoustic Thermometry of the Ocean Climate project (known as ATOC) for Scripps Institution of Oceanography. He received his B.A. from the University of California at Los Angeles in 1974 and his doctorate from the University of California at Santa Cruz in 1978.

Robert Elsner is professor emeritus at the Institute of Marine Science, University of Alaska, Fairbanks, and previously was on the faculty of the Scripps Institution of Oceanography, University of California, San Diego. His scientific interests concern the environmental physiology of animals and humans, including reactions to diving, cold, and altitude. He served on the U.S. Marine Mammal Commission from 1984 to 1991. He received his B.A. at New York University and M.S. and Ph.D. degrees from the University of Washington.

Darlene R. Ketten has a joint appointment as an associate scientist in biology at Woods Hole Oceanographic Institution, Woods Hole, Massachusetts, and assistant professor at Harvard Medical School in otology and laryngology. Her current work is divided between modeling sound reception and hearing mechanisms of marine mammals and developing in vivo microimaging techniques for diagnosing ear trauma and disease, particularly in humans. She received a B.S. / B.A. degree in biology and French from Washington University in 1970, M.S. in biological oceanography from Massachusetts Institute of Technology in 1979, and a Ph.D. in neuroethology and experimental radiology from The Johns Hopkins University in 1984.

Christina Lockyer is currently a senior scientist at the Danish Institute for Fisheries Research, Charlottenlund, working on the problems of interactions between marine mammals and fisheries and specifically the bycatch of harbor porpoises in gillnets. For most of her nearly 30 working years, she has worked in England for the Natural Environment Research Council's Sea Mammal Research Unit in Cambridge while periodically taking leave to work abroad on specific research projects. From 1989 to 1991 she served as president of the Society for Marine Mammalogy and is currently chairman of the European Cetacean Society. She received her B.Sc. in biology from the University of East Anglia, Norwich, England, in 1968, her M.Sc. in zoology from the University of London in 1972, and her D.Sc. in zoology from the University of East Anglia in 1989.

Helene D. Marsh is professor of environmental science at James Cook University, Townsville, Queensland, Australia, where she chairs the School of Tropical Environment Studies and Geography. For more than 25 years she has studied dugong reproduction and life history based on analysis of dugong carcasses collected from aboriginal hunters or killed incidentally in fishing nets. Since the mid-1980s she has been conducting large-scale aerial surveys to determine dugong distribution, abundance, and population trends in northern Australia. She currently is co-chair of the Sirenia Specialist Group of the World Conservation Union—IUCN. She earned her B.S. in zoology in 1968 from the University of Queensland and her doctorate from James Cook University in 1973.

William A. McLellan is a researcher in the Department of Biological Sciences and the Center for Marine Science Research at the University of North Carolina at Wilmington. His research is focused on learning and describing the detailed anatomy that allows cetaceans to survive in a cold, high-drag environment. Other interests include development of health assessment parameters to monitor coastal and nearshore populations of cetaceans. He received his B.A. in marine sciences from the College of the Atlantic, Bar Harbor, Maine, in 1988.

Thomas J. O'Shea is a biologist with the U.S. Geological Survey's Midcontinent Ecological Science Center in Fort Collins, Colorado. Areas of study have focused on bats, cetaceans, sirenians, and other mammals, including research on the occurrence and impacts of environmental contaminants. Field work has been carried out in the United States and various tropical regions abroad, including Kenya, Venezuela, Panama, the Caribbean, and the South Pacific. He is a former member of the U.S. Marine Mammal Commission's Committee of Scientific Advisors and is chair of the interagency Working Group on Marine Mammal Unusual Mortality Events mandated under the Oceans Act of 1992. He has a B.S. in zoology from Colorado State University and his M.S. and doctorate degrees from Northern Arizona University.

Daniel K. Odell is a research biologist at SeaWorld, Inc., Orlando, Florida, research associate at Hubbs-SeaWorld Research Institute, Florida office, and on the adjunct faculty of the Department of Biology, University of Central Florida. He previously served on the graduate faculty at the Rosenstiel School of Marine and Atmospheric Science, University of Miami, Florida. In 1974 he cofounded the manatee carcass salvage network in Florida and he has served as scientific coordinator for the Southeastern U.S. Marine Mammal Stranding Network since its inception in 1977. Research interests include the natural history of the Florida manatee and the bottlenose dolphin as well as mass stranding phenomena. He received his B.S. in wildlife biology in 1967 from Cornell University, his M.A. in zoology from the University of California at Los Angeles in 1970, and his doctorate in biology from the University of California, Los Angeles in 1972.

D. Ann Pabst is associate professor in the Department of Biological Sciences and the Center for Marine Science Research at the University of North Carolina at Wilmington. Research interests focus on locomotor and thermoregulatory functions in cetaceans, and current work has involved investigations of the biomechanics of dolphin skin and blubber and the development of swimming muscles in dolphins. She received a B.S. in zoology from the University of Maryland in 1982 and her doctorate in zoology from Duke University in 1989.

Galen B. Rathbun is a research biologist with the Western Ecological Research Center of the U.S. Geological Survey in California. He previously served as the U.S. Fish and Wildlife Service's project leader for manatee research in Florida, and then as project leader for the sea otter translocation experiment in California. Currently, his research focuses on the behavioral ecology of several declining terrestrial vertebrates in central California. He received his undergraduate degree in zoology from Humboldt State University in Arcata, California. Following service in the Peace Corps, he obtained his doctorate from the University of Nairobi, Kenya, studying the behavioral ecology of elephant shrews.

John E. Reynolds III is professor of marine science and biology at Eckerd College, St. Petersburg, Florida, where he has taught marine mammalogy and comparative anatomy since 1980. His scientific research has primarily involved Florida manatees. He has served as chairman of the U.S. Marine Mammal Commission since 1991 and, before that, was a member of the Commission's Committee of Scientific Advisors. He received his B.A. in biology from Western Maryland College, and his M.S. and doctorate degrees in biological oceanography from the University of Miami's Rosenstiel School of Marine and Atmospheric Science.

Sentiel A. Rommel is a member of the research staff of the Marine Mammal Pathobiology Laboratory of the Florida Marine Research Institute, St. Petersburg, Florida. He also serves as adjunct faculty member at Eckerd College, St. Petersburg, and a research associate in vertebrate zoology at the U.S. National Museum, Smithsonian Institution, Washington, D.C. His recent scientific work has focused on gross anatomy of marine mammals with a special interest in functional morphology. He received his B.S. in physics from the U.S. Naval Academy, Annapolis, Maryland, and his M.S. in electrical engineering and doctorate in biological oceanography from the University of Maine, Orono.

Donald B. Siniff is a professor in the Department of Ecology, Evolution and Behavioral Biology at the University of Minnesota, and has been studying Antarctic seals since 1968. He is currently carrying out studies of the Weddell seal at McMurdo Station, and is participating in the international Antarctic Pack Ice Seal (APIS) program. From 1975 to 1995 he also carried out studies on sea otters in Alaska and California. His research has focused on the population dynamics, life histories, and behavior of large mammals with particular reference to marine mammals. He received his B.S. in fisheries and wildlife from Michigan State University in 1957 and his M.S. in statistics in 1958. He earned his doctorate from the University of Minnesota. He served on the U.S. Marine Mammal Commission from 1975 to 1986 and on the National Research Council's Polar Research Board from 1993 to 1997. He currently is a member of the California sea otter and Steller sea lion recovery teams.

Peter L. Tyack is an associate scientist with the Biology Department of Woods Hole Oceanographic Institution, Woods Hole, Massachusetts, specializing in the acoustic communication of cetaceans. Research topics have included vocal learning and mimicry in the natural communication systems of cetaceans, acoustic structure and social functions of the songs of baleen whales, and development of methods to identify which cetacean in a group is producing a sound. He has served on committees of the National Academy of Sciences to review the potential impact of low-frequency noise on marine mammals. He received his undergraduate degree in biology from Harvard College and his doctorate in animal behavior from Rockefeller University.

Douglas Wartzok is associate vice chancellor for research and dean of the Graduate School at the University of Missouri-St. Louis. He previously held positions at Indiana University-Purdue University, Fort Wayne, Indiana, and The Johns Hopkins University, Baltimore, Maryland. Laboratory work has focused primarily on the visual psychophysics of seals. Field work has involved seals, walruses, and large whales with an emphasis on the development of new technologies to better understand animal behavior. For nine years, he was editor of *Marine Mammal Science,* the journal of the Society for Marine Mammalogy, and is editor of the Special Publications of the Society for Marine Mammalogy. He earned a B.A. in physics and mathematics from Andrews University in 1963, an M.S. in physics from the University of Illinois in 1966, and a Ph.D. in neurophysiology from The Johns Hopkins University in 1971.

Randall S. Wells is a conservation biologist with the Chicago Zoological Society, Brookfield, Illinois, and directs the Center for Marine Mammal and Sea Turtle Research at the Mote Marine Laboratory, Sarasota, Florida. He also serves as adjunct associate professor of ocean sciences at the University of California at Santa Cruz. Current research is directed toward the behavior, ecology, health, and population biology of bottlenose dolphins along the west coast of Florida. He received his B.A. in zoology from the University of South Florida, and his M.S. in zoology from the University of Florida. He obtained his doctorate in biology from the University of California, Santa Cruz, in 1986.

Terrie M. Williams is an associate professor of biology at the University of California at Santa Cruz. A comparative exercise physiologist, she has spent more than 15 years studying the energetics of locomotion in mammals. Current research interests include the evolution of cost-efficient swimming in marine mammals, and the balance between exercise and thermoregulatory and diving responses in marine mammals. She obtained her doctorate in physiology from Rutgers University, New Brunswick, New Jersey, in 1981, and did postdoctoral work at Scripps Institution of Oceanography, La Jolla, California.

Index